C0-AXD-339

ADVANCES IN **SOLAR ENERGY TECHNOLOGY**

(in four volumes)

Volume 2

ADVANCES IN
SOLAR ENERGY TECHNOLOGY

Proceedings of the Biennial Congress of
the International Solar Energy Society
Hamburg, Federal Republic of Germany, 13–18 September 1987

Edited by

W. H. BLOSS and **F. PFISTERER**
Institute of Physical Electronics
University of Stuttgart, Federal Republic of Germany

Volume 2

PERGAMON PRESS
OXFORD · NEW YORK · BEIJING · FRANKFURT
SÃO PAULO · SYDNEY · TOKYO · TORONTO

U.K.	Pergamon Press plc, Headington Hill Hall, Oxford OX3 0BW, England
U.S.A.	Pergamon Press, Inc., Maxwell House, Fairview Park, Elmsford, New York 10523, U.S.A.
PEOPLE'S REPUBLIC OF CHINA	Pergamon Press, Room 4037, Qianmen Hotel, Beijing, People's Republic of China
FEDERAL REPUBLIC OF GERMANY	Pergamon Press GmbH, Hammerweg 6, D-6242 Kronberg, Federal Republic of Germany
BRAZIL	Pergamon Editora Ltda, Rua Eça de Queiros, 346, CEP 04011, Paraiso, São Paulo, Brazil
AUSTRALIA	Pergamon Press Australia Pty Ltd., P.O. Box 544, Potts Point, N.S.W. 2011, Australia
JAPAN	Pergamon Press, 5th Floor, Matsuoka Central Building, 1-7-1 Nishishinjuku, Shinjuku-ku, Tokyo 160, Japan
CANADA	Pergamon Press Canada Ltd., Suite No. 271, 253 College Street, Toronto, Ontario, Canada M5T 1R5

First edition 1988

Library of Congress Cataloging in Publication Data

International Solar Energy Society. Congress (10th: 1987: Hamburg, Germany)
Advances in solar energy technology: proceedings of the Biennial Congress of the International Solar Energy Society, Hamburg, Federal Republic of Germany, 13–18 September 1987/ edited by W. H. Bloss and F. Pfisterer.—1st ed.
p. cm.
Includes index.
1. Solar energy—Congresses. I. Bloss, W. H. (Werner H.), 1930–
II. Pfisterer, F. III. Title.
TJ809.2.I573 1987
621.47—dc19 88–17824

British Library Cataloguing in Publication Data

International Solar Energy Society. *Congress (1987: Hamburg, Germany)*
Advances in solar energy technology. 1. Solar energy. Conversion 1. Title II. Bloss, W. H. (Werner H.)
III. Pfisterer, F.
621.47
ISBN 0–08–034315–5

In order to make this volume available as economically and as rapidly as possible the authors' typescripts have been reproduced in their original form. This method has its typographical limitations but it is hoped that they in no way distract the reader.

Printed in Great Britain by A. Wheaton & Co. Ltd., Exeter

TABLE OF CONTENTS

Volume 1

Volume 2

Volume 3

Volume 4

Contents of Volume 2

Fresh Air Solar Heating System - Seasonal Operating Performance

George O.G. Löf
Gary Cler
Thomas E. Brisbane

Colorado State University
Ft. Collins, CO 80523

ABSTRACT

A solar heating system comprising a 25 m^2 evacuated tube solar collector, 4 m^3 stratified hot water storage tank, liquid-to-air heat exchange coil in a commercial fresh air heating-cooling unit in which heat recovery in a regenerative air-to-air exchanger is employed, and an electric auxiliary heater was operated and monitored continuously through four months of the 1985-86 winter in a residential style building of 200 m^2 floor area. Design improvements over a previously tested system included the control system which provided higher solar collection efficiency, better temperature stratification in storage, higher reflectivity of evacuated tube reflectors, and the supply of fresh air rather than recirculated air without significant thermal and economic penalty.

Average daily solar collection efficiency during four months of operation, January through April, was 42% based on collector aperture area. The portion of the heating requirements met by solar was 53%, and 61% of the hot water was supplied by solar.

An improved reflector increased energy collection 6.8%, and use of a solar intensity sensor resulted in a 16% increase in energy collection compared with conventional delta T control.

KEY WORDS: Solar Heating, Solar Cooling, Evacuated Tube Collectors

INTRODUCTION

In a solar systems development laboratory at Colorado State University, three residential buildings are heated and cooled primarily by solar energy. House I, with a total floor area of 259 m^2 on two levels, is cooled by use of a solid desiccant in a rotating bed, followed by heat exchange and evaporative cooling. The desiccant is regenerated by the return air after heating in an exchanger coil supplied with a solar heated antifreeze solution.

Space heating with this 100 percent fresh air system is accomplished by use of rotating heat recovery exchanger in which supply air is heated by return air and then by solar heated glycol solution. Service hot water is also supplied by heat exchange with hot water in main storage.

During the 1985-86 testing program, the hydronic solar heating and cooling system used the previous year (Karaki, 1985) was modified by installing a desiccant cooling-fresh air heating subsystem manufactured under the trade name, "Energymaster," by American Solar King, Inc. (Venhuizen, 1984; Coellner, 1986). The Sunmaster TRS-81 evacuated-tube all-glass collector was used as the primary source of heat for the system from which conditioned air was distributed to the rooms of the building. A hot water solar storage tank and the DHW subsystem with an in-tank heat exchanger were unchanged from the system used in the 1984-85 program.

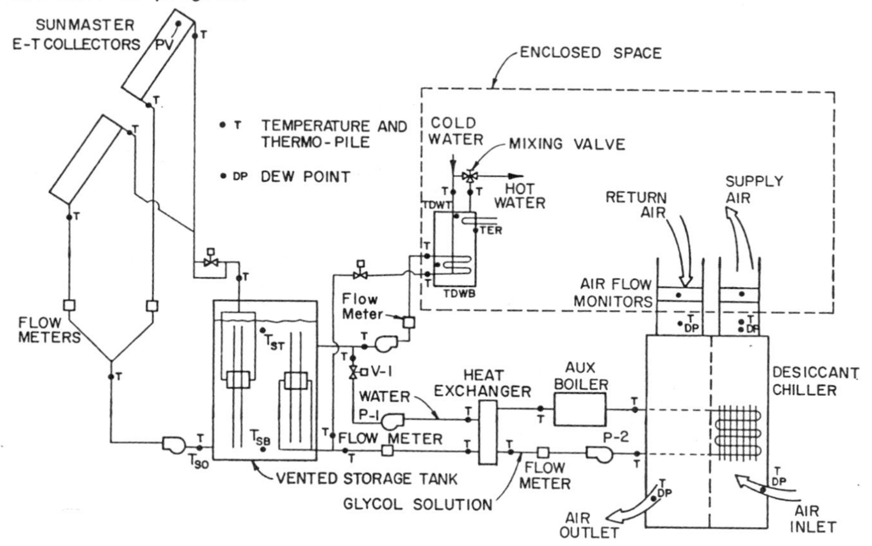

Fig. 1. Schematic diagram of the heating system for the 1985-86 winter season.

HEATING SYSTEM

General Description

The heating system arrangement for the winter season (1985-86) is shown in Fig. 1 (Löf, 1986). Water is heated in an evacuated tube collector in a drain-back system with a large return pipe. Water circulated from the storage tank through the collector is returned to the tank through a stratification manifold. Solar heating is accomplished by circulating water from the storage tank to a heat exchanger through which a glycol solution is circulated to a liquid-to-air heating coil in the air supply side of the ASK Energymaster unit. Heated fresh air is circulated to the rooms, and return air passes through the discharge side of the Energymaster unit where heat is recovered in the regenerative exchanger. Domestic water heating is accomplished in a single tank system with a coil (supplied from main storage) submerged inside the DHW tank. When the solar system is unable to adequately heat the building, auxiliary heating can be provided by a supply of electrically heated glycol solution to the air heating coil in the ASK unit.

Principal Components

Collectors. Each collector tube is 52 mm outside diameter and 1.83 meters in length. There are 8 tubes per module on 152 mm spacing with a modified compound parabolic specular reflector below the tubes. The aperture area of the array is 25.688 m^2. The evacuated-tube assembly consists of three concentric glass tubes, serving as cover, absorber, and vent. The outer surface of the absorber tube has a selective black coating, and the space between the absorber and cover is evacuated. Water flows upward between the vent and absorber tubes and downward by gravity through the vent tube. The collector is designed for use in a drain-back system.

Twelve collector modules were installed in two parallel rows of six each on the west half of the Solar House I roof, as indicated by Fig. 2. To assure collector drainage, the manifold of each module was installed at an adequate slope. Pipe sections between modules are connected with silicone hose and hose clamps.

Collector efficiency is highly dependent on multiple reflections, especially at mid-day, and any improvement in reflectance increases collector performance. In late fall, the reflectors on the upper collector array were replaced with a set having the same curvature and a newly developed reflecting film. The lower array was left to facilitate a comparison of the two reflector surfaces.

Fig. 2. Evacuated tube collector installation on Solar House I roof.

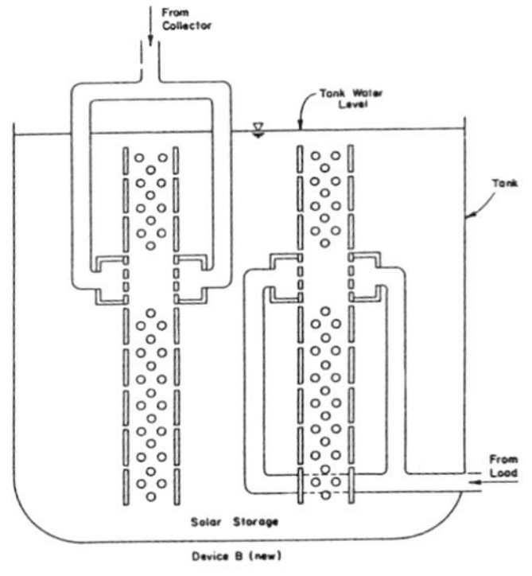

Fig. 3. Stratification manifolds in liquid storage tank.

Hot water storage. A 4000-liter insulated tank containing 3600 liters of water was used for thermal storage during winter operation. Water from storage is circulated through the collector, as well as to the load. Storage temperatures did not exceed 95C (the local boiling point of water) during the winter season, so pressurization of the tank was not required.

Stratification manifolds. The design in Fig. 3 is a modification of a unit investigated by Gari (1983). It consists of an inlet chamber and porous sections attached above and below the chamber. Water enters the inlet chamber horizontally from two sides, and after its momentum is dissipated radially, it leaves the inlet chamber and rises or descends to an appropriate density (temperature) level in the tank. Thermocouples were installed at five vertically aligned positions, 9 cm below the water surface and at 31.6-cm intervals to the fifth level 39 cm above the rounded bottom of the tank.

RESULTS

Solar Energy Collection

Monthly averages of daily energy collection and collection efficiency of the evacuated-tube collectors during the 1985-86 winter are shown in Fig. 4. It is seen that the monthly average corrected daily collection efficiency, including net energy drained back to storage after daily shut-down, ranged from 41 to 52%, and the average for the season was 45%. This performance may be compared with that in the 1984-85 season when the January-April average efficiency was 36%. The increase was due mainly to a better collector control system and to an improved reflector behind one-half of the collector tubes.

Daily collection efficiency was increased by use of photocell control. Figure 4 shows average daily collection of 8.07 MJ/m^2,day and collection efficiency of 45% for the full array. Corresponding figures for the 1984-85 heating season are 6.86 MJ/m^2,day and 36% seasonal collection efficiency (Karaki, 1985). The percentage of solar radiation received during pump operation increased from 69% in '84-'85 to 89% in '85-'86. Although other minor system changes had also been made, improved control is considered the principal factor in increasing energy collection. Higher collectible solar radiation ("solar while collecting") results from the greater duration of pump operation. Occasional heat losses due to circulation of hot storage water through the collector during periods of low solar intensity were more than offset by less frequent collector drainback, decreased pump lock-out by the high temperature switch, and a longer operating day.

Comparison of energy collection and daily efficiency of the lower (unimproved) collector array in 1985-86 with the 1984-85 values shows that collector control by a solar sensor increased energy collection by 18.2% over that with delta T control. Energy collection in the upper array containing the improved reflectors was typically about 6.4% higher than in the lower array of the same area. The efficiency of the upper array was 45.9% compared with 35.7% efficiency the previous year. Energy delivery at comparable radiation levels would be increased 28.6% by the combined effects of improved reflectors and a better control system.

Space Heating

The monthly distribution of the space heating load is shown in Fig. 5. With approximately two-thirds of the season included in the measurements, the measured solar space heating percentage was 53%. The 1985-86 solar heating supply of 162 MJ/day was more than 50% higher than the 105 MJ/day solar supply the previous season.

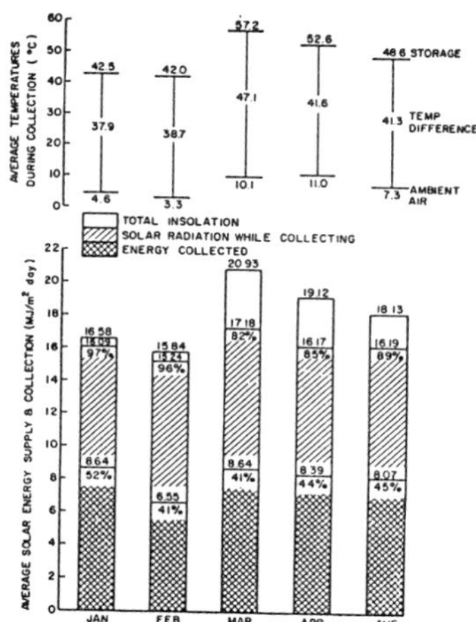

Fig. 4. Monthly average energy collection, CSU Solar House I, Winter 1985-86.

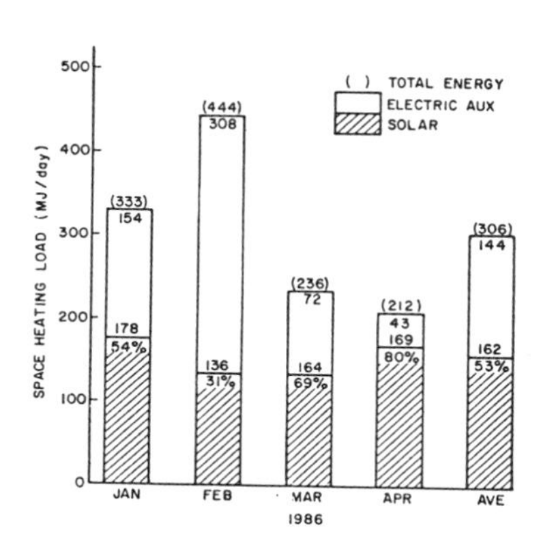

Fig. 5. Space heating load, CSU Solar House I.

Water Heating

During the four winter months starting in January, 60% of the average daily 34.8 MJ DHW load was supplied by solar energy. Average water heating demand in 1984-85 was 41.4 MJ/day, of which solar provided 70%. Lower hot water demand in 1985-86 is the principal reason for the decreased solar use for water heating.

System Energy Flows

A pictorial representation of energy flows in the solar heating system is shown in Fig. 6. The indicated percentages are based on the total incident insolation on the aperture area of the collectors. Uncollectible energy (below threshold) losses of 11% are significantly lower than the 33% uncollectible in 1984-85 with this collector under delta T control. Optical and thermal losses from the collector increased, however, from 33% in '84-'85 (coincidentally, the same percentage as the uncollectible portion) to 44% in '85-'86. Longer periods of operation with photovoltaic control resulted in less uncollectible energy but more losses due to operation when radiation levels near sunrise and sunset were low.

Collector Efficiency Improvement

Time traces of instantaneous collector efficiencies, along with temperatures and solar irradiance are shown in Fig. 7 for a clear day in 1986. In the lower portion of the figure, the points show instantaneous efficiency of the entire collector, including upper and lower arrays, (based on aperture area) as a function of $(T_i-T_a)/I_T$. Also shown is the manufacturer's efficiency curve based on gross collector area (dashed line), and on aperture area (solid line).

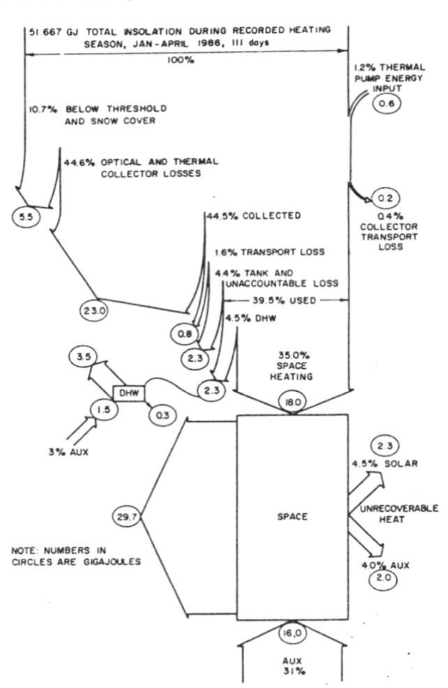

Fig. 6. Energy flows for heating season, CSU Solar House I, 1985-86.

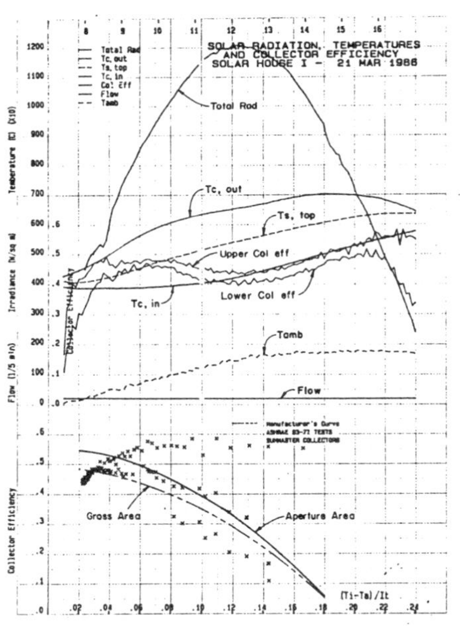

Fig. 7. Solar system performance on a clear winter day, with photocell control of evacuated tube collectors, CSU Solar House I, 21 March, 1986.

In the upper portion of Fig. 7, the collection efficiencies of the upper and lower arrays are individually shown. The benefit of higher reflectivity of the back surface is evident.

CONCLUSIONS

Evaluation of results obtained with the solar heating system and the desiccant cooling system in Colorado State Solar House I during the June 1985 - May 1986 year leads to the following conclusions:

- During the winter and spring months, the Sunmaster TRS-81 evacuated tube collectors operated with an average on-time collection efficiency of 50% and all-day efficiency of 45%, providing heat at an average storage temperature of 48.6C. These efficiencies may be compared with 52% and 36%, respectively, for the same collectors the previous year.
- The replacement of a delta T controller with a photovoltaic controller resulted in an average increase in solar energy collection of 18%, primarily through an increase from 69% to 89% in the portion of total daily radiation received during collector operation.
- The collector array containing improved reflector material collected 6.4% more energy than the array of equal area with unimproved reflectors.
- The percentage of incident radiation that was delivered for use in space heating and hot water was 39%. The fraction of total heat requirements met by solar was 0.54.

ACKNOWLEDGEMENT

The financial support of the Solar Buildings Program in the U.S. Department of Energy and the equipment and advice provided by the Sunmaster Corporation and American Solar King, Inc., are gratefully acknowledged.

REFERENCES

1. Karaki, S., G.O.G. Löf, T.E. Brisbane, D. Weirsma and G. Cler (1985). Performance of Solar Heating and Cooling Systems Employing Evacuated-Tube Collectors, CSU Solar House I, June 1984 - July 1985 Operation. Final Report, CSU Report No. SAN-11927-28, November.

2. Venhuizen, D. (1984). Solar King's cooling gambit. Solar Age, October, 25.

3. Coellner, J.A., D.S. MacIntosh, and M.C. Blanpied (1986). Open Cycle Desiccant Air-Conditioning System and Components Thereof. U.S. Patent 4594860, June 17.

4. Löf, G.O.G., T.E. Brisbane, G. Cler, and S. Beba (1986). Performance of Solar Heating and Cooling Systems. Final Report, CSU Report No. SAN-11927-44, November.

5. Gari, N.H.K. (1983). Stratification Enhancement in Solar Liquid Thermal Storage Tanks, Analysis and Test of Inlet Manifolds. Ph.D. Thesis, Mechanical Engineering Department, Colorado State University.

SOLAR ASSISTED AIR HEATING SYSTEMS IN COMMERCIAL BUILDINGS

N.Fisch, E.Hahne

Institut für Thermodynamik und Wärmetechnik
Universität Stuttgart
Pfaffenwaldring 6,FRG-7000 Stuttgart 80

ABSTRACT

The thermal performance of solar-assisted air-heating systems for its applicability in commercial buildings was investigated with the program "SIMUL". This modular computer program was specifically developed for solar systems with air as the heat transfer medium. The design parameters of a typical office building were determined. With the computer program "HAUS", the hourly heating load of the building was calculated using the weather data for Stuttgart. The thermal perfomance (e.g. solar fraction, coefficient of performance and system efficiency) of systems with and without a heat store was determined.

KEYWORDS

Solar heating system; System simulation; Room heating; Office building; Solar fraction; System efficiency; Coefficient of performance.

INTRODUCTION

In the Federal Republic of Germany, about 40 % of the primary energy is used for room heating. Although air as a heat carrier offers a number of advantages, the application of air-heating systems in dwellings is unimportant in Europe compared to the USA. In commercial buildings, mechanical ventilation and air conditioning by HVAC-systems is most common. For these buildings, solar air-heating (SAH)-systems are very interesting, because on the one hand installations for ventilation already exist and on the other hand heating of such buildings is necessary during the day.

SYSTEM SIMULATION

The simulation of the thermal performance of a SAH-system in commercial buildings has been performed in two steps. In the first step, the hourly heating load of the building is calculated with the program "HAUS" (Fisch, Munder and Hahne 1984) depending on weather conditions, building design data and operation parameters of the heating system. In the second step, the program "SIMUL", developed by Fisch (1984), is used to simulate the thermal behaviour of a SAH-system. The data flow for a simulation cycle is presented schematically in Fig. 1. The results of the simulation are e.g. the monthly and yearly solar fraction, the efficiency and the coefficient of performance of the solar plant.

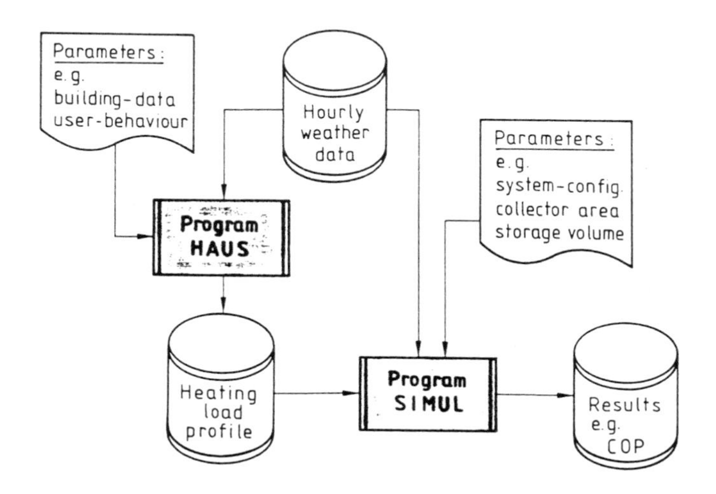

Fig. 1. Scheme of a simulation cycle

The user of SIMUL has a choice between eight different configurations of SAH-systems (SYS01-SYS08). For this paper, two different configurations were investigated. Fig. 2 shows the scheme of a system without a heat store and with an open collector circuit ("SYS01"). The SAH-system with a pebble-bed storage and a closed primary circuit ("SYS05") is presented schematically in Fig. 3.

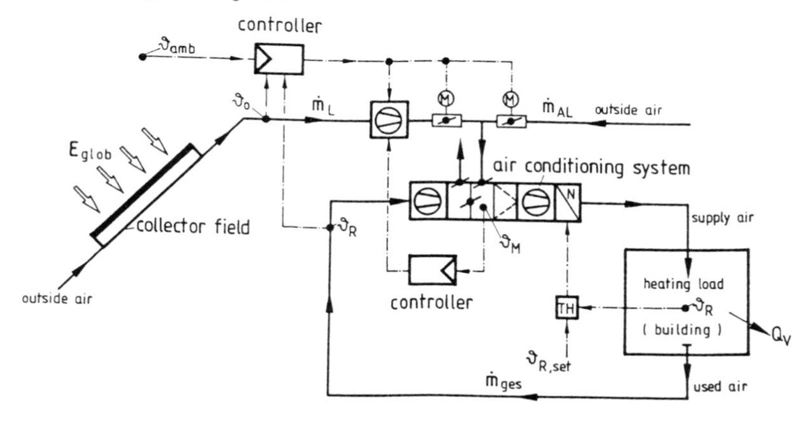

Fig. 2. Scheme of the SAH-system "SYS01"

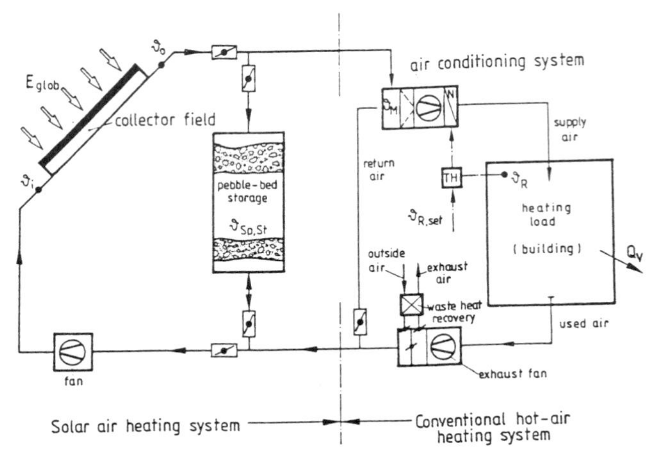

Fig. 3. Scheme of the SAH-system "SYS05"

RESULTS OF THE PARAMETER STUDY

Standard parameters for the investigation

For the simulation, hourly weather data (global and diffus solar radiation, ambient temperature and wind speed) from Stuttgart during 1979 are used. The essential design parameters for the office building and the solar air-heating system are:

Global radiation on horiz.	$924 \frac{kWh}{m^2 y}$	**OFFICE BUILDING**	
Average ambient temp.	$10°C$	Three floors, flat roof	
Degree days	$3326\ DD$	External dimensions	$50x12x10\ m$
SOLAR AIR-HEATING SYSTEM		Floor space	$1800\ m^2$
Collector field area	$400\ m^2$	Volume	$6000\ m^3$
Air flow rate	$25 \frac{kg}{h\ m^2}$	U-Value of the walls	$0.5 \frac{W}{m^2K}$
No. of collector cover plates	2	Window area in % of the wall area;	
Absorptivity of the absorber	0.97	south/west/east/north:	70/15/50/15
Emissivity of the absorber	0.95	U-Value of the windows	$2.4 \frac{W}{m^2K}$
Storage volume	$125\ m^3$	Room temperature	$20°C$
Storage capacity	$140 \frac{MJ}{K}$	Air exchange rate	$4.3 \frac{1}{h}$
Thermal insulation thickness	$0.1\ m$	Internal heat gain	$10\ kW$

Effect of overnight room-temperature set down

The office buildings are fully heated during the normal working hours (7 a.m. to 6 p.m.). In the night and on the weekend, a reduced room temperature (15°C) is common. Compared to residential buildings, the heating load and the solar energy are in a better coincidence. Therefore, a greater solar fraction can be expected for SAH-systems in commercial buildings. Fig. 4 shows the effect of the overnight room-temperature set down time period on the monthly heating load and solar fraction.

For the standard parameters, the yearly heating load is 149 MWh or 83 kWh per m^2 floor area and the solar fraction is 20 % for a system without a heat store ("SYS01") and 30 % for a system with a pebble-bed storage. The heating load increases about 2 % by decreasing the nightly room temperature set down time period by two hours (7 p.m. - 6 a.m.). By increasing the time period (5p.m. - 8a.m.), the heating load is reduced about 10 %. The solar fraction increases with increasing reduced room temperature time period. With a relatively simple SAH-system "SYS01", about 18 to 23 % of the incident solar energy can be directly used (without a heat store) for heating (see Fig. 4).

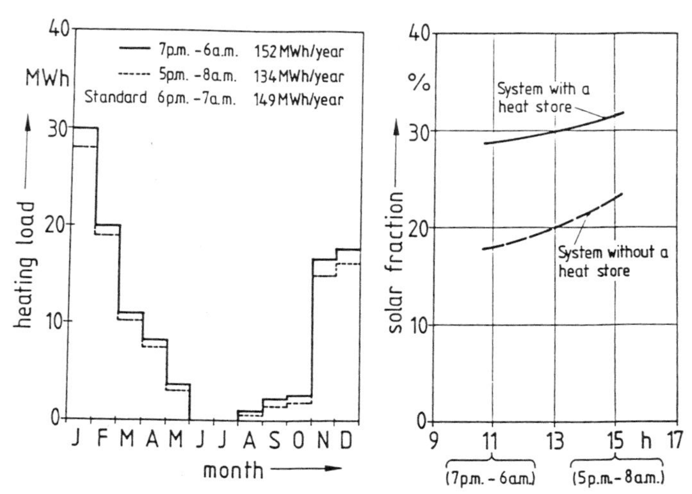

Fig. 4. Effect of the time period for the overnight room temperature set down on the heating load and solar fraction

Effect of the collector field area

Fig. 5 shows the effect of the collector area on various thermal performance factors for a SAH-system with and without a heat store. The growth of the solar fraction decreases with increasing collector area (Fig. 5a). An increase in the collector area from 200 to 400 m^2 brings forth an increase of the solar fraction more than twice as high as an extension from 400 to 600 m^2. The system efficiency (Fig. 5b) and the coefficient of performance (Fig. 5c) are inversely proportional to the size of the collector area. For the standard parameters, the system efficiency for "SYS05" is one third greater compared to "SYS01". The COP for the system with the heat store and the closed collector circuit ("SYS05") is lower compared to the system configuration "SYS01". The COP of "SYS05" for 400 m^2 collector area is about 11; this means that 9 % of the used solar energy is expended to operate the fans. For a collector area of 600 m^2, the electrical auxiliary amounts to 15 % and for 1000 m^2 about 35 % of the used solar energy (see Fig. 5c). For the "energetic" optimization of the collector area, the difference between the used solar energy (Q_{used}) and the electrical auxiliary (W_{el}) are calculated and is shown in Fig. 5d.Considering the total efficiency (circa 33 %) for the generation of electric energy ($\Delta Q = Q_{used} - 3\ W_{el}$) the optimal collector area is about 400 m^2 for "SYS05" and 600 m^2 for "SYS01". In general, the solar fractions of systems in commercial buildings are higher compared to systems in residential houses (Fisch, N. 1984).

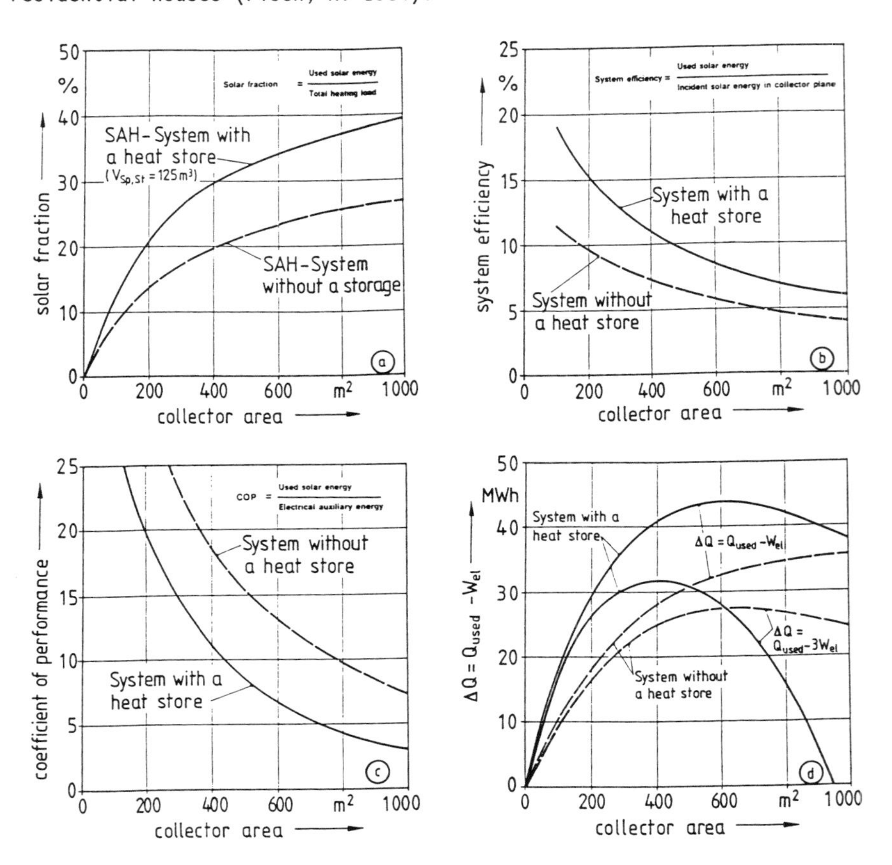

Fig. 5a-5d. Effect of the collector area on the solar fraction system efficiency, coefficient of performance and ΔQ for a SAH-system with and without a heat store

REFERENCES

Fisch, N. J. Munder and E. Hahne (1984). Untersuchung über Maßnahmen zur Reduzierung des Wärmeverbrauchs eines Wohngebäudes. Bauphysik, Heft 2, S. 55-62.

Fisch, N., E. Hahne (1984). Simulation of Solar Air-Heating Systems. Proc. First EC Conference on Solar Heating, p. 208-215, Amsterdam, 30.4.-4.5.1984.

Fisch, N. (1984). Systemuntersuchungen zur Nutzung der Sonnenenergie bei der Beheizung von Wohngebäuden mit Luft als Wärmeträger, Diss. Univ. Stuttgart, ISBN-Nr. 3-922-429-10-6

DEVELOPMENT AND TESTING OF SIMPLE SOLAR AIR HEATING COLLECTORS

M. Reuss, T. Schmalschlaeger, Dr. H. Schulz
Landtechnik Weihenstephan, Technical University of Munich
D-8050 Freising, F.R. Germany

1. ABSTRACT

In the last years solar air heating collectors have gained significance especially in the field of industrial and agricultural applications. In the period from 1982 to 1986 a project of development and testing of simple air heating solar collectors sponsored by the German Ministry of Research and Technology was carried out at our institute. The main topics were:

1. Development and improvement of simple solar air heating collectors for agricultural and industrial applications.
2. Performance measurements under steady- and quasi-steady-state conditions.
3. Material testing was performed with respect to durability of transparent collector glazing. The mechanical stability and the spectral transmission were measured.
4. Real size plants were designed and monitored in practice mainly for drying of hay and other agricultural products.

2. KEYWORDS

Solar air heating collector, air collector, collector test, performance monitoring of systems, test of materials, drying

3. INTRODUCTION

At present most solar collectors are used for water heating - mostly for domestic hot water or heating of swimming pools. However, for applications with air as heat transfer fluid like space heating or drying, there are apparent advantages for using air heating collectors:

High flowrate - hence good heat removal from the absorber, leakage is less critical, no heat exchangers, less problems with corrosion, sedimentation and freezing, simple constructions and therefore cheap solutions are possible.

For different applications in agriculture (e.g. low and medium temperature drying) different types of solar air heating collectors were developed. Among these were uncovered collectors, single and double glazed ones. Samples of these were mounted on a test rig for parallel performance measurements under outdoor conditions. The test installation was built in agreement with the recommendations of the CEC Collector and Systems Testing Group.

4. DESCRIPTION OF THE COLLECTORS

Various types of solar air heating collectors without, with single or double glazing and different types of absorbers were developed.

4.1 Unglazed Collectors

We have developed several unglazed air heating collectors - with roof tiles or corrugated metal sheets, but also other materials are possible. The first one uses roof tiles as an absorber with a polypropylen net inbetween giving a small gap to suck air through. The backside of the rafters is covered with a wood fibre board. At the ridge the air flows between the rafters into a main duct (fig. 1). The construction of the second collector is similar but instead of roof tiles corrugated metal sheets are used (fig. 2). More types were built, for example black plastic tube types, however, will not be mentioned here. Such unglazed collectors are designed for low temperature applications (temp. rise 5 - 10 K over ambient) sufficent for hay drying.

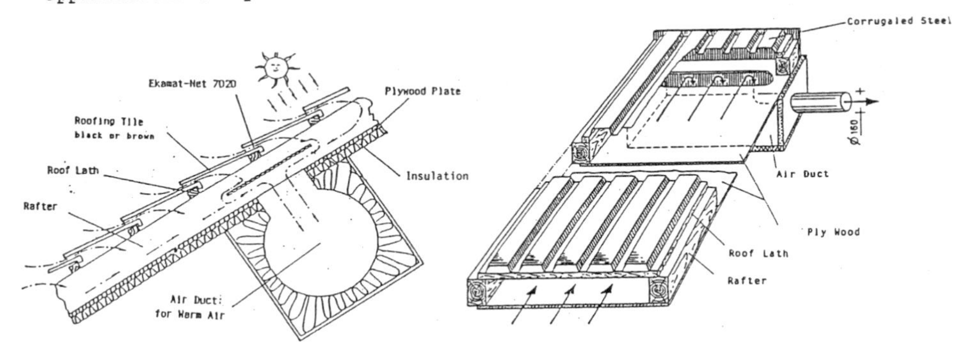

Fig.1 Roof-tile-collector

Fig.2 Collector with corrugated steel absorber

4.2 Single and Double Glazed Collectors

Most of the collectors we designed for higher temperature applications (T > 50K) were covered with transparent plastic sheets. Due to the limited temperature resistance of these materials (acrylic glass, polycarbonate or polyester) we use double glazing. The inner ones made of a temperature resistant plastic foil. If overheating is prevented all these types could be built with single glazings. As an absorber we have used corrugated metal sheets black painted or with selective surface, a copper foil with selective surface and a black painted wood-wool-plate. Air flow is usually below the absorber to prevent dust sedimentation on the absorber surface. A wood-wool-plate or a black polypropylen net is used as a permeable absorber with air flow from underneath to the front. As all these collectors are mainly designed for drying application, ambient air is sucked through the collector and thus leakage is a negligible problem.

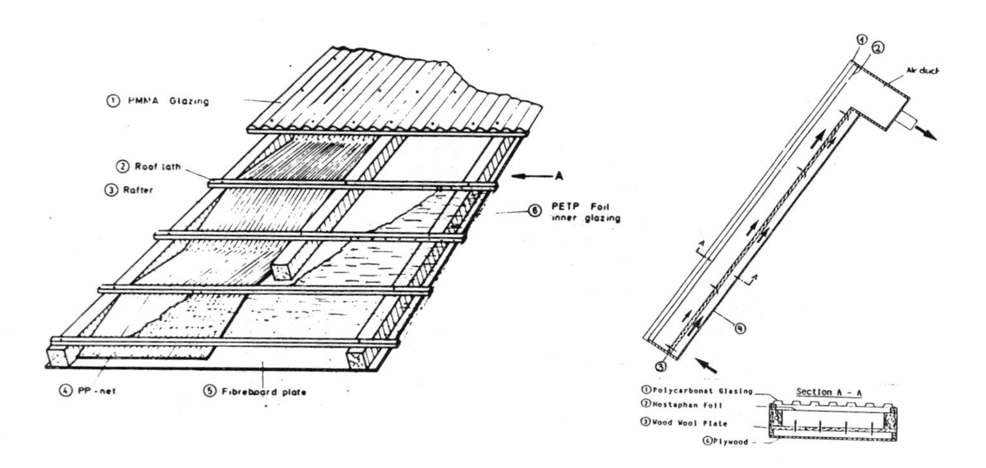

Fig.3 Air collector with permeable plastic net absorber

Fig.4 Air collector with wood-wool-plate absorber

All these collectors are constructed to be built by the user himself or by lokal workshops with lokal available materials. The material costs for unglazed collectors are about 10 - 15 US $\$/m^2$, for single glazed 25 - 35 US $\$/m^2$ and for double glazed ones 35 - 40 US $\$/m^2$. Commercially available solar air heating collectors usually cost more than 100 US $\$/m^2$.

5. PERFORMANCE MEASUREMENTS

For testing of air heating collectors an outdoor test facility for 10 collectors (parallel testing) was built. All collectors were integrated in a roof structure with a slope of 35° facing south. At all collectors the inlet- and outlet-temperature was measured with high accuracy Pt 100 sensors. The flowrate was measured either with small anemometers or orifice plates mounted in measuring tubes. Beside these, meteorological data - global radiation in collector plane, diffuse radiation, ambient temperature, air humidity, wind speed and direction - were registered every 30 sec. and recorded on magnetic tape for further evaluation at the computer centre of the university.

The collectors were measured under quasi-steady-state conditions, which are similiar to steady-state except that inlet temperature is equal ambient. The evaluation was made according to the following equation:

$$n = n0 - UL * T^*$$
$$T^* = (To - Ta) / 2G$$

The optical conversion factor n0, the overall heat loss coefficient UL and the mean daily efficiency of a selection of collectors are listed in table 1.

Fig. 5 View of the air collector test installation

Collector	n0	UL(W/m²K)	n
Unglazed:			
Roof-tile-collector	0.90	48.0	0.34
Corrugated steel collector	0.78	64.0	0.30
Single glazed:			
Alu absorber black	0.76	18.7	0.51
Steel absorber select.	0.76	17.3	0.49
Double glazed:			
Black wood-wool-absorber	0.76	19.2	0.52

Table 1. Performance measurement results

6. MATERIAL TESTING

Some of the used collector glazings were examined for their stability. According to DIN 53455 the tensile strength of weathered and unweathered specimens was measured. The results of a PVC and GUP sample is shown in table 2. The degradation after 3 years weathering is significant for PVC and highly significant for GUP.

Material	Condition	Tensile Strength (N/mm²)	Stand.Dev.
PVC	unweathered	67.2	1.7
	weathered	64.6	2.1
GUP	unweathered	70.2	8.2
	weathered	45.9	9.8

Table 2. Mechanical properties

Beside this the optical properties of different materials were tested. Fig. 6 and fig. 7 show the spectral transmission curves for GUP, PVC and acrylic glass.

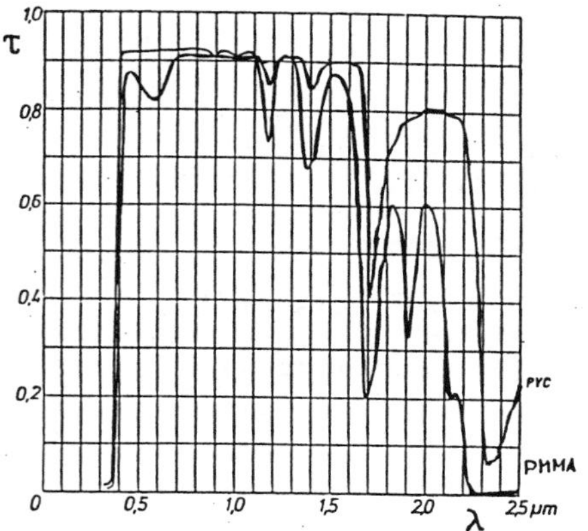

Fig.6 Spectral transmission of GUP
1 new sample
2 3 years weathering in the collector
3 sample from coll. inside (only temp. influence)

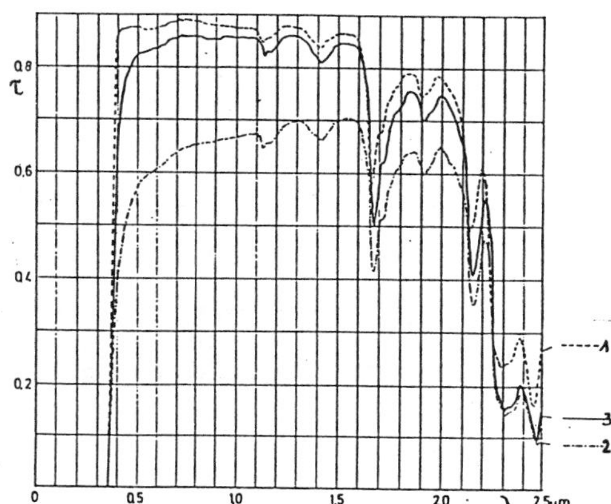

Fig.7 Spectral transmission (PVC, PMMA)

7. CONCLUSIONS

The application of solar air heating collectors for drying of agricultural products is realized in simple and cheap systems. Do-it-yourself has proved to be practical for reduction of costs and improving of performance. Landtechnik Weihenstephan designs could be built by the user himself with local available materials at costs less than 40 US $/m². Although the collectors are rather simple, the performance measurements show encouraging results. The material tests have shown that GUP and PVC are not recommendable for longterm applications, these materials have to be replaced after 4 - 6 years. Polycarbonate and PMMA are more expensive but also more durable. The large number (more than 100) of practical systems with collectors built using the construction plans of Landtechnik Weihenstephan is really encouraging.

8. REFERENCES

M. Reuss, T. Schmalschlaeger, H. Schulz (1986). Ueberpruefung von Einfachluftkollektoren, Final report of a BMFT research project, Freising-Weihenstephan, F.R.Germany

M. Reuss (1986). Simple Solar Collectors for Agricultural Applications, CNRE (FAO) Workshop on "Solar Heating of Animal Houses", Lund, Sweden

J.E. Moon, W.B. Gillet (1981). Draft Recommendations Testing of Air Heating Solar Collectors, University College Cardiff, U.K.

HIGH EFFICIENCY SOLAR FLAT PLATE COLLECTORS WITH CAPILLARY STRUCTURES

R.Digel, N.Fisch, E.Hahne

Institut für Thermodynamik und Wärmetechnik
Universität Stuttgart
Pfaffenwaldring 6, FRG-7000 Stuttgart-80

ABSTRACT

High efficiency flat plate solar collectors with an inner translucent capillary tube structure were tested. The air passes through the capillary tubes and in this way these act as heat exchangers. Modules with glass capillary tubes were examined for their thermal behaviour. The investigation includes the determination of the efficiency curve according to the ASHRAE Standards 93-77 (1978) and the determination of the thermal efficiency, the time constant and the pressure drop as a function of the mass flow. Provided that the inlet temperature is not much higher than the ambient temperature, the collector has a very good thermal efficiency.

KEYWORDS

High efficiency flat plate solar collector, air collector, translucent insulation, capillary structure, glass capillary tubes, collector testing.

INTRODUCTION

Fig. 1 shows the cross section of the tested collector. The field of the translucent capillary tubes absorbs the thermal radiation and is heated up. For the passing air the capillary tubes act as a heat exchanger. Due to the arrangement of inlet and outlet flow there is a decoupling between the "cold" front side and the "hot" rear side of the collector as shown in Fig. 1. In 1982, the first test collector with an inner polycarbonate capillary structure was investigated. This collector however was damaged under non-flow conditions. In 1986 a new module with a capillary structure made of glass was built and tested. In this paper only the results of the collector with the glass capillary structure are presented.

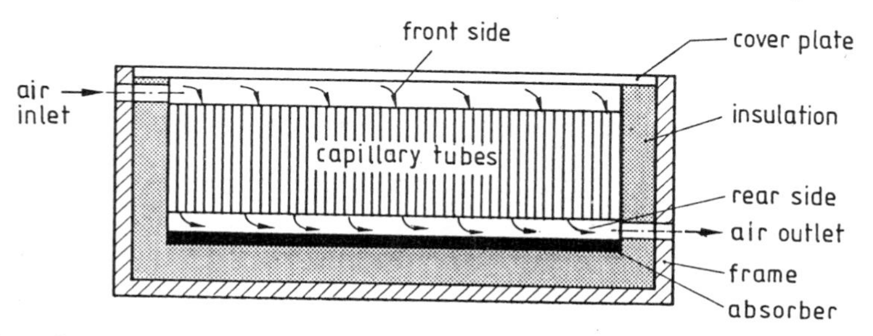

Fig. 1. Cross section of the collector

SPECIFICATION OF THE COLLECTOR

Since a commercially made field of glass capillary tubes was not available at the time, a small test module was constructed by hand with an aperture area of 0.105 m^2. Commercially available glass capillary tubes with a length of 180 mm, a diameter of about 1.7 mm and a ratio of wall thickness to diameter of about 1:28 were used. For the cover plate, a customary glazing of 4 mm thickness and for the absorber, a black coloured aluminum plate were used. The transmittance of the glazing as well as the absorptance of the absorber were not determined. The frame of the collector was made out of plywood. The side walls and the back side are insulated with 200 mm and 100 mm glass wool layers, respectively.

INVESTIGATION OF THE THERMAL PERFORMANCE OF THE COLLECTOR

Outline of the test stand

The collector was investigated on a biaxial resetting test stand. Fig. 2 shows the flow diagram of the open test loop. The collector was tested in compression operation, that means with a small over-pressure. Because of the low air flow, rotameters were used as flowmeters. The collector inlet and outlet temperatures were measured with six thermocouples respectively; these were distributed over the cross section.

Results of the tests

The collector was tested according to the guidelines of ASHRAE Standard 93-77 (1978). The thermal efficiency can be determined using the equation

$$\eta = F_R(\tau_g \cdot \alpha_a) - F_R \cdot U \frac{\vartheta_i - \vartheta_{amb}}{E_{glob}}$$

Within the investigated range of inlet temperatures, the heat transfer loss coefficient U and the heat removal factor F_R are approximately constant, so that the measuring points can be approximated by a line. Fig. 3 shows the thermal efficiency curve of the tested collector. Due to the construction of the collector, the heat losses are much more dependent on the inlet temperature than on the outlet temperature. Therefore the slope of the curve is rather steep.

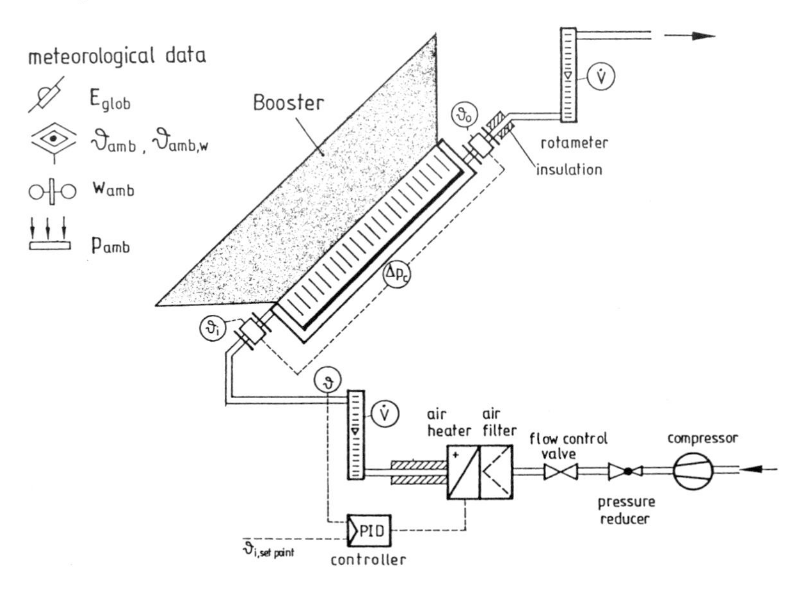

Fig. 2. Flow diagram of the open test loop

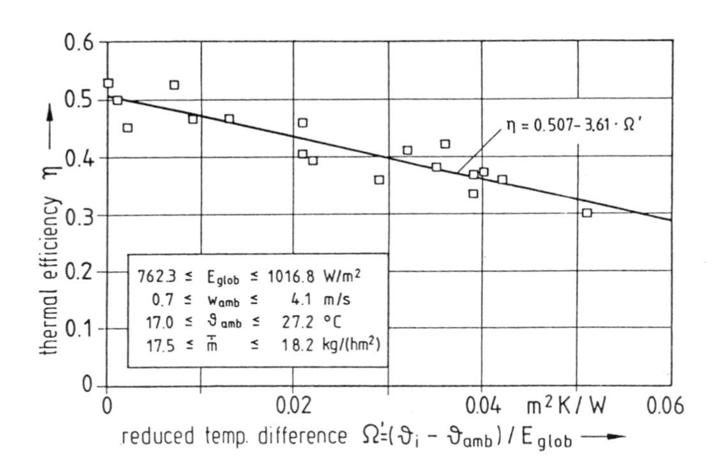

Fig. 3. Thermal efficiency curve according to ASHRAE 93-77

Thermal efficiency as a function of the mass flow

As the collector works well at low inlet temperatures, the efficiency was determined for different mass flows with inlet temperatures near to the ambient temperature (Fig. 4). Another curve gives the increase in outlet temperature at constant inlet temperature and global radiation of 1000 W/m^2 when the mass flow decreases.

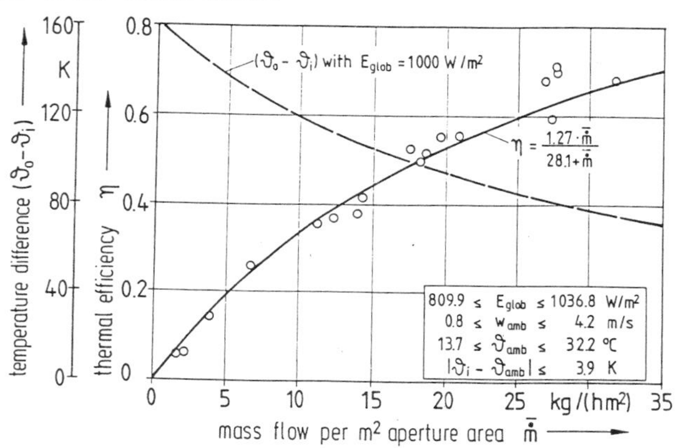

Fig. 4. Efficiency as a function of the mass flow

To reach even higher temperatures, a simple booster was built in the form of a truncated pyramid with a base area of 0.45 m^2 and a height of 300 mm. The top area is the same as the aperture area (0.105 m^2). To be able to compare the results of operation with and without the booster, the mass flow is related to the aperture area, but the calculation of the efficiency is based on the radiation to the base area. Fig. 5 shows the efficiency and the temperature increase in the collector.

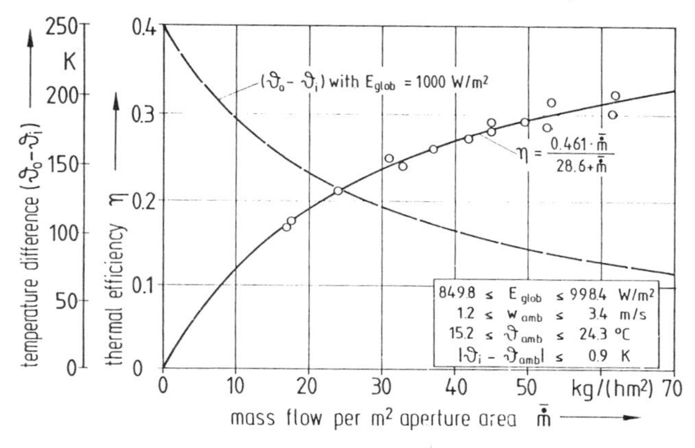

Fig. 5. Efficiency as a function of the mass flow with add on booster

Time constant and pressure drop as a function of the mass flow

The time constant is the governing quantity of the dynamic behaviour of a collector and the pressure drop is a parameter for the blower energy. Fig. 6. shows both the time constant and the pressure drop as a function of the mass flow. The time constant of the collector is rather high, but it seems that shorter capillary tubes are sufficient and with industrial fabrication, it will be possible to reduce the wall thickness of the capillary tubes, so that the thermal mass will be smaller. Because of the low flow velocity inside the capillary tubes, the pressure drop of the collector is low.

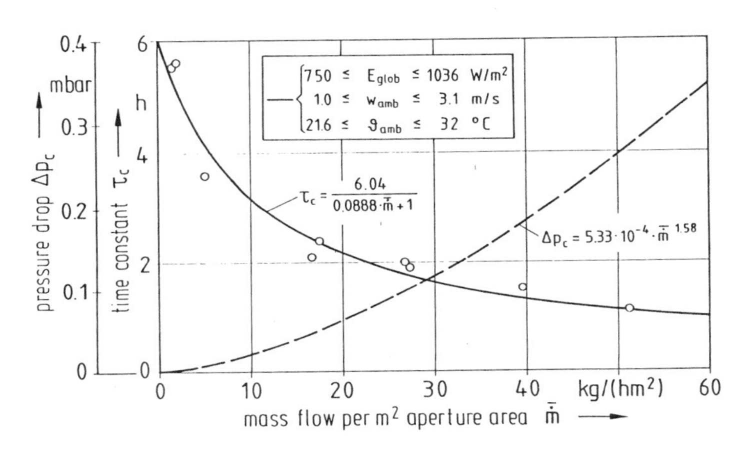

Fig. 6. Time constant and pressure drop as a function of the mass flow

NOMENCLATURE

E_{glob}	global radiation
F_R	solar collector heat removal factor
U	solar collector heat loss transfer coefficient
w_{amb}	ambient wind velocity
p_{amb}	ambient air pressure
$(\tau_g \cdot \alpha_a)$	effective transmittance-absorbtance product
ϑ_{amb}	ambient temperature
$\vartheta_i / \vartheta_o$	collector inlet/outlet temperature

ACKNOWLEDGEMENTS

The investigations have been financed by the Federal Ministry of Technology under 03E-8021-A
The authors greatfully acknowledge this support.

REFERENCES

ASHRAE Standards 93-77 (1978). Method of Testing the Thermal Performance of Solar Collectors, ASHRAE, New York

PERFORMANCE STUDY OF A NEW SOLAR AIR HEATER

M. S. Abdel-Salam[+], I. A. Sakr[++] and W. I. Habashy[+]
+ Mech. Power Dept., Faculty of Eng., Cairo University.
++ Solar Energy Laboratory, N.R.C., Giza, Egypt.

ABSTRACT

Two solar air heaters have been constructed and tested. The first heater is an ordinary (or simple) one, while the second heater is a modified one to increase the heat transfer rates by increasing the turbulence in the stream. This "supper-turbulence" improves the collector thermal performance, and increases the pressure drop along the air passages.

Different experiments were carried out with different air discharges and different tilt angles. The experiments were extended along different seasons. The thermal performance and pressure drops in the two heaters were compared.

KEYWORDS

Solar energy; air heater; super-turbulence; tilt angle; overall efficiency; pressure drop.

NOMENCLATURE

Ap	absorber plate area = 1.746 m^2
C_p	specific heat of air = 1012 j/kg. ^oC
M_a	air flow rate, kg/sec.
m_a	specific air flow rate, kg/(hr.m^2).
(ΔP)	pressure drop across heater, mm H_2O
Q_u	useful energy gain, W
$Q_{u,i}$	ideal energy gain, W
S	Total solar energy impinging upon the collector surface.
t_i	inlet air temperature, ^oC
t_o	outlet air temperature, ^oC.
η_o	collector overall efficiency.

INTRODUCTION

Solar air heating is an old idea. Several types of solar air heaters have been constructed in different ways and several countries. Solar-heated air could be used effectively for drying under controlled conditions.

Most workers, concerned with solar dehydration, concentrated their work on the construction of solar air heaters (e.g. see references at the end of the paper). The collector plates ought to be suitably shaped or profiled to increase both surface area and air turbulence.

The present work is an extension to the work of Abdel-Salam and coworkers (ref. I) who constructed a solar air heater with baffles inside the air passages. These baffles increase both the air-flow turbulence and the pressure drop.

This work introduces a comparison between two types of solar air heaters; an ordinary air heater without baffles, and the new one with baffles which cause super-turbulence in the air stream. The two heaters are tested for a large range of air flow rates, different tilt angles, and a long period of time. The thermal performance as well as the pressure drop are studied for the two heaters.

DESCRIPTION OF APPARATUS

The main parts of the apparatus are two solar collectors fabricated from iron sheets 1x2 meters and 0.8 mm thickness. Fig. 1 shows the configuration of the first (ordinary) air heater, while Fig. 2 shows that of the second (modified) one. An ordinary window glass was used as cover, and polyurethane foam was used as an insulation at the back and both sides of each heater. A detailed local section at the heater sides is shown in Fig. 3. Fig. 4 shows the assembly of both heaters with the fan. All joints between the fan and the system were connected by flexible rubber joints in order to avoid the effect of vibration. Three Tee sections were used to control the air flow rate out of the blower and through each heater.

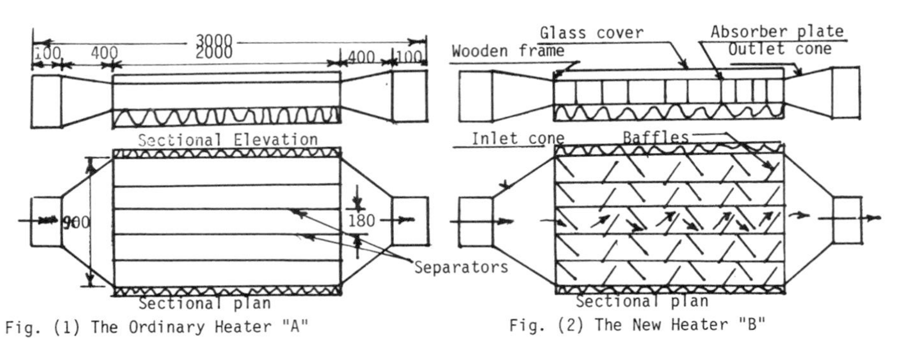

Fig. (1) The Ordinary Heater "A" Fig. (2) The New Heater "B"

The Two Solar Air Heaters Configurations (not to scale) (Dims. in mm.)

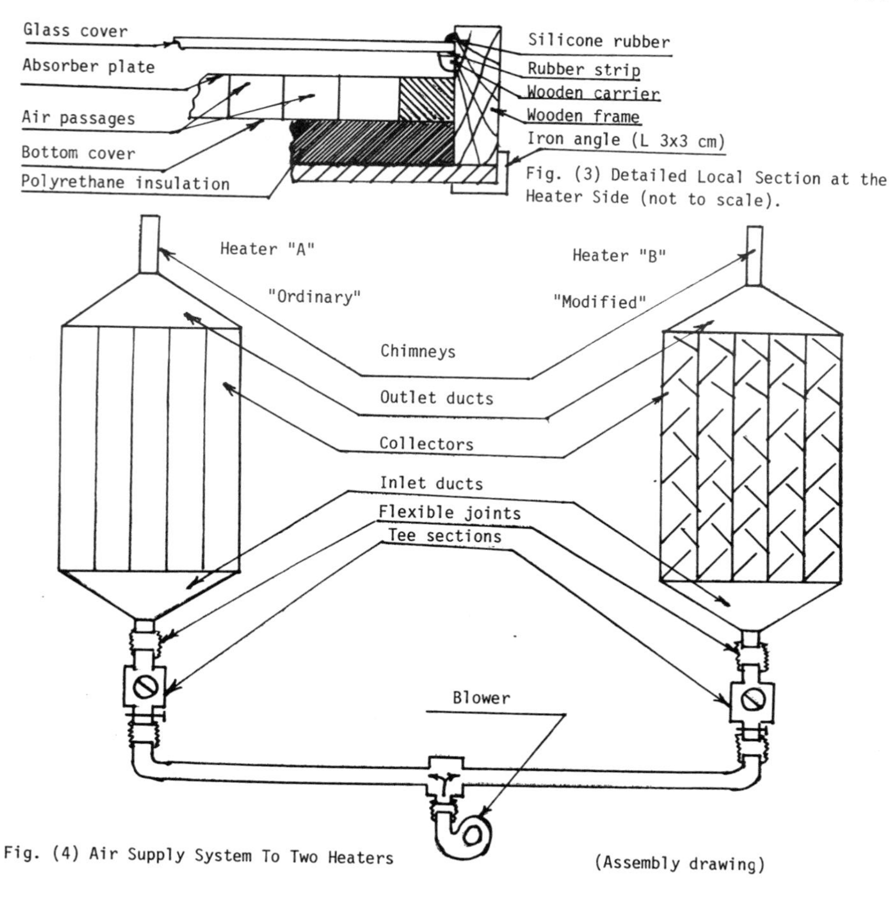

Fig. (3) Detailed Local Section at the Heater Side (not to scale).

Fig. (4) Air Supply System To Two Heaters (Assembly drawing)

METHOD OF CALCULATION

Many quantities had to be calculated in the present work. The following is the method of calculation of the most important quantities.

Useful Energy Gain:

The useful energy gain is the heat added per unit time to the air during the heating process. It can be calculated as follows:

$$Q_u = M_a \, C_p \, (t_o - t_i), \qquad W \qquad (1)$$

Collector Overall Efficiency:

This was calculated as follows:

$$\eta_o = \frac{Q_u}{A_p \, . \, S} \qquad (2)$$

RESULTS AND DISCUSSION

(a) Effect of Air-Flow Rate:

A satisfactory set of data was obtained during the different groups of testing. Clear and non-meteorologically interrupted days were chosen to express the heaters performance. The relation between the overall efficiency of the collector and the air-flow rate is shown in Fig. 5. The overall efficiency increases as the air-flow rate increases. This result is natural, as the temperature level and heat losses are reduced with larger air discharges. It is noticed also, from Fig. 5, that the efficiency of the modified heater (heater B) is always higher than that of the ordinary heater (heater A). The increase depends on the season and the rate of discharge. Thie improvement in efficiency is due to the increasing turbulence in heater B.

In a trial to find out a correlation between the collector efficiency and the air mass flow rate, the following form was found suitable.

$$\eta_o = K.m_a^n \quad \%$$

where:

m_a = air mass flow rate, $kg/hr.m^2$(of absorber surface area).
K & n are constants, found experimentally.

The constants K & n depend on different variables such as the collector design, season and the time of the day. A suitable value of n was found 0.5. The value of K was found for the different seasons for both heaters at noon. It has the same

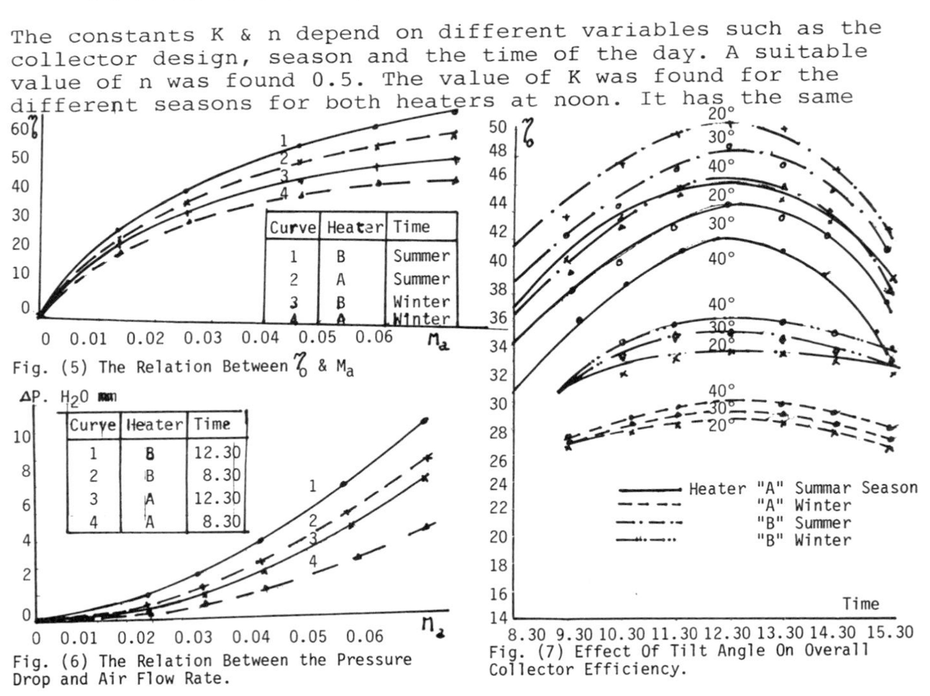

Fig. (5) The Relation Between η_o & M_a

Fig. (6) The Relation Between the Pressure Drop and Air Flow Rate.

Fig. (7) Effect Of Tilt Angle On Overall Collector Efficiency.

value for spring and autumn, but different values at other seasons. Table (1), gives the values of the constant K for both heaters and for different seasons.

Table (1) Values of the constant "K" for both heaters for different seasons.

Season \ Heater	A	B
Summer	0.0535	0.058
Spring & Autumn	0.051	0.0572
Winter	0.042	0.0462

Fig. 6 shows the pressure drop as a function of the air flow rate for the two heaters. The pressure drop increases very much as the air discharge increases. It is noticed also that the baffles in heater B cause an increase in pressure drop ranging between about 60 and 80%. This increase results in an increase of the energy required to drive the air blower.

A correlation between the pressure drop across the heater and the air mass flow rate could be found to take the form

$$(\Delta P) = Z \quad m_a^{n'} \qquad mm\ H_2O$$

where Z & n' are constants depending upon the season, time of experiment and the heater design. The value of n' is always about 2. The value of (Z) at noon was 0.00065 for heater A, and 0.0011 for heater B. The values of m_a are taken in kg/(hr. m^2). In this way it becomes easy to expect the pressure drop when the air-flow rate is known.

(b) Effect of Tilt Angle:

The effect of the collector tilt angle on the overall efficiency is shown in Fig. 7. It is noticed that the optimum tilt angle depends on the season. In summer the best angle is 20°, while the worst one is 40°. In winter the conditions change. The best angle is 40°, while the 20° gives the lowest efficiency. Heater B is more efficient than heater A in all seasons and at all tilt angles. An arrangement of tilt angles is shown in Table 2 to give the tilt angles in the order of decreasing efficiency for the different months of the year.

Table (2): Arrangement of the tilt angle in order of decreasing efficiencies.

Month		Tilt angle	Month		Tilt angle
1	Jan.	40.30.20	7	July	20.30.40
2	Feb.	40.30.20	8	Aug.	20.30.40
3	Mar.	30.40.20	9	Sep.	30.20.40
4	Apr.	30.40.20	10	Oct.	30.20.40
5	May	30.20.40	11	Nov.	30.40.20
6	June	20.30.40	12	Dec.	40.30.20

REFERENCES

(1) Abdel-Salam, M.S. and coworkers, Proc. First Ar. Int. Solar Eng. Conf., Kuwait, Dec. 1983, pp. 145-149.

(2) Duffie, J.A. and Beckman, W.A. (1974). Solar Energy Thermal Process. John Wiley and Sons, New York.

TWO AND THREE DIMENSIONAL ASPECTS IN THE THERMAL BEHAVIOUR OF SOLAR COLLECTORS

A. Oliva, M. Costa, C.D. Pérez Segarra

Departament de Termoenergètica UPC, Secció Termotècnia.
C/Colón 9, 08222 Terrassa, Barcelona (Spain).

ABSTRACT

This work analyses the thermal behaviour of a flat-plate collector, taking into account the non-uniformity of phenomena and of the boundary conditions which characterise it. The study has been done specifically with an air heater with ducts of rectangular section. It has been possible to quantify, using a numerical simulation by finite elements, the influence on the thermal behaviour of such aspects as: flow non-uniformity distribution, the three dimensional nature of the heat transfer in the insulation, the existence of areas of shadow, etc.

KEYWORDS

Flat-plate collector, Air solar collector, Two and three dimensional numerical simulation, Collector efficiency.

INTRODUCTION. CHARACTERISTICS OF THE SOFTWARE USED.

The thermal behaviour of a flat-plate collector is difficult to analyse due to the large number of parameters to be taken into consideration. Therefore simplified models (Duffie and Beckman, 1980) have had to be used which, for the degree of precision sometimes required, are suitable. Nevertheless it is obvious that any improvement in the thermal modelization of a collector will have an effect on its design and on the estimation of its thermal behaviour. The results of a two and three dimensional modelization are presented in this paper. These results cannot be predicted by means of more simplified models.

The predictions have been made on the basis of numerical simulations of the different components (Oliva, Costa, Pérez and Cendra, 1987). As general characteristics we can enumerate:
-In the insulation the discretization is three-dimensional and by finite differences. Criteria of the central difference and an overrelaxation factor are used.

-In the absorber plate and the ducts the discretization is in finite elements, resulting from the subdivision of each one of the components. An overrelaxation factor is used.
-In the fluid the discretization is in the flow direction. Upwind criteria and overrelaxation are used. The evaluation of the heat exchange by forced convection has been calculated using empirical expressions.
-The thermal radiation between the finite elements, in which the exchange surfaces are subdivided, is done by the net-radiation method.
-In the cover the discretization is two-dimensional. An underrelaxation factor is needed.

The solution of the governing equations is by the line by line method (Patankar, 1980).

RESULTS

The tables and the figures shown below illustrate the modelization possibilities of the software used. Likewise they give information about the incidence of different aspects of non-homogeneous surroundings in collector performance, which produce in it a three dimensional behaviour. They also quantify the influence of the geometry, the properties of the materials and the empirical expressions used.

The collector used in this study (Fig. 1) corresponds to the following data:

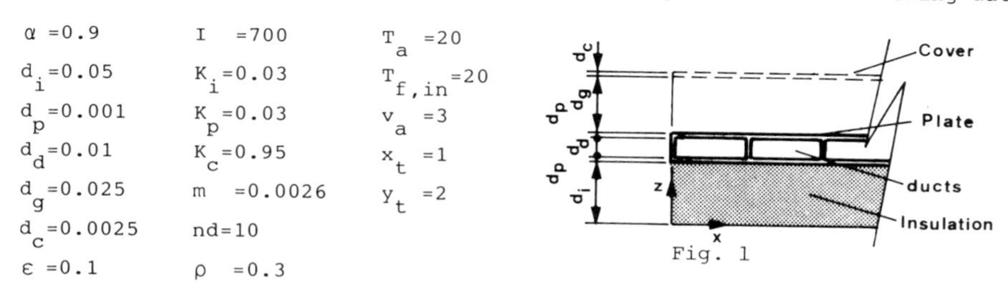

α =0.9	I =700	T_a =20
d_i=0.05	K_i=0.03	$T_{f,in}$=20
d_p=0.001	K_p=0.03	v_a =3
d_d=0.01	K_c=0.95	x_t =1
d_g=0.025	m =0.0026	y_t =2
d_c=0.0025	nd=10	
ε =0.1	ρ =0.3	

Fig. 1

In refering to different cases all parameters except the ones being varied remain constant.

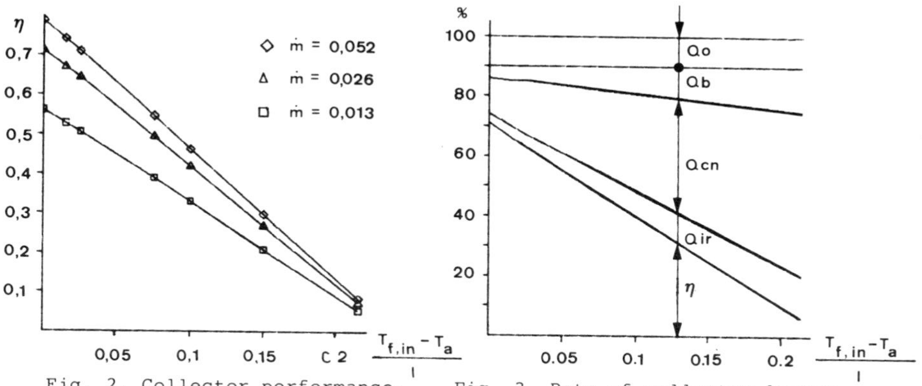

Fig. 2. Collector performance. Fig. 3. Rate of collector losses.

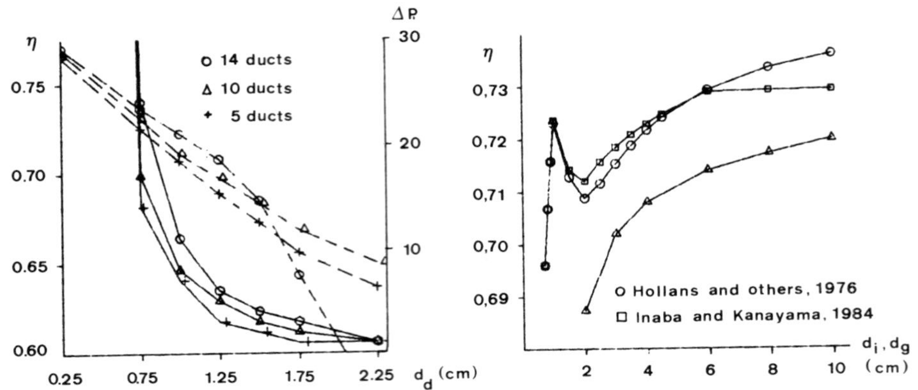

Fig. 4. Efficiency and pressure drop versus height ducts.
— Pressure drop.
-- Efficiency.

Fig. 5. Collector performance versus insulation thickness Δ and versus air gap height using the expressions of two studies.

TABLE 1 Variation of Some Characteristic Parameters of the Collector for Different Grid Densities.

$N_c \times M_c$	3x3	3x3	3x3	3x3	3x3
$N_p \times M_p$	21x21	21x21	21x21	21x21	31x31
M_f	21	21	21	41	31
Z_d	2	2	2	2	2
$N_i \times M_i \times Z_i$	21x21x14	11x11x14	11x11x7	21x21x14	31x31x21
η(%)	71.15	68.43	71.05	71.34	71.18
Qu	1326.54	1275.82	1324.61	1330.09	1327.12
Ql	536.55	525.84	537.18	537.14	535.74
Iterations	12	16	15	11	14
I-Qu-Ql	1.14	62.57	2.43	2.99	1.37

TABLE 2 Variation of Some Characteristic Parameters of the Collector for Different Thermal Conductivities of the Ducts and the Insulation.

	K_p			K_i	
	58	204	384	0.02	0.03
η(%)	70.50	71.59	71.17	72.59	69.80
$T_{f,out}$	70.06	70.54	70.52	71.54	69.56
T_p	69.34	66.75	66.08	67.07	65.15
T_c	30.85	30.02	29.80	30.15	29.59

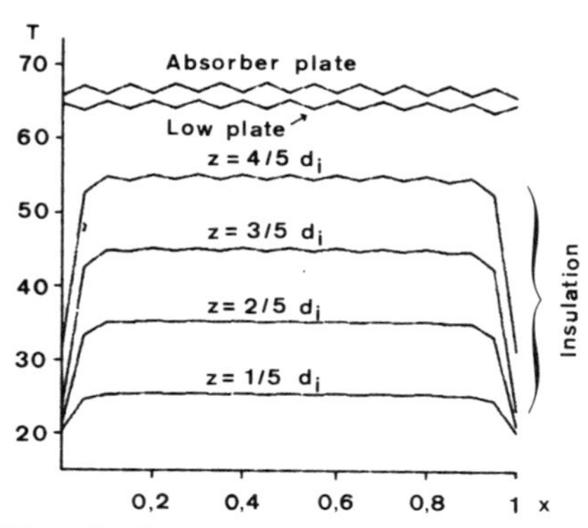

Fig. 6. Temperature profile in the middle of a cross section of the collector.

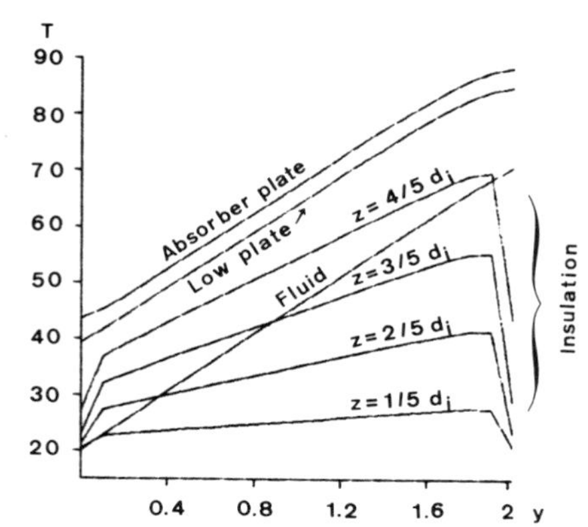

Fig. 7. Temperature profile in the middle of a longitudinal section of the collector.

TABLE 3 Some Characteristic Parameters of the Collector when Some Ducts are Plugged.

	plugged up ducts			
	1,2nd	5,6th	1,2,3rd	5,6,7th
η(%)	68.70	70.30	65.96	69.42
Qu	1218.80	1301.10	1229.00	1294.20
Qcn	1261.90	213.20	298.40	257
Qo	180.80	180.80	180.80	180.80
Qir	65.50	59.60	79.58	64.04
Qb	79.50	71.03	90.80	74.30
			case A	case B

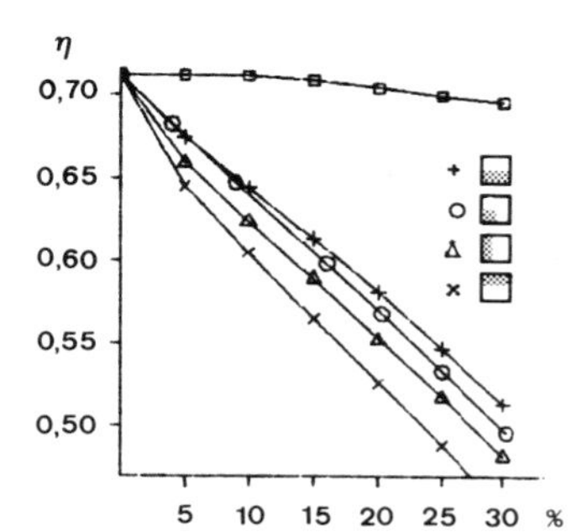

Fig. 8. Collector efficiency versus : □ % lineal variation of m; ×, + , Δ , o % of shadow in the upper part, lower part, one side and lower corner.

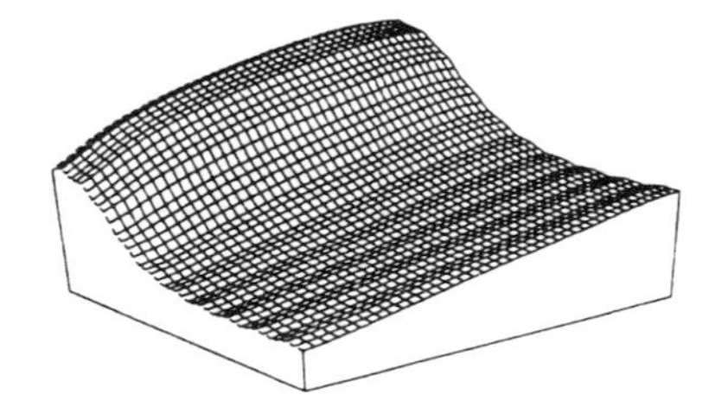

Fig. 9. Absorber plate temperature in case A (Table 3).

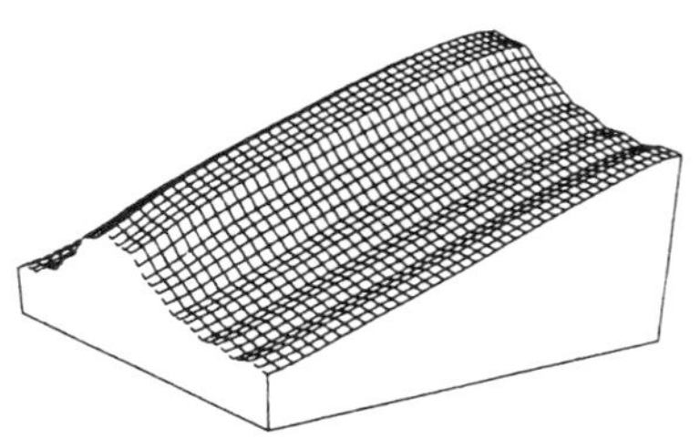

Fig. 10. Absorber plate temperature in case B (Table 3).

CONCLUSIONS

The results presented show the importance in the collector performance of such aspects as: shadows (Fig. 8) and non-uniformity flow distribution (Fig. 8 and Table 3). For the type of collector chosen, the figures indicate the influence on its efficiency of geometry (Figs. 4 and 5) and the physical properties required of the materials used (Table 2). On the other hand the calculating capacity required for the discretization density is within the range of medium-size computers (Table 1). It is obvious that the kind of study carried out in this paper can be applied to most collectors. The study has been done in a steady-state way, a development to transient-state would be recommended.

NOMENCLATURE

α	Solar absorptance.
d	Thickness.
ε	Absborber emittance.
I	Solar energy per unit area.
K	Thermal conductivity.
m	Air mass flow rate.
N,M,Z	Grid density.
η	Collector efficiency.
nd	Ducts number.
P	Pressure drop.
Qb	Back insulation losses.
Qcn	Conv. loss from abs. plate
Qir	Infra. loss from abs. plate.
Ql	Total losses.
Qo	Optical losses.
Qu	Useful energy.
ρ	Ground reflectivity.
T	Temperature (ºC).
V	Wind speed.

Subscripts

a	Ambient.
c	Cover.
d	Ducts.
f	Fluid.
g	Air gap.
i	Insulation.
in	Inlet.
out	Outlet.
p	Plate.
t	Total.

Superscripts

-	Mean value.

Except when it says to the contrary S.I. units have been used.

REFERENCES

Duffie, J.A. and W.A. Beckman (1980). Solar Engineering of Thermal Processes. John Wiley & Sons, New York.

Hollands, K.G.T., T.E. Unny, G.D. Raithby and L. Konicek (1976). Free convection heat transfer across inclined air layers. Journal of heat transfer, 98, 89-193.

Inaba, H. and K. Kanayama (1984). Natural convective heat transfer in an inclined rectangular cavity. JSME, 27, 1702-1708.

Oliva, A., M. Costa, C.D. Pérez and J. Cendra (1987). Modelización bidimensional y tridimensional en colectores solares planos. I Congreso Iberoamericano De Energía Solar, Madrid, Vol. I, 189-206.

Patankar, S.V. (1980). Numerical Heat Transfer and Fluid Flow. McGraw-Hill, Washington.

This work was supported by Comisión Asesora de Investigación Científica y Técnica, Project No. 1937.

EXPERIMENTAL RESULTS OF A GREENHOUSE HEATED BY SOLAR ACTIVE SYSTEM COMPOSED OF AIR COLLECTORS, ROCK BED STORAGE AND HEAT PUMPS.

S. Camporeale*, F. Pacciaroni*, G. Picciurro*. M. Romanazzo**, G. Mennini**, E. Marinelli**

*Biological and Agricultural technologies Division, TECAB Dept., ENEA, CRE Casaccia, P.O.Box 2400, 00100 Rome A.D., Italy

**Instrumentation Laboratory, IMPRO Division, FARE Dept., ENEA, CRE Casaccia, P.O. Box 2400, 00100 Rome A.D., Italy

ABSTRACT

As a part of a project for the use of renewable energy in agriculture, a greenhouse for floriculture production, covering a surface of 537 m2, has been built at the ENEA - CRE Casaccia in Rome.

The greenhouse is heated by an active solar system composed of air solar collectors, a rock bed storage and two heat pumps.

Double layer polymetylmetacrilate is used as covering material and thermal screens made of aluminium and polyester are employed for light control and energy saving.

Climatic, meteorological and process data are collected by an automatic data acquisition system for energetic behavior evaluations.

The solar system has provided satisfactory greenhouse temperatures during the winter and good levels of relative humidity have been recorded.

The quality of the plant production was excellent in shape and foliage.

Further work is required to perform better control of the solar system by mean of low cost microprocessors and to improve heat exchanges in the store and the use of heat pumps.

KEYWORDS

Solar greenhouse, air solar collectors, rock bed storage, active solar system, auxiliary heating heat pumps.

INTRODUCTION

In the last years ENEA (National Committee for Nuclear Research and Alternative Sources) has started a project in order to study new typologies of greenhouse for a more efficient use of alternative sources of energy (solar, geothermal, etc.)

At the CRE Casaccia in Rome was built a greenhouse for floriculture production heated by mean a solar active system composed of air

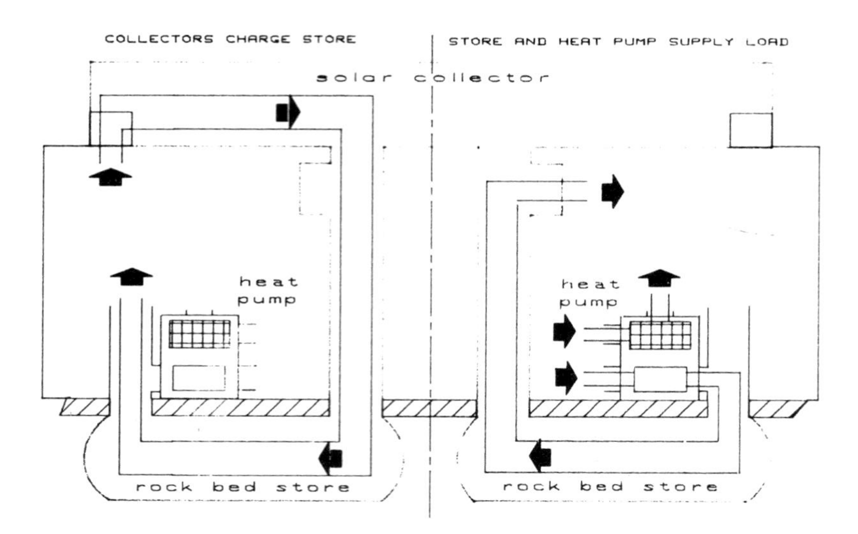

Fig. 1 Air flow circuit diagram of the solar heating system.

collector disposed on the roof and a rock bed store underground. In the greenhouse a normal activity of growing ornamental plants is carried out.

The goals of the experimental activity were two:
- maximizing the efficiency of the various parts of the system;
- verify the availability of microclimatic conditions suitable for commercial production in terms of shape of the plants, time of growing, absence of diseases or desorders.

MATERIALS AND METHODS

Greenhouse characteristics

The greenhouse covers a surface of 537 m2 and is oriented with its long axis in a East-West direction. A modular typology is obtained installing the solar collectors on the roof, so that the cross section is nearly a shed type.

For a high thermal insulation, 16 mm thick double layer polymetylmetacrylate is used as covering. Movable thermal screens made in polyester and aluminium, able to save 40% of energy, are employed; the screens are also used to control internal solar radiation.

The plants are placed on raised benches made of concrete in order to increase the thermal capacity of the grenhouse.

The solar system and its operation

The air solar collector is made of expanded polyuretane and covered by 4mm thick double glass. The total area is 60 m2 and the collectors are

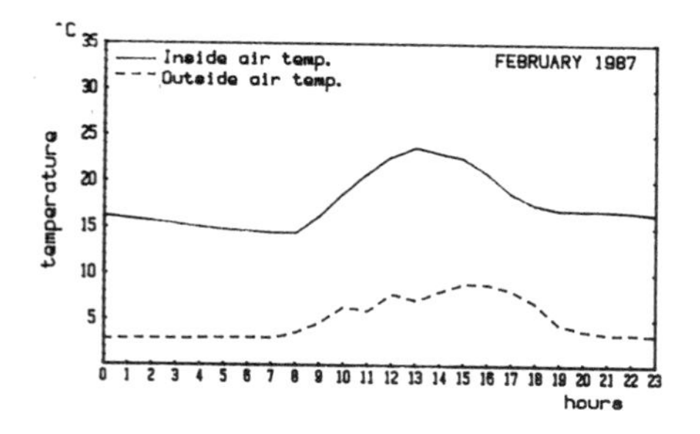

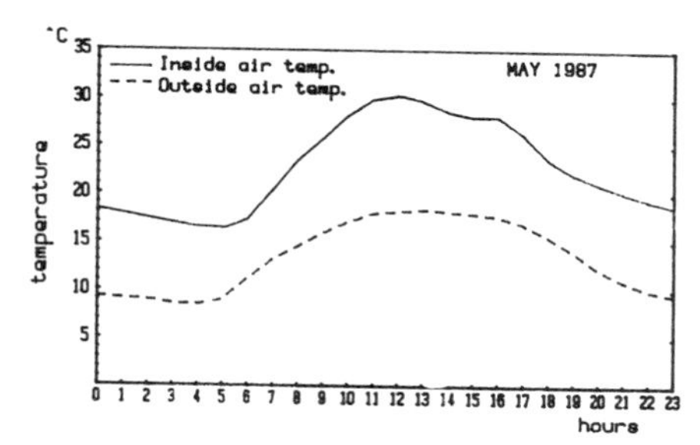

Fig. 2 Hourly mean values of temperature.

disposed in two series of 25 ones; the tilt angle is 60 deg. from horizontal and the orientation is Sud.

The rock bed store has a volume of 60 m3 and is placed underground in two steel tahks.

A schematic view of airflow is shown in fig. 1. The aim of the system is to utilize also the greenhouse as solar collector recovering the extra heat of internal air that should be ventilated to avoid excessive temperature. (Zabeltitz, 1984) During the daytime, two centrifugal blowers force internal air in the solar collectors; and the heated air charges. During the night, when the internal temperature falls below the set point, the internal air is conveyed in the store. An automatic control system based on conventional thermostats is employed.

Two air-air electric heat pumps, with a thermal capacity of 14.5 KW, are connected to the solar system in order to provide auxilary load supply. Greenhouse internal air is heated by the condenser of the heat pump; at the same time, also the evaporator cools the internal air that is blown in the store. In this way the heat pumps extract heat from the store even when the internal air is a little warmer than the store.

Monitoring system

The data monitoring has been finalized to collect data about the energy efficiency of the different subsystems and physiological parameters that affect plant confort. Measurements are executed at time intervals consistent with the dynamic caracteristic of each quantity and time averages are hourly recorded.

Esperiments performed

Three different species of ornamental plants have been grown in the greenhouse: Phylodendron spp., Syngonium spp. and Spatiphillum spp.. The set-points selected were: minimum 15 ^ C for the operation mode of the store alone and 12 ^ C for the heat pumps auxiliary supply; the maximum temperature before ventilation was 28 ^ C.

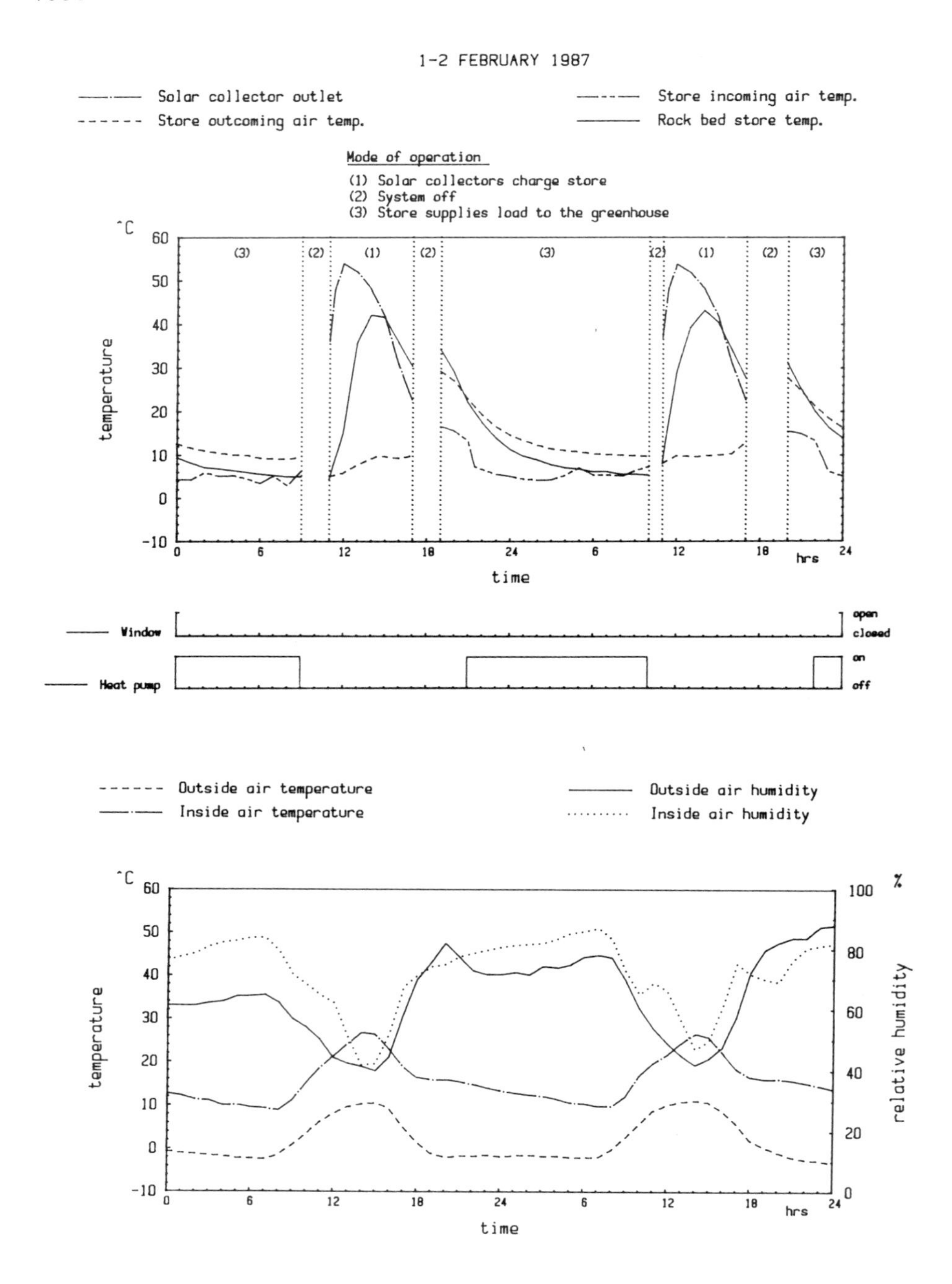

Fig. 3 Performance of the solar heating system in a typical winter day.

RESULTS AND DISCUSSION

Microclimatic results

Good levels of the air temperature and humidity were observed during the whole winter: minimum temperature recorded was about 9 ° C. The hourly mean values of temperature recorded in February and in May (fig. 2) show that temperature was rather comprised in the set points levels. The mean value recorded of the relative humidity in the night was about 80%; this is a good level for ornamental plants and it is an interesting result because, tipically, in solar greenhouses, insulation and poor ventilation adopted to reduce heat losses, induce high humidity and, consequently, plant diseases. The problem didn't appear in the greenhouse under investigation because of two reasons: the first is the optimal ratio between foliage area and the volume of the greenhouse; the second is that the solar system continously moves air in the greenhouse and excessive humidity is condensed at the evaporator of the heat pumps and in the rock bed store.

System performance

In fig. 3 the performances of the solar system are shown for a typical winter day in comparison to internal microclimatic conditions. It appears that thermal capacity of the store is larger then the energetic input coming from the solar system; the store outcoming temperature, raised only up to 10 ° C, points out a high thermal gradient in the rock bed store. Calculations made on load supplyed by the solar system, have shown a low efficiency of the air collector, about 15%, and that the use of thermal screens during the daytime has to be limited. Difficulties appeared in the control of the system performed by the thermostatic equipment so that unnecessary use of heat pumps was observed.

CONCLUSIONS

The solar systems has provide satisfactory levels of temperature and relative humidity during the winter. In order to increase the efficiency of the system and to reduce electric consumption of the heat pumps a more accurate microprocessor control should be necessary. The practical reduction of the efficiency of the solar collectors suggests the use of low cost systems.

REFERENCES

Garzoli, K.V., and G.S.G. Shell (1984). Performance and cost analysis of an australian solar greenhouse. Acta Horticulturae, 148, 723-729.

Picciurro, G., G. Boaga, V. Calderaro and L. Martincigh (1984). Prototype of solar Bioclimatic greenhouse. Acta Hort, 148, 755-769

Zabeltitz, CHR. V., and J. Damrath (1984). Greenhouse heating with sun energy - The greenhouse as collector. Acta Horticulturae, 148, 763-769.

CONVECTIVE SYSTEMS: THE USE OF AIR IN THERMOSYPHONIC FLOW IN SOLAR THERMAL SYSTEMS - A NEW WAY TOWARDS COST-EFFICIENT APPLIANCES -

Michael Grupp*, Bernd Kromer**, Joachim Cieslok***

*Synopsis, Route d'Olmet, F - 34700 Lodève, France
**Institut fuer Umweltphysik d. Universitaet, D - 6900 Heidelberg, W.-Germany
***INCO, Alexanderstr. 69-71, D - 5100 Aachen, W.-Germany

ABSTRACT

The use of air as heat transfer fluid in solar thermal systems has important practical (no problems with leaks, phase transitions, corrosion, absorber conductivity) as well as economical advantages. In the past, the low specific heat and the low heat transfer coefficients caracteristic for air, together with the large sections needed for air ducts, have limited air systems to direct space heating and drying purposes.

Since 1981, research conducted at Synopsis on a new arrangement of absorber and heat exchanger has lifted this restriction and opened the way for applications that had been reserved before to liquid cooled or concentrating designs, such as flat plate collectors for the heating of water, mono-block water heaters, advanced batch-type water heaters, solar cookers and medical sterilizers. The basic physical features of these systems are outlined.

A parallel procedure of thermal simulation and experiment was used to optimize the many different heat transfers involved. Results are presented.

KEYWORDS

Air as heat transfer fluid; air in thermosiphonic flow; two-way air-type collectors; thermal diode; freeze protection; intrinsic selectivity; thermal simulation calculations; flat plate collectors; mono-block water heaters; batch-type water heaters; solar cookers; medical sterilizers.

INTRODUCTION

A qualitative review of the advantages and drawbacks of air as heat transfer fluid in solar thermal energy applications is given in the table below. It can be seen that the advantages are bound to make air-type appliances simple and economical, whereas the drawbacks can result in low efficiency.

TABLE Air as Heat Transfer Fluid

Advantages	Drawbacks
Abundant	Low density
Small leaks uncritical	Low specific heat
No phase changes (freezing, boiling)	Low heat transfer coefficients
Absorber conductivity uncritical	Difficult heat transfer to target
Corrosion uncritical	High top losses in simple designs

Attempts to use air as transfer fluid for indirect applications (e.g. water heating flat plate collectors) are not new: the first patent (Sutter, 1947) is 40 years old; about 15 more patents have been granted later on the subject. To our knowledge, none of these processes has found any lasting application, the main reason being low efficiency of the proposed designs: the air heated by an absorber circulates in direct contact with the glazing which results in high top losses. See Fig. 1.

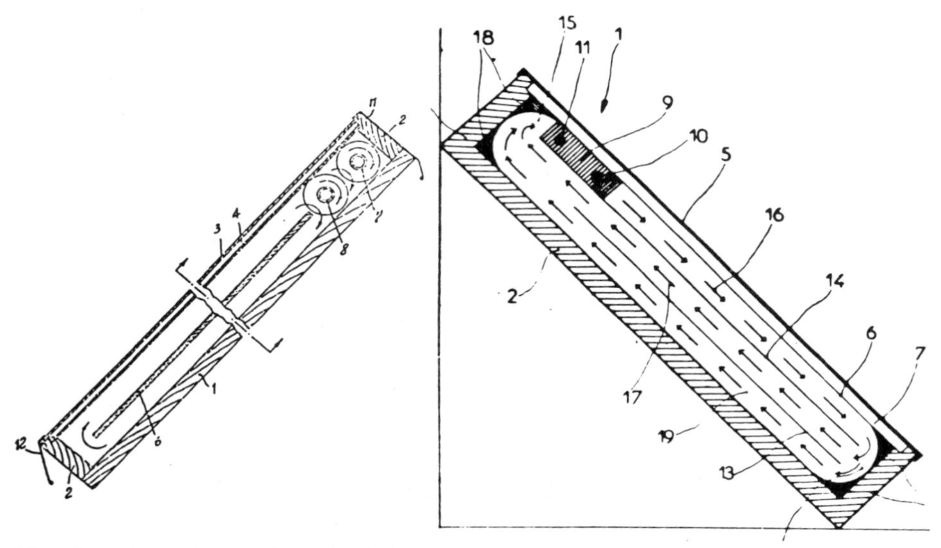

Fig. 1: Comparison of an inefficient and an efficient air-type, water heating collector design. The drawing shown on the left is taken from the patent of Sutter (1947): hot air rises in contact with the glazing, passes through a heat exchanger, cools and returns to the absorber channel through the back of the collector. The drawing on the right is taken from the patent of Grupp (1981, 1982): the hot air rises behind an intermediate glazing, passes through a heat exchanger, cools and returns in contact with the outer glazing.

THE MAIN CHARACTERISTICS OF CONVECTIVE SYSTEMS

The design shown on the right-hand side of Fig. 1 confines the hot air in the back of the collector, behind an intermediate glazing. Heat losses through this intermediate glazing are - at least partly - recovered in the cold-air duct. This set-up has some interesting characteristics:

Thermal efficiency: in-house (Beall and co-workers, 1983) as well as independent tests (Cardi, Gatt, Gschwind, 1984) have shown that the primary circuit does not reduce efficiency noticeably, compared to classical collectors. A pre-series "PU4" collector showed an efficiency at ambient temperature η_o =.76 and a loss coefficient U=6 W/m2K.

Thermal diode: the air flow in the collector stops when solar irradiance stops, which results in very low heat losses (this is important for freeze protection and for average efficiency in changing irradiance conditions). Loss reductions of 75 % were recorded (Cardi, Gatt, Gschwind, 1984). The effect can be further improved by placing the heat exchanger in the top of the hot-air duct.

Intrinsic selectivity: even without a selective absorber, a convective collector shows intrinsic selectivity. Infrared radiation from the absorber is absorbed in the intermediate glazing and - at least partly - recovered in the cold-air duct. Comparative tests on an identical collector with selective (Maxorb) and non-selective absorber (matt-black paint) show (Beall and co-workers, 1983) that the selective absorber improves η_o by 3% (rel.) without changing U significantly. Simulation calculations yielded similar results (see below).

Reduced heat losses at high inlet-outlet temperature differences: since the heat exchanger is placed in a counter-flow position, high inlet-outlet temperature differences mean relatively low air temperatures at the outlet of the heat exchanger, which, in turn, means low heat losses through the glazing. This is important to counterbalance the efficiency losses typical for low coolant flow rates, e.g. in thermosyphonic systems. However, a joint project of the Ecole Nationale Supérieure des Mines and Synopsis (M. Grupp, M. Gschwind, 1987) showed that flow conditions at the collector inlet can significantly alter the efficiency at low flow rates.

DIFFERENT APPLICATIONS

The different applications that convective systems have found so far are shown in Fig. 2 and Fig. 3.

Fig. 2: The different applications of convective systems.
Left: "PU4" convective collectors in a hot water installation.
Right: A prototype of a convective mono-block water heater, combining high system efficiency (60 % at ambient temperature (Cieslok, 1984)) with improved freeze protection.

Fig. 3: The different applications of convective systems.
Left: Solar cooker prototype, family size (see also Grupp and co-workers (1987)).
Right: A solar medical sterilizer in a Sudanese hospital. Water vapor at 121 C, produced by a "PU4" convective collector, is used for sterilization in an autoclave (see also Grupp and co-workers (1987)).

DESIGN OPTIMIZATION BY SIMULATION AND EXPERIMENT

There are many different heat transfers involved in convective systems which corresponds to a great number of free design parameters that have to be optimized (for an example, see Fig. 4). It is evident that optimization by prototype testing alone would be far too slow and not precise enough to solve all problems.

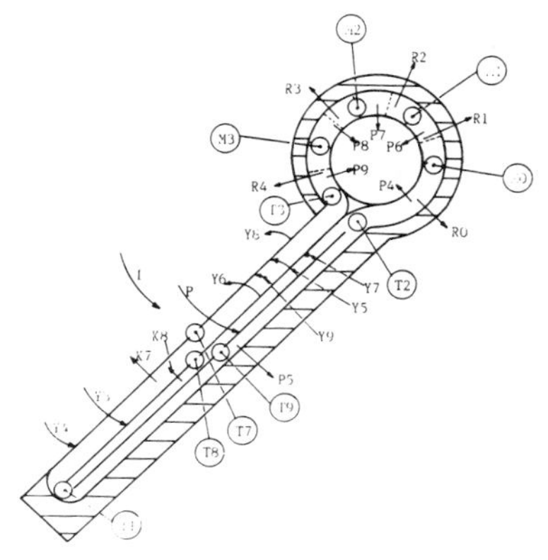

Fig. 4: The main heat transfers in a convective mono-block water heater
The schematic cut shows the radiative (bent arrows) and conductive / convective heat transfers (straight arrows) that have been taken into account in the simulation calculations of one of the versions of the convective mono-block water heater. Cycles indicate simulated temperatures (Cieslok, 1984).

Therefore, a parallel procedure of simulation and experiment has been adopted. In a first step, a crude iterative simulation program, taking into account the most important heat transfers, is written. In a second step, a prototype is built and tested at stable conditions. Apart from irradiance, ambient temperature and efficiency, internal temperatures (see circles in Fig. 4) are recorded. In a third step, the simulation model is used in an attempt to reproduce the experimental conditions. Significant differences between simulation and experiment usually indicate wrong assumptions on particular heat transfers, which are then corrected on a trial and error basis. In order to assess systematical errors of this procedure, the simulation model is used in a "Monte-Carlo"-mode, with distributions attached to the entry parameters.

Once the simulation model is able to reproduce different experimental conditions in a satisfactory way, it is used to test the effect of design changes, such as multiple glazing, different absorbers, effects of reflectors, different geometries and dimensions of heat exchangers, performance under different climatic conditions (e.g. in the tropics), etc. In this way, economic criteria can also be taken into account (e.g. prices of materials, labour costs for production).

An explicit comparison between simulation and experiment for the case of a convective solar cooker can be found in (Grupp and co-workers, 1987).

REFERENCES

Beall, J. and co-workers (1983). Le capteur solaire hybride. Internal Report, Synopsis.

Cardi, J.M., Gatt, P., Gschwind. M. (1984). Mesure des performances thermiques du capteur solaire Synopsis PU4. Ecole Nationale Supérieure des Mines, Centre Energétique.

Cieslok, J. (1984). Der Hybridmonoblock. Thesis, University of Aachen.

Grupp, M. (1982, 1983). French Patent 2 536 156 and European Patent 0 109 344 A1

Grupp, M. and co-workers (1987). Two novel solar appliances for developing countries: 1. Convective solar cooker (CSC) 2. Solar medical sterilizer, ISES SOLAR WORLD CONGRESS 1987, Proceedings.

Grupp, M. and Gschwind, M. (1987). The influence of flow conditions on the performance of flat plate collectors, to be published.

Sutter, J. (1947). Swiss Patent 257348.

STABILITY ANALYSIS OF A THERMOSYPHON LOOP

B.J. Huang and R. Zelaya
Department of Mechanical Engineering
National Taiwan University, Taipei, Taiwan 10764
China

ABSTRACT

A stability analysis of a rectangular thermosyphon loop based on linear system theory was carried out. It was found that the stability of a thermosyphon loop can be shown by a stability diagram in terms of parameters P, Y, and $\Delta Z/L$. Stable and unstable flow solutions are presented to show the instability which might occur in a real solar thermosyphon water heater.

KEYWORDS

Solar energy; thermosyphon system

INTRODUCTION

Due to low installation cost and little maintenance requirement, the thermosyphon solar water heater has become the most successful solar commercial product, with several millions installed all over the world. It has long been found that flow reversal occurs even during the energy collecting periods at daytime [1,2]. This can be attributed to the instability of the thermosyphon water heater. To eliminate this effect, a general design rule was deduced from experiences which states that the storage tank must sit on a level at least 30 cm higher than the collector [3,4]. Unfortunately, the research of the stability in thermosyphon solar water heaters is still not underway. Only a few studies were conducted toward the problems of stability phenomena in some simple-geometry thermosyphon loops.

In the present paper, a rectangular thermosyphon loop which can closely emulate a solar thermosyphon heater was studied. See Fig.1. The loop consists of a partly-heated vertical leg at one side and a partly-cooled vertical leg at the other side. The heating is furnished directly by a constant and uniform heat flux in the heater to simulate the solar irradiation on a collector. The cooling is provided by a cooling jacket with coolant flow which is similar to the heat loss or thermal load from the storage tank in a solar thermosyphon water heater. The understanding of the stability phenomena in this loop thus can provide useful information for the similar phenomena in a real solar thermosyphon heater.

STABILITY ANALYSIS

1. *1-D Modelling*

To facilitate the analysis, viscous dissipation, axial conduction, and heat loss from the loop to the ambient were all neglected. The 1-D approximation and constant properties (except for the use of Boussinesq approximation for the density) were used. It was further assumed that the cooler is composed of a heat exchanger with coolant flow in the upward direction (See Fig.1). The temperature of the coolant was assumed to vary linearly along the flow direction, and, the heat transfer is from the loop fluid having a mean temperature $\bar{T}$ to the coolant having a mean temperature $\bar{T}_c$. Since the wall shear force may be greater than that in a straight pipe, an "effective length" L_{ec} was used which is the sum of the geometric straight length of the loop L and the "equivalent length" L_e accounting for for the friction due to form losses in curved parts, i.e. $L_{ec} = L + L_e$. Therefore, when taking the integral of the momentum equation around the loop, the shear force term τ_w should be taken along the effective length L_{ec}.

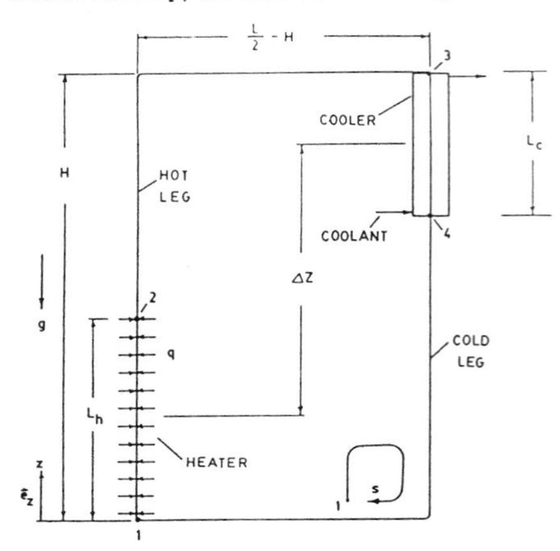

Fig.1 Schematic diagram of a thermosyphon loop.

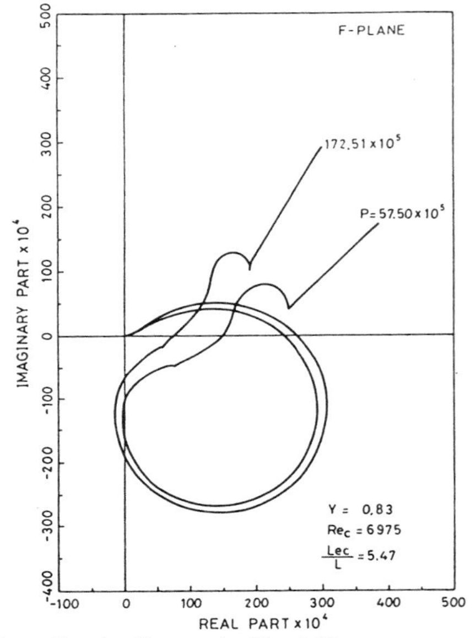

Fig.2 Nyquist diagram for $Y = 0.83$.

Applying a momentum and energy balance to the loop and noting that the shear force relation $\tau_w = f\rho_r V^2$, the nondimensional equations can be obtained (see Nomenclature and Fig.1) :

$$\frac{dv}{d\tau} + 4(L_{ec}/D)fv^2 = St^* \oint \theta dx\, \mathbf{e_z} \cdot \mathbf{e_s}, \tag{1}$$

$$\frac{\partial\theta}{\partial\tau} + v\frac{\partial\theta}{\partial x} = \frac{4St^*}{D/L}, \qquad 0 \le x \le L_h/L, \tag{2}$$

$$= -\frac{4St^*}{D/L}(\theta - \theta_c), \qquad 1/2 \le x \le 1/2 + L_c/L, \tag{3}$$

$$= 0, \qquad \text{otherwise}, \tag{4}$$

$$\frac{L_c/L}{2v_c}\frac{\partial\bar{\theta}_c}{\partial d\tau} + (1+\sigma)\bar{\theta}_c = \sigma\bar{\theta} + \theta_{c,i}, \tag{5}$$

$$St^* = \frac{U}{\rho_r C_p V_r} = \frac{U^3}{qLg\beta(\rho_r C_p)^2}, \qquad V_r = \frac{qLg\beta\rho_r C_p}{U^2} \qquad \sigma = \frac{\pi D L_c U}{2m_c C_p}, \qquad x = s/L.$$

where the subscripts "r" denote the reference states. Assuming that the friction factor f is $f = a/Re^b$ where a and b equal to 8 and 1, respectively, for laminar flow, and .03955 and 0.25 respectively for turbulent flow. The steady state solutions of the above equations (denoted by the subscripts "ss") have been obtained by Huang and Zelaya [5] which give

$$Re_{ss} = \begin{cases} (PY/8)^{1/2}, & \text{for laminar flow,} \\ (PY/0.3955)^{4/11}, & \text{for turbulent flow,} \end{cases} \tag{7}$$

where $Re_{ss} \equiv \rho_r D V_{ss}/\mu$, and P and Y are defined as

$$P \equiv \frac{\rho_r^2 g \beta D^4 q}{\mu^3 C_p}, \qquad Y \equiv \frac{(\Delta Z/L)(L_h/L)}{(D/L)(L_{ec}/L)}.$$

2. *Linearized Approach*

Equations (1) to (5) can be solved to give transient solutions [5]. However, to understand the stability phenomena in the thermosyphon loop, a linear perturbation analysis was employed. Assuming that a small perturbation is imposed upon the loop at a steady state. Mathematically, this can be expressed as, for the velocity and temperature distribution: $v(\tau) = v_{ss} + v'(\tau)$ and $\theta(x,\tau) = \theta_{ss}(x) + \theta'(x,\tau)$, where v_{ss} and θ_{ss} are the steady-state velocity and temperature, v' and θ' are the small disturbances. Substituting the above relations into eqn(1) to (5), subtracting the steady-state equations and then neglecting the second-order terms, we obtained the perturbed equations:

$$\frac{dv'}{d\tau} + \frac{4(2-b)a(L_{ec}/L)v_{ss}^{1-b}}{(D/L)Re^{*b}} v' = St^* \oint \theta' \, dx \, \mathbf{e_z} \cdot \mathbf{e_s}, \tag{8}$$

$$\frac{\partial \theta'}{\partial \tau} + v' \frac{\partial \theta_{ss}}{\partial x} + v_{ss} \frac{\partial \theta'}{\partial x} = \begin{cases} -\dfrac{4St*}{(D/L)}\theta', & \text{cooler;} \\ 0, & \text{otherwise} \end{cases} \tag{9}$$

The linearized equations have the solution forms : $v' = \tilde{v}e^{w\tau}$ and $\theta' = \tilde{\theta}e^{w\tau}$, where $\tilde{v}$ and $\tilde{\theta}$ are the amplitudes and w is a complex growth rate. Substituting the above solutions into equations (8) and (9) and solving for $\tilde{v}$, we obtain the characteristic function $F(s)$:

$$F(s) = \frac{H'(s)}{s} - \frac{s v_{ss}}{St^*} - \frac{4(2-b)a(L_{ec}/L)v_{ss}^{1-b}}{(D/L)St^* Re^{*b}} = 0, \tag{10}$$

where $H'(s)$ is presented in Appendix A, and $s \equiv w/v_{ss}$, $Re^* \equiv \rho_r V_r D/\mu$.

3. *Stability Diagram*

The stability of the loop can be analyzed using the characteristic equation (10) and Nyquist criterion. It can be seen from the Nyquist diagram shown in Fig.2 that the contours do not encircle the origin. However, the contours move closer to the origin as Y or P continue to increase. This indicates that the loop tends to be unstable at low loop friction and high heating rate or high relative elevation of the heater ΔZ. If ΔZ (or Y) and heating rate q (or P) continue to increase or the loop friction continue to decrease (Y increases), the contour starts to encircle the origin and thus the loop becomes unstable, as can be seen from Fig.3. This takes places as $P > 2.07 \times 10^7$and $Y > 2.54$, for the particular loop studied.

To construct a stability diagram for the thermosyphon loop, the Nyquist analysis was performed for different operating conditions and loop configurations in terms of $\Delta Z/L$, P and Y. The use of these three parameters as the characteristics for loop stability can be seen from the coefficients of characteristic equation and the relations of the steady-state solutions [5]. The stability diagram was then constructed using an iterative process and is shown in Fig.4.

The curves shown in Fig.4 partition the stable and unstable regions. It can be seen that for a given heating rate, (fixed P), the higher the height of the heater $\Delta Z/L$, the more unstable the loop. The higher the loop resistance or the shorter the heater length (i.e. lower Y), the more stable the loop since the operating point tends to locate in the stable region. For a given loop design (fixed Y and $\Delta Z/L$), the higher the heating rate (higher P), the more unstable the loop. However, the stability becomes insensitive to the variation of the heating rate if P is large. That is, after P has been increased to a large value, the stability seems depends on the geometric parameter Y only.

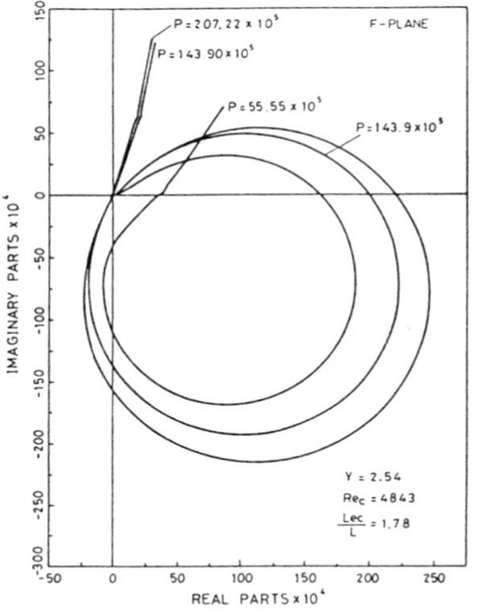

Fig.3 Nyquist diagram for $Y = 2.54$.

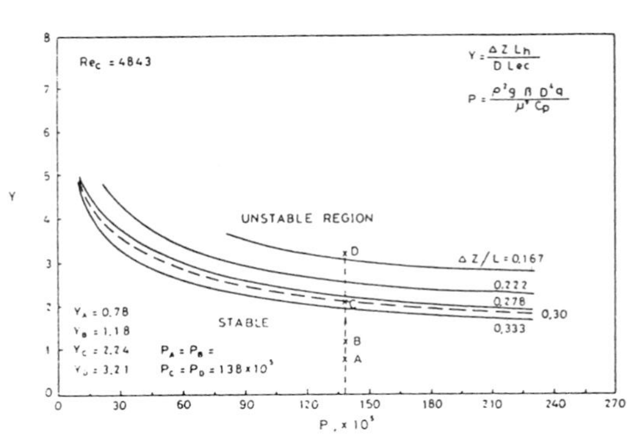

Fig.4 Stability diagram.

4. *Transient Flow Solutions*

To see the stability phenomena, equations (1) to (5) were solved numerically for transient flows. In the present study, only a loop with $\Delta Z/L = 0.3$ was studied and whose stability diagram was drawn as the dotted line in Fig.4. The transient solution was carried out by applying a step change to a initial steady-state condition (point A in Fig.4) and letting it approach to a final states (points B, C and D).

For the transition from A to B, the circulation velocity was found to first oscillate then to be damped out with a final stable velocity, as shown in Fig. 5. This represents a stable operation of the loop since point B lies in the stable region. For the transition from A to C, the loop circulation velocity was found to continue to oscillate as shown in Fig.6. This is the neutrally stable operation since point B lies on the partition line. For the transition from A to D, the loop was found to exhibit a unstable oscillating phenomenon as shown in Fig.7. This shows the instability of the loop since point D lies in the unstable region as predicted by the stability diagram. The above numerical examples were shown to be consistent with the predictions based on the stability diagram which were developed from the linear theory.

STABILITY IN SOLAR THERMOSYPHON WATER HEATER

Since the solar thermosyphon water heaters are very similar to the thermosyphon loop of the present study, the above results indicates that to avoid flow reversal or instability, the relative height between the collector and the tank and the loop friction of the solar water heater should be carefully controlled. That is, increasing the relative height of the tank and use of taller collector (both increasing Y) would increase the thermosyphon flow, but the energy efficiency may not be increased. Instead, the efficiency might be reduced if the loop friction is so low (Y high) that the loop instability occurs. This seems coincides with the experimental evidence found by Ong [6] who showed that the efficiency of the thermosyphon water heater first increases with the relative height of the tank and the collector, then tends to decrease and there seems exists an "optimum height". Ong attributed this to the mixing and thermal stratification effect in the tank. In fact, the loop stability may play an important role.

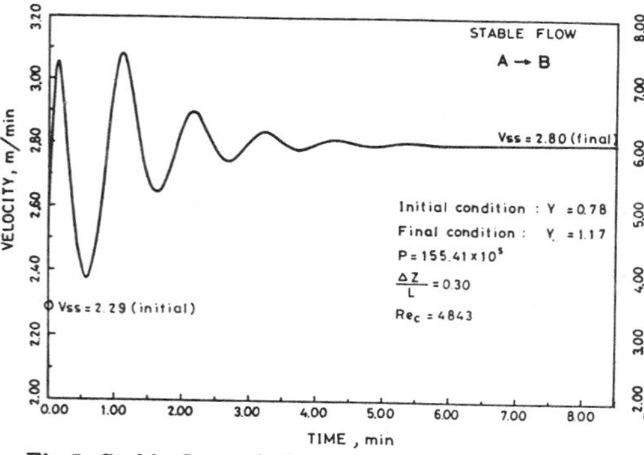

Fig.5 Stable flow solution.

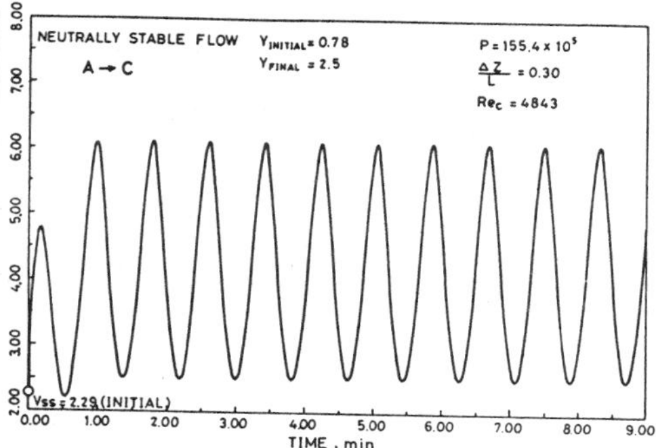

Fig.6 Neutrally stable flow solution.

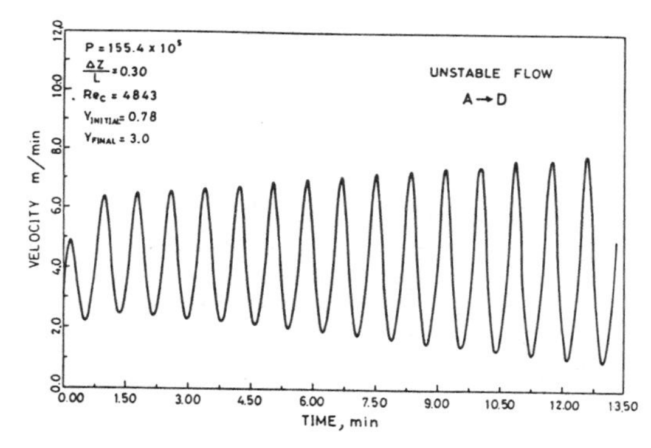

Fig.7 Unstable flow solution.

The flow reversal phenomenon was also observed by Gupta and Garg [1] in an experiment during the period of daytime while solar irradiation was high. Since the water heater used in their experiment was designed with parallel-plate type collector, the loop resistance is expected to be low (Y high) and thus the solar water heater tends to be unstable as predicted by the present study.

CONCLUSIONS

A stability analysis of a rectangular thermosyphon loop based on the linear theory was carried out. The results show that the loop tends to be unstable for higher heating rate, taller heater, lower friction, and higher relative height between the cooler and the heater. It was found in the present study that the stability of a thermosyphon loop can be characterized by a stability diagram which can be shown in terms of two parameters, P and Y, and the relative height between the cooler and the heater $\Delta Z/L$. Several numerical examples for transient flows are presented to show the instability phenomena during transient which might occur in a real solar thermosyphon water heater. The present results imply that reducing the loop friction and increasing the tank elevation in thermosyphon solar water heater design though would improve the collector efficiency most of the time, but will suffer from the risk of occurrence of instability which in turn would reduce the efficiency. Use of taller collector will increase Y and hence reduce the stability of the solar water heater.

Acknowledgement — The present study was supported by the National Science Council of the Republic of China through Grant No. NSC73-0401-E002-08.

NOMENCLATURE

C_p	specific heat, $kJ/kg\,°C$	D	inside diameter of the loop, m
$\mathbf{e_s}$	unit vector in the s direction	$\mathbf{e_z}$	unit vector in the upward direction
g	acceleration of gravity, 9.80 m/s^2	k	thermal conductivity of loop fluid, $W/m\,°C$
m_c	mass flowrate of coolant, kg/s	q	heat flux per unit heated area in heater, W/m^2
Re_c	Reynolds number of coolant flow	s	spatial coordinate running around the loop, m
T	fluid temperature, $°C$	$T_{c,i}$	inlet temperature of coolant, $°C$
t	time, s	U	overall heat transfer coefficient of cooler, $W/m^2\,°C$
V	fluid velocity, m/s	β	thermal expansion coefficient, $°T^{-1}$
μ	viscosity of loop fluid, $kg/m\,s$	ρ	density, kg/m^3
τ	$\equiv V_r t/L =$ dimensionless time	θ	$\equiv (T-T_r)/(q/U)$
$\theta_{c,i}$	$\equiv (T_{c,i}-T_r)/(q/U)$	$\bar{\theta}$	$\equiv (T-T_r)/(q/U)$
$\bar{\theta}_c$	$\equiv (T_c - T_r/(q/U)$		

REFERENCES

1. Gupta, C.L. and Garg, H.P., "System Design in Solar Water Heaters with Natural Circulation". *Solar Energy* , Vol. 12, 163(1968).
2. Ong, K.S., "A finite-difference Method to Evaluate the Thermal Performance of a Solar Water Heater". *Solar Energy,* Vol.16, 137(1974).
3. McVeigh, J.C., "Sun Power". Chapter 9, Pergamon Press, Oxford, (1977).
4. Vaxman, M. and Sokolov, M., "Effects of Connecting Pipes in Tthermosyphonic Solar Systems". *Solar Energy,* Vol.37, pp.323-330(1986).
5. Huang, B.J. and Zelaya, R., "Heat Transfer Behaviour of a Rectangular Thermosyphon Loop". to be published in *ASME J. Heat Transfer* , (1987).
6. Ong, K.S., "An Improved Computer Program for the Thermal Performance of a Solar Water Heater". *Solar Energy* , Vol. 18, 183(1976).

APPENDIX A. $H'(s)$ in equation (10).

$$H'(s) = -(p + \frac{A_2}{s v_{ss}})\left[e^{-sL_h/L} - 1\right] - K_2\left[e^{-sH/L} - e^{-sL_h/L}\right] + \frac{K_3 s}{s + A_2}\left[A_7 e^{-A_8 s} - e^{-(s+A_2)/2}\right]$$
$$+ K_4\left[e^{-(1/2+H/L)s} - e^{-A_8 s}\right] + (A_{10} - \frac{A_2 L_h}{L v_{ss}}), \qquad \text{where} \qquad A_2 = \frac{4St^*}{(D/L)v_{ss}},$$

$$A_3 = A_2(\theta_{3ss} - \bar{\theta}_{css})e^{A_2/2}, \qquad A_7 = e^{-A_2(1/2+L_c/L)}, \qquad A_8 = \frac{1}{2} + \frac{L_c}{L},$$
$$A_{10} = \frac{A_3}{v_{ss}A_2}(A_7 - e^{-A_2/2}), \qquad K_2 = p + \frac{A_2}{s v_{ss}}\left(1 - e^{sL_h/L}\right), \qquad K_3 = K_2 e^{-A_2/2} - \frac{A_3}{s v_{ss}} e^{s/2},$$

$$K_4 = A_7\left[K_3 + \frac{A_3}{s v_{ss}} e^{A_8 s}\right], \qquad p = \frac{A_2 e^{-A_2/2}\left[1 - e^{sL_h/L}\right] + A_3\left[e^{A_8 s} - e^{s/2}\right]}{(s v_{ss}/A_7)e^s - s v_{ss} e^{-A_2/2}}.$$

VALIDATION OF A CLOSE-COUPLED THERMOSIPHON UNIT USING OUTDOOR TEST DATA

*R. Marshall, +S. Griffiths

*Solar Energy Unit, University College Cardiff
+Dept. Mechanical Engineering & Energy Studies, University College Cardiff

Newport Road, Cardiff CF2 1TA, Wales

ABSTRACT

Validation of a close coupled thermosiphon system is considered using outdoor test data for a period of 55 hours. The results show that the model works well for predicting the day-time temperature trajectories but that the overnight reverse flowrate is over estimated although it is a small value, less than 0.5 g/s. The average mid-store temperature error is less than 1.2% (based on the observed 34.9 deg. C temperature swing) with an RMS percentage error of 5.2%.

KEYWORDS

Solar, Thermosiphon Systems, Modelling, Validation, Reverse Flow

INTRODUCTION

A close-coupled thermosiphon unit is, typically, a collector with a horizontal cylindrical tank located immediately above the collector outlet. This arrangement permits pre-fabrication and easy installation. Thus a low installed system cost is achieved. The close coupling (without a non-return valve) results in a loss of performance because reverse thermosiphoning can occur, see Morrison (1986). Closed-coupled thermosiphon units are commonly available in the sunnier parts of the world where a loss in performance is compensated by a generous supply of irradiation. The objective of the work reported herein is to validate our model of a close-coupled thermosiphon unit using outdoor test data.

EXPERIMENTAL SET-UP

A thermosiphon unit was constructed and mounted at our outdoors Durability Trials Station, Cleppa Park, near Cardiff. A view of the experimental set-up is seen in Plate 1. The single glazed collector is a header-and-riser design with 2.1 square metres of selectively coated absorber. The collector had been well tested indoors as part of the indoor validation exercise, see Riddle (1985). The standard collector test indicates values for η_o and U of 0.63 and 4.5 W/m²K respectively. The collector was also tested at reduced flow rates typical of those (0.007 kg/s) occurring in thermosiphonic operation.

For this validation exercise a different storage tank was used. A 53 litre in, approximately, the shape of a right cylinder is mounted horizontally. The tank was lagged with 80 mm of mineral wool insulation and covered by a plastic bag to protect it from any rain. The upriser and downcomer pipes were 22 mm and insulated with 15 mm pre-formed Armaflex insulation.

A calibrated temperature probe was constructed and mounted in the centre of the tank to measure the vertical stratification in the store. The probe consisted of a hollow insulating tube (Bakelite) with 5 Type-K thermocouple beads just penetrating the surface. These data were logged manually for the first day at 5 minute intervals. A microdata 12 channel data logging system was used to monitor the inlet temperature, temperature difference, mid-store temperature, ambient temperature, wind speed, and irradiation on the 42 degree tilted south-facing surface. All but the store temperatures were measured using calibrated resistance thermometers (PT100). The mid-store temperature was a Type-K thermocouple, as noted above. Timing data (minute, hour, and day) were also recorded.

Raw data recorded on audio cassettes in terms of logger units (0-255) were recorded every 30 seconds for a 55 hour period beginning at 10:00 on February 18, 1987. This period was cold and clear with overnight ambient temperatures dropping to -5 deg. C. The data was then processed off-line using a number of computer programs.

No demand load was applied as the tank loss and reverse thermosiphon loss served to deplete the store sufficiently to give a true dynamic operation.

MODELLING, IDENTIFICATION, AND DATA REDUCTION

Modelling

In previous work, Riddle (1985), a general thermosiphon model had been established and validated indoors using a solar simulator. A constant irradiance of about 750 W/m^2 was used and the validation looked at just the charging period. For the present exercise a different store was used and in a non-standard configuration, see Plate 1. The upriser connects to the upper right side of the store. The downcomer connects to the lower left side of the store. Thus, the flow is both across and down. In addition, the vertical separation between upriser and downcomer is quite small, 0.16 m. Thus, the configuration would suggest that strong stratification could not be achieved. In the model, therefore, we used 3 nodes in the store to represent the "partially" stratified store in contrast to the 6 to 10 nodes one would normally use to model a fully stratified store. The store loss parameter, Us W/m^2K, also had to be changed from that used in the indoor testing. Experience at Cardiff and elsewhere with testing thermal storage devices, even when factory-insulated, has shown that a loss coefficient 4 to 5 times the theoretical value is a likely experimental value. For the present case a range in Us between 2.0 and 3.0 W/m^2K was required while the nominal value was estimated to be 0.48 W/m^2K. Our approach was to use the data beginning with the evening of the first day to the morning of the second and "identify" the best estimates for the store and pipe loss coefficients. Values of 1.96 and 2 W/m^2K were established. Thereafter, all the data were used and the full transient response of the system was computed and compared with the experimental results.

Data Reduction

Using the system model, the average error between the predicted and measured was formed for the collector inlet, exit, and mid-store temperatures together with the root-mean-squared error (RMS). In addition, the energy balance on the store was performed. Here, the difference in energy stored must equal the energy in minus the energy out. Energy into the store is taken to be that arising from a positive flowrate through the collector. The energy out of the store is considered in two parts, a) the energy lost from the store, and b) the energy depleted during reverse thermosiphoning, i.e. when the flow rate is negative.

RESULTS

Temperature trajectories. The results are depicted graphically in Fig. 1. The uppermost plot depicts the irradiance. The afternoon of the first day indicates broken passing clouds while the following two days indicated that broken clouds passed by shortly before mid-day. It was bright with the irradiance exceeding 1000 W/m^2 on the inclined surface.

The ambient temperature, 3rd from the top in Fig. 1, indicates a day-time peak of only 4 to 5 deg. C while overnight lows dipped to -5 deg. C. Coupled with the clear night sky there was a possibility that the fluid in the collector may have begun to freeze.

For the first 4 hours the model overpredicts the mid-store temperature. This is thought to arise from an imprecise knowledge of the initial system temperatures. At this time the irradiance drops and the flow rate becomes negative for the first time. At hour 6 (i.e. 16:00 in the afternoon) the irradiance drops below the critical level, @ 420 W/m^2, and the flow rate goes negative for the remainder of the day. At this time the deviation between the model and the experiment increases. The discrepancy indicates that the reverse flow rate of approximately 0.4 g/s predicted by the model is larger (in absolute magnitude) than the true flow rate. We conclude that the model does not predict well the flowrate in the reverse thermosiphoning mode: the friction factors are different from those in the forward mode and their values at such low flow rates are in doubt. In addition, as the fluid temperature in the collector approaches zero, the correlation equation for the density of water, and hence the buoyancy, exhibits larger errors.

Despite the discrepancy in the collector outlet temperature, by the start of the second day, 23.5 hours into the test, both the measured and predicted mid-store temperatures are in close agreement and up to hour 30 (14:00 on the second day) the agreement remains good. The model predicts that the critical irradiance level was reached and that reverse thermosiphoning begins while the experiment indicates a 2 hour stagnant (no-flow) period. The parallel lines during the overnight decay confirms that the loss coefficient, Us, identified from the first days overnight cooldown is a good estimate.

The overnight cooldown at the collector outlet is, again, not well predicted in contrast with the good agreement found for the collector inlet. The transient behaviour at the start of day 2, hour 47, agrees well with the experimental transient and by the end of the test, hour 55, there is but a 1.2 K error between the measured and predicted mid-store temperature. (The store sensor accuracy is 0.5 deg. C.)

In general, the temperature trajectories agree well enough, especially during the day time, to enable good estimates of system performance. Taking the mid-store temperature swing between a low of 38.5 deg. C on day 2 to a high of 73.4 deg. C, there is an average error of 0.45 K between the model and experiment with an RMS error of 1.88 K over the 55 hour test period. Taking the temperature swing as a basis, i.e. 34.9 K, the percentage error and percentage rms error are 1.2% and 5.2% respectively.

Energy transfers: At the end of the 55 hour test period 25.1 MJ was collected 18.3 MJ was lost from the store through the insulation (recall, no demand was imposed), and 0.25 MJ was lost through reverse thermosiphoning. Recall, too, that the model predicted too large a reverse flowrate so that the reverse flow loss is expected to be less than that indicated. Thus, only 1% of the collected energy was lost through reverse thermosiphoning. This loss could be reduced if a suitable, low pressure loss one-way valve can be found.

CONCLUSIONS AND FUTURE WORK

Validation of a close-coupled thermosiphon unit using outdoor test data has been considered. The model was extended to permit reverse thermosiphonic flow. The predicted day-time temperature trajectories agree well with the measured values but the model appears to over-estimate the magnitude of the reverse flow rate although it is still small, less than 0.5 g/s. The model is sufficiently accurate to enable the calculation of long term system performance which is the next task. The major objective still to be achieved is the development of a low pressure loss, passive, non-return valve so that modelling of the reverse flow is not necessary.

REFERENCES

Morrison, G. (1986). Reverse circulation in thermosyphon solar water heaters. Solar Energy, 36, 377-379.

Riddle, D. (1985). Modelling the transient performance of natural circulation solar hot water systems, Proc. 4th Intl. Conf., Numerical Methods in Thermal Problems, 15-18 July 1985, Vol. 4, Part 2, 1029-1039.

Plate 1. View of the test stand

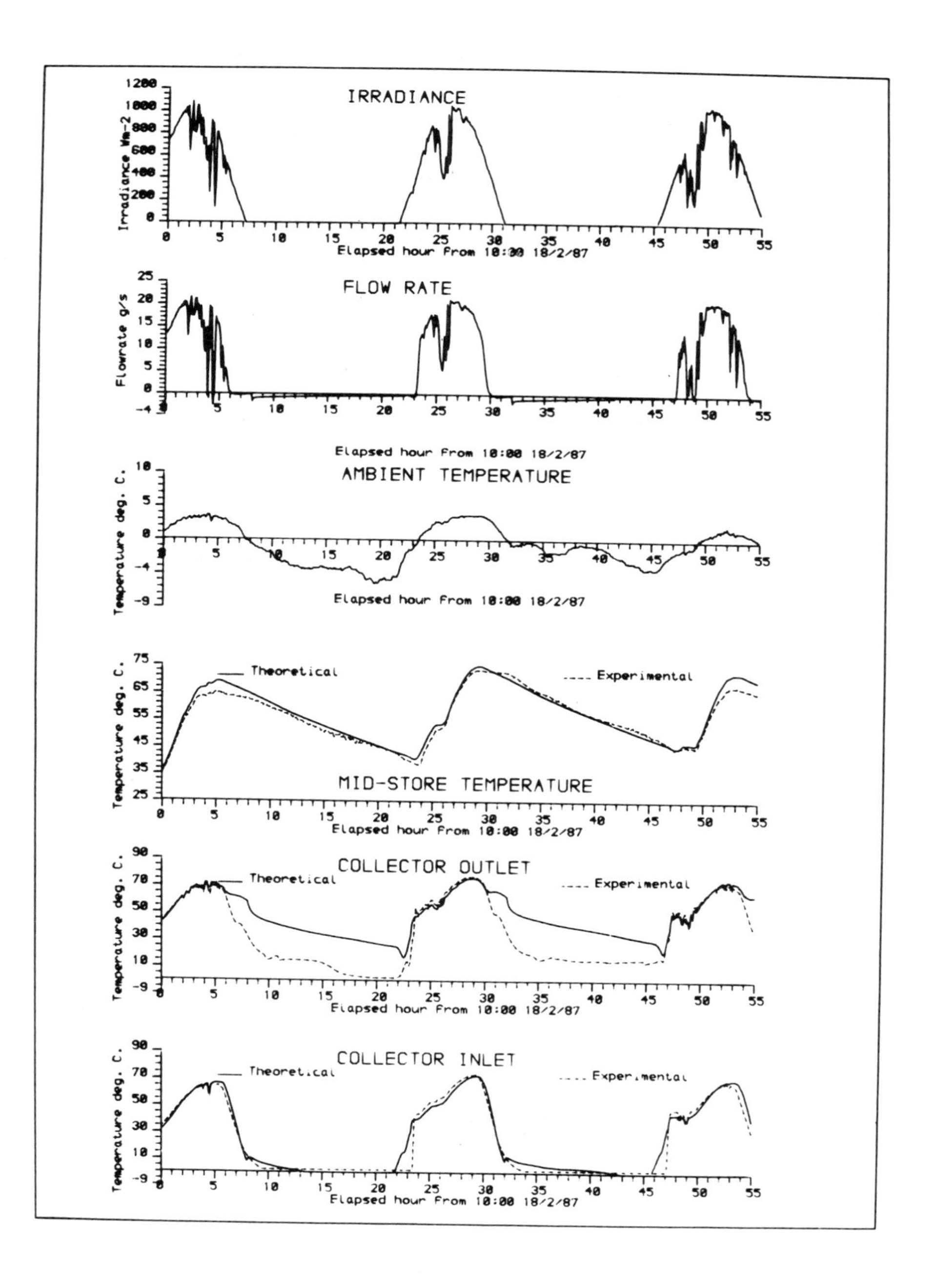

Fig. 1 Comparison of results

CORRELATING THE DAILY PERFORMANCE OF INDIRECT THERMOSYPHON SOLAR-ENERGY WATER-HEATERS

P.A. Hobson, S.N.G. Lo[+], B. Norton [*+], and S.D. Probert

Solar Energy Technology Centre
School of Mechanical Engineering
Cranfield Institute of Technology
Bedford MK43 OAL, UK

[+] Member ISES
[*] Author to whom correspondence should be addressed

ABSTRACT

Thermal performance data was acquired from the long-term monitoring of an indirect thermosyphon solar-energy water heater. This data was used to verify successfully a technique for correlating the daily performance of such systems. The analysis is based on the solution to a simplified transient heat balance on the system.

KEYWORDS

Thermosyphon; water heating; correlations; monitored performance;

ANALYSIS OF SYSTEM PERFORMANCE

A transient heat balance was carried out on a generic solar-energy water heater (see figure 1). The following assumptions were made:-

i) The collector and storage tank both have the same mean temperatures.

ii) The solar radiation intensity and ambient temperature are assumed to remain invariant over the day at their respective mean measured values.

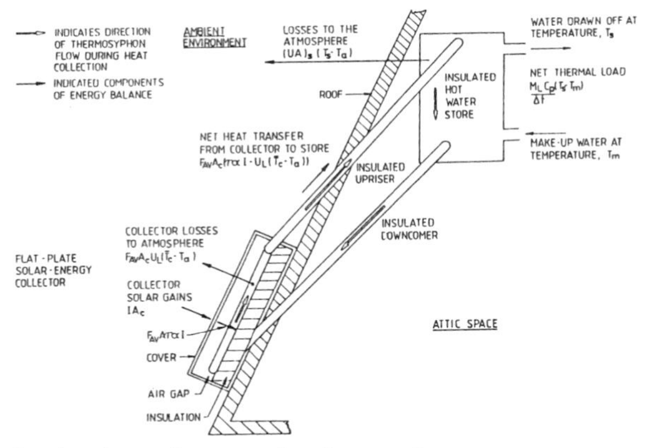

1. Holistic heat transfer mechanisms for a generic thermosyphon solar-energy water heater.

iii) The total draw-off of water from the solar store is assumed to take place as a single event at the end of the insolation period. By introducing as an independent parameter the mass of water drawn off, the analysis is as applicable to 'third tap' as to preheat systems.

iv) Water is drawn off at the mean temperature of the store and replenished at the mains water supply temperature.

v) Tank and pipe losses are considered negligible.

vi) The store is initially at the mains water supply temperature at the onset of the insolation period.

An instantaneous heat balance on the system gives, with assumption (i)

$$M_s\ C_p\ \frac{dT_s}{dt} = F_{AV}\ A_c\ [(\tau\alpha)_e\ \frac{H}{\Delta t} - U_L(T_s - T_a)] \tag{1}$$

where $M_s\ C_p$ is the thermal capacity of the water within the store, T_s is the mean temperature of the store, F_{AV} is the collector efficiency factor evaluated at the mean collector fluid temperature, A_c collector area, $(\tau\alpha)_e$ effective transmittance-absorptance product, H is the total daily insolation incident on the plane of the collector, Δt is the durations of the insolation period, U_L is the overall collector heat loss co-efficient, and T_a is the ambient temperature.

Using assumptions (ii) and (v), equation (1) becomes, after making the substitution $\theta = T_s - T_a$ and rearranging

$$\frac{d\theta}{dt} + \frac{F_{AV}A_cU_L\ \theta}{M_sC_p} = \frac{F_{AV}A_c(\tau\alpha)_e\ H}{M_sC_p\Delta t} \tag{2}$$

The solution to equation 2 using assumption (vi) as a boundary condition gives

$$\theta - \theta_o = [\frac{F_{AV}A_c(\tau\alpha)_eH - F_{AV}A_cU_L\Delta t(T_m-T_a)}{F_{AV}A_cU_L\Delta t}]\ [1-e^{-(\frac{F_{AV}A_cU_L}{M_sC_p})t}] \tag{3}$$

where T_s for θ_o corresponds to the mean diurnal mains temperature.

The difference in the mean store temperature between the beginning and the end of the insolation period is therefore

$$\theta_e-\theta_o = [\frac{F_{AV}A_c(\tau\alpha)_eH - F_{AV}A_cU_L\Delta t(T_m-T_a)}{F_{AV}A_cU_L\Delta t}]\ [1-e^{-(\frac{F_{AV}A_cU_L\Delta t}{M_sC_p})}] \tag{4}$$

From assumptions (iii) and (iv), the solar fraction can be expressed as

$$f = \frac{M_LC_p(\theta_e - \theta_o)}{Q} \tag{5}$$

Substituting equation 4 into equation 5 and rearranging gives

$$\frac{f\,Q}{M_c C_p (T_m - T_a)\left(1 - e^{-\left(\frac{F_{AV} A_c U_L \Delta t}{M_s C_p}\right)}\right)} = \frac{F_{AV} A_c (\tau\alpha)_e H}{F_{AV} A_c U_L \Delta t (T_m - T_a)} - 1 \quad (6)$$

from which three dimensionless groups can be defined as

$$X = \frac{fQ}{M_L C_p (T_a - T_m)} = \frac{\text{TOTAL HEAT DELIVERED BY SYSTEM}}{\text{CHANGE IN INTERNAL ENERGY OF WATER DRAWN-OFF WHEN RAISED FROM MAINS TO AMBIENT TEMPERATURE}} \quad (7)$$

$$Y = \frac{F_{AV} A_c (\tau\alpha)_e H}{M_s C_p \,(T_a - T_m)} = \frac{\text{TOTAL DAILY INSOLATION ABSORBED}}{\text{CHANGE IN INTERNAL ENERGY OF STORE WHEN RAISED FROM MAINS TO AMBIENT TEMPERATURE}} \quad (8)$$

$$Z = \frac{F_{AV} A_c U_L \Delta t}{M_s C_p} = \frac{\text{TOTAL COLLECTOR HEAT LOSS COEFFICIENT}}{\text{HEAT CAPACITY OF STORE}} \quad (9)$$

So in terms of the dimensionless groups defined, equation 6 becomes

$$\frac{X}{1 - e^{-Z}} = \frac{Y}{Z} + 1 \quad (10)$$

Equation 10 therefore indicates a linear relationship between X/(1-exp(-Z)) and Y/Z.

APPLYING THE ANALYSIS TO THE EXPERIMENTAL RESULTS

Three basically-identical thermosyphon solar-energy water-heaters were retro-fitted to three adjacent occupied houses in Wharley End, Bedfordshire, U.K. (Norton et al, 1985).
Data for 250 complete days automated monitoring of an indirect thermosyphon solar-energy water-heater is considered here. The values of the mean ambient and mains water supply temperature and the total insolation and mass of solar heated water drawn off from the store were calculated from the available data in order to determine the daily values of the dimensionless groups X, Y and Z defined by equations 7, 8 and 9 respectively. The daily solar fraction, f, was determined from

$$f = \frac{\Sigma[M_L C_p (\bar{T}_s - T_m)]}{Q} \quad (11)$$

where $\bar{T}_s$ is the mean temperature of all the axially-placed thermocouples in the store contained within a distance from the top of the store corresponding to the volume of water withdrawn.

A plot of X/(1-exp(-Z)) against Y/Z for the 250 days monitored, is shown in figure 2. A strongly linear relationship is observed with a correlation coefficient of 0.96. The apparent scatter of the data points about the origin in figure 3 is due to slight seasonal variations in the functional relationship between the dimensionless parameters. This can be seen in figure 3 where for data monitored over the month of September the scatter is

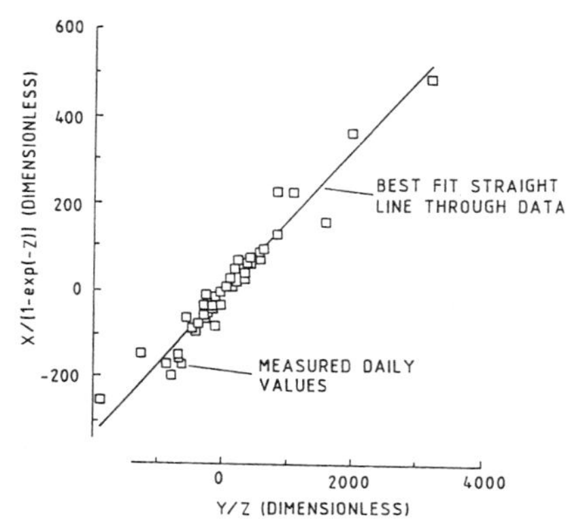

2. X/(1-exp(-Z)) and Y/Z correlated over a period of 250 days

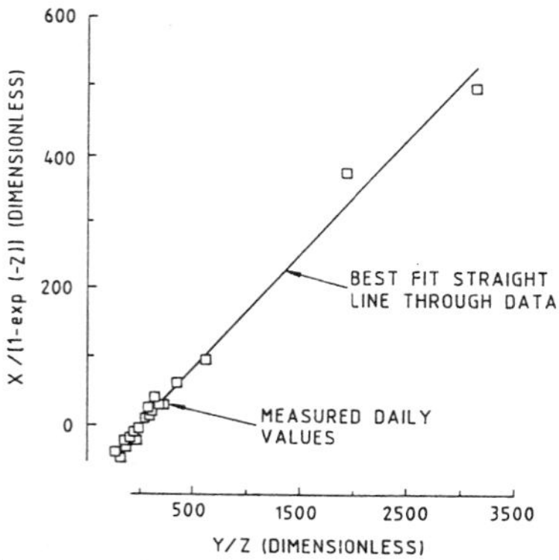

3. A graph of X/(1-exp(-Z)) versus Y/Z over the month of September

minimal about the origin. The correlation coefficient for this month is 0.99 and is typical of the values obtained for individual months. There is evidence of gradual seasonal variations in the gradients of the correlations. If a series of daily measured values were to be correlated in order to establish a single mean characteristic thermal performance curve similar to that shown in figure 2, for a system, a wide range of values of X, Y and Z would have to be correlated.

USING THE CORRELATION AS A DESIGN TOOL

Inspection of the dimensionless groups X, Y and Z indicate that the linear relationship (shown graphically in figure 2) between the grouped parameters obtained using data measured from the particular thermosyphon solar-energy water heater described, is applicable to other systems. However, in addition to the exogeneous parameters $F_{AV}(\tau\alpha)_e$, $F_{AV}U_L$, A_c and M_s which define the system configuration and thermal characteristics, the performances of buoyancy-driven systems also depend on other factors such as the position of the store relative to the collector. The influence of these other factors may be encapsulated in the relationship between the store and collector temperatures, that is the validity, or otherwise of assumption (i). An initial investigation indicated a constant temperature difference between the daily mean collector (T_c) and store (T_s) temperature over the insolation period. Figure 4 shows a plot of the daily T_c against T_s for the month of August indicating a linear relationship. The best straight line fit through the data gave

$$T_c - 1.00T_s = 1.41 \quad (12)$$

with a correlation coefficient of 0.97. Rearranging equation 12 gives

$$T_s - T_c = 1.41$$

Thus more generally

$$T_s - T_c = \beta \quad (13)$$

Introducing into the analysis such a constant temperature difference β, between the store and collector results in the dimensionless group Y being redefined as,

$$Y = \frac{F_{AV}\ A_c\ [(\tau\alpha)\ H + U_L\ \beta\ \Delta t]}{M_s\ C_P\ (T_a - T_m)} \quad (14)$$

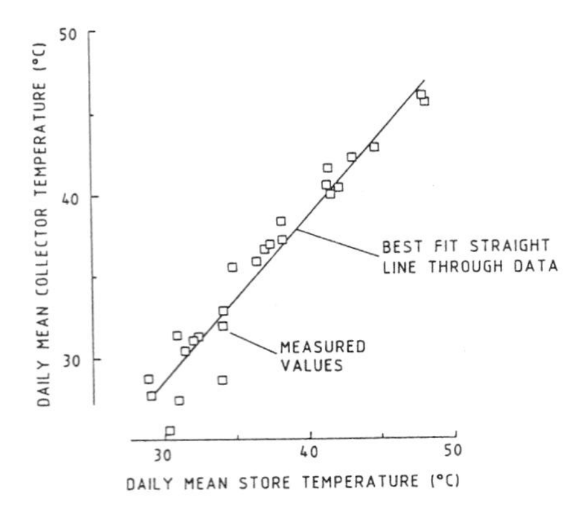

4. Correlation of the mean collector and hot-water store temperatures, each averaged over single periods of insolation.

Using this modified value of Y, would result in a correlation with a more-universal application potentially. The performance of any system can then be determined from several daily measurements of β. Experimentally-determined values of β, would enable X, Z and the modified Y parameter to be used in predicting the daily performance of the system under the applied conditions using a universal correlation. Alternatively, daily values of β could be generated using a high-level simulation model (Hobson et al, 1987) and correlated against the height of the store relative to that of the collector.

CONCLUSIONS

An analysis has been developed enabling the long-term monitored performance of an indirectly-heated thermosyphon solar-energy water-heater to be correlated, on a daily basis, with the thermal demands made on the system and the prevalent meteorological conditions. The correlation exhibits a high degree of linearity over a wide range of climatic conditions, and draw-off patterns.

Acknowledgements

This study was supported financially by the Science and Engineering Research Council, Swindon, U.K. and by the Commission of the European Communities, Brussels, Belgium.

REFERENCES

P.A. Hobson, B. Norton and S.D. Probert, (1987). Improved Simulation Model of Thermosyphon Solar-Energy Water-Heaters, Proceedings ASME Winter Annual Meeting, International Symposium on Natural Circulation, Boston, USA. November.

B. Norton, P.D. Fleming, S.N.G. Lo and S.D. Probert, (1985). Data Acquisition from Thermosyphon Solar-Energy Water Heaters for a Terrace of Three Dwellings, Proc. of UK-ISES Workshop on Solar-Energy and Building Design, Birmingham, UK. April.

MATCHING COLLECTOR'S AZIMUTHAL ORIENTATION AND ENERGY DEMAND PROFILE FOR THERMOSYPHONIC SYSTEMS

by

M. SOKOLOV and M. VAXMAN
Department of Fluid Mechanics and Heat Transfer
Faculty of Engineering
Tel-Aviv University
Ramat Aviv 69978, Israel.

ABSTRACT

Numerical simulation for different collector's azimuthal orientation and discharge times is performed to investigate the output water temperature. The results of this study indicate that for a specific thermal load demand, a preferred direction for a given hour and season can be found to reach an optimal temperature.

KEYWORDS

Thermosyphonic system; Collector's azimuthal direction.

INTRODUCTION

Most investigations of solar thermosyphonic systems deal with the study of systems in which the collectors face south and the performance of the systems were studied without thermal load [1, 2, 3]. This is the correct method when total energy collection is to be maximized or when demand is made in the evening hours. However, there are many cases in which the collectors cannot be placed facing south (e.g., due to shadowing, lack of space, etc.) or when energy demand is not in the evening hours. In these cases, it is likely that the collectors will not be placed facing a southerly direction in order to improve the system performance.

The purpose of this study is to investigate the performance of a thermosyphonic system with a load of 30 liters of water. The temperature of the load with various collector azimuthal directions and time dependent demand is checked.

Two modes of load time dependency are investigated:

a) Single load mode, where a load is withdrawn at a given time from a system that started at uniform temperature of 20°C in the morning. This mode simulates the energy available after the system was completely drained the day before.
b) Periodic load mode, where a load is made each day at the same time so that a steady state is achieved.

SIMULATION PROCEDURE

The thermosyphonic system was simulated for the following conditions:

a) Days of simulation: 21 June, 21 September, 21 December.
b) Direction of collectors: S, S-E, E, E-N, N, N-W, W, W-S.
c) Latitude 32N.

Each system was composed (Figure 1) of a tank of 150 liters of water and two collectors $1.5m^2$ each. The energy and momentum equations, and the system dimensions are similar to those of [4].

The total solar radiation on a tilted surface was estimated following [5]. Thus:

$$I_T = Z[\sin\theta_z \sin S \cos(\gamma_s - \gamma) + \cos\theta_z \cos S + 1/2\,(1 + \cos S)\,ZC + 1/2\,Z(1 - \cos S)\,(\cos\theta_z + C) \quad (7)$$

where

$$Z = G\exp(-B/\cos\theta_z) \quad (8)$$

and S is the tilt angle of the collector, θ_z - solar zenith angle, γ_s - sun azimuth, γ - collector's azimuth and ρ - surrounding reflectivity.

The constant G, B, C are:

For June 21,	$G=1088.3\ w/m^2$	B=0.205	C=0.134
For September 21,	$G=1151.2\ w/m^2$	B=0.177	C=0.092
For December 21,	$G=1233.2\ w/m^2$	B=0.142	C=0.057

The ambient temperature is given by:

$$T_a = F_1 + F_2 \sin[\pi(t-6)/12] \quad (9)$$

where F_1 and F_2 are given by:

$$F_1 = 7 + 3\sin[\pi(n-90)/365)] \quad (10)$$
$$F_2 = 9 + 5\sin[\pi(n-90)/365)] \quad (11)$$

The set of energy and momentum equations were solved numerically by the finite difference method, using backward differentiation. For this purpose each component length (in the flow direction) was divided into 20 equal parts. The time interval was 5 seconds.

RESULTS AND DISCUSSION

The main parameter used in this work to compare the various systems is the mixed cup temperature of the load. As mentioned earlier the load consists of 30 liters of water withdrawn from the top of the tank at different hours.

In calculating the performance of thermosyphonic systems without thermal load (such as total efficiency, average temperature in the tank etc.) the performance is mainly affected by the amount of solar energy which has been accumulated in the system during its operation, but the temperature of the thermal load is also strongly affected by the local solar radiation (the solar radiation near the time of discharge). For example, in the afternoon hours, when solar radiation decreases, the water passing through the collector and entering the top of the tank is cooler than the water that passes through the collector at noon. Therefore, the water withdrawn in the afternoon hours (when the solar radiation is strong enough to cause direct flow in the system) is cooler than the water withdrawn earlier.

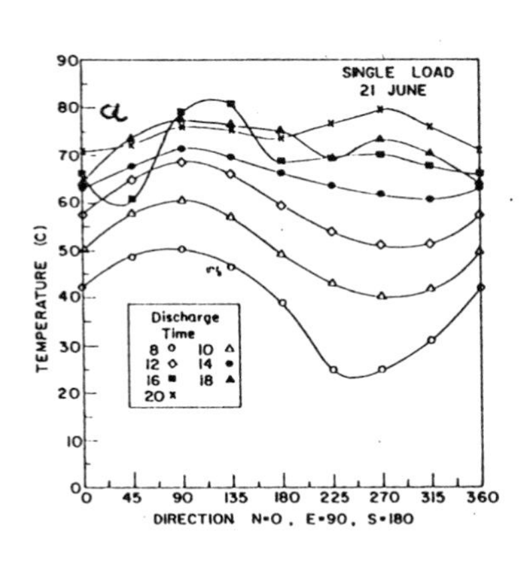

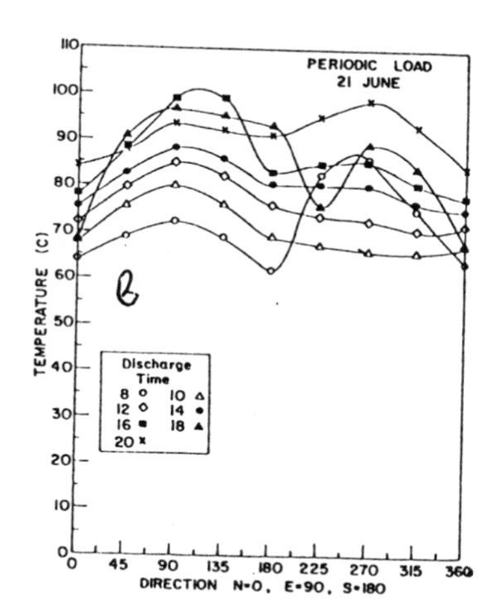

Fig.1 Temperature vis. direction for June

In the evening hours the reverse flow in the system (even weak) causes the lower layers of water in the tank to flow to the top of the tank and so the load water is hotter.

Figures 1(a), 2(a) and 3(a) exhibit the mixed cup temperatures (for June, September and December respectively) for various collector's azimuthal angles with the hour of the load as a parameter, for a single load operating mode.

The clear result of figure 1(a) is that if the load is required between 8 a.m. - 2 p.m. the east system will yield best energy output. Thus, if a single load is required at noon the east facing system will provide it at 68°C while the south and west facing system will yield only 63°C and 51°C respectively. When load hours move towards the evening the west system may have an advantage because the solar radiation is received near the demand time.

It is interesting to compare the performance of systems directed to the east of the south ("eastern system") to those directed to the west of the south ("western systems"). At 4 p.m. the "western systems" still collect energy and their temperature is still below the maximum. At the same time, weak solar radiation reaches the "eastern systems" and the reverse flow takes place causing lower water layers in the tank to reach the discharge point with temperatures higher than those of the "western systems".

For "eastern systems" the discharge temperature is higher at 6 p.m. than at 8 p.m. while the "western systems" discharge temperature is lower. The reason for this phenomena is that after 6 p.m. the "eastern systems" are well beyond solar radiation time and their temperatures are already decreasing while for "western systems" at 6 p.m. the solar radiation is weak and the discharge temperatures are low as long as the direct flow occurs. Later on the radiation is weak enough to cause reverse flow before 8 p.m., which brings the hot layers of water to the top of the tank causing higher discharge temperature than those at 6 p.m.

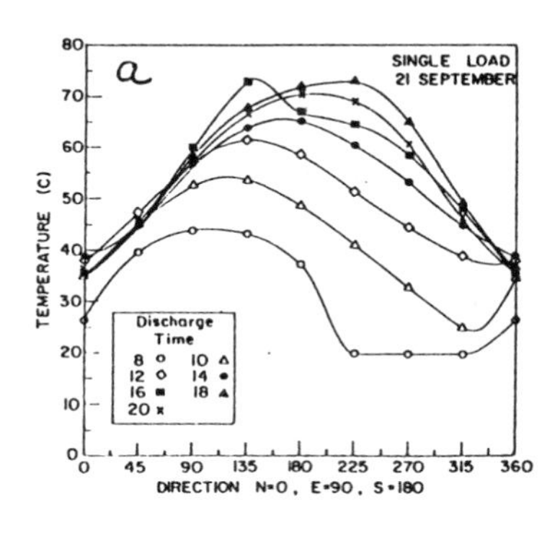

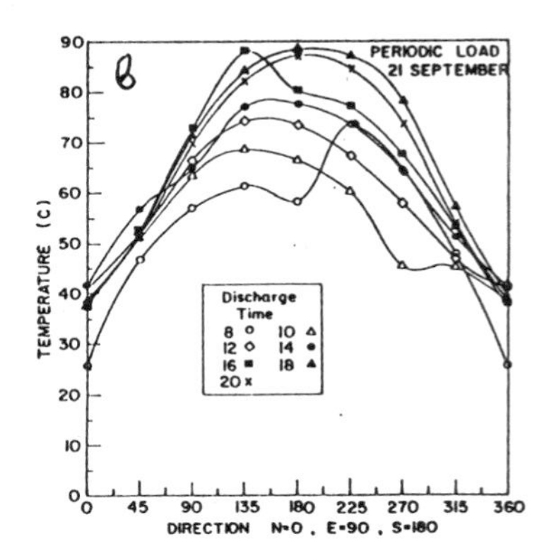

Fig.2 Temperature vis. direction for September

In Figure 2(a), for September, one can see that at 8 a.m. the preferred orientation is east, while between 10 a.m. and 12 p.m. the preferred orientation moves toward the south. At 4 p.m. the preferred orientation is southeast. This can be explained by the fact that the reverse flow begins earlier in the system, and as a result, the temperature of the discharge water rises.

In Figure 3(a), for December, it may be seen that the preferred direction of the collectors between 8 a.m. and 10 a.m. is south-easterly. For the remaining hours considered, the preferred direction is south. This excludes 4 p.m., for which the preferred direction is south-east. Here, as in the previous cases, the explanation is that of reverse flow occurring at that time.

In Figures 1(b), 2(b) and 3(b), the temperatures versus the direction, according to periodic load mode operation of the system, are shown. Generally, the temperatures are considerably higher, as the liquid in the system is hotter at the start of the day, having warmed up during the preceding days.

Fig 1(b) describes the temperature distribution for June. For 8 a.m. the discharge temperatures are higher for the west side than for the east. The explanation for this is that for the east side, at this hour, there is a weak thermosyphonic flow with weak solar radiation. This situation causes the upper layers of the water to be colder than the lower layers. In contrast, in a "western system" the upper layers are warmer, due to the slow reverse flow at night.

From 10 a.m. through 6 p.m. the preferred side is the east, while for 8 p.m., the west side is preferred.

In Figure 2(b) the temperature distribution for September may be seen. In this month, at 8 a.m., the preferred side is south-west and from 10 a.m. - 4 p.m., the preferred side is south-east.

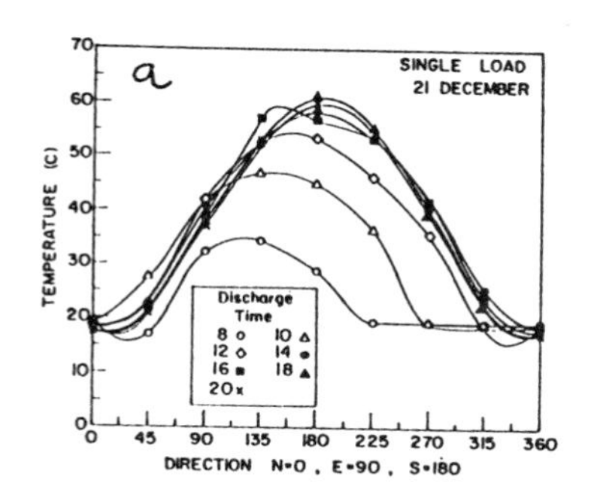

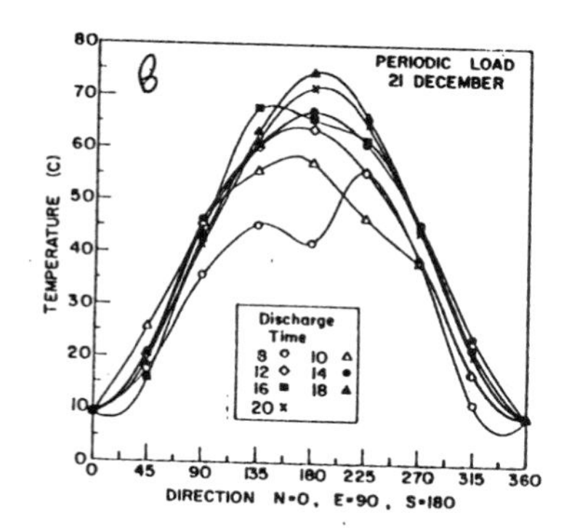

Fig.3 Temperature vis. direction for December

Figure 3(b) shows the distribution of temperatures for December. For most of the hours, the preferred side is south, with the exception of 8 a.m. and 4 p.m. The previously offered explanations are applicable to these results as well.

REFERENCES

1. J. Close, (1962). The performance of solar water heaters with natural circulation. Solar Energy 6, 33-40.
2. K.S. Ong, (1976). An improved computer program for the thermal performance of a solar water heater. Solar Energy 18, 183-191.
3. J. Huang, (1980). Similarity theory of solar water heater with natural circulation. Solar Energy, 25, 105-116.
4. M. Vaxman, M. Sokolov, (1986). Effects of connecting pipes in thermosyphonic solar systems. Solar Energy, 37, 323-330.
5. E.E. Anderson, (1983). Fundamentals of Solar Energy Conversion. Addison-Wesley Publishing Co.

The BUBBLE PUMPED SOLAR DOMESTIC HOT WATER HEATING SYSTEM known as THE SOL-PERPETUA SYSTEM

Wilfred Sorensen, President
Bubble Action Pumps Ltd.

4 Cataraqui Street
Kingston, Ontario, Canada
Postal Code K7K 1Z7 tel.(613) 5424045

Key Words:
bubble pump, two-phase flow, thermo-siphon
self-pumping, self-controlling

Brief Description:

The bubble pump is a two-phase thermo-siphon device which is admirably suited to liquid filled solar collectors because it makes them self-pumping and self-controlling. The bubble pump moves the heated liquid from the collector down to the heat exchanger at the storage tanks, then returns it to the collector to be reheated. Motors, mechanical pumps, controls and wiring are eliminated from all systems where the storage tank must be located at a lower level than the collectors such as in frost prone climates and also wherever the weight or appearance of the storage tank is undesirable on the roof.

Operation:

In the bubble pump device a liquid is caused to circulate in a closed loop system by the driving power of vapour bubbles rising in a bubbling tube. The bubbles are condensed at the top of the bubbling tube by losing heat to the cooler liquid returning from the heat exchanger (see diagram). The bubbles are generated by the heat source, and since the system is free of noncondensible gases, the bubbling action will begin as soon as heat is added to the liquid no matter what its temperature. Hence, the bubble pump will work within a wide range of temperatures.

Bubble Action Pumps Ltd. has developed a device for moving heated liquids downwards from the point of heat input against its normal desire to move upwards. The motive power is the gentle force of rising vapour bubbles. We call it the "Sol-Perpetua" system. The trademark shown on the right will be affixed to solar water heating systems built in accordance with Bubble Action Pumps' engineering know-how and patents.

The many advantages of the Sol-Perpetua system are as follows:

1. No electric power is used; it's passive but has the advantages normally associated with active systems.
2. No pump and no motor to wear out.
3. No controls to malfunction.
4. No wiring needed whatsoever.
5. No parasitic power loss. Instead, the solar powered pump actually adds energy to the system.
6. Operation is quiet.
7. There is no drain down and no expansion tank.
8. Pumping action starts as the collector(s) and pump become warm and stops as they cool off. Rate of pumping varies as the rate of insolation. It cannot pump at night to cause loss of energy.
9. Sol-Perpetua can be used with virtually any type of collector including evacuated tube collectors.
10. One pump can serve two or more panels even if they have differing orientations. For example, a panel facing east will flow in the morning while there is no flow in a panel facing west. In the evening, the reverse will happen. This quality will appeal to home-owners who do not have a south facing roof.
11. No tank on the roof to look ugly, cause structural problems or spring a leak. Tank can be 15m or more below collector.
12. Heat exchanger at tank is convective type requiring no motor or controls. Tank remains stratified as only hot water is delivered to it. Water delivery is approximately 50 C even on cloudy days.
13. Heat exchanger can be inside tank or on the outside according to code requirements. A second tank is not needed.
14. Working fluid is freeze protected as required down to -50. Seldom needs changing.

15. System is sealed after all corrosion causing gasses have been removed by pumping or by boiling. Interior passages remain clean.
16. Bubbling action is visible even from the ground. The condition of the system is immediately apparant and

IT'S FUN TO WATCH !!!

LIMITATIONS

1. Plumber must be able to make an airtight piping system.
2. Good performance of collector array depends on a suitably sized heat exchanger (as for any system). Bubble Action Pumps Ltd. will carry the appropriate sized heat exchangers.
3. Top of bubble pump must be at least 6m above top of tank.

PERFORMANCE

The performance of a Sol-Perpetua system will depend on location, quality of collector used, the heat exchanger, the tank size, and the competance of the installer. The bubble pump simply pumps as dictated by the rate of insolation and the temperature of the water in the bottom of the tank. There are as yet no quantitative means of evaluating the bubble pump.

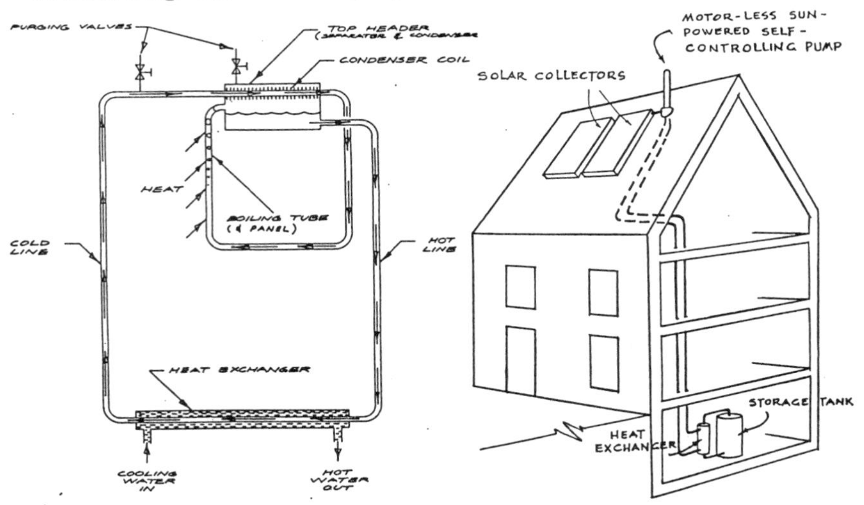

FIGURE 1: THE BUBBLE PUMP

THE BUBBLE PUMPED D.H.W. SYSTEM

Simplest possible solar circuit

Svend Erik Mikkelsen and Simon Furbo

Thermal Insulation Laboratory
Technical University of Denmark
DK-2800 Lyngby, Denmark

Abstract

Different simple designs of the solar circuit will be presented and the performances of the systems will be compared. These are based on calculations, partly verified by experiments and measurements.

One of the results is that with the heat exchanger located low in the storage tank (containing domestic hot water), back flow during night time is not important since no heat can be extracted from the storage. Futhermore the control of the pump can be simplified to a lightmeter or even a clock.

Keywords

Solar DHW system. Simple control. Back flow. Simple system. Solar circuit. Solar collector loop. Heat exchanger. Air escape. Open expansion. Light meter.

Background

Experiences from practical use of solar systems show that most problems are connected to the solar circuit.

The problems can be very different, fx.:

Back flow is cooling the tank during nighttime

The sensor for starting the pump is too slow and the system is boiling before the pump starts.

The air that is seperated from the liquid in the collector (temperature is rising and pressure is dropping), is not removed through an air escape valve. At last the circulation stops and the systems is boiling.

The pressure loss through the collector is too small and the flow distribution is unequal. Maybe the flow is zero in part of the system.

The pump is placed so the air is trapped in the pump.

The one way valve is not working.

The best way to solve all these problems is to design a system so that the problems can not arise. On the Thermal Insulation Laboratory we have for some years worked on simplifying the solar systems, among that especially the solar collector loop and the storage system.

The figure shows different designs for the solar circiut:

1) Traditional system in Denmark. Pressurized system. Control of the pump by a differential thermostat. One way valve to prevent back flow during the night.
The solar circuit is shown with 3 different storage systems to make it possible to compare the different systems. In the system with two heat exchangers the top one is for pree-heating and is connected to a boiler. It is not possible to compare the performance of this system directly to one without this heat exchanger, because of the different heat losses involved.
We have good experienses with the system with two heat exchangers.

2) Traditional system with natural circulation. This is in many ways superior to the other systems and therefore widely used. The only problem is that the tank has to be placed at a higher level than the solar collector. This does not fit to Danish building codes and tradition and it give problems with freezing.

3) Here a one way valve or magnet valve is controlling the back flow.
The control system is a light meter that starts the pump when the solar irradiance is above 100 w/m^2. Or the pump is running between 8 am and 4 pm, when the collector is facing south.
The control system ensures that the pump is running when it is possible to collect energy. But if the storage temperature is high the pump will run at times where no energy can be collected. In fact the tank will be cooled. That is why the

height of the heat exchanger is important. Only the water below the level of the heat exchanger will be cooled, and because the heat conduction in the water is low, the temperature in the top of the tank is not very much influenced by the cooling of the bottom.
The system is futhermore with open expansion and a combined expansion and air escape container (1 liter) is placed in top of the system by the collector. This implies that the collector, the pipings and the heat exchanger is made of corrosion proof material, f.eks. copper.
The pump is placed on a vertical pipe pumping upwards. This means that no air can accumulate in the pump. As the system is open the temperature will not be much higher than 100 C (boiling), and the pump can be placed before the heat exchanger where the temperature is highest. This can give problems if the system is pressurized.

4) Identical to system no 3, but the one way valve is omitted.
This means that back flow will take place during the night resulting in a cooling of the tank.
The higher the heat exchanger is placed the more energy can be drawn from the tank.
The calculated performance of this system is based on the assumption that back flow is taking place when the temperature of the tank is higher than the temperature of the air surrounding the collector, unless the temperature in the tank is below 4 C. Then the heat convection from the water to the heat exchanger is dropping rapidly. Experiments verifies this, the temperature in the bottom of the tank surrounding the heat exchanger never drops below 4 C.
It is possible that the loss calculated in this way is exaggerated. Maybe the flow, for different reasons, will be very low in many cases.

5) Identical to system no 4, but in this system there is a heat exchanger in the top connected to a boiler. Or it could be an electrical heater. The top of the tank is then heated all year, and only one DHW tank is installed.
This lay-out is very carefully tested and widely used in danish solar systems today.

Performance of the different systems

The figure shows the calculated yearly performance. For system no 1 the performance has been fixed to 100 for all 3 tank systems, A B and C. The performances are then to be compared vertically in the figure.
Regarding the comparison of the 3 systems (spiral coiled heat exchanger, mantle heat exchanger and system with two heat exchangers) we refer to the paper "Is low flow operation an advantage for solar heating systems?" from this same conference.

The performance of system B is only slightly less than the performance of system A, becaurse the stratification in the tank is very efficient. This is verified by experiments.

Otherwise the calculations are based on -

solar collecter area 5 m². South. Selective absorber, one cover.

Storage tank. 200 liter in system A and C. 300 liters in system B.

Hot water consumption. 150 liters per day, heated from 10 to 50 C.

Conclusions and recommendations.

Systems with natural circulation is superior if it is possible to use it.

When the heat exchanger is placed in the bottom third of the tank and the pump is controlled by a simple light meter the performance is reduced with only 2% compared to more sophisticated control systems.

If futher the one way valve is omitted, the performance is redused with 8%.

If now the heat exchanger is taking up the bottom 1/6 of the tank the performance is reduced with about 5%. Still compated with the sophisticated system with a one way valve and a differenceial termostate control system.

If then the top of the tank is heated all year (only one DHW tank in the house), the performance is reduced with 10%.

But if now the heat exchanger is taking up only 1/12 of the tank, the performance is redused with 7%.

The conclusion is that the height of the heat exchanger is not critical when discussing simple control. But if there is no one way valve and especially if the top of the tank is heated, the height of the heat exchanger is important.

The very simple system with the simple control, no check valve and with two heat exchangers (only one DHW tank in the house), can be used when the height of the heat exchanger for the solar collectors is about 1/10 or less of the height of the tank.

The performance will then be slightly below the theoretical optimum. But the advantages with such a simple system (price, relyability, maintanance) is much more important in our opinion.

Ref. 1: Mikkelsen, S. E., "Output and experience with 31 solar domestic hot water systems in Denmark", Thermal Insulation Laboratory, Technical University of Denmark, report no. 86-1, february 1986.

SYSTEM	control	FLOW / PIPES	One way valve?	Expansion	Air escape	Performance compare vertical! H/E	A	B	C
1	Differential thermostat	1,0 l/min per m² ϕ16 mm	yes	Pressurized	Valve at collector	1, 3, 6, 12	=100	=100	=100
2	Natural circulation	Variable ϕ22m	no	Open	Expansion tank				>100
3	Light meter I>100 W/m²	0,5 ϕ10mm	yes	Open	same	1	1)		
						3	98		
						6	99		
						12	99		
4	same	same	no	Open	same	1	65		
						3	92		
						6	95		
						12	96		
5	same	same	no	Open	same	1		2)	
						3		80	
						6		90	
						12		93	

1) Totally dependent on flow in heat exchanger. 2) Very small.

Explanations

H/E Height of storage divided by height of heat exchanger.

A Heat exchanger (spiral coiled) connected to solar collextors.

B Heat exchanger in top for post-heating, connected all year.

C Mantle heat exchanger connected to solar collectors.

Performance of system 1 fixed to 100 for A, B and C.

Compare performances vertically in the figure.

Pump

Control

One way valve

Safety valve

Expansion (pressure)

Expansion (open)

DEVELOPMENT OF A LOW-COST SOLAR COLLECTOR SYSTEM

Henk Muis
Energy Unlimited (Energie Anders)
Leeuwenstraat 9-11, 3011 AL Rotterdam, Netherlands

ABSTRACT

In this paper a solar hot water system is discussed, based on a once-through principle and built of plastic materials. The aim was to produce a system which would be economically feasible in Dutch climatic conditions. Results of both calculations and measurements are given. These indicate that the proposed system is indeed feasible, both technically and economically.

KEYWORDS

solar, collector, low-cost, hot water, plastic

INTRODUCTION

Some years ago developments started to design and build a prototype of a small low-cost solar hot water system, capable of producing an average of 60 liters of hot water a day. Most of the necessary research was done in our firm and was sponsored by the Dutch government within the framework of the National Solar Energy Research Programme.

The starting point was the observation that a complete solar domestic hot water system (DHW), in order to be economically feasible, should cost no more than the energy saved in 10 years.

In Dutch circumstances the price for buyers therefore must not exceed Dfl. 1.000,-- ($500,--, 1987).

As a consequence, the system had to be very simple, lightweight and easily transportable. This makes it suitable for the do-it-yourself sector, avoiding the cost of installation which makes up for almost 50% of the total cost. Also cheap materials had to be used and there would have to be the possibility of mass production. Therefore most of the components were made of plastic material. When the system had been designed, calculations were done concerning its efficiency and economic feasibility. Also a prototype was built, using standard components.

On this prototype measurements have been taken, both in outdoor circumstances and under an artificial sun. Calculations and measurements proved that the system should be feasible.
Another item of research was the development of the plastic components especially the absorber. After completion of the research a manufacturer has taken up the design.

SYSTEM

In order to achieve a simple system the number of components has been reduced as compared to conventional solar systems.
No electronic equipment has been used. The only connection which has to be made is the water supply. This enables a quick installation by consumers themselves.
In Fig. 1 the configuration is shown.

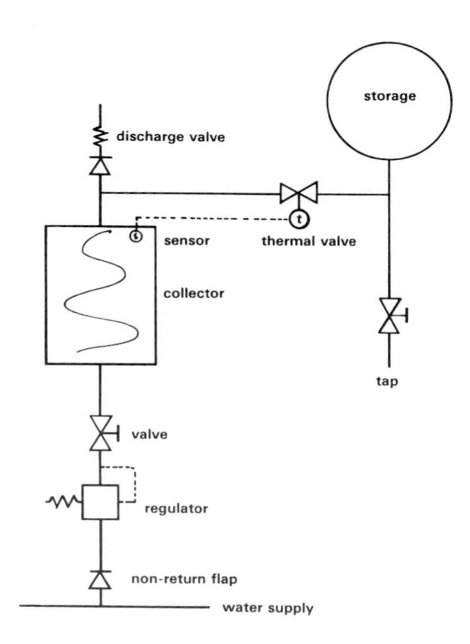

Fig. 1. Schematic lay-out of the solar hot water system

The most important feature of this system is the fact that the water only passes through the collector once, driven by the water pressure, which eliminates the need for a pump. The flow is controlled by a thermal valve, opening when the water inside the collector has reached a pre-set value. The water is stored in an insulated container.
Figure 2 shows the complete system, and the arrangement of its components.

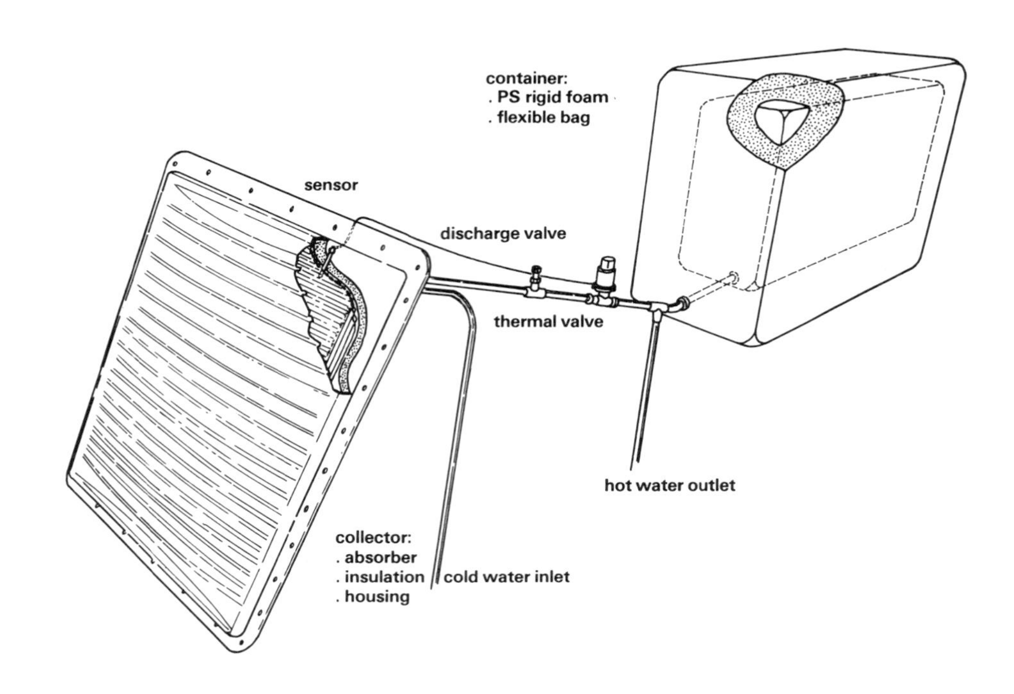

Fig. 2. The complete system

RESULTS

Calculations on a computermodel of the system and measurements on a prototype system showed that the DHW's efficiency is comparable to conventional systems. The yield of hot water was calculated by simulating the system using the data of sunshine in the Netherlands in the year 1962. From the available 1072 kilowatthours per square meter 494 kilowatthours were converted into hot water, approximately 10 cubic meters of 50 degrees Celsius. The efficiency of the system, averaged over the year, is thus 0.45.

In the experiment under controlled circumstances in the Dutch solar test facility, the collector was illuminated with 785 Watts per square meter. Every 6 minutes 2.5 liters of hot water were produced and transported to the storage. In Fig. 3 some of the registered parameters are shown. These results agreed with the calculations.

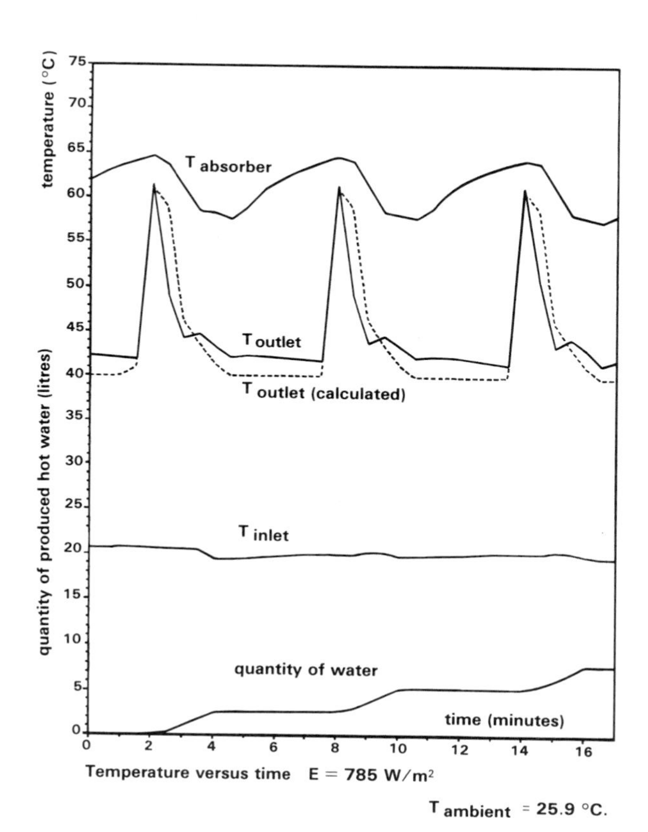

Fig. 3. Temperature and produced hot water versus time in solar test facility

PLASTIC ABSORBER

Next to the development and testing of the once-through system research was done concerning an all-plastic collector.

Materials were identified which could satisfy the rather severe conditions of an absorber both in an operating and non-operating mode. During stagnation the temperature of the absorber could reach a temperature (in extreme conditions) of 155 C.
In winter the contents of the absorber could freeze. Apart from this the absorber material must be heat-sealable, non-toxic and durable.
Furthermore it should be possible to increase the absorption index by blackening.
An extensive survey was made of available plastic films and only some fluoropolymers proved to be useful. Small-scale absorbers have been made, using a heat-sealing method to create parallel channels. To scale these up to life-size absorbers, suitable for mass-production, more research is needed.

CONCLUSION

Results sofar have indicated that it is indeed possible to manufacture a solar hot water system for less than $500,-- per square meter. This can be achieved by using a very simple system and plastic materials. The system proved to operate satisfactorily. More research is needed for the plastic absorber.
Recently a manufacturer has taken up the design in order to market the system, starting in the recreational sector.

THEORETICAL STUDY OF A NEW SIMPLE DESIGN FOR SOLAR FLAT PLATE COLLECTOR

I.A. Sakr* A. Hegazy**

* Solar Energy Lab. National Research Centre, Giza-EGYPT.
**Mech. Power Dept., Faculty of Engg., Menoufia Univ.

ABSTRACT

In this paper two new Variants of Flat Plate Collector are presented. A mathematical model for analyzing these two variants is proposed. The results obtained by the proposed mathematical model are discussed and compared with those of the conventional Flat Plate Collector.

KEYWORDS

Flat Plate Collector; beam radiation, diffused radiation, absorptivity, reflectivity, re-reflected, power factor, tube mean heat flux.

INTRODUCTION

Solar energy is presently being increasingly collected by flat-plate collector (FPC) and used at moderately low temperatures; less than 100°C. FPCs have the advantages, of being able to absorb both direct and diffused sun radiation, and they need a little maintenance.

There are two main objectives in the thermal design of a solar FPC:

1. Absorb as much as possible of the available solar energy (high optical efficiency),
2. Transfer as much of this received energy as possible into water (low thermal losses).

The first item can be easily realized by using a transparent cover having high transmissivity and a selective coating for the absorbing surface with high absorptivity for solar radiation. To fulfill the second item, it necessitates using high technology for manufacturing the collector absorbing surface. This results in, high price of the FPC.

In this work, two new designs of FPC are theoretically and mathematically studied. In these designs, the construction of FPC is simple, and thereby simple techniques are required for manufacturing the absorbing surface and in turn , a low price FPC is obtained.

INVESTIGATED SIMPLE FPC SHAPE

Two new variants of FPC are shown in Fig. 1 "a & b". The FPC in Variant I (Fig. 1a) consists of the collector risers; i.e the absorbing element (1) suspended in the enclosure (2) which has a front side (3) of a transparent material like glass, while the back side (4) is covered with a specular reflecting material such as Aluminum Foil. The risers have an outer diameter d and are arranged such that their axes are longitudinally and apart from each other by equal spacing t. The risers are located at a distance h_T above the reflecting surface (4). Variant II of such FPC is quite similar to variant I, but its absorbing tubes (risers) are provided with one side fin (5) of width w_F.

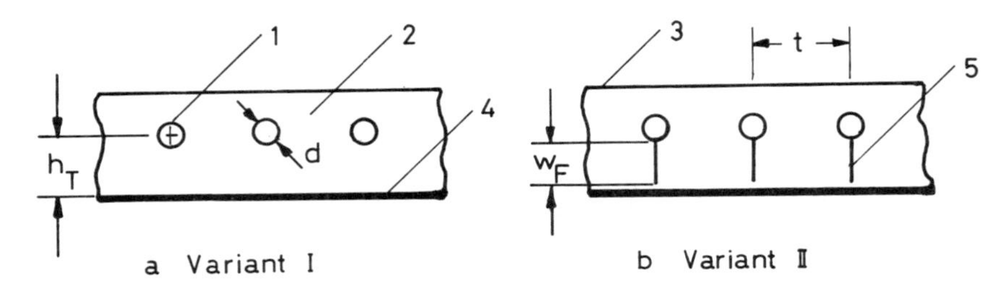

Fig. 1 Investigated FPC Variants
1 absorber element 2 enclosure 3 transparent cover 4 reflecting surface
5 fins.

The incident solar radiation (Fig. 2) goes thru the glass. Part of it impinges directly upon the absorbing surface, while the rest goes thru the spacing between the absorbing surface elements (tubes or tubes and fins) where it is reflected by the reflecting surface back to the absorbing surface and the glass. For the mathematical analysis in this work, it is assumed that the reflection is specular although it might be mixed of specular and diffused one.

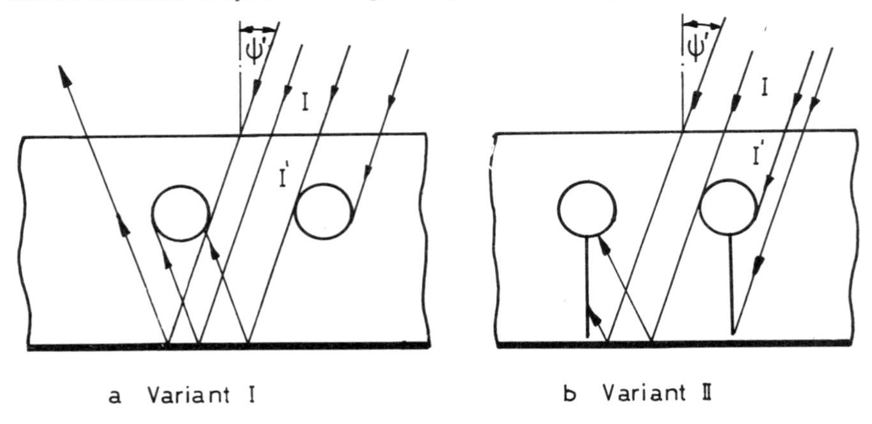

Fig. 2 Incidence of sun's radiation on the absorbing surface.

The ratio of absorbed and reflected energy of the incident solar radiation on the absorbing surface depends upon the absorbing surface coating and the incident angle. The reflected radiation might go out thru the glass as loss or it might be directed to the reflecting surface or to another element of the absorbing surface.

It is clear, that the solar radiation absorbed, reflected and re-reflected exchanges between the different parts of FPC under these three modes. In the first one, the absorbing surface absorbs the sun radiation incident directly on it and it is denoted as direct absorbed radiation. In the second mode it absorbs the reflected radiation by the reflecting surface behind the absorbing surface. This absorbed energy is denoted by reflected absorbed radiation. For the third mode, the absorbing surface absorbs the re-reflected solar radiation and this absorbed energy is denoted by multi re-reflected radiation.

DETERMINATION OF THE ABSORBED SOLAR RADIATIVE ENERGY

To get the performance of the new design FPC (Variants I and II in Fig. 1. a & b) at different conditions, it is necessary to determine the values of the three modes of absorbed energies mentioned above. The determination of these energies requires pursuing the incident solar rays on the FPC aperture. The mathematical model described here is used for calculating the values of these energies caused by the beam radiation. Some assumptions have been made for simplification of the mathematical model:

1. The length of the FPC is so long, that the effect of the collector sides on heat-exchange is neglegible.
2. All reflections taking place in FPC are specular.
3. Transmissivity of the collector cover and absorptivity of the absorbing surface coating are dependent on the incident angle.

The sun intensity I penetrates the collector cover and will be

$$I' = I\ \tau(\psi') \tag{1}$$

where $\tau(\psi)$ is the transmissivity of the collector cover. The sun's rays impinge on a riser tube as shown in Fig. 2a. The area of the tube exposed to direct radiation subtended between angles φ_{d_1} and φ_{d_2} (see Fig. 3).

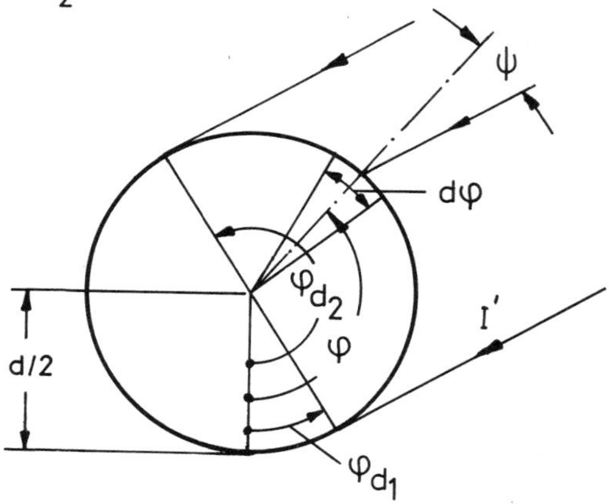

Fig. 3 Details of direct radiation absorbed by a riser tube

The power absorbed Q_d of the direct incident solar radiation is given for a tube unit length by

$$Q_d = \frac{I d}{2} \int_{\varphi_{d_1}}^{\varphi_{d_2}} \tau(\psi')\ \alpha(\psi)\ \cos\psi\ d\varphi \tag{2}$$

where ψ is the incident angle. Being able to determine $\varphi_{d_1}, \varphi_{d_2}, \psi'$ and ψ, eqn. (2) can be numerically solved to give the value of Q_d. This is carried out simply by using vector analysis and simple geometry.
By analogy with the previous derivation one can get similar formulae for the reflected and multi re-reflected absorbed power Q_r and Q_{mr}:

$$Q_r = \frac{I d}{2} \int_{\varphi_{r_1}}^{\varphi_{r_2}} \tau(\psi')\ \alpha(\psi)\ \rho\ \cos\psi\ d\varphi \tag{3}$$

and

$$Q_{mr} = \frac{d}{2} \int_{\varphi_{mr_1}}^{\varphi_{mr_2}} I''(\varphi)\ \alpha''(\varphi)\ CF\ d\varphi \qquad (4)$$

In equ. (3) ρ is the reflectivity of the reflecting surface. In the last equation the intensity of radiation I'' (φ) and the absorptivity $\alpha''(\varphi)$ at a point (φ) on the tube are variable and depend on the angle φ. They are resulted due to reflection of radiation from different orts on the absorbing and reflecting surfaces. CF is the cosine factor for the different radiation directions.

Power Factor ζ is representing the ratio of power absorbed to the power incident on the collector aperture. Thus:

$$\zeta_d = \frac{d}{2\cos\psi'\ t} \int_{\varphi_{d_1}}^{\varphi_{d_2}} \tau(\psi')\ \alpha(\psi) \cos\psi\ d\varphi \qquad (5),$$

$$\zeta_r = \frac{d}{2\cos\psi'\ t} \int_{\varphi_{r_1}}^{\varphi_{r_2}} \tau(\psi')\ \alpha(\psi)\ \rho\ \cos\psi\ d\varphi \qquad (6)$$

and

$$\zeta_{mr} = \frac{d}{2\,I\cos\psi'\ t} \int_{\varphi_{mr_1}}^{\varphi_{mr_2}} I''(\varphi)\ \alpha''(\varphi)\ CF\ d\varphi \qquad (7)$$

Given equations (5), (6) and (7), it follows for the absorption efficiency η_N of the new design that:

$$\eta_N = \zeta_d + \zeta_r + \zeta_{mr} \qquad (8)$$

The tube mean heat flux q_{N_m} of the FPC is given in consideration of equations (2),(3) &(4) by

$$q_{N_m} = \frac{Q_d + Q_r + Q_{mr}}{\pi d} \qquad (9)$$

This is primary indication of the water temperature that can be attained.

EFFFECT OF FPC GEOMETRY ON THE POWER ABSORBED

Analysing eqns. (2) thro' (9) one could conclude the results in figure 4. It shows for both Variants I and II the power factors $\zeta_d, \zeta_r, \zeta_{mr}$ and the absorption efficiencies η_N, η_C of the two variants as well as of the conventional FPC versus the tube spacing ratio t/d for altitude angle $\Theta = 60°$ and azimuth angle $\gamma = 60°$. The figure shows also the tube mean heat fluxes q_{N_m} & q_{C_m} of the two variants and the conventional FPC.

It is obvious from the left hand side diagrams of Variant I that, the absorption efficiency decreases with increasing value of t/d. q_{Nm} increases sharply with the increase of t/d till about t/d = 3, then the rate of increase becomes very slow. So, to have satisfactory high absorption efficiency and possibly high water temperature for Variant I and at the same time to reduce the collector costs the ratio of t/d has to be near the value 3.

The diagrams in the right hand side of Fig. 4 for Variant II reveal that the absorption efficiency and the tube mean heat flux of Variant II is higher than that of the conventional FPC in the range t/d = 1 to 6.5.

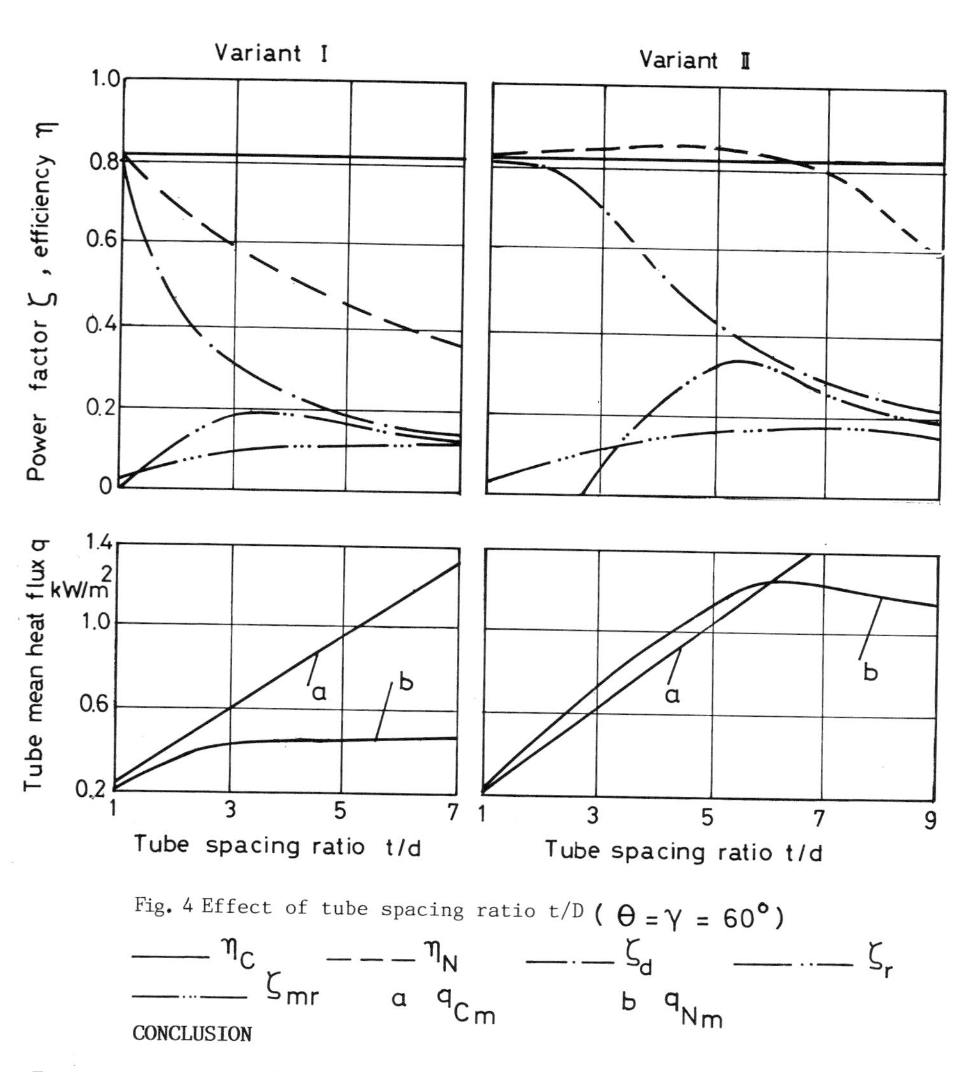

Fig. 4 Effect of tube spacing ratio t/D ($\Theta = \gamma = 60°$)

——— η_C – – – η_N —·— ζ_d —··— ζ_r

—···— ζ_{mr} a q_{Cm} b q_{Nm}

CONCLUSION

The present paper proposed mathematical analysis of two new designs of FPC.
A theoretical study has been carried out to reveal the effect of the main geometrical parameters of the two design variants on the energy absorbed by the absorbing surface. For Variant I the tube spacing ratio t/d has to reach the value 3. The tube height ratio h_T/d should have value of 1 and 1.5, while for Variant II t/d has to reach value 4 and 5 and h_T/d between 3.5 and 5.

REFERENCES

1) Duffie, J.A., and W.A. Beckman: Solar Engineering of Thermal Processes. John Wiley & Sons Inc., New York, 1980.

THERMAL PERFORMANCE OF SIMPLE UNCOVERED FLAT-PLATE COLLECTORS UNDER SYRIAN CLIMATIC CONDITIONS

Jamal Alomar, L.I. Kiss, M.Sharif

Technical University Budapest
Institute of Thermal and Systems Engineering

ABSTRACT

Simple uncovered roof collectors acting also as night-sky radiators and heat exchangers with the ambient air offer a simple and cheap solution to maintain human comfort in buildings under Syrian climatic conditions. A computerized design method has been worked out to predict the heating and cooling performance of roof collectors. The algorithm is based on a finite-difference type discretized numerical model of the thermal behaviour of the collector.

INTRODUCTION

Most of artificial cooling equipment are using closed thermodynamic cycles using high-value energies like electricity or gas for their operation. Production of these devices needs a relatively high level of manufacturing technology, their maintenance needs qualified personnel. The utilization of the cooling potentials of "natural" heat sinks as night-sky, ambient air or ground to avoid overheating due to high solar gain promises a solution adequate to several developing countries. Simple uncovered roof absorber/radiator arrays can produce not only cold water during night to charge a cool-storage tank, but they can produce hot water for domestic use. At the same time daytime use of roof absorbers helps the reduction of heating loads on the building. The contribution will be to the reduction of cooling demands will be significant only if the hot water demand is high or the re-cooling of hot water overproduction can be solved by using natural air or ground heat sinks.

The aim of present work is to create a fast flexible procedure to analyze cooling and heating performance of roof collector arrays and to perform of a comparative study of different-storage system variants.

METHOD

As uncovered absorber/radiators arrays are in direct thermal connection with ambient air and sky having different temperatures the concept of overall heat loss coefficients between absorber plate and ambient air must be replaced by a detailed description of heat transfer mechanism. Also instead of the classical approach (see for example in Ref. [1] using closed, explicit form functions to describe temperature change of heat-carrying medium, the discretized model of [2] and [3] was applied.

The main model assumptions are as follows:

1. In the absorber plane temperature variations in the direction perpendicular to the flow are taken into account by using average temperatures determined by fin efficiencies.
2. Along the flow direction of the medium, only its convective enthalpy flux takes part in heat transfer – the heat conduction in the absorber is neglected.
3. The low temperature emission and absorption of solar radiation is characterized by two average spectral properties, directional behaviour of absorber surface is diffuse-type.

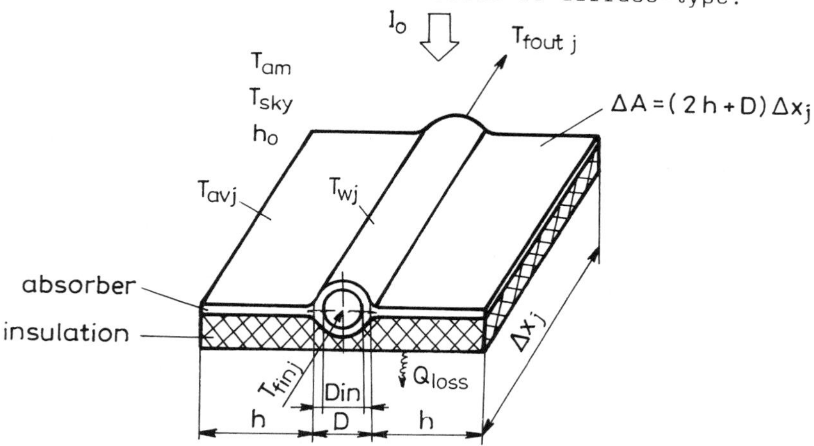

Fig.1. Element of absorber/radiator

Along the flow direction the collector is lumped into discrete parts (fig.1.) and for each of the heat transport in the direction perpendicular to the base plane is described by a nodal type heat balance equation as follows:

$$I_o \cdot a \Delta x_j (2h+D) = h_{in} D_{in} \pi \cdot \Delta x_j (T_{wj} - T_{fav_j}) + 2h\, K_1 \cdot \Delta x_j (T_{av_j} - T_o) + D\pi/2\; K_1 \Delta x_j (T_{wj} - T_o) + h_{1j} \cdot 2h \Delta x_j (T_{av} - T_o) + h_{sj} \Delta x_j \, D\pi/2 (T_{wj} - T_s) \qquad (1)$$

where: $K_1 = h_o + K$

$$h_{1j} = \varepsilon \cdot \sigma_o (T_{av_j}^4 - T_s^4)/(T_{av} - T_o)$$

$$h_{sj} = \varepsilon \cdot \sigma_o (T_{wj} + T_s) \cdot (T_{wj}^2 + T_s^4)$$

In the jth discrete part of ΔA_j surface the heat-carrying medium is warmed up by

$$\Delta T_{fj} = \frac{\dot{Q}_j}{\dot{m}\ c_p} \qquad (2)$$

where $\dot{Q}_j$ can be calculated using the resultant tube wall temperature from (1) and either a predicted average fluid temperature in the jth section or the fluid outlet temperature from the (j-1)th part:

$$\dot{Q}_j = h_{inj}\ \Delta A_j (T_{wj} - T_{fj}) \qquad (3)$$

The algorithm is advancing along flow direction step-by-step in each step solving the nonlinear heat balance Equation (1). Special subroutines compute necessary coeficients for Equation (1). A solar processor determines the total solar intensity on receiving surface depending on geographical location, calendar day, collector tilt and orientation, total and diffuse horizontal radiation data. Measured solar radiation data, air temperatures, humidites, wind velocities for each month in Hungary and Syria are stored on magnetic tape for easy creation of time series driving the simulation procedure. Clear sky and cloudy days can be combined in an arbitrary sequence.

Thermal conductance and fin efficiencies in (1) are also determined by their own subroutines. Equivalent sky temperature can be computed from Swinbank formula (Ref. [1]).

The only important result of collector performance computation is the fluid outlet temperature. This temperature together with the mass flow rate gives the coupling of the collector to the other parts of a whole system as heat-exchanger, storage tank.

SYSTEM ASPECTS

As the amount of water which can be chilled by nocturnal radiation from the roof of a given house is usually not enough to ensure total conditioning, special attention should be paid to proper selection of storage size, to combination with hot water supply, ground storage etc. Several variants are recently under study, one of them intended to reach nearly an energy-autonomous operation mode is illustrated by fig. 2.

To analyse the performance of such a system, long-term simulation is necessary.

The simulation procedure has a hierarchic structure it consist of various part models as

- Meteorological data processor
- Cooling load simulator
- Control system behaviour simulator
- Mass flow network model
- Collector/radiator model
- Storage simulator
- Heat exchanger subprogram.

The procedure can be used for design purpose indirectly: starting with intuitively given size of components the systems performance in form of temperature history at storage tank outlet is determined.

PERFORMANCE CHARACTERISTICS OF ROOF ABSORBERS/RADIATORS

As uncovered collectors are in thermal contact with heat sources and sinks at various temperature levels the traditional colector efficiency definition does not give proper information about the thermal performance. In the traditional collector efficiency definition only solar irradiation serves as base of comparison for the heat flow rate represented by the heat carrying medium, if ambient air is warmer then inlet fluid, efficiency will be higher than 100 percents. In night time operation the original definition cannot be applied as the denominator is zero. For nocturnal radiators a dimensionless temperature ratio can be introduced to characterize their performance

$$\theta = \frac{T_{fout} - T_{fin}}{T_{eq} - T_{fin}} \tag{4}$$

where T_{eq} denotes the equilibrium temperature of the absorber/radiator plate without fluid circulating inside. T_{eq} is affected by T_s , T_{am}, emissivity of plate, wind velocity, position and is an equivalent or resultant heat sink temperature of sky and air. θ defined by Equation (4) gives the ratio about the utilization of the cooling potential of equivalent heat sink.

RESULTS

According to the above mentioned the fluid outlet temperature satisfactorily characterizes the behaviour of uncovered roof absorber/radiators. As example figure 2 is given. To study the effect of mass flow rate on outlet temperature, figure 3 is presented.

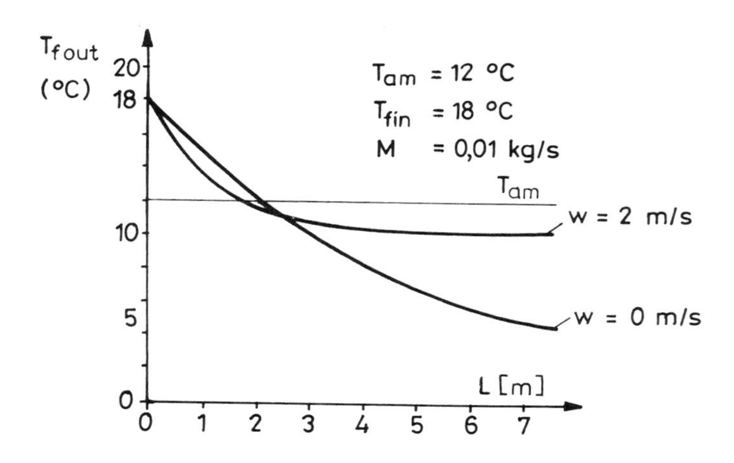

Fig. 2. Characteristic of absorber/radiator (Night time)

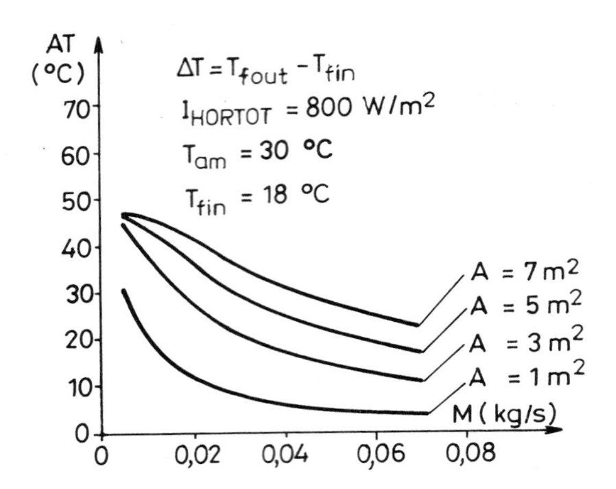

Fig.3. Effect of mass flow rate on outlet temperature (Day time)

NOMENCLATURE

ΔA : element collector area
a : absorptance of absorber
c_p : specific heat of fluid
D : tube diameter
D_{in} : inside tube diameter
h_o : heat transfer coefficient between air absorber
h : distance between each 2. tubes
h_1 : heat transfer coefficient at x-div. through the plate
h_3 : heat transfer coefficient through the tube wall
h_{in} : heat transfer coefficient between fluid and tube wall
I_o : solar radiation intensity
K : conductive heat transfer
m : fluid mass flow rate
Q : heat transported by fluid
T_o : ambient temperature
T_{av} : plate average temperature
T_{fav} : fluid average temperature
T_{fin} : fluid inlet temperature
T_{faut} : fluid outlet temperature
T_s : sky temperature
T_w : tube wall temperature
w : wind velocity
ε : emissivity of the plate for solar spectrum
π : constant
σ_o : Stefan Boltzmann constant

REFERENCES

[1] Duffie, J.A., Beckman, W.A.: Solar Energy Thermal Processes, Wiley New York (1974)
[2] Kiss, L.I.: Simulation of the Thermal Characteristics of Flat--Plate Solar Collectors. Report of TOKAI University, Tokyo 1980
[3] Imre, L., Kiss, L.I.: Analysis of the Transient Thermal in "Numerical Methods in Heat Transfer" Vol. II.Wiley, 1983. pp. 395-446.

ROOF SPACE COLLECTOR

S. Østergaard Jensen and V. Korsgaard

Thermal Insulation Laboratory
Technical University of Denmark
Building 118, DK-2800 Lyngby, Denmark

ABSTRACT

The ratio between output and cost of solar heating systems can be improved by using simple but cheap collectors. The roof space collector is such a simple and cheap collector. With a roof space collector covering the south side of the roof 50% of the total energy demand for space heating and hot water production for a normal Danish house and 75% of this demand for a low-enbergy house can be covered under Danish weather conditions.

KEYWORDS

Roof space collector, simple and cheap collector, rock bed storage, air based solar heating system, high solar fraction.

BACKGROUND

The main problem in utilizing solar energy in Denmark is that there is only very little sunshine during the heating season. The total solar irradiation in December is only 400 Wh/m^2 per day, while the total solar irradiation in June is practically 6000 Wh/m^2 per day.

It is in principle possible to cover the demand for heat 100% by using seasonal storages and collecting heat during the summer and use it during the winter. However, seasonal storages for a single or a few houses are far from being cost effective today.

Another solution is to improve the ratio between the performance and the cost of the systems and in that way make large collector areas possible. You can either develop more efficient solar collectors, even if they are more expensive, or you can develop cheaper and therefore maybe less efficient collectors. With small temperature differences between the absorber and the ambient temperature (0-20°C) a simple collector is almost as efficient as more complicated collectors. To cover a major part of the heating demand during the winter months it is necessary to have a large collector area. The ratio between the output and the cost is thus best for a simple but cheap collector if the primary demand is to cover the heat demand (at 20°C) during the winter months. A very simple, cheap and large solar collector has been developed when making the roof space collector.

ROOF SPACE COLLECTOR

An experimental house with a roof space collector, financed by the Danish Ministry of Energy and the Danish Board of Technology, has been built at the outdoor test area of the Thermal Insulation Laboratory - see Fig. 1 and 2. The experimental house is 70 m^2 and insulated according to the Danish standards. The roof space collector is 44 m^2 and is connected to a rock bed storage of 2.7 m^3. The tilt of the roof is 30°. This is not the optimum tilt, but it is the most common tilt today in Denmark. The most important marked for roof space collectors will, therefore, be found when renewing the roofs of existing houses. If the roof space collector is profitable with a tilt of 30°, it will also be profitable at higher and better tilts.

Fig. 1. The experimental house with the roof space collector.

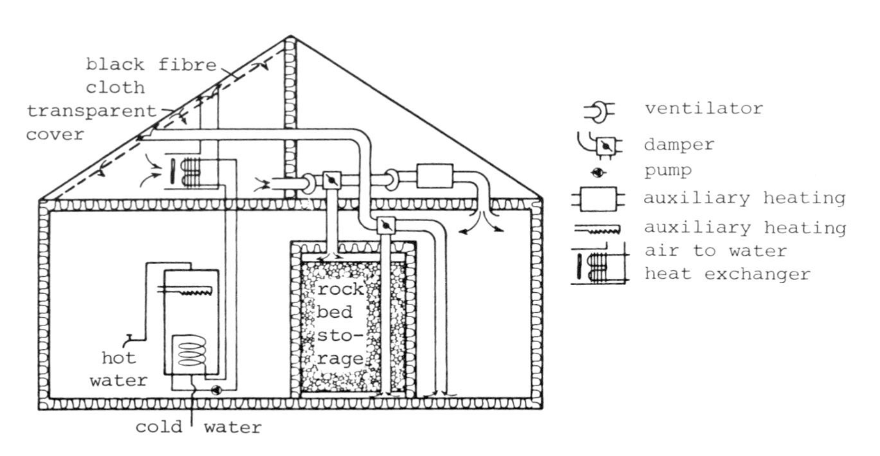

Fig. 2. A diagram of the experimental house.

As seen from Fig. 1, the roof space is divided into two separat collectors with different covers. Both covers consist of acrylate. The cover to the right has only one layer, while the cover to the left has two layers - a ribbed plate with double wall. In this way it will be possible to evaluate the influence of the covers on the efficiency of the roof space collector. (Fig. 2). A black fibre cloth is mounted between the cover and the roof space. This causes a higher heat transfer between absorber and air, than if the walls and the floor of the roof space were used as absorber. The space between the cover and the fibre cloth will be "colder" than the roof space, which gives a lower heat loss through the cover, and in that way a lower overall heat loss from the collector.

As shown in Fig. 2 there is an air to water heat exchanger in one of the two collectors. This heat exchanger supplies the domestic hot water tank with heat. Electrical auxiliary heating has been installed both in the water tank and in the air channel to the room. The operating modes for the solar heating system is shown in Fig. 3. The heating system maintains a room temperature of 19-21°C. The solar heat has first priority. The auxiliary heating is only switched on when the storage can no longer cover the heat demand.

Measurements

Measurements on the house started in the middle of February 1987 and are still going on. On the basis of the measurements a computer based programme (European Modelling Group Programme 2 - EMGP2 (Dutré 1985) has been validated (Jensen, 1987a, 1987b). The model has been validated to simulate the performance of the rock bed storage with a very good result, e.g. the difference between the measured and the calculated energy output from the storage lies below 5%.

Until now only very few measuring points can be used to determine the efficiency of the collectors. The efficiency equation based on these measurements is:

$$\eta = 0.43 - 4.1(T_i - T_a)/G \qquad (1)$$

where T_i is the inlet temperature, T_a the ambient temperature and G the global solar irradiation on the collector. The mean flow through the collector was 1560 m^3/h. The efficiency based on the mean collector temperatur will thus be:

$$\eta = 0.53 - 5.0\ (T_m - T_a)/G \qquad (2)$$

We had expected a higher starting efficiency and also a somewhat higher heat loss coefficient. This is also indicated by previous measurements, which have shown a higher overall efficiency. Further measurements will clarify this. There is not detected much difference in the efficiency of the two collector types but it seems that the collector with only one cover is slightly better than the collector with 2 covers. This is probably due to the higher transparency, which more than compensates for the lower insulation value of the single cover.

Simulations

EMGP2 has been used to give a first estimate of the performance for a house with a roof space collector. (Jensen, 1987b). The results from the valida-

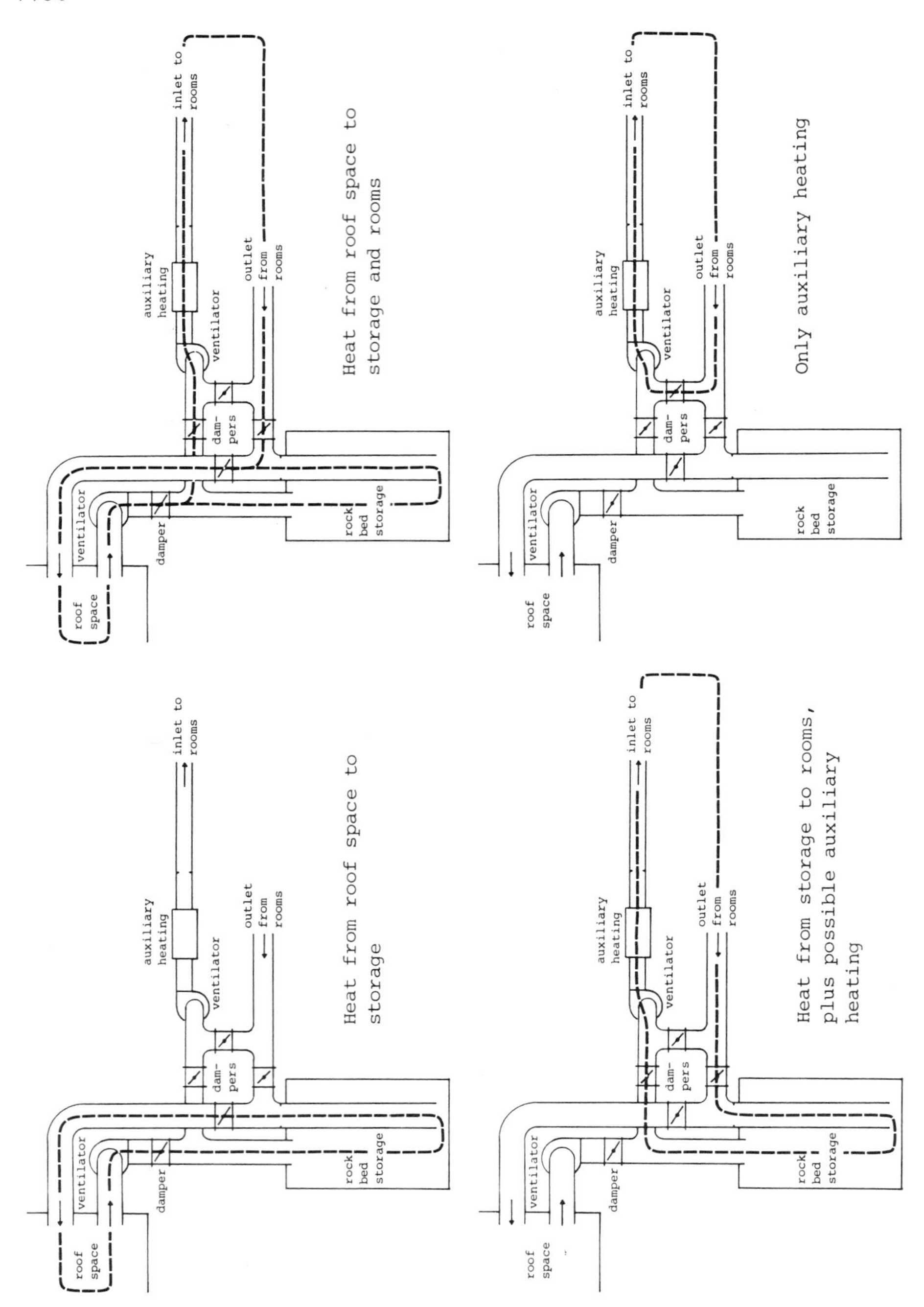

Fig. 3. Operation modes for the solar heating system.

tions have been used for the simulations: System parameters for the storage, ratio between collector area and storage volume, efficiency equation (2) for the collector, etc. To simplify these first calculations only the space heating was covered by the roof space collector while the domestic hot water was produced in a traditional water based solar heating system with a collector area of 5 m^2.

Figure 4 shows the results obtained from simulations with EMGP2. The performance has been calculated for two houses: 1) a house insulated according to the Danish standards - space heating demand 13,500 kWh/year and 2) a low-energy house - 6,200 kWh/year. The hot water demand was 2,500 kWh/year. The reference heating system was a traditional Danish water based central heating system with an oil fired boiler, which has an efficiency of 0.85 and a heat loss of 350 W. The calculations were furthermore made for two tilts of the roof: a) 30° - total collector area = 85 m^2 and b) 45° - total collector area = 105 m^2. The orientation of the house was south. The volume of the storage was 0.06 m^3 per m^2 collector area. The electricity to the ventilators is considered an extra energy demand for the houses with roof space collectors.

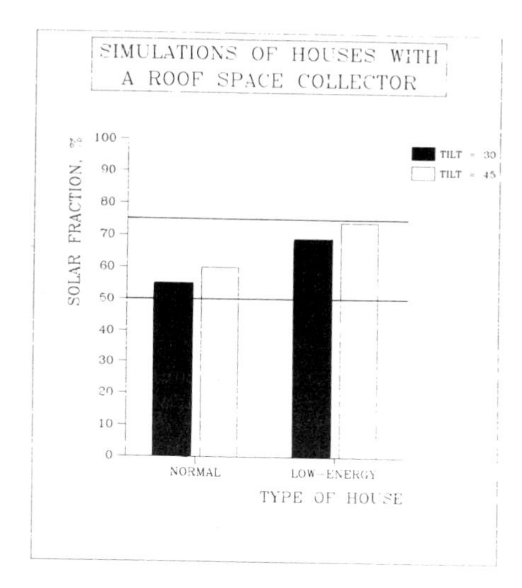

Fig. 4. The results obtained from simulations with EMGP2.

CONCLUSION

Although the investigations are not finished and the above mentioned calculations are a bit uncertain it seems reasonable to conclude that it is possible to cover 50% of the total energy demand for space heating and hot water production for a normal Danish house and 75% of this demand for a low-energy house under Danish weather conditions by using a roof space collector.

Economical calculations have not yet been performed but especially for new houses the roof space collector will be cost effective also because the traditional Danish water based central heating system is more expensive than an air based central heating system.

REFERENCES

Dutré, W.L. (1985). An European Transient Model for Solar Systems. Solar Energy R&D in the European Community - series A, Volume 5. D. Reidel Publishing Company for the CEC.

Jensen, S.Ø. (1987). Validation of models for simulation of solar heating systems. A paper to the ISES Solar World Congress (1987), Hamburg.

Jensen, S.Ø. (1987). Roof space collector - validation of and simulation with EMGP2. Thermal Insulation Laboratory. Technical University of Denmark. Report no. 87-15. Not yet published.

DHW HEATING BY MEANS OF A SELECTIVE ENERGY ROOF

R.M. Lazzarin
Istituto di Fisica Tecnica ed Impianti Termotecnici
University of Bari, via Re David, 200
70125 BARI, Italy
P.Romagnoni, L.Schibuola
Istituto di Fisica Tecnica
University of Padova, via Marzolo, 9
35100 PADOVA, Italy

ABSTRACT

The DHW heating by means of a selective energy roof was experimented. Two identical plants were connected to different receiving surfaces, made of selective steel or copper plate. The hot water was taken from the storage tanks according to the RAND profile to simulate a real working. The system revealed an appreciable efficiency particularly for the selective tiles, touching 20% on a daily basis (october) at temperatures as high as 40°C. It can be an interesting alternative to the traditional solar collectors for the advantages of lower cost and easy installation.

KEYWORDS

Energy roof; DHW heating; selective surfaces; selective steel.

INTRODUCTION

An important obstacle to a widespread utilization of solar collectors is the difficulty and the related cost of their installation. Even a strong reduction in the production cost is minimized by this almost fixed additional cost. One possible solution is the utilization of an energy roof, i.e. a structure with all the functions of a roof, but possessing integrated channels in order to let flowing a fluid, collecting energy. The installation cost may be very low, sharing the cost with the one related to the building roof. The aesthetic result is very good and the collecting surface occupies but the area of the roof. Moreover the shape of the roof can be by no means modified.
Of course an energy roof suffers by the strong effects of not disposing a transparent screen. Two main disadvantages are expected: thermal losses should increase strongly due to the higher convection and the plate is no more protected against direct weather threats. As far as the former disadvantage is concerned, in effect till now the utilization of the energy roofs was limited to the coupling to heat pumps with low thermal levels (Lazzarin, 1986). In order to reach the suitable levels for DHW the tiles were selectively treated. The reduction of the radiative component of the

losses was considered sufficient as the convective loss was not so high if computed by recently proposed formulae (Sparrow, 1977; Ramsey, 1980). As regards the latter disadvantage, it appears particularly redoubtable for a directly exposed selective surface. That is the reason why the experimentation started testing the behaviour of various selective surfaces (16 different types) before and after three months exposure to the atmosphere. Two treatments demonstrated excellent resistance to weathering: one on stainless steel, the other on copper. Though stainless steel has a low thermal conductivity, it was chosen for the experiments as it was produced in plates of a size suitable to form the tiles. Instead the copper surface was provided in coils too narrow to allow the manufacturing of the tiles and it was moreover too thin in thickness. Of course success in experiments could lead to a manufacture on purpose of the copper coils.
In the frame of the National Project PFE2 an experimental plant was built up to test selective roofs in DHW heating.

DESCRIPTION OF THE EXPERIMENTAL PLANT

The experiments have been carried out on two identical DHW heating plants connected to the following energy roofs:
ROOF1: the tiles, made of stainless steel, are obtained from a selective plate; the tiles are set on a base of copper tiles in order to improve the thermal conductivity of the system;
ROOF2: the tiles are formed of a copper coil.
The roofs were supported by a wooden structure and located on a terrace of the Metalux factory at Musile di Piave near Venice. The tile profile is the usual patented one of TOM Srl (fig.1). The tiles were directly riveted on the wood. The surfaces are facing South with a 38.4° tilt. A sketch of one roof can be seen on fig.2 with the main sizes; the whole area is 16.5 m^2.
Each roof is connected to an identical hydraulic circuit, formed of a storage tank with internal coil heat exchanger, whose capacity is 300 liters, circulation pump and the other usual fittings of a simple DHW solar plant (fig.3). The load is simulated by means of a solenoid valve at the outlet of the tank. This valve is activated by the control system for different periods each hour as a function of time in order to realize the RAND profile of demand. The tank is not supplemented directly by the mains but by an open constant level tank above the thermal room. Therefore the head remained constant as well as the flow through the valve.
The system is monitored by a data logger Delphin Control 300, connected to a PC-IBM. This provides:
- on/off of the pumps according to the usual differential control;
- on/off of the solenoid valves according to the intervals suggested by the RAND profile;
- data logging and saving on minidisks.

The surveys are still proceeding. Some first results related to two different period in October 1986 and June 1987 are here reported.

EXPERIMENTAL RESULTS

Fig.4 illustrates a typical daily trend for solar radiation intensity unto the roofs, outside air temperature and wind velocity in October. Figs. 5, 6 report for the same day inlet/outlet temperatures and energy roof efficiency for ROOF1 and ROOF2 respectively. The demand was kept at about 300 liter each day with the already specified distribution. The efficiency is satisfactory for both the roofs, being higher than 20% at midday, even if the temperatures are rather high. In fact the outlet temperature overcomes

often 40°C. This limits the daily obtainable efficiency, since in the afternoon the water in the tank is already too hot, because the RAND profile implies the highest request in the evening. The DHW was anyhow supplied at temperatures always higher than 35°C.
It must be underlined a particular feature of the feeding system. As the open tank which restores the tanks is set on the factory roof, the water is heated by the sun so that the temperature varies during the day from a starting value of about 11°C to more than 20°C at noon. This penalizes solar collection which cannot provide this low temperature heating.

CONCLUSIONS

The efficiencies are on the whole satifactory: see in fig.7 the comparison of the daily efficiencies of the two roofs during some days. Fig.8 reports daily solar radiation on the roof and average diurnal ambient temperature The efficiencies are of course the ratios of the daily useful collected energy and the daily solar radiation on the roof. The copper roof always overcomes 10% and it is sometimes above 15%. The selective roof is always above 15% and once it reaches 20%. A 300 liter demand was always supplemented with water at a temperature between 35 and 40°C which is acceptable for direct utilization.
The higher efficiency of the selective roof seems to justify the treatment whose cost is about 10-15% of the final cost of the installed tile. Of course these first results must be confirmed by a longer experimentation particularly in winter with lower outside temperatures and a stronger wind.
The selective energy roof seems to be an interesting alternative to the traditional solar collectors: the efficiency is about half, perhaps 1/3, but the installed cost can be estimated 5 times lower. At the same time the various locating, installation, maintenance and aesthtic problems find a common solution.

REFERENCES

Lazzarin, R.M. and Schibuola, L. (1986). Energy roof as heat pump source/sink, Int. J. of Refrg., 9, 108-117.
Ramsey, J.W. and Charmchi, M. (1980). Variances in solar collector performance predictions due to different methods of evaluating wind transfer coefficients, J. of Heat Transfer, 102, 766-768.
Sparrow, E.M. and Tien, K.K. (1977). Forced convection heat transfer at an inclined and yawed square plate: application to solar collector, J. of Heat Transfer, 99, 507-512.

ACKNOWLEDGEMENTS

The work was funded by the National Project PFE2, Contract 84.02811.59 CNR. We are grateful to Metalux SpA and TOM Srl of Musile di Piave for the essential technical support during experimentation.

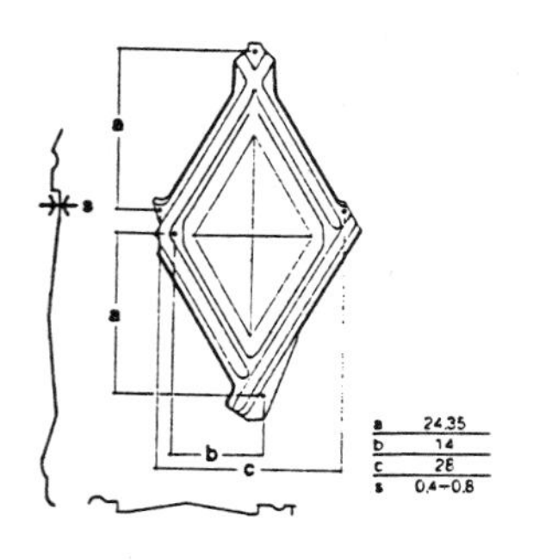

Fig.1 A view of the tile.

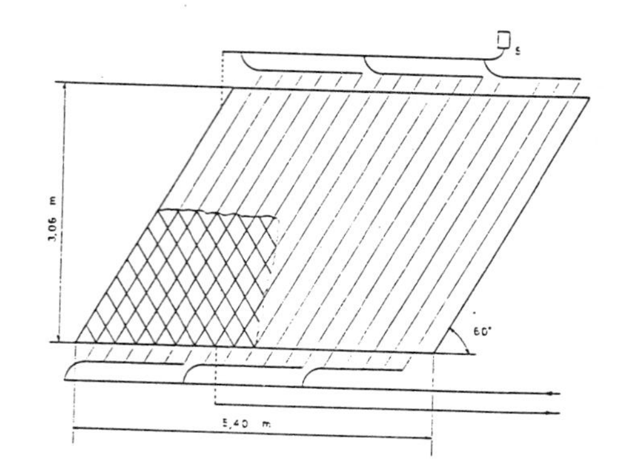

Fig.2 Sketch of one of the energy roofs.

Fig.3 A scheme of the plant.

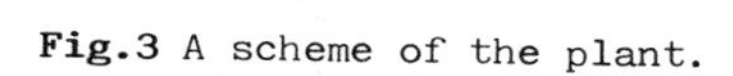

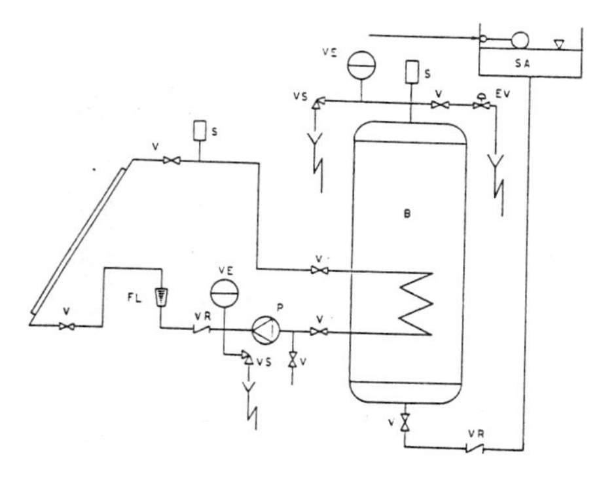

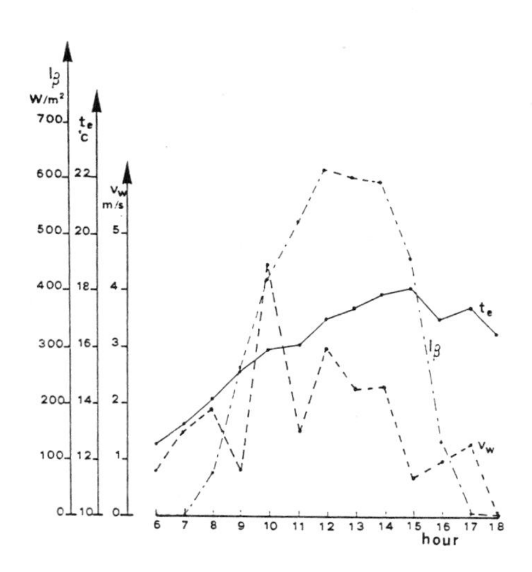

Fig.4 Solar radiation intensity on the roof (I_β), wind velocity ($\mathbf{v_w}$) and outside air temperature ($\mathbf{t_e}$), October 16th.

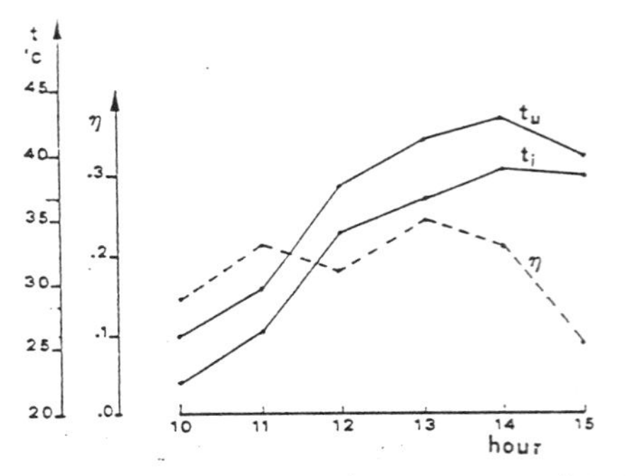

Fig.5 Efficiency (η), inlet (t_i) and outlet (t_u) temperatures of the roof one during October 16th.

Fig.6 Efficiency (η), inlet (t_i) and outlet (t_u) temperatures of the roof two during October 16th.

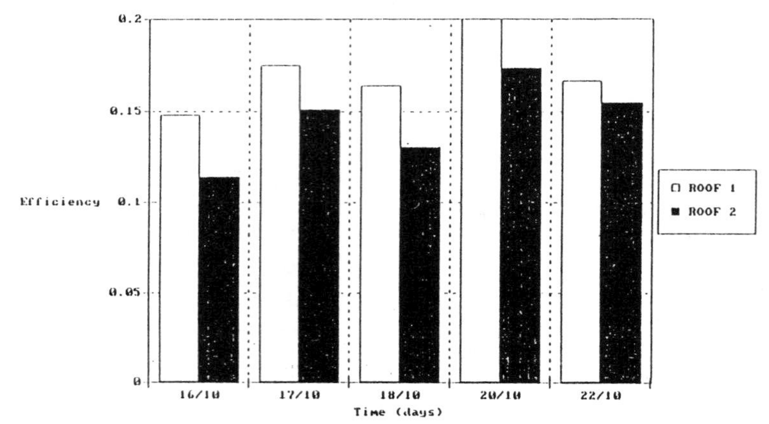

Fig.7 Daily efficiencies for roofs one and two during some days.

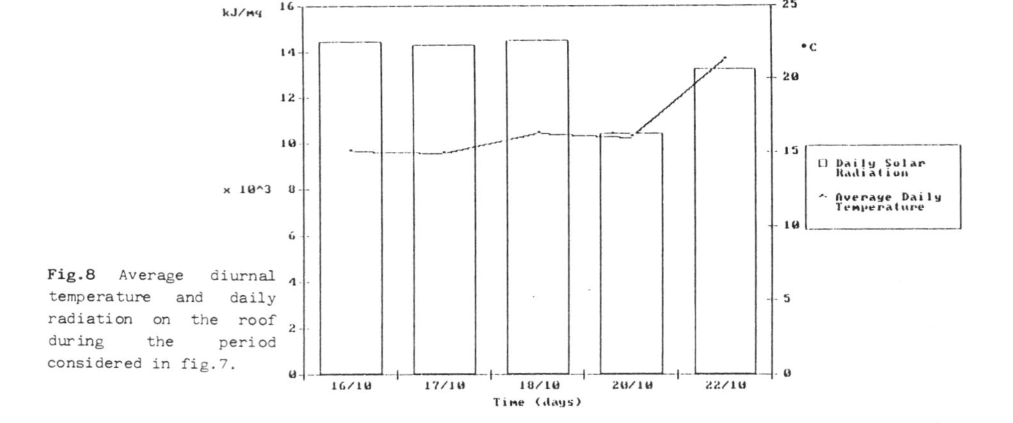

Fig.8 Average diurnal temperature and daily radiation on the roof during the period considered in fig.7.

FIELD TEST WITH WATER FILM SOLAR COLLECTORS

Monika Söderlund
Department of Water Resources Engineering (WREL)
Luleå University of Technology
S-951 87 LULEÅ
SWEDEN

ABSTRACT

Water film solar collectors are asphalt and roof surfaces used as solar heat absorbers. Research into this field was started in 1983 at WREL, Luleå University of Technology. The performance of the collector has been studied by modelling of field tests which have been carried out.

During the summer of 1987 the water film solar collector has been tested in a solar energy system. The collector consisted of a sheet-metal roof surface and the extracted solar heat was used for charging of a borehole heat store. During the test period measurements of climatic data and heat output were made. In this paper the technical feasibility of the energy system is discussed and some of the test data are presented.

KEYWORD

Solar heat, field test, low temperature, energy wells, ground heat, recharging.

INTRODUCTION

Asphalt and roof surfaces are heated by the sun to relatively high temperatures during the summer. These surfaces can be used as solar collectors by spreading and collecting a thin water film flowing over the surfaces. In Fig. 1 a principle outline of water film solar collectors in an energy system is shown.

The water film solar collector has been studied since 1983 at Luleå University of Technology. The investigation started with a field test (Nordell and Söderlund, 1984) and the performance of the collector has further been simulated with the governing mass- and heat transfer equations (Söderlund, 1985). The water film solar collector is suitable for low temperature applications. Temperatures close to the ambient temperature entail small losses to the surroundings. Using these circumstances the calculated energy output in Stockholm during June to August is 460 kWh/m^2.

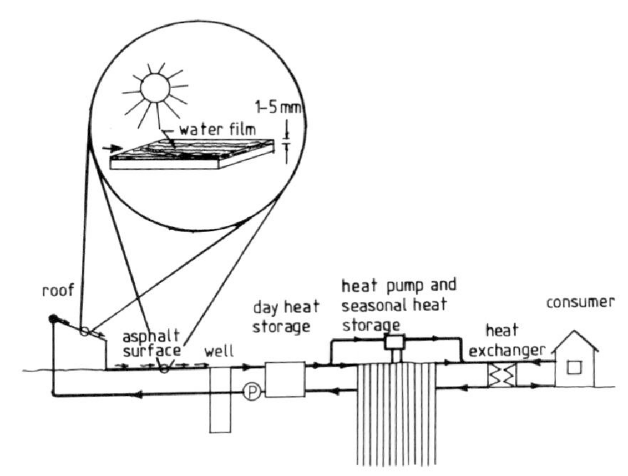

Fig. 1. Principle outline of water film solar collectors in an energy system.

The feasibility of the water film solar collector for recharging of energy well systems has been simulated. In an energy well system the ground is used as a heat source to a heat pump. As a consequence of the heat extraction the bedrock is cooled in the long run and therefore the heat output from the wells is reduced. By recharging the energy wells the heat efficiency of the system is maintained.

During the summer of 1987 a pilot plant for heat storage has been charged with solar heat. The pilot plant has previously been used for seasonal heat storage studies (Nordell et.al., 1983). In this paper the technical feasibility of the solar energy system and some field measurements are presented. A more detailed evaluation is going to be made during this autumn and will be reported to the Swedish Council for Building Research in December 1987.

THE SOLAR ENERGY SYSTEM

The Trial Installation

The heat store consists of 19 boreholes, 52 mm in diameter, and 21 m deep. The upper 6 m pass through earth and the lower 15 m are drilled in rock. A liner protects the borehole where it passes through the earth layers. The store volume is 400 m^3 of crystalline rock. The spacing between the boreholes is 1.3 m. In Fig. 2 a plan and section of the store is shown.

The store was charged with solar heat during the summer of 1987. The main features of the installation and a view of the general appearance are shown in Fig.3.

The circulation system within the store was open and consisted of six groups of holes, each containing three boreholes, connected in parallel with the six groups.

The Instrumentation System

Climatic data such as wind speed, total global radiation, relative humidity and air temperature were recorded by a datalogger every 15 minutes. The water flow rate and the input/output temperatures of the flowing water were measured at the boundaries of the surface.

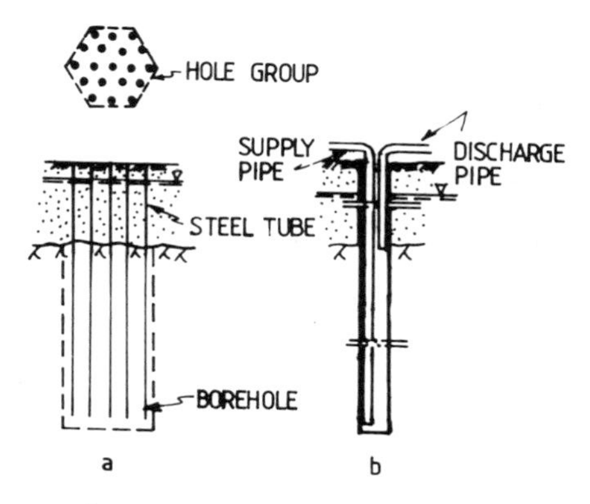

Fig. 2. a/ Plan and section through the heat store. b/ Section through a borehole.

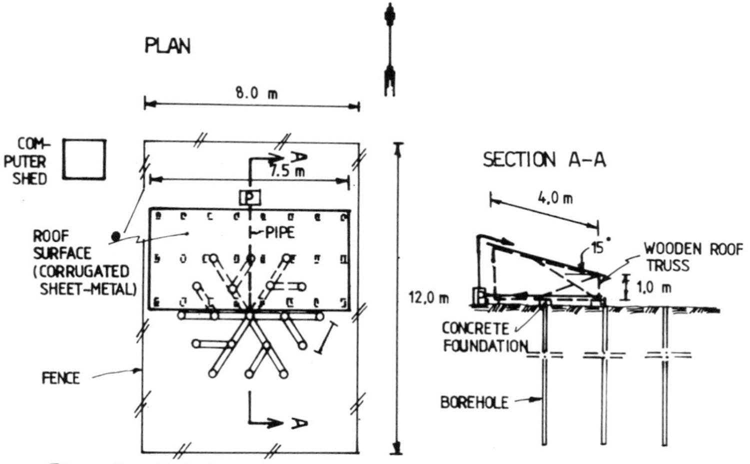

Fig. 3. Trial installation for the solar energy system.

FIELD TEST

The trials started on June 22, 1987 and continued until August 18, 1987. Statistically this summer has been more rainy and colder than usual. The average temperature in June was 11.5 °C (normally 12.4 °C) and the precipitation has been 95 mm (normally 46 mm). In July the average temperature has been 14.5 °C (normally 16 °C) and the precipitation has been 69 mm (normally 47 mm).

The iron content in the groundwater in this area is high, 4.9 mg/litre. When the water gets in contact with the oxygen in the air, some of the iron is precipitated. In a full scale heating system the collector and store circuits ought to be separated by a heat exchanger. Because of the iron precipitations a couple of operation interruptions were necessary. Fig. 4 shows inlet, outlet water temperatures and water flow rate during July and August. The charging period was between 7 a.m. and 7 p.m. every day. The highest temperature difference between inlet and outlet is 7 °C and the lowest one is 1 °C.

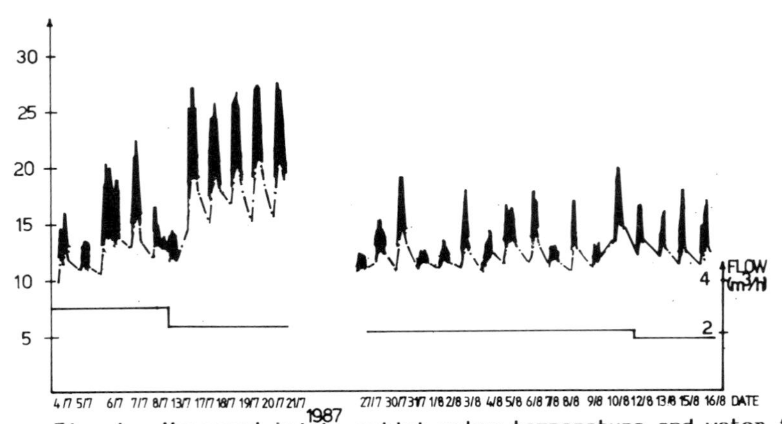

Fig. 4 Measured inlet, outlet water temperature and water flow rate during July and August 1987.

In Fig. 5 the charged energy and the heat input per borehole metre to the wells are shown. The charged energy during the period was 2340 kWh and the heat input varies between 5 W/m and 75 W/m. This means an energy output of 80 kWh/m^2 roof surface and an average heat input to the wells of 24 W/m during daytime.

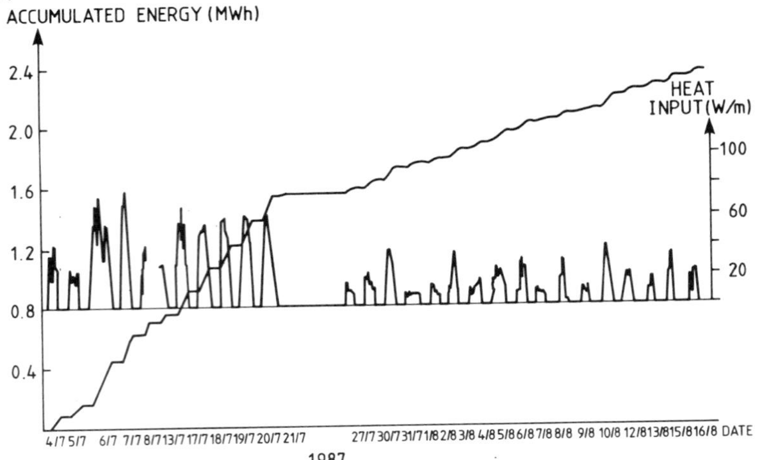

Fig. 5 Accumulated charged energy and heat input during July and August 1987

The calculated average bedrock temperatures for the heat input presented above (i.e. a pulse of 24 W/m during 12 hours each day) are shown in Fig. 6. Consequently, during the charging period the bedrock temperature is increased by approximately 8 °C. Because of the pulsation in charging during the day the actual bedrock temperature fluctuats. The fluctuations during 2 days in the middle of the charging period are shown in the Fig. 6.

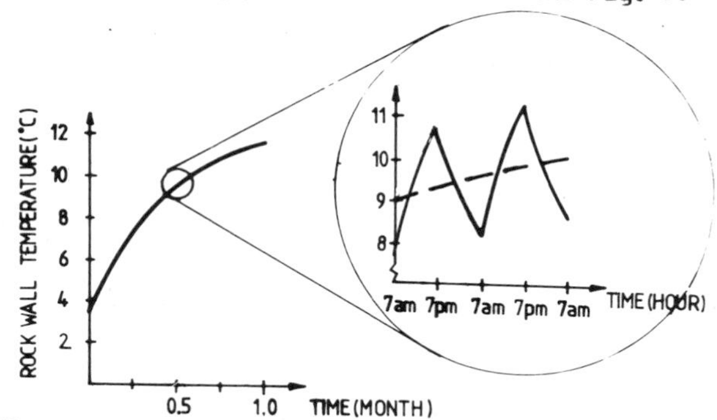

Fig. 6. Average bedrock temperatures and temperature fluctuations with a charging input of 24 W/m during daytime.

FURTHER WORK

The results from the field test will be further evaluated during this autumn. The evaporation process contributes to a large part of the heat losses from the collector (see Söderlund 1987), therefore it is necessary to determine this quantity carefully. The evaporation rate has been measured during the field test and an evaporation model for the water film solar collector will be made. It should be of interest to decrease the amount of evaporated water and consequently increase the outlet temperature from the collector. A couple of small laboratory experiments indicate that this is possible by adding an additive to the water. The continued evaluation includes a study of the effect of additives. Also a thorough estimate of costs for the solar heat charging system will be made.

REFERENCES

Nordell B. et.al. (1986). A borehole heat store in rock. Pilot trials in Luleå and preliminary design of a full-scale installation. D6:1983. Swedish Council for Building Research.

Nordell B., Söderlund M. (1984). Heat collection from asphalt and roof surfaces. Field test with surface solar collectors. Dept of Water Resources Eng., Luleå. Technical Report 1984:58T (in Swedish).

Söderlund M. (1985). Surface Solar Collectors. Paper presented at the 7th Miami Int. Conf. on Alternative Energy Sources.

Söderlund M. (1987). Water Film Solar Collectors. Solar heat from asphalt and roof surfaces. Dept of Water Resources Eng., Luleå University of Technology. Licentiate thesis 1987:07L.

Testing the efficiency of a Do-it-yourself Solar Collector System

Andreas Wagner

Wagner & Co GmbH
Sonnen-Energie-Technik
Zimmermannstr. 1
D-3550 Marburg

Abstract:

Since 1982 the portion of D.I.Y. solar systems of all in Germany installed solar collectors has grown continuously up to about 25%. There are some successful systems with high technical standard, little consumption of material and low cost. The performance of 14 complete solar installations was tested under identical conditions in 1985/86 by the German Ministry of Science and Technology (BMFT) in Munich. Tested were: Vacuum and flat plate collectors with thermosiphon and forced circulation. Among the industrially produced collectors the D.I.Y. system showed high efficiency. The remarkable result: low cost does not mean low efficiency!

Keywords:

- Do-it-yourself (D.I.Y.) Solarsystem
- Low Cost Collector Kits
- Collector Test 1986
- Heat Leakage of Solar Storages
- Relative Energy Expenditure "R"

Philosophy of Do-it-yourself Solarsystems

About 25% of all solar systems in Germany are not installed by professionals but by home-owners themselves. This fact can be easily explained by the growing tendency towards "Do-it-yourself". Apart from that there are even more aspects which encourage the non-professional to install his own working solar system:

economical aspects

- savings of mounting costs
- simple technology
- no special tools necessary
- maintenance can be carried out by owner himself

personal motives

- compensation for daily work
- urge to take part actively in the conservation of the environment
- interest in technology

Solar systems consist of large units. Ready-made collectors cause relatively high freight and packing costs. Unit-construction systems are therefore suitable as well for professional mounting, because the additional expenditure for mounting is relatively low. Moreover, certain prejudices could be met with in the meantime, for Do-it-yourself systems obtain partly the same or even higher results than industrially made solar collectors as you will see from the following:

Description of the different components of the solar collector:

The collector:

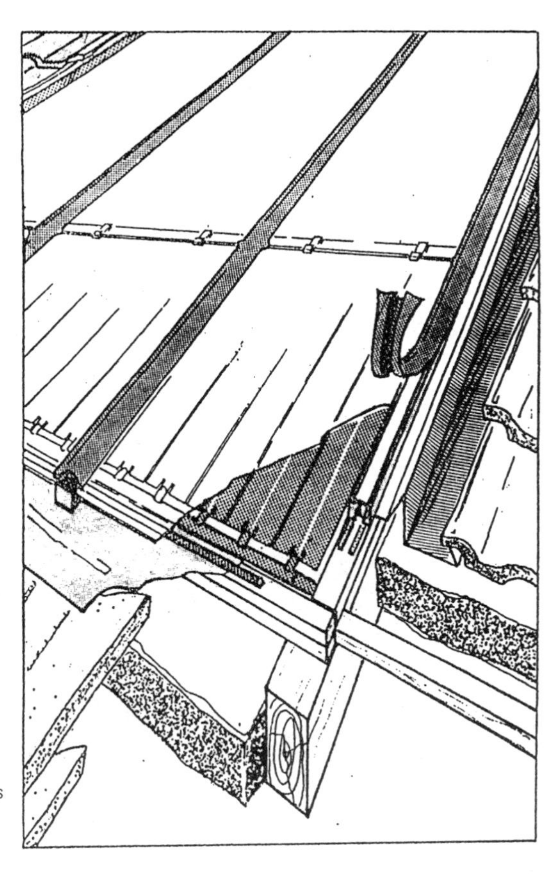

Low costs by:

- standard insulation
- standard glass measurements
- standard glazing profiles
- easy roof-fitting
- no collector surround
- Sunstrip Do-it-yourself absorber

High efficiency by

- thick insulation
- low heat-loss through frame
- no pipes on roof-top
- selective sunstrip aluminium fins with a copper pipe

The circulation system:

Copper tubing with long life expectancy, i.e. no problems with combining materials.

Use of normal heating armatures or low cost safety armatures for thermosiphon circulation.

The storage tank

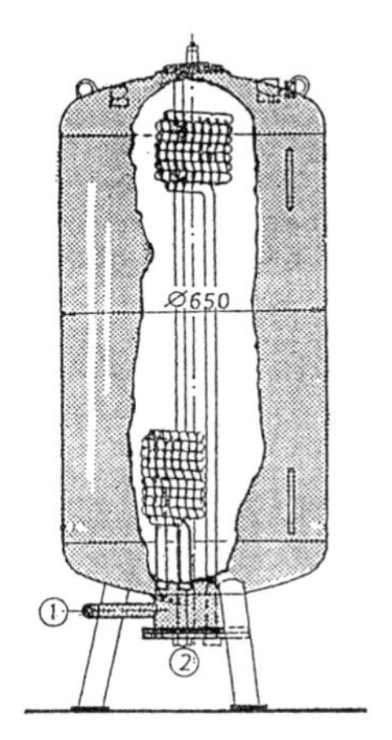

Low costs by:

- enameled steel
- standard storage capacities with a great number of variations
- standard heat exchangers from a large range

High efficiency by:

- very low heat leakage by a special way of tubing
- thick insulation (up to 15 cm)
- a good heat layering
- large heat exchanger surface

Results of the test of solar collectors in 1985/86

The German ministry of Science and Technology (BMFT) carried out comparisons of 14 complete solar systems under identical test conditions in Munich 1985/86.

In the test 200 litres of warm water were drawn daily from each solar system. Construction and regulation were up to the test's participants.

To represent the data on efficiency levels a new concept, that of **R** = relative energy expenditure, was introduced. If **R** is depicted in graph in relation to E, the average daily insolation (kWh/m² d), the ensuing characteristic curve enables us to predict how large a proportion of energy may be covered by solar energy in various locations.

Performance of the storage

The efficiency of a solar storage can be represented by the special stray power $q = \frac{W}{K}$ or by $R_o = \frac{\text{added energy}}{\text{real energy}}$

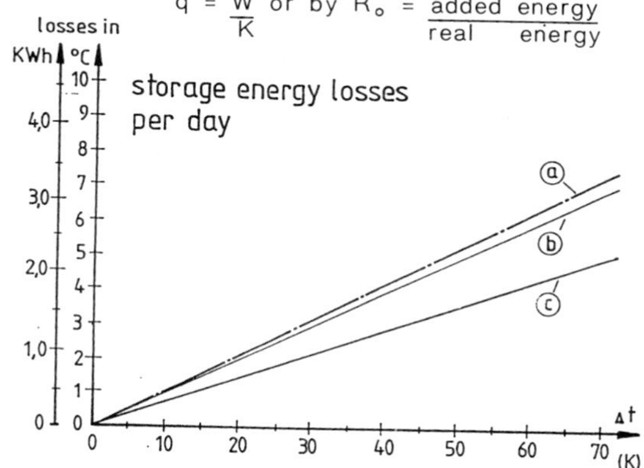

In figure 1 you can see the energy losses of the storage. The better the insulation of the storage, the more even the curve.

a = the tested EMS-storage of Wagner & Co
b = the best storages in the test
c = the new storage ECO made by Wagner & Co

These values characterise the construction of the storage. With R_o = 1,22 and q = 2,0 W/K the Wagner & Co EMS tank obtained a good result in the test. Nevertheless Wagner & Co created a new solar storage type **ECO.** ECo was planned according to the results achieved in the test carried out by the BMFT. Special connection technology of pipes and fittings reduces heat losses to a minimum. Heat losses were lowered again by about 50% (own measurements) down to q = 1,4 W/K.

High performance of the "Do-it-yourself" sunstrip collector

The main difference of the Wagner & Co solar system in comparison to conventional solar heating systems is due to the "Do-it-yourself" sunstrip collector. The performance can be perceived in the efficiency diagram.

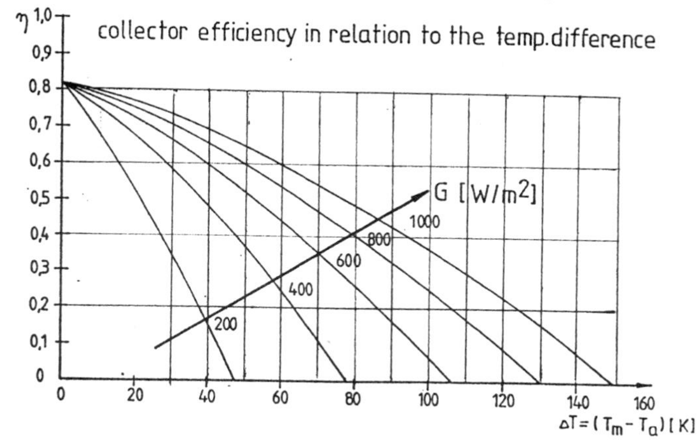

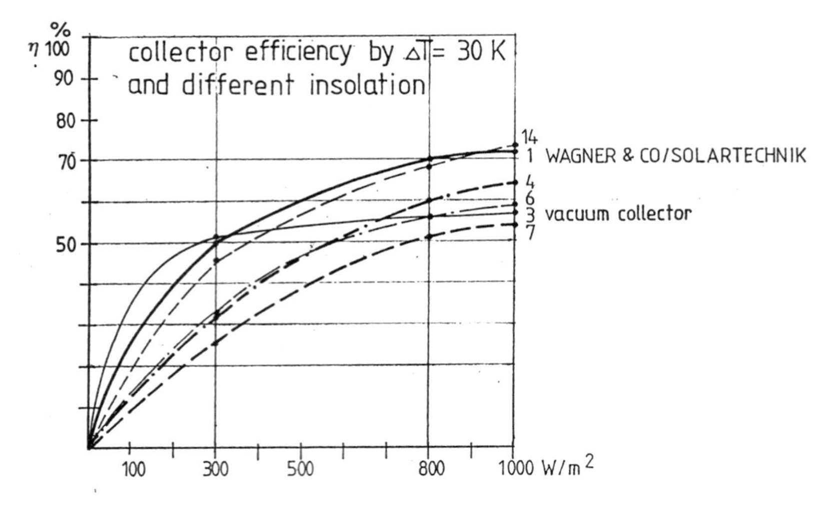

At a difference in temperature of e.g. 30 K and in an operating range between approx. 300 and 800 W, our system achieved the highest values concerning the performance of the collectors in the test.
The following aspects are mainly decisive for these high values of performance:

a) a low value of K (=3,2 W/m^{2}K) because of the small frame losses and the good insulation of the back side. This is due to the fact that in contrast to collector cases our collectors consist of a closed surface.
b) a high η_o value. This means: Absorption (α = 0,95) and light permeability of the glass cover plane are comparatively elevated.
c) a high value of emission with ϵ = 0,15.
d) a good transition of heat from the surface of absorption to the sunstrip tube.

Performance of the whole unit

The annual average insolation rate for Munich is E = 3,2 kWh/m^2 d. In regard to this insolation the test resulted in a value for R (added energy/real energy) = 0,38. That is equivalent to a solar cover rate of 69%. This value shows that it is possible to reach high levels of efficiency even with "Do-it-yourself" solar collectors. If we consider the better efficiency of the new ECO solar storage, it is possible to reach a yearly solar cover rate of about 74%.

DEVELOPMENT AND TESTING OF TWO HYBRID-COLLECTORS

M. Reuss, T. Schmalschlaeger, Dr. H. Schulz
Landtechnik Weihenstephan, Technical University of Munich
D-8050 Freising, F.R. Germany

1. ABSTRACT

Air heating collectors, which are used for solar drying of hay, grain, herbs and other agricultural products do in our middle European climate offer operating only a short period a year. To increase the operation time, we have developed two new types of collectors, the so called hybrid-collectors, for heating water and air. They have two main advantages:

- Multiple usage of the usually big collector areas increases the time of operation and improves the economy
- The cost of such collectors are rather low

On our test facility we have built two types of hybrid-collectors and carried out measurements. The values of o and UL are determined under steady state conditions. Also the heat capacity, an important characteristic, was measured with different methods. The thermal performance as water collectors is comparable to that of industrial collectors. Also the air collectors perform quite well, but have still a potential of improvment. At least the collectors can work as water to air heat exchangers. Also for this operation mode we determined the characteristic performance data.

2. KEYWORDS

Air-collector, water-collector, hybrid-collector, collector testing, collector-characteristic, heat capacity measurement, water/air heat exchanger.

3. INTRODUCTION

Within the framework of a project of the German Ministry of Research and Technology (BMFT) on testing of simple solar air heating collectors two hybrid-collectors for air and/or water heating were developed and tested. Such collectors have especially in agricultural applications two main advantages:

- Multiple usage of the usually big collector areas increases the time of operation and improves by this the economy.
- The costs of such collectors are rather low-especially in the case of self built systems

There are three main modes of operation:

1) Air collector operation only:
   The collector is heating up air for example for drying purposes. In this low temperature application for example ambient air is heated up with a typical temperature raise of 20-30 K at a flowrate of 50-100 kg/hm²2.
2) Water collector operation only:
   The collector is heating water for space heating, heating of stables, greenhouses or domestic hot water. The typical temperature raise under sunny weather conditions is about 15 K.
3) Hybrid collector operation:
   The collector is heating water and air at once. Hot air is directly supplied to the drying system, the hot water is delivered to a storage tank. In the case of oversized systems heat peaks during noon could be stored in the water tank.

Beside these modes a hybrid collector could also be operated as a water to air heat exchanger for example for drying during night time using the exess heat stored during daytime.

4. DESCRIPTION OF THE COLLECTORS AND TEST INSTALLATION

4.1 Collectors

A scheme of these hybrid collectors are seen in fig. 1 and fig. 2. The first collector (I) has an absorber for water heating made of stainless steel with a selective surface (=0.90, =0.16). The air flow of this collector is over the absorber. The backside insulation is made with a 50 mm thick polyurethane plate. This collector has a double glazing - the outer one is made of a corrugated polycarbonat sheet, the inner one is a high transparent PETP (polyester) foil.

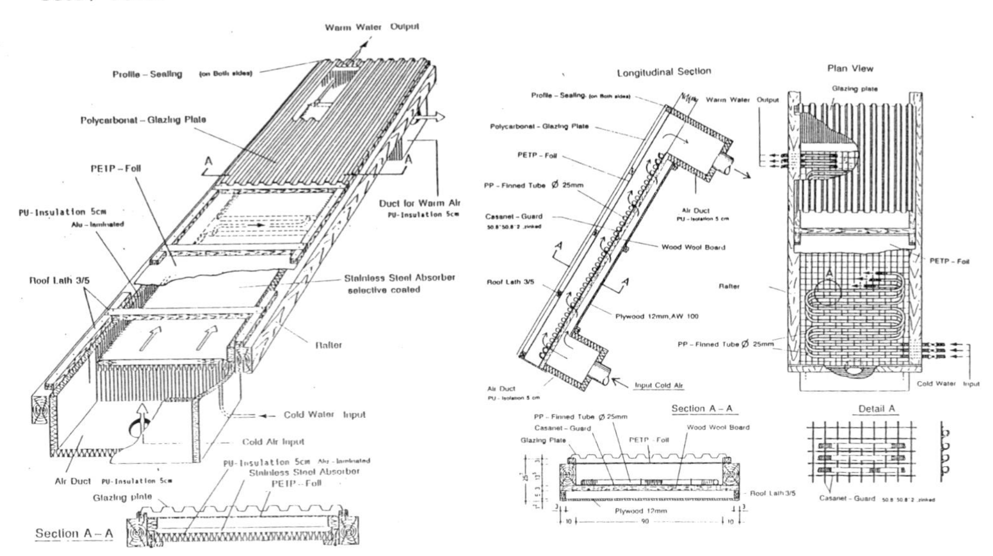

Fig. 1 Hybrid collector (I) with stainless steel absorber

Fig. 2 Hybrid collector (II) with PP-plastic tubes as water absorber

The second collector (II) developed uses a black painted porous wood wool board as absorber for air heating. The air is sucked from the back to the front through this board removing the absorbed energy. The air duct on the back has a hight of 50 mm. On the wood wool board flexible black plastic tubes made of polypropylene (PP) were fixed in a serpentine way. The inner diameter of these tubes is 20 - 25 mm with a thickness of the wall of 0.9 mm. Through these tubes the water is circulated and heated up. This collector has double glazing of the same structure as collector (I).

4.2. Test Installation

The performance measurements were done according to the recommendations of the Collector and Systems Testing Group of the Cominission of the European Communities (DG XII).

The collectors were tested in parallel mounted on a rig with a slope of 35° facing south. A series of meteorological data like global radiation, diffuse radiation, ambient temperature air humidity, wind speed and wind direction were registered. Beside these the collector inlet and outlet temperatures and the massflowrate were measured in both the water and the air circuite. The inlet temperatures were kept constant by a control unit, the massflowrate was controlled manually. The flowrate measurement was done with orifice plates in order to reach the required accuracy.

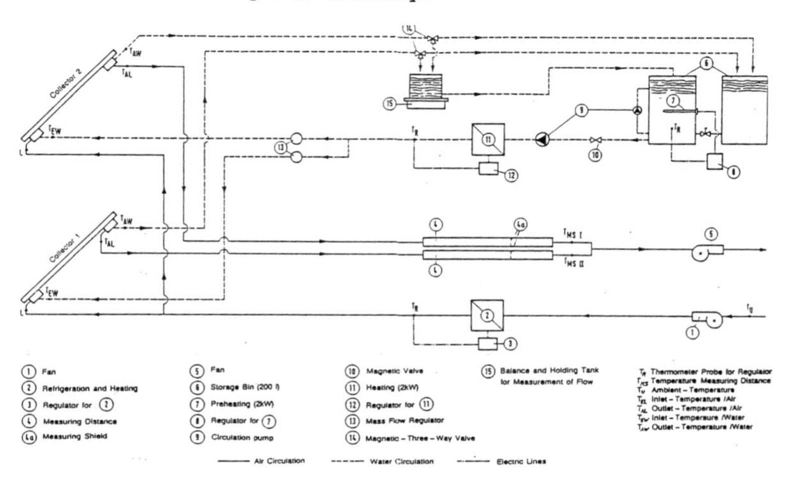

Fig. 3: Scheme of the Test Installation

5. RESULTS OF THE PERFORMANCE MEASUREMENTS

The performance measurement were carried out under steady-state- conditions arround noon when the variation of the solar flux was less than 50 W/m²2 and that of the ambient temperature was less than 0.5 K. The results are plotted in the following fig. 4 and fig. 5 and concluded in table 1.

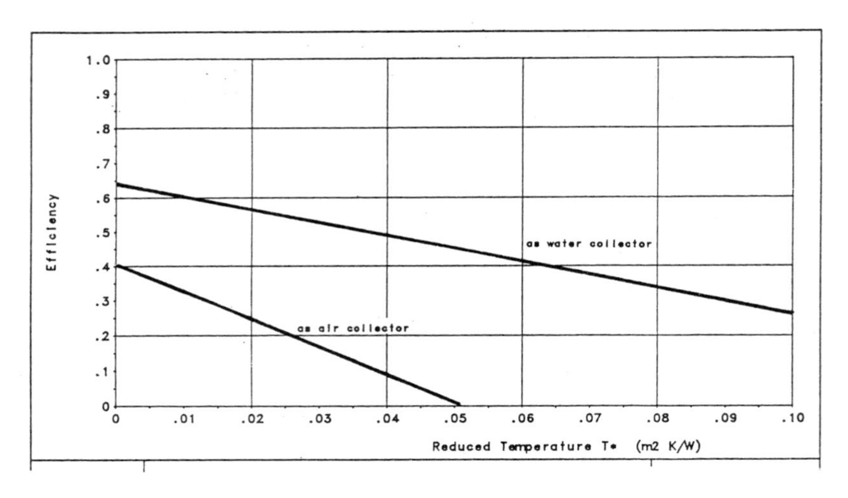

Fig. 4: Efficiency curves of hybrid-collector I

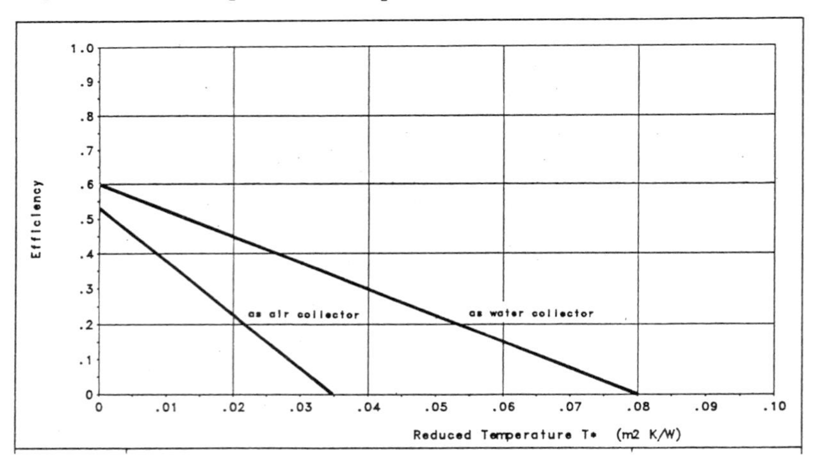

Fig. 5: Efficiency curves of hybrid-collector II

Collector	Oper.Mode	n0	UL ($W/m^2 2K$)
I	air coll.	0.407	7.7
	water coll.	0.637	3.7
II	air coll.	0.529	15.3
	water coll.	0.593	7.4

Table 1: Performance measurements results

Beside these standard test results the heat capacity of such collectors is a further important characteristic, as it influences the response time. The heat capacity of a collector depends strongly on the used materials and in our case, as the collector was built mainly of wooden materials it is also dependent on the humidity of these materials and of the ambient air.

The heat capacity measurements were carried out after three different methods found in literature with good agreement between each other. Under our conditions the most practical methode was to increase continuously and then decrease the inlet temperature during night time. The difference between the mean fluid temperature and ambient is drawn against time. The heat capacity is then calculated from an energy balance made at two points of the same temperature difference, one in the inclining part and one in the declining part of the temperature difference curve. Heat capacity determination in air collector operation mode was done without water. The results are given in table 2.

Collector	oper.mode	CK (kJ/K)
I	air coll.	12.0
	water coll.	60.0
II	air coll.	50.3
	water coll.	158.3

Table 2: Results of heat capacity measurements.

6. WATER/AIR HEAT EXCHANGER OPERATION

As already mentioned the hybrid collectors could also be operated as heat exchangers (water/air). This could be useful in drying plants for agricultural products (low temperature drying), when excess heat stored during daytime could be use during the night.

For qualification of this heat exchanger operation a temperature exchange efficiency (see Fig. 6 and 7) is defined in the following way

$$nT = (mL*cpL*(ToL - TiL))/(mW\ cpw\ Tiw - mL\ cpL*(TiL)$$

nT	temperature exchange efficiency	m	massflow rate (kg/s)
Cp	spec. heat capacity (kJ/kg K)	To	inlet temperature (°C)
Ta	outlet temperature (°C)		
	Indices : L for air, W for water		

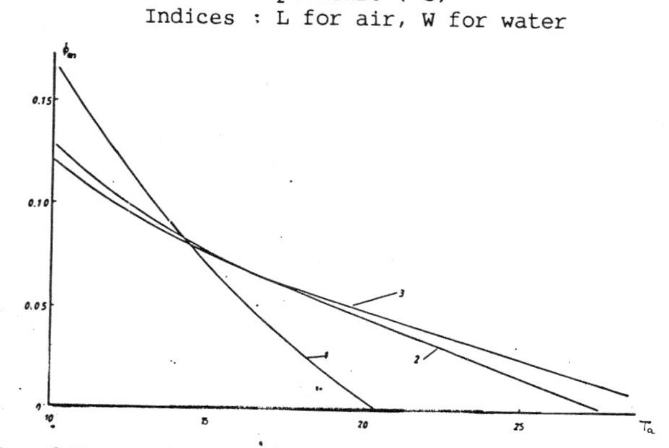

Fig. 6 Temperature exchange efficiency of hybrid collector I

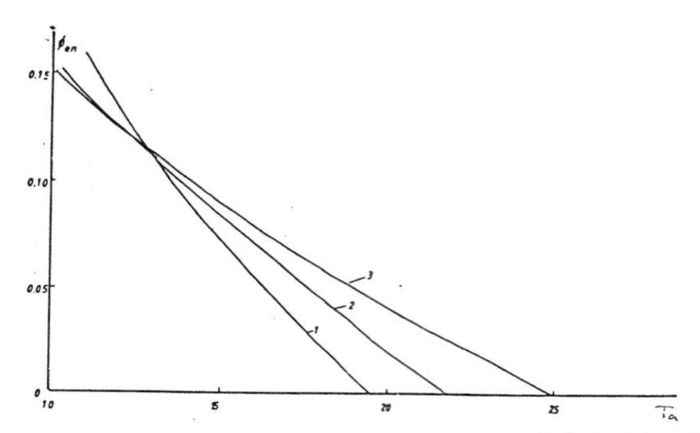

Fig. 7 Temperature exchange efficiency of hybrid collector II

7. CONCLUSIONS

Solar drying plants for agricultural products have usually a quite big area of air heating collectors (more than 50 m^2), which are only used for several weeks a year. This fact influences the possible investment and the economy of such a plant. An improvement of this situation could be given by the development of a multiple usage or hybrid collector for air and/or water heating, especially in the case of selfbuilt collectors. Performance measurements on two hybrid collectors at our collector test installation have shown encouraging results, even if there is still a potential of further development and improvement. These systems could not only be used as solar collectors but also as water/air heat exchangers for example in drying plant during night time. It is planned to use such a hybrid collector in a big solar drying plant for vegetables (demonstration project of DG XVII, CEC) in Spain.

8. REFERENCES

Reuss, M., Schmalschlaeger, T. and Schulz, H. (1986). Überprüfung von Einfach-Luftkollektoren luftkollektortest II

Schmalschlaeger, T. (1985). Vergleich zweier Hybridkollektoren für Wasser und Luft. Journal Sonnenenergie & Wärmepumpe, 4/86, 32-35

Aranovitch, E. and Ledet, M. (1976). Methods for the determination of the effective heat capacity of a solar collector. Report of Ispra Establishment, Commission ofthe European Communities

Bougard, J. (1978). Integral method for the determination of the apparent heat capacity of a solar collector. Laboratoire de Thermodynamique, Mons, Belique

Schoelkopf, W. and Sizmann, R. (1982). Measurement of characteristic solar collector parameters in transient operation. Ludwig-Maximilians- Universität München

Moon, J.E. and Gillet, W.B. (1983). Recommendations for European solar Collector Test methods, Commission of the European Communities.

THE EUROPEAN SOLAR STORAGE TESTING GROUP

H.A.L. van Dijk

TNO Institute of Applied Physics
(Technisch Physische Dienst TNO-TH)
P.O. Box 155 , 2600 AD DELFT
The Netherlands

ABSTRACT

This paper describes the background, aim, scope and current activities of the European Solar Storage Testing Group (SSTG).
The SSTG was established by the Commission of the European Communities in the beginning of 1982, as one of the concerted actions of the energy R & D Programme of the CEC. The one and a half year of co-ordinated efforts by six participating countries resulted in the CEC publication 'Recommendations for European Solar Storage Test Methods' (EUR 9620, November 1984).
A follow-up of the concerted action in storage testing was started in the course of 1986 with Denmark, France, Germany, United Kingdom and the Netherlands as participating countries, co-ordinated by the TNO Institute of Applied Physics in Delft, the Netherlands. The project will be completed within a period of three years from the starting date.
The aim of the project is to develop a set of test procedures for short term thermal energy storage systems for solar applications. The test procedures will comprise both a simple test method for the basic characterization of storage systems and an extensive test method, resulting in a detailed characterization of storage systems. The results of the extensive test should yield parameter values for a calculation model of the storage subjected to the tests.
The development of such a general model is a special task within the research project. The joint action programme consists of three successive series of tests aimed at evaluation and improvement of the test procedures an (a) general storage model(s). A number of subtasks have been defined and distributed to deal with specific questions.

KEYWORDS

Test procedures; storage system; model parameters; short term storage; standardized test method; European Commission; thermal characteristics; model development; latent heat storage materials.

1. INTRODUCTION

The European Solar Storage Testing Group (SSTG) was established in the beginning of 1982 as a Concerted Action within the second four year Energy R & D programme (1979-1983) of the Commission of European Communities. The aim of the SSTG, co-ordinated by the TPD, was to draw up recommendations for solar storage test procedures.
The contract ended in 1983 (Van Galen) with the preparation of the CEC publication 'Recommendations for European Solar Storage Test Methods' (Van Galen and Van den Brink, 1984) the result of one and a half year of co-ordinated efforts by participants from six European countries.
Within the new EC Non-Nuclear Energy R & D Programme (1985-1988) a follow-up of the concerted action on storage testing was started in the course of 1986, to investigate a number of topics more in detail. The current SSTG project is again co-ordinated by TPD.
Participants in this project are:

S. Furbo	Technical University of Denmark, Denmark.
T. Vest Hansen	Danish Solar Energy Testing Laboratory, Denmark.
P. Achard	Ecole Nationale Supérieure des Mines de Paris (jointly with CSTB), France.
J. Sohns	University of Stuttgart, Federal Republic of Germany.
R.H. Marshall	University College Cardiff, United Kingdom.
J. van der Linden	TNO Institute of Applied Physics, The Netherlands.

The project's duration will be three years from its start.

2. AIM

The aim of the project is to develop a set of test procedures for short term thermal storage systems. The emphasis with respect to the kind of systems and the selected conditions for the tests is on applications in solar energy systems for space heating and/or domestic hot water.
Commercial storage systems often use water as storage material. Nevertheless also so-called latent heat storage systems are taken into consideration as systems which might be commercialized in the near future.
The aim is to develop procedures both for a simple test method for the basic characterization of storage systems and for an extensive test method, resulting in detailed characterization of the storage systems. The results of the extensive test should yield parameter values for a calculation model of the storage subjected to the tests.

3. TEST PRINCIPLES

In the Recommendations, published in 1984 (Van Galen, Van den Brink) procedures are specified to determine the following quantities:
- the storage capacity as a function of temperature;
- the storage capacity over the design temperature range;
- the heat loss rate at finite flow rate;
- the heat loss rate at zero flow rate;
- the heat exchanger effectiveness and the potential for stratification in the flow direction;
- the storage efficiency.

For this purpose heat is supplied to or extracted from the store during each individual test by means of a fluid (usually water) with a certain flow rate and temperature.
The flow rate and temperature are controlled in a specific way depending on the aim of the test.
Examples are presented in Fig. 1 and Fig. 2.

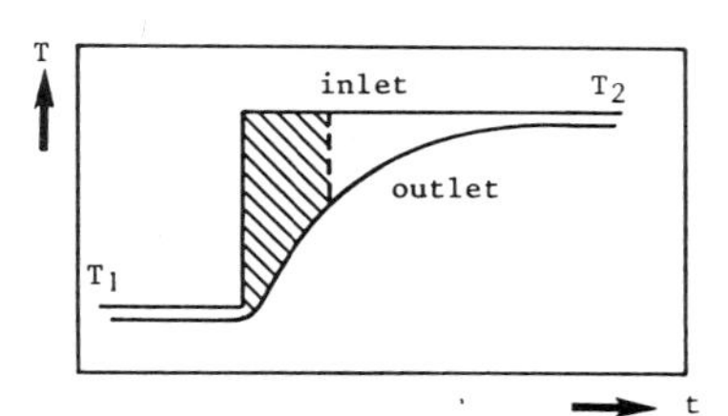

Fig. 1. The storage capacity and storage efficiency are determined by measuring the response on a step change in inlet temperature.
Efficiency: heat content stored within a certain time.
Capacity : total amount of heat stored.

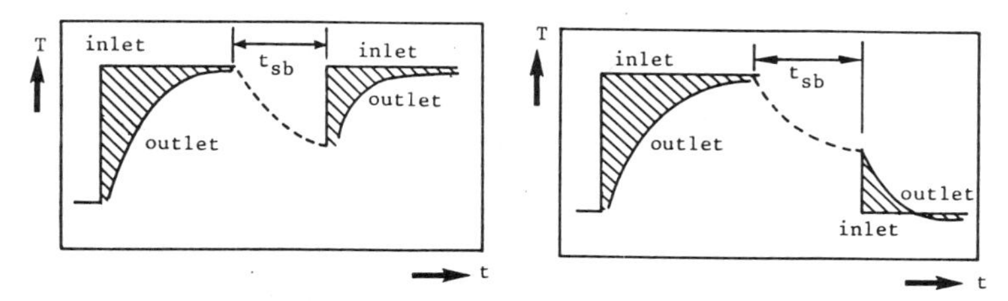

Fig. 2. Various alternatives are under examination for the measurement of the heat loss during stand-by (t_{sb}); e.g. by recharging or by discharging to a situation with well-known heat content.

In the current SSTG project the possibilities, accuracies and limitations of these and other tests are investigated in order to end with a complete package of test procedures which meet the requirements defined above.

4. WORKPLAN

4.1. Joint actions

The joint action programme is set-up around three series of tests which serve to develop and improve the test procedures and to investigate the applicability of the test results for the derivation of parameter values for a computer model of the store.
The test series are:

- Test series A (1987):
  A Round Robin test on a selected commercially available sensible heat storage system (see Fig. 3).
- Test series B (early 1988):
  Test on different types of commercially available heat storage systems.
- Test series C (late 1988):
  Final test on various types of heat storage systems.
  Test series C is followed by a dynamic test for model validation.

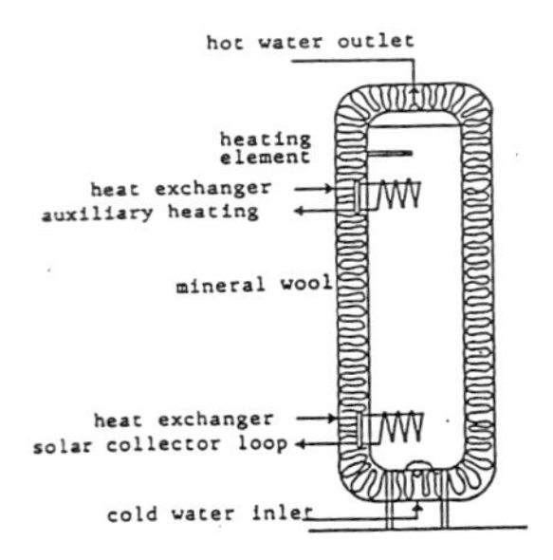

Fig. 3. The storage system for the first series of tests.

4.2. Model development

Simultaneously, two participants, UK and France, with supporting effort from the Netherlands, are involved in a special task: the development of one or a few general computer models which can be used to adequately simulate the performance of a thermal storage system.
In this context different possibilities are being investigated. Towards the end of the project the then available models will be evaluated and one of the models will be selected and subjected to validation in a dynamic test. One of the specific requirements of this model is that is must be compatible with the European programme EMGP2 for solar energy systems.
In Cardiff a model of a 4 port storage system is being developed. The model parameters are to be found by fitting the calculated response with the measured temperature curve.
In the meantime at Valbonne progress has been made with the development of a model identification mode. A specific algorithm computes the coefficients which describe the relation between input and output variables and specified state variables.

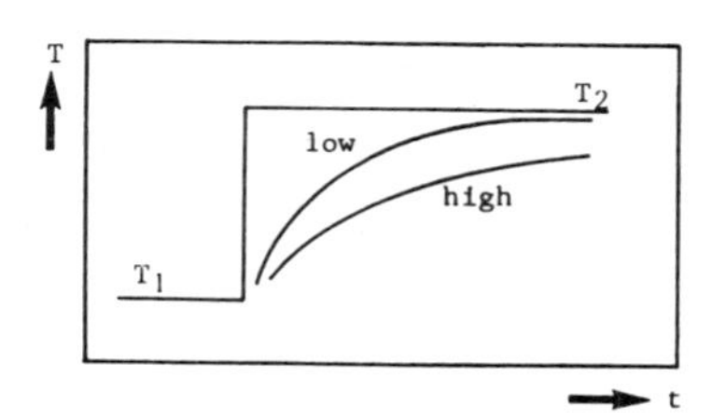

a. High resp. low heat exchanger performance.

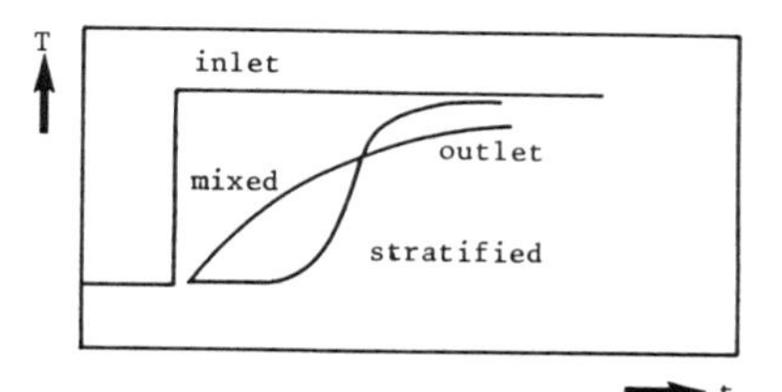

b. Mixed resp. thermally stratified storage.

Fig. 4. Typical step response curves showing - at least qualitatively - the effect of heat exchanger performance and the degree of thermal stratification.

In this computation, the coefficients (organised in matrices) are found by minimizing the difference between calculated and measured output variables. The number of state variables can be chosen (e.g. two: one fluid and one store temperature, etc.) and the coefficients can be forced to a fixed value or a linear relation with other coefficients. In this way in fact different models are created, from which the most appropriate one has to be selected. This method does not require neat stepwise changes in inlet temperature.

The first results both on a fully mixed store and a stratified store are quite promising. Nevertheless, a number of problem areas have been identified which demand further investigation.

Simultaneously a method is under development for a graphical analysis of the reponse of the storage exit temperature on a step change in inlet temperature. The method could either be used as a first rough characterization of the storage system or as a means to derive the main storage parameters in selected, i.c. simply to describe, systems.

Whatever the outcome of the investigations, it is clear that the step response curve already reveals a number of characteristics (Fig. 4).

4.3. Subtasks

Eleven subtasks have been defined and distributed among the participants. Each subtask is carried out by a selected number of participants. Each of the subtasks deals with a specific question, either in relation to the development of test procedures or the general storage model, or in relation to possible extensions of the application area.

The following subtasks have been distributed:

- Development of a general method to process measured data.
- Definition of a steady-state criterion for heat loss tests.
- Investigation of the influence of a fluctuation in ambient temperature on the measurement accuracy.
- Investigation of possible new or modified heat loss tests.
  Comparison of heat loss test methods.
  Heat loss distribution over the storage surface.
- Investigation and prescription of the required accuracies of the test results for modelling of solar energy systems.

- Definition of stratification.
- Evaluation of 'black box' test approach by using internal measurements.
- Definition of a dynamic test for model validation.
- Identification of necessary modifications of test procedures for larger short term heat storage systems.
- Identification of necessary modifications of the test procedures for low and high temperatures.
- Test methods for storage materials and transfer fluids.

5. EXPECTED RESULTS

The aim is to conclude the project with the publication of the following final results:
- final recommendations for test procedures for the simple and the extensive tests;
- description of a general method (Fortran package) to process the measured data, including manual;
- description of (a general) storage model(s) for EMGP2, including a clear indication of the possibilities and limitations;
- a list of failures for solar heat storage systems and experienced ways to possibly prevent them.

6. REFERENCES

Galen, E. van (1983).
The European Solar Storage Testing Group, proceedings of the EC Contractors meeting, held in Brussels, 1-3 June 1983.
Project A: Solar Energy Applications to dwellings. D. Reidel Publishing Company, Dordrecht, the Netherlands.

Galen, E. van, and G.J. van den Brink (1984).
Recommendations for European Solar Storage Test Methods (sensible and latent heat storage devices), EUR 9620.

THERMAL STRATIFICATION WITHIN A HOT WATER STORAGE TANK

W.J. Spaulding and C.K. Rush

Department of Mechanical Engineering
Queen's University
Kingston, Ontario, Canada

ABSTRACT

This research represented an attempt to improve upon previously developed thermal stratification devices. The device developed and tested consisted of a polycarbonate pipe vertically oriented within a hot water storage tank. The pipe had holes located at its upper and lower ends and was small enough in diameter to fit through the one inch pipe thread hole located at the top of a water storage tank. When hot water entered the tank the inlet flow-rate was fixed. When cool water entered the tank the inlet flow-rate was varied according to information received from a flow direction detector located in the upper holes to minimize the flow of water through those holes. The destratifying effect of plume entrainment was thus reduced. The results demonstrated that this devise was able to generate thermal stratification within a hot water storage tank over a wide variety of test conditions.

KEYWORDS

Thermal stratification; hot water storage; energy storage; active solar; plume entrainment; inlet jet mixing.

INTRODUCTION

Solar domestic hot-water heating system performance can be improved dramatically by inducing thermal stratification within the storage vessel (Hollands, 1986; Sharp, 1979). The present project attempted to advance the development of a thermal stratifier by broadening the range of conditions under which such a device would work (Spaulding, 1986).

In a water storage tank there are many factors that can lead to a well mixed condition, including plume entrainment and inlet jet mixing (Hollands, 1986). Plume entrainment occurs when an inlet stream enters a tank at a location where the tank water and the inlet water do not match in temperature. Inlet jet mixing involves a stream of fluid entering a body of water in the form of a jet. This effect is characterized by the modified Richardson Number (buoyancy forces divided by inertial forces) (Sliwinski, 1978).

This study concentrated on developing a device that eliminated plume entrainment.

Loehrke and Gari (1977, 1979, 1982) attempted to solve the destratifying plume entrainment problem by enclosing the incoming stream of water in a pipe. Their pipe was oriented vertically in the tank and was perforated with many holes which made it possible for the incoming stream to enter the tank at any elevation. Their device was made up of an inlet chamber and a porous section. Central to the successful operation of this device was the pressure difference between the interior of the stratifier and the tank. Equation 1 describes this relationship.

$$\frac{d(P_S-P_T)}{dz} = \frac{g}{g_C}(\rho_T-\rho_S) + \frac{2f\mathring{m}^2}{DA^2\rho_S} + \frac{1}{A^2}\frac{d}{dz}\left(\frac{\mathring{m}^2}{\rho_S}\right) \quad (1)$$

$$f = \frac{\pi^2}{32}\;\frac{gD^5\rho_S(\rho_S-\rho_T)}{g_C\;\mathring{m}^2} \quad (2)$$

Equation 2 describes the friction factor, provided by internal resistance elements, necessary to enable the stratifier to deliver water to the desired elevation and is a direct result of equation 1.

PRESENT WORK

A simplification was made to the Loehrke and Gari stratifier by the elimination of the inlet chamber, thus making the upper end of the porous section the top of the stratifier. The stratifier tested was further simplified by opening holes at the top and bottom only.

In order to obtain the stratifier operation desired, equation 2 must be satisfied by physically linking two of the variables and setting the others constant. Linking of the flow friction to the temperature difference was tried but after many attempts this line of work was halted. Inlet flow-rate and temperature difference were then linked, controlling the flow-rate while keeping the friction factor more or less constant. Control was based on the temperature of the incoming water relative to that of the tank and whether or not water was being expelled or entrained through the stratifier's top holes. Fig. 1 illustrates this stratifier.

Ten thermocouples were evenly spaced, vertically, within the 1549 mm high, 152 mm diameter acrylic, insulated tank. While cool water was entering the tank direction of water flow through the top holes was sensed by a flow direction detector inserted into one of the top holes. When the inlet flow-rate was too

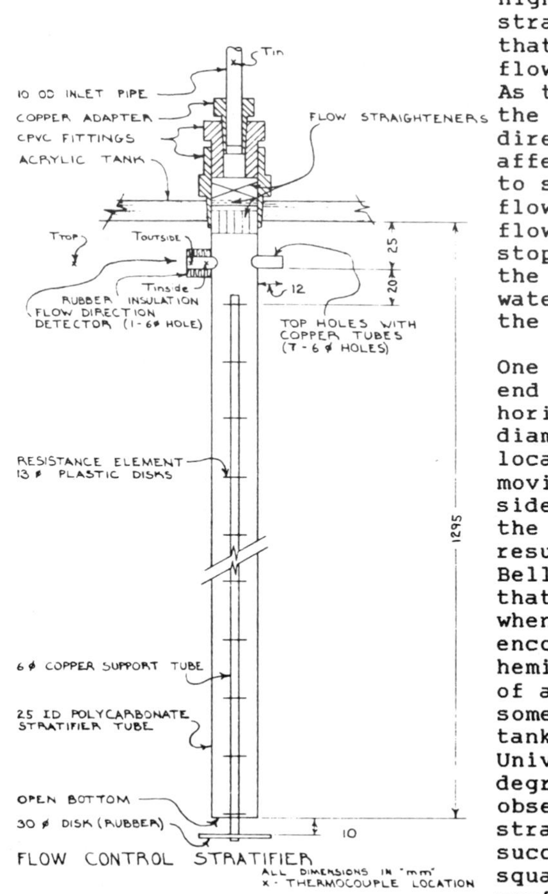

Fig. 1 Flow control stratifier

high, pressure in the stratifier was greater than that in the tank and water flowed out of the top holes. As the water flowed outward, the thermocouples in the direction detector tube were affected and a signal was sent to slightly decrease the inlet flow-rate. Conversely water flow-rate was increased to stop in-flow of water through the top holes. While hot water was entering the tank the inlet flow-rate was fixed.

One centimeter below the lower end of the stratifier, a horizontally oriented, 3 cm diameter, rubber disk was located so that downward moving flow would be deflected sideways toward the walls of the tank. This was done as a result of work by Cole and Bellinger (1982) indicating that upward mixing could occur when a downward moving flow encountered the hemispherically shaped bottom of a standard water tank. In some early work using such tanks at the Queen's University Solar Lab a degradation in performance was observed when an open ended stratifier that had been successfully tested in a square bottomed tank failed to perform as well when installed in a tank with a hemispherical bottom.

ANALYSIS AND RESULTS

A determination of how well the stratifier performed was obtained by considering three quantities: 1) the flow-rate of water expelled or entrained by the stratifier; 2) the degree of stratification within the tank, and; 3) the modified Richardson Number of stratifier expelled flows.

An indirect method was used to determine the flow-rate of expelled or entrained water by calculating the rate of energy change within a tank element and eliminating from it the effects of heat transfer to the surrounding air and water flowing down the inside of the stratifier.

The "Degree of Stratification" or DOS was defined by Sliwinski (1978) as the average rate of change of temperature with respect to the change in height at the thermocline region of the tank. An arbitrary distinction of 5 °C temperature range over the entire height of the tank was chosen to distinguish between a stratified and non-stratified condition.

In addition to the calculated determination of stratifier performance, examination of temperature versus elevation, and temperature versus time plots of experimental data provided an alternate perspective. A temperature versus elevation plot only is presented here.

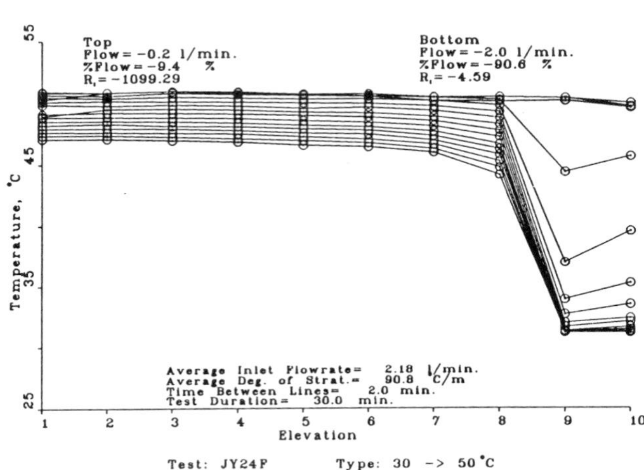

Fig. 2 Temperature vs. elevation

Results from a typical test are presented in Fig. 2 in which cold water entered a hot tank. Stratification was generated in all tests over a wide variety of test conditions. The tests indicated that the destratifying effects of plume entrainment could be virtually eliminated. Cycling of the flow-rate in this test inevitably led to slight entrainment and expulsion of water at the top holes resulting in very slow degradation of the stratified condition. Tests in which hot water entered a cold tank demonstrated that all the incoming hot water was expelled through the top holes until the tank thermocline was within a few centimeters of the bottom holes.

Some upward mixing took place during the test shown as a consequence of water expulsion through the bottom holes despite a calculated bottom modified Richardson number of 4.59. This indicates that jet mixing does play a role in destratification at modified Richardson Numbers above 0.24 when vertical wall interaction is possible. Many more tests are required before this result can be confirmed because no direct measure of expelled flow-rate was made.

CONCLUSIONS

Several advances were made during this study beyond the work of Loehrke and Gari and the work done at the Queen's University Solar Lab (Rush, 1983, 1984; Spaulding, 1986). The stratifier changed from a two part device with inlet section and porous section as conceived by Loehrke and Gari, to a device with a single porous section. This represented a significant simplification both in physical application and in theoretical description, without degradation of performance. The development of equation 2 served to clarify the restrictions placed on the design of a stratifier of this kind.

NOMENCLATURE

A cross-sectional area of stratifier
D inside diameter of stratifier
f friction factor
g gravitational acceleration
g_c conversion factor
$\mathring{m}$ mass flow-rate through stratifier
P pressure
T temperature
V velocity
z vertical distance
ρ density of water

Subscripts
s stratifier
t tank

REFERENCES

Cole, R.L., and Bellinger, F.O., Thermally Stratified Tanks. ASHRAE Transactions, 1982, Vol. 88, Pt. 2.

Gari, H.N., and Loehrke, R.I., A Controlled Buoyant Jet for Enhancing Stratification in a Liquid Storage Tank. Journal of Fluids Engineering, December, 1982, Vol. 102, pg. 475.

Hollands, K.G.T., and Lightstone, M.F., Low-Flow, Stratified-Tank Solar Water Heating Systems. SESCI Conference Proceedings, Winnipeg, Manitoba, June 22-26, 1986.

Loehrke, R.I., Gari, H.N., and Holzer, J.C., Thermal Stratification Enhancement for Solar Energy Applications. Technical Report HT-TS792, Mech. Eng. Dept., Colorado State University, June 1977.

Loehrke, R.I., et. al., Stratification Enhancement in Liquid Thermal Storage Tanks. Journal of Energy, May-June, 1979, Vol. 3, No.3, pg. 129.

Rush, C.K., Technical Report on the Development of a Low Flow Solar Collector for Domestic Hot Water. NRC Contract Report, September, 1983.

Rush, C.K., Stratified Storage Tanks for Solar Water Heaters. Proceedings of the Annual Meeting of the Solar Energy Society of Canada, Inc. (SESCI), Calgary, 1984.

Sharp, M. K., and Loehrke, R. I., Stratified Thermal Storage in Residential Solar Energy Applications. Journal of Energy, 1979, Vol. 3, No. 2, pg. 106.

Sliwinski, B.J., Mech, A.R., and Shih, T.A., Stratification in Thermal Storage During Charging. 6th International Heat Transfer Conference, Toronto, 1978, Vol. 4, pg. 149-154

Spaulding, W.J., and Rush, C.K., Dynamic Resistance Elements in a Tank Stratifier. SESCI Conference Proceedings, Winnipeg, Manitoba, June 22-26, 1986.

SOLAR THERMAL ACTIVE SYSTEMS: FROM CLOSED FORM MODELS TO SIMPLE SIZING RULES FOR COLLECTOR CIRCUITS

Bernard BOURGES, Consultant
6, rue de l'Armor, 35760 St-Grégoire (France)

Jérome ADNOT, Centre d'Energétique, Ecole des Mines
60, Bd St-Michel 75006 Paris (France)

ABSTRACT

A simple closed-form model of thermal solar active systems (Space or Water heating) is presented. It deals both with fully-mixed storage and perfectly stratified storage and enables a better understanding of physical phenomena occurring within the system (e.g. low collector flow-rate effects). Direct application of this model makes it possible to establish partial sizing rules for the system components: storage capacity, storage insulation, pipe insulation, heat-exchanger area, collector flow-rate.

KEYWORDS

Solar heating; solar Domestic Hot Water; design; closed-form model; stratification; solar collector; storage; heat-exchanger; pipe loss.

INTRODUCTION

Active solar thermal system operation can still be improved; this involves physical phenomena introduced by "system effects" which are as yet insufficiently known. The so-called "low collector flow-rate" discussion over these last few years (Van Koppen, 1979; Fanney and Klein, 1986; Hollands and co-w., 1986) pointed out this fact. Detailed simulation programmes contributed in a large way to this improvement (Wuestling, 1985). Simpler models, e.g. closed-form, can also provide essential information. This concerns basis phenomena, like stratification, but has also practical consequences through various sizing rules for system components. This is a brief presentation of this type of simple model and some of its applications.

CLOSED-FORM MODEL

This model has been developed by Bourges and Adnot (1986). It is based on a number of usual linear thermal assumptions for each component: solar collector, heat-exchanger, storage heat loss, etc. The space-averaged storage temperature T is governed by the differential equation

$$M\ c\ dT\ /\ dt = A\ F_r'(n_0\ I - U_1\ (T_b - T_a)\ ) - A_s U_s\ (T - T_e) - L \qquad (1)$$

The right-hand side of this equation includes three terms: collector energy output (controlled and therefore never negative); storage heat loss; energy taken from storage to meet the load. In the fully-mixed storage case, the bottom storage temperature, Tb, is simply the average temperature T. If the storage is perfectly stratified, Tb may be computed during one storage cycle from the initial conditions within the tank. The energy taken from the store may be computed in several ways according to the energy extraction type (Directly from the store or at collector outlet), but it is generally characterized by a draw-off flow-rate, m2.
The solution of eqn (1) in the "mixed case" involves the "system heat storage correction factor", introduced by Phillips (1981). This term is representative of daily system efficiency and characterizes the drawback of limited storage capacity as compared to an infinite capacity. In the most simple case (constant irradiance during collector operating time td, no draw-off, no losses), it takes the form:

$$G_{cst} = (\ 1 - EXP\ (\ - A\ F_r'\ U_1\ t_d\ /\ Mc\ )\ )\ .\ Mc\ /\ (\ A\ F_r'\ U_1\ t_d\ ) \qquad (2)$$

This solution may be generalized to provide a stratified system for which an equivalent mixed system may be defined by introducing two stratification coefficients K1 and K2 (Bourges and Adnot, 1986) (Table 1).
The first one applies as a multiplicative factor both to Fr' and m2; the second one to storage heat loss. The following basic assumption is required: the storage water passes through the collector at least once during collector operating time (one storage cycle at least). Theoretically a whole number of cycles or passes is needed, but that has no effect for more than two cycles.

Table 1 Stratification coefficients vs $az = (1 - AF_r'U_1/m_1c_1)\ (1-m_2/m_1)$

	$A_sU_s\ /\ m_1c_1z = 0$					$A_sU_s\ /\ m_1c_1z = 0.25$				
az	1.0	0.8	0.6	0.4	0.2	1.0	0.8	0.6	0.4	0.2
K_1	1.00	1.12	1.28	1.53	2.02	1.00	1.11	1.26	1.50	1.95
K_2	1.00	1.00	1.02	1.07	1.20	1.00	1.00	1.02	1.07	1.20

STRATIFICATION EFFECTS

It is well known that for thermosyphon or direct systems the collector flow-rate has no effect on daily efficiency if no draw-off occurs during the operating period (Gordon and Zarmi, 1981): the stratification coefficient K1 exactly compensates the decrease in the heat-removal factor Fr. If water is drawn-off from the tank during the day, an optimal collector flow-rate is observed; it corresponds to a single-pass system (one storage cycle during the collector operating time) (Wuestling, 1985; Fanney and Klein, 1986). This result can be illustrated with the solution to eqn 1 (constant irradiance) (Fig. 1): the larger the draw-off, the more important the performance improvement due to the low flow-rate as compared to an infinite rate.

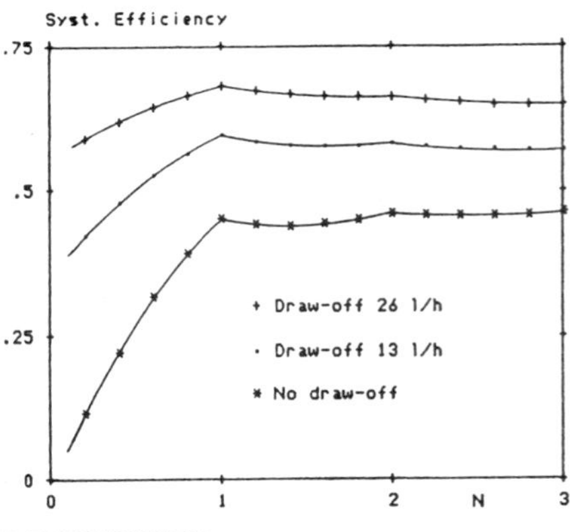

Fig. 1: Daily system efficiency of a stratified system as a function of number of cycles (direct syst.)

STORAGE SIZING: MAXIMAL DAILY TEMPERATURE

The system heat storage factor (eqn 2) makes it possible to calculate the storage energy content at the end of the daily collecting period, and therefore the storage temperature if the solar irradiance is constant (no draw-off). It has been generalized (Bourges and Adnot, 1986) to cover other irradiance profiles, like a sine-shaped radiation (more realistic). Values are given in Table 2: they are a bit higher than for constant irradiance although the operating time is shorter, but differences are negligible as long as the operating period is less than half of characteristic time. This factor makes it possible to compute the maximal storage temperature for a given day with maximal irradiation Ht (and no-draw-off) from the initial temperature T0

$$E_s = A\ F_r'\ G\ n_0\ \emptyset\ H_t \tag{3}$$

Daily Utilizability, Ø, is computed at the threshold irradiance defined by initial temperature T0.

Table 2 System heat storage correction factor vs $x = Mc\ /(A\ F_r'\ U_l\ t_d)$ for constant (G_{cst}) and sine-shaped (G_{sin}) radiation

x	0.20	0.50	0.75	1.00	1.50	2.00	3.00	5.00
G_{sin}	.271	.493	.596	.664	.749	.800	.857	.910
G_{cst}	.199	.432	.552	.632	.730	.787	.850	.906

STORAGE AND PIPE INSULATION

Pipe and duct heat loss can be integrated in the collector characteristics, as proposed by Beckman (Duffie and Beckman, 1980). These formulas can be slightly modified and generalized by considering inlet and outlet pipes (with respective heat loss coefficients: AinUin, AoutUout) as heat-exchangers (with their own effectiveness). The modified collector loop characteristics can then be computed as follows

$$U_l^* = U_l + (A_{in}U_{in} + A_{out}U_{out})\ /\ A \tag{4}$$

$$n_0^* = n_0 \cdot \mathrm{EXP}\,(- A_{out}U_{out} / m_0c_0) \cdot F_r / F_r^* \qquad (5)$$

The effect of pipe heat loss on long-term system performance may be approximated for a given typical day by eqn 3. It is more important for stratified systems, with low collector flow-rate.

As stated by table 1, storage heat loss is slightly enhanced by stratification (due to higher temperatures obtained within the storage), but mainly has a negative effect on the stratification (Kl).

HEAT-EXCHANGER SIZING

Most of the systems include an external or immersed heat-exchanger between collector and storage. Sizing of the heat-exchanger is somewhat critical. This theory provides some useful information.

Immersed coils generally imply a mixed tank. Losses caused by the heat-exchanger are taken into account by the usual collector-heat-exchanger efficiency factor Fr' whith the limiting value for an infinite flow-rate

$$F_x = 1 / (1 + A\,U_l / A_eU_e) \qquad (6)$$

Its variations are represented by fig.2: The lesser of the losses in output from the flow-rate (Fr) and the heat-exchanger (Fx) may be disregarded in practice. At a given flow-rate, a minimum heat-exchanger size is needed but a precise heat transfer coefficient computation is not required. This is an interesting result, if we keep in mind the poor accuracy of such heat-transfer calculation (Feiereisen, 1982).

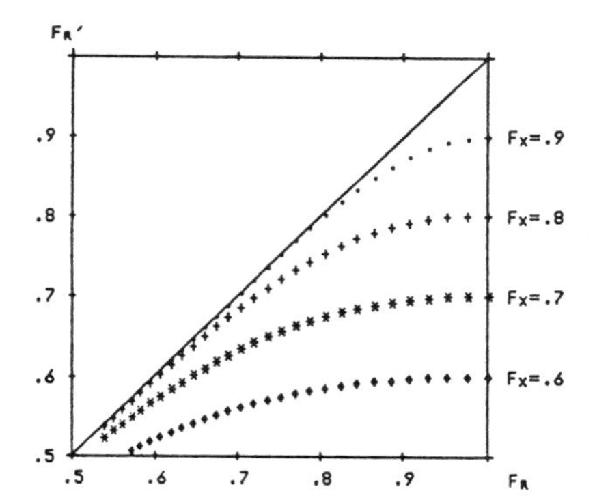

Fig.2: Collector heat-exchanger correction factor for immersed coils

External heat-exchangers are concerned both with F'r (like immersed coils) and stratification coefficients. The parameter to be considered is their respective product. Fig. 3 shows that for a given primary loop (collector + heat-exch.) there is an optimal flow-rate ratio on both sides of the heat-exchanger for which Fx (eqn 6) is determined again. As for direct systems, flow-rates can be "forgotten", provided they are in a pre-determined ratio, equal to Fx: low flow-rates are compensated by a high degree of stratification (at least one storage cycle is still required). But the optimum flow-rate cannot be determined unless the daily draw-off profile is known.

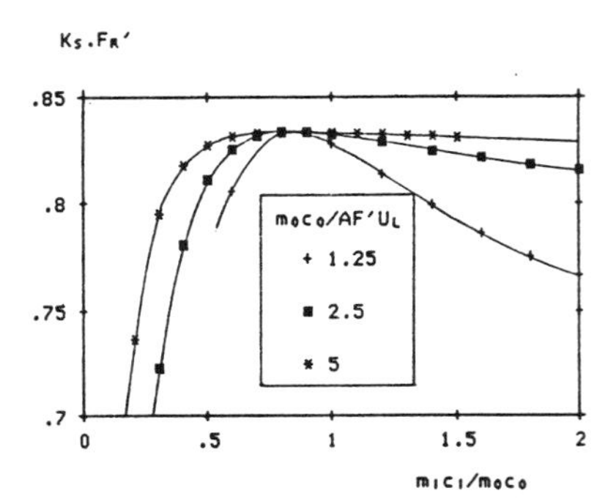

Fig.3: Optimum flow-rate ratio for an external heat-exchanger

REFERENCES

BOURGES, B., and J.ADNOT (1986). A generalized closed-form model for solar hot water systems. (to be published in SOLAR ENERGY)

DUFFIE, J.A, and W.BECKMAN (1980). Solar Eng. of thermal processes. WILEY.

FANNEY, A.H., and S.A.KLEIN (1986). Thermal performance comparison for solar hot water systems subjected to various collector array flow-rates. INTERSOL 85, 538-543, Pergamon Press.

FEIEREISEN, T.J., and co-w. (1982). Heat transfer from immersed coils. ASME 82-WA/Sol-18.

GORDON, J.M., and Y.ZARMI (1981). Thermosyphon systems: single vs multi-pass. SOLAR ENERGY, 27, 441-442.

HOLLANDS, K.G.T., D.R.RICHMOND and D.R.MANDLESTAM (1986). Re-engineering domestic hot water systems for low flow. INTERSOL 85, 544-548, Pergamon Press

PHILLIPS, W.F. (1981). Integrated performance of liquid-based solar heating systems. SOLAR ENERGY, 26, 287-295.

PHILLIPS, W.F., and R.N.DAVE (1982). Effects of stratification on the performance of liquid-based solar heating systems. SOLAR ENERGY, 29, 111-120.

VAN KOPPEN, C.W.F, and co-w. (1979). Proceedings ISES biennal meeting, Atlanta, 2, 576-580, Pergamon Press.

WUESTLING, M.D., S.A.KLEIN and J.A.DUFFIE (1985). Promising control alternatives for solar water heating systems. ASME J. of Solar Energy Engineering, 107, 215-221.

NOMENCLATURE

A_eU_e	Heat transfer coefficient	m_2	Draw-off flow-rate
H_t	Daily global irradiation	t_d	Collector operating time
m_0	Collector flow-rate	T_m	Mains cold water temperature
m_1	Heat-exch./storage flow-rate	T_0	Storage initial temperature

HEAT TRANSFER FROM FINNED AND SMOOTH TUBE HEAT EXCHANGER COILS IN HOT WATER STORES

R. Kübler, M.Bierer, E. Hahne

Institut für Thermodynamik und Wärmetechnik
Universität Stuttgart
Pfaffenwaldring 6, FRG - 7000 Stuttgart 80
Federal Republic of Germany

ABSTRACT

Heat transfer capacity rates $(UA)_{hx}$ of finned and smooth tube heat exchanger coils in hot water stores have been studied experimentally. Using heat transfer correlations for single horizontal tubes from literature, $(UA)_{hx}$ was calculated and compared with the measurement results. Good agreement was found for the smooth tubes. For the heat transfer coefficient on the outside of the finned tubes new correlations were determined. A simplified correlation of $(UA)_{hx}$ for use in simulation programs is presented.

KEYWORDS

Hot water storage, heat exchanger, natural convection heat transfer, heat transfer capacity rate, finned tubes, smooth tubes

INTRODUCTION

Today most small-scale hot water stores are used for domestic hot water heating and mostly equipped with internal heat exchangers to provide separation between the heat transfer fluid and drinking water. For system design or simulation purposes, it is necessary to know the heat transfer capacity rate $(UA)_{hx}$ of the heat exchanger (see Fig.1).

It has been shown earlier (Kübler 1987) that $(UA)_{hx}$ is dependent on heat transfer fluid flow rate and thermal properties, storage temperature and heat exchanger inlet temperature and finally on the geometry of the heat exchanger.

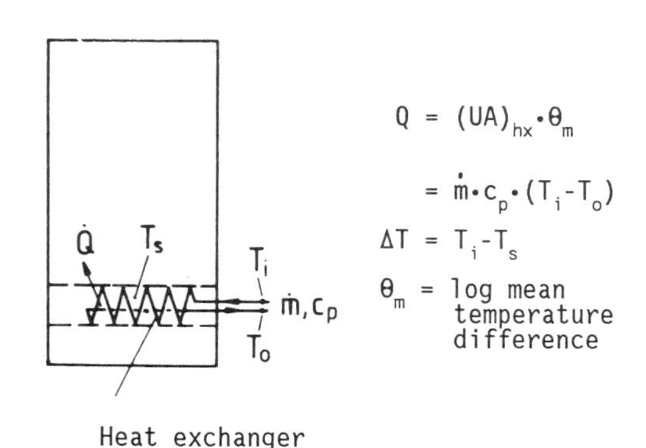

Fig. 1. Basic equations and notation

Fig. 2. Heat exchangers

In this report an improved calculation model, new correlations for the outside heat transfer on the finned tubes and simplified equations for the heat transfer capacity rate are presented. The heat exchangers investigated are shown in Fig. 2; their relevant data are summarized in Table 1.

Measurements on all heat exchanger coils were carried out with the axis horizontal (as indicated in Fig.1). On heat exchangers No. 1 (the smooth tube) and No. 3 also measurements with the axis vertical were carried out, using both flow directions, from top to bottom and vice versa. Experimental conditions covered a range of $150<\dot{m}<1000$ kg/h, $15 < T_s < 90$ °C and $3 < \Delta T < 25$K.

TABLE 1 Data of the Heat Exchangers

Heat Exchanger			1	2	3	4
Tube Type			smooth	finned	finned	finned
Orientation of Axis			h + v	h	h + v	h
Outside Area	A_{out}	[m^2]	0.283	0.962	1.224	2.532
Inside Area	A_{in}	[m^2]	0.245	0.169	0.237	0.490
Coil Diameter	D_w	[m]	0.127	0.115	0.123	0.137
Inside Tube Diameter	d_{in}	[mm]	13.0	10.5	12.5	16.5
Outside Tube Diameter	d_{out}	[mm]	15.0	12.5	14.5	18.5
Fin Diameter	d_r	[mm]	-	19.5	21.5	25.5
Fin Distance	b	[mm]	-	1.9	1.9	1.9
Tube Length	l	[m]	6.0	5.13	6.04	9.45

h = horizontal , v = vertical

THE CALCULATION MODEL

To account for changing U along the heat exchanger (mainly due to the natural convection heat transfer on the outside of the tubes), the model calculates the heat transfer capacity rate for small segments (one turn of the coil) and forms a weighted sum (eqn. (1)).

$$(UA)_{hx} = \frac{1}{\theta_m} \cdot \sum_{j=1}^{N} Q_j = \frac{1}{\theta_m} \cdot \sum_{j=1}^{N} (UA)_{hx,j} \cdot (T_{f,j}-T_s) \qquad (1)$$

where θ_m is the log-mean temperature difference (for the whole heat exchanger) between heat transfer fluid and the surrounding storage water. The heat transfer capacity rate of each segment $(UA)_{hx,j}$ can be calculated using eqn. (2), if inside heat transfer coefficient h_{in} and outside heat transfer h_{out} are known.

$$(UA)_{hx,j} = \frac{1}{\dfrac{1}{(h_{in} \cdot A_{in})_j} + \dfrac{S^*}{\lambda} + \dfrac{1}{(h_{out} \cdot A_{out})_j}} \qquad (2)$$

S^* = shape factor for thermal conduction

Heat Transfer Inside the Tubes

Heat transfer coefficients in tubes under different conditions have been studied and reported extensively. For the calculation of the inside heat transfer coefficient an equation by Gnielinski (1986) for helically coiled tubes was applied. It takes into account the heat transfer augmentation due to the curvature.

Heat Transfer on the Outside of the Tubes

Heat transfer correlations from literature are only available for single smooth and finned tubes and for vertical arrays of horizontal tubes. While the heat exchanger with vertical axis is similar to the case of arrays of horizontal tubes, no information is available for coils with horizontal axies for both tube types and for arrays of horizontal finned tubes.

Smooth tubes. Churchill and Chu (1975) have given an equation for the heat transfer from horizontal smooth tubes covering a large range of Rayleigh numbers with one single equation (eqn. (3)).

$$Nu_{d,out} = \left(0.60 + \frac{0.387 \cdot (Ra_{d,out})^{1/6}}{\left[1 + (0.559/Pr)^{9/16} \right]^{8/27}} \right)^2 = \frac{h_{out} \cdot d_{out}}{\lambda} \qquad (3)$$

Finned tubes. Henderson and Caolo (1983) have reported an equation for the heat transfer from single, horizontal finned tubes (eqn. (4)). Both Nusselt and Rayleigh numbers are based on the fin to fin distance b.

$$Nu_b = C \cdot (Ra_b \cdot b/d_r)^m = h_{out} \cdot b/\lambda \qquad (4)$$

where $C = 0.0912$ and $m = 0.536$ for $Ra_b \cdot (b/d_r) < 10^{2.25}$
and $C = 0.3365$ and $m = 0.285$ for $Ra_b \cdot (b/d_r) > 10^{2.25}$

RESULTS

The heat transfer capacity rate of all four heat exchangers with the axis horizontal and the two heat exchangers with the axis vertical were calculated with the model, using eqns. (1),(2),(3),(4) and the equation given by Gnielinski (1986) for the inside heat transfer coefficient. Then calculated and measured values of $(UA)_{hx}$ were compared.

Smooth tube heat exchanger

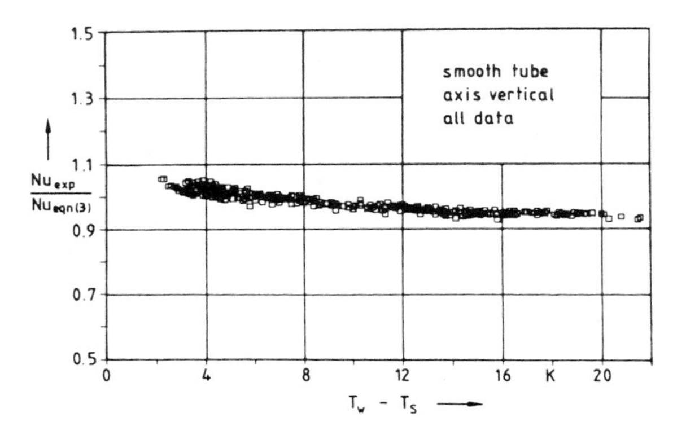

Agreement between model calculation, using the correlation for the single horizontal tube (eqn. (3)), and measurement is excellent at low temperature differences $\Delta T = T_i - T_s$ between heat exchanger inlet and storage water. Only at high ΔT (=20K) the model predicts slighthy higher values

<- Fig. 3 Ratio of measured to calculated (eqn.(4)) Nusselt numbers for the smooth tube heat exchanger (axis vertical)

than the measurement. In Fig. 4 the ratio of measured Nusselt number to the Nusselt number from eqn. (4) is plotted versus the temperature difference between outside tube wall and storage water (please note the difference to the ΔT in Fig. 1 , where $\Delta T=T_i-T_s$). At low temperature difference the measured values are about 5% higher, at high temperature difference about 5% lower than the values from eqn. (3). This result is remarkable in view of work published by different authors on heat transfer from arrays of horizontal tubes. Marsters (1972), Sparrow and Niethammer (1981), Tokura and co-workers (1983) have reported measurement results on arrays of up to nine horizontal tubes at different spacing (vertical center to center distance of the cylinders).

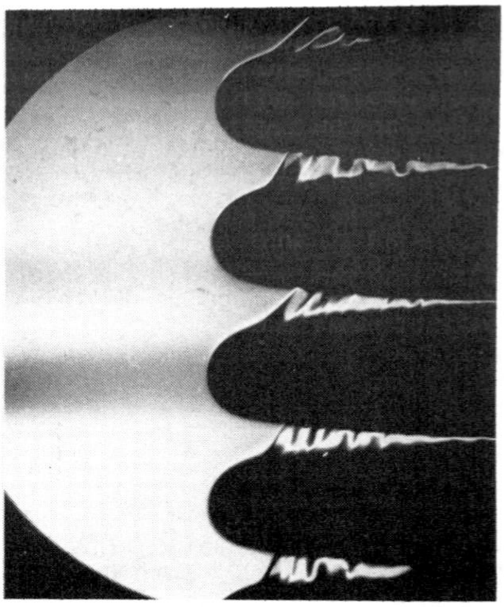

Fig. 4. Convection flows

Their results show a decrease in heat transfer at tube spacings of less than two diameters (which we have also here) of around 20% at the second and up to 50% at the further downstream located tubes compared to a single tube. Our values are very close to the single tube results. Schlieren optical investigations (Fig. 4.) show, that at low temperature differences the convection flows are directed towards the coil axis and the tubes do not influence each other. At higher temperature difference eddies start to form on the lower turns and flow around the upper ones, leading to a slight decrease in heat transfer.

Influence of axis orientation. Both flow directions in the heat exchanger with the axis vertical deliver exactly the same heat transfer capacity rate, if the slight stratification along the heat exchanger axis is considered in the calculation of the log mean temperature difference θ_m. Mounting the coil horizontally changes the heat transfer capacity rate by about 5% compared to the vertical axis. The outside heat transfer coefficient can, with very good accuracy, be calculated using eqn. (3).

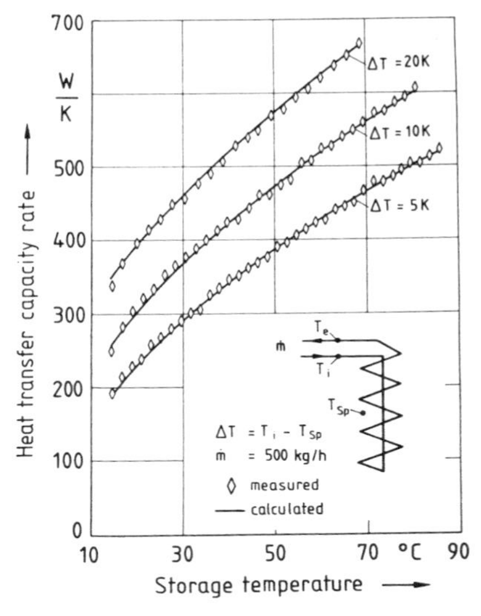

Fig. 5. $(UA)_{hx}$ of heat exchanger 3, vertical axis ($\dot{m}$ = 500 kg/h, ΔT=10 K)

Finned Tube Heat Exchanger

The comparison between model calculations using eqn. (4) and the experimental data showed up to about 20% difference in the outside heat transfer coefficient at the finned tube. An new correlation was therefore determined (eqn.(5)) for the outside heat transfer coefficient from the experimental data. Regression was carried out for heat exchanger 3.

The comparison between measured values and model results is shown in Fig. 5. for the case of the vertical axis. Due to the additional constant, only one equation was needed to cover the whole range.

$$Nu_{b,out} = C_1 + C_2\,(Ra_b \cdot b/d_r)^{0.27} \qquad 10^{1.2} < Ra_b \cdot (b/d_r) < 10^{2.8} \tag{5}$$

where for horizontal axis: $C_1 = -0.76$ $C_2 = 0.57$
for vertical axis: $C_1 = -0.55$ $C_2 = 0.48$

Regression for heat exchangers No. 2 and No. 4 was not yet possible, as the data basis, in particular the range of operation conditions, was too small. Using eqn.(5) with the coefficients for the horizontal axis, however, leads to very good agreement between measured and model results with deviations less than 10%.

SIMPLIFIED CORRELATIONS

A dimensional analysis shows, that the heat transfer capacity rate can be described with an equation in exponential form (eqn. (6)) as function of the heat exchanger inlet temperature T_i.

$$(UA)_{hx} = a \cdot T_i^{\,c} \tag{6}$$

For heat exchanger 3 the following coefficients were determined:

axis vertical

$a = (15 + 0.02 \cdot \dot{m} + 0.5 \cdot \Delta T)$

$c = 0.65$

axis horizontal

$a = (13 + 0.02\,\dot{m} + 0.55\,\Delta T)$

$c = 0.67$

where $\dot{m}$ = mass flow rate in kg/h $\Delta T = T_i - T_s$

Eqn. (6) represents the experimental data with less than 15% deviation in the range of $250 < \dot{m} < 1000$ kg/h, $5 < \Delta T < 20$ K and $20 < T_i < 90$ °C. Similar regression for the smooth tube heat exchanger leads as yet to higher deviations over the whole range. The method will be developed further with the aim to determine the coefficients of eqn. (6) from a small number of measurements of $(UA)_{hx}$ at different operation conditions.

ACKNOWLEDGEMENT

This work was financed by the Federal Minister for Research and Technology under 03 E 8021 A. The authors gratefully acknowledge this support.

REFERENCES

Churchill, S.W., and Chu, H.H.S. (1975). Correlating equations for laminar and turbulent free convection from a horizontal cylinder. *Int. J. Heat and Mass Transfer, 18*, 1049-1053.

Gnielinski, V. (1986). Heat transfer and pressure drop in helically coiled tubes. *Heat Transfer 1986*, Proceedings of the 8th International Heat Transfer Conference, San Francisco, USA (1986), 2847-2584.

Henderson, J.B., and Caolo A.C. (1983). Optimization of radial finned tube heat exchangers for use in solar thermal storage systems. *DOE Report 25247-t1*, Department of Energy.

Kübler, R. (1987). Heat transfer capacity rate of heat exchangers in hot water stores. *Short term water heat storage*, Proceedings of the IEA energy conservation through energy annex IV workshop, Utrecht, The Netherlands (1987).

Marsters, G.F. (1972). Arrays of heated horizontal cylinders in natural convection. *Int. J. of Heat and Mass Transfer, 15/1*, 921-933.

Tokura, J., Saito, H., Kishinami, K., and Muramoto, K. (1983). An experimental study of free convection heat transfer from a horizontal cylinder in a vertical array set in free space between parallel walls. *Journal of Heat Transfer*, *105*, 102-107

A MODEL OF HEAT STORAGE IN PHASE-CHANGE MATERIALS FOR ACTIVE SOLAR SYSTEMS

H. Visser*,***, D. van Hattem**, G.C.J. Bart* and C.J. Hoogendoorn*

* Delft University of Technology, Applied Physics Department, P.O. Box 5046, 2600 GA Delft, The Netherlands

** Centre of the European Communities-Joint Research Centre, Heat Transfer Division, 21020 Ispra (Varese), Italy

*** Now with TNO-Institute of Applied Physics, Delft, P.O. Box 155, 2600 AD Delft, The Netherlands

ABSTRACT

A new subroutine on phase-change heat storage has been developped as a component in the TRNSYS transient simulation program. It is based on the solution of the energy conservation equation with the enthalpy method. To show the possibilities of this subroutine it has been applied in the description of a solar energy domestic hot water system. A comparison has been made between differently sized latent heat storage vessels and sensible heat storage in a water tank with different degrees of stratification. In the domestic hot water system no advantage could be noticed in the application of a latent heat storage. Nevertheless this subroutine can be a valuable tool and in fact will be used in the design of sophisticated solar energy applications as there are solar driven LiBr-airconditioning systems or solar driven refrigerators (NH_3-cycle).

KEYWORDS

phase-change materials (PCM), latent-heat storage (LHS), solar energy, transient simulation program (TRNSYS).

INTRODUCTION

The purpose of this work is to develop a component model that can be used as part of the TRNSYS Transient Simulation Program, a software package that has been designed to describe the performance of solar energy systems. At the CEC-Joint Research Centre in Ispra (Italy) effort is put in the development of a solar driven LiBr-airconditioning system. This kind of systems demands heat within a narrow temperature range. This heat will be supplied from a latent heat storage system as sensible heat storage systems are not adequate here. As the TRNSYS-program does not provide a latent heat storage system one has to make a subroutine on its own in order to be able to simulate the system. It was chosen to adapt a program that earlier was developped at the Delft University of Technology. As an example of the capabilities of this program a number of test runs have been made to compare the behaviour of a sensible heat storage with that of a latent heat storage device in a solar energy domestic hot water system. See Fig. 1.

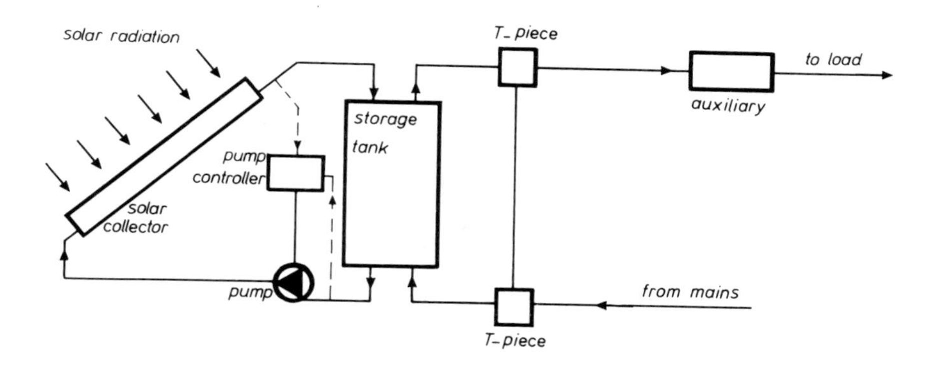

Fig. 1. The simulated domestic hot water system.

THE STORAGE SYSTEM AND ITS NUMERICAL DESCRIPTION

The storage vessel consists of a number of closed cylindrical pipes filled with the phase change medium. See Fig. 2. These pipes are surrounded by a heat transfer fluid. The flow in the vessel is controlled by the flow conditions in the total solar system. With two T-pieces of tubing the vessel can be connected to the collector loop and the load loop. A second model, consisting of a cylindrical pipe embedded in the phase-change material, has also been developped. It can be found in the original report (Visser, 1985).

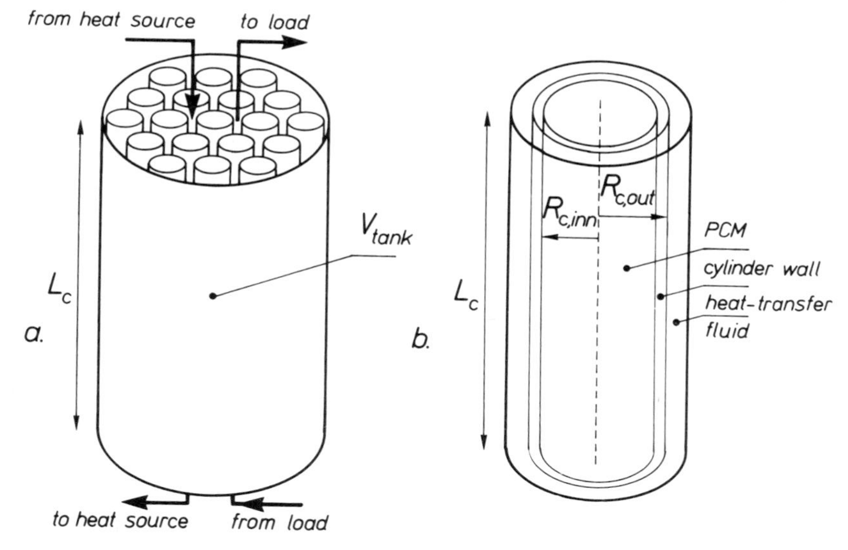

Fig. 2. The storage vessel and its numerical model.

In the numerical model of this vessel it is assumed that all pipes filled with the phase-change material behave identically. So the complete vessel can

be considered as a parallel connection of separate tubes. The total heat loss of the vessel will then be equally distributed over the tubes. In this way the threedimensional problem is reduced to a twodimensional one.
The program comprises the solution of the heat balance equation for the phase-change material. Use is made of the enthalpy method (Voller and Cross, 1980) to solve these equations. Only conduction has been taken into account. The thermal conductivity of the liquid can eventually be enhanced if convection plays a role. The phase transition range has been modelled according to Fig. 3.

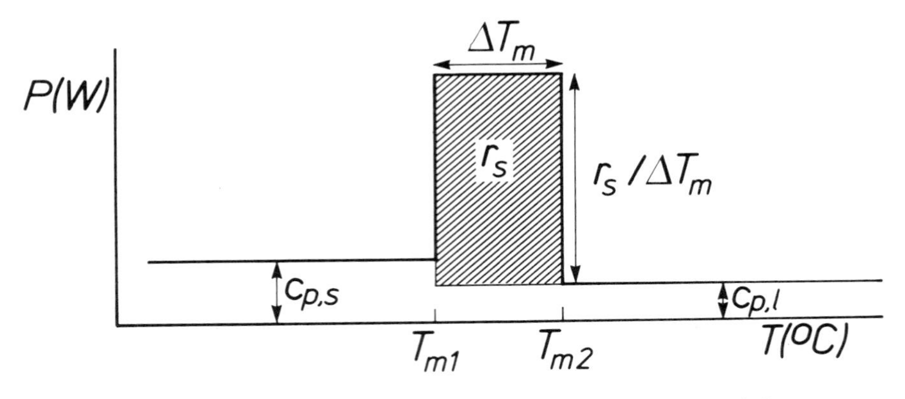

Fig. 3. Specific heat curve of phase change material.

It can be shown that for an arbitrary shaped specific heat curve there exists an equivalent rectangular one with only slight deviations in its step response behaviour.
As the influence of the pipe wall cannot be neglected on forehand a separate control volume has been attributed to it.
In the heat balance for the heat transfer fluid conduction in the flow direction can be neglected compared to the heat transfer by convection. Depending on the flow regime (laminar, turbulent, forced/free convection) of the heat transfer fluid empirical relations have been used to obtain values for the heat transfer coefficient from fluid to pipe wall. A measured heat loss coefficient has been assumed between the heat transfer fluid and the surrounding world.
In the TRNSYS-program a timestep is used that is synchronized with the weather data. Within the subroutine we describe here the time step preferably is synchronized with the time needed for the flow to travel over the distance between two grid points (control volumes). So this time step is flow dependent. This decrepancy is solved by chosing the greatest time step that is smaller than the preferable internal one and that fits a whole number of times within the TRNSYS time step. In between each internal time step the heat balance of the heat transfer fluid has to be adjusted then.

PARAMETERS USED IN SIMULATIONS

The sensible heat storage vessel did have a volume of 0.15 m^3 and a height of 0.7 m. It was filled with water with an initial temperature of 10 °C. The heat loss coefficient to the surroundings was set to 0.4 W/m^2K. In the simulations numbered 1 to 3 the number of nodes was varied in the sequence 1, 3, 10 to simulate different degrees of stratification. If the number of nodes equals 1 the tank is fully mixed (one temperature), for the numbers of nodes

being 10 the tank can be considered fully stratified.
A latent heat storage tank simulation has been carried out in eight different situations. First we will mention the properties that were not altered during these simulations. These are the density $\rho = 900$ kg/m^3 and the specific heat in the solid and liquid range $c_{ps} = c_{pl} = 2$ kJ/kgK of the phase change material. And for the pipe wall $\rho_w = 900$ kg/m^3, $c_{pw} = 2$ kJ/kgK and the thermal conductivity $\lambda_w = 0.277$ W/mK. The properties of the heat transfer fluid (water) were set: $\rho_f = 1000$ kg/m^3, $c_{pf} = 4190$ kJ/kgK , $\lambda_f = 0.616$ W/mK and the vicosity $\mu_f = 0.75\ 10^{-3}$ kg/ms. The initial temperature and heat loss coefficient were chosen equal to those of the sensible heat storage tanks.
Simulation no. 4 is used as a reference. Its properties are given hereafter. An equal thermal conductivity of the phase-change material in solid, liquid and transition range $\lambda_s = \lambda_l = \lambda_t = 0.286$ W/mK. The transition range lies between the temperatures $T_{ml} = 30$ °C and $T_{m2} = 35$ °C, with a latent heat $r_s = 150$ kJ/kg (see figure 1). The volume of the storage tank equals 0.15 m^3 and its surface equals 1.58 m^2. In it are 37 cylinders containing the phase-change material with a lenght of 0.7 m, an inner radius of 0.035 m and an outer radius of 0.037 m. In simulation no. 5 the transition range has been shifted with $T_{ml} = 20$ °C and $T_{m2} = 25$ °C. In simulation no. 6 $T_{ml} = 40$ °C and $T_{m2} = 45$ °C. In the 7th simulation the thermal conductivity of the phase-change material is set to $\lambda_s = 0.143$ W/mK, $\lambda_l = 0.556$ W/mK and $\lambda_t = 0.277$ W/mK. In simulation no. 8 the only difference with simulation no. 4 is a latent heat value $r_s = 300$ kJ/kg. In simulation 9 the number of cylinders has been increased upto 91 and consequently the inner radius is lowered to 0.022 m and the outer radius to 0.0235 m. In simulation no. 10 the number of cylinders has been decreased to 19 with an inner radius of 0.050 m and a outer radius of 0.052 m. In the 11th and last simulation the tank volume is lowered to 0.12 m^3 and consequently its surface area to 1.36 m^2. Now the lenght of the cylinders turns out to be 0.65 m with an inner radius of 0.032 m and an outer radius of 0.0335 m.
The daily domestic hot water demand has been set to 15 kg between 6 a.m. and 7 a.m., 30 kg between 7 a.m. and 8 a.m., 15 kg between 8 a.m. and 9 a.m., 7.5 kg between 11 a.m. and 12 a.m., 15 kg between 12 a.m. and 1 p.m., 7.5 kg between 1 p.m. and 2 p.m., 15 kg between 4 p.m. and 5 p.m., 30 kg between 5 p.m. and 6 p.m. and 15 kg between 6 p.m. and 7 p.m.. Outside these hours there has been no demand. The delivery temperature of the water has been supposed to be 55 °C. With a cold water temperature of 10 °C the total demand in the simulated one week period amounts to 198.0 MJ. From the weather data a total solar insolation of 243.4 MJ has been found on the collector area of 3.0 m^2 for this period.

PERFORMANCE OF SIMULATED DHW-SYSTEMS

The results of the aforementioned simulations have been collected in Table 1.

Table 1. Performance of DHW heat storage systems.

Simulation number	1	2	3	4	5	6	7	8	9	10	11
Collector efficiency	.466	.497	.510	.489	.486	.484	.483	.506	.487	.482	.469
Storage efficiency	.740	.780	.794	.775	.765	.789	.776	.707	.791	.771	.808
Stored fraction	.218	.197	.187	.210	.223	.195	.212	.282	.194	.214	.176
Solar contribution	.423	.477	.498	.466	.457	.469	.461	.440	.473	.459	.466

In this table the collector efficiency has been determined as the fraction of the solar insolation that net has been transferred into the storage system. The storage-efficiency has been defined as the fraction of input energy of the storage vessel that has been transported into the load. Part of this input energy still remains in the vessel, it has been called the stored fraction. The solar energy transported into the load only can fullfill partly the demand, this has been expressed as solar contribution.

CONCLUSIONS

The collector efficiency for our PCM-heat storage is hardly affected by measures like enhancing the heat transfer from heat transfer fluid to the phase change material. In this respect the proposed storage vessel cannot be further improved. With the water vessel we see, as expected, a distinct influence of the degree of stratification on the collector efficiency. So an other conclusion is that, with respect to the collector efficiency, stratification in the PCM-vessel is optimal. In the PCM simulations the only clear effect is that of the heat capacity of the storage. With a higher heat capacity (simulation no. 8) the collector efficiency is increased somewhat as with a lower heat capacity (simulation no. 11) it is decreased. To raise the collector efficiency to a better acceptable level in these simulations the solar collector itself should be improved first.
The storage efficiency effects are more difficult to understand. The energy named stored fraction is not yet lost but still can be of some use and in our simulations also cannot be neglected. Nevertheless we will focus attention mainly to the storage efficiency of the storage system as it is a more positive property to put the energy through as soon as possible in order to have more capacity available in case of a supply of solar heat.
For the water tank again it is the degree of stratification that determines the storage efficiency. It is rather easy to see that this effect will even be more pronounced in case of a bigger storage volume. In case of a PCM-storage more parameters are important.
From simulations no. 4, 5 and 6 it can be concluded that it is advantageous to have the latent heat stored at a temperature level just above the demanded temperature. This can easily be understood as this increases the chance on periods without auxiliary heating.The other differences that we see are due to the two-dimensional nature of stratification. The energy is transported by the heat transfer fluid in an axial direction and when stored it is spread out in the radial direction. This effect can clearly be seen by the pipes with small diameters behaving better than the bigger ones. In simulation no. 8 the more available latent heat, in between the periods of demand, sorps the sensible heat of the liquid even more. With the small storage tank, simulation no. 11, the mean storage temperature becomes higher giving an enhanced heat transfer to the hot water system. The net result is an equal solar contribution compared with simulation no. 4.
So as a general conclusion we can say that, especially with the latent heat storage system, it seems that measures to improve collector efficiency respectively the storage efficiency are antagonistic.
However we expect this routine to be a valuable tool in the design of the LiBr-airconditioning system.

REFERENCES

Visser, H., (1985). Energy storage in phase change materials - Development of a component model compatible with the TRNSYS - transient simulation program. Final report, contract no. 2462-84-09 ED ISP NL CEC JRC Ispra.

Voller, V. and M. Cross (1980). Accurate solutions of moving boundary problems using the enthalpy method. Int. J. Heat Mass Transfer, Vol. 24, 545-556.

LATENT HEAT THERMAL ENERGY RESERVOIR USED AS A FLOW-THROUGH WATER HEATER

Malatidis, N.A.
Institut für Thermodynamik und Wärmetechnik der Universität Stuttgart, Pfaffenwaldring 6,
7000 Stuttgart 80, FRG

ABSTRACT

In this paper a latent heat thermal storage (LHTS) flow-through heater system is presented. It consists of a special heat exchanger and an integrated storage material.
The heat exchanger is characterized by the ability to simultaneously charge and discharge from two different heat sources and sinks in the storage system.
The experiments with a prototype have demonstrated its function and performance.

A modification, the LHTS - flat solar collector-compact system, is characterized by the storage and use of solar energy by means of a phase change material in a highly efficient heat exchangers as follows:

- use of one side of the heat exchanger / storage container as absorber
- transfer of the absorbed energy to the phase change material through fins
- use of the stored heat by heating water at constant temperature.

KEYWORDS

Solar energy utilization, latent heat storage, special heat exchanger, phase change material, flow-through heating effect, latent heat reservoir with a solar collector as a compact system.

INTRODUCTION

The amount of sunshine varies very much with the time of year, and the maximum available sunshine occurs when there is little or no energy demand. This makes it necessary to use efficient energy storage systems. Without heat storage, there is no economically feasible application of solar systems, at least not in Central Europe.

LATENT HEAT THERMAL ENERGY STORAGE (LHTES)

The essential characteristic of LHTES is that absorption and supply of heat occur while the storage material changes its phase at constant temperature. This is the case with the inorganic eutecticum $Mg(NO_3)_2 \cdot 6\ H_2O + NH_4NO_3$ at a eutectic temperature $\vartheta_E = 50\ ^\circ C$. This material is compared with water, of which the sensible heat is shown in Fig. 1.
The advantages of the phase change storage material (PCSM) are:
- high energy storage capacity
- lower heat losses to the surroundings
- energy supply at nearly constant temperature
- use of heat at a relatively low temperature level.

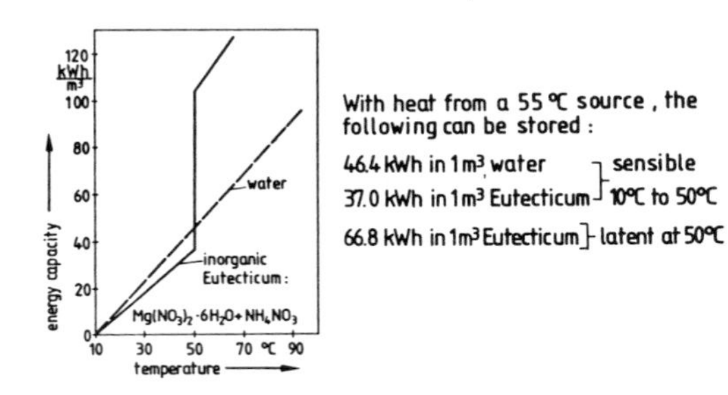

Fig.1: Temperature dependence of the energy capacity of a Eutecticum (phase-change material) compared to water.

STORAGE AND USE OF LATENT HEAT

The storage and use of latent heat reguires a heat transfer system which exchanges thermal energy by a PCSM (Abhat and co-workers, 1981; Abhat, Aboul-Enein, Malatidis, 1982). The problems with storing and using solar energy efficiently are:
- selection of PCSMs which must meet certain criteria concerning their thermophysical properties
- design and production of heat exchangers which allows charging and discharging in short times (Malatidis, 1984; Hahne, 1986).

It is therefore required to use a special heat exchanger design which forms an integrated storage unit combining storage mass and heat exchanger. It is taken for granted that both components are balanced in their physical, chemical and electrochemical properties.

LATENT HEAT THERMAL ENERGY STORAGE (LHTES) SYSTEM

In this paper a LHTES-System for use as a water heater is presented. The experiments made to test its function and performance as a prototype system are described by Malatidis und Erhorn, 1986.
Furthermore, there is a presentation of a promising further development of this LHTES-system which can be regarded as a modification and can be defined as a compact system integrating LHTES with a solar collector.

Phase Change Storage Material (PCSM)

The working temperature is given by the melting and solidification point of the PCSM. The melting and solidification behaviour and the physical and corrosion properties of the PCSM are given as the essential criteria for the heat exchanger design. Selecting the PCSM with its working temperature and its heat capacity determines the field of application and the use of the storage systems. Those systems with working temperatures near 50 °C ,the presented eutecticum has ϑ_E = 50 °C, can be applied for solar usage in Central and Southern Europe.

The Heat Exchanger

The design of the heat exchanger is very important for the storage and use of latent heat. It is desirable to combine optimal volumetric storage capacity with a geometry which allows continuous uniform charging and discharging. The geometry should not lead to essential local superheating of the system when charging or to essential subcooling of the melted PCSM when discharging. Fig.2 is a schematic diagram of the prototype heat exchanger integrating LHTES and flow-through heating. It is a flat container, which surrounds equidistant fins.
The inner surfaces of the larger sides of the container have a fixed metallic connection with the edges of the fins while the outer surfaces have channels in a double-walled bottom for the flow of a fluid (Fig. 3). The container is thermally insulated from the surroundings. The inorganic eutecticum presented here (ϑ_E = 50 °C) is the filling with which the system operates. The channels on the larger sides of the container are used as the charging and discharging system. Such a heat exchanger concept operates simultaneously as a charging and discharging system from two different heat sources and sinks.

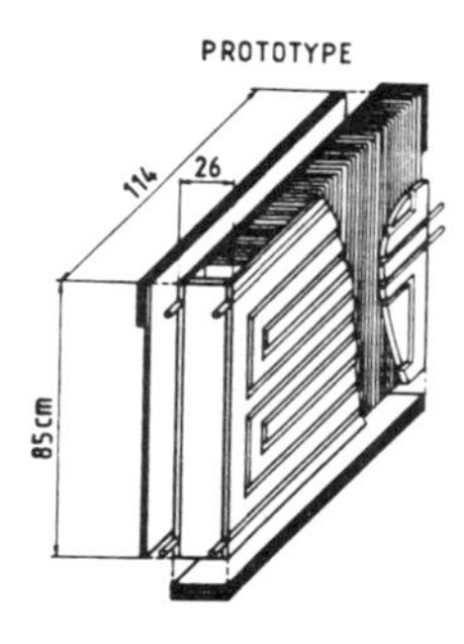

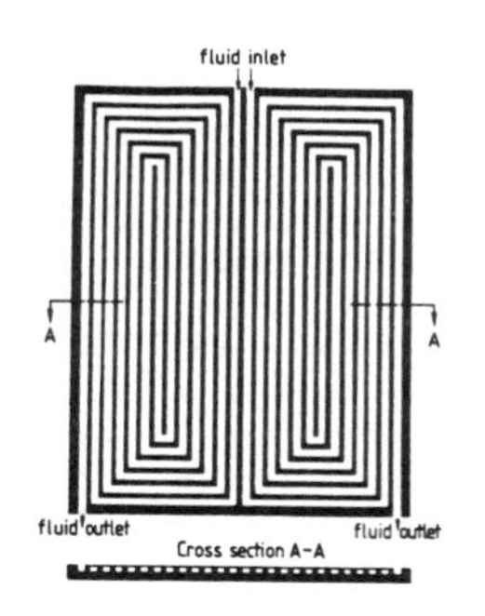

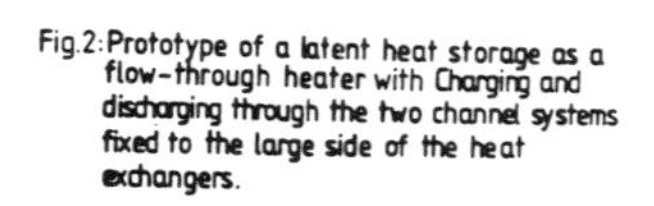
Fig.2: Prototype of a latent heat storage as a flow-through heater with Charging and discharging through the two channel systems fixed to the large side of the heat exchangers.

Fig.3: Diagram of the labyrinthine arrangement of the channel systems fixed to the large sides of the heat exchanger.

EXPERIMENTS AND RESULTS

Heating and melting processes in a closed loop, and cooling and solidification processes in an open loop, at defined flow rates and driving temperature potentials were conducted to investigate the optimal operational properties of the prototype. When heating in a closed loop, the channels of both larger outer surfaces of the container are connected to a setup which controls the temperature drop of the water supply and water return and as such keeps the water supply temperature constant. When cooling, the channels are directly connected to the ($\vartheta \approx 10\ ^\circ C$) tap water supply. A cooling and solidification process is shown in Fig. 4. It is shown that the stored latent heat of $Q_{ges.}$ = 6.6 kWh is discharged as the heat of crystallisation at different heat output levels during two time intervalls. The solidification curve of the eutecticum has a continuous linear shape in the phase change range at different loads. When the water flow rate is raised during the second time interval, the system responds with a step-like decrease of the temperature difference, however with a considerable increase in the heat output. By taking the measured properties into account, the function and performance of the given system is demonstrated.

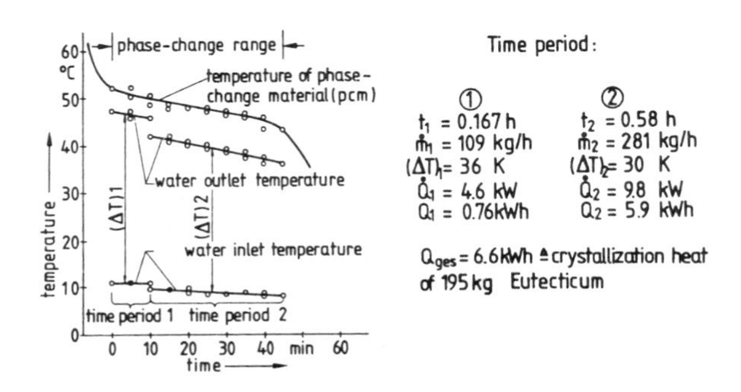

Fig.4: Discharge mode of the prototype latent heat storage used as flow-through heater.

LATENT HEAT STORAGE WITH FLAT SOLAR COLLECTOR AS COMPACT SYSTEM

Figs. 5 and 6 show a modification of the presented heat exchanger geometry which is a compact system comprising heat exchanger and flat solar collector. For this compact system, on one large side of the container, the false bottom side with its channels is missing / deleted. Instead of this surface, there is a flat solar collector. Various types of solar collectors can be used. The heat exchanger is operated by the eutecticum.
The LHTS with flat solar collector, used as a roof mounted element, is able to

- receive solar radiation and transfer it into heat,
- transfer this heat for starage in the PCM by means of the absorber surface and fins,
- discharge the stored (latent) heat by means of the labyrinthine channel system such that the system operates as a flow-through water heater,
- transfer the supplementary heat from the absorber surface through the fins directly to the consumer.

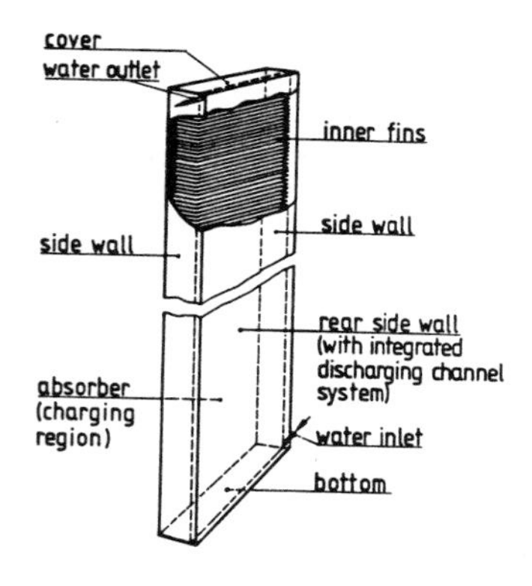

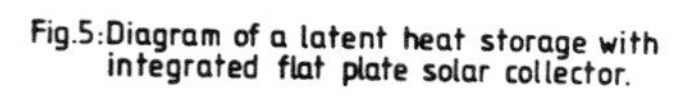
Fig.5: Diagram of a latent heat storage with integrated flat plate solar collector.

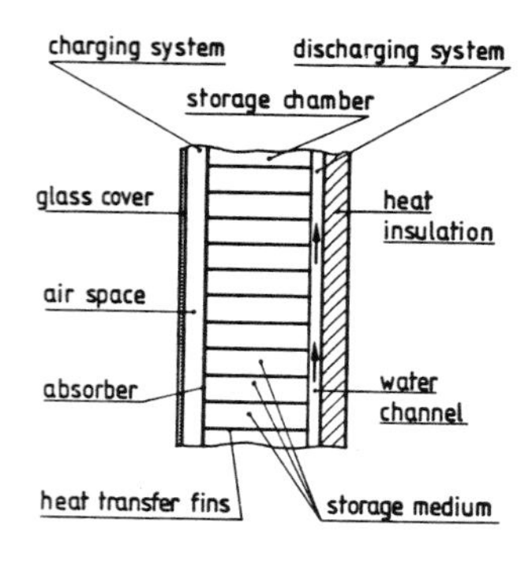

Fig.6: Fundamental design of the latent heat storage with integrated solar collector as a flow-through heater to supply warm water.

With the modular design of the LHTS-flat solar collector system with a 2 m² absorber surface, the heat output of the system is dependent on the solar radiation at the place of operation. The mean solar radiation (Atlas, 1984) is important for the depth of the heat exchanger / storage container.
From the experiments with the LHTS flow-through water-system operating with this modification, i.e. compact system with flat solar collector, it can be deduced: In areas which have higher solar radiation (Southern Europe) during the entire year, this system may be useful during the entire year. In Central Europe the use might be restricted to spring, summer and autumn.

REFERENCES

- Abhat, A.; Heine, D.; Heinisch, M.; Malatidis, N.A. und Neuer, G.: Entwicklung modularer Wärmeüberträger mit integriertem Latentwärmespeicher. Forschungsbericht Nr. BMFT-FB-T-81-050, Bonn (1981).
- Abhat, A.; Aboul-Enein, S. und Malatidis, N.A.: Latentwärmespeicher - Bestimmung der Eigenschaften von Speichermedien und Entwicklung eines neuartigen Wärmeübertragungssystems. Forschungsbericht Nr. BMFT-FB-T-016, Bonn (1982).
- Malatidis, N.A.: Latente Thermospeicher als Bauteile zur Solarenergienutzung. Forschungsbericht SA 01/83 des IBP, Stuttgart (1984).
- Hahne, E.: THERMAL ENERGY STORAGE SOME VIEWS ON SOME PROBLEMS. Proc. of the Eighth International Heat Transfer Conference. San Francisco 1986, H.Vol.1, pp. 279-292.
- Malatidis, N.A.; Erhorn, H.: Experimentelle Untersuchungen von latenten Wärmespeichern - Prototyp und Modelle. Forschungsbericht SA 8/85 des IBP, Stuttgart (1986).
- Atlas über die Sonnenstrahlung Europas, Band II: Global und Diffusstrahlung auf vertikale und geneigte Oberflächen. Verlag TÜV Rheinland, Köln (1984).

SIMULATION OF PHASE CHANGE ENERGY STORAGE FOR SOLAR SPACE HEATING

D. Laing** L.F. Jesch* L. Jankovic* K. Fellague*

** Fakultät für Maschinebau, Universität Karlsruhe, W. Germany

* Solar Energy Laboratory, Dept. Mechanical Engineering
University of Birmingham, Birmingham B15 2TT, UK

ABSTRACT

A well chosen phase change material (PCM) can increase the storage capacity in a space heating system if the PCM is always kept in the transition phase. Simulations show that a good control system will bring better results than using only sensible heat storage in water. The preconditions for a good control system are the results of a simulation which was developed during the last three years in two laboratories by several research workers. The simulation is being validated by monitoring an installation in Birmingham UK

KEYWORDS

Phase change material; transitional phase; enthalpy; control strategy; forced stratification; stepwise optimisations

INTRODUCTION

The usefulness of phase change material (PCM) based energy storage for solar space heating depends largely on whether the high heat capacity of such materials can be utilised in operation. As collector exit temperatures in most systems vary with changing radiation intensities according to Fellague and coworkers (1986) the PCM could be in any one of the three phases: solid, transitional or liquid.

The heat capacity of popular PCMs in the solid and in the liquid phases is lower than that of water. This results in the PCM based solar energy storage being uneconomical unless the system is able to operate most of the time in the high heat capacity transitional phase.

To maximise the high heat capacity of the transitional phase it is necessary to be able to assess the response of the heat store to solar energy input and load demand. This requires a reliable simulation of both the physical changes in the PCM and of the control strategy.

Two physical models were used in simulation. The first was developed by

Laing (1985) and further improved by Fellague (1986) in The Solar Energy Laboratory in Birmingham UK. The other was developed at the TNO in Holland by de Jong and Hoogendoorn (1986) and adapted for TRNSYS The results of these simulations are being compared with measurements made on an active solar space heating installation using PCM storage in the Solar Demonstration House in the Bournville Solar Village, Birmingham UK. This house is being monitored as part of the demonstration programmme of the Commission of the European Communities.

SYSTEM DESCRIPTION

The active space heating system consists of 16 m^2 selectively coated solar collectors connected to two PCM store tanks and to coils imbedded in the floors and in the ceiling. An electric heater provides auxiliary energy and an excess heat dump tank is used for surplus energy.

Each PCM tank is filled with 285 kg water and 680 kg PCM which is in one metre long, 38 mm diameter sealed plastic tubes. The PCM has a phase change temperature of 31 ^{o}C temperature. There are 2x234 tubes in each tank, fairly tightly packed. The rest of the tank volume is filled with water which is used as the heat transfer fluid.

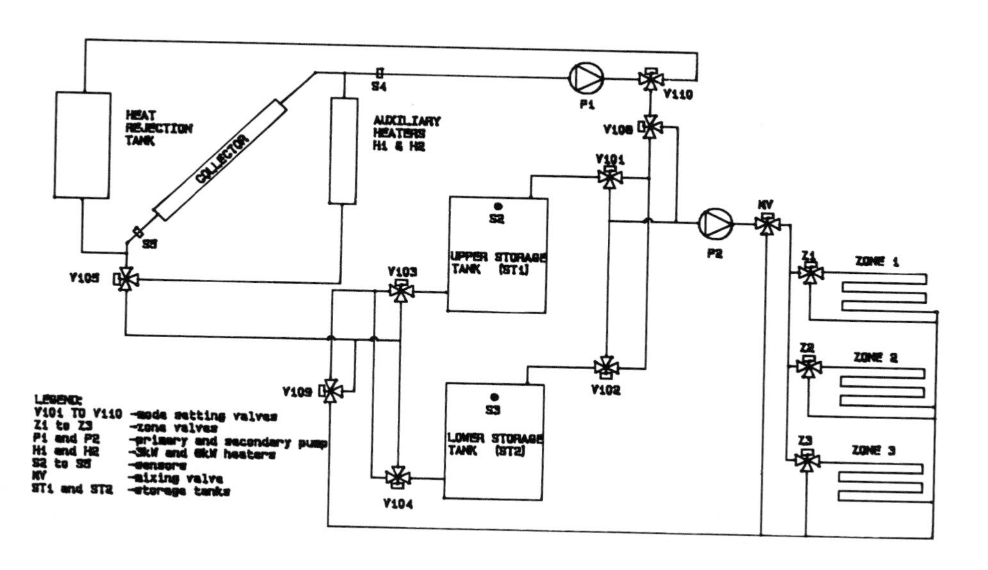

Fig. 1 Primary and secondary circuits

MATHEMATICAL MODELLING

Based on tests carried out by the manufacturer a three part model of the temperature vs enthalpy relationship was created to simulate the behaviour of the store while the material is in solid, transitional and liquid phase as a consequence of varying energy inputs and outputs.

The actual c_p value of the PCM is determined at every 6 minutes by using the monitored temperature and the temperature - enthalpy model. Thus the enthalpy of the store is calculated for each time step and these enthalpy values are compared with those at the interface between phases. Once the momentary enthalpy is known the condition of the PCM can be characterised by chosing the c_p value of the solid, transition or liquid phase. For the calculation of the c_p value at the transition phase of the PCM a temperature difference of 5°C between the solid-transition and transition-liquid phase onset temperatures is taken as illustrated in Fig. 2. This temperature difference was observed during production tests.

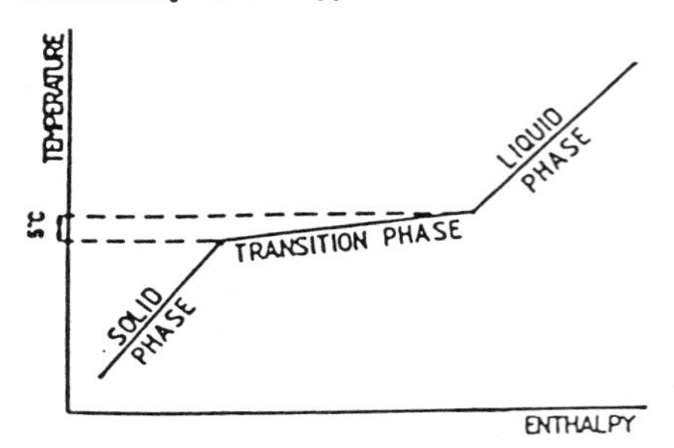

Fig. 2 Temperature vs enthalpy

Thus the effective heat capacity value for the transition phase is obtained through dividing the latent heat by the temperature difference given by the two ends of the slope of the temperature-enthalpy curve in the phase change condition

$$\text{latent heat} / 5^{\circ}\text{C} = 37.24 \text{ kJ/kg,}^{\circ}\text{C}$$

which is 8.9 times greater than that of water.

Finally an equivalent c_p value for the combined water and PCM store, proportional to their mass is calculated and used as a single heat capacity value for the energy balance.

CONTROL STRATEGY

The control system is based on a computer which receives input signals from solarimeters, temperature sensors, flowmeters and status indicators which produces opto-isolated output signals through interfaces to operate valves and pumps. The control strategy is stored in the computer memory and is readily changed by user input or software if required.

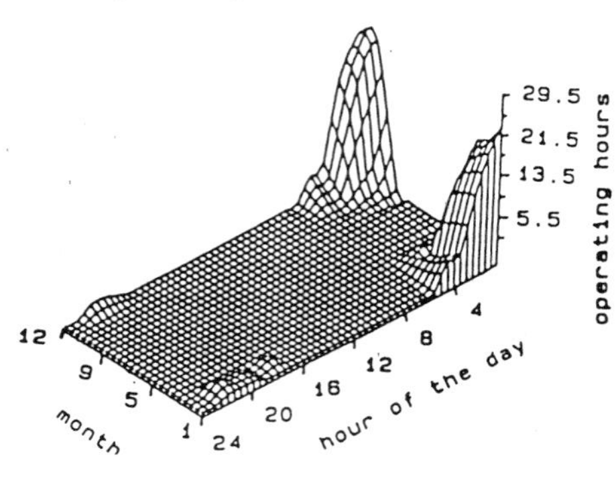

Fig. 3 Axionometric view of the operating time of the auxiliary heater

The aim of the control strategy was the necessity to guarantee the specified room temperatures at all times at minimum cost. This meant that the auxiliary energy, electricity, should only be accessed during the night at the off-peak rate.

A series of simulations (Laing 1985) have shown that by selecting a 9 kW electric heater it is possible to avoid daytime use of electricity for heating. Fig. 3 shows the bin data for the temporal distribution of auxiliary heater use when the 9 kW unit size is employed. The heater is only used during reduced rate night time periods and mostly in winter.

With the preferential use of the two tanks, auxiliary input preferring one and output to the load preferring the other, the two tanks are usually at different temperature levels. This forced stratification can be utilised (Jesch 1979) for increased system efficiency and solar fraction.

SIMULATION

Passive solar analysis

Before the full scale active solar heating simulation runs were made the design was subjected to passive solar analysis by using F chart. The analysis was based on monthly average data and the parameters listed below.

Room temperature	$18^{o}C < T_{room} < 23.6^{o}$
Solar gain	Direct only
UA value	157 W/^{o}C
Thermal capacitance	27 MJ/^{o}C
Air change	0.6 per hour

Results showed that no auxiliary heating was required between May and September and the yearly solar fraction of just below 0.4 was predicted.

Active dynamic simulations

Further to the passive analysis full year dynamic simulations using hourly meteorological data were performed to test system response to a variety of design considerations. Single variable parametric stepwise optimisations were used to establish design specifications which were used to determine optimal configurations. This resulted in an optimum collector flow rate of 0.01 kg/m^2 and an optimum collector area of 30 m^2.

Results for collector area size reflected a short term (1 - 4 days) energy storage solution. When interseasonal or long term (100 - 400 days) energy storage was simulated it showed very large collector areas and store sizes, beyond the economic feasibility of this type of solution for the problem. These findings were echoing the reports of Duffie and Braun (1984)

In the simulation tests the assumed temperature band width of the phase change was also varied to establish the importance of the 5 ^{o}C test value provided by the manufacturer. The simulation showed little difference in the yearly solar fraction.

Temperature band with	^{o}C	2.5	5.0	10.0
Solar Fraction	SF	0.528	0.528	0.513

Perhaps the most revealing result came from the simulation of various concentrations of PCM to water ratios. At the optimum the solar fraction was 0.528 for the standard case which was not much higher than the water only case (0.525) in which the same tank volume was filled with water but without PCM. This is connected to the fact that the PCM has lower heat capacities than water both below and above the phase change temperature band. It is also believed to be the result of the inability of the control system to keep the store temperature inside a very narrow temperature band around the phase change temperature of 31 ^{o}C. In reality the behaviour of the PCM may not be as narrowly strict as theory predicts and the widening of the phase change temperature band will increase the efficiency of the PCM store. Only monitored tests will tell.

Reports on the economic feasibility of thermal storage (Shelpuk 1976) show scepticism. Morrison and Abdel Khalik (1978) have reported better results when PCM was used in connection with air collectors. Jurinak (1979) has investigated a variety of proposed solutions. The accuracy of the mathematical models when they describe the behaviour of the PCM near the phase change temperature has still not been ascertained.

CONCLUSIONS

From the series of simulations to establish the optimal configuration of design parameters for the space heating system several points became clear.

The installed system has a collector surface area which is too small for providing more than about half the total load by solar. This limitation is imposed by architectural design which was constrained by low overall construction cost which necessitated a small roof area.

The store size represented by the two tanks of water and PCM mixture is nearly correct as it is and the auxiliary heater with a capacity of 9 kW can top up the store with sufficient quantity of energy from low tariff off peak electricity to last for the next day under any condition.

The relatively small increase in thermal performance of PCM based heat stores over sensible heat stores containing water can only be improved by changes in control strategy. The operating results can be predicted more accuretly by the use of better mathematical models which describe the physical events during phase transition more realistically.

REFERENCES

Braun, J.E. and J.A. Duffie, (1984) Long term energy storage in solar systems. Long term energy storage: an overview. UK-ISES Conf., London

de Jong and Hoogendoorn (1986) TNO, Delft, The Netherlands

Fellague, A (1986) Thermal analysis of a solar heated house M.Sc. Thesis University of Birmingham, Birmingham, UK

Jesch L.F. and A. Zekios (1979) Adaptative control for a multiple tank, multiple temperature energy storage system. Sun II Proc. ISES Solar World Congress, Vol. 1, 852 - 856, Atlanta

Jesch, L.F. (1984) A community of active and passive solar houses at Bournville Village in Birmingham UK. Proc. DGS 5th International Solar Forum, pp 1051 - 1068, Berlin

Jurinak, J.J. (1979) On the performance of an air based solar heating system utilizing phase change energy storage, M.S. Thesis, University of Wisconsin-Madison

Laing, D. (1985) Simulation of an active solar space heating and energy storage, Solar Energy Laboratory, University of Birmingham

Monsen, W.A., S.A. Klein and W.A. Beckman, (1981) Prediction of direct gain solar heating system performance, Solar Energy Vol. 27, 143- 147

Morrison, D. and S. Adel Khalik, (1978) Effect of phase change energy storage on the performace of air based and liquid based solar heating systems. Solar Energy Vol 20, 57-67

Shelpuk, B., P. Roy and M. Crouthamel, (1976) Technical and economic feasibility of thermal storage -Final report COO/2591-76/1 RCA Advanced Technology Laboratories, Camden, N.J.

LOW TEMPERATURE THERMAL ENERGY STORAGE FOR SOLAR HEATING APPLICATIONS

PRELIMINARY INVESTIGATIONS

S.M. Hasnain and B.M. Gibbs
Department of Fuel and Energy, The University of Leeds,
Leeds, LS2 9JT, United Kingdom

ABSTRACT

A novel latent heat thermal energy storage concept, using technical grade paraffin and sodium thiosulphate, is under investigation. An experimental perspex cell has been developed. A material holding chamber (20 mm wide, 40 mm high, 112 mm long) with a flat metal plate was used. Various types of metal plates were tested. A method is also established for the thermal cycling tests.

The preliminary investigations have shown that segregation of sodium thiosulphate lowers the heat of fusion and solidification along with phase transition temperature. Thermal analysis measurements show the effects of using different metal plates on temperature distribution in phase change materials. Repeated thermal cycling with paraffin shows a few degree differences in solidus and liquidus points. It also reveals the effects of using extended surfaces on transition temperature as compared to the first cycle.

KEY WORDS

Thermal energy storage; Heat storage materials; DSC measurement; Thermal cycling; Aluminium honeycomb; Paraffin and sodium thiosulphate.

INTRODUCTION

An exact knowledge of the thermophysical properties of phase change materi- als (pcm) is important in the design and operation of a latent thermal energy storage system, particularly for inorganic salt hydrates. Abhat (1980) stressed the importance of the determination of thermophysical properties of pcm, due to the fact that there are many discrepancies in the manufacturer's data and measured values. Table 1 lists thermophysical properties and cost data on selected pcm's. It was the intention of this investigation to (1) determine the actual latent heat of fusion, melting and freezing range of the selected pcm's; (2) examine the temperature dis- tribution and effects of material of construction on temperature profiles in pcm; (3) examine the correlations between the phase change temperatures and the number of thermal cycles by considering different container's geom- etry and material of construction.

TABLE. 1. Literature and Manufacture's data of thermophysical properties on selected technical grade phase change materials.

Heat Storage Materials	Melting Point [C]	Density [KG/m³]	Specific Heat [KJ/KG] [KJ/dm]	Latent Heat of Fusion [KJ/KG]	Latent Heat of Fusion [KJ/KG]	Thermalconductivity [W/M.C]	Average Cost 1987 [Sterling per KG]
$CaCl_2.6H_2O$	29	1.63	1.34 2.31	170	286	--	2.1
$Na_2HPO_4.12H_2O$	36	1.52	1.69 1.94	280	422	0.512	2.9
$Na_2S_2O_3.5H_2O$	48	1.73 s 1.66 l	1.45 2.39	200	344	0.465	1.7
Paraffin	49 *	0.91 s 0.78 l	2.89 2.1	210	145	0.20	2.4

s = solid l= liquid * congealing point

EXPERIMENTAL SETUP AND PROCEDURE

Measurement Techniques.
To obtain the precise values of phase transition temperature during the melting and freezing of test substances, the fol-lowing conceivable measurement techniques were employed. A method was also developed for repeated heating and cooling cycles.

Differential Scanning Calorimeter (DSC). The experiments were carried out on about 3 mg test substance using hermatically sealed pans in the Perkin-Elmer DSC model-4.

Thermal Analysis (TA). This technique provides the knowledge of the temperature distribution in pcm. It also involves the determination of heating and cooling curves, by considering large amounts (ca 50 g) of test substance as compared to the DSC technique. For thermal analysis a rectangular perspex cell with flat metal plate (thickness 1 mm) was designed and developed. Fig. 1 shows a schematic drawing of the experimental apparatus with water flow configuration, it also gives the information about the instrumentation of k-type thermocouples in the system. The internal dimensions of the pcm holding chamber were 20 mm wide x 40 mm high x 112 mm long. The floor of the storage chamber was a replaceable metal plate (stainless steel/copper) used to provide heating/cooling to the test substance from below. Above the pcm level an air space was provided in order to allow the expansion of the pcm.

Thermal cycling. Previous investigations (Heine and Abhat, 1978) have revealed that the influence of thermal cycling on the pcm characteristics must be measured experimentally. Since, in case of solar heating, short term heat storage system undergoes a daily charge/discharge cycle, therefore it is necessary to evaluate the effect of thermal cycle by running the tests in accordance with the practical applications. This has not been the case in previous studies. Nevertheless, to obtain the actual practical picture of pcm, after a number of repeated heating and cooling cycles, a method has been developed. In this method one glass and two stainless steel (one inch diameter) tubes were used. One out of these two stainless steel tubes contained an aluminium honeycomb structure. All tubes were filled half way with known amounts of pcm, and k-type thermocouples were instrumented in the middle of the medium. These tubes were then placed in a thermostatically controlled water bath and its temperature was adjusted

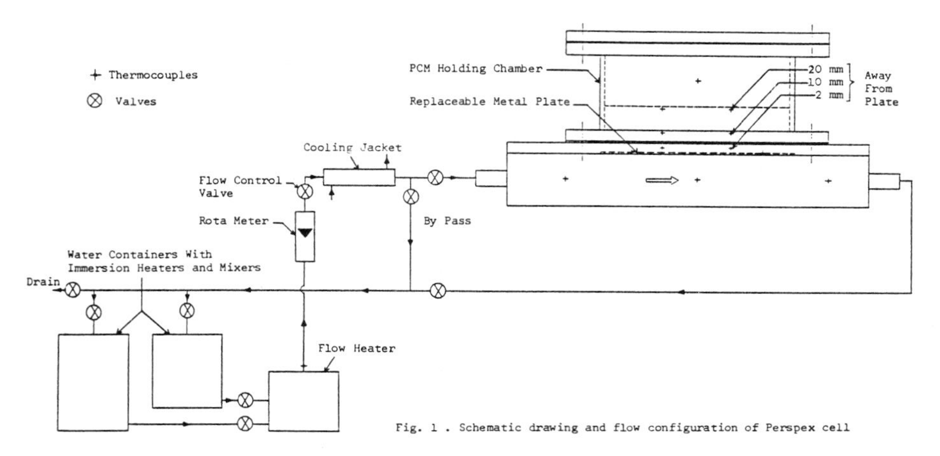

Fig. 1 . Schematic drawing and flow configuration of Perspex cell

at 15°C above the melting point of pcm. During cooling the water bath heater was switched off and cooling was achieved by natural convection in an undisturbed water bath.

A total of 150 cycles (one thermal cycle per day) were carried out on paraffin and the degradation of the storage substance was checked.

RESULTS AND DISCUSSION

Table 2 reveals some results obtained from DSC measurements with technical grade paraffin. It exhibits the thermal cycling of paraffin and confirms primary and secondary phase transitions, namely, solid to liquid and solid to solid respectively. The primary peak shows about 75% of the total heat content during the melting/freezing process. All endothermic/exothermic thermograms are similar, even up to 15 alternate thermal cycles.

Results obtained from the DSC technique, using sodium thiosulphate, are summarized in Table 3. The yielded thermograms (Fig. 2) show a single endothermic peak during the first heating process. No exothermic peak was observed during the first freezing process, even up to -30°C. Similarly, heating sodium thiosulphate back to 60°C, did not reveal any endothermic peak. However, a sharp crystallization peak was found during the second

TABLE. 2. Results from the DSC measurement on technical grade paraffin. Temperature range ofmeasurement : 7 °C to 60 °C.

Process	Total Number of Thermal Cycles Performed	Solid to Solid Phase Transition (Average)		Solid to Liquid Phase Transition (Average)		Total Heat of Fusion & Crstallization [KJ/KG]	Average Primary Peak Temperature [C]
		Temperature range [C]	Related Heat Content [KJ/KG]	Temperature range [C]	Related Heat Content [KJ/KG]		
Fusion Cycle	15	18.85 to 44.2	48.67 ± 1.1	44.20 to 52.85	131.95 ± 0.32	180.62	49.65 * 44.2
Crstallization Cycle	15	52.85 to 42.60	133.3 ±1.35	42.60 to 18.85	47.20 ± 1.0	180.50	49.2 ** 42.6

* Onset melting temperature
** Onset freezing temperature

TABLE. 3. Results from the DSC measurement on technical grade sodiumthiosulphate pentahydrate. Temperature range of measurements: 60 °C to -36 °C.

Number of cycles	1	2	3	4	5
Peak temperature [C] (During melting)	51.2	*	35.05	34.55	34.45
(During freezing)	*	-23.50	-23.35	-23.65	-23.4
Onset Melting point [C]	49.05	*	33.25	31.85	31.85
Onset Cold Crystallization point	*	-22.05	-23.05	-22.85	-23.0
Latent Heat of Fusion [KJ/KG]	211	*	116	122	120
Solid to Liquid transition temperature range [C]	46.85 to 55.15	*	31.05 to 43.55	30.75 to 43.15	30.75 to 43.15
Heat of Solidification [KJ/KG]	*	110	115	120	120
Liquid to Solid transition temperature range [C]	*	-21.25 to -24.85	-22.65 to -26.15	-22.35 to -25.95	-22.70 to -26.30

* Data evaluation not possible

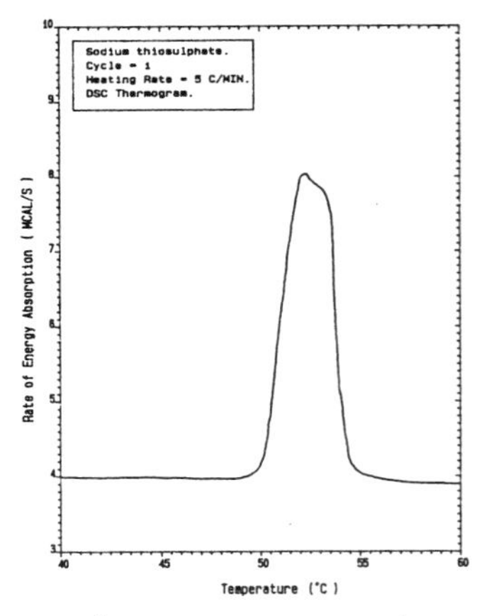

Fig. 2. Fusion behavior of technical grade Sodium thiosulphate pentahydrate.

freezing process. On reaching the third thermal cycle and onwards, two broad endothermic peaks at lower temperatures, and a sharp exothermic peak (Fig. 3) at nearly the previous cold freezing temperature appeared. Fig. 4 provides the evidence for a decomposition/segregation of sodium thiosulphate as a broad endothermic peak. This could be due to the surface free energy of the imperfect smaller crystals formed during cold crystallization. It is also evident that the heat of fusion/solidification is lower (50-60%) than the very first heating process.

The DSC results on paraffin showed that the measured values regarding melting/freezing points and related heat contents are lower than the manufacturer/literature data. This effect has also been reported by Abhat (1978). However, in the case of sodium thiosulphate, the melting point was found about 1°C higher than the manufacturer's data.

During thermal analysis on paraffin, the solidification times, as a function of the solid layer thickness, were measured (Fig. 5). This figure is based on the time to reach a temperature of 0.5°C less than its melting point. It also reveals a longer solidification time with stainless steel plate as compared to copper. This figure also shows that the solid-liquid interface propagates in a direction which, in general, is not perpendicular to the direction of fluid flow in the channel. As a consequence, the first row of thermocouples, nearest the entrance regime, shows solidification temperature prior to the second row, which resides closer to the existing regime.

Figure 6 shows the response of a paraffin specimen to thermal cycling, over a typical seven hour period. The temperature gradients between the centre of paraffin, as a function of the cycle temperature can be observed. The freezing range following thermal cycling is found to be 46-46.8°C. The solidus point is now evident at 46°C as compared to 43.3°C before cycling. The freezing range, on the other hand, has narrowed down to 0.8°C from 1.5°C. It was also found that metallic surfaces increase the solidus and liquidus points.

CONCLUDING REMARKS

(1) A comprehensive knowledge of melting and freezing characteristics of heat storage materials, and their ability to undergo thermal cycling is essential for assessing the short term performance of a latent heat storage system. (2) Technical grade paraffin shows two phase transitions, (a) solid to solid phase, (b) solid to liquid phase. Our results show differences in thermophysical properties of this paraffin as compared to manufac-

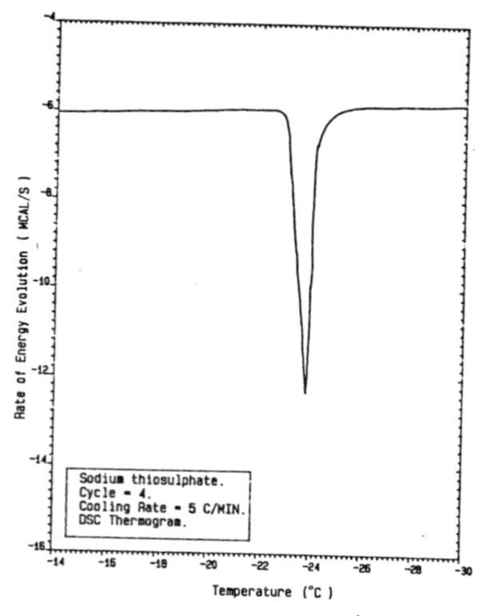

Fig. 3. Crystallization behavior of technical grade Sodium thiosulphate pentahydrate.

Fig. 4. Fusion behavior of technical grade Sodium thiosulphate pentahydrate.

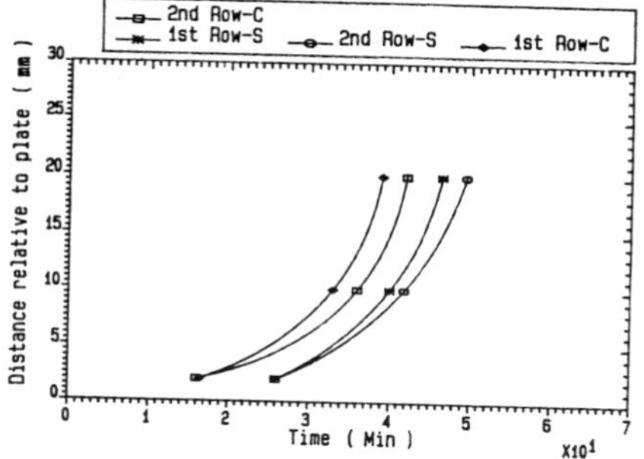

Fig. 5. Solidified layer thickness as function of time by using copper (c) and stainless steel (s) plates as heat exchange surfaces.

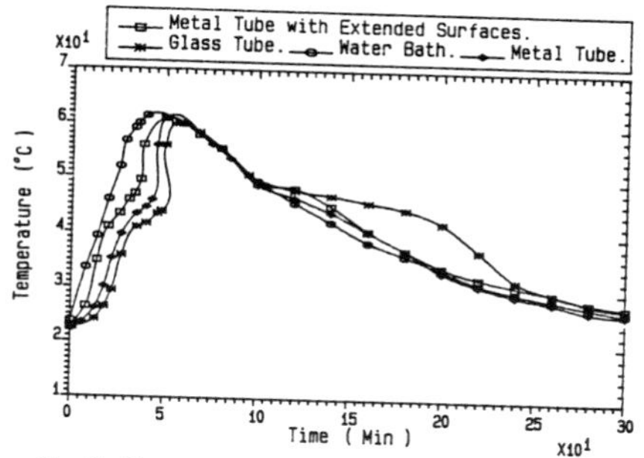

Fig. 6. The time - temperature history of one thermal cycle with Paraffin in different tubes.

turer's data. However, sodium thiosulphate segregated during the first thermal cycle shows fusion peaks at lower temperature in subsequent cycles. Heat of fusion/solidification is also effected due to cold crystallization. (3) In thermal analysis measurements, melting/freezing temperature profiles depend on the heating/cooling rates. (4) Repeated thermal cycling with paraffin showed a few degree differences in solidus and liquidus points, and effects of using extended surfaces on phase change transition temperatures.

ACKNOWLEDGEMENT

One of the authors (S.M. Hasnain) wishes to acknowledge the support of his postgraduate studies by the Ministry of Science & Technology, Government of Pakistan.

REFERENCES

Abhat, A. (1980). Short Term Thermal Energy Storage. Revue Phys. Appl., 15, 477-501.

Heine, D. and Abhat, A. (1978). Investigation of Physical and Chemical Properties of Phase Change Materials for Space Heating/Cooling Applications. Proc. 77, ISEC, New Dehli.

THERMAL OXIDATIVE AGING OF ORGANIC LATENT-HEAT THERMAL STORAGE MATERIALS

K. Hayakawa, H. Taoda, M. Tazawa, and H. Yamakita

Government Industrial Research Institute, Nagoya
Hirate-cho, Kita-ku, Nagoya, Japan

ABSTRACT

In order to estimate the extent of thermal oxidative aging of organic latent-heat thermal storage materials, the thermal deterioration of paraffin and stearic acid derivatives was investigated by applying the following heating-cooling cycles in air, which simulated the daily temperature fluctuation by insolation, over 6 months: 8-hour holding at 150°C as the highest temperature in a day followed by 5-hour holding at 30°C as the lowest temperature.
The resultant decrease in heat of fusion or crystallization, that in weight, and so forth were discussed in relation with the progress of oxidation.
The thermal stability of those materials differed markedly with their substituents. The addition of radical scavengers brought the remarkable suppression on the progress of thermal deterioration of those materials in open atmosphere.

KEYWORDS

Thermal oxidative aging; thermal deterioration; heating-cooling cycle; heat-cycling; organic latent-heat thermal storage material; paraffin; stearic acid derivatives; polyethylene.

INTRODUCTION

The introduction of direct heat exchange technique between heat transfer medium and insoluble organic phase change materials(PCM's) into the solar heat-collecting system will considerably eliminate the economical as well as technological difficulties inherent to such a system.

We have already reported that the organic polymeric materials as the cross-linked and/or surface coated polyethylene and poly(ethylene oxide) could be successfully utilized for the above purpose, because we could get rid of undesirable flowing and self-adhesion problem accompanied with freezing and thawing of them in the heat transfer medium (Hayakawa and others, 1986; Taoda and others, 1980).

Low molecular organic crystalline materials, as paraffin and higher fatty acids, have been shown to be also useful for the same purpose by applying some kind of device on them, though they cannot be crosslinked (Feldman, Shapiro, and Fazio, 1985). Moreover, organic PCM's were superior to salt hydrates in respect of no appreciable supercooling or phase separation.

For the systematization of direct heat exchange by using these organic PCM's, however, the thermal stability in respect of heat storage capacity, especially in air circumstances, should be of utmost importance (Taoda and others, 1985, 1986).

In this paper the thermal stability of these low molecular PCM's as stearic acid and related substances was estimated by applying the heating-cooling cycles to them which simulated the temperature change by daily insolation, and the results were discussed in comparison with those on high polymeric analogues.

EXPERIMENTAL

All materials for experiment were reagent grade and used without further purification. Three grams of PCM's in glass tubes were submitted to the following simplified and accelerated heating-cooling cycles in open atmosphere over a period of 6 months: that is, 8-hour holding at 150°C as the highest temperature followed by 5-hour holding at 30°C as the lowest temperature with transit time in a day, by taking into account of the melting temperature range of the concerned PCM's and possible temperature rise by heat transfer media. As the temperature dependency of diffusion coefficient of oxygen seemed to be large enough in such materials, almost the saturated amount of oxygen was thought to exist in the specimens heated and molten at 150°C (E_d: ca. 10 kcal/mol, ΔH_s: ca. 0 kcal/mol)(Brandrup and Immergut, 1975).

The melting temperature range and the heats of fusion and crystallization were determined by using a Perkin-Elmer DSC-1B differential scanning calorimeter at a rate of 10°C/min. The peak temperature of endothermic thermogram was taken as the representative of melting range, and the heat of fusion and that of crystallization were calculated from the area under the thermogram, each point in table and figures being an average of at least three or more determinations.

The activation energy of thermal oxidative decomposition was calculated from thermogravimetric analysis accompanied with the derivative thermogravimetry. Oxygen content was obtained from C, H, and N contents measured by a CHN-Corder. The functional groups formed on PCM's by heat-cycling were determined by IR and H-NMR measurements.

RESULT AND DISCUSSION

Table 1 shows the overall change of physical properties and so on of the concerned PCM's brought by 6-months of 150-30°C heat-cycling.

The thermal stability of polyethylenes was excellent, but the amorphous surface layers were formed which contained 11-13 percent oxygen at the end of 6 months. Stearamide and Pb stearate belonged to rather stable sort of PCM against heat, whereas paraffin, stearyl alcohol, stearylamine, stearyltrimethylammonium chloride and other metallic salts of stearic acid lost most of their heat-storing capacities in 6-months. It is very difficult to deduce any conclusion

TABLE 1 Change of Physical Properties by 150-30°C Heat-Cycling

Material	mp.(°C)[1*]	Heat of fusion (cal/g)	Wt. decrease (%)	Act. en. of dec.(kcal/mol)[2*]
Paraffin(mp. 58-60°C)	61 --→ 51	48 ---→ 9	6.1	25 --→ 20
Stearic acid	70 --→ 57	50 --→ 11	9.5	28 --→ 20
" + Inh.(3 wt. %)[3*]	71 --→ 69	49 --→ 42	1.5	
Stearyl alcohol	60 --→ 40	56 ---→ 3	14.8	
Stearylamine	54 --→ 26	57 ---→ 9	1.9	
Stearamide	100 --→ 67	39 --→ 20	1.9	22 --→ 18
Stearyltrimethyl-ammonium chloride	94 --→ 19	23 ---→ 2	10.6	
Na stearate	133 -→ 125	14 ---→ 7	20.7[4*]	
Ca stearate	119 -→ 100	25 ---→ 3	5.0	
Pb stearate	106 -→ 103	29 --→ 22	2.8	
Al stearate(mono)	122 ---→ -	3 ---→ 0	5.0	
HDPE	135 -→ 131	49 --→ 49	-0.2	70 --→ 71
LDPE	107 -→ 103	15 --→ 13	-2.3	67 --→ 44

1* A peak temperature in DSC thermogram. When multiple peaks were present in a thermogram, the maximum one was shown in the above table.
2* At the end of 3-month heat-cycling.
3* Inh. = 2,2'-methylenebis(6-t-butyl-p-cresol).
4* Original specimen contained water.

about the role of substituent on the thermal stability. But, the substance which possesses the higher melting point could be more stable against heat-cycling.

Figures 1 and 2 show the decrease in weight and that in heat of fusion with the lapse of heat-cycling time on paraffin and some of stearic acid derivatives. In the 6-month heat-cycled paraffin and stearic acid, high content of ketone carbonyl group was revealed and also a presence of peracid and/or perester group was suspected. And in 6-month heat-cycled stearyl alcohol, ketone group and also the conjugated ketone group were detected, while original hydroxyl group completely dissappeared.

Although the chemical changes brought by heat-cycling in air were much the same in these substances of similar chemical composition, and surely the feasibility to oxidation of the end group may affect the results, other factors than substituent seemed to play a big role, as was exemplified by the difference in thermal stability between paraffin and polyethylene.

The principal nature of oxidation, however, looked to be more complicated, because even in polyethylenes the feature of oxidative deterioration was distinctly differed between HDPE and LDPE. Red-browning rapidly proceeded from the earlier stage of heat-cycling only on the superficial part touching air on LDPE and often the glass tubes containing LDPE sample were broken during heat-cycling, whereas homogeneous browning proceeded gradually throughout the whole body in HDPE. Thus phenomenology of oxidation is left unsolved in these PCM's.

Addition of radical scavengers as Ionol remarkably suppressed the thermal degradation of PCM's during heat-cycling. Figure 3 indicates that about 3 percent by weight of 2,2'-methylenebis(6-t-butyl-p-cresol) greatly improved the thermal stability of stearic acid.

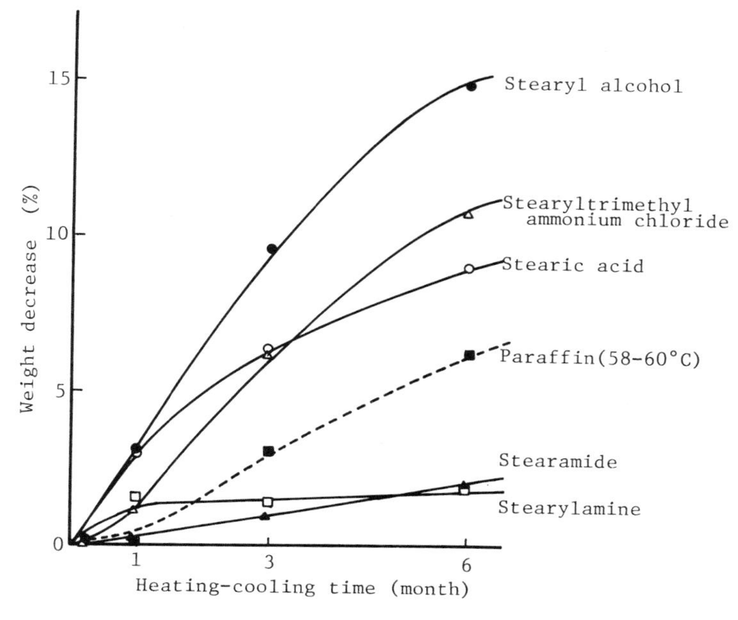

Fig. 1. Weight decrease of PCM's by heating-cooling cycles(150-30°C).

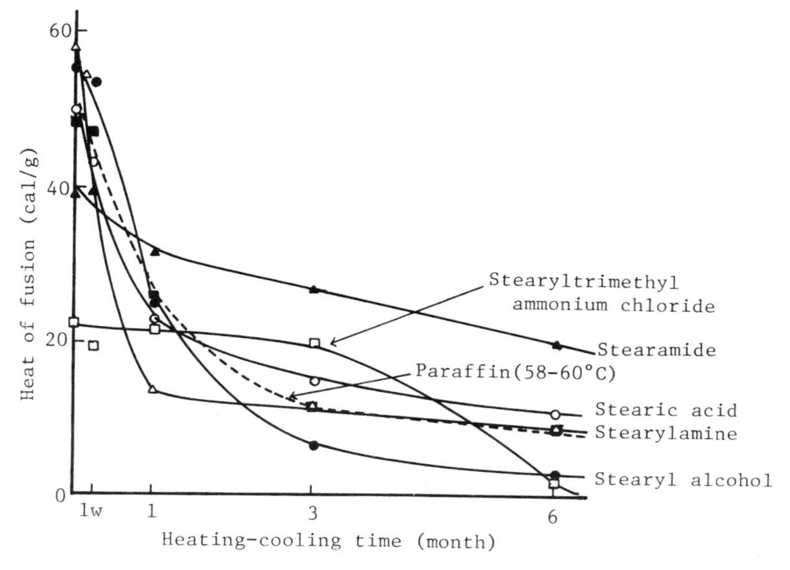

Fig. 2. Lowering in heat of fusion of PCM's with duration of heating-cooling cycles as measured by DSC.

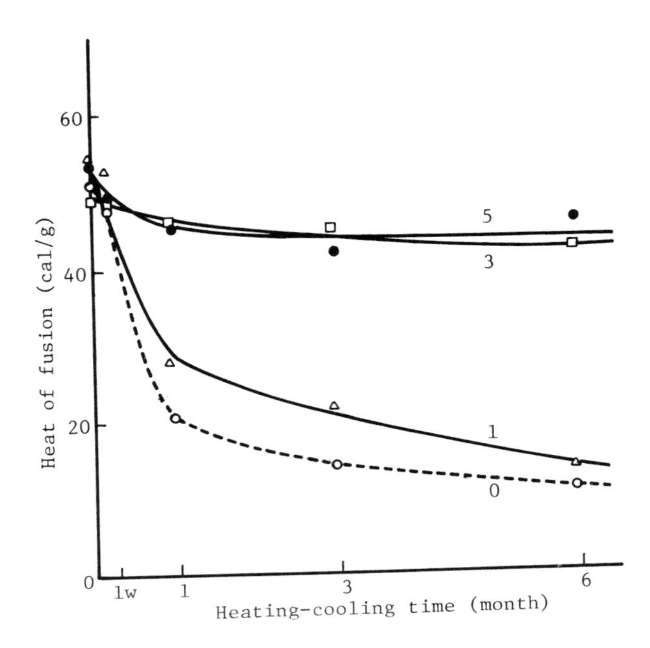

Fig. 3. Suppressing effect of 2,2'-methylenebis(6-t-butyl-p-cresol) on thermal deterioration of stearic acid. The numbers in the figure show the weight percentage of the additive.

When we assume that the heat transfer medium as water touches these PCM's directly at even as high as 90°C, sufficiently long service-life could be expected on these organic PCM's from the above-mentioned data by taking the activation energy of thermal decomposition shown in Table 1 into account.

In conclusion, the organic PCM's were considered to be endurable for long period of the practical heat-cycling even in the direct heat exchange device, if due concern was paid for their usage.

REFERENCES

Brandrup, J., and E. H. Immergut (1975). Polymer Handbook, 2nd Ed., Wiley.

Feldman, D., M. M. Shapiro, and P. Fazio (1985). Polym. Eng. Sci., 25, 406-411.

Hayakawa, K., H. Taoda, M. Tazawa, H. Yamakita, and J. Shimizu (1986). Abs. of 40th Semiannual Conf. of Govt. Ind. Res. Inst., Nagoya, 14-15.

Taoda, H., K. Hayakawa, T. Asahina, and M. Kosaka (1980). Reports of Govt. Ind. Res. Inst., Nagoya, 29, 31-41.

Taoda, H., K. Hayakawa, K. Kawase, and M. Kosaka (1985). Kobunshi Ronbunshu, 42, 151-158.

Taoda, H., K. Hayakawa, K. Kawase, and M. Kosaka (1986). Kobunshi Ronbunshu, 43, 353-359.

STUDY ON COMPOUND PCM SOLAR HEATING SYSTEM FOR TEHRAN

A. Kaabi-nejadian, Y. Nakajima

Dept. of Architecture Kogakuin Univ.
24-2, 1-chome, Nishishinjuku,
Shinjuku-ku, Tokyo, Japan

ABSTRACT

The operation of a novel solar-energized long term storage system has been studied using computer simulation. The system uses a phase change material (PCM) consisting CaCl2, 6H2O for short-term heat storage; crushed-rocks sensible heat storage bin with enclosed hot water tank and heat pump system for medium-term heat storage, and soil for long-term heat storage. Calculations indicate that this system will provide a higher heat gain than a sigle bin system studied earlier. The single bin system was studied analytically and theoritically, comparing the use of a PCM and crushed-rocks, and included a heat pump operating from the same bin. Due to analysis and design method of experiment, the effect of six variance factors on system performance for higher PCM efficiency was studied.

KEYWORDS

Short-term heat storage; medium-term heat storage; long-term heat storage; PCM.

INTRODUCTION

The single bin studies explored the relative advantages of using encapsulated PCM storage elements as compared to crushed rocks. And the high input temperature heat pump arrangement was found to be ineffective in evaluating the usefulness of the PCM for two reasons. Firstly, it was difficult to distinguish between actual heat gain from the PCM and heat produced independently by the heat pump. Thus the potential heat storage efficiency expected from the use of PCM as compared to the use of crushed rocks could not be confirmed. Secondly, contrary to one's expectation, the amount of heat gained from the soil decreased rather than increased when PCM was substituted for crushed rocks, despite the fact that PCM is known to have good short-term heat storage capacity. An examination of the above results led to a further development of heat storage concepts, namely a system including the use of both high and medium temperature PCM's for short-term heat storage, and the separation of the heat pump and main hot water tank from the PCM material and fan as shown in Fig. 1.

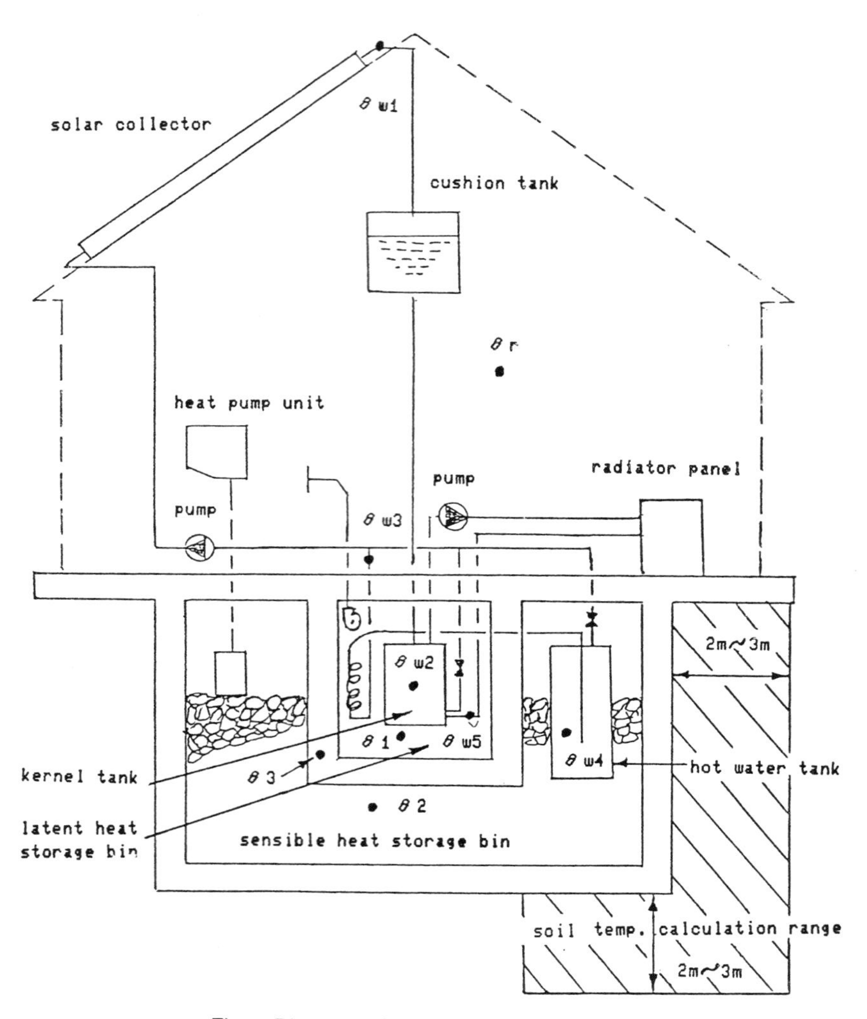

Fig. 1 Diagram of the system simulations

To determine the temperature distribution in the soil a rectangular model was used, the explicit method of finite differencing was considered because it includs temperature at a number of nodal points over a rectangular region and because the finite difference approximation of two dimensional unsteady state heat conduction equations can be used for both the cylindrical and rectangular coordinates.

There are a maximum of 8 unknown terms in the improved model system (locations shown by dots in Fig. 1). Depending on the scheduled system operation and previous temperature, each unknown term was specified as required and N linear equations were then calculated.

CALCULATION METHOD

The variance factors of the system (solar collector surface area, PCM weight, fan air capacity, fan operation time, heat pump operation time and radiator panel operation time) were arranged in orthogonal arrays L27(3) with 3 levels of the 6 variance factors. Then the effects of the coefficient of system performance, the coefficient of heat pump performance, solar collector efficiency, and the ratio of heat gain of the soil (the amount of heat gain of soil to the heat gain of the collector) on system performance is examined by computer simulation.

RESULT of CALCULATION

The results of analysis of variance (ANOVA) show that significant effects are obtained for scheduled heat pump operation time with regard to the amount of heat gain of the soil and for the solar collector surface area variance factor with regard to solar collector efficiency.
Figures 2 and 3 are derived from the results of ANOVA and the variance factor is shown for the corresponding 3 levels.

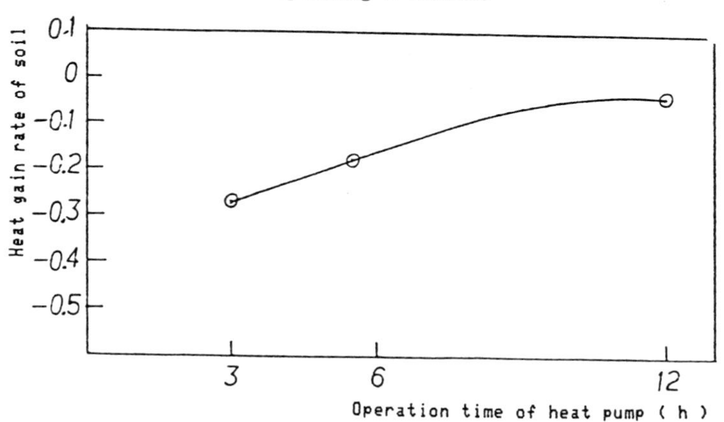

Fig. 2 Effect of operation time of heat pump on heat gain rate of soil.

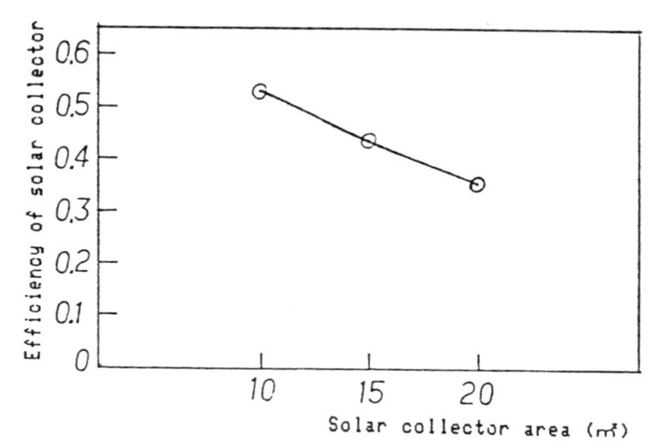

Fig. 3 Effect of solar collector area on solar collector efficiency.

With regards to weather data for direct normal solar radiation, diffuse solar radiation and outside air temperature, model No. 26 was selected from orthogonal arrays L27 for comparison and considered for the date of February 7, a sunny day. Temperature change of the heat storage bin for the above date is illustrated by Fig. 4. Fig. 4 shows the conditions for model No. 26 at the second fixed level of scheduled heat pump operation time. Heat gain and solar radiation variation are shown in Fig. 5.

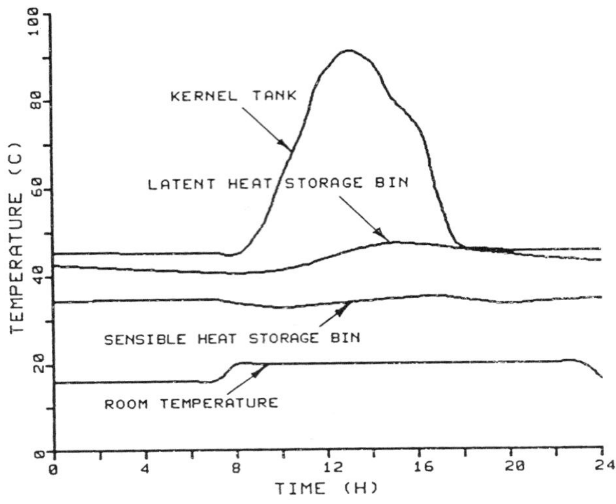

Fig. 4 Temperature variation of kernel tank, heat storage bin, with operation of heat pump.

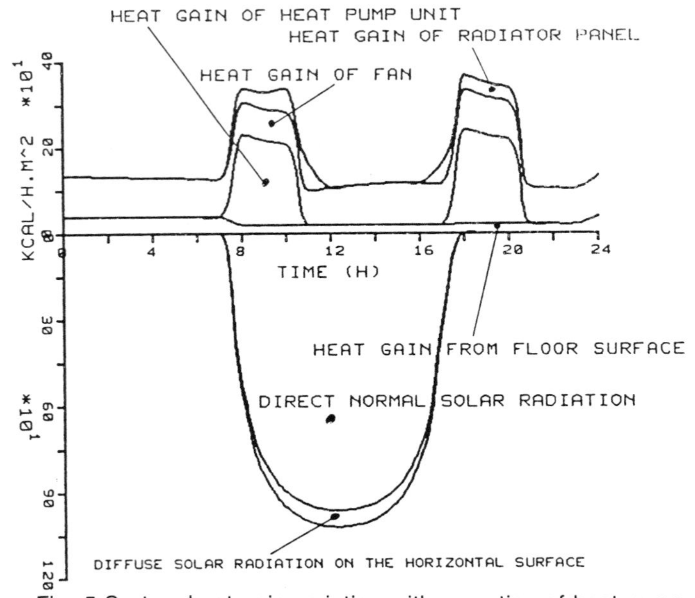

Fig. 5 System heat gain variation with operation of heat pump

CONCLUSION

Analysis of variance shows that interaction between the fan and heat pump unit is avoided by the improved PCM filled long-term underground heat storage for the solar heating system. The effect of heat pump operation time is confirmed to provide significant heat gain from the soil through long-term underground heat storage.

REFERENCES

1) Y. Nakajima et al., Research of long-term underground solar energy storage and heat pump system. Part(1)-(7), summaries of technical papers of annual meeting of AIJ, 1981-1984.

2) Y. Nakajima et al., Design and analysis for solar residence with long-term latent heat storage system through underground. Part(7)-(10), summaries of technical papers of JSES, 1981-1984.

3) Y. Nakajima et al., Study of thermal peculiarity of ground thermal storage material. Part(1)-(6), summaries of technical papers of AIJ, 1980-1985.

4) A. Abhat, Low temperature latent heat thermal energy storage: Heat storage materials. Solar Energy 30, 313-331 (1983).

5) N.A. Malatidis and A. Abhat, Investigation of the thermophysical behaviour of calcium chloride 6-hydrate for use as heat storage material in latent heat stores(in German). Forschung Ingenieur-Wesen 48, 1-26 (1982).

6) V.A. Costello, S.S. Melsheimer and D.D. Edie, Heat transfer and calorimetric studies of a direct contact-latent heat energy storage system. Thermal storage and heat transfer in solar energy system. Presented at ASME Meeting, San Francisco, Calif. 10-15 Dec. 1978.

7) J. Shelton, Underground storage of heat in solar heating system. Solar Energy 17, 137-143 (1975).

8) M. Telkes, Solar energy storage. ASHRAE J. 16, 38(1974).

SIMPLE METHOD FOR ASSESSING THE THERMAL PERFORMANCE OF PCM PANELS.

Ch. Charach and Y. Zarmi, A. Zemel
Applied Solar Calculations Unit
Jacob Blaustein Institute for Desert Research
Ben Gurion University of the Negev
Sede Boqer Campus, 84993, Israel.

Abstract

A simple analytic method for assessing the performance of PCM panels as passive solar energy storage devices is described. Expressions for an energy-efficient choice of the thickness and melting temperature are derived for a given operational scenario.

KEYWORDS: Passive solar systems, latent heat storage, phase change materials.

Introduction

The performance of passive systems for latent heat storage of solar energy is characterized by their storage capacity as well as by the rates of energy charge and release. These characteristics depend on three main groups of parameters: i) time-dependent operation conditions such as solar flux, ambient and room temperatures: ii) thermophysical properties of the phase-change material (PCM) in use; iii) size and geometry of the system. Design of such latent heat storage devices should therefore be treated as an optimal selection of the material and geometry to match best a given operation scenario.

This work describes the methodology for the energy efficient design of a simple storage device, a planar PCM panel, operating at a diurnal charging-discharging mode by melting the PCM during sunshine hours and releasing the accumulated energy as the PCM refreezes at night.

Design Strategy

During the charging stage the panel is exposed to a solar heat flux Q(t) and ambient temperature $T_a(t)$, whereas at nights, when the discharge process occurs, it is brought to thermal contact with room air at a constant comfort temperature T_r. The back side of the panel is insulated during sunshine hours in order to increase the thermal efficiency of the charging process. The insulation is removed at the discharge stage, doubling in this way the effective energy release area to compensate for the smaller heat flux that drives this stage. The resulting system is schematically shown in Fig. 1.

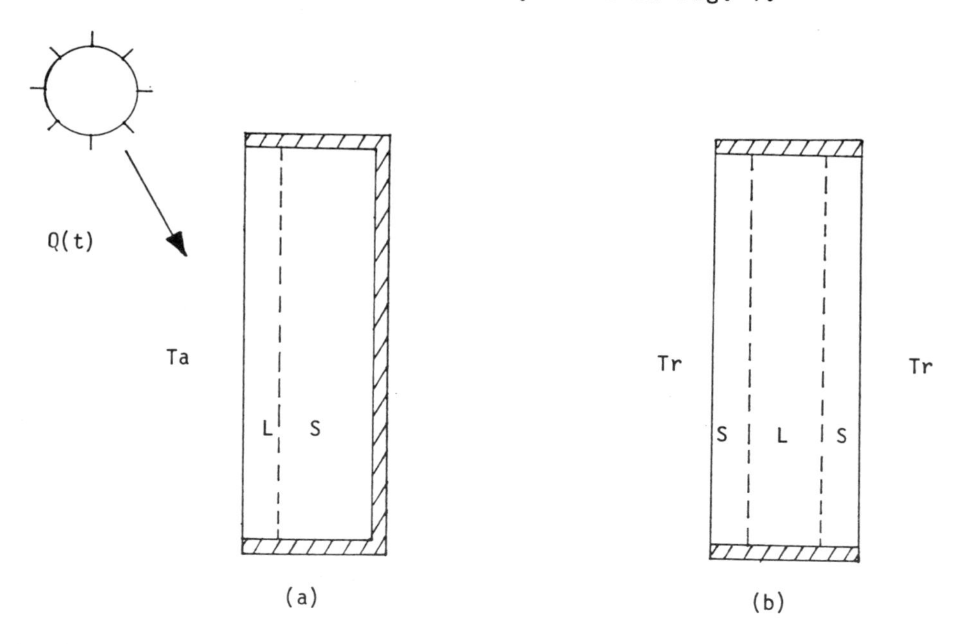

Fig 1: PCM panel for diurnal solar energy storage. a) charging b) discharging mode, S and L refer to solid and liquid phases.

The design of the system concerns two quantities: the thickness of the panel and the melting temperature of the PCM. The latter parameter can be controlled for a number of commonly used materials by introducing various additives, with minor effects on the other thermophysical properties of the material.

The heat storage capacity is, in principle, proportional to the thickness of the slab. However, for very thick panels the amount of solar radiation accumulated during the charging stage may not suffice to melt the slab completely. An upper bound for a useful slab thickness can, therefore, be expressed as:

$$d \leq d_{max} = (1/\rho L) \int_0^{t_m} Q(t)dt \equiv t_m\langle Q\rangle/(\rho L) \qquad (1)$$

Here ρ is the PCM density, t_m is the melting time, L is the latent heat and $\langle Q\rangle$ is the average incoming solar flux. Due to thermal losses from the surface of the slab the actual penetration depth of the emerging liquid phase is, in general, smaller than d_{max}. Its precise value is determined by the melting and ambient temperatures and by the film coefficient h_1, as shown below.

The performance of the system is very sensitive to the selection of the melting temperature, T_m. For enchanced discharge rates large values for the temperature difference (T_m-T_r) are required. Too low values of T_m may result in incomplete freezing at the end of the night with a dual negative effect: poor energy delivery during that night, and reduced energy collection in the subsequent day (since the system begins at a partially charged state). On the other hand, high values of T_m are associated with increased thermal losses during the charging stage. These considerations suggest that an energy efficient design may be achieved by requiring that all the energy collected at the charging stage should be released at the discharge stage. This "efficiency condition" can be used to determine the optimal melting temperature and the corresponding penetration depth, which is adopted, in turn, as the slab thickness. Thicker panels do not melt completely and represent a waste of material. Moreover, at night the unmelted side hinders heat transfer from the slab to room air. If the slab is too thin, it may melt too early and heat up excessively, losing more heat to ambient during the day.

Theoretical Considerations

The analysis is restricted to PCM's in which the heat transfer is conduction dominated. We further assume that the thermo-physical parameters do not vary through the operating temperature range and neglect the difference between the solid and liquid densities.

The ration c/L (deg^{-1}) of specific to latent heat is typically of the order 10^{-2} for materials commonly used for latent heat storage. For melting the ratio of sensible to latent heat is characterized by the flux Stefan number $St_q = d \langle Q \rangle / (\alpha_\ell \rho L)$ (α_ℓ is the heat diffusivity of the liquid). For solar fluxes this number is of the order 10^{-1}. The smallness of these parameters entails important consequences [1-5]. First, most of the energy stored in the panel is in the form of latent heat, the sensible component adding only a minor contribution. Furthermore, the thermal response time is much shorter than the melting time. Thus, the small effect of the sensible heat is even further reduced, as the system relaxes rapidly to the melting temperature after transition between the charging - discharging modes [3-4]. Therefore, one can consider only the latent component in the first approximation. The shortness of the thermal relaxation time has an additional aspect. For a small constant flux, the liquid temperature is known to be quasi-steady, (i.e. linear in the spatial coordinate). This feature remains valid also for slowly varying boundary conditions [2,5]. A recent analysis [5] indicates that the quasi-steady approximation is useful for much more general time dependence of the flux, since the liquid temperature relaxes rapidly to the linear profile even after very fast changes in the flux. These arguments are valid also for the freezing stage, since the corresponding Stefan number is also much smaller than unity. In summary, the general complex problem is considerably simplified by the following approximations:

a) The heat transfer is conduction-dominated.
b) The difference between solid and liquid densities is neglected.
c) Overheating and subcooling are neglected.
d) The temperature profile is quasi-steady and follows closely the variations in ambient conditions.

Analysis

The melting process is described as follows:

$$c_1(\partial T_1/\partial t) = k_1 (\partial^2 T_1/\partial x^2), \qquad 0 \le x \le \delta \tag{2a}$$

$$-k_1(\partial T_1/\partial x) \big|_{x=0} = Q(t) - h_1\left[T_1(0,t) - T_a(t)\right] \tag{2b}$$

$$T_1 \big|_{x=\delta} = T_m \tag{2c}$$

$$(d\delta/dt) = -k_1(\partial T_1/\partial x) \big|_{x=\delta} \tag{2d}$$

$$\delta(0) = 0; \quad T \big|_{x>\delta} = T_m \tag{2e}$$

Here $\delta(t)$ is the position of the interface between the two phases, the subscript "1" refers to the liquid and k is the thermal conductivity. In the quasi-steady approximation the liquid temperature is a linear function of x, with slope and intercept determined by the boundary conditions (2b) and (2c):

$$T_1(x,t) = T_m + \frac{Q(t) - h_1 (T_m - T_a)}{k_1 + h_1(t)}\left[\delta(t) - x\right] \tag{3}$$

The interface position can now de derived (in this approximation) using eqs. (2d), (2e) and (3):

$$\delta(t) + [h_1/(2k_1)]\delta^2(t) = \left[\int_0^t Q(t)dt - h_1\int_0^t (T_m - T_a(t))dt\right]/(\rho L). \tag{4}$$

Therefore, the maximal liquid penetration depth d is given by

$$d + (h_1/(2k_1)d^2 = t_m\left[\langle Q\rangle - h_1(T_m - \langle T_a\rangle)\right]/\rho L \tag{5}$$

where $\langle Q\rangle$ is the average flux defined by eq. (1) and $\langle T_a\rangle$ is the average ambient temperature

$$T_a = \left[\int_0^{t_m} T_a(t)dt\right]/t_m \tag{6}$$

The derivation of eqs. (4-5) is equivalent to replacing the total heat balance by the heat balance at the interface. In this approximation the total energy stored in the melting stage turns out to be just the latent heat content:

$$E = -k_1\left[\int_0^{t_m} (\partial T/\partial x) \big|_{x=0} dt\right] = \rho L d \tag{7}$$

A similar analysis for the freezing stage yields

$$d^2 + 4dk_s/h_2 = 8t_f k_s(T_m - T_r)/(\rho L) \tag{8}$$

Here the subscript "s" stands for for solid, t_f is the full freezing time and h_2 is the film coefficient for the heat transfer between the solid PCM and the room. The "efficiency condition" implies that the d values given by eqs. (5) and (8) are the same, and the melting and freezing time complete the charge-discharge cycle:

$$t_m + t_f = 24 \text{ hours} \tag{9}$$

where t_m is taken equal to the duration of sunshine hours. The problem is most conviently described in terms of the following dimensionless quantities:

$$St_s = c_s(T_m - T_r)/L; \quad St' = c_s(T_r - \langle T_a \rangle)/L; \quad t' = \alpha_s t/X^2;$$

$$Bi = d/X; \quad u = k_1/k_s; \quad v = h_1/h_2; \quad b = t_m/t_f; \tag{10}$$

$$t'_m = \alpha_s t_m/X^2; \qquad q = \langle Q \rangle X/(\alpha_s \rho L),$$

where $X = k_s/h_2$. In terms of these quantities, we wish to determine the Biot number Bi, (or equivalently the slab thickness d) and the Stefan number St_s, (or equivalently the melting temperature T_m), as functions of the given parameters q, St, u, v, b and t_m. To this end, we rewrite eqs. (5), (8) and (9) in their dimensionless form

$$(Bi)^2 + 2(u/v)Bi = 2(u/v)qt'_m - 2ut'_m(St_s + St') \tag{11}$$

$$(Bi)^2 + 4Bi = 8t'_m St_s/b \tag{12}$$

$$b/(b+1) = t_m \text{ (hours) } /24 \tag{13}$$

Solving for Bi and St_s one finds

$$Bi = -C/2 + \left[D + C^2/4\right]^{1/2} \tag{14}$$

$$St_s = b\left[Bi^2 + 4Bi\right]/(8t_m) \tag{15}$$

where $C = 4u(2+bv)/[v(4+bu)]$ and $D = 8ut'_m(q-vSt')/[v \cdot (4+bu)]$. As a typical example we have considered a panel filled with $CaCl_2.6H_2O$ operating inside a room with $T_a=T_r=18°C$ and $h_1=h_2=5W/m^2deg$. For $t_m=10$ hours and $\langle Q \rangle \cong 290\ W/m^2$, the value derived for the thickness is 2.1cm and the melting temperature is optimal at 32°C. The ratio of stored latent heat to the incident solar energy is about 65%, providing some 6700 kJ/m^2 at the discharge stage.

References

(1) A.D. Solomon, Some aspects of the computer simulation of conduction heat transfer and phase change processes, Oak Ridge National Laboratory report ORNL/CSD-100 (1982).

(2) L.N. Gutman, On the problem of heat transfer in Heat Mass Transfer, 29 921 (1986).

(3) Ch. Charach and P.B. Kahn, Solidification in finite bodies with prescribed heat flux: bounds for the freezing time and removed energy, Int. J. Heat Mass Transfer, 30 233 (1987).

(4) P.B. Kahn and Y. Zarmi, Consistent perturbation expansions for phase change material in finite domain, J. Appl. Phys., 61, 5018 (1987).

(5) Ch. Charach, Y. Zarmi and A. Zemel, A new perturbation method for planar phase change processes with time dependent boundary conditions, to J. Appl. Phys., in press.

Phase Change Storage in Solar Pipes

M. Sokolov and Y. Keizman
Faculty of Engineering, Tel-Aviv University
Dept. of Fluid Mechanics and Heat Transfer
Tel-Aviv
ISRAEL.

Abstract

In order to lower the price of solar energy absorption, the possibility of storing the energy with phase change materials is investigated. A solar pipe is composed of two concentric pipes such that the space between them is filled with phase changing materials (P.C.M.'s). Solar radiation is absorbed by the outer pipe's surface and transmitted to the P.C.M. Both sensible and latent heat of the P.C.M. is used for the heat storage. Water passing in the inner pipe absorbes the heat from the P.C.M. in order to deliver the energy demand. The use of such arrangements for both collection and storage of solar energy is investigated.

KEYWORDS
Phase change, Solar pipe, Energy storage, Moving Boundary.

Introduction

Materials which undergo a change of phase are being intensely researched for applications in the field of thermal energy storage and transfer. The high energy density and desirable thermal characteristics of these phase change materials (P.C.M.'s) make them attractive candidates for use in areas such as solar energy collection and peak load leveling. Due to the inherent temperature nonlinearity of the melt-line, the analytical solutions are relatively few in number and require restrictive boundary or initial conditions (1-6).

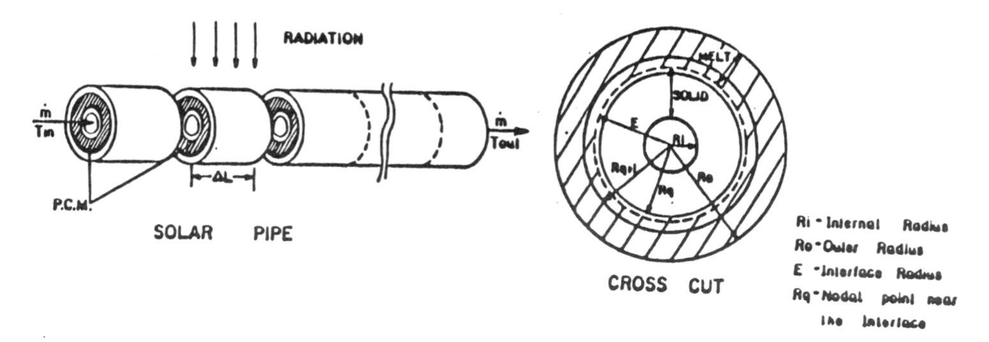

The method presented in this paper was evolved initially by Forster (7) and refined by Murray and Landis (8) and Olson and Springer (9).

In this work an explicit finite-difference model is used for the solution of heat transfer coupled with phase change in the solar pipe.

Thermal performance evaluation for the cylindrical P.C.M. is difficult due to the interior moving boundary because its position is not known a priori and must be found as part of the solution.

At the first stage of the full solution, the following simplifying assumptions were made:

1) Both geometry and boundary conditions are axisymmetric.
2) All physical properties of each of the phases are independent of temperature.
3) Phase change occurs at a single-temperature.
4) Heat transfer through convection in the liquid phase is neglected.
5) The density of the liquid is equal to the density of the solid $\rho=\rho_\ell=\rho_s$.

Mathematical Statement of the Problem.

Consider melting in the r direction of a solid contained between the walls of a long hollow cylinder of inner and outer radii r_i and r_o respectively, as a result of a uniform heat flux entering at the outer surface of the cylinder. Prior to the commencement of melting, temperature distribution within the solid can be determined by conventional heat transfer analysis. After the commencement of melting, the analysis becomes more involved because the problem entails a solid-liquid boundary which moves in the negative r direction. The mathematical formulations of the problem of melting are as follows:

For the solid and liquid regions (in dimensionless form):

$$\frac{\partial \theta_s}{\partial \bar{t}} = B_s \left[\frac{\partial^2 \theta_s}{\partial \bar{r}^2} + \frac{1}{1+\bar{r}} \frac{\partial \theta_s}{\partial \bar{r}} \right]$$

$$\frac{\partial \theta_\ell}{\partial \bar{t}} = B_\ell \left[\frac{\partial^2 \theta_\ell}{\partial \bar{r}^2} + \frac{1}{1+\bar{r}} \frac{\partial \theta_\ell}{\partial \bar{r}} \right]$$

$$\frac{d\bar{\varepsilon}}{d\bar{t}} = \frac{\partial \theta_s}{\partial \bar{r}} - K \frac{\partial \theta_\ell}{\partial \bar{r}}$$

where the dimensionless quantities are defined as:

$$\bar{r} = \frac{r-r_i}{r_i} \qquad K = \frac{k_\ell}{k_s} \qquad B_\ell = k_\ell \left(\frac{L}{k_s\, C\, T_F}\right)$$

$$\bar{t} = \frac{T_F\, k_s}{\rho_\ell\, r_i^2}\, t \qquad \bar{\varepsilon} = \frac{\varepsilon-r_i}{r_i} \qquad B_s = k_s \left(\frac{L}{k_s\, C\, T_F}\right)$$

Results

The computer simulation was monitored for different groups of parameters and different boundary conditions of the P.C.M. for the axisymmetric configuration.

a) Boundary conditions of the first kind - fixed temperatures at the boundaries:

$\bar{r} = 1.0 \qquad \theta_\ell = \theta_{a_2} \qquad \bar{t} > 0$

$\bar{r} = 0.0 \qquad \theta_s = \theta_{a_1} \qquad \bar{t} > 0$

$\theta_s(\bar{r}) = 0.0 \qquad \bar{t} = 0$

Figures 1, 1a, describe the temperature profiles and the interface velocity during the melting process.

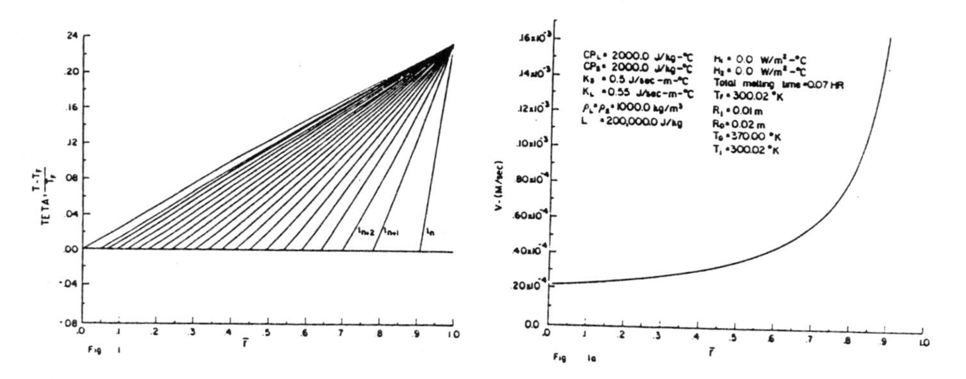

Fig 1 Fig 1a

b) Boundary conditions of the third kind at the outer radius and of the first kind at the inner radius:

$$\bar{r} = 0.0 \qquad \theta_s = \theta_{a_1} \qquad \bar{t} > 0$$

$$\bar{r} = 1.0 \qquad \frac{\partial \theta_\ell}{\partial \bar{r}} = \frac{h_2 r_i}{k_\ell}\theta_{a_2} - \frac{h_2 r_i}{k_\ell}\theta_\ell + \frac{f(t)\cdot r_i}{k_\ell \cdot T_F} \qquad \bar{t} > 0$$

$f(t)$ - radiation

$$\theta_s(\bar{r}) = 0 \qquad \bar{t} = 0$$

Figures 2, 2a describe the temperature profiles and the interface velocity during the melting process.

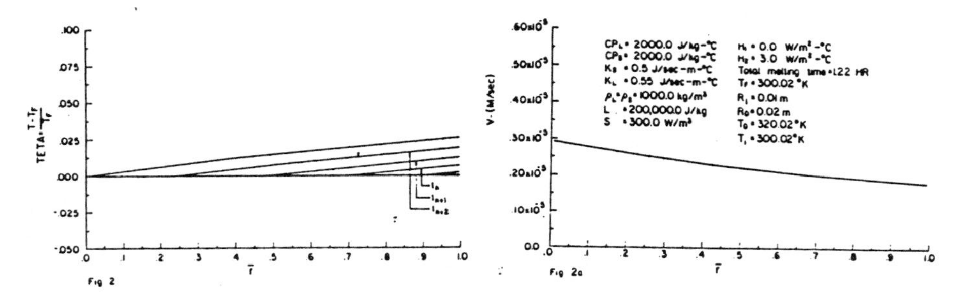

Fig 2 Fig 2a

c) Boundary conditions of the third kind at the outer and inner radius:

$$\bar{r} = 0.0 \qquad \frac{\partial \theta_s}{\partial \bar{r}} = \frac{r_i h_1}{k_s} (\theta_s - \theta_{a_1}) \qquad \bar{t} > 0$$

$$\bar{r} = 1.0 \qquad \frac{\partial \theta_\ell}{\partial \bar{r}} = \frac{h_2 r_i}{k_\ell} \theta_{a_2} - \frac{h_2 r_i}{k_\ell} \theta_\ell + \frac{f(t) r_i}{k_\ell T_F} \qquad \bar{t} > 0$$

$$\theta_s(\bar{r}) = f(\bar{r}) \qquad \bar{t} = 0$$

Figures 3, 3a, describe the temperature profiles and interface velocity during the melting process.

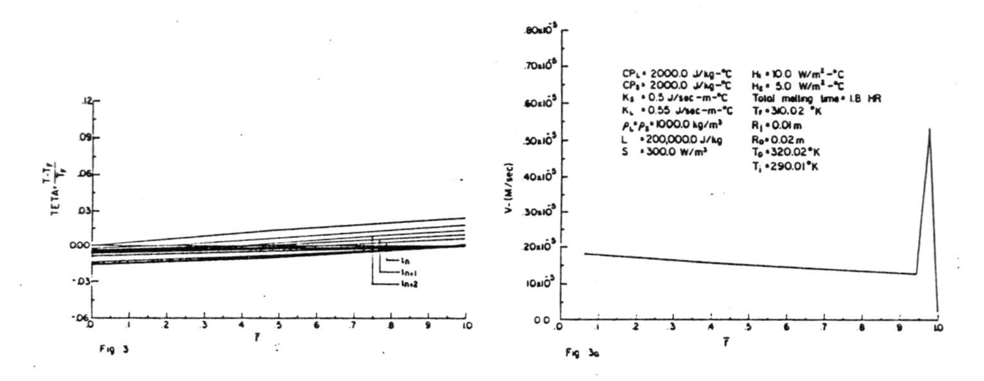

Discussion

From the results of the axisymmetric problem one can get a good estimation of the melting time and of the total energy stored in the solar pipe. For example, the case that is shown in Fig.(3, 3a):

$$St_o = \frac{C(T_o - T_F)}{L} = \frac{2000(320.02 - 310.02)}{200000} = 0.1$$

$$St_i = \frac{C(T_i - T_F)}{L} = \frac{2000(290.01 - 310.02)}{200000} = -0.1$$

The total energy stored in the solar pipe is (for 1m)

$$E = \rho \times \pi (R_o^2 - R_i^2) \times L \times 1 \qquad = \pi \times 1000 (0.02^2 - 0.01^2)\ 200000$$

$$= 188 \times 10^3 \text{ J/m} \approx 45 \frac{\text{Kcal}}{\text{m}}$$

One can get 45 Kcal/m at 37°C as latent heat and the whole process takes approximately 1.8 hours. Therefore, for 1000 Kcal/day, 20 1m pipes can be used and the whole collector area will be approximately 1m^2.

In the calculation a solar flux of $s = 300\ \text{w/m}^2$ was used, which is approximately one half of the solar radiation during a summer day, due to the axisymmetric needs of the calculation.

Conclusions

It may be concluded that the solar pipe has a potential future as the needs of energy storage are growing. Future developments will concentrate on non-axisymmetric problems with dependent solar radiation in order to obtain the efficiency of the solar pipe.

Nomenclature

C - specific heat
h - heat transfer coefficient
k - thermal conductivity
K - conductivity ratio
L - latent heat
r - radial coordinates
T - temperature
t - time
B - material property
ε - fusion front position in radial direction
ρ - density
θ - temperature = $\frac{(T-T_F)}{T_F}$
T_F - fusion temperature

Subscripts

ℓ - liquid phase
s - solid phase
q - radial space step containing fusion front

References

1. T. Goodman, "The heat balance integral and its application to problem involving change of phase", Trans. Am. Soc. Mech. Eng., 80, 335 (1958).
2. G.H. Meyer, "Multidimensional Stefan problems", SIAM J. Numer. Anal. Vol.10, No.3, June 1973.
3. G. Poots, "On the applications of integral-methods to the solution of problems involving the solidification of liquids initially at fusion temperature", Int. J. Heat and Mass Transfer, Vol.5, 1962, pp.525-531.
4. A.D. Solomon, "Melt time and heat flux for a simple PCM body", Solar Energy, Vol.22, 1979, pp.251-257.
5. B.A. Boley and J.M. Lederman, "Axisymmetric melting or solidification of circular cylinders", Int. J. Heat and Mass Transfer, 13, 413 (1970).
6. L.M. Jiji and S. Weinbaum, "Perturbation solutions for melting or freezing in annular regions initially not at the fusion temperature", Int. J. Heat and Mass Transfer, Vol.21, (1977), pp.581-592.
7. C.A. Forster, "Finite-difference approach to some heat conduction problems involving change of state", Report of the English Electric Company, Luton, England, (1954).
8. W.D. Murray and F. Landis, "Numerical and machine solutions of transient heat conduction problems involving melting or freezing", Trans. Am. Soc. Mech. Eng., 106 (1959).
9. G.S. Springer and D. Olson, "Method of solution of axisymmetric solidification and melting problems", ASM Paper 62-WA-246.

EXPERIMENTAL INVESTIGATION AND NUMERICAL SIMULATION OF A LATENT HEAT EXCHANGER

Jialin Sun C. K. Rush
Department of Mechanical Engineering
Queen's University Kingston, Ontario
Canada K7L 3N6

ABSTRACT

In this study, numerical and experimental investigations of a capsule-type latent heat exchanger are considered. The storage medium was a mixture of paraffin wax and aluminum shavings which occupied three percent of the mixture's volume. The improvement in the thermal conductivity of the mixture caused by the present of the aluminum shavings was examined experimentally. The thermal performance of a capsule-type latent heat exchanger was evaluated numerically. The results are presented in the form of NTU-effectiveness charts. The design parameters are presented in terms of the Biot number and the Stefan number.

KEYWORDS

Improvement of thermal conductivity of phase change material; Thermal performance of a capsule-type latent exchanger

INTRODUCTION

Paraffin waxes with a range of melting points from 45 °C to 55 °C have reasonable heats of fusion and suitable melting point for solar energy systems. Proper design of a latent heat exchanger using paraffin wax must consider the low thermal conductivity. Adding aluminum shavings increases the thermal conductivity of the wax because the aluminum shavings provide high thermal conductivity paths through the wax. A study was undertaken to investigatethe effective thermal conductivity of a wax-aluminum mixture and the thermal performance of a capsule-type latent heat exchanger.

EXPERIMENTAL APPARATUS

The experimental apparatus consisted of a single pass vertical heat exchanger. The PCM filled cylinders were firmly held in place as shown in Fig. 1. A schematic drawing of the apparatus for the present experimental study is given in Fig. 2.

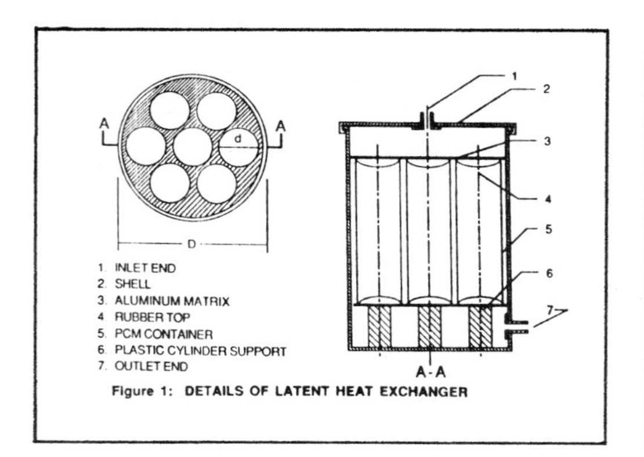

Figure 1: DETAILS OF LATENT HEAT EXCHANGER

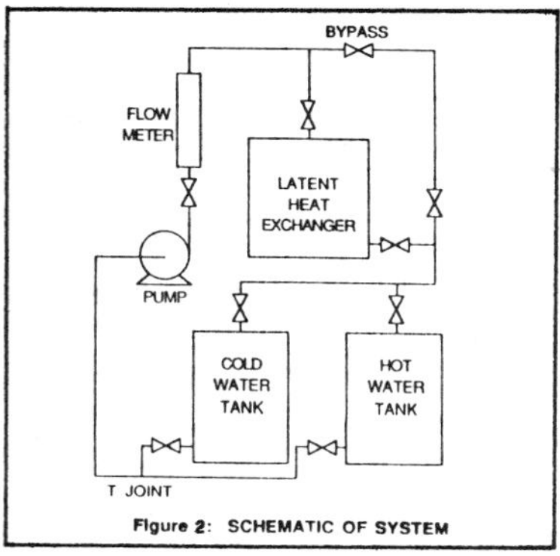

Figure 2: SCHEMATIC OF SYSTEM

NUMERICAL ANALYSIS

To simplify the complexity of the present work, the heat transfer rate for external hexagonal flow is obtained from an equivalent annulus model as shown in Fig. 3. Two dimensional convective heat transfer governing equations were solved using an explicit finite-difference technique. The energy equation for the capsule is the one dimensional heat conduction equation shown in Fig. 4. The enthalpy model, based on the method of weak solution, is used in the present work. Using cylindrical coordinates, the Fourier differential equations in each phase were solved using an explicit finite-different technique.

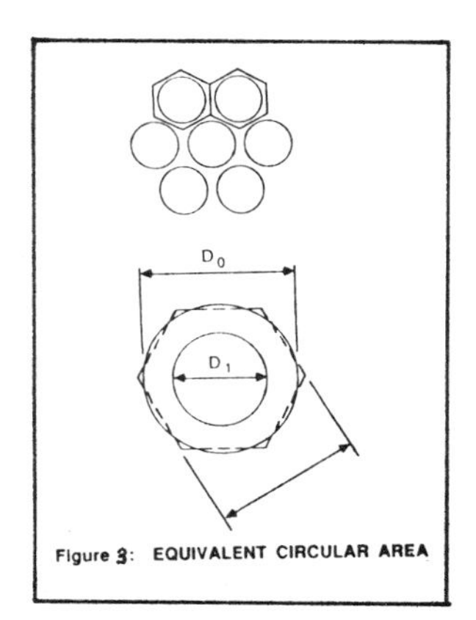

Figure 3: EQUIVALENT CIRCULAR AREA

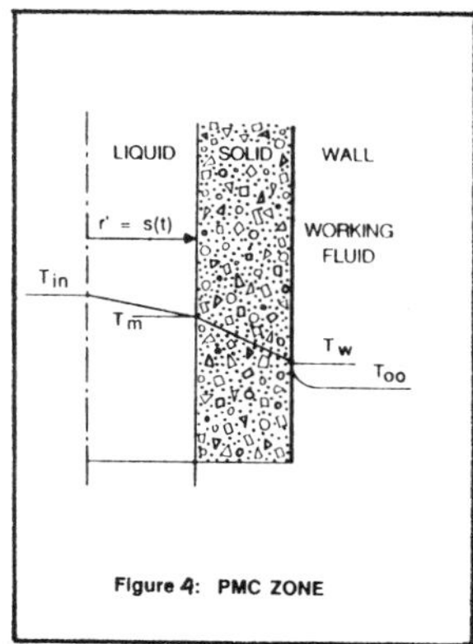

Figure 4: PMC ZONE

Figure 5: ALUMINUM-WAX MIXTURE

RESULTS AND DISCUSSIONS

EFFECTIVE PROPERTIES OF WAX-ALUMINUM MIXTURE

Aluminum shavings were uniformly distributed in the wax as shown in Fig. 5. As indicated in Fig. 6, the curves represent the temperatures at the center of the pure wax and the wax-aluminum mixture containers. The calculation of equivalent thermal resistance of the mixture may be found in ref. 6. By adding the 3% volume fraction of the aluminum shavings, the effective thermal conductivity increased by approximately 58% over that of solid wax. As a result of the improved thermal conductivity, the heat transfer rate was increased as shown in Fig. 7.

THERMAL PERFORMANCE OF THE LATENT HEAT EXCHANGER

Dimensionless parameters are introduced as follows:

Dimensionless heat transfer rate

$$Q' = \frac{1}{Bi}\left(1 + \frac{1}{Ste}\right)\frac{dF}{d\tau}$$

Biot number

$$Bi = \frac{D\,h}{k}$$

Stafen number

$$Ste = \frac{C_p(T_i - T_\infty)}{\lambda}$$

Number of transfer unit

$$NTU = \frac{hS\pi DL}{C_{pm}} = -\int \frac{dF}{Q'F(1+1/Ste)}$$

Effectiveness

$$\epsilon = 1 - \frac{(T_{in} - T_{\infty o})}{(T_{in} - T_{\infty i})}$$

An additional dimensionless parameter is the average frozen fraction of the entire heat exchanger, F', which gives the total amount of PCM that has been frozen at any time, $F' = \int Fdx/L$.

NTU-Effectiveness chart

It presents the variation of effectiveness and average frozen fraction of the heat exchanger with time as displayed in Fig. 8. One important characteristic of the latent heat exchanger which should be noted is that the freezing of PCM will be completed over the length from the inlet to a certain x after a certain time. The heat transfer from this portion is then the only contribution of heat capacity. In effect, the active heat transfer surface is reduced if sensible heat is also released. As a result, the effectiveness for this portion is zero and NTU is reduced. Therefore, the actual NTU will be the designed NTU minus the NTU from the point where the effectiveness is zero. The total amount of heat in PCM will be released when the actual NTU is reduced to zero.

Effect of Biot number

The marked effect of the Biot number on the heat flux and frozen fraction is noted. For low Biot numbers, e.g. 5, the rate at which the heat flux decreases is much less than that at higher Biot numbers, e.g. 50. This is shown in Fig.9. The physical meaning of this is as follows: If the heat flux at time zero is the design output of the heat exchanger, then the lower Biot number causes this output to remain relatively steady with time. The freezing occurs rapidly over the entire length of heat exchanger. From the definition of the Biot number, it is seen that increasing the thermal conductivity will reduce the Biot number. Therefore, selecting a PCM with a high thermal conductivity or using effective methods to increase the thermal conductivity of the PCM will greatly improve the thermal performance of latent heat exchangers.

Effect of Stefan number

The interpretation of the Stefan number is the ratio of the sensible heat to the latent heat of the PCM. The variation in the heat flux with Stefan number is illustrated in Fig. 10.The result illustrates that freezing and the release of heat occur more quickly at larger Stefan numbers. It also illustrates that the rate of heat flux decrease for both Stefan numbers is almost the same when the total freezing time is taken into account. This indicates that the influence of the Stefan number cannot be ignored.

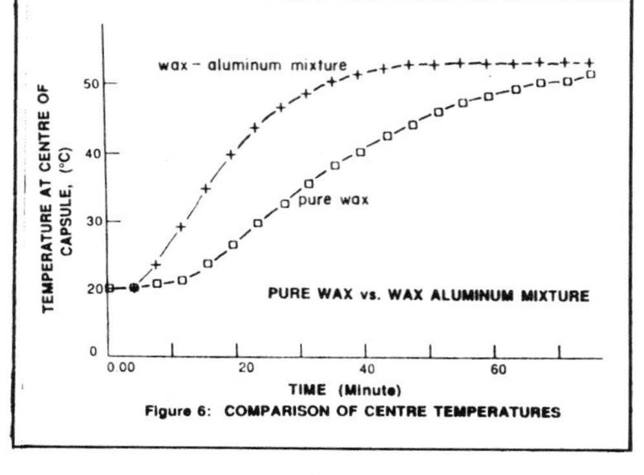

Figure 6: COMPARISON OF CENTRE TEMPERATURES

CONCLUSIONS AND RECOMMENDATIONS

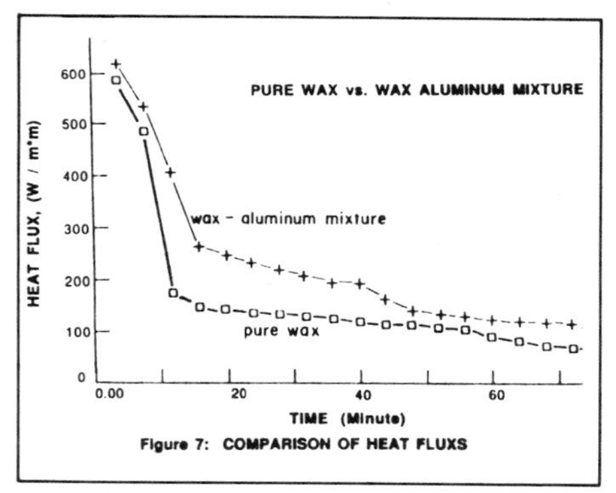

Figure 7: COMPARISON OF HEAT FLUXS

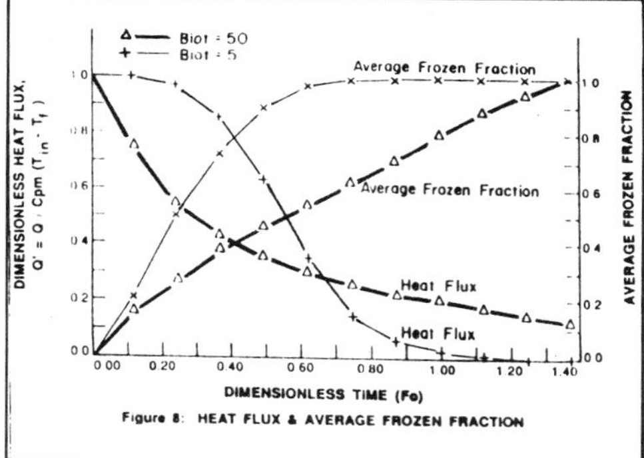

Figure 8: HEAT FLUX & AVERAGE FROZEN FRACTION

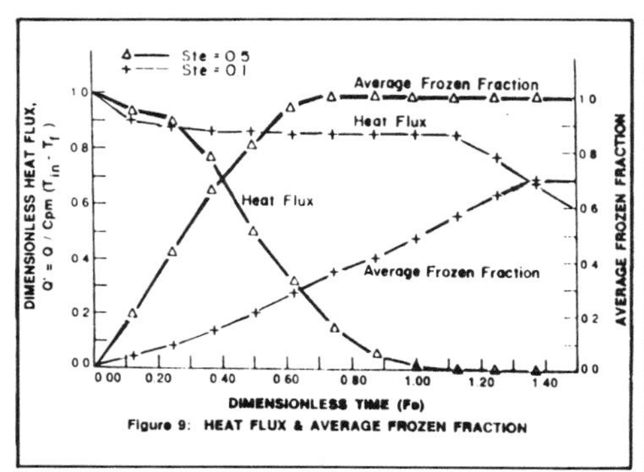

Figure 9: HEAT FLUX & AVERAGE FROZEN FRACTION

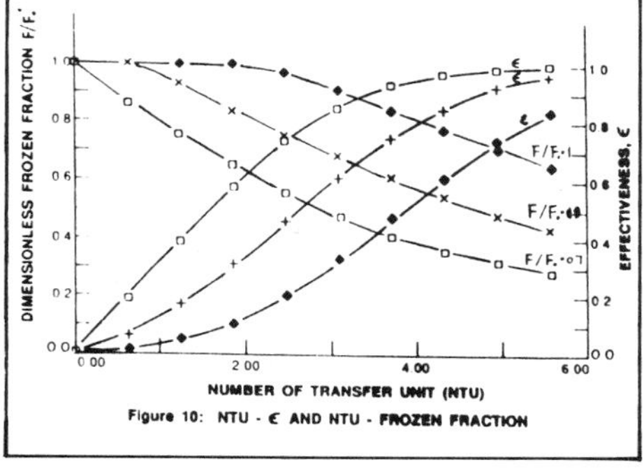

Figure 10: NTU - ∈ AND NTU - FROZEN FRACTION

CONCLUSIONS

1. The thermal conductivity of the paraffin wax is improved by adding the aluminum shavings. The addition of aluminum shavings increases the instantaneous heat flux as well.
2. The marked effect of the Biot number on the effectiveness of a latent heat exchanger indicates the importance of improving the thermal conductivity of the PCM.
3. The influence of the Stefan number cannot be ignored. Heat is released more rapidly at larger Stefan numbers. This shows the contribution of the sensible heat cannot be ignored.

RECOMMENDATIONS

1. The thermal conductivity of the PCM can be improved by increasing the cross-sectional volume fraction of the high thermal conductivity material in the PCM.
2. The NTU-effectiveness method used in the present study requires an experimental examination. The results obtained from this method should only be used to give an indication of the thermal performance of a latent heat exchanger.

ACKNOWLEDGMENT

The authors wish to acknowledge the assistance of Mr. S.J. Harrison in the preparation of this paper. This paper is a contribution of the School of Graduate Studies and Research and the Department of Mechanical Engineering of Queen's University.

REFERENCES

1. Mori, A. and Araki, K. "Methods for Analysis of the Moving Boundary Surface Problem." Int. Chem. Eng. Vol. 16 1976. pp. 734
2. Voller, V. and Cross, M. "Accurate Solution of Moving Boundary Problems Using the Enthalpy Method." J. Heat and Mass Transfer. Vol. 24 1981. pp. 545
3. Bonacina, C. et al. "Numerical Solution of Phase-change Problems." J. Heat Transfer. Vol. 16, 1973. pp. 1825
4. Goodrich, L. E. "Efficient Numerical Technique for One-dimensional Thermal Problems with Phase Change." J. Heat and Mass Transfer. Vol. 21, 1978. pp. 615
5. Shamsundar, N. and Srinivasan, R. "Effectiveness-NTU Chart for Heat Recovery From Latent Heat Storage Unit." J. Solar Energy Eng. Vol. 102 1980, pp. 263
6. Sun J. Master thesis, Department of Mechanical Engineering, Queen's University. Kingston, Canada 1986

NOMENCLATURE

Q' dimensionless heat transfer rate $Q/hA(T_{in}-T_{\infty})$
h heat transfer coefficient
A capsule surface area
T_{in} PCM initial temperature
F frozen fraction in radial plane of a capsule
τ dimensionless time $kt/\rho CD$
S number of capsules
L length of capsules
$T_{\infty i}$ working fluid inlet temperature
$T_{\infty o}$ working fluid outlet temperature

A SUMMARY OF CENTRAL SOLAR HEATING PLANTS WITH SEASONAL STORAGE

Charles A. Bankston
CBY Associates, Inc.
Washington, DC, USA

ABSTRACT

This paper is a summary of a review of the literature pertaininig to the research and development on the use of seasonal energy storage in solar heating systems. More than 30 operating solar heating systems utilizing seasonal storage in a dozen countries have been identified and characterized. The important findings of both national and international studies on the technical and economic merits of seasonal storage for solar heating were also reviewed.

Successful systems have been designed, built, and operated and some approaches are nearing commercial status. A variety of thermal energy storage techniques and collector types have been tested, but many potentially viable configurations have not been studied in depth. Most of the systems that have been built and tested to date are not large enough to show significant economies of scale, but larger plants employing more than 100,000 m^2 collector arrays are planned.

KEYWORDS

CCHPSS (Central Solar Heating With Seasonal Storage); District Heating; IEA (International Energy Agency); Seasonal Storage; Solar Heating.

INTRODUCTION

In Northern Europe, Scandinavia and parts of North America there is little hope of significant utilization of solar energy for building heating without resorting to long-term storage. The past decade has, therefore, seen some important advances in the cost-effective, long-term storage of low-temperature thermal energy and the combination of these storage technologies with solar energy collectors to provide heating for buildings.

This paper is a summary of a review of the literature pertaining to the research and development on the use of seasonal energy storage in solar heating systems. The full review (Bankston, 1987) will be published in Volume 4 of the American Solar Energy Society's Advances in Solar Energy. It covers international activities, developments, and advances in the utilization of large-scale seasonal storage of thermal energy to enhance the performance and economy of solar heating systems.

The Concept and Supporting Rationale

The capacity of the seasonal storage subsystem must be large enough to absorb all or most of the solar collectors' summer excess output without overheating. The ratio of storage volume to collector area is much higher in a seasonal storage system than in a diurnal storage system. Seasonal storage systems also must be large in an absolute sense to be efficient and cost-effective. It is necessary to resort to large-scale construction techniques, rather than manufacturing techniques, to achieve acceptable performance and economies.

Most of the seasonal storage techniques that have been evaluated involve some form of in situ construction using aquifers, earth coils, drilled rock, water pits, or rock caverns. The common element among these techniques is that they become more cost-effective as the storage volume increases.

The primary benefit of seasonal storage for a solar heating system is it allows the collector subsystem to operate at or near its peak efficiency throughout the year. As a result, collectors in seasonal storage systems can deliver two or even three times as much heat as those in diurnal systems.

MAJOR PROJECTS AND STUDIES

Significant work on Central Solar Heating Plants With Seasonal Storage (CSHPSS) has occurred in at least 12 countries. In addition, the International Energy Agency (IEA) has conducted studies that transcend national boundaries. In the full review (Bankston, 1987) the more important projects and studies are discussed by country. For each entry, the purpose of the work, the scope of the study or project, and what has been learned is reported.

Thirty-two seasonal storage projects and experiments are summarized in Table 1. Most of the listed projects have significant solar energy input. Some are included because they represent especially interesting storage concepts. The table lists the important characteristics of the storage, collector, load, and energy conversion systems.

CURRENT STATUS SUMMARY

Knowledge about CSHPSS has increased dramatically in the past ten years. However, the numbers of potential configurations, components, and site conditions are so great that only a few of the many possibilities have been evaluated. Some of the important findings noted in the full review data are summarized below.

System Configurations

Economic studies show that the temperature requirement of the load/delivery system is the most critical parameter in determining economic viability; but total load and load profile also are important. Systems with low-temperature storage and heat pumps generally are most economical--especially in countries with low-cost electricity. Large-scale systems that meet the demand load directly from stratified storage are gaining favor because they can provide a high solar fraction and enhanced energy security.

The Heating Load Delivery System

The economic viability of CSHPSS is sensitive to the required delivery and return temperatures. These temperatures determine the size and efficiency of the collector and storage subsystems as well as the cost of the distribution network and end use delivery equipment. The trend in modern district heating systems is to reduce the delivery temperature so that less expensive materials and simpler installation techniques may be used.

Thermal Energy Storage Systems

Research and field trials have been conducted on aquifers, earth and rock duct systems, in-ground water pits and underground rock caverns. All options appear to be technically viable under the proper geotechnical conditions.

The current estimates of the cost of seasonal storage in Sweden range from less than 0.005 to 0.015 $/kWh for aquifers, ducts in earth, and bore holes in rock; from 0.01 to 0.025 $/kWh for rock caverns; from 0.018 to 0.035 $/kWh for water pits; and from 0.022 to 0.04 $/kWh for steel tanks. These costs have been converted at 7.5 SEK/$US and assume a 15 year economic life and a 6 percent real discount rate, but they include only the cost of construction of the storage unit. Costs of interconnecting piping, controls, operation and maintenance are not included.

Solar Collectors

Most CSHPSS experiments have used glazed or unglazed flat plate collectors, evacuated collectors, or parabolic troughs. A few systems have used natural collectors such as lakes, rivers and fields. The primary determinant of the appropriate collector technology is the delivery temperature required by the load or the maximum temperature in the storage system.

Heat Pumps

The lowest cost CSHPSS systems in many locations are those that employ heat pumps in conjunction with low-temperature distribution systems, low-temperature storage, and unglazed or natural collectors. These systems are usually limited to a solar fraction of 0.7 to 0.8. They are most attractive where the cost of electricity is low relative to the cost of fossil energy.

CONCLUSIONS

Even the limited data reported in Table 1 supports the conclusions that there is considerable interest in CSHPSS and that some of the early small systems and experiments are working well. The more comprehensive review also indicates that large systems, such as Lyckebo and larger plants being planned, are nearly cost competitive with conventional and other renewable energy sources.

REFERENCES

Bankston, C. A. (1987). The Status and Potential of Central Solar Heating Plants with Seasonal Storage: An International Report. In K. W. Boer (Ed.), Advances in Solar Energy, Volume 4. American Solar Energy Society, Inc., Boulder, CO. p. 405.

TABLE 1 Characteristics of CSHPSS Projects and Experiments Throughout the World

Country	Location/ Name	Latitude	Year Built	STORAGE Type	Volume m3	Maximum Temp, C	Minimum Temp, C	Capacity MWh/yr	Annual Eff'ncy	COLLECTOR Type	Area m2	Collector Temp, C	Annual Eff'ncy	HEAT PUMP Type	No. of Units	Tot. Capacity, kW	Annual C.O.P.
Austria	Kranebitten	47	1982	Tubes/Earth	72000	10	5			Unglazed	416			Elect.	1	350 th	2.8
Canada	Scarborough	44	1985	Aquifer	250000	50	3	3100		Evacuated	1300			Elect.			
Denmark	Lyngby	56	1983	Pit	500	95	48	27		(non-solar)							
Finland	Kerava	60	1983	Pit & Duct/Rock	11000	70	10	250		Flat plate	1100		0.27	Elect.	1	240 el	2.7
France	Toulouse-Blagna	43	1980	Tank/Concrete	200					Flat plate	110						
France	Aulnay-Sous-Boi	49	1983	Aquifer	85000	14	4	700	NA	Unglazed	1275			Elect.	3	660 el	3.9
France	Saint Quentin	49	1986	Aquifer	15000	50				Unglazed	300				2	180 el	
France	Cormontreuil	49	1986	Duct/Chalk	15000					Flat plate	110						
Germany, FRG	Stuttgart	48	1985	Aquifer/Pit	1050	35				Unglazed	211	45		Elect.	1	66 th	3.4
Italy	Treviglio	46	1982	Duct/Earth	43000	30	5	591	0.80	Flat plate	2000		0.35	Elect.	12	720 th	3.9
Italy (CEC)	Ispra	46	1981	Duct/Earth	3000	35	15	33	NA	Flat plate	180		0.30	Elect.	1	32 el	
Japan	Yonezawa	38	1982	Aquifer		26	2		0.50	Unglazed	240			Diesel			2.1
Netherlands	Groningen	53	1984	Duct/Clay	23000	60	30	649	0.68	Evacuated	2400						
Netherlands	Bunnik	52	1985	Aquifer	35000	26	14	310	0.79	Flat plate	450	25	0.60	Gas	1	120 th	1.9
Sweden	Studsvik	58	1979	Earth Pit	640	70	30	20		CPC	120		0.47				
Sweden	Ingelstad	57	1979	Tank	5000	95	40	300		Parabolic	1300		0.32				
Sweden	Sunclay	57	1980	Duct/Clay	87000	20	12	654	0.94	Unglazed	1500	<18	0.39	Diesel	4	140 el	
Sweden	Lambohov	58	1980	Rock Pit	10000	70	5	750	0.63	Flat plate	2900	45	0.43	Elect.	2	185 el	
Sweden	Gullspang	59	1982	Tunnel	10000	20	2	209		Lake, summer				Elect.			
Sweden	Lyckebo	60	1983	Rock Cavern	100000	90	40	5500	0.74	Flat plate	4300	62	0.29				
Sweden	Lulea	66	1983	Duct/Rock	100000	65	30	2000	0.60	(non-solar)				Elect.	2	400 el	
Sweden	Vallentuna	59		Lake sediments	1200000	20	2	20500		Lake, summer		20		Elect.	1	8000 el	3.0
Sweden	Kopparberg	59		Aband. Mine	240000	20	3	4745		Lake, summer		20		Elect.			
Sweden	Klippan	56		Aquifer/sand	800000	14	5	8352		River		5–14		Elect.	1	2000 el	
Sweden	Harryda	58		Tubes/Earth	18000	14	2	203		Unglazed	550			Elect.	1	200 el	
Sweden	Suncourt	59		Duct/Rock	30000	16	6	198	0.85	Atrium		20		Elect.			4.0
Sweden	Falun	61		Aquifer/gravel	700000	20	6	1368		Lake, summer		20		Elect.	1	400 el	
Switzerland	Cort.-Neuchatel	47	1981	Duct/Earth	4500	25	5	60	0.68	Flat plate	320			Elect.	1	50 th	3.3
Switzerland	Lavigny-Vaud	47	1981	Duct/Sand	250	40	15	3		Flat plate	50						
Switzerland	Vaulruz	47	1982	Duct/Earth	3500	40	5	170	0.84	Flat plate	520		0.39	Elect.	1	160 th	3.0
Switzerland	Meyrin-Geneva	47	1986	Duct/Earth	20000	30	0	564		Flat plate	550			Gas	1	210 el	
U.S.A.	Hatfield	42	1983	Duct/Earth	1926	50	20	33		Flat plate	171			Elect.	27	470 el	3.5

TABLE 1 Characteristics of CSHPSS Projects and Experiments Throughout the World (cont.)

Country	Location/ Name	Latitude	Year Built	Type	LOAD Building Area, m2	Annual Load, MWh	Delivery Maximum	Temps, C Minimum	PERFORMANCE Fractional Contributions Solar	HP Input	Aux.
Austria	Kranebitten	47	1982	Milt. Barracks	65000	1370			0.66	0.32	0.02
Canada	Scarborough	44	1985	Office Bldg	30000	1800					
Denmark	Lyngby	56	1983	Heat Exchanger							
Finland	Kerava	60	1983	Resid./44 units	4400	550	60	45	0.50		
France	Toulouse-Blagna	43	1980								
France	Aulnay-Sous-Boi	49	1983	Resid./225 units		2473	50	25	0.66	0.24	0.10
France	Saint Quentin	49	1986	School	2500						
France	Cormontreuil	49	1986	Function Hall	5200						
Germany, FRG	Stuttgart	48	1985	Office Bldg	1375	175	60	25	0.73		
Italy	Treviglio	46	1982	Apart./102 unit	9200	1235	45	30			
Italy (CEC)	Ispra	46	1981	Simulated							
Japan	Yonezawa	38	1982	Snow Melt & Bldg							
Netherlands	Groningen	53	1984	Resid./96 units	10000	1200	43		0.66		0.34
Netherlands	Bunnik	52	1985	Office Bldg	10000	660	50	40	0.30		
Sweden	Studsvik	58	1979	Office Bldg	500	23	30	30	1.00		
Sweden	Ingelstad	57	1979	Resid./52 units	6500	940	80	50	0.50		
Sweden	Sunclay	57	1980	School	15000	1650	55	45	0.54		0.50
Sweden	Lambohov	58	1980	Resid./55 units	7000	940	55	25	0.70	0.30	
Sweden	Gullspang	59	1982	Office Bldg.							
Sweden	Lyckebo	60	1983	Resid./550 unit	45000	7000	70	55			
Sweden	Lulea	66	1983	Univ. Bldg		2000					
Sweden	Vallentuna	59		Distr. Heating							
Sweden	Kopparberg	59		Distr. Heating		4200					
Sweden	Klippan	56		Distr. Heating		16000					
Sweden	Harryda	58		School							
Sweden	Suncourt	59		Apart./40 units	3500	320		50			
Sweden	Falun	61		School							
Switzerland	Cort.-Neuchatel	47	1981	Resid./12 units		190	60		0.40		
Switzerland	Lavigny-Vaud	47	1981								
Switzerland	Vaulruz	47	1982	Adm. Bldg & Garage	3200	341	50		0.46	0.19	0.35
Switzerland	Meyrin-Geneva	47	1986	Industr. Bldg	30000	1000					
U.S.A.	Hatfield	42	1983	School	4025	293	90	60			

MONITORING RESULTS OF THE GRONINGEN - CSHPSS

A.J.Th.M. Wijsman and J. Havinga
TNO Institute of Applied Physics
P.O. Box 155, 2600 AD DELFT
The Netherlands

ABSTRACT

In Groningen, the Netherlands, a group of 96 solar houses is connected to a central seasonal heat storage reservoir in the subsoil. This Central Solar Heating Plant with Seasonal Storage (CSHPSS) has been in full operation since the end of 1984. The objective of the project was to gather practical experience with the actual realisation and operation of such a system, and to gather data for model validation. A detailed monitoring programme has been carried out over a period of more than two years. This paper describes the monitoring programme and gives the monitoring results.

KEYWORDS

Solar houses, evacuated tubular collectors, seasonal heat storage in the soil, vertical heat exchanger tubes, short-term heat store, monitoring programme, monitoring results.

Photo 1: Aerial view of the Groningen CSHPSS

INTRODUCTION

One of the possible applications for solar energy in the Netherlands is to use it for space-heating in houses. However, in this application solar heat input and heat demand of the house do not occur simultaneously. To solve this problem, the heat collected during the summer will have to be stored till winter. A feasibility study showed that, under Dutch climatic conditions, seasonal storage of heat can give a twice as high solar contribution of the solar energy system as compared to short term heat storage only. Seasonal heat storage in the subsoil was found to be a good option.
In Groningen such a system was built. In 1981 a start was made with the design of the system and during the second half of 1982 the seasonal heat store was built, followed by a field test during 1983. On the basis of the test results it was decided to carry out the next phase: construction of the houses and their coupling to the seasonal heat store. Since the end of 1984 the system has been in full operation. A detailed monitoring programme was carried out from February 1985 till the end of January 1987.
This project was financed from the Dutch National Solar Energy Programme and by the Commission of the Eur pean Communities.

SYSTEM DESCRIPTION

The system consists of 96 solar houses grouped around the seasonal heat store in the soil and the central boiler house (see Fig. 1).

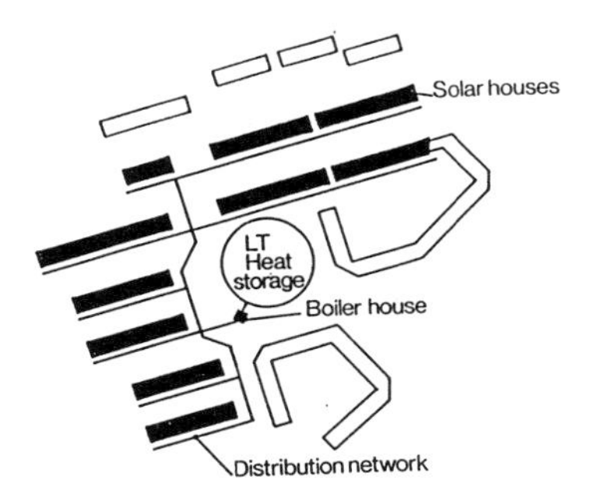

Fig. 1. Schematic representation of the Groningen system.

There are 9 blocks of houses with each a common collector roof. The solar collectors applied are evacuated tubular collectors (type Philips VTR-261). The total collector area installed is about 2400 m² (25 m² per house).
The seasonal heat store has a cylindrical shape (diameter 39 m, depth 20 m, total volume 23.000 m³). Only the top of the store is insulated. Heat exchange between the soil and the transfer medium is effected with flexible polybutene tubes which have been inserted vertically into the soil up to a depth of 20 m. In the centre of the heat store a 100 m³ - water tank is located which acts as a short-term (daily) heat store.
Heat is withdrawn from the system for space heating and preheating of domestic hot water. The houses are equipped with a low temperature heating system (radiators with a large area). There is a central gasboiler which comes in when the heat delivery from the solar energy system is insufficient. In each house there is a back-up system which heats up the preheated tapwater to the required tap temperature of 60 °C.

DESIGN BACKGROUND

For the design work on the system a computer simulation model of this CSHPSS-plant has been used. After the field test (1983) the model for the seasonal heat store has been modified.
The solar contribution of the system after 3 or more years of operation is expected to be 732 MWh, which is 63% of the total load. In Fig. 2 a Sankey diagram is given.

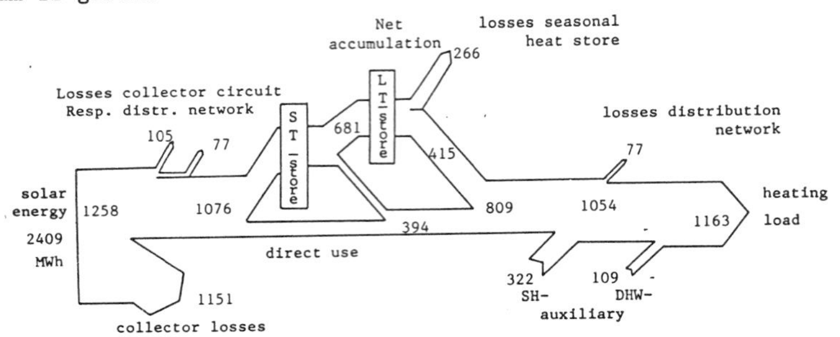

Fig. 2. Sankey diagram of the Groningen system (pumping energy not included) - DESIGN.

The net solar contribution is 310 kWh per m² of solar collector. The annual efficiency of the collector roofs is 48%. The seasonal heat store will operate between 30 and 60 °C. The electricity consumption of this system (for pumps) is 77.000 kWh, i.e. 800 kWh per house.

MONITORING PROGRAMME

After the system came into operation, a detailed monitoring programme was started. First a diagnosis of the entire system was made: characteristics of components were determined, flowrates and controls were checked. After that the actual monitoring started which lasted two years. A long monitoring period is essential for systems with seasonal heat storage: the effect of a summer with less solar input will not show until the next spring in the form a higher auxiliary heat supply.
The monitoring programme consisted of measurements on individual houses, on blocks of houses, of central measurements and of detailed measurements on the seasonal heat store:

- in 6 individual houses the heat consumption for both DHW-preheating and space heating was measured. Also, room temperatures were recorded as well as the electricity and gas consumption.
- in each of the 9 blocks of houses both the heat gain of the solar collector roofs and the total heat delivery for space heating and DHW-preheating was measured.
- centrally the heat input/output for both the short term and seasonal heat store were measured. Also, meteorological data were gathered (solar radiation, ambient temperature and other).
- in and around the seasonal heat store, soil temperatures have been measured in many spots. Also measurements on heat losses through the top insulation, on ground water movement, ground water level and on horizontal/vertical deformations of the soil store were done.

Data are collected by an on-site intelligent datalogger/computer-combination. Daily review-tables with hourly data are generated to check the operation of the system. After first processing, data are stored on-site on magnetic tape. From Delft there is a direct telephone connection with the data-acquisition system.

MONITORING RESULTS

The system has been monitored for two years: February 1, 1985 till January 31, 1987. In Fig. 3 the Sankey diagram is given for the first year of monitoring.

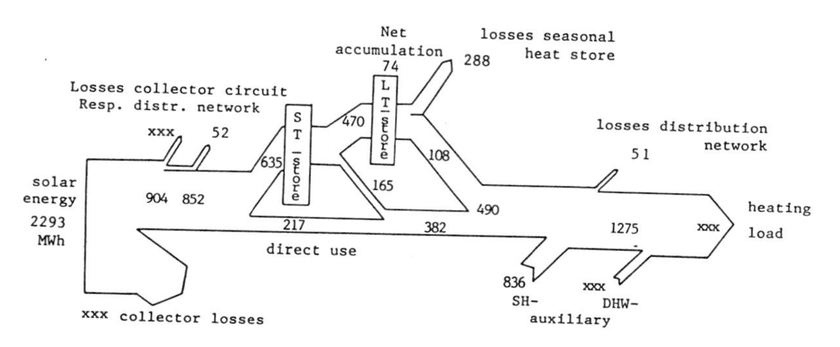

Fig. 3. Sankey diagram of the Groningen system (pumping energy not included) - first year.

The solar contribution of the system was 439 MWh: 217 MWh via direct use, 165 MWh via short term storage, 108 MWh via seasonal storage and minus the heat losses of the distribution network towards the houses (77 MWh). The annual efficiency of the collector roofs was 39%. The net heat accumulation in the seasonal heat store was 74 MWh, the heat losses of this store were 288 MWh.
In Fig. 4. the Sankey diagram is given for the second year of monitoring.

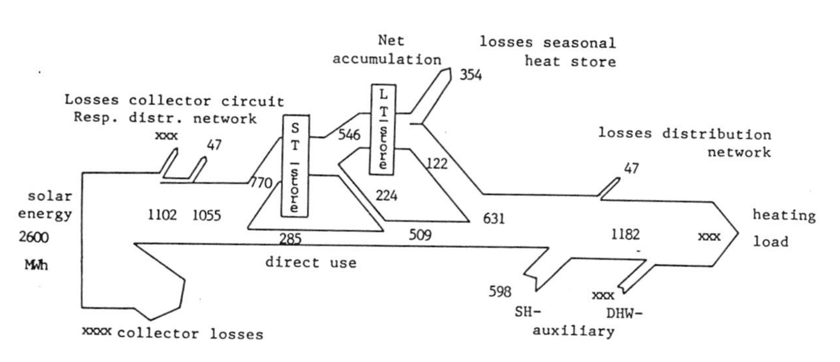

Fig. 4. Sankey diagram of the Groningen system (pumping energy not included) - second year.

The solar contribution of the system was 584 MWh: 285 MWh via direct use, 224 MWh via short term storage, 122 MWh via seasonal storage and minus the heat losses of the distribution network (47 MWh). The annual efficiency of the collector roofs was 42%. It was observed, that the several blocks showed differences in performance: 39-45%.

The net heat accumulation of the seasonal heat store was 70 MWh (so a "steady" year cycle has not been reached), the heat losses of the store 354 MWh.

In Fig. 5. the mean temperature in the seasonal heat store is shown for the two years of monitoring.

As an example Fig. 6. shows the temperature distribution in and around the seasonal heat store at the end of the summer of the second year (date: October 1, 1986).

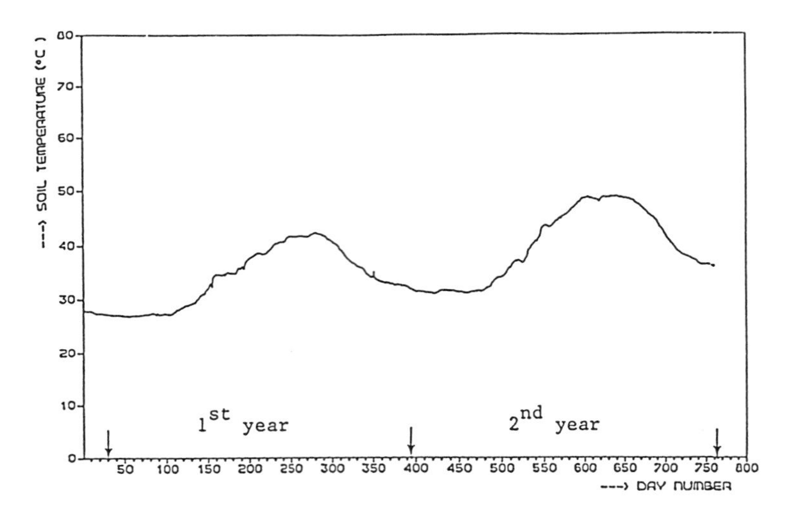

Fig. 5. Mean temperature in the seasonal heat store during the two years of monitoring: Feb. 1, 1985 - Jan. 31, 1987.

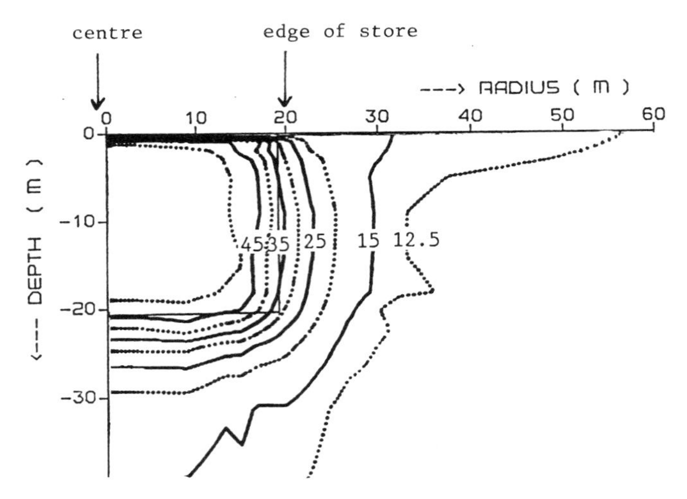

Fig. 6. Temperature distribution in and around the seasonal heat store. Date: Oct. 1, 1986

The electricity consumption of the CSHPSS-plant was measured to be 112.000 kWh annually (i.e. 1150 kWh per house).
Finally it was observed that the space heating system in the houses did not function optimally. This led to some complaints about too low temperatures in the houses and heat consumption for space heating in summertime. The heat consumption in summertime reduced the input to the seasonal heat store and so its seasonal heat storage function.

DISCUSSION OF THE RESULTS

In the first year the solar contribution of the CSHPSS plant was 439 MWh, the second year 584 MWh; for the third year a contribution of 640 MWh is expected because the net heat accumulation in the seasonal heat store goes to zero. So compared to the design (732 MWh) the solar contribution is 15% lower.
Main reasons for this lower solar contribution are the lower efficiency of the solar collector roofs and the higher heat losses of the seasonal heat store.
In the first year the efficiency of the collector roofs was 39% and the second year 42%. The lower efficiency during the first year was caused by the fact that in the spring of that year part of the collector roofs was out of operation because of frost damage. According to the design 48% was expected; so the measured efficiency in the second year was 13% lower. An additional study yielded the causes for the lower performance.
The heat losses of the seasonal heat store are about 1.6 times higher than expected. It was found from the heat flow measurements through the top insulation that both the top losses and the losses to the surrounding soil are that rate higher. For the top losses the cause is obvious (the second insulation layer with Argex clay grains does not function properly). What causes the higher losses to the surroundings is not clear yet (the ground water movement may be higher than expected).
The electricity consumption of the CSHPSS-plant is higher than expected: 112.000 kWh instead of 77.000 kWh. This is mainly caused by electricity consumption of other apparatus (valves, control equipment and others) than by extra consumption of the pumps. Finally the space heating system in the houses has been adapted to solve the problems of too low temperatures and of heat consumption for space heating in summertime.

RESUME

During the two years of monitoring on the Groningen plant a lot of experience was gained with the operation of the system. During the second year the solar contribution of the plant was 584 MWh, which corresponds to about 800 m^3 of natural gas per house, the electricity consumption of the plant was 112.000 kWh (1150 kWh per house). Expected was 1000 m^3 of gas savings per house and an electricity consumption of 800 kWh.
During the next 3 years the monitoring of the system will be continued on a low level.

REFERENCES

Wijsman, A.J.TH.M. and J. Havinga (1985)
The Groningen project: 96 solar houses with seasonal heat storage in the soil. Proceedings 'Intersol 85' Congress, Montreal, Canada.

SOLAR ENERGY FOR DISTRICT HEATING OF NEW HOUSING AREAS

by

Simon Furbo, P.h.D. Poul E. Kristensen, B.Sc.
Planum International, Consulting Engineers FIDIC
Havnegade 41, DK-1058 Copenhagen

SUMMARY

Calculations show that the supply temperature for a new district heating grid in a new housing area can be reduced from 60°C to 45°C and still be sufficiently high for supplying the houses with all the required energy for space heating and domestic hot water if low temperature radiators and special designed hot water tanks are installed in the houses.

For a supply temperature of the district heating grid of 45°C a mean return temperature in the new housing area of 20°C can be achieved if the above mentioned low temperature radiators and hot water tanks are used in all the houses. Such low supply and return temperatures are very attractive from the point of view of a solar heating plant.

SYSTEM LAY-OUT

The principle of the system taken in calculation is scematically shown in figure 1. The radiators and the hot water tank are connected in parallel resulting in the greatest possible temperature decrease of the fluid passing the installations. Radiator thermostatic valves on each radiator controls the volume flow rate in the radiators.

The hot water tank has a built-in heat exchanger spiral and an electric heating element. The volumetric flow rate in the heat exchanger spiral is controlled by the temperature inside the tank by means of a valve with a temperature sensor.

The house taken in consideration is a 120 m^2 single family house which is well-insulated. The design heat loss (20°C indoor and -12°C outdoor) is 3700 W and the total heat demand for space heating in the Danish Test Reference Year is 6280 kWh.

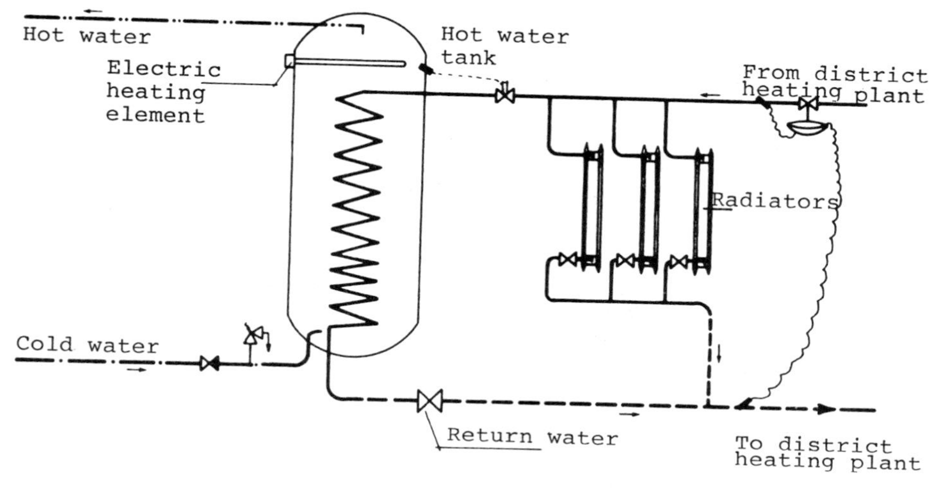

Figure 1. Schematical illustration of the principle of the radiator system and the hot water tank heated by the fluid of the district heating system.

The data for the hot water tank used in the calculations are given in table 1.

Hot water tank	
Form	Cylindrical
Material	Steel
Volume	200 l
Diameter	412 mm
Height	1500 mm
Insulation	
Material	mineral wool
Thickness	5 cm
Thermal bridge	1 W/k at the bottom of the tank
Ambient temperature	20°C
Heat exchanger spiral	
Dimension	10/12 mm copper pipe
Length	20 m evenly located at the lowest 80% of the tank
Constant supply temperature	45°C
Electric heating element	
Power	500 W
Placement	Upper 1/5 of the tank
Thermostate temperature	40°C
Control system of the volume flow rate of the fluid passing the heat exchanger spiral	
Location of temperature sensor:	upper 1/5 of the tank
Volume flow rate	0,5 l/min. if sensor temperature < 38°C 0 l/min. if sensor temperature > 41°C

Table 1. Data for the hot water tank.

The hot water consumption taken in consideration is 210 l/day with a cold water temperature of 10°C and a hot water temperature of 40°C, corresponding to 2650 kWh/year.

The space heating demand of the house is calculated every half hour in the year. By means of the model developed the annual energy consumption of the electric heating element, the annual thermal loss of the hot water tank and the mean return temperature of the fluid is calculated.

RESULTS

It is possible to obtain a low mean return temperature, if the system is designed and controlled in the right way. Figure 2 shows the mean return temperature of the fluid passing the heat eschanger spiral of the hot water tank and the daily thermal loss of the tank and the daily energy consumption of the electric heating element in the tank as a function of the location of the temperature sensor. In order to obtain a low return temperature it is important to locate the temperature sensor in the upper part of the tank. A mean return temperature of 16°C is achieved with the sensor located in the upper 1/5 of the tank.

The results of the calculations for the total system for different radiator sizes are shown in figure 2. For increasing radiator size the mean return temperature decreases.

A dimensioning radiator power of 10.000 W is sufficiently to cover all the space heating demand of the house during all year. A mean return temperature as low as 20°C is achieved with a dimensiong radiator power of 20.000 W and the radiators and hot water tank connected in parallel.

CONCLUSION

Calculations show that new dwellings in a district heating system can be heated with a supply temperature as low as 45°C and that a mean return temperature as low as 20°C can be achieved without a considerable increase of the cost of investment of the house.

The extra costs for oversized radiators and specially designed hot water tank is preliminary calculated to be 6.500 Dcrs. for each dwelling. The reference costs are calculated for a traditional installation designed for 60°C flow temperature.

Larger groups of new houses therefore are attractive from the point of view of a solar heating plant connected to a district heating grid. It is therefore recommended to carry out detailed investigations in order to elucidate the suitablility of solar heating plants for new housing areas.

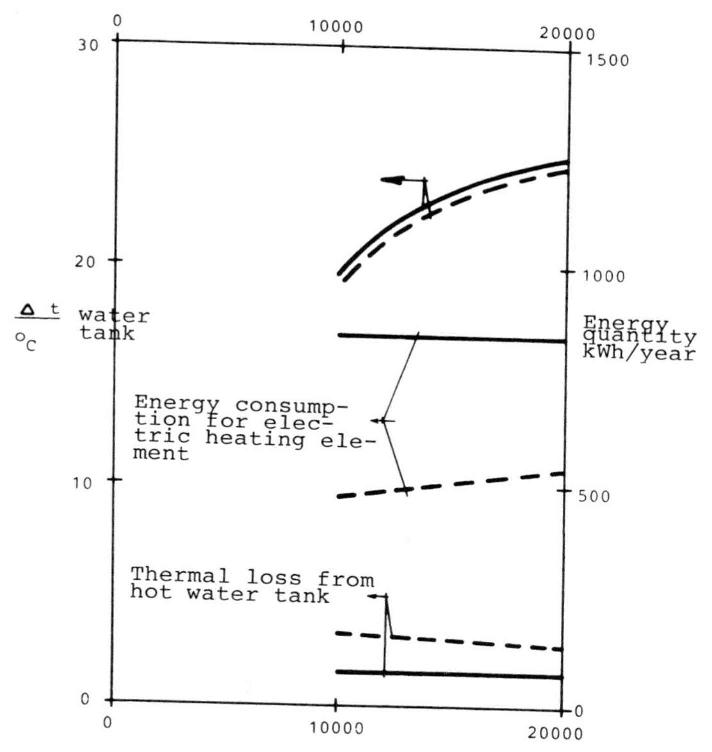

Figure 2 Mean temperature decrease of the fluid passing the house, the annual thermal loss from the hot water tank, the energy consumption for the electric heating element as a function of the radiator size and the system configuration.

——Radiators and hot water tank connected in parallel.

----Radiators and hot water tank connected in series.

FLAT PLATE OR VACUUM TUBE COLLECTORS FOR SWEDISH SOLAR DISTRICT HEATING

B. Karlsson*, C. Brunström*
Swedish State Power Board
Alvkarleby Laboratory, Sweden

ABSTRACT

The annual performance of vacuum tube, concentrating and well insulated flat plate collectors have been monitored during 4 years of parallel operation. Typically the long vacuum tube delivers 440 kWh/m², the concentrating collector 350 kWh/m², the flat plate collector 330 kWh/m² and the short vacuum tube 300 kWh/m² at an operating temperature of 65°C. Furnishing of the flat plate with a low-iron glass will increase its heat production to around 350 kWh/m². The combination of relatively high performance and efficient installation in large arrays makes the large area flat plate to the most cost-effective alternative for swedish solar district heating.

KEY WORDS:

Large collectors, vacuum tubes, concentrating collectors, district heating.

INTRODUCTION AND BACKGROUND

Solar district heating in combination with seasonal storage is considered as the most promising application of solar energy in Sweden. This technology allows to design a solar-system where 80 percent of the annual load is supplied. These systems requires solar collectors which are easily installed in large arrays and with high efficiencies at temperatures between 60 and 90°C. This demand on the temperature leaves a choise between vacuum tube, concentrating or well insulated flat plate collectors. In an earlier paper (Karlsson and Hultmark 1985) it was shown that the annual production of heat of a large collector, equipped with convection suppressing teflon films, was comparable with the production from the advanced collectors. In the present paper the result from a continued long term testing is presented, with the long vacuum tube collector VTR 361 from Philips included in the assembly of tested collectors.

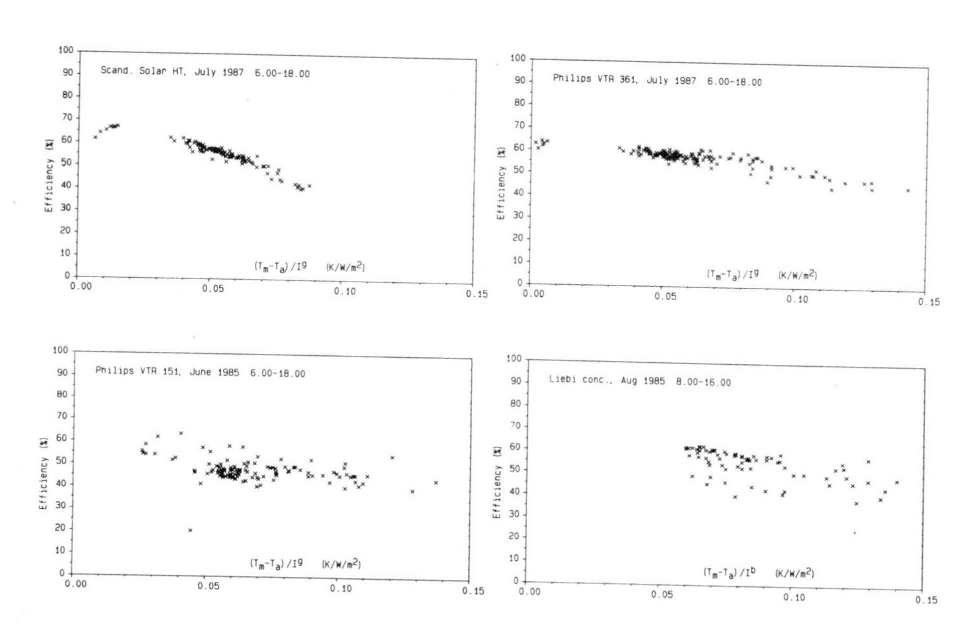

Fig. 1 a-d. Efficiencies of collectors during all hours of complete operation.

COLLECTORS AND METHODS OF TESTING

The following collectors were tested:

Scandinavian Solar HT (η_o =0.71, $F'U_L$ =2.8 W/m K) single glass large 2x6 m² flat plate collector with two films of teflon between the high-iron glass and the Sun-strip absorber. This collector, with a high transmitting low-iron glass, is installed in the large Lyckebo system (Brunström 1986). The Lyckebo collector will accordingly have a 0.04 units higher optical efficiency. The ratio of area used for calculation of efficiency to the transparent area was 0.965.

Polisolar-Liebi (η_o =0.68, $F'U_L$ =1.0 W/m K) linear parabolic collector, tracking around an axis tilted 45, was originally installed in the Ingelstad solar heating plant (Dalenbäck, Jilar 1986). The ratio between the width of the parabola and the diameter of the absorber was 40 to 1. The ratio between used area and total reflector aperture was 0.955.

Philips vacuum tube VTR 151 (η_o = 0.57, $F'U_L$ = 1.6 W/m K) of a tube length of 1160 mm and with 19 tubes in a module of a width of 1420 mm. Thus giving a ratio of 1.25 between defined collector area and effective absorber area.

Philips vacuum tube VTR 361 with a ripple reflector (η_o = 0.65, $F'U_L$ = 1.0 W/m K) of a tube length of 1760 mm and with 14 tubes in a module of a reflector- width of 1550 mm, giving a ratio of 1.84 between reflecting and absorbing areas.

The areas of all collectors except VTR 151 were defined by the effective optical area so the loss free conversion factor could be written $\eta_o = F'\alpha\tau$ for the flat and $\eta_o = R\tau\alpha$ for the concentrating collector. This means that relatively small areas were assumed and accordingly higher efficiency were obtained. The collectors were tilted 45° from the horizontal and pointed towards south. The collectors were installed on a platform on the laboratory roof with separate piping and controll systems for each collector. The temperatures were controlled by adjustment of the flow through the collectors and the flow of cooling water through the heat-exchanger. The global radiation was monitored on the horizontal and on the collector plane, further the the global and the diffuse irradiation were measured in the tracking plane of the concentrating collector.

TABLE1 ANNUAL HEAT PRODUCTION OF COLLECTORS

	1984		1985		1986		1987	
	E(kWh)	T(°C)	E(kWh)	T(°C)	E(kWh)	T(°C)	E(kWh)	T(°C)
SSHT- Fl.pl	328	67	302	75	335	59	332	60
Ph VTR-151 vac	296	61	301	67	-		-	
Ph VTR-361 vac	-		-		-		439	59
P-L Conc	321	63	363	70	353	58	-	
Global-45°	958		1005		962		976	
Beam-tracking	807		904		850		-	

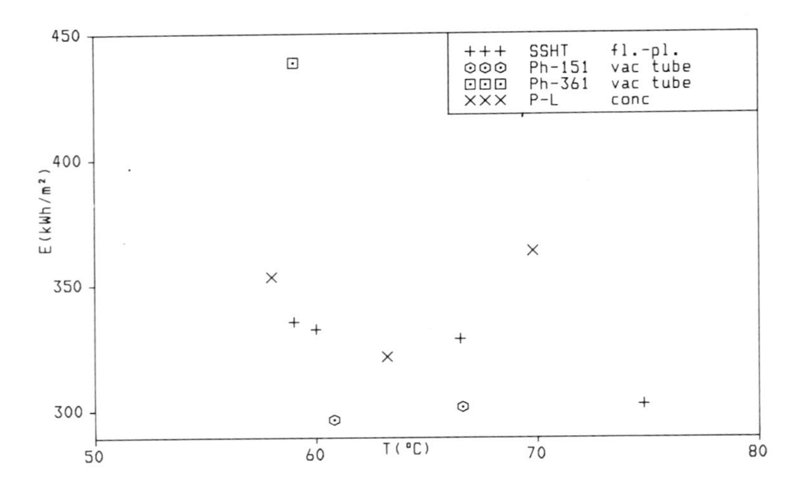

Fig. 2. Annual heat production during the years 84-87 at given operating temperatures.

Hourly mean values of meteorolocical and collector performance data were stored. Data points for determination of the linearized efficiency parameters were selected from hours of irradiances above 900 W/m. The required spread of data was obtained by varying the operating temperatures for a limited number of hours. The parameters obtained for the vacuum tubes are in good agreement with earlier measurements (Grijs and de Vaan 1983).

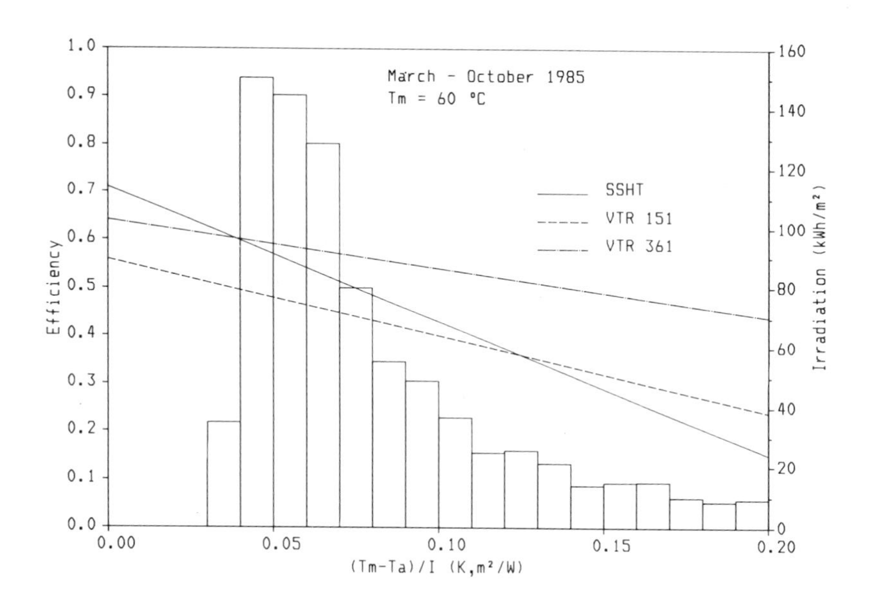

Fig.3. Distribution of annual irradiation and efficiency of stationary collectors.

RESULTS

The figure 1 a-d illustrates the long term performances of the collectors during all hours of complete operation. The very good performance of the long vacuum tube is shown by its very little spread of data up to low irradiances. The absence of data points for the flat collector at low irradiances is explained by its very few hours of complete operation at these intensities. The tracking collector shows spread of data for early and late hours, when the paraboles give internal shading.

The table 1 and the figure 2 concludes the results of four years of collector testing and irradiation measurements. The long tube produced 438.5 kWh/m², which to our knowledge is the highest reported value from a swedish system. The concentrating collector delivered slightly more heat than the flat plate, apart from the year of 1984 with an extremely low amount of direct radiation

ANALYSIS AND CONCLUSIONS

The last figure shows the annual distribution of irradiation in the efficiency diagram at an assumed operating temperature of 60°C. In this figure is drawn the linearized efficiency curves derived from hours of high intensities. The flat plate has a higher efficiency than the short vacuum tube during hours containing 70 percent of the irradiation, while the long tube has a higher efficiency than the flat plate for hours of 81 percent of irradiation. This figure illustratively explains the relative difference in annual performance between the collectors.

An detailed analysis (Brunström and Karlsson 1986) showed that the plane of the flat collectors during 1985 received 794 kWh/m² of irradiances exceeding 300 W/m during 1273 hours at an mean ambient temperature of 14°C, simultaneously the tracking collector received 739 kWh/m² of beam radiation of intensities above 300 W/m² during 1207 hours. This means that the plane and the concentrating collectors operate with acces to almost the same amount of useable irradiation. The performance of the flat collector could easily be improved by introduction of a low-iron glass, which should increase η_o by 0.04 units. A high efficient concentrating collector would be described by $\eta_o = 0.75$ and F'U = 0.5 W/m²,K. These upgraded versions of flat plate and concentrating collectors would have produced around 350 kWh/m² and 435 kWh/m² respectively at 65°C during 1985 as compared with the 438 KWh/m² delivered by the long vacuum tubes.

However, the 25 percent increase of produced heat can hardly pay the extra costs associated with the complexity of installation of the tubes and the complexity of both installation and operation of the paraboles. Therefore the future efforts for making the solar district heating cost effective should be put on technology for increasing the efficiency of operation and decreasing the cost of production of the large flat plate collectors.

REFERENCES

Brunström, C., Merkell, A-L and Larsson, M. (1986). The Lyckebo project - thermal performance of system and storage. Proc. North Sun 86. Copenhagen, Denmark.

Dalenbäck, J-O and Jilar, T. (1986). Ingelstad - A Solar heating plant with seasonal storage. Proc. North Sun 86, Copenhagen, Denmark. pp 39-49.

Brunström, C., Karlsson, B. and Larsson, M. (1986). Climatic limitations and collector performance in the middle of Sweden. Proc. North Sun 86, Copenhagen, Denmark. pp 161-166.

Grijs, J.C. and de Vaan, R.L.C. (1985). Requirements for testing evacuated tubular collectors with heat pipes. Proc. Solar World Congress, Perth, Australia. V2,pp. 1061-1070.

Karlsson, B. and Hultmark, G. (1985). The superiority of flat plate collectors for district heating. Proc. Intersol 85, Montreal, Canada. pp. 1003-1007.

THE SUN TOWN PROJECT - SWEDISH PLANS FOR THE BIGGEST SEASONAL STORAGE PLANT IN THE WORLD

Torbjörn Jilar

Department of Building Services Engineering
Chalmers University of Technology
S-412 96 Gothenburg
Sweden

ABSTRACT

The initial project phase, a pre-study, has recently been carried out concerning technology and economics for a very big solar heating plant planned to be in operation around 1990 in the town of Kungälv in the west of Sweden. The essential aim of the project is to present a feasible system concept featuring substantial solar heat coverage for the entire building stock of the town.

About 50 % of the total building space is planned to be heated by the central solar heting plant. The annual energy requirement is 52 GWh including residential houses for 6000 inhabitants and 115 000 m^2 of buildings for commerce, municipial service and industry. The preliminary outlined plant design employs 116 000 m^2 of high-temperature flat plate solar collectors and 380 000 m^3 of waterfilled rock caverns for seasonal heat storage. The plant is connected to an extensive district heating network. The design objective is to meet 75 % of the annual heating requirement by solar energy and the rest by fossil fuelling. The calculated cost of the heat produced by the plant is SEK 0,42/kWh (1 SEK ≈ 0,16 US$).

KEYWORDS

Solar heated town, System design, Economics.

INTRODUCTION

This paper briefly describes the technology and the economics for a very big solar heating plant planned to be in operation around 1990 in the town of Kungälv in Sweden. The essential aim of the project is to exhibit technically feasible and cost attractive system concepts featuring substantial solar heat coverage for the entire building stock of the town.

Since the middle of the 1970's there has been significant progress in the Swedish development of large-scale solar heating technology. At present, four Swedish solar heating plants incorporating seasonal heat storage are in operation. The comprehensive experiences gained have been utilized for the planning of the Kungälv plant. Since a couple of years plans are developed to introduce district heating on a wide base in the town of Kungälv. Swedish district heating technology is far developed and it is of great universal application to combine with large-scale solar heating technology. Concerning the national potential a number in order of 100-150 plants of the size described in this paper would meet around 10 % of the annual energy requirement for space heating and domestic hot water of the entire Swedish building stock.

SYSTEM DESIGN AND PERFORMANCE

At present about 90 % of the heating demand for the buildings in Kungälv is covered by heat from oil-fired boiler plants. At the beginning of the 1990's about 50 % of the total building space is planned to be heated by the central solar heating plant. The buildings have been calculated as requiring 52 GWh/year for space heating and domestic hot water production. About 60 % of the energy is required for the heating of residential houses for 6000 inhabitants. The rest of the energy is required for the heating of buildings for commerce, municipal service and light industry. A number of around 15 multy-building and 30 single-building, now existing heating centrals will be taken out of operation and rebuilt into service units connected to an extensive district heating network supplied by heat from the central plant.

An image of the plant concept and the town of Kungälv is shown in Fig. 1. The plant will probably be located to a place quite near the central part of the town where suitable land area for solar collectors and rock material for underground heat storage is available.

The preliminary outlined operating principle is quite simple. The design objective is to meet 75 % of the annual heating requirement by solar energy. The rest of the energy requirement is expected to be met by heat from cogeneration and top fossil fuelling (Fig. 2). Rock caverns will be used for heat storage. The water temperature in the store ranges between approximately 45 ^{0}C and 95 ^{0}C. A temperature swing in the order of 50 ^{0}C is possible through use of high-temperature solar collectors in combination with low district heating system temperatures.

Concerning the choice of a suitable type of solar collector experiences gained through Swedish developments in large-scale solar heating tehcnology since the 1970's is of interest. Rapid developments in the field of high efficient flat plate collectors have taken place. Thermal performance standards have been improved and collector array costs have been lowered to a great extent. High efficient flat plate collectors are suitable for use at 70 ^{0}C, the intended operating temperature in the Kungälv installation. It appears possible to design for around 360 kWh/m^2 of annual thermal yield. The choice of a suitable type of heat store is based upon Swedish experiences in the main concerning underground building technology and science in geology and hydraulics. Blasted rock caverns have been compared to boulder-filled rock caverns, deep vertical ducts and ducts combined with rock tunnels. Essential aspects treated are structural mechanics, hydro chemistry, thermal performace and economics. The best alternative appears to be blasted rock caverns essentially owing to good abilities to maintain thermal stratification, small thermal mass, good heat storage capacity and no severe hydro-chemical problems. The total heat storage capacity demand is 22 GWh corresponding to 380000 m^3 of water-filled caverns. The preliminary store configuration is 4 parallell cavern units having a width of 20 m and a height of 30 m. The building technology is very well known from the field of crude oil storage.

ECONOMICS

The calculated cost of the heat produced by the Kungälv plant is SEK 0,42/ /kWh including solar heat and supplementary heat from natural gas fuelling. The heat cost is obtained for a solar energy coverage of 75 % and included is the central plant cost and the district heating system cost (figur 3). The cost presented here in 1987 terms corresponds to a solar collector system cost of SEK 1200/m^2 and a heat store cost of SEK 100/m^3 (1 SEK ≈ ≈ 0,16 US $). The additional cost for ancillary piping and quipment for connecting to the heat store and the distribution system represents about 5 % of the total collector and heat store cost. Even the boiler plant represents an additional relative cost in the order of 5 %. The heat cost in this context is based on a 4 % real rate of interest and an economic lifetime of 40 years for the heat store, 20 years for the collectors and 15 years for the rest of the plant. The total plant cost amounts to SEK 246 million out of which about 18 % corresponds to the district heating system cost. The heat cost for the solar heating plant has been compared to the heat cost for the presently used alternative featuring many individual boiler plants for oil and other alternatives featuring one central plant for combined fuels like the one for the solar heating plant (figure 3).

Today the solar heat concept appears to be less competitive compared to the fuel use concepts. However, the global risk for near future fuel price increases and local problems concerning the availability of wood-chips and industrial waste heat should somehow be taken into account in cost analyses of this kind.

REFERENCES

Claesson, T., G. Hultmark, and T. Jilar (1986). Solar heat with seasonal storage in Kungälv - A pre-study summary. The Swedish Council for Building Research, T9:1986, in Swedish.

Dalenbäck, J-O. (1987). Large-scale Swedish solar heating technology - System design and rating. The Swedish Council for Building Research, document for publication in 1987.

Jilar, T. (1985). Large-scale solar heating technology. The Swedish Council for Building Research, Document D2:1985.

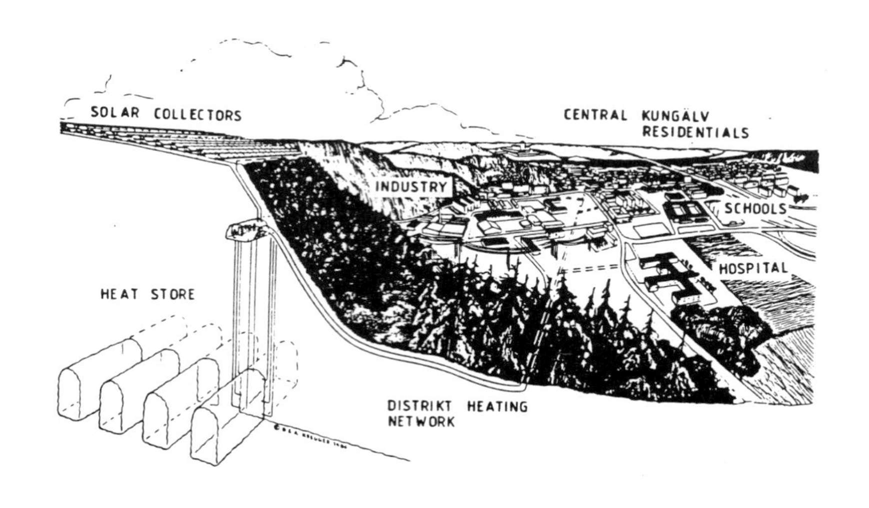

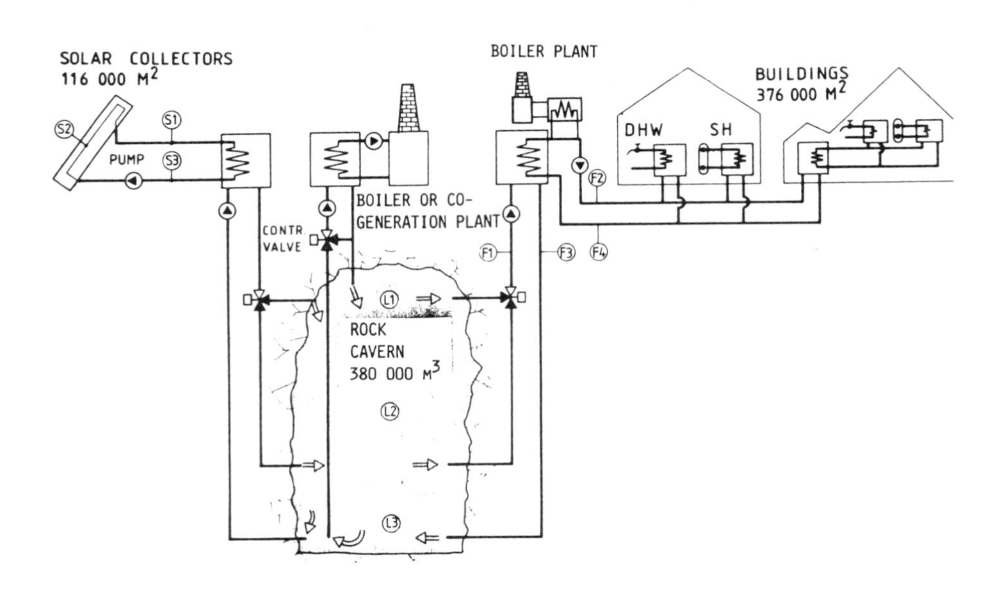

Fig. 1. The planned central solar heating plant in the town of Kungälv, Sweden. Top: A view over the town and the plant site. Bottom: System principle.

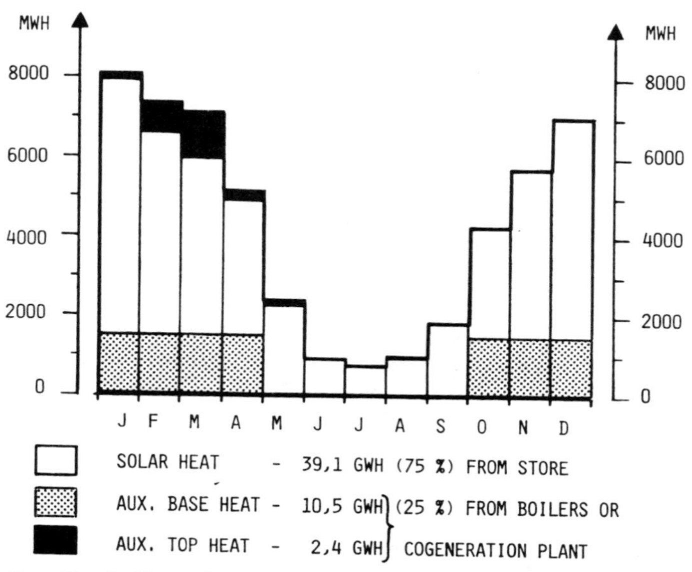

Fig. 2. Plant thermal performance 5 years after operation start. Distribution of solar and supplementary heat.

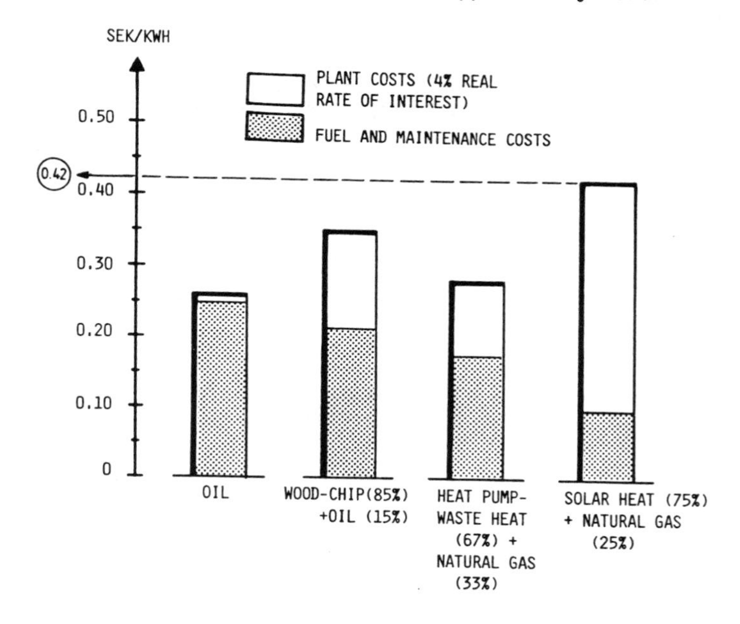

Fig. 3. Heat cost analysis for the Kungälv plant 1987. Heat production cost for the central solar plant compared to a central plant for oil and wood-chips or waste heat and natural gas and to individual boiler plants for oil.

SWEDISH SOLAR HEATING WITH SEASONAL STORAGE - ECONOMICAL PROSPECTS TODAY

Jan-Olof Dalenbäck, Torbjörn Jilar

Department of Building Services Engineering
Chalmers University of Technology
S-412 96 Gothenburg
Sweden

ABSTRACT

One feasible Swedish application for solar heating is group and district residential heating systems. If seasonal storage systems are to be competitive to fossil fuelling the specific solar collector array cost must not exceed SEK 800 - 1000/m^2 (1 SEK = 0,16 US $). This corresponds to solar heat costs in the order of SEK 0,30/kWh for large solar heating plants with storage in uninsulated rock caverns. For smaller solar heating plants the specific heat store cost must not exceed SEK 200 - 300/m^3 for having solar heat costs in the order of SEK 0,40/kWh. Consequently, the most important issues today are to lower collector array costs about 30 % from present levels around SEK 1200-1400/m^2 and to develop new, low-cost designs of insulated heat stores for smaller plants.

KYEWORDS

Solar group heating, investment costs, heat cost targets.

INTRODUCTION

This paper outlines economical prerequisites if solar heating plants with seasonal storage are to be competitive to other forms of heat production for residential heating in northern climates.

Since the middle of the 1970's there has been significant progress in the Swedish development of large-scale solar heating technology. An interesting application for high-temperature solar heating systems is the far developed Swedish group and district heating technology.

Smaller systems, i.e. those supplying a few hundred residential units, are known as group heating plants. Larger systems for some thousands of residential units are known as district heating plants. The following guideline values are based on practical plant experiences and comprehensive computer calculations made for new plant designs. Large modular flat plate solar collectors and water heat stores are suggested in these applications.

TABLE 1

Type of plant	Collector surface area m^2	Type of heat store	Useful solar heat kWh/m^2, year
Group heating plant	5.000-15.000	Insulated pit	Approx. 310
District heating plant	over 30.000	Uninsulated rock cavern	Approx. 330

For residential heating a maximum solar heat coverage of 70-80 % of the annual heat load is feasible from an economic point of view. The required heat storage capacity corresponds to 3-4 m^3 of store volume per m^2 of collector area. The required volume is smaller if for instance the heat load distribution over the year is very uniform or if the store temperature swing is high.

SOLAR COLLECTOR SYSTEM COSTS

In recent years a series of larger solar heating plants has been built for research purposes. True costs as well as performance results are known in detail. Rapid developments in the field of hign efficient flat plate collectors have taken place. Thermal performance standards have been improved and collector array costs have been lowered to a great extent. Fig. 1 shows heat costs specifically for collector arrays. The investment cost shown is the total cost for the array in operation expressed in 1987 price levels. For recent plant designs collector array costs are in the order of SEK 1200-1400/m^2 (1 SEK = 0,16 US$).

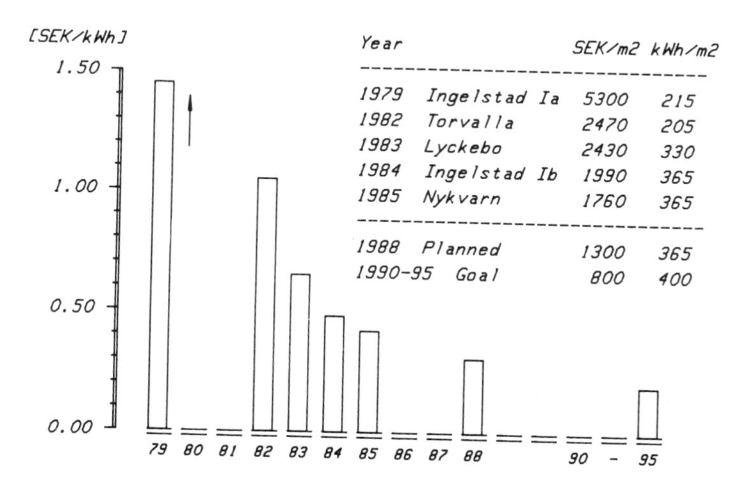

Fig. 1. Specific heat costs for solar collector arrays at 70 °C operating temperature. Real rate of interest 4 %, depreciation time 20 years.

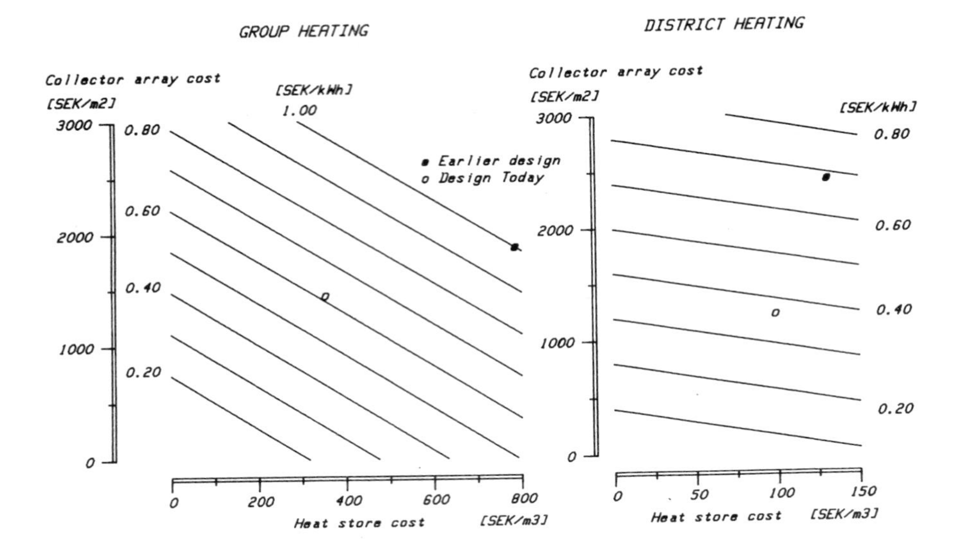

Fig. 2. Solar heat costs versus collector array and heat store investment costs (3,5 m^3 store/m^2 collector). Real rate of interest 4 %, depreciation time 20 years for collectors and 40 years for heat stores.

The results have been achieved by a combination of several factors. For instance the area of the panels has been heavily increased and panel mountings and connections have been simplified. Fig. 1 also shows postulated costs for installations possible in the near future. Assessments say that is should be possible to manufacture collector arrays on an industrial basis for total costs not exceeding SEK 800/m^2. This cost target is assumed to be met if production volumes in the order of 300.000 m^2/year can be assured.

HEAT STORE COSTS

For large, blasted rock caverns the building technology is very well known from the field of crude oil storage. The investment cost is in the order of SEK 100/m^3. Larger insulated heat stores for seasonal storage built so far, are restricted to one concrete tank and one rock pit. Both were built in the late 1970's at costs in the order of SEK 800/m^3 in 1987 price levels. At the present feasibility studies on new designs show that it is possible to build insulated, excavated pits for costs in the order of SEK 350/m^3 using quite conventional building methods. The studies indicate that it should be possible further to reduce the cost using new methods and closely controlled contracts.

SOLAR HEAT COSTS

The cost of heat from solar heating plants is determined primarily by the investment costs of the collector array and heat store. Cost diagrams have been obtained (Fig. 2) in which one axis represents the cost of the solar collector array plus piping system, expressed in SEK/m² of collector area and the other axis represents the cost of the heat store in SEK/m³ of store volume. Corresponding costs for the collectors and store for given costs of heat can then be read off for group heating plants and district heating plants, provided that the numerical values of real rate of interest, depreciation time and maintenance costs for the collectors and heat store are constant (Fig. 2), and using the values of net useful heat as shown in the table above.

It should also be assumed that the ratio of store volume to collector surface area is fixed. The ratio used when preparing Fig. 2 was 3.5, which is intended to reflect the fact that costings should primarily relate to the situation in existing heating systems with only moderately low return temperatures and with relatively unevenly destributed heat demand characteristics.

In Fig. 2 the cost development from the earlier designs to the present day designs is illustrated. Today both of the plant concepts appear to be less competitive to more conventional solutions for residential heat supply. During a couple of years in the near future conventional group heating probably will cost about SEK 0,40/kWh and district heating about SEK 0,30/ /kWh. Obviously collector and heat store costs still below the designs today, are necessary. Realistic collector costs are in the order of SEK 800-1000/m² (about 30 % below present) if the market for solar heating expands. For the related cases the following cost targets now can be stated:

• Group heating plants (200-600 flats)	Collector array costs SEK 800/m² resp. 1000/m² corresponds to max. heat store costs SEK 300/m³ resp. 200/m³ (about 15-40 % below present).
• District heating plants (> 1200 flats)	Collector array cost SEK 1000/m² corresponds to max. heat store cost SEK 100/m³ (present level).

REFERENCES

Dalenbäck, J-O. (1987). Swedish solar heating with seasonal storage - Design studies. Chalmers University of Technology. Paper submitted to ISES'87.

Hultmark, G. (1987). Development of the MEGA-technology for large scale solar heating. Scandinavian Solar AB. Paper submitted to ISES'87.

Jilar, T. (1987). The sun town project - Swedish plans for the biggest seasonal storage plant in the world. Chalmers University of Technology. Paper submitted to ISES'87.

Development of the Mega-technology for large scale solar heating

Göran Hultmark
AB Andersson & Hultmark
Box 24135, 400 22 Göteborg, Sweden

At the beginning of the 80's, it was obvious that the solar collector fields that so far had been built were too expensive both with respect to cost of solar collectors and, even more, installation costs.
A development was therefore started in 1981 with a long ground based on site build collector. From this collector, a five year development of what now is called the Mega-technology has been carried out.
The development is a combination of theoretical studies and practical construction work.

The practical work has resulted in solar collector installations such as Torvalla 1 700 m2, Lyckebo 4 300 m2, Ingelstad 1 600 m2, Nykvarn 4 000 m2 and the construction Kronhjorten 500 m2.

A lot of questions have had to be answered during this period. The size of the solar collector has increased from the normal 1-2 m2 over, in one period, 20-30 m2 and down to the last considerations of 12.5-13 m2.
Different types of convection supression barriers such as one or several teflon films, pleated films, special acrylic constructions and cellulosa-film have been tested.
Absorbers with very low flow rates have been constructed resulting in possibilities to start running collector fields of serveral thousand square meters without any regulation or air-outlet.
Considerations regarding how much work should be done on the site and how much shall be done in factories resulted in a very high degree of factory work even as regards large modul solar collectors.
The piping system has been constructed in three different ways. It started with a traditional way of piping, continued with a piping system, used in Lyckebo, where the pipes followed the solar collector lines and ended with the last development where all flows are passed inside the collectors resulting in a minimum of piping on the field.
Great efforts have been made to minimize the material required in the collector and in the total installation. All these considerations have resulted in lower cost for solar energy in large-scale solar heating.

The total cost for an installation has been reduced from 0.21 ECU/kWh in Ingelstad 1979 to 0.12 ECU/kWh in Torvalla, 0.048 ECU/kWh in Lyckebo, 0.41 ECU/Kwh in reconstruction of Ingelstad and 0.034 ECY/kWh in Nykvarn 1985.

SWEDISH SOLAR HEATING WITH SEASONAL STORAGE - DESIGN STUDIES

Jan-Olof Dalenbäck

Department of Building Services Engineering
Chalmers University of Technology
S-412 96 Gothenburg Sweden

ABSTRACT

This paper summerises the Swedish research program for solar heating systems with seasonal heat stores. Existing plants are described together with the background to new plants, the designs and ratings of which are directly based on experience from and research related to, the existing plants. Based on present-day knowledge, these solar heating systems should be designed and rated to provide an energy coverage of about 70-80 % of the heat requirement of the load. They should be realized using flat plate collectors in large moduls, with a basically simple design incorporating separate heat storage and discharge circuits.

KEYWORDS

Solar heating, seasonal heat store, system design, system rating, simulations.

INTRODUCTION

A pure solar heating plant with a seasonal heat store is one in which solar collectors produce heat at a sufficiently high temperature to allow it to be used directly for space-heating and domestic hot water purposes. As the times of year at which heat is required on the one hand and insolation levels are high on the other do not coincide, a heat store is needed to store surplus heat from the summer to the autumn and winter. Heat is stored at a sufficiently high temperature to allow it subsequently to be used directly for heating purposes.

The main feature that must therefore be required of a pure solar heating system with seasonal heat storage is that it must be possible to store and abstract heat at a high, directly useable temperature level, i.e. 50-60 ^{0}C. In these respects, there is little experience from the use of ground heat stores, which indicates that under present conditions work should be concentrated on water heat stores, which can be regarded almost as established technology.

Several solar heating plants incorporating seasonal heat stores have been built in Sweden for experimental purposes. In order to achieve the maximum benefit in terms of knowledge and experience relating to different types of plants, a mix of designs and sizes has been built, with performance and other aspects being monitored closely. The experience obtained from these plants, together with that from other larger plants, has resulted in the accumulation in Sweden today of internationally unique experience, firmly founded in reality.

EXISTING PLANTS

TABLE 1 summarises the main components of four solar heating plants with seasonal heat stores in which water is used as the storage medium. The connected annual loads vary from about 40 MWh in Studsvik to more than 8000 MWh in Lyckebo. The solar heat coverage for each system is the intended design coverage.

TABLE 1 Existing plants

Plant name	Collector type	Heat store	Solar heat coverage
Studsvik	120 m^2, partly concentrating	640 m^3, insulated excavated pit	100 %
Ingelstad Ia	1320 m^2, concentrating parabolic	5000 m^3, insulated concrete tank	50 %
Ib	1425 m^2, flat plate, high-temperature	5000 m^3, insulated concrete tank	50 %
Lambohov	2700 m^2, flat plate, roof-integrated	10 000 m^3, insulated rock pit	85 %
Lyckebo[1)]	28 800 m^2, flat plate, high-temperature	105 000 m^3, uninsulated rock cavern	100 %

Schematic diagrams of the existing systems are shown in Fig. 1. It can be seen that some plants have separate charging and discharging circuits, while in others it is possible to bypass the store and supply heat from the collectors directly to the load. System design in which constant and variable flowrates through the collector and storage circuits are combined with heat supply to the store at one or more levels are represented, as are designs with abstraction from one or more levels.

---

1) The full solar collector array has not yet been installed, but is partly simulated by an electric boiler.

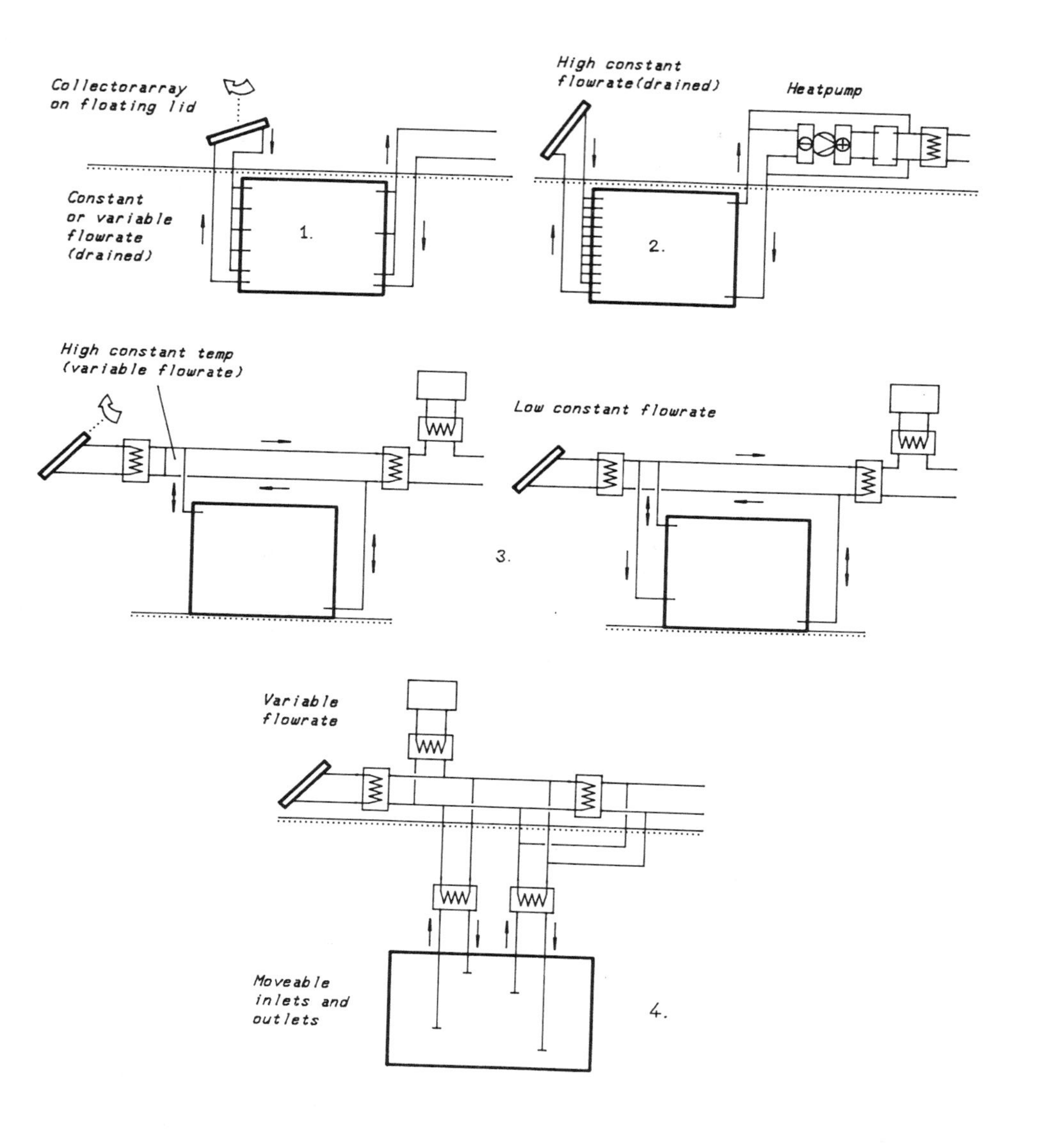

Fig. 1. Schematic drawings of existing systems
1. Studsvik, 2. Lambohov, 3. Ingelstad
4. Lyckebo.

Figure 2 is showing a comparison of the measured annual heat balances in the plants. The quantities of solar heat and supplementary heat have been related to the magnitude of the heat load, which has made it possible to compare the measured degree of energy coverage with the intended design coverage. It is also possible to see the magnitude of the heat losses from the store in proportion to the heat load. In making such a comparison, it is important to bear in mind the relative sizes of the systems. The Ingelstad and Lambohov plants, for example, are about ten times larger than the Studsvik plant, while Lyckebo is larger again by a further factor of ten.

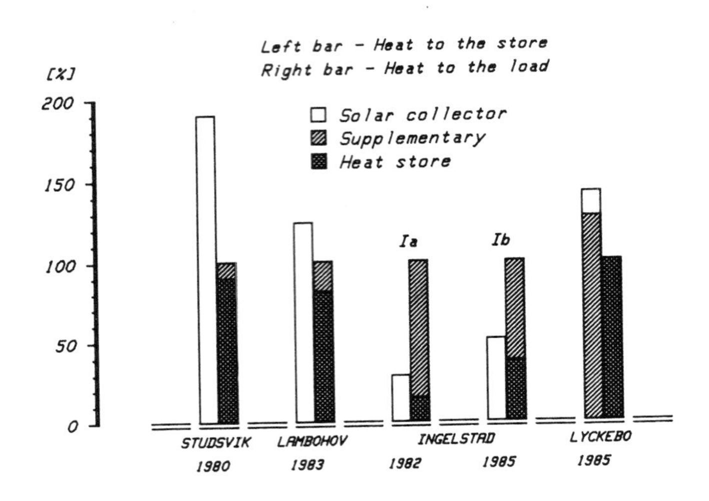

Fig. 2. Measured annual heat balances

Summarising, it can be seen that as-measured degree of heat coverage is close to the design coverage for all installations except Ingelstad Ia. The concentrating collectors were replaced with flat plate high-temperature collectors, which brought the yield up to the expected value. Heat losses from the stores, however, differ considerably from the design losses for Lambohov (insulated heat store) and Lyckebo.

To a greater or lesser degree, design calculations have been made for all existing plants before building them. These calculations were made without any possibility of comparison with actual results. The performance measurements that have been made together with new computer simulations of performance of the existing plants and of similar plants, make it possible to isolate the effects of particular system design features, choice of collector types and store types and rating of these elements.

Computerised simulation models, the results of which have been confirmed by comparison with measurements from existing plants, have been used exclusively during the work of the feasibility studies carried out for new projects in recent years. The results of these feasibility studies are therefore much more closely linked to reality than was previously the case.

SYSTEM DESIGN AND RATING - NEW PLANTS

Based on present-day knowledge, solar heating systems with seasonal heat stores should be designed and rated to provide an energy coverage of about 70-80 % of the heat requirement of the connected load. They should be realised using flat plate solar collectors in large modules, with a basically simple design incorparating separate heat storage and discharge circuits, as shown in Fig. 3. The solar collector and the storage circuit should be designed for a low constant flow rate.

When roughing out the initial design of solar group heating plants for operation in southern Sweden, the following guide values can be used to provide a 70-80 % energy coverage of the connected load by means of solar heat.

> 6 m^2 of solar collector surface area per kW of nominal connected load power demand, which requires a ground surface area of 15 m^2;
>
> 18 m^3 of store volume of water per kW of nominal connected load power demand.

Smaller systems, i.e. those intended to supply heat loads of the order of 100-3000 kW, require the use of insulated heat stores ranging in size from 2000 to 60 000 m^3 of water. Larger systems, intended to supply loads of 5 MW or more can utilise water-filled uninsulated rock caverns, in size of 100 000 m^3 or more, as their heat stores.

A new pilot plant for a smaller installation, Ingelstad II, having an insulated heat store has been built and a feasibility study for a large system in Kungälv is under way.

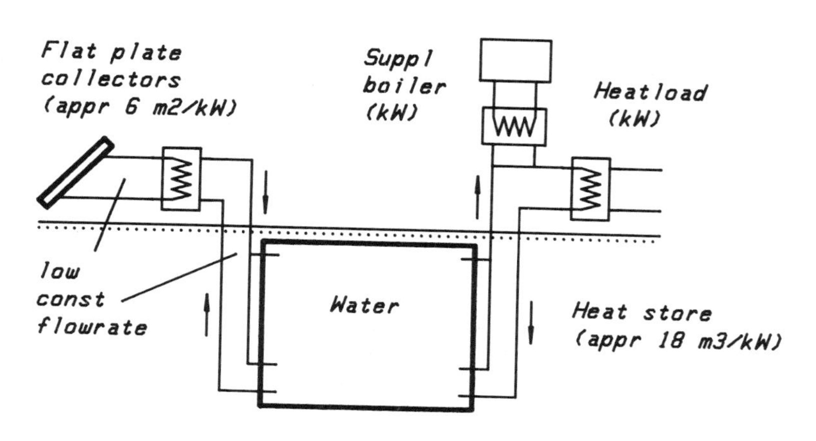

Fig. 3. Schematic drawing of new systems

REFERENCES

Dalenbäck, J-O, (1987). Large-scale Swedish Solar Heating Technology. System Design and Rating. Swedish Council for Building Research, DX: 1987.

THE LYCKEBO PROJECT - HEAT LOSSES FROM THE ROCK CAVERN STORAGE

C. Brunström*, B. Eftring**, and J. Claesson**

*Swedish State Power Board, Alvkarleby Laboratory, S-810 71 Älvkarleby, Sweden
**Dep. of Mathematical Physics, Lund Institute of Technology, Box 118, S-221 00 Lund, Sweden

ABSTRACT

A hypothesis explaining the high heat losses from the Lyckebo rock cavern heat storage is presented. A convective heat flow through a crack system from the cavern top via a tunnel surrounding the cavern and back to the storage bottom through an expansion duct is suggested. It is shown that a water flow of around 2 m^3/h is possible in the actual system. Computer simulations of cavern heat losses and cavern temperatures taking this type of convection into account give results in excellent agreement with measurements. The hypothesis is also supported by unexpectedly high water temperatures measured in the expansion tunnel system and by measurements of rock temperatures and deformations close to the top of the cavern.

KEYWORDS

Solar heating; seasonal storage; rock cavern; heat losses.

INTRODUCTION

The Lyckebo solar district heating plant has been in operation since summer 1983. The system is designed to supply an annual heat load of 8 GWh by solar heat seasonally stored in a rock cavern storage. Details of system design and economy as well as performance data have been presented in several reports, Brunström (1986a, 1986b, 1986c, 1987). The evaluation showed a reliable system function, a collector field output at designed values and a very good storage temperature stratification.

The Lyckebo rock cavern storage has the shape of a ring with a height of 30 m, a width of 18 m and a diameter of 75 m. The cavern roof is situated 30 m below the rock surface. The central rock pillar has a diameter of 39 m. The storage volume is 100 000 m^3 and the cavern walls are not insulated. The cavern heat losses measured for three annual cycles are considerably higher than expected, Fig. 1. Furthermore, losses seem not to decrease as expected when the rocks surrounding the cavern gradually becomes warmer. Marked in the figure are measured annual heat losses compared with the calculated conductive losses based on actual storage temperatures. The measured annual mean storage temperatures are indicated. The solid line represents the annual decrease of losses as expected from the predesign calculations at a mean storage temperature of 65°C. The difference between the line and the marked calculated losses is explained by the difference in mean storage temperatures.

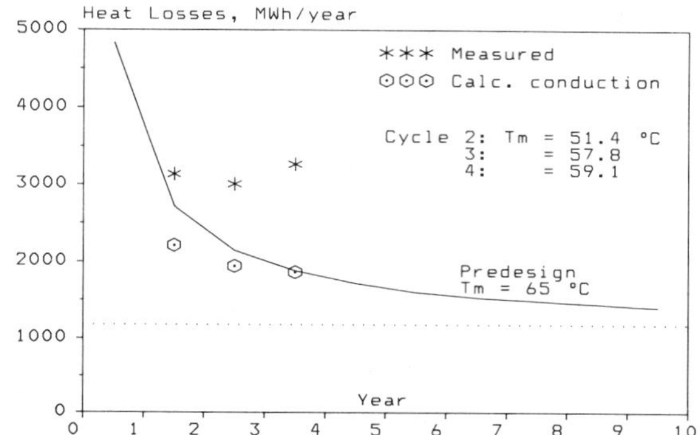

Fig. 1. Measured, calculated and predesign heat losses versus time for the Lyckebo rock cavern storage.

In this report different convective loss mechanisms occuring in rock caverns are discussed. A probable explanation for the high losses in the Lyckebo system, supported by water temperature measurements in the cavern, in a temporary access tunnel close to the storage and in the surrounding rock, is presented. Finally computer simulations of storage heat losses, taking both conduction and convection into account, are presented and compared with measured losses.

CONVECTIVE HEAT LOSSES

The extra heat losses above the expected ones due to heat conduction in the rock can, as far as we can judge, only be explained by convective losses due to moving water in cracks and cavities outside the cavern. Many flow patterns are conceivable. There may be a regional groundwater flow in permeable rock and, in particular, in shallow layers above the bedrock. Another possibility is convection cells above the rock cavern or outside its walls. But the permeability of the rock is very low. The regional gradient of the water table is small, and the more permeable top-soil layers lie 25 m above the top of the rock cavern. These flow types can not, in our judgement, explain the extra heat loss.

There is no net loss of water from the cavern. The convective flow must take place in a closed path, where warm water leaves the top of the cavern and colder water returns at lower levels. The driving force is the density difference between the warm water in the cavern and colder water along the flow paths in the ground. The largest driving force, when the whole height of the cavern is involved, becomes around 0.5 m water head in the present case with cavern temperatures from 80 to 30°C and ground temperatures down to 5°C. The convective heat loss corresponds to a steady water flow of 2 m^3/hour, if the temperature of the water that leaves the cavern is 50°C higher than the return temperature. There is probably a flow in cracks in the immediate vicinity of the cavern, which results in an extra heat loss. However, the flow paths must extend further away from the cavern in order to cause a substantial heat loss. But the flow resistance increases with the length of the crack flow paths. We must look for a closed flow path that extends sufficiently far away from the cavern and has a sufficiently low total flow resistance.

Figure 2 illustrates the tentativ explanation. A major part of the flow path is the tunnel used during construction of the cavern. The entrance of the tunnel to the cavern was thoroughly sealed by 1 m of concrete and refilled with soil for the next 9 m. However, there is an 'expansion' or by-pass duct with a diameter of 0.15 m from the bottom of the cavern to the open tunnel. The flow path through duct and tunnel is closed by a presumed crack system, probably from the top of the cavern out to the tunnel. There is secondary bouyancy flow in the tunnel above the cracks. The warm tunnel with the diameter of 5 m and the length of 300 m loses heat by heat conduction to the surrounding ground.

A cruical point in this hypothesis is that the crack system must be sufficiently open to let 2 m^3/hour pass with a driving water head somewhere below 0.5 m. According to Schlichling (1968) the water flow through a rectangular crack is proportional to the crack width cubed. For example a rectangular crack with the area 50 m, the width 1 mm and 0.5 m water head gives a water flow Q=2 m^3/hour. A few cracks with a width of 0.1 mm would give a flow that is 100-1000 times too small.

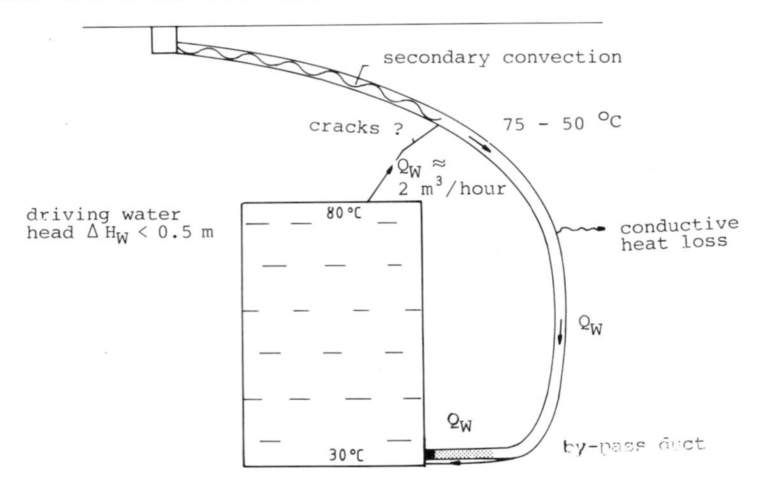

Fig. 2. Tentative explanation of the convective heat loss.

WATER TEMPERATURES

The explanation of the extra cavern heat losses discussed in the previous section suggests a convective flow path through the tunnel surrounding the cavern. A secondary buoyancy flow above the cracks should increase the water temperature in this part of the tunnel. This is indead also what has occurred. It has not been possible to registrate the temperature through the complete tunnel. But measurements have been made close to the tunnel entrance near the ground surface. In February 1987 this resulted in a temperature as high as 48°C. The expected temperature for the same month according to conductive loss calculations is close to the undisturbed ground temperature.

A computer model for rock cavern storage systems (Eftring, 1983), has been used to analyse losses and rock temperatures in the Lyckebo system. The model has been improved with possibilities to inject and extract heat at arbitrary water temperature levels. It is also possible to simulate a convective heat flow where warm water leaves the top of the cavern and colder water is returned at lower levels.

A detailed analysis of the storage heat balance was performed for the period Sep. 15, 1985 to Oct. 31, 1985. Figure 3 shows development of measured and calculated mean storage temperature with time during this period. The calculations was performed based on measured daily flows and temperatures of injected and extracted heat. Results with only conductive and with conductive and convective heat losses taken into account are shown. The measured storage water temperature profile and simulated rock temperatures on Sep. 15 were used as a starting point for all calculations.

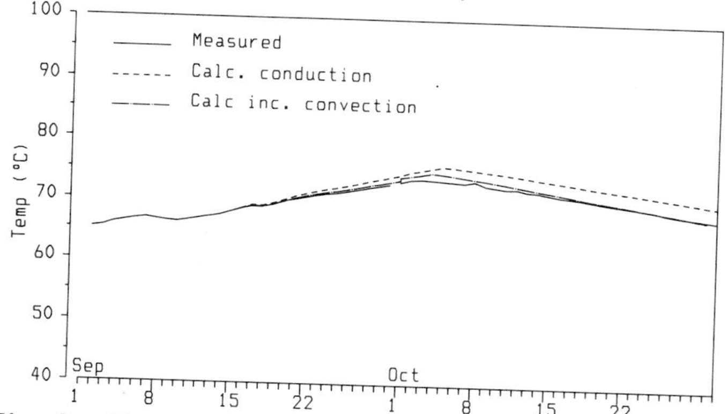

Fig. 3. Time development of measured and calculated mean storage temperature.

The difference between measured and conductive heat losses during the studied period was about 280 MWh. This represents at $\Delta T = 55^oC$, a convective heat flow of about 4 m^3/h, which was used during the calculations. The convective heat flow was assumed to leave at the top of the cavern and to return close to the bottom at a temperature of 28^oC. The annual difference between measured and conductive heat losses represent a mean convective flow of about 2 m^3/h. The difference varies, however, considerably with actual heat content of the storage which was relatively large during the studied period.The development of storage water temperature profiles between Sep. 15 and Oct. 15, 1985 as measured and as calculates with and without convective heat losses are shown in Fig. 4. The calculated storage temperature development, with convection according to the hypothesis taken into account, shows an excellent agreement with measurements.

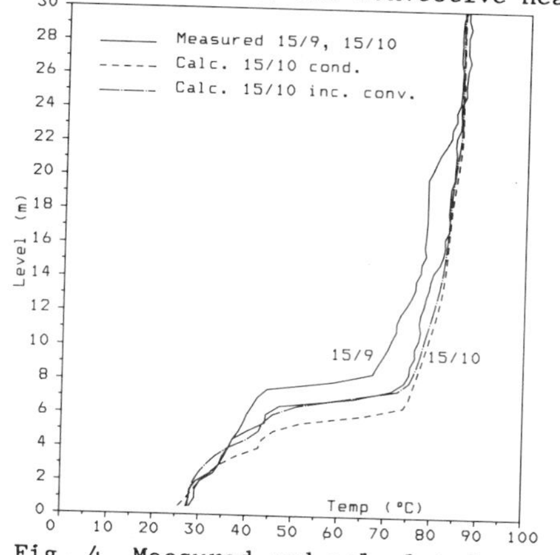

Fig. 4. Measured and calculated storage temperature profiles.

Measurements of rock temperatures have been performed during the spring 1987. Four bore-holes in the vicinity of the cavern have been used. One vertical bore-hole above the cavern top, one vertical through the pillar and finally two inclined bore-holes through the pillar. Registrations of deformations have been performed several times since 1983 in the two vertical bore-holes. No bore-hole for measurements is available in the region of primary interest for the convective loss mechanism discussed in this paper. Comparisons between measured and calculated rock temperatures indicate an increased temperature level in the pillar and especially in a region close to the top of the cavern. Very large rock deformations have been observed in the vertical bore-holes in a zone just above the cavern. These deformations have increased continuousely with time since 1983. The measurements show that a crack system exists close to the cavern top. It is quite conceivable that this system extends out to the tunnel.

CONCLUSIONS

The explanation for the high heat losses in the Lyckebo rock cavern storage suggested in this paper is supported by a comparison between measurements and computer simulations of storage heat losses and storage temperatures when convection is taken into account. It is shown that the measured storage temperature development can be reproduced by calculations if a convective heat flow where warm water leaves the top of the cavern and colder water is returned at lower levels is assumed. Indirectly the hypothesis is supported by high water temperatures observed close to the entrance of the tunnel surrounding the cavern. Furthermore, measurements of rock temperatures and deformations show that a crack system exists in a region just above the cavern top. This system might be the same system as presumed for the convective heat flow.

A final test of the presented hypothesis should be an accurate registration of the temperature in the tunnel system. Hopefully this can be realized during 1987. We are presently testing the hypothesis against all available field data. The convective heat loss should, if the presented explanation is the correct one, be stopped by blocking the flow path at any point. The simplest point is at the by-pass duct. A major diffuculty is the high temperture in the tunnel (over 50°C), which may prohibit the use of a diver. We are also considering the possibility of pumping water between the tunnel and the bottom of the cavern in such a way that the driving water head is counter-balanced.

REFERENCES

Brunström, C. (1986). Lyckebo central solar heating plant. Results of first year of operation. Alvkarleby Laboratory, UL-FUD-A 86:1.

Brunström, C., Merkell, A-L. and Larsson, M. (1986). The Lyckebo project - thermal performance of system and storage. Proc. North Sun '86, Copenhagen, Denmark.

Brunström, C., Larsson, M. and Hillström, C-G. (1986). The Lyckeby project - economy and measured performance. Alvkarleby Laboratory, UL-FUD-B 86:5.

Brunström, C. and Hillström, C-G. (1987). The Lyckebo project - results of system evaluation, final report. To be published.

Efring, B. (1983). Stratified Storage Temperature Model. Manual for Computer Code. University of Lund, Nov. 1983.

Schlichtling. (1968). Boandary Layer Theory, McGraw-Hill, pp 77.

MINSUN SIMULATION OF THE LYCKEBO PLANT

Håkan Walletun
Studsvik Energiteknik AB
S - 61182 Nykøping

ABSTRACT
This report describes the results obtained from simulation and economic optimization of the solar distric heating plant in Lyckebo, 13 km north of Uppsala, Sweden. The work has been performed using the computer program MINSUN, which was developed within the 'IEA Task VII of the Solar Heating and Cooling programme', ref. 1.

The work has been concentrated to questions concerning the optimal portion of solar energy compared with the price of energy, variations in climate from year to year, and different operating conditions. Another question which has been considered is the optimization of the collector surface for the solar plant.

The results show that the optimum solar energy fraction is 80-85 % with current (1986) energy prices. The optimal collector surface area and storage volume for the plant are approximately 25 000 m² and 100 000 m³ respectively. Further, the calculations have shown that a large propertion of the auxiliary energy can be reduced if the operational strategy can be improved, for example by reducing the return temperature of the distric heating system. The energy price for the optimized plant was then 0.40 SEK/kWh (6 US cent/kWh), calculated using 1986 cost level.

KEYWORDS
Solar energy; Seasonal storage; District heating; Simulation; Optimization; MINSUN; IEA Task VII.

1. INTRODUCTION

The solar heating plant in Lyckebo, 13 km north of Uppsala, Sweden, was built in order to develop a large scale district heating system mainly based on solar energy. The plant has been operational since the summer of 1983, and the first phase of the evaluation covers the period 1984-1985, ref. 2.

The plant currently supplies approximately 550 dwellings in a newly built area with energy for heating and water. On average the annual energy requirement is about 8 GWh. The system comprise a solar collector unit of about 4 320 m² flat plate highly efficiency solar collector and an underground cavern storage of about 100 000m³ for seasonal storage, Figure 1.

The solar collector area corresponds to 15 % of the area which would be necessary to supply the entire heating needs. The remainder of the solar collector area has been simulated by a 6 MW electric boiler, which can also be used for additional heating. A planned expansion of the unit supply 100 % of the demand is currently under discussion, and this work should be considered to be part of the preliminary study.

2. SIMULATION

2.1 Assumptions

The external assumptions for these system simulations using MINSUN are the technical and economic performance characteristics and actual data. The technical performance characteristics and known parameters have been used in the set-up which describes the entire district heating plant. This includes everything from insulation thickness on culverts to the thermal conductivity of the rock and annual heating period.

The economic input comprises known facts from the plant performance as well as normal depreciation, interest rates and suchlike.

2.2 Parameter variation

Three different cases have been studied.
Case 1 has considered which effects are to be expected under different climatic conditions. As realistic a set-up as possible has been assigned to the system and has been used for all of the simulations in Case 1. The different climatic conditions used were Lyckebo 1984, Lyckebo 1985 and Copenhagen 1978. The first two are the results from measurements carried out in Lyckebo whereas the Copenahgen climate ia a reference which has often been used in IEA Task VII. 1985 was a relatively normal year; 1984 was mild but with unusually little sunshine. The Copenhagen climate is typical for Southern Scandinavia, see Table 1 and Figure 2.

The results of the simulation give good correspondence with regard to the actual heating requirements and the energy stored underground in the cavern. Some of the simulation results are shown in Table 2 and Figure 3. Despite a fairly large variation in the climate the optimal solar fraction for the Lyckebo plant is 80-85 %, whereas about 90 % is optimal for

Copenhagen.

The aim with Case 2 was primarily to determine the optimal collector area for an expansion of the existing plant. The Lyckebo 1985 climate was used with a normal heat load of 8.2 GWh. The cost assumed for solar collectors was 1200 SEK/m^2 (approx. 180 US$/$m^2$). The results show that the optimal collector area is about 25 000 m^2, se Figure 4. The plant could then produce heat at a price of about 0.40 SEK/kWh (approx. 6 US cent/kWh) in 1986 figures.

Case 3 has studied different ways of changing the operational conditions and thus attempt to improve the economics and energy production. The Lyckebo 1985 climate has been used with an extremely high load: about 9.1 GWh. The high load was used in an attempt to exaggerate the effects of changes in operating strategy.

It was found that the most interesting measure is to try and reduce the return temperature in the district heating system, see Figure 5. Four different calculations have been performed. The first three show the proportions of energy from storage and auxiliary production if the return temperature is reduce from 45 to 41°C and further to 37°C . The need for extra production of energy is reduced from 14 % to 9 %. The fourth calculation has been performed for comparison and describes a normal district heating system with higher temperatures than those used in Lyckebo. It should be noted that the proportion of auxiliary produced energy required can be halved by improving the situation.

3. CONCLUSIONS

Despite the fact that the Swedish climate is very variable the optimal solar fraction for this type of plant has been shown to be about 80-85%, at current energy prices.

The optimal collector area is about 25 000 m^2 for an optimal storage capacity of about 100 000 m^3, assuming a collector cost of 1 200 SEK/m^2 (approx. 180 US$/$m^2$).

Optimal operational strategy (primarily reducing the return temperature in the district heating network) can reduce the propertion of auxiliary produced energy required by up to 50 % .

The resultant energy price for the optimized plant is about 0.40 SEK/kWh (approx. 6 US cent/kWh).

The simulated results are in good agreement with the measured values.

The MINSUN program is very useful for such studies.

REFERENCES

1. V Chant, J Hickling and R Håkansson
The MINSUN simulation and optimization program application and users guide, September 1985.
Document # CENSOL2
National Research Council, Ottawa, Canada.

2. H Walletun, P Holst and H Zinko
Performance and operating results for the solar district heating plant at Lyckebo, Uppsala, Sweden.
STUDSVIK Report EI-85/46 (1985).
Studsvik Energy, 611 82 Nyköping, Sweden.

TABLES

Table 1.	Lyckebo 1984	Lyckebo 1985	Copenhag 1978
Global horizontal irradiation (kWh/m^2)	798	874	1018
Yealy mean temperature (^{o}C)	8.2	5.4	8.1

Table 2.	Lyckebo 1984	Lyckebo 1985	Copenhag 1978
Collector supply (MWh)	8585	9866	10315
System load (MWh)	7418	9060	7198
Collector supply/System load	116%	109%	143%

FIGURES

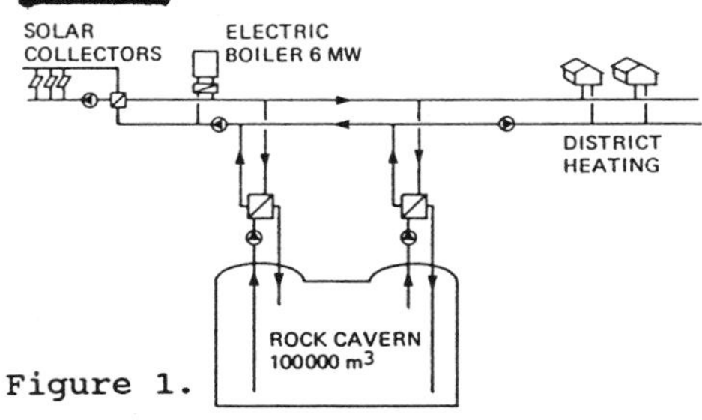

Figure 1.

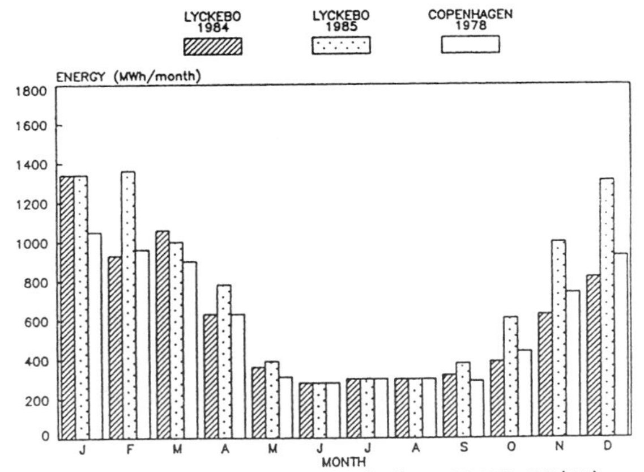

Figure 2.

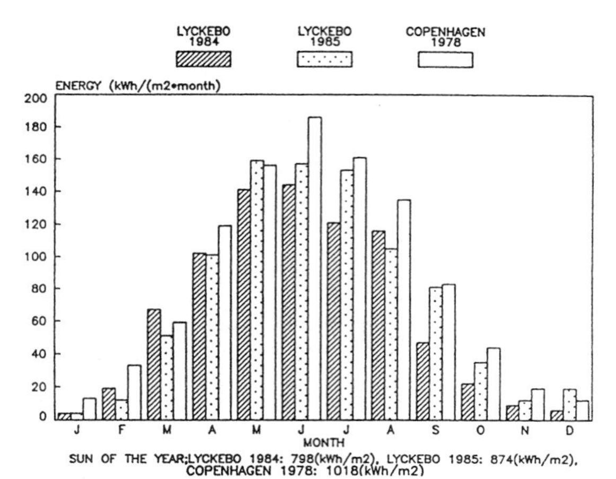

Figure 3.

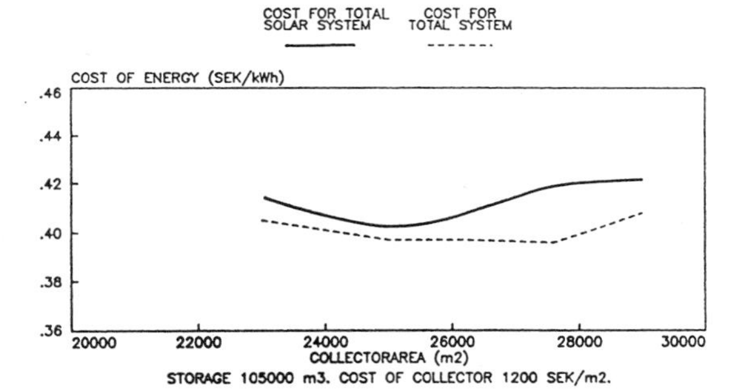

Figure 4.

LYCKEBO 1985

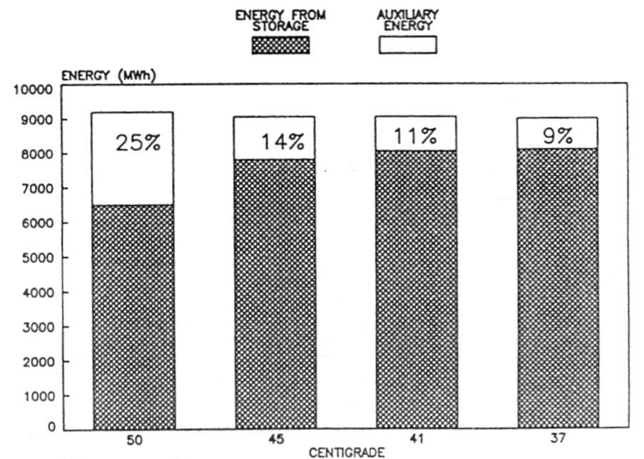

Figure 5. RETURN TEMPERATURE IN DHW-SYSTEM

THREE-DIMENSIONAL FINITE ELEMENT MODEL FOR OPTIMIZATION OF SEASONAL HEAT STORAGE IN BURIED TANKS

M. Alabiso*, V. Brignoli*, L. Castellano**,
F. Parrini* and R. Viadana*

* ENEL (Italian Electricity Generating Board) C.R.T.N., Milan, Italy
** MATEC S.r.l., Milan, Italy

ABSTRACT

This paper deals with the Computer Code SERBIN developed at ENEL-CRTN to analyze the thermal performance of seasonal heat storage in a buried tank under various operating conditions. The presented results are related to an interseasonal combined solar-heat pump system used for the space conditioning of an ENEL commercial building in Turin. The final release of the computer code will consent to solve the complete set of mass, momentum and energy conservation equations, in the framework of the FEM. At present only the solution of the heat transfer problem has been implemented. The effects of the natural and forced convection are simulated by suitably changing the positions of the inlet and outlet flow. In spite of this limitation, reasonable suggestions have been obtained to evaluate, from the economic point of view, the optimum ground insulation of the storage tank.

KEYWORDS

Solar heating systems; computer simulation; mathematical models; three-dimensional calculations; energy storage; heat losses; energy conservation; thermal storage tank.

INTRODUCTION

Many solarized plants have been accomplished and monitorized by ENEL-CRTN to study the rational exploitation of the solar energy for space conditioning purposes, /1/, /2/. This paper deals with an interseasonal solar-heat pump system planned for a commercial building in Turin (North Italy), Fig. 1. The energy exchanges between the storage tank and the other components are reported in Fig. 2. The tank consists of two twin cylinders connected in the shape of the number 8. The characteristics of the basic storage unit are shown in Fig. 3 and Table 1.

MATHEMATICAL MODEL

Owing to the effects of both natural and forced convection phenomena, the distributions of temperature and velocity in the tank are strictly interdipendent. For the complete version of the Computer Code SERBIN, the development of two different ways of proceeding is foreseen. The first one requires the numerical solution of the Navier-Stokes equations, written according to the Boussinesq approximation, /3/. In the second one, the velocity field is simply computed by interpolation of the available experimental data in such a way as to satisfy the continuity equation. At present, the energy equation is written as:

$$\rho C_p \frac{\partial T}{\partial t} = \frac{\partial}{\partial x}\left(K_x \frac{\partial T}{\partial x}\right) + \frac{\partial}{\partial y}\left(K_y \frac{\partial T}{\partial y}\right) + \frac{\partial}{\partial z}\left(K_z \frac{\partial T}{\partial z}\right) + Q \qquad (1)$$

where the convection terms have been dropped out and C_p indicates specific heat, (K_x, K_y, K_z) thermal conductivity, t time, T temperature, (x,y,z) cartesian co-ordinates, ρ density. The source term Q is the algebric sum of the following contributions: heat supplied from solar collectors, heat supplied from air conditioning, heat directly removed for space heating and heat pump. The numerical model is based on the use of the FEM according to the Bubnov-Galerkin formulation, /4/. The implemented boundary conditions can be summarized as the usual form, /4/:

$$K_x \frac{\partial T}{\partial x} n_x + K_y \frac{\partial T}{\partial y} n_y + K_z \frac{\partial T}{\partial z} n_z + \Phi + \alpha (T-T_{ref}) = 0 \qquad (2)$$

$$T(\underline{x},t) = \text{prescrived (time dependent) temperature} \qquad (3)$$

when Φ indicates the prescrived heat flux and α the convective heat transfer coefficient. The discrete model consists of 20-nodes (finite) elements. The elements where the source term Q is supposed to be different from zero can change in time in order to simulate, to a certain extent, the effect of the mixing due to the convective motion.

RESULTS AND CONCLUSIONS

A portion of the discrete model is shown in Fig. 4. The overall model consists of 273 elements and 1674 nodes. Figs. 5a-6b and Table 2 summarize the main results of a preliminary analysis of the heat storage in the cylindric tank described in Fig. 3 and Table 1. The labels (a) and (b) are used to distinguish the case (a) in which the bottom is completely coated with a layer of insulating material from the case (b) in which the bottom is assumed without any insulation. Apart from the obvious conclusion that the tank works better in the case a), the following remarks seem to be particularly important:

- the heat loss at the top wall is almost independent from the insulation of the bottom;
- the heat loss at the lateral wall in the case a) is greather than in the case b): this is a consequence of the greather temperature difference between the soil and the water in the tank;
- in spite of the fact that the position of the source term has been changed, during the computation, in order to avoid instable configurations, Figs 5a - 5b show that after a certain time the onset of

the natural convection is fated: this is a consequence of the boundary conditions at the top wall.

Two main conclusions can be drawn:

- the heat conduction model is an useful tool for the qualitative analysis of the behaviour of the system;
- the correct evaluation of the efficiency of the storage process requires a detailed analysis of the effects of both the forced and natural convection.

AKNOWLEDGMENTS

The authors wish to thank D. Borgese, head of ENEL-CRTN, for his helpful suggestions and his continued encourangements of this work.

REFERENCES

/1/ Parrini, F. et al. (1984). Description and experimental long range evaluation of a solar electric plant installed in a residential building in Southern Italy. Proceedings of the First EC-Conference on Solar Heating, Amsterdam.

/2/ Paganuzzi, A. et al. (1982). Studio di un impianto ad energia solare e pompa di calore per la climatizzazione della nuova sede di Zona Torino Nord. ENEL-DSR-CRTN, G16/82/02.

/3/ Zienkiwicz, O.C. (1977). The Finite Element Method. McGraw-Hill.

/4/ Gartling, D. K., and E.B. Becker. (1976). The Finite Element Analysis of Viscous, Incompressible Fluid Flow. Comp. Math. Appl. Mech. Engeen., 8, pp. 51-138.

TABLE 1. Characteristics of Heat Storage Unit.

- Max storage design temperature		70	°C
- Tank dimensions	- Depth	5.6	m
	- Radius	5.12	m
- Insulation thickness	- Walls	0.2	m
	- Top	0.3	m
	- Bottom	0.2	m
- Concrete thickness	- Walls	0.45	m
	- Top	0.35	m
	- Bottom	0.5	m
- Insulation thermal conductivity	- Walls	$3.3 \cdot 10^{-2}$	$W \cdot m^{-1} \cdot °C^{-1}$
	- Top	$3.68 \cdot 10^{-2}$	$W \cdot m^{-1} \cdot °C^{-1}$
	- Bottom	$3.68 \cdot 10^{-2}$	$W \cdot m^{-1} \cdot °C^{-1}$
- Concrete thermal conductivity	- Walls	0.5	$W \cdot m^{-1} \cdot °C^{-1}$
	- Top	0.5	$W \cdot m^{-1} \cdot °C^{-1}$
	- Bottom	0.7	$W \cdot m^{-1} \cdot °C^{-1}$
- Soil thermal conductivity		1	$W \cdot m^{-1} \cdot °C^{-1}$

TABLE 2. Heat Balance After a Year Simulation.

	Total energy storaged into the tank (kW·day)	Total energy lost through boundary (kW·day)	Total energy supplied into the tank (kW·day)	Average temperature of tank (°C)
Tank completely insulated	1301.7 (E_0=452.34)	-574.5	1424.2	34.5
Tank without insulation on the bottom	1193.5 (E_0=452.34)	-683.4	1424.2	31.7

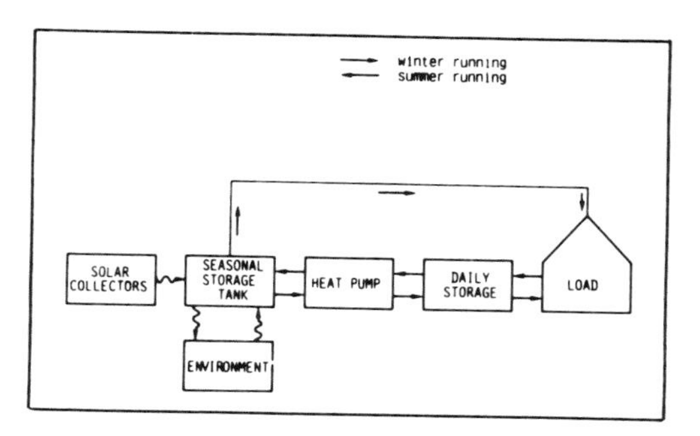

Fig. 1. Scheme of the plant.

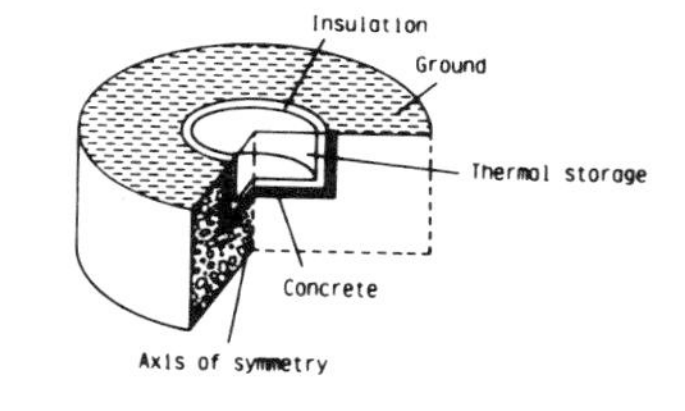

Fig. 3. Schematic rapresentation of the tank.

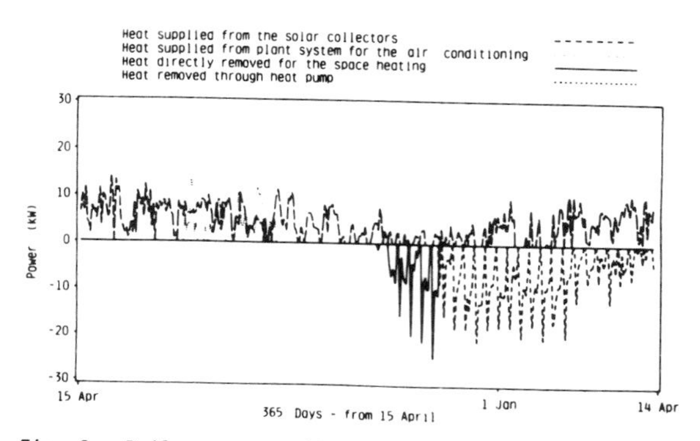

Fig. 2. Daily energy collection and energy extraction.

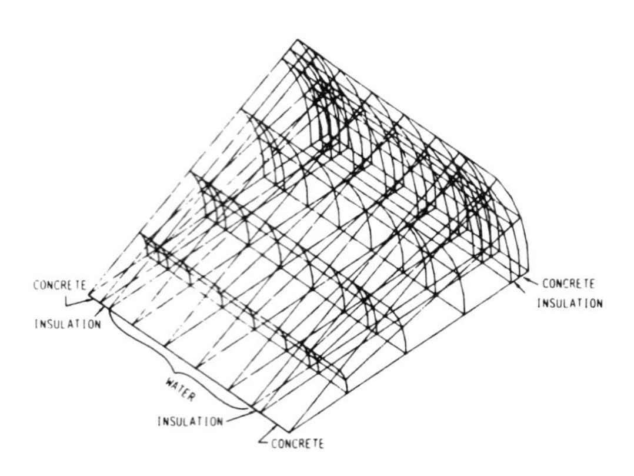

Fig. 4. Portion of the discrete FEM model of the tank with water.

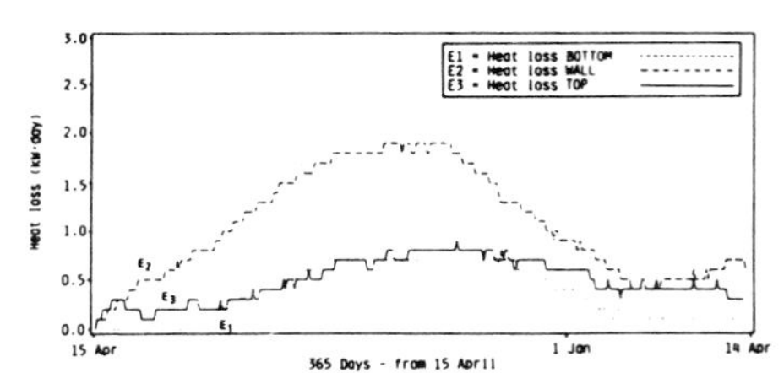

Fig. 5a. Heat losses for tank completely insulated.

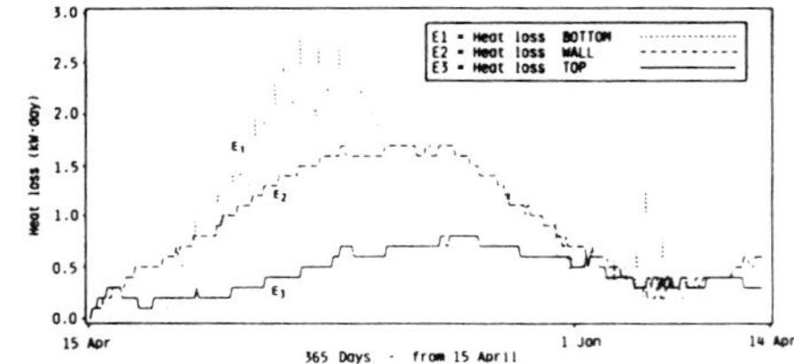

Fig. 5b. Heat losses for tank without insulation on the bottom.

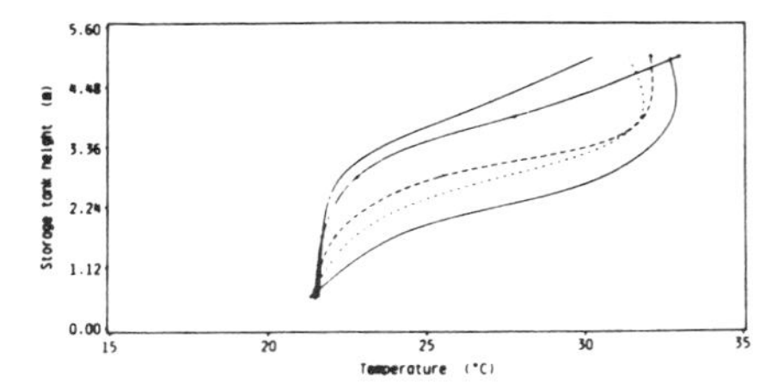

Fig. 6a. Vertical distribution of the horizontally averaged temperature in tank completely insulation at different time (5 days interval).

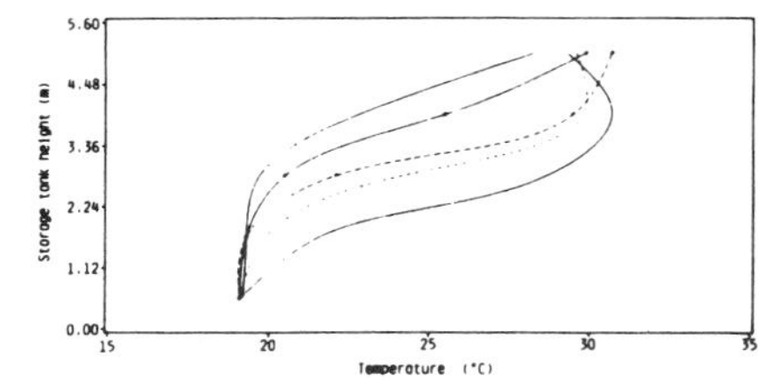

Fig. 6b. Vertical distribution of the horizontally averaged temperature in tank without insulation on the bottom at different time (5 days interval).

A HYBRID THERMAL ENERGY STORAGE DESIGN

P.D. Lund

Helsinki University of Technology
Department of Technical Physics
SF-02150 Espoo 15, Finland

ABSTRACT

The thermal performance of the seasonal heat energy storage system in Kerava solar village is discussed. The hybrid thermal energy storage employed in Kerava comprises a subsurface water storage surrounded by vertical heat exchanger pipes in the rock. The measured yearly heat recovery efficiency of the storage system has been of the order of 80%. To increase the capacity for storage of summertime solar heat, an improved storage operational strategy has been suggested which could increase the yearly solar fraction by 6-9%-units.

KEYWORDS

Seasonal storage solar heating; hybrid thermal energy storage; system simulation; CSHPSS

INTRODUCTION

Large-scale exploitation of new heat sources at northern latitudes often necessitate long-term thermal energy storage. In solar heating systems, a high solar fraction requirement is not obtainable without seasonal storage. Most of the potential seasonal storage concept rely on subsurface techniques such as aquifers, pits, caverns, or pipes in the ground.

A most interesting recent storage concept is the so-called hybrid thermal energy storage in which the seasonal and diurnal storage functions are accomplished by different storage techniques. A schematic of a hybrid heat storage is shown in Fig. 1 (Lund, 1985). In the center of the storage system, there is a rapid-response water storage working as a buffer between the storage system and heat production/consumption. The seasonal storage function is mainly accomplished by vertical pipes in the ground. The pipe storage has in general a very slow response, but the cost per cu.m. is low.

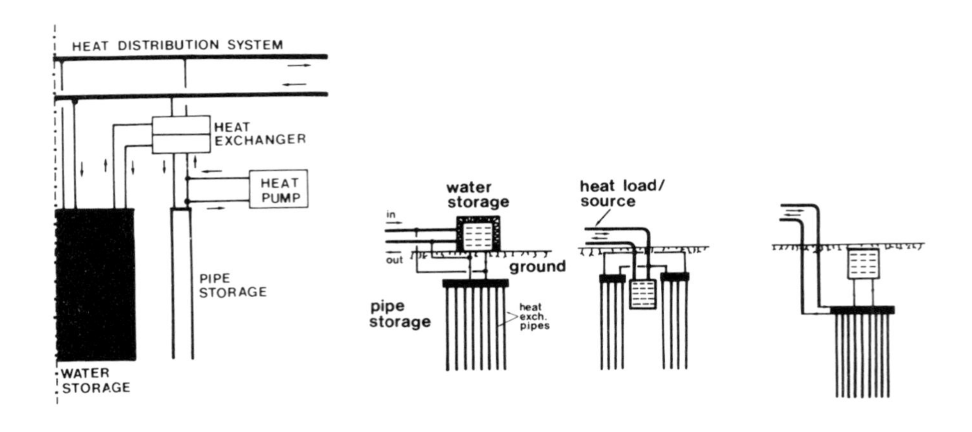

Fig.1. Principle of a hybrid thermal energy storage having both a diurnal and seasonal function.

The pipe storage may also operate at high temperatures without a heat pump when constructed into stable bedrock. The short-term buffer tank may be an ordinary on-ground tank, or alternatively, a subsurface construction. For subsurface applications, storage insulation would in most cases be unnecessary.

At Kerava solar village near Helsinki, Finland (60°N), where the northmost seasonal storage solar heating experiment in the world has been going on since 1983, a variant of the hybrid thermal storage is being tested and monitored.

This paper shortly describes the experiences obtained so far with the storage subsystem in Kerava solar village. Based on these findings, improved storage control strategies are discussed to increase the overall system performance.

KERAVA STORAGE SYSTEM

Kerava solar village (KSV) is a full-scale central solar heating system with seasonal storage (CSHPSS) and employs also a heat pump. The 44 residences have a yearly load of 550 MWh and the total solar collector area is 1,100 m^2. A full description of the system principle is given by Lund (1987).

A schematic of the storage configuration is shown in Fig. 2. The storage comprises a 1,500 m^3 rock water pit surrounded by a conical vertical pipe storage in rock with a volume of 11,000 m^3. The pipes are located on two circles: the innermost have 18 pipes and the outer 36 each 25 m in length. Both storage types have here a seasonal storage function -usually the buried water storage would only serve as a buffer tank to compensate the large inertia of the pipe storage in ground.

The boreholes (heat exchanger pipes) in Kerava have been employed so far only passively, i.e. the pipes have recovered heat losses from the water storage through the heat pump. Active charging of the rock has not taken place. Originally only the outer borehole circle was intended to be used for discharging and the inner for charging; in 1984 the circles were connected in series during discharge to improve the heat recovery.

The main findings of the measurements are shown in Table 1.

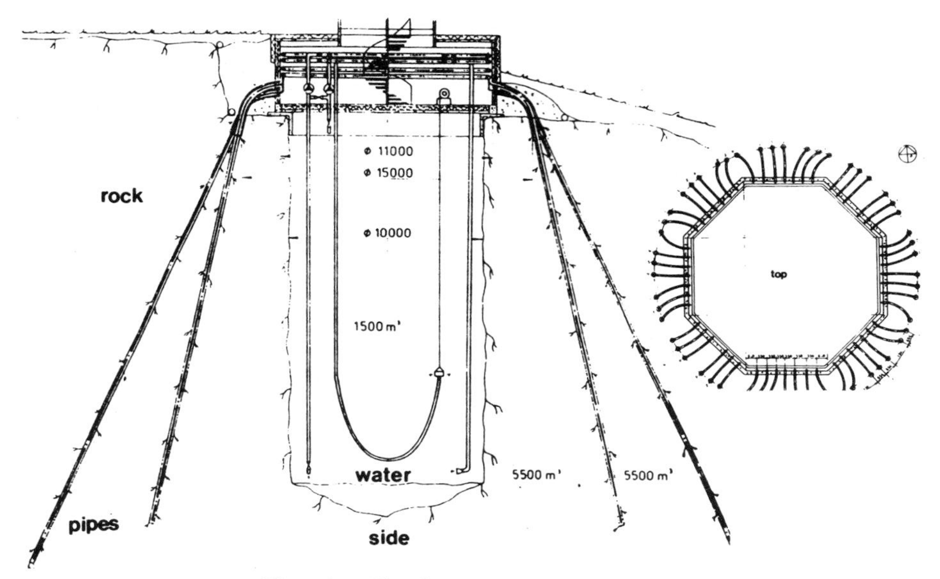

Fig. 2. The Kerava storage system.

The pipes have in average recovered about 25% of the heat losses from the water storage. This figure could be still higher, because malfunctions in the heat pump's operation has occasionally stopped the discharge. The power effect of the pipes has been in average about 20 W/m, or, 4-5 W/Km.

Even though the present storage configuration in the light of the long-term measurements has worked well, its capacity has not been adequate to store all the summer-time solar heat available from the Kerava collectors. In practice, the water storage which is the primary storage medium becomes full already in late May leading to a major dissipation of summertime solar heat. Therefore, it is of large interest to investigate the use of the pipes for charging the rock thus increasing the overall useful storage capacity. In later chapters, the effect of storing the access heat in the rock is being assessed in more detail. Actually, the storage has been recently revised so that the active charging is already possible.

MATHEMATICAL ANALYSIS

Transient computer simulations were accomplished to study the performance of the KSV hybrid storage in greater detail as also to assess the effects

TABLE 1 Measured Storage Performance in Kerava Solar Village

Year	Water storage losses (MWh)	Water storage η_s without pipes (%)	Recovered by pipes (MWh)
1983+)	140	63	3
1984	134	80	51
1985	179	76	60

+) June-December

of using the pipes both for heat injection and extraction. It is worth noting that there also exists a certain thermal interaction between the pipes and the water storage, e.g. if heat extraction through the pipes is high then the losses of the water storage could increase.

Model

The analyses are based on a semi-numerical model in which the pipe response is described by an analytical expression (Lund, 1987):

$$\phi(t) = F_p \, \Delta T \, e^{-\tau(t)} \qquad (1)$$

where F_p = flow factor,
ΔT = temperature factor,
$\tau(t)$ = time factor.

The macroscopic behaviour of the ground is given by the heat conduction equation which is solved numerically by the FDM-method:

$$\vec{\nabla}^2 T_g - \frac{1}{\alpha}\frac{\partial T_g}{\partial t} = -\frac{1}{\lambda} q'(\vec{r},t) \qquad (2)$$

where the source term q' is given by eqn(1).

The temperature of the water storage is obtained from a balance equation, or alternatively, described by a periodic function.

The mathematical model for the hybrid storage is incorporated in the Fortran program HYBTESS also available for a IBM PC (Lund, 1987):

Simulations

The simulations cover the most interesting cases encountered in a hybrid thermal storage and the results are shown in Table 2. The figures for the pipe storage represent maximum energy values which cannot be exceeded by the present system configuration due to physical limitations of the rock storage.

TABLE 2 Results of Simulations

Case	Losses from water storage (MWh)	Injection (pipes) (MWh)	Extraction (pipes) (MWh)	Total benefit (MWh)
1. Measured (8/84-8/85)	161	-	60	
2. No injection Extraction both circles	169	-	62	
3. Injection inner circle Extraction both circles $\Delta T=0.5^oC$+)	134	88	86	+53
4. Injection inner circle Extraction both circles $\Delta T=5^oC$	127	83	57	+34
5. Injection both circles Extraction both circles $\Delta T=2^oC$	133	204	163	+131
6. Injection both circles Extraction both circles $\Delta T=5^oC$	128	194	130	+104

+) heat extracted only if fluid temperature minus local ground temperature around the pipes is less than ΔT

Firstly, the model was tested against measured performance between a period from 8/1984-8/85. During this period system operation was in general very stable and thus suitable for model verification. The agreement between simulated and measured energy flows (cf. measured and case 2 in Table 2) is satisfactory. The water storage temperature is here well described by a sinus curve ($T_{max}=48^oC$, $T_{min}=17^oC$).

It is estimated that during the summertime some 90 MWh of solar heat (=30-35% of all collected solar energy) is presently lost in KSV due to a limited water storage capasity. Actually, this is almost exactly the amount of injected energy that case 3 and 4 represent and thus the overall benefit of the active use of the pipes is given by these figures. The net benefit thus obtained would rise the yearly solar fraction to 27-31% compared to the present 21%. Combined with a more optimal overall system control, a 45% solar fraction could be obtained in KSV.

REFERENCES

Lund, P.D. (1987). Fundamentals of thermal processes in a hybrid thermal energy storage. Helsinki University of Technology. Report TKK-F-A609 (1987).

Lund, P.D. and M.B. Östman (1985). A numerical model for seasonal storage of solar heat in the ground by vertical pipes. Solar Energy 34,351-366.

Solar Assisted Low Temperature Heating System with Seasonal Storage

M. Hornberger, N. Fisch, E. Hahne

Institut für Thermodynamik und Wärmetechnik
Universität Stuttgart
Pfaffenwaldring 6
D-7000 Stuttgart 80

ABSTRACT

Results of the first operational year of a solar assisted low temperature heating system with a heat pump and a seasonal store are presented. Simulations with the computer code MINSUN show good agreement with measured values. Parametric studies with variations of store volume and absorber area have been performed.

KEYWORDS

Seasonal storage, solar heating system, low temperature heating system, heat pump system, absorbers, system simulation.

1. INTRODUCTION

The University of Stuttgart solar heating plant is the first solar assisted heating system with seasonal storage in the FRG. Since April 1986, it has been in operation with a gravel-water long term store (Fisch, Giebe, Hornberger, 1987). The investigation of the overall system performance as well as the investigation of the performance of individual components in the system operation are of interest. Simulations of the first operational year have been done with the computer code "MINSUN". In this report results of the first operational year as well as results of MINSUN Simulations are presented.

2. SYSTEM DESCRIPTION

The scheme of the overall system is shown in Fig. 1. It consists of:

- 211 m^2 absorbers
- electric heat pump (66 kWh_{th})
- 1050 m^3 long-term heat store
- low temperature distribution system (50/40°C)

- cogeneration plant at the University of Stuttgart

Three different absorber types are installed on four absorber sheds. Heat from the absorbers can be delivered either into the long term store or directly to the evaporator of the heat pump. To cover the load, heat can be discharged by the heat pump or directly via three different heat exchanger systems. The one used in the first operational year is a horizontal coil system (Fisch, Giebe, Hornberger, 1987). Heat from the heat pump is stored in a buffer store (1 m^3), from where it is supplied to the low temperature distribution system which heats the eastern part of an office building. Additional energy can be delivered from the cogeneration plant.

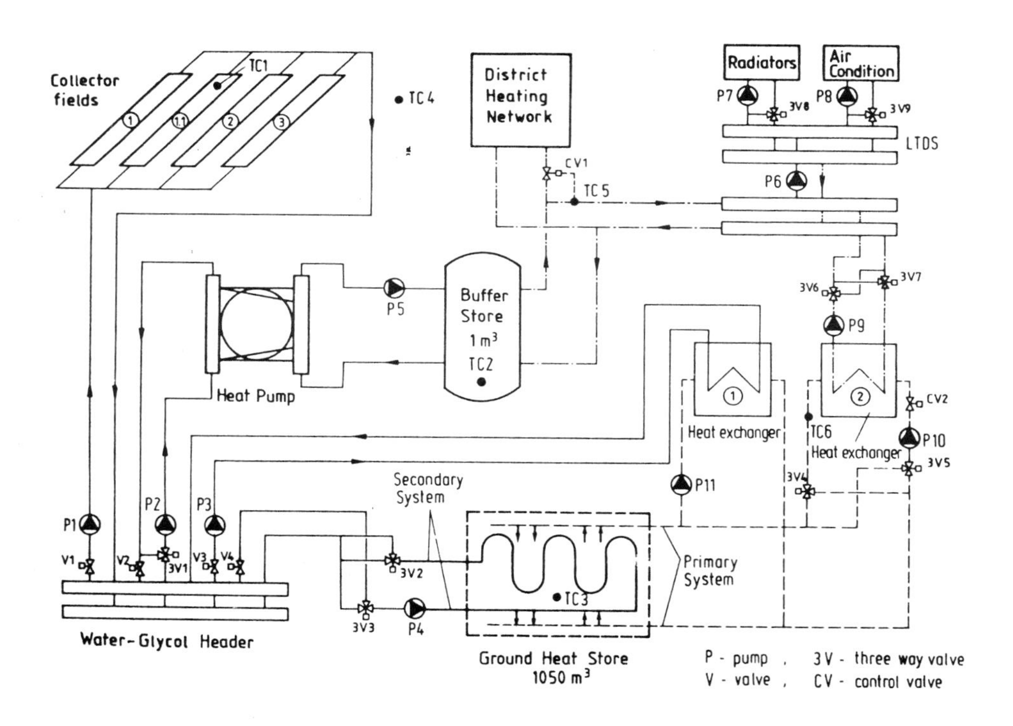

Fig. 1. Scheme of the solar heating system.

2.1 Control Strategy

Several programs are available to control the system operation. From April 1986 to April 1987 a program with the following strategy was used: If there is a heat demand from the office building, heat is delivered by the heat source with the higher temperature (either absorber field or long term store). If no heat is necessary, the long term store is charged by the absorber field as long as the absorber temperature (TC1, see Fig.1) is about 7 K higher than the store temperature (TC3). Heat from the cogeneration plant is supplied if the heat pump can not cover the load by itself.

3. RESULTS OF THE FIRST OPERATIONAL YEAR

The results of the system operation (18.4.1986 - 17.4.1987) are shown in Fig. 2. Of the incident solar energy 32% has been used by the absorbers, 10.6% of the used heat has been supplied directly to the heat pump's evaporator, and 89.4% has been charged into the long term store. Only 8% of the building's heat consumption had to be delivered by the cogeneration plant. The thermal performance factors of the solar heating system are:

absorber field effectiveness	0.32
seasonal performance factor of the heat pump	2.76
storage effectiveness	0.82
solar fraction	0.6

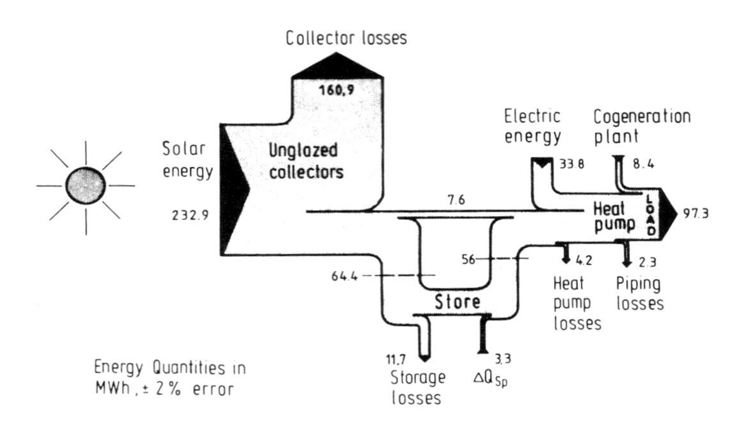

Fig. 2. Heat balance of the first operational year for the overall heating system.

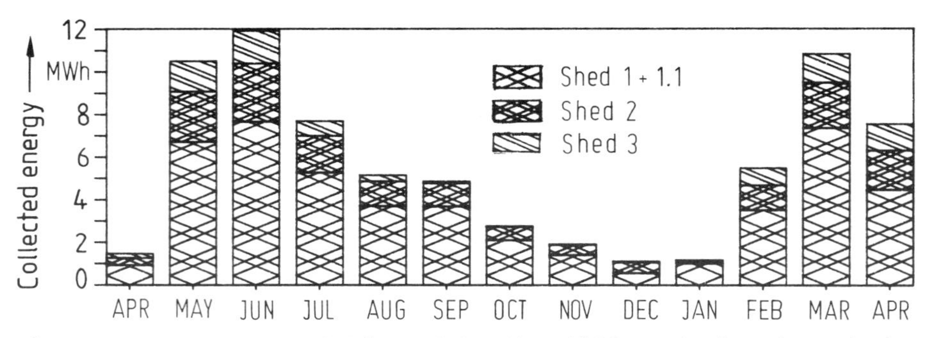

Fig. 3. Monthly heat delivered by the different absorber sheds.

Fig. 3 shows the monthly heat delivered by the absorber sheds. Shed 1 and shed 1.1 had an effectiveness of about 39%, shed 2 and shed 3 of 23.9% and 19.3% respectively. Shed 2 and shed 3 are constructed for a convective use of heat from the environment. As there was only in February and March a reasonably long time period with absorber fluid temperatures below ambient temperature, these two absorber types had a low effectiveness during the whole year. Considerable quantities of used heat at low solar irradiation ($E_{glob,K}$ < 100 W/m^2) have only been obtained in February and March (see Fig. 4). Detailed results of the thermal performance of the seasonal store are described by Giebe and Hahne (1987), those of the heat pump and the heating load by Fisch and Hornberger (1987).

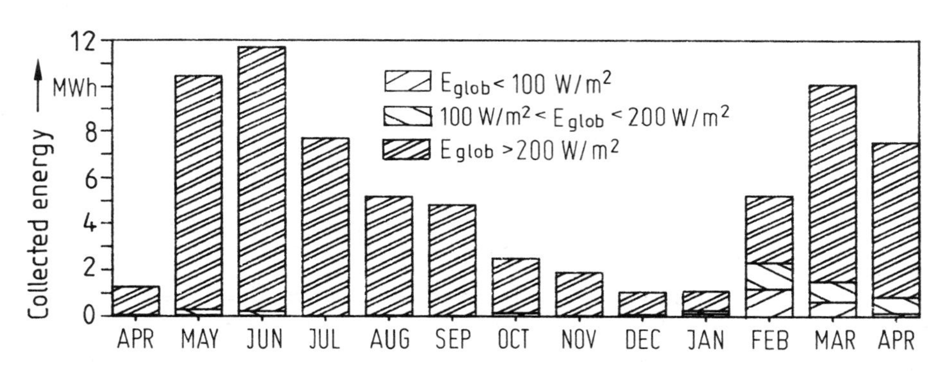

Fig. 4. Monthly heat delivered by the entire absorber field for different solar irradiation.

4. MINSUN SIMULATION RESULTS

4.1 Results from the first Operational Year

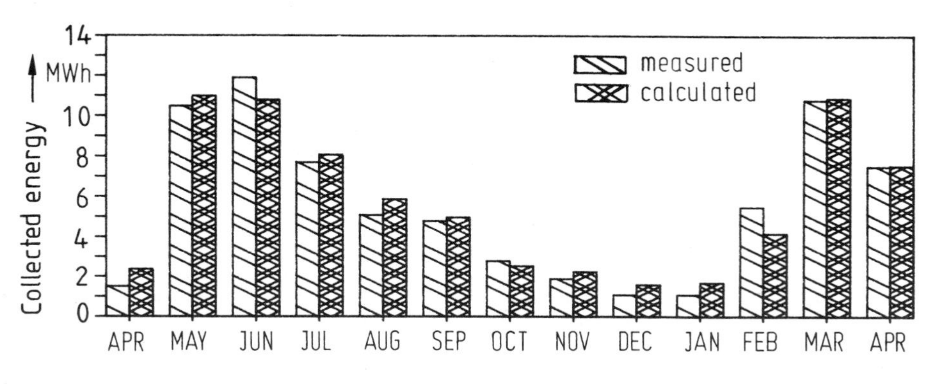

a: Solar energy collected.

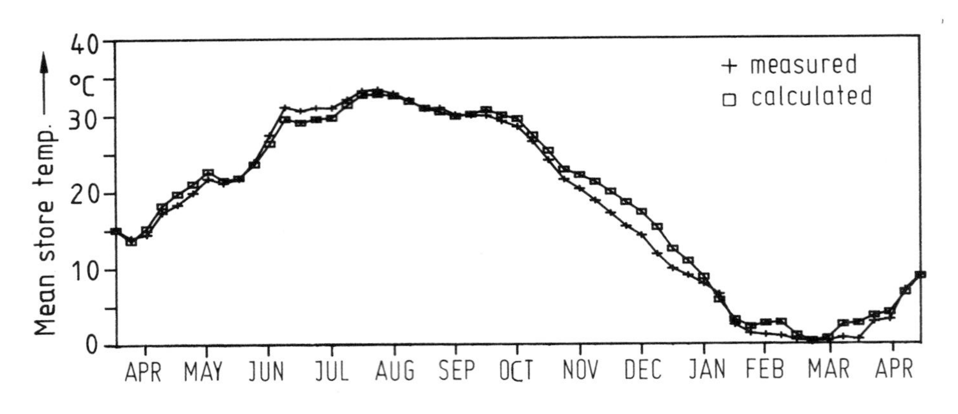

b: Mean store temperature.

Fig. 5. Comparison between monthly values of the MINSUN simulation results and the measured performance.

With measured weather data from Stuttgart, a MINSUN simulation has been performed for the first operational year. Comparisons between measured and calculated results are shown in Fig. 5a and 5b. Good agreement between measured and calculated performance is obtained, and the calculated mean store temperature differs only by 2 K from the measured value. Details about theoretical and experimental results are given by Fisch and Hornberger (1987).

4.2 Parametric Studies

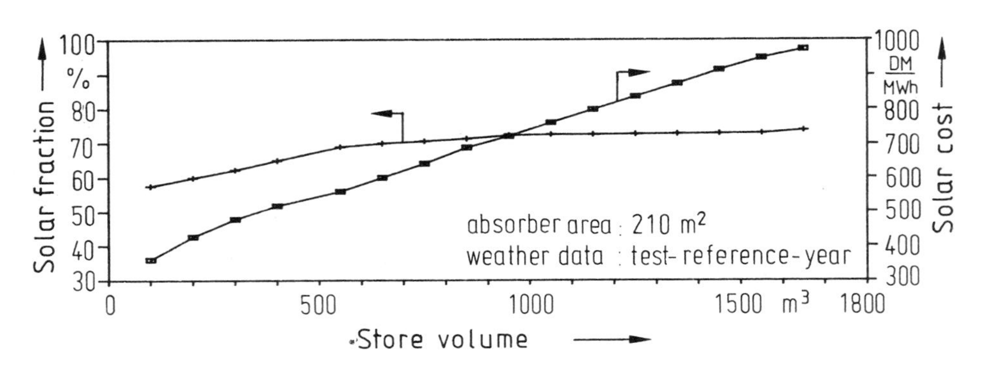

Fig. 6. Effects of store volume on solar cost and solar fracion.

The effects of store volume and absorber area on solar cost and solar fraction were investigated in a parametric study (see Fig. 6 and 7). A store volume of more then 800 m^3 as well as an absorber area of more then 250 m^2 results only in a large increase of solar costs, while the solar fraction remains nearly constant.

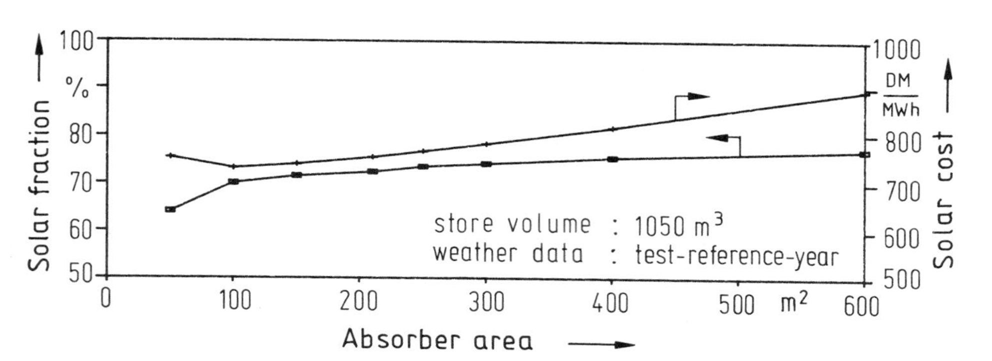

Fig. 7. Effects of absorber area on solar cost and solar fraction.

5. ACKNOWLEDGEMENTS

The investigations have been financed by the Federal Ministry of Technology (BMFT) under 03E 8187A; the authors gratefully acknowledge this support.

6. REFERENCES

Fisch, N., R. Giebe, and M. Hornberger (1987). National Report 1986, prepared for the IEA-Task VII June 1987 Meeting in Helsinki. Universität Stuttgart.

Fisch,N., and M. Hornberger (1987). Evaluation Report of the Stuttgart University Project, prepared for the IEA-Task VII June 1987 Meeting in Helsinky. Universität Stuttgart.

Giebe,R., N.Fisch, and E. Hahne (1987). Operational results of a seasonal heat storage project.ISES Solar World Congress, Hamburg 1987, Proceedings.

OPERATIONAL RESULTS OF A SEASONAL HEAT STORE

R.Giebe ,N.Fisch ,E.Hahne

Institut für Thermodynamik und Wärmetechnik
Universität Stuttgart

ABSTRACT

Results after one year of operation of an artificial (man-made) long term store (V=1050 m^3) used with a low temperature heating system are presented. The maximum average store-temperature, reached in summer, was 33°C, the lowest in winter was 0°C. During the first long term cycle , 82% of the stored heat was used for heating. The chemical and biological quality of the storage water has been monitored from the beginning of the storage operation. The results of the analyses show no alarming changes in water quality.

KEYWORDS

Artificial storage tank; long term cycle; monthly heat balance; temperature distribution; storage losses; water quality.

INTRODUCTION

The aquifer storage pilot plant at the University of Stuttgart is the first artificial (man-made) seasonal heat storage project in the FRG. After one year of construction, the store has been in operation since December 1985. First loading and unloading tests with heat from the university power-plant were performed to examine its correct function and to investigate short-term thermal behaviour of the store (Fisch and Giebe, 1986; Giebe, Kübler and others 1986). During two long-term cycles (starting in April 1986, ending in April 1988) the thermal behaviour of the store and the surrounding soil as well as water quality will be studied. The store is a part of the heating system for an office building. Unglazed solar collectors (absorbers) of 211 m^2 deliver heat to the store which supplies an electric heat pump (66 kW_{th}). Hornberger, Fisch and Hahne (1987) are presenting the results of the system-performance. The results of the first long-term cycle (Apr.1986-Apr.1987) are presented in this paper.

STORE DESIGN

The store has the shape of a truncated cone with a volume of 1050 m³; it is filled with gravel of 8-32 mm grain size and 37% porosity and flooded with water. A 2.5 mm thick high-density polyethylene liner prevents water leakage. The store is insulated only at the top with a 0.9 m thick layer of lava stone.
In this pilot-project, various loading and unloading systems will be investigated : two systems for direct water exchange will be used for short-term experiments and a system of coiled polyethylene pipes for indirect loading and unloading will be used for the long-term cycles. The design of the store is presented in detail by Fisch and Giebe (1986).

RESEARCH PROGRAM

The aim of the research program is the investigation of two heat storage applications used in practice : a) long-term heat storage in connection with heat-pumps and low temperature heating systems and b) short-term heat storage in connection with cogeneration-plants and district heating systems.
The following thermal investigations are the main part of the research program : a) thermal stratification in the store and temperature distribution in the surrounding soil, b) heat losses, storage efficiency and c) heat transfer in water-flooded gravel beds.
Water quality analyses are performed 4-6 times per year to monitore chemical and biological changes in the storage water.
This project is the German contribution of two cooperation programs in the International Energy Agency (IEA), where several storage concepts are evaluated.

OPERATIONAL RESULTS

Thermal Results

During the first operational cycle, heat was exchanged only by the polyethylene coils. The monthly heat balance of the store as obtained from the measurements is presented in Fig. 1.

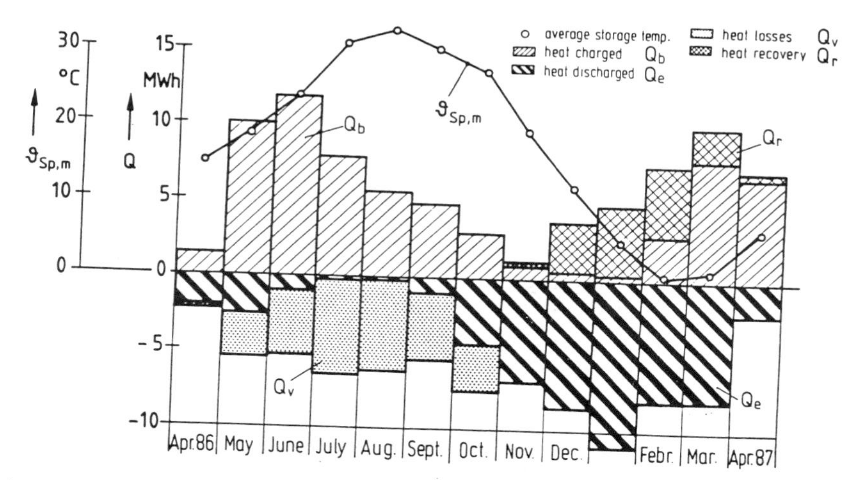

Fig. 1. Monthly heat balance and monthly store temperature.

From the middle of April 1986 until July 1986, the average monthly storage temperature increased from 15°C to 33°C. The amount of heat fed to the store in May and June was greater than in the typical summer months of July and August. With increasing storage temperatures, the absorber efficiency decreased and the storage losses increased. The heat losses (Q_v) of the store increased considerably from May to August; in August they were even slightly greater than the supply of heat (Q_b). At the beginning of the heating period (end of September), heat was discharged (Q_e) from the store. The store cooled down below the temperature of the surrounding soil, whereby heat (Q_r) was recovered. During this period the heat content of the store changed by -3.3 MWh, 64.4 MWh were charged and 56 MWh were discharged; the heat losses amounted to 27.4 MWh and the heat gain from the soil amounted to 15.8 MWh. Considering the heat gain, the net heat losses amounted to about 18% of the stored heat, thus 82% of the stored heat was used for heating. Figures 2 and 3 show the isothermal lines in a cross-section of the store and the surrounding soil at the beginning (6.10.1986) and in the middle (15.2.1987) of the heating season, respectively. During the summer period, the entire storage medium had temperatures above 30°C. At the beginning of the heating period, the lower storage layers cooled down below 30°C due to heat losses to the surrounding soil. The soil temperature 5 m below the storage bottom increased by about 2K from 12 to 14°C.

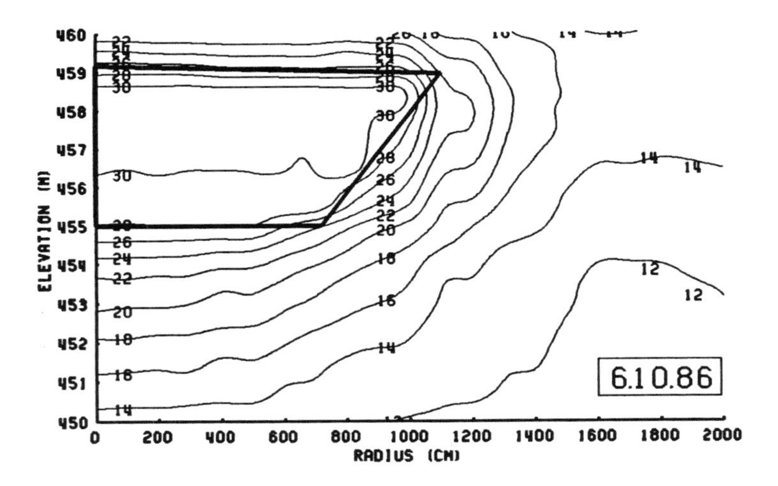

Fig. 2. Isothermal lines at the beginning of the heating period.

In February (15.2.1987) the store temperature reached its minimum of 0°C (Fig. 3); and part of the storage water was frozen. The temperature gradients at the sidewall and bottom indicate the heat gain from the surrounding soil. Due to the large heat supply from the absorbers in March 1987 (s.Fig. 1) the store temperature increases slightly until the end of the heating period (April 1987).

The monthly losses or heat gains of the storage walls -bottom, sidewall and cover- are shown in Fig. 4. The maximum heat loss through the bottom and sidewall occurred in July, the maximum heat gain in January and February. The maximum heat losses through the cover occurred approximately 2.5 months later than the maximum losses through the bottom and sidewall.

Though the cover area is larger than the sidewall area, the lava insulation layer on top, helps to keep the maximum heat losses below those of the sidewall. The heat loss through the bottom is rather small compared to the sidewall-loss.

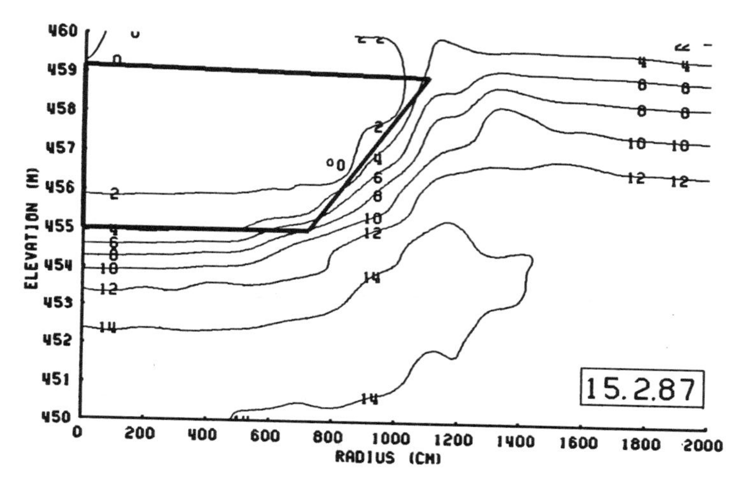

Fig. 3. Isothermal lines in the middle of the heating period

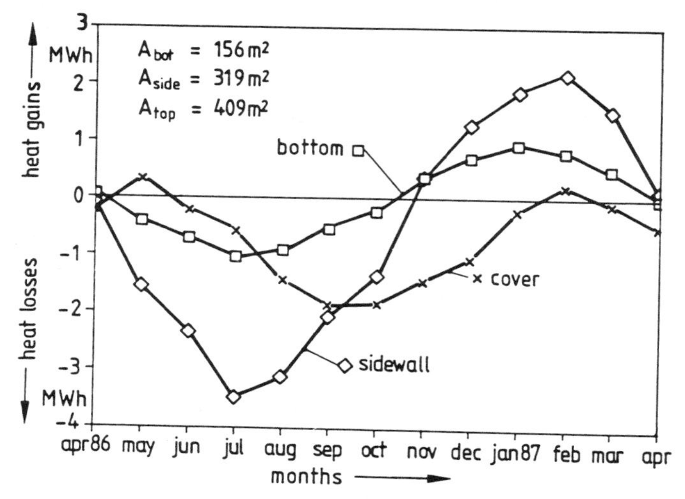

Fig. 4. Monthly storage losses and heat gains at the bottom, sidewall and cover.

Chemical and Biological Water Quality

Five water samples were taken from the store to determine the chemical water quality. On October 10, 1985 the store was filled with tap water. Substances with large concentration changes are presented in Fig. 5.

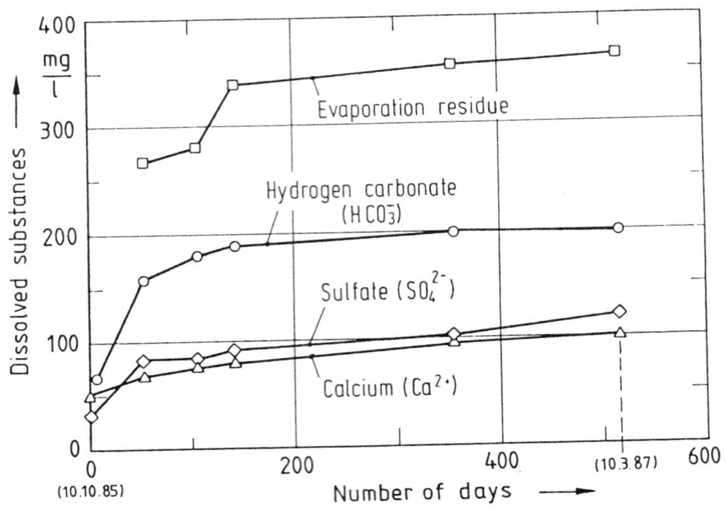

Fig. 5. Water quality changes

The evaporation residue, which characterizes the sum of all substances dissolved in water, increased. Limestone, $Ca(HCO_3)_2$, is found in gravel; the Ca^{2+} and the HCO_3^- concentration increased considerably. Calcium sulfate, $CaSO_4$, existing in the gravel, also dissolved into the water and contributes to the increase of Ca^{2+} concentration. The bacteria count of seven water samples was analysed in a separate water biology program. The values for the colony forming units vary between 3500 and 33000 units per ml. An increase of the number of bacteria with time is not observed. A noticeable bacteriological contamination has not occured.

ACKNOWLEDGEMENT

The investigations have been financed by the Federal Ministry of Technology (BMFT) under 03E 8187A; the authors gratefully acknowledge this support.

REFERENCES

Fisch, N., and R. Giebe (1986). National report 1985 of the Stuttgart university project. International Energy Agency (IEA) Task VII, Programme : Central Solar Heating Plants with Seasonal Storage (CSHPSS).

Giebe, R., R. Kübler, N. Fisch, J. Sohns and E. Hahne (1986). Man-Made Aquifer Thermal Energy Storage (ATES) Pilot Project. Proc. NORTH-Sun Conference, 58-64. Copenhagen, 10.-12.6. 1986.

Hornberger, M., N. Fisch and E. Hahne (1987). Solar-Assisted Low Temperature Heating System with Seasonal Storage. Proc.ISES Solar World Congress. Hamburg, 13.-18.9. 1987

SOLAR ENERGY STORAGE IN AQUIFERS

J.T. Pytlinski
Center for Energy and Environment Research
University of Puerto Rico
G.P.O. Box 3682
San Juan, Puerto Rico 00936

ABSTRACT

Aquifers are considered to be compact underground rock formations containing groundwater in very large quantities (on the order 10^6-10^{10} m^3). Their physical properties are characterized by porosity and permeability. They consist of rocks of sedimentary origin such as limestone, dolomite, siltstones, sandstones, and conglomerates. Aquifers store thermal energy in both the rock confining water and in the water itself. The thermal energy storage capacity of aquifers is a function of the rock type, porosity and temperature variation. An estimate is that average storage capacity could be considered to be in the range of 480 kcal/m^3°C. Access to the aquifer energy storage could be through a single well, pairs of wells, or multiple wells system. Aquifer thermal energy storage is compatible with residential and commercial heating and cooling systems, and often is considered for seasonal energy storage by using waste and solar generated heat.

KEYWORDS

Energy storage; solar energy storage; solar energy storage in aquifers.

INTRODUCTION

Aquifers are underground cavities or units of permeable rocks and unconsolidated sediments that contain or conduct water. Most aquifers are of sedimentary origin and either lie flat or gently slope within a particular location. Aquifers range in depth from as little as 2m to several kilometers below ground surface. The usefulness of a particular aquifer to accept, retain, and release injected hot water is dependent on several hydraulic properties of the aquifer, and hydraulic tests should be performed during the initial phase of a feasibility study.

The groundwater to be used as a thermal energy storage medium should fulfill several requirements such as not being subject to precipation of solute ions over the temperature range of interest, not reacting chemically with the piping of the heat exchanger or well screen, not transporting and depositing particulate material, and not supporting the growth of bacteria. Water with total dissolved solids of less than 500 mg/l is desirable for heat storage in aquifers. Aquifer confining materials least likely to react with stored water are siliceous sandstones, quartz sands, negoliths, and metamorphic rocks. Limestone, dolomite, shale and basalt may qualify on the basis of porosity and permeability but they contain reactive minerals and intrusive rocks that may yield alkalines, alkaline earths, and iron by hydrolysis of feldspathic and ferromagnesian minerals. Thus, knowing the composition of rocks confining the aquifer is an important factor in the overall decision making and the system design.

SOLAR HEAT STORAGE IN AQUIFERS

The aquifer thermal energy storage system operates in three modes: 1) water withdrawal/energy conditioning/injection; 2) water storage; and 3) water withdrawal/utilization/reinjection. The system normally uses two or more wells which reach the aquifer, and which are connected by insulated pipes to the load. After use, the cold or hot water withdrawn through the supply well is reinjected back to the aquifer through a reinjection well located some distance away. Water withdrawn from the aquifer is used directly or circulated through a heat exchanger to heat or chill a dedicated fluid in the distribution system.

The possible applications of aquifer stored solar heat are shown in Fig. 1. Solar energy is collected year-round or seasonally and stored in the aquifer. The aquifer's large storage eliminates or minimizes the need for a backup system. It is believed that a decrease of 60 to 70 percent in collector area can be achieved by using aquifer solar heat storage. By using a suitable aquifer, a lower capital cost can be attained than the cost of short term storage presently utilized in most solar systems.

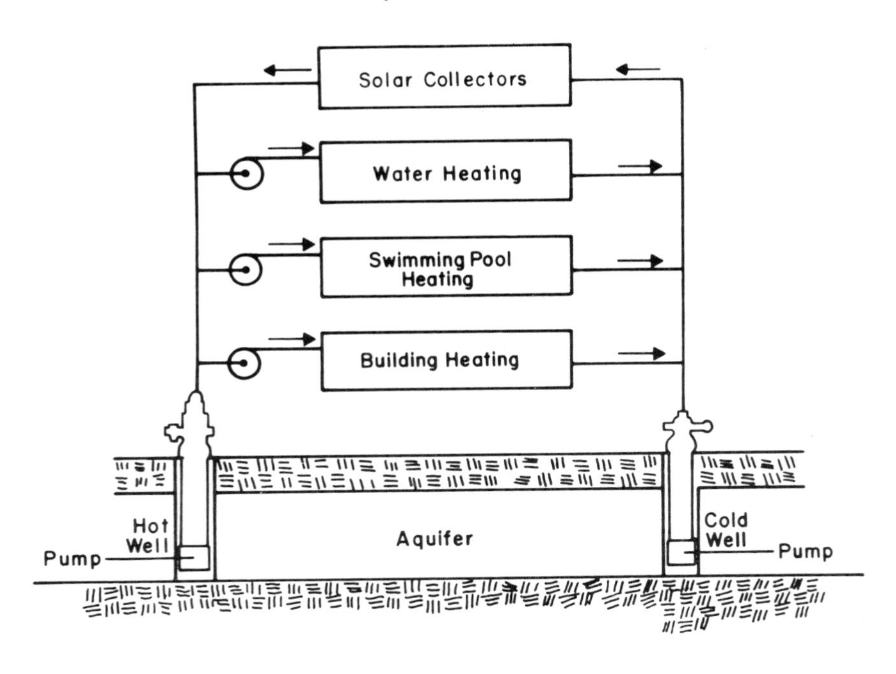

Fig. 1. Possible applications of aquifer stored solar heat

BASIC CALCULATIONS

When the aquifer properties such as porosity, permeability, thickness, physical size, natural flow direction and velocity, and the thermal requirements are known, the storage system can be determined in the following way (Schaetzle and others, 1980):

a) The thermal capacity of the aquifer can be calculated taking into account that thermal energy is stored in both the water and rock in each unit volume of aquifer. The thermal capacity of the aquifer is:

$$q = (\rho \times C_p)_{rock} \times (1 - \phi) + (\rho \times C_p)_{water} \times \phi \quad (1)$$

where;

ρ = Specific density
C_p = Specific heat

ϕ = Porosity of the aquifer rock; values from 0.01 to 0.5 are used.

b) Volume of the thermal storage can be calculated as:

$$V = Q/(q \times \Delta T \times \eta_A) \quad (2)$$

where

Q = Thermal storage load
ΔT = Temperature difference between the hot and cold wells
η_A = Aquifer efficiency

c) When seasonal energy storage is utilized, the seasonal heating and/or cooling load can be used as the thermal storage value. When seasonal storage of solar heat is planned, the amount of heat collected can be calculated as:

$$Q_s = I \times A_c \times \eta_c \quad (3)$$

where

I = Seasonal solar insolation
A_c = Solar collectors area

η_c = Efficiency of solar collectors

Substituting Q_s by a heating or cooling load, one can calculate the approximate area of solar collectors needed to collect the amount of energy required.

d) Aquifer horizontal area can be calculated now if the aquifer thickness, b, is known:

$$A = V/b \quad (4)$$

e) Well spacing required for one well pair is calculated as:

$$R = \sqrt{A/1.05} \qquad (5)$$

Several well pairs are usually required when the water flow rate is not sufficient to meet peak loads. Detailed calculations of aquifer thermal energy storage are presented by W. J. Schaetzle and others (1980).

The aquifer efficiency , η_A , can be determined from Fig. 2 (Schaetzle and others, 1980).

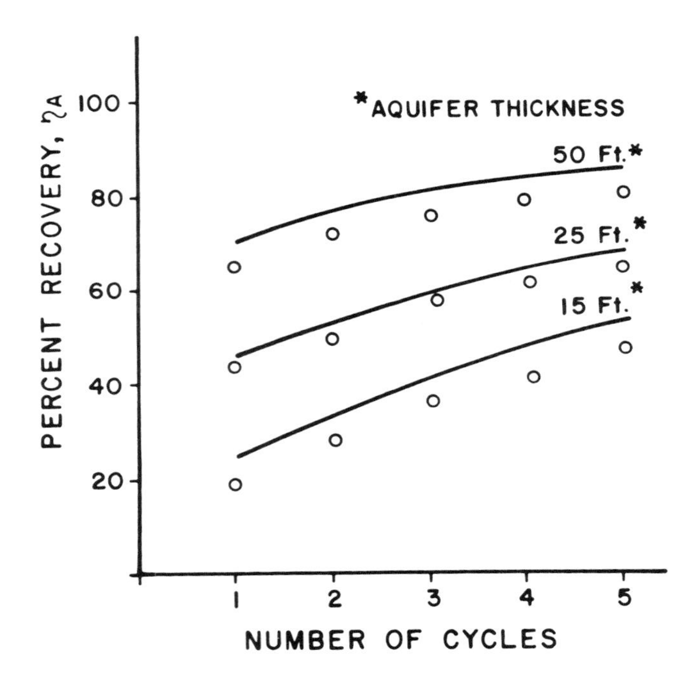

Fig. 2. Aquifer efficiency versus number of recovery cycles

RESEARCH AND DEMONSTRATION RESULTS

Experimental work on energy storage in aquifers has been done in Europe, the United States, Canada, Japan, and China. Some of this work has already led to practical applications.

In a field experiment conducted in Campuget, France (Iris and others 1980), 20,000m^3 of water was stored at 33°C in a sand and gravel type aquifer. The hot water was produced by a heat pump. No well clogging was observed, but the hot water was contaminated by bacteria. It was suggested that a chlorination of the water might solve this problem. During the approximately nine-month period of the experiment, the hot water temperature dropped from 33°C to 14.5°C. Calculations showed that this heat loss occurred mainly at the limits of the storage (35 percent) and at the aquifer surface (8,000^2)

toward the atmosphere (33 percent) because the hot water level in the aquifer was only 1.5m below the ground level. These heat losses contributed toward a low heat recovery of 18.5 percent. The conclusion was drawn that to have a good heat recovery, the aquifer should be located more than 10m below ground level, if not, surface thermal insulation is necessary to reduce heat losses. Even in this situation, the temperature of storage must be much higher than the temperature of water needed for winter use. Field experiments on the use of aquifers have been performed in Mobil, Alabama, St. Paul, Minnesota, and Stonybrook, New York, in the United States (Chin Fu Tsang and others, 1980; Allen and Raymond, 1982). Experiments have been conducted in the United States to store not only medium (50°C) but also higher temperature water (150°C) in aquifers for industrial uses.

The confined aquifer in Mobil, Alabama, U.S.A., is located between 24 and 34m below the ground surface (Chin Fu Tsang and others 1980). The surrounding aquifer ground is composed of a medium sand containing approximately 15 percent silt and clay by weight. Containment above and below is composed of different types of clay. By using a pair of wells, the first water injection/storage/recovery cycle lasted six months. The water from the aquifer was raised in temperature from 20°C to 55°C by passing it through an oil-fired boiler and then reinjecting it into the aquifer. Approximately 54,784m^3 of hot water were injected. After a storage period of 51 days, some 53,345m^3 of water were produced at a temperature of 33°C, giving a 65 percent thermal energy recovery rate. The second cycle lasted 64 days. The injection rate varied from 587 l/min to 814 l/min resulting in an injection of 58,010m^3 of water at 55°C. During production, the pumping rate of water at 33°C was 836 l/min. The second cycle energy recovery was 76 percent confirming the computer prediction of an increase in energy recovery with a multiple injection/storage/recovery cycle. The major aquifer thermal energy storage problem encountered in Mobil was clogging at the injection/production well.

In the period of 1983-1985, a field experiment was performed in Puerto Rico on the capacity of a limestone aquifer to receive, store and release injected freshwater (Whitesides and others, 1985). Although the water was not heated, this experiment raised the question of the feasibility of limestone aquifer use for thermal energy storage in Puerto Rico. A site in north-central Puerto Rico about 72.4 km , west of San Juan, the island capital, was used. An irrigation well was connected to an aquifer containing saltwater with a chloride concentration of about 930 mg/l to conduct an experiment on the feasibility of injecting and recovering excess water runoff from the nearby river Rio Grande de Arecibo. The pipe which was used on the injection-recovery well was 0.25m in diameter and descended to a depth of 52 meters. The first 30 meters were within the predominantly clay aluvium. The remaining 21 meters were open into a highly permeable limestone aquifer (the transmissivity value was estimated at 5,109 to 7,432m^3/day) situated along the north coast of Puerto Rico. A #50 slot galvanized well screen was installed to screen the injection recovery well water. Tests were conducted in the area to determine the aquifer's hydraulic properties. A chloride mass balance was applied to determine the efficiency of the aquifer to retain and release the injected water. Average injection rates varied from 995 l/min to 1726 l/min. The pumping rate for recovery of the injected water was usually about 380 l/min. A total of five injection and recovery tests were conducted with the injection time ranging from 23 hours to 241 hours. The specific conductance, water temperature, and chloride concentration were measured during the duration of each test. No well plugging occurred during injection or recovery. The experiment showed that recovery rates ranged from 5 to 26 percent of the fresh water injected. This low recovery rate put in question the suitability of the limestone aquifer at this site for thermal energy storage. Since ground water is withdrawn along the north coast limestone aquifers in Puerto Rico for agricultural, industrial, and population uses, the potential use of these aquifers for thermal energy storage creates an environmental issue which needs to be resolved.

In Falun, Sweden, an aquifer of 800,000m^3 is used to store sea water heat for space heating in a school and in Klippen, Sweden an aquifer of 700,000m^3 is used to store solar heat for district heating (Soderlund 1985).

CONCLUSIONS

Seasonal heat storage in aquifers is possible and is already used in several countries. Obtaining geological survey data by drilling test wells is a necessary initial step in determining aquifers'locations and matching them with end use. Aquifer thermal energy storage is site specific. A feasibility study of a selected aquifer site from the point of view of its geophysical properties is required. The chemical properties of an aquifer's water and the composition of rocks confining the aquifer should also be known. Hydrological modeling of the aquifer's thermal energy storage before proceeding with practical applications could help to make preliminary predictions of the aquifer's performance as a solar energy storage. Given the right conditions, aquifers can store very large amounts of hot or cold water for long periods of time at low cost.

REFERENCES

1. Allen, R.D., and J.R. Raymond (December 13-15, 1982), Progress in Seasonal Thermal Energy Storage, Proceedings of the 5th Miami International Conference on Alternative Energy Sources, Miami Beach, Florida, U.S.A.

2. Chin Fu Tsang, F.J. Moltz, and D. Parr (August 18-22,1980), Experimental and Theoretical Studies of Thermal Energy Storage in Aquifers, Proceedings of the 15th Intersociety Energy Conversion Engineering Conference, Seattle, Washington, U.S.A., pp. 1244-1248.

3. Iris, P., G. de Marsily and Y. Cormary (August 18-22, 1980), Heat Storage in a Phreatic Unconfined Aquifer, the Campuget Experiment Proceedings of the 15th Intersociety Energy Conversion Engineering Conference Seattle, Washington, U.S.A., pp. 1259-1264.

4. Schaetzle, W.J., C.E. Bretl, D.M. Grubbs and M.S. Seppanen, (1980), Thermal Energy Storage in Aquifers, Pergamon Press, New York.

5. Soderlund, M., University of Lulea, Sweden,(December, 1985), Private Communication.

6. Whitesides, D.V., V. Quinones-Aponte, and A. Zack (May 5-8, 1985), Estimating the Capacity of a Salty Limestone Aquifer in Puerto Rico to Receive, Store, and Release Injected Freshwater Using Chloride Mass Balance, Proceedings of the International Symposium on Tropical Hydrology and 2nd Caribbean Islands Water Resources Congress San Juan, Puerto Rico, pp. 50-55.

A MULTIPURPOSE 3-D COMPUTER SIMULATION MODEL FOR ATES APPLICATIONS

M.T.Kangas and P.D.Lund

Helsinki University of Technology,
Department of Technical Physics
SF- 02150 Espoo, Finland

ABSTRACT

Computer model applications for aquifer thermal energy storage (ATES) systems along with modelling methodology are described. The base case system employing just groundwater as heat source for a heat pump would be sufficient to provide nearly one half of the yearly heat load in a community heating system in a strict northern climate. The non-purchased energy fraction can still be increased by 15-25 per cent by interseasonal storage of heat injected in the summertime from a low grade heat source, e.g. lake surface waters or simple solar collector designs.

KEYWORDS

Aquifer heat storage; seasonal storage; computer simulation; community heating systems.

INTRODUCTION

Energy storage provides a means for matching the different time dependence of energy production and consumption. Storage is essential when utilizing intermittent renewable energy sources such as solar energy, but it may also enhance the economy of traditional energy sources. In cold climates, a large fraction of energy consumption is in the form of heat. E.g. in Finland, space heating of building comprises about one fourth of the total energy consumption. Storage of energy in the form of heat can thus produce significant savings. Thermal energy storage is relatively simple, and many different methods for it have been devised.

Aquifers, which are geological formations containing groundwater, offer a simple and potential way of storing heat for long periods of time. The storage medium in aquifers consists of water saturated gravel or sand. Being a natural formation, the storage volume already exists, and all that one has to do is to provide some means of transferring heat into and from the storage. Large storage volumes are also available, typical aquifer sizes ranging from hundreds of thousands to millions of cubic meters. In Finland, the total groundwater resources are estimated at 25 to 70 km^3.

STORAGE PRINCIPLE

With no natural groundwater flow, both charge and recovery function of the storage can be accomplished by a single well drilled into the water containing formations (Tsang, 1983). However, in case of rapid groundwater flow, which is typical for Finnish geological conditions, heat is transferred away from charging area by convection, and recovery can not be accomplished through the same well as injection. In general, two groups of wells must be used, one for injection and another for heat recovery (Fig. 1). The distance between the injection and recovery areas depends on the desired storage time and on the rate of groundwater flow. Ideally, the stored heat moves with the groundwater and reaches the discharge area at the time of recovery.

Because of losses in energy quality, multi-well systems are best suited for low temperature storage. Low temperatures are also environmentally more benign, causing less chemical and biological changes. This is especially important in Finland, where groundwater is in great extent also used as drinking water for communities.

SIMULATION MODEL

In the type of energy systems studied, storage is an essential component, and the dynamics of it must be carefully accounted for. Special care has therefore been devoted to the simulation of aquifer storage, and a separate independent model for it has been developed. Because an efficient recovery is possible only if the stored water reaches the recovery area at a right time, accurate knowledge of rate and direction of groundwater flow is needed. The location of recovery wells must also be carefully considered. In the model, versatile input routines for both the aquifer parameters and well placing are provided.

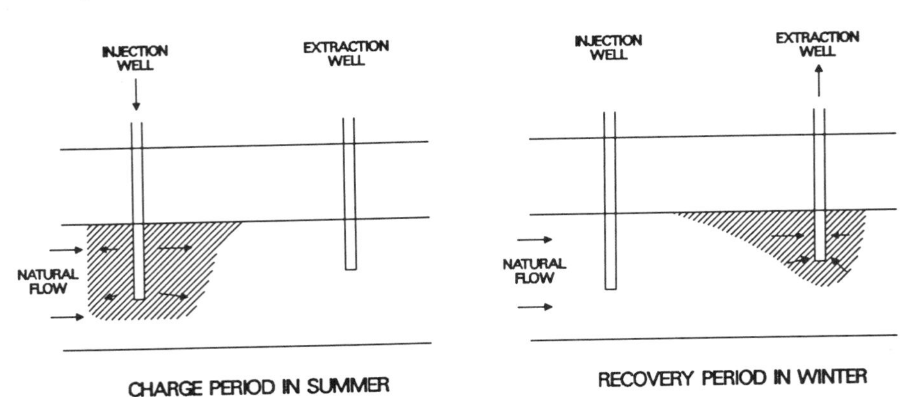

Fig. 1. The principle of a multi-well aquifer heat storage

For system studies, the model has been enlargened to include subprograms describing energy production and consumption. In addition to accurate subsystem simulation, interactions between subsystems are carefully simulated to ensure proper description of system dynamics.

STORAGE SIMULATION

The flow of water and heat through porous media is described by:

$$\vec{\nabla}\cdot(\frac{k}{\mu}(\vec{\nabla}p - \rho\vec{g})) = -Q \quad , \tag{1}$$

$$(\rho c)_g \frac{\partial T}{\partial t} = \vec{\nabla}\cdot(\lambda\vec{\nabla}T) - (\rho c)_f \vec{q}\cdot\vec{\nabla}T + H \quad , \tag{2}$$

where T is the temperature, p the pressure, $\vec{q}$ the flux of fluid (m^3/m^2s), $\rho(T)$ and $\mu(T)$ the dynamic viscosity and density of water, $(\rho c)_g$ and $(\rho c)_f$ the heat capacities per unit volume of the fluid saturated soil and the fluid itself, λ the aquifer heat conductivity, H and Q terms representing a source or a sink, and $\vec{g}$ the acceleration of gravity. Permeability $\bar{\bar{k}}$ is in general a tensor quantity. First of these equations describes groundwater flow, and the second heat transfer. Heat transfer equation is coupled to fluid flow equation by convective terms, and, on the other hand, flow equation to heat transfer equation through ρ and μ.

The above equations are solved in three dimensions by a FDM computer program THETA. The program consists of 2500 FORTRAN instructions and requires 450 kbytes of core memory. Simulation times for typical esker aquifers are 5-10 minutes per year.

THETA has a special menu input to enable arbitrary well placement in the ground. Also, highly anisotropic, vertically heterogeneous aquifer properties can be included. The program output consists flow and temperature fields in user defined horizontal or vertical cross sections as well as temperatures and energy balances of individual wells. Separate routines for drawing temperature contour lines and flow field in the aquifer are also provided. THETA has been succesfully verified against measured data from three major field experiments, two of them in Finland, and one, the SPEOS experiment, in Switzerland (Kangas, 1987).

SYSTEM SIMULATION

In the latest version of THETA (version 3.0), the aquifer model has been combined with various subroutines for energy production and consumption subsystems. These include two kinds of solar collectors, heat pump, boiler, buffer storage, and heat load. Heat source or sink with arbitrary input pattern can also be defined.

The heat load subsystem comprises the space heating and hot water demand of a community heating system. Heating load is calculated on the basis of hourly weather data and insulative properties of houses. For heat pumps, the actual power and temperatures are calculated with the known values for operating point and with prevailing temperature values and power demand given by simulation. Electric boiler can be used to raise temperature further to desired level. Buffer water storage with two sets of input/output connections is available for smoothening temperature variations. Solar collectors as well as auxiliary heat source can be used either directly through heat pump or through aquifer.

Subsystems can be connected in miscellanous ways to form the desired energy system. Connection is accomplished with connection matrixes CLINK and AQLINK, so that

$$CLINK(i,j) = 1$$
$$AQLINK(k,m) = n$$

means that the output of system i is connected to the input of system j, and that system k is connected through well number m to aquifer for storage input (n=+1) or output (n=-1). Fig. 2 shows a typical system configuration for the model.

SIMULATION RESULTS

The base system was chosen in accordance with an actual community heating system using groundwater heat pump, built recently in southern Finland. Input data system design was thus readily available. The base case system consists of 200 houses, and system configuration is that shown in Fig. 2. Heat pump with condenser power of 435 kW is used to supply heat, with additional heating by electric boiler if necessary. The natural groundwater temperature is 7 °C, and minimum allowable output from heat pump 2 °C. Groundwater flow rate was 600 metres/year, other aquifer properties also being those of a typical Finnish esker. In the heat storage applications, the distance between charging and recovery areas is 300 metres. The studies were limited to low temperature applications, which are most appropriate for Finnish geological conditions.

Figure 3a shows the non-purchased fration (F) for the base system. Aquifer was used as a heat source for the heat pump in three ways: no heat input, a 25 °C heat pulse input lasting four months annually (May - August), and a lake water input with sinusoidally variable temperature. The temperature of pulse input was chosen to give stored energy about equal to that of lake water input. Heat pulse can be seen to be the best alternative of these, giving a non-purchased fraction of 56 per cent, exeeding the value with no heat injection by 15-25 per cent. However, the differences between the two storage loading alternatives are small. With pulse input, the medium temperature is higher, but the duration of input shorter than with lake water. During the storage time, the pulse is effectively smeared out, reducing the difference between the two loading schemes.

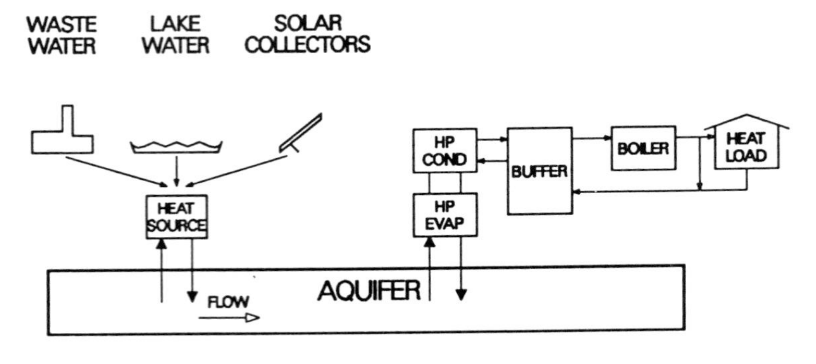

Fig. 2. Typical system configuration.

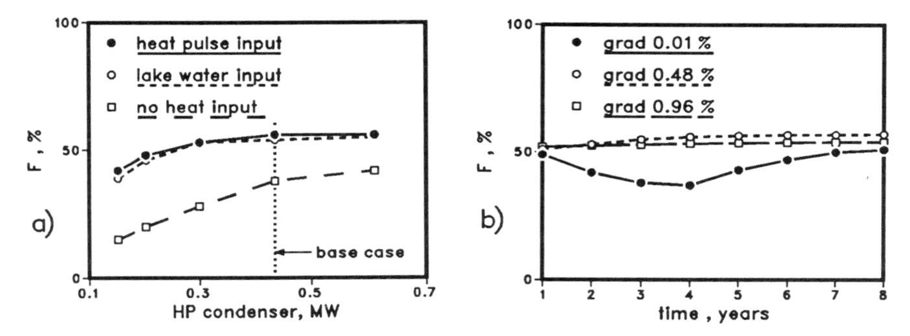

Fig. 3. Non-purchased fraction of energy for some ATES systems.

In Fig. 3a can also be seen the effect of heat pump condenser power on F. The non-purchased fraction first increases with condenser power, and then saturates as the limits of heat pump performance are reached. The maximum of F is determined by the COP of heat pump.

The effect of groundwater flow rate on system performance can be seen in Fig. 3b, which shows F as a function of time with different hydraulic gradients. For the first year, the non-purchased fraction increases with increasing gradient.The differences are small, however, because the main front of stored heat has not yet reached recovery wells. After the first year, F increases slowly as steady state is approached, with high gradient giving slightly inferior performance. With a small gradient and thus a small flow, the performance is totally different. The non-purchased fraction first reduces rapidly, and only after the fourth year begins to rise again. This is caused by cold water from heat pump evaporator output, which, with a small groundwater flow, penetrates in the upstream direction to the evaporator input wells, reducing heat pump input temperature and system performance. After the fourth year, the stored energy reaches recovery area causing an increase in the non-purchased fraction.

On the whole, substantial energy quality losses in the storage can be observed. Temperature fronts tilt rapidly causing the mixing of cold and warmer water. The pulse-like structure of storage temperature is evident for the first years but is later smeared out as steady state is approached. For low temperature applications, aquifer thus acts mainly as a buffer storage. With higher groundwater flow rate, storage behaviour decipted in Fig. 1 is approcahed.

REFERENCES

Kangas M.T., Lund P.D. (1987). The Simulation of Aquifer Thermal Energy Storage Systems. Nordic Workshop on Computational Methods of Seasonal Storage Solar Heating Systems, June 12-13, 1987 (ed. P.Lund). Helsinki University of Technology, Department of Technical Physics, Report TKK-F-A612.

Tsang C.F. (1983). Aquifer Storage Simulation - In Theory and Practice. Proceedings of an International Conference on Subsurface Heat Storage in Theory and Practice. Stockholm, June 6-8, 1983.

INFLUENCE OF A PIPING ON HEAT COLLECTING RATE OF DISTRIBUTED COLLECTOR SYSTEM

I. Tsuda, T. Tanaka and T. Tani

Electrotechnical Laboratory
1-1-4, Umezono, Sakura-Mura, Niihari-Gun,
Ibaraki, Japan

ABSTRACT

Heat collecting characteristics of collector system are influenced by collector performance and heat losses from pipe system connecting collectors. Although, it is important to improve collector performance, effect of piping heat loss on heat collecting characteristics of a collector system is emphasized in this paper. Because it is reported that the heat loss from the pipes during operation is greater than its value designed. We, therefore, made experimental study to clarify its effect. It is shown from the experimental results that it is important to make the total length of the pipes as short as possible from the viewpoint of the improvement of heat collecting characteristics.

KEYWORDS

Solar thermal system;Residence time;Thermal response time;Influence of heat capacity.

INTRODUCTION

Various types of solar thermal systems were constructed in the world to utilize effectively solar thermal energy. To obtain a large amount of the heat from the sun in these systems, a great number of collectors or heliostats were used because of the low density of solar radiation. Above all a piping which comprises a collector system is needed in a distributed collector system(DCS). It is often reported from the operation results of the DCS that heat loss from the pipe system is greater than its value estimated at the design stage and that the performance of collector system is decreased remarkably by the pipe heat loss.
Solar total energy system is promoted as a consistency of solar thermal power generation system research in Sunshine Project, which is one of national energy projects in Japan. Solar total energy experimental facility (STEEF) was constructed at Tsukuba Second Research Center, Agency of Industrial Science and Technology, and the operation has been being made (Tanaka and others, 1982; Tsuda and others, 1983; Tsuda, Tanaka, Tani, 1985;

Tanaka, Tsuda, Tani, 1986). The experimental result indicated that characteristics of the STEEF is influenced remarkably by the piping heat loss in the collector system which is called the low temperature loop (LTL). Therefore this experimental study was carried out by using the LTL. In order to clarify an influence of the piping on a collecting characteristics, two conditions having different routes of piping were chosen. This report gives the major experimental results.

OUTLINE OF PIPING HEAT LOSS EXPERIMENT

16 units of improved reverse flat plate collector (IRFP), which is a collector of non-concentrating non-tracking type, are used in the LTL. The LTL consists of 4 branches. As shown in Fig. 1, 4 units of the IRFP were connected in parallel in each branch. The IRFP is composed of 3 units of evacuated tube collector (ETC) and 1 unit of reverse flat plate collector (RFP). The parallel arrangement of the IRFP is determined so as to attain easily the desired temperature by one unit of the IRFP i.e., desired temperature is 120°C, under the condition of inlet temperature 60°C, thermal fluid flow rate of the LTL 1.25ton/h, and insolation 0.65kW/m^2.
However, at the start time of the operation of the STEEF, the outlet temperature of the LTL rises very slowly because the header pipes installed at the inlet and outlet of the IRFP in each branch are very long. Accordingly the piping route of the branch is reconstructed to make a thermal response time of the outlet temperature faster.
Thermal response time is defined by the following equation (Wright,1981),

$$\tau_{th} = \frac{C_f+C_p}{C_f}\tau_r \qquad (1),$$

where τ_{th}:thermal response time in the piping (sec), τ_r:residence time of thermal fluid in piping (sec) =L/v (L:length of piping, v:velocity of thermal fluid), C_f:heat capacity of thermal fluid in the pipe per unit length (J/m), C_p:heat capacity of the pipe per unit length (J/m). This equation can be applied to the case when $C_f/(C_p+C_f)<0.7$ and when heat capacity of the thermal insulation material is ignored. The condition expressed in equation (1) is nearly satisfied in the piping of the LTL.
It is shown from equation (1) that the in-

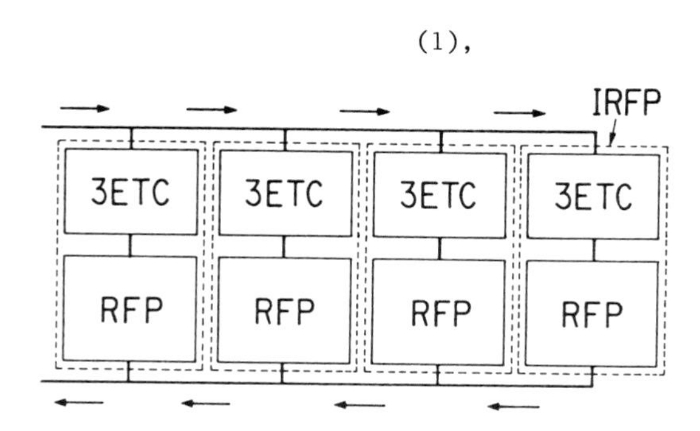

Fig. 1. 1 branch of the LTL before the reconstruction.

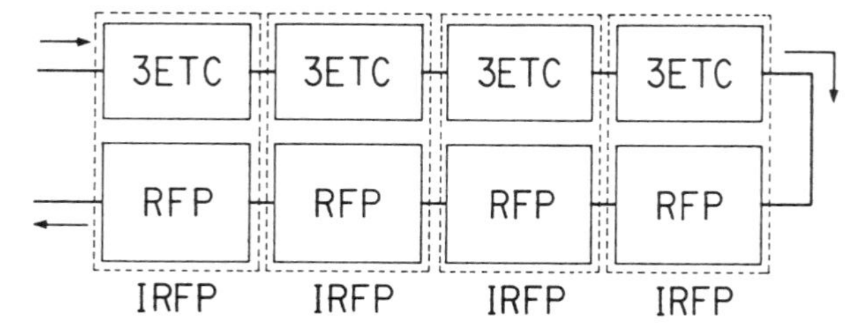

Fig. 2. 1 branch of the LTL after the reconstruction.

Table 1 characteristics of Collector and Pipe with Reconstruction

	Collector		Pipe System			Total
	ETC	RFP	Inlet Header	Outlet Header	Other	
Heat Capacity of Thermal Fluid (kJ/°C)	6.47 (5.29)	9.31 (6.61)	--- ($2.04X10^3$)	--- ($5.19X10^2$)	44.7 (48.6)	$4.98X10^2$ ($2.97X10^3$)
Heat Capacity of Pipe (kJ/°C)	7.01 (6.76)	41.9 (41.2)	--- ($1.06X10^3$)	--- ($4.13X10^2$)	28.7 (30.3)	$1.04X10^2$ ($2.49X10^3$)
Residence Time (min.)+	0.83 (2.74)	1.15 (3.29)	--- (68.5)	--- (16.0)	1.42 (1.56)	16.0 (97.6)
Thermal Response Time (min.)+	1.72 (6.24)	5.88 (24.0)	--- ($1.04X10^2$)	--- (29.0)	2.33 (2.53)	46.5 ($1.78X10^2$)
Operating Temperature(°C)*	80 (70)	120 (110)	--- (40)	--- (125)	90 (80)	--- ---
Warming up Energy (MJ)	0.86 (0.65)	5.33 (4.51)	--- (76.8)	--- (99.8)	5.36 (5.03)	$1.32X10^2$ ($2.85X10^2$)
Collecting Area (m^2)	1.70 (1.70)	7.19 (7.19)	--- ---	--- ---	--- ---	196 (196)

+:flow rate is 1 m^3/h.
*:at noon in the clear day.
() before the reconstruction.

crease of fluid flow rate makes thermal response time shorter. However, it does not contribute to the rise of the outlet temperature of collector system because the outlet temperature of the collectors becomes low with the increase of the flow rate. Therefore, to make τ_{th} small, it is necessary to make the values of the ratio of the heat capacity and τ_r small. To attain this purpose, the piping of the LTL was exchanged by another pipe which has a different kind of material, and was reconstructed to shorten the total length of the piping.

Piping route after the reconstruction is shown in Fig. 2. 4 units of the IRFP are arranged in series. The header pipes installed at the inlet and outlet of the IRFP is taken off in this configuration.

Absorber tube in the ETC and the RFP before reconstruction were connected in series. But the connection was changed into parallel not to increase pressure drop.

Total length of the piping before the reconstruction was about 500 m, and that after the reconstruction is about 230 m.

EXPERIMENTAL RESULTS AND DISCUSSION

Operating time of the STEEF is from 8 a.m. to 4 p.m. in solar time. The thermal fluid flow rate is controlled at constant flow rate until outlet temperature attains set value. Then flow rate is controlled by a PID controller.

Characteristics of collecting rate per day before and after the reconstruction is shown in Fig. 3. The horizontal axis indicates the total amount of the incident solar radiation on collector area. The vertical axis gives the heat gain in collector system. It can, therefore, be observed in Fig. 3 that the threshold insolation rate per day decreases from 500kWh/day to 300kWh/day. The slope is almost the same before and after the reconstruction.

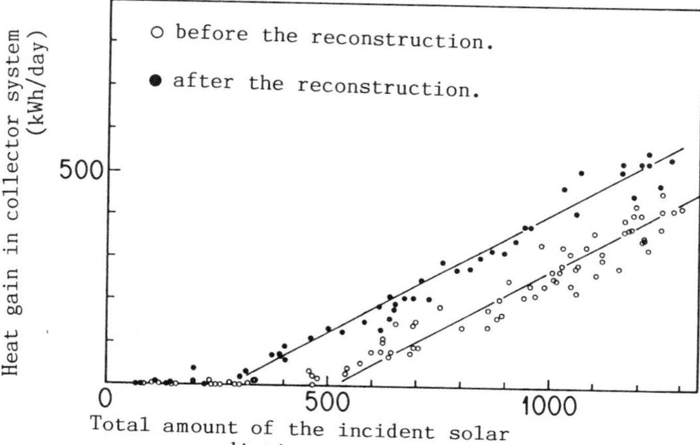

Fig. 3. Change of the collecting rate as the reconstruction.

It is shown from Table 1 that the heat energy required for warming up of the LTL was decreased by the reconstruction. That is, the heat energy required to warm up from ambient temperature (in this case 15°) to operating temperature in the LTL is changed from 285MJ to 132MJ.
Heat loss was measured near the solar noon under the condition when the insolation is stable. The heat loss per day of the header pipes before the reconstruction become about 108MJ at the outlet side and 82.8MJ at the inlet side from the result of calculation for clear day. This heat loss is decreased by removing the header pipes after the reconstruction.
A sum of the decrease of the heat loss and the decrease of the heat energy required to warm up from ambient temperature to operating temperature is equivalent to the increase of the collecting rate per day after the reconstruction. Therefore, the increase of the rate is caused by the decrease of the heat loss of the piping and the heat energy required for warming up of the piping of the LTL with the reconstruction.
It is recommended to make the length of the piping as short as possible in collector system. It contributes to an improvement of a collecting characteristics in collector system.

CONCLUSIONS

The following conclusions are obtained by the experimental results before and after the reconstruction of the LTL.
(1) Collecting rate per day increases by the decrease of the heat loss of piping and the decrease of the heat energy required for warming up with the decrease of the heat capacity of piping.
(2) It is recommended to make the length of the piping as short as possible in collector system. It contributes to an improvement of a collecting characteristics in collector system.
(3) Reducing heat energy required for warming up of the piping is essential for an improvement of performance. Furthermore, collecting rate per day increases by reducing the heat capacity of collectors, because it can

reduce the warming up energy.

REFERENCE

Tanaka, T., T. Tani, S. Sawata, K. Sakuta and I. Tsuda (1982). Development of Solar Total Energy System for Industrial Sectors. 17th, IECEC, 829256, pp. 1556-1561.
Tsuda, I., T. Tanaka, S. Sawata and T. Tani (1983). Operating Characteristics of Solar Total Energy Experimental Facility. 18th, IECEC, 839325 pp. 1990-1995.
Tsuda, I., T. Tanaka and T. Tani (1985). Transient Response of Collector Subsystem in Solar Total Energy Experimental Facility. TASE, pp. 131-136.
Tanaka, T., I. Tsuda and T. Tani (1986). Development of Solar Total Energy System in Japan. Solar Energy, Vol. 37, No. 1, pp. 55-63
Wright,J.D.(1981). Analytical Dynamic Modeling of Line-Focus Solar Collector. ASME Journal of Solar Energy Engineering, Vol. 103 pp. 244-250.

TEMPERATURE AND FLOW DISTRIBUTIONS IN A COLLECTOR BANK

Wasim Y.Saman Laith T. Fadhl

Solar Energy Research Center
P.O. Box 13026,Jaderiyah,Baghdad, Iraq.

ABSTRACT

The paper describes an experimental study on a bank of ten tube in plate type collectors,connected in parallel. Extensive temperature and pressure measurements were carried out at the points of connection between collectors as well as at the risers of some collectors. The effects of flow rate and positions of the bank inlet and outlet connections were studied. Both velocity and temperature results indicate that the risers nearest the outlet port carry the maximum flow while the central portion between the inlet and outlet have the lowest velocities and highest temperatures. It is evident from the results that,at low flow rates,more energy is collected when the inlet and outlet are kept at the same end of the bank;while less pumping power is required for this arrangement.

KEY WORDS

Flat plate collector;solar collector pressure;solar collector temperature; collector bank connection.

INTRODUCTION

Available methods for predicting the performance of flat plate collector systems usually assume that the flow is equally divided between the parallel risers. As velocities and hence hydraulic losses are variable along the headers,the velocity and temperature distributions in the risers will also be affected causing some detrimental effect on the collector efficiency (Saman and Mohamed,1985). From this stand point,the case of a bank of collectors,connected in parallel,requires particular attention as this connection method leads to a large number of parallel risers. This case has been studied by Dunkle and Davey (1970) who concluded,on the basis of temperature measurements,that the number of parallel risers should be limited to 16. A study by Soin (1984) showed that the risers in the central collectors may carry less than 10 per cent of the fluid flowing through the risers near the connection points. This was obtained for banks of 5 and 10 collectors each consisting of 12 risers.

The present study aims at obtaining detailed experimental results of flow distribution,energy collection and pressure losses in a bank of 10 collectors connected in parallel. Four alternative connection arrangements of the inlet and outlet were examined at various flow rates with emphasis on the low flow range.

EXPERIMENTAL

The collectors used were single glazed. Each consisted of 7 risers of length 64 cm and diameter 14 mm with headers of 27 mm diameter. The total effective collector area was 11.4 square metres. The bank was inclined at 33 degrees to the horizontal (the latitude of the city of Baghdad) facing due south.

The temperature of the inlet water was kept constant (at 30 or 35°C $\mp$1°C) by means of hot and cold water tanks. Thermocouples and pressure tappings were fixed at all points of connection between collectors at the top and bottom headers. In addition,attention was focused on the first,sixth and last collectors. The temperature and pressure were measured at the top and bottom of the first,fourth and last risers of these three collectors. Inverted U tube differential manometers,inclined at 40 degrees to the horizontal,were used to measure the pressure differences along the headers and across the risers. The temperatures were measured by means of Cu Cn thermocouples and electronic thermometer via suitable selector switches. The following arrangements of inlet and outlet location were examined:

1- The inlet and outlet at the same end of the collector bank.
2- The inlet port at the begining of the bottom header and the outlet port at the end of the top header.
3- The inlet port at the middle of the bottom header and the outlet port at the middle of the top header.
4- The inlet port at the middle of the bottom header and the outlet ports at both ends of the top headers.

The flow rate was kept constant at all tests and a range of 100-400 litres per hour was examined

Two experimental procedures were used for evaluating the thermal performance.

1- Constant insolation experiments which were carried out at noon of clear days.
2- Daily performance experiments,in which hourly readings were taken throughout the day.

RESULTS AND DISCUSSION

All results obtained confirmed the considerable differences in temperatures and flow distribution that existed at all flow rates and connection methods (see Fadhl,1987 for details). Taking the lowest flow rate of 8.8 litre/hr per unit collector area when the outlet is at the same end as the inlet,the highest riser exit temperature (77.1^{o}C) was recorded at the last riser of the 6th collector; while the minimum temperature (62.9^{o}C) was found at the first riser of the first collector. Under similar flow and solar insolation,with the outlet port at the end of the top header,the outlet temperature of the last header of the sixth collector remained high at 68.7^{o}C just below the absolute maximum (69.5^{o}C) which was recorded at the seventh riser of the first collector. The minimum riser outlet temperature (59.9^{o}C) was located at the first riser of the last collector. The overall water temperature rise was 35.7^{o}C for the first connection arrangement and 28.1^{o}C for the second arrangement giving instantaneous efficiencies of 33.5 and 26.7 per cent respectively. The temperature trends remain at higher flow rates. However,as the flow per square metre of collector area reaches 35 litre/hr,the efficiencies become 58.7 per cent for the first case and 60.5 per cent for the second. Figures 1 and 2 show the temperature rise in the nine risers under consideration plotted as a function of flow rate through the collector bank for the two connection arrangements being considered. It may be inferred from Fig. 1 that at the higher flow rates the first collector carries a considerably higher proportion of the flow than the middle and end collectors which produce largely similar temperature rises little affected by the change in flow rate.

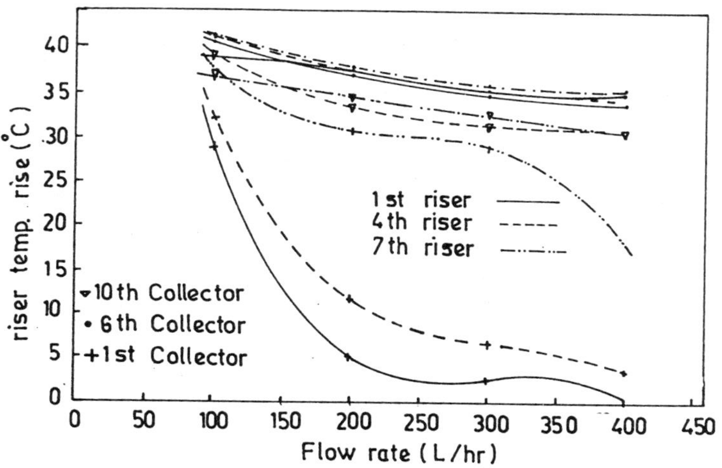

Fig.1 Temperature rise in some risers (1st connection method date 24/9/86)

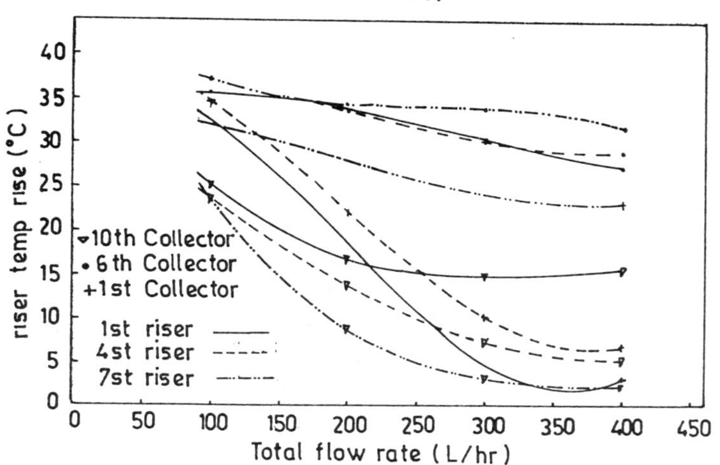

Fig.2 Temperature rise in some risers (2nd connection method date 23/9/86)

When the connection is at opposite ends (Fig.2),low flow rates result in the last collector carrying a large portion of the flow. As the flow increases,only the first and last few risers of the whole bank have the markedly low temperature rise corresponding to a large share of the flow with the temperature rise increasing as we move nearer to the middle section of the bank.

It may be noted that the third and fourth connection methods are respectively similar to the first and second methods but with half the number of risers in parallel. Both the third and fourth methods produced considerably less variation of the temperature differences in their risers in comparsion with the first and second methods. For the experiments of 17.5 litre/hr.m^2 collector area,the noon temperature rise ranged between 3-36°C and 6-34°C for the first and second cases;the corresponding values for the third and fourth cases were 21-37°C and 14-27°C respectively.

The variation of collected energy and total pressure loss in the collector bank with the water flow rate using the first two connection methods is shown in Fig.3. For the range of flow rates considered,it is evident that connecting the outlet port at the same end as the inlet produces more collected energy with less pumping power in comparison with the more common connection of the inlet and outlet at opposite ends. However as the flow rate goes above 30 litre/hr per unit collector area the second connection arrangement produces more energy but with considerably higher pumping power.

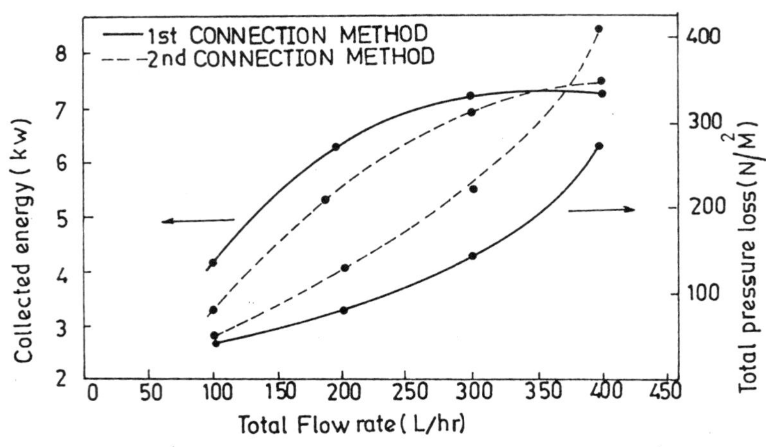

Fig.3 Variation of collected energy and pressure loss with flow rate.

In spite of the difficulties encountered in measuring the small differences in pressure across the risers especially at the low rates considered,it was possible to locate those risers having relatively high flow velocities (the flow being laminar). Figure 4 is a plot of the velocity at particular risers divided by the average velocity,assuming equal flow distribution through all risers,versus the flow rate per unit collector area. It may be observed from this plot that a few risers carry considerably more than the average riser flow rate especially when the water flow rate through the collector bank is increased. The velocity results are in general agreement with the temperature measurements discussed above.

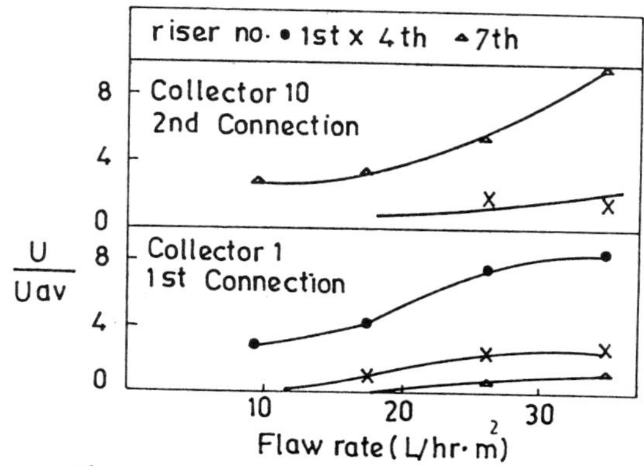

Fig.4 Velocity distribution in some risers

CONCLUSION

In the light of this study it may be concluded that in a system containing a large number of parallel risers,the risers nearest to the flow exit carry a high proportion of the flow . When the inlet and outlet are at the same side of the collector bank,as opposed to being at opposite ends,lower pressure losses are always incurred and higher efficiency is obtained for flow rates lower that 30 litre/hr per square metre of the collector area. The situation of the inlet port at the middle of the lower header,which is comparable to using half the number of parallel headers,reduces the flow maldistribution considerably.

REFERENCES

Dunkle,R.V. and Davey,E.T. (1970) Flow distribution on solar absorber bank,Int. Solar Energy Society Conference,Melbourne.

Fadhl,L.T. (1987) Flow in flat plate collectors M.Sc. thesis University of Technology,Baghdad.

Saman,W.Y. and Mohamed,A.A. (1985) Effect of hydraulic losses on velocity and temperature distribution in flat plate collectors. Conference of Int. Solar Energy Society,Montreal.

Soin,R.S. (1984) Flow distribution in large solar collector array and methods for balancing. Renewable Energy Sources: international progress,Part A.

DYNAMIC OPTIMIZATION OF CONTROL FOR DESIGNING THE OPTIMAL GEOMETRY OF A SEASONAL GROUND STORAGE

A.R. Logtenberg and A.G.E.P. van Delft

Eindhoven University of Technology, System and Control Engineering Group,
Building W&S 1.35, P.O.Box 513, 5600 MB EINDHOVEN, Netherlands

ABSTRACT

In this paper we describe a method of evaluating the performance of a seasonal ground storage of sensible heat with an arbitrary geometry by comparing it with a theoretical optimal model.

KEYWORDS

Dynamic Optimization, Seasonal Storage of Heat.

INTRODUCTION

In general for space heating systems based on the application of active solar energy the supply of and demand for energy are virtually in phase opposition. A way to bridge this gap is to store the solar heat gained in matter. For this purpose large water tanks, water containing soil layers (so-called aquifers, in which the hot medium is injected into the soil) or ground storage by means of an heat exchanger (closed circuit) can be used.

Like a number of other European countries the Netherlands has started to introduce storage systems of this kind: e.g. an aquifer system for an office building at Bunnik and a soil heat-exchanger at Groningen.

Over the last years within the group of Thermal Energy Systems (TES) we concentrated our attention on closed circuit systems (soil heat-exchanger). A line of investigation is aimed at solving the problem which pipe geometry in the ground is most suitable for a given situation.

Various studies on systems with a seasonal storage of sensible heat have been carried out in our group, consisting in both simulation and optimization studies. In all cases however, one started from a fixed ground geometry, usually being a system of coaxial tubes which were imagined to be arranged in concentric rings, or radially from a single point at the ground surface. Besides use was made of a so-called 'open system approach', in which the number of vertical tubes was increased unlimitedly, thus transforming into a flow density distribution in stead of a number of discrete liquid flows.

Extensive literature search for approaches of tackling the problem of determining the optimal geometry of a soil heat-exchanger did not lead to results. In practice one chooses either for

coaxial- or loops of pipes. Next one decides for a certain position of the pipes (mostly vertical). Finally it is assumed that the pipes themselves or the concentric pipe rings are arranged to be equidistant. What remains for optimization is the distance (one parameter) between the pipes or pipe rings as well as the the pipe diameter, thickness and material properties such as the coefficient of heat conduction.
Unfortunately it was not possible to achieve the goals as described in the preliminary abstract. In particular we were not able to carry out the dynamic optimization of the control strategies for practical pipe configurations. We therefore have to restrict ourselves to presenting the results accomplished so far, whereby the problem solving approach will be in the focus of attention.

OPTIMAL CHARGE/DISCHARGE STRATEGY OF A GROUND STORAGE

Main characteristic of the adopted approach is that we first of all look for the theoretical optimal result to be achieved with a soil heat-exchanger. This implies that we concentrate on the boundaries set by thermodynamics and for the moment refrain from considering the technical constraints, such as e.g. the fact that in practice only a limited number of pipes can be used and that these should be put into the ground rectilinearly by preference.

Anticipating the results it is interesting to note that the theoretical optimal control of a soil heat-exchager gives rise to a temperature stratification in the soil analogous to the results obtained with short-term storage vessels.
First we will state the assumptions made:
- the soil is homogeneous and isotropic.
- the possibility of heat transfer through convection (ground water) is not considered; the heat transfer in the ground is exclusively described by a constant heat capacity and heat-conduction coefficient.
- the top-layer of the soil is perfectly isolated.
- at a large distance from the heat-exchanger the temperature level is constant.
- one year is divided into a charging and a discharging period.

Recognizing the highly variable nature of the weather conditions the heat supply is assumed to be variable per unit of time, both in quantity and in temperature level. In the case of the usage of a solar collector this is no superfluous sophistication of the problem definition. However, it is at all times possible to choose the time interval in such a way that the temperature may assumed to be constant during that time interval, see Fig. 1.

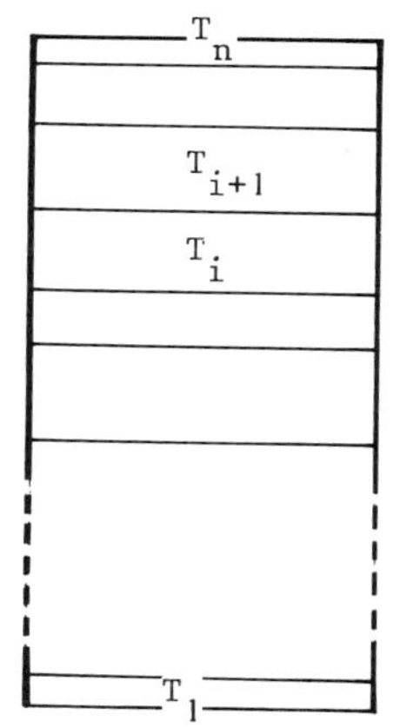

Fig. 1 Supply register. Daily/monthly solar collector gain.
$T_{i+1} > T_i$ for $1 \leq i \leq n-1$

The soil is thought to be a grid of sufficiently small discrete elements, in which the temperature may assumed to be constant for a sufficiently short time interval.

Given a particular subdivision of supply medium and soil storage volume into basic elements we now suppose that we are able to combine an arbitrary element from the supply register with an arbitrary soil element. In this context combining means the operation of complete temperature equalization within a negligible short time. After every combination it is possible to combine the element from the supply register (whose temperature has decreased) once again with another soil element, until it is impossible to transfer more heat to the soil.

At present the question is according to what strategy the n elements from the supply register should be combined with m soil elements in order to achieve a maximum heat transfer. The answer reads: a supply and a soil element having temperatures of T1 and T2 respectively are allowed to be combined, provided that both in supply register and soil no temperature between T1 and T2 can be found. This procedure should be iterated until no further combinations are possible. At this point maximum heat transfer is achieved. The proof for this statement is rather complex and is ommited here. It can be shown that by application of this strategy used for this first combination has at all times the highest final temperature; the soil element used for the second combination has the but one highest temperature and so on.
Starting from a soil with an uniform temperature this implies that the temperature differences between the soil elements will become increasingly larger as the charging proceeds.

Remembering that in practice the soil elements are in thermal contact, giving rise to thermal conduction, through which the temperature stratification degrades and heat is lost, it is intuitively clear that in a spherical body of matter heat can be stored for a longer time than in the same quantity of matter of an arbitrarily different shape. Hence it is obvious to consider a spherical volume of soil and to divide this in concentric sphere segments, which can be seen as the soil elements referred to above, see Fig. 2.

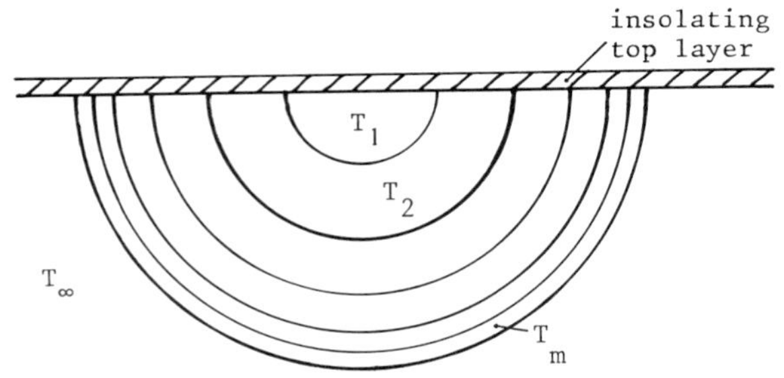

Fig. 2 Concentric sphere segments model. All segments are of equal volume.

Elaborating the consequences of the charging strategy described here and applied to this concentric sphere model we can prove that:
1) a monotonous decreasing temperature profile will develop from the center to the edge;
2) the center soil element assumes the highest possible temperature at a given heat supply;
3) the total heat content of the storage is the maximum attainable one for a given heat supply.

Then it can be shown that even if temperature equalization due to conduction is involved in the consideration, the total heat content of the soil elements for all $t>0$ is a maximum when the theoretical optimal charging strategy is applied ($t=0$ is time at which the charging proces starts).
Note that thermal conduction only takes place in the radial direction, so that it makes no difference at all if we consider a hemispherical ground body which is thermally perfectly insolated at the top.

Beyond that, as far as discharging is concerned minimal heat loss can be achieved if exactly the same strategy of heat exchange is applied.

DYNAMIC OPTIMAL CONTROL OF A COLLECTOR-GROUND STORAGE-USER SYSTEM

R. Wijnen (student) has determined the dynamic optimal control of a combination of a collector, ground storage and a user, starting from a charging and discharging period of 6 months each, the weather conditions being assumed to be constant for each month.
The optimization criterion is the minimization of the auxilliary heating over a one year period. The quantity to be optimized during the charging period was the collector flow. During discharge the flow was assumed to be constant.
The reference system is characterized by a storage radius of 22 meter, a collector area of 1250 m2 and a collector heat loss factor of 2 W/m2.
The dynamic optimal control strategy during the charge period is given in Fig. 3.

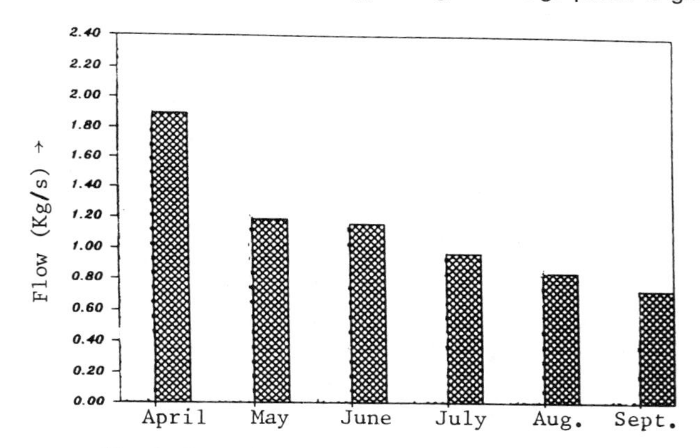

Fig. 3 Dynamic optimal control of the collector flow.

Initially the flow is relatively high, so that the temperature is low. However the medium temperature gradually rises. Obviously the control prevents large temperature differences between the medium and soil to occur. For the same system we also have determined the instantaneous optimal control, i.e. the control which transfers the maximum quantity of heat during every time step, without consideration of the temperature level. The control curve is plotted in Fig. 4.

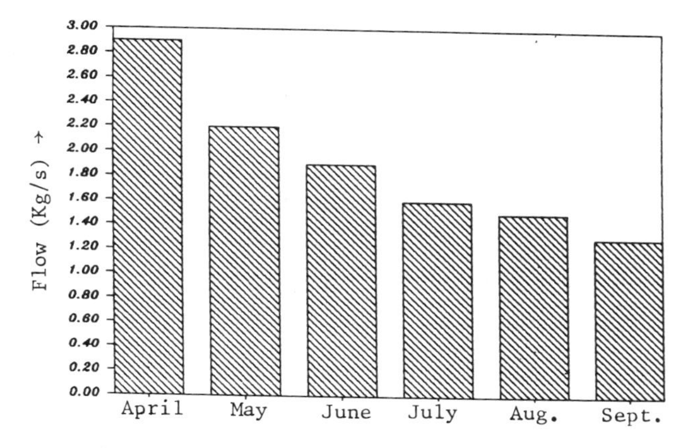

Fig. 4 Instantaneous optimal control of the collector flow.

The comparison between the energy flows (in units of 10**10 Joules) shows that for a heat demand of 204 the instantaneous optimal control requires an auxiallary heating of 55 whereas the dynamic optimal control requires 38 units.

PROSPECTS

The theoretical optimal way of using a quantity of matter for heat storage purposes has been described above. It is clear that in practice this optimum can only be approximated. Hence the model is valuable in so far as it gives a criterion by which practical configurations can be evaluated. Consequently our next task is to determine the dynamic optimal control strategies of several obvious configurations in which coaxial pipes are used, see Fig. 5.

Fig. 5 Examples of ground geometries to be evaluated.

After the comparison with the theoretical optimal model we should be able to judge whether or not it is advantageous to look for those configurations which best satisfy this optimum.
We hope to inform you about the outcome of this investigation on another occasion.

DYNAMICS OF A VERTICAL HEAT EXCHANGER PIPE IN GROUND

A. Keinänen, P.D. Lund

Helsinki University of Technology
Department of Technical Physics
SF-02150 Espoo 15, Finland

ABSTRACT

The thermal behaviour of a vertical heat exchanger pipe in a solid material has been investigated with experimental and computational methods. A major task has been the testing of a new semi-analytical model under controlled conditions. Comparisons to full scale systems are also discussed.

KEYWORDS

thermal energy storage; duct storage; CSHPSS; computer simulation

INTRODUCTION

Exploitation of new energy sources causes a seasonal mismatch between energy consumption and production. To compensate this phase change, effective energy storage means must be used to reach a high fraction of non-purchased energy. A very promising approach to cost-effective seasonal storage is a duct storage, consisting of vertical heat exchanger pipes in ground. The positive experiences gained from long-term field experiments have proved that a duct storage is simple and cheap to construct and the storage efficiency is satisfactory.

The basic problem encountered with duct storage systems is their slow response to outside fluctuations: the large thermal inertia of ground results in poor transient behaviour. This problem may, however, be overcome by using attenuation, e.g. a water buffer, between the pipe storage and the remaining heating system. To enable an optimal operational strategy for the two storage components, it is important to predict the time-dependent thermal response of a single borehole.

This paper concentrates on investigating the thermal behaviour of a single borehole by a new semi-analytical model (Lund, 1987a). Two laboratory borehole models were constructed to study the heat transfer conditions at the fluid-rock boundary. Especially the effect of variations in mass flow rate to the overall thermal resistance between fluid and rock were examined.

MODEL

The heat conduction equation describing heat transfer processes in a duct storage is solved by dividing the problem into two parts: the transient heat flows in the viscinity of the pipes are uncoupled from the macroscale slow thermal processes (Claesson, Hellström, 1981).

To enable a requisite coupling between these two processes, a separate local temperature field, θ, just around the pipes is defined. When accompanied with a short-term storage, levelling out high frequency fluctuations, the discharge and charge conditions of the pipe storage are almost stationary. Thus the dynamics of a single pipe is described by the following first-order differential equation:

$$d^2(\rho c)_g \frac{\partial \overline{\theta}}{\partial t} = \frac{\overline{T}_f - \overline{\theta}}{R} = \phi \tag{1}$$

where R is average thermal resistance between fluid and ground, $\overline{\theta}$ is average local ground temperature, $\overline{T}_f$ is average fluid temperature along the pipe and ϕ is power rate.

If we denote $\tau = d^2(\rho c)_g R$ and suppose constant fluid inlet temperature, the solution of Eq. (1) is

$$\overline{\theta}(t) = \overline{\theta}_0 e^{-f(t)} + \overline{T}_{f,0}(1-e^{-f(t)}), \tag{2}$$

where

$$f(t) = \int_0^t \frac{1}{\tau(t)}\, dt . \tag{3}$$

τ is the storage time constant and has a dimension of time (s).

As expected, Eq. (2) shows an exponentially decaying power rate for constant mass flow conditions. The overall thermal resistance, R, is a sum of resistances caused by the surface between fluid and pipe wall the thermal conductivity of the pipe wall, the possible air gap between the pipe wall and the thermal conductivity of the ground itself. These terms are discussed in detail by Claesson and co-workers (1985) and Lund (1987a).

LABORATORY EXPERIMENTS

Let r_1 be the radius of the local region of a borehole. Then, for symmetry reasons

$$\left(\frac{\partial \theta}{\partial r}\right)_{r=r_1} = 0 . \tag{4}$$

Consequently, local thermal processes in a large mesh can be reduced to a single borehole provided that the single borehole model is properly insulated to satisfy the boundary condition.

From Fig. 2 we see that with laminar flow the influence of mass flow rate to surface resistance is considerable. When the transition region is entered, the resistance saturates to an almost constant value.

The measured results were compared to results where the thermal resistance was determined from the commonly employed correlations between the Nusselt and the Reynolds number. A good agreement was found, especially in the transition region. The values of the Nusselt numbers with laminar flow tend to be greater than usually, which depends on a greater scale factor (D/L) of the model.

The next step was to repeat the measurements in case the water was substituted with a solid material, gypsum, which resembles the real conditions. The radial distribution of the temperature field was measured with thermocouples cast in a radial line in the middle of the cylinder. The average temperature of the solid was determined by integrating the temperatures over the radius of the local region and dividing with the length of radius. Because the average ground temperature as a function of time was accordingly in hand, the time constant could easily be obtained by fitting Eq.(2) to the curve $\tau(t)$.

It was found that the contribution of the surface resistance to the overall resistance is minor. The governing component in these measurements was unexpectedly the air gap between the pipe and gypsum. A 1 mm thick air gap accounted for more than 60% of the total resistance while about one third was due to the solid medium.

Since the ratio of all dimensional parameters were kept constant during the model construction, the findings can be applied to full scale heat exchanger pipes. As the dimensionless parameters for full scale pipes remain the same, the only component to be changed during scaling is the resistance due to the air gap. The resistance of an air gap is inversely proportional to the radius of the pipe. For a pipe whose dimensions are tenfold compared to the laboratory model, the resistance caused drops to one tenth. For full-scale lined pipes, the same 1 mm air mass would account for about 25% while the rock resistance would be about 50% of the total. For unlined pipes, the surface resistance would be about 35% at most and the rock resistance would cover the rest.

In laboratory experiments, the storage time constants were around five hours. From the definition of τ we see that in real systems the time constant increases with the area of the local region. In Kerava solar village, a Finnish full-scale seasonal storage heating experiment, the estimated time constant for continuous discharge is about 50 days.

In a pipe, most of the thermal transfer normally occurs at the bottom of the pipe, so that at startup the temperature field will be axially dependent, i.e. is strongly stratified. If the pipe is lined and there is air between the pipe and the ground, destratification can be encountered as observed in the laboratory experiment: after one hour's discharge the temperature in the upper end of the model was 1-2 C colder than the bottom. Thus, to avoid unexpected thermal performance in a pipe storage, open circulation systems are preferred, whenever possible.

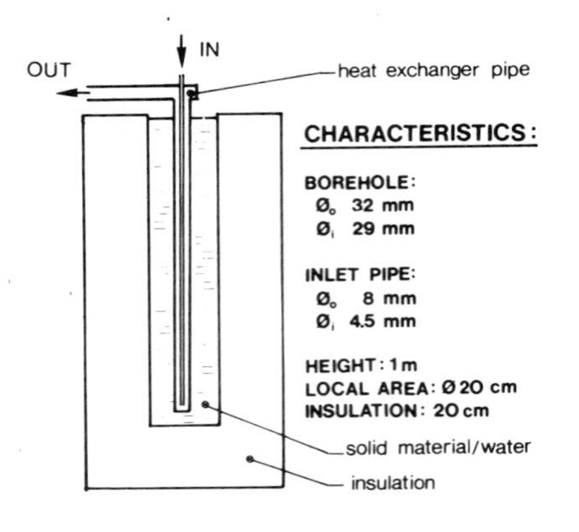

Fig 1. Principle of a model used in laboratory measurements

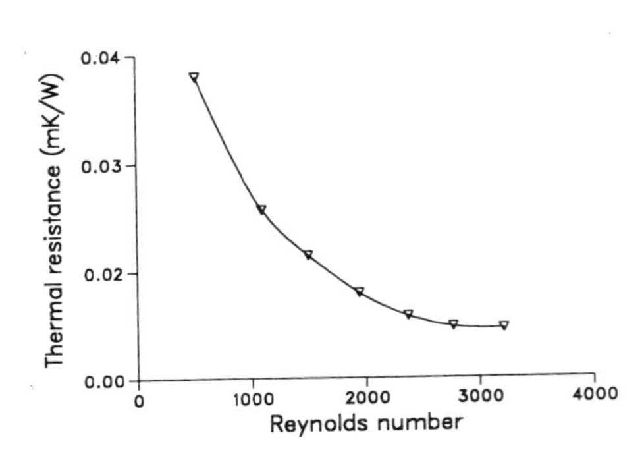

Fig. 2. Measured surface thermal resistance as a function of Reynolds number

A schematic of the constructed model for studying the surface resistance is shown in Fig. 1. The scale was 10:1 and the Reynolds number was dimensioned to 500-4000. The water around the pipe was mixed continuously to eliminate the conductive resistance of the surrounding material. The total measured resistance consisted then of the resistances caused by the borehole pipe wall and the surface resistance.

The surrounding water, or local pipe storage region, was discharged from 30 °C with a constant mass flow and a ΔT of 25 °C. For separate measurements, the mass flow rate could be varied between 0-9 l/min.

The average temperature was recorded as a function of time and the time constant was defined from the temperatures obtained. As all material parameters were known, the surface resistance could be calculated from the results. Measured results are listed in Table 1 and shown in Fig. 2.

TABLE 1 Measured and Calculated Values of Surface Thermal Resistance

m (l/min)	Re	τ (min)	R_f (mK/W)	Nu	h_w (W/m^{2}K)
1.35	516	73.6	0.03797	14.71	524
2.86	1094	50.0	0.02570	21.73	774
3.91	1495	41.8	0.02144	26.05	928
5.07	1939	35.0	0.01791	31.18	1111
6.19	2367	30.8	0.01572	35.52	1265
7.24	2768	28.8	0.01469	38.01	1354
8.40	3212	28.5	0.01453	38.43	1369

DISCUSSION

The laboratory experiments demonstrated that the new analytic model can be used to predict the time dependence of the power rates in pipe storage. The time constant of the storage is dependent on the material properties of the ground and the flow conditions inside the pipe, so the preliminary designs of duct storages can also be accomplished by the method. A comparison has been made for a hypothetical TES against a numerical model (Lund, 1987b). The results are shown in Fig. 3. Though the analytic model seems to overestimate slightly the pipe power effect during the startup and decays faster than the numerical approach, the agreement is satisfactory.

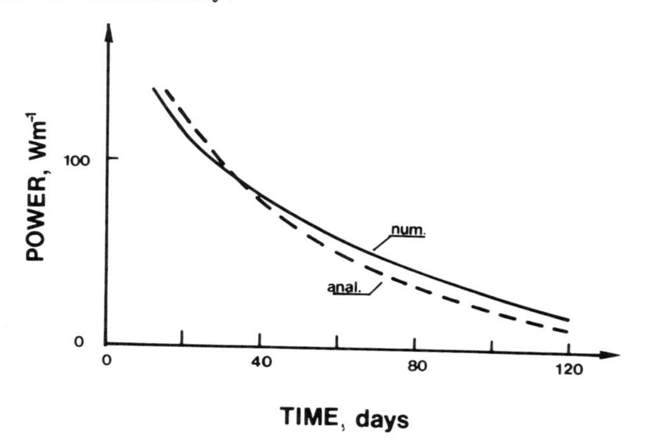

Fig. 3. Comparison of the analytical and numerical heat exchanger pipe model. Constant fluid inlet temperature and flow rate.

Normally the mass flow rates during charge and discharge periods are not constant, but vary with time. For example, in Kerava solar village charging and discharging occur only during daytime, so that the mass flow rate is a step function, step length being eight hours. According to the dynamic character of the model, the time dependence is easily included in the model. In another paper, presented at this conference (Lund, 1987b) results of simulations for the Kerava system are presented. Even then a good agreement between the measured and calculated results was shown.

REFERENCES

Lund, P.D. (1987). Fundamentals of thermal processes in a hybrid thermal energy storage. Helsinki University of Technology, Report TKK-F-A609.

Lund, P.D. (1987). A hybrid thermal energy storage design. Paper in the ISES Solar World Congress 1987. Hamburg, September 13-18.

Claesson J. and Hellström G. (1981). Model studies of duct storage systems. Paper in the IEA-Conference in Berlin.

Claesson J. and co-workers (1985). Markvärme, En handbok om termiska analyser, del II Värmelager. Byggforskningsrådet.

CENTRAL HEATING AND COOLING PLANT FOR YEAR-ROUND AIR CONDITIONING AND HOT WATER USING SOLAR ASSISTED HEAT PUMP

Enrique Alaiz

COBRA, S.A.
General Moscardó, 3
28020-MADRID
SPAIN

ABSTRACT

This paper describes a central heating and cooling plant operation, designed and constructed to supply year round air conditioning and hot water for a 20.000 m² residential, academic and office complex distributed in five buildings in Hoyo de Manzanares, Madrid. It also includes an analysis of the expected energy output and consumption.

The system consists basically of a 375 kW water-to-water solar assisted heat pump. The low temperature solar system is composed of 960 flat plate, selective solar copper collectors with a total absorption area of 2045 m², two storage tanks with capacities of 700 m³ and 100 m³ and absorption chiller units.

In a working operation while in the heating mode, the auxiliary energy required for heating is derived from electric heaters which are connected during the night and the heat is stored in the 700 m³ tank.

The analysis of the energy consumed and that supplied by solar systems and nighttime electrical energy storage shows the scarce utilization made of the daytime electrical energy as well as the high percentage of solar energy available.

The results obtained in the first two years of starting up and regulation of the system confirm the energy previsions and show the high efficiency of the solar assisted pump.

KEYWORDS

Solar heating and cooling; heat pump; absorption chiller; 2045 m² flat plate collectors; nighttime electrical storage; microprocessor control; security programs; automatic maintenance.

DESCRIPTION OF THE OPERATION MODE

The operation of the system is carried out through several different program options which can be executed simultaneously. These programs are:

- Solar energy collection and storage
  - Summer
  - Winter
- Heating
- Nighttime electrical storage
- Cooling
- Domestic hot water
- Heat dissipation. Security and automatic maintenance

SOLAR ENERGY COLLECTION AND STORAGE (Circuit B1-B2)

In winter solar energy is collected and stored for heating and domestic hot water supply. These processes are simultaneously carried out whenever the collectors produce useful heat at a temperature which exceeds the temperature of either one of the two tanks. In summer this solar radiation is collected and stored in the 100 m^3 tank for the production of domestic hot water and for chilled water production in the absorption unit at temperatures above 70ºC.

In winter the energy collected is stored in both tanks, priority being given to the low temperature tank of 700 m^3, until it reaches the maximum admission temperature of the heat pump evaporator (23ºC).

After that, solar energy is stored in the 100 m^3 hot water tank to be used directly for heating, until the tank reaches its maximum admissible inlet temperature in the fan coils and climatizers of the building (65ºC). When this temperature is reached, the storage process continues in the 700 m^3 tank.

This process is controlled by means of circulation pumps Bl and B2 and two 3-way valves that direct the flow to each of the storage tanks.

As long as the 700 m^3 tank reaches temperatures over 23ºC the inlet temperature of the water in the heat pump evaporator is controlled by the 3-way valves installed for this purpose.

HEATING (Circuit B3, B7 and B13)

The 700 m^3 tank is the low temperature heat source for the heat pump. The condenser's heat is delivered to the distribution circuit where the 100 m^3 tank is located. The water from this tank passes through the condenser, where it is heated, then circulates through auxiliary electric heaters, where occasionally it is overheated to match transitional demand. This heated water is distributed to the fan coils of the buildings, by means of a distribution pipe ring, returning to the 100 m^3 storage tank. The flow rate through this ring is variable according to the demand. The heat pump has a ten-step graduated power control which gives flexibility and the necessary capacity to continuously match the variable heat demand of the building.

Two electronic thermostats, in response to a maximum temperature in the condenser and/or maximum temperature in the evaporators are in charge of sequentially connecting or disconnecting the ten steps.

The electric heaters consist of eight groups of electric resistors distributed in eleven vessels which are connected sequentially by a control which checks the temperature in the heated water for buildings at fixed time intervals. Heaters are used during the night when the electric rates are reduced. The quantity of heat delivered in the storage tank is controlled in function of the temperatures of the tanks and the daily average demand of heat.

The nominal heating capacity of the heat pump is 1.150.000 kcal/h. The maximum combined heating power of the plant is 2.150.000 kcal/h. The circulating pumps B7 and B13 are in charge of the heating flow.

COOLING (Circuit B11, B7, B10, B14, B9)

The necessary chilled water for the air conditioning systems of the buildings is delivered first by the absorption chiller, whenever the solar energy is capable of supplying water at a temperature over 70ºC to the unit.

In addition to the absorption chiller, the heat pump is used as an auxiliary system for the production of chilled water with a total cooling capacity of 300 tons.

The cooling capacity of the absorption chiller unit installed is 33 tons. This power capacity will be increased in the future when the solar-assisted absorption system is proved efficient.

DOMESTIC HOT WATER (Circuit B4, B5)

The production of domestic hot water is carried out by a flat plate parallel heat exchanger. The primary of these heat exchangers is fed with the water of the 100 m^3 tank. One electric heater is used as an auxiliary heating system. The capacity of this hot water system is 8500 liter/hour of water at 50ºC.

NIGHTTIME ELECTRICAL STORAGE (Circuit B2)

When solar energy is not available day after day due to weather conditions, or heating demand is over the capacity of the solar energy supply during several days, the 700 m^3 storage tank reaches its lowest temperature for the heat pump (10ºC). Under these conditions the nighttime electrical storage program will start connecting the electric heaters during the night when low electric rates are in effect.

SECURITY AND AUTOMATIC MAINTENANCE. HEAT DISSIPATION

The automat that controls the system actuates several different security programs, for example:

- Start-up of the reserve pump (in case of main pump failure)
- Valve position check
- Alarms (in case of malfunction, etc.)

Another program establishes a heat dissipation circuit through the cooling tower as a security measure for the tanks, flat plate solar collectors and in several hot fluid pipes (Circuit B2).

Still another group of programs permits an automatic system check of the proper functioning of each of the components in the entire system.

ENERGY BALANCE

This diagram shows the breakdown of the energy inputs used to satisfy the heating load. To determine the solar energy inputs the simulation program of the Ø F-Chart method has been used. A daily temperature model for the tanks was used to calculate the heat pump working conditions.

Due to the difficulty of a rigorous simulation of the total system, several simplifications were necessary and hence the end results must be considered approximate. However, these previsions are starting to be confirmed by the data available after the first year of start-up and regulation of the system.

CONCLUSIONS

This solar energy assisted heat pump system is able to supply 80% of the total energy demand in combination with the electric nighttime heat for the five building military complex in Hoyo de Manzanares, Madrid. It is interesting to point out the low daytime electric heat input estimated to be 5%, which together with another 20% of electric energy absorbed by the compressor of the heat pump makes a total of 25% of daytime energy consumption compared to conventional systems, which accounts for the low energy cost of the installation.

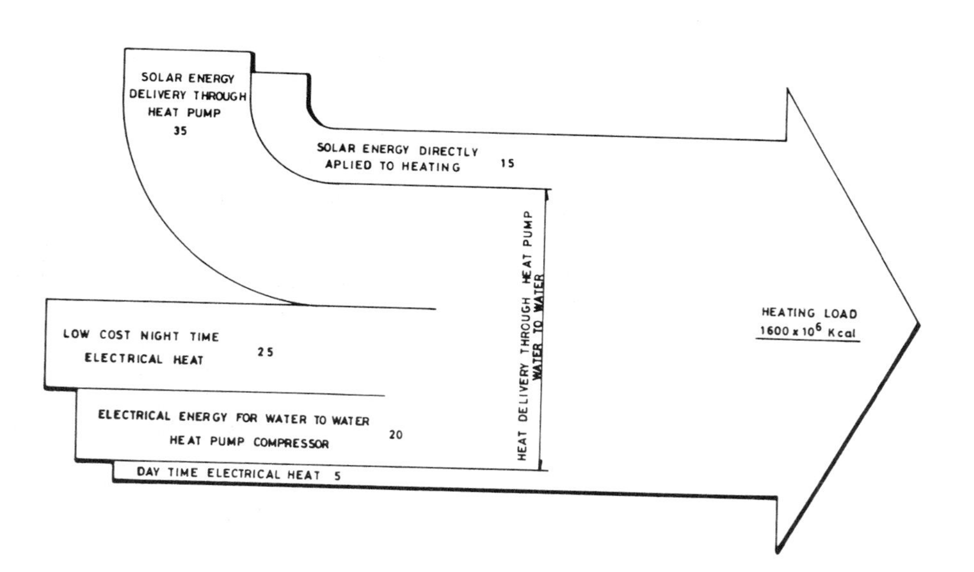

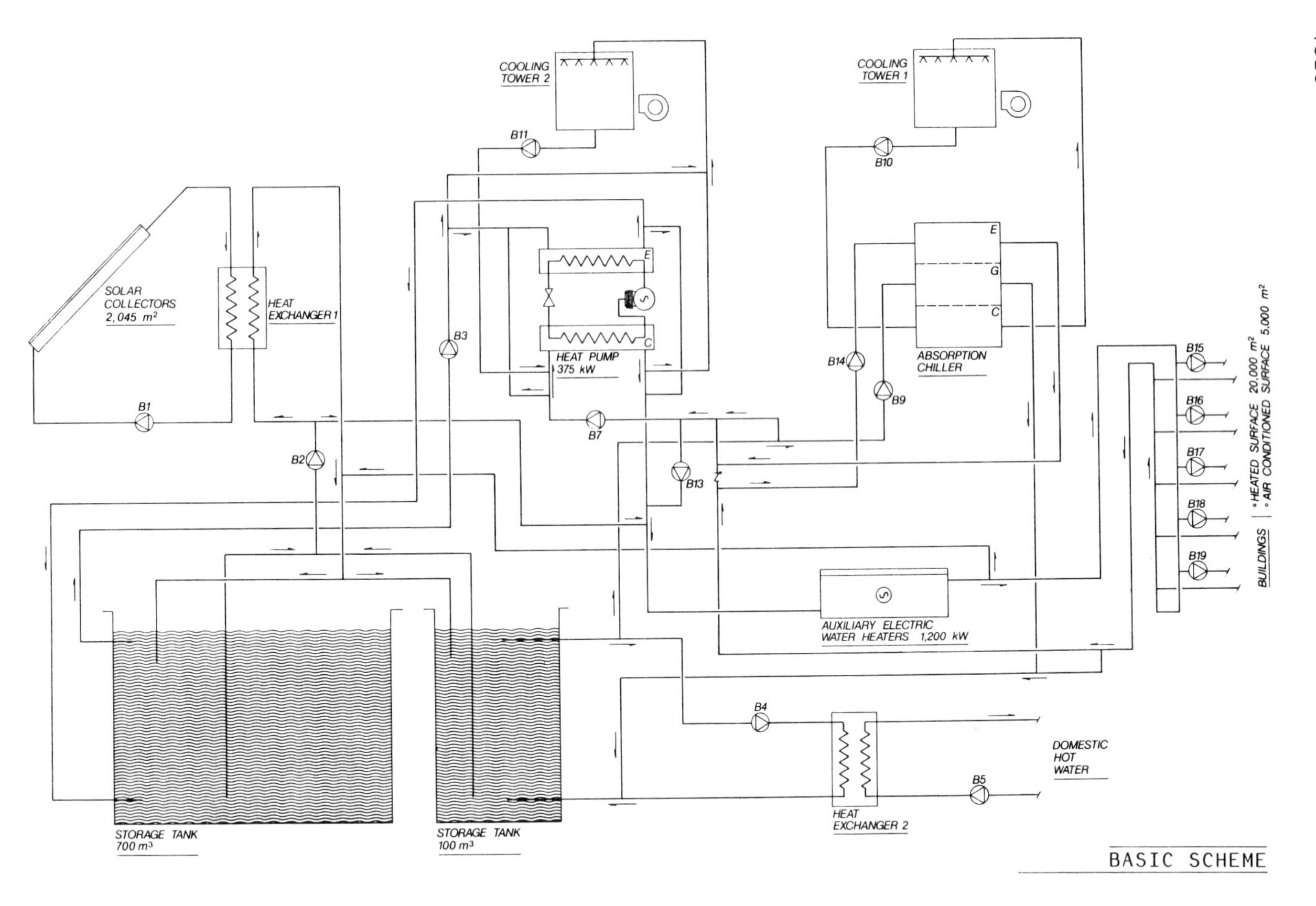
COOLING TOWER 2
COOLING TOWER 1
B11
B10
SOLAR COLLECTORS 2,045 m²
HEAT EXCHANGER 1
E
G
C
B3
HEAT PUMP 375 kW
B14
ABSORPTION CHILLER
B9
B1
B7
B2
B13
BUILDINGS
• HEATED SURFACE 20,000 m²
• AIR CONDITIONED SURFACE 5,000 m²
B15
B16
B17
B18
B19
AUXILIARY ELECTRIC WATER HEATERS 1,200 kW
B4
DOMESTIC HOT WATER
B5
HEAT EXCHANGER 2
STORAGE TANK 700 m³
STORAGE TANK 100 m³
BASIC SCHEME

SIMULATION PROGRAMME INVESTIGATES: WHICH TYPE OF COLLECTOR IS BEST FOR SELFSUFFICIENT SOLAR SPACE HEATING IN MIDDLE EUROPE ?

Ulrich Radons
German Section of ISES
Beethovenstr. 11, 8068 Pfaffenhofen, BRD

ABSTRACT

Costly mistakes and serious backsteps in the reputation of the Solar idea are provoked if applications of Solar Systems and their components are not selected with proper foresight. In account of the changing dayly and seasonly weather and the complexity of the systems it is essential to optimise systems in a simulation programme before they are built.

In search of the most cost efficient way to realise selfsufficient solar space heating for middleeuropean conditions such a simulation programme was used. It was investigated, which type of collector would be best: A flat plate collector with a wide sampling characteristic but high losses or a concentrating type with a narrow sampling characteristic but low losses.

It was found that the effect of sampling characteristic on annual heating capacity was only 12 %. The effect of loss was found to be large. Reducing the loss coefficient from 3 W/m²deg as common for good flat plate collectors down to 0,3 W/m²deg as possible for concentrating collectors the heating capacity will increase by 75 %.

The investigation showed that a nontracking reflective trough collector will do better than a flat plate collector, better even than the expensive vacuumpipe collector.

KEYWORDS

Solar house; computer simulation; optimal collector; reflective trough collector; vacuum pipe collector; nontracking; energy autarc; self-sufficient; space heating; survey.

INTRODUCTION

Dwindling resources and increasing pollution ask for new ways in all aspects of energy application. In order to find the optimal way to heat space singulary solar it was decided to test solar house components and their cooperation in computer simulation. Amongst other features it was to be investigated which type of collector would be best.

COLLECTOR PERFORMANCE

Good collectors need high acceptance of solar radiation and low losses. The commonly used flat plate collectors have good acceptance but high losses. Low losses are offered by concentrating reflective trough collectors. These collectors are cheap and simple if they are nontracking. But will they collect enough of the direct and diffuse radiation in spite of their narrow sampling characteristic ?

To find out their characteristic and how it can be simulated a collector was build and tested. The concentration ratio was 1.7 which was thought to be small enough for good sampling. The orientation of the absorber axis was put horizontal. The collector, as superimposed in Fig. 1, is more shallow than the designs presented by Rabel, O´Gallagher and Winston (1980) and allows the incorporation of an insulating vacuum tube. For this design a loss coefficient of 0,43 W/m²deg applies. There is no double reflection and for a given concentration rate the reflector rim is closer to the absorber. This allows better diffuse pick up.

The predicted sampling efficiency η' is shown in Fig. 1 approximated by a cosine rule. The maximum sampling efficiency is η_s . The sampling window α_s is defined as the width of the characteristic halfway between maximum and minimum sampling efficiency.

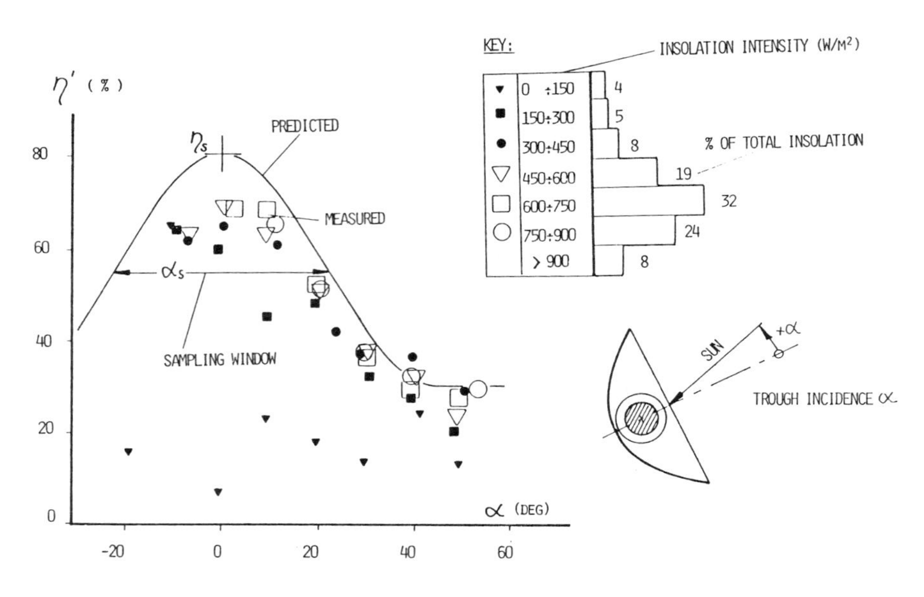

Fig. 1 Sampling Efficiency η' measured for different Insolation Intensities.

Tests were done over a year under real weather conditions in Pfaffenhofen BRD. Measurements of radiation intensity and sampling efficiency were taken continuously. The caloric bomb principle was used with an integrating time of one hour. All the points taken in September/October 1985 were averaged for different ranges of insolation intensity and trough incidence. They are plotted in Fig. 1. They are all reasonably close to the prediction except for the very low intensities where the diffuse radiation is dominating and poorly directioned. As shown by the Key of Fig. 1 these points represent only 4 % of the total radiation and can be neglected. The points above 150 W/m²K show good sampling efficiency. This proves that almost all the radiation coming down was fit to be focussed. The fact that the prediction was not quite met was possibly due to impurities and buckling of the mirror. So the predicted values were taken for the simulation procedure with some extra allowance, for the loss of performance at lower insolation intensity.

SIMULATION SYSTEM OF SOLAR HOUSE

Figure 2 shows on the right a summary of what the computer programme does and on the left the simulated solar house and its basic components. The collector array is arranged at a steep tilt to give optimum performance in winter (hight of sun, snow not to settle). The heat flows are indicated by arrows. For calculating the way of sun and the radiation incident onto the collector some of Revfeim's (1982) formulas were used. The heat gained in the absorbers can either be used directly for heating or to heat up the heat storage positioned in the ground. This will be handled by control criteria set to minimize losses. The programme can handle storage in one phase media and in phase change mixtures. Melting temperature, melting energy and specific heat of the mixture can be selected. The hot water storage is integrated in the heat storage so that all hot water heat losses will be caught and used for heating.

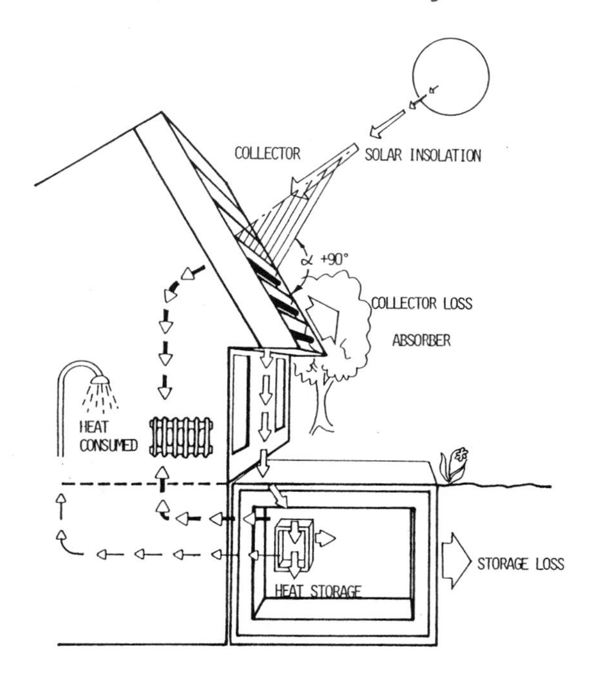

SIMULATION PROGRAMME

INPUT

GEOGRAFIC position of house and angular arrangement of collector

WEATHER of the year: hourly global radiation, day/night temperatures

HEAT DEMAND criteria of house: heat loss coefficient, heat capacity, hot water usage, desired room/water temperatures

COLLECTOR characteristics for sampling and loss, size of collector

HEAT STORAGE characteristic for storage and loss, size of heat storage

CONTROL CRITERIA

GENERATION

WAY OF SUN relative to collector

COLLECTOR OUTPUT from way of sun, weather

HEAT MANAGEMENT for:
collector: to storage or heating, loss
storage : input, output, loss
house : heating, heat reluctance
hot water: heating, usage, loss to storage

OUTPUT

MINIMUM STORAGE TEMPERATURE reached in the course of the year

MONTHLY HEAT: collected, consumed, stored, lost.

Fig. 2 Solar House and its Simulation Programme

SIMULATIONS AND RESULTS

The simulations were aimed at heating a house which would conventionelly use 2000 ltr light oil per year. The location is Pfaffenhofen BRD at a latitude of 48.5°. The orientation of the roof is 35° SSW, the tilt is 64.5°. The collector area was fixed at 80 m² and the heat storage at 40 m³. A salt/water mixture crystallizing at 48 °C was used as storage medium (Furbo, 1982).

The simulation procedure always started at the end of September with the heat storage full. Minimum storage temperatures would be reached in February/ March and the storage would refill again towards April/June. The heat demand criteria were always adjusted to meet a minimum storage temperature of 30 °C. From these the annual heating capacity was calculated.

For investigation of the sampling characteristic the sampling window α_s was varied between 28° and ∞ . The other data were kept as predicted. The results in Fig. 3 show that the effect is very small, which means that a narrow sampling characteristic is hardly a handicap to the collector. The

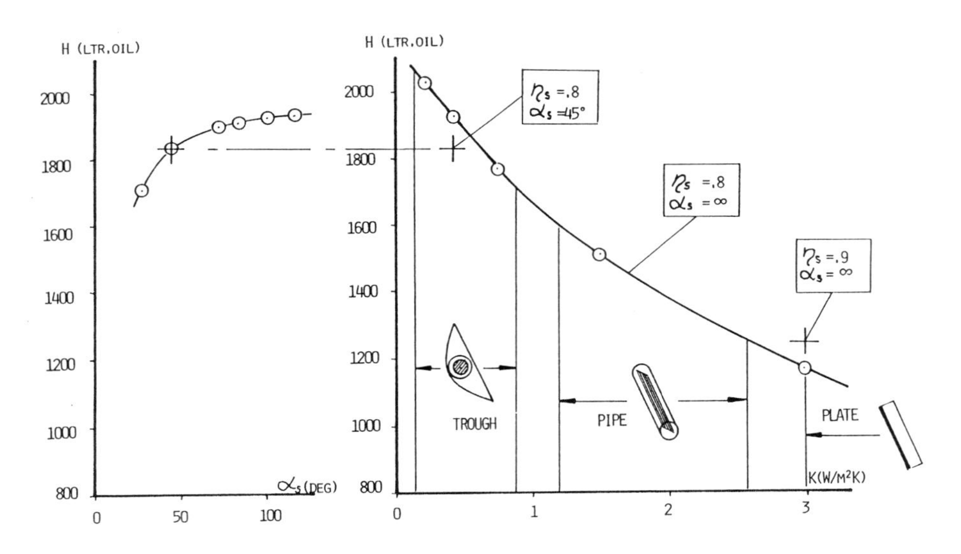

Fig. 3 Effect of Sampling window α_s and Collector Loss Coefficient K on annual Heating Capacity H

reason is that in critical wintertime, when full performance of the collector is essential, the sun will shine mainly in a narrow band of trough incidence. In Summer, when the system goes almost idle, bad performance does not matter.

The effect of collector loss was researched at a flat sampling characteristic ($\alpha_s = \infty$). The results (Fig. 3) show a very pronounced improvement at low losses in the range reserved to the reflective trough collectors.

Even with allowance for the sampling characteristic, performance is excellent for the reflective trough. The vacuum pipe collector does not reach its class, even with no corrections done to allow for its unfavourable orientation of absorber axis. In the range of flat plate collectors performance is quite poor. This does not change much if η_s is raised to 0,9 as a benefit for better, because of direct optics.

DISCUSSION

The computer simulation demonstrated that, at the task of singularly solar space heating in the center of Europe, reflective trough collectors give far better performance than any other collectors. This baffling fact gets understood, having in mind, that most of the socalled diffuse radiation is directioned well enough to be focussed at low concentration. At concentration below and around 1.7 the troughs don´t have to be tracked. This is because in the critical wintertime the sun keeps in a quite narrow band of trough incidence.

Reflective trough collectors are not as simple as flat plate collectors, so are they cheap and practical ? They are not so bad: because of their small absorbers and lightweight mirrors there is less weight involved than in a flat plate collector. At mass production the more complicated design should not matter so much. Considering the good optical sampling performance at 1.7 concentration it should be worthwhile to further increase concentration and save further material. Narrower sampling windows could be compensated by simple tracking facilities if necessary.

Computer simulation of solar systems is a powerful tool to select promising items from extensive snick snack. In the course of the simulation exercise a lot of fascinating features got tried out and dismissed. It would have cost a lot of money and time and reputation if they all would have been tried out in reality.

REFERENCES

Rabl, A.; O´Gallagher, J.; Winston, R. (1980)
Design and Test of nonevacuated solar collectors with compound parabolic concentrators.
Solar Energy 25 335-351

Revfeim,K.J.A. (1982)
Simplified Relationships for estimating solar radiation incident on any flat surface.
Solar Energy 28 509-517

Furbo, S. (1982)
Heat Storage Units using salt hydrate.
Sunworld 6 134-142

DESIGN AND DEVELOPMENT OF A NEW GENERATION OF EVACUATED COMPOUND PARABOLIC CONCENTRATORS

Joseph J. O'Gallagher and Roland Winston
The Department of Physics and the Enrico Fermi Institute
The University of Chicago, Chicago, IL 60637, U.S.A.

William Schertz
Argonne National Laboratory, Argonne, IL 60439, U.S.A.

ABSTRACT

The integrated Compound Parabolic Concentrator (ICPC) evacuated tubular collector has the potential to become the general purpose solar thermal collector for the future. To date it has remained a highly successful experimental concept which has demonstrated that very efficient high temperature performance from fixed non-tracking concentrators is a technical reality. Several initiatives are now underway with the objectives of overcoming the remaining technical and practical barriers. One of these uses a large diameter (125mm) borosilicate glass tube with a specially configured internal shaped mirror CPC coupled to a U-tube absorber. Another is scaled to be compatible with the use of presently available fluorescent glass tubing and production tooling.

KEYWORDS

Evacuated tube, CPC, Heating, Cooling, High-Temperature.

INTRODUCTION

The combination of non-tracking Compound Parabolic Concentrator (CPC) reflectors (1) with evacuated absorbers and spectrally selective coatings (2,3) brings together three of the most significant developments in solar collector technology of this century. The incorporation of a high reflectivity stationary concentrator (using nonimaging CPC type optics) directly inside the vacuum enclosure itself is shown for two alternative methods of implementation in Fig. 1. The concentration substantially reduces heat losses and results in significantly improved high temperature performance. The integrated design enables the concentration to be done while maintaining high optical efficiency which ensures that excellent low and moderate temperature performance is also achieved. The characteristically long lifetimes typical of properly designed evacuated collectors are retained as well.

Several years ago, about one hundred prototype tubes based on this concept were fabricated and a test panel successfully demonstrated superior high temperature performance over more than three years of testing at the University of Chicago (4). The technical goal of the present research is to develop an advanced fully stationary (fixed mount) solar collector capable of delivering thermal energy at a temperature of 175°C with an annual efficiency of 50% while at the same time having a performance and cost which will be competitive with existing collectors at domestic hot water temperatures.

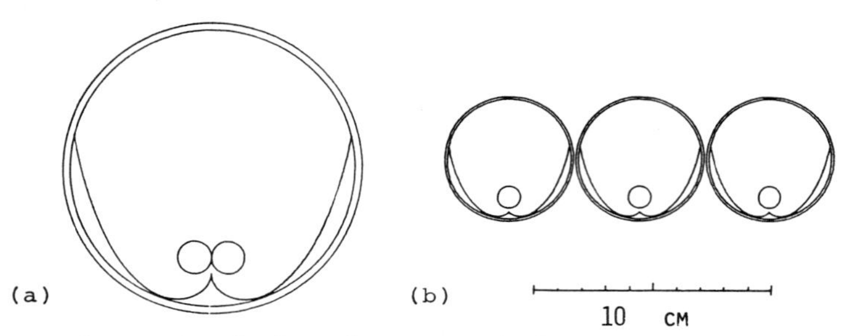

Fig. 1. Two alternative designs for a practical manufacturable advanced integrated CPC (ICPC): a) A lower concentration version (1.3x) using a large tube and a U-Tube absorber which minimizes fluid inventory, b) A high concentration version (1.7x) scaled to standard fluorescent tube diameters with a circular absorber tube and a counter-flow feeder tube inside.

UNIQUE POTENTIAL OF INTEGRATED CPC COLLECTORS

1) System Flexibility and Performance: In Fig. 2, the calculated peak performance for a 1.6X integrated CPC, based on a model derived from the earlier pre-prototype measurements, is compared with other conventional solar thermal collectors. CPC's are the only simple and efficient method for delivering solar thermal energy in the temperature range from 100°C to about 300° C without tracking. They are also a very efficient source of low temperature (30°C to 100°C) heat with a performance which is virtually independent of both delivery and ambient temperatures. The annual energy delivery, calculated from hour-by-hour simulations based on Typical Meterological Year (TMY) data for several locations in the United States, is shown in Fig. 3. This demonstrates the excellent performance of CPC's relative to all other collector types in a wide range of climates. Thus CPCs provide the most universally adaptable source for any and all solar thermal energy systems.

2) Innovative Technology: CPCs embody a completely new optical technology, invented less than 15 years ago. The first commercial CPCs simply combined the new collecting optics with existing dewar type evacuated absorbers. However it has become clear that these designs are not optimized with respect to either performance or economics of fabrication. The concept remains a highly successful experiment. The development of a real operational collector subsystem incorporating the technical features of the preprototype still requires a concerted research effort with the full participation of a potential manufacturer.

3) Potential low cost: Recent research on varieties of integrated evacuated CPC's suggests that these types of collectors can be fabricated and deployed at a cost equal to or less than that for current flat plate collectors. Inexpensive glass tubing, already produced in very large quantities for fluorescent lamps can avoid prohibitive tooling costs. Light-weight, inexpensive reflector substrates and materials are being developed. New designs with a center-flow absorber will eliminate any need for expensive bellows. There are promising new concepts for glass-metal seals with potential low costs. Finally we note that there is an economic advantage to moderate concentrations which reduces the cost/aperture for absorber material and coating.

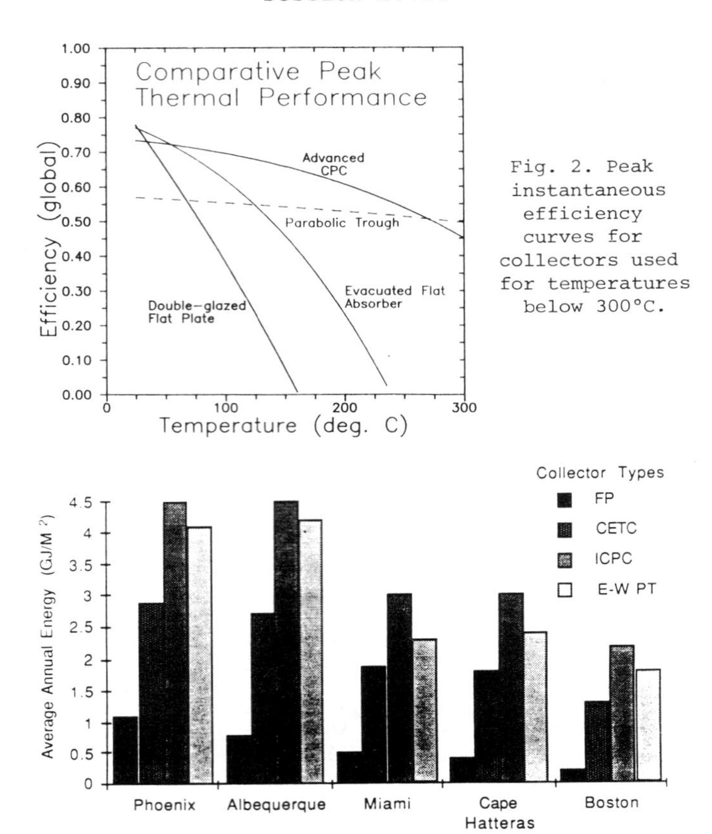

Fig. 2. Peak instantaneous efficiency curves for collectors used for temperatures below 300°C.

Fig. 3. Comparison of the annual energy delivery characteristics for a flat plate (FP), contemporary evacuated tube collector with external reflectors (CETC), the integrated CPC and an East-West Tracking Parabolic Trough (E-W PT).

CURRENT STATUS OF INTEGRATED CPC TECHNOLOGY

1) Tube design: The optical and thermal design parameters are well understood. The pre-prototype panel of tubes, in which the CPC was formed by shaping the outer glass tube, maintained a peak efficiency of more than 50% at 200°C over several years of testing(4). A variety of trade-off studies for a manufacturable prototype definition have been carried out, but a final design needs to be optimized subject to mechanical fabrication requirements. A shaped metal concentrator reflector insert, preferred at present to reduce tooling requirements, is under development. Practical methods for optical alignment of concentrator and absorber, thermal isolation of absorber, etc., alternative heat extraction procedures, and suitable glass to metal seals need to be developed further.

2) Collector: The methods for assembly and connection of an appropriate number of tubes to form a full collector have an important impact on system design and performance (5), (in particular pumping requirements and parasitic thermal

losses) and need further study. Configurations for integrating the concentrating tubes into building components (walls, roofs, etc.) have been suggested and need further development.

3) Array: Techniques appropriate for minimizing collection and distribution losses in large arrays of other collector types are not likely to be applicable for ICPCs operating at moderate to high temperatures. Therefore, development of innovative array layout configurations and heat transport systems will be necessary.

4) Materials: Candidate materials with the properties necessary to produce an ICPC with the excellent performance characteristics shown in Fig. 2 are available today. The most critical materials are those used for the reflector surface and the absorber selective coating. Several possible choices in each area are under review and evaluation at The University of Chicago.

5) Supporting Sybsystems: This is another area requiring considerable development. No advanced CPC high temperature system has ever been built. Particular attention needs to be paid to control strategies and subsystems, heat transfer fluid transport and storage, and stagnation protection.

POTENTIAL APPLICATIONS

General Comments: If the solar thermal source is ever to be effective for widespread use in the temperature range from 100°C to 300°C, for both small scale and diffuse climate applications, it will have to be done with CPCs. An advanced CPC collector will provide an efficient source of high temperature solar thermal energy with an inexpensive and uncomplicated system. This can be as competitive or more competitive with natural gas and fossil fuels for any application at these temperatures as flat plate collectors are at domestic hot water temperatures. Specific examples:

-<u>Conventional low temperature applications</u>: The performance of the Advanced CPC is competitive with existing collector technologies even at temperatures below 100°C (Fig. 1). It's potential low cost will make it quite attractive for well established solar applications, such as domestic hot water heating and conventional solar space heating and serve as a market entry opportunity for early production.

- <u>Improved space heating systems</u>: The relatively constant high efficiency of the Advanced CPC below 100°C means that space heating systems can be designed to take advantage of delivery temperatures somewhat higher than are practical with conventional collectors. This can mean less stringent requirements on heat-exchangers, storage volumes, system control strategies, etc. and result in less expensive and better performing systems.

-<u>District heating applications</u>: The flat efficiency characteristic of the Advanced CPC will make it an ideal collector for district heating applications and/or charging large seasonal storage capacities since its performance will not vary significantly with demand temperature or ambient temperature.

-<u>Cooling</u>: The ability of an Advanced CPC to provide an efficient source of solar heat at temperatures between 110°C and 175°C, makes viable several cooling technologies which are otherwise only marginal. Since tracking parabolic troughs are not likely to be practical in residential scale systems or effective in many environments where cooling is nevertheless required, the development of an Advanced CPC is absolutely necessary if space cooling is ever to become a reality. Possible cooling systems include double-effect or multistage regenerative absorption cycle chillers, desiccant/reactants inside the tube itself which might make an integrated refrigeration cycle possible, and high temperature desiccant systems with significantly improved performance at elevated temperatures.

-Steam: Advanced CPC collectors can be designed to generate steam in situ. Systems for the transport and use of hot steam are very well developed and familiar to most building engineers and installers. The potential for this application needs to be quantitatively explored.

-Process heat: There are a great variety of end use applications for process heat used by the chemical, food and material processing industries. Much of the demand lies at temperatures for which flat plate and conventional evacuated collectors are too inefficient and in environments for which tracking troughs are impractical and/or ineffective. Many of these would be ideal loads for an Advanced CPC.

-Co-generation: The development of an Advanced CPC will provide a fixed, non-tracking collector capable of driving systems combining solar heating, solar cooling and solar electrical power generation. The elimination of the tracking requirement will greatly expand the number of potential sites which may be suitable for such systems.

DEVELOPMENT PLAN

The performance and range of applications outlined above provide strong arguments for a concerted effort to develop practical collectors and systems utilizing these concepts. Presently two different paths are being pursued.

The first of these involves an international effort to design, fabricate, test and model the performance of a new manufacturable ICPC. The participants include The University of Chicago and Colorado State University in the U.S., an evacuated collector manufacturer in France, the DFVLR in Germany and The University of Geneva in Switzerland. The proposed design, shown in Fig. 1a), uses a large diameter (125mm) borosilicate glass tube with a specially configured internal shaped mirror CPC coupled to a U-tube absorber. Performance projections in a variety of systems applications using the computer design tools developed by the International Energy Agency (IEA) task on evacuated collectors are being used to optimize the optical and thermal design. Another effort involves the participation of a major U.S. manufacturer in a joint project with Chicago and Argonne National Laboratory to develop ICPC components scaled to be compatible with the use of presently available fluorescent glass tubing and production tooling. A profile of the baseline design is shown in Fig. 1b).

The major activity will be a hardware development effort, in cooperation with one or both manufacturers, to fabricate and test a substantial number of prototype evacuated CPC tubes of the designs in Fig. 1. This will be accompanied by analytical modeling to define collector and array performance specifications and to design a moderate size prototype array. At the same time systems studies will be carried out to identify the most appropriate high performance chiller technology for a prototype experiment.

ACKNOWLEDGEMENT

This work was supported by U.S. Department of Energy grant DE FG02 85SF15753.

REFERENCES

1) W.T. Welford and R. Winston, Optics of Nonimaging Conc., Acad.Press,N.Y.1978.

2) J.D. Garrison, Solar Energy 23, 93 (1979).

3) B. Seraphin and A. Sievers in Topics in Applied Physics 31: Solar Energy Conversion, ed. B. Seraphin, Springer-Verlag, N.Y. (1979).

4) K.Snail, J. O'Gallagher, R. Winston, Solar Energy 33, 441 (1984).

5) J. O'Gallagher, R. Winston, Proc. IEA Workshop, San Diego, CA, p.487, 1984.

Thermal Analysis And Operational Limits Of Compound Parabolic Concentrators In Two-Phase Flows With Saturated Exit States

Ahmed Y. El-Assy

Department of Mechanical Engineering (Power and Automotive)

Faculty of Engineering, Ain Shams University

Cairo, Egypt

Abstract: *Compound Parabolic Concentrators (CPC's) in two-phase flows are studied and their thermal performance is analyzed. New parameters identifying the CPC optical and thermal performance are defined. The limits of operation of CPC's in two-phase flows for saturated fluid exit states are identified as a function of the solar flux and the CPC working parameters such as the working pressure and the coolant mass flow rate. The effect of the CPC concentration ratio on the thermal performance and the operational limits is studied. It is shown that CPC's in two-phase flows have greater thermal efficiency than similar CPC's in single-phase flows under same working conditions. It is also shown that the working pressure can be increased as the concentration ratio is increased thus, increasing the thermal conversion efficiency.*

1 Introduction

The Compound Parabolic Concentrator (CPC) was invented by Roland Winston in the mid-1960's but it was not until 1974 that the suitability of the CPC for solar energy collection was recognized in the USA [1]. Sponsered by the U.S. Department of Energy after the oil embargo in 1973, extensive research for the development of the CPC took place at Argonne National Laboratories. Winston and Hinterberger [2] developed the theory of light collection within the framework of geometrical optics. Giugler et al [3] and Thodos and Winston [4] developed principles of cylindrical concentrators and concepts of CPC's used for solar thermal power generation. Once the principles were established, Rabl, et al [5] worked on practical design considerations and their effects on the CPC performance. Collares-Pereira, O'Gallagher, Rabl and Winston [6] and [7] studied the effect of evacuated receivers which enabled them to build a CPC operating at $300°C$. Also they studied the relation between high temperature performance and optimum concentration ratio [8]. Their results show that with evacuated receiver tubes, at $300°C$, reasonable efficiency is feasible for fixed collectors. On the other hand, at lower temperatures, at about $100°C$, calculations indicated that even non-evacuated CPC collectors with proper design can operate with acceptable efficiency, surpassing that of flat plate collectors [9] and [10]. Hsieh [11] designed a system using CPC collectors to produce industrial process steam. He also performed a comprehensive thermal analysis of a CPC using the Hottel-Whillier-Woertz-Bliss formalism [1]. His predicted results compared well with experiments. El-Assy and Clark [12] performed a detailed analysis of the optical and thermal performance of CPC's working under non-boiling, boiling and superheated conditions. In this study [12], the authors developed a new formulation for the optical efficiency, η_o, the generalized heat removal factor, $\mathcal{F}_S$ and the generalized overall thermal loss coefficient, U_L. Also El-Assy [13] studied a Rankine cycle using a CPC working under superheated conditions for power generation. The economic viability of such a Rankine cycle is also discussed in a study by El-Assy [14].

The objective of the present work is to analyze the thermal performance of a CPC working under boiling conditions with saturated exit states. The operational limits of the saturated exit state conditions are presented as a function of the CPC working parameters such as the coolant mass flow rate, the solar flux, the properties of the coolant used, etc., and the concentration ratio of the CPC.

2 CPC Thermal Performance Parameters

The overall thermal efficiency of the CPC, η, is shown to be

$$\eta = \mathcal{F}_S \left[\eta_o - \frac{U_L(T_1 - T_a)}{IR} \right] \tag{1}$$

where $\mathcal{F}_S$ is the generalized heat removal factor, η_o is the optical efficiency, U_L is the overall thermal loss coefficient for a collector operating with multiphase flows, w/m^2K, T_1 and T_a are the inlet fluid temperature and the ambient temperature, K, and I is the total incident irradiation, w/m^2. R is the CPC concentration ratio defined as the ratio of the aperture area, A_c, to the receiver area, A_r, m^2. The optical efficiency, η_o, is formulated [12] as follows

$$\eta_o = \frac{A(\frac{C}{R}+1)}{(C+1)} \tag{2}$$

where A is a constant which involves the optical properties of the CPC materials and C is the ratio of the diffuse components to the direct components of the solar radiation. It is to be noted that in the absence of the diffuse components, C is equal to zero and the optical efficiency becomes constant independent of the concentration ratio, R. The formulation for $\mathcal{F}_S$ and U_L [12] for the general case where the coolant through the CPC channels undergoes non-boiling, boiling and superheating conditions are

$$\mathcal{F}_S = \frac{F'}{a_R}(1-e^{-a_R z^*}) + F'_B(1-z^*-z^{**})\,e^{-a_{Rb}z^*} + \frac{F'_S}{a_{SR}}\left(1-e^{-a_{SR}z^{**}}\right)e^{-a_{Rs}z^*} \tag{3}$$

$$U_L = \frac{1}{\mathcal{F}_S}\left[\frac{F'}{a_R}\left(1-e^{-a_R z^*}\right)U_{NB} + F'_B(1-z^*-z^{**})\,e^{-a_{Rb}z^*}U_B + \frac{F'_S}{a_{SR}}\left(1-e^{-a_{SR}z^{**}}\right)e^{-a_{Rs}z^*}U_S\right] \tag{4}$$

respectively. For the purpose of this study where the coolant exit states are limited to the saturated conditions, that is, no superheated conditions occur and the dimensionless superheating channel length, z^{**}, is zero, equations (3) and (4) for the saturated exit states become

$$\mathcal{F}_S = \frac{F'}{a_R}(1-e^{-a_R z^*}) + F'_B(1-z^*-z^{**})\,e^{-a_{Rb}z^*} \tag{5}$$

$$U_L = \frac{1}{\mathcal{F}_S}\left[\frac{F'}{a_R}\left(1-e^{-a_R z^*}\right)U_{NB} + F'_B(1-z^*)\,e^{-a_{Rb}z^*}U_B\right] \tag{6}$$

where U_{NB} and U_B are the thermal loss coefficients for the non-boiling and boiling regions through the channels, respectively, F' is the non-boiling collector efficiency factor defined [12] as

$$F' = \frac{\frac{1}{U_{NB}}}{\frac{1}{U_{NB}} + \frac{R_o \ln\frac{R_o}{R_i}}{k_w} + \frac{R_o}{h_l R_i}} \tag{7}$$

where k_w is the thermal conductivity of the receiver wall material, w/mK and h_l is the convective heat transfer coefficient, w/m^2K and F'_B is a similar factor for the boiling region defined as

$$F'_B = \frac{\frac{1}{U_B}}{\frac{1}{U_B} + \frac{R_o \ln \frac{R_o}{R_i}}{k_w} + \frac{R_o}{h_b R_i}} \tag{8}$$

where h_b is the two-phase boiling heat transfer coefficient. The dimensionless capacitance rate for non-boiling conditions, a_R, is defined as

$$a_R = \frac{F' U_{NB}}{(w/A_c)\, c_{pl} R} \tag{9}$$

where w/A_c is the coolant flow rate per unit collector area, kg/sm^2, and c_{pl} is the liquid phase specific heat, J/kgK. The dimensionless capacitance rate for boiling conditions, a_{Rb}, is defined as

$$a_{Rb} = \frac{F'_b U_B}{(w/A_c)\, c_{pl} R} \tag{10}$$

where F'_b defined as

$$F'_b = \frac{\frac{1}{U_B}}{\frac{1}{U_B} + \frac{R_o \ln \frac{R_o}{R_i}}{k_w} + \frac{R_o}{h_l R_i}} \tag{11}$$

The dimensionless non-boiling length, z^*, in equations (3) through (6) is defined [13] as

$$z^* = \frac{1}{a_R} \ln \frac{\eta_o - U_{NB} \frac{T_1 - T_a}{IR}}{\eta_o - U_{NB} \frac{T_{sat} - T_a}{IR}} \tag{12}$$

3 Saturated Exit Operational Limits

The limits of operation of a CPC having saturated exit states are shown in figure 1. The emphasis in this section is to identify those limits analytically. At small solar flux, especially for large inlet sub-cooling, $T_{sat} - T_1$, boiling will not occur and the collector will operate under single-phase, non-boiling conditions. These states are shown in figure 1 as the line A-A for z^*=1.0. The collector thermal efficiency for non-boiling conditions is obtained from equations (1) through (3) in which the value of z^* in equations (2) and (3) is unity. These states are illustrated in figure 1 by the line A-A corresponding to z^*=1.0. Another limiting case is that for saturated liquid inlet conditions. Since the fluid enters the collector channels as a saturated liquid boiling occurs from the collector inlet. The entire collector length is then under boiling conditions. This is the fully boiling condition shown in figure 1 by the line F-F. For these states there will be a limiting value of the solar flux, I^{**}, at which the fluid quality at the collector exit, x_e, is unity and the fluid at the exit is in an incipient superheated state. This is expressed as follows from a thermal analysis of the channel flow

$$wh_{fg} = F'_B A_c \left[I^{**}\eta_o - U_B \frac{(T_{sat} - T_a)}{R} \right] \tag{13}$$

Rearranging and introducing equation (9), equation (13) is rewitten as

$$I^{**} = \frac{h_{fg} U_{NB}}{a_R (F'_B/F') c_{pl} R \eta_o} + \frac{U_B (T_{sat} - T_a)}{R \eta_o} \tag{14}$$

Introducing a new dimensionless boiling capacitance rate, a_{RB}, defined as

$$a_{RB} = a \frac{F'_B}{F'} \frac{U_B}{U_{NB}} \tag{15}$$

the critical solar flux , I^{**}, for saturated vapor exit state is

$$I^{**} = \frac{U_B (T_{sat} - T_a)}{\eta_o R} \left[1 + \frac{1}{a_{RB}} \frac{h_{fg}}{c_{pl} (T_{sat} - T_a)} \right] \tag{16}$$

The thermal efficiency corresponding to the solar flux I^{**} determines the upper limit of the efficiency for the fully boiling condition. The lower limit of the efficiency for the fully boiling condition is the intersection of the limit F-F with the fully non-boiling line A-A, figure 1. The reduced temperature at which this intersection takes place is designated as $\left(\frac{\Delta T}{I}\right)^*$ and is derived as follows. For the fully non-boiling conditions, that is, at z^*=1.0, $\mathcal{F}_S$, equation (5), becomes F_R, that is,

$$F_R = \frac{F'}{a_R} (1 - e^{-a_R}) \tag{17}$$

For the fully boiling conditions, that is, at z^*=0.0, $\mathcal{F}_S$ is F'_B. Since the value of the efficiency at the intersection of the limits F-F and A-A is identical for the conditions of z^*=1.0 and z^*=0.0, it is clear that

$$F_R \left[\eta_o - U_{NB} \left(\frac{\Delta T}{IR} \right)^* \right] = F'_B \left[\eta_o - U_B \left(\frac{\Delta T}{IR} \right)^* \right] \tag{18}$$

Hence, the reduced temperature becomes

$$\left(\frac{\Delta T}{I} \right)^* = \frac{R \eta_o}{U_{NB}} \frac{[1 - F'_B/F_R]}{[1 - a_{RB}/(1 - e^{-a_R})]} \tag{19}$$

If the solar radiation intensity is greater than I^{**}, superheated exit states will occur. Because boiling conditions enhance the collector thermal performance by increasing its efficiency compared to the non-boiling conditions efficiency at the same coolant flow rate [12] and [13] and because superheating increases the thermal losses, thus decreasing the collector efficiency, there exist a point for each $I > I^{**}$ at which the collector efficiency has a deflection point (point B in figure 1). Each point B is actually the collector operating conditions at which the exit fluid is in an incipient for superheated exit state. The locus of points B is, therefore, the upper limit of operation for saturated exit states. Using equation (13) and recognizing that, in general, the collector area needed to provide an amount of useful energy equal to wh_{fg} is $A_c(1 - z^*)$ equation (13) is written as

$$wh_{fg} = F'_B A_c (1 - z^*) \left[I\eta_o - U_B \frac{(T_{sat} - T_a)}{R} \right] \tag{20}$$

Also it is shown [12] that

$$\frac{IR\eta_o - U_B (T_{sat} - T_a)}{IR\eta_o - U_B (T_1 - T_a)} = e^{-a_{Rb} z^*} \tag{21}$$

Substituting equation (21) into (20) and rearranging, the locus of points B is determined by

$$\frac{T_1 - T_a}{I} = \frac{R\eta_o}{U_B} - \frac{h_{fg}}{I a_{RB} (1 - z^*) c_{pl} e^{-a_{Rb} z^*}} \tag{22}$$

4 Results and Discussion

For an R-11 mass flow rate per unit collector area, w/A_c, of 0.002 kg/sm^2 and a working pressure of 0.32 MPa (T_{sat}=60°C) the CPC thermal efficiency is shown in figures 2 through 6 at different solar fluxes for concentration ratios, R, of 1, 2, 3, 4 and 5, respectively. The performance of the CPC of R=1, figure 2, is shown to be as follows: at solar fluxes equal to 300 and 400 w/m^2, boiling occurs enhancing the performance of the CPC compared to the non-boiling conditions represented by the line A-A, figure 2, and the exit conditions are saturated with an exit quality, x_e, less than unity. The inlet temperature being T_{sat}, the exit quality, x_e, is exactly unity when the solar flux reaches the value I^{**}, equation (16), shown to be equal to 488 w/m^2 for the conditions represented in figure 2. As the solar flux is increased to values greater than I^{**}, superheated conditions will occur at reduced temperatures, $(T_1 - T_a)/I$, identified as points B determined by equation (22). As seen from figure 2, the reduced temperature B decreases as the solar flux is increased. The CPC efficiency drops at reduced temperatures greater than that determined by point B at a specified solar flux as the superheated conditions increase the thermal losses. Furthermore, as seen from figure 2, at solar fluxes greater or equal to 800 w/m^2, the CPC efficiency drops as the CPC channels are mostly under superheated conditions.

As the concentration ratio is increased, the critical solar flux, I^{**}, decreases considerably and the coolant undergoes superheated conditions even at the lowest used solar flux value of 300 w/m^2, figures 3 through 6. Hence increasing the concentration ratio the fluid working pressure and/or the mass flow rate are increased to maintain the CPC efficiency from dropping due to the excessive superheat.

Figure 7 represents the CPC thermal efficiency at a mass flow rate per unit collector area of 0.004 kg/sm^2 for a concentration ratio of 3. Comparing figure 7 to figure 4, that is for R=3, it is clear that increasing the mass flow rate, at the same solar flux, the efficiency is increased as superheated conditions occur at relatively higher reduced temperatures.

Figure 8 represents the thermal efficiency for the same moderate concentration ratio of 3 for w/A_c of 0.002 kg/sm^2 at a working pressure of 0.7 MPa (T_{sat}=90°C). It is shown that increasing the working pressure the thermal losses are increased and the CPC efficiency decreases slightly compared to the conditions where w/A_c=0.002 kg/sm^2 and T_{sat}=60°C (figure 4). This slight decrease takes place regardless the solar flux. Considering point A, figure 8, for I=400 w/m^2 at $(T_1 - T_a)/I$=0.1 Km^2/w, the CPC thermal efficiency, η, is shown to be 0.643. Point A, figure 8, shows superheated exit conditions as it is beyond point B (x_e=1) in the direction of the increase of the reduced temperature. Figure 9 represents the CPC thermal efficiency for R=3, w/A_c=0.004 kg/sm^2 and T_{sat}=90°C. At point A, figure 9, for I=400 w/m^2 at $(T_1 - T_a)/I$=0.1 Km^2/w, the thermal efficiency is 0.705 and exit conditions are saturated, $x_e < 1$, as point A, figure 9, is before point B where x_e=1. The comparison of the values of points A in figures 8 and 9 explicitly shows that the CPC thermal efficiency for saturated exit states is greater than the CPC thermal efficiency for superheated exit states at the same solar flux. It is to be emphasized that the saturated exit states for I=400 w/m^2 in figure 9 were obtained by increasing the mass flow rate, w/A_c, to 0.004 kg/sm^2 compared to the superheated exit states obtained at w/A_c=0.002 kg/sm^2 in figure 8.

Finally, to illustrate the effect of increasing the concentration ratio, R, figure 10 represents the CPC thermal efficiency for w/A_c=0.004 kg/sm^2 and T_{sat}=90°C. At point A, figure 10, for I=400 w/m^2 at $(T_1 - T_a)/I$=0.1 Km^2/w, the efficiency is 0.713 which is compared to the efficiency of point

A, figure 9, of 0.705, the increase of the thermal efficiency as R is increased is explicitly demonstrated.

Acknowledgements

The author wishes to acknowledge the helpful support, the orientation and the guidance of Dr. John A. Clark, Professor of Mechanical Engineering, Department of Mechanical Engineering and Applied Mechanics, The University of Michigan, Ann Arbor.

References

1. Hsieh, C.K.: Thermal Analysis of CPC Collectors. Solar Energy, vol 27, pp 19-29, 1981

2. Winston, R. and Hinterberger, H.: Principles of Cylindrical Concentrators for Solar Energy. Solar Energy, vol 17, p 255, 1975

3. Giugler, R. et al: Compound Parabolic Concentrators for Solar Thermal Power. Argonne National Lab., Rep ANL-75-52, 1975

4. Thodos, G. and Winston, R.: Development and Demonstration of Compound Parabolic Concentrators for Solar Thermal Power Generation: Argonne National Lab., Rep ANL-76-71, 1976

5. Rabl, A., Goodman, N.B. and Winston, R.: Practical Design Considerations for CPC Solar Collectors. Solar Energy, vol 22, p 373, 1979

6. O'Gallagher, J., Collares-Pereira, M., Rabl, A. and Winston, R.: A Compound Parabolic Concentrator for Operation at 300°c. Proceedings of the 1978 annual meeting, American Section of ISES, Denver, n 1, p 885, 1978

7. Winston, R. and O'Gallagher, J.: Non-Imaging Concentrators for Solar Thermal Energy. Rep DOE/ET/20236-18, 1980

8. Collares-Pereira, M. et al: High Temperature Solar Collector of Optimal Concentration. Proceedings of the International Solar Energy Society meeting, New Delhi, 1978

9. Collares-Pereira, M. et al: Design and Performance Characteristics of Compound Parabolic Concentrator with Evacuated and non-Evacuated Receivers. Proceedings of the ISES congress, Atlanta, GA, n 2, p 1295, 1979

10. Rabl, A.: Optical and Thermal Properties of Compound Parabolic Concentrators. Solar Energy, vol 18, p 497-511, 1976

11. Hsieh, C.K: Design of a System Using CPC Collectors to Collect Solar Energy and to Produce Industrial Process Steam. Argonne National Lab.,

Rep ANL-79-102, 1979

12. El-Assy, A.Y. and Clark, J.A.: A Thermal-Optical Analysis of A Compound Parabolic Concentrator for Single and Multiphase Flows, Including Superheat, Warme-und Stoffubertragung, vol 21, pp 189-198, 1987

13. El-Assy, A.Y.: Technical and Economic Analysis of The Thermal Performance of A Solar Boiling Concentrator For Power Generation. Ph.D. Thesis, The University of Michigan, Ann Arbor, 1986

14. El-Assy, A.Y.: Economic Evaluation of A Solar Powered Rankine Cycle, Manuscript submitted to Solar Energy, July 1987

Dr. Ahmed Youssef El-Assy, Assistant Professor, Mechanical Engineering Department (Power and Automotive), Faculty of Engineering, Ain Shams University, Abbassia, Cairo, Egypt.

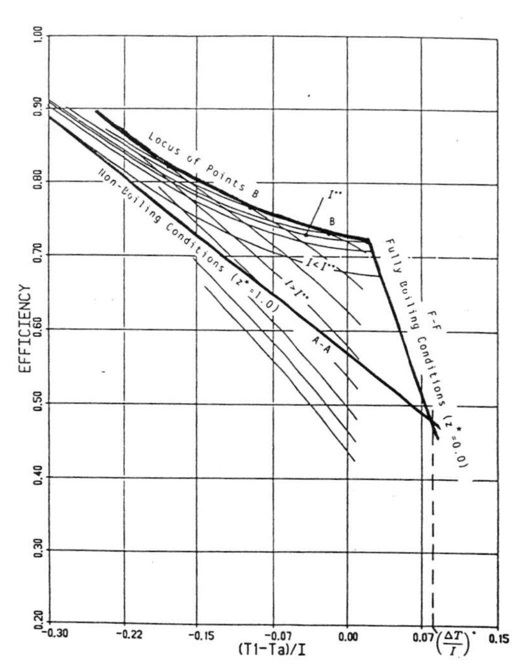

Figure 1: Illustration for the CPC operational limits

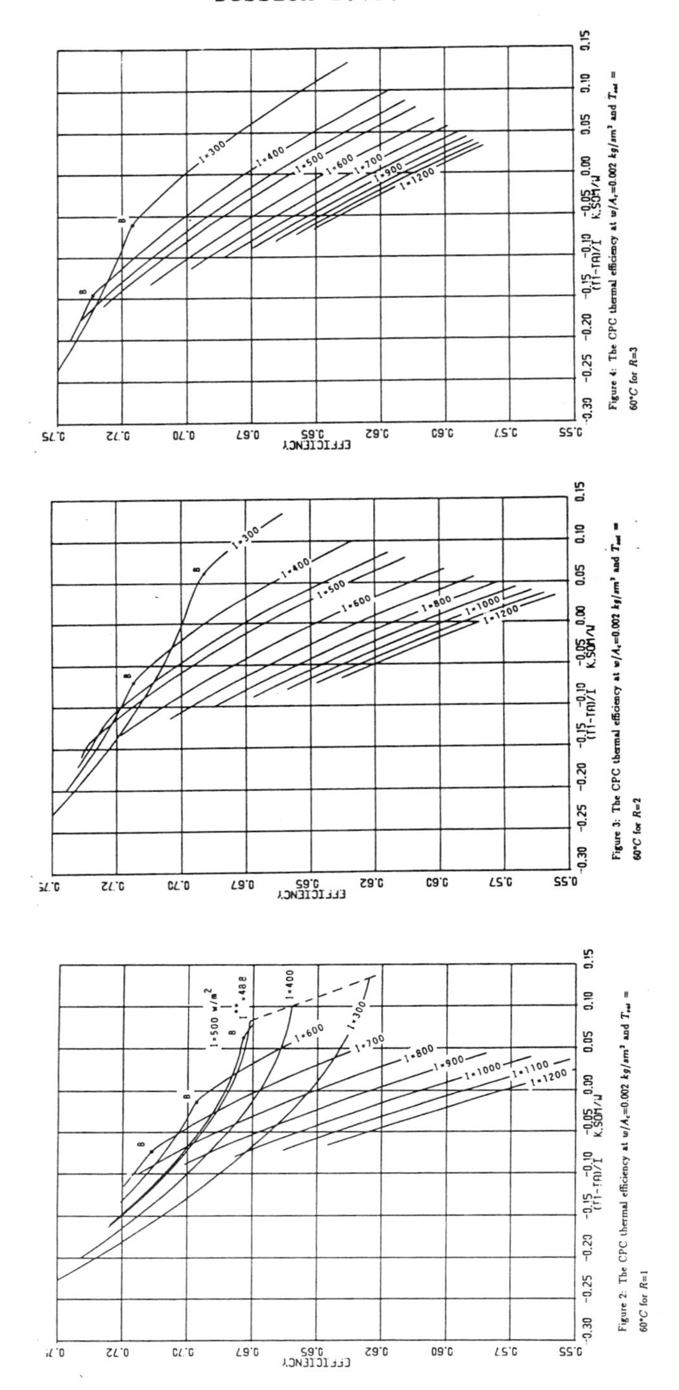

Figure 4: The CPC thermal efficiency at w/A_c=0.002 kg/sm^2 and T_{ref} = 60°C for R=3

Figure 3: The CPC thermal efficiency at w/A_c=0.002 kg/sm^2 and T_{ref} = 60°C for R=2

Figure 2: The CPC thermal efficiency at w/A_c=0.002 kg/sm^2 and T_{ref} = 60°C for R=1

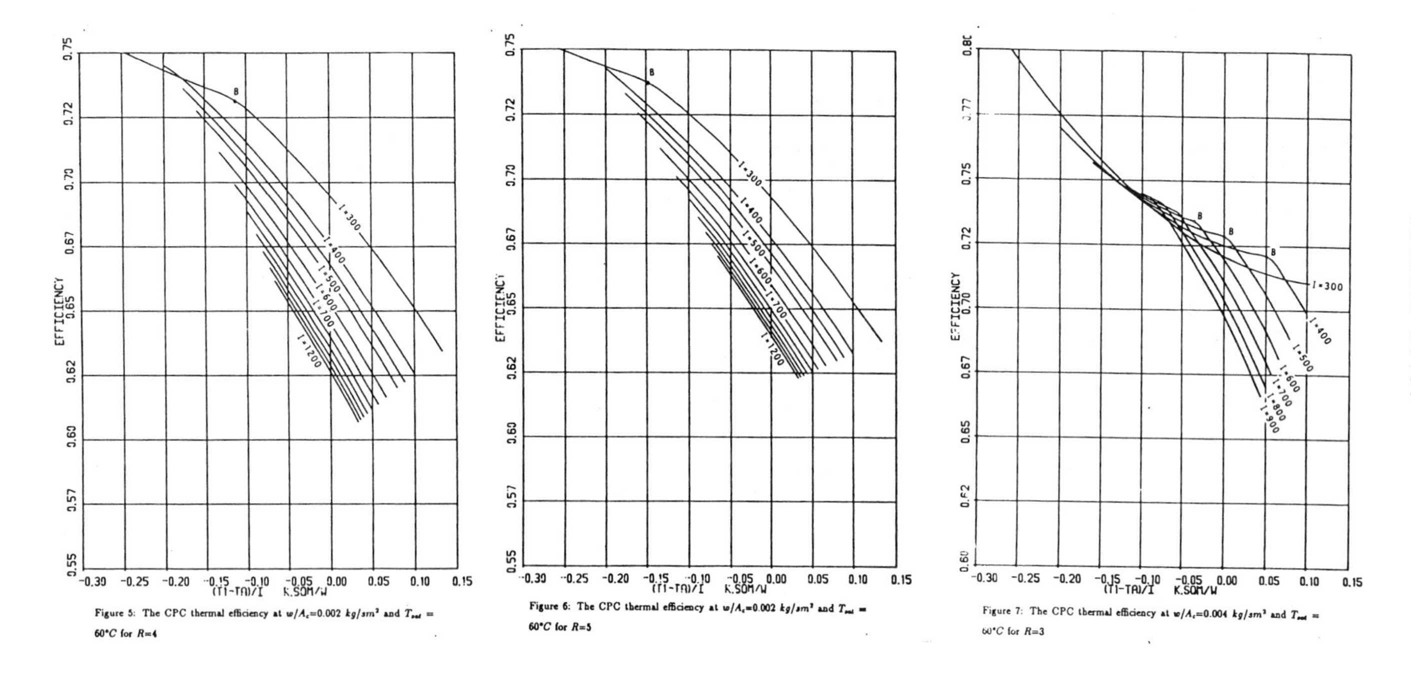

Figure 5: The CPC thermal efficiency at w/A_c=0.002 kg/sm^2 and T_{out} = 60°C for R=4

Figure 6: The CPC thermal efficiency at w/A_c=0.002 kg/sm^2 and T_{out} = 60°C for R=5

Figure 7: The CPC thermal efficiency at w/A_c=0.004 kg/sm^2 and T_{out} = 60°C for R=3

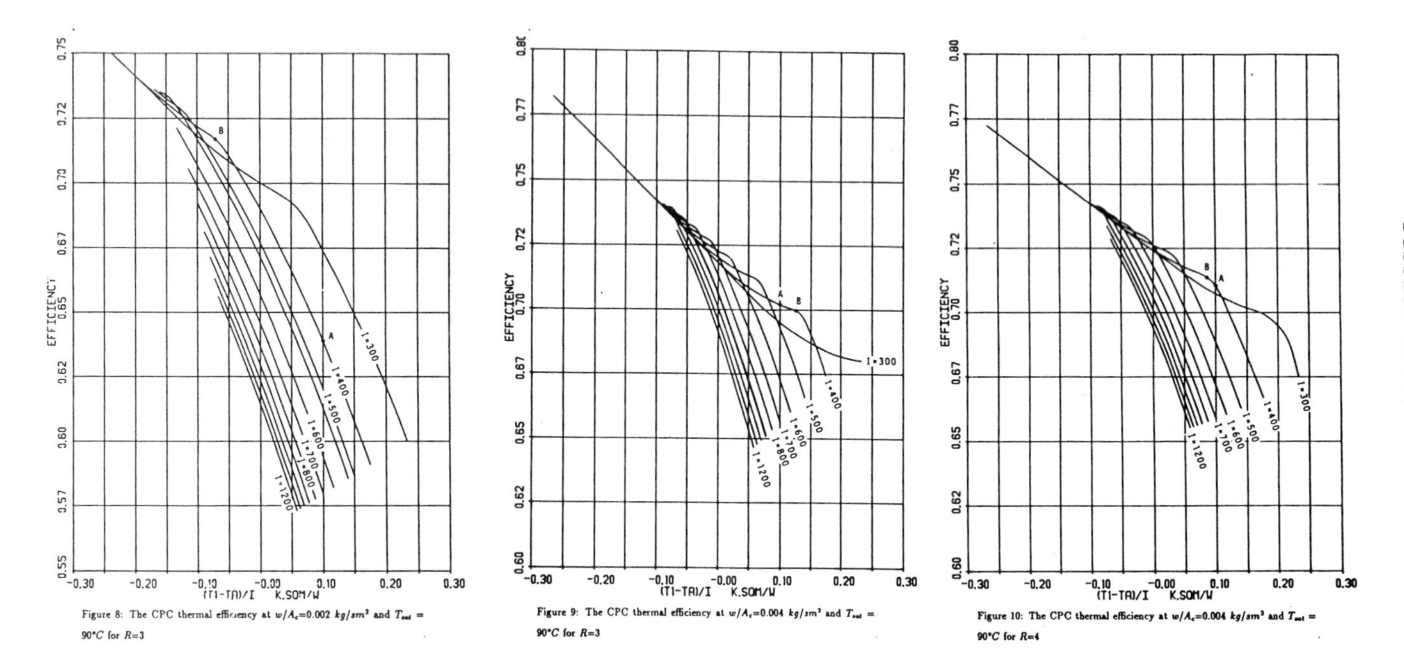

Figure 8: The CPC thermal efficiency at w/A_c=0.002 kg/sm^2 and T_{out} = 90°C for R=3

Figure 9: The CPC thermal efficiency at w/A_c=0.004 kg/sm^2 and T_{out} = 90°C for R=3

Figure 10: The CPC thermal efficiency at w/A_c=0.004 kg/sm^2 and T_{out} = 90°C for R=4

DEVELOPMENT OF AN EFFICIENT AND INEXPENSIVE EVACUATED TURBULAR COLLECTOR WITH A CONCENTRATING REFLECTOR

P. Kofod and J. Hvid

Thermal Insulation Laboratory
Technical University of Denmark
Building 118, DK-2800 Lyngby, Denmark

ABSTRACT

In this project a stationary evacuated turbular collector with concentrating reflectors for high temperature operation was designed and tested. It proved a very good performance at temperatures beyond 70°C above ambient compared to flat plate collectors, but a reduced performance at lower temperatures.

KEYWORDS

Evacuated CPC-collector, process heat, solar power, solar cooling.

INTRODUCTION

The purpose of the project is to develop a stationary inexpensive solar collector for process heat, particularly for cooking, cooling and power generation in developing countries and for industrial processes.

Only concentrating collectors yield a high performance at the temperature levels needed for these purposes. Among the concentrating collectors the compound parabolic concentrator (CPC) provides the largest concentration ratio for any given acceptance angle. The relation between the acceptance angle V and the concentration ratio C is given as

$$C = \frac{1}{\sin V}$$

In a stationary east-west aligned CPC-collector the acceptance angle must be at least 35°, giving a concentration ratio of 1.74.

In order to further reduce the heat loss coefficient an evacuated turbular receiver can be used.

Design of the Collector

A collector module consisting of three reflector troughs was designed and manufactured. A Cortec evacuated tube with a flat plate absorber was chosen as receiver. The reflector troughs were made of anodized aluminium sheets, shaped by six frames (Fig. 1).

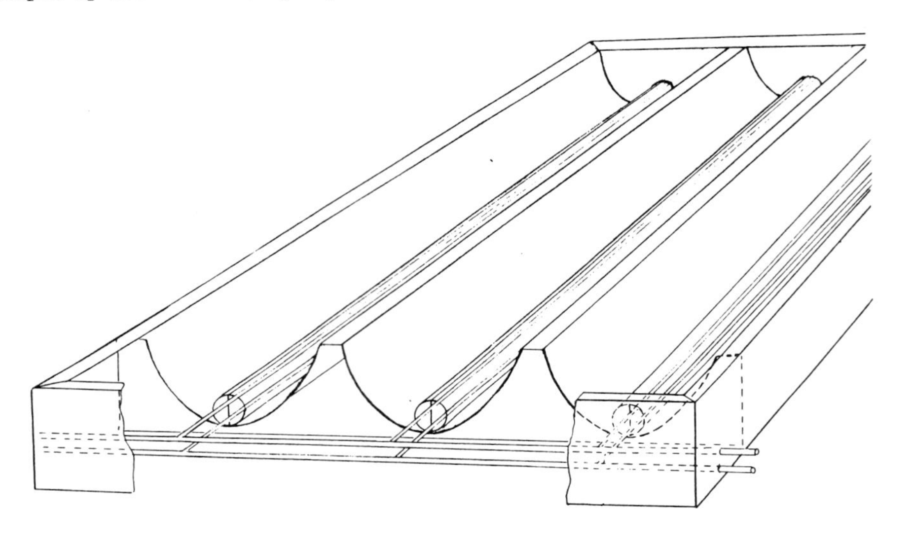

Fig. 1.

The module was designed to produce steam directly in the absorber. Circulation is established by the buoyancy of the steam, so no circulation pump is needed.

Collector Test

Three different test procedures were applied in order to determine the performance of the collector:

1) The optical efficiency was determined by connecting the three evacuated tubes in series, passing water through the tubes at ambient temperature and measuring the flow and the temperature difference.
2) The collector efficiency at temperatures up to 200°C was determined by measuring the steam flow rate from the collector at varying insolation levels.
3) The collector efficiency at higher temperatures was determined by measuring the stagnation temperatures at varying insolation levels.

The performance of the collector cannot be represented unambiguously as a function of the global insolation I_g, since part of the diffuse radiation will not be accepted due to the reduced acceptance angle.

Therefore the parameter I_g is replaced by the reduced insolation parameter

$$I_{red} = F_A \cdot I_g$$

where F_A is the fraction of the insolation with an incidence angle less than 35°. The magnitude of this fraction most of all depends on the haziness of the sky, and was found to be 0.8 on a very hazy day and 0.9 on a clear day.

The optical efficiency C_o given as

$$C_o = \frac{q_o}{I_{red} \cdot A}$$

where q_o is the output at $\Delta T = 0°C$, was found to be 0.56, which is less than calculated. The main reason for that discrepancy is supposed to be that the plane shape of the absorber provides a considerable increase in optical losses, compared to a cylindrical absorber. The use of a cylindrical absorber is expected to provide an increase in optical efficiency to approx. 0.62.

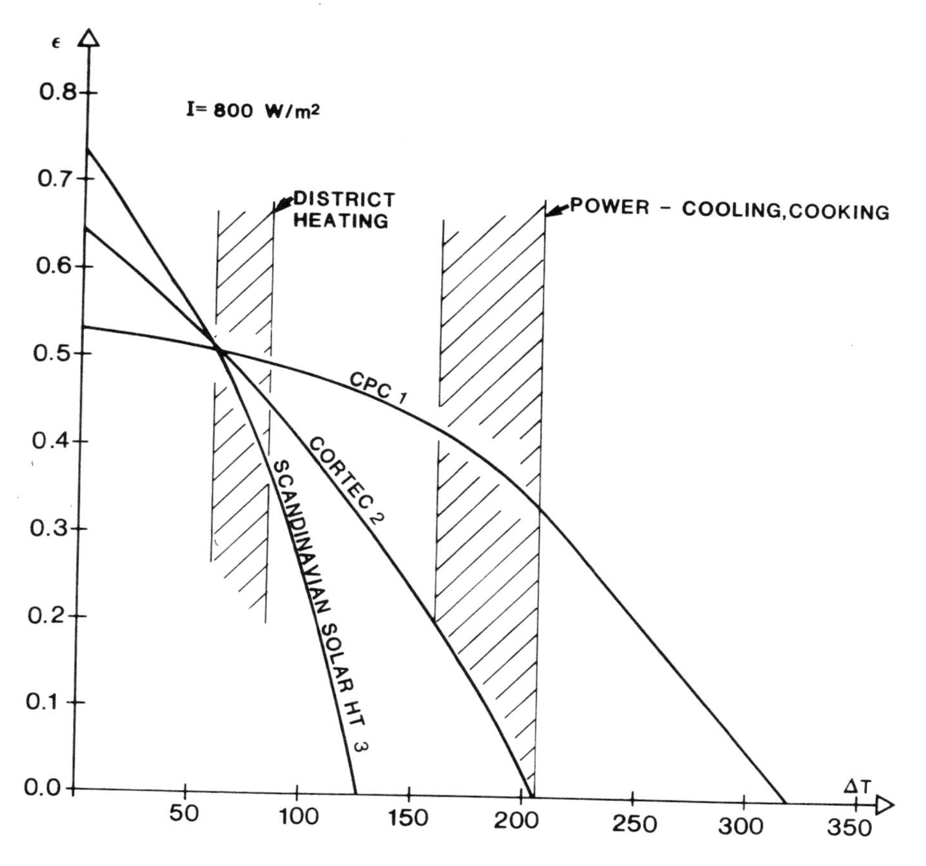

Fig. 2.

Fig. 2 represents the performance of the concentrating collector (1) compared with a nonconcentrating Cortec-cellector (2) and a flat plate collector (3) at I_g = 800 Wm^2 and F_A = 0.95.

As it is seen the concentrating collector outperforms the two other collector types at higher temperatures than 70°C above ambient. In the range of temperatures of cooling and power, the concentrating collector yields more than twice as much as any of the other collector types.

Power and Cooling

The concentrating collector has been used as a steam generator for a steam pump. A collector area of approx. 27 m^2 supplied a very simple steam pump with energy. The system performance at a steam pressure of 6 bars was found to be 2%. This efficiency is supposed to be increased by improving the steam engine.

At present a solar absorption refrigerator for cooling of medicine, using a concentrating collector, is being constructed. It will be tested in India during 1988-89.

CONCLUSION

The concentrating solar collector developed has proved a very good performance at temperatures from 100°C to 250°C, and it could probably find wide application for a series of purposes like cooling of medicine and milk, cooking and baking, pumping and desalination of drinking water in developing countries, and for industrial processes.

The collector can be produced by simple and inexpensive manufacturing methods, and the production price is expected to be less than 250$ per m^2 in large production series.

A HEAT PIPE COMPARISON FOR A CPC SOLAR COLLECTOR APPLICATION

M. Collares-Pereira*, F. Mendes*, O. Brost**, M. Groll** and S. Roesler**

* DER/LNETI - Estr.Paço do Lumiar, 1699 LISBOA Codex, Portugal
**IKE/Universitat Stuttgart - 7000 STUTTGART 80, F.R.G.

ABSTRACT

In the present paper, the use of heat pipes as means to extract energy from an integrated CPC evacuated tubular collector is investigated and the first test results for three different heat pipes are presented. The choice of the experimental parameters is justified in terms of operation of the complete system which is to be built in the near future. It is shown that a longitudinally grooved heat pipe performs best, taking into account the low heat load and low fill ratios for different temperatures, inclination angles, and start-up conditions.

KEYWORDS

Two-phase closed thermosyphon; grooved heat pipe; fill ratio; start-up; overall thermal resistence; vacuum solar collector;CPC

INTRODUCTION

The use of heat pipes as heat transfer devices has been discussed in the literature [1,2] and many applications have been found, including solar collectors. Some advantages over the classic solutions for solar energy applications are [3] : i) supression of freezing and corrosion problems; ii) thermal diode effect ; iii)simpler collector assembly, mounting and maintenance, owing to the separation of absorber and heat transfer circuits, also resulting in lower pressure losses, lower pumping power and simpler control strategies. CPC type collectors combined with heat pipes result in a reduction of the number of heat pipes per unit aperture area, and in the present case of a specially formed vacuum tube (Fig.1), only one metal-glass seal is necessary [4]. We tested a longitudinally grooved heat pipe, which seemed to be more convenient for the present application, and compared it with a smooth thermosyphon and a circumferentially threaded thermosyphon.

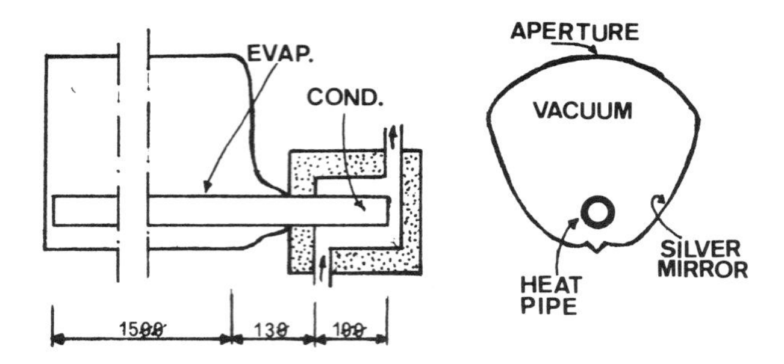

Fig. 1. CPC profile on a specially formed vacuum tube with heat pipe.

HEAT PIPE THERMAL RESISTANCE

Setting zero the vapour flow thermal resistance and neglecting further the liquid-vapour thermal resistance and the axial thermal conduction [2], the total resistance is :

$$R_{hp} = R_{pe} + R_{we} + R_{wc} + R_{pc}$$

where R_{pe} and R_{pc} are the external wall thermal resistances at the evaporator and condenser sections. R_{we} and R_{wc} are the liquid saturated wick thermal resistances, based on an effective thermal conductivity of the wick at the evaporator, KE_e, and condenser, KE_c [1]. In the case of the two-phase closed thermosyphon, R_{we} and R_{wc} represent the liquid film thermal resistances, and are based on the inside convection heat transfer coefficient at the evaporator, H_e, and condenser, H_c. H_c is evaluated using the falling film condensation theory of Nusselt and H_e is evaluated taking into account not only the heat transfer through the falling film, H_{film} (evaluated as for H_c) but also the heat transfer correlations for boiling in a liquid pool, H_{pool}, [5]. R_{we} is then obtained as :

$$R_{we} = (1-F)\,[1/(A_e\, H_{film})] + F\,[1/(A_e\, H_{pool})]$$

where A_e is the evaporator external wall area and F is the fill ratio (% of evaporator volume filled with working fluid).

EXPERIMENTAL WORK

Description of the Experiments

The goal of the present heat pipe testing program (parameters given in TABLE 1) was to establish the usefulness of heat pipes as solar energy heat transfer devices, in connection with CPC (1.3 x) shaped vacuum tubes, with the possibility of E-W or N-S operation - inclination angles $< 5°$ and $\approx 38°$(latitude of Lisbon) respectively. The tests must also reproduce other conditions encountered by the heat pipe in normal operation, namely: a) an amount of power to be transferred, from 0 to Q_{max}(=80 W) ; b) variation of the power to be transferred, resul

TABLE 1 Description of Heat Pipes

	Circ.Thr.Th.	Smooth Th.	Groov.H.P.
Fill Ratio [%]	4.4/10/20/25/30	20	13/20
Evap.Length [m]	1.23	1.0	1.4
Adia.Length [m]	.2	.1	.065
Cond.Length [m]	1.05	.2	.24
Out.Diam. [m]	.02	.02	.02
Ins.Diam. [m]	.017	.017	.015
Groove Width[m]	–	–	.00063
Groove Depth[m]	–	–	.001
Fin width [m]	–	–	.001

ting from variable incident solar radiation with slow and fast changes during the course of the day, which has been simulated by gradually or suddenly increasing the power from 0 to 80 W or 100 W, respectively. The combination of water as working fluid with copper as the container material has been chosen, owing to the good durability results already otained [6]. Low fill ratios, F, have been investigated because of safety considerations. In fact, an evacuated tube may have to survive stagnation conditions, e.g. due to pump failure. The resulting temperature can exceed the critical temperature of the water, and if the tube has enough water to feed its continuous evaporation, the pressure will increase exponentially. With a low mass of water, however, the evaporation will cease before reaching a high level pressure and after that it will increase only linearly with temperature, according to the ideal gas law, finally reaching a sustainable value at the stagnation temperature. With the low fill ratios, the dry-out appears as a natural heat pipe performance limit, but it is expected that the small power to be transferred may balance the tendency to dry out, and this is also investigated.

Heat input to the heat pipes was provided by an electrical tape wrapped around the evaporator and insulated with foam. Heat extraction at the condenser was achieved by circulation of water at the desired temperature. The power was adjusted by a transformer, and the temperature of 12 thermocouples (10 for the smooth tube), attached to the external wall of the tubes, was continuously monitored. Thermocouples 1 and 2 (T1,T2) were always diametrically opposed at the bottom of the evaporator.

Results and Discussion

Circumferentially threaded thermosyphon. With the very low fill ratio initially chosen, F=4.4 %, a dry-out situation appeared for 2° inclination angle, resulting from the very small amount of water and consequently insufficient fluid return from the condenser. The same happened with F=10 % ; only with 20 % fill ratio did the pipe work without dry-out for all the aforementioned operating conditions. However, an overheated liquid pool is formed, resulting in temperature oscillations. The overall thermal resistance ($R_{hp}=(T_e-T_c)/Q$) decreased for lower inclinations, higher powers and higher operating temperatu

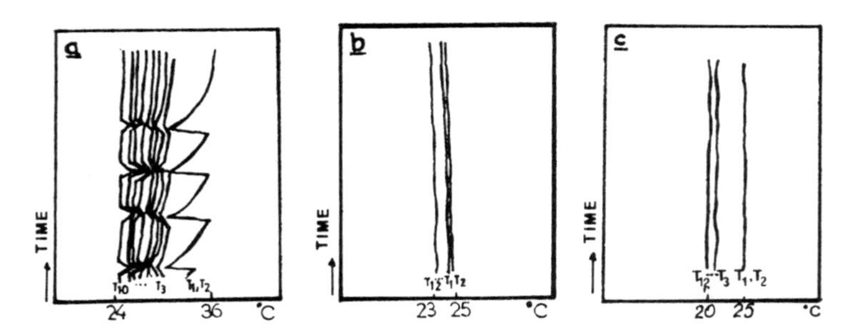

Fig. 2. Transient temperature profiles (F=20 %, β=2°, Q=100 W) of the smooth therm.(a), threaded therm.(b) and grooved wick heat pipe (c).

res, because in all these cases the stable and overheated pool was destroyed. In Fig.3., the theoretical model is compared with some of the experimental results, showing better agreement for higher inclinations, which agree with [5].

Smooth thermosyphon. With the experience gained with the threaded thermosyphon, only a 20 % fill ratio was tested. The thermosyphon operated without dry-out, for all the required operating conditions. However its temperature difference, $T_e - T_c$, was greater than that of the threaded thermosyphon, and the operation at low temperature, Fig. 2, shows a very different transient temperature profile, which probably results from a poor water distribution on the inside perimeter.

Grooved wick heat pipe. We started with 13 % fill ratio, corresponding to a liquid volume equal to the groove volume. At low temperature and zero inclination a dry-out situation appeared. On increasing the inclination angle to 1°, the heat pipe showed very good perfomance with the smallest temperature difference observed during the tests. No problem occurred at an inclination angle of 38°, but the formation of a bottom pool increased the temperature difference, $T_e - T_c$. An increase of the fill ratio to 20 % resulted in a greater temperature difference, without any significant advantage. A comparison between experimental results and the proposed theoretical model is shown in Fig. 3. The solid line is the prediction result of the model obtained with a high value for R_{we}, according to the evaluation of KE_e in reference [1]. It is seen that the model grossly overestimates the temperature difference, $(T_e - T_c)$. In fact in reference [1] the calculation assumes no pool, which is not our case, and moreover, the fins and grooves can be expected to behave differently, since the fins inside the pool act as nucleation inducers, spraying liquid on the evaporator walls not covered by the pool and changing drastically the assumptions made in [1] for the evaluation of KE_e. If however, it is assumed that the resistance in the evaporator, outside of the pool, is calculated as the resistance of the condenser (broken line in Fig.3), a better agreement between results and model is obtained.

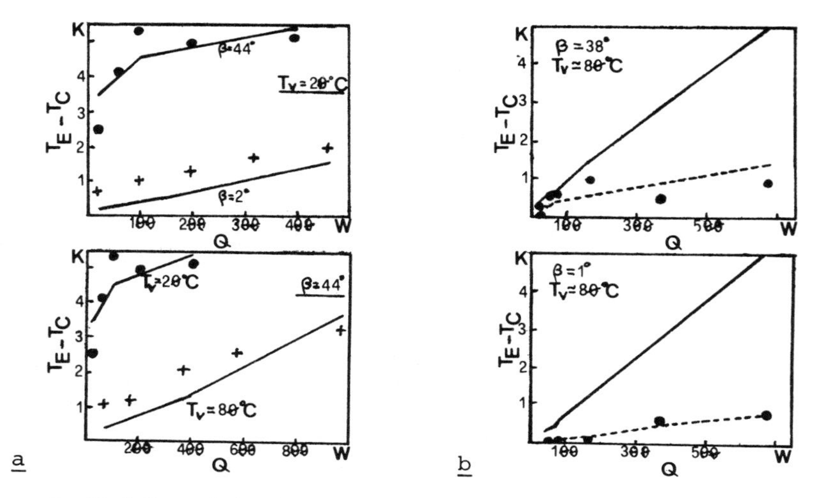

Fig. 3. Model and test results (F=20 %) of the circunferentially threaded thermosyphon(a), and grooved wick heat pipe(b).

CONCLUSIONS

The longitudinally grooved heat pipe was found to perform better than the two thermosyphons tested. It worked without dry-out at low fill ratio and with a small temperature difference along the heat pipe. It might be worthwhile to try decreasing the fill ratio further or to use a tube with small grooves. For near horizontal operation, an inclination angle of 1° seems to be a good solution because it doesn't noticeably penalize the collected energy of the E-W oriented collectors, and it doesn't permit a reverse mode operation. There seems to be reasonable agreement between the simple theoretical model proposed and the results obtained. However, in the case of the longitudinally grooved heat pipe, the model seems to require special assumptions for the heat exchange mechanism in the evaporator and this is to be studied in more detail.

REFERENCES

[1] - Chi S.C. (1976).Heat Pipes, Theory and Pratice. Hemisphere Publishing Co.

[2] - Dunn P., Reay D.A. (1978). Heat Pipes. Pergamon Press Ltd.

[3] - Akyurt M. (1984). Development of Heat Pipes for Solar Water Heaters. Solar Energy, vol.32, Nº5.

[4] - Collares-Pereira M., Mendes F. (1984). Tubo de Calor para Extraccao da Energia Captada por Colector do tipo CPC. Proc.2nd Iberic Congress of Solar Energy, Lisboa.

[5] - Shiraishi M., Kikuchi K., Yamanishi T.(1981).Investigation of Heat Transfer Characteristics of a Two Phase Closed Thermosyphon. Proc.IVth Int. Heat Pipe Conf., London.

[6] - Groll M., Heine D., Spendel T. (1983). Heat Recovery Units Employing Reflux Heat Pipes as Components.IKE, Stuttgart

DEVELOPMENT OF MIRRORS FOR SOLAR ENERGY[1]

R. Almanza, M. Mazari and F. Muñoz
Instituto de Ingeniería, Universidad Nacional Autónoma de México
Ciudad Universitaría, 04510, México, D.F. México

ABSTRACT

In this paper are presented the advances obtained in first surface mirrors using SiO, SiO_2 and Si_2O_3 films as front surface over an aluminum film thermally evaporated over a glass or plastic substrate. The size of the mirrors is .30 x .30 m for SiO and Si_2O_3 with 0.003 m substrate thickness. The SiO_2 mirrors were obtained by evaporating Pirex glass with an electron gun in a small evaporator.
In order to obtain a higher reflectance in the SiO mirrors, the SiO was evaporated in an atmosphere of oxygen at pressure 10^{-4} Torr to produce a coating of Si_2O_3. The highest measured specular reflectance of these mirrors was 0.86.

KEYWORDS

Mirrors, solar energy, first surface mirrors, front-surface mirrors, reflectors, second surface mirrors.

INTRODUCTION

If one wishes to use solar energy in thermal devices at temperatures at above 100°C, it is necessary to concentrate it with mirrors over absorbers in order to increase the irradiance flux.
Mirrors are used in different types of concentrators such as parabolic troughs, parabolic dishes, spherical bowls and heliostats. The mirrors in each of these concentrators have to comply with the following characteristics:

a) High optical performance that includes reflectance higher than 0.85, specularity of the order of 2 milliradians and good geometrical configuration.
b) Infrequent maintenance in the field
c) Long life (at least 10 years)
d) Low initial cost, and
e) Resistance to ultraviolet radiation, pollutants and environmental degradation.

1 Work in colaboration with the Physics Institute, UNAM

Most mirrors employ either silver or aluminum films as reflecting surfaces. The aluminum mirrors are more resistant to degradation, while the silver ones must be protected from physical and chemical deterioration. Both types of mirrors have been used in different solar energy applications but, after a few years of use, the reflectors lose their reflectance properties mainly due to corrosion.
Research and development has been done on second surface mirrors using plastics and glasses as substrates and in first surface mirrors using plastics as the front surface (Benson, 1984; Bieg, 1980; Czanderna, 1985; Dennis, 1980; Goodyear, 1980; Griffin, 1980; Jacob, 1983; Marion, 1980; Pitts,1984; Schissel, 1985). However for first surface mirrors using glass films as the front surface, not enough research has been done.
This paper presents advances obtained in first surface mirrors using SiO, Si_2O_3 and SiO_2 as the front surfaces. In the case of SiO, the coatings were deposited using thermal evaporation in a vacuum. An electron gun was used to evaporate boro-silicate (Pyrex, 80.5% SiO_2) and silica (Vycor, 96% SiO_2) on aluminum or silver films deposited previously in glass substrates or polymethylmethacrylate (PMMA)

SELECTION OF SUBSTRATES

The substrates selected were of two types. One is commercial glass (soda lime) with iron oxide that reduces the transmittance due to absorption of the long visible wave length of the light, but is the only one that is produced in Mexico. In some other countries glass with low iron content is produced. It is known as low iron glass with which higher reflectances can be obtained in second surface mirrors.
The other types of substrates selected were plastic: PMMA (acrylic), and poliester resin reinforced with fiberglass.
All the substrates measured 0.3 x 0.3 m with 0.003 m thickness, except for the SiO_2 films that were obtained on a small evaporator with an electron gun.

PREPARATION TECHNIQUES

The most widely used technique for depositing reflecting films is by evaporation in high vacuum (between 10^{-5} and 10^{-6} Torr). The equipment used for the experiments carried out in this investigation was a 1.1 m diameter evaporator (Fig. 1)
Cleaning of the substrates was first effected with soap for a rough degreasing. Then solvent cleaning with isopropyl alcohol was carried out. As a final cleaning, a glow-discharge in an oxygen or argon plasma was applied under the following conditions: 3.5 kV with 220 ma. Fig. 2 shows the circular aluminum glow-discharge cathode consisting of a 0.02 m-wide aluminum ring. A glow discharge produced under these conditions during 15 minutes was found to be sufficient to clean the glass substrates before applying the aluminum coating. However, for acrylic substrates, cleaning by this method was not totally satisfactory.
For the size of substrates employed, only one tungsten filament was necessary in order to get an aluminum film with uniform thickness. After evaporation of the aluminum, the SiO was evaporated also by thermal means using a tantalum crucible made out of a tube with 0.003 m external diameter and 0.04 m long. Along the lenght of the tube, 3 small holes of 0.0005 m diameter were drilled in order to allow passage of the SiO vapors. The distance between the substrate and the evaporation source was about 0.25 m.

Fig. 1 Evaporator 1.1 m diameter

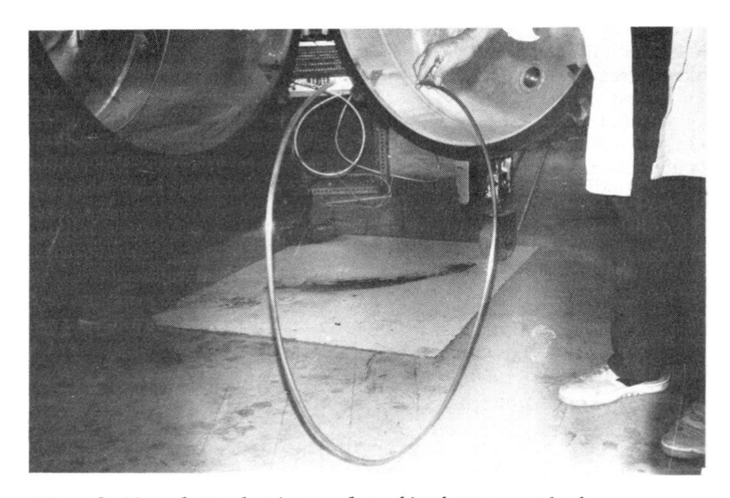

Fig. 2 Circular aluminum glow-discharge cathode

However, higher reflectance is obtained if the SiO is changed to Si_2O_3 (Drummeter, 1964; Hass, 1982). Vacuum systems make it possible to deposit films in the presence of oxygen for reactive evaporation. In order to produce films with negligible absorptance in the visible range, it is neces-- sary to use low deposition rates (4 Å /sec) and at an oxygen pressure of about 10^{-4} Torr (Hass, 1984).

REFLECTANCE OF FIRST AND SECOND SURFACE MIRRORS

For the different first surface mirrors the specular reflectance is given in table 1. The reflectance was measured with a solar spectrum reflectometer

(SSR) from Devices & Services Co. The main characteristics of this device are: resolution 0.001 units, repeatability ± 0.003 units, accuracy ± 2%

TABLE 1. First Surface Mirrors.

Mirror	Evaporation (front surface)	Reflectance
SiO	Thermal	.79
Si_2O_3	Thermal	.86
SiO_2	Electro Gun	.86
Al.	(without front surface)	.89

Also the following second surface mirrors were obtained in acrylic and glass as second surface. Table 2 shows the results

TABLE 2. Second Surface Mirrors

Mirror	Reflectance
Glass	.75
Acrylic	.67
Fiber glass	.65

Each of the mirrors listed in table 2 was protected with epoxic resins in order to decrease the corrosion. The last one was made by evaporation of aluminum over the fiberglass and protected with acrylic .003 m thick.

CONCLUSIONS AND RECOMMENDATIONS

The aluminum first surface mirrors can give a high reflectance necessary to get good efficiency in concentrators that use the beam component of solar irradiance. However, it is necessary to continue the R and D of this subject.

The silver first surface mirrors can be made with higher reflectance, however, they need more protective films, because of the poor adhesion between substrate and Ag, and between Ag and front surface. Therefore, it is necessary to study this alternative more carefully.

Outdoor tests are also necessary to know how the environment can affect the reflectance; these tests are in progress in our laboratory.

ACKNOWLEDGEMENT

E. Zironi, G. Mendoza and F. Calderon made some test in the laboratory; R. Chicurel reviewed the final version

REFERENCES

Benson B. A., (1984). Acrylic solar energy reflecting film 3 M brand ECP-244 (formerly FEK-244). Progress in Solar Energy, ASES

Bieg K.W. and K.B. Wischamann (1980). Plasma-polymerized organosilanes as protective coatings for solar front-surface mirrors. Solar Energy Materials Vol. 3, No. 1,2.

Czanderna, A.W., K.Masterson, and T. M.Thomas, (1985). Silver / Glass Mirrors for Solar Thermal Systems. SERI / sp - 271-22 93.

Dennis W. E., and J.B. Mc Gree (1980). Silicon resins for protection of first surface reflectors. Solar Energy Materials Vol. 3 No. 1,2.

Drummeter, L.F., and G. Hass (1964). Solar Absorptance and Thermal Emittance of Evaporated Coatings. In G. Hass and R.E. Thun (Ed), Physics of Thin Films, Vol. 2, Academic Press, New York.
Goodyear, J.K., and V.L. Lindberg (1980). Low absorption float glass for back surface solar reflectors. Solar Energy Materials Vol. 3 No. 1,2
Griffin, R.N (1980). Thin film solar reflectors. Solar Energy Materials Vol. 3 No. 1,2
Hass, G., J.B. Heaney, and W.R. Hunter (1982) Reflectance and Preparation of front Surface Mirrors for Use at Various Angles of Incidence from the Ultraviolet to the far Infrared. In G. Hass, M.H. Francombe and J.L.Vossen (Ed.). Physics of Thin Films. Vol. 12, Academic Press.
Jacob, J.W. (1983) Solar Energy - Mirrors and Panels. Solar World Congress Vol. 3 Szokolay S.V. Pergamon Press
Marion, R.H. (1980). The use of thin glass reflectors for solar concentrators. Solar Energy Materials Vol. 3 No. 1,2.
Pitts, J.R., T.M. Thomas, and A.W. Czanderna (1984), Surface analysis of silver mirrors made from organometallic solutions. Solar Energy Materials Vol. 11
Richmond, J.C. (1984). Measurement Techniques for Evaluating Solar Reflector Materials. NTIS-PB85-1194469.
Schissel, P., H.H. Neidlinger, and A.W. Czanderna AW (1985) Silvered polymer reflectors. Solar thermal Research Program Annual Conference, Denver Co, U.S.A.

A DUAL FLOW FORCE BALANCED ROTATING JOINT FOR HIGH TEMPERATURES AND PRESSURES

R.E. Whelan
Department of Engineering Physics
Research School of Physical Sciences
Australian National University
G.P.O. Box 4
CANBERRA A.C.T. 2601 AUSTRALIA

ABSTRACT

The requirement to transfer high temperature fluid at high pressure between relatively rotating systems cannot be met by the elastic deformation of specially formed flexible metal hoses or tubes nor can industrial rotary unions provide satisfactory solutions due to temperature limitations, heat loss and leakage.

The development by the Department of Engineering Physics of elastomer sealed rotating joints that have operated successfully in the field since 1983 with superheated steam up to 500°C, overcame the temperature limitation and did not require thermal insulation.

Recognition and application of those elements that have contributed to the success of the rotating joint has led to the evolution of a new design that has fewer moving parts and should be easily maintained in the field. This new design, the **Dual Flow Force Balanced Rotary Joint** is well suited to the new series of large solar concentrators ($\approx$300m^2) where steam is the working fluid and the heat engine is mechanically decoupled from the concentrator structure.

The validity of the design principles involving controlled temperature gradients to determine seal positions, and interfluid heat transfer are confirmed by experiment.

KEY WORDS

High temperature steam connections; reduced heat loss; large tracking concentrators; experimentally verified design.

GENESIS

During the design and development of the experimental solar power station (Kaneff, 1983) at White Cliffs, New South Wales, by the Department of Engineering Physics at the Australian National University, and funded jointly by the N.S.W. Energy Authority, a need was established for a device to connect the receivers of the fourteen 20m^2 concentrators to the station manifolds.

Devices were required to connect feed water at 60°C and superheated steam at 550°C, both at a maximum pressure of 70 atmospheres with minimum heat loss and leakage.

The concentrator tracking mount was of two axes configuration with rotations of ±45°elevation and ±180° in azimuth and two fluid paths were to be provided per axis. The major factors in the selection of a suitable device were:

- fluid compatability
- temperature rating
- pressure rating (and its interaction with temperature)
- tolerance of thermal cycling
- heat loss (and hence steam quality)
- rotational limitations

After examining without success available rotary joints and other flexible connections (coiled tubes, braided metal hoses) and testing special purpose devices based on modified steam values, it was decided to develop a design "in house", based on seal positioning concepts proposed by Carden and Whelan, and reported by Whelan in 1986.

Design Principles and Testing

The rotary joint developed in the Department (Whelan, 1986) met all the requirements and additionally provided the transmission for both steam and water in one unit.

The principal feature of the design is the use of an elastomer seal ('O' ring) in a system of high temperature and pressure fluids in such a position as to allow sealing action to be maintained through rotation.

The positioning depends on a designed temperature gradient (Fig. 1) in the interfluid material (the seal carrying element), established by designed heat flow between the steam and feed water channels contained in a single element. The heat flow is determined by the thermal cross section of the interfluid material, the temperature difference and the materials' thermal properties.

Additionally, the interfluid material is chosen to be suitable for the construction of a containment system that has as the basis of its design a symmetry about the axis of rotation.

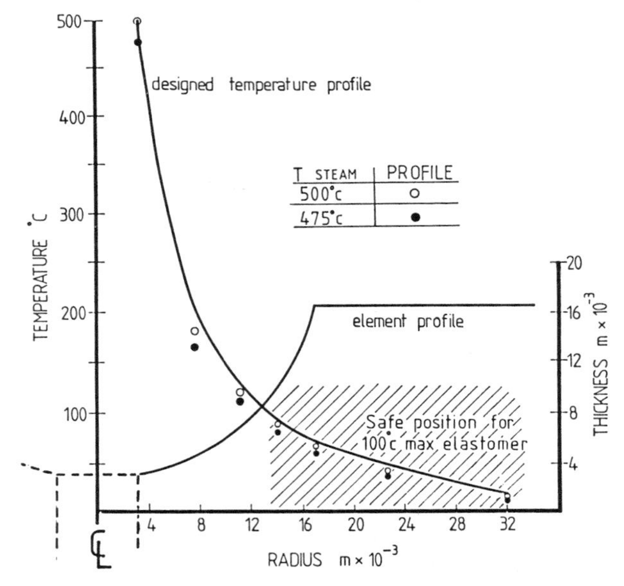

Fig. 1: Confirmation of designed temperature gradient

As a consequence of the dual flow configuration the elastomer seal has only the pressure difference between the water and the steam to seal against. This is significantly less than the pressure difference across the high temperature seal in a conventional rotary joint and so tolerances are less stringent.

The design approach is to approximate the heat flow and temperature gradient by applying Fourier's law of conduction for one dimensional flow and is generally expressed, for circular section, by the equation

$$Q = -KA \frac{dt}{dr} \quad (1)$$

where Q is the heat flow rate, K a constant of proportionality, the thermal conductivity, A is the area of the section, and (dt/dr) the radial temperature gradient.

This design approach has been verified by experiments that use radially positioned thermocouples for determining the temperature gradient and both water and steam calorimetry for the heat flow (Fig. 2). Heat flow (loss) for other devices has also been determined.

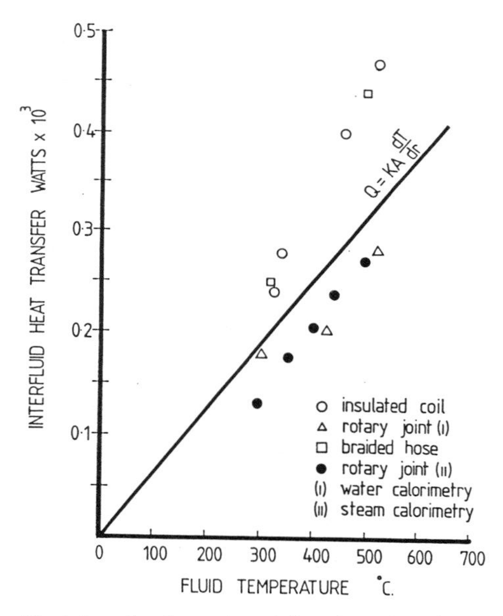

Fig. 2: Heat transfer for rotary joint, heat loss for coil and hose

Twenty eight rotary joints of this design have been in operation at White Cliffs since 1983 and have operated at steam temperatures of approx. 450°C. Some failures have occurred in the elastomer seals due to incorrect alignment and steam injection into the water passage ways during tests.

EVOLUTION

The current trends towards large tracking solar concentrations of up to 300m^2 in area have generated new interest in the problem of connecting the steam generator to the engines with minimum heat loss. This is of more significance now that the engines are not carried on the concentrator frame (Power Kinetics Inc., 1986).

The application of the design approach used in the successful joint for White Cliffs has led to a new configuration that overcomes some of the limitations encountered when that model is increased in size to match the new concentrators.

The essential feature of a temperature gradient determining the seal position remains, but to overcome predicted high interfluid heat transfer rates, a number of seal carrying element profiles have been considered (Fig. 3). The favoured shape is the cone because of its inherent strength to withstand pressure generated forces, its ease of manufacture and flexiblity of design parameters (length/diameter ratio).

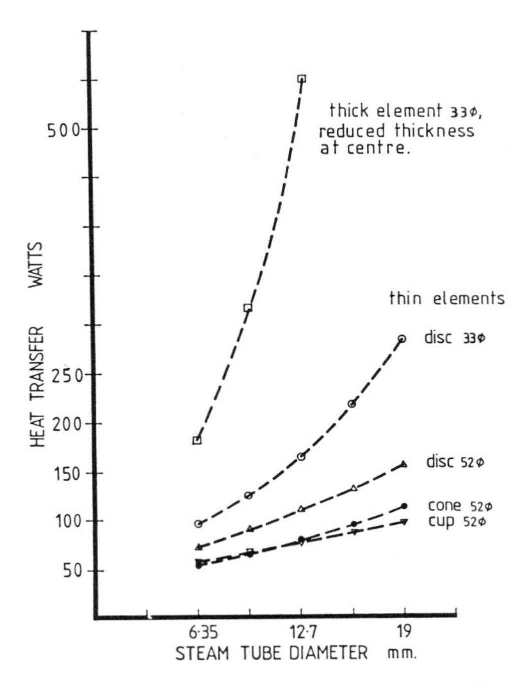

Fig. 3: Heat transfer for various element profiles

The major difference however is not in the shape but in the doubling of seal carrying elements to four (Fig. 4).

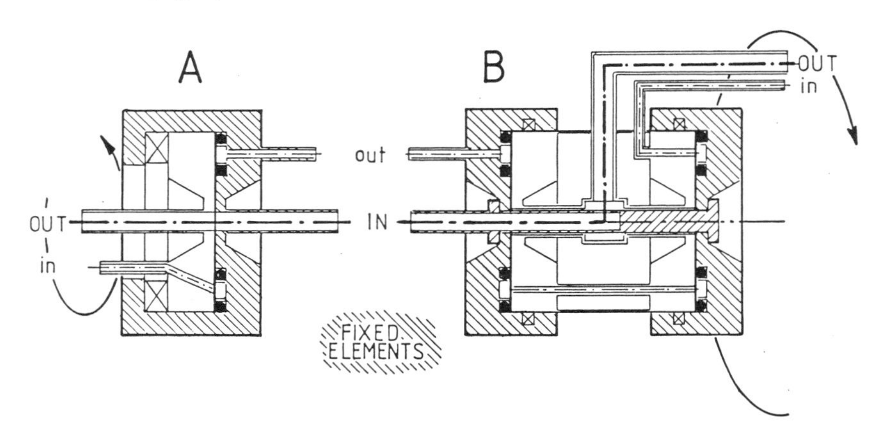

Fig. 4: Cross section showing flow path for 2 element (A), and 4 element force balanced (B), rotary joints

The forces associated with the steam pressure acting on the seal element area exceed 30,000 newtons for pipe sizes above 12mm diameter and the thrust bearing is therefore an essential but expensive component within the original rotary joint design. With the four element or force balanced design, the centre or rotating section is subjected to equal and opposite axis forces, and apart from simple plain bearing for radial location, no expensive bearings are required.

Interfluid Heat Transfer

In conventional connections heat lost through poor insulation to the air is lost from the system and represents wasted collection area. The dual flow system provides a minimum loss arrangement because heat transferred from the steam to the seal elements to the feedwater passes to the reciever as preheat. This is not without disadvantages in that the receiver must run at a higher temperature to maintain the manifold temperature. Fortunately, this problem diminishes with scale. For example, a $20m^2$ concentrator output, passing through a joint with transfer rate of 310 watts degrades from 550°C to 517°C at 56 atmospheres, whereas at the same pressure a $280m^2$ concentrator output passing through a joint of Q=650w will degrate from 550°C to 545°C.

This is the equivalent of approximately five metres of conventionally insulated pipe.

CONSEQUENCE

With reduced interfluid heat flow, fewer components, and predicable characteristics, the **Dual Flow Force Balanced Rotary Joint** will soon undergo in–field testing for reliability and will be installed with confidence based on the performance of its predecessor.

ACKNOWLEDGEMENTS

The author wishes to thank Professor S. Kaneff for his continuous support and his encouragement for the development of ideas; Dr H. Williamson for assistance in finding the appropriate integrals for heat transfer calculations, and especially the author wishes to thank his daughter, Roslyn Whelan, for her contribution to the graphical presentation of results and the associated poster drawings.

REFERENCES

Kaneff, S., "The White Cliffs Solar Power Station", *Proc. Solar World Congress*, Vol.4, Perth, Australia, August 1983.

Power Kinetics Inc., "Definitive Design of the Small Community Programm at Molokai, Hawaii", U.S. Department of Energy Contract DEFCO4–85–AL–32887, Troy, N.Y., 1985.

Whelan, R.E., "Rotating Joints for High Temperature Fluids and Their Application for Tracking Solar Concentrators", *Proc. 5th A.N.Z.A.A.S. A.I.S.T. Conf. Science Technology*, Sydney, July 1986.

CAMPBELLTOWN HOSPITAL SOLAR STEAM PROJECT

Dr. D.R. Mills, School of Physics,
Univ. of Sydney, NSW, AUSTRALIA
and
Mr. Alan Moore, HOSPLAN, Private Bag 5,
ROSELLE, 2039 AUSTRALIA.

ABSTRACT

This paper describes a demonstration solar steam boiler plant now being installed at Campbelltown Dirstrict Hospital on the outskirts of Sydney, Australia. The collector technology uses evacuated tubes and periodically adjusted concentrating reflectors, and is intended to eventually compete with bulk industrial fossil fuel.

KEYWORDS

Solar steam, solar boiler, solar thermal, medium temperature solar thermal, solar industrial process steam, process steam.

INTRODUCTION

A solar industrial process steam demonstration plant is under construction on a hospital site near Sydney, Australia. The plant is being funded by the National Energy Research, Development and Demonstration Program (NERDDP) of the Department of Primary Industry and Energy and is being administered by the Hospital Planning Advisory Centre (HOSPLAN) of the New South Wales State Department of Health, who will continue to monitor plant performance throughout its lifetime. The collector technology was developed by the University of Sydney Department of Applied Physics, who will oversee research and development on the site during the grant period of two years.

The eventual object of the development program is to produce a solar collector system which will generate industrial process steam at a cost competitive with, or cheaper than, that generated by inexpensive bulk industrial LPG and natural gas now available to Australian industry. Australian industrial fuel is relatively inexpensive by world standards.

The function of this initial project is to technically demonstrate the compatability of the solar steam boiler with a conventionally fuelled system, and to monitor system performance. Should this first phase be successful, then a considerably larger economic demonstration would be attempted at a later date.

COLLECTOR TECHNOLOGY

Periodically Adjusted Collector (PAC) technology has been recently advocated [Mills, 1986] as the most cost effective means of using evacuated tube technology to produce medium temperature thermal energy. In this type of solar collector, an adjustable linear parabolic trough mirror of geometric concentration between 3 and 6 is used with a fixed evacuated tubular absorber. Continuous tracking is not necessary, but about 10-15 adjustments per year are required to maintain close to optimal performance.

The aim of the collector is maximum cost-effectiveness, not maximum efficiency; the chosen concentration ratio for this project is about 4.4x, whereas the most efficient concentration ratio would be approximately 2x for this output temperature range (100-200°C). The primary aim of the concentrating mirror is to reduce the number of evacuated tubes and associated plumbing as a function of energy collected. The collector mirror is of a shallow parabolic design to reduce mirror area, a significant capital cost.

The use of low concentration optics allows a considerable relaxation in mirror formation and alignment accuracy, and a reduction in support structure rigidity. The fixed heat transfer pipework in a single plane is comparatively inexpensive and trouble-free. Evacuated tubes are simply slipped over the steel pipework, and may be replaced easily if damaged.

While original prototypes [Mills et al., 1981] used pumped circulating heat transfer oil and a heat exchanger, the present system uses direct boiling of water inside steel tubes (Fig. 1), and no heat exchanger. Our earlier 5x prototype [Mills, 1984, 1985], first erected in 1983, is a stand-alone steam two-phase thermosyphon system and is still operational, being used in a solar steam pump development program by a local Sydney firm. The CDH plant requires a pump to lift evaporant up from the present reservoir to the roof, but the array plumbing design is versatile and is used unchanged.

THE CAMPBELLTOWN DISTRICT HOSPITAL ARRAY

Campbelltown D.H. is a modern hospital with a large flat, unshaded roof area and an up-to-date computerised energy monitoring and management system.

The hospital uses steam in two separate buildings. The larger tower block contains a steam dishwashing facility supplied by an 80kW high voltage electric boiler, the steam being used is a combined drying and sterilisation facility which also services medical centres in the surrounding district. This is presently supplied by two 80kW boilers identical to that in the tower block.

The 196 m^2 project array [Fig 2.] will only supply steam to the dishwashing facility, initially with a peak rating of 55kW using BA finish 316 stainless steel mirrors, and later with a peak of 73kW using a protected and anodised aluminium mirror lining planned for later in the project. The array will use modules of 14 mirrors and tubes each. Each mirror has an aperture area of 0.6 m^2. Mirrors are ganged together and adjusted every few weeks by hand, according to a predetermined schedule.

A possible and inexpensive heat dump path for surplus steam exists in a

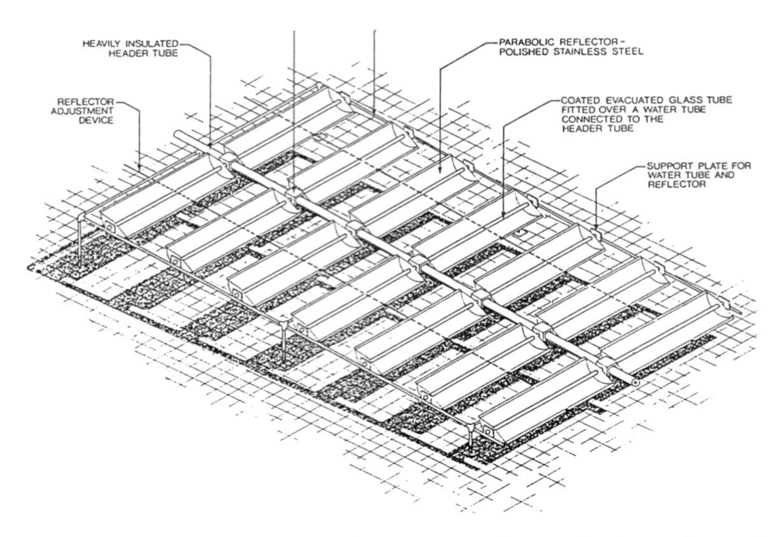

Fig. 1 C.P.H. array module. Boiling occurs in metal pipework enveloped by evacuated tubes.

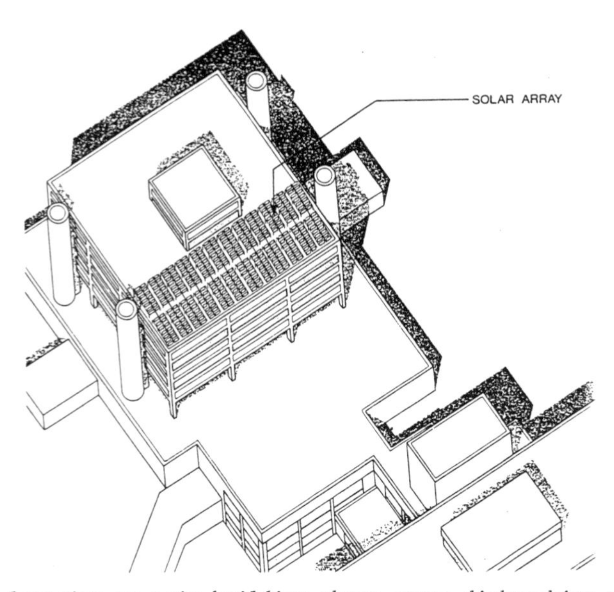

Fig. 2 Array location on main building above steam dishwashing plant.

separate natural gas powered hot water system, but this is probably not required with a 365 day a year daytime load pattern and an undersizd system. On the adjacent roof, there is sufficient space to install a 400 m^2 array sufficient to meet the peak load of the sterilisers, but this would require additional funding to install.

The general plan of the array is shown in Fig. 2. Evaporant is pumped up to a small buffer reservoir on the roof, which is connected by a manifold to a line of 12 parallel modules, each module being connected in series to another for a total of 24 modules. The water level in the reservoir is controlled to ensure that the entire plumbing system is flooded. Steam produced in the array is dried in a steam trap and then directly injected into the main steam line of the hospital through a non-return valve.

The system operates as a fuel saver and includes no thermal storage, with backup being provided by the existing boiler system. However, in situations where conventional fuel is very expensive, thermal storage could raise the annual fraction of energy supplied by solar energy.

The annual thermal energy supplied to the project site by the stainless steel version array is expected to be 250GJ, and by the later aluminium version to be 390GJ. Figures 3 and 4 show calculated annual estimates of performance as a function of clearness index for the latitude of Sydney, 34^0 S. It is evident that annual performance is highly dependent upon available direct beam radiation and that the technology is particularly suited to sites as sunny as Sydney or better. The technology would clearly not be viable in Northern Europe except at very high commercial fuel costs.

In addition, particular attention should be paid to minimisation of losses from system heat transfer pipework in all climates. Effective low loss pipework has been developed by the Department of Applied Physics, but conventional pipework will be used initially on the project site.

COST-EFFECTIVENESS

The small 196 m^2 project array is not intended as a cost-effectiveness demonstration, and its installed cost will be about US$43,000, about twice the projected installed cost of US$21,000-25,000 in large scale production at a rate of 100,000 m^2 per annum. The large volume ex-factory price after 4 years of production has been estimated at US$64 per m^2.

On the CDH site, the cost of conventional fuel (electricity) is high, at about US$12-16 per GJ, so that under continuous operating conditions the array could expect to save US$3,000-6,200 per annum on straight cost per GJ saved basis. However, actual savings may be greater at the site because the hospital presently incurs large penalty rates for exceeding electricity supply demand maxima, and the solar contribution may reduce penalties applying to the entire electrical bill. There seems every likelihood that even the first project array may prove cost-effective against the expensive electricity now used. Improved low loss pipework could raise savings, to between $5,300 and $7,000.

Detailed discounted cash flow analyses have been carried out recently, based upon both stainless steel and aluminium versions of the collector. The stainless steel version delivered a life-cycle energy cost of US$2.50-$6.60 (depending upon economic assumptions) per GJ, compared to between US$3.60-$6.80 LPG for a sunny inland climate similar to Parkes NSW, while the aluminium version in the poorer Sydney climate would be US$3.20-$6.00 per GJ, which is cost-competitive against natural gas in that city.

SUMMARY

A prototype periodically adjusted evacuated tube solar boiler array is being installed at Campbelltown District Hospital near Sydney. This simple type of array is intended, in large scale production, to compete with cheap conventional industrial fossil fuel in Australia, and would be suitable for installation in most countries having an annual average clearness index greater than 0.5. The tehnology could be viable in other areas at higher fuel prices.

REFERENCES

1. Mills, D.R. (1986). Relative Cost-Effectiveness of Periodically Adjusted Solar Collectors Using Evacuated Absorber Tubes. Solar Energy 36, No. 4, pp 323-331.

2. Mills, D. Harting, E. and Giutronich (1981). Simple High Efficiency Non-tracking Thermal Concentrator for Temperatures up to 250°C. Solar World Forum, p.1723, Proc. Ises Conf. Brighton.

3. Mills, D. (1984). Periodically Adjusted Solar Collectors for Medium Temperature Applications. ISES ANZ Section Annual Meeting. Auckland.

4. Mills, D., (1985) Periodically Adjusted Solar Collectors for Medium Temperature Applications. Intersol Congress, Montreal.

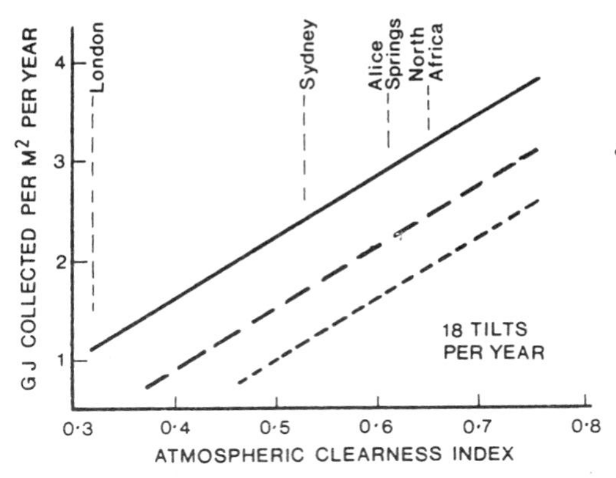

Fig 3. Graph of annual performance — stainless steel mirror.

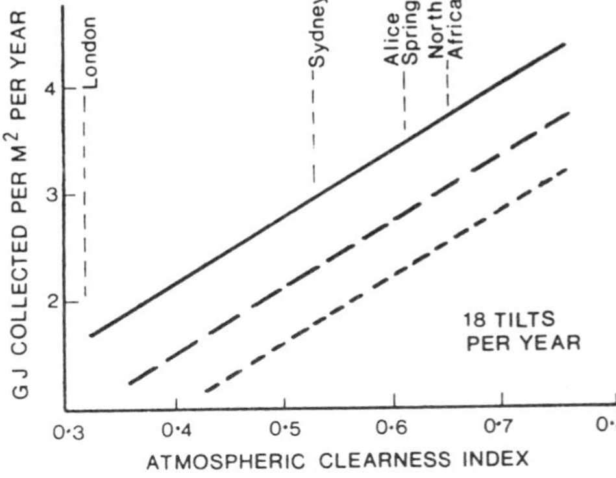

Fig. 4. Graph of annual performance — aluminium mirror.

The array uses some low loss pipework, and will lie between the extremes.

——— Optical collection only
— — — Low loss system pipework (150°C)
- - - - - Conventional pipework (150°C)

THE DEGREES OF FREEDOM IN THERMAL SOLAR PLANTS

A. ADELL, Laboratoire de Chimie-Physique, Université des Sciences et Techniques du Langueduc, 34060, Montpellier, France

ABSTRACT

In order to achieve the thermodynamic optimization of thermal solar plants only some particular parameters are required. The thermodynamic degree of freedom is the number of independent decision variables necessary to the thermodynamic optimization. It has been shown that these variables are generally distinct from those which correspond to the thermodynamic closure of the problem and from those which characterize the equivalent ideal Carnot machine. A certain amount of prudence is then recommended when discussing the maximum performances available from thermal solar plants.

KEYWORDS

Thermal solar plants, thermodynamic optimization, degree of freedom, maximum performances, Carnot machine.

INTRODUCTION

The optimization of thermal solar plants is necessary in order to improve their competitivity and to promote their industrial development. In order to render this optimization possible, we must clearly set up the particular problem we want to treat and especially we must enumerate carefully all the variables that have to be optimized. In this report we have focussed our attention on the manner of proceeding to this enumeration.
The optimization problems have been presented in a previous report (Adell, 1987). They generally consist of searching for the best value of the parameters characterizing the machine and the conditions of its working, in order to satisfy certain predetermined criteria. The present study is limited to the *thermodynamic optimization problems* which consist of maximizing only the energetic production of the apparatus. In the case of a solar apparatus, the maximum energetic production for a given solar irradiation and collecting area, is obtained when the total solar efficiency is at its maximum. The total solar efficiency involves the efficiency of the collecting system and the efficiency of the isolated thermal machine. Generally these two efficiencies can not be optimized separately because of the interdependence of certain variables on which they depend. So, among the set of variables, the independent ones which can be optimized while satisfying to the objectives of the technical specifications of the project, must be determined. In a first step, this optimization is made independently of economic considerations. The variables of the optimization must be *primary variables*, that is variables which allow us to answer essential questions linked to energetic problems and concern as well as the realization of the machine as its working. These variables are the *thermodynamic decision variables* and their number is the *thermodynamic degree of freedom.* It is obvious that these variables have to be enumerate before the resolution and the discussion of any optimization problems.
One of the principal difficulties occurs when, in order to discuss the maximum

performances accessible to a thermal machine, the ideal Carnot machine is refered to. Now as it will be shown, the Carnot machine does not have the same degree of freedom as the real corresponding machine. Besides, the degree of freedom does not correspond to the number of independent variables which allow the complete closure of the problem from a thermodynamic viewpoint; this may raise some confusion in the discussions.
Felli (1983) has entered upon the subject in the case of absorption refrigerators.

DEGREE OF FREEDOM IN DITHERMAL MACHINES: IDEAL AND REAL CYCLES

The study of the maximum energetic performances accessible to thermal machines working thanks to the flow of a pure fluid in a closed cycle can be done by comparison with the performances of the corresponding Carnot machine which is the machine working with the same temperature sources and which only involves monothermal or adiabatic evolutions. The performances are maximum if the machine undergoes completely reversibles transformations. This implies that there can be no external or internal dissipations of exergy, and requires that the evolutions with heat exchanges be isothermal and without internal creation of entropy, and that the evolutions without heat exchange be isentropic. The only thermal machine that can have a reversible working is thus the Carnot machine.

The principal quantities which allow us to characterize the energetic objectives are: the temperature of heat sources (T_i), the amount of heat exchanged with each source (Q_i) and the total mechanical work (W). These quantities are the energetic variables of the system. The fundamental *objective function* linked to the thermodynamic optimization criterion is the exergetic efficiency which has an intrinsic physical meaning contrary to the energetic efficiency which depends on the heat exchanges temperature. To reckon this objective function no other variables than the above ones, are needed.
We shall consider that the *quality factors* that are not linked to the systematic decreases of exergy which result of the different technological choices in the real machine, are not true decision variables and do not intervene in the calculation of the thermodynamic degree of freedom (the quality factors are the physical parameters which are the cause of the exergy degradations in the machine). The thermodynamic optimization of the quality factors is a quite trivial partial optimization: they become optimum when their value corresponds to an infinite quality.

Ideal Cycle: Carnot Reversible Machine

Number of variables corresponding to the thermodynamic closure of the problem.
If the nature of the working fluid is known, this implies the knowledge, at least formally, of the fundamental thermodynamic equation of the fluid in the retained representation, for instance: $g = g(T,P)$. This equation implicitly contains a quasi infinity of coefficients some of which intervene in the evaluation of the quality factors linked to the fluid: specific heat, viscosity... .
The four equilibrium states concerning the Carnot cycle are defined by the eight parameters (T_i, s_i); only four of them result independent when taking into account the particular conditions of the cycle: $T_1 = T_2$, $T_3 = T_4$, $s_1 = s_4$ and $s_2 = s_3$.
When making the energy balance, we must distinguish the calorific transfers "q" from the mechanical transfers "w":

$$q_1 = \int_1^2 T\,ds = T(s_2 - s_1) \;, \quad q_2 = 0 \quad , \; q_3 = \int_3^4 T\,ds = T(s_4 - s_3) \;, \quad q_4 = 0$$

$$w_1 = \int_1^2 (vdP)_T \quad , \; w_2 = \int_2^3 (vdP)_s \;, \; w_3 = \int_3^4 (vdP)_T \quad , \; w_4 = \int_4^1 (vdP)_s$$

where the small letters stand for quantities linked to the unit of mass of the working fluid.
In permanent regimes the knowledge of the instantaneous energetic quantities requires the knowledge of the mass flow rate "$\dot{m}$", and the knowledge of the total energetic quantities requires the knowledge of the duration "Δt" of the

working; while in intermittent cyclic regimes, the knowledge of the total energetic quantities can be obtained by the knowledge of the total cycled fluid mass"Δm".

The number of variables corresponding to the complete thermodynamic closure of the dithermal Carnot machine, is then 7, for instance:

fluid nature, T_{hot}, T_{cold}, $s_1(t)$, $s_3(t)$, $\dot{m}(t)$ and Δt or

fluid nature, T_{hot}, T_{cold}, $s_1(m)$, $s_3(m)$, $\dot{m}_i(m)$ and Δm .

where $m = m(t)$ is the mass of the cycled fluid and $\dot{m}_i(m) = \dot{m}_i(t)$ is the mass flow rate in the element "i" of the machine.

Degree of freedom in energetic problems. The working of the Carnot reversible cycle is perfect from an energetic viewpoint; therefore no improvement is to be expected, so that the thermodynamic optimization makes no sense. To comply with the energetic objectives or constraints that have been assigned to the machine, it is necessary to know the value of a minimum number of variables. If we are interested in the time-global balance of the working, these energetic variables are: T_{hot}, T_{cold}, Q_{hot}, Q_{cold} and W .

The conditions imposed by the fundamental laws of Thermodynamics bring in two relations: $\int dh_{cycle} = 0 \rightarrow Q_{hot} + Q_{cold} = W$; $\int ds_{cycle} = 0 \rightarrow Q_{hot}/T_{hot} + Q_{cold}/T_{col} = 0$

The number of independent global energetic variables is then 3, for instance: T_{cold}, T_{hot} and W . These three energetic variables are also *primary decision variables*. If we are interested in the instantaneous energetic quantities, the knowledge of the duration Δt of the working is necessary.

In the ideal Carnot machine no irreversibility is implied so that no quality factor intervenes. The four energetic variables are also the four thermodynamic decision variables and thus the degree of freedom is 4. Parameters like pression, mass flow rate, specific mass... do not directly intervene in energetic balances: they are *technical variables* that are only taken into account when one wishes to treat technical problems. In the same manner, the choice of the nature of the working fluid does involve any consequences for the energetic efficiencies in the case of the Carnot reversible machine.

Real Cycle: Dry Vapor Compression Machine

Because of the great technological complicacy of Carnot machines -- all their elements must exchange mechanical work in very special conditions of reversibility so that each element must include a moving part -- the machines which have actually been built, definitely work in a different way.

The differences between the real machines and the Carnot ones, stand on two points: on the one hand the imperfections of the quality of the components of the real machines (elements or working fluid) involve *occasional degradations of exergy*; on the other hand the necessity of building up real machines with a little sophisticated technology, entails some specific constraints which bring about *systematic degradations of exergy*.

For instance let us consider the special case of a dithermal refrigerator working by compression of vapor in a dry regime. Such a machine has only one motive element: the compressor. Suppose that the working of the machine is quasi perfect, that is it does not involve any occasional dissipations of exergy. Then the irreversibilities come only from the systematic dissipations of exergy at the level of the desuperheating of vapors (external irreversibility) and at the level of the adiabatic expansion in the throttling device (internal irreversibility).

Number of variables corresponding to the thermodynamic closure of the problem

The knowledge of the fundamental thermodynamic equation of the working fluid involves that only two intensive thermodynamic variables are necessary to set up the chart of the cycle. As for the Carnot machine the instantaneous or total energetic quantities are deduced from the chart thanks to the mass flow rate $\dot{m}(t)$ and the duration Δt for permanent cycle or thanks to $\dot{m}_i(m)$ and Δm for intermittent cycles. Indeed, all the energetic powers are evaluable for each evolution, since all these evolutions are reversible in an internal manner, except for the expansion which does not involve any exchange of energy. The number of variables which corresponds to the thermodynamic closure is then 5, for instance: fluid nature, T_{cond}, $T_{ev}(t)$, $\dot{m}(t)$ and Δt .

Degree of freedom corresponding to energetic problems. The energetic variables of the quasi-perfect real machine are: T_{amb}, T_{cold}, Q_{amb}, Q_{cold}, W and Δt. Because of the systematic irreversibilities, only the first law of thermodynamic links these variables directly together. So, there are 5 independent energetic variables, for instance: T_{amb}, T_{cold}, Q_{amb}, Q_{cold} and Δt.
Then, the problem consists of indicating what are the primary decision variables that allow to specify the characteristics of a quasi-perfect real machine the working of which correspond to an imposed value of these five energetic variables. The knowledge of the fluid nature and of the mass flow rate allows the calculation of the systematic irreversibility powers. These two new variables permit the introduction of the two first fundamental laws of Thermodynamics and leave the total number of independent variables unchanged. Thus the thermodynamic degree of freedom is 5 and corresponds to the case of the complete thermodynamic closure, for instance: fluid nature, T_{amb}, T_{cold}, $\dot{m}$ and Δt .
Similarly to the case of the Carnot machine, these five thermodynamic decision variables do not allow us to accede to the totality of the primary decision variables of the machine; for instance the mass flow rate depends in fact upon two other decision variables: the volumetric ratio and the rotation speed of the compressor.
In order to characterize the optimum performances accessible to real machines, an over-simplified method is sometimes envisaged. This method consists of taking into account the constraints imposed by the nature of the fluid to the latent heats exchanged during the changes of phase, while neglecting the systematic irreversibilities. In the case of a compression refrigerator with a permanent flow, it follows:

$$\dot{Q}_{amb}/\dot{m} \simeq \Delta h_{cond}(T_{amb}) \quad \text{and} \quad \dot{Q}_{cold}/\dot{m} \simeq \Delta h_{ev}(T_{cold})$$

where Δh are the massic latent enthalpies. This situation which can offer some interest if we wish to know the order of magnitude of the maximum powers accessible to the unit of mass of the working fluid, must be envisaged very prudently, since in this case the exergetic efficiency of the machine can become higher than 1 which shows its total unrealizability!

OPTIMIZATION PROBLEMS IN SOLAR COLLECTORS

In the case of a flat solar collector the phenomenological equation giving the collected power $\dot{Q}_c$ at a given time, may be written in the form (Cooper, 1975) (Adell, 1984) : $\dot{Q}_c = (h\,E(t) - h'\,T_c + h'')\,S$

where S is the collecting area, $E(t)$ is the global solar irradiation of the unit area and T_c is the collecting temperature. h, h' ,h" are three phenomenological coefficients which depend on the quality factors and on the environmental and utilization parameters of the collector. The three coefficients h, h' and h" do not constitute true thermodynamic decision variables, since for a collector of infinite quality, they respectively take the value 1, 0 and 0.
The energetic characteristics of a solar collector are given by the four quantities: $T_c(t)$, $E(t)$, S and $Q_c = \int_{one\ day} \dot{Q}_c\,dt$ then the thermodynamic degree of freedom of a solar collector is 3.
The adaptation between a solar collector and a thermal machine in perfect conditions of energetic coupling, leads to: $T_c(t) = T_{hot}(t)$ and $Q_c = Q_{hot}$
If the coupling is done with the help of a heat accumulating device, it genarally follows: $T_c = T_{hot} = \text{constant}$; $\dot{Q}_c(t) \neq \dot{Q}_{hot}(t)$

with once again if the coupling is perfect: $Q_c = Q_{hot}$.

So that the degree of freedom of the complete solar apparatus is higher one unity than that of the thermal machine alone.
Practically the solar irradiation E(t) of the collector is seldom a directly accessible experimental datum. The local meteorological stations generally give the ground irradiation or the direct normal intensity. The passage from one type of data to another, needs the knowledge of the collector tilt. This tilt angle constitutes then a new decision variable. However we have shown that it is possible to proceed to a *partial thermodynamic optimization* of the orientation angle of a flat solar collector, independently of the characteristics of the collector (Adell, 1982). This thermodynamic optimization is made once and for all, as a function of the place and of the days provided for the utilization, so that the tilt angle no longer constitutes a decision variable.

CONCLUSION

In order to determine the best energetic conditions for a thermal solar plant, we have sought what are the independent primary decision variables that permit us to answer such elementary questions as "how to fabricate the machine and how to make it operating?". These variables have been called thermodynamic decision variables because they permit to proceed to the thermodynamic optimization and their number is the degree of freedom.
We have shown that the thermodynamic decision variables are generally distinct from the variables that correspond to the thermodynamic closure of the problem allowing the knowledge of all the thermo-energetic quantities. We have also shown that the ideal Carnot machines the characteristics of which are sometimes utilized to discuss the maximum performances accessible to real machines, have a degree of freedom lower than real machines. Thus a certain amount of prudence is recommended when discussing optimization problems.
The notion of "quality factors" has been introduced for defining some parameters of the system. These quality factors are responsible for the degradations of energy in the machine. We have specified that it is only the quality factors that are influenced by the technological choices of the apparatus, that play a real part in the thermodynamic optimization and that may intervene in the calculation of the degree of freedom. The knowledge of the thermodynamic decision variables permits us to proceed to the thermodynamic optimization of thermal solar plants which can be made within various scenarios corresponding to various technological choices.

REFERENCES

Adell, A. (1982). Determination of the optimum tilt of a flat solar collector in function of latitude and local climatic data. Rev. Phys. Appl.,17, 569-76.
Adell, A. (1984). Determination des performances d'un capteur solaire plant à générateur solide incorporé utilisé dans la réfrigération solaire. Rev. Int. d'Héliotechniques. 1er semestre, 52-59.
Adell, A. (1987). Optimization problems in thermal solar plants. I.S.E.S. Congress, Hamburg.
Cooper, P.I. (1975). Symp. Applications of Solar Energy Research and Development in Australia I.S.E.S. A.N.Z. Section,Melbourne. 1-11.
Felli, M. and G. Galli (1983). The degrees of freedom of absorption systems. Proceedings of the XVIth International Refrigeration Congress, 560-566, Paris.

A RANGE OF HIGH PERFORMANCE UNIFLOW RECIPROCATING STEAM ENGINES POWERED BY SOLAR, BIOMASS AND OTHER SOURCES OF STEAM

S. Kaneff, E.K. Inall, R.E. Whelan

Department of Engineering Physics
Research School of Physical Sciences
Institute of Advanced Studies
The Australian National University
Canberra, A.C.T., Australia.

ABSTRACT

A medium level technology for producing robust high performance steam engines is described, which results in engines in the range of a few kW to megawatts, being based on converting certain diesel engine blocks to steam operation. Such engines have operated in systems with outputs of 25 kW_e and 50 kW_e, efficiently and reliably (in the former case since 1982) with less required attention than for diesel engines.

The engines may be employed in various applications, including the use of steam generated from solar energy, biomass (including crop residues), waste heat, and the burning of municipal wastes to produce mechanical and electrical outputs (and process heat) for electricity generation in remote areas, rural village power supplies, mechanical drives for water pumping, co-generation and many others.

Scope exists for further improvement of the already high efficiency by the use of higher quality steam made possible by the use of ceramic components. Simple units may be produced for about $1000/kW for sizes around 50 kW and less for a production run. A target of $250/kW is being pursued and is expected to be met following further development.

INTRODUCTION

In recent years much effort has been directed to developing, on the one hand, high efficiency heat engines (for example employing Stirling, Brayton and other cycles) for utilizing, very efficiently, the high quality heat from concentrating solar collectors; and on the other, to produce very inexpensive low efficiency steam and other engines which can be built in Third World workshops and supplied by energy from biomass. The former objectives have not as yet been realised because problems of performance, especially reliability and cost effectiveness, have not been adequately overcome; the latter also for various reasons have not resulted in that technology reaching significance.

But there is another approach which employs medium level technology and is based on mass-produced readily-available components, supplemented by some special items, to produce steam engines with heat to mechanical-work conversion efficiencies of over 20%, robust, reliable, able to be maintained by those with automotive and agricultural experience, and having the potential for cost-effectiveness for many applications.

Such engines have resulted from our development of the White Cliffs Solar Thermal Power Station (Kaneff (1983),(1987)), where a unit has operated for many thousands of hours giving electricity supply reliably. Other units are currently working in USA in Troy NY and at Albuquerque at the Sandia test facility, preparatory to operation on Molokai (Hawaii). Further units are to be used for a rural village power supply in Fiji and for the utilisation of crop wastes in Australia.

ENGINE DETAILS

Engines have been designed already for 40 kW mech., 75 kW mech. and 200 kW mech; but much larger units can be produced, all based on the conversion of diesel engines to steam operation, this approach being perceived as being very appropriate with respect to costs and spare part provision. Diesel engine blocks also tend to have very robust construction.

To indicate some details and properties, the following apply to the engine used on the White Cliffs solar power station which employs a field of collectors to provide steam to power the prime mover, the nett output being 25 kW elec. The engine starts automatically when the steam from the 14 paraboloidal collectors is hotter than 180°C and the pressure reaches 28 kg/sq cm. The full supply of steam being produced by the array is used when ever the engine is running. When the steam supply drops, due to lack of sunshine, the engine stops and the starting circuits are reset to monitor the steam and await the starting conditions. Water drains from the steam lines via a steam trap. The engine is not sensitive to water in the steam and it operates reliably over a wide range of steam conditions giving partial output from wet low temperature steam when the sun is attenuated and a maximum output of 34 kW when the insolation is 1 kW/m^2. The collectors then supply 50.4 gm of steam per second at a pressure of 42 kg/cm^2 and a temperature of 415°C.

The engine exhausts into an evacuated condenser via a vortex chamber which collects most of the engine lubricant and water droplets. However, the steam carries some oil to the condenser and the condensate then contains a dispersion of oil droplets which must be removed before the water is recirculated through the solar collectors. A process using little power was devised to do this and return as much as possible of the oil to the engine.

THE ENGINE WITH PISTON OPERATED VALVES (POV ENGINE)

Most of the engine is made from parts of two diesel engines which are on the market. The general form of the engine is shown in Fig. 1. Steam is supplied to a chamber in the head of each cylinder and the engine is started by a standard electric motor. As a piston approaches top dead centre the pins in its crown lift the three ball valves from their seats and steam enters the cylinder until the valves seat again past dead centre. The steam expands while applying pressure to the piston until the piston exposes the normal exhaust ports in the cylinder liner which was made for a two stroke diesel engine. The cylinders, cylinder heads, valve seats and steam chambers, that is, the conversion components, can be produced by relatively simple workshop techniques from cast iron, mild steel and stainless steel.

All parts of the engine are inexpensive, do not require special machine tools to fabricate, and two men could rebuild the engine with replacement parts between sunset and sunrise should that ever prove to be necessary.

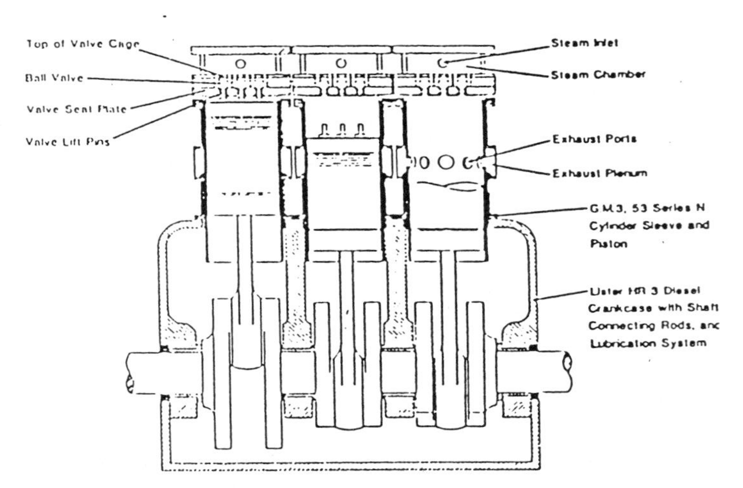

Fig. 1: Valve and piston arrangement in the POV engine

The White Cliffs engine configuration is:

Bore	98.4mm
Stroke	114.3mm
No. of Cylinders	3
Maximum Steam Pressure	70 kg/cm^2 (abs)(6.9 MPa)
Maximum Steam Temperature	450C
Condenser Pressure	0.25 kg/cm^2 (abs)(24.5 KPa)
Expansion Ratio (Adjustable)	1:25 (used)
Lubrication	as in Lister engine
Lubricant	Specially selected
Measured Efficiency (Steam Press 42 kg/cm^2, temp. 415°C)	21.9%

Results of measurements of efficiency made by staff of the N.S.W. Energy Authority are shown in Fig. 2. This shows that the efficiency drops slowly with reduction of super heat until saturated steam occurs, after which wet steam causes a rapid reduction in efficiency.

SEPARATION OF OIL FROM THE CONDENSATE

Fig. 3 shows the interconnections between the components of the steam system. The exhaust steam line enters tangentially the upper end of a cylindrical vortex chamber. Oil and water droplets are stopped by a gauze sleeve on the inner surface and drain to the bottom of the chamber, the steam passes up through stainless steel wool held in a cylinder, which is attached to the top plate of the vortex chamber. The smallest oil drops pass into the condenser with the steam and the condensate carries these as a very fine suspension to compartment 1 of the feedwater tank via the vacuum pump. This dispersion of oil in the condensate can not be removed by conventional filters or the centrifuge, and special means must be employed.

The filter pump draws water from compartment 1 of the feedwater tank and delivers it up to a filter system mounted in compartment 2 of the feedwater tank. The oil coalesces in the filters and the partly cleared water passes over a weir into compartment 1 from where it is recirculated through the

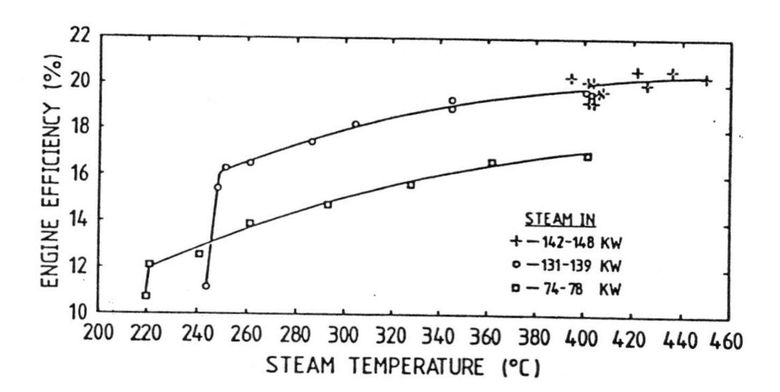

Fig. 2: Variation of the steam engine efficiency with temperature for three different power levels. The sharp drop in efficiency occurs when the steam is wet. The solar collectors produce 170 kW of steam at more than 350C when insolation is 1 kW/m^2. Efficiency increases with throughput.

filters. The feedwater enters compartment 3 by passing through a further filter in the bottom of compartment 2. Beads of oil form on the internal filters and float to the surface in compartment 2 from where they are carried over the weir into compartment 1.

Level switches in the vortex chamber control the operation of the oil-water pump which delivers the oil-water mixture to the centrifuge via the skimmer pump. Oil from the centrifuge is returned to the engine tank and water returns to compartment 1. Oil also collects on the surface of the water in compartment 1 which varies in level depending upon the amount of water or steam in the return line from the array. A special skimmer, which floats to the level needed to skim at the rate determined by the flow into the skimmer pump, is used to collect the oil with water. The skimmer pump delivers this mixture to the centrifuge. About 2 ml/s of oil is recovered from the engine exhaust with the complete condensate treatment. The oil is washed and cleaned by this process, leaving solids in the various filters and the centrifuge chamber.

The condenser is cooled by water from a tank holding 100000 litres of rain water collected from a nearby dam.

When the engine is stationary a motorised bypass valve is open and water or steam from the solar collectors passes to the condenser until the steam conditions are correct for the engine to start. Some water can collect in the steam line to the engine. When the engine is to start and the throttle opens, water and steam arrive at the engine. When this enters the cylinders the starter motor cannot crank the engine if the water can only escape via the inlet valves. A motor operated drain valve is fitted with a port to each cylinder. In some configurations this is not necessary. Interlocks prevent the starter functioning until the drain valve is open. When it is, the water can flow to the evacuated exhaust line to the condenser, through ports which limit the amount of steam lost. As soon as the engine starts the drain valve is closed by a signal from a speed measuring instrument. When the engine is running any water in the steam from the solar collectors is diverted via a steam trap into exhaust lines.

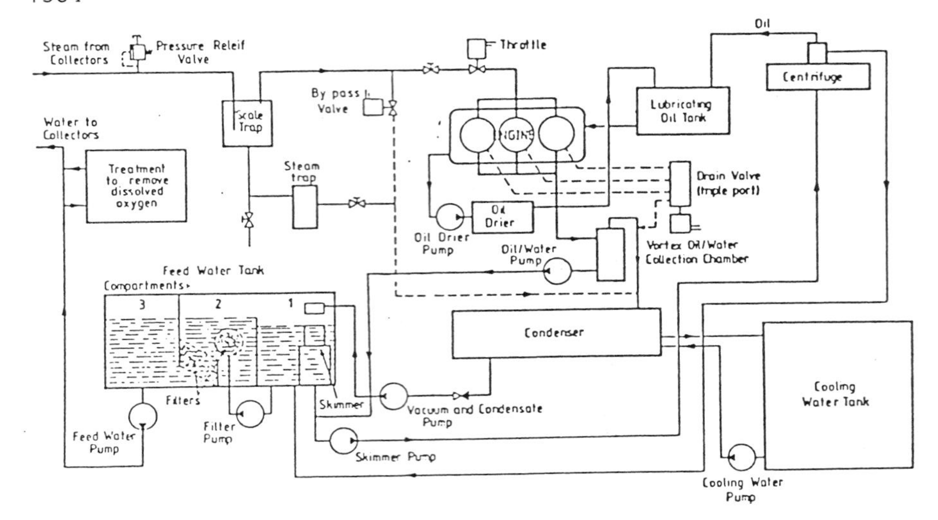

Fig. 3: The interconnection of the steam engine and its auxiliary plant.

CONCLUSIONS

Efficient, robust engines can be built using parts from currently available diesel engines. The highest temperatures at which such engines can operate are limited by the properties of the materials used for the valves and seats, mainly. Current limits are set at 450°C, but ongoing research to employ ceramic components give potential temperatures of over 800°C. A target efficiency of up to 30% may eventually be achieved.

The engine can be serviced or repaired by any person familiar with the usual internal combustion engine used in vehicles and could be used in any location where superheated steam could be produced by any means including the combustion of waste material or garbage.

Initial cost indications for the first units produced, on a one-off basis, suggest - excluding burner and boiler but including generator and controls - some \$Aust.1000/kW_e for sizes of 25-50 kW_e, falling to about \$A500/kW_e for a small production run. Current simplifications and improvements are expected to lower these costs further, target cost being \$250/kW_e. System sizes up to 2 MW are considered practicable.

REFERENCES

Kaneff, S. (1983). The White Cliffs Solar Power Station, Proc. Solar World Congress, ISES, Perth, Pergamon Press 1984, Vol.4, pp.2722-2776.

Kaneff, S. (1987a). The White Cliffs Solar Thermal Electric Power Systems, Proc. Joint ASME-JSME-JSES Conference, Honolulu March, Solar Engineering 1987, pp.815-820.

Kaneff, S. (1987b). A Viable Role for Solar Thermal Power in Remote Areas. Proc. I.E. Aust Engineering Conference, Darwin, May, pp.192-198.

50 KW SOLAR CONCENTRATORS WITH STIRLING ENGINE, RESULTS OF THE SIX MONTHS OF OPERATION IN SUADI ARABIA

Ali Al-Rubaian, KACST, John Hanson, DFVLR
King Abulaziz City For Science and Techlology
Solar Energy Program, P.O. Box 6086
Riyadh 11442, Saudi Arabia.

ABSTRACT

This paper discribes the results of the first 900 hours of operation by the two 50 kW Dish with Stirling engine units in Riyadh, Saudi Arabia. The units were constructed, installed and are being operated as a joint project between the Federal Ministry for Research and Technology (BMFT) for West Germany and King Abdulaziz City for Science and Technology (KACST) for Saudi Arabia. The Design and construction was led by Schlaich and Partner, Stuttgart who developed the novel 17m diameter membrane concentrator and who were responsible for the system design and integration incorporating a control system by MBB, and the 50 kW Stirling engine generator units by USAB of Malmo, Sweden. Construction ended in spring 1986 and a programme of operation and evaluation was begun under the leadership of KACST with participation by DFVLR and Schlaich and Partner.

The two dishes have demonstrated how effectively the system operates as a solar generator, starting to deliver useful AC power when insolation exceeds 350 W/m^2. Over all the hours operated so far the useful outputs have averaged 20.5 kWh per dish per running hours for 6.5 hours per day. The present actual performances are around 20% above this overall average and the improvements are possible.

Figure 1. 50 KW Solar Concentrator

Testing has indicated room for improvement of the receiver performance particularly to minimise wind effects and a need for further work on controlling the membrane focus. Results of evaluation of reflectivity and cleaning procedures are also presented.

INTRODUCTION

The Saudi Solar Village which is located 45 KM northwest of Riyadh the capital, was selected as a testing site for the two 50 kW solar concentrators with stirling engines. The solar village is known to be a solar testing site because it include among its solar testing facilities a 350-kW concentrating photovoltaic power plant the world's first and one of the largest operating PV systems.

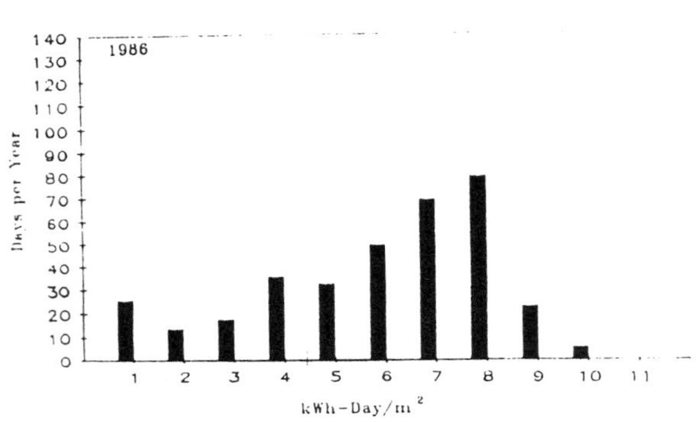

Fig.2 Distribuation of Direct Normal Insolation

A complete weather and solar radiation data for the site are being recorded since 1981. The feasibility of a using the site for solar thermal application is illustrated by figure 2 which demonstrates the relatively high frequency of days with good Direct Normal Insolation (DNI) throughout the year.

After inauguration at the end of April 1986 the two solar dishes began to operate on a preliminary basis according to the initial operating plan to concentrate on routine electrical generation and establish operating procedures, requirements and to build up an effective operating establishment.

Numerous unforeseen problems were encountered, technical, organisational and site related. The result was a virtual close down until the most important problems were solved and operation was effectively resumed in August 1986.

Since August 1986 routine operation has been achieved and consistent record maintained.

Both dishes were operated daily for six days a week with some interruptions untill mid of October when dish 2 operation had to be terminated as the engine could no longer be started, while dish 1 continue a routine power generation and some special testing. The scope of these testing is however, very limited due to an almost complete lack of recording equipment for measurement data. In the absence of calibration equipment the available data must not be interpreted as absolutely accurate.

SYSTEM DESCRIPTION

Figure 3 shows the major subsystems of the 50 kW solar concentrator in a simplified block diagram which consists of the following:

a) Reflector (Dish): This subsystem consists of a two flexible 0.5 mm thick membranes mounted on 17m diameter ring. Between the two membranes a vacuum system is installed to maintain the parabolic shape of the concentrator by stablizing the pressure of the inside space. on the top side very thin square glass mirrors are bonded. The whole struture is suspended in a rotating stand made of steel tubes, which runs on rails, by means of two motors.

b) Receiver: Is the mean of converting solar energy provided by the reflecting mirrors to a thermal energy. The receiver consists of 96 tubes bent to conical involute shape. The outer diameter of tubes is 6 mm and the inner diameter is 4 mm.

c) Stirling Engine: A 4-cylinder double acting stirling engine with coaxially arranged cylinders in parallel. The engine convert the thermal energy to shaft work by using hydrogen gas as a working fluid

d) Generator: A synchronous generator which is located in the focal plan and directly coupled to the stirling engine and has production capacity of 54.5 KWe with efficiency of 91%.

e) Stirling Control System: It is independent subsystem which control and regulates the stirling engine, generator operation and starting, shutdown sequences according to the data gathered from the receiver, stirling engine and generator. It receives a starting (weak up) signal from the dish control and sends to it a detrack signal in case of any fault in the engine, generator or receiver.

f) Dish Control System: It is the master control system which is responsible for the overall dish tracking mechanism. It is based on a computing algorithm for defining the sun position and two high precision angle step signal transmitters which transmit the current position of the dish to the control system, thus an accurate positioning of the dish with respect to the sun can be achieved by the azimuth and elevation motors three operation mode, manual semi-automatic or automatic.

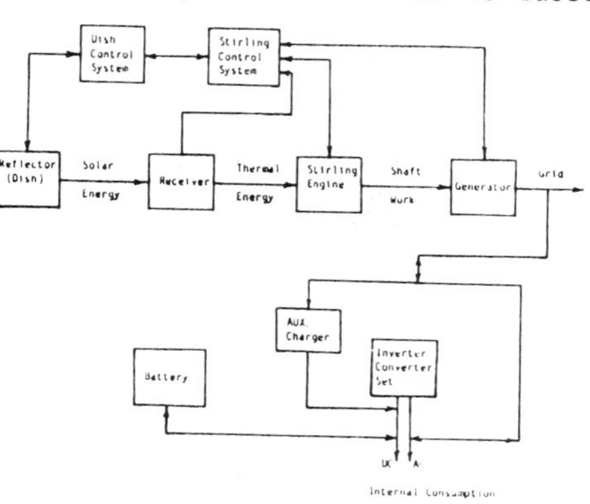

Fig. 3 Simplified Block Diagram of 50KW Solar Concentrator with Stirling Engine

g) Battery Subsystem: Consists of a 110 led acid cells connected in series with a total of 1000 Ah storage capacity and it is being charged either by the inverter/converter set or by the auxiliary charger.

PERFORMANCE AND OPERATION

Operation Overview

The vital statistics of the operation so for are given in Table 1; and illustrated in Figure 4 & 5 where

Gross output: KWh out from dish unit after internal consumption
Net output : KWh out from TDSA site to grid
Input : Radiation incident on the $227m^2$ projected area of the total not the actual reflective area
Hours : Hours synchronised to grid
Dish day : A day when a dish achieved synchronisation (1 per dish)

Table 1 Operation Summary

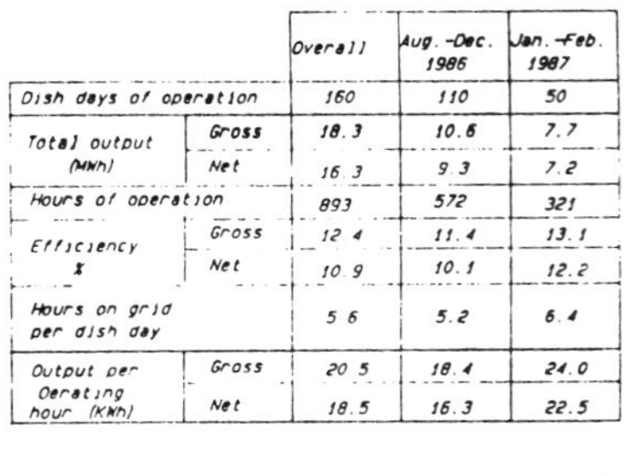

		Overall	Aug.-Dec. 1986	Jan.-Feb. 1987
Dish days of operation		160	110	50
Total output (MWh)	Gross	18.3	10.6	7.7
	Net	16.3	9.3	7.2
Hours of operation		893	572	321
Efficiency %	Gross	12.4	11.4	13.1
	Net	10.9	10.1	12.2
Hours on grid per dish day		5.6	5.2	6.4
Output per Oerating hour (KWh)	Gross	20.5	18.4	24.0
	Net	18.5	16.3	22.5

The figure illustrate the improving results being achieved as problems are understood and dealt with and the organisation improved. It must be emphasised that all the above figures refer to overall average figures covering all the operating hours from morning to night. They must not be used as a basis for comparison with spot performance measurements made in other plants.

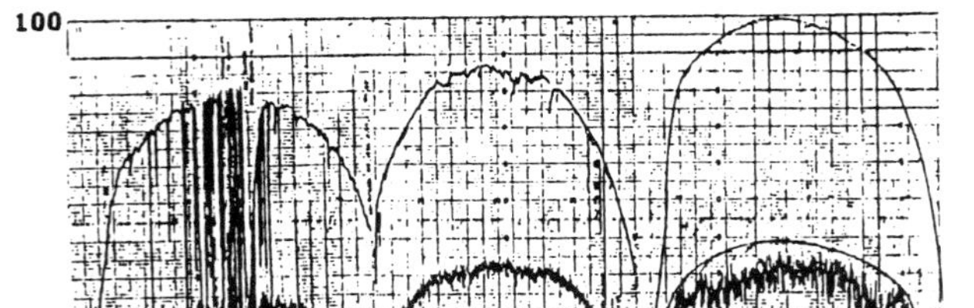

Fig. 4 Typical Daily Characteristics

Some of the technical problems which has occured during this operation period gave the site staff a good experience and key lessons in the plant operation. Below is a summary of these problems:

a) Stirling engine: Rings in the Hydrogen compressor of engine on dish 1 failed in Nov. 1986 and repairs were done by the site crew without difficulty. A

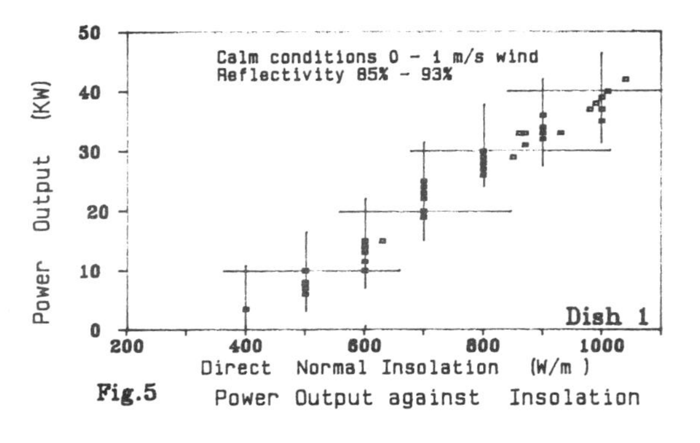

Fig.5 Power Output against Insolation

second failure was reported after approximately 450 operation hour after the fisrt one. In the second fialure the piston cap and the cylinder bottom were damaged.

The engine on dish 2 suffered serious mechanical damages before inauguration and unrectified consiquences of this event led to operation close down in October.

The lubrication systems cause problems with start up procedure in winter when the oil temperature below the minimum which is 14°C.

b) Dish Structures and Mechanism: The under presure pumping equipment has proved effective so far. Membrane under pressure has never fallen out of the control range of the DC Pumps.

c) System Control: The system software is not yet fully developed and various problems have been exposed. Tracking accuracy is one of the major problems, a continuous offset adjustment is required.

This must not be allowed to abscure the clear demonstration that Dish/stirling concept is a very effective unit for solar electric power generation.

System Performance

The characteristics at three typical days are shown in Fig. 4, where the scale for insolation is 1:10 and for power 1:1. Fig. 4-I is for cloudy conditions, Fig. 4-II is for clear and calm day, and Fig. 4-III is for a windy conditions. The conversion efficiency of the system can be deduced from Fig. 5, and it is on the average 16% for 1kW/m² of insolation. 36 test points distributed on the reflector surface were selected for reflectivity measurements. The history of activities is summarised in Fig. 6. Fig. 7 illustrate the severe effect of wind, when the average wind speed is 3.9 m/sec the degradation in the output is about 15% of no wind condition.

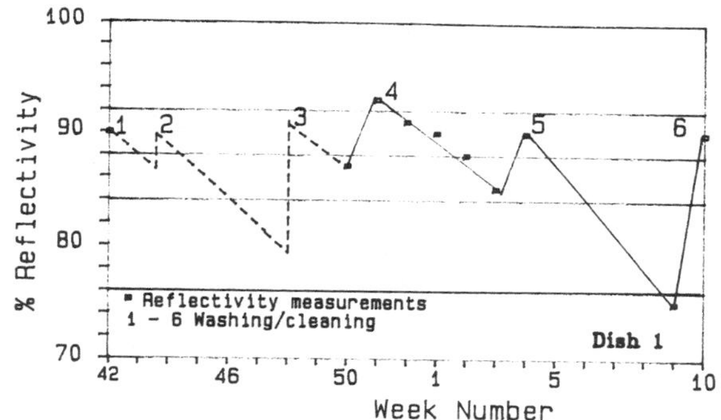

Fig.6 Cleaning and Reflectivity measurement

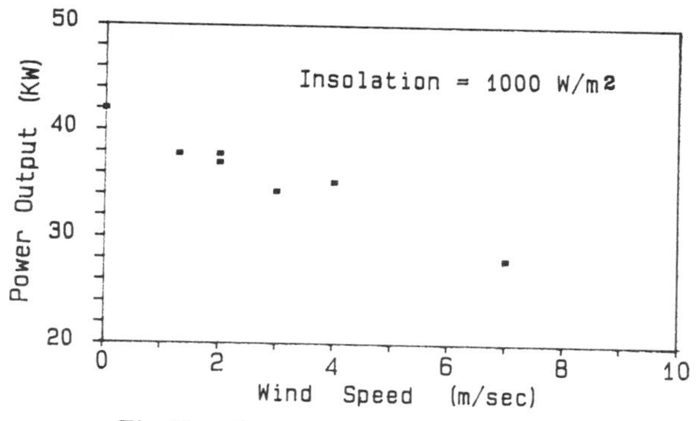

Fig.7 Power against Wind Speed

THE STM4-120 STIRLING ENGINE FOR SOLAR APPLICATIONS

R.J. Meijer

Stirling Thermal Motors, Inc.
2841 Boardwalk
Ann Arbor, MI 48104 U.S.A.

Abstract

The STM4-120 Stirling engine currently under development at Stirling Thermal Motors, Inc. in Ann Arbor, Michigan, USA, has been specifically optimized for use in solar energy applications. This engine has a power output of 25 kW at 1800 rpm, with an efficiency of 40% from heat input to shaft output.

Sealing and power control problems which plagued previous Stirling engines have been eliminated by using a simple variable swashplate piston stroke control. The use of a sodium heat pipe system to bring heat into the engine has not only solved earlier problems with integrated heater heats, but has also made this engine very versatile and easy to use for many applications.

This paper briefly describes the engine and the sodium solar receiver and presents some preliminary test bench measuring results.

Introduction

In recent years many different authors have published a wide variety of papers about the Stirling engine and its unique properties. There are, however, no Stirling engines in commercial production.

It is relatively easy to make a small, running, Stirling engine as a sort of model or toy. These models can show all of the beautiful qualities of the Stirling engine, and it is fascinating to see how one of them placed in the focus of a small parabolic mirror can run on solar energy.

The difficulty arises when one needs a more powerful Stirling engine, particularly one comparable in size to an internal combustion engine, where the heat load must be increased 10, or even 100, times. At this point one encounters the complexities of Stirling engine theory in regard to thermo- and aerodynamics and suitable mechanisms.

Background

At the time that N.V. Philips started the investigations on the hot air engine, later called the Stirling engine, in 1938, the specific weight of the hot air engine was about 1500 kg/hp. In the 40 years Philips was involved with the Stirling cycle, the specific weight was reduced roughly to 3 kg/hp. This was possible because of the basic research Philips applied to the cycle and drive mechanism.

Philips had several licensees, including General Motors and Ford, but none of them was patient enough to reduce all this knowledge to a practical commercial engine. Philips' last licensee was Ford Motor Company. In 1972 Ford obtained a world-wide exclusive license from Philips for the Stirling engine, covering virtually all applications, including automotive. Philips Research Labs designed and built four 175 hp engines for the Ford Torino automobile. Two of these were actually installed. In 1976 the two Stirling-powered Torinos and an older Stirling bus from Philips were successfully demonstrated in Dearborn, Michigan for three days. Unfortunately, by this time the entire auto industry was in trouble. Ford terminated its Stirling engine activities in 1978 to make manpower available for short-term technological problems. A year later, Philips, discouraged by the loss of such a major licensee, stopped all work on the Stirling engine.

At this time Dr. Roelf Meijer, who had headed the Stirling development program at Philips since 1947, retired from Philips and formed his own company, Stirling Thermal Motors.

Stirling Thermal Motors

Stirling Thermal Motors, Inc. (STM), located in Ann Arbor, Michigan, U.S.A., was organized in June 1979, and was founded for the specific purpose of developing commercial Stirling engines. STM became a licensee of N.V. Philips in November of 1979. This gave the company access to all Philips' Stirling know-how, patents and simulation computer codes.

STM was convinced that the time was ripe for commercialization of the Stirling engine. All the ingredients were present for a simple, inexpensive, and reliable engine with long life.

General Approach, Basic Applications

STM's approach to commercialization of its STM4-120 Stirling engine is based on the conclusion that competition with existing internal combustion engines should be avoided, at least in the beginning. Rather, markets should be found where the IC engine cannot be used and where use of the Stirling engine would be economical and would make use of the unique properties of Stirling engines. Of the many possible applications, particular attention was given to four:

1) Solar energy conversion
2) Biomass energy conversion
3) Prime mover for heat-driven heat pump
4) Engines for underwater applications.

Technical Approach

A Stirling engine suitable for commercialization must be simple, reliable, inexpensive, and must have a long service life. None of these requirement should be allowed to have an adverse effect on the performance of the engine.

Several years of designing, consulting with suppliers, price calculations and component testing led to STM's new version of the Stirling engine. The STM4-120 (4 cylinders, 120 cm^3 swept volume per piston) is equipped with variable angle swashplate drive.

The engine was meant for use in solar applications from the very beginning, and was, therefore, optimized for its highest efficiency at 25 kW at a speed of 1800 rpm. This makes it suitable for use with an 11-meter parabolic concentrator. Calculated performance curves are shown in Fig. 1.

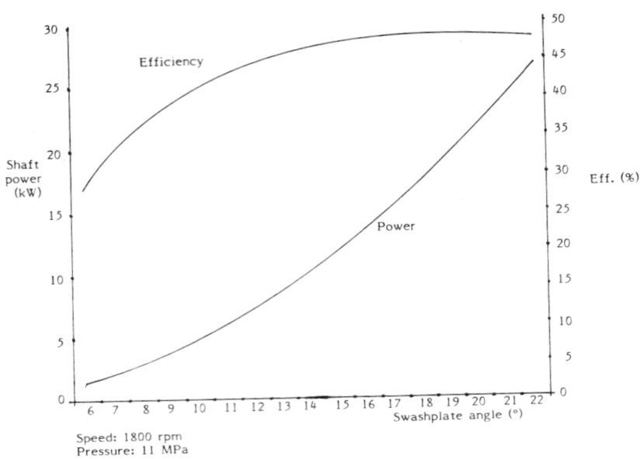

Fig. 1 - Power and Efficeincy of the STM4-120 Stirling Engine

The design of the STM4-120 has also overcome the three most important obstacles in the way of mass production that plagued earlier Stirling engines:

1) Integrated heater head - The use of heat pipe technology allowed STM to make the four heaters as small as the coolers and, thus, fully suitable for mass production.

2) Power control system - The invention of variable angle swashplate power control completely eliminated the problems presented by the sophisticated and complicated power control systems used previously. Varying the angle of the swashplate changes the stroke of the pistons, thereby controlling the power. Fig. 2 is a photograph of the variable angle swashplate from the STM4-120 in two positions.

Fig. 2 - Variable Angle Swashplate From the STM4-120 Shown in Two Positions

3) Reciprocating rod seals - Variable swashplate power control eliminates sealing problems by allowing the drive to be enclosed in a pressurized crankcase. For electrical generation applications, such as solar, the generator can also be enclosed in the crankcase (or the rotating shaft can be sealed with a commercially available seal.

These three technological breakthroughs were integrated into the design of the new STM4-120, shown as a layout drawing in Fig. 3. Fig. 4 is a photograph of the engine without heat pipe connections to a heat source.

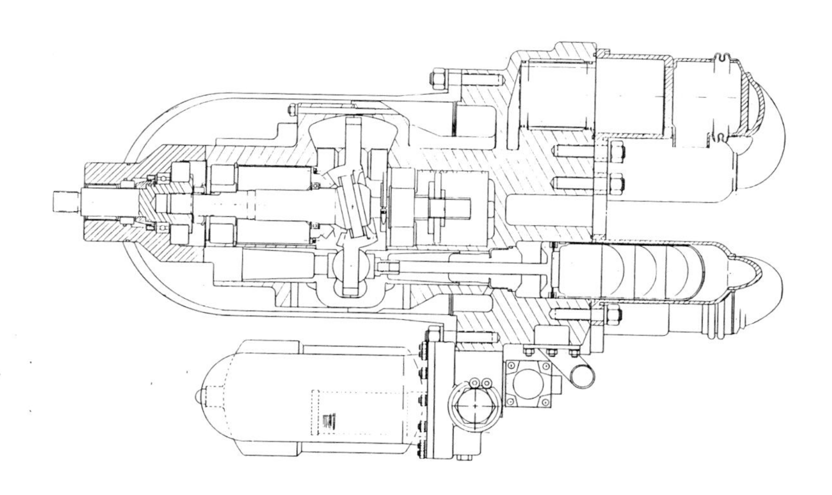

Fig. 3 - Layout Drawing of the STM4-120 Stirling Engine

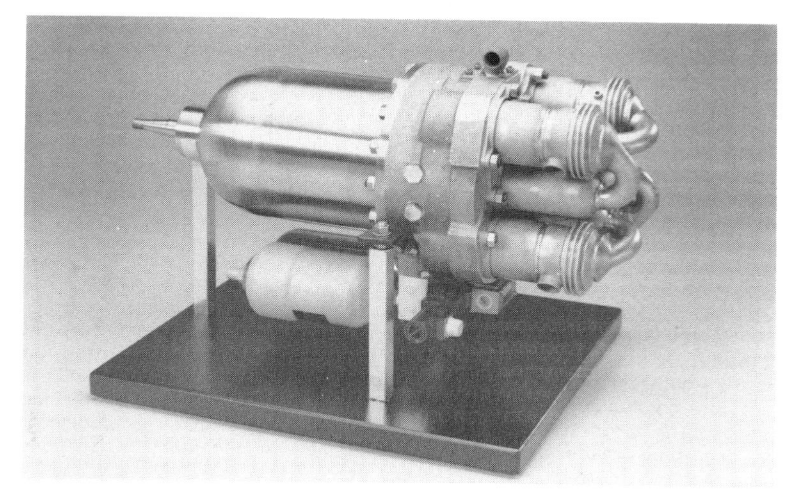

Fig. 4 - The STM4-120 Without Heat Pipe Connections

Heat Pipe Technology

A sodium heat pipe is the most effective heat transporting system known. Condensing sodium inside a heat pipe guarantees an absolutely uniform temperature distribution.

The use of a heat pipe makes the STM4-120 easily adaptable to different applications and heat sources. The same engine can be used with virtually any heat source simply by changing its heat exchanger.

Fig. 5 is a photograph of an STM4-120 model with permanent magnet generator enclosed in the pressurized crankcase, and with a solar heat pipe receiver that STM developed in cooperation with Sandia National Laboratories and the Department of Energy.

Fig. 5 - Model of the STM4-120 with Solar Heat Pipe Receiver

Testing the STM4-120

The test program for the STM4-120 is divided into two parts, a cold motoring test and a performance test in which the engine is heated with a gas-fired heat pipe.

Testing the variable swashplate drive under load conditions with the pressurized crankcase and rotating shaft seal is the most important test run for the engine. Here STM must show that the STM4-120 does, in fact, solve the sealing and power control problems which have always prevented the mass production of Stirling engines. This test started a year ago with the engine equipped with dummy cylinders. After debugging, the engine was put into endurance testing. Preliminary results of the engine performance test have been very encouraging, and the testing will be finished by the end of 1987.

Conclusion

The new STM4-120 Stirling engine, developed and currently being tested by Stirling Thermal Motors, Inc., solves the three major problems that have prevented mass production of Stirling engines. Both the troublesome reciprocating rod seals and the complex and cumbersome power control system were eliminated by the use of a variable swashplate drive contained inside the pressurized crankcase. Indirect heating via heat pipes made an integrated heater head unnecessary.

The use of heat pipe technology results in a very versatile engine especially suitable for solar application. The testing program underway at STM strongly indicates that the STM4-120 will be the first successful Stirling engine for commercial applications.

References

1. Norman D. Potsma, Ford Motor Company, & Rob van Giesel and Frits Reinink, N.V. Philips, The Stirling Engine for Passenger Car Applications, Paper 730648, SAE, Chicago, Illinois, June 1973.

2. R.J. Meijer, The Philips Hot Gas Engine with Rhombic Drive Mechanism, Philips Technical Review 20, 245-262, 1958/9, No. 9.

3. R.J. Meijer and B. Ziph, A Variable Displacement Automotive Power Train, Proceedings, 5th International Automotive Propulsion Systems Symposium, Dearborn, Michigan, April 1980.

4. B. Ziph and R.J. Meijer, Variable Stroke Power Control for Stirling Engines, Paper 810088, SAE, Detroit, Michigan, February 1981.

ELECTRIC GENERATORS FOR SMALL SOLAR-THERMAL SYSTEMS

Dr. Otto J.M. Smith
Professor

ELECTRICAL ENGINEERING AND COMPUTER SCIENCES DEPARTMENT
UNIVERSITY OF CALIFORNIA
BERKELEY, CALIFORNIA, 94720, USA

ABSTRACT

Induction machines are suitable electric generators for small solar-thermal systems using a Rankine-cycle turbine. Th turbine shaft is coupled to a 2-pole induction machine which feeds into an alternating current public utility.

Remote solar installations are often serviced by only single-phase public utilities. Three-phase machines, however, ar more efficient than single-phase machines, cost less, and have less vibration. A three-phase induction generator coupled to turbine can be operated at balanced voltages and currents, and feed single-phase power into a single-phase line using thi new winding connection with capacitors.

KEYWORDS

Generator, Induction-Generator, Phase-Converter, Rankine-Turbine, Capacitor-Excitation, Wye-delta, Three-Phase Single-Phase.

INTRODUCTION

The system being considered is shown in Figure 1. A solar collector delivers pressurized water at a temperature T into a buffer tank. The water is pumped from the buffer tank through a boiler and preheater in series, and the reduced tem perature water is fed back into the bottom of the buffer tank. The ORC working fluid is refrigerant 113 which is preheate and boiled and delivered to the turbine at temperature T_2. The turbine exhaust is cooled in the regenerator and condensed i the condenser and can be stored in a liquid working fluid storage not shown. The feed pump compressed the ORC liquid t the pressure needed by the turbine, and the high pressure liquid is heated successively in the regenerator and the preheater In the boiler, it changes phase and exits to the turbine at temperature T_2. The condenser uses an external cooling water sup ply.

INDUCTION GENERATOR

The turbine shaft is coupled to the ac induction generator. It is desirable to have a high turbine speed, and therefore th generator should have only two poles. Induction motors are much more commonly available than induction generators, an therefore this paper will assume that a commercially available motor is used for the generator. Further, three-phase motor are more efficient than single-phase motors, and therefore a three-phase machine will be used.

Figure 2 shows the reconnection of a conventional motor which had the following name-plate full-load characteristics:

TABLE I

Phases	3	Voltage, volts	230.0/115.0
Poles	2	Current, Amperes,	47.43/94.86
Shaft power, kw	15.0	Electrical Input, KWe	16.91

Efficiency	0.887	Electrical Input, KVA	18.895
Power Factor	0.895	Double-Wye Winding	6 coils
Power Factor Angle Θ	26.5		

IOTOR CIRCUIT

This was a dual-voltage machine, so that at 115-volts, the two wyes were in parallel. These were six separate coils, vo in each phase. All 12 leads were brought out. In a conventional motor connection, A, B, and D would be as shown in igure 2, but for the lower voltage, terminals AA, BB, and CC would all be connected to N.

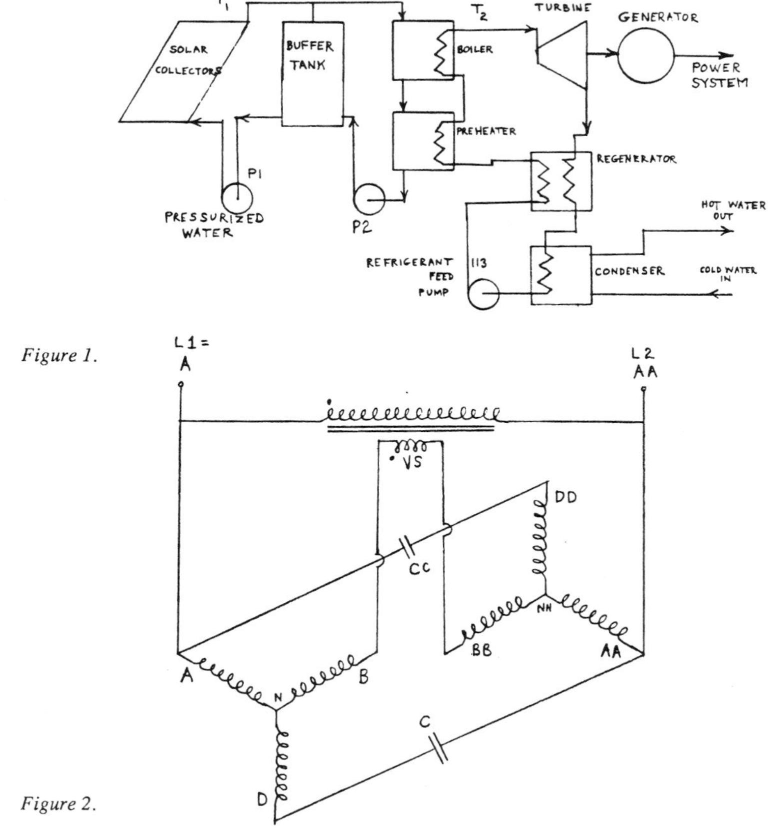

Figure 1.

Figure 2.

ENERATOR CIRCUIT.

By disconnecting these latter three windings, and connecting the opposite ends together to NN, the new wye nfiguration of AA, BB, DD, shown in Figure 2 is realized. In this case, A, B, D is a 115 volt, 7.5 KW machine, and AA, 3, DD is a second 115 volt, 7.5 KW machine.

Terminal B is connected to terminal BB through the secondary of a step down transformer, so that the voltage VS parates the two. Each wye machine is therefore operating at less than rated voltage of 115 volts.

EFFICIENCY TESTS

I have made laboratory tests on an induction motor operated as an induction generator, and measured the maximum efficiency to occur at 15 percent less than rated voltage. This is because the flux in the generator is higher than the flux in the motor for the same terminal voltage. The higher flux causes larger iron hysteresis losses and eddy current losses and larger magnetizing currents. The larger magnetizing currents flowing through the windings causes larger copper losses.

One of the goals of Figure 2 is to reduce the wye voltage to about (0.86)(115) = 98.9 volts. Another goal of Figure 2 is to excite windings D and DD with capacitor currents that are exactly the full load current at exactly the correct phase angle. When these goals are achieved, each wye will be a balanced three-phase machine.

PHASOR DIAGRAM

The phasor diagram for this balanced condition is shown in Figure 3. The power supply from L1 to L2 is 230 volts, so that the voltages V_{AB} plus V_{BB-AA} plus VS equals 230 volts.

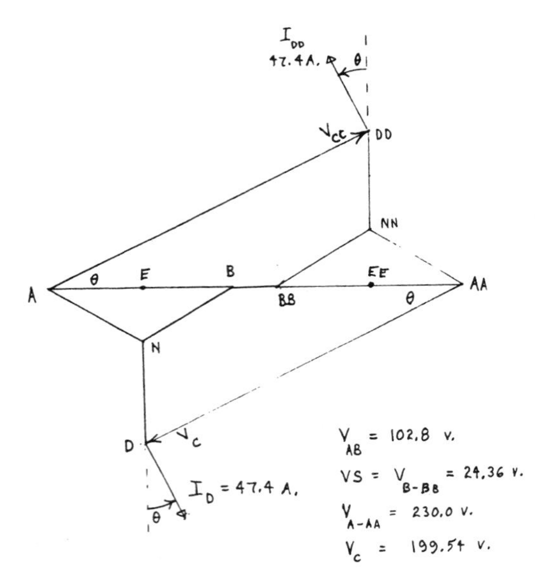

Figure 3.

$$2V_{AB} + VS = 230.0$$

$$VS = 230 - 2V_{AB}.$$

The voltage drop V_{CC} from A to DD causes a current to flow through capacitor CC which leads this voltage drop by 90°. This is the current I_{DD} which enters winding DD. This current should be 47.4 amperes with a power factor angle of Θ with respect to the voltage V_{DD-NN}. The projection of the current I_{DD} on the voltage drop from DD to NN is negative, which means that the winding is operating as a generator, not as a motor.

Since I_{DD} is perpendicular to V_{CC} and V_{DD-NN} is perpendicular to V_{AB} the angle of V_{CC} with respect to V_{AB} is also Θ. We can therefore solve for all of the voltages.

$$V_{E-D} = \sqrt{3}(V_{AB}/2) \tag{3}$$

$$V_{E-AA} = 230 - V_{AB}/2 \tag{4}$$

$$\tan\Theta = V_{ED}/V_{E-AA} \tag{5}$$

$$\tan\Theta = \sqrt{3}V_{AB}/(2\times 230 - V_{AB}) \tag{6}$$

$$2\times 230\tan\Theta = V_{AB}(\sqrt{3} + \tan\Theta) \tag{7}$$

$$V_{AB} = 230(2\tan\Theta)/(\sqrt{3} + \tan\Theta) \tag{8}$$

For this case where Θ = 26.5°,

$$V_{AB} = 230(0.447) = 102.8\text{volts} \tag{9}$$

m (2),

$$VS = 24.36\text{volts} \tag{10}$$

m (3)

$$V_{ED} = 89.0V$$

Matching the phase angle perfectly did not reduce the wye voltage quite as much as desired, but this is not important, ce the variation in efficiency with voltage near the maximum efficiency value is very small. To summarize the results n Figure (3), the transformer should be 230 volts/24.36 volts, 47.4 amperes. Each capacitor has a voltage of

$$V_c^2 = V_{ED}^2 + (230 - V_{AB}/2)^2 \tag{12}$$

$$V_c = \sqrt{(89^2 + 178.6^2)} = 199.54\text{volts}. \tag{13}$$

reactance is

$$X_c = V_c/I_c = 4.2097\text{ohms} \tag{14}$$

capacitance is

$$C = I_c/\omega V_c \tag{15}$$

60 Hertz, this is 630 microfarads. For 50 Hertz, ω is 314.2, and C is 756.1 microfarads. Using the circuit of Figure 2, electrical power delivered is

$$P_e = 2 \times V_{AB} \times I \times \cos\Theta \times \sqrt{3} \tag{16}$$

$$P_e = 15.11\ KW_e \tag{17}$$

It is interesting to draw a temporary conclusion that the ac electrical power delivered in kilowatts as a generator, using Figure 2 circuit, is almost exactly equal to the motor nameplate shaft power.

The circuit of Figure 2 is suitable for any 6-winding motor whose power factor is larger than 0.87, that is, Θ is less 1 29.5 degrees.

-SITE ADJUSTMENT

When a complete system is installed, there will be deviations from the design values computed. The voltage VS might different due to the use of an available standard low cost transformer. The supply voltage from the power company might different. In many developing countries it is below nominal. In California, it is often above nominal. The turbine torque ld be either more or less than expected, due to unknown turbine efficiencies. My recommendation is to operate the com-te system with a wattmeter in the power line, and to adjust the capacitance values to maximum the measured wattage.

ALL GENERATOR

Motors smaller than the 15 KW one given in Table I will have lower power factors and larger values of Θ. A circuit able for Θ of 30 degrees and a power factor of 0.866 is given in Figure 4. The motor for this circuit is given in Table II.

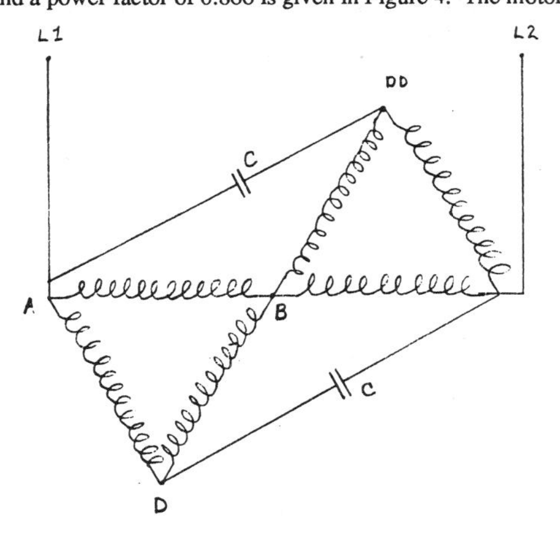

Figure 4.

TABLE II
Motor Ratings

Phases	3	Voltage, volts	460/230
Poles	2	Current, Amperes,	9.98/19.92
Shaft power, kw	6.0	Electrical Input, KWe	6.873
Efficiency	0.887	Electrical Input, KVA	7.936
Power Factor	0.873	Double-Wye Winding	6 coils
Power Factor Angle Θ	30.0°		

Each of the six coils in the double-wye motor has a voltage of $230/\sqrt{3}$ which is 132.8 volts. When one set of 3 coils is reconnected into a delta configuration, and the other set of 3 coils is also reconnected into a delta configuration, each delta should be supplied with 132.8 volts to operate as a motor at rated design voltage. In Figure 4, I have connected the two deltas in series across 230 volts, so that each one receives only 115 volts. The flux is therefore reduced by the ratio 115/132.8 which is 0.866, a very desirable reduction which sets the machine to near maximum efficiency. The voltage across each capacitor has a phase of thirty degrees with respect to the line voltage.

$$V_c \cos 30 = (230)/(3/4) \tag{18}$$

$$V_c = 199.2 \text{volts} \tag{19}$$

Each winding in the original wye connection was rated to carry 9.96 amperes. In a delta connection, the terminal current is $\sqrt{3}$ times the winding current, so the I_c in Figure 4 should be 17.25 amperes. Each capacitor should have a reactance of 199.2/17.25 = 11.55 ohms. The value of each capacitance should be

$$C = I_c/(\omega V_c) = 0.0866/\omega \tag{20}$$

For a 60 Hertz system, C is 229.7 microfarads. For a 50 Hertz system, C is 276 microfarads. The electrical power delivered is

$$P_e = 2(115) \cdot 17.25 \cdot 0.866 \cdot \sqrt{3} = 3(115)(17.25) = 5950 \textit{watts}. \tag{21}$$

In this case, the generator power in kilowatts is one percent less than the shaft power as a motor.

For very poor power factor machines, usually of low kilowatt rating, a third capacitor can be connected between points D and DD in Figure 4 to adjust the winding currents to exactly balanced 3-phase voltages and balanced 3-phase currents.

There is an experimental method of adjusting the capacitors. If both C are too large in Figure 4, then the voltages V_{AD} and V_{DB} will both be larger than V_{AB}. If the third capacitor is too small, V_{BD} will be larger than V_{AD}. If the third capacitor is too large, V_{AD} will be larger than V_{BD}. For a quick calculation, C is given by

$$C = I_{MOTOR\,(230)} \times 10^6/(\omega \cdot V_{MOTOR\,(230)}) \text{microfarads}. \tag{22}$$

Each capacitor volt-amperes equals $P_e/\sqrt{3}$ or 3435 volt-amperes, approximately half as much as the original motor ac watts.

SUMMARY

A conventional three-phase motor can be used as a 3-phase induction generator by reconnecting the windings to reduce the voltage per coil, and to produce a configuration with four terminals. Two of the terminals are connected to the single phase power line, and the other two terminals are excited by capacitor currents. Each capacitor is connected from an unexcited phase to a power line which produces the maximum possible voltage across the capacitor. Remote systems with only single-phase can therefore use high-efficiency three-phase generators.

REFERENCE

"Three-phase Induction Generator for Single-Phase Line," by Otto J.M. Smith, AIEE-ASME Joint Power Generation Conference, Portland, Oregon, October 21-23, 1986.

PERFORMANCE STUDY OF SOLAR STILLS

M.A. Sharif, L.I. Kiss

Technical University Budapest
Institute of Thermal and Systems Engineering

ABSTRACT

The daily fresh-water production of a solar still is determined by a numerical simulation procedure applied to a desktop computer. The mathematical model can take into account all the significant parameters in solar still design as various material properties, changing meteorological conditions, geometry etc. The results meet experimental daily yields with a satisfactory accuracy, that promises reliable application for practical design purposes.

INTRODUCTION

In several regions the increasing demand for potable water can be met only by desalination of brackish or seawater. In developing countries desalination technologies with large, concentrated units can not grow fast due to high investment costs. Desalination technologies with low cost installation, simple operation, low maintenance can be competitive in case of distributed demand, in household applications.

The aim of work is to study the performance and economical aspects of solar stills with various design and materials. The project includes the following parts:

1. A computerized numerical method has been worked out to support design work. The mathematical model is based on the numerical solution of the nodal type set of energy balance equations; the equivalent conductances for radiation, convection, filtration, evaporation etc. as well as moist air properties and mass transfer coefficients are calculated by special subroutines.
2. Experimental work to identify unknown parameters in mathematical model as well as for the direct testing of performance under real conditions.
3. Development of design variants for solar still which meet the requirement of serial production, reliable operation, low-maintenance.

MODEL

The basic heat and mass transfer processes in the solar still /Fig.1/ are modelled by nodal equations of the cover and basin. The driving force of mass transfer within the solar still is the temperature difference between the water in the basin and that of the cover. These temperatures reach their equilibrium values when the corresponding net energy fluxes due to radiation, convection, conduction and condensation/evaporation are equal to zero. The effect of mass transfer on the thermal equilibrium of the nodes is taken into account by forming equivalent heat transfer coefficients, and summing them to the convective heat transfer coefficients.
The corresponding equations for cover and basin are:

$$I_o a_1 A_1 + h_{12} A_1 /T_2 - T_1/ - h_o A_1 /T_1 - T_o/ - \varepsilon_1 \sigma_o A_1 /T_1^4 - T_s^4/ = 0 \qquad /1/$$

$$I_o t_1 a_2 A_2 - h_{12} A_2 /T_2 - T_1/ - \varepsilon_{12} . \sigma_o A_2 /T_2^4 - T_1^4/ + k A_2 /T_2 - T_o/ = 0 \qquad /2/$$

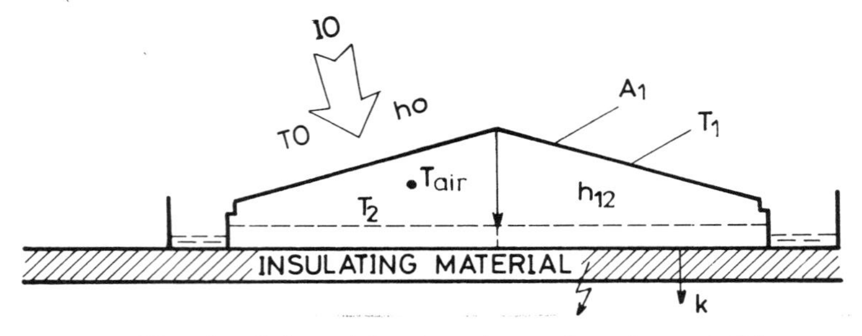

Fig.1. Schematic drawing of solar still

The nonlinear set of equations for T_1 and T_2 is solved by an iterative method.
The heat transfer coefficient h_{12} consists of two parts corresponding to convection and evaporation/condensation heat transport

$$h_{12} = h_{12/conv./} + h_{12/evap./} \qquad /3/$$

Free convection heat transfer inside the enclosure is calculated by introducing a weighted average temperature representing the state of air above the basin /Fig.1/. Then from Ref[3,4] from basin to air

$$h_{2aconv} = 1,32 \; /\frac{T}{L}/^{[0,25]} \quad /laminar/ \qquad /4/$$

$$h_{2aconv} = 1,43. /\; T/^{\frac{1}{3}} \quad /turbulent/ \qquad /5/$$

and for heat transfer from air to cover:

$$h_{a1} = 1,32 \; /\frac{T}{L}/^{[0,25]} \quad /laminar/ \qquad /6/$$

$$h_{a1} = 1,43 \; / \; T/^{\frac{1}{3}} \quad /turbulent/ \qquad /7/$$

The overall convection heat transfer coefficient from basin water surface to cover is then

$$h_{12,conv.} = \frac{1}{h_{2a}A_2 + h_{a1}A_1} \qquad /8/$$

To follow the energy transfer due to evaporation-condensation we adopted the approach of Ref [5], assuming the mass transfer coefficients to be proportional to the convective heat transfer coefficients.
The energy transported by the mass flow at evaporation

$$\dot{Q}_2 = \dot{m}_2 r_2 \qquad /9/$$

where

$$\dot{m}_2 = \sigma_2 A_2 \, /x_2 - x_{air}/ \qquad /10/$$

analogously for condensation on cover

$$\dot{Q}_1 = \dot{m}_1 r_1 \qquad /11/$$

and

$$\dot{m}_1 = \sigma_1 A_1 \, /x_{air} - x_1/ \qquad /11a/$$

From the Lewis equation $\sigma = \frac{h}{Le \, c_p}$ /12/

If the still is perfectly tight against mass transfer by filtration, then in steady-state $\dot{m}_1 = \dot{m}_2$, which is the yield of the still.
The x_1, x_2, x_a moisture contents are calculated using the corresponding temperatures of the basin, cover and air, using formulae

$$x = 0.622 \frac{p_w}{p - p_w} \qquad /13/$$

assuming all three points lying on the saturation line, so

$$p_w = e^{/25.14 - \frac{5096}{T}/} \qquad /14/$$

where we applied an approximation for the vapour tension curve. The equivalent evaporation and condensation heat transfer coefficients are formed by using the same temperature differences for driving forces as they are at convection:

$$h_{2a,evap.} = \frac{\dot{Q}_2}{A_2 /T_2 - T_a/} \qquad /15/$$

$$h_{a1,evap.} = \frac{\dot{Q}_1}{A_1 /T_a - T_1/} \qquad /16/$$

$$h_{12,evap.} = \frac{}{A_2 h_{2a,evap.} + A_1 h_{a1,evap.}} \qquad /17/$$

In calculation of sky radiation, free or forced convection heat transfer coefficients from outer surface of the still the approach and formulae from References [1], [2] have been applied:

$$T_{sky} = 0.0552x/T_O/ \qquad \text{Ref [1]} \qquad /18/$$

$$h_O = 5,7 + r,8\ w \quad 1 \quad \text{if } 0 < w < 5 \ \text{m/s} \qquad /19/$$

$$h_O = 7,6x/w/^{0.8} \quad 1 \quad \text{if } w > 5 \ \text{m/s} \qquad /20/$$

As the heat transfer coefficients are depending on temperatures T_1, T_2 indirectly, multiple iteration procedure was applied. The architecture of the program is modular, making possible to replace modules computing the different parameters by others if changes in geometry are to be taken into account.

SIMULATION

The program for solving the mathematical model has been worked out for desktop computers. For a daily performance study the hourly meteorological data are the input data for the main program. The program was tested by using meteorological data for Budapest, as for those conditions experimental data of solar stills of various design are available for the sake of comparison.

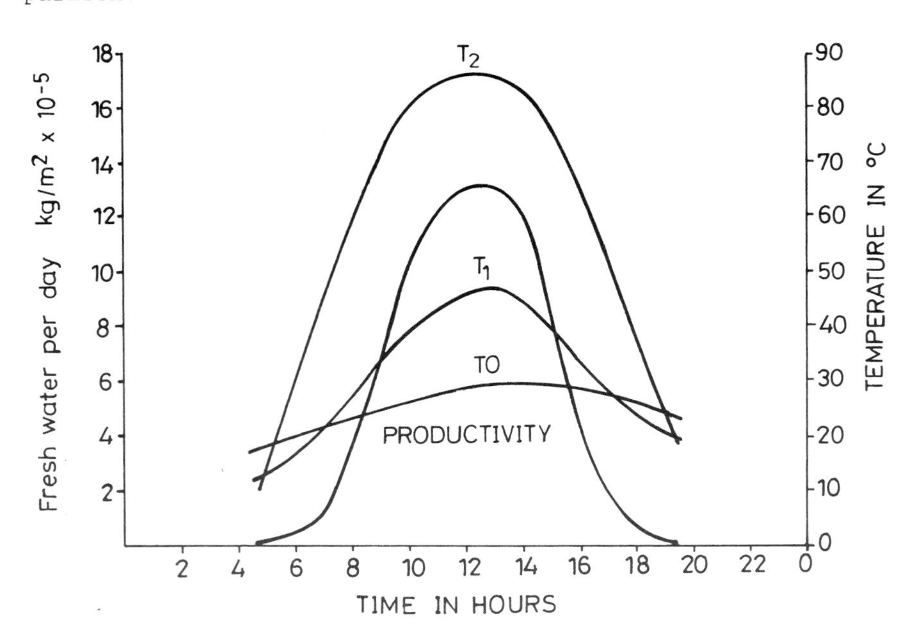

Fig.2. Productivity, temperatures for Budapest 17. July

RESULTS

To illustrate the output of the program in Figure 2. the temperature histories of cover and basin together with the fresh water production profile are presented. The case corresponds to Hungarian geographic and meteorological conditions, for a simple glass covered solar still without any heat recovery.

SUMMARY

A computer code has been developed and applied for the study of the operation of solar stills. The method makes easy to analyze the influence of various materials, sizes, geometry, changes in transparency of cover or absorptivity of basin, as well as it can quickly predict fresh water yield under various climatic and geographic conditions.

NOMENCLATURE

a_1, a_2 absorptivity of cover and basin respectively
A_1, A_2 area of cover and basin respectively
h_o heat transfer coefficient on outer surface of cover
h_{12} heat transfer coefficient between water and cover
h_{1a}, h_{2a} free convection heat transfer coefficient from air to cover and from basin to air respectively
I_o solar intensity on the plane of the cover
k conductive heat transfer coefficient
l basin length
m_1, m_2 mass of water-vapour from air to cover from water to air respectively
P total pressure of moist air
p_w water vapour's partial pressure within air
q_1, q_2 energy transported during condensation, evaporation respectively
r heat of evaporation of water
T_o ambient temperature ; T_1 cover temperature
T_2 water temperature ; T_s sky temperature
w wind velocity
x_1, x_{air}, x_2 absolute moisture content at T_1, T_{air}, T_2 resp.
$\varepsilon_1, \varepsilon_2$ emissivities of cover and water respectively
σ_o Stefan-Boltzmann constant
σ mass transfer coefficient.

REFERENCES

1 S.A. Duffie, W.A. Beckman, /Solar energy thermal processes, Wiley, New York /1974/.
2 I.F. Kreider, F. Kreith, Solar energy handbook, McGraw Hill /1981/.
3 I. Holman, Heat Transfer, McGraw Hill, Kogakusha /1976/.
4 M. Ozisik, Basic Heat Transfer, McGraw Hill /1977/.
5 Tasnádi, Cs., Calculation of heat transport with material transfer by analogy to heat exchangers dimensioning, /in Hungarian, English translation BNWL-TR-218/, Energia és Atomtechnika 25 /6/, pp.252-258 /1972/.

A STUDY FOR A SOLAR STILL COLLECTOR

M.R.Leeson

and H.Boutebila

Mechanical Engineering Department
Loughborough University of Technology,
Loughborough, England LE11 3LY

ABSTRACT

A solar collector consisting of a thin film, inclined ,free flow, flat plate has been studied both theoretically and experimentally. As the objective is distillation applications, a condensing system was introduced in the experimental rig.
The theoretical model considered both the liquid and vapour phases and a two dimensional flow analysis was carried out in the vertical plane. From this study it has been shown that the significant parameters of the two phase flow are,film thickness, liquid flow rate, collector length and inclination and solar radiation and plate heat flux.

The small scale experimental unit of one square metre in area with glass cover, indicated that atmospheric wind speed and vapour leakage are also of importance. Experimental studies were carried out under both laboratory and direct outside solar conditions at Loughborough University (Latitude 52° 46'). These tests confirmed the significance of the parameters identified in the theoretical study.

KEYWORDS

Solar Desalination Still Twophase Flow Irrigation Inclined

INTRODUCTION

Desalination of sea water for the production of drinking water has received increasing attention as industrialization of countries has proceeded. Solar distillation is a specialised application which in the case of the simple Basin Still imitates the natural cycle in one unit. The size and capital cost is relatively high in relation to the output, and with a high maintenance level to prevent loss of vapour through leakage is regarded as a last resort for the supply of drinking water in small, remote low technology communities with no alternative. The economic development of communities requires increasing supplies of water. If irrigation water even on a small scale can be provided in arid areas it has been demonstrated that Nature is able to develop the ecological system multiplying the effective water supply many times over.

OBJECTIVE

This study is aimed at developing design information for large stills with improved efficiency and without increasing the level of technology excessively. Here the work so far covered relates to a model unit of one square metre which was used to simulate various positions in a full sized unit.

PREVIOUS WORK

The basin still has received considerable attention both theoretically and experimentally and there are numerous examples described in the literature. Good design requires, low heat capacity of the still, low angle of glass inclination (10-20 deg) and good insulation below the still.

THEORETICAL STUDY

In this study a collector is considered inclined to maximise heat absorption. This inclination enables the evaporating fluid to flow freely downward. The evaporated vapour which condenses on the inner surface of the cover as well as the condensing plate is collected in a trough similar to basin still arrangements.

Many theoretical analyses have been reported of investigations into the fluid flow and heat transfer of this type of collector for varied applications. Collier (1979) considered a system for absorption refrigeration and assumed steady flow and based his analysis on the energy equation. Peng and Howell (1982) attempted to improve the analysis with both mass and heat balance equations, but assumed constant liquid mass flow rate and heat capacity which is not valid for the present study.

For this analysis we commenced with the continuity, momentum and energy equations for the liquid and the interface region. In a previous similar study by Spindler(1981) the initial equations were simplified. With these equations and making the assumptions that the physical properties are constant for the working temperature range, the flow is laminar and heat transfer is by conduction, and, from the conclusions of Cooper that 70% of the transmitted solar radiation is absorbed by the plate; performance predictions were made using a computer programme. Most of these predictions were confirmed during the experimental work.

EXPERIMENTAL EQUIPMENT

The still was divided into several parts; the still itself, with adjustable inclination, storage tanks at either end of the collector, and a circulating pump, and distillate collector. Figure 1 shows a cross section of the experimental still module which consists of a black base plate of aluminium with a 6mm wooden insulation plate, a 6mm glass cover, top and bottom tanks and the condensate collection trough.

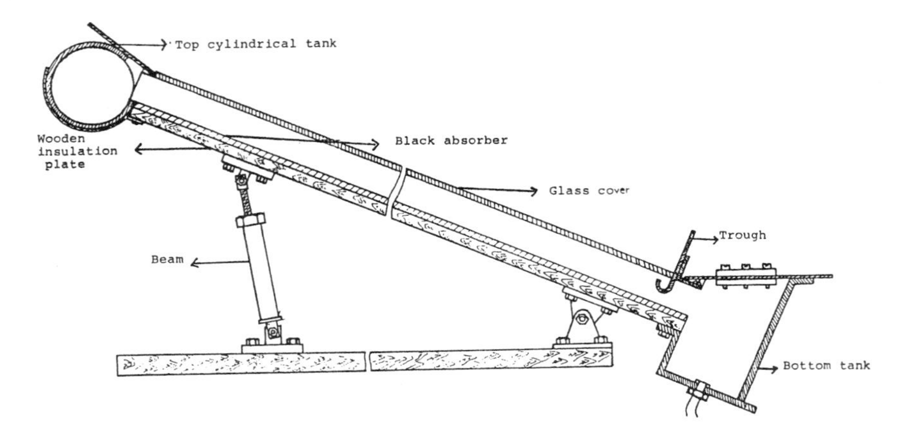

Fig.1. A cross section of the free flow flat plate solar still

In the laboratory although artificial lighting was considered the radiation levels expected were simulated more accurately by heaters under the base plate which could be set to run at various temperature levels or heat inputs. The laboratory still was fitted with flow and temperature measuring equipment which is seen in the photograph below.

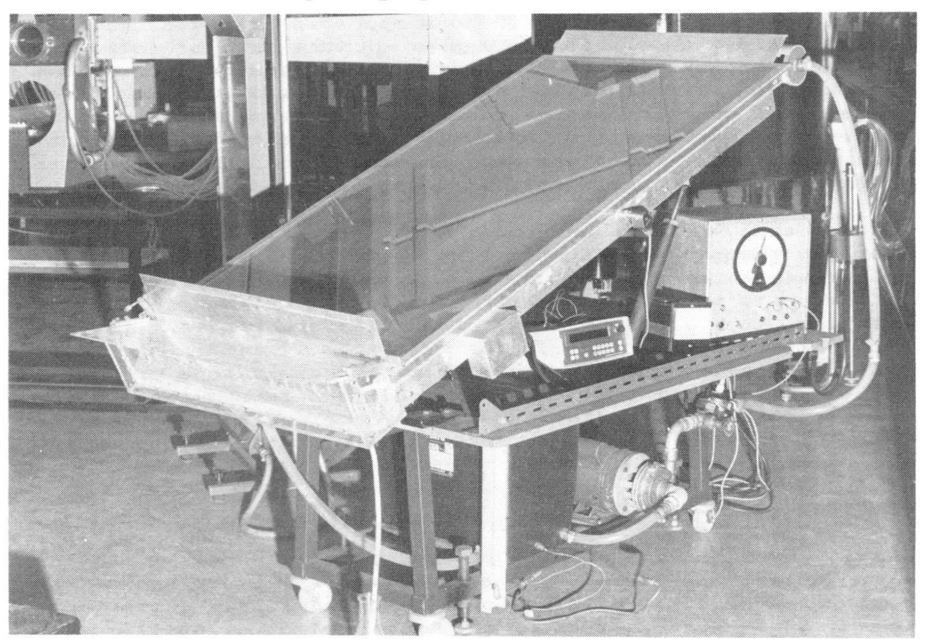

The equipment under outdoor conditions used solar radiation for heating in the designed manner with similar instrumentation, augmented by the weather station equipment which was in the same enclosure where the experimental still was installed for the tests. The weather station equipment included a solarimeter (Casella) and a data logger through which it was possible to transfer data for analysis by computer. In the outdoor tests the brine was allowed to circulate to reach evaporation temperature.

EXPERIMENTAL TESTS

Outdoors

Since the still module was run with recirculation to optimum conditions when operating under solar exterior conditions full simulation of a large still was not possible. From the outdoor experiments it was concluded that to improve efficiency, the still base should be well insulated, wettability of the black base surface should be given careful attention and a vapour tight cover will reduce the considerable effect of wind speed on still performance. This was marked at this high latitude of 52° 46'.

Indoors

With the indoor experimental heating system the temperature of the plate could be controlled and simulation of the whole still carried out.

EXPERIMENTAL RESULTS

The experiments on solar radiation demonstrated clearly the predictions of the theoretical study. In the test to examine the effect of liquid mass flow rate variation on distillate production with a plate inclination of 20 deg. and temperature of 50 °C the experimental results were in agreement with the theoretical model showing that as the flow increases the distillate output decreases.

For variation in still inclination tests showed the distillate production reached a peak value at 20 deg. inclination which is in agreement with previous studies although the teoretical study suggested that production would be higher at 10 deg. inclination. This difference is attributable to dropwise condensation which was clearly observed and produces a significant reduction in output.

Ambient temperature variations do not correspond exactly with variations in distillate output. Indicating that other factors such as wind speed are related.

CONCLUSIONS

It was found from the theoretical study that a relationship exists between liquid and vapour velocities due to the interactive effect between the two phases and this was clearly observed in the experimental still.

Variations in liquid film thickness down the still has a significant effect on the general performance of the still and particularly in the case of a long large scale plant should not be neglected.

From the outdoor experiments it was concluded that to improve efficiency the still base should be well insulated, wettability of the black base surface should be given careful attention and vapour tightness of the cover will reduce the effect of wind speed on still performance.

FUTURE STUDY

The next stage should be a 30 m length at inclination of 1 in 3 (approximately 20 deg) which would be suitable for a steep hillside.

REFERENCES

Collier, R.K.(1979) 'The analysis and simulation of an open cycle absorption refrigeration system', Solar Energy V23, 357-366

Chingshiang, Peng and J.R.Howell (1982) 'Analysis of open inclined surface solar regenerators for absorption cooling applicatins, comparison between numerical and analytical models', Solar Energy v28, No 3, 265-268

Johannsen A and Grossman G (1983) 'Performance simulation of regenerating type solar collectors' Solar Energy v30, No 2, 87,92 .

Vaxman, M. and Sokolov M. (1985) 'Analysis of free flow solar collector', Solar Energy, v35, No 3, 287-290,

Spindler, B. (1981) 'Equations gouvernant l'ecoulement plan d'un film liquid avec flux de chaleur a la paroi et changement de phase a l'interface' Commissariat a l'energie atomique, France, Rapport CEA, R5061,

Cooper, P.J. (1973) 'The maximum efficiency od single effect solar stills', Solar Energy, v15, 205-217

Boutebila, H. (1987) 'A free flat plate solar still' M Phil thesis, Loughborough University of Technology

AN ECONOMIC FOIL-BASED SOLAR STILL FOR SEAWATER DESALINATION - DESIGN CONSEPT AND OPERATING RESULTS

P. H. Koske, N.P. Schmidt

Institut für Angewandte Physik, Universität Kiel, Kiel, Germany

ABSTRACT

A new concept for an economic solar still is presented which is characterized by two main features:

- the still consists of a tube like system made of plastic foils, thus reducing the specific costs of investment for the plant,

- the still is operated with active air ventilation, either using a fan, driven by photovoltaic panels as part of the system, or by thermoconvection, thus improving the efficiency of the plant.

The classical solar stills or "greenhouse" stills consist of rectangular basins with tilted glass covers in appropriate frames. They produce 2 - 3 l of fresh water per m^2 and day at costs of about 100 DM/m^2 plant area. Thus a production unit with an overall capacity of 1 m^3/d fresh water amounts to 40 000 DM (400 m^2 plant area x 100 DM/m^2), which is too expensive to be installed and used for instance in small communities of the Third World.

The new concept described in the present paper is an attempt to improve this adversely economic situation considerably.

KEYWORDS

Solar still; greenhouse distillation; solar seawater desalination; economic solar still concept.

INTRODUCTION

Solar stills or "greenhouse" stills have been used in the past frequently in a wide variety of configurations for the distillation of fresh water from seawater. Usually a rectangular basin made for instance of concrete and painted black inside holds the seawater and is covered air tight with glass in appropriate frames. These covers are slightly tilted for two reasons: for a better passage of the incident solar radiation through the glass cover without too much reflection losses at the outer glass surface and for the better drainage of the condensed water at the inner glass surface and its collection in a drain channel.

With these particulars the main operating principle of a classical solar still has been indicated. The seawater in the concrete basin is heated up by the solar radiation which passes through the transparent glass covers and is absorbed by the water, the black bottom, and the black walls. According to the higher water temperature the partial pressure of water vapor in the atmosphere, the humidity in the air-phase above the seawater layer, is increased. The temperature of the glass cover above the humid air on the other hand is mainly determined by the outside air temperature, which is considerably lower than the inside temperature of the system. Therefore the water vapor condenses at the inner, relatively cool surface of the glass and, because of its tilted design, flows down into a drain channel along the lower part ot the cover.

There are several parameters that are of importance to this simple process scheme. So the efficiency of the process is determined by the vapor transport from the warm seawater surface as evaporator to the cooler glass cover which acts as condenser. This transport is activated by thermoconvection and therefore depends on the temperature difference between the two surfaces which again is determined by the flow of latent heat from the water to the glass and its thermal conductivity.

The incident solar radiation as heat source for the evaporation is influenced by the transparency of the glass cover which in turn is decreased by condensation and droplet formation on the inner glass surface.

The condenser temperature finally, the temperature of the outside glass surface, depends strongly on the exchange of air in the neighbourhood of the glass cover, as this air acts as final sink for the heat of condensation.

The efficiency of a solar still is the amount of product water actually collected from the plant relative to the water that could be evaporated theoretically by the irradiated solar energy. As can be seen from this short discussion the classical solar stills are by no means optimized process plants. Because of the above mentioned various terms of influence the efficiency is variable dependent on the momentary combination of process parameters.

It seems reasonable therefore to consider variants with better controllable and defined process parameters in order to increase the product output and to establish more constant operating conditions. The presently proposed new concept is one attempt to realize such conditions.

THE TUBE CONCEPT WITH ACTIVE AIR VENTILATION

The cross section of the solar still is shown in two variants in Fig. 1. Both consist of 3 layers of plastic foil which form a tube like structure. The bottom layer is made out of black foil, is shaped like a trough by a corresponding shaping of the ground, and acts as plastic lined basin for the seawater. The necessary cover for this trough consists of 2 layers of trans-

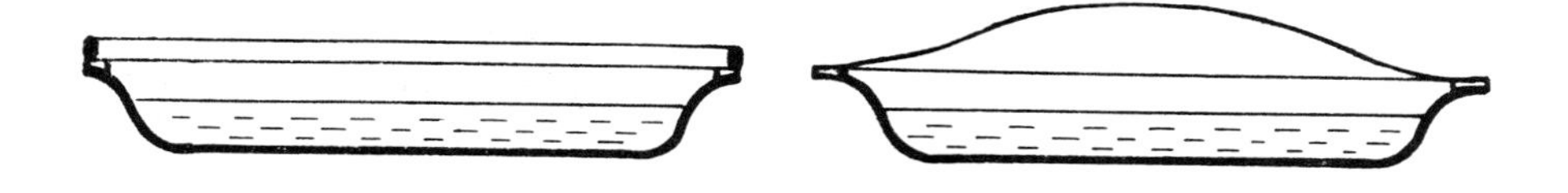

Fig. 1: Cross section of the foil-based tubular solar still; 2 variants for the cover: frame supported and air-pressurized

parent foil kept apart by air pressure in between them. Both parts, the black bottom element and the transparent cover are joint together air tight thus forming a channel filled with humid air.

In contrary to the classical solar still the 2 layers cover with pressurized air acts as insolating element thus preventing condensation of water at the inner surface of the cover. Condensation is accomplished outside of the tube by using a fan at one side (air inlet) and a water cooled condenser at the other side (air outlet) of the channel. The water saturated air is conducted through the external condenser, the temperature of which is determined by the temperature of the seawater which is used as operating medium for the condenser. The condensed water is collected in a product storage tank.

Integral part of the above described system is a photovoltaic solar energy converter which is used as power source for the fan as well as for the seawater pump supplying seawater to the external condenser. Whereas the fan operates with a constant power of 5 W and is 12 V/DC operated the power consumption of the pump depends on the hydrostatic height which has to be provided for the particular condenser arrangement. Here the appropriate combination of water flow through and temperature increase in the condenser has to be optimized with regard to energy consumption: the higher the condenser is placed above sea level, the lower the flow rate of cooling water should be adjusted with the result of a slightly elevated mean condenser temperature.

An alternative to the air ventilation by a photovoltaic-activated fan is the application of thermoconvection. At the outlet end of the solar still an upward tilted part of the foil-based tube connects the horizontal air-channel with the condenser. Because of the higher temperature in this connecting element the humid air is moved upward and supplied to the condenser again, without the help of a fan. In this alternative the auxiliary power from photovoltaics is limited to the condenser pump.

After this short description of the new concept for solar stills a few theoretical considerations shall be discussed before in the final part the first results of field runs are presented.

PROCESS FUNDAMENTALS

Whereas with classical solar stills the process cannot be controlled from outside by an operator the situation is different with the tube concept in this paper. Referring to the simplified process scheme of Fig. 2 the following equation hold:

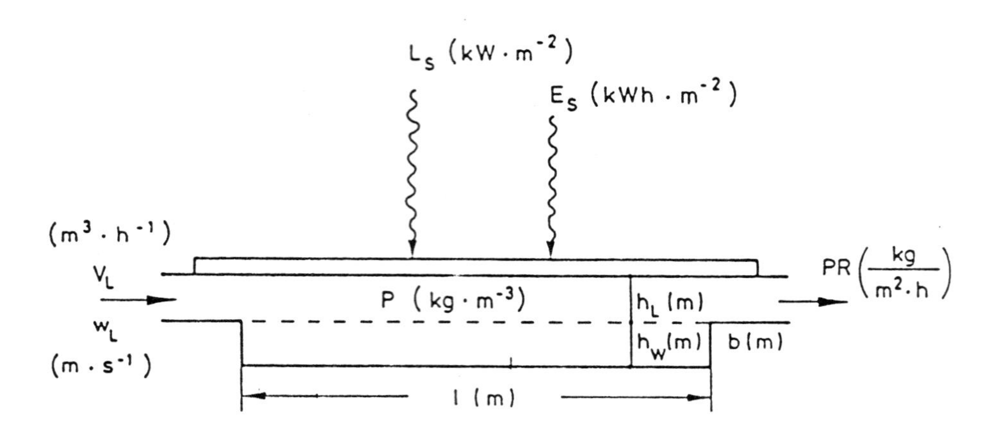

Fig. 2: Simplified process scheme of the tubular solar still

Energy balance: $L_S \cdot F \cdot \eta = PR \cdot H_v$ (1)

with
- L_S: irradiated power (kW/m^2)
- F : irradiated plant area (m^2)
- η : radiation absorption efficiency
- PR: water production rate (kg/h)
- H_v: enthalpy of evaporation (kWh/kg)

Mass balance: $V_L \cdot P = PR$ (2)

with
- V_L: volume flow of air (m^3/h)
- P : water content in air (kg/m^3)

The water content P in saturated humid air can be approximated within the temperature range from 20°C to 70°C by the simple semi-empirical equation

$$P = k \cdot \exp(-H_v/RT) \quad (3)$$

with
- $k = 7 \cdot 10^5 \ kg/m^3$
- $H_v = 0.660$ kWh/kg
- $R = 1.28 \cdot 10^{-4}$ kWh/kg · k

Combining equations (1) and (2) and using equation (3) and the relations

$$V_L = w_L \cdot h_L \cdot b \quad \text{and} \quad F = l \cdot b$$

with
- w_L : flow velocity of air (m/s)
- h_L : height of air channel (m)
- b : width of air channel (m)
- l : length of air channel (m)

the air velocity in the system can be calculated according to

$$w_L = \frac{L_S \cdot \eta \cdot l}{H_v \cdot h_L \cdot k \cdot 3600} \cdot \exp(H_v/RT) \tag{4}$$

As can be seen from equation (4) the flow velocity of air in the tube-system depends besides thermodynamic constants or plant dimensions on the irradiated power L_S. Therefore steady-state operation of the solar still with constant temperatures in the seawater phase requires air velocities that follow the daily variation of the radiated power for optimal performance.

EXPERIMENTAL RESULTS

After extensive test runs with artificial sunlight in the technology center of Kiel University the solar stills were operated under natural field conditions in Kotor, Yugoslavia, in close cooperation with the Biological Institute Kotor and the Physics Department of Titograd University. This cooperation was promoted by the International Bureau of the GKSS Research Center Geesthacht and funded within the German/Yugoslavian Bilateral Research Program.

Two results of such field experiments are shown in fig. 3.

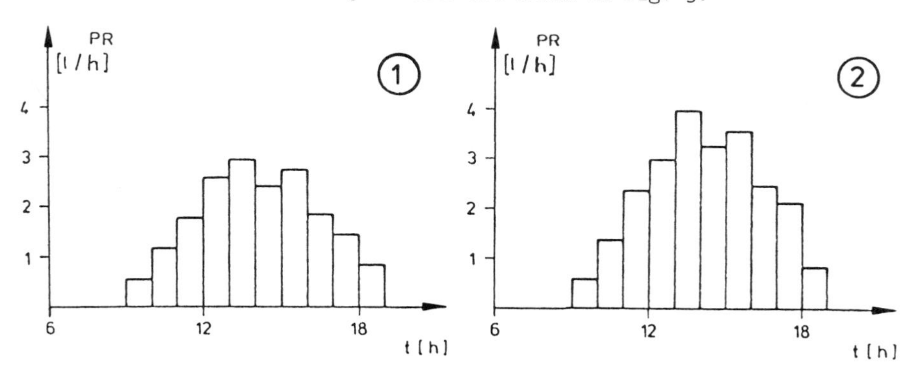

Fig. 3: Typical results of 2 distillation runs on June 13, 1987 in Kotor, Yugoslavia (42° N);
1) Fan operated plant with 5 m^2 area
2) Thermoconvection plant with 10 m^2 area

In this figure the hourly collected product water is shown for 2 different plants, for a fan operated plant with 5 m^2 active area and a thermoconvection plant with two times 5 m^2 active area conbined connected to the same condenser.

As can be seen from the graphs the production of distilled water follows the irradiated energy with peak values between 13 an 14 hours. The somewhat reduced production between 14 and 15 hours is due to partly cloud coverage. The specific production for the fan operated plant 1 amounts to 4 l/m^2 d whereas the corresponding value for plant 2 (thermoconvection) is 2.5 l/m^2d, which is lower then expected and probably due to the non-optimized thermoconvective air flow. Further work is in progress.

CONCLUSIONS

It has been shown in field experiments that the new foil-based tubular solar still with active air ventilation can be operated as anticipated. The specific costs of investment are low (less than 20 DM per squaremeter plant area) and the specific production rate can be optimized to values between 4 and 5 l/m^2 d. It is expected that the operational variant of thermoconvective air flow will produce comparable results to the fan operated system.

YANBU SOLAR-POWERED SEAWATER DESALINATION PILOT PLANT-PROBLEMS ENCOUNTERED AND EXPERIENCE GAINED

Dr. Ghassan Hamad

King Abdulaziz City for Science and Technology
P.O. Box 6086, Riyadh 11442, Saudi Arabia

ABSTRACT

The solar-powered seawater desalination pilot plant at Yanbu combines state-of-the art solar thermal power technology with a freeze desalination technique to produce 200 cubic meters of product water in a day.

The data collected during the past two years of operation of the plant are currently being analyzed and technically documented. Operation and test results have indicated the potential of the plant as well as key problem areas within the broader perspective of research and development for solar thermal and desalination technologies.

The PKI (Power Kinetics Inc.) point focus dish collectors concentrate the sun's radiation onto thermal cavity receivers, and the energy collected is turned into shaft power to produce refrigeration required for CBI's (Chicago Bridge and Iron Company) freeze desalination process. The paper describes briefly the integration of the two technologies earlier cited, significant experience, and the valuable knowledge gained to advance the development of both the technologies.

KEYWORDS

Fresnel concept point focus dish collector; freeze desalination.

INTRODUCTION

Background

The desalination project, which is located on the shore of the Red Sea at

Yanbu, is a three-phase project namely: preliminary design, construction, operation and data collection. All those phases concluded recently.

Integration of Two Technologies

The cost of solar equipment in a solar-powered desalination plant is about half the total cost of the plant. Minimum product water cost is achieved when energy consumption is minimized by selecting appropriate technolgies both on the solar side and the desalting side, and integrating the two in the most effective manner. This led to the consideration of several solar desalination technologies. Analyses and optimizations led to the most desirable combination of solar power and desalination technologies.

The Yanbu pilot plant uses a distributed array of two-axis tracking, point focus collectors. This choice permits high efficiency energy collection, using modular capacity for flexibility of design. Desalination capability is provided by indirect contact heat transfer freeze desalination process.

DESCRIPTION OF THE PLANT

Fig. 1 Solar desalination plant

Figure 1 is a photograph of the pilot plant. The plant consists of two major systems, the first being the energy power system (Fig. 2), and the second being an independent desalination system (Fig. 3). The energy system consists of four smaller subsystems - solar energy collection, auxiliary energy supply, energy storage, and energy delivery. Solar energy is converted to thermal by PKI collectors which use a fresnel concept. The collector field employs 18 units, with each unit capable of a nominal 50 kW thermal output at 338°C outlet temperature. The receiver is a monotube cavity design mounted at the end of receiver boom and located at the concentrator focal point. A heat transfer oil (SYLTHERM 800) is heated while passing through the receiver, and delivers its heat into the molten salt (PARTHERM 290) of the storage subsystem. The back-up energy system (oil heater) is designed to provide full thermal energy for plant operation in the absence of solar energy. A steam generator produces steam from the molten salt's heat energy to drive a steam engine, which in turn drives the compressor of the primary vapor compression refrigeration system. An absorption refrigeration system recovers energy remaining in the engine exhaust steam and produces additional refrigeration. Ice is produced in the freeze heat exchangers, where brine flows down the exchanger tubes, and heat is removed by the freezer shell's ammonia refrigerant. The absorption system cools and partially condenses the ammonia refrigerant. Ice crystals which are formed in the exchanger tubes have still brine adhering to their surfaces, and they are washed by fresh water and melted to become product water.

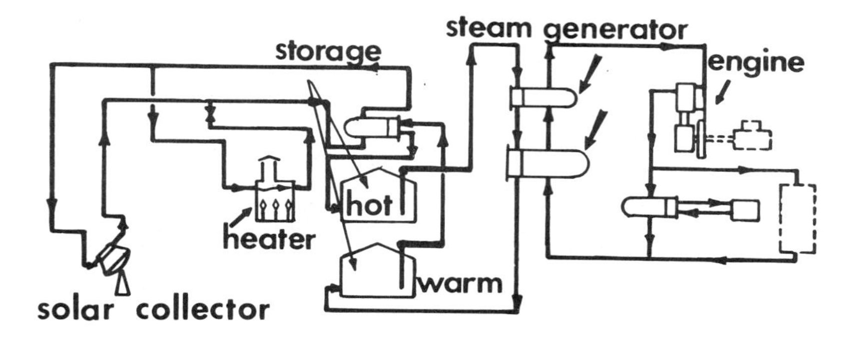

Fig. 2 Process flow diagram of the solar power system

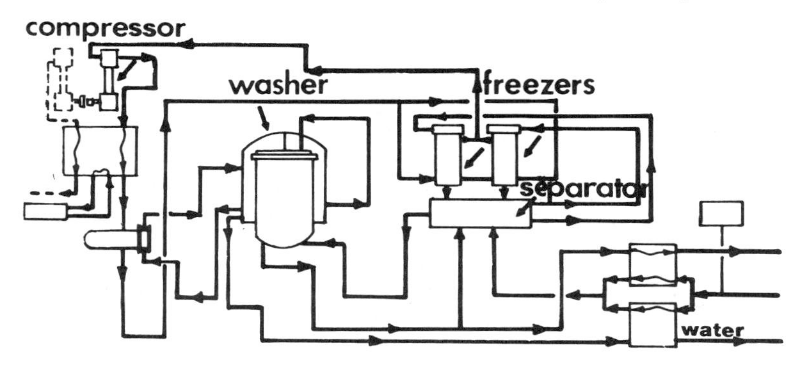

Fig. 3 Process flow diagram of the freeze desalination system

UNIQUE FEATURES OF THE PLANT

The PKI collector design employs fresnel concept silvered glass reflective surfaces which have high reflectivity and durability; incorporates a drive mechanism which has a rapid focus-defocus capability without excessive power consumption; and tracking controls based on dependable feed-back concept.

The desalination process is designed to produce crystalline water from an impure aqueous solution via indirect contact heat transfer. The process allows bulk crystallization without degradation of the heat transfer surface. Among the potential benefits of this process are: no scheduled replacement of components, broad acceptibility of feed stream quality, and reduced corrosion due to low operating temperatures.

Efficient energy utilization is achieved by using high grade heat to produce compression refrigeration and low grade heat to produce absorption refrigeration.

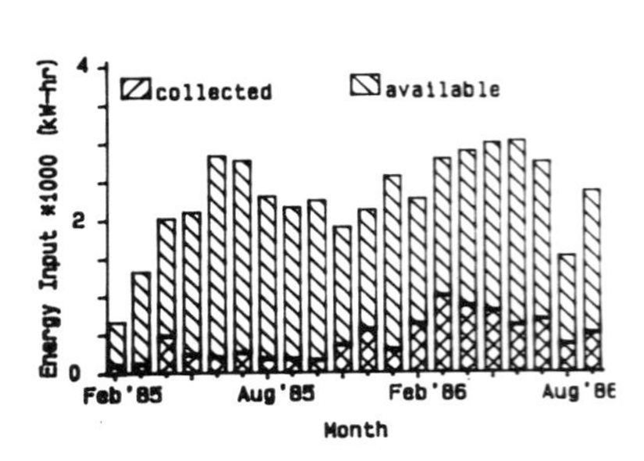

Fig. 4 Collected solar energy

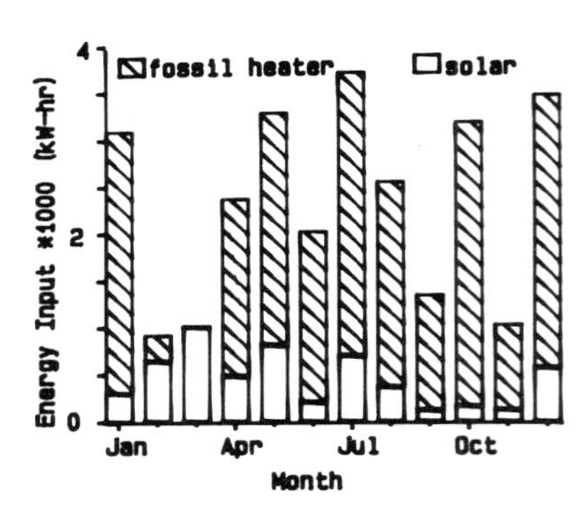

Fig. 5 Energy supplied to plant

PROBLEMS ENCOUNTERED AND EXPERIENCE GAINED

Energy Power Subsystem

Available solar energy and collected thermal energy quantities are shown in Fig. 4. It can be noticed that the collectors performance is substantially better in 1986 than that of 1985. This rise is due to increased operating experience and components developments.

Energy available from solar was not enough to operate the plant for 24 hours continuously. Figure 5 shows that much of the energy that needed for the desalination was provided by the heater. Three factors were behind the extensive use of the heater are: the solar energy available from the sub-system was not designed to operate the system continuously for a full day, the performance of the collector field was less than expected, and the energy demand for the desalination system was more than planned for. The later two factors will be described in this section. Energy collection and energy conversion performance of the plant is summarized in Fig. 6. The collector subsystem captured 19% of the daily solar energy available. More than 90% of the collected thermal energy was available to storage, and 74% of the collected energy was available to the refrigeration cycle. Thus, the energy power system efficiency was 14%.

Collector performance dropped significantly at midday, though performance should peak with insolation at midday. This is attributed to the fact that a constant solar geometry is not maintained by the collectors.

The collectors employ shadowband sensors for feedback control of the drive motors. The diffuse insolation at Yanbu widens the dead band of the sensor control, and this affects precise tracking. On the other hand, if the shadowband limits are tightened it can result in continual overshoot, which is not desirable. Retrofit of "sunglasses" to the shadowband sensors improved diffuse tracking. The lack of precise tracking also caused the melting of the aluminum flux traps which are mounted on receivers' cavities.

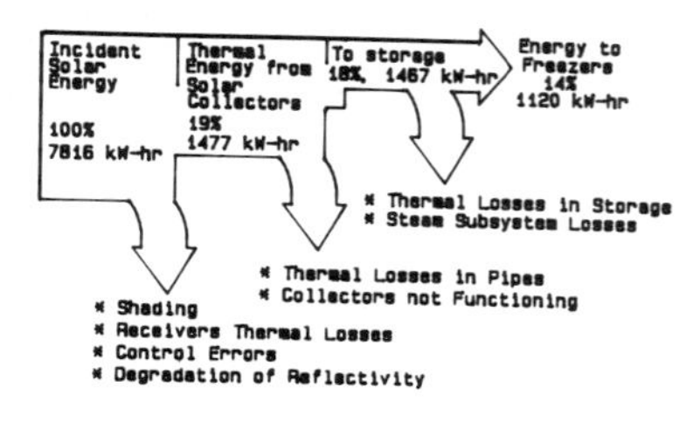

Fig. 6 Daily solar energy flow

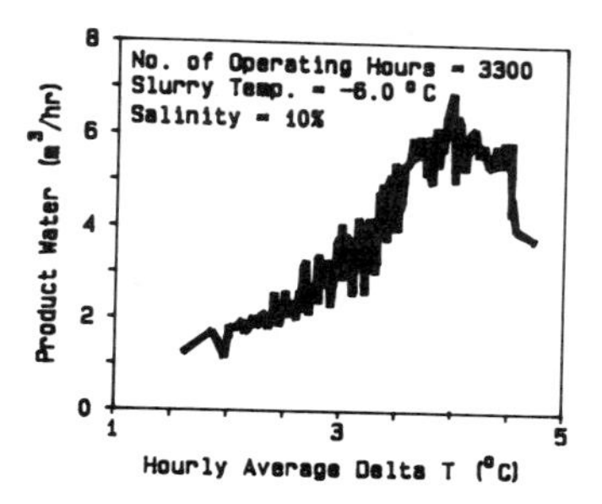

Fig. 7 Desalination performance

Edge corrosion was observed on the mirror tiles of the collectors. Humid environment and the presence of sulfurdioxide in the atmosphere are considered to be the main reasons for this problem.

Thermal Storage

The durability and availability of the dual-tank (hot and warm) thermal storage system was excellent in comparison to other plants subsystems. The storage medium Partherm 290 salt, which is a mixture of nitrite and nitrate with a melting point of 142°C, has maintained its physical properties within the design limits and its chemical composition with little contaminants throughout the operating period.

Freeze Desalination Subsystem

Freeze separation of water from the salt is based on the principle of freezing or crystallization of water molecules from the brine when the solution is brought to a thermodynamic state in which the component with a higher freezing point separates itself by crystallizing within a saline solution, which becomes increasingly concentrated and subsequently has a progressively lower freezing point. The freezing point is a function of salinity that is controlled by the relative flows of seawater reject brine.

Operation at various salinities showed that the optimum value is 10%. Lower salinity reduces freezer tubes resistance to icing while higher values precipitates sodium sulphate decahydrate from brine and makes plant operation unstable. Figure 7 shows water production for 10% salinity and as a function of delta T (temperature difference between brine and ammonia refrigerant). It can be noticed that optimum production is obtained at delta T of 3.6°C.

The primary cause of freezers thermal performance degradation was found to be the compressor lubricating oil migrating to the shell and the scaling of the tube walls by the seawater salts when brine circulation was stopped.

PERFORMANCE CHARACTERISTICS OF THE DUAL-TANK THERMAL STORAGE SUBSYSTEM OF THE SOLAR-POWERED WATER DESALINATION PILOT PLANT AT YANBU

A. R. Al-Ibrahim*, H. A. Gari**, and M. C. Gupta*

*Solar Program, KACST, P.O. BOX 6086, Riyadh 11442, K.S.A
**Department of Mechanical Engineering, King Abdulaziz University, P.O. Box 1540, Jeddah 21413, K.S.A.

ABSTRACT

The solar energy powered desalination plant at Yanbu in the Kingdom of Saudi Arabia is an unique experimental pilot plant integrating the state-of-the-art solar collector technology with an indirect contact heat transfer freeze desalination process. The plant was operated for about two years, and presently a detailed and critical analyses of the various subsystems are being made in order to assess the potential of such a system.

The plant consists of four major subsystems: collection of solar energy, storage, energy delivery, and water desalination. This paper describes the dual-tank thermal storage subsystem, and presents its performance characteristics. A computer program was developed to compute the overall heat transfer coefficient and effectiveness of the oil/salt heat exchanger. The paper presents an analysis of the heat loss from the dual-tank system and of the efficiency of the storage subsystem. The paper concludes with a note on the choice of the molten salt as a storage fluid.

KEYWORDS

Dual-tank thermal storage; heat loss to ambient; thermal efficiency; Syltherm and Partherm heat transfer fluids.

INTRODUCTION

Background

The solar-powered seawater desalination pilot plant is a research project in the program, named SOLERAS, under the auspices of the Saudi Arabia-United States Joint Commission on Economic Cooperation. The primary objective of this project is to advance the development of solar powered desalination as a visible and cost - competitive energy alternative, and to facilitate the developed technology transfer between the two countries. During its period of operation for about two years, it has provided valuable information for the design and cost analysis of a commercial plant. The pilot plant has two major systems: solar energy power and freeze desalination. The thermal storage is a crucial component of the solar power system, and its performance is the subject matter of this paper.

Thermal energy storage is required to provide the thermal capacitance effect to match the fluctuating energy collection to the energy demands. This improves both system operability and utility. Operability is improved by the addition of thermal storage between the solar collector field and the energy delivery subsystem, which eliminates transient effects of one upon the other. Utility is imporved by providing enough storage for continous operation.

Thermal energy can be stored as sensible heat in a solid or liquid medium, as latent heat of phase change, or in a reversible chemical reaction. The solar-powered desalination pilot plant employs sensible heat storage in a molten salt. This type of storage uses the internal energy of the material for heat storage. The amount of energy stored is dependent on the mass and heat capacity of the storage medium for a given temperature defferential during the heat transfer process. High specific heat and density would be obvious requirements. Additional requirements are negligible corrosion and resistance to chemical or physical changes resulting from thermal cycling.

Selection of Storage System and Storage Medium

Two different salt storage system concepts were considered for the pilot plant: thermocline, where the high and low temperature salts are stored stratified in the same tank, and a dual-tank, where the hot and warm salts are stored in separate tanks. A comparison of the relative merits of the two systems led to the choice of a dual-tank system.

SYSTEM DESCRIPTION

The process flow diagram of the solar power system in the desalination pilot plant (Fig.1) shows the position of the thermal storage tanks in relation to the other subsystems. The storage system process is a closed loop containing the hot and warm salt tanks, oil/salt heat exchangers, the steam generator, and the superheater. Salt from the warm storage tank is pumped to the oil/salt heat exchanger, where it is heated by the high temperature oil from the solar collectors field and/or the fossil fired heater (backup system). The hot salt is stored in the hot salt storage tank. On demand the hot salt is pumped to the steam superheater and steam generator to provide steam to the power plant. The warm salt leaving the steam generator is stored in the warm salt storage tank.

The two halves of molten salt system operate independently: warm salt is heated and stored hot at whatever rate the energy collection subsystem is capable of providing energy, and hot salt is used and restored warm at whatever rate is required for power generation at the time. Thus as thermal energy supply fluctuates, the levels of stored hot and warm salts also vary, opposite to each other, while remaining near their design temperatures.

The storage tanks are constructed of 6.35 mm (0.25 inch) thick carbon steel. The tanks have 0.305 m (12 inches) of mineral wool blanket insulation on the sides and top of the tanks, and 0.254 m (10 inches) of load bearing cellular glass insulation at the bottom of the tank. The operating temperature of the hot salt tank is about 377^{0}C and of the warm salt tank is about 225^{0}C. A nitrogen cover gas is provded to each tank to minimize exposure of the molten salt to air.

The oil/salt heat exchanger has four identical units connected in series. All of the four units are shell and tube counter flow single pass type of heat exchangers. Dow-Corning Syltherm 800 is the shell side fluid and Partherm 290 molten salt is the tube side fluid. In each unit there are 155 tubes, each 3/4 inch outside diameter, 14 feet long, with a 3/8 inch diameter core buster inserted in each tube. The core busters decrease the cross-sectional area of the tube, increase flow velocity into the turbulent region, thereby enhancing the heat transfer rate. Area density, which is the ratio of heat transfer area to the volume of the heat exchanger is about 26 m^{-1}.

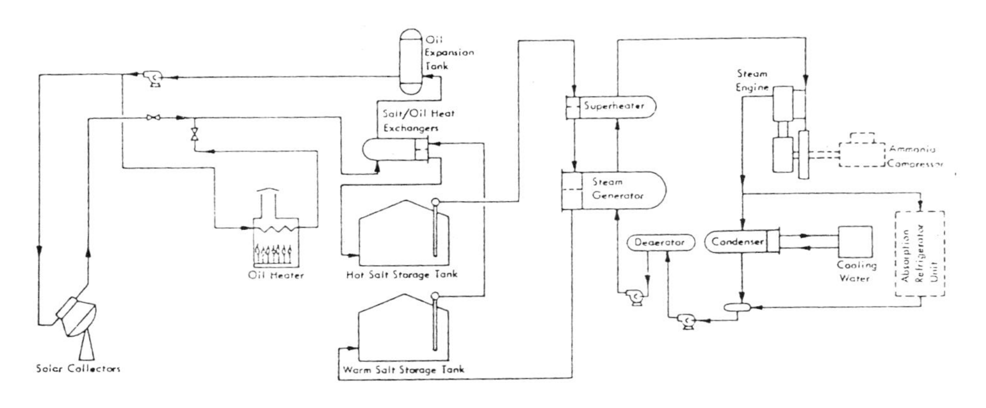

Fig. 1. Process flow diagram of the pilot plant (solar power system)

EXPERIMENTATION

Figure 1 shows the closed loop molten salt system integrated with the closed loop oil system (energy collection) and the closed steam power system (energy delivery) to provide all primary and secondary refrigeration energy needs for operation of the freeze desalination process. The two tank thermal storage system completely isolates the energy collection system from the energy delivery system. The warm salt pump can deliver warm salt from the warm salt storage tank to the oil/salt heat exchanger at a rate that it will allow accepting energy from the solar collector field at rates up to 1100 kW. The hot salt pump can deliver hot salt to the steam generator with an energy delivery rate of 600 kW. Normally when full, the hot salt storage tank has a thermal capacity of about 6000 kWhr above a base temperature of 225°C. 225°C is used as the reference temperature, since the temperature of the salt returning to the warm storage tank ranges between 220°C and 235°C, and normally averages about 225°C . Thermal energy below 225°C is not of any use in the present system. When there is no energy being drawn from the hot salt storage tank by the steam generator, the solar collector field and/or the fossel fired heater can fill the storage tank from the 20 percent level in about 5 hours. The level of the hot salt storage tank is not allowed to go below 20 percent in order to allow ample time to start up the fossil fired heater and to insure that there will not be an interruption in the delivery of hot salt energy to the steam generator. When energy is being delivered to the steam generator at the rate of 600 kW, it takes 8 to 12 hours to fill the hot salt tank. Once the hot salt tank is filled the energy collection subsystem is turned off for the next 8 to 10 hours while the energy in the hot salt tank is used.

The storage cycle duration varied from 16 to 24 hours. The temperatures and the levels of the molten salt in the two tanks were recorded every hour, and heat loss computed from this data. It may be mentioned here that the pilot plant was designed and set to operate at specified temperatures and flow rates, and hence there was not much flexibility in varying the parametric values during experimentation.

PERFORMANCE CHARACTERISTICS

Oil/Salt Heat Exchanger

The variables effecting the performance of a heat exchanger are the mass flow rates, specific heats, inlet and outlet temperatures of the hot and cold fluids, and the overall heat transfer coefficient. The performance of a heat exchanger is characterized by its effectiveness, defined by the ratio of the actual heat transferred to the maximum possible heat transfer. Fig. 2 shows the overall heat transfer coefficient, logarithmic mean temperature difference, and salt flow as a function of time on a typical operating day. As the hot salt temperature is to be maintained at 377 °C. The oil and salt flow rates vary accordingly, and as a consequence the overall heat transfer coefficient and the logarithmic mean temperature difference also vary. Fig. 3 shows the effectiveness as a function of time.

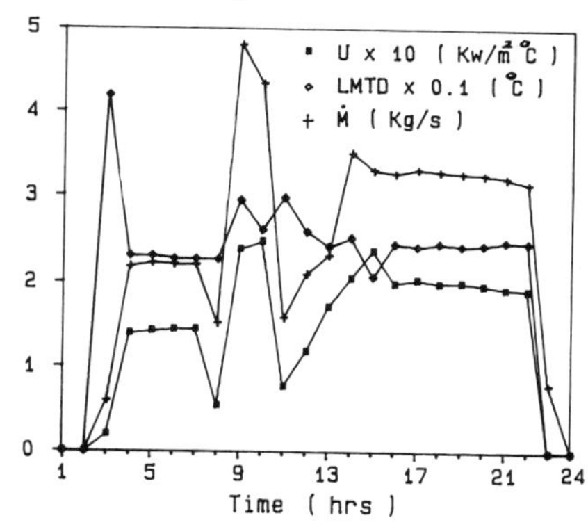

Fig. 2. U LMTD and M as a function of time

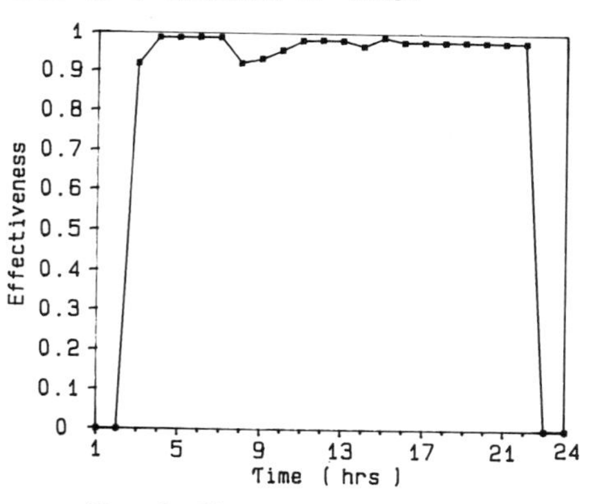

Fig. 3. Effectiveness as a function of time

Heat Loss from the Storage Tanks

For a nonstratified tank, an energy balance on the tank can be used to predict its temperature as a function of time.

$$M\,C_p\,\frac{\partial T_s}{\partial \tau} = Q_{in} - Q_{out} - (UA)_s\,(T_s - T_{amb})$$

$$Q_{in} = Q_{out} = 0$$

For finite increment in time,

$$\Delta T_s = -\frac{(UA)_s}{MC_p}\,(T_s - T_{amb})\,\Delta\tau$$

$$\text{or } T_{s2} - T_{s1} = -\frac{(UA)_s}{MC_p}\,(T_s - T_{amb})\,(\tau_2 - \tau_1)$$

T_{s1} and T_{s2} are the temperatures at times τ_1 and τ_2 respectively

The combined thermal resistance of the hot and warm salt tanks are 62.81 and 72.19 °C/kW respectively. Mean ambient temperature is assumed to be 25 °C. Figures 4 and 5 show the recorded temperatures of the hot and warm salt tanks compared to the calculated temperatures. The recorded temperatures compare favourably with the theoretically predicted temperatures.

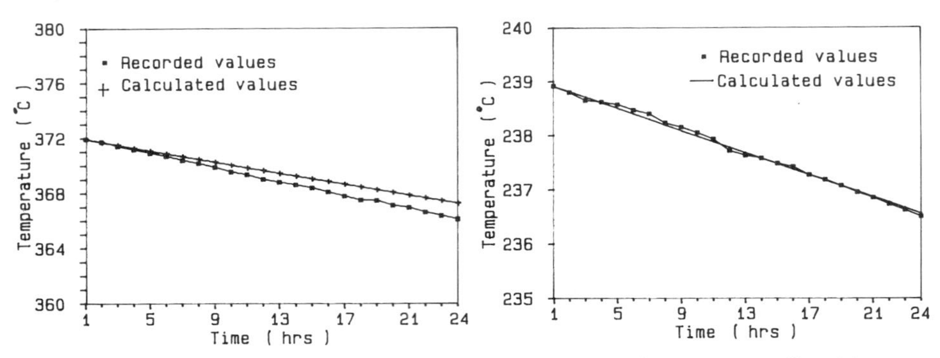

Fig. 4. Temperature - Time history of the hot tank

Fig. 5. Temperature - Time history of the warm tank

Thermal Storage Efficiency

The thermal storage efficiency is defined as the ratio of the energy delivered by the salt to the steam cycle to the energy transferred by the oil to the salt. Considering the week of September 18, 1986 as typical operating days, the average thermal storage efficiency for the five days is calculated to be 91.7 %. The effectiveness of the oil-to-salt heat exchanger is 0.97, and this gives an efficiency of 88.95 % for the storage subsystem.

Heat Transfer Fluids:

The two heat transfer fluids, Dow Corning's Syltherm 800 and Park Chemical's Partherm 290, have performed in a trouble free manner, retaining their physical and chemical properties. Syltherm 800 is a modified dimethylsiloxane polymer, low viscosity liquid, and noncorrosive. Partherm 290 (KNO_3 $NaNO_3$, $NaNo_2$) has a maximum operating temperature of 593 °C, and a negligible vapor pressure.

CONCLUSION

The energy storage subsystem comprising molten salt tanks and oil-to-salt heat exchangers has given a good performance. The storage efficiency is about 90 percent. The subsystem availability approached almost 100 %.

REFERENCES

Al-Ibrahim A.R. (1987). The performance of a thermal storage subsystem. Project report for B.S.Degree in Mechanical Engineering, King Abdulaziz University, Jeddah, Saudi Arabia.

Andrepont, J.S. and T.K. Shah (1986). Problem assessment and lessons learned Solar energy water desalination engineering test facility, MRI/SOLERAS Report, Midwest Research Institute, Kansas City, MO 64110, USA.

Zimmerman, J.C., M.C. Gupta, and G.F. Hamad (1987). Solar-powered water desalination project at Yanbu, MRI/SOLERAS Report, Midwest Research Institute, Kansas City, MO 64110, USA.

ON THE NOCTURNAL PRODUCTION OF A CONVENTIONAL SOLAR STILL USING SOLAR PRE-HEATED WATER

Ernani Sartori

Lab. de Energia Solar, Univ. Fed. da Paraíba, 58059 João Pessoa PB, Brazil

ABSTRACT

A conventional roof-type solar still working in the nocturnal period is analysed theoretically and experimentally. During the night solar heated water stored in an insulated tank was used. The nocturnal experimental behavior of the solar still was simulated using the mathematical model due to Dunkle (1961) and Morse and Read (1968) with the necessary modifications for the nocturnal case. Three liters per square meter can be obtained for the still working 11 hours at night with continuous feeding of heated water. The theoretical water and glass temperatures of the simulated still agree well with experimental values. The simulated distilled water production was, in general, 8% higher than that measured. The comparison of the theoretical with the corresponding experimental nocturnal data had never been done before. A brief cost analysis was made based on the experimental apparatus. This analysis showed that a system formed by a conventional still with pre-heated water working 24 hours a day does not offer a competitive production compared to a series of similar stills of an equivalent cost working during the diurnal period. An analogous system could be useful when we have waste thermal energy, as is the case in the Northeast of Brazil in the sugar cane alcohol distilleries.

KEYWORDS

Conventional solar still; nocturnal production of distilled water; continuous feeding of solar pre-heated water.

INTRODUCTION

Over the years several theoretical and experimental researchs have been made in order to understand better the processes and parameters involved in solar distillation and obviously to improve the efficiency and production of solar stills. Although great progress have been reached due to these constant efforts dedicated to the study of the conventional solar still (CSS) it is observed, however, that its application can only be considered economically feasible in extreme cases. Such situation justifies the several attempts of modification and study of the diverse solar still types which have been made until to date.

One of the reasons that compete to restrict the distilled water production of the CSS is its installation, generally of elevated cost, which remains nearly useless during the night. One way to surpass this difficulty would be to heat the water to be distilled at night. Tleimat and Howe (1966) simulating the industrial waste heat investigated the utilization of pre-heated water in a tubular solar distiller. They concluded that the distiller production could be increased 2 to 3 times using such method. Identically Proctor (1973) studied a CSS working with water pre-heated by industrial waste heat. That distiller produced 3.18 times more than a CSS of equal area without pre-heated water. In this paper a study of a CSS working in the nocturnal period due to solar pre-heated water is made. The production of a CSS using pre-heated water working 24 hours a day is compared to a series of similar stills of an equivalent cost, working during the diurnal period. The model due to Dunkle (1961) and Morse and Read (1968) with the modifications for the nocturnal case is used and tested on the abbility of producing theoretical temperatures and production curves equivalent to the experimental ones.

THE CSS MATHEMATICAL MODEL

The basic operation of a CSS can be summarized as follows. The solar radiation which is transmitted through the glass is partially absorbed by the water and bottom of the distiller after some reflections at the water level and bottom, heating up the water to a temperature superior than that of the transparent cover. Thereafter the water exchanges heat with the cover through simultaneous heat transfer processes by radiation, convection of the humid air, and heat and mass transfer by evaporation. In contact with the glass the water vapor condenses and runs down the cover at which end the condensate is collected. Heat can also be added to the still by heated water flow. Figure 1 shows a schematic view of the CSS and the main thermal processes involved.

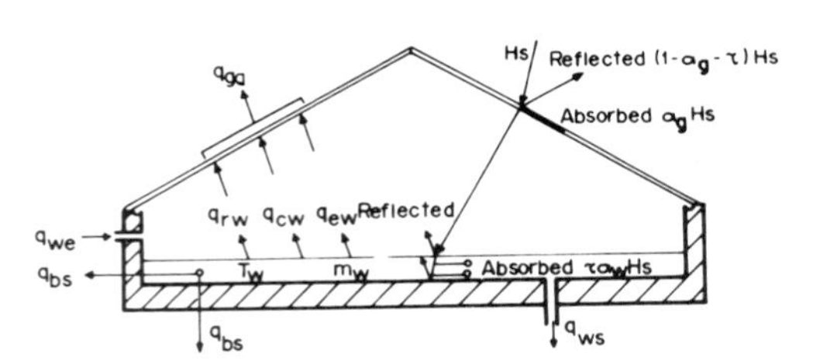

Fig. 1 Energy balance in the solar still

The steady state heat and mass transfer relations that govern the operation of a solar still have been stablished by Dunkle (1961) to which Morse and Read (1968) included a thermal storage term to allow for transient conditions. This equation can also receive the terms which represent the input and output energy of water of the still. So, an energy balance for the still is formulated

$$\alpha_g H_s + \alpha_w \tau H_s + q_{we} = q_{ga} + q_{bs} + q_{ws} + C_{st}(dt_w/d\theta) \tag{1}$$

where α_g and α_w are the glass and water absorptivities, τ is the glass transmissivity, H_s is the global solar radiation incident on a horizontal plane, t_w is the water temperature, θ is time and C_{st} represents the system thermal capacity which is function of the thermal capacity of the water, the glass and of the other structural parts of the distiller. The heat transfer between the water and the glass is given by the heat transfer rates by radiation, convection and evaporation whereas the heat flux to the ambient is the sum of such quantities plus the solar energy absorbed by the glass, i.e.

$$q_{ga} = q_{rw} + q_{cw} + q_{ew} + \alpha_g H_s \tag{2}$$

Using Dunkle's relations (1961) the internal heat transfer rates can be ex - pressed in $kJm^{-2}hr^{-1}$ as

$$q_{rw} = 0.9\sigma[(t_w + 273)^4 - (t_g + 273)^4] \tag{3}$$

$$q_{cw} = 3.182[t_w - t_g + (P_w - P_g/2009 - P_w)(t_w + 273)]^{1/3}(t_w - t_g) \tag{4}$$

$$q_{ew} = 2.918x10^{-3}[t_w - t_g + (P_w - P_g/2009-P_w)(t_w + 273)]^{1/3}(P_w - P_g)h_w \tag{5}$$

where h_w is the latent heat of vaporization of water, t_g is the glass temper- ature and P_w and P_g are the water vapor partial pressures at the water and glass temperatures, respectively, and are represented by (Keenan and Keyes, 1936)

$$P = 165807.59x10^{-[\frac{y}{T}(\frac{a + by + cy^3}{1 + dy})]} \tag{6}$$

being P in mmHg, T in degrees Kelvin, y=647.27-T, a=3.2437814, $b=5.86826x10^{-3}$, $c=1.1702379x10^{-8}$, $d=2.1878462x10^{-3}$. The latent heat of vaporization of water can be expressed in $kJkg^{-1}$ by (Cooper, 1970)

$$h_w = 3160.7 - 2.411(t_w + 273) \tag{7}$$

The heat dissipation from the glass to the surroundings in $kJm^{-2}hr^{-1}$ is given by

$$q_{ga} = 0.9\sigma[(t_g + 273)^4 - (t_{sky} + 273)^4] + h_{ca}(t_g - t_a) \tag{8}$$

where h_{ca} is the convective heat transfer coefficient and t_a is the ambient temperature. The input and output energies of water are respectively repre - sented by

$$q_{w(e,s)} = \dot{m}_{(e,s)}c_w(T_{(e,s)} - 273) \tag{9}$$

being c_w the specific heat of water, $\dot{m}_{(e,s)}$ the input and output flow rates and $T_{(e,s)}$ their respective temperatures. The heat loss from the still base and sides to the ambient is

$$q_{bs} = k_{bs}(t_w - t_a) \tag{10}$$

where k_{bs}, for a soil elevated still, is obtained through the over-all heat transfer coefficient for a multilayer wall. On the other hand, the theoretical yield rates can be calculated through the expression

$$Y = \int_0^\theta q_{ew} d\theta / h_w \tag{11}$$

The above set of equations allow us to elaborate a computer program in order to simulate the operation of a CSS with pre-heated water, or not.

THE THEORETICAL SIMULATION

In order to evaluate the validity of the mathematical model employed, simula- tions were performed with the aim at reproducing theoretically the experimental results. Therefore, in the computer program hourly values of H_s, t_a, $\dot{m}_e$, $\dot{m}_s$, T_e, T_s and fixed others such as α_w, α_g, τ, C_{st}, k_{bs} were introduced. When eqns (3-10) and the values of the above parameters are substituted in eqns (1-2), two equations which are functions of t_w and t_g result. After using the Runge-Kutta and the secant methods for solving this set of equations, hourly values of t_w and t_g and as a consequence hourly values of q_{ew}, q_{rw}, q_{cw}, q_{ga}, etc, besides still efficiency and production were obtained.

THE EXPERIMENTAL SYSTEM AND THE TESTS

The experimental system is formed by a one square meter flat plate solar collector, a 75 liters thermal reservoir and a one square meter CSS. The CSS was made with glass fiber and the cover inclination is 20 degrees. A glass cover of 3 mm of thickness was used. The still was elevated 0.4 m from the soil and glass wool was used for its thermal insulation. Silicone rubber was utilized for the sealing of the parts. Basically three types of tests were performed: A) Solar pre-heated water flow during the first hour of the test and subsequently the still operated with the water thermal inertia; B) Continuous feeding of solar pre-heated water during all the test duration and without withdrawal of water; C) The same of Test A but with withdrawal of all water after 2 hours and repeating Test A again. For comparison, the still also operated in the diurnal period from 7 AM to 5 PM without pre-heated water, with batch feeding and using 2 cm of water thickness. The hourly recorded parameters during the tests were: global solar radiation, ambient temperature, wind velocity, still production, input and output water volumes, the temperatures of the water, of the glass and of the thermal insulation, thermal storage tank temperatures, inlet and outlet solar collector temperatures. The system temperatures were measured with the aid of copper-constantan thermocouples and recorded by a Datalogger 1000.

RESULTS AND CONCLUSIONS

The results here presented are samplings of the tests performed during four months of experiments. Figure 2 to 4 present the simulated and the experimental curves of t_w and t_g as well as nocturnal productions of distilled water according to the three types of tests described above. It is observed the good agreement between the simulated and the experimental t_w and t_g curves as well as between the simulated and the experimental still yield curves. It can also be seen that there is a tendency for the increasing of still production with the increasing of water temperature. This agrees with Tleimat and Howe (1966) results. After all tests have been made an average of 8% between the theoretical and experimental still productions was obtained. The average experimental yield was 216 $m\ell m^{-2}hr^{-1}$ for Test A, 279 $m\ell m^{-2}hr^{-1}$ for Test B and 300 $m\ell m^{-2}hr^{-1}$ for Test C. Although the production of Test C is close to the production of Test B, the latter is more indicated for nocturnal distillation because its execution form is simpler than that of Test C, which may to require an automatic system for its operation. The continuous feeding was also recommended by Tleimat and Howe (1966). The average diurnal production of the solar still was 5 liters per square meter.

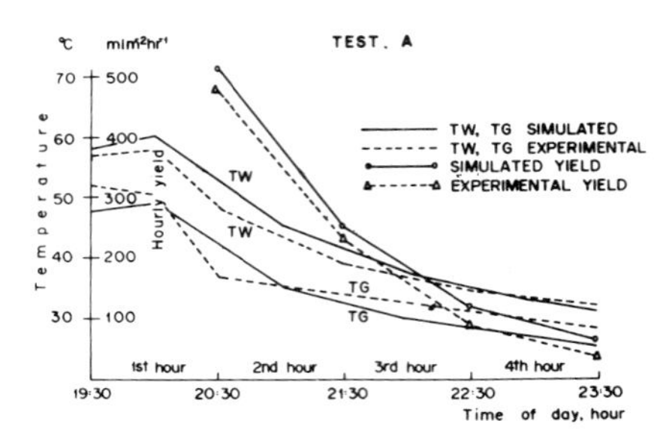

Fig. 2 Initial feeding

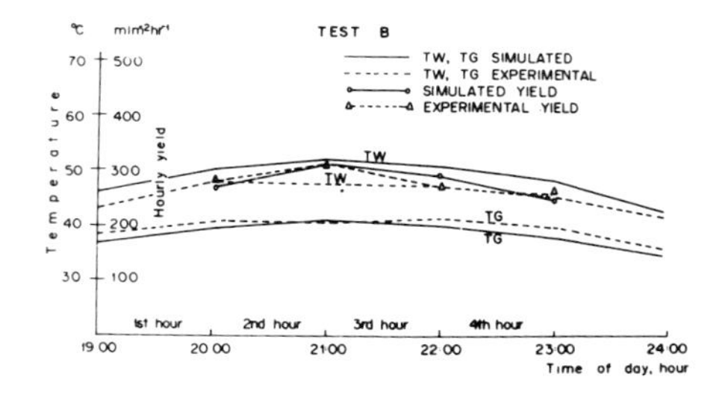

Fig. 3 Continuous feeding

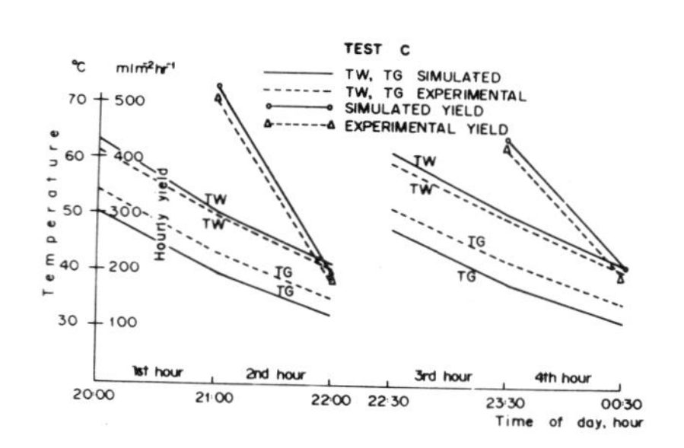

Fig. 4 Intermittent feeding

A brief cost analysis was made based on the experimental apparatus. Such analysis showed that the system cost (solar collector, water storage tank and solar still) to the solar still cost ratio is approximately equal to 3. So, if the CSS area were increased with the fraction 2 utilized in excess for the nocturnal production, the diurnal production would be elevated to 10 ℓm^{-2}. Considering 11 hours of still operation during the night and with the average production of Test B, it can be verified that the total nocturnal production can rise to 3 ℓm^{-2}. Comparing 10 ℓm^{-2} with 3 ℓm^{-2} we can see that the system formed by a CSS and a solar pre-heated water system working 24 hours a day does not offer a competitive production compared to a series of similar stills of an equivalent cost working during the diurnal period.

It is finally pointed out that an analogous system could be useful when we have waste thermal energy. This is the situation found in the Northeast of Brazil in the sugar cane alcohol distilleries.

REFERENCES

Cooper, P.I. (1970). The Transient Analysis of Glass Covered Solar Stills, PhD Thesis. University of Western Australia. p. 103.

Dunkle, R.V. (1961). Solar water distillation: the roof type still and a multiple effect diffusion still. ASME 1961 International Heat Transfer Conference, 895-902.

Keenan, J.H. and F.G. Keyes (1936). Thermodynamic Properties of Steam, Wiley, New York. p. 14.

Morse, R.N. and W.R.W. Read (1968). A rational basis for the engineering development of a solar still. Solar Energy, 12, 5-17.

Proctor, D. (1973). The use of waste heat in a solar still. Solar Energy, 14, 433-449.

Tleimat, B.W. and E.D. Howe (1966). Nocturnal production of solar distillers. Solar Energy, 10, 61-66.

ACKNOWLEDGEMENT

The author acknowledges Dr. E. F. Jaguaribe for his valuable comments on the development of the present work.

SOME COST-EFFECTIVE COMMERCIAL APPLICATIONS FOR SOLAR HOT WATER SYSTEMS

Robert K. Swartman and Peter T. Swartman
Solcan Ltd
London, Canada, N6A 4B7

ABSTRACT

A review of several particularly attractive commercial applications for solar domestic water includes nursing homes, car washes, fitness centres, campgrounds and laundromats. The domestic or service hot water load is quite substantial in these various applications which ensures the solar system can be cost-effective. Energy figures are based on design and actual monitoring results.

KEYWORDS

Solar energy; domestic hot water; commercial applications; hot water loads.

INTRODUCTION

We are continually seeking applications for solar thermal energy which are cost-effective in terms of the cost of the displaced energy. The challenge is to match the thermal load to the availability of the solar energy. Excellent applications exist for solar heated hot water in nursing homes, car washes, fitness centres, campgrounds and laundromats. The domestic or service hot water load is substantial in these various applications. Solcan Ltd has designed, built and installed a variety of solar systems ranging from 4 m^2 to 225 m^2, over 5,000 m^2 of collector area. The system types have included thermosiphon, closed loop (with propylene glycol), drainback and draindown. Our most successful systems have been closed loop; however, thermosiphon systems are the most attractive because of their simplicity. Even in our other designs, simplicity is always a major factor, along with reliability and dependability. The cost is sometimes high but we believe that reliability and quality cannot be compomised.

REVIEW OF APPLICATIONS

Applications for solar heated domestic or service hot water are varied. The common factor is that they use a large quantity of hot water on a regular basis. It is preferable if the load is concentrated during daylight hours to minimize storage requirements. Further, it is desirable to utilize the hot water on days when it is most available; for example, a car wash is more likely used on a sunny day than on a rainy or even a cloudy day.

The applications which we have identified as those with the most potential because of their hot water usage include:

Nursing homes - usually have a dining facility, laundry as well as a requirement for warm baths. They present a year-round use, the management are intersted in long term savings and, if possible, financing the cost of the project over the payback period. The hot water load is predicted on the number of patients/beds, number of meals and the amount of clothes and dish washing.

Car washes - many use cold water with appropriate detergent but if the chill could be taken off the water a better wash would result. Some use propane or electricity to warm the water slightly but the temperature is not significant. The hot water load is based on the number of cars washed per day based on seasonal experience. Some car washes have as many as 200,000 vehicles per year, concentrated more in the cold months and on sunny days.

Fitness centres - are characterized by large number of patrons using facilities during the day and having a shower. Many centres also have an indoor pool which is a good application, particularly in preheating the make-up water which is required daily by health regulations. The hot water load is based on the number of patrons plus pool heating plus special applications such as dish or clothes washing.

Campgrounds - are usually a seasonal application, i.e. usually June, July and August in Canada. Some seasonal applications extend from April to October and some operate throughout the year, for cross country and alpine skiing. The best applications are the washroom buildings, but other applications include swimming pools and hot tubs. The hot water load for the campground washrooms is based on the number of campers. In seasonal applications thermosiphon systems are particuarly appropriate. The energy load for pools is based mostly on the surface area, the temperature of the pool, and the atmospheric temperature. There are other losses but the greatest are radiation and evaporative heat losses which can be reduced by using a pool blanket.

Laundromats - use hot water continually. Some laundromats cater more to tourists in the summertime so are particularly good applications with the greater hot water load when the sun's energy is the most available. The hot water load is based on the number of washers with a factor accounting for seasonal use.

In summary, the various applications described above are excellent uses for solar heated hot water because they are all characterized by substantial hot water loads.

DESCRIPTION OF SYSTEMS

Most of the solar systems which we have installed are simple in design, dependable and reliable in operation, and have a life expectancy of many decades.

We have designed/built various kinds of systems including: closed loop, drainback, draindown and thermosiphon.

The operation of an active solar thermal system can be simulated using a computer. A particularly useful version for us is WATSUN (developed by the University of Waterloo) because it uses hourly weather data, then produces monthly totals of energy gained. If the hot water load has been over- estimated for the WATSUN simulation the predicted energy will also be much greater than the measured energy. Overestimating the hot water load will result in providing too large an area of collectors unless the solar fraction is kept low; we have found that it is generally not economical to provide more than about 20% solar fraction when there is an auxiliary fuel and the application needs a minimum temperature. Table 1 shows for a variety of applications the energy predicted and measured.

We have found system costs total between \$400 - \$500/m^2. Each collector produces between 2.0 and 3.0 GJ/yr.m^2. Assuming 2.8 GJ/yr.m^2 an installed cost of \$450/m^2 gives an energy cost of \$160/annual GJ. We have investigated combining the solar collectors and the support system in an integrated mega-collector which shows promise to reduce the installed cost for a commercial system by about 15%.

We have found that as the flow is reduced through the collectors the temperature rise increases which, in turn, increases the water temperature.

When the water flow is low enough and its temperature is high the storage tank will be stratified and the system efficiency will be much higher. Reducing the flow rate also reduces the size of pumps and piping which, in turn, reduces the system cost. A typical system in 1983 cost \$58,850 and was predicted to produce 180 GJ/yr. A similar system in 1987 cost \$39,440 and is predicted to produce 220 GJ/yr, an improvement of 18% in energy and reduction of \$19,440 in cost, \$327/GJ down to \$179/GJ. This, while inflation has been increasing at approximately 5% per year over the intervening four years. This indicates the strides we have made in reducing costs. Further cost reductions will be possible when a market develops and sales volumes can be increased. As we have reduced the costs of installation and operation we have, at the same time, striven to increase energy output.

At the present we are hampered by the public perception that solar is too expensive compared to the conventional fuels eg. gas and electricity; these other energy sources have artificial price levels which can be influenced by government intervention. Worst of all is we are comparing a renewable resource like solar energy to non-renewable sources like oil, gas, coal and some forms of electricity. Electricity from hydraulic sources is renewable but the sites are limited. We have been championing the Canadian government for a "level playing field" so we can compete fairly with oil, gas, and nuclear electricity. Those energy forms all have vested interests and financial incentives with more influence than solar. Over time as our oil and gas resources become depleted and as we do more to protect our environment, we will have to use more renewable resources such as solar energy.

ACKNOWLEDMENT

The assistance of Energy, Mines and Resources Canada in the presentation of this paper is sincerly acknowledged.

Table 1

Application	Project	Area (m2)	Energy Output	System Type	Installed Cost	Energy Cost ($/GJ)
Nursing Homes	Leamington	66	210 GJ	CL	$37,800	180
	Essex N. H.	135	321 GJ	CL	$54,000	168
	Gables N. H.	44	120 GJ	CL	$26,000	216
	Craigwiel	36	95 GJ	CL	$17,261	181
Car Washes	Leamington	53	176 GJ	CL	$32,000	181
	Oxford Dodge	19	49 GJ	DD	$6,316	128
	Zurich	40	124 GJ	CL	$23,950	193
Fitness Centre	Fitness Forum	240	695 GJ	CL	$125,550	180
	Thames Hall	178	540 GJ	CL	$99,225	183
	Town & Country	18	59 GJ	CL	$8,225	139
Campgrounds	Bingeman Park	269	489 GJ	POOL	$47,615	97
	Nor-Halton	44.4	89 GJ	POOL	$8,000	89
	Spring Valley	81.4	122 GJ	POOL	$11,000	90
	UTRCA	59.2	83 GJ	POOL	$9,318	112
	Braemer Valley	37	52 GJ	POOL	$4,186	80
	Holiday Beach	46.8	79 GJ	POOL	$5,586	70
	Erie Woods	37	52 GJ	POOL	$4,445	85
Laundromats	West Street	52.8	163 GJ	CL	$29,000	177
	Bagot & Earl	64	64 GJ	DB	$11,200	175
Other	Sussex Centre	225	730 GJ	CL	$137,000	187
	ROM	72	220 GJ	CL	$39,440	179
	Sudbury Downs	88	228 GJ	CL	$39,500	173

* CL - Closed Loop DD - Draindown DB - Drainback PL - Pool

DIRECT EVAPORATING FLAT PLATE AND TUBULATOR COLLECTORS FOR PROCESS STEAM PRODUCTION

V. Heinzel, J. Holzinger

Institute of Reactor Technology
University of Karlsruhe
Germany

ABSTRACT

Direct evaporating collectors were developed for process steam generation. Within the temperature range up to 150°C most of the process steam is used for applications like food sterilization, textile industry etc. Evaporation within the collector encompass two phase flow, which is well known for flow instabilities. Design guidelines for flat plate as well as for tubular collectors were elaborated to master two phase natural convection within the collector. They allow to exploit the advantages of direct evaporation, good heat transfer and allow to eliminate the steam generator.

KEYWORDS

Process steam, direct evaporating collectors, two phase flow instabilities, natural convection.

INTRODUCTION

Direct evaporating solar collectors were developed for process steam generation. Process steam is used for industrial applications like food sterilization largely in the temperature range up to about 150°C. Direct evaporation exploits the better heat transfer within the collector and allow to eliminate the steam generator. Both reduce the temperature difference between receiver and consumer and therewith foster the temperature dependent collector efficiency.

Elimination of the steam generator presents a cost reduction with growing importance at decreasing plant power mainly below 100 kW. Circulating the coolant through the collector by gravity supplements saving in costs by eliminating the pump.

However, direct evaporation means two phase flow and two phase flow is afflicted with flow instabilities like steam blockages and flow pattern changes. Instabilities may cause vibrations harmful to the collector and as second consequence the steam is unsteadily released to the consumer.

In horizontal tubes two phase stratified or wavy flow pattern with poor heat transfer coefficients are prevented by a sufficiently high velocity of the fluid. Natural convection, therefore, not only has to replace evaporated water but also has to maintain a sufficient circulation through the collector.

FLAT PLATE COLLECTORS

Development of flat plate collectors for direct or internal evaporation start from the standard design. That means, for heat removal several parallel tubes are attached to the receiver, which are in this case boiler tubes. They are provided with water from a lower header and feed a steam-water-mixture to an upper header. A riser connects the upper header with an external steam drum which separates the mixture. The steam is released to the consumer, whereas the water flows back through the downcomer to the lower header. Condensate is added at the downcomer.

In heated pallel ducts with two phase flow steam blockages occurre if a flow rate reduction increases the steam production resulting in a acceleration pressure loss higher than the also caused decrease of friction pressure loss. For remedy the friction pressure loss has to be increased by a higher throughput. In case of loops with natural convection the buoyancy and therewith the velocity increases with the height difference between heat scource and sink, here the upper header and steam drum.

Buoyancy and pressure loss calculation showed that a riser height of 0.5 m suffices to preclude steam blockages. For reason of confirmation a glass modell with 4 parallel boiler tubes provided with electric heater was built up (Fig. 1). Indeed steam blockages did not appear, but geyser like pulsations with almost constant periods were found. Periods up to 50 sec were observed (Fig. 2). They decreased with lower water lever in the steam drum, boiler tube diameter and higher power.

The steam pulses started with a rapid increase of the velocity. The accelerations shaked the experimental device. At the end of the pulses all tubes were streamed through with the colder water of the drum. Then a reheating phase followed and the velocity decreased to zero.

At the end of the reheating phase a first steam bubble was formed in the riser. The bubble reduced the geodetic height in the riser, and further bubbles were formed below. Within less than 1 sec the flow pattern in the riser changed from one phase flow to bubble, slug and film flow. Finally the riser contained steam only. According to this observations a computer program was written. Pulse frequency, velocity and steam flow rate could be calculated correctly. By means of this code and the experiments collector designs were developed, to avoid those pulsation.

The first solution is an orifice, installed at the steam exit of the drum. This orifice causes a pressure difference, which increases, if the steam production rate grows at start-off of a pulse. As repurcussion the loss of geodetic height in the riser is balanced and the pulse doesn't continue. The pressure differences at the orifice necessary to suppress pulsations completely are plotted in Figure 3.

The second solution is an enlarged header, which acts as internal steam-water separator. Only steam leaves the collector whereas the water returns directly to the lower header. Again glass models were used to investigate the two phase flow stability. A collector with 12 parallel boiler tubes and an internal steam-water-separator was manufactured and tested in open air. The receiver had an area of 2 m^2. The aperture was enlarged by two flat mirrors to 4.2 m^2. During several days at noon time the collector produced steam at a continuous rate corresponding a power of 1.8 kW.

TUBULAR COLLECTORS

A tubular collector consists of a glass tube enveloping the receiver to which a tube system for heat removal is attached. Largely the tubular collectors are installed horizontally and are provided with booster mirrors. Direct evaporating tubular collectors may be connected to a common header according to the experiences of the flat plate collector or each collector tube is equipped with a steam-water separator in an annexe.

The annexe shown in Fig. 4 and 5 in a compact version consists of a riser with an overflow at the upper end and a downcomer. At the overflow steam is separated and released through the steam exit, whereas water flows through the downcomer and return-tube back to the boiler tube. Before entering the boiler tube condensate is added. Riser and downcomer dimensions are laid out so that weight difference of their content propel the circulation through the boiler tube.

Compact steam-water separators are developed for collectors with vented glass tubes (see Fig. 4) as well as for vacuum tube collectors type Corning (see Fig. 5). The vented tube collectors are installed with parabolic boosters. Each collector has an aperture of 1 m^2. A receiver plate was attached to the boiler tube and housed in a vented glass tube. The water return tube is installed coaxially inside the boiler tube. Collector tube, booster and steam-water separator form one unit and are turned for tracking together. The separator can be deflected from vertical position to $\pm$ 45° at still steady steam flow rate. Instabilities or pulsations are not observed. A group of 8 collectors are installed for open air test. They provided steam with 100°C and atmospheric pressure equivalent a power of 2.5 kW respectively an efficiency of 35 %. The design allows steam temperatures up to 130°C and with minor changes up to 150°C.

The vacuum tube collectors are shown in Fig. 7. The upper leg of the U-tube for heat removal guides the steam water mixture to the separator. Through the lower leg water coming from the separator returns. For tracking the separator can be inclined till $\pm$ 45° without risking puls free operation. The booster aperture is 0.7 m^2. A group of 8 collectors is under preparation to be installed for open air test.

Both test group are envisaged to be installed in Turkey next year (financial support from BMFT, Federal Ministry for Research and Technology).

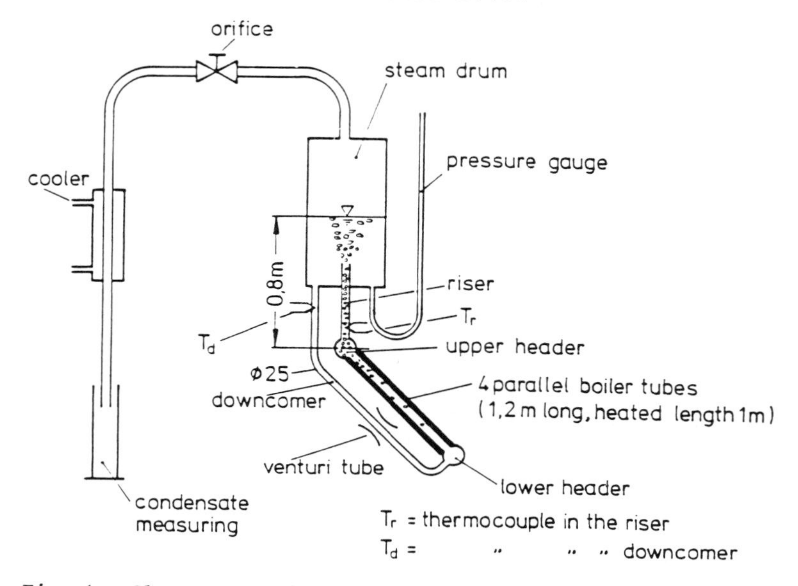

Fig. 1: Glass model simulating a flat plate collector

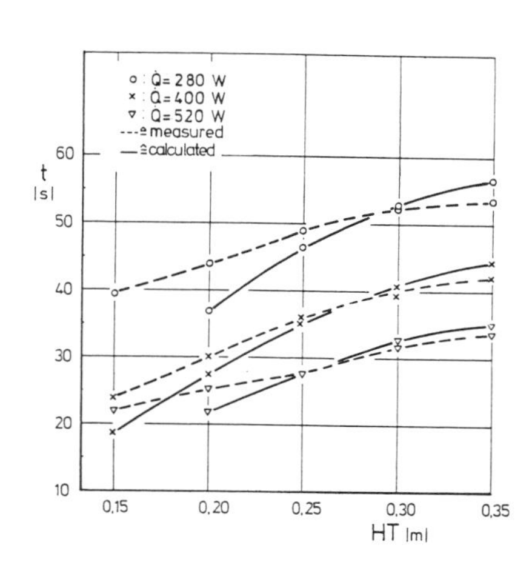

Fig. 2 Pulsation period depending on the water level HT in the steam drum and the power Q fed to the boiler tube

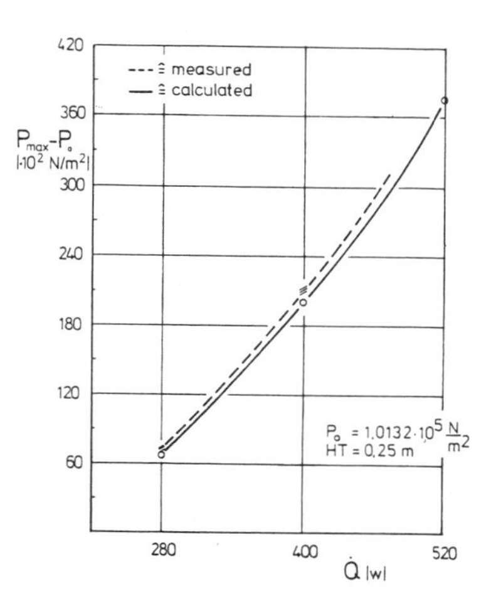

Fig. 3. Pressure difference at the orifice necessary to suppress pulsation depending on the power Q (HT = water level in the steam drum)

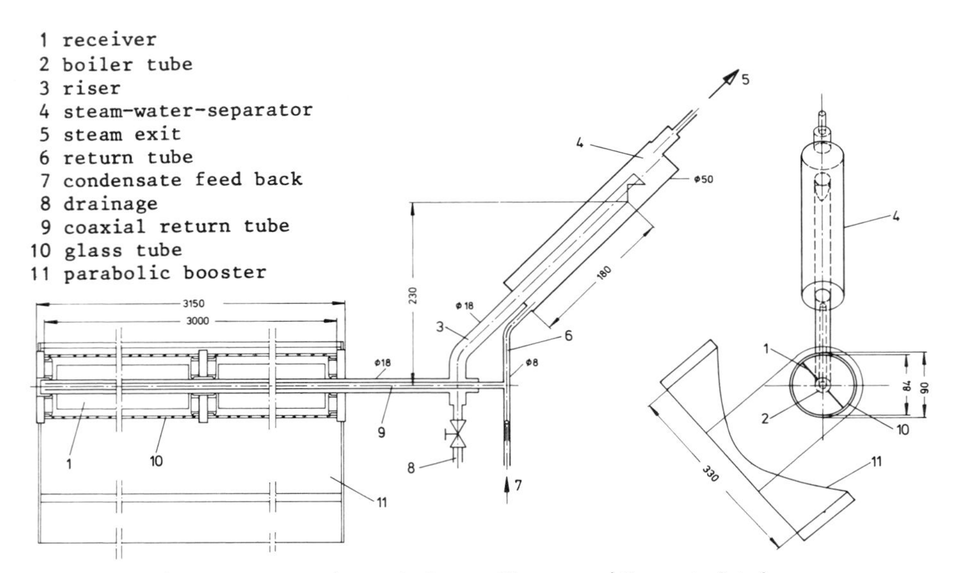

Fig. 4: Direct evaporating tubular collector with vented tube

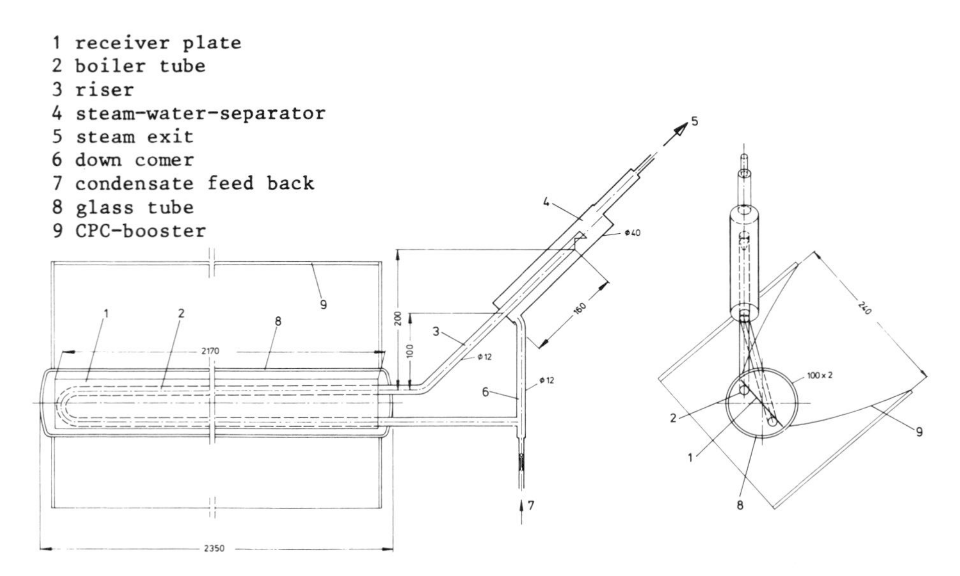

Fig. 5: Direct evaporating tubular collector with vacuum tube, type 'Corning'

ANALYSIS AND STUDIES ON THE MOVING-TYPE SOLAR ENERGY HEATING SYSTEM FOR ROAD PITCH

Shao Shu Ren, Hu Xian Yun, Xia Qing Wei

Ezhou Highway Maintenance Branch, Hubei, China

ABSTRACT

According to the highway's character of long distance, of large area and of mobility under the construction, from the point of view of economizing on energy and raw materials, the author in the paper makes analysis and studies on necessity and feasibility by using a kind of moving-type solar energy heating system for road pitch which is easy to be installed and removed in the course of workers' moving. Meanwhile, it also makes technical explanation on how to apply combination of multiple solar concentrating modes to heating pitch.

KEYWORDS

moving-type; pitch; green-house; pond; tracker; dehydration

INTRODUCTION

Pitch is one of the major materials for building superior and sub-superior roads. It is lacking under the influence of world-wide petroleum crisis. In some places and countries, the cost of pitch roads is as high as that of cement concrete roads. Even so, for pitch roads are rather plain, less noise and comfortable when driving on, it is used on a large scale in the road construction. Some engineers in highway construction think pitch is a rather good material in application to heavy-duty transportation roads construction. Especially in developing countries, most experts of road construction think in the area where the state of road surface is not good, because of heavy traffic, wider and plainer roads with pitch paved on, transportation efficiency and capacity will raise twice as much, but wear and tear of vehicles as well as transportation cost will reduce greatly. Therefore, in order to raise the standard of highway construction, it is an important way to use pitch roads.To meet the requirements of pitch road construction,pitch must be dehydrated and heated,the temperature ranges within 160°C-180°C(excluding emulsified pitch).To heat pitch,it must not only cost a lot of conventional energy resource, but also bring about pollution to surroundings.So from 1981,we started to study the solar energy heating system for road pitch and have made a desirable improvement.To see Fig.1,four main kinds of solar energy heating systems used for highway construction at present in our country.

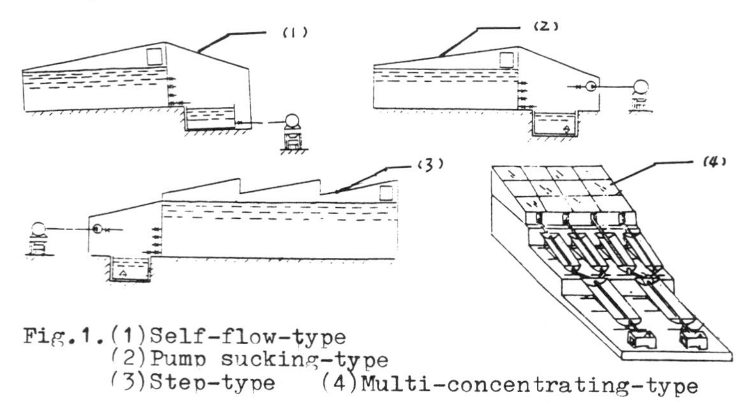

Fig.1.(1)Self-flow-type
(2)Pump sucking-type
(3)Step-type (4)Multi-concentrating-type

ANALYSIS OF THE SYSTEM

In the systems shown above, Green-house Ponds consist of brickstone structure which is fixed,it is invested only once.The following principles will be considered when it is designed.
1.The solar energy heating system for road pitch is applied to a limited area.The transport line of the pitch and stone-materials is short.
2. The slope of topography should be considered.
3. To select slope facing south where ground water level is low.
4.The system should be located in the centre of the area where many roads are scattered.
5.The operating radius of the system is less than 30 kilometres.
The above principles have limited the operating extent of the system and the located place of the system.Although the system built according to the principles can get remarkable economical result,it confines operating area.We know that roads as loading-body of transport vehicles are extending according to the requirement,their length is variable.But,the system for road pitch of the road construction is not easy to remove.So in order to guarantee the need of the road construction,the heating system must be increased,as a result,it should cost a lot of funds,materials and land.In fact,after the roads are built and available for use, the area of the road maintenance and repairing is not large, generally,only occupying one tenth of the new-built roads or less. From this point of view,the fixed system would be surplus and its using efficiency will reduce to one tenth or even lower.However, the funds for maintenance and repairing of the systems can't decrease greatly.If the solar heating system is not fixed install,

but movable,and its fundamental function does not change,we can give full play to the system.On the basis of this circumstance, the moving-type solar energy heating system for road pitch is born.

THE STRUCTURE OF THE MOVING-TYPE SOLAR ENERGY HEATING SYSTEM AND SOME TECHNICAL EXPLANATION

To see the Fig.2 of the system:

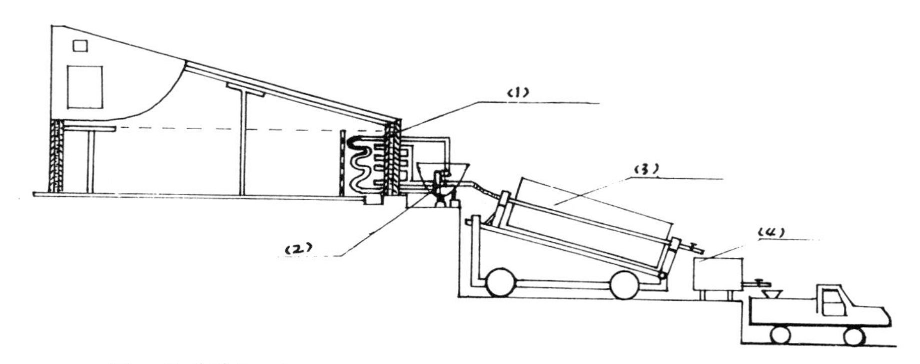

Fig.2.(1)Moving-type Solar Energy Pitch Pond
(2)Pre-heating Exchanger(CPC)
(3)Parabolic Tracker Collector
(4)Pitch Tank Heated Electrically

The Pitch Pond is the major part of the system, which has two functions:1)Used as pitch storage pond, 2)Used as pre-heating pond For the convenience of installation and dismantling,the wall of the pond consists of combined structure which is made of metal. The wall body is of certain rigid intensity.This knowledge is on the basis ofthe following conditions:
Pitch is in solid state at normal temperature, and semi-fluid state at 60°C.In practical observation and measurement,we found the changing state of the pitch stored in the green-house pond. The data as follows:
Time: At 12 o'clock(at noon),September 4,1983
Radiation intensity:cal.1.28/cm.min.
When the depth of the pitch oil surface is 15cm,the maximum temp. is 78°C(to account up to down).When it is 15-25cm,the average temp. is 50°C and when it is 25-50cm,the average temp. is 40°C or so.When it is below 50cm,the temp. reduces step by step until at normal temp.According to the circumstances shown above,strong cohesion exists in the molecules,it reduces the pressure to the wall of the green-house pond,so the wall intensity can be guaranteed. After absorbing the solar energy,green-house effect in the pond takes place,the thermal insulation of the pond wall is very important. The wall of the fixed solar energy heating pitch pond consists of brick-stone structure mainly, the wall of moving-type pitch pond consists of metal plates. To see the Fig.3(1),(2) below:

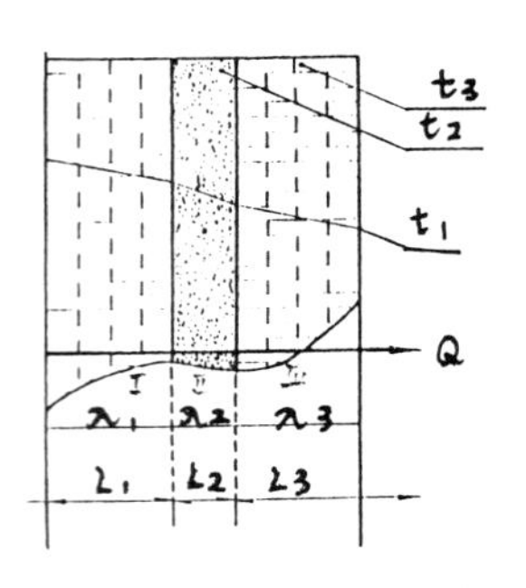

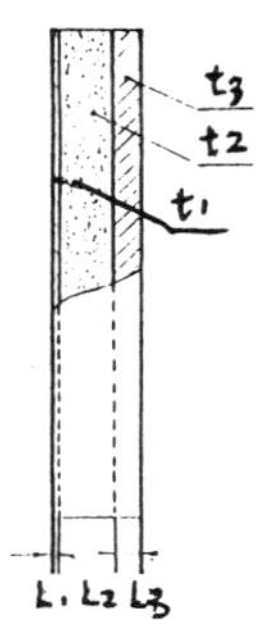

Fig.3(1)Fixed Wall Fig.3(2)Moving-type Wall

Fig.3(1) shows the wall has multi-layers and is flat.If $\lambda_1, \lambda_2, \lambda_3$ is thermal conductivity of each layer respectively,and L_1, L_2, L_3 is thickness of each layer,the design will do according to some formulas of heat exchanging theory. Fig.3(2) shows the wall is moving-type,combined structure,interior wall is made of steel plates,in intermediate layer are thermal insulation plates materials,zinc-plated plates are in outside.If temp. in the pond is available(The average temp. from May to September is at 65°C or so),and the ambient temp. outside is also available,since the resistance of thermal conductivity(inner wall of the metal)is small.Namely,the temp. of inner wall of the plates can be considered as that of pitch in the pond.If λ (thermal conductivity of insulation materials) and a_2 (heat-release coefficient outside) are known,to determine q (heat loss of flat wall) first, then δ (optimistic depth of thermal insulation materials) can be determined easily.To see following formula:

$$q = \frac{t_n - t_u}{1/a_2 + \delta/\lambda} \ \mathrm{kcal/m^2 \cdot h}, \quad \text{then}, \delta = \frac{\lambda(t_n - t_u)}{q} - \frac{\lambda}{a_2} \ \mathrm{m}$$

The moving-type solar energy heating system for road pitch is multi-concentrating type(combined).To heat road pitch by using solar energy in multi-concentrating forms is determined by the physical character of the pitch itself.It is not only the need of collecting solar energy but also the requirement in the course of heating pitch.It is pointed out above, that at normal temp. the pitch is in solid state, at 60°C,it begins to soften and then becomes semi-fluid state, because of green-house effect,the temp. of the pitch in the pond can't rise further, but focussing equipment can do.Since the equatorial system is parallel with earth-axis, so long as the changing range of declination angle is available,the focussing equipment can join the outlet of the green-house pond, as a result of dropping, the flowing resistance can be made weaker, the temp. of the flowing pitch can be raised. The fact can not be negnected that pitch contains about 5% water moisture in the course of transfer and storage.The standard of containing water moisture is from 0.2%-2%,according to different grade of pitch.Practical calculation shows that when water moisture reaches 5% , the energy cost for dehydrating only 3% of water moisture amounts to 29% of the total energy cost!It is not wise of us to dehydrate pitch by using solar energy.Experiment proves that the optimum temp. is from 60°C to 90°C when pitch is

separated from water moisture. When dehydrated in the green-house pond, the water moisture remains in different temp. fields because of different temp. in each layer. Equatorial system uses two dimensional focussing, focussing line temp. is higher than 100°C ,so, the water moisture is vaporized while it cost a lot of energy with the result that it would reduce the effective heating efficiency of the concentrators. Other means should be taken hereby. Focussing ratio of pre-heating exchanger (CPC)can be limited 3 or so,it is put in east-west direction, parallel to the green-house and normal to the equatorial system.It utilizes space without occupying ground area.If it is used as "heater" a heating exchanger (medium oil in it) is put outlet of the green-house pond,then, different temp. fields can change partly The water sinks downwards, then exhausted through exhausting valve. The dehydrated pitch flows into the equatorial system through compound parabolic concentrating pipes(each layer separated).It is possible to recover "remained heat".In order to ensure the quality, a relevant electrical heater put outlet of equatorial system is necessary. Of sun-tracking ,the problem was resolved in 1982, it has stable property and result is good.
In conclusion,the moving-type solar energy heating system for road pitch consists of the following parts.

1. Green-house Pitch Pond-steel plates,combined install
2. CPC collector(including heat exchanger,unitized install)
3. Cylindrical Parabolic Tracker Collector-systematic construction
4. Electrical Heater
5. Attached Electric Systematic generator set

The equipment can be removed by using vehicles with trailers. The design capacity of the system is determined according to the output each shift. The pitch temp. can reach 160°C-180°C.

SOLAR SEA-WATER DESALINATION AND THE TECHNICAL AND ECONOMICAL FEASIBILITY OF SOLAR POND POWERED DISTILLATION PLANTS

M. Posnansky

Atlantis Energy Ltd., Thunstrasse 43a, 3005 Berne/Switzerland

ABSTRACT

Desalination is an important and interesting application for the use of solar radiation as a source of undepletable energy. After almost a decade of research and development including the installation and testing of various smaller pilot systems, our solar desalination technology - among others - is now becoming available on a commercial level.

The paper discusses the evolution of the technology both of the desalination- and the collector-subsystems as a result of the technical and economical constraints associated with the utilization of solar energy, a highly fluctuating energy source of low surface density.

Performance data is presented in particular for the coupling of a selfregulating MSF unit with a solar pond energy collection and storage system, both inhouse developments. The performance and layout data was obtained from computer simulation and experimental results with a small sized solar pond and desalination subsystem in Switzerland.

The economy assessment, which is presented for Middle East climate conditions, clearly demonstrates that solar desalination already becomes competitive for medium sized installations at remote locations. Potential further cost reductions particularly through upscaling may well lead to the use of desalinated water for agricultural applications one day.

INTRODUCTION

Solar energy is an undepletable source of energy which is available in abundancy on this planet: The present world-wide energy consumption only represents a fraction of less than 1 % of 1 ‰ of the incoming solar radiation.

It is not a coincidence that fresh water is lacking most where the highest solar radiation exists: The incident solar radiation, particularly in desert arid areas averages 5 ÷ 6 millions kWhrs/sqkm day.

Solar desalination is an important and most interesting application for the utilization of solar energy: The future depletion of conventional fuel resources shall ultimately lead to a widespread harnessing of solar energy. The use of solar energy at an important scale in near term however, will only take place if desalinated water can be produced reliably at a cost which is competitive to the cost of water produced with existing systems consuming fuel.

Should it be possible to significantly reduce the cost of water to conventional alternatives wide agricultural usage can be envisaged - thus leading to a tremendous potential for solar sea-water desalination (unthinkable using depletable fuel resources).

SOLAR DESALINATION METHODS AND THE BASIC REQUIREMENTS FOR THE DEVELOPMENT OF AN APPROPRIATE TECHNOLOGY

In principle all presently known desalination methods (including cogeneration schemes) can be applied for the use of solar energy since solar radiation is a high grade source of energy that can be transformed into low, medium and high temperature thermal energy or electricity to power the respective systems.

In practice the utilization of the highly fluctuating energy source of low surface density requires a technology which

- can cope with strong variations of energy supply (clouding, day/night and seasonal cycles)
- provides an extremely low cost storage and solar conversion system, feasible to be installed over very large areas with very low maintenance requirements
- ensure a high reliability and effectiveness under severe environmental conditions (storm, dust) over extended periods of time (15 - 20 years life cycle)

THE ATLANTIS SOLAR DESALINATION TECHNOLOGY

Atlantis engaged into the development of energy efficient thermal distillation desalination equipment to be powered by a low temperature solar conversion and storage system almost 10 years ago. This decision was based on the conviction that low temperature heat could be collected and stored at significantly lower cost than high temperature heat or electricity.

As a result of the R + D efforts, special multistage evaporative desalination plants capable of operating under the required variable heat input conditions were conceived. The successful development and implementation of a solar pond energy collection and storage system to work in conjunction with such equipment demonstrates now that above decision proves to be still valid.

EXAMPLES OF REALIZED SOLAR DESALINATION PLANTS WHICH ILLUSTRATE VARIOUS STAGES OF TECHNOLOGY DEVELOPMENT

Delivery of the world's first solar desalination plant (1979 Bari; Agip). With this plant Atlantis gained practical experience which led to the development of the "Autoflash" desalination technology which allows an inherently stable operation of multistage plants under strongly variable conditions of heat supply.

First Atlantis "Autoflash" unit delivered to the Kuwait Institute for Scientific Research (1980, KISR). The plant has been successfully working for several years thanks to its selfregulating ability for operation under variable conditions. A break-through was achieved with respect to the development of simple reliable and energy efficient multistage distillation equipment for solar and waste heat applications.

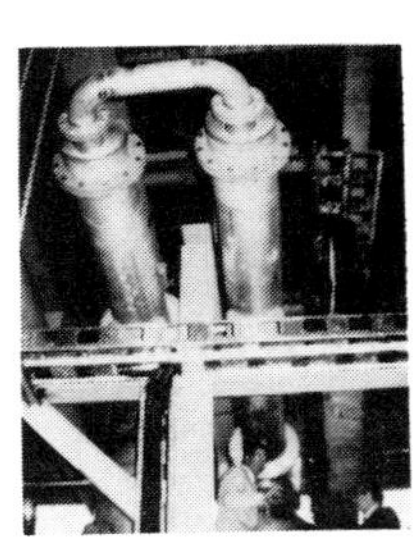

Upscaled version of the Atlantis "Autoflash" unit. This particular installation was designed for "recycling of waste waters to regain valuable protein substances" powered by industrial waste heat (1986, Germany). "Autoflash" unit sizes of 500 - 1000 m3/d have matured from the experiences gained.

Solar laboratory completely designed and delivered by Atlantis to Qatar (1985) Various solar collector systems (1), (2) developed by Atlantis were installed to power a 22 stage vertical Atlantis "Autoflash" unit (3). For the first time a solar pond energy collection and storage system (4) was implemented here. In addition photovoltaic generators (5), hybrid system (2) for supply of electricity into the grid were delivered.

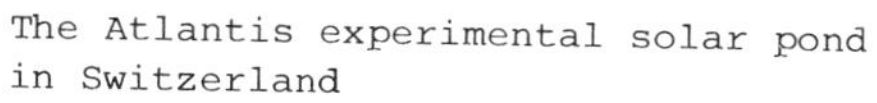

The Atlantis experimental solar pond in Switzerland

Pilot solar pond established on the Canary Islands

THE ATLANTIS SOLAR POND ENERGY COLLECTION AND STORAGE SYSTEM POWERED DESALINATION PLANT - THE MOST PROMISING OPTION

General Description

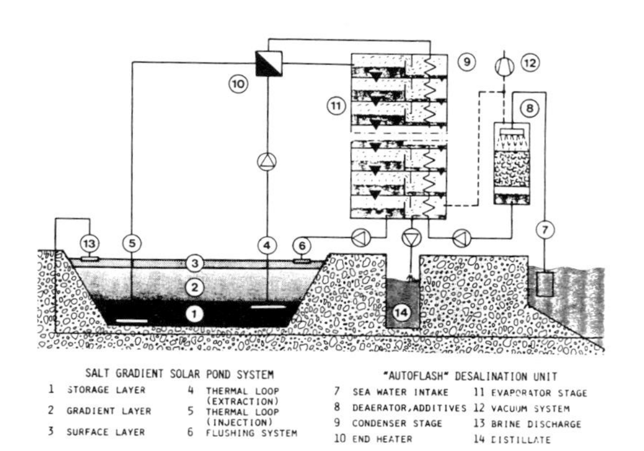

A solar pond consists of 3 layers of saline water:

- A top layer of low salinity of approx. 0,3 m thickness. This layer is designed to protect the lower layer from environmental influences (sea-water is generally used for replacement of evaporation losses and surface cleaning).

- An intermediate layer of approx. 1,2 m thickness - the so called salt gradient layer - where the salinity increases with depth. Due to the increasing density, convection currents are suppressed. Heat losses therefore only occur through heat conduction. This layer acts as a very good "transparent" insulation.

- A lower bottom or storage layer with the highest concentration. In this zone, the actually useful radiation (up to 30 % of the total received solar energy) is absorbed. The water, being "trapped" by its high density is heated up to the desired temperature. Depending on the thickness of the storage zone vast amounts of thermal energy can be stored at temperatures up to 80 - 90 °C. The heat accumulated in this storage zone is extracted much like from a conventional thermal storage vessel and fed to the endheater of the desalination plant.

Main Advantages of the Solar Pond System

- o Low investment costs per installed collection area.
- o Thermal storage is incorporated into the collector.
- o Diffuse radiation ("cloudy days") is fully used.
- o Very large surfaces or "solar lakes" can be built thus large scale desalination becomes possible.
- o Heat can be extracted at individual points from the storage zone in contrast to other solar collector systems where the heat must be collected and transported over long distances requiring expensive piping, insulation, armatures, pumps and controls.
- o The pond system is "selfcleaning" (dust sinks to the bottom and surface pollution is washed away).

Some Drawbacks

- o The yearly average collector efficiency is lower than for other systems meaning that for equal thermal output 2 ÷ 3 times more collection surface (not ground area) is required.
- o Cost effectiveness is only achieved with large pond areas (>20'000 m2) and under suitable geological conditions.
- o Installation of the pond is limited to locations where saline water is available (coastal areas, inland locations with saline ground water).

Brief Summary of the Solar Pond Technology Development of Atlantis

In 1980 the company started the development of the solar pond technology in cooperation with the Swiss Federal Institute of Technology. Following initial theoretical studies and laboratory developments, a solar pond with a size of 500 m2 was built in Switzerland. Methods for the establishment and maintenance of the salt gradient were developed. This year, the pond is entering its 4th year of operation. The pond is equipped with a data acquisition system in order to measure the behaviour under Swiss climatological conditions. In parallel, a dynamic computer model was developed in order to simulate and predict the behaviour of a given pond under a given climate.

The comparison of the performance calculated with the dynamic model with the actually measured performance (the actual measured Swiss climate being an input) showed a very good agreement. As a consequence, the model can be used to simulate the behaviour of a given pond under the climate conditions of a particular location.

The combination of the dynamic model for the pond with a computer based model describing the "Autoflash" desalination plant, made it possible to simulate the seasonal and yearly output of desalinated water and also to optimize a complete solar pond powered distillation plant (see Fig. 1).

Based on those results, the economical data presented here were established for Middle East climate conditions (see Fig. 2).

The first potable water with a solar pond system was produced in 1985 with an experimental solar pond of 1'500 m2 built by Atlantis in cooperation with IDTC for the solar laboratory in Qatar. The pond was coupled to an Atlantis "Autoflash" 22-stage desalination unit with a capacity of 20 m3/day and a specific heat consumption of 70 kWh/m3. Just recently Atlantis build up a pilot pond on the Canary Islands.

The experience gained so far shows that:

- Further development is necessary on our side with respect to the maintenance of the pond in order to keep the critical water layers sufficiently clean over extended periods of operation.
- In order to reliably ensure proper overall system behaviour, execution and control of important operation, maintenance and management functions must be automated using computerized equipment.
- Preliminary system optimization suggests to operate the desalination plant at a temperature of 75 ÷ 80 °C in summer and to lower it to 55 ÷ 60 °C in winter. The flexibility of the desalination plant for operating at variable temperatures proves to be of great value in this respect.

COST COMPARISON OF THE ATLANTIS SOLAR POND POWERED DESALINATION OPTION WITH FUEL CONSUMING DESALINATION SYSTEMS (RO, VC, MSF/ME)

In the attached table the water cost comparison between the reverse osmosis process, the fuel fired MSF/ME plants and the solar pond system is shown for production capacities of 1'000 m3/day and 10'000 m3/day. The investment costs for the required solar pond of approx. 800'000 m2 has been estimated based on our own analysis and experiences as well as from literature. Investment costs (indicative) for the desalination plants are based on commercial data obtained for the various systems.

CONCLUSIONS

Solar sea-water desalination using the "solar pond energy collection and storage system" as a heat source to power specially conceived multistage distillation plants can compete with conventional desalination methods (RO, VC, MSF, ME) which use depleting fuel resources.

Large areas are required for solar energy collection, but this should not be a limiting factor in arid or desert locations.

A potential for obtaining rather favourable water costs exists and thus irrigation for agricultural applications may become a reality one day.

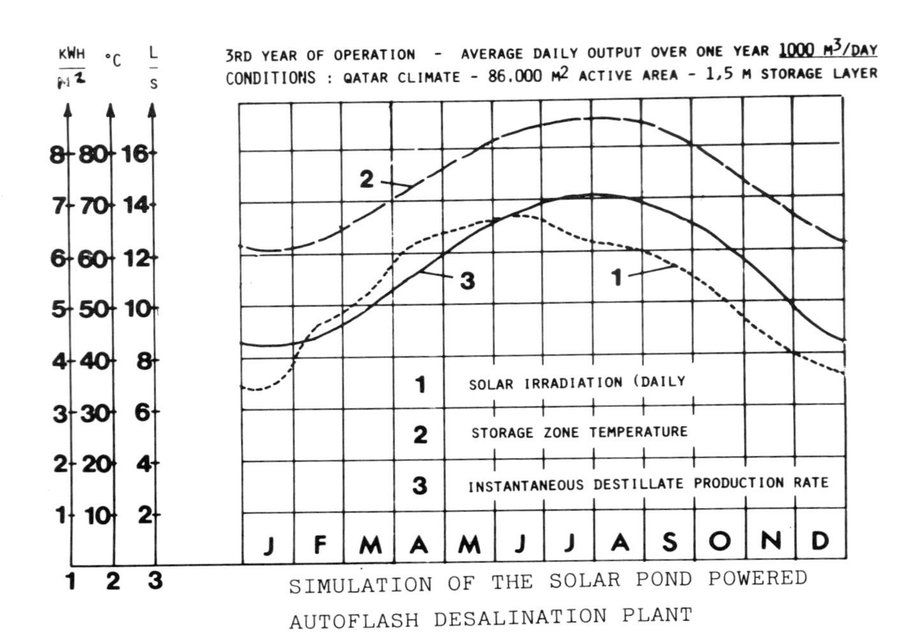

SIMULATION OF THE SOLAR POND POWERED AUTOFLASH DESALINATION PLANT

COST COMPARISON OF THE SOLAR POND POWERED DESALINATION SYSTEM WITH FUEL CONSUMING DESALINATION PLANTS

Plant capacities: 1'000 and 10'000 m3/day (330 days of operation per year); amortization period: 15 years

PLANT TYPE / COST ITEM	REVERSE OSMOSIS - 8 kWh_{el}/m3 -		MSF/ME - HEAVY FUEL FIRED - 70 kWh_{th}/m3 ≙ 7 kg/m3 -		ATLANTIS SOLAR POND SYSTEM - 70 kWh_{th}/m3 -	
	1'000 m3/d	10'000 m3/d	1'000 m3/d	10'000 m3/d	80'000 m2 1'000 m3/d	800'000 m2 10'000 m3/d
A) 15 years cost (x 1'000$)						
o Energy costs	5'940 (15 cts/kWh)	59'400 (15 cts/kWh)	5'700 (25 $/barrel)	57'000 (25 $/barrel)	1'340' (15 cts/kWh for auxiliary electricity)	13'400
o Investment costs (installed, indicative)	3'000	18'000	3'100	18'000	3'800 (desal) 2'560 (pond) ($ 32/m2)	23'000 16'000 ($ 20/m2)***
o Interests (7 %, 15 years am.)	1'875	11'700	1'937	11'700	4'134	25'300
o Operation, maintenance spare parts, chemicals	2'400**	14'000**	2'000**	12'000**	2'400**	14'000**
o Membrane replacement (1 % per month)	800	6'000	---	---	---	---
Total expenditures 1)	14'015	109'100	12'737	98'700	14'234	91'700
B) Water generating costs ($/m3)	2.8	2.18	2.55	1.97	2.85	1.84

* plant located in place with favourable insolation conditions: 5,5 kWh/m2/day average

** tentative values (are variable depending on local conditions)

***pond prices according to literature (3) could be as low as 9 $/m2

1) present value calculation not applied

COST EFFECTIVE SOLAR AIR HEATERS FOR DRYING AGRICULTURAL PRODUCTS

C.J. van der Leun, J.H. Boumans

ECOFYS Research and Consultancy Cooperation
Nachtegaalstraat 60 bis, 3581 AL Utrecht
The Netherlands

ABSTRACT

The drying of agricultural products requires heat at low temperatures. This process is therefore well suited for the use of solar air heaters. Due to the fact that agricultural drying takes place during a limited part of the year, the investment costs of the collector should be kept low. This is possible, because the temperatures needed for the drying process can be achieved with simple collectors. ECOFYS has evaluated two projects in this field.
The performance in practice of a solar air heater, used mainly for the drying of onions, was studied. The collector in this case has an absorber consisting of a porous, fibrous material. This allows air to be drawn through the absorber surface, ensuring a good transfer of heat. The efficiency of this collector was found to be approx. 54 %. The investment costs for this collector are low: \$ 60 per m^2, including air ducts and assembly.
In a second project, ECOFYS has designed a low-cost solar air heater for the drying of cereals. The collector in this case has no cover. The absorber consists of the corrugated roofing itself. The air flows between the roofing and a layer of insulating material. The collector has been installed this year (1987). Expected efficiency is approx. 30 %. System investment costs are only \$ 30 per m^2.

KEYWORDS

Solar air heaters; agricultural drying; low-cost collectors.

INTRODUCTION

Many agricultural products have to be dried quickly after the harvest to maintain quality during storage. Typically the moisture content has to be reduced by 4 to 8 % in a matter of days. This can be done with outside air, the drying capacity of which can be substantially improved by heating it to 25 or 30 °C. The demand for heat at such low temperatures makes this process suitable for the use of solar air heaters.
Due to the fact that agricultural drying in a moderate climate takes place during a limited part of the year, the investment costs of the collector should

be kept low. This is possible, because the temperatures needed for the drying process can be achieved with simple collectors.

ECOFYS has evaluated two projects in this field. In this article the design and performance of the systems studied will be discussed and an economic evaluation will be made.

PERFORMANCE OF A SOLAR AIR HEATER WITH A POROUS ABSORBER

In this project the object was to assess the performance in practice of an existing solar air heater, used mainly for the drying of onions. The commercially available flat collector, situated on the roof of a storehouse, is of the matrix type: see Fig. 1. The absorber consists of a porous, fibrous material. This allows air to be drawn through the absorber surface, ensuring a good transfer of heat and an even distribution of the air flow over the absorber surface.

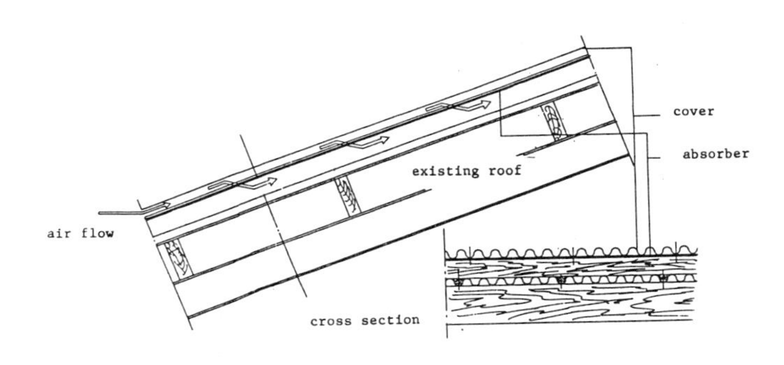

Fig. 1. Flat collector with porous absorber

The collector studied was the first of this type to have been installed in the Netherlands (1985). In 1986, three more have been installed. The absorber is fixed on top of the existing roof. The cover consists of corrugated polyester, reinforced by fibreglass, and has a transmittance of 0.80, according to the manufacturer.
A total of 240 m² of the roof of a storehouse is used for the collector. Unfortunately, this area had been distributed evenly over the two sides of a gabled roof, with a slope of 20°, thus exposing half of the area towards the northeast.

An important parameter in the design of solar air heaters is the flow rate in the system. A high flow rate has the advantage of reducing losses in the collector by keeping down the temperature of the absorber and of the air in the collector. Moreover, drying processes have a higher efficiency at high flow rates. The energy consumption of the fan, however, increases rapidly with the flow rate, as the pressure drop over the collector, the air ducts and the produce rises.

The system studied had been designed for a flow rate of 0.04 kg/s per m^2. The fan used is of the axial flow type with a flow rate of 7 to 9 m^3/s; it has an electric motor with a power of 4.0 kW.
Due to errors in the design of the duct system, however, the resistance of the duct system was too high, reducing the flow rate considerably. The pressure drop over the air ducts, now more than 50 % of the total pressure drop over the solar air heater and the produce, could be reduced by over 60 % by a simple improvement in the design.

Performance

During the drying season of onions (September/October) measurements of meteorological variables (global radiation, temperature, wind speed and direction) and of temperatures and flow rates inside the solar air heater were carried out. The efficiency of the collector was found to be approx. 54 % at a flow rate of 0.025 kg/s per m^2 and 30 % at 0.010 kg/s per m^2.

Economic evaluation

The investment costs for this type of collector are very low: approx. Dfl. 95 ($ 48) per m^2, including on-site assembly. In the case studied the cost of the air duct system was Dfl. 4300 ($ 2150). Taking into account that a larger fan is needed, the total investment cost for the solar system amounts to Dfl. 125 ($ 63) per m^2. Government subsidies, up to 40 % in the Netherlands, can further reduce the costs.
The 240 m^2 collector in the system studied has a mean efficiency of 42 %, yielding a heat output of 50 kW at a global radiation of 500 W/m^2. The fan uses 4 kW.
In a better optimized system a 120 m^2 collector with an efficiency of 54 % can provide a heat output of 32 kW at a global radiation of 500 W/m^2. In such a system the fan will use no more than 2 kW.
Application of this type of solar air heater is economically viable if the purchase of a conventional oil fired air heater can be avoided. In this case a pay-out time of 4 to 5 years is possible, even at the present low fuel prices.

DESIGN OF A LOW-COST SOLAR AIR HEATER

ECOFYS has designed a low-cost solar air heater for the drying of cereals. The collector in this case has no cover. The absorber consists of the dark corrugated steel roofing itself; the absorptance is 0.92. The air flows through a space, 4 cm wide, between the roofing and a layer of insulating material: see Fig. 2.

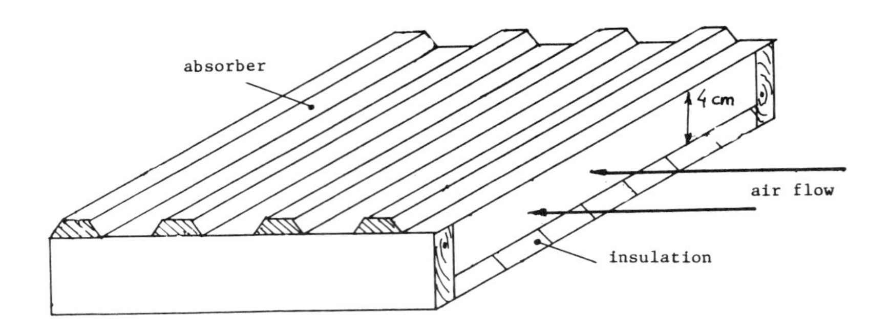

Fig. 2. The collector

The efficiency of a collector of this type strongly depends on the wind speed, due to convective cooling of the absorber. The effect of the wind speed on the collector efficiency at a flow rate of 0.015 kg/s per m² is shown in Fig. 3.a. The efficiency rises with increasing collector air flow rate. The calculated relationship is shown in Fig. 3.b, for a number of values of the wind speed.

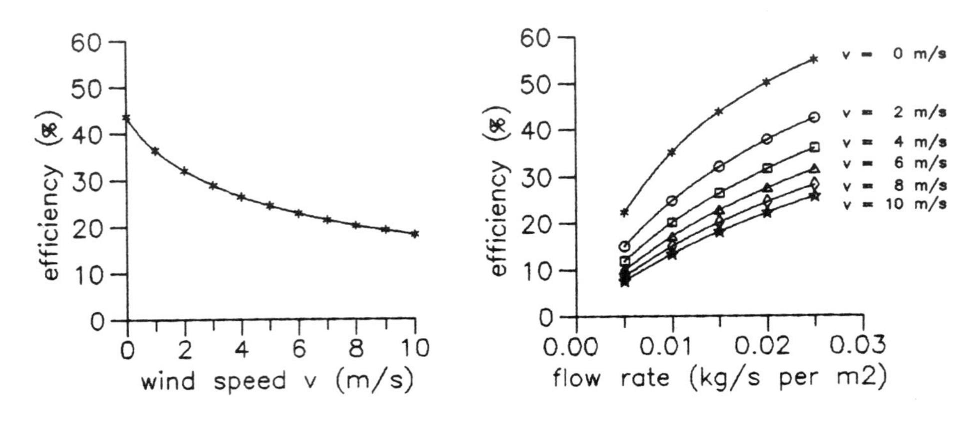

Fig. 3. Effect of wind speed v and collector air flow rate on collector efficiency

Since the electricity consumption of the fan rises strongly with increasing air flow rate, too, an optimization is required, taking into account the electricity costs of the fan and the drying capacity of the air from the system. A computer simulation over the drying season (July, August, September) showed the optimal flow rate to be 0.010 - 0.015 kg/s per m², under local wind speed conditions. The pressure drop over the solar air heater then amounts to approx. 40 % of the pressure drop over the grain.

Performance

Recently this solar air heater has been constructed. The collector has been installed on a tilted roof with a slope of 30°, exposed to the south. The total area is 350 m², which should be sufficient for a drying capacity of approx. 200 metric tons of grain per month.
At the moment measurements are being carried out. Preliminary results show that the efficiency of the system is 20 - 40 %, depending on the flow rate, in accordance with the calculated efficiency.

Economic evaluation

The investment costs for this type of collector are still lower than those of the collector described above: the total investment cost for the solar system amounts to Dfl. 60 ($ 30) per m², including air ducts and assembly.
A breakdown of the costs is given below. Government subsidies have not been taken into account.

		Dfl.	Dfl/m²	$/m²
Collector:	insulating material	6000	17.1	8.6
	remaining materials	2000	5.7	2.9
Air ducts:	materials	1000	2.9	1.4
System:	materials	9000	25.7	12.9
	on-site assembly	12000	34.3	17.1
	Total	21000	60.0	30.0

A 350 m² collector with an efficiency of 30 % has a heat output of 53 kW at a global radiation of 500 W/m². In a well designed system of this size the fan will use approx. 3 kW.

CONCLUSIONS

- In designing solar air heaters for drying purposes, an optimization is required, taking into account the electricity costs of the fan and the drying capacity of the air from the system. The optimization can be done by means of a computer simulation over the drying season.
- In order to minimize pressure losses, a careful design of the air ducts is necessary.
- System investment costs of $ 30 - 60 per m² are attainable.
- If the purchase of a conventional oil fired air heater can be avoided, reasonable pay-out times can be achieved at the current oil prices.

ACKNOWLEDGEMENT

This work has been done with support of the Dutch Project Management Office for Energy Research (PEO).

DRYING OF MEDICINAL PLANTS IN A GREENHOUSE TYPE SOLAR DRYER IN YUGOSLAVIA

J. Müller*, M. Tesić** and J. Kisgeći**
* Institute for Agricultural Engineering
Hohenheim University
Federal Republic of Germany
** University of Novi Sad (Yugoslavia)

A solar dryer for medicinal plants was developed on the basis of a plastic foil covered greenhouse. This dryer was tested in Yugoslavia in an area where medicinal plants are produced in large scale.
To decrease the investment costs of the drying unit the solar air heater is incorporated into the roof of a greenhouse construction. The solar air heater consists of a transparent plastic cover foil, a black textile fabric and an air-bubble foil, which avoids heat losses on the back. The foils as well as the textile fabric are connected onto the construction by a special weather strip fixing system. To enlarge the area of the solar air heater, the roof is extended on the southern side almost to the ground.

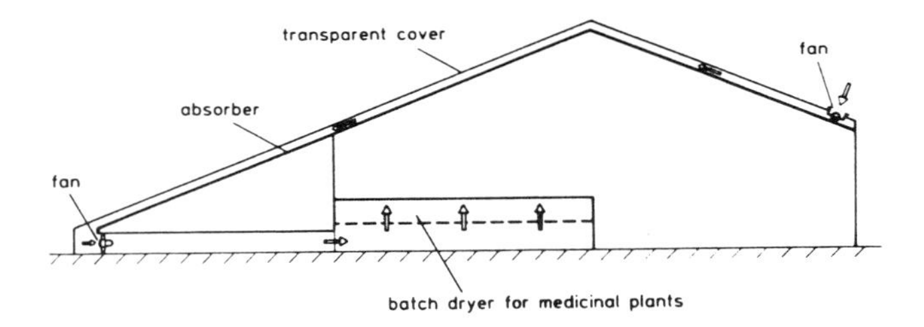

One of the major advantage of this solar air heater design is the low air resistance. Therefore only a small fan is required to direct the drying air through the collector. A supplementary fan forces the heated air through a batch dryer containing the drugs.
The pilot equipment, 15 m in width, 16 m in length and 4,5 m in hight, has a capacity of 3 tons of peppermint and 1,7 tons of camomile respectively.
The modular design allows simultaneous drying of various kinds of drugs and adaption of the system to altering yields. Compared to natural drying in the shadow the solar dryer reduces the drying time to almost 50%. By keeping the temperature on a permissible level the contents of ethereal oils is well preserved.
The building also makes some space for further processing of the drugs. In times the dryer being out of operation the building can be used as barn, machine shed or after removing of the absorber even as a conventional greenhouse for nursing medicinal plants.

DEVELOPMENT OF A MULTI-PURPOSE SOLAR CROP DRYER FOR ARID ZONES

K. Lutz and W. Mühlbauer*

* Inst. for Agr. Eng., Hohenheim Univ., Stuttgart, FRG

A B S T R A C T

A multi purpose solar crop dryer was developed for drying various agricultural products such as fruits, vegetables, medicinal plants etc. The new developed system consists of a small fan, a solar air heater and a tunnel dryer. The simple design allows the production either by farmers themselves using cheap and locally available materials or by small scale industries. Due to the low investment, the solar dryer is predestinated for application on small farms in lower developed countries.

Depending on the crop to be dried and the size of the dryer 100 to 1000 kg of fresh material can be dried within one to seven days to safe storage conditions. The solar dryer was successfully tested in Greece, Yugoslavia, Egypt, Ethiopia and Saudia Arabia drying grapes, dates, onions, peppers and several medicinal plants.

K E Y W O R D S

Solar drying, tunnel dryer, air heater, drying time, raisins,

INTRODUCTION

Sun drying is still the most common method to preserve agricultural products in most tropical and subtropical countries. Due to the lack of sufficient preservation methods, farmers have to spread the crop to be dried in thin layers on paved grounds or on mats where they are exposed to sun and wind.

Considerable losses may occur during natural sun drying due to various influences such as rodents, birds, insects, rain, storm and microorganisms. The quality of the dried products may also be lowered significantly. Overdrying, contamination by dust and insect infestation are typical for natural sun drying. The resulting decrease in product quality causes the products in question not to be marketable on domestic or international markets.

In the following a newly developed solar dryer which is based on a different principle as the solar dryers presented in the literature (Anonymous, 1975, Pablo,I.S., 1978, Szulmayer, W., 1971) will be described. Instead of being forced through a bulk, the air is directed over the crop which is spread in a thin layer, in order to reduce air resistance significantly (Lutz, K., W. Mühlbauer, 1987). This reduces the power consumption of the fan to an amount which can be covered either by batteries or solar cells.

OBJECTIVES

General objective of the investigation described in this paper is the development and testing of a solar tunnel dryer with integrated collector. Specific objectives included:

- Design of a low-cost solar drying system which can be assembled by farmers or craftsman using simple tools and relatively cheap materials.

- Development of a fastening system for the transparent cover foil which allows easy mounting and replacement of the foil.

- Determination of the influence of the weather conditions on drying time, energy consumption and uniformity in moisture content of the dried product in comparison to natural sun drying.

- Evaluation of product quality with respect to colour, taste, texture, insect infestation and contamination with foreign material.

- Economic evaluation of the solar dring system.

EQUIPMENT

The solar dryer consists of a small radial flow fan, a solar collector, and a tunnel dryer in which the crop is spread in a thin layer, Fig. 1. Heat is generated by absorption of solar energy on the absorber of the collector as well as on the crop itself.

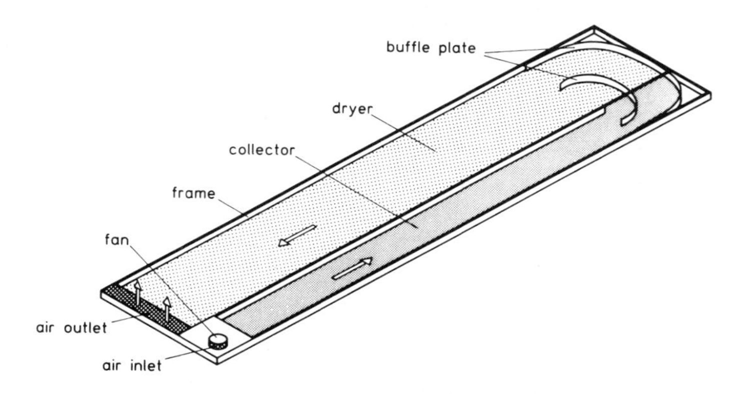

Fig. 1. Solar tunnel dryer with integrated collector

The frames of the collector as well as of the tunnel dryer are fixed on the ground. Both components are covered with a transparent foil. Black plastic materials are layed out between the walls of collector and dryer. To reduce conductive heat losses on the back side of the collector a heat insulator is installed underneath the absorber foil.

The drying air is forced by the fan through the collector, heated up by converting solar radiation, turned by 180 degrees and conducted into the tunnel dryer. To achieve the required uniform air distribution over the cross section of the dryer, buffle plates were installed at the reversing section. Wire mash at the air inlet as well as at the air outlet prevent the crop from insect infestation. During rain the air outlet can be closed with a flap. Operating the fan with closed flap prevents remoistening of the crop as well as it causes water flowing down the vaulted cover foil.

The solar dryer is 20.0 m in length and 2.0 m in width. The collector has approximately the same length and is 1.0 m in width. The holding capacity depends mainly on the bulk density of the crop and ranges from 100 kg when drying medicinal plants up to 1000 kg when drying grapes.
The high efficient fan with backward curved blades is driven by a 100 W AC-motor. PE-EVA foil is used as cover material. Block wooven glas fabric as absorber.

The air-bubble foil is provided with a weather strip fastening profile, which is welded to it along the long sides. The foil is fastened into the frame of the collector and dryer by pulling the weather strip into a PVC-profile which is fixed on top of the side walls, Fig. 2.

At the end wall of collector and tunnel dryer the foil is squeezed between frame and an angular steel. Since the cover foil of the tunnel dryer has to be removed for daily control, loading or unloading the crop, a weather strip fastening profile is fixed on a water tube. Thus the cover can be rolled up by only one man within less than one minute. After closing the cover the tube can be clamped underneath the profile of the frame as shown in Figure 5. The cover foil of collector and dryer can be tightened if necessary by distance holders which are installed every two meters. These distance holders also consist of water tubes. To prevent corrosion all metallic parts are galvanized.

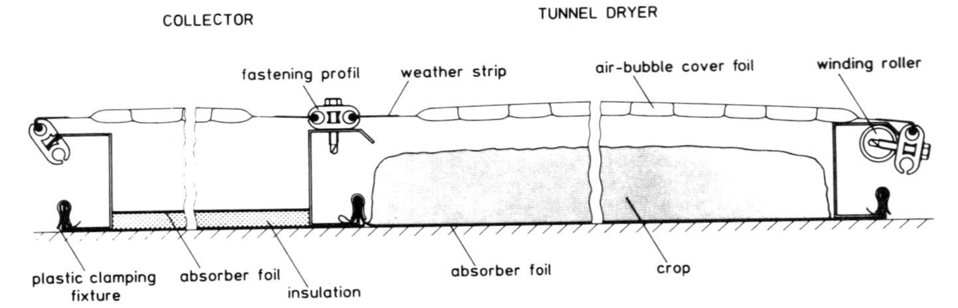

Fig. 2. Cross section of collector and dryer

If the crop which has to be dried is blanched or dipped into a chemical solution before drying it is recommended to spread the crop on a black perforated material which allows the drainage of the surplus of water.

If the solar dryer is assembled by farmers themselves, of course cheaper materials can be used for the construction. Single PE-foil can be used as transparent cover, black stones or pebbles as absorber and rice husks or dry leaves as insulation material.

EXPERIMENTS

In the following, experiments are described exemplarily for the drying of grapes. The tests were carried out in Greece in 1984 and 1985. Similar investigations were conducted in Yugoslavia drying medicinal plants in 1986. Furthermore on-farm tests were carried out in Ethiopia, Egypt and Saudi Arabia drying various fruits and vegetables.

Instrumentation

Sensors were arranged for continuous measuring of direct and diffuse radiation, temperature and relative humidity of the ambient air, inlet and outlet air of collector and dryer, and the wind velocity. To measure temperature, humidity and static pressure of the drying air along the length of collector and dryer respectively, additional sensors were placed at various points. All data were registered in ten minutes intervals by data logger and recorded on disc.

The weight loss of the grapes was determined by weighing 15 samples which were located at various points in the drying area.

Material

The drying tests were carried out with freshly harvested Thompson seedless sultana grapes. Depending on the stage of maturity of the grapes the initial moisture content ranged between 74 and 78% w.b. and the acidity between 3.5 and 8.0% (Eissen, W., 1984). The raisins have to be dried to a final moisture content of at least 18% w.b. To increase the drying rate the grapes were pretreated in a chemical solution.

Results and Discussion

Collector efficiency. The reference method served to test the thermal performance of the collector used for heating the drying air. Fig. 3 shows the effectiveness of the use of an air-bubble foil as cover material in comparison to the use of a single PE-foil cover. The plots indicate that the air-bubble foil has certain advantages at the relatively high operating temperatures which are needed when the collector is used in grape drying.

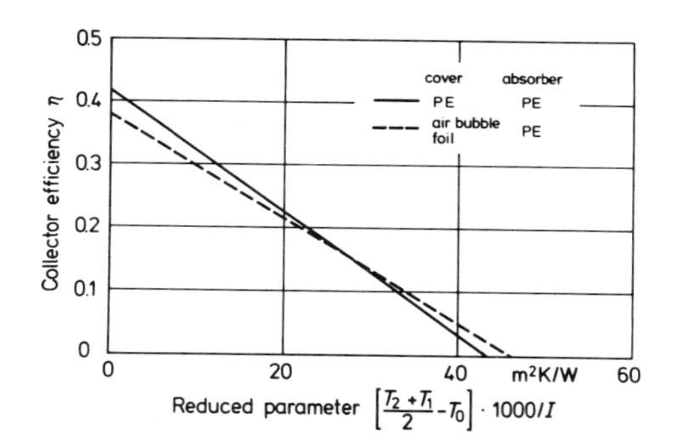

Fig. 3. Influence of the cover material on collector efficiency versus normalized temperature rise

Operating conditions. To prevent undesired discolouration and spoilage of the product the dryer has to be operated continuously during the first two days. During this period a relatively high air-flow rate of approximately 1200 m^3/h equivalent to an air velocity of 3 m/s inside the tunnel dryer is required to remove the water which evaporates at a high rate during the first phase of the process. Tests have shown that enzymatic reactions during this phase, which occur when the fluid is stagnant, can only be prevented by running the fan throughout the night. After two days of drying the air-flow rate can be reduced to 600 m^3/h and ventilation is not longer necessary during the night.

Fig. 4 shows the drying air temperature profile in collector and tunnel dryer. The plots indicate that the temperature seems to rise almost constant with only a slight decrease at the air outlet.

This tendency which is characteristic for the solar tunnel dryer with integrated collector can be explained as follows.

As the grapes absorbs 82 to 85 percent of the incoming solar radiation, the reduction in temperature as a result of the evaporation is almost compensated by converting solar energy into thermal energy.

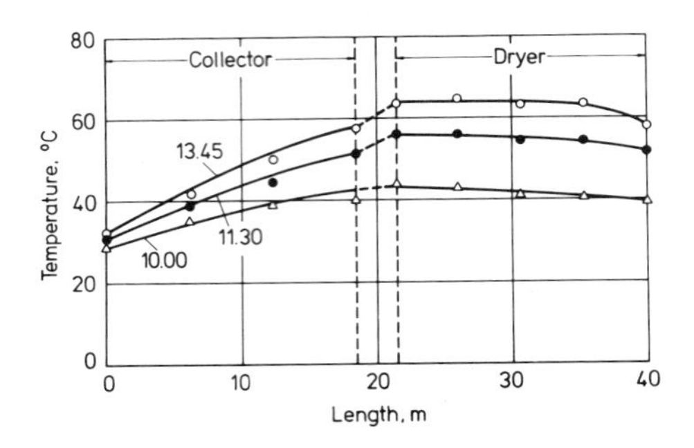

Fig. 4. Temperature profiles in collector and dryer (Air-flow rate: 600 m^3/h)

Drying time. Depending on the weather conditions, solar drying takes 4 to 7 days. Fig. 5 exemplarly shows the decrease of the moisture content of grapes dried in the solar tunnel dryer in comparison to natural sun dried grapes.

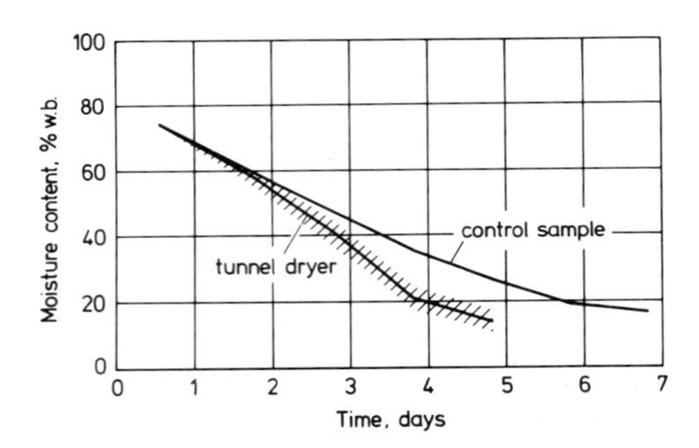

Fig. 5. Drying behaviour of solar and natural sun dried grapes

Especially at the end of the drying season the reduction of the drying time is of great importance. In order to be dried to safe storage conditions with natural drying methods, drying has to be completed when the humid season starts.

Energy costs. An energy source for driving the fan is essential for operating the solar drying system, because the evaporated water cannot be removed by natural convection or by wind (Eissen, W. a. W. Mühlbauer, 1983). The tests have shown that only an energy source of 80 to 250 W is necessary to force 1200 m^3 air per hr through collector and dryer. Such low energy requirement results from an extremely low pressure drop in collector and dryer of less than 10 dPa.

<u>Quality</u>. All tests have shown that solar drying leads to improved quality compared to the traditional drying methods. As the grapes do not come into contact with insects during drying, chemical treatment with insecticides is not required before storage.

<u>Economics</u>. The following cost calculation is based on a sales price of 1250 US$ which is achieved in serial production. Assuming a constant interest rate during the life cycle of 10 percent, a life span of 5 years for the materials and of 10 years for the fan leads to capital annuities of 330 US$. The yearly operating costs of 75 US$ include the labour for mounting and dismounting the solar system, the energy costs for operating the fan and also the costs for renewing the absorber foil. In this calculation operating costs such as those for dipping materials, the labour required for dipping, loading and unloading the dryer, and as for supervision are not included because there is no significant difference between the solar drying system and the traditional methods.

The benefits of the solar dryer are in the form of increased revenue as a result of improved quality and reduced mass losses. Assuming that 2000 kg of raisins can be dried annually, 10 percent mass losses occur during natural sun drying, and the better quality of the solar dried raisins affects the sales price by 0. 2 US$ per kg, an increase revenue of 600 US$ can be calculated. Based on the annuities of expenditure and the additional revenue a pay-back period of less than three years can be expected. Assuming do-it-yourself assembling the pay-back period can be reduced to less than one year.

This demonstrates the profitability of the solar drying system and leads to the conclusion that the introduction of solar tunnel dryers seems to be a way of increasing the income of raisin producing farmers in Greece.

CONCLUSIONS

Compared to the traditional sun drying methods the newly developed solar tunnel dryer with integrated collector shows the following advantages:

- The drying time can be reduced significantly by converting solar energy into thermal energy in collector and tunnel dryer.

- During drying the crop is completely protected against rain, dust, insects and animals which leads to the desired reduction of the mass losses and also improves the quality of the dried product.

- The simple system enables manufacturing either by small scale industries or by farmers using locally available materials.

- A cost-benefit analyses shows that the pay-back period ranges from 1 to 3 years taking into account the reduction in mass losses and the improved quality of the product.

- The multi-purpose solar dryer can be used for drying various crops such as vegetables, fruits, medicinal plants, herbs and spices.
- In comparison to the additional earnings the electricity costs for operating the fan are negligible.

At present the use of the solar dryer is limited to electrified areas. Yet, the very low power consumption of the fan offers the possibility to use batteries or solar cells as energy source. The solar dryer with integrated collector was originally developed for use in arid zones. With a few modifications it also can be used in humid tropical countries.

REFERENCES

Anonymous (1975). A survey of solar agricultural dryers. Technical Report T 99, Brace Research Institute. McGill. Univ. Quebec, 151 pp.

Eissen, W. (1984). Trocknung von Trauben mit Solarenergie. Research Report T 84 - 089, Federal Ministry of Research and Technology, Bonn.

Eissen, W., W. Mühlbauer (1983). Development of Low-cost solar grape dryers. Proceedings of the International Workshop on Solar Drying and Rural Development, Bordeaux, 263-272

Lutz, K., W. Mühlbauer (1986). Solar tunnel dryer with integrated collector. Drying Technology, Vol. 4, Nr. 4, Marcel Dekker, INC., New York and Basel, 583-603

Pablo, I.S. (1978). The practicality of solar drying of tropical fruits and marine products for income generation in rural areas. Proceedings of the Solar Drying Workshop, Manila, 263-272

Szulmayer, W. (1971). From sun-drying to solar dehydration. Food Technology Australia, 23, 23-25

AIR COLLECTORS FOR HAY DRYING

Effect of collector cover material, collector size and air flow rate on the efficiency and the pressure drop

V.Kyburz, J.Keller
Swiss Federal Institute for Reactor Research
CH-5303 Wuerenlingen / Switzerland

ABSTRACT

Air collectors for hay drying are more and more installed on Swiss farms. They are composed of the roof cover used anyway and plywood boards nailed at the bottom side of the rafters. Ambient air is sucked through the duct in between, heated up and blown into the haystack. To establish a collector design basis for agricultural consultants, the performance of four collector types covered with different materials has been investigated. Four collector modules of 10 m^2 each have been built, each of them being provided with movable bottom boards enabling a varation of the air duct height.

For specific air flow rates of 100, 150 and 200 $m^3/h\ m^2$ and for duct heights between 4 and 10 cm the collector efficiency and the pressure drop have been measured. From these results the thermal performance of collectors with sizes and duct heights used in practice was calculated. The efficiency conversion from the collector module to the whole roof is based on a model involving the invariance of the heat transfer coefficients. The air pressure drop is obtained using the well known laws for pressure drops in tubes.

KEYWORDS

Air heating; solar drying; hay drying; air collectors; air collectors in agricultural applications; thermal performance of air collectors; modelling; design basis

INTRODUCTION

Heating air by means of solar air collectors is of particular interest in those cases where the heated air is directly used (i.e. without heat exchangers or stores). A promising field of application is the drying of hay and other agricultural goods, the temperature of the heated air being of the order of the ambient temperature. Because of the low temperature level these installations require relative simple collector roofs. However, the performance of theses roofs depends significantly on parameters like roof cover material, collector size, air duct height, air flow rate etc. Therefore it is important to design the collector properly by determining the optimum values of these quantities.

In order to test experimentally the thermal performance of such air collectors, a twin outdoor test facility has been erected which allows the testing either in an open loop or in a closed loop mode. By using two completely independent test loops it is possible to test two collectors in parallel, using one as a reference in order to correct the test data for meteorological changes. The main features of the test facility are:

- collector area : up to 10 m^2
- air flow rate : 80 - 2000 m^3 / h
- inlet temperature : 0 - 50 ^0C

Four collector modules of 10 m^2 (L = 5 m, W = 2 m) each have been built (see Fig. 1). Each of them is equipped with movable bottom boards in order to enable a variation of the duct height H:

- corrugated, brown coloured roof cover "Eternit" (opaque), air stream across the corrugation (type 1). Rem.: "Eternit" is a kind of fibercement
- corrugated, brown coloured roof cover "Eternit" (opaque), air stream along the corrugation (type 2)
- corrugated, brown coloured aluminum roof cover (opaque), air stream across the corrugation (type 3)
- corrugated polycarbonate roof cover (transparent), air stream across the corrugation (type 4)

This program was financially supported by the Commission for the Promotion of the Scientific Research (KWF), the National Energy Research Foundation (NEFF) and the Swiss Federal Institute for Reactor Research (EIR).

PERFORMANCE OF THE COLLECTOR MODULES USING OUTDOOR EXPERIMENTS

Outdoor Experiments

Solar air collectors for hay drying are operating in the open loop mode, i.e. ambient air is sucked into the collector. Thus the collector inlet temperature T_i is equal to the ambient temperature T_a. The basis of the test procedure is the well known collector equation

$$(1) \quad \eta = F_R \, \alpha \, \tau - F_R \, U_L \, [(T_i - T_a)/G] \; ,$$

α and τ being the absorptance of the absorber and the transmittance of the transparent cover, respectively, and G being the irradiance. With $T_i = T_a$, Eq. (1) reduces to

$$(2) \quad \eta = F_R \, \alpha \, \tau$$

It can be shown that the heat removal factor F_R of a collector of vanishing thermal inertia is only a function of the involved heat transfer coefficients (i.e. of collector design and air velocity v) and of the air flow rate. From Eqn. (2) it follows that the efficiency η is expected to be independent of irradiance. Thus a test under clear sky conditions may be performed during the whole day and not only during a short period around noon. Based on temperature difference, flowrate and irradiance the collector efficiency is calculated over the whole day. Fig. 2 shows the efficiency of two modules during a clear day. Due to the thermal inertia of the collector the efficiency is a slightly increasing function of time. In order to get a steady state efficiency η for the test day, each measured (non steady state) efficiency $\eta_{meas}(t)$ is corrected according to

$$(3) \quad \eta = \eta_{meas}(t) + \tau_{coll} \, [d\dot{q}_{meas}(t) \, / \, dt] \, / \, G(t) \quad ,$$

$\dot{q}_{meas}(t) = \eta_{meas}(t) \, G(t)$ being the measured thermal power per unit collector area. The time constant τ_{coll} is to be varied until the linear least square fit yields a time-independent function, i.e. the efficiency has a constant, representative value (except at early morning and late afternoon time). Values for τ_{coll} were found to be of the order of 20 to 40 minutes.

Fig. 1 : Collector modules on the outdoor test facility for air collectors

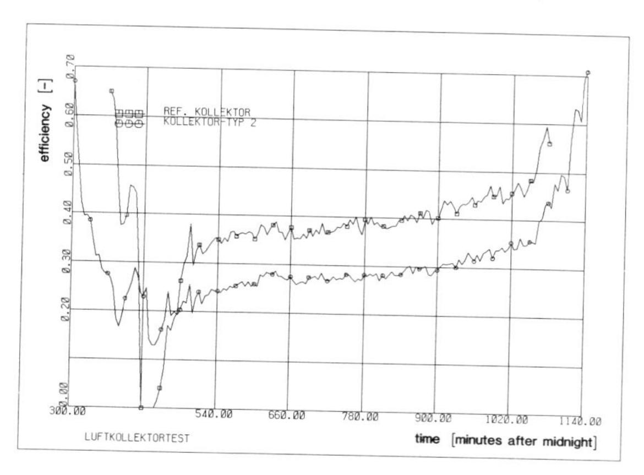

Fig. 2 : Example of an efficiency plot (2 collector modules)

Applying this procedure, the collector efficiency η was determined for 3 different specific flow rates $\dot{w}$ (around 100, 150 and 200 $m^3 \, / \, h \, m^2$) and 3 different duct heights H (4, 6 and 10 cm). For the same set of parameter values the pressure drop Δp was measured.

Data Representation by Analytical Functions

Any analytical function representing the efficiency as a function of w must yield $\eta = 0$ for $\dot{w} = 0$ and η = constant for $\dot{w} = \infty$, reflecting the behaviour of F_R. In general the efficiency may be expressed as follows:

$$(4) \quad \eta = \alpha' \; \tau \; \{1 - \exp[-\kappa(\dot{w},H) \; w]\}$$

For opaque collector covers ($\tau = 1$) the absorptance α is to be corrected for the heat conductivity of the roof cover as follows:

$$(5) \quad \alpha' = F_\lambda \; \alpha$$

$$(6) \quad F_\lambda = 1 \; / \; (1 + U_t \; d/\lambda_d)$$

The correction factor F_λ depends on the thickness d of the cover, on its heat conductivity λ_d and on the top loss coefficient U_t. For transparent collectors, $\alpha' = \alpha$ holds. The coefficient κ in Eqn. (4) turned out to depend on specific flow rate $\dot{w}$ as well as on duct height H.

In a similar way the pressure drop may be expressed as a square function according to the friction laws in tubes:

$$(7) \quad \Delta p = \lambda(\dot{w},H) \; [L/d_h] \; \rho \; v^2/2$$

where d_h is the hydraulic diameter of the collector and $\lambda(\dot{w},H)$ is a coefficient analogous to $\kappa(\dot{w},H)$.

THERMAL AND HYDRAULIC PERFORMANCE OF WHOLE COLLECTOR ROOFS

To convert the test data of the collector modules into the performance data of whole collector roofs, a calculation procedure based on similarity has been developed.

Efficiency. From Eqn. (2) it is obvious that two collectors of a given type, but different sizes have the same efficiency only if their coefficients F_R are equal. Thus for a given specific flow rate w each heat transfer coefficient involved must have the same value for both the collector roof in practice (index 1) and the collector module unter test (index 2). Assuming the relation $Nu \propto Re^{0.8}$, the ratio of the duct heights H_1 and H_2 of collector 1 an 2 can be calculated according to

$$(8) \quad H_1 \; / \; H_2 = (L_1 \; / \; L_2)^n$$

If one states that the collector is modeled as two parallel plates, then $n = 0.75$. If it is modelled as a rectangular tube, then $n = 0.80$. This small difference is quite astonishing for the ratio of the characteristic lengths used in the two models is of the order of 50 to 100.

Conversely, the efficiency of a roof of given size and duct height may be calculated by determining the correspondent duct height of the tested collector module. By using Eqn. (4) and the experimentally determined function $\kappa(\dot{w},H)$, the efficiency is obtained.

Pressure drop. Two collectors 1 and 3 are similar if $Re_1 = Re_3$ and $L_3 \; / \; L_1 = H_3 \; / \; H_1 = B_3 \; / \; B_1 = l_v$. For the ratio of the correspondent pressure drops Eqn. (9) holds:

$$(9) \quad \Delta p_1 \; / \; \Delta p_3 = l_v$$

The ratio of the specific flow rates is calculated by

$$(10) \quad \dot{w}_1 / \dot{w}_3 = l_v$$

To calculate the pressure drop Δp_1 of a collector 1 of given dimensions L_1, B_1 and H_1, the duct height H_3 and the specific flow rate $\dot{w}_3$ of the similar test collector 3 are determined using Eqn. (10) (L_3 = 5 m). The pressure drop Δp_3 has been determined by experiment as a function of H_3 and $\dot{w}_3$ according to Eqn. (7). Hence the pressure drop Δp_1 of the collector in practice is then given by Eqn. (9). It has to be mentioned that the duct height H_3 of the "similar" test collector module is different from the duct height H_2 of the test collector module with the same efficiency as the collector roof in practice.

RESULTS AND CONCLUSIONS

By using the transformation laws discussed above, efficiency and pressure drop diagrams for arbitrary collector dimensions were calculated. Fig. 3 shows an example of the efficiency and pressure drop diagrams which may be used by agricultural consultants in order to design a solar hay drying installation. The hatched area in the efficiency diagram is a forbidden region, where the pressure drop is above an allowed limit of 150 Pa.

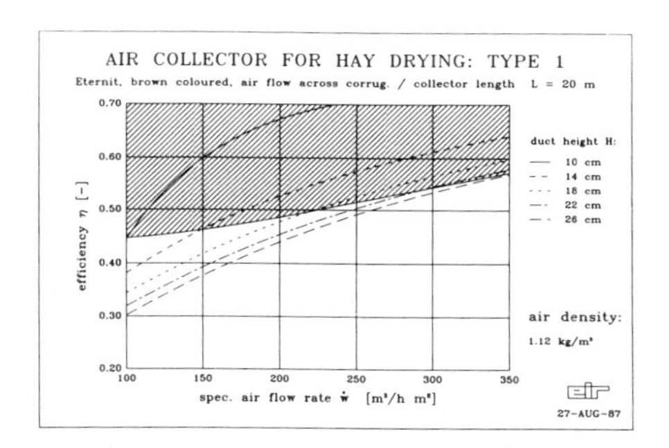

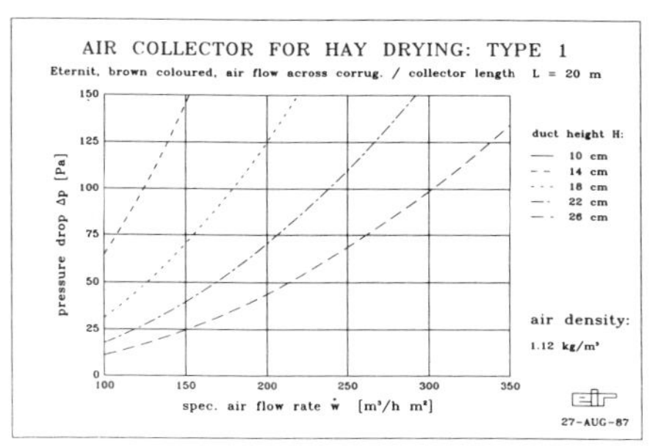

Fig. 3 : Efficiency and pressure drop curves for the "Eternit" covered collector (type 1). L = 20 m.

The results show that the efficiency is an increasing function of the specific flow rate as well as of the collector length L and a decreasing function of duct height H. As an example, for opaque roof covers (i.e. "Eternit" or aluminum) of L = 20 m at specific flow rates around 200 $m^3/h\ m^2$ (recommended in Switzerland) the efficiency amounts to 40-55%, H varying between 26 and 14 cm. The collector type with the transparent cover shows, for the same range of duct heights, a higher efficiency range of about 55-65%. By increasing the duct height by, e.g., 4 cm the efficiency of the opaque types are decreased by 3-5% whereas in the case of the transparent type the decrease is only 1-2%.

The results of the above model have been compared with the performance of a real collector roof measured in 1980 (Eidg. Forschungsanstalt für Betriebswirtschaft und Landtechnik, Tänikon, Switzerland). This collector is provided with a transparent cover (i.e. type 4) and the collector area is 250 m^2 (L = 20 m, B = 12.5 m). The average specific flow rate over the time of operation was 206 m^3 / h m^2 leading to an average efficiency of 0.62. The value predicted by the model is 0.61, which is in very good agreement with the in situ measurements.

SOLAR DRYING WITH A NEW HYBRID COLLECTOR

M. Reuss, T. Schmalschlaeger, Dr. H. Schulz
Landtechnik Weihenstephan, Technical University of Munich
D-8050 Freising, F.R. Germany

1. ABSTRACT

This paper describes the concept and the design of a solar drying plant for vegetables on an industrial scale, which will be built in 1988 in the south of Spain. The project is sponsored by DG XVII of the European Community as a demonstration project.

The innovation in this project is mainly the use of solar energy in the production of dried vegetables on an industrial scale and in an improvement of the drying techniques by introducing a multistage drying process, which is charcteristic for a fully controlled drying process. 3 different types of solar collectors are used. For the predrying (1. stage; 80°C) an air collector with selectiv absorber is used. The main drying (2. stage, 60°C) is like the predrying bivalent heated - solar and conventional with an air collector with black aluminium absorber. For the 3rd stage (40-60°C) only solar energy is used. For this a new hybridcollector for air and water-heating is used, which was developed at the Institute of Agricultural Engineering of the Technical University of Munich. This collector provides the warm air for the 3rd stage of the drying process and heats up parallel or alternative, depending on the heat demand of the drying prozess and the meteorology, a 100 m^3 water tank. The stored energy covers the hot water demand of the factory and provides the drying during night with solar energy. This is a shorttime storage for 24-48h.

The expected amount of energy savings is about 7250 MWh/a, about 1/2 of this is due to the use of solar energy the rest due the heat recovery system and multistage drying. The primary energy demand will be reduced by 2/3's compared to a conventional plant.

The capital return period is calculated to be about 10 years, which is about half of the expected lifetime of 20 years of the complete installation.

2. KEYWORDS

Industrial drying, solar drying, multistage drying, air collectors hybrid--collectors for air and water heating, heat recovery, shortterm storage.

3. INTRODUCTION

Conservation and distribution of food from superfluity to penury areas could really help to solve the problems of human nourishment. Modern drying techniques, especially solar drying, is a procedure of great importance for developing countries. Here we have to keep the nature products from deteriorating by rotting without consuming scarce foreign currencies. Nearly all kinds of fruits and vegetables could be dried. Fast withdrawal of moisture from the product, reducing time between harvest and drying determine quality and appearance of the product. The main procedure used in vegetable drying is thermal drying in one or more stages. The ideal case of a drying process, which takes care of the quality of the product, is a fully controlled drying process. There is a big market for these high quality products especially in Europe and USA.

Based on this situation a plant for drying of vegetables on an industrial scale was designed and will be built nearby Malaga in Spain. This paper will therefore describe the concept and design of the installation and its components. The whole project will be sponsored by DG XVII as a demonstration project.

4. DESCRIPTION OF THE PLANT

The purpose of this project is the demonstration of a multistage drying process with the use of solar energy on an industrial scale. The plant will consist of a solar system with air heating and hybrid collectors, a drier, a heat recovery system and a conventional heating system.

4.1 Drying System

There are two different methods of drying:

- mechanical drying, like pressing etc.
- thermal drying, the water of the product is evaporated, transported to the surface and transfered to the ambient air by thermal energy

The ideal case would be a fully controlled drying process. That means the temperature and humidity of the drying air is controlled according to the momentary temperature and humidity of the product. Experiments have shown that this ideal situation could be nearly approached by a multistage drying process.

In this project 3 steps are used:

- pre-drying 80°C; 1-3 h drying time
- main-drying 60°C; 1-4 h drying time, final product humidity 10-15 %
- final-drying 40-60°C to a final product humidity of 5-6 %

For the first 2 stages tray-driers are used with a bivalent heating system-- solar air heating collectors and a backup system fired with heavy fuel. The 3rd stage is performed in special waggon driers heated monovalent by solar energy. This is possible, as after the 2nd stage the vegetables have a humidity of 10-15 % and could so be stored for some hours without losses of mass and quality even under bad weather conditions.

The trays in the pre-and main driers are moving in steps from the top to the bottom, the air flow goes in the opposit direction (counterflow). The contents of two pre-driers is filled into one main drier, and the contents of one or two main-driers is filled into one final-drier-waggon.

This third stage, requiring the highest amount of energy per kg water to be evaporated but at a rather low temperature level of 40-60°C could be done monovalent with solar energy. The massflowrate is the highest one of the whole drying process. Due to the low temperature level and the low humidity heat recovery and recirculation as in the first two stages is not possible. These drying waggons are connected to a fixed air duct till the product reaches his final humidity.

4.2 Solar System

The whole solar heating system utilizes three different typs of collectors.

- air collector with selective absorber - 2000 m^2
- air collector with black painted absorber - 920 m^2
- hybrid collector with selective absorber - 1400 m^2

The collector will be integrated in the roof of the factory building. The standard type air collector has a metal-frame-backside with an incorporated PU-foam-plate as insulation. Corrugated steel (selective coated) or aluminium (black painted) sheets mounted on this frame serve as absorber. As glazing corrugated PMMA-sheets are used. Fig. 1 gives an overview of the collector construction and in table 1 data measured at our test facility are given.

The hybrid collector for air and water heating (fig. 2) has also an absorber made of stainless steel with selective surface. The double glazing is made of an PETP-foil (inner) and also PMMA- sheets (outer).

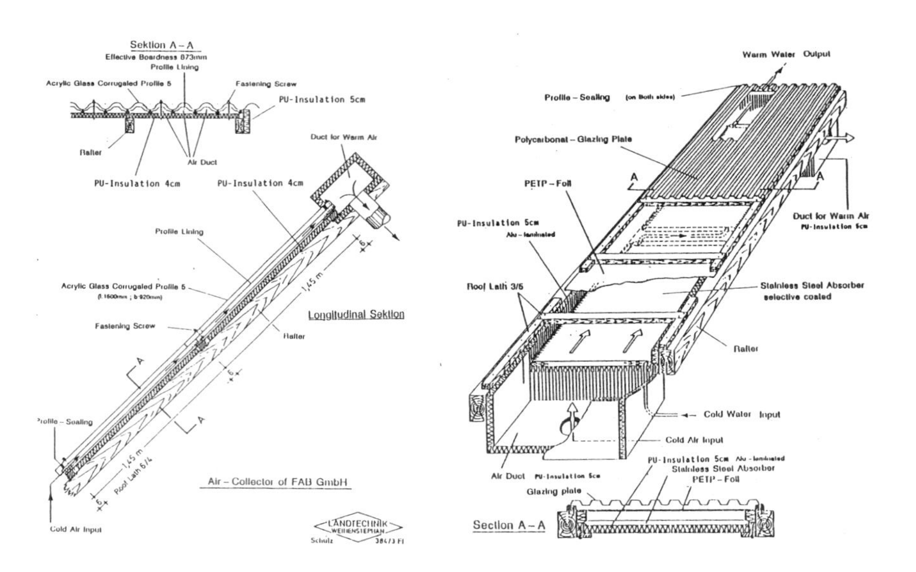

Fig.1 FAB solar air collector

Fig.2 Hybrid collector

collector	flowrate (kg/hm²)	no	UL (W/m²K)
Air coll. black	70	0.76	9.4
Air coll. select.	70	0.76	8.7
hybrid coll.			
air operation	70	0.41	7.8
water operation	50	0.64	3.7

Table 1. Collector performance data

Two water tanks of 50 m³ each are used as heat storage (day to night). These tanks were well insulated with a layer of 200 mm PU-foam. The maximum operating pressure is 2.5 bar, the max. temperature is 95°C, operating temperature will be 30-70°C.

4.3 Capacity of the Plant

This plant is designed to dry about 10000 t/a fresh vegetables respectively 660 t/a dry product. The starting humidity is 80-95 %, the final is 5-6 %. The specific energy demand per kg dry product is 25-55 MJ/kg. The mean daily operation time is about 22 h, which yields a yearly time of operation of 5500 h/a.

F1	= Area of collector 1 (sel. absorb.)	2000	m²
F2	= Area of collector 2 (black absorb.)	900	m²
F3	= Area of collector 3 (sel. absorb.)	1400	m²
QS1	= Insolated energy coll 1 (apr-feb)	2970	MWh/a
QS2	= Insolated energy coll 2 (apr-feb)	1456	MWh/a
QS3	= Insolated energy coll 3 (apr-feb)	2079	MWh/a
QK1	= Useful energy from coll 1	1188	MWh/a
QK2	= Useful energy from coll 2	728	MWh/a
QKL	= Useful energy from coll 3 (air)	312	MWh/a
QKW	= Useful energy from coll 3 (water)	624	MWh/a
QSS	= Useful energy from the storage	530	MWh/a
QVS	= Heatlosses from the storage	94	MWh/a
n1	= Mean daily efficiency coll 1	0.40	
n2	= Mean daily efficiency coll 2	0.50	
n3	= Mean daily efficiency coll 3 (mixed)	0.45	
QWR	= Energy from heat recovery	1991	MWh/a
QWV	= Energy in the exhausted air	5992	MWh/a
EWT	= Power of the heat exchangers	308	kW
QHÖ	= Energy of the heavy fuel	4492	MWh/a
QHN	= Useful energy from the heavy fuel	3234	MWh/a
QHV	= Heatlosses of the heating system	1258	MWh/a
nHT	= Effic. thermal oil heating syst.	0.72	
nHD	= Effic. vapor heating system	0.63	
QTB	= Used primary energy (heavy fuel)	396	t/a
QTN	= Energy demand of the drier	7983	MWh/a
QTA	= Energy leaving the drier	7983	MWh/a
MT	= Mass dried products	660	t/a
MN	= Mass fresh products	10000	t/a
TB	= Time of operation of the plant	5410	h/a

Table 2. Energy balance - data

With meteorological data (mean values over 10 years), the available data from collector tests and requirements on capacity of the user an energy balance was made for the solar drying system. The plant consists of 3 different collectors, 2 air heating and 1 hybrid collector, a heat recovery system of air to air heat exchangers, a water storage tank, backup heating system with thermal oil as heat transfer fluid, tray driers for pre- and main drying and waggon drier for the third stage.

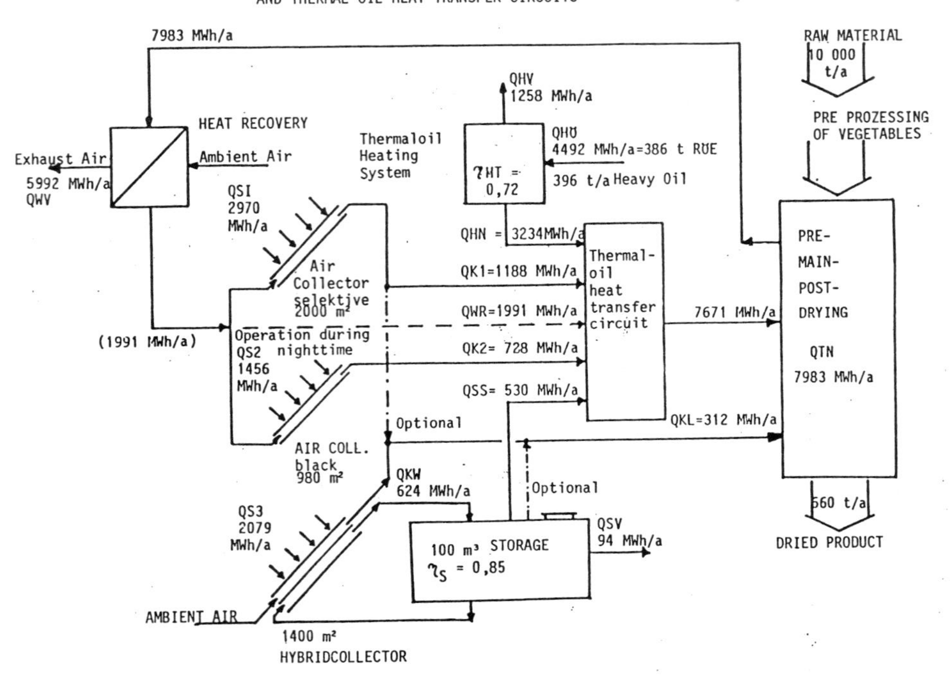

Fig. 3 Energy balance

4.4 Economical Considerations

The savings of primary energy of this plant in comparison to a conventional plant with thermaloil as heat transfer fluid is about 7750 MWh/a (heavy fuel) with a value of almost 100 000.- ECU/a (case 1); compared to a conventional plant with steam as heat transfer fluid to the driers the savings are 120 000.- ECU/a (case 2). With cost for operation and maintenance of 30 000.- ECU/a a capital return of 13.6 year in the first case and 9.1 years in the second case is calculated. In further projects a reduction of investment of 10 % is expected.

5. CONCLUSIONS

This project should demonstrate the drying of vegetables on an industrial scale with improved drying process mainly with the ecologically benificial utilization of solar energy. A non-conventional innovative 3 stage drying process is used, where each stage has it's own specific required temperature and flowrate. An energy balance of one year shows, that a plant like this designed for 660 t/a of dried product has an energy demand of almost 8000 MWh/a. About 3400 MWh/a are delivered from solar energy. The rest is delivered by heat recovery and backup system.

The solar energy is used with two big air collector fields - 2000 m^2 with selective coating and 980 m^2 with black painted absorber. 1400 m^2 of a novel hybrid collector for air and/or water heating will supply the third stage of the drying process solely with solar energy. The solar fraction will be 65 % of the total energy demand.

A simple economical estimation shows that the time to return the investment will be in the order of 10 years. Hopefully the plant, which will be built in 1988, will validate rove all the design considerations and help to reduce energy demand of agriculture and the food processing industry.

6. REFERENCES

M. Reuss, E. Thome; Solar Drying of Vegetables with a New Hybrid- Collector for Air and Water Heating; proposal for a demonstration project of DG XVII of CEC; 1986
Landtechnik Weihenstephan, Technical University of Munich, 8050 Freising; Deutsche Babcock Anlagen AG, 4150 Krefeld- Uerdingen; F.R. Germany

PERFORMANCE MONITORING ON A SOLAR DRYING PLANT FOR HERBS

M. Reuss, T. Schmalschlaeger, Dr. H. Schulz
Landtechnik Weihenstephan, Technical University of Munich
D-8050 Freising, F.R. Germany

1. ABSTRACT

Drying of agricultural products with solar air heating collectors is one of the main applications in agriculture.

A solar drying plant for herbs with an air collector area of 530 m^2 was built on a farm north of Munich. Ambient air is preheated in the collector, then heated up to the required temperature with the conventional heating system and led into 5 - peforated conveyor - drier. Additionally a heat recovery and air recirculation system was installed.

The different products to be dried require different temperature levels. On nice days about 50 % of the insulated energy could be delivered to the drying system. Depending on the temperature level (50 - 100°C) the solar fraction varies between 40 - 20 %. In 1986 about 6000 l fuel could be saved by the solar system.

The payback period of the solar system is about 14 years, the energy payback time only 1.5 years.

2. KEYWORDS

Solar air heating collector, solar drying of agricultural products performance of air collectors.

3. INTRODUCTION

One of the most important and promissing from an economical point of view applications of solar energy in agriculture is the field of drying. Especially because favourable agreement between the supply of solar radiation and the heat demand for drying exists. In a project, partly sponsored by the EC and partly by the German Ministry of Research and Technology, a solar drying plant for herbs was monitored. This plant was built in 1981 on a farm north of Munich in a herb growing region. There are several different leafy products like peppermint, parsley, majoram, chervil and spinach to be dried with the plant. Due to this fact the installation is almost used to capacity during the drying season from May to November. This long period of use has a favourable effect to the economy of the whole system.

The monitoring programme should give an overview over all energy flows in the plant and the performance of the air collector field. The experience gained from this work will be introduced to the planning of further solar drying plants, carried out at our institute or by the agricultural extension service.

4. DESCRIPTION OF THE PLANT

The total installation consists of four main parts- the drier itself, the conventional heating system, the heat recovery and the roof integrated solar air heating collector field.

The solar collector, integrated in the roof, double glazed air heating collector with air permeable absorber. The area is about 530 m² with an inclination of 27° facing south Fig. 1 shows a scheme of the collector. The back of the roof structure is covered with wood fibre boards. An air permeable black polypropylen net mounted inbetween the rafters serves as an absorber. The air entering the collector at the eaves is passing through the absorber several times on the way to the collecting duct just below the ridge. The collector has double glazing, the inner one is made of a temperature resistant PETP-foil (polyester), as outer one corrugated PMMA sheets (acrylic glass) are used. This covering is expected to maintain performance over a long period of about 20 years. At the beginning the collector was covered with a GUP sheet (polyester), which degrated very fast and has to be replaced after 5 years.

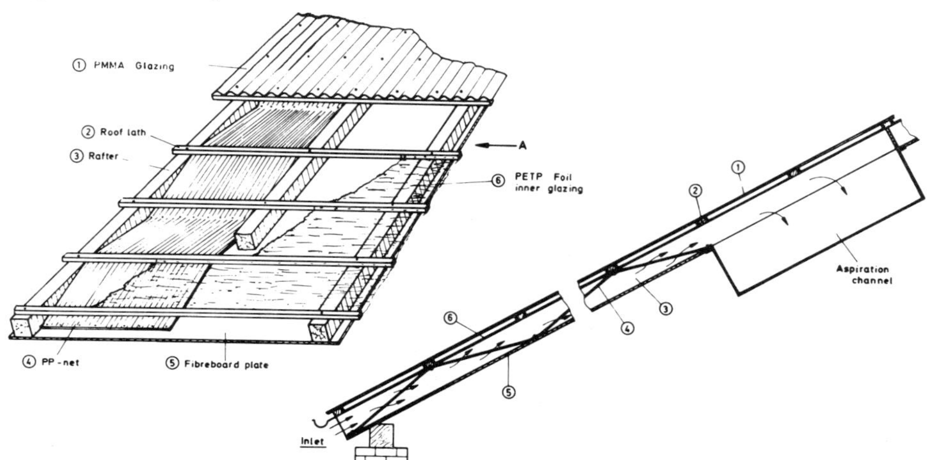

Fig. 1 Scheme of the air collector

From the collecting duct at the ridge the solar heated air is led to the conventional heating system consisting of two furnaces (650 kW, 350 kW) for controlled final heating to the required drying temperature (45-100°C depending on the product). Then the air is blown into the drier with 5 mesh type which are

- useful energy from the solar collector
- useful energy from the backup system
- useful energy from the heat recovery and recirculation systems

The following figures show as an example the results of a day in september 1986, when peppermint was dried. In fig.3 the meteorological data are plotted against time, in fig. 4 the different useful energies are seen as an hourly distribution. During noon more the 50 % of the whole demand is delivered from the solar system. The amount of energy from heat recovery and recirculation is rather low. The daily sum of global radiation measured on this day was 5.75 kWh/m², the useful energy from the collector is 3.19 kWh/m², this results in a daily collector efficiency of 55 %. The solar fraction at this day was due to long operation time in the evenning only 40 %.

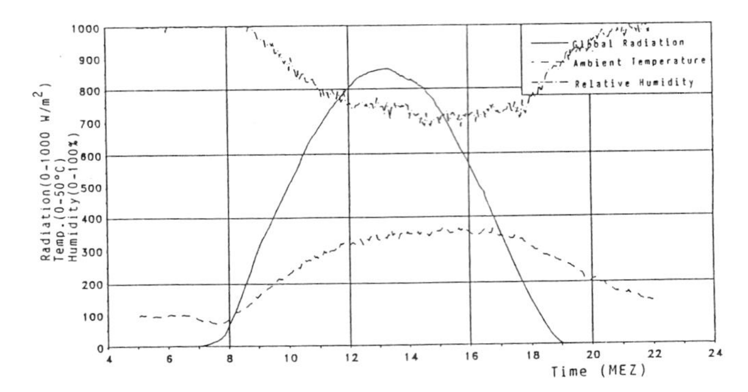

Fig. 3 Meteorological data of a sunny day

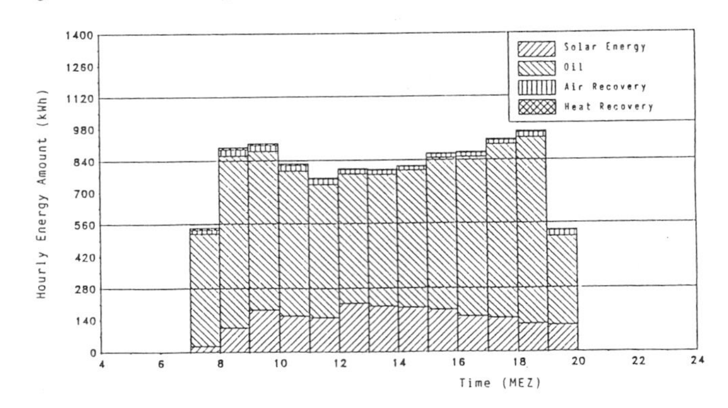

Fig. 4 Useful energies of this sunny day (drying product: peppermint)

The yearly consumption of fuel in 1986 was 19000 l which give a useful energy of 153 MWh/a. The savings with this plant were about 7000 l fuel. This gives a yearly solar fraction 28 %, which was caused be the high amount of products to be dried at higher temperatures (parsely, spinach).

conveyors. In order to conserve energy heat is recovered from the exhausted wet warm, transfered via a water circuite and introduced before the conventional heating system. This part of the heat recovery works during higher temperature drying (< 70°C). The second part is recirculation of semi-saturated wet air from one exhaust chimney to the air heater (350 kW) supplying the upper conveyor.

The total air flowrate is about 30 000 kg/h, the air velocity in collector itself is about 1.8 m/s. At this flowrate the collector heats ambient air under optimal meteorological conditions up to maximum 60°C, which is the maximum drying temperature for peppermint. Under these conditions peppermint could be dried solely with solar energy.

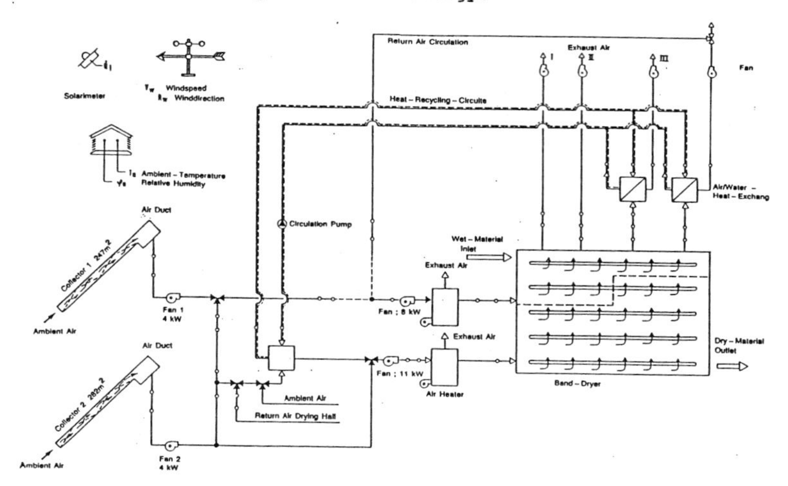

Fig. 2 Scheme of the plant

5. MEASUREMENTS AND RESULTS

At different relevant locations in the plant sensors for measuring temperatures, air flowrates, air humidity, electric power for operating the system (fans) and the fuel consumption were installed. Additionally meteorological data like global radiation at the collector slope, wind speed and direction, ambient temperature and humidity are registered.

For temperature measurement Pt 100 sensors were used, preselected and calibrated before installing, with an accuracy of 0.2 K. The air humidity measurement was done with a Rotronic sensor, which has a polymere as sensitive element. Air flowrate measurement in big air ducts of existing installations is one of the most complicated problems, if reasonable accuracy is required. In this plant small anemometers were used either mounted in the centre of the duct or with a mechanical device scanning the cross-section. The accuracy of flowrate measurement reached in this plant is about 5 %.

All these data were measured every 60 and 180 sec respectively with a computer controlled datalogger and recorded on magnetic cassetts for further processing at the computer centre of the instute.

These data were used to calculate the different energy flows in the plant,

With data of specific energy input of materials the collector and duct system was analysed. The power demand to operate the system is in plants with air as heat transfer fluid rather high (16 kW for fans). Nevertheless this gives an energetic amortisation of the plant of about 1.5 a, when the performance of the conventional heating system of 75 % is taken into account.

The monetary amortisation of 14 years is rather long compared to the energetic one. This is due to the fact, that the air ducts were rather expensive, the collector glazing has to be replaced after 5 year of operation because of degradation, and non efficient performance of the drying system itself.

6. CONCLUSIONS

The solar drying system has performed quite well during the last years. After replacing the degradated glazing of the collector, the performance has really improved. Daily collector efficiencies of more than 50 % could be measured at flowrates of about 50 kg/m^2h. The yearly savings in 1986 were about 6000 l fuel.

The energetic amortisation period of 1.5 years is really encouraging. The quite long monetary amortisation time of 14 years could be reduced by improving the operation of the plant, which means using as much as possible nice and sunny days for drying. This is of course not a simple task in our climate.

The heat recovery and air recirculation performs not as expected. By improving the conventional part of this system the solar fraction could be higher and the amortisation period shorter.

These results will be introduced in planning of further similar solar assisted drying.

7. REFERENCES

M. Reuss, W. Schoelkopf; Analysis and Comparison of Air and Water Heating Systems, 1984 Proceedings of the Congress on Energy Economics and Management in Industry; Albufeira, Portugal

M. Reuss, K. Meuren, S. Vogt; The Performance Measuring Data of an Air Collector System for Drying Herbs, 1984; Proceedings of 1st E.C. Conference on Solar Heating; Amsterdam, Netherlands.

H. Edelmann; Untersuchungen an einer solarunterstützten Trocknungsanlage, 1987; Diplomarbeit an der Landtechnik Weihenstephan, Freising, F.R. Germany

H.J. Wagner; Energieaufwand zum Bau und Betrieb ausgewählter Energieversorgungstechnologien; 1978, KFA Jülich Jül-1561; F.R. Germany

H. Edelmann, T. Schmalschläger, 1986 530 m^2 Luftkollektor zur Kräutertrocknung in Ismaning, Journal Sonnenenergie und Wärmepumpe, 4/86, F.R. Germany.

SOLAR DRYING OF PEAT

Timo Järvinen, Senior Research Engineer
TECHNICAL RESEARCH CENTRE OF FINLAND
DOMESTIC FUEL LABORATORY
P.O.Box 221, 40101 JYVÄSKYLÄ
Finland

ABSTRACT

Peat production is based on the solar drying of peat on the open fields of bogs. In the Domestic Fuel Laboratory of Technical Research Centre of Finland has been erected a solar simulation chamber for peat production to make operational instructions, planning figures and basic knowledge of drying more accurate. In different climatic and field conditions drying models have been made for both milled and sod peat and also studied basic heat and mass transfer phenomena. The test facility has shown clearly that the only possibility for better understanding and control of solar peat drying is to simulate natural conditions on a laboratory scale.

KEYWORDS

Solar peat drying; Modelling of peat drying; Peat production; Peat production control; Simulation of peat production; Solar drying simulation; Climatic simulation chamber.

INTRODUCTION

Peat production is based on the solar drying of peat on the open fields of bogs, which are specially prepared (drained and levelled) for peat harvesting. The total bog area in Finland is approx. 10 million hectares: in a natural state 3.5 mill. hectares including 0.7 mill. hectares preservation bogs, 0.1 mill. hectares are used as power plant reservoirs, 0.7 mill. hectares are in agricultural use, 5.6 mill. hectares are used for forestry and 0.1 mill hectares are reserved at the moment for the peat industry. So far now 0.05 million hectares have been prepared for peat production. Production figures increased rapidly during the 1970's and last year peat production reached the level of 7 million tons, of which 6.5 million tons were used as fuel.

RESEARCH INTO SOLAR DRYING

In order to make operational instructions, planning figures and also the basic knowledge of solar drying for peat production more accurate, a solar simulation chamber for conventional sod and milled peat production has been erected in the Domestic Fuel Laboratory of Technical Research Centre of Finland (Fig. 1). The chamber consists of a peat block with a certain ground water level and climatic conditions for drying. The radiation is as similar to the natural sunlight spectrum as is possible within reasonable costs limits and its maximum heat output is approx. 2 000 W/m^2. The temperature and humidity are also controlled and programmed. The whole test facility is controlled by a PID-controller and a microcomputer, and information is gathered and handled automatically.

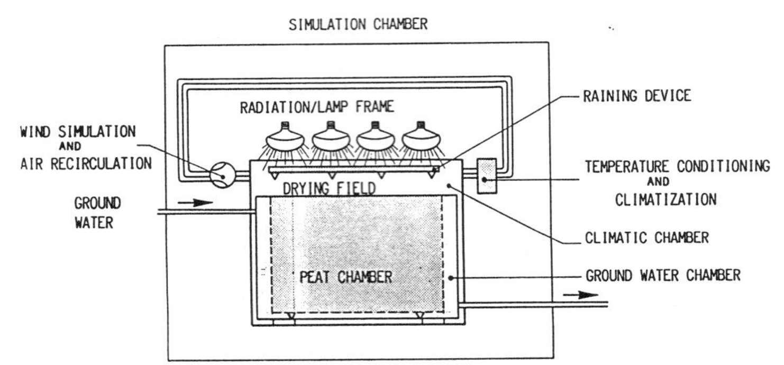

Fig. 1. Diagram of the simulation chamber.

The climatic conditions in the chamber can be kept very stable if needed. This has been proved with experiments when evaporation has been measured with an evaporimeter placed in the chamber in stable conditions (Fig. 2). Also the drying curves are repeatable. This can be seen from experiments in which the drying tests has been repeated under the same conditions with the same thicknesss (or field load kg DS/m^2) of milled peat layer using peat as equally as possible (Fig. 3).

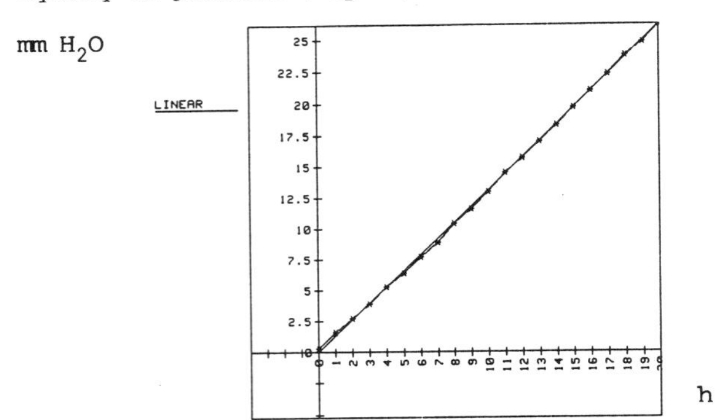

Fig. 2. The measured cumulative evaporation with evaporimeter under the chamber stable conditions.

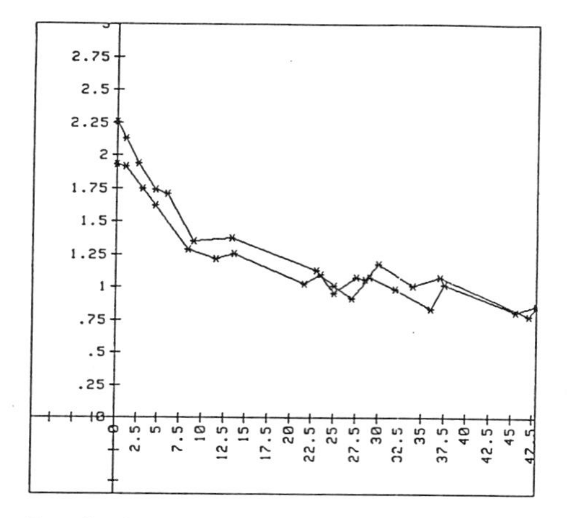

Fig. 3. Two drying curves of milled peat layer under the same (stable) conditions.

MODELLING OF DRYING

Solar peat drying can be described using general heat and mass transfer equations. Those relations give reliable information for instance for a production control system only in the case when all parameters affecting with drying are well known. Unfortunately this isn't the case of solar peat drying on open fields. That is why it is necessary to determine drying curves and models under different conditions according to weather and field (climatic and peat deposit conditions) measurements. Up until now drying models have been derived for extraordinary good stable conditions with two radiation levels and for good changing (day and night) conditions with the chosen ground water level and on a certain peat block "drying field" (Fig. 4 and 5). The mathematical equations best fitted for the models are at the beginning of drying the linear form $y = A \cdot x + B$ and after a certain time depending on layer properties the exponential form: $y = A \cdot e^{B \cdot x}$. The experiments for determining drying curves for peat samples have shown very clearly, that peat layer, of whatever type it is, also transfers heat to lower lying bog layers and (ground) water if reflection of radiation to the atmosphere is excluded, which is also a very important factor affecting solar peat drying. This means it is important to somehow determine and also control the heat transfer properties of a peat layer and its products to be dried together with its environment.

kg H_2O/kg DS

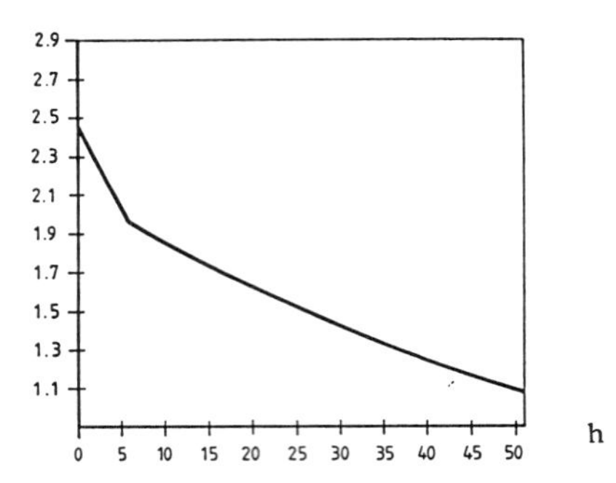

Fig. 4. The drying model (curve) of milled peat as a function of time under extraordinarily good conditions with radiation level 552 W/m^2.

kg H_2O/kg DS

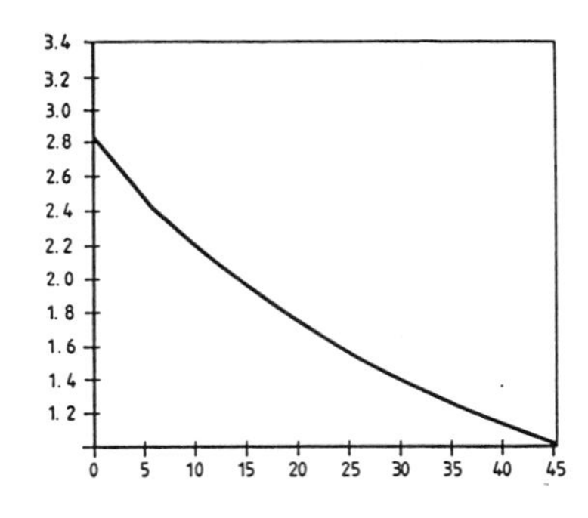

h

Fig. 5. Drying model of milled peat layer as a function of time under changing (day and night) conditions.

THE COMPARISON OF RESULTS ON LABORATORY SCALE WITH DRYING ON OPEN FIELDS

The most important factor affecting solar peat drying is the temperature gradient, which is formed in the bog down to the ground water, if reflection of radiation (albedo) is excluded. If the temperatures of simulator and "native" peat fields are compared, it can be found that the simulators' temperatures are a little higher (2 - 3 ^{0}C) in upper layer of the simulator bog than in the field. The temperature of ground water is on the same level in both cases and also very close together at a depth of 30 cm. This would suggest that heat transfer is more effective in the simulator than in the field and that the drying of the peat layer is slower than under same natural climatic conditions. On the other hand so far the drying conditions in the chamber have been extraordinary good and cannot be directly compared with field conditions.

DISCUSSION

The test facility has shown very clearly that the only possibility for better understanding and control of solar peat drying is to simulate natural conditions on a controlled (laboratory) scale. The use of a simulation chamber enables one to get more accurate information and data for production control and machinery planning. Also it gives new prospects for the utilization of solar energy. The most important things in creating simulation facilities are to achieve radiation, heat and mass transfer phenomena that are as natural and controlled as is reasonable within certain cost limits. The experiments have proved that the conditions in the chamber are stable enough and can also be controlled according to the programmed climatic data which have been measured in the field. Also the drying curves are repeatable. The only question is how well the temperature gradient of the chamber "bog" describes the corresponding situation in the "native" field and also if it differs to some extent, what is its effect on the drying of a certain peat layer in the chamber compared with solar drying of the same layer on a field. This work is still going on.

IMPROVED TECHNOLOGY FOR SOLAR DRYING OF VINE FRUITS

R.J. Fuller*, D.R. Kaye, and M.J. Schache
(Project Supervisors: G. Redding & R.W.G. Macdonald)
* Dept. of Agriculture and Rural Affairs,
Ag. Eng. Centre, Werribee 3030, Australia.

ABSTRACT

Simple solar technology has been incorporated into a drying rack, similar to those used by 90% of Australian dried fruit producers. Drying times have been reduced by 3 to 4 days, and final moisture content levels of 13.5% were achieved on the rack itself, indicating that the on-ground finishing process can be eliminated. Final dried fruit quality was also consistently high.

KEYWORDS

Dried vine fruit; drying rack; unidirectional rockpile; corrugated iron solar air collectors.

INTRODUCTION

The production of dried vine fruit (dvf) is the major horticultural enterprise of the Sunraysia region of Australia. The area produces over 80 percent of the national output of dvf which in 1986 totalled 93,000 tonnes. 60% of the crop is exported and international competition is strong, principally from the US, Greece and Turkey.

In the present method of production, fresh fruit is spread out on wire netting suspended in long open sided racks and drying is effected by the natural potential of ambient air. A moisture content level (w.b.) of 13.5% is required but drying down to this level is rarely achieved on the racks. Consequently the fruit is shaken off the racks at approximately 16.5% M.C. and is then sundried on ground sheets. Drying may take anywhere between 10 and 30 days.

Improving quality, reducing drying times and eliminating on-ground finishing would all bring benefits to the industry and this project has investigated ways of improving the existing technology.

EXPERIMENTAL DESIGN AND OPERATION

A 36 metre drying rack of standard design, oriented north-south, has been divided into 4 sections. Three of these have been modified as described below.

Curtain - Only System

Roll-up polyethylene curtains have been mounted below the roof of the rack. These are raised and lowered manually at hours 0900 and 1900 respectively once the grapes have reached 25% M.C., the level where the danger of re-absorption of moisture may occur. Using the curtains any earlier would prevent night-time moisture loss which occurs in the earlier stages of drying.

Rockpile System

This system employs polyethylene curtains as the Curtain-Only system but in addition heat is supplied at night by a rockpile built into one bay of the rack. The rack roof above the storage and adjacent bays has been modified to create 2 simple solar air heaters, one glazed (Fig. 1) and the other unglazed (Fig. 2), with a total area of $32m^2$. $4.5m^3$ of 20mm nominal dia. washed basalt is used for the storage medium. Perforated PVC pipes at ground level have been used as discharge ducts, the air being forced through the system by a 200 W centrifugal fan at a design flow rate of $0.16m^3\ s^{-1}$. The rockpile-collector system operates in a unidirectional mode with air always being drawn along the collectors, forced down through the rockpile and out of the ducting. Control of the system is by a 2-stage time clock. Heat collection begins at hour 1030 and ceases at hour 1630, and heat delivery begins at midnight and ceases at hour 0800, with the objective of the peak temperature front emerging from the rockpile during the coldest part of the morning.

Collector - Only System

This system uses a $13m^2$ glazed solar collector (Fig. 1) but with a nominal gap of 50mm between glazing and absorber plate. Heat collection during the day is controlled by a time clock which activates a 250 W centrifugal fan between hours 1030 and 1630. Heat is transferred directly to the rack space by perforated PVC pipes located at ground level at a design air flow rate of $0.26m^3\ s^{-1}$. The rack space is enclosed during the day by polyethylene curtains but a flap level with the top netting tiers is left open to allow the moist air to exhaust. At night this flap is closed manually to fully enclose the bays as in the other two systems.

INSTRUMENTATION AND MEASUREMENT

A 30 Channel Hewlett Packard computer and logger has been used to record temperature and climatic data. Dry and wet bulb temperatures have been measured in the upper and lower tiers of the 4 sections, on the adjacent lawn area and roof of the rack. Dry bulb sensors have been located in the outlet plenums of the various collectors and in the top and bottom layers of the rockpile. 50 trays containing similar quantities of grapes have been weighed in the morning and evening to determine moisture uptake and loss. Random sampling of 100 berries has been used as the method of determining moisture content at various time intervals. Quality assessment was performed by commercial graders at a local packing shed.

RESULTS AND DISCUSSION

2 trials of the systems were conducted during the 1987 drying season and the performance of the three modified sections has been compared against the unmodified section of the rack. Fruit quality, moisture content levels and microclimate differences in the sections have been compared.

Fruit Quality

Fruit quality is measured commercially by a crown grading system. The higher the number of the crown, the better the quality of the fruit and the more uniform its colour. 5 crown is regarded as top quality fruit. Further quality is designated by the classification 'dark' or 'light', a higher price being paid for light coloured fruit. The commercial quality gradings of the fruit produced in two trials are shown in Table 1.

	Control	Collector-Only System	Rockpile System	Curtain-Only System
Trial 1	5L	5L	5L	5L
Trial 2	N.A.[a]	4L[b]	4L[b]	4L[b]

Table 1. Quality Gradings of Dvf Produced in 1987 Trials

a. The fruit from the control bays was not graded. It became very dark, could not be dried down to the required moisture content level on the ground and was not considered to be commercially viable.

b. Final quality was reduced by the inclusion of some poor fresh fruit. When this was removed by hand, 5 crown light gradings were achieved.

It can be seen from the above that top quality dried fruit is consistently achievable with the modified systems provided that good quality fresh fruit is initially available.

Moisture Content Levels

Prime objectives of the project are a reduction in overall drying time and drying down to the 13.5% moisture content on the rack itself. The final moisture content levels achieved in the two 1987 trials are shown in Table 2.

	Control	Collector-Only System	Rockpile System	Curtain-Only System
Trial 1	15.5	13.5	14.5	15.0
Trial 2	16.5	13.5	14.0	15.5

Table 2 Final Moisture Content Levels Achieved in the 1987 Trials

The implications of the above results are :-

(i) that even late in the drying season, moisture content levels of 13.5% can be achieved on the rack, thus eliminating the on-ground finishing process.

(ii) When translated into time savings, the best modification could save at least 3 to 4 days, being the estimated time required to reduce the control from 16.5% to 13.5% by sun drying. In poor drying conditions the time saving would be greater.

(iii) That the Collector-Only System is the most promising of the three systems.

Effect of Curtains

The curtains did retard moisture uptake, but the addition of heat at night from the rockpile almost eliminated it. However, moderately strong wind caused the curtains to lift away from the rack, thus allowing excessive infiltration of ambient air. The effectiveness of the curtains could be improved if a more positive sealing system is devised.

Effect of Solar Systems

The effect of the solar systems on the enclosed rack spaces was analysed during the second trial period when the climatic conditions were the least favourable. Daytime temperatures within the Collector-Only system were usually 3°C higher than those in the control for at least 8 hours each day, with peak collector outlet temperatures 6 to 10°C above ambient achieved regularly on the 11 days investigated. The discharge of heat into the Rockpile system caused temperature rises of approximately 5°C above the control (Fig. 3) with corresponding reductions of 20 percentage points or more in relative humidity levels (Fig. 4). Temperature levels of 40 to 50°C were regularly achieved in the top section of the storage with the peak temperature front emerging at approximately hour 0600 (Fig. 5). The glazed collector was superior in terms of temperature rises achieved to the unglazed unit, with differences of 10°C or more on several occasions. (Fig. 6).

CONCLUSIONS AND FUTURE WORK

The systems described are to be evaluated over a 2 year period, and a second trial will be used to confirm the initial results. The results from the first year's experimentation indicate that simple improvements can be made to the existing technology so that better quality dried fruit, reductions in drying time and the elimination of on-ground finishing are all possible.

Improvements to the curtains are to be made to reduce infiltration. This should further reduce moisture uptake and improve the utilization of heat delivered. A full economic evaluation will be necessary to determine which of the systems will offer dvf growers the best option for improving their racks.

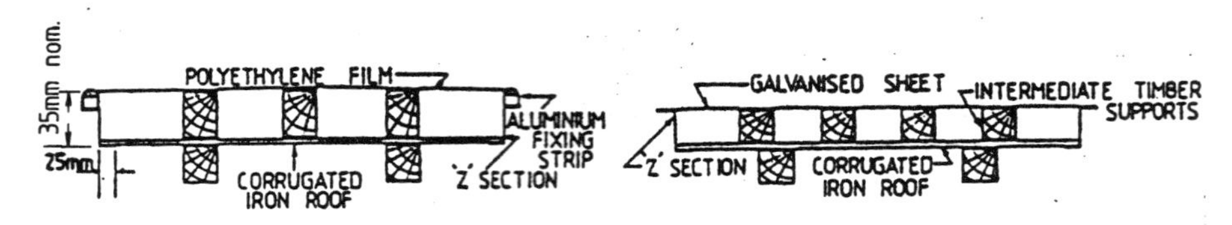

FIG.1. GLAZED COLLECTOR SECTION

FIG.2. UNGLAZED COLLECTOR SECTION

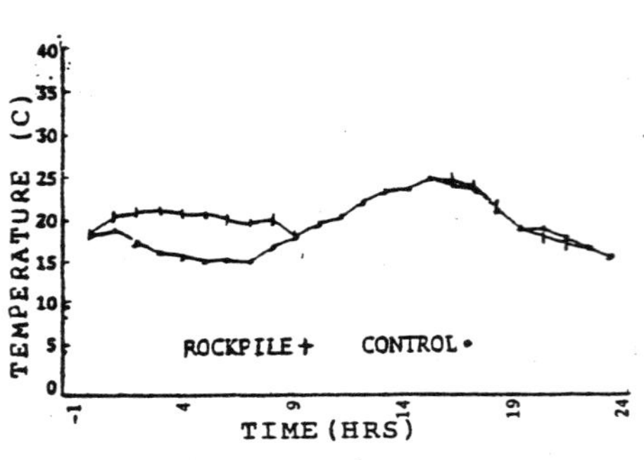

FIG.3. COMPARISON OF DRY BULB TEMPERATURE LEVELS IN ROCKPILE SYSTEM AND CONTROL

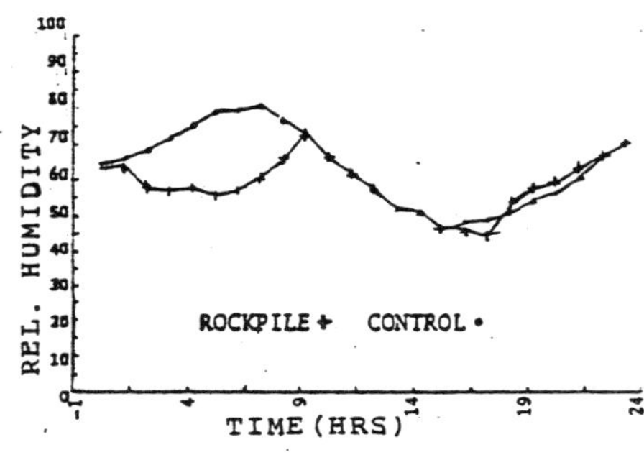

FIG.4. COMPARISON OF RELATIVE HUMIDITY LEVELS IN ROCKPILE SYSTEM AND CONTROL

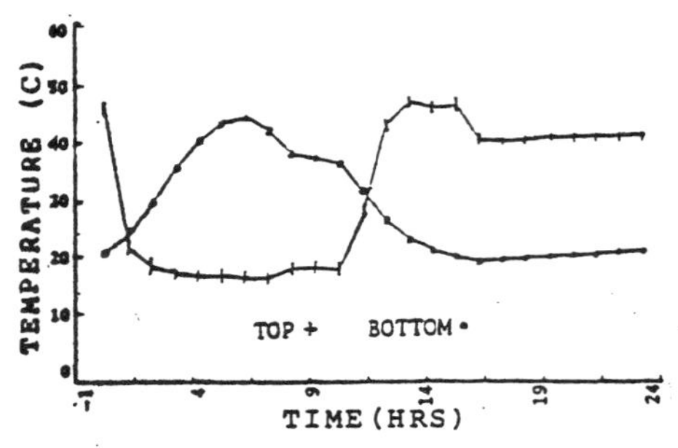

FIG.5. CHARGE AND DISCHARGE CYCLES IN ROCKPILE

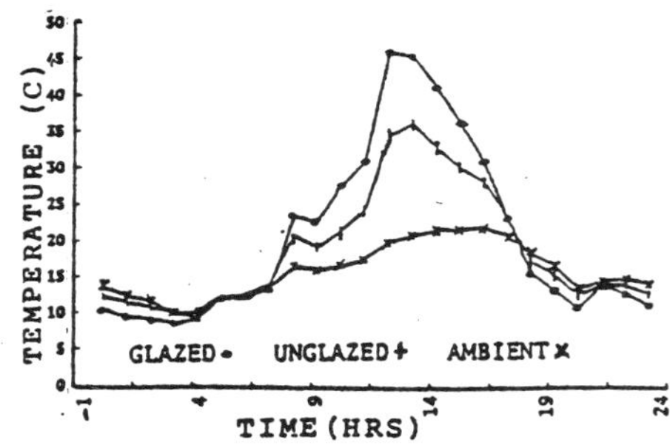

FIG.6. GLAZED AND UNGLAZED COLLECTOR COMPARISON

DOUBLE-DIFFUSIVE INTERFACE BEHAVIOUR: GROWTH AND EROSION IN SOLAR PONDS.

B.P. Marett

Department of Mechanical and Industrial Engineering,
University of Melbourne,
Parkville, Victoria, Australia. 3052.

ABSTRACT

A model of the surface mixed layer has been developed which portrays the main features of double-diffusive and wind-shear induced mixing. The stability and depth of the boundary region are shown to be important in limiting density gradient erosion. Some practical implications for solar pond operation are discussed.

KEYWORDS

solar ponds, surface mixed layer, density gradients, mixed layer model.

INTRODUCTION

To optimise the thermal efficiency and daily operation of solar ponds it is important to minimise the depth of the surface mixed layer. The major influences which contribute to the deepening of the surface mixed layer are the surface fluxes of thermal energy and salt and the turbulent kinetic energy associated with wind energy and water inflow.

It is the purpose of this paper to study density gradient stability in the upper region of salt-gradient solar ponds, which are subject to both double-diffusive (thermal energy and salt) convection and wind driven convection. In solar ponds, a strong stable density gradient (or step) normally exists at the lower boundary of the surface mixed layer. (Figure 1) This gradient opposes further layer deepening, and practical operational experience suggests that the rate of erosion diminishes when some limiting density gradient is approached. In a salt gradient solar pond, the boundary thickness (defined as where eddy diffusion is observed) can be a significant fraction of the main density gradient zone depth, and should therefore be included in any model. Eddy diffusion plays an important part in gradient erosion, at least in solar ponds subject to strong wind regimes. At the windy Laverton site, the solar ponds' boundary thickness can be up to one-fifth the total diffusive interface thickness.

The presence of a strong boundary density gradient, even during prolonged periods without significant wind, suggests that the stable salinity gradient dominates mixed layer deepening at Laverton. During, and for some time following a period of strong winds, the boundary moves downard, creating a stronger boundary density gradient as it does. When the wind and eddy motion subside, the boundary salinity gradient no longer has a flux sufficient to maintain its magnitude and so it weakens. Under certain circumstances, the boundary region may then grow, at a rate consistent with eddy diffusion rates. The mechanism of this process is not yet clear.

In the presence of low levels of kinetic energy in the surface mixed layer, salt diffuses upward across the boundary region creating a weak gradient. This gradient may strengthen in time, weakening the original boundary gradient. At higher levels of kinetic energy this gradient will be eroded.

Solar ponds are sometimes subject to internal waves generated by high insulation levels and wind-wave action. These mechanisms, and possibly others, enhance the molecular diffusion of the salt and thermal energy. The time-averaged diffusion coefficients in a solar pond are often several times larger than those imposed by the molecular limit.

THEORY

To describe the processes eroding a density gradient in a general manner, some account must be made of double-diffusive convection and wind-induced convection. To date, these two mechanisms have been described separately. Models accounting for turbulent kinetic energy (wind induced) erosion ignore the contribution of the thermal energy flux, and double-diffusive convection studies assume no other destabilising kinetic energy input apart from the thermal energy flux. The following approach has been developed to identify the conditions under which each approach is valid.

In this analysis, the density is described in terms of the saline (S) and thermal energy (T) contributions to the density profile. In the gradient region each density component contribution may be described by:

$$\rho_{T\ (S)} = \rho_r + \overline{\rho} N_{T\ (S)}^2 z/g \tag{1}$$

where $N_T{}^2(S) = -\ (g/\overline{\rho})\ (d\rho_{T(S)}/dz)$ are the initial gradients expressed in the usual buoyancy frequency notation.

Consider an interface which has been eroded by both double-diffusive and external kinetic energy sources. A mixed layer of initial depth d with initial density steps $\Delta\rho_{oT}$ and $\Delta\rho_{oS}$, is deepened to a depth h, with B_T and B_S representing the net thermal and saline buoyancy fluxes respectively, at the water surface. Upon completion of energy and salt balances, the density steps at the boundary can be expressed by:

$$\Delta\rho_T = (-\tfrac{1}{2}(h^2 - d^2)N_T{}^2 + B_T t + dg\ \Delta\rho_{oT}/\overline{\rho})\overline{\rho}/(hg) \tag{2}$$

$$\Delta\rho_S = (\tfrac{1}{2}(h^2 - d^2)N_S{}^2 + B_S t + dg\ \Delta\rho_{oS}/\overline{\rho})\overline{\rho}/(hg) \tag{3}$$

To inter-relate these equations, a boundary region stability parameter is defined in terms of the ratio of the opposing density steps. Linden [2] has used this approach to describe mixed layer development in double-diffusive systems. If the boundary region has considerable thickness as in actual solar ponds, then this ratio may also be taken to be the ratio of the boundary gradients of the two diffusing components, T and S.

$$k = -\Delta\rho_T/\Delta\rho_S = -(d\rho_T/dz)/(d\rho_S/dz) \tag{4}$$

In double-diffusive systems the value of k may fall in the range of $1/3 <= k <=1$, the lower bound being attained when all the kinetic energy of the buoyancy driven convection is used to mix the layer, and increase the potential energy of the system. [3] The upper bound is the case observed in purely double-diffusive systems (within experimental error) and implies that the boundary region of the interface is of marginal stability, and that molecular diffusion is the means of maintaining boundary and interface stability. Note that in wind-induced convection, a differen t lower bound will apply as explained in the Discussion section of this paper.

By combining the last three expressions, the mixed layer depth of a single layer can be described by:

$$h = [\ 2(kN_S^2 + N_T^2)^{-1}[\ (B_T - kB_S)t - gd(\Delta\rho_o/\bar{\rho})\ ] + d^2\ ]^{\frac{1}{2}} \tag{5}$$

Turner [3] suggested the following mechanism for the formation and growth of mixed layers. Diffusion of the faster component T ahead of the growing layer produces a boundary layer which ultimately becomes unstable when a critical Rayleigh number is exceeded. Successive layers can form in this way. Linden [2] incorporated a critical Rayleigh number expression into the last equation and experimentally confirmed the validity of this approach for a double-diffusive system.

For layer growth due to turbulent kinetic energy input, boundary shearing, direct intrusive mixing and other mechanisms are present. These mechanisms are interdependent and are too complex to separate and examine individually. Without a priori knowledge of how k varies in time and depth in such circumstances, it is more useful to adopt other experimental evidence of layer deepening. Firstly (1) and (2) are combined in the form of the net density step $\Delta\rho$. With the additional definitions of the net density gradient $N^2 = N_S^2 + N_T^2$, and the net buoyancy flux at the upper surface of the mixed layer, $B = B_S - B_T$:

$$gh(\Delta\rho/\bar{\rho}) = \tfrac{1}{2}(h^2 - d^2)N^2 - Bt + gd(\Delta\rho_o/\bar{\rho}) \tag{6}$$

Following [4] the velocity of the mixed layer lower boundary due to predominantly shear flow, may be written in terms of an overall Richardson number ($Ri = gh\Delta\rho(\rho u_*^2)^{-1}$). The velocity u_* is taken to be the surface friction velocity. The velocity of the boundary can then be described in terms of both diffusive components:

$$\begin{aligned} u_E &= dh/dt = c\ u_*\ Ri^{-1} \\ &= c\ u_*^3[\tfrac{1}{2}(h^2 - d^2)N^2 - Bt + gd(\Delta\rho_o/\bar{\rho})]^{-1} \end{aligned} \tag{7}$$

where c is a constant. If the kinetic energy input term u_*^2, and the fluxes are assumed to be constant in time, then this equation can be integrated and rearranged to yield the expression:

$$\begin{aligned} t = {} & [\tfrac{1}{2}N^2B^{-1}(d^2+2Ld+2L^2) - \tfrac{1}{2}d^2N^2B^{-1} + dgB^{-1}(\Delta\rho_o/\bar{\rho})]\exp[(h-d)/L] \\ & - \tfrac{1}{2}N^2B^{-1}(h^2+2Lh+2L^2) + \tfrac{1}{2}N^2d^2B^{-1} - gdB^{-1}(\Delta\rho_o/\bar{\rho}) \end{aligned} \tag{8}$$

where $L = cu_*^3/B$ is a measure of the influence of buoyancy in shear flows and is known as the Monin-Obukov length. [5] The ratio of h/L is directly proportional to the ratio of the power due to buoyancy to the power of kinetic energy. Generally, $h/L << 1$ in solar ponds at windy sites and (8) can be simplified and expressed in terms of the depth;

$$h = -cu_*^3B^{-1} + [\, -2BN^{-2}t + c^2\, u_*^6\, B^{-2} + 2cu_*^3B^{-1}d + d^2]^{\frac{1}{2}} \tag{9}$$

For dominant buoyancy driven mixing $h/L << 1$, and (8) may be simplified with the depth being described by:

$$h = -cu_*^3B^{-1} + [\, -2BN^{-2}t - c^2{}_*u^6\, B^{-2} - 2gdN^{-2}(\Delta\rho_o/\overline{\rho}) + d^2]^{\frac{1}{2}} \tag{10}$$

In the case of purely double-diffusive mixing, that is in the limit $u_* \rightarrow 0$, equation (5) (with k=1) is returned. In general, equation 8 must be solved numerically for h. (Figure 2).

DISCUSSION AND CONCLUSION

The deepening of a wind mixed surface layer has similar features to double-diffusive layer development (c.f. (5) & (9)) as described by Turner [1]. The dependence of the depth on time and gradient strength are particularly interesting. In the presence of shear and buoyancy the depth increases proportionally with the square root of time, as it does when buoyancy only is present. With only the shear mechanism operative, it can be shown that the depth increases proportionally with the cube root of time. [4] Figure 2 illustrates a relatively sharp transition between wind mixed and buoyancy driven convective deepening. The friction velocity contours suggest that relatively low average windspeeds will often lead to the wind mixing mechanism dominating.

The mixed layer depth has been shown to be inversely proportional to the square root of the density gradient. This result is particularly significant as it applies to double-diffusive and wind shear induced mixing. Furthermore, it can be shown [6] that penetration length of simple convective plumes has the same gradient proportionality. Briggs [7] (cited by Turner [5]) showed that for a wide range of observed plumes, the penetration length was inversely proportional to the density gradient raised to the power three-eights. The boundary density gradient, is therefore dependent on the interface density gradient as well as acting to minimise further gradient erosion. In fact, a thick boundary density gradient will buffer the weaker interface gradient from some of the mechanisms of mixing. This suggests that gradient erosion in practice could be minimised by introducing a strong density gradient in the upper region of the interface.

It is clear that k is a complicated function of the kinetic energy of the mixed layer. Note from (5) above, that $k > -N_T^2/N_S^2$, and if this limit is approached as it may be initially during wind mixing, then the depth would increase rapidly. In practice it is found that the boundary gradient (or step) approaches a maximum value after which time erosion proceeds more slowly. This fact and (8) imply that k never reaches this limit. Empirical evidence of the variation of k under field conditions is currently being collected at the Laverton field site. [6] It is expected that average k values will help provide a better understanding of gradient erosion.

REFERENCES

1. B.P. Marett, R.W.G. Macdonald, W.W.S. Charters, and D.R. Kaye, Recent

developments at the Laverton solar ponds, ISES Solar World Congress, Hamburg. (1987).

2. P.F. Linden, The formation and destruction of fine-structure by double-diffusive processes, Deep-Sea Research 23, 895 (1976).

3. J.S. Turner, The behaviour of a stable salinity gradient heated from below, J. Fluid Mech. 33, 183 (1968).

4. H. Kato and O.M. Phillips, On the penetration of a turbulent layer into stratified fluid, J. Fluid Mech. 37, 643 (1969).

5. J.S. Turner, Buoyancy Effects in Fluids, Cambridge Press (1973).

6. B.P. Marett, Thesis in preparation, University of Melbourne (1987).

7. G.A. Briggs, Plume rise. U.S. Atomic Energy Commission Critical Review Series. (1969).

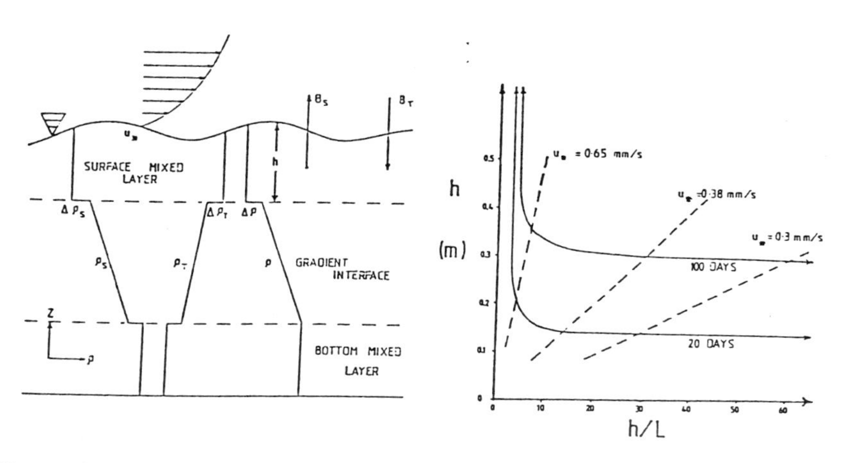

Figure 1 Solar pond density profiles

Figure 2 Mixed layer depth as a function of layer stability and time, (d=0, $B=-1.39\times10^{-8}m^2s^{-3}$, $N^2=2.89s^{-2}$, c=2.5)

RECENT DEVELOPMENTS AT THE LAVERTON SOLAR PONDS

B.P. Marett, R.W.G. Macdonald, W.W.S. Charters and D.R. Kaye

Department of Mechanical and Industrial Engineering
University of Melbourne,
Parkville, Victoria, Australia. 3052.

ABSTRACT

Recent work at the Laverton solar ponds has proceeded in three main areas. After several years of operational experience, a generalised instrument controller has been developed to automate the monitoring of solar ponds. A method for clarifying bitterns has also been developed. Furthermore, a ring-wave suppression system has been tested and the preliminary results suggest that it is very effective.

KEYWORDS

Solar ponds; bitterns; instrumentation; rings; wave suppression; convection suppression; water clarity.

INTRODUCTION

Salt-gradient solar ponds can provide thermal energy for electricity generation or for industrial process heat. A number of studies have been conducted in Australia to determine the technical feasibility of solar ponds and the size and nature of the market. For example, in a recent study [1] solely devoted to the prospects for solar pond electricity production, it was concluded that some 29 to 65 potentially useful pond sites existed in Western Australia, South Australia, Queensland and the Northern Territory. Further sites were identified in regions without a sufficient demand for electrical energy. In a recent paper [2], the present authors cautiously forwarded the figure of 2,730 x 10^{12} J/yr (70,000 kl/yr Diesel fuel) as being the potential displacement fuel value of Diesel fuel displaced by electricity generating solar ponds, and a corresponding figure of 5,000 x 10^{12} J/yr for low grade process heat. The latter figure was based on 5% of the total solar potential replacement for industrial thermal energy usage at temperatures <100°C. These figures are derived from an independent assessment of the studies referenced in [2].

A considerable potential exists for solar ponds in Australia. The technical feasibility of pond operation in harsh climatic conditions has been confirmed following several years of pond research and development in Australia.

The Laverton Solar Pond Project has been concerned with three main aspects of solar pond operation. Attention has been placed on the use of bitterns as the source of salt and water. Bitterns is concentrated seawater, and the waste concentrate of the solar salt manufacturing industry. It is also found underneath natural salt lakes in central Australia. Another aspect of research has been the development of solar pond instrumentation for both research and commercial solar ponds. In conjunction with these two areas of study, some basic research in pond stability has also been conducted. [3]

THE USE AND CLARIFICATION OF BITTERNS

The major reasons for the use of bitterns are the low cost and its high density which enables a very stable density gradient to be maintained in a solar pond. The most common salt used in solar ponds is sodium chloride. However, this salt can only provide marginal stability for solar ponds operating at high temperatures. Very little operating experience with bitterns has been accrued anywhere.

The Laverton bitterns is discolored by the presence of various algae. Many chemical techniques exist to clarify natural waters but none are directly applicable to bitterns. Extensive chemical tests have been conducted during the period of this project. It has been concluded from the tests conducted that standard chemical water treatments are inadequate to effectively clean the water. [4]. Only very high concentrations of some chemicals (eg Ultrion, Catoleum Pty. Ltd.), provided near to acceptable levels of water clarity. (Figure 1) At such high concentrations, these treatments were not considered to be economic.

Following the failure of the chemical tests, filtration experiments were undertaken. Microscopic investigations showed that the algae grow up to approximately twenty microns in diameter, but a large fraction are below the filtration limits of common sand (15 micron) and diatomaceous earth (8 micron) filters. Various multiple bed filtration systems were tested but without success. Laboratory scale filtration using high pressure filtration through 1μ and 5μ (absolute) filters increased water clarity. However the high operating pressure led to fragmentation of the contaminants which allowed further material to pass through the filter. (Figure 1)

A commercial ultra-filtration unit with a rating of 0.2 micron absolute, has also been investigated. This unit, provided by Amicon Australia, consists of hollow glass straws set in a polycarbonate canister casing. Water flows through the straws and the tangential pressure on the glass walls leads to filtrate production on the outside of the walls. The concentrate is returned to the raw bitterns storage. Material which collects on the glass walls is sheared off by the flow rate, by reversing the flow direction, or by back flushing across the glass walls. This filtration unit has been found to be very effective in removing all colored material and producing crystal clear water. No other water clarification technique investigated has provided similar results. In Figure 1 ultra filtration trial #1 represents filtrate produced by the filters under normal operating pressures. Trial #2 filtrate was produced by the filters under higher pressures.

Field tests with four filter canisters (each of approx 3m^3 effective surface area) were undertaken at Laverton, Victoria. The canisters were joined in parallel with the flow directed by solenoid valves. A 5m^3 sand filter was used as a prefilter. Electronic control gear, designed and built in-house, closes and opens the valves to reverse the flow direction every twenty minutes. The first series of experiments were designed to optimise operating pressures. Filtrate flow rate and quality were shown not to be very sensitive to the operating pressure range available. In the second series of experiments, the filtrate flow rate was shown to be inversely proportional to bitterns concentration, at least over the specific gravity range 1.18-1.32. An average filtrate flow rate of 10 l per minute was maintained by the filter system for a period of several days of continuous operation. Raw bitterns of specific gravity 1.18 was used during this period. No deterioration in performance was observed. This filtrate flow rate is quite low when compared to those used in sand and multi-bed filtration systems. However, under continuous operation it is believed that this ultrafiltration system could economically clean bitterns for a solar pond. Further system performance data are available. [4].

DENSITY GRADIENT MAINTENANCE

Erosion of the density gradient must be minimised if optimal thermal performance of the pond is to be achieved. Wind induced mixing is the major source of gradient erosion at windy locations. The Laverton site is well situated to study wind induced gradient erosion. Originally it was proposed to compare two identically constructed solar ponds for the testing of a ring wave suppression system [4]. One pond was to be covered with the rings in a hexagonal close packed arrangement, the other pond was to be left uncovered. The two pond plan was abandoned due to a lack of funds and man-power, as well as a potential shortage of raw bitterns. Bitterns, the salt source for the salt gradient, is only available after the salt harvest once yearly.

A quasi-steady state approach was adopted to study the effect of the wave suppressors. Long term data has been collected on the rate of erosion of a wind exposed gradient at Laverton. Observations have also been made for the pond covered with rings. The resulting data are not strictly quantitatively comparable since the time periods studied for the ponds differed. An average surface convective zone depth of 0.80m resulted from wind induced mixing on a pond with no wind or wave suppression devices. For the pond with the rings a depth of 0.50m would be appropriate under similar conditions for the same time period. This data applies to ponds of similar gradients and surface salinities. If such a difference in performance is indeed representative, and could be sustained, then the thermal efficiency of a pond with rings can be shown to be substantially increased.

Recent theoretical work [3] confirms the observation that a reduction in the kinetic energy of the surface mixed layer will slow, the rate of gradient erosion. Further, increasing the strength of the gradient near the surface mixed layer, will also reduce the rate of gradient erosion. Theoretical and experimental work in these areas is continuing.

INSTRUMENT CONTROL

One of the major objectives of the Laverton Project has been the development of suitable instrumentation for solar ponds. An instrument controller has been designed and built to provide cost effective and time efficient collection of data. It is believed that a commercial demand exists for a versatile instrument controller. Many tasks performed by the Laverton controller are of a general nature, and therefore are directly relevant to a multipurpose commercial controller.

For the Laverton Solar Pond Project, a reliable and accurate controller was required to assist in the automatic measurement of water clarity, acidity, salinity and temperature. Measurements were required throughout the depth of the stratified solar pond. A sampling probe was mounted vertically on a motor driven pulley assembly. The instrument controller was designed to direct the probe's movement vertically within a few millimetres resolution. Once positioned, the controller starts a pump to move fluid from the probe to the inline instruments. Readings from these instruments are then collected by the controller and sent to a Hewlett-Packard 3497 data logger.

A commercial controller could be made in modules, each module performing specialised tasks. The basic electronic module is the motor control unit which provides a link between TTL logic levels and the 12 volt, 4 amp supply required by the motor. This module allows direct motor control via an external data logger.

Another basic module provides a link between the motor module and the data bus of a micro or mainframe computer. The present computer control module contains a HPIB bus controller. This allows the control of other instruments and data logging units as well as the down-loading of data. If a micro or mainframe computer is not available, or a very portable controller unit is required, then a third module would be made available. This module contains a microprocessor, mass storage, and an external panel of push button controls with a display. Data collected with this module can be downloaded to a computer for analysis at a later date.

The present Laverton controller unit is equivalent to an assembly of the three modules described here. This unit has proven to be reliable over a period of two years.

DISCUSSION AND CONCLUSION

Bitterns is readily available at many potential solar pond sites in Australia. However if its clarity is not high it must be clarified before building the density gradient. The ultra-filtration method described here has the advantage of being portable as it is compact and light, but has the disadvantage of low filtration rates. It is believed that this system will prove to be economically feasible in many situations.

The ring-wave suppression system has been shown to be effective in reducing gradient erosion. Whilst a direct quantitative comparison has not been possible, the experimental evidence is interpreted as being very encouraging.

In commercial solar ponds, the instrumentation and data collection and analysis should be minimised. The instrument controller that has been developed at the University of Melbourne simplifies instrumentation requirements and time consuming data collection. A major advantage of the controller is its low production cost. The three modules together would cost approximately A$1300 (approx. US$900) to manufacture. As the final stage of this development, a commercial instrument manufacturer has been approached to evaluate and market the controller. This company has not actively marketed the instrument but intends to alert the appropriate customers to the product.

REFERENCES

1. National Energy Research, Development and Demonstration Program. Report Number 15. Feasibility of establishing a national solar pond research facility. Department Resources and Energy Canberra (1986).

2. B.P. Marett, W.W.S. Charters, and R.W.G. Macdonald. A review of advances in Australian solar pond technology. ISES ANZ Sect. Annual Conf. (1986).

3. B.P. Marett, Double-Diffusive Interface Behaviour: Growth and Erosion in solar ponds. ISES Solar World Congress Hamburg (1987).

4. R.W.G. Macdonald, W.W.S. Charters, B.P. Marett, D.R. Kaye 1986-87 Solar Pond Project Report (Final). Victorian Solar Energy Council (1987).

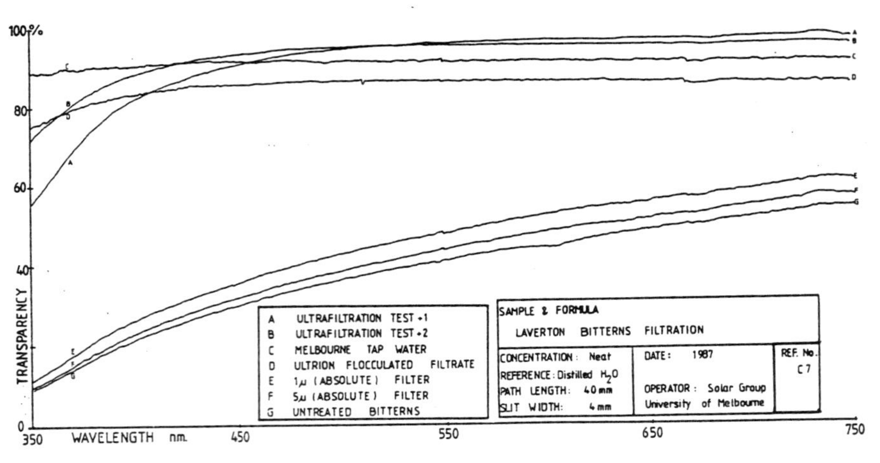

Figure 1. Filtrate clarity as a function of wavelength.

GRADUAL INSTALLATION OF THE NONCONVECTIVE ZONE OF A SOLAR POND

R. R. Johnson*, C. H. Huang*, J. A. Edwards*, and P. Lowrey**

*Applied Energy Research Laboratory, North Carolina State University, School of Engineering, USA

**San Diego State University, San Diego, California, USA

ABSTRACT

The objective of this paper is to show that the initial heating of a solar pond can be accelerated if the nonconvective zone (NCZ) of a solar pond is installed gradually. Comparative simulations are presented for a solar pond in which the NCZ is either installed immediately or the NCZ is installed gradually over the time it takes for the storage zone (SZ) to reach its operational temperature. The rate of thickening is accounted for by a convective velocity term in the equations

KEYWORDS

Solar ponds; Nonconvective zone.

INTRODUCTION

Salt gradient, or non-convecting solar ponds, work by neutralizing natural upward convection. In solar ponds the salinity increases with depth. Consequently the density increases with depth even after sunlight has heated the lower layers. Tabor and Weinberger (1982) give an excellent and comprehensive review of solar ponds.

Solar ponds develop, and are analyzed, in terms of four zones. There is a thin, upper convective zone (UCZ) which is homogenized in salinity, temperature and density. Beneath the UCZ is the non-convecting zone (NCZ), which contains the salt gradient which suppresses convection. Beneath the NCZ is the storage zone (SZ). Underlying the pond is the ground zone (GZ). Heat is lost from the storage zone into the ground zone.

The NCZ of a solar pond is usually constructed over a period of a few days. Subsequently the pond is allowed to heat until the SZ has risen to its operational temperature. Weinberger (1964) showed, however, that there is a relationship between the temperature in the storage zone and the NCZ thickness at which SZ heat absorption is maximized. This feature has been recognized by Batty, Riley and Bhise (1986) in a paper on seasonally varying the thickness of the NCZ. It is also true during the initial heating of a pond. The analysis presented in this paper considers the construction of the NCZ over the full period it takes for the SZ to come to operational temperature, such that the rate of heat absorption may be optimized.

ANALYTICAL MODEL

The properties density ρ, conductivity k, and specific heat C, are assumed to not change with temperature. Also the radiation absorbtion function h(z) is assumed to be an exponential function with two terms. Tybout (1967) gives an equation corresponding to #3 seawater with the form:

$$h(z) = 0.58\, e^{-0.96733z} + 0.42 e^{-15.8466z} \qquad (1)$$

A simple model is considered. The UCZ is assumed to be at a constant temperature T_a = 15°C. The NCZ is modeled as pseudo-steady state heat transfer in which the temperature pro-file goes from one steady state condition to another. The SZ, by contrast, is modeled as being unsteady. The temperature is assumed uniform but is changing because of the rate of energy stored. The ground zone is taken as pseudo-steady state conduction in a thick slab.

Consider the non-convective zone NCZ. If the rate of heat stored in the NCZ is assumed small compared to the rate of heat stored in the SZ, then the unsteady term may be set to zero. The resulting differential equation is:

$$k \frac{d^2T}{dz^2} = H_o \frac{dh}{dz} - \rho C_p V \frac{dT}{dz} \qquad (2)$$

The terms in equation (2) can be identified as being associated with conduction, radiation and convection. Lowery and Johnson (1986) introduced the convection term for solar ponds in a paper concerning the benefits of decanting hot water from the top of the SZ, and returning the same water to the bottom of the SZ after extracting heat. Here it is not the velocity in the SZ that is invoved, but rather an implied velocity because of the thickening of the NCZ with time. The inclusion of this term is an extension of the an analysis previously reported by Lowery, Johnson and Huang (1987).

Integrating twice from ℓ_1 to z (as a differential equation), taking the heat transfer into the storage zone as q_{SZ}, and evaluating at z = $\ell_2 + \ell_1$, results in the following expression for T_{SZ}:

$$T_{SZ} = T + \frac{q_{SZ}}{\rho CV} \left(e^{-\frac{\rho CV \ell_2}{k}} - 1 \right) + \frac{H_0 h_2 g_2}{h_2 - \rho CV g_2} \left(e^{-\frac{\rho CV \ell_2}{k}} \right) \qquad (3)$$

The function g is given as the integral of (h/k) from z=ℓ_1 to z.

The temperature in the SZ is assumed uniform and the same as the temperature at the bottom of the NCZ. An energy balance on all of the SZ may be written as

$$\rho \ell_3 \, C_p \, \frac{dT_{SZ}}{dt} = q_{SZ} - q_{GZ} \quad (4)$$

For convenience the heat transfer in the ground zone GZ is modeled as pseudo-steady conduction through a slab of thickness ℓ_4.

$$q_{GZ} = \frac{k_g}{\ell_4} \left(T_{SZ} - T_a\right) \quad (5)$$

With these basic equations consider two strategies for installing the NCZ; case 1 where the NCZ is assumed to have a constant thickness and V=0, and case 2 where the NCZ is installed gradually and the NCZ thickness ℓ_2 varies with time and $V=d\ell_2/dt$. In case 2, the rate q_{SZ} is optimized when the temperature gradient at the bottom of the NCZ is zero (Weinberger, 1964), or when q_{SZ} = Ho h_2.

The equations are most easily solved numerically with a marching procedure. For a specified value of ℓ_2, the value of T_{SZ} can be found from equation (3). Knowing the previous and current values of T_{SZ}, the time interval to reach the new temperature may be calculated from equation (4).

RESULTS AND DISCUSSION

The results presented here are for variations of a nominal pond which has an UCZ thickness of 0.3m, a NCZ thickness of 1.4 m, a SZ thickness of 2.5m, and a GZ thickness of 5.0m. These dimensions are similar to those planned for large, deep solar ponds for power generation. The net radiation at the surface is taken as H_o = 215 W/m². The thermal conductivity and specific heat of the brine in the NCZ is set at k = 0.59 W/m°C and C = 3800 J/kg°C respectively. The conductivity of the ground k_g = 1.28 W/m°C. The results assume constant temperature T_a = 15°C at the top of the NCZ and at the bottom of the GZ. The operational condition is achieved when the temperature gradient at the bottom of the NCZ is zero for a desired NCZ thickness. The reference case corresponds to making the NCZ thickness equal to the desired final NCZ thickness during all heating. The comparative case

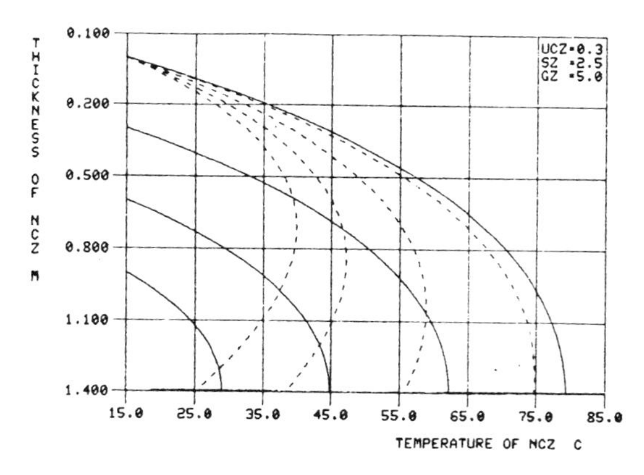

Fig. 1. Comparative change in temperature profiles with time.

allows the NCZ to gradually thicken, at a rate that optimizes heat stored, until the final desired value is reached.

Figure 1 shows a comparative schematic of the change in temperature profiles in the NCZ for the two installation strategies. Dashed lines represent case 1, conventional NCZ heating. Solid lines represent case 2, gradually thickening NCZ. The end conditions for both strategies are the same and the question is the time it takes to reach that end condition. For the case of variable NCZ thickness shown in figure it can be seen that the temperature gradient at the bottom of the NCZ is always zero.

Figure 2 contains a plot of the rise in SZ temperature with time for the case where the desired thickness of the NCZ is 1.4 m. The gradually installed NCZ takes less time to reach the final condition. The rate at which the NCZ has to be thickened can be seen in the corresponding time/thickness plot in Fig. 3.

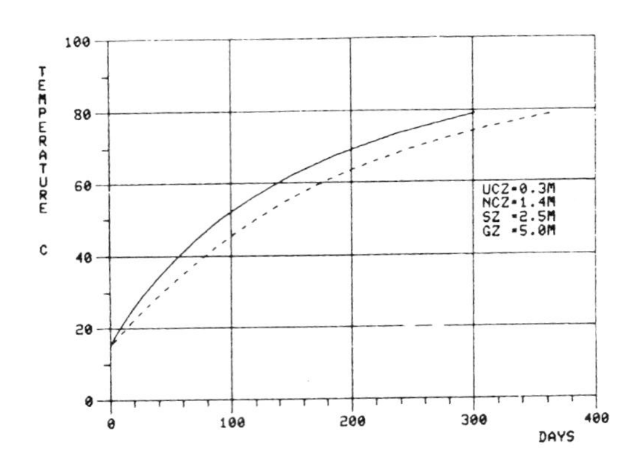

Fig. 2. Temperature rise in the SZ as a function of time.

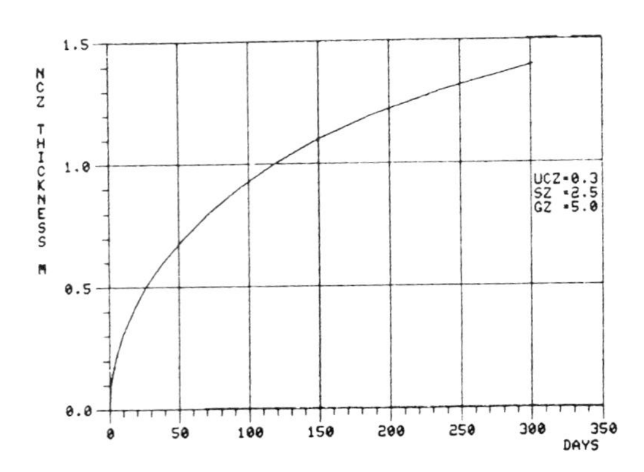

Fig. 3. Rate of thickening of the NCZ.

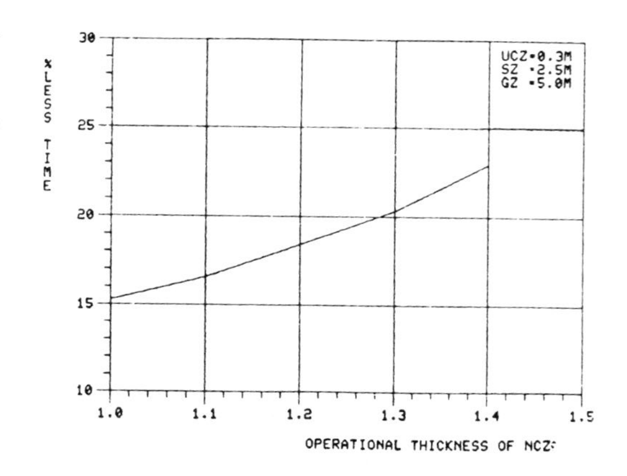

Fig. 4. Reduction in time to reach operating condition.

The time reduction to be expected from the gradual installation of the NCZ is plotted in fig. 4 for variation in the oprational temperature. For a NCZ thickness of 1.4m and an UCZ of 0.3m, the percentage savings of time is 23% . It is also possible that the process of gradually installing the NCZ may be used to limit the growth of an UCZ during heating, which in turn could be expected to make still a larger reduction in time.

CONCLUSIONS

A strategy is suggested for reducing the time it takes a solar pond to initially heat-up to its operational temperature. The idea is to gradually install the NCZ over a period of months rather than a few days. The heat gain in the SZ may be optimized by installing the NCZ at a rate that always maintains a zero temperature gradient at the bottom of the NCZ. Results indicate that there is a reduction in time taken to heat a pond by the new strategy. Calculations for a pond with a UCZ of 0.3m, NCZ of 1.4m, SZ of 2.5m and a GZ of 5.0m show that the expected reduction in time is 23% for a pond with average clarity.

REFERENCES

Batty, J.C., Riley, J.P. and Bhise, N.K. (1986). Optimum Thickness of the Nonconvective Zone in Salt Gradient Solar Ponds, Solar Energy, 36, pp. 15-20.

Lowrey, P., and Johnson R. (1986). Simulation of a Solar Pond Using Upward Flow Through the Storage Zone. ASME J. of Solar Energy Engineering, 108, pp. 325-331.

Lowrey, P., Johnson, R. R., and Huang, C. P. (1987). Optimizing the rate of installation of the NCZ of a solar pond. Proceedings of the International Congress on Solar Ponds, Cuernavaca, Mexico.

Tabor, H. and Weinberger, Z. (1982). In Kreith and Kreider (Eds.), Solar Energy Handbook, Chp. 10.

Tybout, R. A. (1967). A Recursive Alternative to Weinberger's Model of the Solar Pond, Solar Energy, 11, pp. 109-111.

Weinberger, Z.(1964). The physics of the solar pond. Solar Energy, 8, 45-46.

STEADY-STATE ANALYSIS OF THE RISING SOLAR POND

John R. Hull* and Carl E. Nielsen**
*Argonne National Laboratory, Argonne, IL 60439 U.S.A.
**Dept. of Physics, The Ohio State University, Columbus, OH 43210 U.S.A.

ABSTRACT

The rising solar pond is a configuration in which brine is injected into the pond so as to give all or part of the gradient an upward velocity. This alters the salinity profile, steepening it adjacent to the upper gradient zone boundary as required to prevent gradient erosion. This paper presents an analytic expression for the salinity profiles in a rising pond, calculated for temperature dependent solutal diffusivity, and for particular cases it evaluates the upper boundary gradient obtained and the internal stability of the gradient zone.

KEYWORDS

solar pond, salinity gradient solar pond, rising pond, gradient stability

INTRODUCTION

The salinity gradient solar pond requires a salinity profile configured to insure the internal stability of the gradient and to prevent boundary erosion. Because the natural equilibrium profile usually provides a salinity gradient at the upper gradient boundary that is too small to prevent boundary erosion, the surface zone will grow continuously downward at the rate required to create the necessary boundary gradient, becoming deeper at the expense of the gradient zone (Nielsen, 1982). In the rising pond (Assaf, 1976, Nielsen, 1979, 1980) the gradient zone is moved upward at this same rate, thus creating the increased boundary gradient necessary to hold the upper gradient boundary stationary. There is no corresponding difficulty at the lower boundary, because in a pond properly designed and operated to give optimum thermal performance the temperature gradient is close to zero at the lower boundary, and the salinity gradient required there is correspondingly small. With a natural diffusional equilibrium profile the upper boundary erosion limits the total temperature difference possible to at best about 45°C in NaCl ponds. We analyze two variations, including combinations, of the rising pond concept: (1) concentrated brine injected into the lower zone, and (2) brine injected into the gradient zone interior.

SALINITY PROFILES

We define a vertical coordinate system such that $z = 0$ is the bottom of the gradient zone and $z = d$ is the top. We assume that surface and lower zone salinities are maintained constant, the boundaries remain at fixed levels, and effects at side walls may be ignored. Mass transport rate Γ in a diffusive medium moving upward with speed v is given by

$$\Gamma = Sv - \kappa_S dS/dz \; , \tag{1}$$

where S is salinity and κ_S is the solutal diffusivity. Equation (1) applies at any level inside the gradient or at the gradient boundaries. If subscripts u and ℓ refer to values of the quantities at upper and lower boundaries, respectively, then with injection into the lower zone so that v is uniform throughout the gradient,

$$\Gamma_u = S_u v - (\kappa_S dS/dz)_u \; , \quad \text{and} \quad \Gamma_\ell = S_\ell v - (\kappa_S dS/dz)_\ell \; . \tag{2a,b}$$

In steady state, total mass flux Γ is constant and the same everywhere. The quantity $-\kappa_S dS/dz$ is always positive because dS/dz is always negative. In a solar pond κ_S is typically two to three times smaller at the upper boundary than at the lower boundary. This variation of κ_S makes the steady-state salinity gradient larger at the top of the gradient zone even without the influence of an upward v. When v is given a nonzero positive (upward) value, the mass motion term is smaller at the upper boundary than at the lower boundary, because $S_u < S_\ell$, and it follows that diffusion transport must become larger at the upper boundary. This requires a further increase in the gradient $(dS/dz)_u$ at the top of the gradient zone.

If we assume a temperature profile of the form $T = T_\ell - az^2$, which is a fair approximation to the real temperature profile when heat is extracted at the optimum depth, and if we approximate the diffusion coefficient variation by a linear dependence upon temperature, then we can write for the diffusion coefficient at any level

$$\kappa_S(z) = \kappa_\ell(1 - b^2 z^2) \; , \tag{3}$$

in which κ_ℓ is the value at the maximum temperature, i.e., at the lower gradient boundary. Substituting Eq. (3) into Eq. (1) gives

$$\frac{dS}{S-\Gamma/v} = \frac{v}{\kappa_\ell} \frac{dz}{1-b^2 z^2} \; . \tag{4}$$

Integrated over the interval from 0 to z, this yields the solution

$$\ln\left[\frac{S(z)-\Gamma/v}{S_\ell-\Gamma/v}\right] = V \ln[Z(z)] \; , \tag{5}$$

$$V = v/(2b\kappa_\ell) \; , \quad \text{and} \quad Z(z) = (1+bz)/(1-bz) \; .$$

Solving for S(z), we obtain

$$S(z) = \Gamma/v + (S_\ell - \Gamma/v)[Z(z)]^V \; . \tag{6}$$

The values of Γ are related to the upper and lower boundary gradients through Eqs. (2a,b), and there is a whole family of possible salinity profiles depending upon the boundary gradient values. Only certain ranges of boundary gradients are possible of course.

One may wish to specify the upper boundary gradient to correspond to that required to maintain the boundary at a fixed level, according to pond experience at a given site. Putting z equal to gradient thickness d in Eq. (5) gives a relation between S_u, S_ℓ, Γ, and v. The quantities Γ and $(dS/dz)_\ell$ are eliminated from the set of Eqs. (2a,b) and (5), leaving an equation relating $(dS/dz)_u$ to v,

$$(dS/dz)_u = -(v/\kappa_u)(S_\ell - S_u)[Z(d)]^V/\{[Z(d)]^V - 1\} . \tag{7}$$

Fig. 1 gives salinity profiles calculated for constant κ_S and zero upward v, and for variable κ_S and several values of upward v. The gradient thickness is 1.6 m, and S_ℓ and S_u are 280 kg/m^3 and 40 kg/m^3 (at 20°C), corresponding to about 24% and 4% by weight. Temperatures are 85°C at z = 0 and 25°C at the top of the gradient, at z = 1.6 m. The corresponding diffusion coefficients are κ_ℓ = 3.9x10^{-4} m^2/day at the lower boundary and κ_u = 1.3x10^{-4} m^2/day at the upper boundary. In Fig. 1, v = 15 cm/yr results in $(dS/dz)_u$ = -836 kg/m^4, as compared with v = 0, for which $(dS/dz)_u$ = -320 kg/m^4.

STABILITY ANALYSIS

Given fixed salinities at the upper and lower boundaries, a strengthening of the gradient adjacent to the upper boundary necessarily entails a weakening somewhere else. In a rising pond the salinity profile is weakened near the bottom of the gradient, and internal instability may occur there. It can be shown mathematically (Weinberger, 1965), that to a very good approximation in the interior of the gradient the gradient will become unstable unless

$$|dT/dz| < [(Pr+\tau)(\beta/\alpha)/(Pr+1)]|dS/dz| , \tag{8}$$

where Pr is the Prandtl number, $\tau = \kappa_S/\kappa_T$, κ_T is the thermal diffusivity, β is the solutal expansion coefficient, and α is the thermal expansion coefficient. Equation (8) applies everywhere within the gradient zone. In practice, due to sidewall heating and other phenomena, the pond is often operated with a margin of safety so that the temperature gradient is one-half that given by Eq. (8).

Using the thermophysical properties (compilation of Elwell et al. (1977)) for NaCl, and evaluating the parameters together in Eq. (8) indicates a strong temperature dependence, but a much weaker salinity dependence. To simplify the analysis, in the calculational examples we shall assume only a temperature dependence and use the values for a fixed salinity of 10%. Salinity gradients corresponding to the salinity profiles shown in Fig. 1 are plotted in Fig. 2, together with the salinity gradient for marginal stability gradient (M1P) calculated from Eq. (8) and also the gradient with a safety margin of 2 (M2P). If the salinity gradients are larger than the marginal values, then the gradients are stable. Thus, all curves to the right of the marginal curves in Fig. 2 are stable. Rise velocities up to 10 cm/yr give profiles that are stable with respect to the M2P marginal curve. A velocity of 15 cm/yr gives a profile that is stable against the M1P criterion but not against M2P. If instability does occur, the most likely region is the lower-middle part of the gradient.

INTERIOR INJECTION

It has been shown experimentally (Nielsen, 1983) that brine injection into the gradient zone a short distance below the upper boundary can be used to increase the boundary gradient enough even to make the boundary move upward. We may extend the analysis to include both a lower zone input producing upward velocity v and an interior input of salinity S_1 producing additional upward velocity v_1 above the injection level z_1. With interior injection the mass transport equations become

$$\Gamma_\ell = Sv - \kappa_S dS/dz \ , \tag{9}$$

$$\Gamma_u = S(v + v_1) - \kappa_S dS/dz \ , \tag{10}$$

$$\Gamma_u = \Gamma_\ell + S_1 v_1 \ . \tag{11}$$

The solution follows the same procedure as described above. With a single interior injection in a rising solar pond the four parameters v, v_1, z_1 and S_1 determine the salinity profile. Table 1 gives the salinity gradient at the upper boundary for a few selected parameter values.

Table 1. Salinity gradient at upper boundary for selected injection parameters for the case S_ℓ = 280 kg/m^3, S_u = 40 kg/m^3, d = 1.6 m, T_ℓ = 85°C, T_u = 25°C, and an optimized temperature profile.

Entry	v cm/yr	v_1 cm/yr	z_1 m	S_1 kg/m^3	$-(dS/dz)_u$ kg/m^4
1	0	0			320
2	5	0			463
3	15	0			836
4	0	5	0.5	280	468
5	0	5	1.4	280	527
6	0	15	0.5	280	846
7	0	15	1.4	280	958
8	5	10	1.4	280	910
9	10	5	1.4	280	869

We see that for a given input rate the boundary gradient is increased more as the input level is raised from the lower zone toward the gradient upper boundary (entries 2, 4, 5 and 3, 6, 7). Thus when two inputs are used simultaneously the input into the gradient has proportionately more effect than input into the lower zone (entries 8 and 9). In practice the choice of v and v_1 will involve both salt recycling considerations and the internal stability of the gradient profile produced.

DISCUSSION

An analytical expression for the steady-state salinity profile in a rising solar pond shows that a profile can be established using NaCl that is stable in the interior of the gradient and has an upper boundary gradient large enough to resist erosion with a bottom temperature of 85°C and a surface temperature of 25°C. In order to shape the profile to give the best margin of safety for stability throughout the gradient, it is likely to be desirable to use injection into the gradient itself as well as into the lower zone. It is necessary to operate the pond with fairly close to the optimum temperature profile, having small or zero temperature gradient at the lower gradient boundary, to avoid internal instability and lower boundary erosion. Salt recycling cost will be a relevant consideration in determining the injection pattern used, but details of salt management are outside the scope of the present paper.

ACKNOWLEDGEMENTS

Part of this work was sponsored by the U.S. Dept. of Energy under contract No. W-31-109-ENG-38. One of us (C.N.) thanks the Dept. of Physics of the Ohio State University for continuing support.

REFERENCES

Assaf, G. (1976). The Dead Sea: a scheme for a solar lake, Solar Energy 18, 293-299.

Elwell, D. L., T. H. Short and P. C. Badger (1977). Stability criteria for solar (thermal-saline) ponds, in Proc. Am. Sec. Int. Solar Energy Soc., Orlando, 16.29-16.33.

Nielsen, C. E. (1982). Surface zone behavior in solar ponds, ASME Paper 82-WA/Sol-25.

Nielsen, C. E. (1983). Experience with heat extraction and zone boundary motion, in Proc. Am. Solar Energy Soc., Minneapolis, 405-410.

Nielsen, C. E. (1979). Control of gradient zone boundaries, in Proc. Int. Solar Energy Soc., Atlanta, 1010-1014.

Nielsen, C. E. (1980). Nonconvective salt-gradient solar ponds, in Solar Energy Technology Handbook, ed. W. C. Dickinson, and P. N. Cheremisinoff, Marcel Dekker, New York.

Weinberger, H. (1965). Physics of the solar pond, Solar Energy, 8, 45.

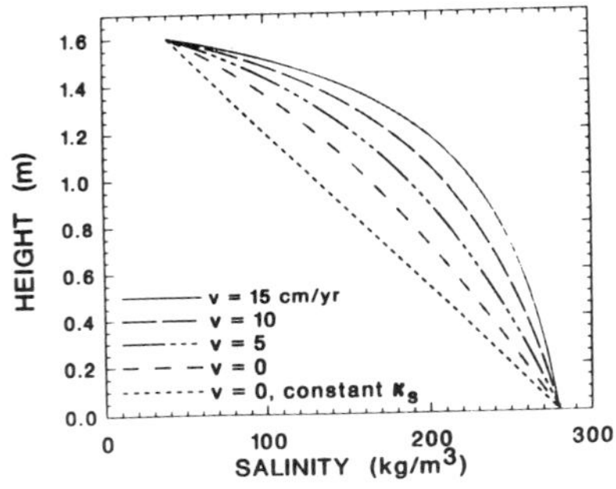

Fig. 1. Calculated salinity profiles for a rising solar pond.

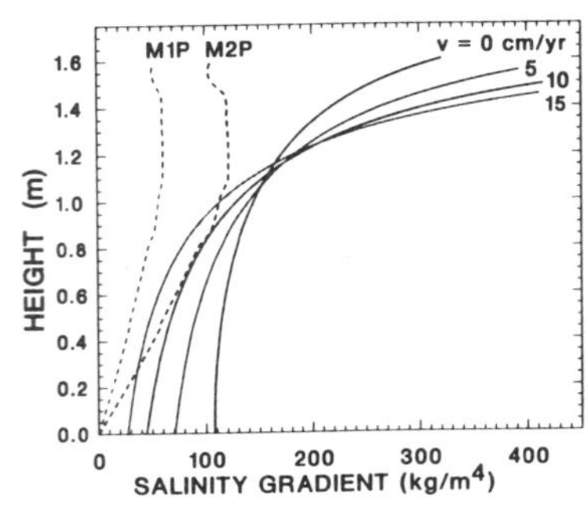

Fig. 2. Comparison of calculated salinity gradients for different rise velocities (solid curves) with marginally stable salinity gradients (dashed) for parabolic temperature profile: M1P - marginal curve, M2P - curve M1P with safety factor of 2. Boundary conditions are the same as in Fig. 1.

SOLAR PONDS FOR NORTHERN LATITUDES

G. K. Abdul Sada and L. F. Jesch

Solar Energy Laboratory, Department of Mechanical Engineering
University of Birmingham, Birmingham B15 2TT, UK

ABSTRACT

The salt gradient solar pond is a body of saline water in which the concentration of salt increases with depth. It is used for the collection and seasonal storage of solar energy. The pond has three identifiable layers: the upper convective zone UCZ, the non-convective zone NCZ and the lower convective zone LCZ which is used for high temperature storage. This temperature does not fluctuate with changes in ambient temperature.

A computer model was made for Birmingham, UK (52.4 N) with a cold/moderate climate and frequent changes in the weather. This model simulates the dynamic behaviour of the pond. Hourly data are used of solar radiation, ambient air temperature and wind speed measured in the Solar Energy Laboratory for the past ten years. The governing heat balance equations of the pond are solved numerically, taking into account heat losses from the surface, side walls and bottom. The response of the pond to solar gain and heat losses is followed for three years. The study shows that solar ponds could be used in high latitudes for space heating.

KEYWORDS

Solar pond; earth pit; vertical temperature profiles; seasonal storage

INTRODUCTION

The solar pond has been developed for hot regions where high solar radiation is predominent (Joshi,1984, and Zangrando 1977). In the countries of high northern latitudes, where the solar radiation is less, several theoretical studies have been made. In the UK Bryant (1977), and Hawlader (1981), in Finland Lund (1982) have reported new ideas and feasibility estimates.

In this study the thermal behaviour of a pond at high latitude and cold climate has been analysed by solving the energy balance eguations numerically, taking into account heat losses from the surface, side walls and bottom.

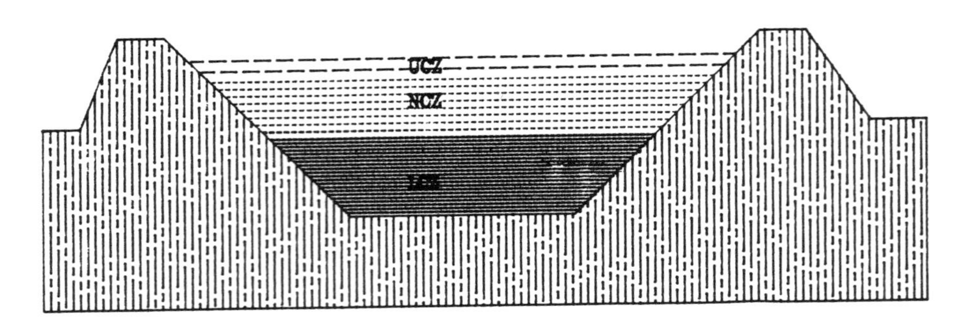

Fig 1 Zones of the salt gradient solar pond:
UCZ: upper convective zone
NCZ: non convective zone
LCZ: lower convective zone (storage)

ECONOMICAL EXCAVATION OF A SOLAR POND

An earth pit with side slopes up to 45° is one of the most economical construction,(Short 1979) for large commercial solar ponds. There are three ways of excavating the pond as listed in Table 1. The first one is to excavte it to the desired depth. Thus the soil has to be taken from the bank of the pond. Therfore the cost of the pond includes the cost of transportion and excavation. The second one excavates the pit to about one half of the desired depth and removes the soil to form the raised bank above the level of the surrouding ground. The third way uses no excavation and requires to transport the soil to build the pond. The second type is most economical.

Table 1 shows pond construction types

CASE	DESCRIPTION
I	The surface of the pond is at the same level as the land surface 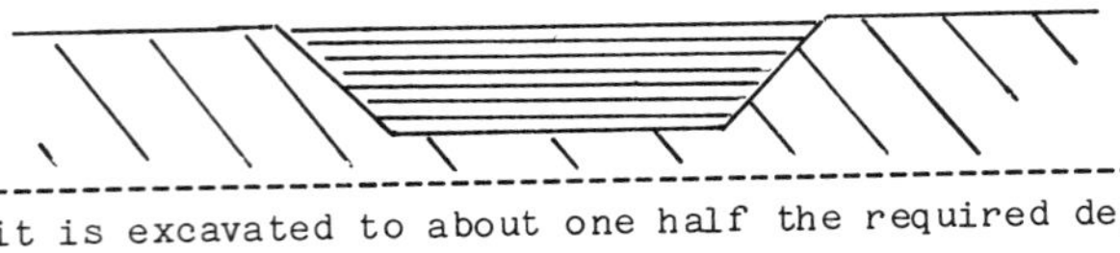
II	The pit is excavated to about one half the required depth 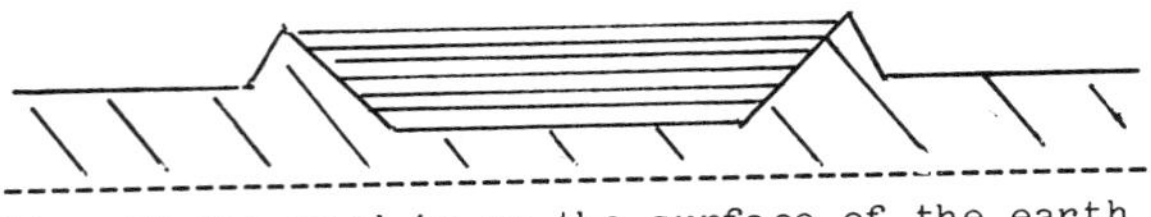
III	The bottom of the pond is on the surface of the earth 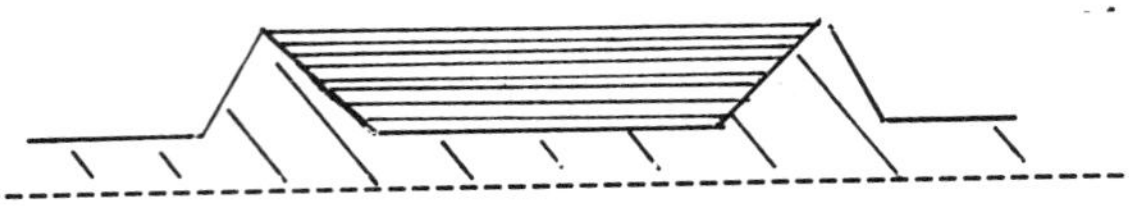

SIMULATION RESULTS

The simulation results were tested with experimental data from a solar pond at Sussex UK (Unsworth 1983). Fig. 2 shows the variation in the storage temperature, solar radiation incident on horizontial surface and ambient temperature with time for three years of simulation.

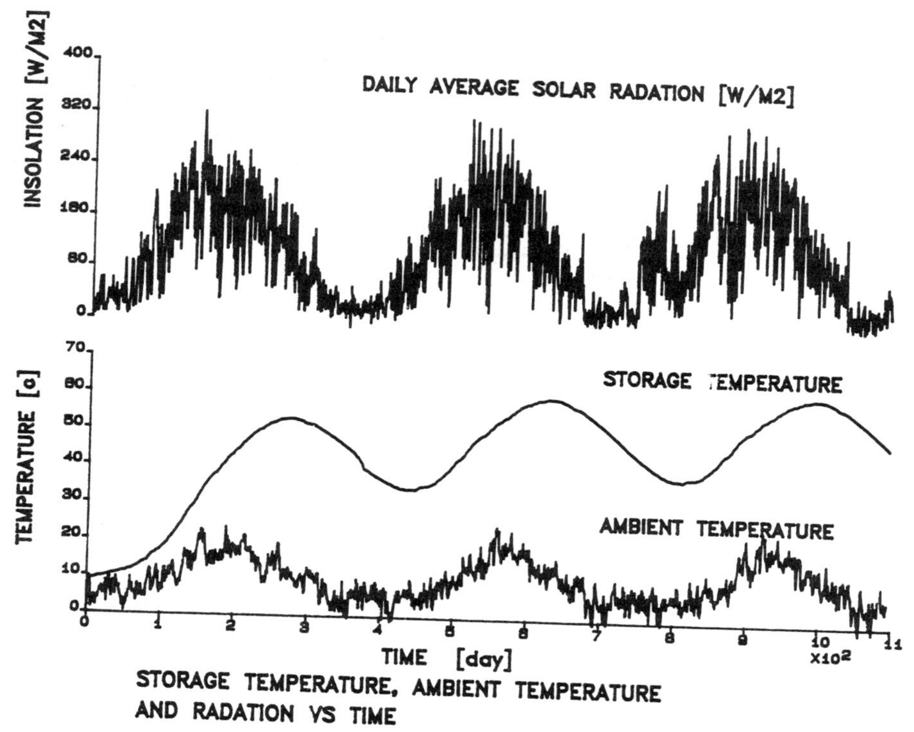

Fig. 2: Solar radiation, storage and ambient temperatures vs. time

The three dimensional plot in Fig. 3 shows the rate of useful energy available to be extracted from the solar pond. It is based on bin data for 24 hours daily and for 365 days a year. In Birmingham most energy is only availabe from April to October.

The development of vertical temperature profiles through the depth of the pond were calculated for a three year period from January 1982. These are shown in figures 4a, 4b and 4c. In each year only the months of April, July and September are given here. It is seen from Fig. 4c that the pond approched the steady state as the highest temperatures reached in September have not increased much after the second year.

The experimental results of Unsworth (1983) indicating the vertical temperature profile are given in Fig. 5 for comparison. It is reasonably close to our calculated values for Birmingham.

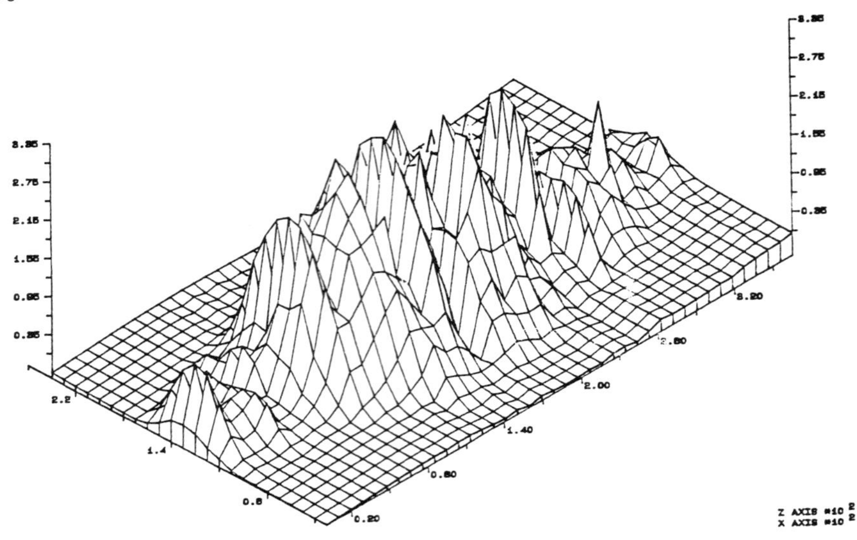

Fig. 3: Useful energy extractable from a solar pond in Birmingham, UK

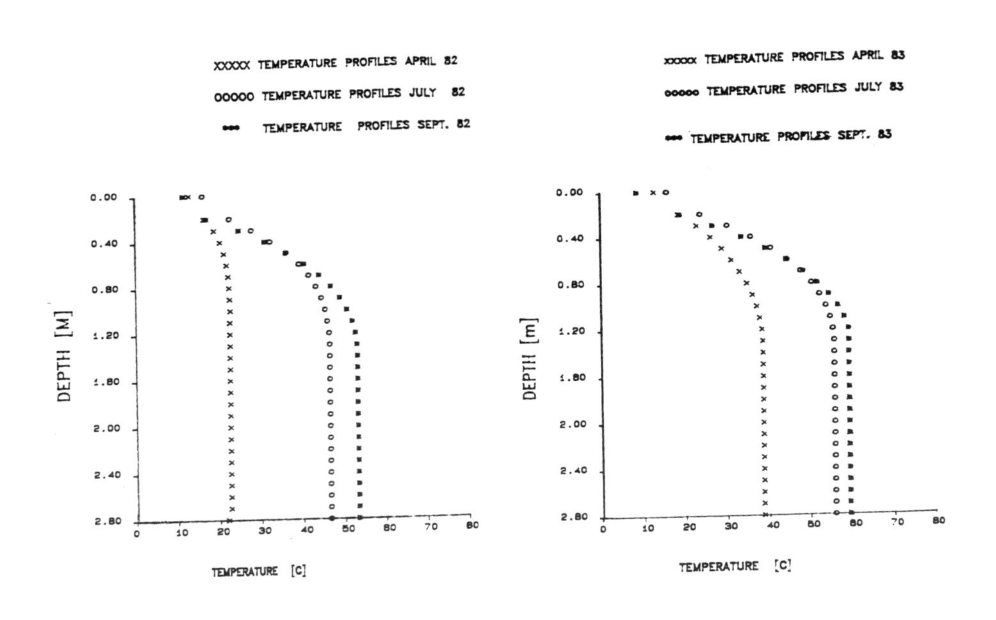

Fig. 4a: 1st Year temperature profile

Fig. 4b: 2nd year temperature profile

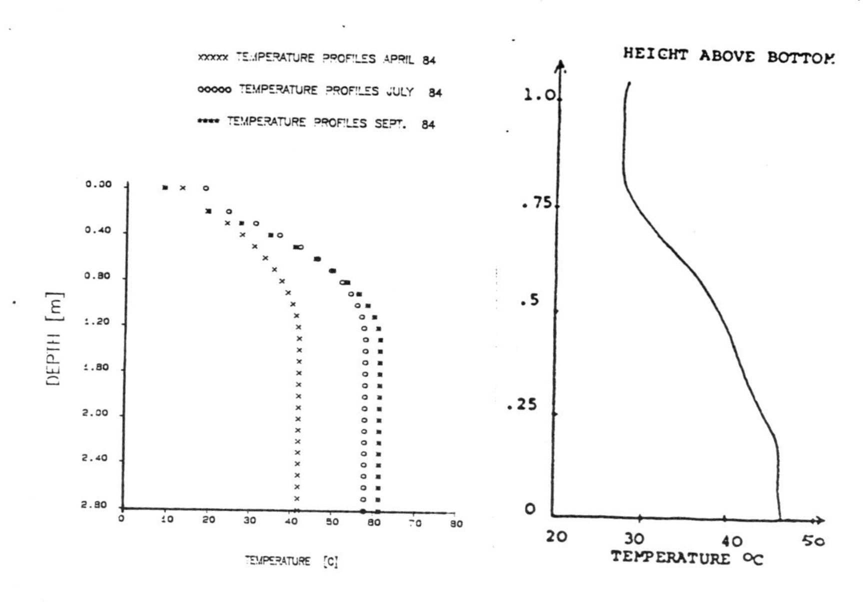

Fig.4c: 3rd year temperature profiles Fig.5: Unsworth (1984) exp. profile

CONCLUSION

The study shows that solar ponds at high latitudes can be used for space heating but not for power production at the present time.

REFERENCES

Bryant H.C. and I.Colbeck (1977): Solar Pond For London, Solar Energy 19, p.321

Lund P.D. and J.T.Routti (1982): Feasibility of solar pond heating for northern cold climates , Report TKK-F-A491

Short, T.H. (1979): The development and demonstration of a solar pond for heating greenhouses, Sun II, v2, p 1021

Unsworth P.J., N. Al-Saleh and V. Philips (1984): Improving solar pond efficiency by heat extraction from the non-convective zone, pp 44-55, UK ISES C35, London

Veena Joshi, V.V.N.Kishore and K. S. Rao (1984). A Digital Simulation of the non-convecting Solar Pond for Indian Conditions. Renewable Energy Sources International Progress, Part A, Edited by T.N.Veziroglu

Zangrando H. and H.C. Bryant (1977): Heat extraction from salt gradient solar ponds. Proceedings of the international conference on alternative energy sources, Miami Beach, Florida.

COMPUTER SIMULATION OF AN INTERNAL HEAT EXTRACTION SYSTEM FOR SALT GRADIENT SOLAR PONDS

M. Metwally* and A. Abouhabsa*

* Faculty of Engineering & Tech. at Mattaria (HELWAN Univ.)
P.O. 11718 Cairo/EGYPT.

ABSTRACT

In this study, a computer model for a heat extraction system for a solar pond to be built in Cairo zone was developed. The heat extraction system simulates grain drying, in which the temperature of the ambient air is raised when it passes through a liquid-to-air heat exchanger that is coupled to an in-pond polyethylene-tube heat exchanger. The model was used to predict the optimum size and performance characteristics of the involved heat exchangers for a variety of pond conditions and heat loads applied for Egypt climatology.

KEYWORDS

Solar applications; solar ponds; pond simulation; heat extraction; in-pond heat exchanger; Egypt climatology.

INTRODUCTION

Salt gradient solar ponds are effective solar energy collectors which provide simple, economical and reliable long-term energy. This energy is suitable for many applications including industrial process heat, district heating and, when combined with suitable conversion system, production of electricity.

An atractive application for the use of solar ponds in Egypt is providing heat for grain drying processes. The proposed heat extraction system consists of in-pond polyethylene tube heat exchanger that is coupled to a conventional fin-tube liquid-to-air heat exchanger, to produce the heated air. Figure 1, shows a schematic of the proposed extraction system.

The objective of this study is to develop a computer simulation procedure to provide a complete analysis of pond and heat extraction system performance.

SOLAR POND PERFORMANCE

The model chosen for the analysis of pond heating up and heat extraction processes is shown in Fig. 2. It consists of three layers: The surface layer of thickness l_S, has a low and uniform salt concentation and a constant temperature which is assumed to be equal to the ambient air temperature; a nonconvective gradient layer of thickness l_G, in which the salinity increases with the depth and it serves as insulating layer; and the bottom convective layer of thickness l_B, in which the salt concentration and the temperature are assumed constant. It serves as heat collection and storage medium.

Considering the bottom layer as a thermodynamic system and applying the energy conservation law yields :

$$dT_P/dt = (1/m\ C)\ [(dq_I/dt) - (dq_E/dt) - (dq_L/dt)] \quad (1)$$

Where :

dq_I/dt = The rate of solar energy absorbed at the bottom layer. dq_E/dt = The heat extraction rate. dq_L/dt = The heat loss rate. dT_p/dt = Estimated pond temperature rate. mc = heat capacity.

The Runge-Kutta method was used for solving the linear differential eqn(1) in order to predict the temperature T_P of the convective bottom layer (pond temperature) over the course of the year. By using small time increments in the itrative computer program the external driving functions dq_I, dq_E and dq_L were accurately followed for any geometrical and physical parameters of the solar pond model. The computer model is assumed to be in steady state when the temperature T_P at the end of the year equals to within 0.1 % the temperature T_P at the beginning of the year.

Both hourly and daily computation increments were applied. It has been found that computations using daily average insolation agreed with those using hourly insolation data. As a result the pond simulation presented here was performed over time steps of a day thus permitting its day-to-day transient behaviour to be predicted with a minimum of computational effort.

The values of the mean daily global solar radiation were assumed to be constant over the month and were computed from the measured values listed in Table 1, for Cairo zone.

TABLE 1 : Mean Daily Monthly Global Radiation for Cairo [KWh/m²/day].

Jan.	Feb.	Mar.	Apr.	May	June	July	Aug.	Sep.	Oct.	Nov.	Dec.
3.29	4.31	5.35	6.47	7.30	7.81	7.57	7.04	6.13	5.02	3.69	3.13

The solar energy absorbed at the bottom layer of the pond over an entire day is given by :

$$q_I = H\tau(\alpha_{eff}, z) \quad (2)$$

Where H is the monthly mean daily total radiation, and the transmittance τ is function of the monthly mean daily effective solar altitude angle α_{eff} and the vertical depth z.

For the transmittance function τ, the following simple correlation as proposed by Reddy (1986) was selected :

$$\tau(\alpha, z) = 0.08\ (1-R)\ \ln\ [90\ \cos\theta_R/z\ ] \quad (3)$$

Where R is the Fresnel coefficient of reflactance at the surface of the pond; θ_R is the angle of refraction at the surface of the pond; z the vertical depth from the surface of the pond to the upper surface boundary of the bottom layer $z = l_S + l_G$.

The heat loss from the solar pond bottom layer was simulated over the course of the year using measured data of the monthly mean daily average of the ambient temperature and the ground temperature for Cairo zone (EMA, 1976-1982). The total heat lost from the bottom storage layer was computed as the sum of heat loss through top, bottom, and sides of the bottom layer.

$$q_L = q_{TOP} + q_{BOT} + q_{SW} \quad (4)$$

PERFORMANCE OF THE IN-POND HEAT EXCHANGER

The design work for the in-pond heat exchanger is concerned with the following two major phases:
- Estimation of the heat exchanger size for the maximum heat extraction rate.
- Estimation of the performance characteristics of the designed heat exchanger for the varying pond temperatures and heat extraction rates over the course of the year.

In the first phase of design the maximum heat extraction rate q_E and the respective pond temperature T_P are determined from the solar pond computer program. The rate of heat extraction can be expressed in form :

$$q_E = \pi\ d_o\ L\ U.LMTD \quad (5)$$

Where d_o and L denote the external diameter and the total required length of tubes respectively, the LMTD is the logarithmic mean temperature difference LMTD = $(T_{wo} - T_{wi})/\ln\ [(T_P - T_{wo})]$ and the overall heat transfer coefficient U is given by :

$$1/U = [(d_o/d_I h_i) + (d_o/2\ k).\ln\ (d_o/d_i) + (1/h_o)] \quad (6)$$

The heat transfer coefficient inside the tubes h_i can be aproximated quite well by the following correlation for liquids being heated under turbulent flow conditions in long, straight, smooth circular pipes :

$$Nu = 0.023\ Re^{0.8}\ Pr^{0.4} \quad (7)$$

The heat transfer coefficient outside the tubes is limited by the natural convection between the outer tube walls and the brine of the bottom layer. For natural convection the Nusselt number is given in the following form :

$$Nu = c\ (Gr.Pr)^n = cRa^n \quad (8)$$

Where c,n are constants depending on configuration and art of the flow. For horizontal submerged cylinders, the characteristic dimensions for natural convective heat transfer is the outer diameter d_o. The Nusselt number can be calculated from the following relations as recommended by Karlekar (1977):

$$Nu = 0.53\ Ra^{1/4} \qquad \text{for } 10^4 < Ra < 10^9$$
$$Nu = 0.13\ Ra^{1/3} \qquad \text{for } 10^8 < Ra < 10^{12} \quad (9)$$

The fluid properties like viscosity, thermal conductivity, etc..., entering into the above equations are evaluated at a temperature that is the arithmatic mean of the wall temperature and the ambient fluid temperature.

The required total heat transfer area of the heat exchanger was calculated according to the following steps:

- Calculate the LMTD from the terminal temperatures T_P, T_{wi} and T_{wo}.
- Calculate the appropriate rate of flow for the heat exchange fluid from $m = q_E/[C\ (T_{wo} - T_{wi})]$. The number of parallel pipes may be obtained by estimating the flow velocity inside the tubes, which has an upper limit given by the maximum admisable value of the pressure losses.
- Calculate the overall heat transfer coefficient U.
- Calculate the required heat transfer area A.

The performance of the in-pond heat exchanger can be estimated from eqn (5) after substituting for LMTD and simplifying we obtain

$$T_{wo} = T_P - (T_P - T_{wi})/EXP(UA/mC) \quad (10)$$

Equation (10), was used for evaluating the heat exchanger performance for different heat extraction rates and pond temperatures according to the following steps :

- Assume T_{wo}.
- Calculate LMTD and U.
- Estimate T_{wo} from eqn (10) and compare with the assumption.
- Repeat until satisfactory agreement is obtained then calculate q_E.

RESULTS AND DISCUSSION

Computations were performed for a proposed solar pond to be built in a Cairo suburb having a square surface area of 10000 m² and a total depth of 3.0 m. The pond shall be used for providing hot air for grain drier. The recommended allowable drying air temperature is about 65 °C throughout the year (Cimbria, 1986).

To study the efect of the gradient layer thickness on pond temperature , three program runs were made, in which the thickness of the gradient layer was changed from 0.2 to 2.2 m with an increment of 0.2 m. For each run, the pond was assumed to be loaded at a level 5%, 7.5%, and 10% of the monthly mean daily insolation. For all simulations, the overall depth of the solar pond remained constant with a constant surface layer thickness of 0.3 m.

The results-presented in Fig. 3, indicate that an optimum thickness of the gradient layer for maximum pond temperature is near 0.8 - 1.2 m depending on pond loading.

The effect of different pond loadings are illustrated in Fig. 4. The results indicate that increasing the loading rate reduces the temperature of the bottom layer considerably. Fig. 4 shows that a variable heat extraction rate at 7.5% of the mean daily insolation maintains the temperature in

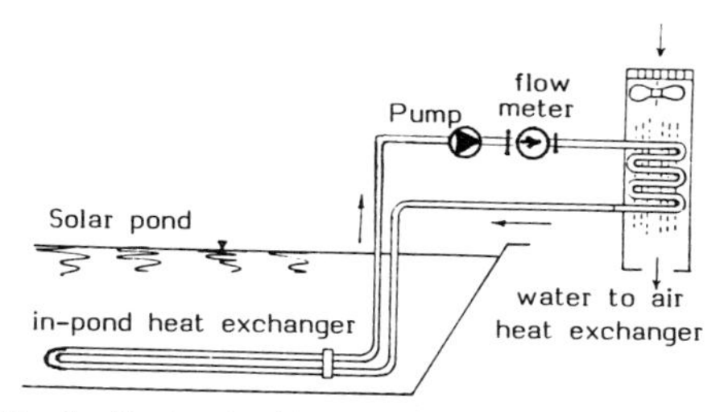

Fig. 1. Heat extraction system of SGSP, using submerged polyethylene H.E.

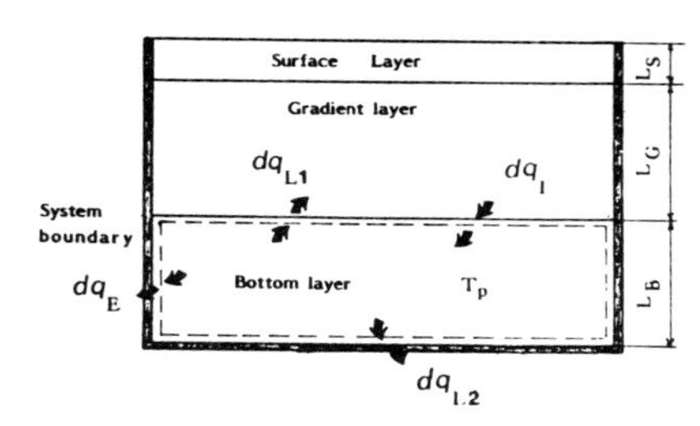

Fig. 2. Solar pond model and system boundaries.

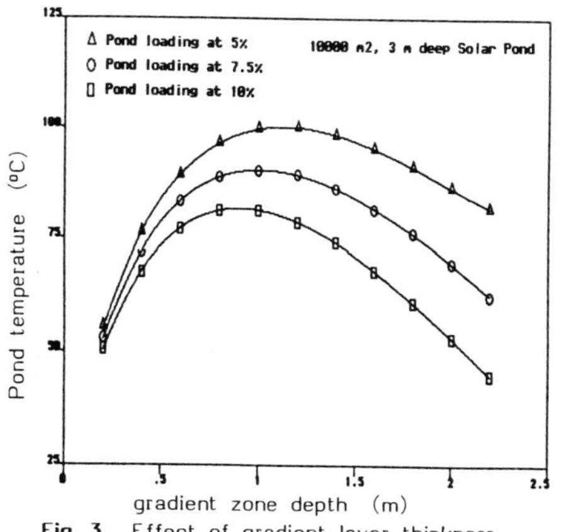

Fig. 3. Effect of gradient layer thickness.

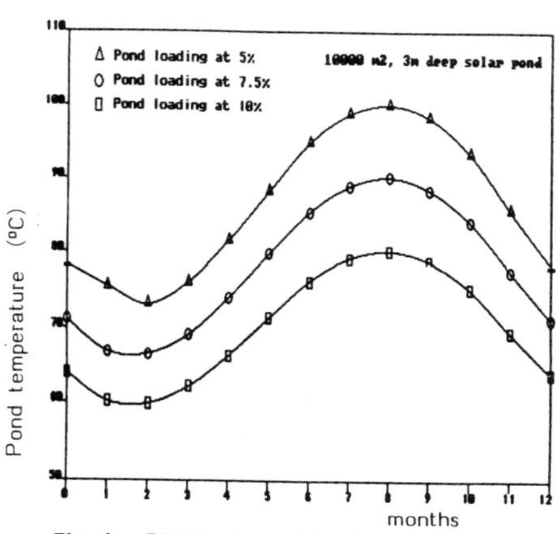

Fig. 4. Effect of pond loading rate.

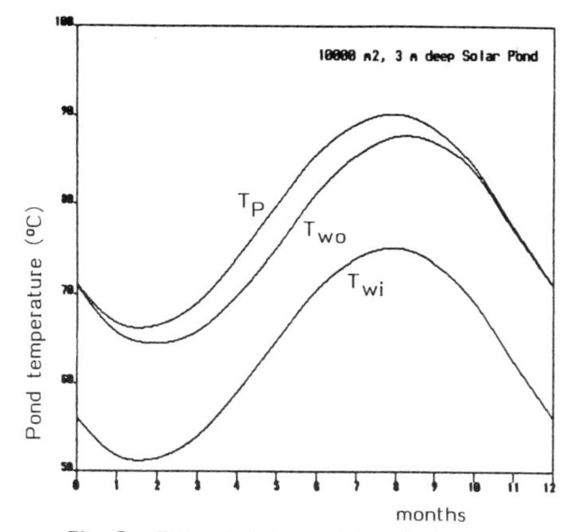

Fig. 5. Estimated in-pond terminal temperatures.

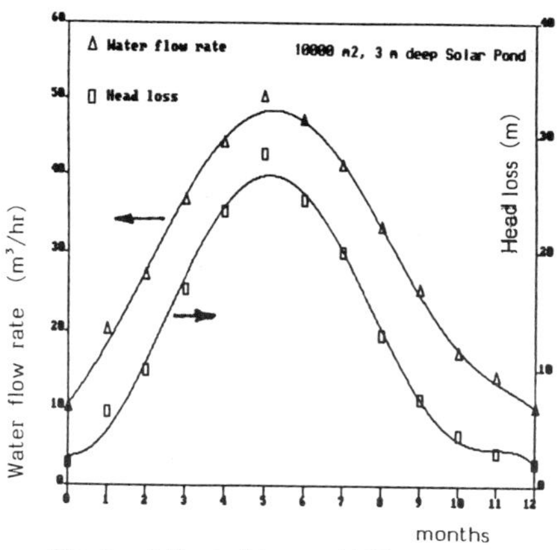

Fig. 6. Estimated in-pond H.E. water flow rate & head loss.

the bottom layer high enough, so that a good performance of the associated heat exchangers for producing drying air at 65 °C is secured. For pond loading at 7.5% of the mean daily insolation the bottom layer temperature reached a peak of 90 °C in September and was close to the drying air temperature during February and March.

As a result a gradient layer thickness of 1.1 m and a heat extraction rate at 7.5% of the mean daily insolation is recommended. This yields a monthly variable pond loading from a maximum value of 5800 KWh/day in June to a minimum value of 1700 KWh/day in January.

An in-pond submerged polyethylene-tube heat exchanger as described by Sabetta (1985) and Metwally (1985) was designed to carry out the maximum pond heat extraction loading rate of 600 KW (assuming 10 working hours per day), where the number of tubes was selected so that the power consumed for pumping water through the in-pond heat exchanger is not to exceed 1% of the pond loading.

The design characteristics are given as follows : Brine temp. (80 °C), water outlet & inlet temperatures (75 and 65 °C), no. of tubes (26), tube O.D. (25 mm) and tube thick. (2.1 mm), which gave the following "results" : Overall tube length (6566 m), rate of flow (52 m^3/hr), head loss (31 m), power consumption (5.9 KW), overall heat transfer area (515 m^2) and overall heat transferc coefficient (127 w/m^2 °C).

The performance characteristics of the in-pond heat exchanger have been computed for the monthly varying pond loading over the course of the year. The resulted water terminal temperatures at inlet T_{wi} and outlet T_{wo} are shown in Fig. 5. The computed water outlet temperature reached a maximum value of 87.6 °C in September and a minimum value of 64.5 °C in March.

Simulation of the varying pond loading was performed through controlling the water flow rate pumped into the in-pond heat exchanger. Figure 6, shows the required water flow rate throughout the year. Also the variation of the head loss in the tubes around the year were estimated as shown in Fig. 6, thus yields a good selection of the required pumping unit.

CONCLUSIONS

(1) A simple computer program was developed for simulation of the performance of solar ponds and associated heat extraction system. **(2)** Based on computational results it appears that a variable heat extraction rate at about 7.5 % of the mean daily solar irradiance is realistic loading for the proposed 10000 m^2 (3 m depth)solar pond, to produce hot air at 65 °C for grain drying. **(3)** A higher loading results a lower pond temperature, which does not fullfil the recommended drying air temperature.

REFERENCES

Cimbria, (1986). Grain dryers guide. Cimbria Unigrain ltd., A/S 3700 Thisted, Denmark.

Egyptian Meteorological Authority (EMA), (1976-1982). EMA Yearly reports, Cairo/Egypt.

Karlekar, B.V. and R.M. Desmond, (1977). Engineering heat transfer. West publishing company.

Metwally, M.N. (1985). Solar ponds as a source of heat and power generation in Egypt. FRCU project no. 830217. Report no. 7. Fac. of Eng. & Tech. at Mattaria, Cairo/Egypt.

Reddy, T.A., S. Jumpa and G.Y. Sannier, (1986). Effective daily mean position of the sun for solar ponds. Solar Energy, vol. 37, no. 1.

Sabetta, F., M.' Pacetti and P. Principi (1985). An internal heat extraction system for solar ponds. Solar Energy, vol. 34, no. 4/5.

TRANSMISSION RADIATION IN SOLAR PONDS

S. Perez, A. Macias, A. Hernandez, A. Gomez

Dpto. Ing. Procesos, E.T.S.I.I., Univ. Politec.
de Canarias, Tafira, 35017, Las Palmas, España

ABSTRACT

In this paper an analysis of transmissions spectra for different salt solutions is made. Base solutions are: NaCl, $MgCl_2$, KCl,and $MgSO_4$. They were studied in binary and ternary way. The transmission spectra has been measured in the range 300-900 nm

KEYWORDS

Solar pond; transmission spectra; attenuation factor.

INTRODUCTION

In order to obtain a good collecting efficiency a sufficient transparency of the pond is essential. The attenuation of solar radiation depends on the quality of water, and on the of presence of salt and suspended particles. The main purpose of this report is to determine the attenuation factors of the $NaCl-MgCl_2-H_2O$, $MgCl_2-KCl-H_2O$, $KCl-NaCl-H_2O$, $MgSO_4-MgCl_2-H_2O$, systems, as well as those corresponding the NaCl, $MgCl_2$, $MgSO_4$, solutions. Transmissions spectra for differents salt solutions have been measured.

EXPERIMENTAL AND THEORETICAL APPROACH

The transmission spectra were measured with a spectrofotometer VIS-UV Shimadzu UV-260 in the range 300-900 nm. using quartz-glass sample containers of 5 cm. length. Utilized salts are part of the contents of sea water in the Atlantic Ocean, Canary Islands. Taking into account the positive and negative ions analysis and also hypothetic composition of the water, (Rodier, 1971) we are in position to known the ocean own salts. Ions in disolution do not exist in a combined way and the group under shape based on their reciprocal affinities. An analysis of the salts is shown on TABLA 1.

TABLE 1 Quantitative Analysis of the Atlantic Ocean, Canary Island

NaCl	29.25(gr/l)	79.61(%)
$MgCl_2$	3.59 "	9.77 "
KCl	0.85 "	2.31 "
$CaSO_4$	1.10 "	2.99 "
$MgSO_4$	1.95 "	5.31 "

Concentrations of the systems to the studied have to be such that they correspond to the unsaturated disolution zone on each system phase diagram, which have been determined from available data in literatura (Brunisholz,1963; Damilano,1961,1962; Ughfoot 1946) of the eight systems which have been studied, two of them are represented on figure 1. On each phase diagram the criteria, within the no saturation zone, of keeping a proportion amoung the components of each system 9:1, 3:2, 2:3, and 1:9, three points being chosen for the system eight differents in transmission spectra.

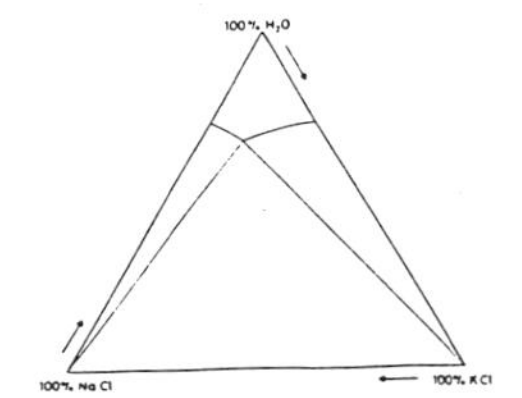

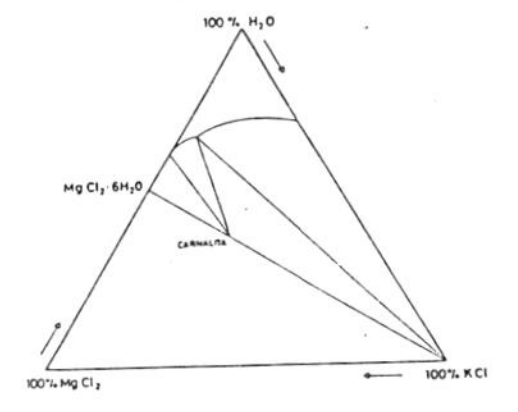

Fig. 1 Phase diagrams of the $NaCl-KCl-H_2O$, and $MgCl_2-KCl-H_2O$ systems, 25ºC

For each sample transmission spectra has been measured in the spectrofotometer, forty graphics being obtained as shown on fig. 2. If the spectra distribution at the surface is known, one may use thin data as input into the solar pond model. One assumes the spectral distribution to have the same form as the standard solar irradiance curve,(Thekaekara, 1967). A more practical form of presenting the transmission properties of the solutions is to give the relative attenuation of the solar spectrum in the solution. We define an attenuation factor (Lund, 1984),which is transmitted through the sample solution.

$$\tau = \frac{\int_0^\infty \tau_s(\lambda)\ I(\lambda)\ T(\lambda)\ d\lambda}{\int_0^\infty I(\lambda)\ d\lambda} = \frac{\sum_i \tau_{s,i} I_i\ T_i\ \Delta\lambda}{\sum_i I_i\ \Delta\lambda} \qquad (1)$$

where $\tau\ (\lambda)$ is the measured transmission spectrum, $I(\lambda)$ is the solar radiation spectrum, and $T(\lambda)$ is the transmission at air-water interface which we assume the unit.

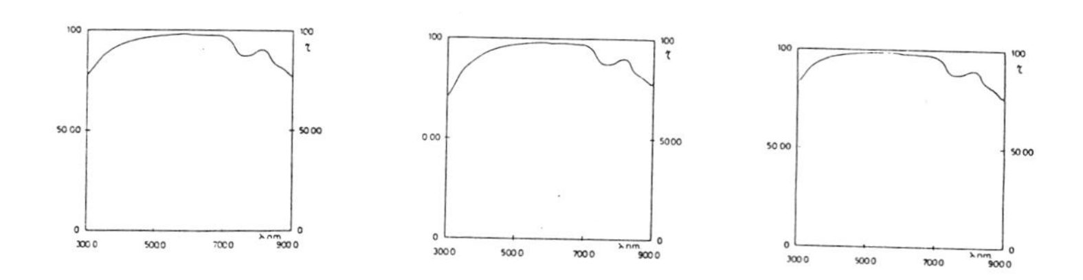

Fig. 2 Transmission spectra measured for a) $NaCl-MgCl_2-H_2O$, b) $MgCl_2-KCl-H_2O$, c) $MgCl_2-MgSO_4-H_2O$

RESULTS AND CONCLUSIONS

We may combine the data in Fig. 2 and the standar solar irradiance curve to get the total intensity of a direct solar beam integrated over all wavelengths. With Hull (1986) criteria, the spectrum is divided into 24 nm bins with the first of 40 bins centered at 325 nm and the last bin centered at 900 nm. With eqn. 1, the attenuation factor is determined, as shown on Table 3.

TABLE 2 Transmission Spectra Measurement in the Range 300-900 nm.

	τ		τ
$MgSO_4$ (02.75%) $MgCl_2$ (25.50%)	0.682	NaCl (02.50%) $MgCl_2$ (22.50%)	0.677
NaCl (02.50%) KCl (21.50%)	0.696	$MgCl_2$ (25.00%)	0.677
		NaCl (22.00%)	0.698
$MgCl_2$ (25.00%) KCl (03.25%)	0.679	KCl (20.00%)	0.692
		$MgSO_4$ (12.00%)	0.695

The attenuation factor of radiation in salt solutions has been investigated. Following solutions were studied $MgSO_4-MgCl_2$, $MgCl_2$-KCl, NaCl-KCl, NaCl-$MgCl_2$, $MgCl_2$, $MgSO_4$, NaCl and KCl. On found that the absortion of pure sea water was practically identical with that of distilled water (Clarke, 1939).

REFERENCES

Brunisholz, G., and M. Dodmer (1963). Helv. Chim. Acta, 46, 7, 2566-74.
Clarke, G. L., and H. R. James (1939). Laboratory analysis of the selective absorption of light by sea water. Optical Soc. Amer. Jour., 29, 43-55.
Damilano, J., and J.Baabor (1961). Bol. Soc. Chilena Quim., 11, Mes 1-2, 3-7.
Damilano, J., and J. Baabor (1962). Bol. Soc. Chilena Quim., 12, Mes 1, 7-16.
Hull, J. R. (1979). Physics of the Solar Pond, NTIS, Ames Laboratory, Iowa State University, 36-37.
Lighfoot, W.J., and C.F. Prutton (1946). J. Am. Chem. Soc., 68, 101-102.
Rodier, J. (1971). L'Analyse Chimique et Physycochimie de L'Eau, Dunod, Paris, p. 373.
Thekaekara, M. P. (1967). Solar Energy, 9, 7.

THE MEASUREMENT OF RADIATION TRANSMISSION IN A SALT-GRADIENT SOLAR POND

A.A. Green, A.L.M. Joyce and M. Collares-Pereira

Departamento de Energias Renováveis
L.N.E.T.I., Estrada do Paço do Lumiar, 22
1699 LISBOA Codex Portugal

ABSTRACT

A spectrophotometric technique is described for determining the spectral absorption coefficients of solar pond samples. Spectral transmittance measurements made on pure water are found to be in good agreement with published data, validating the technique and providing an indication of its accuracy. A radiation transmission profile deduced from measurements made on slightly turbid, unchlorinated water from an outdoor swimming pool is compared with direct broadband transmittance measurements made in the pool using a specially-adapted, submersible pyranometer. Discrepancies between the two sets of results are discussed. A simple numerical model of the absorption of beam radiation in a solar pond is used to deduce a radiation transmission profile from spectral transmittance measurements made on samples obtained at various depths in a solar pond in Portugal. The profile is compared with broadband transmittance measurements made in the pond using the submersible pyranometer. It is concluded that the spectrophotometric and pyranometric techniques are complementary in this application, with the former able to produce reliable data for the first metre below the surface, and the latter more suitable and sufficiently accurate for routinely measuring the radiation reaching the storage zone.

KEYWORDS

Solar pond; solar radiation; transmission profile; submersible pyranometer; spectrophotometer; spectral absorption coefficients.

INTRODUCTION

The use of a pyranometer to make broadband transmittance measurements in a solar pond involves several difficulties, including large errors caused by refractive effects. These problems may be avoided by collecting samples at various depths in the pond and employing a spectrophotometric technique to determine their spectral absorption coefficients, from which the radiation transmission profile of the pond can then be deduced. Such an approach has been developed to aid the interpretation of pyranometric measurements made in the salt-gradient solar pond at Pégões in Portugal.

TECHNIQUE

Precision sample cells of 40 mm and 5 mm optical pathlengths are located respectively in front of the sample and reference beam entrance ports of a 60 mm diameter integrating sphere in a Perkin-Elmer Lambda 9 UV/VIS/NIR spectrophotometer, using the holders provided. The cells were matched by comparing their transmission spectra, in order to obtain a pair with almost identical spectral transmittance curves in the 330-1400 nm wavelength range. Errors due to small differences in spectral transmittance between them are minimised by running a background correction scan with them empty and clean, prior to each sample scan. With the same liquid sample filling both cells, the spectrophotometer directly measures its spectral transmittance over the 35 mm pathlength difference between the two cells. The spectral absorption coefficients of the liquid are then calculated from the transmittance data. A detailed description of this technique is given by Green and Joyce (1987). Absorption coefficients of greater accuracy could be obtained using a longer cell in front of the sample beam entrance port, but cells of 50 mm optical pathlength are not readily available in Portugal, and anything longer than that cannot be accommodated in the integrating sphere compartment.

RESULTS

The foregoing technique has been validated using degassed, distilled water. The measured spectral transmittance curve for the 350-1350 nm wavelength range is compared in Fig. 1 with values deduced from optical constants data for pure water (Hale and Querry, 1973). It can be seen that corresponding spectral transmittance values differ at most by 0.02, and those which are higher than 0.9 differ by less than 0.005.

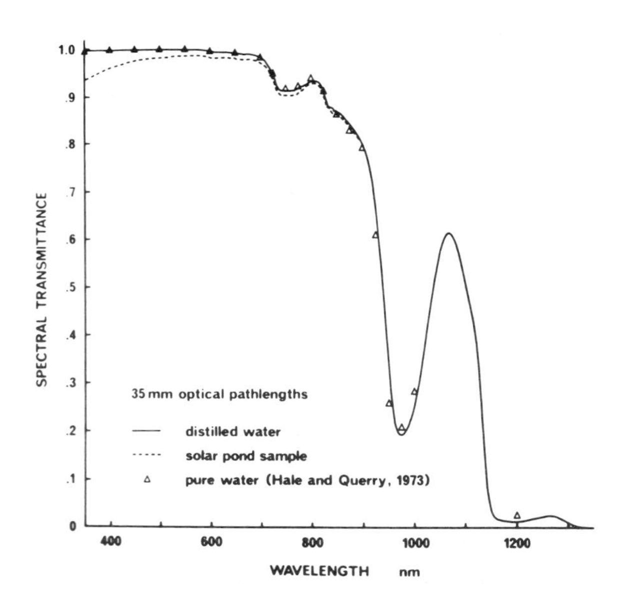

Fig. 1. Spectral transmittance curves for distilled water and a solar pond sample, compared with values deduced from optical constants data for pure water.

The "selected ordinate" procedure of Lind, Pettit and Masterson (1980) is used to derive broadband transmission profiles from spectral transmittance data. This approach requires that the spectral absorption coefficients of the liquid be determined at 100 selected wavelengths corresponding to the 100 equal-energy intervals of a representative AM 1.5 solar spectrum; the wavelength corresponding to each interval being that which subdivides it into equal-energy halves. When applying this method to aqueous solutions, the absorption coefficients corresponding to the 20 selected wavelengths greater than 1400 nm can be justifiably treated as being infinite, because radiation of those wavelengths is so strongly attenuated in water that it penetrates no further than a centimetre below the surface. Transmittance values corresponding to those wavelengths are therefore assumed to be zero. The absorption coefficients are used to calculate, for each of several pathlengths in the range 0.1-2.0 m, the spectral transmittance of the liquid at the 80 selected wavelengths less than 1400 nm. The broadband transmittance over each pathlength, for the representative AM 1.5 solar spectrum, is then obtained by averaging the 80 calculated and 20 zero spectral transmittance values for that pathlength.

Transmission profiles derived in this manner from the spectral data for pure water and from spectrophotometric measurements made on slightly turbid, unchlorinated water from an outdoor swimming pool are compared in Fig. 2 with a transmission profile for clear sea water calculated using the equation of Rabl and Nielsen (1975). The latter profile is evidently unrepresentative of the somewhat turbid water typical of dusty locations. Also presented in Fig. 2 are direct broadband transmittance measurements made in the outdoor swimming pool using a specially-adapted, submersible pyranometer devised by Collares-Pereira and Joyce (1986). The discrepancies between these and the calculated transmission profile decrease with increasing depth, suggesting that the errors to which the submerged pyranometer is subject diminish as the transmitted radiation becomes more diffuse. Further measurements will be made to confirm this. The pyranometric errors due to refractive effects probably depend on the angular distribution of the transmitted radiation, which in turn depends on both the angular distribution of the radiation at the surface and the turbidity of the water. The preliminary results do not indicate a simple correction procedure for pyranometric measurements made in clear water.

The calculation of the transmission profile of a solar pond is complicated by the change in composition, and hence spectral absorption coefficients, of the pond water with depth. Spectrophotometric measurements are made on samples collected near the centre of the pond, at 0.4 m intervals between the surface and a depth of 2 m. These are used to compute, for each of the 80 selected wavelengths less than 1400 nm, the least-square cubic regression curve of the corresponding absorption coefficient on depth. Cubic equations are utilised in order to accommodate different forms of variation with depth. The regression curves are used to compute, for each of several depths in the range 0.1-2.0 m, the spectral transmittance of the pond at the 80 selected wavelengths less than 1400 nm. This is done using a simple numerical model of the absorption of beam radiation propagating at a constant angle to the vertical in a solar pond (Green and Joyce, 1987). As before, the broadband transmittance at each depth is obtained by averaging the 80 computed and 20 zero spectral transmittance values for that depth.

The measured spectral transmittance curve of distilled water is compared in Fig. 1 with that of a sample collected 0.1 m below the surface of the solar pond at Pégões. The organic and particulate matter and dissolved salts in the pond evidently have little effect on the spectral transmittance of the

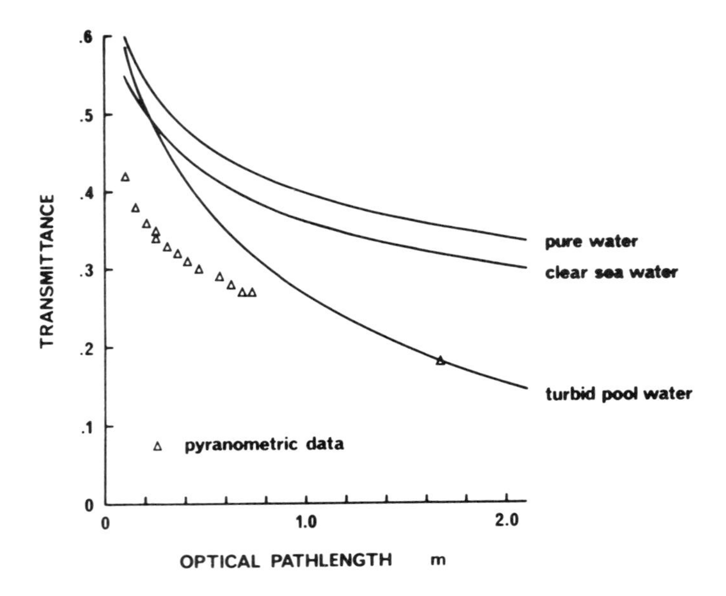

Fig. 2. Transmission profiles deduced from spectral data for pure water, sea water and slightly turbid water from an outdoor swimming pool, compared with pyranometric measurements made in the swimming pool.

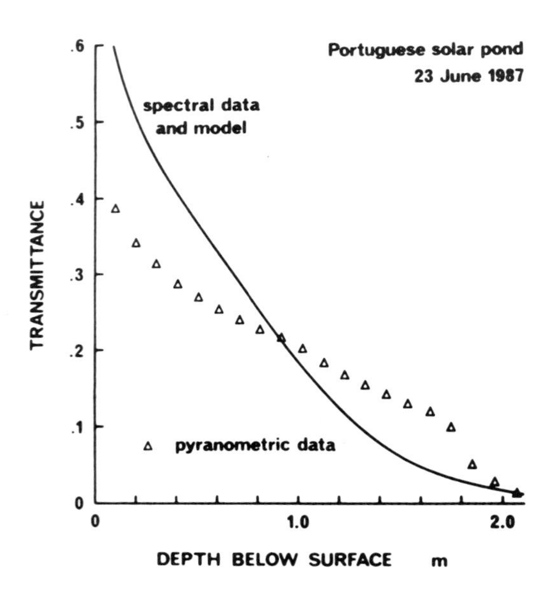

Fig. 3. Pyranometric measurements made in the Portuguese solar pond, compared with the transmission profile computed from spectral transmittance measurements made on samples of the pond water.

water at wavelengths greater than 900 nm. The transmission profile of the pond, computed from such spectrophotometric data, is compared in Fig. 3 with broadband transmittance measurements made in the pond using the submersible pyranometer. It can be seen that the pyranometric measurements and computed transmission profile converge more rapidly with increasing depth than was the case for the less turbid swimming pool water, suggesting that the pyranometric measurements made at depths greater than 1 m may be fairly reliable, since the transmitted radiation is almost entirely diffuse there. The accuracy of the computed transmission profile, on the other hand, is expected to deteriorate with increasing depth for a number of reasons. Small errors in the values of absorption coefficients deduced from 35 mm pathlength spectral transmittance measurements become significant when those coefficients are used to calculate transmittances over much longer pathlengths. Ray bending due to a variation in the refractive index of the pond water with density and salt concentration gives rise to errors which also increase with depth. Furthermore, the rapid fading of an appreciable dark tint in samples from deeper than about 1.5 m, after they have been exposed to air, results in underestimation of their *in situ* spectral absorption coefficients. It is possible that these sources of error are largely responsible for the discrepancies between the two sets of results for depths greater than 1 m.

CONCLUSION

The spectrophotometric technique described in this paper was conceived as a means of determining a simple procedure for correcting refractive errors in pyranometric measurements made in solar ponds. Comparisons of transmittance data for the Portuguese salt-gradient pond, obtained using both techniques, indicate that those errors are strongly dependent on the angular distribution of the radiation incident on the submerged pyranometer, being greatest just below the surface of the pond but relatively unimportant in the storage zone where the transmitted radiation is mostly diffuse. Generally the user needs to measure only the radiation that reaches the storage zone. The preliminary results reported here suggest that this can be done with sufficient accuracy using the submersible pyranometer. The more complicated spectrophotometric technique can therefore be reserved for special purposes such as accurately determining the transmission profile above the storage zone.

REFERENCES

Collares-Pereira, M. and A. Joyce (1986). Salt gradient solar pond for a greenhouse heating application. Proceedings of the Ninth Biennial Congress of the International Solar Energy Society "Intersol 85", 1515-1520, Pergamon.

Green, A.A. and A.L.M. Joyce (1987). A spectrophotometric method for determining the transmission of radiation in a solar pond. Proceedings of the CEC Workshop on Optical Property Measurement Techniques, Ispra, Italy.

Hale, G.M. and M.R. Querry (1973). Optical constants of water in the 200 nm to 200 μm wavelength region. Applied Optics, 12, 555-563.

Lind, M.A., R.B. Pettit and K.D. Masterson (1980). The sensitivity of solar transmittance, reflectance and absorptance to selected averaging procedures and solar irradiance distributions. Trans. ASME Journal of Solar Energy Engineering, 102, 34-40.

Rabl, A. and C.E. Nielsen (1975). Solar ponds for space heating. Solar Energy, 17, 1-12.

THERMAL PERFORMANCE OF SMALL SCALE EXPERIMENTAL SALT GRADIENT SOLAR POND IN CAIRO/EGYPT

M. Metwally*, H. Heikal* and A. Abouhabsa*

* Faculty of Engineering & Tech. at Mattaria (HELWAN Univ.)
P.O. 11718 Cairo/EGYPT.

ABSTRACT

Salt gradient solar ponds (SGSP) are an effective method for solar energy collection and thermal energy storage with a lower cost per unit area. In order to produce suitable design data for large ponds and to obtain an estimate of the effect of the various design and operating parameters on the performance of such ponds, a small 4 m² - 1 m deep outdoor SGSP was built and tested under climatic conditions of Cairo zone (30 °N) in Egypt.

KEYWORDS

Solar applications; experimental solar ponds; heat extraction systems; polyethylene heat exchanger; Cairo climatology.

INTRODUCTION

Two major methods may be used for heat extraction from solar ponds: in the first one a heat exchanger is submerged into the lower convective zone, while in second design the hot brine is withdrawn from the lower convective layer, passed through an external heat exchanger and then returned to the bottom of the pond. The first method was applied in this study by utilizing a cheap and simple in-pond heat exchanger made out of reinforced polyethylene tube which is preferable to metal pipes because of the abscence of corrosion or erosion.

DESCRIPTION OF THE EXPERIMENTAL SGSP

The small experimental SGSP was constructed in the premises of the Faculty of Eng. & Tech. at Mattaria, Cairo, Egypt, which consists of welded steel octagonal plane with aparture surface area of 4.25 m² and 1 m depth (Fig. 1). The internal pond surface was coated by 1.5 mm thick reinforced fiberglass sheets with an isophthalic polyester resin. All outside walls and bottom surfaces were thermally insulated. An air blower (7.5 HP) was installed as a source of air stream via air diffuser at variable speeds for wind effect study (not included).

Initially twelve tungsten-halogen lamps (500 watts each) were used to accelerate heating of the pond with about 600 W/m² as intensity of infrared radiation (0.7 - 2.8 µm) at a height of 80 cm from pond surface. Black matt metallic cover cap was used above the lamps to keep all radiation directed to pond surface. The electric current was stabilized at 220 volt to fix the level of lamp illumination. This artificial heating was applied for 20 days in continous operations, then the pond was opened to the direct exposure of solar radiation.

Filling and Washing System

Filling the pond was accomplished by applying the "salinity redistribution method", while the pond was filled partially at first time with high salinity brine (raw washed sodium chloride local salt of LE 50/ton was used), to 2/3 of the pond depth. Fresh water was then pumped through a diffuser of 7.5 cm diameter (closed circular bottom) with 2.5 mm circumferntial space, with peripheral exist velocity of about 6 cm/s ($\sim$ 2 lit/min) to avoid agitation, starting just over the depth of desired bottom zone (50 cm). The diffuser was subsequently risen to the surface in upward vertical discrete steps of 2.5 cm for each 5 cm increase in water level up to the pond surface. The same diffuser was used to wash the pond surface to make-up for evaporation ($\sim$ 10 mm/day) as well as to clean it from dirt using a discharge of 10-15 lit/day.

Heat Extraction System

An in-pond heat exchanger cross linked polyethylene tube was submerged just below the interface between the convective and the salt gradient regions, in spiral round shape (16/20 mm diameters, with 12 m total length) as shown in Fig. 2. The polyethylene tube is easily bent and has relatively good thermal characteristics (0,38 W/m °C thermal conductivity, 2.2 kj/kg °C specific heat and 2.10^{-4} $°C^{-1}$ as coefficient of expansion up to 100 °C), combined with good resistance at moderate temperature up to 100 °C (Sabetta, 1985). Its density is about 940 kg/m^3 which is lower than the brine density of about 1150 kg/m^3 at 60 °C. That makes the heat exchangers when filled with fresh water float in the convective zone, which can be easily affixed to the bottom.

The rate of fresh water flow circulation was 1 lit/min during 4 hours/day at temperature difference Δt_{io} = 6 °C between inlet and exist of the external heat exchanger (corresponding to 10% of daily incident solar irradiance on the pond). This fresh water flow rate was regulated by means of a circulating pump (Solar Maxi, 5130-27 SMC, 80 Watt) of variable speed. The operation of the pump is controlled by means of adjustable "differential temperature control unit" (Commodore 103/180, central heating circulator, SMC, ranging Δt_{io} from 2 to 12 °C), which is connected to two temperature sensors (hot and warm heat exchanger water sides).

Pond and Meteorology Monitoring Instruments

The following equipments were used :

(a) Twenty type T stainless steel sheathed in-pond thermocouples were used spaced at 5 cms immersed in the pond centre. Also five thermocouples were used in the wall and bottom insulation, and four thermocouples used in the external heat exchanger.

(b) Two flow meters were used, one for filling and washing rates, and the second was for circulating fresh water flow rates through the in-pond and the external heat exchanger.

(c) Conductivity electronic transmitter (870 EC electrodeless-Foxboro), accompanied with PX-sensor, which has a calibrated indicator reading salinity (in % wt.) directly.

(d) An Epply black and white pyranometer (intensity of global solar radiation), a precision linear thermistor (ambient temp.), humidity probe (relative humidity), and micro response anemometer (wind spead) from WEATHERTRONIC. All of these equipments were equiped with signal conditioning modules.

(e) Data Acquisition System (DAS), including "Data logger" (Digistrip II,Kaye Instruments, 48 Channels) and an IBM computer.

The meteorological, pond and heat extraction system data were collected, manipulated and analysed by the "DAS".

TESTING PROCEDURE AND ANALYSIS OF RESULTS

(a) Mean daily data of solar irradiance, ambient temperature, relative humidity and wind speed were measured hourly by the local meteorological station in the solar laboratory of the faculty and are represented in Fig. 3, for one year.

(b) The brine salinity was measured once a day; the convective zone had been stabilized in few days at 50 cm, while gradient zone moved from 30-40 cm and surface layer moved from 10-20 cm; as indicated in Fig. 4, during the heating period.

Fig. 1. General top view of the experimental (4 m² - 1 m deep) solar pond, installed in Fac. of Eng. (Helwan Univ., Cairo - EGYPT).

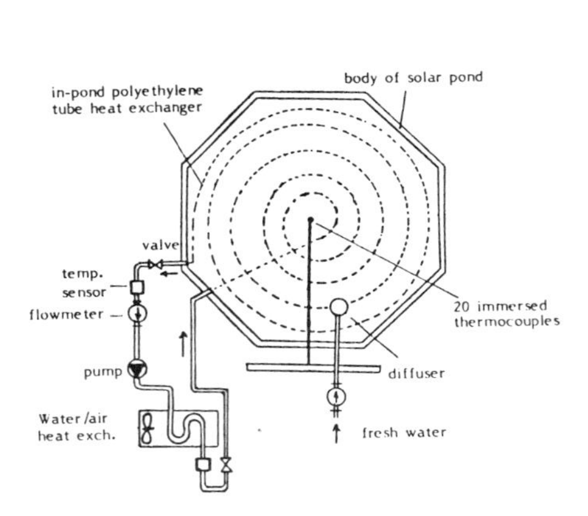

Fig. 2. Schematic of heat extraction system from the exp. SGSP.

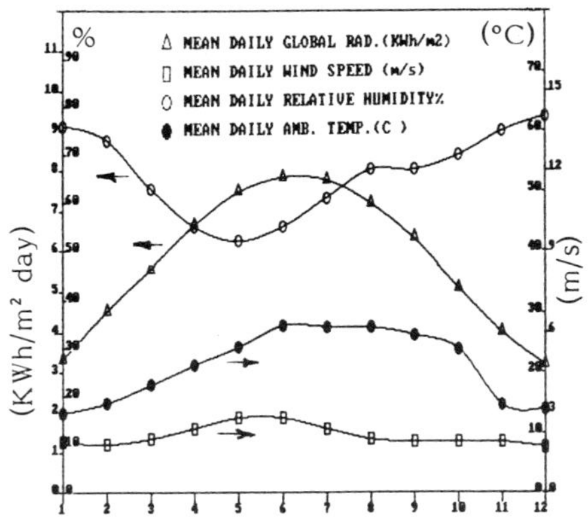

Fig. 3. Climatological conditions of Cairo zone (30 °N).

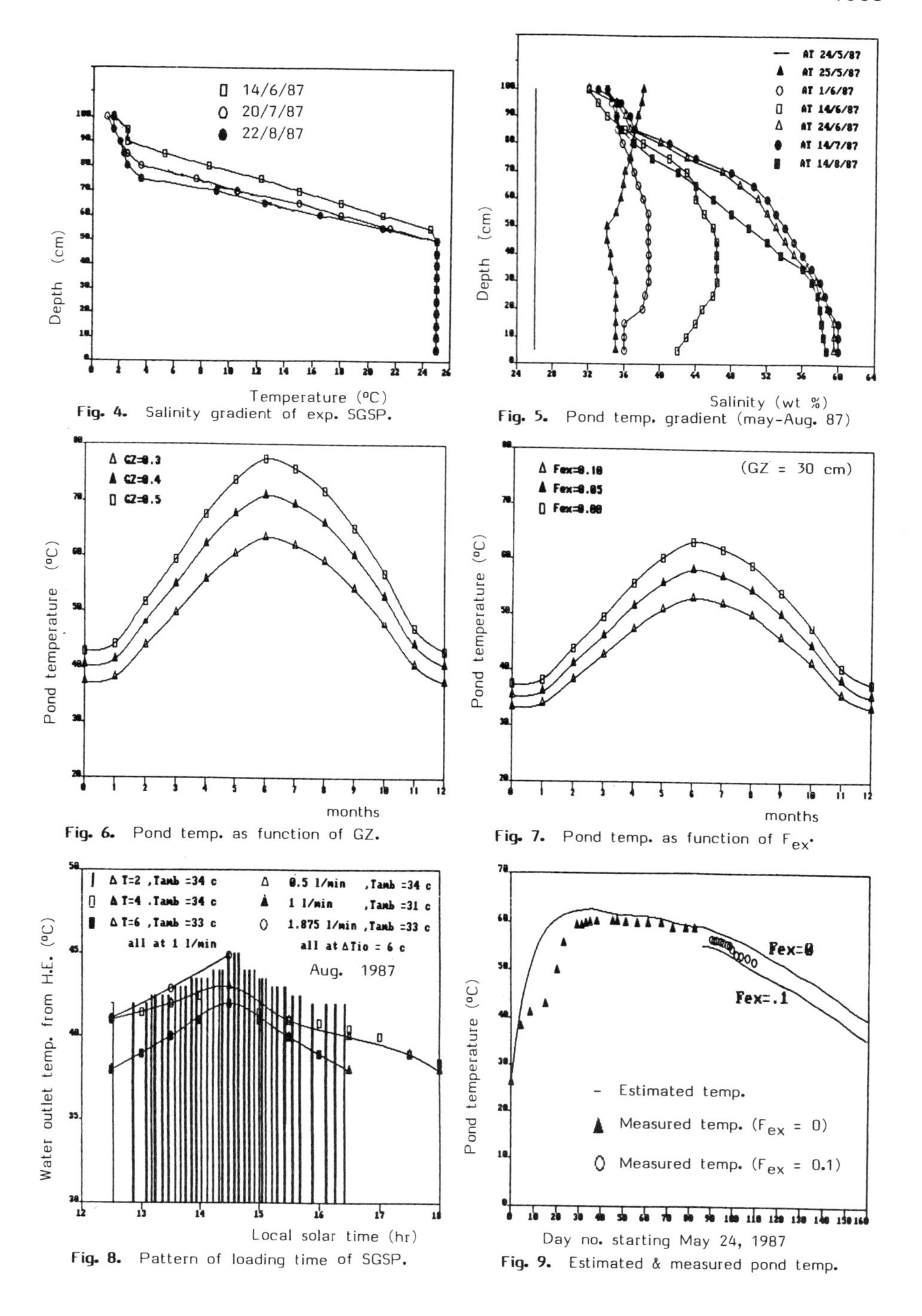

Fig. 4. Salinity gradient of exp. SGSP.

Fig. 5. Pond temp. gradient (may-Aug. 87)

Fig. 6. Pond temp. as function of GZ.

Fig. 7. Pond temp. as function of F_{ex}.

Fig. 8. Pattern of loading time of SGSP.

Fig. 9. Estimated & measured pond temp.

(c) In-pond temperatures were measured hourly and it was found that maximum daily pond temperature (bottom zone) usually occurs around 3 PM. Therefore we are presenting an historical temperature gradient in the pond at this time for subsequent days, the first period (24 May to 14 June) represent the effect of the artificial heating 24 hrs/day (increase of 20 °C/20 days) as all of this heat was transfered by conduction (IR-radiation absorbed at the surface). It is noted from Fig. 5, that on the 14 th June there is a negative gradient slope of temperature from the middepth towards the bottom wall, which was the reason behind the cancelation of the artificial heating.

(d) Natural exposure to solar radiation started in June 15. Three months of monitoring indicated that the maximum hourly bottom temperature occurs also around 3 PM, and max. daily existe in June as shown in Fig. 5, with subsequent decrease in the following months. This figure shows that the general configuration of the temperature gradient for the three zones follows the expected shape and thus loading we found afterwards should yield reliable results.

(e) A computational program referred to Metwally (1985, 1987) had been proposed for hourly and daily calculation of pond temperature (T_P) in the bottom zone, taken in consideration the shadow effect of side walls (Lund, 1984). The effect of varying the depth of gradient zone (GZ) from 0,3-0,5 meter on T_P was presented in Fig. 6, all round the year, which represents a limitation of max. T_P of about 62 °C for GZ = 0,3 m (actual present stabilized case). Generally the increase of GZ provide the ability to increase T_P all over the year. Figure 7, represent by the same program the monthly variation of T_P as function of the percentage of heat extracted (F_{ex}) referred to the incident monthly irradiance in the pond. It can be deduced that if F_{ex} exceeds 10%, T_P will drop less than 40 °C (mean of the year) thus causing problems in loading. Hence an F_{ex} of 10% should not be exceeded for small SGSP in Egyptian conditions.

(f) Figure 8, represents the loading effect on varying the circulation flow rate of fresh water (m_w) through the external heat exchanger and its temperature difference (Δt_{io}) on the pattern of daily loading time at F_{ex} = 10 % (1.4 Kwh/day in August). It was found that at constant Δt_{io} = 6 °C and varying m_w, it affects only on leading period, but it had continous operation. For constant m_w and low values of Δt_{io} = 2 °C, we had an intermittent loading period.

(g) Figure 9, illustrates an excellent coincidence between the theoretical estimation presented by Metwally (1987) and the experimental curves of the present work representing the heating process (F_{ex} = o) of bottom zone for natural solar radiation exposure (about 30 days after starting heating). First 2 weeks of heat extraction (F_{ex} = 10%) is also presented in the same Fig. 9. Despite the fact that the loading period was short, it agrees quite well with the trend of the theoretical estimation.

CONCLUSIONS

(1) A simple installation and low cost of tube heat exchanger was proposed, which can be applied for heat extraction system from any size of solar ponds. Monitoring of pond heating history and computational program assure the feasibility of large ponds application for heating purposes in Egypt. **(2)** Extraction factor must be within the limit of 10% of daily incident irradiance under the climatology conditions of Egypt to obtain acceptable pond temperatures. **(3)** In order to obtain a continous loading period, the temperature difference Δt_{io} of 4-8 °C was found necessary. **(4)** At a temperature difference of 2 °C, any flow rate, however, small, caused an intermittent loading pattern. The intermittency, none the less, was found insensitive to flow rate.

REFERENCES

Lund, P. and J. Routti (1984). Feasibility of solar pond heating for northern cold climates. Solar Energy, Vol. 33, No. 2.

Metwally, M.N. (1985). Salt gradient solar ponds as source of heat and power generation in Egypt. FRCU project no. 830217, report no. 7. Fac. of Eng. & Tech. at Mattaria, Cairo/Egypt.

Metwally, M.N. and A.R. Abouhabsa (1987). Computer simulation of an internal heat extraction system for salt gradient solar ponds. ISES World Congress in Hamburg.

Sabetta F., M. Pacetti and P. Principi (1985). An internal heat extraction system for solar ponds. Solar Energy, Vol. 34, No. 4/5.

RESEARCH ON MEMBRANE VISCOSITY STABILIZED SOLAR POND

M. Taga*, T. Matsumoto** and T. Ochi**

*Faculty of Science and Technology, Kinki Univ.,
Higashi-Osaka City, Osaka, 577 Japan
**Osaka Prefectural Technical College,
Neyagawa City, Osaka, 572 Japan

ABSTRACT

This paper described on the basic experiments and the calculations for the optimum design of the membrane viscosity stabilized solar pond. In the basic experiments, the stability criterion for the onset of convection was measured on the several layer of polymer solution which is used as the thickener of its pond. And the theoretical calculations of the pond temperature were carried out by solving the equation of heat balance at each element in the two-dimentional model pond. As a result of these studies it was concluded that the optimum conditions for the pond are 0.5 m of total thichness of the insulating layer and 7 sheets of film.

KEYWORDS

Solar pond; gel pond; membrane solar pond; viscosity solar pond; solar energy store; convection stability criterion.

INTRODUCTION

Most of the research on solar ponds has been confined to the salt gradient solar pond (S.G.S.P.). However, the S.G.S.P.has a number of difficulties: it causes environmental polution in the event of a salt leakage, and needs continually maintenance of the non-convective salt gradient zone.
In order to eliminate the above problems a new type of solar pond using a transparent polymer gel as the non-convective zone was proposed by Shaffer (1978). The polymer gel has low thermal conductivity and is used at a state of near-solid, so that it will not convect. And the polymer gel is not harmful to the environment, in that it is a useful method for supplying domestic heat or small-scale energy.
The first experimental study on the gel pond was performed to prove its effectiveness by Wilkins and others (1981) at the University of New-Mexico. And, many thickeners which can be used for the non-convective layer have been examined over a wide range, and also special techniques for the cross-linked polymer gels to form permanent gels have been developed by Drumheller

and others (1975). However, on the gel techniques, still research and development may be needed to learn how to make uniform gel on a large scale. Accordingly, at the present technical level, it may be recommended that the non-convective layer is composed by the viscous polymer solution rather than the near-solid gel.

On the viscosity stabilized solar pond (V.S.S.P.) which uses polymer solution as the non-convective zone, the authers (1985) have already reported their results which involve a discussion on the physical properties of polymer thickener, experiments and theoretical calculations of the model pond.

In the above V.S.S.P., the polymer solution must be of high viscosity which exeeds a certain concentration for suppressing convection. And the larger the values of concentration the smaller the light transmittance becomes, then the collection efficiency decreases. But on the other hand, since the values of critical concentration for suppressing convection becomes smaller with the decrease of the layer thickness, if the polymer layer is partitioned by a number of transparent membranes a thinner polymer solution can be utilized. Therefore the transmittance of the polymer layer itself increases. However, in such a membrane viscosity solar pond (M.V.S.P.), the light transmittance through the insulatiting polymer layer to the bottom water layer is reduced significantly with the number of membranes. Therefore, in the case of design in the M.V.S.P., the determination of adequate thickness of insulating layer and the number of membranes is very important. This paper describes the basic experiments on the polymer solution and theoretical calculations for determining these optimum values.

EXPERIMENTS ON THE CRITICAL VALUES FOR THE ONSET OF CONVECTION

First of all, as the basic experiments, the stability criterion for the onset of convection was measured on a layer of polymer solution confined between two rigid horizontal surfaces with heating from below according to the method of Tien and others (1969).

A typical set of experimental data is shown in Fig.1. For a given series of experiments, 8 to 13 sets of readings of Q vs. ΔT were obtained. The temperature difference ΔT from a few degrees up to 24 °C was maintained, corresponding to a power input up to 240 W. The stable time required for each reading was from a few hours up to 60 hrs. The two lines at both regions of conduction and convection are obtained by drawing best-fit lines visually

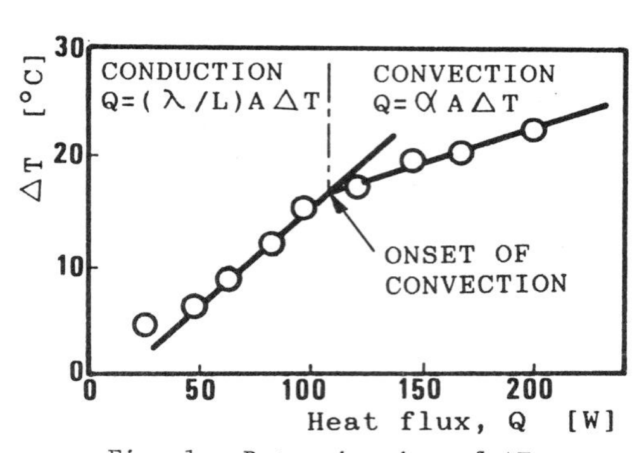

Fig. 1. Determination of ΔT_{cr} (0.5 % Polymer, L = 100 mm)

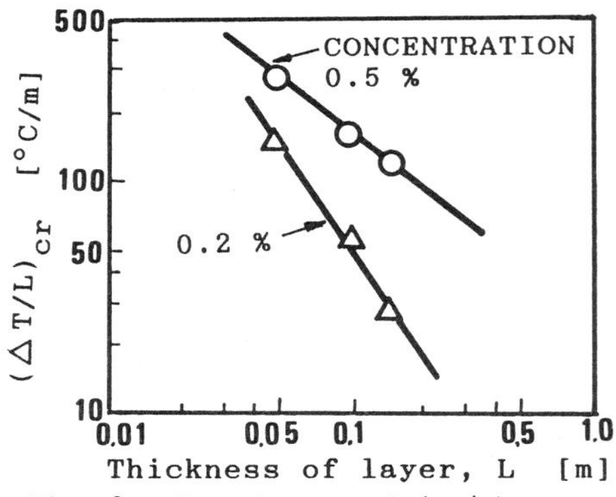

Fig. 2. Experiments of $(\Delta T/L)_{cr}$

through the data points, and the intersection points of these lines give the values of ΔT_{cr} at the onset of convection. The experimental results in the cases of 0.2 % and 0.5 % weight concentration is shown in Fig.2. Temperature differences of the criterion $(\Delta T/L)_{cr}$ [°C/m] to any thickness of polymer layer L [m] were expressed in the following formula; $(\Delta T/L)_{cr} = cL^m$, where, c = 1.47, m = -1.53 at 0.2 % and c = 25.6, m = - 0.81 at 0.5 % concentration.

THEORETICAL CALCULATIONS

Physical Model and Governing Equations. The increase of temperature in the bottom water layer was calculated numerically by the physical model of Fig.3. Hourly temperature in each element in Fig.3 ($t_1 \sim t_A$) changes satisfying the following heat balance equations (1), (2) and (3) during every pitch hour.

$$t_1 = (H_1 + Q_R + Q_C + Q_E + Q_1) / (G_W c_W) + t'_1 \qquad [\,°C\,] \qquad (1)$$

$$t_n = (H_n + Q_{n-1} + Q_n) / (G_P c_P) + t'_n \qquad [\,°C\,] \qquad (2)$$

$$t_A = (H_A + Q_n + Q_G) / (G_W c_W) + t'_A \qquad [\,°C\,] \qquad (3)$$

where, t'_1, t'_n, t'_A = temperature at a hour before one pitch [°C],

$G_W c_W$ = heat capacity of the water layer [kJ/m²h°C],

$G_P c_P$ = heat capacity of each polymer layer [kJ/m²h°C].

The values of radiation loss Q_R, convection heat loss Q_C and evaporation loss Q_E were calculated by the usual equations[1] using typical climate data of Osaka[2]. The absorption energy into the each element H_n [kJ/m²h] was calculated by the following equation.

$$H_n = H_0 (1 - r_W)(1 - r_S)(1 - r_P)^{n-2} \eta_W \eta_{P,2} \cdots \eta_{P,n-1} T_M^{n-1} [\,1 - (1 - r_P) \eta_{P,n} T_M] \qquad (4)$$

Here the values of incident solar energy H_0 were given by standard data[2] of

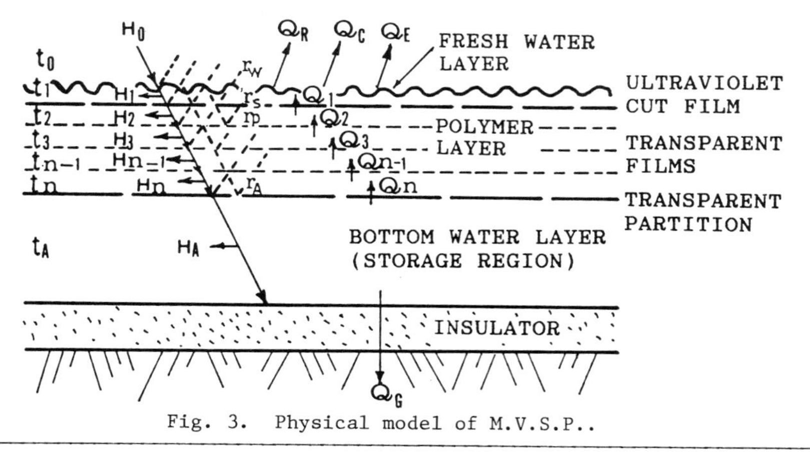

Fig. 3. Physical model of M.V.S.P..

[1]Japan Solar Energy Soc. (1985). Solar Energy Utilization Handbook, 47-76

[2]Japan Solar Energy Soc. (1978). The Basis and Applications of Solar Energy, 29.

total solar radiation for every month. Reflective losses r_W, r_S, r_P and transmittances of fresh water layer, each polymer layer and each film η_W, η_P and T_M were modified by monthly data of light path coefficients K_L which take into consideration the solar altitude and direction. The values of transmittance of polymer layer η_P were given by the data of Makino and others (1987) after determining the critical concentration owing to the previous mentioned experimental data of Fig.2. Conduction heat fluxes $Q_2 \sim Q_{n-1}$ and the heat fluxes from the water layer Q_1, Q_n and Q_G were calculated respectively by the formula $Q = K_J \Delta t$ using the experimental coefficient of overall heat transmission K_J.

Results. Temperature of the bottom water layer t_A were calculated on various number of films (4 ~ 9 sheets) ranging in polymer thickness of (0.3 ~ 1.1 m). Fig.4 shows an example of the results calculated under the conditions which are described in the figure. The temperature of the bottom water layer reached its maximum value, $t_{A,max} = 72$ °C, on September for both the first and following years.

Where the same total thickness of polymer layer exists, the critical concentration for suppressing convection decreases with the increase of the number of membrane films, therefore the light transmittance of the polymer layer itself increases. But an increase in the number of films causes an increase of reflective loss and the absorption loss of films. Therefore, there are an optimum number of films to obtain the maximum temperature in these conditions. Fig.5 shows the values of $t_{A,max}$, ε_{max} and $Q_{e,max}$ in several thicknesses L. Where ε_{max} is the maximum extraction exergy which is defined by the following equation,

$$\varepsilon_{max} = Q_{e,max}[(T - T_0) - (T_0 \ln(T / T_0))] \quad [\mathrm{kJ/year\,m^2}] \qquad (5)$$

where, $Q_{e,max}$ = maximum extraction energy [kJ/year m^2],

$T = t_{A,max} + 273.15$ [K], $T_0 = 25 + 273.15$ [K].

The maximum value of $t_{A,max}$ was obtained in the case of total thickness is equal to 0.9 m and the number of films is equal to 8. However, the values of the extraction load $Q_{e,max}$ which are 20 % of the solar energy that reaches the bottom water layer, decrease with the total thickness. And the values of ε_{max} show the maximum when the thickness of the polymer layer equals to 0.5 m . Therefore, if also the cost which increases with the total thickness is considered, optimum conditions for the pond of this type are 0.5 m of total polymer layer and 7 sheets of film.

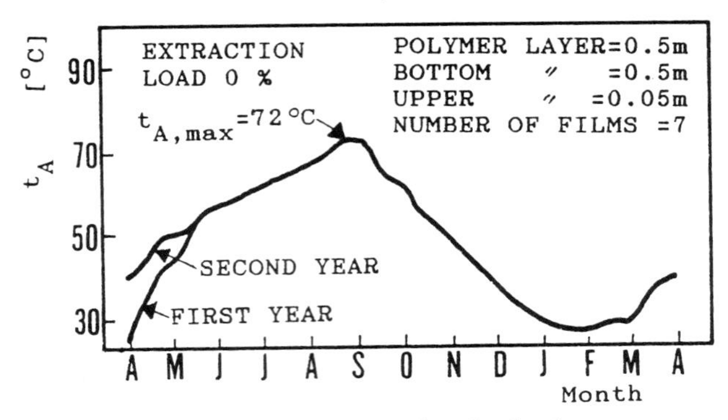

Fig. 4. A result of calculation.

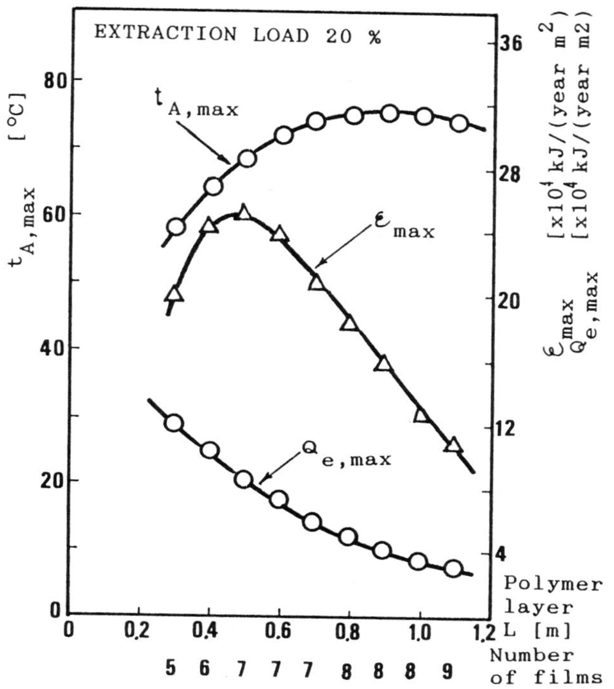

Fig. 5. Calculation of optimum condition.

ACKNOWLEDGMENTS

We would like to express our appreciation to Sanyo Chemical Industries, Ltd. for preparing the polymer. And the data of transmittance of the polymer solution were taken mostly by Dr. T. Makino and others at Kyoto University.

REFERENCES

Drumheller, K. and others (1975). Comparison of solar pond concepts for electrical power generation. Battelle Pacific Northwest Laboratories. BNWL - 1951, 3.23 - 4.18 .

Makino, T., Maerefat, M. and Kunitomo, T. (1978). Measurement of optical and thermophysical properties of polymer gels for a viscosity-stabilized solar pond. Proc. of the ASME - JSME Solar Energy Conf. Vol.1, pp.172-177.

Shaffer, L. H. (1978). Viscosity stabilized solar ponds. Proc. Int. Solar Energy Soc., New Delhi, 1171 - 1175.

Taga, M., Matsumoto, T. and Ochi, T. (1985). Experimental studies on viscosity stabilized solar pond. Proc. Int. Conf. Solar and Wind Energy Applications, Beijin, A.6 - 11.

Tien, C., Tsuei, H. S. and Sun, Z. S. (1969). Thermal instability of a horizontal layer of non-Newtonian fluid heated from below. Int. J. Heat Mass Transf., 12, 1173 - 1178.

Wilkins, E. S., Yang, E. and Kim, C. (1981). The gel pond. Int. Soc. Energy Conversion Eng. Conf., 2, 1726 - 1731.

EFFECT OF THE GAP SPACING ON THE SHALLOW SOLAR POND PERFORMANCE

H. M. ALI

Mechanical Engineering Department,
Amirkabir University of Technology,
Hafez Ave, Tehran, Iran.

ABSTRACT

It is found that for shallow solar pond where the gap spcing is large (300 mm), the convective heat loss is somewhat lower that for a conventional collector with a shorter gap spacing (20 mm), Garg (1, 82) . But howmuch the total heat transfer coefficient is lower and what is the effect of the gap spacing on the performance of the shallow solar pond, these were not mentioned. In this study the effect of the gap spacing on the shallow solar pond performance is studied. For this purpose a computer programme is constructed and is experimentally tested using a shallow solar pond of 6.6 m^2 area. It is concluded from the results that the effect of the gap spacing on the top loss coefficient is independent of the temperature difference between the upper film of the water bag and the glazing of the shallow solar pond. The results also show that the change in the gap spacing of the shallow solar pond has not an important effect on the pond performance. Hence the large gap spacing in the conventional design shallow solar pond (300 mm) cannot be considered as a difference with the flat plate collector design(20 mm).

INTRODUCTION

The shallow solar pond (SSP) is a water bag inwhich the lower layer is a black plastic film and the upper layer is a clear plastic film. The depth of the water in the bag is in the range of 4-15 Cm. The bag is coverd by a clear plastic layer to reduce the heat losses from the water. Higher temper ature up to 100c in summer has been obtained but the typical peak temperatures range is between 60c to 40c in summer and winter respectively. The annual efficiency of the SSP with a water depth in the range of 7.5-10 cm is about 50%., Garg (1982).

The typical value of the total heat transfer coefficient of (SSP)is about 7.3 W5/m^2. K, (2) which is nearly equal to the thermal losses from a flat-plate collector of on cover and with non - selective absorber surface. The convective heat losses from the water in the bag of SSP consists a significant part of the total thermal losses. So that the space between the water bag and the upper glaxing is used as a resistance to reduce the upper convective and radiative losses. This resistance depends on the gap of the spacing as well as on the physical properties of the air and the temperature difference between the two transparent plastic layers. Hollands (1976) studied the effect of the gap spacing between the absorber and the cover on the free convection heat transfer coefficent across inclined air layer. Tht heat losses due to the conduction and convection in the plate spacing was found to be decreases with an increase in plate gap spacing to reach a minimum when the convection was initiated, then the losses increased until reaches a maximum, and then decreases gradually with further increase in plate gap spacing. The temperature difference between the two surfaces was greatly influencing the gap spacings at which the maximum and minimum losses occure.

Buchberg(1976), studied the effect of gap spacing on a flat-plate collector performance. The gap spacing was in the range of 9.52-50.8mm and one operating condition was done for single, double, selective and non selective flat plate collectors. The effect of gap spacing was significant when selective black absorber is used.

Duffie et al., (3)predicted the effect of gap spacing on the top loss coefficient, for gap spacing range 0-40mm and for selective, nonselective, one and two cover - flat-plate collectors. The top loss coefficient for 25mm gap spacing was suggested to be use for other gap spacings with little erros as long as gap spacing greater than 15 mm is used. In thier study the effect of the gap spacing on the flat-plzte collector performance was not done.

The convective heat losses in flat-plate collectors has been studied also by Agarwal (1984), and a simple correlation approach has been adopted to predict the heat transfer coefficient for the flat plate collector spacing of tilt angle range 0-90°. Also the gap spacing effect on the flat plat plate collector performance was not performed.

Garge(1982) found that for SSP where the gap spacing is large(300mm) the convective heat loss is somewhat lower than for a convectional collector with a shorter gap spacing (20mm). But howmuch the heat transfere cefficient is lower and what is the total effect of the gap spacing on the performance of SSP were not mentioned. In this study the effect of the gap spacing on the performance of the shallow solar pond is studied, ie, can the large gap spacing in SSP be considered as a useful characteristic in developing the system performance. To do this study SSP of 6.6m2 area was constructed to calibrate the mathematical model.

E EXPERIMENT

The SSP was construeted of 3 2.2m dimensions and its base and side walls are constructed form concrete. The galzing of the shallow solar pond is polythelyne plastic film which is placed over Iron arches where thier ends are fixed inside the concrete curbing. The endges of the plastic glazing is made air sealed. The water bag of 2.3 1.5 m dimensions is made from two PVC films, the lower black and the upper clear films.

ANALYSIS

The effect of the gap spacing on the convective heat transfer coefficient between the glazing and the upper film of the water bag is given by:

$$h_{p-c} = N_u \, Ka/L \qquad (1)$$

Where h_{p-c} is the convective heat transfer coefficient, N_u is Nusselt number, K_a is air thermal conductivity and L is gap spacing.

The Nusselt number is calcutated from Hollands (1976) coerelation (8) which is valid for Rayliegh number in the range 1708-108 , and is expected to be valid in the limit that Rayliegh number approaches infinity

$$N_u = 1+1.44 \left[1- 1708/R_a \cos \phi\right]^{*} \left[1-(\sin 1.8\,\phi)^{1.6} 1708/R_a \cos \phi\right]$$

$$+\left[\left(\frac{R_a \cos\phi}{5830}\right)^{1/3} -1\right]^{*} \quad (2)$$

Where R_a is Rayliegh number, ϕ is the tilt angle and the star brackets become zero when negetive. for SSP where $\phi = 0.$, the correlation becomes:

$$N_u = 1 + 1.44 \left[1 - 1708/R_a\right]^{*} + \left[\left(Ra/5830\right)^{1/3} - 1\right]^{*} \quad (3)$$

Rayliegh number is given by(7):

$$R_a = 2g(T_w - T_c)\, L^3\, pr/ \nu^2 (T_w + T_c) \quad (4)$$

Where g is acceleration gravity, T_w is daily average upper film temperature of the water gag which is assumed equal to the water temperature, T_c is daily average glaxing temperature which its value is assumed, p_r is prandtle number and ν is kinematic viscocity.

The effect of the gap spacing on the top heat loss coefficient is predicted from calculating the top heat loss coefficient:

$$U_t = \left[(h_{p-c} + h_{r,p-c})^{-1} + (h_a + h_{r,c-s}\right]^{-1} \quad (5)$$

where U_t is top heat loss coefficient, $h_{r,p-c}$ is radiation heat transfer coefficient between the glaxing and the upper film of the water bag, h_a is wind heat transfer coefficient and $h_{r,c-s}$ is radiation heat transfer coefficient between the glaxing and the sky.

$h_{r,p-c}$ is given by (3):

$$h_{r,p-c} = \sigma(T_w^2 + T_c^2)(T_w + T_c)/(1/\epsilon_w + 1/\epsilon_c - 1) \quad (6)$$

where σ is stefan - Bottxman, ϵ_w is emissivity of the upper film of the water bag and ϵ_c is emissivity of the glaxing.

h_a and $h_{r,c-s}$ are given by Duffie (1980):

$$h_a = 5.7 + 3.8\,V$$

$$h_{r,c-s} = \epsilon_c\, \sigma\, (T_c^2 + T_s^2)(T_c + T_s)(T_c - T_s)/(T_c - T_a) \quad (7)$$

where Vis daily average wind speed and T_s is daily average sky temperature which is given by Duffie(1980):

$$T_s = 0.0552 \quad T_a^{1.5} \quad (8)$$

Since the daily average top glaxing temperature is not known experimentally, hence it is first assumed as it is mentioned before andt then is calculated to findout the accuracy of the assumed value using the following equation:

$$T_c = T_w - U_t (T_w - T_a)/(h_{p-c} + h_{r\,p-c}) \quad (9)$$

The effect of the gap spacing on SSP performance is predicted by calculating the houry water temperature for the total heat loss coefficients of maximum and miniumum gap spacing.

The water temperature is assumed equal to the temperature of the upper and lower films of the water bag. Hence the water bag is considered as the absorber plate in the flat plate collector. Also the heat losss to the ground is assumed steady state in one dimension, Kreith (1978). The energy balance equation SSPis given by:

$$M_w C_w \frac{dT_w}{dt} = A_c \left[\tau\alpha(t)\, S(t) - U_t(T_w(t) - T_a(t) - U_b(T_w(t) - T_g)\right] \quad (10)$$

Where M_w is water mass, C_w is water specific heat, T_w is hourly water temperature, t is time, A_c is the surface area of the water bag, $\tau\alpha(t)$ is hourly optical efficiency, S(t) is hourly total solar radiation falling on the pond, $T_a(t)$ is hourly ambient temperature, U_b is ground heat loss coefficient, and T_d is ground water temperature at some distance below the ground and it is assumed equal to the yearly average ambient temperature. The water temerature is given by:

$$T_w = T_{wi} + \frac{t}{w_w h_w}\left[\tau\alpha(t)S(t) - U_t(T_w(t) - T_a(t) - U_b(T_w(t) - T_a)\right] \quad (11)$$

RESULTS AND DISCUSSION

The effect of the gap spacing on the top loss coefficient U_t is shown in figure 1. The maximum top loss coefficient after the minimum is at around 20 mm gap spacing, then the top loss coefficient reduces very slowly. The reason for the very slowly reduction in the top loss coefficent is because that the variable resistance due to the change in the gap spacing is important compared to the other resistances only when the gap spacing is very smal, ie, around 20mm.

The top loss coefficient is affected by the temperature difference between the upper film of the water bag and the glaxing of SSP, as it does with the flat plate collector. This temperature difference is changable according to the weather conditions, hence two typical summer and winter conditionswhen SSP performance is experimentally studied, are used to predict the effect of the gap spacing on the top loss coefficent with changing the weather conditions. Figure 1 shows the change in the top loss coefficient with the gap spacing is nearly the same for the two different conditions. The results are also found the same for other conditionswwhich are not shown in fig. 1. It can be concluded from these results that the effect of the gap spacing on the top loss coefficent is independent on the weather condition, ie, the change in the top loss coefficient due to change in the resistance of the gap spacing is nearly the same at any outside condition. Figure 2 shows the experimental and the theoretical temperature of the water in SSP lar pond, and good agreement between the results is ssen from the figure.

Figures-3-4 show the water temperature of SSP in 20 mm and 300mm gap spacing cases, and for the two typical days. I t is seen from the figures that the maximum difference is in around 3C. It is concluded from the results that the change of the gap spacing of SSP has not an important effect on the performance of the shallow solar pond. Hence the large gap spacing in conventional design SSP (300 mm) can not be considered as major difference with the flat plate collector design(20 mm).

REFERENCES

Agrawal V. K. (1984). Convective Heat losses in flat plate collectors, Energy Research, Vol. 8, pp. 297-301.

Buchberg, I. H. Catton and D. K. Edwards(1976), Natural Convection in Enclosed Spaces-Areview of Application to Solar Energy Collection, Journal of Heat Transfer (Transfactions of theASME), No. 89. pp. 182-188.

Clark, A. F., and W. C. Dickison, (1980), Shallow Solar pond, in Solar Energy Technology Handbook, Chap. 12, pp. 377, Marcel Dekker, New York.

Duffie, J. A. and W. A. Beckman(1980), Solar Engineering of Thermal processes, John Wiley and Sons, Inc.

Garg, H. P., B. Bandyopahyay, U. Rani and D. S. Hrishikesan(1982).

Shallow Solar pond: State-of - the -Art, Energy conversion and Manageme-nt, Vol. 22, No. 2, pp. 117-131.
Hollands, K. G. T., T. E. Unny, G. D. Raithby and L. Konicek, (1976). Free convective Heat Transfer Across Inclined Air Layrs, Journal of Heat Transfer (Transactions of ASME), No. 89, pp. 189-193.
Kreith, F., and J.F.Kreider (1978) Principles of Solar Engineering , Hemisphere Publishing Corporation.

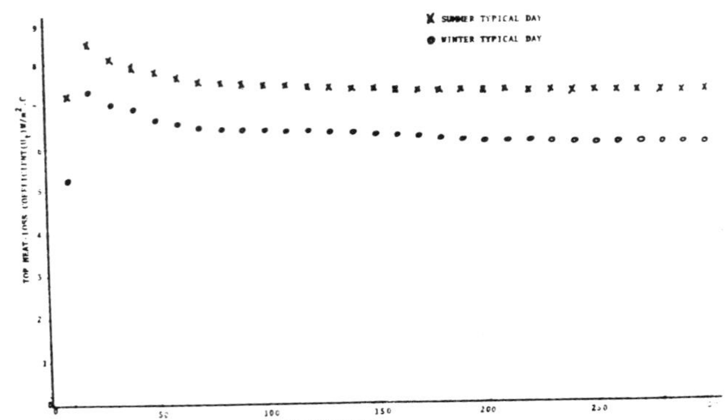

Fig.1.Gap spacing-top loss coefficint relationship

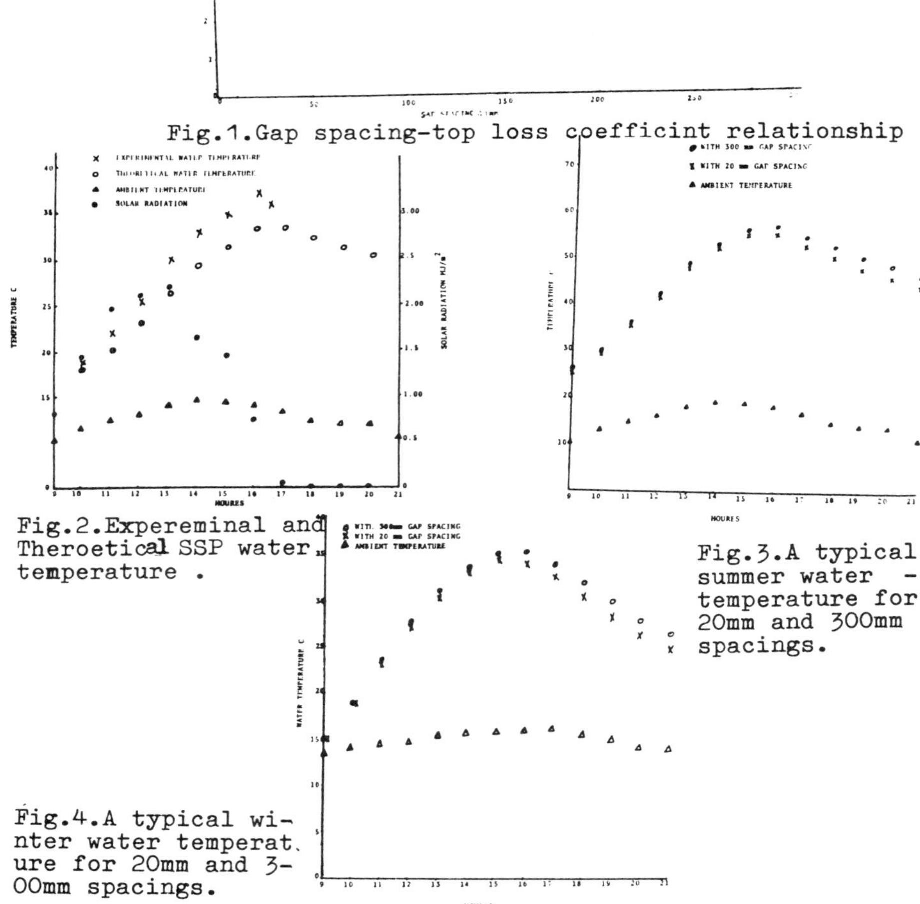

Fig.2.Experiminal and Theroetical SSP water temperature .

Fig.3.A typical summer water - temperature for 20mm and 300mm spacings.

Fig.4.A typical winter water temperature for 20mm and 300mm spacings.

CONVECTIVE LAYERS GENERATED BY SIDE WALLS IN SOLAR PONDS

A. Akbarzadeh

Department of Mechanical and Production Engineering

R.M.I.T.
Melbourne Vic 3001, AUSTRALIA

ABSTRACT

In this paper the results of a series of laboratory experiments simulating the possible instabilities in density-gradient solar ponds which can be induced by absorption of sunlight on the sun facing wall are reported. It was observed that under simulated solar pond conditions, convective layers with thickness between 10 - 20 mm can be generated. The front of these layers advance with a speed of 0.2 - 0.3 m/h. It has also been observed that these layers disturb regions far ahead of them by horizontal flows.

KEYWORDS

Solar pond; density gradient; convective layers, jet flows.

INTRODUCTION

The existence of a stable density gradient layer is an essential element in the operation of salt gradient solar ponds. Experimental evidences indicate that sloping walls can be a source of instabilities in this region(Nielser, 1976; Collins, 1983). Akbarzadeh (1984) studied the effect of sloping walls on salt concentration profile in salt gradient solar ponds. In his analysis it was shown that the salt concentration gradient and therefore the salt flux in the upper region of the non-convective layer is lower than its value in the bottom of this region. This conclusion results from the fact that in the presence of sloping walls the surface area of the pond increases with height. This effect was confirmed by Srinivasan and Guha (1987) in a solar pond with a top surface area 240m^2 and 45° wall angle.

In this paper effects of sloping walls in creating horizontal convecting layers caused by absorption of solar radiation in salt gradient solar ponds are discussed.

HORIZONTAL LAYERS INDUCED BY WALL HEATING

Because of the absorption of solar radiation, the presence of walls in solar ponds can create horizontal gradients of temperature which in turn result in horizontal gradients in density. These gradients would then create a variation of pressure in the horizontal direction which in turn may result in horizontal flows. Generally speaking, the water at the surface of the heated wall becomes warmer and therefore lighter than the interior part of the pond at the corresponding level. Therefore the water separates from the wall and rises towards the upper levels which have lower densities, to find its equilibrium position. When this level is reached then because of its kinetic energy the fluid travels horizontally. At the same time water from the interior part of the pond moves toward the heated wall to compensate for the previous flow and thus a horizantal convective layer is formed. The behaviour of a salt-stratified fluid adjacent to a heated inclined plane has been studied theoretically and in the laboratory by Chen and others (1971) who showed that the onset of double-diffusive instability and the growth of the resultant layers into the body of the fluid is described by the natural length scale

$$h= \alpha.\Delta T/(1/\rho.\partial\rho/\partial z) \qquad (1)$$

and the Rayleigh number, based on h and the reduced gravity $g.\cos\theta$

$$R= g.\cos\theta.\alpha\Delta Th^3/(k\nu) \qquad (2)$$

In the above relation α is the coefficient of thermal expansion of the fluid, ΔT is the temperature difference between the heated wall and the interior of the fluid, ρ the density of the fluid, z the vertical co-ordinate (positive upward), and θ is the angle between the normal to the slope and the horizontal. κ and ν are the fluid's thermal diffusivity and kinematic viscosity respectively. In experiments with a 45^0 wall Chen found the range of R = 18000-26000 for the critical Rayleigh number separating quiescent and convective regions.

The problem of layering due to the presence of a heated wall with a uniform heat flux in a stratified fluid was first considered by Narusawa and Suzukawa (1981). As a criteria for the onset of convection they defined a non-dimensional number

$$\Pi_3=- \alpha\ (q/K)\ /\ \beta\ (dS/dz) \qquad (3)$$

where q is the applied heat flux at the wall, K is the thermal conductivity of the fluid, β is the coefficient of expansion due to salinity, and S is the salt concentration. From a series of experiments they found the critical values of Π_3 for different solutes having different relative diffusivities. For the case of common salt Π_3 was found to assume a value of 0.28.

For the solar pond application, the square of the buoyancy frequency, $N^2 = -(g/\rho)(d\rho/dz)$, is around $1.4S^{-1}$ and, from (1) for a temperature difference between heated slope and ambient fluid of 4K, the length scale h is 9 mm, in accord with the observations in the laboratory experiments mentioned later. From (2), the Rayleigh number for 45^0 slope would be 40,000. A typical heat flux at the slope would be about 200 Wm^{-2}, so from (3) $\Pi_3 \cong 0.75$. By both measures then, there should be strongly convective layers adjacent to the solar heated slope.

THE EXPERIMENT

The laboratory tank in which the experiments were conducted was made of 3mm perspex sheets. The tank was 300 mm high and 100 mm deep and 228 mm long at the bottom. It had a vertical wall on one side and a 45° sloped wall on the other side. A 0.25 mm sheet of brass was placed inside the tank along the sloping wall. Depending on the voltage applied to the brass the desired flux of heat could be generated to simulate absorbtion of solar energy by the wall in a solar pond. Fluid motion was made visible by means of shadowgraphy. Events were recorded by photographs using potassium permanganate dye streaks as a tracer. The linear density profile was set up in the tank having a salt concentration gradient of 250 $kgm^{-3}m^{-1}$. This is the gradient which can exist in a solar pond with a 1m thick gradient layer and assuming that the top of the pond is fresh water and the bottom heat storage is close to saturation point with common salt. The applied electric voltage to the heating plate was such that it created a flux of heat equal to 200 Wm^{-2}.

After 23 minutes of heating, small convective cells started appearing on the heated plate. The activities within the cells were intitally very weak. After 32 minutes convective cells had covered the entire wall and they had propagated about 25mm into the main body of the fluid. The temperature of the plate was about 4K above the water temperature in the interior body of the fluid. A photograph which was taken at this time is shown here (Photograph1) The cells which now were becoming long enough to be called layers showed a slight downward slope. Apart from thick top and bottom convective layers, the layers in the middle region of the wall were initially 7 - 10 mm thick. Merging of thinner layers soon produced layers with thicknesses between 10 - 15 mm. With the passage of time the layers advanced further in the tank and after 3 hours of wall heating, convective layers covered the entire tank. The average advancement speed of the front of the layers was about0.2 m h^{-1}. The Reynolds number using the average thickness of the cell as a characteristic dimension was about 0.6. In the interior of a convective layer it was observed that the flow near the top is away from the heated wall and it is towards the wall near the bottom, assuming a zero somewhere in between. By injection of dye the maximum speed of the flow in the layers was found to be about 0.5 m h^{-1}.

An interesting phenomenon was observed during the course of the experiments. It was noticed that there were horizontal flows in the intial salt gradient layer far ahead of the advancing fronts of the convective layers. In order to visualise the effect dye was introduced to produce a vertical column in the undisturbed ambient fluid. As soon as a column of dye was introduced it began to deform. Photograph 2 clearly indicates that there was a horizontal flow in the region far ahead of the advancing front of the convective layers generated at the wall. The reason for the existence of the above phenomenon can be explained as follows. When a convecting cell is generated at the wall the structure of density which originally varies linearly with height becomes almost uniform across the thickness of the cell. This in turn changes the state of hydrostatic pressure within the cell with respect to the ambient. This means that some net horizontal forces will be present at the interface (the front separating the advancing convective layers from the non-convective region). These horizontal forces are then responsible for the horizontal shear flow (or jets) ahead of the advancing fronts. Attempt was made to quantify the speed of the induced horizontal shear flow by comparing photographs which were taken at different times. It was found that the speed of the horizontal flow at that location was around 0.1 m h^{-1} which is almost half the speed of the advancing front of the convective layer.

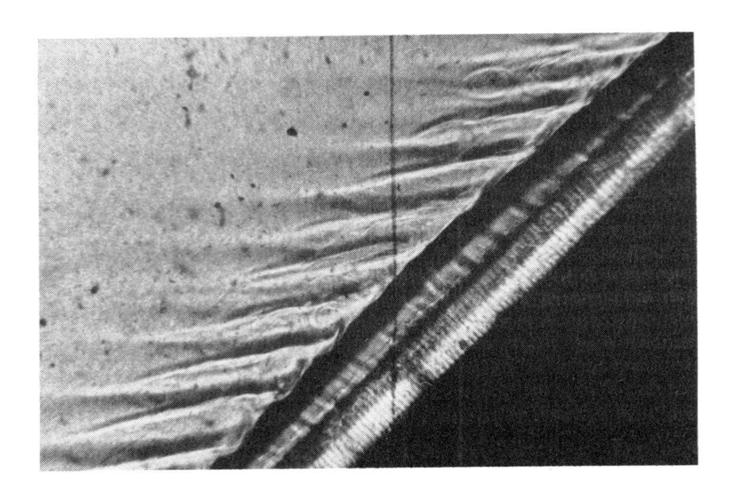

Photograph 1. This photograph was taken 32 minutes after the start of wall heating (9 minutes after the onset of instability). At this stage convecting cells are developing into small layers which have propagated 20-30 mm into the body of the fluid. The layers have a slight downward slope.

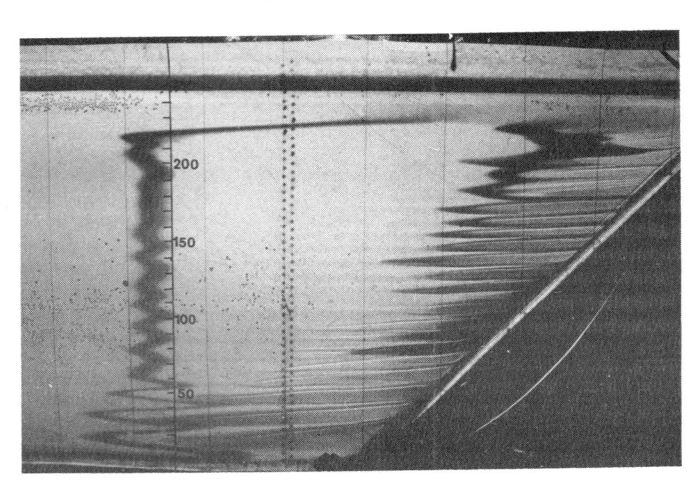

Photograph 2. This photograph was taken 7 minutes after the introduction of vertical column of dye in the left part of the tank. The deformation of the dye column indicates the flow field ahead of the advancing fronts of the convective layers.

EFFECT OF CONVECTIVE LAYERS ON UPWARD TRANSPORT OF SALT

In order to quantify the rate of enhanced upward salt transport resulting from wall heating, a series of measurements on the vertical salt concentration profiles were made. Comparison of these profiles (before the onset of the convective layers and the time that the layers had covered the tank) provided a measure of the degree of enhancement in the upward transport caused by the wall heating. The potential energy of the system calcuted from the salinity profiles was then compared with the anticipated rate of increase of potential energy of the system due to molecular diffusion alone. The comparison could then give a measure of the relative increase of salt transport in the former case compared with the latter. The results indicated that under the conditions of this experiment, the rate of increase of potential energy, in the presence of convective layers was 4 times the rate of increase of potential energy under the condition of no convection.

FURTHER REMARKS

In the laboratory model it was shown experimentally that the presence of an inclined wall(θ=45°) heated at a rate of 200Wm^{-2} and a stratified fluid with an initial density gradient of 0.182^{-1}, will generate horizontal convective layers. The above condition is very similar to the actual case in salt-gradient solar ponds. Also in locations where solar ponds are feasible the flux of heat generated at the surface of the sun-facing wall (45° wall angle with respect to horizon) and at a depth of 1 or less, will be in excess of 200 Wm^{-2}, provided the water is clear. Therefore the similarity between the experimental conditions in this investigation and the actual situation indicates that convective layers can be generated on the walls of typical solar ponds. Examination of the existing literature on the instabilities induced by wall heating in stratified fluids in relation with the present experimental investigation indicates that :

* Under the density gradient conditions in typical salt-gradient solar ponds a temperature difference of 2.5 - 3.5K between the surface of the wall and the interior of the pond will generate convective layers.
* Further, for the non-dimensional number, $\Pi_3=-\alpha(q/K)/\beta\,(dS/dz)_2$ to exceed its critical value of 0.28 for common salt, a flux of only 80Wm^{-2} is required to be present at the wall. If the heat losses due to conduction through the walls are not high, the heat flux generated at the surface of the sun-facing wall over the entire thickness of the density gradient layer generally can be more than this value.
* From the lab experiments, the layers generated in solar ponds will have an initial thickness of 7 - 12mm. The speed of advancement of the front of the convective layers is of the order of 3 mm min^{-1} (~0.2 mh^{-1}). Of course one should not expect that the convective cells can continue to grow indefinitely. At some point the side-wall buoyancy-induced flow is balanced by the shear dissipation of convection.
* It was shown experimentally that the existence of the convective layers under the conditions of the experiments conducted, increased the upward salt transport by a factor of about 4 times. This is likely to be an upper bound for a solar pond, being applicable only over the fraction of the area covered by the layers and possibly the upstream jets.
* Because of the large value of density-gradient ratio due to salt concentration and temperature in a typical solar pond, the mere presence of a sloping boundary (without any heating or cooling of the wall) cannot generate convective layers (see Linden, 1977). Walls in solar ponds can produce a significant layering only if they are heated by solar radiation or possibly cooled

by conduction of heat through the ground.

REFERENCES

Akbarzadeh, A. (1984). Effects of sloping walls on salt concentration profile in a solar pond. Solar Energy, 33, 137 - 141.

Chen, C. F. Briggs, D.G., and R. Wirtz. (1971). Stability of thermal convection in salinity gradient due to lateral heating. International Journal of Heat Transfer, 14, 57-66.

Collins, R.B. (1983). Alice Springs solar pond project, Final Report to NERDDC, Australia.

Linden, P.F. and J.E. Weber, (1977). The formation of layers in a double diffusive system with a sloping boundary. Journal of Fluid Mechanics, 81, 757-773.

Narusawa, U. and Suzukawa Y. (1981). Experimental study of double-diffusive cellular convection due to a uniform lateral heat flux. Journal of Fluid Mechanics, 113, 387-405.

Nielsen, C. E. (1976). Experience with a prototype solar pond for space heating. International Solar Energy Conference Proceeding. Winnepeg, pp. 182-196.

Srinivasan, J. and A. Guha (1987). Concentration profile in the gradient zone of small solar ponds. Solar Energy, 38, No. 2, 135-136.

THERMAL ENERGY COLLECTOR AND STORAGE WITH RADIATIVELY HEATED WALLS

M. Rommel, V. Wittwer and A. Goetzberger

Fraunhofer Institut für Solare Energiesysteme
Oltmannsstr. 22, D-7800 Freiburg, West-Germany

ABSTRACT

Experimental results of a small storage cube with transparently insulated walls are given. In 1986 the storage was insulated with silica aerogel pellets, in 1987 with a honeycomb structure from polycarbonate. First results of simulation calculations of bigger storages are reported.

KEYWORDS

Thermal energy storage, transparent insulation materials.

INTRODUCTION

Due to the improvements of transparent insulation materials completely new concepts of thermal energy storage can be thought of.
Conventional storages, insulated with opaque materials, do always have losses. If the heat conductivity of the material is very low and the insulation is very thick the losses may be rather small, but inevitably there always will be losses.

If, on the other hand, a storage tank is considered, which has its walls insulated with a transparent insulation material it is possible to have no losses at all, because the losses can be compensated by the radiative energy gain during the daytime.

Such a transparently insulated storage can also be regarded as a big integrated storage collector. When the energy gains during the daytime exceed the losses, the surplus energy will heat up the storage tank. Therefore it is not only a storage but a collector and storage, a complete solar system.

In order to investigate the prospects of this simple but very effective principle /1/, we have set up an initial experiment.

CONSTRUCTION OF A SMALL CUBICAL WATER STORAGE TANK WITH TRANSPARENTLY INSULATED WALLS

This cube is sketched in fig. 1. The edge length of the aluminium tank is 50 cm only and the total water volume is therefore not more than approximately 120 litres. It is placed on a 15 cm thick insulating base of polyurethane. On the top, there are heat conducting fins, because the tank is not completely filled with water. If necessary, excess water is drained through an overflow pipe. The tank is at atmospheric pressure. A selectively absorbing film (α=.9, ε=.1) is applied to the outer walls of the tank, which then act as absorbers. The outermost cover is made of low iron glass panes.

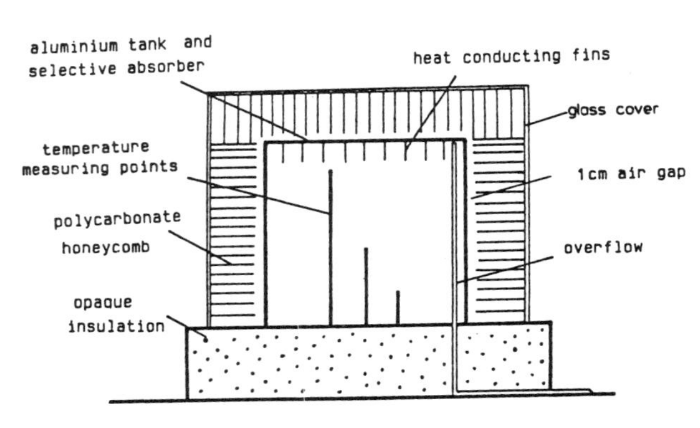

Fig. 1: Sketch of small cubical water storage with transparently insulated walls.

In the space between the absorber and the glass cover, two different transparent insulation materials have been tested so far, as will be described in the following chapters.

The storage water temperature is measured continuously at three levels, top, middle and bottom. Neither is water drawn, nor is heat extracted from the storage, it is simply operated under stagnation conditions.
The solar irradiation is measured by 5 pyranometers facing to the N, S, E, W and upwards.

RESULTS FROM 1986; TRANSPARENT INSULATION WITH SILICA AEROGEL PELLETS

From January to December 1986 we used a 2.5 cm thick layer of silica aerogel pellets /2/. The pellets had diameters of 3 to 5 mm and were simply filled into the space between the absorber and the glass cover. Fig. 2 shows the measured daily mean storage temperatures for the whole year. It can be seen that even during winter the temperature of the storage did not fall below +9°C . The whole storage was boiling for some hours during the best period of the year. The yearly mean temperature was 42.1°C.

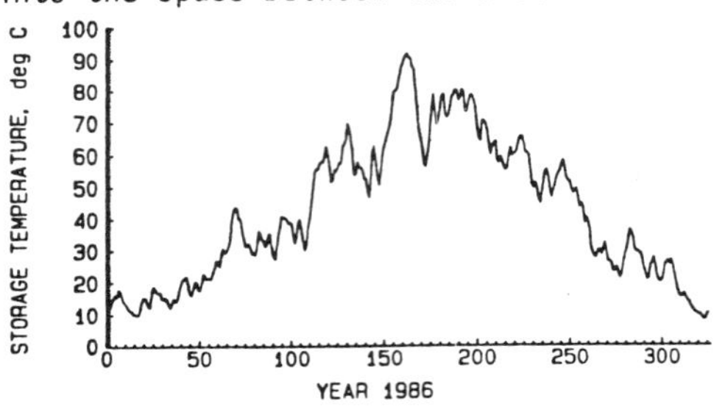

Fig. 2: Measured storage temperatures

Fig. 3 shows a diagram which is similar to an efficiency curve of a solar collector. It originates from the integration of the energy balance equation of the storage tank for

24 hours:

$$\frac{m\,c\,(T_1-T_0)}{A\,\bar{G}\,\Delta t} = \tau\alpha - U\,\frac{(\bar{T}_{st} - \bar{T}_a)}{\bar{G}}$$

The left hand side, which is the energy change of the storage medium within 24 h devided by the total radiative energy on the absorber area within 24 h, is plotted on the y-axis.

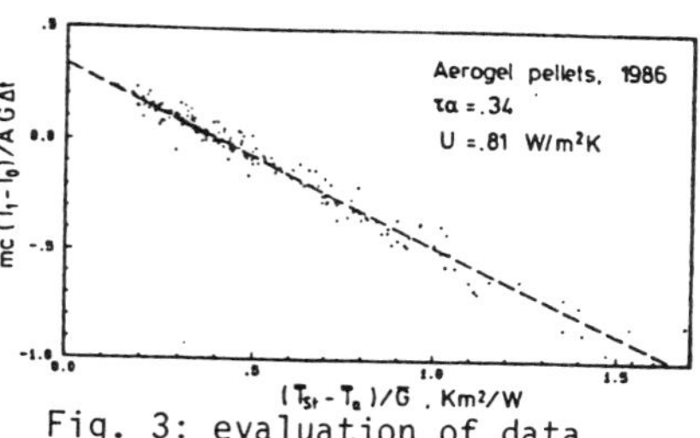

Fig. 3: evaluation of data

RESULTS FROM 1987; TRANSPARENT INSULATION WITH POLYCARBONATE HONEYCOMBS

At the end of January 1987, the aerogel insulation was replaced by a 14 cm polycarbonate honeycomb structure and an air gap of 1 cm was introduced between the honeycomb and the absorber, as indicated in fig. 1. From the known properties of the honeycombs /3/, values of $\tau\alpha$ =.6 and U=1 W/m²K approximately had been expected. Fig 4 shows the weekly mean values, which have been measured up to now. From the beginning up to day number 117 (April 27th), the U-values determined were varying strongly between 1.2 and 1.8 W/m²K. It turned out that this was caused by convection which could take place because the honeycomb had not been mounted flush to the glass cover, so that free air channels were formed. After having re-aranged the honeycombs, the measured U-value is now indeed 1.07 W/m²K.

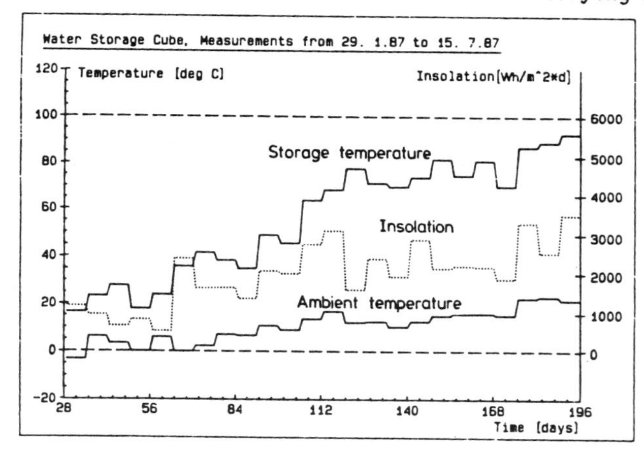

Fig. 4: Measurements from 1987. Up to day no. 117 additional convection losses occured.

The drastic effect on the storage temperature associated with this problem - and more generally with a good insulation material with low heat losses - can be seen by comparing the temperatures from two periodes having the same irradiation. For example, around day number 100 and day number 155 the irradiation was approximately 2000 Wh/m²·day but the storage temperature was 50°C and 75°C respectively.
Up to August 20th, the storage temperature had reached 100°C on 58 days in 1987 already.

STORAGE OF HEAT WITHOUT LOSSES

To a first approximation, the stagnation temperature difference, more precisely the temperature difference (T_{st} - T_a) at which the losses are compensated by the radiative gains, can be calculated from $G \cdot \tau\alpha / U$, when long-term effects are neglected. This relation is shown in fig. 5 for the two insulation materials that have been investigated. The factor $\tau\alpha / U$ describes the effectiveness of a given transparent insulation material when it is used for a transparently insulated storage tank. It can be seen that a storage tank which is insulated like the cube in 1987 can store heat at a temperature of approximately 60-70°C without losses from March to September under Central European climatic conditions. This is an important conclusion with respect to short term storages such as storage of the waste heat from factories over a weekend.

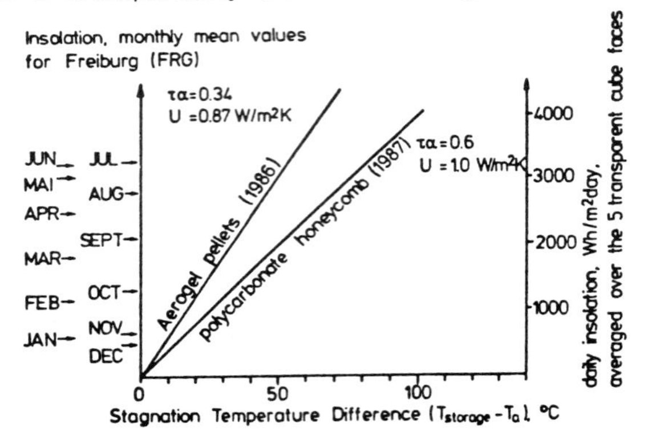

Fig. 5: Heat storage without losses

MORE DETAILED SIMULATIONS

We have started to develop a more detailed simulation program to study transparently insulated storage tanks. For example, the shape of the storage, its orientation and the angular dependence of the transmission as well as the temperature dependence of the heat loss coefficient of the transparent insulation material are taken into account. In order to give an outlook on larger storage volumes, fig. 6 shows first simulation results for two storage tanks, both transparently insulated with 10 cm polycarbonate honeycombs on 5 walls. Storage A has a base of 10m x 10m and a height of 2m, whereas storage B has a base of 38m x 38m and a height of 21m.

The gometric data for storage B are chosen similar to those of the storage tank in Wolfsburg (FRG), which is the largest free-standing storage tank with steel walls (and opaque insulation, of course), that has been built in West-Germany. The technical limit for this kind of storage tank is a cylindrical construction, approximately 100m in diameter and 13m high.

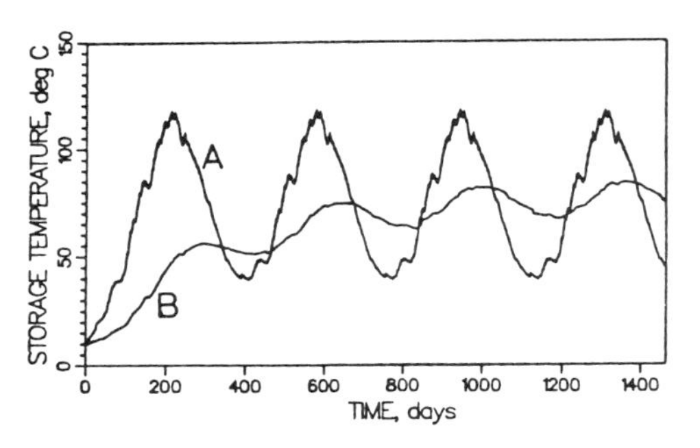

Fig. 6: Simulation results for larger storages

The figure shows that for these very large storage volumes the temperature only varies very smoothly (between 67 and 83°C) from winter to summer because of the high thermal inertia. Even for the smaller storage A, the lowest temperature in winter is 40°C.

If these storage tanks were operated as solar systems delivering energy at, for example, 40°C, approximately 61 MWh/year could be extracted from storage tank A and 1400 MWh/year from storage tank B. That would mean system efficiencies of 38% and 41% respectively.

CONCLUSION

A small water storage cube has been set up to study collector and storage systems experimentally. A detailed simulation program is beeing developed now to investigate possible applications of transparently insulated storage tanks as solar systems, or as components of solar systems, for example in conjunction with heat pumps.

ACKNOWLEDGEMENTS

This work was supported by the German Ministry for Research and Development (BMFT) under contract No.: 03 3500 3A.
We thank Chr. Hansen for evaluating most of the measurements.

REFERENCES

/1/ A. Goetzberger. Seasonal Storage of Thermal Energy with Radiatively Heated Storage Walls. International Journal of Solar Energy, 2 (1984)

/2/ J. Fricke (Ed.), Aerogels, Springer, Berlin (1986)

/3/ A. Pflüger, W. Platzer, V. Wittwer. Transparent Insulation Systems Composed of Different Materials, these conference proceedings.

OBTENTION OF KCl FROM "SILVINITA" IN SOLAR PONDS

J. Doria, M.C. de Andrés, J. Yuguero, C. Armenta, M. Collares-Pereira*

Grupo de Energía Solar. Facultad de Ciencias Físicas.
Universidad Complutense. 28040 Madrid (SPAIN)
*LNETI. Estrada do Paço do Lumiar. 1699 Lisboa (PORTUGAL)

ABSTRACT

Solar ponds can be used to obtain KCl from "silvinita" by differential solution, since in the binary solution KCl-NaCl-water, KCl solubility increases and the NaCl decreases when the temperature rises. On the other hand, the solar pond of "silvinita" remains stable.

KEYWORDS

Solar ponds, differential solubility method, separation of salts.

INTRODUCTION

Solar pond, system used in photothermical conversion of radiation solar, are studied, not only from their manifold and complexes phisical phenomena point on view, but from their useful applications: heat supply (Rabl and others, 1975) Collares-Pereira and others, 1985), power electric generation (Tabor, 1983; Doria and others, 1983), electrochemical conversion (Doria and others, 1985), chemical products purification (Lesino and others, 1980; Mangussi and others, 1987; Doria and others 1987), etc. The last one has been studied due to its very interesting features. This kind of applications can be mainly limited by: salt solubility, solar radiation transmitance in solution and the required temperatures in the process. Obtention of KCl from a "silvinita" Solar pond by using the differential solubility method has been the main goal of this paper.

Spanish "silvinitas" have the next average composition:

K_2O	15.3 %
Na^+	30.4 %
Ca^{++}..........	0.3 %
Mg^{++}..........	0.1 %
$SO_4^=$	0.8 %
Cl^-	55.0 %
Insolubles....	0.4 %
Others........	0.1 %

As KCl and NaCl forms the large proportion of "silvinita", these two salts become the controllers of the solar pond behaviour. First tests were made in KCl-NaCl-water solution; later tests in spanish "silvinita" solution.

LABORATORY TESTS

I. Checking the KCl-NaCl-water system.

NaCl and KCl show an increasing solubility in water as temperature rises (Fig. 1). The former is more soluble below 30ºC and the latter above this temperature.

The results of first laboratory tests in NaCl-KCl-water solution, at different invariant points, are presented in Table I. An invariant point represents the solution-solid phase equilibrium of salt at a given temperature.

TABLE I. Solubility and density NaCl-KCl-water system.*

	Single salt solution		KCl-NaCl-Water system			
T(ºC)	NaCl	KCl	NaCl	KCl	NaCl+KCl	$\rho(g/cm^3)$
0	35.7	27.6	31.8	10.3	41.3	-
20	36.0	34.0	30.4	14.8	45.2	1.230
40	36.6	40.0	29.3	19.6	48.9	1.234
60	37.3	45.5	28.4	24.6	53.0	1.241
80	38.4	51.1	27.7	30.0	57.7	1.243
100	39.8	56.7	27.5	35.3	62.8	1.246

*Solubity is given in grams of salt per 100 g of water.

The analysis of Table I show:

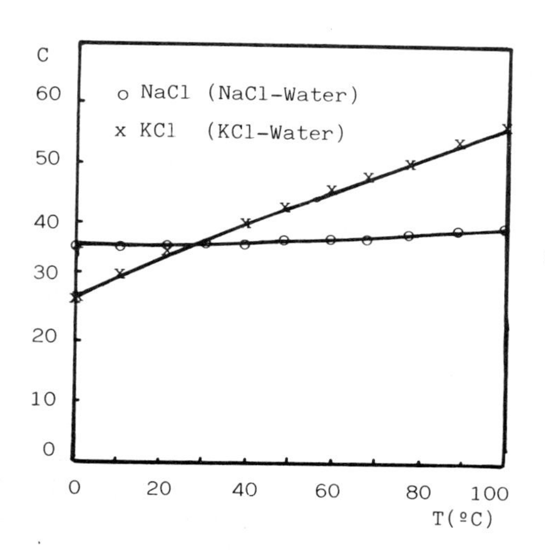

Fig. 1. Solubility curves for NaCl(o) and KCl(x) in water

1) KCl solubility in NaCl-KCl-water solution increases with temperature, while NaCl decreases. This odd characteristic is the basis foundation for the obtention of a very pure KCl, because only KCl precipitates when cooling a hot solution from an invariant point.

2) Density of KCl-NaCl-water system, at the invariant points, increases when temperature rises. As a result of this, a solar pond with salt excess at the bottom would be abble to remain stable, since the final state of spontaneous evolution would be the equilibrium at the invariant points throughout all the temperature gradient zone.

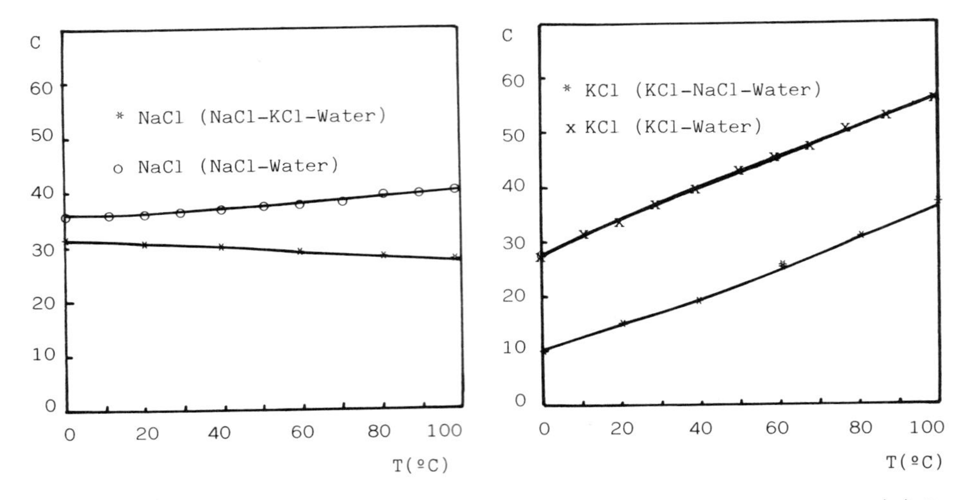

Fig. 2. NaCl solubility curves. (o) In water; (*)KCl-NaCl-water

Fig. 3. KCl solubility curves. (x) In water; (*) KCl-NaCl-water.

II. Spanish "silvinitas" tests.

II.1. Solubility tests. It is well Known that solubility processes involve surface phenomena. as the "silvinita" appears forming part of rocks, any method to exploit KCl requires a preliminary trituration process up to get a final particles size of an equivalent diameter of 1 mm.

Experiments were made by flowing water through a "silvinita" layer of 4.5 cm of thickness (size of particles : 1 mm of diameter). Table II shows the results of these tests. As may be seen, the invariant point is achieved in low time (15 min) for very small flowing current (10^{-6} m/s).

TABLE II. Solubility tests with spanish "silivinitas"

$V(cm^3)$	t(min)	Flux(cm^3/s)	Vel.(m/s)	Density	Temperature
60	15.20	0.0657	9.38×10^{-6}	1.232	25
120	29.25	0.0684	9.66×10^{-4}	1.234	34
180	47.47	0.0633	9.04×10^{-4}	1.234	34

II.2 Pouring off test. Solid particles (clays) can reduce solar radiation in a high density "silvinita" solution. Different density samples solution showed that no longer than 6 hours was neccesary to draw off all the solid particles. So, clays particles turbidity will never represent a problem, using "silvinita" like spanish type, without organic wastes.

II.3. Crystalliization test. Solid wastes of crystallization from a cooled solution of "silvinita" were dryed up to constant weight. Results agreed with expected values from Table I. Chemical analysis shows very low proportion of NaCl and other salts. These impurityes can be imputed to the adsorber solutions by the crystal.

SOLAR POND OF "SILVINITA

A "silvinita" solar pond will be avalaible, according to the results presented before. This type of pond should be made creating a mineral reserve at the bottom. Once the solution at the bottom has reached the previosuly optimum density determined by level radiation and climate local conditions, the hot solution can be extracted and then cooled and precipitated. Cooling could be made by removing the excess of change phase and sensible enthalpy either to the atmosphere, or by the use of a heat pump, whose cooler was the solution, and the heater the lower convective zone in the solar pond. This method requires more sophisticated technology though higher performance could be obtained.

Density tests "in situ" are absolutely neccesary. The technique to check it has revealed as complicated. The device proposed in Fig. 4 try to solve this problem. It is based on the compensating forces between electromagnetic and hydrostatic effects. The scale must be calibrated round the checking point to obtain a good precision (0.001 g/cm^3)

OTHER APPLICATIONS

Obtention of $NaNO_3$ from "sabach" has been studied using an analogue process as described before in a previous work (Doria and others,1987) The most important drawback this process has been the very poor richness in $NaNO_3$ the "sabach" On the other hand, surface saltpetre measures contains high organic wastes and a big proportion of non transparent particles, increasing the turbidity of the pond, and reducing solar beam transmision. The way of avoiding it is purifying the solution by chemical processes.

$NaNO_3$ can be used as a raw material in the solar pond of "silvinita" to obtain KNO_3 according to the chemical reaction:

$$NaNO_3 + KCl \longrightarrow KNO_3 + NaCl$$

Chemical system is formed by two reciprocal couples $(Na^+,K^+)/(NO_3^+,Cl^-)$ and water. The Jaencke diagrams are very useful to represent this complex system.

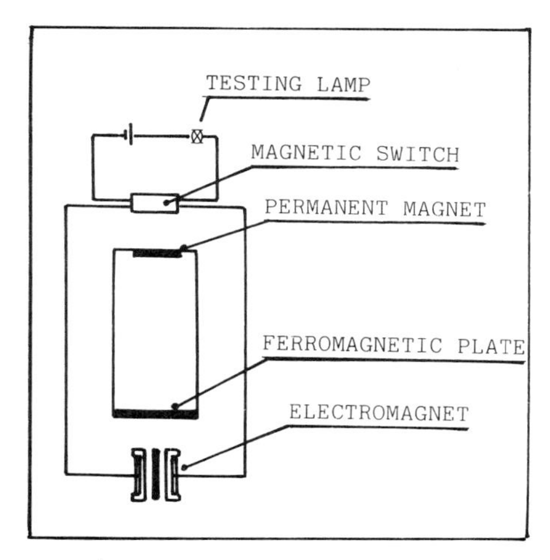

Fig. 4. Device to density measurement

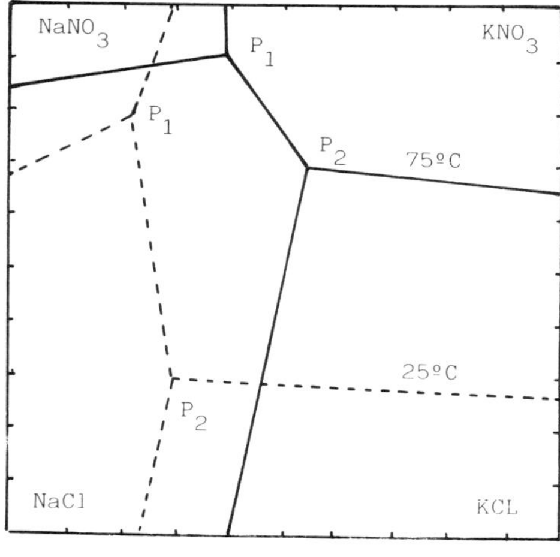

Fig. 5. Jaencke diagram (25º y 75ºC)

CONCLUSIONS

High concentration gradient can be established in a "silvinita" solar pond (or "silvinita and $NaNO_3$). This type of pond could be remain stable.

KCl obtained by crystallization (in a cooling process) of hot "silvinita" solution depends more on cycle thermal gap than on hot solution temperature. Therefore, when the excess of enthalpy is removed to the atmosphere, efficiciency could be practically constant all over

KCl purity can reach 99%.

Cheaper costs can be obtained due the cold solution must be recycled.

ACKNOWLEDGEMENT

To D. José Moya, Director of "Potasas Section", Explosivos Riotinto S.A. by its collaboration on realitation of this paper.

REFERENCES

Collares-Pereira, M., Joyce,A. (1985). A salt gradient solar pond for greenhouses applications. Anales de Física, Serie B, Vol. 79,3 261-271.

Doria, J.,de Andrés, M.C.,Armenta, C. (1983). Modificaciones de un estanque solar para su aprovechamiento térmico y electroquímico. Fundamentos. Anales de Física, Serie B Vol. 79,3 261-271

Doria, J., de Andrés, M.C.,Armenta, C.(1986) Electrochemical solar ponds. INTERSOl 85, Montreal, Canadá, Proc. Pergamon Press, Vol. 3 1500-1504

Doria, J., de Andrés, M.C., Yuguero, J., Alonso,A.,Collares-Pereira, M.(1987) Aplicación de los estanques solares a: I) Purificación de $NaNO_3$, II) Obtención de KNO_3. Proc. III Congreso Iberico, I Congreso Iberoamericano de Energía Solar ISES 87, Madrid, Vol. I, 249-260

Lesino, G.,Sarvia, J., Mangussi, J.,Galli,G. (1980) Incorporación de las popozas solares al proceso Industrial de producción de sulfato de sodio". Actas de la sexta Reunión de ASADES.

Mangussi, J., Saravia, L. (1987). Disolucion de sales en el fondo de una poza solar". Proc. III Cogreso Ibérico, I Congreso Iberoamericano de Energía Solar ISES 87, Madrid, Vol. I 249-267

Rabl, A., Nielsen,C.E. (1975) Solar ponds for space heating. Solar Energy Vol. 17, 1-12

Tabor, H. (1983) Solar ponds. Review Article. Solar Energy Vol. 27 181-194

STUDY OF KAOLINITE AND MONTMORILLONITE CLAYS AS LINER FOR SOLAR PONDS

R. Almanza, F. Muñoz, G. Segura and A. Martínez

Instituto de Ingeniería, Universidad Nacional Autónoma de México
Ciudad Universitaria, 04510 México, D.F. México

ABSTRACT

A technical option for liners for solar ponds is the use of compacted clays. This aspect is under study in a small laboratory model used to measure thermal and mechanical properties of these materials as well as the ion exchange. CH-type clays are under study. The first one was a kaolinite type that gave a thermal conductivity of the order of 1.04 W/m°C for a temperature range of 50 - 60°C when hot water was used; the vertical permeability was on the average 2.2×10^{-7} cm/sec. For the same clay, these properties were obtained with a hot NaCl brine (50°C): the thermal conductivity was 0.89 w/m°C and the vertical permeability was 1.4×10^{-7} cm/sec.
The second clay is under study, it is a montmorillonite. This clay gave a thermal conductivity of the same order, but the vertical permeability is lower, about 9×10^{-8} cm/sec, when hot water was used. The next step is to use hot brine.
The cations and anions are being analysed in the laboratory in order to know the ion exchange during the diffusion of the salt through the clay.

KEYWORDS

Solar ponds, clays, liners.

INTRODUCTION

The use of CH clays as liners for ponds has been initially developed in Mexico mainly for a 3 km^2 cooling pond of a power plant (Auvinet, 1980,1986) and other smaller ponds. This option is interesting and cheaper than membrane liners in big ponds, since the latter ones have many problems during installation as well as during use , as discussed by Auvinet (1986).
In order to determine whether clay liners can be used for solar ponds, it is necessary to study their mechanical and thermal properties as well as their chemical interaction between the brine and the soil components in use, mainly due to ion exchange.

MECHANICAL PROPERTIES

The techniques of soil mechanics used to classify a soil are well known as discussed by Lambe, (1951 ; 1969) and Smith (1982).Tests were conducted in the laboratory in order to determine the liquid limit (LL) as well as the plastic limit (PL), and thus to establish the plasticity index of the soil, defined as

$$Ip = LL - PL \qquad (1)$$

The LL corresponding to the Casagrande test (Lambe, 1951) for the Kaolinite clay was 69.9% of moisture content. The value of PL that was obtained in the laboratory following the technique describe by Lambe (1951)as well as the value of Ip were 32.8% and 37.1 % , respectively. Using these data it is possible to localize the soil according to the plasticity chart (Lambe, 1969). Using such chart, this clay is an inorganic compound of the CH type of high plasticity. This soil is found in the industrial area in the North of Queretaro City, Mexico (Ponce, 1976).
The same parameters were measured for montmorillonite clay. The LL had 189% moisture content, the PL 36.1% and the Ip 152.9 %. This soil is also localized in the plasticity chart as a CH clay of high plasticity and is found in Nuevo Leon State, Mexico.
The permeability of these soils can be reduced by compaction methods. In order to get the humidity necessary to obtain a maximum impermeability, it is a standard practice to follow the Proctor compaction method (Lambe 1951, 1969). The data in laboratory test for the kaolinite showed that the suitable humidity to get the lowest permeability was 36%.With the same method was found that the suitable humidity for the montmorillonite clay was 30%. With these data a prototipe was built with which one could measure some parameters like the vertical permeability and the thermal conductivity. The model is shown in Fig. 1. It is an asbestos cylinder with a diameter of 1.44 m with a drain in its base. A fiber glass sheet was stuck inside the cylinder in order to prevent water filtering through the asbestos.
It was formed stacking different layers of soil material, the lower one being a 0.10 m gravel and the one over that a 0.20 m sand layer. The next layer is the clay that was compacted and saturated according to the data obtained in the laboratory. Over the clay layer there is a sand layer. The object of the latter is to work as ballast bed, since, if the clay were to be in direct contact with the liquid brine, then the compacted layer would lose its compaction in the upper zones. In this model it is possible to measure the vertical permeability of the clays, applying the Darcy's law to the data obtained:

$$Q = XiA \qquad (2)$$

Q = the rate of flow, m^3/sec
X = vertical permeability, m/sec
i = hydraulic gradient
A = area, m^2

THERMAL PROPERTIES

The main thermal property in the underground of a solar pond is the thermal conductivity of the soil enabling one to know the heat loss to the ground. In the literature there exist several empirical formulas (Kersten, 1949; Smith, 1984; Van Rooyen, 1957; Woodside, 1958) that give values of thermal conductivity for dry clays as well as for humidity clays; however this formulas have only been verified to temperatures up to 21°C and with a moisture content over 5%. If the empirical formulas from Kersten (1949) and Van Rooyen (1957) are applied to the kaolinite clay with 36% moisture, then a value of 1.21 W/m°C is obtained. This value can be used at least as reference to compare it with values at higher temperatures.
In the protoype shown in Fig. 1 it is possible to measure the thermal con-

ductivity at different temperatures using the Fourier equation

$$q = -KA \frac{dT}{dx} \quad (3)$$

where

q = rate of heat transferred, W
K = thermal conductivity, W/m°C
A = area, m^2
T = temperature, °C
x = distance, m

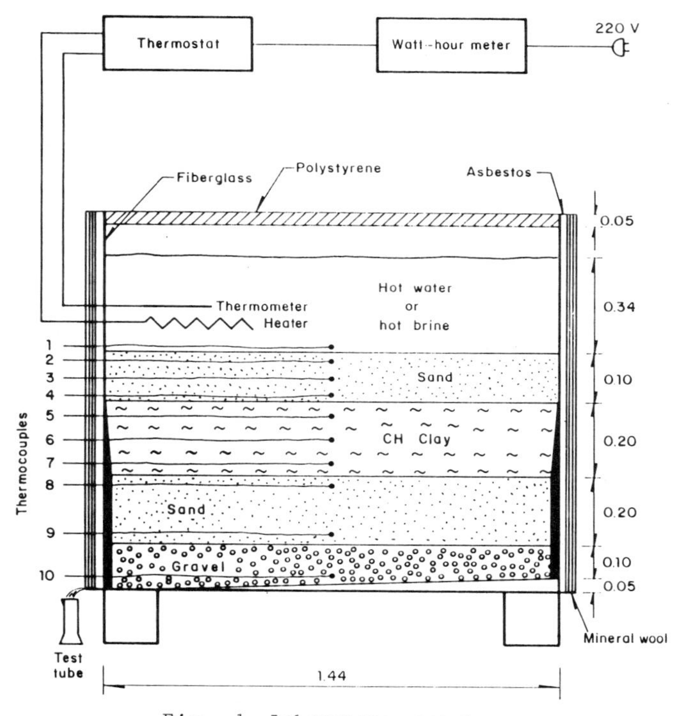

Fig. 1 Laboratory prototype

TESTS AND RESULTS

The first set of tests were carried out with the kaolinite clay using hot water at 50° and 60°C; the thermal conductivity was 1.03 w/m°C and 1.05 W/m°C, respectively. The corresponding permeability at these temperatures were 2.24×10^{-7} and 2.15×10^{-7} cm/sec. After these measurements a clay sample was taken in order to analyse the possible changes in their mechanical properties. These tests were carried out to know if the clay under a thermal gradient would suffer changes in its internal structure. The measurements show that no changes in mechanical properties ocurred. The next set of tests were carried out with saturated NaCl brine at 50°C. The thermal conductivity during the first month was 0.89 W/m°C and after 4 months was 0.68 W/m°C. The vertical permeability on the average had a value of 1.47×10^{-7} cm/sec. If these values are compared with the values obtained with hot water at the same temperature, they show appreciable differences.

In order to determine if some changes were induced in the clay mechanical properties, a sample was analysed. It showed that the liquid limit was reduced from 70 to 60% and therefore the plasticity index changed.
The second set of tests started recently and were carried out with the montmorillonite clay. Using hot water at 50°C, the thermal conductivity was of the same order of the kaolinite, however the vertical permeability is about 9×10^{-8} cm/sec.
The ions identified in the effluent, coming from the kaolinite with hot brine, are: Ca^{++}, Mg^{++}, Na^{+}, K^{+}, Cl^{-} and $SO_4^{=}$. The cations were identified by means of an atomic absorption spectrophotometer and flame photometry.

MATHEMATICAL MODEL

A mathematical model has been developed for the solar pond using an experimental transmittance function (Almanza, 1985). In order to have a more realistic simulation using clay liners, it is necessary to involve other equations that consider the heat and mass transfer as well as the ion exchange between the soil liner and the brine. These equations are being solved and have been presented previously (Almanza, 1987).

ACKNOWLEDGMENTS

Discussions with M. Mazari and G. Auvinet were helpful in clarifying some points in the paper. A. Sanchez, G. Mendoza and C. Juarez made some tests in laboratory. The chemistry School (UNAM) gave help for the analysis of the effluents.

REFERENCES

Almanza R., and J. Lara (1985). Simulation of Solar Ponds with an Experimental Transmittance Function. Transactions of the ASME, Journal of Solar Energy Engineering Vol. 107 No. 4
Almanza R., F. Muñoz, G. Segura, and A. Martínez. (1987). Study of CH-Type Clays as Liners for Solar Ponds. Proceedings of the International Progress in Solar Ponds Conference, Cuernavaca, Morelos, México. (In Press).
Auvinet G., and G. Hiriart (1980). An artificial cooling pond for the Rio Escondido coal fired power plant. Symposium on surface water impoundments, ASCE, USA.
Auvinet G., and R. Esquivel (1986). Impermeability of artificial ponds. LIMUSA and Soc. Mex. de Suelos. (In Spanish).
Kersten M.S. (1949). Thermal properties of soils. Engineering Experiment Station Bolletin No. 28, University of Minnesota
Lambe T.W. (1951). Soil Testing. John Wiley
Lambe T. W., and R.V. Whitman. (1969). Soil Mechanics. John Wiley.
Ponce J.A. (1976). Soils of Queretaro, Qro. Reunion Nacional de Mecanica de de Suelos , tomo 1. (In Spanish)
Smith G.N. (1982). Elements of Soil Mechanics for Civil and Mining Enginner, Granada
Smith W.O. (1948).The thermal conductivity of dry soil. Soil Science Vol.53 No. 6
Van Rooyen M., and H.F. Winterkorn (1957). Theoretical and practical aspects of the thermal conductivity of soils and similar granular systems. Highway , Research Board, Bolletin No. 168
Woodside W. (1958). Calculation of thermal conductivity of porous media. Can. J. Phys. Vol. 36

SOLAR POND FOR HEATING ANAEROBIC DIGESTERS. FINAL REPORT

Li Shensheng, Lu Huanmin and Song Kehui
Department of Physics, Beijing Teachers College,
China

ABSTRACT

A small experimental solar pond is used for heating two small anaerobic digesters and with other two of the same size and physical conditions for comparison. The experimental results of the raise of digestion temperature and the increase of biogas production rate are presented and discussed.

KEYWORDS

Solar pond; anaerobic digester; digestion temperature; biogas production rate.

INTRODUCTION

Using biogas to provide thermal and electrical energy usages is an important way for solving the shortage of conventional energy resources in the vast rural areas of China. It is well known that the production rate of biogas is strongly affected by the digestion temperature, and the most suitable digestion temperature is about 30-35°C. Since in the wide countrysides of China, especially in the northern rural areas, the temperature of most anaerobic digesters, which are built underground or semi-underground, is much lower than 30°C, even lower than 20°C during about half a year, and there is nearly no biogas to yield in severe winter.

Due to its long-term and large capacity of thermal storage, using solar pond for heating anaerobic digesters will appreciably raise its digestion temperature and thus increase its biogas production rate, and will make the digesters to operate pretty well even in winter.

The theoretical analysis and the numerical results of calculation for a solar pond heating anaerobic digesters in Beijing area in China, is presented and discussed earlier(1)(2)(3). At present, we will present the experimental results of using a small solar pond for heating two underground anaerobic digesters, during the period from September 1984 to August 1986.

EXPERIMENTAL RESULTS

The small experimental solar pond was built in summer, 1984. Its dimensions were 3.5 * 3.5 * 1.2 m, using 1.5 tons of NaCl to form a convective solar pond, with the thickness of lower convective zone 0.3-0.4 m, and that of upper convective layer 0.1-0.2 m.

A polypropylene heat exchanger was set in the pond at the depth of 0.24 m above the bottom. It was connected with pumps, valves, flow meters, pressure meters and the heat exchangers set in the digesters to form a closed loop, in which the fresh water as working medium was cycling.

Two anaerobic digesters were insulated with 0.05 m polyethylene and heated by the solar pond, with other two of the same size and physical conditions for comparison.

The experimental results during following four periods are reported:

1. September 11-23, 1984:

During this period, heat was extracted for about 10 hours each day except September 16-18, and a total amount of 141 MJ was extracted from the solar pond. The average thermal utilization efficiency η of the pond, viz. the ratio of the amount of heat extracted from the pond Q to the amount of solar irradiation upon the surface of the pond Q_0, was about 10%. The digestion temperatures were raised to 27-28°C from 23°C, i.e., 4-5°C higher than that of the unheated digesters, and the biogas production rate increased to about 240%.

For 16-18 September, we had not extracted heat from the pond, the temperature of which almost unchanged, while the temperatures of the two heated digesters dropped significantly. On September 22, it was overcast and rainy whole day, but we had extracted an amount of heat about 12 MJ and maintained the digestion temperatures still over 27°C. It was evidently, in spite of the bad weather, the heating by solar pond can also raise the digestion temperature and maintain it rather steady without large fluctuations in a short period. This character would be of great importance and advantage to anaerobic digestion, and it would be rather difficult to achieve by any other means of solar thermal utilizations.

2. November 26 - December 4, 1984:

During this period, the solar insolation and ambient temperature both decreased significantly. On the other hand, the surface of the solar pond was shadowed by 1/3 due to the restriction of experimental place. Hence we provided heat to only one digester and with another one for comparison. The digestion temperature raised only by 2-3°C, and no biogas produced. The average efficiency of the pond was also about 10%.

3. April 5 - May 9, 1985:

During this period, the heat was extracted from the pond and provided to two digesters all the day. The digestion temperatures raised again by 4-5°C, and the biogas production rates increased to about 240%. From April 30 to May 3, we had stopped the heat extraction, the temperatures of the two heated digesters dropped significantly, while the temperature of the pond raised continuously. The average efficiency of the pond still maintained as 10%.

4. July 10 - August 2, 1986:

During this period, we had just provided heat to one digester, and with another one for comparison. The raise of temperature reached 5-7°C. In spite of the rainy weather during this period, the temperatures of the solar pond and the heated digesters were both rather steady, only slightly affected by the weather. The fluctuation of the pond temperature were generally 2-3°C, and the variations of temperature in adjacent days were but only about 1°C. Evidently, these aspects were very profitable for anaerobic digestion. During this period, the increase of biogas production rate was only 160%, because the temperatures of both heated and unheated digesters were all in the neighbourhood of 30°C. The thermal utilization efficiency of the pond was still about 10%.

All the experimental results of above four periods are summerized in the following table:

Duration	Q_0 (MJ)	Q (MJ)	η (%)	ΔT (C)	ΔR/R
84,9,12-22	1467	141.0	9.68	4-5	1.4
84,11,26-12,4	434.5	42.3	9.73	2-3	---
85,4,9-5,9	3755	373.6	9.95	4-5	1.4
86,7,10-8,2	4526	436.8	9.65	5-7	0.6

Where R is the biogas production rate of the unheated digester and ΔR is the difference of biogas production rates of the heated and the unheated digesters.

CONCLUSION

It is shown by the experimental results that the heating of anaerobic digesters by solar pond is definitely technically available. Not only the effect of temperature raise is evident, but also it is rather steady. Therefore, it is not only possible to increase the biogas production rate in spring, summer and autumn appreciably, but also there is the possibility to make the anaerobic digesters to operate pretty well even in winter. Because of the effective area of the solar pond we used is too small, and the physical condition is not good, so the experimental results are not satisfactory. But we do expect, if we can enlarge the effective area of the solar pond appreciably, then both the temperature of the pond and the temperature difference with the ambient surroundings would be raised evidently, so the digestion temperature and the biogas production rate can also increase effectively, hence there exists the possibility to solve the problem that the digesters cannot produce biogas during winter time in most areas of China. Therefore, it is an encouraging approach to release the shortage of conventional energy resources in the vast rural areas in China. We are also in confidence that the above results would be of great importance and interest to many developing countries and districts, where the insolation and salt resources are rich but lack of conventional energy resources, and this approach would be a valuable choice by these countries and districts.

REFERENCES

(1) Ma Wenqi and Li Shensheng, Acta Energiae Solaris Sinica, 5(4), 1984.

(2) Lu Huanmin and Li Shensheng, ibid., 4(3), 1983.

(3) Lu Huanmin, Song Kehui and Li Shensheng, Proceedings of the 1985 International Conference on Solar and Wind Energy Applications, 1985.

PROCESS STEAM GENERATION BY TEMPERATURE BOOSTING OF HEAT FROM SOLAR PONDS

Gershon Grossman and Khaled Gommed

Faculty of Mechanical Engineering
Technion-Israel Institute of Technology
Haifa 32000, Israel

ABSTRACT

Solar ponds provide inexpensive means for collecting and storing solar heat at temperatures below 100°C. The most common application of this heat to date has been the generation of electric power. Due to the thermodynamic limitation set by the relatively low source temperature, the overall efficiency of the electricity generation process is of the order of 1%. The present paper is concerned with the use of heat from solar ponds to generate process steam, which may be performed at relatively high efficiency. The use of an absorption heat transformer makes it possible to produce low-grade steam from the pond's heat at a COP of about 0.5. The paper discusses the operation of the heat transformation process and provides quantitative results on efficiencies and temperatures of the output steam.

KEYWORDS

Solar ponds; process steam; absorption; heat transformer.

INTRODUCTION

Solar ponds provide inexpensive means for collecting and storing solar energy at temperatures below 100°C. The use of a salt gradient to inhibit natural convection in a body of water had been observed in natural ponds as early as 1902 and was proposed over thirty years ago as a method for producing a significant temperature rise in artificial ponds due to solar radiation trapping at the bottom. Since the late 1950s, solar ponds have been studied and experimented with in a number of countries. Fundamental, technological, and economic aspects have been investigated.

An excellent survey of the state-of-the-art of solar ponds was given by Tabor[1], who has been one of the pioneer researchers in this field. Additional analytical and experimental work has been done since the survey was published, with a number of interesting results presented at the recent international conference[2]. Noteworthy is the progress report on the world's largest (5 MWe) solar pond at Bet Ha'Arava, Israel.[3] Typically, solar ponds have achieved collection/storage efficiencies (solar to stored heat) of 18-20% at favorable locations such as the Dead Sea.

The most common application of solar ponds to-date has been the generation of electric power by means of a Rankine cycle-based heat engine operating between the high and low temperatures of the pond. Typical annual pond output temperatures based on the experience at Bet Ha'Arava[3] are 85°C (185°F) and 30°C (86°F) at the bottom and top, respectively. The Carnot efficiency associated with a heat source at 85°C and a heat sink at 30°C is 15%. Due to this second law limitation, the actual electric generation efficiency has been on the order of 5%, bringing the total system efficiency (solar to electric power) to about 1%. The amount of heat rejected in the condenser is about 19 times greater than the useful power generated, and the top surface of the pond is often insufficient to reject all that heat at the required rate.[3] The low overall efficiency makes the economics, at present energy prices, non-competitive for common applications.

The present work is concerned with another application of solar ponds--the generation of process steam. While the latter is a commodity less universally required than electricity, it presents an attractive alternative with favorable economics where energy-intensive industrial plants and solar ponds may be located adjacently.

A heat-actuated heat pump is employed, which utilizes the heat of the pond for temperature boosting as well as for a source of power. Among the different heat pump concepts for temperature boosting of low-grade heat[4], perhaps the most promising is an absorption heat transformer. It operates among heat reservoirs at three different temperatures: the low temperature of the pond's top providing a heat sink; the intermediate temperature of the pond's bottom providing a source of low-grade heat; and the high temperature of the output process steam. In a simple, single-stage system, for each unit of heat taken from the bottom of the pond, approximately half a unit is temperature-boosted to provide process steam and another half rejected to the sink. This reduces considerably the size of the sink required in comparison to the electric power generation application, raising the possibility of using the top surface of the pond for this purpose with no extra cooling pond. The heat transformer does not require any electric or mechanical power for its operation except for small, auxiliary equipment.

THE ABSORPTION HEAT TRANSFORMER

The absorption heat transformer may be regarded as an inverse version of the more common absorption heat pump or chiller. While the basic thermodynamic cycle has been known since the beginning of the century, intensive research and development as well as applications have begun to develop only less than a decade ago, mainly for waste heat utilization[5-8].

Several versions of the absorption heat transformer cycle may be used to achieve different degrees of temperature boosting under various operating conditions. To illustrate the principle of the cycle, a schematic diagram of the simplest, single-stage system is described in Fig. 1. The system requires a pair of working fluids--an absorbent and an absorbate. (The latter is often referred to as "refrigerant" in absorption cooling applications.) In the present study, a well-established fluid pair, lithium bromide/water, is considered. The advantages of this material combination for this application over others are many[5], but a detailed discussion is beyond the scope of this article.

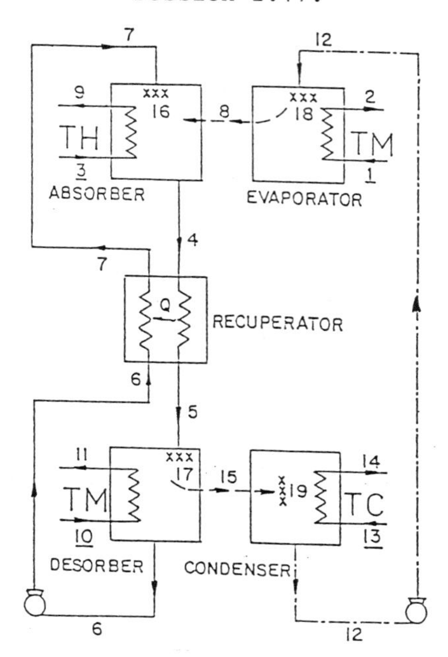

Fig. 1. Schematic description of a single-stage absorption heat transformer

In Fig. 1, the three temperature levels of the heat sink at the top of the pond, heat source at the bottom of the pond, and process heat output have been labeled T_C, T_M, and T_H, respectively. Absorbate water at state point 12 enters the evaporator where heat input from the bottom of the pond is provided by the stream 1-2. The absorbate heats up to the boiling point (state 18) and evaporates. The vapor leaving at state 8 enters the absorber where it is absorbed by a concentrated LiBr/H_2O solution entering at state 7 and leaving more dilute at state 4. The absorption process, which includes an adiabatic step 7 -> 16, releases heat 16 -> 4 at a temperature higher than T_M, ideally by an amount known as the "temperature lift" of the working fluid pair. This heat is transferred to a stream of water entering at state 3 and leaving as process steam at state 9. The diluted LiBr/H_2O solution (state 4) passes through a recuperative heat exchanger into the desorber, where it is reconcentrated by evaporating water from it through the application of additional pond heat at the temperature level T_M, brought by the stream 10-11. For effective desorption at this temperature level, a low-pressure sink for the desorbed water vapor is provided by the condenser, cooled to the temperature level T_C by the stream 13-14 from the top of the pond. The reconcentrated LiBr/H_2O solution leaving the desorber at state 6 returns to the absorber at state 7, via the recuperator. The water condensate (state 12) returns to the evaporator.

Many industrial applications require process heat temperatures greater than what may be obtained from the solar pond by a single-stage heat transformer. There are several possibilities for staging heat transformers for the purpose of obtaining a larger temperature boost[9]. In preferred configuration for a double-stage system[5], a second absorber/evaporator stage is added to the system of Fig. 1. The output stream from the first stage absorber serves as an input to the added evaporator and is temperature-boosted further in the second-stage absorber. Both evaporator/absorber stages are served by a single desorber/condenser system to regenerate their $LiBr/H_2O$ solution.

SYSTEM PERFORMANCE CALCULATIONS

In order to evaluate the capability of the system solar pond/heat transformer to generate process heat, detailed calculations were carried out using typical pond temperatures and realistic heat transformer design and operating conditions. Both the single-stage and double-stage heat transformer options were considered.

The calculations were performed using a computer code developed for simulation of absorption systems in a flexible and modular form[10], which makes it possible to investigate various cycle configurations with different working fluids. The code is based on unit subroutines containing the governing equations for the system's components. Those are linked together by a main program according to the user's specifications to form the complete system. The equations are solved simultaneously, and fluid properties are taken from a property database.

Table 1 lists the design parameters for the single-stage heat transformer described schematically in Fig. 1. Typical pond temperatures were selected for the design condition heat source and heat sink, and a typical low-pressure process steam temperature was specified for the output. Flowrates were selected so as to yield 100 kW of process heat. UA values (overall heat transfer coefficient times area) were specified based on practical experience to allow for reasonable cost. The resulting heat transfer effectiveness of all the heat exchangers vary in the range 0.70 : 0.80. Based on these input parameters, the program calculates the unknown temperatures, flowrates, concentrations, enthalpies, and pressures at all the 19 state points in the system as well as the heat duties of the absorber, evaporator, desorber, condenser, and recuperator. Calculated values of the strong and weak solution concentrations, heating capacity, and COP at the design point are also presented in Table 1.

Figure 2 describes the coefficient of performance (COP) and the relative capacity of process heat delivered at the absorber ($\overline{Q}_A$) as functions of the temperature of the heating water from the bottom of the pond, for different cooling water temperatures. $\overline{Q}_A$ is normalized with respect to the design point capacity. The steam delivered is at 120°C in all cases, and the flowrates and UAs are fixed at the design value. A black dot indicates the design condition. It is evident that the higher the source temperature and the lower the sink temperature, the better the performance. The heating capacity increases almost linearly with the source temperature. The COP increases initially, then levels off and shows a slight drop at large values of the heat supply temperature. The level COP value is slightly lower than 0.5. This is explained by the fact that under effective operating conditions, for every kilogram of absorbate circulating in the system, the absorber rejects the heat of absorption of that kg, and the evaporator and desorber receive the heat of evaporation and desorption of that kg, respectively. Since the heats of evaporation, absorption, and desorption are approximately equal to each other, the system output is approximately half the input. Raising the heat supply temperature does not help gain in COP but rather increases the losses due to the imperfect heat exchange in the recuperator. It is also evident from Fig. 2 that there exists a minimum

Table 1. Design condition for single-stage heat transformer

1. Temperatures:	
Heating water inlet to evaporator and desorber	- 85°C
Cooling water inlet to condenser	- 30°C
Steam outlet from absorber (saturated) at 2.0 Bar)	- 120°C
2. Flowrates:	
Heating water in desorber	4.80 kg/sec
Heating water in evaporator	5.00 kg/sec
Cooling water in condenser	5.10 kg/sec
Strong solution from desorber to absorbers	0.55 kg/sec
3. UA (overall heat transfer coefficient times area):	
Absorber	18.5 kW/°C
Desorber	32.0 kW/°C
Recuperator	3.5 kW/°C
Condenser	34.6 kW/°C
Evaporator	34.5 kW/°C
4. Calculated parameters:	
Weak solution concentration (C_L)	56.59% LiBr
Strong solution concentration (C_H)	61.00% LiBr
Heating Capacity (Q_A)	100.00 kW
Coefficient of Performance (COP)	0.483

value of heat supply temperature below which the heat transformer cannot work at given cooling and output temperatures.

Performance calculations similar to the above were carried out for a double-stage heat transformer. Figure 3 describes performance curves for both the single-stage and double-stage heat transformers, showing the COP as a function of steam delivery temperature, for fixed pond temperatures (T_M = 85°C, T_C = 30°C). The ideal performance curve obtainable with a Carnot heat-activated heat pump is plotted along for comparison. The Carnot COP may be calculated as follows: Assuming a reversible heat engine operating in place of the desorber/condenser between two heat reservoirs at temperatures T_M and T_C and supplying work to a reversible heat pump operating in place of the absorber/evaporator between two heat reservoirs at T_H and T_M, the following expression is obtained:

$$(COP)_{Carnot} = \frac{T_H(T_M - T_C)}{T_M(T_H - T_C)} \quad (1)$$

The performance curves in Fig. 3 show clearly the trade-off between COP and output temperatures, as well as the limits on both, for the single-stage and double-stage heat transformers. The desirable operating point for both systems is in the knee of the performance curve, where the maximum approach to ideal performance is reached.

COMPARISON OF APPLICATIONS

Having calculated the performance of the system solar pond/heat transformer, it is now possible to compare between the electric power and process steam generation alternatives (Table 2). In both cases it is assumed that 100 kWh of solar radiation reaching the pond are converted, at a typical seasonal collection efficiency [1], to 20 kWh of heat stored at the bottom at a temperature about 85°C. Under alternative 1, using a Rankine heat engine, the 20 kWh of heat are converted at a typical 5% efficiency to 1 kWh of

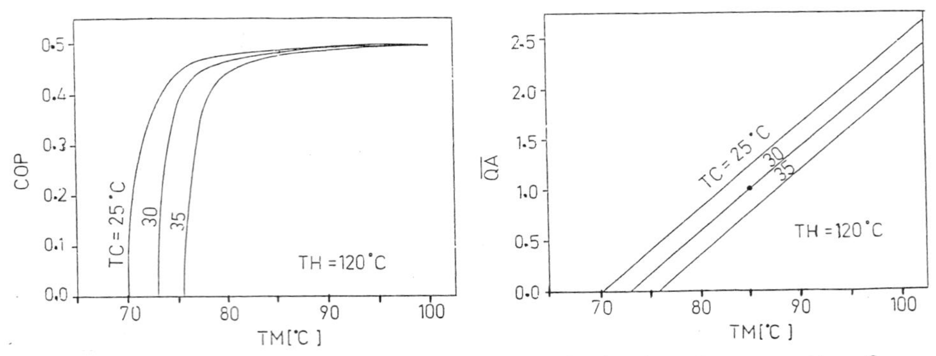

Fig. 2. Coefficient of Performance and relative heating capacity of single-stage heat transformer as a function of heat source temperature for different cooling temperatures

Table 2. Comparison of methods for utilizing pond heat

	Alternative	Solar input (kWh)	Pond heat at 85°C (kWh)	Useful output (kWh)	Rejected heat (kWh)	Output value (US cents)
1	electric power	100	20	1	19	1-12
2	process steam	100	20	10	10	20-35

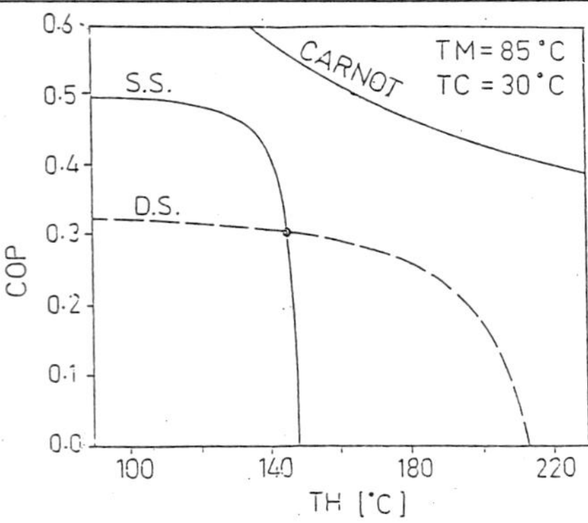

Fig. 3. Performance curves of single-stage and double-stage heat transformers showing Coefficient of Performance as a function of output temperature for fixed heat source and heat sink temperatures. Ideal Carnot performance curve is given according to Eq. 1

electric power. Alternative 2 employs a single-stage absorption heat transformer to convert the 20 kWh of pond heat, at a typical COP of 0.5, to 10 kWh of process steam. Table 2 also shows the quantities of heat that must be rejected under each alternative.

Looking one step beyond the thermodynamics of the different conversion methods, one may try to assign an economic value to their useful output. Based on current energy prices, the cost of 1 kWh of electricity generally varies in the range 4-12 US cents and that of process heat in the range 2-

3.5 US cents. It is evident from Table 2 that the output of Alternative 3 is of the highest value.

CONCLUSION

A study of the application of solar ponds for producing process heat has shown it to be a viable alternative to electric power generation. Absorption heat transformers utilizing the pond's heat itself as a source of power, can boost the temperature of this heat to produce process steam at temperatures about 120°C in a single-stage mode, and about 150°C in a double-stage mode, at coefficients of performance about 0.5 and 0.3, respectively. The overall conversion efficiency of solar radiation to steam is about 10%--ten times greater than for electric power. The heat rejection requirements are also reduced considerably, making it possible to use the top surface of the pond for a heat sink with no extra cooling area. The economic value of the process heat produced from a given solar input is 1.5 to 8 times greater than that of electric power.

REFERENCES

1. H. Tabor, "Solar Ponds -- Review Article," Solar Energy **27**, 181-194 (1981).

2. J. M. Huacuz, ed., Proceedings of the Conference on International Progress in Solar Ponds, Cuernavaca, Mexico, March 1987 (in publication).

3. H. Z. Tabor and B. Doron, "The Beit Ha'Arava 5 MW(e) Solar Pond Power Plant -- Progress Report," Proceedings of the Conference on International Progress in Solar Ponds, Cuernavaca, Mexico, March 1987 (in publication).

4. H. Perez-Blanco, "Heat Pump Concepts for Industrial Use of Waste Heat," ORNL/TM-7655, Oak Ridge National Laboratory, Oak Ridge, Tennessee, 1981.

5. G. Grossman and H. Perez-Blanco, "Conceptual Design and Performance Analysis of Absorption Heat Pumps for Waste Heat Utilization," International Journal of Refrigeration **5**, 361-370 (1982).

6. W. R. Huntley, "Performance Test Results of an Absorption Heat Pump Using Low Temperature (60°C) Industrial Waste Heat," Proceedings of the 18th Intersociety Energy Conservation Engineering Conference, (IECEC), **4**, 1921-1926, Orlando, Florida, August 1983.

7. H. Bokelmann, "Industrial Heat Recovery With Heat Transformers--Practical Applications and Development of Advanced Systems," Proceedings of the IEA Heat Pump Conference, 287-301, Orlando, Florida, April 1987.

8. K. Mashimo, "Overview of Heat Transformers in Japan," Proceedings of the IEA Heat Pump Conference, 271-285, Orlando, Florida, April 1987.

9. G. Grossman, "Multistage Absorption Heat Transformers for Industrial Applications," ASHRAE Transactions **91**, Part 2b, 2047-2061 (1985).

10. G. Grossman, K. Gommed, and G. Gadoth, "A Computer Model for Simulation of Absorption Systems in Flexible and Modular Form," ASHRAE paper No. NT-87-29-2, presented at the Summer Annual Meeting, Nashville, Tennessee, 1987.

WHITE CLIFFS - THE FIRST SEVEN YEARS

S. Kaneff

Department of Engineering Physics,
Research School of Physical Sciences,
Institute of Advanced Studies
The Australian National University,
Canberra ACT, Australia.

ABSTRACT

White Cliffs is supplied by a stand-alone solar thermal electric system based on paraboloidal dish-generated steam driving a high performance uniflow reciprocating engine, driving an ac/dc unit and battery. The paper reports various performance and characteristic features of the solar station as revealed over the years and discusses the many novel features which have been developed and have wider application, including non-solar uses. Overall the project has demonstrated that solar thermal dish/engine systems can more than compete with diesel based power in inland Australia and that much larger grid-connected units are feasible and can be cost effective. Apart from continuing to provide power for White Cliffs, the station acts as a test bed for new ideas which can enhance the performance of next generation systems.

INTRODUCTION

To ascertain the feasibility of using solar thermal power based on paraboloidal systems, the New South Wales Government commissioned the Australian National University in 1979 to develop a family of small systems and to determine their viability for use in inland Australia. The prototype unit (25 kW_e), was to be large enough to be non trivial and small enough to limit the required resources. Potential for cost effectiveness was seen to be a key requirement. The units were expected to run automatically and unattended. As in principle the approach ought to apply to systems up to several megawatts, features of such units should be present in the first experimental system - use of solar arrays, full scale protection, control, etc.

Overall generation costs of production units were to be demonstrably lower than diesel electric costs. Although the prototype unit was in every real sense an experimental laboratory system, the system would need to operate in the field from the start, providing useful power reliably to a small inland community, which was eventually chosen to be the small opal mining township of White Cliffs, 1100km West of Sydney. The system was completed in December 1981, checked out, commissioned and run for a year on dummy load, then connected to the township in November 1983, since which time solar power, with diesel back-up, has been supplied to White Cliffs on a continuous stand-alone basis with a reliability matching that of the Western New South Wales grid. Maintenance and operation has been provided very satisfactorily by local inhabitants on an "as required" basis.

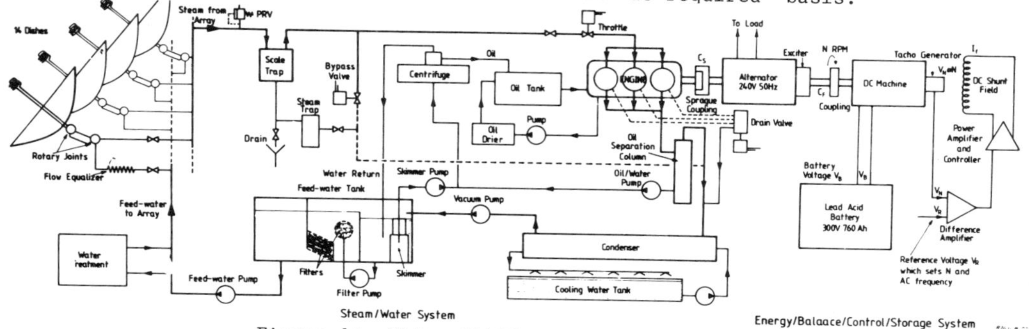

Figure 1: White Cliffs Functional Diagram

SYSTEM DESCRIPTION

The White Cliffs installation has been reported elsewhere (eg. Kaneff (1983) (1984)(1987)). Figure 1 shows the system functional diagram.

The station comprises:

- 14 dish modular solar array;
- water/steam system - feedwater/tank; pumps; fluid lines to each dish via 2 rotary joints and flow equalizer capillaries; receivers; insulated lines to engine room; pressure relief valve; scale trap; steam trap; motorized bypass, throttle and drain valves; 3-cylinder reciprocating uniflow engine; oil separation vortec chamber; water-cooled condenser; and vacuum pump.
- oil water treatment, to remove oil from the exhaust steam and condensate.
- engine lubrication system, comprising separate oil tank, sump, engine oil pump, filter, oil drier and oil pressure and level protection.
- mechanical/electrical conversion, battery storage, shaft torque-balance, frequency control. The ac and dc machines are mechanically coupled to the engine via a free-wheel coupling, allowing continuous electrical machine rotation and intermittent steam engine operation. Digression from 50 Hz causes the dc machine field to adjust operation to restore set frequency in accordance with excess or deficit of energy from the engine. At night or in cloudy periods, the ac load is supplied by the battery via the dc machine (the engine being stationary).
- diesel back-up system
- overall station control system - initiates and directs the daily solar start sequence, ensures that all auxiliaries are functioning, causes the engine to start when steam quality is adequate (usually a few minutes after dishes have tracked); oversees various protection alarms and closes down appropriate sub-systems on malfunctioning; stops solar operation at the end of the day, parks the array and closes down auxiliaries; prevents complete battery discharge by bringing in the stand by diesel unit, switching off the diesel at a set battery voltage. The station is designed to run without operator intervention.

PERFORMANCE AND OPERATING EXPERIENCE

Kaneff (1984), (1987) reports many aspects of performance and operation of the White Cliffs system. Of interest here is the long term performance which depends on climatic factors, system effectiveness and reliability, and successful operating strategies. Tables I and II give an overview of the respective solar, diesel system performances over the full years 1984-1986.

Table I: White Cliffs 1 Solar Generation Costs

Solar		1984	1985	1986	Prospective
Equipment & Expendable Materials	$	2659	2048	3467	1000
Local Contractors	$	15312	12296	9196	5000
Communications	$	810	922	541	100
Travel	$	1334	625	384	-
Total O&M Costs	$	20115	15891	13588	6100
Solar Station Availability		70%	93%	86%	100%
Insolation-Total kWh/m²		2283	2091	2120	2380
Insolation - Total above 700 W/m² kWh/m²		1828	1588	1557	1900
Nett electrical energy generated (Solar generated less auxiliaries) kWh (1)		22195	25187	19620	38000
Costs of generated energy:					
O&M Costs ¢/kWh		91	63	69	16
10% Capital Costs ¢/kWh (2)		141	124	159	82
Total costs ¢/kWh		232	187	228	98
Nett energy to town (including through battery) kWh (1)		10193	12461	10140	30000
Costs of energy to town:					
O&M costs ¢/kWh		197	128	134	20
10% Capital Costs ¢/kWh (2)		306	250	308	104
Total costs ¢/kWh		503	378	442	124

(1) In 1986, continuing diesel problems prevented solar system from operating at times; waiting for replacement parts (commercial) to arrive also put solar supply out of action for 46 days on items not expected to be troublesome. Output was consequently low.
(2) On Capital Cost of $312,000 (1981).

Since this is an experimental unit, built on a one-off basis, conventional commercial cost assessments and comparisons are hardly appropriate; but they are included as a means for indicating that commercial viability relative to diesel/ electric power is now practicable. Tables I and II show the operation and maintenance (O&M) costs /kWh generated for the solar system lower than those from the diesel.

Table I shows a general improvement in performance from 1984 to 1985 due to better reliability, even though insolation fell. Output in 1986 fell further because of less insolation, but mainly because the system was out of action for reasons which involved the auxiliaries and/or commercial components which should not have broken down.

Table II - White Cliffs I Diesel Generation Costs

	1984	1985	1986
Diesel	$	$	$
Contracted Maintenance & Equipment Replacement	11570	11748	12300
Local Operators	2365	2860	4872
Fuel	8649	10948	14017
Solar Station Operators	760	1675	3397
Total O&M Costs	23344	27231	34585
Nett electrical energy generated kWh (3)	27820	33967	36011
Costs of energy generated:			
O&M costs ¢/kWh	84	80	96
10% Capital Cost ¢/kWh (4)	16	13	15
Total Costs ¢/kWh	100	93	111
Nett energy to town kWh	10947	17667	29270
O&M Costs ¢/kWh	213	154	118
10% Capital Costs ¢/kWh(5)	41	26	19
Total Costs ¢/kWh	254	180	137

(3) Diesel system had continuing problems in 1986, requiring replacement of engine.
(4) On Capital Cost $44,500 (1981)
(5) On Capital Cost of $54,500 (1986 modifications)

PROBLEMS

Over the period 1982-present, neither the solar array, electrical generation/control/storage, nor electronic control systems have required significant attention. Dish integrity and operation has been maintained in all weather conditions; reflectivity has deteriorated less than 3%. The diesel-block-converted uniflow reciprocating engine proved the most demanding and time consuming item; however the engine is now a well developed unit, requiring little attention and its application is clearly far wider - for example powered from waste heat or from biomass.

One area still requiring attention, a situation likely to continue for a long time, involves the performance and lifetime of the solar receivers or absorbers; as these age in the presence of very traumatic heat cycles.

COSTS - ACTUAL AND POTENTIAL

Operation and maintenance costs over the period 1984-86, given in Table I, are considerably less than for the corresponding diesel O&M. While the installed cost for White Cliffs is some 7 times that of the diesel, capital costs must prevail in any overall cost assessment. However, the next generation solar system, currently being built, is targeted at $A 100,000 for a 50 kW_e single dish/engine system (on a single unit production run), a capital cost which is far more competitive; when combined with better reliability and performance, this allows an overall generation cost of 20¢/kWh to be targeted. A production run of such units would have more attractive installed and generation costs.

LESSONS FROM WHITE CLIFFS

- With proper attention to level of technological sophistication and detailed design, it is quite practicable, cost-effective and competitive now to supply sunny areas reliably with paraboloidal dish solar thermal power and for local people to operate and maintain such systems.
- There is great scope and potential for producing more economical systems which should have more-widespread viability, including on-grid connections.
- White Cliffs I has been an effective starting point for further development. Even as a first generation one-off system it compares well

with its associated diesel generator operating under the same conditions.

. There is an economy of size which acts to reduce overall costs substantially (as in many engineering systems).

. We were responsible for the conception, design, development, construction, installation, commissioning and later the operation, maintenance and updating of the station; this close and continuing association, we believe, has been a major factor in success and in gaining and applying lessons for next generation systems.

IMPROVING WHITE CLIFFS I

The following additions or changes are being implemented or are in progress to improve performance and reliability to allow the present system to come nearer to the 'prospective' values in Table I:

. General simplification of systems and reduction of auxiliary demands.
. Addition of an automatic oil-fired superheater to improve steam quality at times of low insolation to allow utilising more of the available solar energy (already working).
. Improved feedwater flow control to each receiver to balance their outputs more effectively.
. Better oil/water separation.
. Installation of air cooled condensation in place of water cooling (to cope with water shortages).
. use of improved receivers designed for longer life and better performance.

Having demonstrated viability (real and potential), the station is continuing to provide valuable experience and data useful in the development of next generation systems, as well as serving the valuable function of providing power to White Cliffs. Advantage is being taken to run the system as an experimental test bed to 'develop' and check out new ideas and approaches; to optimise the present system, and to investigate longer term problems which only time can bring to light.

ACKNOWLEDGEMENTS

Funding support of the New South Wales Government through the Energy Authority of NSW is gratefully acknowledged. Many colleagues have contributed to these developments, especially E.K. Inall, R.E. Whelan, P.O. Carden, H.P. Cantor, O.M. Williams, K. Thomas.

REFERENCES

Kaneff, S. (1983a). The White Cliffs Solar Power Station. Proc. Solar World Congress, ISES, Perth, August 1983, Pergamon Press 1984, Vol.4, pp.2772-2776.

Kaneff, S. (1983b). On the Practical and Potential Viability of Paraboloidal Dish Solar Thermal Electric Power Systems. Proc. Solar World Congress, ISES, Perth, August 1983, Pergamon Press 1984, Vol.4, pp.2767-2771.

Kaneff, S. (1984). Fourteen Dish Community Electric System, White Cliffs; Collector System Performance. Proc. International Energy Agency Workshop, "Design and Performance of Large Solar Thermal Arrays", San Diego, June 1984. SERI SP-271-2664, March 1985. pp.328-356.

Kaneff, S. (1987a). The White Cliffs Solar Thermal Power Systems. Proc. Joint ASME-JSME-JSES, Honolulu, March 1987. The American Society of Mechanical Engineers, Solar Engineering - 1987. Vol.2, pp.815-820.

Kaneff, S. (1987b). A Viable Role for Solar Thermal Power in Remote areas. Proc. Annual Engineering Conference, Institution of Engineers, Australia, Darwin, May 11-15 1987. pp.192-198

LESSONS LEARNED FROM SOLAR THERMO-ELECTRIC PILOT PLANTS EXPERIMENTATION

François PHARABOD

Agence Française pour la Maîtrise de l'Energie
27 rue Louis Vicat, 75015 PARIS - FRANCE -

ABSTRACT

Valuable experience has been gained in the world over the last ten years on various concepts of thermo-electric power plants. Large size units are now connected to the grid in California. Lessons learned from the experimentation of pilot plants show that appropriate technology is now available. Care must be taken in order to minimize auxiliary consumption and operation costs. Commercial plant costs are in the competitive range for some countries but room for improvements exits.

KEYWORDS

Solar power plant ; thermal conversion ; electric generation ; parabolic trough ; parabolic dish ; central receiver.

INTRODUCTION

The thermo-electric conversion of solar energy requires a series of elementary transformations, which leads to complex but performant installations. A wide variety of collector arrays have been tested over the last ten years : low temperature solar ponds and flat plate collectors, distributed collectors for medium temperatures, central receiver plants for high temperatures.

The efficiency of thermo-electric conversion mainly depends on the optical quality of the collector system, on the thermal losses of the receiver and on the efficiency of the thermodynamic cycle. Of all these partial efficiencies, that of the cycle is the lowest, although this is not specific to solar energy.

The global efficiency criterion, which is the ratio of the electrical power obtained to the power of the incident solar radiation on the collecting surface, is not the only technical element which must be taken into account. Daily and seasonal variations of the insolation mean that the energy produced by the system over an entire year's operation must be computed. The installation is then characterised by a daily or annual energy efficiency. Analysis in terms of energy reveals the need for storage, and leads to an in-depth study of thermal losses at transitory ratings and during non-operating periods.

Finally, the economic criterion can also lead to the acceptance of physical transformations with weak performances, especially if the collector is a very cheap one.

LESSONS LEARNED

The pilot plants which have been built cover a very broad range of output power from a few dozen of kilowatts to more than 10 MW. In most cases, the experimental installations were followed by commercial realisations, some units reaching 30 MW (Table 1). More than 100 MW are now connected to the grid in California and several plants are in construction or active design.

The diversity of the technological solutions explored for the thermo-electric conversion of solar energy is remarkable. Although the heliostats which are built today may have similar concepts, over twenty very different models have been tested, from very light heliostats inside plastic bubbles to concrete ones capable of operating under any wind conditions. The reflecting surfaces have also been the subject of very far-reaching exploration, bearing both on the reflecting material (silvered glass, protected aluminium or aluminised polyester) and on the immediate backing of this reflector. A wide variety of parabolic troughs and parabolic dishes has been tested. The thermal receivers, fluids and machines which can be used in the broad range of temperatures and powers concerned have also been very considerably developed.

The experience acquired with both components and systems is now considerable. However, experiments on some installations must still be continued before a detailed assessment can be made.

At present, the lessons learned from installations with distributed collectors can be summarised as follows :

- Availability of modern stationary systems and tracking collector arrays has been quite high ;
- When reflecting surfaces are being constructed, special care must be taken over the edges of the mirrors or the joins between layers in order to avoid corrosion attacks ;
- The absorbant coatings and the thermal protection of receivers must be made with very great care ;
- The use of certain thermal fluids can lead to risks of inflammation ;
- Failures of subcomponents and materials are many and varied, non-solar components fail more frequently than might be expected ;
- The management of the thermal loops and the storage must be simplified, with care being taken to reduce the volumes without circulation and the parasitic thermal inertias and losses ;
- Thermal losses that are much higher than expected appear to be the principal reason for disagreement between measurements and simulation ;
- Performance at nominal point generally correspond closely to forecast ; however, auxiliary consumptions and losses on startup and overnight are a drawback to annual production.

In a similar manner, experiments made with central receiver plants has revealed :

- The satisfactory behaviour of most fields of heliostats and their high availability, although they are sensitive to lightening, and this fact must be taken into account when electrical circuits are being built ;
- The interest of the solution involving a thin mirror glued to a glass backing to avoid corrosion through condensation, which can occur with mirrors which are glued to self-standing modules ;
- The advantage of monophasic fluid, molten salt or sodium receivers over water-steam receivers, because they can be set in operation extremely rapidly and they follow the variations of the flux ;
- The satisfactory behaviour of storage with two tanks, and flexibility which these provide for the running of the installations, since they separate the function of solar energy collection from the production of electricity ;
- The considerable auxiliary consumptions which are due to the complexity and dimensions of the systems implemented, and which are a drawback to the annual production of energy, although performances for power are attained without difficulty when the insolation comply with the forecasts ;
- The need to reduce energies for heating the circuits after shutdown by insulating the circuits and their components extremely carefully, and by adopting an appropriate operating method ;
- The advanced monitoring and control techniques implemented have proved to be well-adapted to central receiver plants, and could be used in conventional plants.

RECOMMENDATIONS

All this information, completed by the comparative study of various designs which can be anticipated, leads to the following recommendations for future projects :

- Learn as much as possible about meteorology at the specific site chosen for the construction of a new plant (direct insolation data, but also wind data, temperature data, lightening frequency, etc) ;
- Use analysis and simulation methods capable of accurately predicting the performance of the system under time varying and off-design conditions ;
- In order to reduce operating and non-operating heat losses, decrease pipe length and size, simplify the piping layout and increase insulation (isolate supports from direct connection to the pipe, insulate all fittings to the extent economiccaly feasible) ;
- Examine the conventional equipment very carefully from the standpoint of the thermal cycling capability required ;
- Regarding central receiver systems, design as large a size as possible, consistent with the constraints of the tower height, the heliostats image size and the atmospheric attenuation ;
- An auxiliary gaz boiler or a diesel engine is often requested to meet utilities' standards ;
- The design should keep personnel to a minimum and use modern automation and reliable hardware ;
- And, of course, attention must be paid to prior efforts made by the world's solar community, so that mistakes are not repeated.

Research and development work is still needed to simplify systems layout and to adapt components to plants which are specifically more often at shut-down than operating. Finding a way to mass produce the collectors at low cost remains a key issue for manufacturers.

Unit sizing is an important parameter of the reduction in available kWh cost price and, in particular, in increase of the plant's yearly productibility. As far as storage and electric power generating system are centralised, the relative importance of auxiliary consumptions decreases when the size increases. At last, it must be borne in mind that the construction of a solar plant must be tailored to requirements. The layout of a project should account for consumer's energy requirements (peak power, load distribution, grid characteristics) as well as the site insolation. Productibility increases quicker than insolation, due to the auto-consumption limitation effect which has been evidenced.

CONCLUSION

The design and construction of numerous pilot installations and the experiments made with these are an essential contribution to research on an energy diversification which uses renewable energies.

We now have reliable scientific data which confirm the emergence of thermo-electric conversion. Economic assessments, which are still incomplete, are hindered by the prototype nature of the installations : components built in small quantities, complementary experimental equipment, increased operation and maintenance costs. The scale effects which began with the design and construction of commercial plants are encouraging. They must be made specific and validated for other plants.

Any improvement in performance characteristics and cost reduction must be considered as progress toward accomplishing the ultimate goal for all solar work : cost effectivness. The cost price of solar plants makes them less competitive in view of the prevailing energy conditions, at least when viewed at short term. The need to look for more remote future or for consequences of a serious crisis occuring in electric generation leads one to maintain program priming potentialities. The security in solar thermal power plants has been demonstrated ; their soundless operation without polluting effects meets the environmental requirements.

The work accomplished by the world's solar community makes it possible to firmly seat the bases of the future development of thermo-electric conversion installations and the systems derived from these for the production of heat.

REFERENCES

Bankston, C.A. (1984). Design and performance of large solar thermal collector arrays. Proc. IEA Workshop, San Diego.

Becker, M. (1986). Solar thermal central receiver systems. Proc. DFVLR Workshop, Konstanz.

Kearney, D.W.,and H.W. Price (1987). Overview of the SEGS plants. Proc. Solar 87 Conference of ASES, Portland.

Lopez, C.W. (1987). Solar One power production phase. Proc. solar thermal technology conference, Albuquerque.

Pharabod, F. (1987). Experience gained in solar thermal energy systems. Proc. IEE Conference on energy options, Reading.

PLANT TYPE	LOCATION (NAME)	OBJECTIVE & DATE	APPLICATION	OUTPUT (kWe)	COLLECTING AREA (m2)	GROSS EFFICIENCY* (%)
Parabolic trough	Coolidge AR, USA	Demonstration 1979	Electricity Irrigation	200	2 140	9
	Kramer Junction (SEGS III & IV) CA, USA	Commercial 1986 (Luz Solar)	Electricity	2 x 30 000	2 x 204 000	15 (inculding gas boiler)
Parabolic dish	Shenandoah (STEP) GA, USA	Demonstration 1982	Electricity Steam Chiller	400	4 500	14
	Warner Springs (Solar Plant) CA, USA	Commercial 1984 (La Jet)	Electricity	5 000	29 900	16
Central receiver	Barstow (Solar One) CA, USA	Research 1982	Electricity	10 000	72 000	16
	Targasonne (Themis) France	Research 1983	Electricity	2 500	10 700	19

*** Overall gross efficiency at design point**

Table 1 : Sample of large thermal collector arrays

THREE YEARS EXPLOITATION OF THE 100 kW AJACCIO SOLAR POWER PLANT

G. Simonnot, A. Louche, Y. Decanini

CNRS UA 877, Université de Corse, Vignola, 20000 Ajaccio, France

ABSTRACT

Running a solar plant in real true scale conditions during more than three years gave us the opportunity to improve the ways of conducting the production. Moreover, and in the case of a continuous constant generation of electricity, some significant knowledge was brought about the relative sizes of the various sub-systems of the plant, that bring us to consider the necessity of hybrid systems.

KEYWORDS

Solar Power Plant. Energy systems dynamic operating.

INTRODUCTION

The Ajaccio-Vignola Solar Power plant (1) produces electricity through a thermodynamic conversion of collected solar radiation. Heat is obtained at a temperature level of 250°C, and converted into mechanical/electrical energy via two turbo-alternators, each of these having a power output of 50 kVA. The working fluid is a fluorinert FC 75, and is operated in a Rankine cycle. Heat transfer and thermal storage are carried out using sensible heat of a thermofluid (Gilotherm PW).

The power plant is fitted with a linear solar concentrator, using an East/West segmented cylindrical mirror. The fixed mirror is focusing onto a boiler moving around the axis of the cylinder (COSS type). A fraction only of the direct radiation incoming on the site is collected ; this part, called here "usable irradiation", takes into account cosine effect, and various effects of shadows on mirror and boiler.

RESULTS OF THREE YEARS EXPLOITATION

After having analyzed the various sub-systems behaviour (2)(3) we proceeded to the running of this plant according to two different production schemes.

The first one was designed to use a control of the production based on the monitoring of the temperature value (250°C) at the output of the boiler ; the storage tank is utilized either for starting operations, or to introduce a delay between heat production and electricity production.

In the second operation scheme, full priority is given to the total efficiency value : the storage is by-passed and thermodynamic loops are started directly from the heat produced at the boiler output.

Figures 1 and 2 show the production, respectively in both the schemes of operation. Weekly, mean values of the electricity daily produced are reported as a function of the mean value of daily direct usable irradiation.

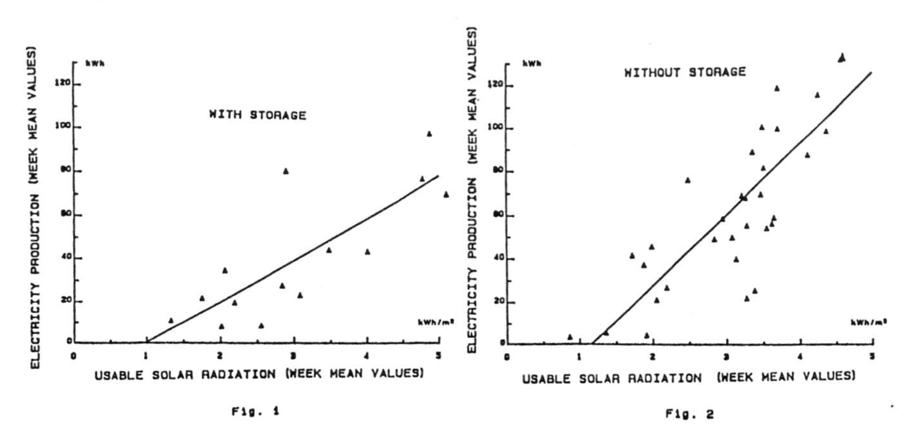

Fig. 1

Fig. 2

Comparing the slopes Fig. 1 and 2, the advantage of the second scheme is obvious.

OPTIMIZATION OF COLLECTOR AREA AND STORAGE TANK CAPACITY. HYBRID SYSTEMS

From experience given by running the solar plant, we tried to seize the opportunity to specify some significant parameters in designing such a system.

For example, to come up to such an expectation of energy, covered with a given rate, it is possible to size the storage tank capacity, and then to determine the corresponding collector area in given climatological conditions, as well as collector theoretic efficiency.

1. Covering expectation of energy and collector efficiency

Analyzing the irradiation data measured on the site of Vignola, a threshold was determined corresponding to the mean value of the daily energy collected which insure a normal running of the plant. Days are qualified "fine weather day" when the value of usable irradiation is above the threshold, and "bad weather day" when below the threshold.

To specify the needs satisfaction and the collector efficiency, a classical power system such as Vignola plant is considered as requested to produce a constant electrical power value. The storage is sized to feed the thermodynamic loops during n_s days without sun (bad weather days), and the collector area is large enough to charge the storage during one fine weather day.

Then, a computer simulation is made, starting from the available data concerning the site of Vignola (3 years). Let be "satisfaction" S_a, the ratio

$$S_a = \frac{\text{Number of operation days}}{\text{Total number of days}}$$

An operation day is so called when the plant is able to produce the requested power.

The collector efficiency corresponds to :

$$E = \frac{\text{Number of operation days}}{\text{Ideal number of operation days}}$$

Ideal number is the number which would be reached starting from a regular favorable succession of fine an bad days. This ideal number is a characteristic of the adaptation of the storage capacity -and correlatively of the collector area- to the climatological conditions on the site (here, Vignola).

Calculating S_a and E for different numbers of operation days, curves are obtained as presented in Fig. 3.

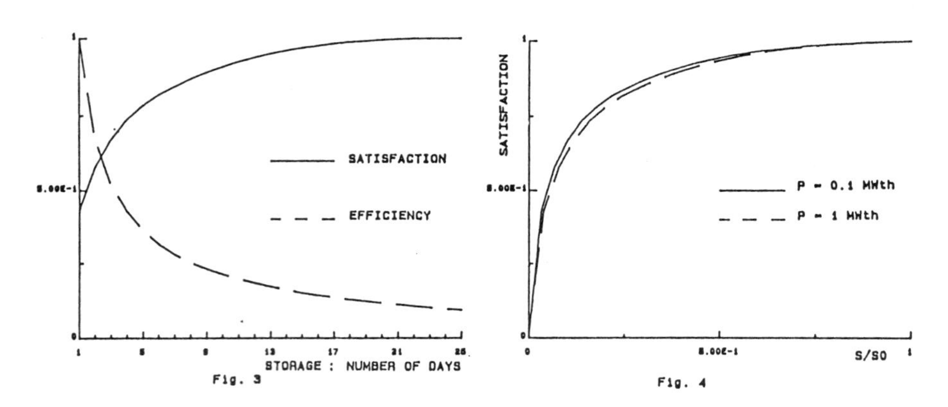

Fig. 3 Fig. 4

It can be seen that, on the Vignola site, satisfaction reaches the unity value when the storage capacity is 22 days. In the same time, collector efficiency decreases sharply when the storage capacity is increased.

In brief, a given value of the demand corresponds to one storage capacity value, one collector area value, and one collector efficiency.

2. Collector area determination

We start from the assumption that the energy decrease (dE) in the storage tank, during a (dt) time, corresponds to the energy P produced by the plant, plus heat losses of the tank during the same time. This is expressed by :

$$dE = - Pdt - k\ E(t)\ dt$$

where E(t) is the energy stored in the tank at the (t) time, and k a constant term depending upon the size of the tank.

Assuming P as constant, this equation, easily integrated gives :

$$E(t) = (E_o + P/k)\ e^{-kt} - P/k$$

At the time t = 0, the storage tank is full and contains a quantity E_o of energy.

In the case of Vignola plant, the storage is considered as empty when 27 % of the total stored energy is still contained (unusable heat). To run during t_o days, E_o has to verify :

$$0.27\ E_o = (E_o + P/k)\ e^{-kt_o} - P/k$$

The term k is as small as the storage is large, that is k is depending upon E_o . The former equation has no analytic solutions and must be numerically integrated using values observed with the Vignola plant.

To find the collector area roughly corresponding to the possibility of filling the storage tank during one day, considering that the same storage will run the system during t_o days, the value obtained for E_o has to be divided by the productivity of one square meter of collector. In our case : 0.8 kWh/m².

Let S_o be the area which permits to come up to the complete needs, and S, the area corresponding to a given rate of satisfaction. A graph is plotted (Fig. 4) giving the satisfaction rate as a function of S/S_o for a given power. It has to be noted that these curves are weakly depending on the power.

For a 100 thermal kilowatts, and for a satisfaction rate of 1 (22 days storage), we need a collector area of the order of 130 000 m², which is a nonsense. Moreover, collector efficiency would be near by 0.1 (Fig. 3).

For a satisfaction rate of 0.9, with the same power, the storage will insure only 10 days ; the collector efficiency will be 0.2 and the collector area will be 37 % of the previous value, that is approximately 48 000 m² (Fig. 4).

In both the cases described above, the collector efficiency is rather bad. When we want this parameter be near by 1, the storage capacity will be of one day, and the collector area around 4 000 m². Two advantages will be gathered : small area and large efficiency of the collector. Then, the satisfaction rate will be 43 %. The complement of energy (in our case 57 % of the needs) has to be produced by a complementary source (gaz, biomass, ...). So we are conducted to the concept of hybrid systems.

In fact, it can be considered that the proportion between conventional and solar sources will be determined by economical considerations, but in any case, these hybrid systems will insure, as we showed, a significant valorization of solar energy.

REFERENCES

1. Bacconet, E., J. Bliaux, and M. Dancette (1983). Systèmes à concentration (COSS) : Centrale de Vignola. Cahiers de l'AFEDES, N° 3. Ed. EETI.

2. Simonnot, G., A. Louche, Y. Decanini, and G. Augé (Avril 1987). Prévisions de la productivité à long terme des systèmes héliothermiques : analyse des performances des sous-systèmes de la Centrale d'Ajaccio-Vignola en vue d'une gestion optimisée. J.I.T.H., Lyon.

3. Simonnot, G., C. Riolacci, and G. Peri (Avril 1985). Bilans et résultats sur l'exploitation de la Centrale héliothermique de Vignola. XXIIIe Rencontre Internationale COMPLES, Séville, Espagne, Congress Proceedings Univ. Seville, pp. 557-570.

A NEW APPROACH FOR DESIGNING A SOLAR CONCENTRATING SYSTEM

A. Louche*, L. Aiache**, G. Simonnot* and G. Peri*

*CNRS UA 877, Université de Corse, Vignola, 20000 Ajaccio, France
**Université d'Aix-Marseille III, Saint-Jérôme, 13013 Marseille, France

ABSTRACT

A new approach, based on the entropy of information from the general theory of system, is used to optimize a specific collector using n reflecting flats, oriented East/West and partially tracking the sun.

KEYWORDS

Entropy. Energy. Concentration. Optimization. Modelization.

INTRODUCTION

The general theory of systems has been recently applied to solar concentrating systems (1) and appears as a powerful tool in designing such systems.

One of the problems arising in focusing reflectors design is to determine the optimum concentration ratio versus physical and technical constraints (manufacture and tracking tolerances) (2). For many years, our team has been working in the field of focusing devices using flat reflecting structures (3)

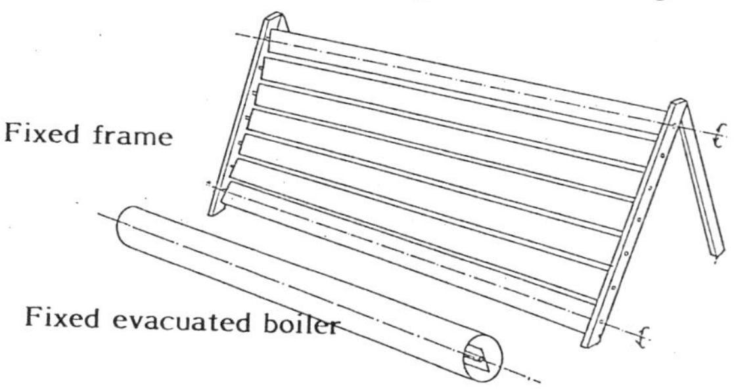

Fig 1. Solar concentrating system

We develop an approach, based on the information theory, to optimize a specific collector using n reflecting flats, oriented East/West, and partially tracking the sun by rotation around this axis (Fig. 1). The system produces an optimized concentration onto a receiver in view of heliothermal conversion.

THE APPROACH

It is obvious that the various systems of solar conversion (physical, chemical, biological, ...) are systems with dynamical behaviour ; most of them are very sensitive to manufacture and, eventually, tracking tolerances.

As presented by Aiache (4) we define an entropy of information which is a function, or a class of functions, that can describe the indetermination, encountered by an observator who wants to characterize a particular element in a whole, as a function of a collective variable and constraints imposed to the model. These constraints are more or less general.

As the entropy, as defined above, is always positive, the definition is compatible with the Shannon's one, and explains the entropy of the absolute and applied metrology, as described by Brillouin.

Let us consider now a collector using n reflecting flats, oriented East/West and partially tracking the sun by rotation around this axis. As n increases, the system concentration increases, but the tolerance (as well mechanical as optical) decreases. The designer's problem is to specify the optimal number of reflecting flats.

For each flat reflecting surface that constitutes an element of the solar adaptor, the tolerance can be measured or estimated as the pointing error, i.e. the difference between the ideal position and the true position of the normal. Then, tolerance is defined for each mirror by Ω the solid angle measuring the whole space corresponding to an authorized position of its central normal.

The measure (considered in the mathematical sense) of the configuration constituted by n reflectors, abiding by these tolerances, is the product of their solid angles.

Then, giving to each surface a partial entropy, we obtain a logarithmic formulation :

$$I = \sum_{1}^{n} \log \Omega_i = \log \prod_{1}^{n} (\Omega_i)$$

Introducing the relative weight ω_i defined as $\omega_i = (\Omega_i / \Sigma \Omega_j)$, and the conditions I (0) = I (1) = 0, and I symetric function towards ω_i, we have the Shannon's formulation. Moreover, if we keep the absolute weight, we obtain the Boltzmann's formulation.

Then, the Shannon's formulation permits to determine when adding up a supplementary reflecting flat is no longer necessary to improve the reflector aperture.

FIRST RESULTS

Considering the solar conversion chain, as represented in Fig. 2, the system optimization will be obtained by the knowledge of numerous parameters, concerning the boiler and the adaptor.

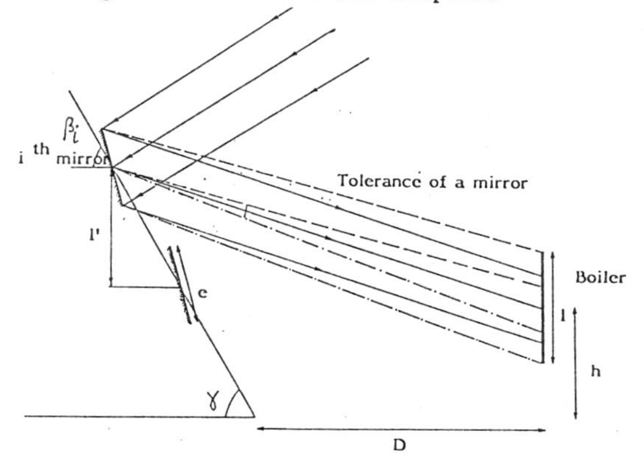

Fig 2. Physical parameters and principle

The mathematical study of this problem shows that the system is described by seven physical parameters : height and width of the boiler, distance from the boiler to the adaptor, angle between the frame and the horizontal plane, width of the flat reflecting surface, distance between two consecutive flats and number of flats. Obviously, we have also solar parameters as declination and hour angle.

The partial entropy of a reflector, as defined above, can be expressed as a function of these seven parameters. The relative weight of the ith reflector is defined by its tolerance divided by the sum of tolerances. Then, if we only use the tolerance, we have an entropy function as defined by Boltzmann, and if we use the relative weight, we obtain an entropy which is a Shannon's one.

On Fig. 3, we can observe the first results obtained with the Shannon's definition. The entropic function of the adaptor is expressed as a function of the number of reflecting flats. We can see that beyond 14 reflectors, we find a step and the entropy does not increase. So, to determine the exact number of necessary reflectors, we need more informations about all the other parameters.

CONCLUSION

At that time, we have demonstrated that the concept of information entropy seems to be very useful to optimize solar devices. We are just beginning the comparison of theoretical results and experimental data. Our aim is to take into account other criteria such as energy distribution on the focal spot, technical constraints and economical factors.

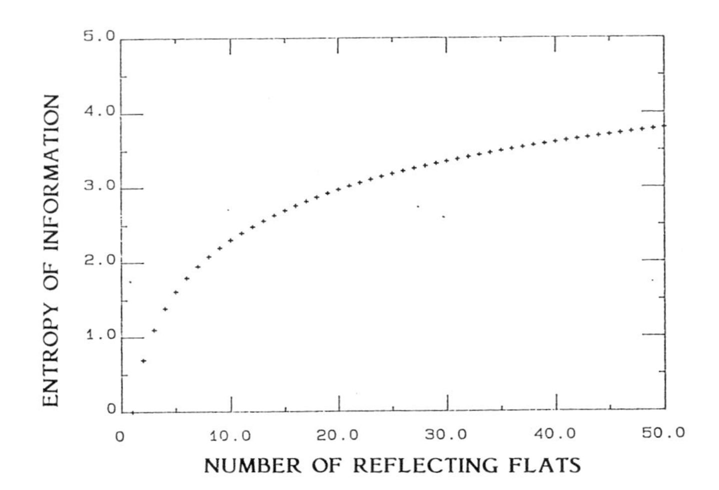

Fig 3. Theoretical results

REFERENCES

1. Aiache, L., P. Gallet, and B. Imbert (1981). Réflecteurs à facettes, structures et champ de miroirs. XXe Rencontre Internationale COMPLES, Rabat, Maroc, Proceedings, 72-77.

2. Aiache, and L., B. Imbert (1985). Réflecteurs à facettes planes et à support sphérique. XXIIIe Rencontre Internationale COMPLES, Séville, Espagne, Proceedings, 463-469.

3. Balbi J.H., N. Balbi, G. Girolami, P. Orenga, G. Simonnot, A. Louche and G. Peri (1986). Modélisation du champ de capteurs de la Centrale de Vignola. Revue Physique Appliquée, 21, 169-180.

4. Aiache L., and R. Zerouali (1985). Une nouvelle axiomatique de l'information, application à la thermodynamique, la géométrie et la métrologie. Analyse de systèmes, Vol. XI-4, 3-39.

THE THERMO-HELIO-ENERGY-KW (THEK) PARABOLIC DISH PROGRAM

M. AUDIBERT, R. PASQUETTI, J. DESAUTEL

Laboratoire des Systèmes Energétiques et Transferts
Thermiques de l'Université de Provence. U.A.CNRS 1168
13397 MARSEILLE cedex 13. FRANCE.

ABSTRACT

This paper gives a brief description and the main measured performances on the parabolic collectors and the experimental solar process heat production sytem developped during the french THEK program.

KEYWORDS

SOLAR ENERGY, THERMAL ENERGY, POWER PLANT, DISTRIBUTED COLLECTORS, PARABOLIC DISH

INTRODUCTION

The THEK Program started by Centre National de la Recherche Scientifique (CNRS/ PIRSEM) in 1975, aimed to develop distributed collectors solar thermal power plants to produce thermal, mechanical or electrical energy within the range of few tens of kW to some MW, at a temperature up to 300°C.

From 1978, this program has been supported by Agence Française pour la Maîtrise de l'Energie (AFME). It involved three phases :

- Phase 1 : Design, construction and experiments on two laboratory prototype collectors.
- Phase 2 : Industrialization of collectors.
- Phase 3 : Design, construction and experiments on an experimental installation named THEK 2, featuring 8 collectors, to produce process heat .

In the first part of this paper we will give a brief description of the THEK collectors and their main characteristics as measured on 4 industrial prototypes. In the second part we will present the THEK 2 Experimental installation and the main results of the experiments conducted on this solar thermal system.

THE THEK SOLAR COLLECTOR.

The THEK solar collector is a parabolic dish collector using a set of about 750 flat triangular mirrors, bonded on a fiber glass-epoxy surface, in order to approximate a parabolic concentrator. This parabolic concentrator is supported by a tubular structure, 5 meters high, and can be rotated around a vertical and an horizontal axis in order to point toward the sun. The two movements are obtained with hydraulic rams.

The reciever, at the focus of the concentrator, is a mono-tubular black painted plane or cavity receiver.

Each collector is equipped with a micro-processor in charge of the orientation of the concentrator and of the safety of the collector in case of excessive output temperature or.

Four industrial prototypes were realized and tested at the Centre d'Essais Solaires de Saint-Chamas, near Marseille. Two of them were made by the BERTIN Cie associated with CREUSOT-LOIRE and the two others by the Société Européenne de Propulsion (SEP).

Table 1 resumes the main specifications given to the manufacturers and the corresponding results of measures made on the prototypes.

Table 1. Main characteristics of the THEK collector.

CHARACTERISTICS	SPECIFIED	MEASURED
Concentrator surface area	50 m2	48.6
Concentration ratio	100<C<300	238
Nominal head losses	1 bar	0.9
Horizontal movement	East : 120°	125
	West : 120°	127
Vertical movement	7 to 90°	5 to 90°
Horizontal angular speed	360°/hour	720
Vertical angular speed	360°/hour	Downward: 828
		Upward : 1260
Efficiency at the receiver output	74%	73%
Efficiency at the collector output	68%	62%

These results show that the prototype collectors fullfilled all the specifications excepted for the heat losses along the tubes between the receiver and the base of the collector. During the 800 hours of operation on these 4 collectors, we observed only minor failures despite severe wind conditions on certain days (wind speed up to 160 km/h).

THE THEK 2 EXPERIMENTAL POWER PLANT

It appeared, from different studies conducted during the first phase of the program, that the production of process heat was certainly, in the present state of the art, the most interesting application for distributed collectors solar thermal systems. The main reasons were that this type of application needs relatively small installations (few thousands of square meters of collectors) and that the direct use of the heat produced by the collectors, without any other energy transformation, is leading to high overall efficiency systems.

It also appeared that average installations, taking into account the needs of different industries (food, textile, paper ....), should have a size of about 100 collectors of the THEK type and produce saturated steam at temperatures ranging between 120°C and 200°C.

Optimization studies of installations of this size have been conducted in 1980 and the THEK 2 experimental power plant represents a module, featuring 8 collectors, of such an installation to produce steam at 180°C-10 bars.

The schematic of the THEK 2 experimental power plant is given in Fig.1. The heat transfer fluid, in the collectors, is pressurized water. The average input temperature is 200 °C and the maximum output temperature is 260°C. The heat collected in the transfer fluid is delivered to the storage fluid through a heat exchanger.

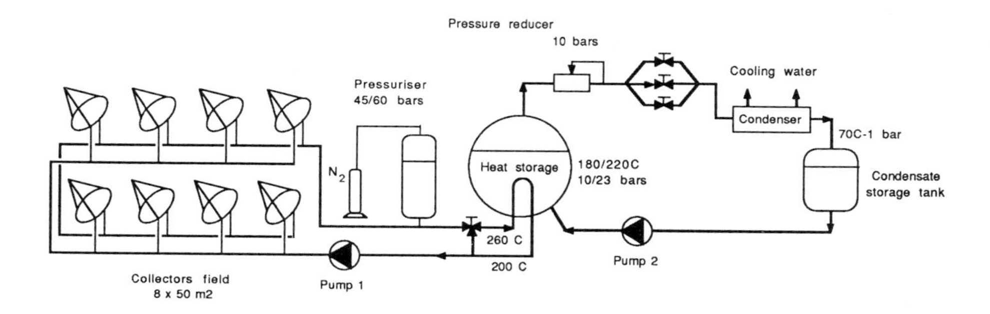

Fig. 1. Schematic diagram of THEK2.

The storage fluid is saturated water at a temperature varying between 180°C-10 bars and 220°C -23 bars. The capacity of the storage tank (8 m3) corresponds to 2 hours of nominal production.

The steam at 180°C-10 bars is obtained from the steam in the upper part of the storage tank through an expander. A set of valves and a condenser allow to simulate the steam consumption of the process. The condensates are recuperated in a tank and pumped back to the storage tank. The nominal power of this installation is 200 kW.

The control system of THEK 2 consists of a central minicomputer communicating with the microprocessors of each collector through a serial RS232 C current loop and with a data acquisition system. The link with the microprocessors allows to get data from the collectors (status, position, temperature) and to send commands as necessary. The link with the data acquisition system allows to get data from the various subsystems other than the collectors and to send signals to the control actuators.

A complete cycle of control operations is executed every 5 seconds by the central minicomputer. The minicomputer terminal provides input-output control and monitor capability to the operator. In case of improper operation, the control system initiates appropriate reactions and edits a message on the printer to alert the operator.

MEASURED PERFORMANCES OF THEK 2 COLLECTORS FIELD

During the long period experiments conducted on the THEK 2 experimental power plant, an important degradation (up to 25%) of the performances appeared rapidly, due to dust deposit on the mirrors, and no economical mean was avelaible to clean the mirrors periodically. Consequently, the exploitation of the results was difficult and it is not possible, within the frame of this paper, to report on these long period tests.

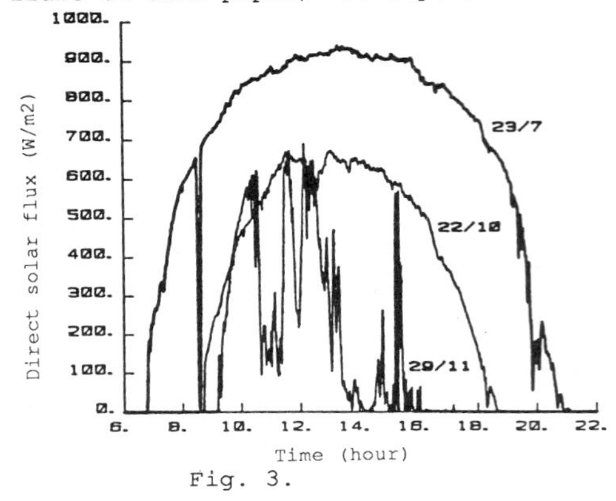

Fig. 3.

We will present hereafter the results of measurements performed during three typical days with clean mirrors : a clear summer day, a hazy fall day and a cloudy day in november.

Figure 3. represents the solar flux during these three days and Table 2. gives the energy balance for the complete day at different levels of the power plant.

If we compare the results obtained on july 23 and october 22 we can see that:

- due to the lower fluid temperature on october 22, the efficiencies at the receiver output level and at the collector output level are higher for this day than for july 23.
- the percentage of the input energy transferred to the storage tank is higher on july 23, because the heat capacity of the installation is the same in both cases and the higher the input energy the lower its influence on the global thermal balance.

Table 2.

	JULY 23	OCTOBER 22	NOVEMBER 29
Operating time	13.5 h	7.25 h	7.5 h
Average fluid temperature	210 °C	150 °C	160 °C
Collected energy (Ec)	3774 kWh	1541kWh	476 kWh
Efficiency at receivers level in % of Ec	65%	68%	43%
Efficiency at collectors level in % of Ec	60%	62%	27%
Energy transfered into the storage tank (% of Ec)	55.5%	54.5%	6%
Electricity consumption of the concentrators	13.4 kWh	6 kWh	10 kWh
Electricity consumption of the auxiliaries	82 kWh	26 kWh	14 kWh
Thermal energy equivalent to the electricity consumption (% of Ec)	7.6%	6.2%	15%
Global efficiency (% of Ec)	47.4%	48.3%	-9%

- the thermal losses between the reciever and the base of the collectors are still too important as noticed on the prototype collectors.
- the relatively higher electricity consumption on july 23 is mainly

due to the primary pump which was running at a higher speed in order to maintain the output temperature under the limit value.

- the global efficiency of about 48% is a promizing result which could be improved up to 55% by reducing the thermal losses in the collectors and the electricity consumption of the pump.

The thermal losses in the field piping -which do not appear directly in this table-, represent only 3% of the input energy on the concentrators. This is due to the use of the same thermal glass-wool insulation for the feed and return pipes and represents a very good result, for these losses were considered as a major limitation for this type of installation.

The results of november 29 show the influence of the thermal capacity of the collectors on their efficiency (43% at the receivers output level compared to 27% at the collectors output level). The low value of the energy transferred to the storage tank is due to the fact that the 3 ways valve was not operating at that time and the fluid was flowing through the heat exchanger even when the output temperature at the collectors was lower than the storage tank tamperature.

CONCLUSIONS

The THEK program has proven the technical feasability of distributed parabolic dishes solar thermal power plants. The measured global efficiency of about 48% on sunny days can be improved up to 55%. The long term operations showed that it is necessary to dispose of an efficient mean to clean the mirrors, the natural cleaning during rainy days is not sufficient.

In the present state of the art, this type of installation is not yet competitive with conventional energy production systems. The cost of the kWh is between five to ten times too high. An important technological effort is still necessary, not only on the collectors but also on the different subsystems, in order to reach the competitiveness.

The THEK 2 experimental power plant

MODELING OF A THEK-TYPE PARABOLIC DISH COLLECTOR EXPERIMENTAL VALIDATION USING INFRARED THERMOGRAPHY

A. ARGIRIOU, R. PASQUETTI, F. PAPINI, A. ARCONADA, M. AUDIBERT

Laboratoire des Systèmes Energétiques et Transferts
Thermiques de l'Université de Provence. U.A.CNRS 1168
13397 MARSEILLE cedex 13. FRANCE.

ABSTRACT

This paper presents some results concerning the validation by IR thermography of two softwares, developed in the frame of the french THEK program, for the determination of brightness concentration distribution and for the dynamic simulation of solar absorbers.

KEYWORDS

Solar energy; infrared thermography; concentration measurement; concentration calculation; dynamic heat transfer simulation; parabolic concentrator.

INTRODUCTION

In the frame of THEK (Thermo Hélio Energy kW) french program (Audibert, 1981), various optical and thermal models of the experimental collectors have been developped. Some of these models were experimentally validated using infrared thermography. In this paper we present :

- the theoretical and experimental procedure for the determination of the brightness concentration distribution
- the dynamic modeling and measurement of the surface temperature of the receiver.

The THEK-type collector has 50 m^2 of reflective area and 4.8 m focal length. The reflector is a paraboloidal dish, approximated by 864 triangular equilateral (0.37 m) plane mirrors. The receiver consists of a stainless steel multitube heat absorber coil.

The camera used is an AGA Thermovision 780 infrared imaging system, sensitive in the 3 to 5.6 µm (SW) or in the 8 to 14 µm (LW) spectral bands. The camera is fitted with a 7° lens and produces a video signal. This signal is applied to an analog to digital converter (ADC), which converts samples of the video voltage to 8 bits or 256 levels. There are 128 points digitized along each line and there are 64 lines generated. Thus a digitized thermogram contains 64 × 128 pixels. The data are stored in real time in the hard disk of a HP 1000 F minicomputer.

BRIGHTNESS CONCENTRATION CALCULATION

Two softwares for the brightness concentration calculation in the focal zone of concentrators were developed in our laboratory (Pasquetti, 1983, 1984). The first is adapted to concentrators of any shape and concavity, assuming a single reflexion of incident rays. The reflector is assimilated to an intensity source and the brightness concentation is calculated by numerical integration, on its surface, of the flux reaching the considered point on the receiver. The second is specific to concentrators constituted by plane polygonal elementary mirrors, thus the calculation time is largely decreased. These softwares take into account the non-uniformity of the solar brightness distribution (Jose, 1957); the modeling of a real concentrator is possible because all the characteristic parameters of the elementary mirrors as well as their position can be altered with errors following Gaussian distributions. We used this second software to calculate the brightness concentration distribution of the THEK-type collector.

BRIGHTNESS CONCENTRATION MEASUREMENT

The method consists to view a target placed in the focal zone of the reflector, with the infrared camera (Argiriou, 1987a). The camera provides thermal images (thermograms), representing the total flux distribution on the target. Two thermograms are needed: one when the collector is almost pointed so that no direct solar radiation impinges on the target and a second when the collector is pointed. The difference between these thermograms gives the radiation distribution of the solar image. From this difference, relative values for the brightness concentration can be deduced. Absolute values yield from the energy conservation equation between the reflector and the target.

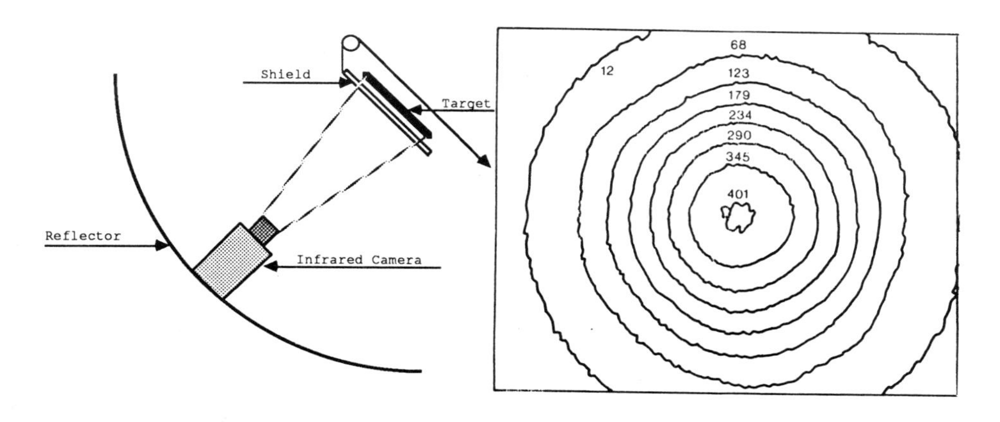

Fig. 1. Experimental set-up.

Fig. 2. Concentration distribution.

A schematic diagram of the experimental set-up is shown in figure 1. The infrared camera is placed in the vertex of the reflector, viewing a target mounted on the receiver plane. The SW scanner is used. The target is a 3 10^{-3} m thick stainless steel disc, normal to the optical axis of the camera, provided with a diffuse coating. A shield is used in order to avoid heating of the target during collector pointing. The experimental procedure is the following: first we film the target, the collector being not pointed. Then we start pointing, the target being shielded. When the collector is pointed, the target is uncovered rapidly and filmed for about 4.5 s. Figure 2 represents measured brightness concentration isolines in the focal zone. A good agreement can be obtained considering a mean value of 0,45° and a standart deviation of 0,25° for the slope error distribution of the elementary mirrors.

RECEIVER MODELING

For the dynamic modeling of multitube heat absorber coils having a flat or cavity shape, we developed a software calculating the temperature field of the receiver (Pasquetti, 1986). Firstly the brightness concentration distribution along the coil is calculated, taking into account the eventual cavity effects. Then the energy conservation equation in the fluid and the non-linear conduction equation in the metal are integrated for every step of time. These two equations are coupled by the fluid film temperature field. It must be also mentioned that a combined heat transfer problem (conductive and radiative) has to be solved numerically:the infrared flux density in the cavity of the receiver depends on its surface temperature field. Also local radiative heat transfer phenomena, between two tubes in contact, are taken into account, using opto-geometrical coefficients, similar to emittance or absorptance factors.

EXPERIMENTAL VALIDATION

The model was validated for a plane receiver, with step change experiments (Argiriou 1987b). This kind of experiment needs good stability of the external conditions because the collector response must depend only on the input variable. Two experiments were conducted, a solar flux step and a flow rate step. For these experiments the LW scanner of the camera has been used. We will limit our presentation to the results of the solar flux step. The receiver must initially be in a thermal steady state, the concentrator being not pointed. The receiver is shielded during pointing. When the concentrator is pointed, the shield is taken off rapidly and the scanning is initiated. The experimental conditions were:direct solar radiation 0 $W.m^{-2}$ before the step and 742 $W.m^{-2}$ afterwards, wind speed 4,4 $m.s^{-1}$, fluid flow rate 569 $l.h^{-1}$, inlet fluid temperature 181,7 °C and ambient temperature 24,8 °C.

Figure 3 shows the thermal profile along the horizontal diameter of the receiver at four different instants (t=1, 3, 13.1, 45.5 s after the step). The solid lines correspond to the experimental thermograms and the dotted lines to the simulated ones. A good agreement is found. The difference observed at the center of the profiles is due to the presence of a cylindric box at the center of the coil; the heat transfer is not very well described at this region. The slight dissymmetry of the experimental profiles is due to the displacement of the solar image on the receiver, because the automatic tracking is out of function during scanning. Figure 4 shows the variations of the calculated temperatures along the absorber pipes for the fluid (curve 1), metal's front external (2) and internal (3) side, metal's back external (4) side, for the instants mentioned above. The back side of the receiver is isolated. The difference between the curves (2) and (3) shows the radial gradient in the cross section of the tube and the difference between (2) and (4) shows the angular gradient. Figure 5 illustrates the calculated temperature field in two cross sections of the coil for the steady states before and after the flux step. The x-axis represents the polar angle of the section and the y-axis the thickness of the tube, in order to facilitate the presentation of the isotherms. For every section, maximal and minimal metal temperatures and fluid temperature are specified. Isotherms are multiples of 10 K. Temperatures are decreasing as the polar angle increases from 0° to 180°.

CONCLUSION

In this paper, relative to the thermal behaviour of a THEK-type collector, we compare theoretical calculations and experimental results obtained by infrared thermography. The infrared thermograms and the numerical calculations allow the determination of the temperature gradients in the metal and the surface temperature field on the receiver. This determination is usefull for aging studies of the absorbing coatings or of the heat transfer fluid; it is also interesting for mecanical resistance studies of the coil tubes. The infrared thermography can thus be considered as an efficient non destructive method for the study of solar devices. The softwares which were validated for a THEK-type concentrator, can also be used for many types of solar concentrating collectors.

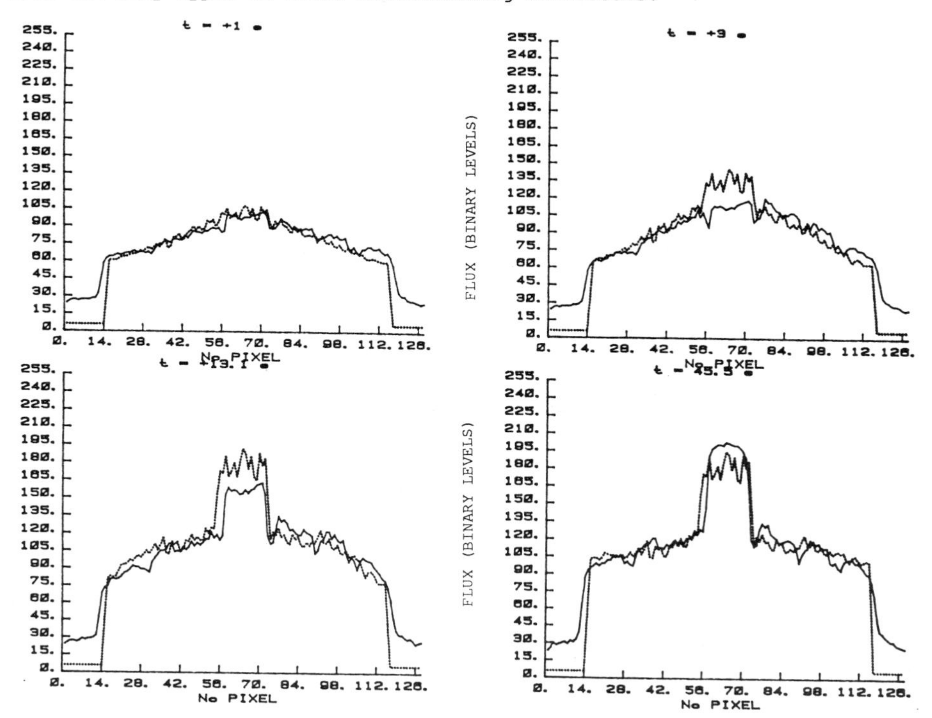

Fig. 3. Thermal profiles on the horizontal diameter of the receiver.

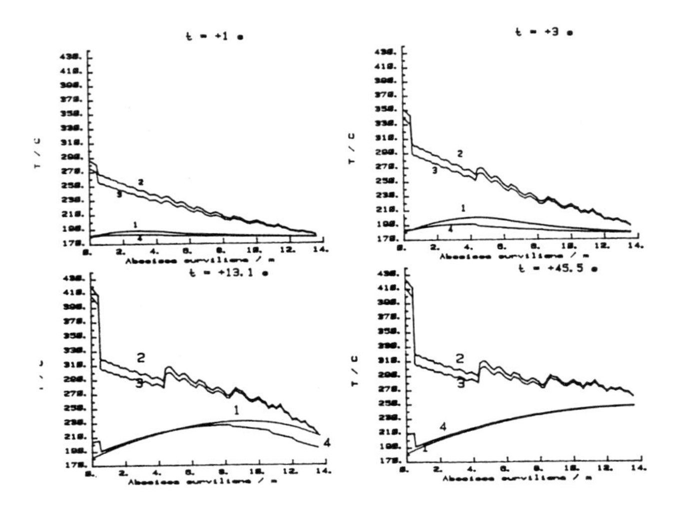

Fig. 4. Temperature variation along the receiver tubes.

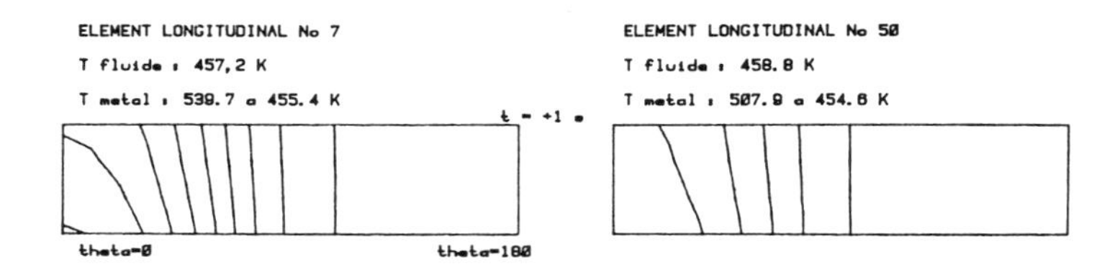

SOLAR FLUX STEP CHANGE

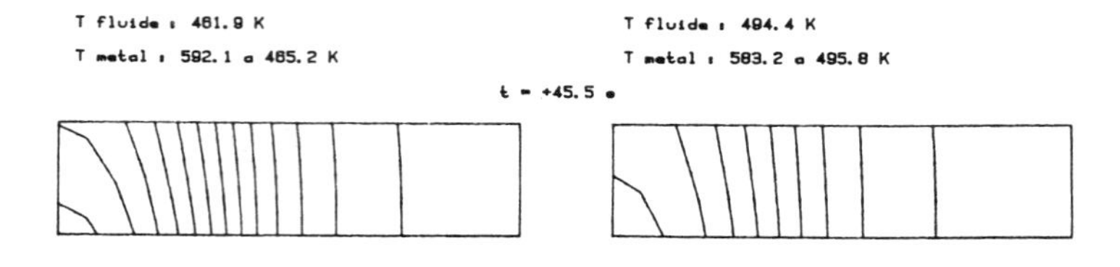

Fig. 5. Isotherms in two sections of the receiver tube.

REFERENCES

Audibert, M. (1981). Rev. Gén. Therm. Fr., 234-235, 509.
Pasquetti, R. (1983). Thèse d'Etat, Université de Provence, Marseille.
Pasquetti, R. (1984). Revue Phys. Appl., 19, 455-460.
Jose, P. D. (1957). Solar Energy, 1, 4.
Argiriou, A. and others (1987). Submitted to Solar Energy.
Pasquetti, R. (1986). Revue Phys. Appl., 21, 785-794.
Argiriou, A. (1987). Thèse de l'Université de Provence, Marseille.

FLUX DISTRIBUTION AND INTERCEPT FACTORS IN THE FOCAL REGION OF A FACETED PARABOLOIDAL DISH CONCENTRATOR

N.D. Kaushika* and S. Kaneff

Department of Engineering Physics,
Research School of Physical Sciences,
Institute of Advanced Studies
The Australian National University,
Canberra, Australia.

*On leave from Centre for Energy Studies,
Indian Institute of Technology, Delhi,
New Delhi, India.

ABSTRACT

Illumination of the full moon has been adapted for photographic determination of spatial distribution of energy in the focal region of a faceted paraboloidal dish concentrator. The observed flux distribution is shown to be represented by the composition of two Gaussian distributions whose peaks coincide in the focal plane and are displaced apart in other parallel planes. The expression for the intercept factor as a function of aperture size is derived. Sample computational results are presented. The procedure outlined is particularly important in design of receivers/absorbers for point focussing systems whose optical characteristics and manufacturing tolerances are difficult to be modelled mathematically.

INTRODUCTION

Parabolic dish solar collector systems have received steadily increasing attention during this decade as a viable means for high temperature heat production for electricity generation and industrial processes. Such systems consist of two-axis sun-tracking dish concentrators with a receiver (usually a cavity-receiver) mounted at their focus. Evaluation of their optical/thermal performance requires knowledge of the energy flux distribution in the focal plane and an expression of the intercept factor as a function of receiver aperture size. Over the years many theories have been advanced to assess the flux distribution in the focal plane (See for example Wen et al (1980) for a review). In general, realistic optical systems are very complex and statistical in nature. Approximations and assumptions are, therefore, inevitably made for mathematical manageability. An approximation often used in this connection was devised by Duff and Lameiro (1974) and it assumes the flux distribution and related variables in the focal plane as of Gaussian distribution.

In recent years the use of surface receivers and modified cavity receivers has been suggested for application in low cost dishes of short focal length and rather imperfect optics (Schmidt et al (1983), Kaneff (1983)). Evaluation of systems with these receivers requires knowledge of flux distributions and intercept factors in the whole focal region and not just in the focal plane. Theoretically, such a determination is cumbersome and does not account for the factors resulting from manufacturing tolerances. In the following, we present a procedure and results of a relatively simple method for the determination of flux distribution and intercept factors in the focal region of a dish concentrator.

EXPERIMENTAL DETERMINATION OF FLUX DISTRIBUTION

A suitable experimental procedure (Thomas and Whelan (1981)) is similar to that of Hisada et al (1957) used for solar furnace measurements. It is based on the fact that the apparent diameter (diameter/distance from earth) of the moon is about 31.1', almost equal to the sun's (whose intense radiations would produce very high temperatures in the focal region). Illumination by the moon may therefore be adapted for the determination of flux distribution in the focal region. The determination, however, embraces an assumption that the surface brightness of the moon is proportional to that of the sun.

Measurements are made on the night of full moon. In the present example, the paraboloidal dish collector is 5m diameter, 19.8m² aperture area and 1.808m focal length (Kaneff, 1983). The dish shell is of 6mm thick fibreglass, has a rim angle of 70° and is rim supported. The reflector is formed by shaped 100mm x 100mm plane glass mirror segments. The dishes are integrated in a steel frame modular unit employing altitude and azimuth tracking driven by a control unit normally acting in response to sun sensor signals. For the measurements, a translucent mylar sheet (thickness = .006cm) was placed in the focal plane where the moon's image was photographed with a calibrated camera using various exposure times. The brightness of the moon's image at the periphery of each pattern is assumed proportional to the reciprocal of the exposure time. This yielded contours of constant brightness. Fig. 1 depicts such contours. Similar measurements were made in other planes parallel to the focal plane but at different axial distances from it. The relative intensity variations obtained from these measurements are shown in Fig. 2.

INTERCEPT FACTOR: FORMULATIONS

From the experimental measurements plotted in Fig. 2, it is apparent that the focal flux distribution may be represented by a composition of two Gaussian distributions. In a plane parallel to the focal plane (x-z plane) and at a distance y_o we have:

$$I(x,y_o) = \frac{I_o'}{2} e^{-c_1 y_o^2}[e^{-k(x+c_2 y_o)^2} + e^{-k(x\ c_2 y_o)^2}] \quad (1)$$

where I_o', k, c_1 and c_2 are constants obtainable from the experimental curves.

$$I(x,y_o) = \frac{I_o}{2} [e^{-k(x+c)^2} + e^{-k(x-c)^2}] \quad (2)$$

where $I_o = I_o' e^{-c_1 y_o^2}$ and $c = c_2 y_o$.
If dE is the energy received between circles of radii x and x + dx in the plane parallel to the focal plane and distant y_o from it, then

$$dE = 2\pi x dx I(x,y_o) = \frac{I_o}{2} [2\pi x e^{-k(x+c)^2} dx + 2\pi x e^{-k(x-c)^2} dx]$$

Put $x = \sqrt{A/\pi}$ or $\pi x^2 = A$

or $2\pi x dx = dA$

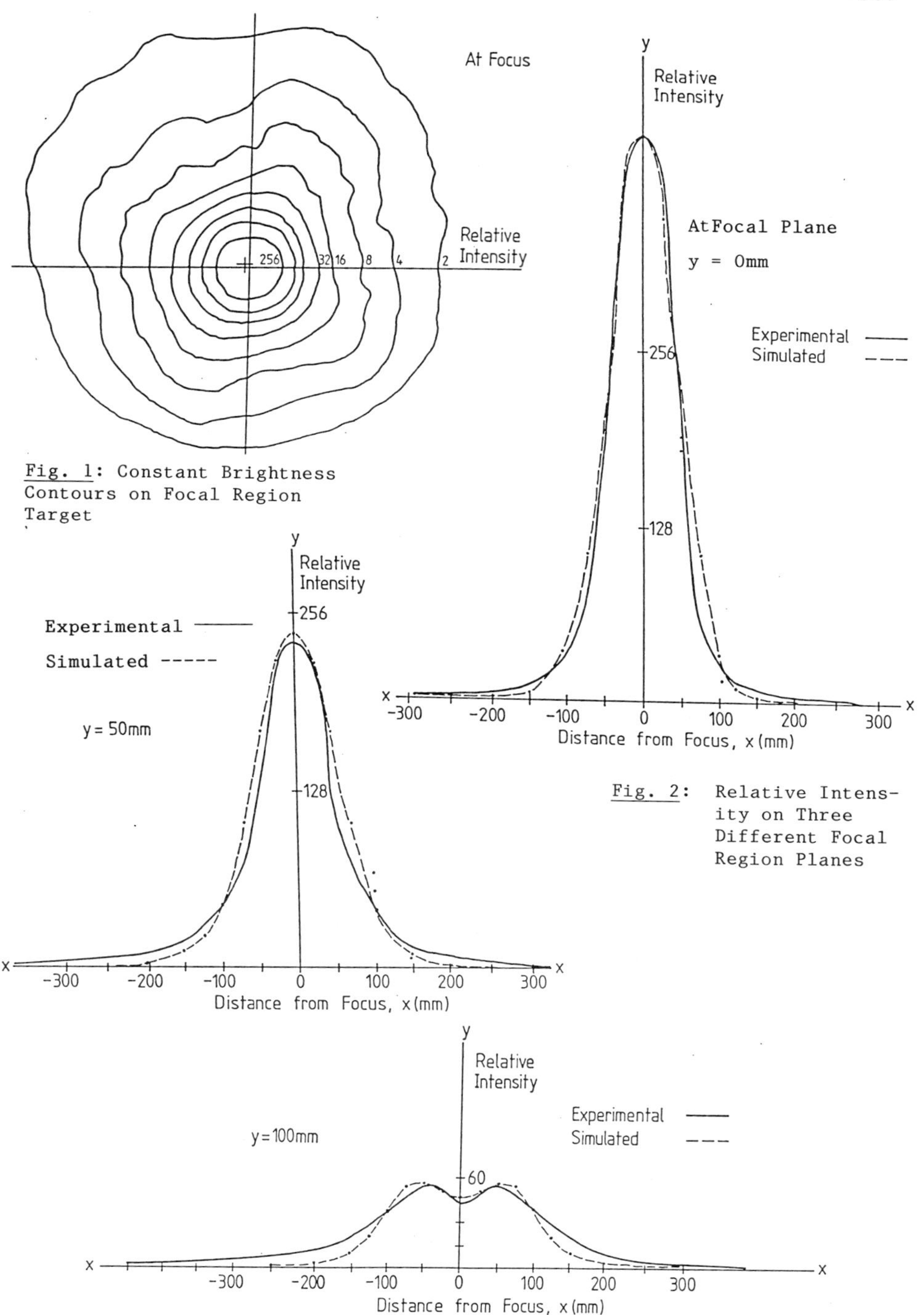

Fig. 1: Constant Brightness Contours on Focal Region Target

Fig. 2: Relative Intensity on Three Different Focal Region Planes

$$dE = \frac{I_o}{2} [e^{-k(\sqrt{A/\pi} - c)^2} dA + e^{-k(\sqrt{A/\pi} - c)^2} dA] \tag{3}$$

Intercept factor is defined as

$$\phi(x_o, y_o) = \frac{{}_0\int^{A_o} dE}{{}_0\int^{\infty} dE(y_o=0)} = \frac{E_o}{E_{to}} \tag{4}$$

where $A_o = 2\pi x_o$

and $E_o = \frac{I_o}{2} [{}_0\int^{A_o} e^{-k(\sqrt{A/\pi} + c)^2} dA + {}_0\int^{A_o} e^{-k(\sqrt{A/\pi} - c)^2} dA]$

Substituting $\sqrt{A/\pi} + c = y$ and $\sqrt{A/\pi} - c = z$ we have

$$E_o = \frac{\pi I'_o}{2k} [{}_c\int^{c+\sqrt{A_o/\pi}} e^{-ky^2} ydy - c \, {}_c\int^{c+\sqrt{A_o/\pi}} e^{-ky^2} dy$$

$$+ {}_c\int^{-c+\sqrt{A_o/\pi}} e^{-kz^2} zdz + \underline{c}_{-c}\int^{-c+\sqrt{A_o/\pi}} d^{-kz^2} dz]$$

$$\text{or } E_o = \frac{\pi I_o}{2k} [2e^{-kc^2} - e^{-k(x_o+c)^2} - e^{-k(x_o-c)^2} + 2kcD_o] \tag{5}$$

$$\text{where } D_o = {}_{-c}\int^{-c+x_o} e^{-kz^2} dz - {}_c\int^{c+x_o} e^{-ky^2} dy \tag{6}$$

$$D_o = \frac{1}{\sqrt{k}} \cdot \frac{\sqrt{\pi}}{2} \{2\phi(c\sqrt{k}) - \phi[(c-x_o)\sqrt{k}] - \phi[(c+x_o)\sqrt{k}]\} \tag{7}$$

Also $E_\infty = {}_0\int^{\infty} dE(x_o, o) = {}_0\int^{\infty} I_o \, e^{-kx^2} 2\pi x dx = \frac{\pi I_o}{k}$

$$\phi_o(x_o, y_o) = \tfrac{1}{2} e^{-c \, y_o^2} [2e^{-kc^2} - e^{-k(x_o+c)^2} - e^{-k(x_o-c)^2} + 2kcD_o] \tag{8}$$

where D_o is given by (7)

RESULTS OF COMPUTATIONS

The assumed expression of focal flux distribution given by equation (1), could be fitted into the observed variations as shown in Fig. 2. The values of corresponding parameters are as follows.

$I_o = 212$
$k = .00024 \text{ (mm)}^{-2}$
$c = .000125 \text{ (mm)}^{-2}$
$c_2 = .60$

Using these parameters and equations (7) and (8), $\phi_o(x_o, y_o)$ was calculated for different values of x_o and y_o. The results are illustrated in Fig. 3.

DISCUSSION

For a dish concentrator, focal flux distribution information is usually required for design of the receiver/absorber. One approach is to perform slope error measurements with optical or contact probe methods (eg. Grilikhes (1968) and Krasilovskii and co-workers (1978)) and to obtain slope error statistics which may be used to compute focal flux distribution by numerical methods which are generally expensive. An alternative approach is to adopt the method outlined in the present paper, which seems easier and more meaningful.

In physical terms, representation of flux distribution by the composition of two Gaussian functions seems logical. Petit (1977) used the sum of two normal distributions to describe the angular distribution of light reflected from a metal or polymeric mirror. In the present case, reflection

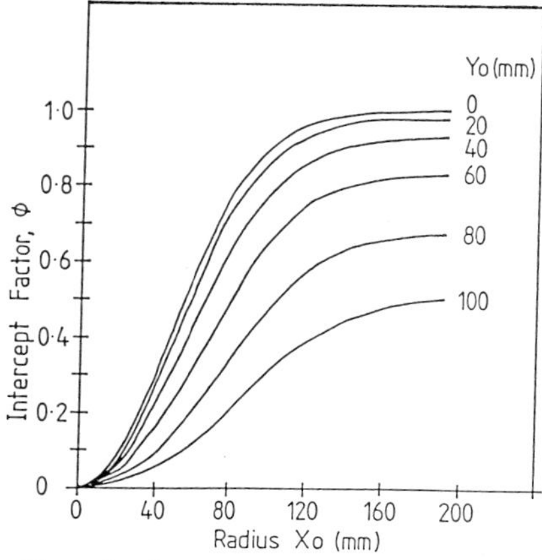

Fig. 3: Intercept factor vs. Radius

is from flat glass mirror segments and the reflected beam profile can be adequately represented by a Gaussian distribution. However, according to the concept of cone optics, the solar radiation reflected from each element of a dish would be a cone of rays which would cast an elliptical image on the focal plane. The flux mapping due to each hemisphere of dish would be the superposition of numerous elliptical images of various orientations and sizes. The mapping due to the other half of the dish would be identical and would overlap the first, in the focal plane. In other parallel planes the two component mappings would be displaced from each other.

ACKNOWLEDGEMENTS

Helpful discussions with R.E. Whelan on the techniques and results of the moonshot assessment of focal region distribution is acknowledged.

REFERENCES

Duff, W.S. and Lameiro, G.F. (1974). A performance comparison method for solar concentrators. ASME paper 74-WA/Sol. 4.

Grilikhes, V.A. (1968). Experimental investigation of the distribution of irradiance in the near focal region of a paraboloidal concentrator. Geliotekhnika, 4, p.31-36.

Hisada, T., Mii, H., Noguchi, C., Noguchi, T., Hukuo, N. and Mizuno, M. (1957). Concentration of the solar radiation in a solar furnace. Jour. Solar Energy Sci. and Engg. 1, No.4, p.14.

Kaneff, S. (1983). The design of viable paraboloidal collectors for high quality heat production. Proc. Solar World Congress, Perth. Pergamon Press (1984). Vol.4, pp.2762-2766.

Krasilovskii, V.I., Tarnezhevskii, B.V. and Yver'yanovich, E.V. (1978). Facility with sectioned photo receiver and laser radiator for determining solar radiation concentrator accuracy characteristics. Geliotekhnika, 14, No.1, p.30-35.

Petit, R.B. (1977). Characterization of the reflected beam profile of solar mirror materials. Solar Energy, 19, p.733.

Schmidt, G., Schmid, P., Zewen, H., and Moustafa, S. (1983). Development of a point focussing collector farm system. Solar Energy 31 (4), p.299.

Thomas, K. and Whelan, R.E. (1981). Assessing Solar Flux Distribution in the Focal Region of a Paraboloidal Dish by Moonshots. Department of Engineering Physics Research Report EP-RR-44, February 1981. The Australian National University.

Wen, L., Huang, L., Poon, P. and Carley, W. (1980). Comparative study of solar optics for paraboloidal concentrators. J. Solar Energy Engg., 102, p.305.

COMPARATIVE OPTICAL BEHAVIOURS OF PRINCIPAL LINE-AXIS CONCENTRATING COLLECTOR-TYPES

D.E. Prapas, B. Norton° and S.D. Probert

Solar Energy Technology Centre, School of Mechanical Engineering, Cranfield Institute of Technology, Cranfield, Bedford MK43 OAL

° Author to whom correspondence should be addressed

ABSTRACT

A detailed optical analysis of parabolic-trough solar-energy collectors, obtained by a ray-tracing technique, is presented. The fraction of the diffuse insolation that can be harnessed by this type of concentrating collector has been found to be significant for concentration ratios less than ten. The effect on the collector performance of the angular distribution of the diffuse insolation has been demonstrated to be very weak. The acceptance angle and the exploitable part of the diffuse insolation for a generalised parabolic-trough collector have been demonstrated to be both lower than those for a corresponding compound parabolic concentrating collector.

KEYWORDS

Concentrating solar collectors; CPC; PTC; optical behaviour; diffuse insolation distribution.

1. INTRODUCTION

Two types of line-axis concentrating solar-energy collectors are in common use today, namely: (i) the Compound Parabolic Concentrating (CPC) collector; and (ii) the Parabolic-Trough Concentrating (henceforth referred to as the PTC) collector (see Fig.1). The temperatures achieved by low concentration ratio PTC collectors are comparable with those obtained with CPC collectors. The optics of CPC collectors, with respect to the amount of insolation they

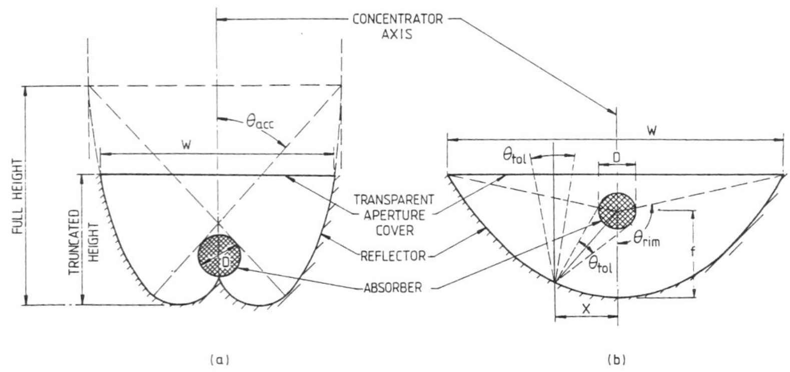

Fig. 1. Schematic cross-sections of (a) a compound parabolic concentrator (CPC); and (b) a parabolic-trough concentrator (PTC).

harness, have been investigated extensively [1],[2]. Cachorro and Casanova [3] studied the optical efficiency of a semi-static PTC collector. Evans [4] investigated the effect of non-uniformity of the insolation over the solar-disk on the performance of the PTC collector. A study by Mills [5] on the cost-effectiveness of periodically-adjusted PTC collectors included the contribution of the diffuse insolation (not accounted for by [3,4]), but assumed it to have an isotropic distribution.

In the present study a computer-based ray-tracing analysis has been used to reveal the individual contributions, of both the direct and the diffuse components of the insolation, to the flux reaching the absorber of a PTC collector.

2. ANALYSIS

A line-focus CPC collector is characterised by its acceptance half-angle θ_{acc} (see Fig.1a). This angle determines the maximum attainable concentration ratio, which is given [1] by

$$C_{max} = 1/\sin\theta_{acc} \quad (1)$$

In a real application, with a tubular absorber, the concentration ratio is expressed [1] as

$$C = W/\pi D \quad (2)$$

The value given by equation (2) is lower than that given by eqn (1) because of (i) truncation of the concentrator top, undertaken normally to reduce the capital cost [2]; and (ii) oversizing of the absorber's diameter, to allow for optical scatter introduced by imperfections arising during manufacture and operation. Absorbers of non-circular cross-section may also be employed.

The concentration ratio for a PTC collector (see Fig. 1b) is also given by eqn (2). The finite diameter of the absorber allows some additional rays, not parallel to the concentrator axis, to reach the absorber. This can be expressed by a local tolerance angle, θ_{tol} (see Fig.1b.), whose value varies according to the position at which a particular light ray is incident on the reflector's surface. A mean tolerance angle, θ_m, is defined as the average of the local tolerance angles across half the aperture width, W/2: θ_m can also be interpreted as the mean acceptance angle of the parabolic concentrator, namely

$$\theta_{acc,PTC} \equiv \theta_m \quad (3)$$

The mean acceptance angle as expressed by eqn (3), quantifies the average optical behaviour of a PTC collector and is, thus, comparable with the acceptance angle of a CPC collector.

Unlike that achievable with flat-plate collectors, only a fraction of the diffuse insolation is exploitable by concentrating collectors. For a CPC collector, this is given [1] as

$$g_{D,CPC} = 1/C \quad (4)$$

For a PTC collector, the absorber can "view" itself on the reflector. Thus, the exploitable part of the diffuse insolation for a PTC collector, $g_{D,CPC}$, is less than that suggested by eqn (4). The total insolation, I_u absorbed profitably can be expressed as

$$I_u = (\tau\ \rho\ \alpha)\ \gamma\ I_{eff} \quad (5)$$

where $(\tau\ \rho\ \alpha)$ is the transmittance-reflectance-absorptance product for a concentrating collector, γ is the intercept factor [6] accounting for the optical losses occurring in a real PTC due to optical errors, and I_{eff} represents the effective insolation at the concentrator's aperture [7], which is given by

$$I_{eff} = \beta_B\ I_B + \beta_D\ g_{D,CPC}\ I_D \tag{6}$$

In eqn (6), β_B and β_D are correction coefficients accounting for the parts of the direct and the diffuse insolation, I_B and I_D respectively, that reach the absorber directly, i.e. without the participation of the reflector. The values of these coefficients slightly exceed unity [7]. The optical efficiency η_{opt} of a PTC collector is defined as the ratio between the insolation I_u absorbed by the absorber and the total hemispherical insolation on the plane of the collector, I_{tot}, i.e.

$$\eta_{opt} = I_u/I_{tot} \tag{7}$$

3. RESULTS

The mean acceptance angle and the exploitable function of diffuse insolation obtained for a PTC collector, as predicted with a computer-based ray-tracing technique, were contrasted with those for a comparable CPC collector. The latter was assumed to be truncated in such a manner that its maximum concentration ratio (see eqn (1)) is reduced by a factor of 0.3. It can be seen from Fig. 2 that a CPC collector exploits a greater part of the available diffuse insolation compared with a PTC, although this advantage diminishes as the concentration ratio increases. A CPC collector also maintains its superiority with respect to the acceptance angle.

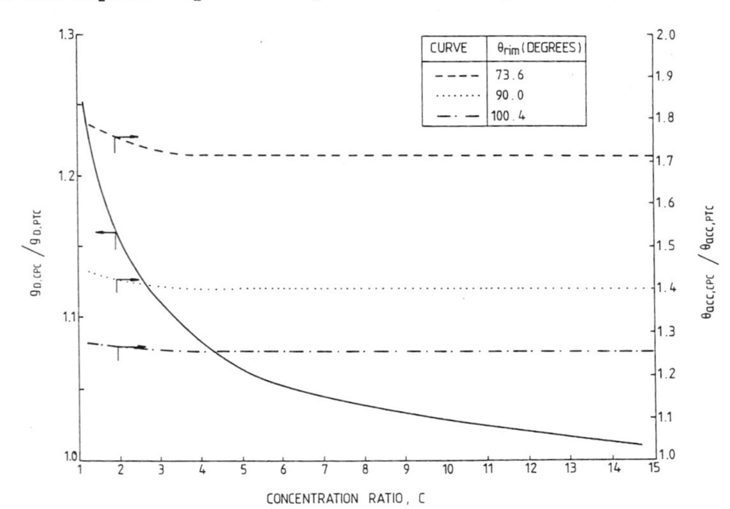

Fig. 2. Comparison of the behaviours of PTC and CPC collectors on the bases of (i) the part of the diffuse insolation they can exploit (solid curve); and (ii) their mean acceptance angles (broken curves).

It is informative to consider how the following particular skyward angular distributions of the diffuse insolation affect the performance of a PTC solar-energy collector: (i) isotropic, (ii) cosine, or (iii) hybrid Gaussian. The hybrid Gaussian distribution combines an isotropic background with a circumsolar Gaussian part. This distribution is more realistic for a tracking system than both the isotropic model (which underestimates the insolation intensities at incidence angles near zero) or the cosine model (which underestimates the intensities at large incidence angles). The results presented in Fig. 3 derived for the following set of data: $(\tau \rho \alpha) = 0.79$, $\gamma = 1$ and $\theta_{rim} = 90°$. The hybrid Gaussian distribution yields invariably a higher optical efficiency whereas the efficiency curve for the isotropic distribution can be regarded as the lower limit. The difference in the optical efficiency for these two cases, although more pronounced for the lower concentration ratio and for higher I_D/I_{tot} ratios, is small: it is, for example, less than 2.5% for a concentrating ratio of 3 and less than 1.1% for a concentration ratio of 10.

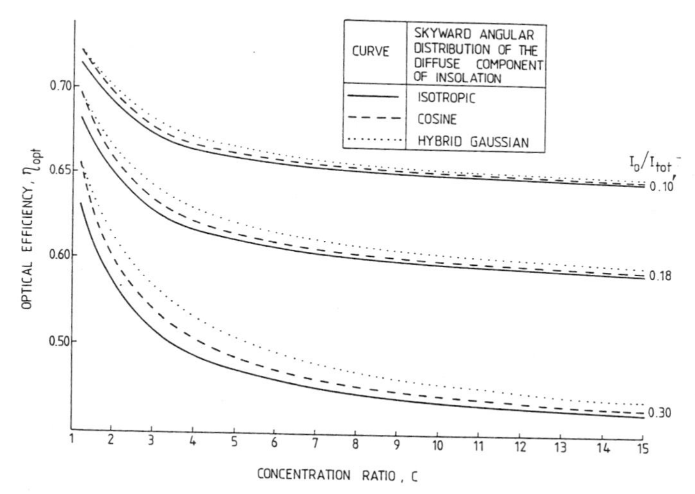

Fig. 3. Effect of skyward distribution of the diffuse insolation on the optical efficiency of a PTC collector.

4. DISCUSSION AND CONCLUSIONS

An analysis of the optical behaviours of PTC collectors, based on a ray-tracing technique, has been undertaken. It has been demonstrated that the skyward angular distribution of diffuse insolation does not affect significantly the performance of such a collector. Thus, for most engineering design purposes, it is reasonble to assume an isotropic distribution, which has been demonstrated to provide a lower performance limit.

A comparison, performed on the basis of the acceptance angle and the exploitable fraction of the diffuse insolation of PTC and CPC collectors, revealed the latter to be the more efficient design. However, the ranges of concentration ratios found in commercially-manufactured collectors of the two types are not the same: CPC collector designers take advantage of the possibility of constructing stationary concentrators, which require concentration ratios less than 2 [2]. By contrast, for effective operation, PTC collectors should track the Sun. Commercially-manufactured PTC solar-energy collectors may be divided into two types:- (i) those with a concentration ratio of the order of 40 can be regarded as devices which are very accurate optically; and (ii) those with a concentration ratio in a range of 3→10, which can afford to be of moderate optical accuracy [6].

The exploitable part of the diffuse insolation becomes negligible for high concentration ratios. Thus PTC collectors can be divided into two categories: those with low concentration ratios (i.e. with $C \leq 10$) for which a significant amount of diffuse insolation can be collected, and (ii) those with high concentration ratios (i.e. with $C > 10$), for which the amount of the diffuse insolation collected is negligible.

REFERENCES

1. Rabl, A. (1976). Comparison of solar concentrators, Solar Energy 18, 93-111.
2. Carvalho, M.J., Collares-Pereira, M., Gordon, J.M. and Rabl, A. (1985). Truncation of CPC solar collectors and its effect on energy collection, Solar Energy 35, 393-399.
3. Cachorro, V.E. and Casanova, J.L. (1986). Optical efficiency of semi-static cylindrical-parabolic concentrator, Solar Energy 36, 147-149.
4. Evans, D.L. (1977). On the performance of cylindrical parabolic solar collectors with flat absorbers, Solar Energy 19, 379-385.
5. Mills, D.R. (1986). Relative cost-effectiveness of periodically adjusted solar collectors using evacuated absorber tubes, Solar Energy 36, 323-331.
6. Rabl, A. and Bendt, P. (1982). Effect of circumsolar radiation on performance of focusing collectors, ASME Journal of Solar Energy Engineering, 104, 237-250.
7. Prapas, D.E., Norton, B. and Probert, S.D. (1987). Optics of parabolic trough solar-energy collectors possessing small concentration ratios, Solar Energy, In Press.

PERFORMANCE BENEFITS FROM NONIMAGING SECONDARY CONCENTRATORS IN POINT FOCUS DISH SOLAR THERMAL APPLICATIONS

R. Winston and J. O'Gallagher

The Enrico Fermi Institute
The University of Chicago
Chicago, Illinois 60637, U.S.A.

ABSTRACT

A performance model based on a Monte Carlo ray-trace technique is used to evaluate the benefits associated with the addition of a nonimaging , maximally concentrating secondary, to a conventional paraboloidal solar dish. The model shows that for a primary dish with a focal length to diameter ratio of 0.6, a characteristic slope error of 5 milliradians, and a cavity receiver operated at temperature of 1000°C, the optimized efficiency with a secondary is 0.70 compared to 0.59 for the primary alone. The relative performance advantage provided by a secondary decreases somewhat if either the temperature or the primary slope error are decreased, however, it remains significant at all temperatures above 400°C, even in the "high performance limit" of slope errors < 2 milliradians. In this limit, there is an alternative secondary optimization strategy, for which the off-track tolerance can be doubled relative to a single stage configuration while maintaining the identical efficiency. The relative performance when the primary is spheroidal rather than paraboloidal shows that the optimum performance with a secondary is much better than the best performance with no secondary. This may have important implications for the use of primaries utilizing stretched-membrane technologies.

KEYWORDS

Paraboloidal dish, secondary concentrator, nonimaging, slope error, trumpet.

INTRODUCTION

Nonimaging secondaries (1,2), such as the well known Compound Parabolic Concentrator (CPC), deployed in the focal zone of image forming primary concentrators, offer the optical designer an additional degree of freedom unavailable with any conventional approach. For a given angular acceptance, such two-stage configurations have the capability of delivering an additional factor of two to four in system concentration. In general, this is accomplished without doing anything to the primary. This unique capability can be used either to increase concentration or angular acceptance. For solar thermal applications these parameters are related respectively to increased performance (thermal efficiency) or relaxed optical tolerances (mirror slope error, angular tracking, etc.).

Several alternative configurations are possible including the Compound Elliptical Concentrator (CEC), and the flow-line or "trumpet" concentrator in which the reflector is a hyperboloid of revolution. The latter has been developed extensively, both analytically and experimentally (1). In addition, several versions of a secondary 5X CPC, were fabricated and tested at the Advanced Component Test Facility at Georgia Technical Institute (2). The motivation for the present work is to provide a quantitative basis for the systematic evaluation of the potential performance benefits associated with such two-stage designs over a wide range of design configurations.

THE PERFORMANCE MODEL

We use a methodology patterned after an approach originally introduced by Jaffe (3), however, we have incorporated two very important new features.

1) We have used Monte Carlo ray trace calculations to determine the focal plane distributions, which in turn are used to evaluate the trade-off between energy intercepted by apertures of different sizes (related to the geometric concentration ratio) and receiver heat losses. Our approach not only provides a better representation of the actual focal plane distributions, but it allows treatment of non-gaussian optical error distributions, deviations from axial symmetry (e.g. off-track bias) and non-paraboloidal contours.

2) We model the secondary efficiency to take account of variations in throughput as a function of the fraction of energy it intercepts. Jaffe simply assumed a constant secondary efficiency, typically between 0.9 and 0.95, independent of the relative size of the secondary compared to the scale of the focal plane distribution. This is unrealistic, particulary for gaussian distributions, since it artificially introduces a loss in the two stage system, whether or not the secondary is doing anything.

The details of the model are reported elsewhere (4) and will not be presented here. The basic procedure can be understood as follows. For a particular set of optical parameters a large number of randomly chosen rays, governed by a gaussian or other probability distribution, are traced to determine the detailed focal plane distributions. These in turn are then used to determine the intercept factors as a function of (variable) target receiver aperture sizes. For each value of target radius r_1, the corresponding heat loss and the net thermal efficiency are determined as a function of receiver operating temperature. These results are then tabulated and used to determine the maximum efficiency at each temperature resulting from the trade-off between intercept factor and heat loss.

The effect of a nonimaging secondary is included in the above analysis in a parallel set of calculations incorporating two modifications. 1) For each value of target radius, a second value of the relative heat loss is calculated, corresponding to a receiver aperture reduced in area by a factor equal to the secondary concentration ratio. 2) The effective net optical efficiency is reduced by a factor η_2, to account for losses due to the secondary. An analytical relationship is used which effectively approximates the average number of reflections in the secondary by the fraction intercepted by the secondary and is a very good approximation in the case of the trumpet. Detailed raytracing of a number of selected configurations has shown that the approximation used is accurate to a fraction of one percent and is more than adequate for the kind of trade-off analysis done here.

For purposes of illustration, we have chosen a baseline reference design with the parameters defined and listed in Table 1. We selected nominal values for the key quantities which are characteristic of what we think might be a typical two stage configuration. In particular we chose a focal length to diameter ratio F/D = 0.6, an operating temperature T = 1000°C, and a characteristic primary slope error of 5 milliradians.

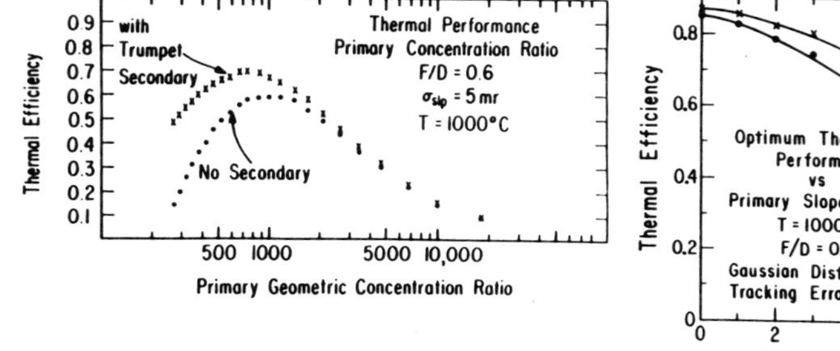

Fig. 1. Illustration of the optimization procedure for the baseline case. The combined effect of optical and thermal losses, as a function of effective focal plane target size (primary concentration ratio), is shown with and without a secondary.

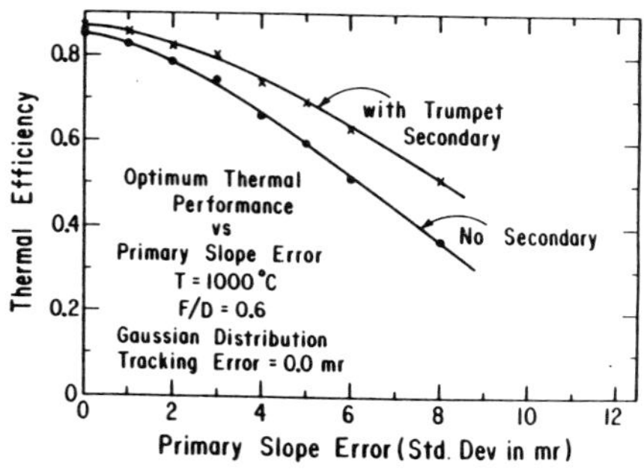

Fig. 2. Illustration of the performance improvement provided by a trumpet secondary for the baseline case when the primary slope error is varied while the temperature is held constant.

The basic optimization procedure is illustrated for the baseline case in Figure 1. This is a plot of optimum efficiencies, with and without a secondary, as a function of primary geometric concentration ratio. The maximum value reached by each curve occurs when the loss in intercept with increasing concentration is precisely balanced by the corresponding reduction in heat losses. For our selected baseline illustration the optimum thermal efficiency without a secondary occurs at (primary) concentration ratio of about 1000 X and yields a thermal efficiency of 0.59. With a secondary the optimum occurs at a somewhat lower primary concentration ratio of about 750X (and thus a larger optical intercept) but at a combined two-stage net concentration ratio of about 1400 X. The resulting lower heat losses and higher optical intercept yield a thermal efficiency of 0.70 or a relative increase in energy delivered of more than 18%. In the next section we use this model to evaluate these benefits as a function of some other system parameters.

SUMMARY OF RESULTS

The optimized thermal conversion efficiencies, determined by the procedures described in the section above, are shown in Figure 2 as a function of primary characteristic gaussian slope error while temperature is held constant and in Figure 3 as a function of receiver operating temperature with slope error held constant. The other baseline values are all held constant at their values in Table I. As the slope error is reduced below its value for the baseline, the optimized efficiency for both designs improves, with the two stage always remaining somewhat above the single stage.

TABLE 1

Parameters for Baseline Thermal Concentrator Performance Analysis

Dish Diameter	11 Meters
Focal Length	6.6 Meters
Focal Ratio (F/D)	0.6
Rim Angle	45.24°
Primary slope error*	5 mr
Specularity error*	1.5 mr
Sun size*	2.73 mr
Shape of Angular Incidence Distribution	Gaussian
Primary Reflectivity	0.9
Secondary Reflectivity	0.95
Receiver Operating Temperature	1000°C
Absorbtivity of cavity aperture	1.0
Emissivity of cavity aperture	1.0
Aperture convective loss coefficient	16.0 W/M^2°C
Ambient Temperature	20°C
Direct Normal Insolation	800 W/M^2

* Standard Deviation of the one-dimensional projection of the corresponding two dimensional circular normal distribution.

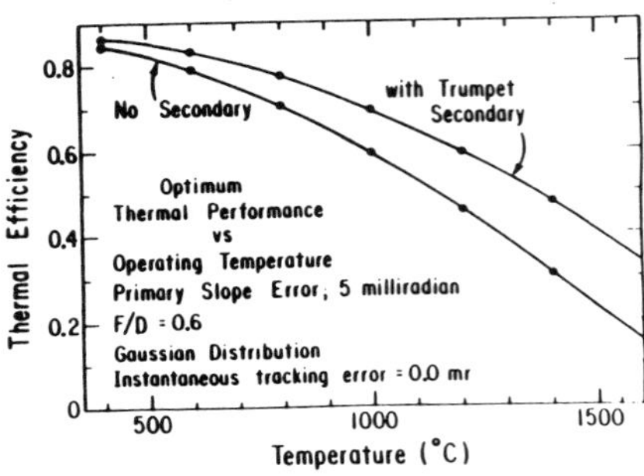

Fig. 3. Illustration of the performance improvement provided by a non-imaging trumpet secondary as a function of design temperature.

Fig. 4. Thermal performance, with and without secondaries, versus tracking bias error in the "high performance limit" of very small slope error.

The single most important result of these comparative optimizations is that the efficiency of the two-stage is always greater than the single stage if all the other design parameters are the same.

In the limit of small slope errors and/or low receiver operating temperatures, when the relative efficiency gain with a secondary becomes marginal there are other optical benefits from a secondary which do not show up on an efficiency graph such as Figs. 2 or 3. In particular we refer to increased circumsolar collection and relaxed tolerance for off-track errors. For illustration, the efficiency for a single-stage concentrator having an aperture optimized for a 1 mr slope error, is shown in Fig. 4 as a function of pointing direction as it is moved systematically off the sun. Similar curves are shown for two-stage systems with secondaries of two different sizes. It is clear that in this high performance limit, where the efficiency of the single stage is already quite high, the properties of the secondary can be utilized to great advantage for increasing the angular acceptance.

AN INNOVATIVE APPLICATION

For some time we have been suggesting that the optimum application for secondaries may lie in their combination with potentially inexpensive approximations to a paraboloidal figure, rather than with a true paraboloid (2,4). One example of such a configuration is a spheroidal mirror. We determined the optimized performance for the spherical analog of our baseline case for effective values of F/D between 0.6 and 1.2 (4). The results are plotted in Fig. 5. These preliminary results appear quite encouraging. The two-stage performance levels off at a value about 35% better than the best single stage. The corresponding optimized performance for a single stage at moderate to long focal ratios is dramatically better than the spheroid alone and even exceeds the performance of true paraboloid at short focal lengths. We suggest that this may have possible applications with stretched membrane primaries and may offer a very attractive path to economical delivery of solar thermal energy.

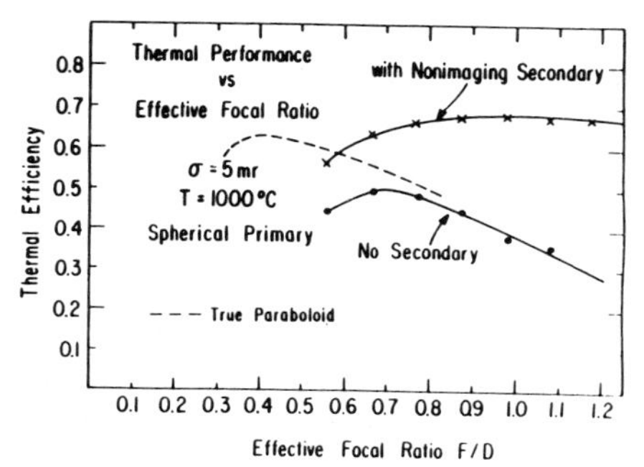

Fig. 5. Thermal performance (for the baseline optical errors and operating temperature) for single and two-stage concentrators using a spheroidal primary, as a function of focal ratio. The corresponding optimized performance for a single stage paraboloidal contour is shown for reference.

ACKNOWLEDGEMENT

This work was supported in part by the U.S. Department of Energy and The Solar Energy Research Institute under subcontract DOE SERI XK-4-04070-03.

REFERENCES

1) J. O'Gallagher and R. Winston, Test of a 'Trumpet' Secondary Concentrator with a Paraboloidal Dish Primary, Solar Energy, 36, 6, (1986).

2) J. O'Gallagher, R. Winston, D. Suresh, and C.T. Brown, Design and Test of an Optimized Secondary Concentrator with Potential Cost Benefits for Solar Energy Conversion, Energy, 12, 217, (1987).

3) L.D. Jaffe, Optimization of Dish Solar Collectors with and without Secondary Concentrators, JPL Technical Report No. DOE-JPL-1060-57. U.S. Department of Energy, May 1982.

4) J. O'Gallagher and and R. Winston, Performance and Cost Benefits Associated with Nonimaging Secondary Concentrators Used in Point-Focus Dish Solar Thermal Applications. SERI Subcontract Report STR-253-3113, August, 1987.

INVOLUTE FOCONE AS SECOND STAGE CONCENTRATOR

R.V.G.Menon*, G.Inayatullah** and P.K.Rohatgi***

* College of Engineering, Trivandrum, India
** Engineering Department, Mu'tah University, Jordan
*** Materials Engineering, University of Wisconsin, USA

ABSTRACT

The use of second stage concentrators in solar concentrating systems is examined. The Involute Focone, obtained by revolving a Trombe-Meinel Cusp about its central axis, is suggested for this application, and is compared with the CPC, CEC and the Trumpet. A new criterion, the Coefficient of Advantage, is defined as a measure of the advantage derived from the use of the secondary mirror. An approximate analytical expression for COA is derived and checked against numerical results. Finally, the results of an experiment, for determining the improvement in heat gain consequent to the introduction of a second stage involute focone, are reported.

KEYWORDS

Solar; concentrator; second stage; involute focone; Trombe-Meinel Cusp.

INTRODUCTION

The heat flux concentration achieved in the focal region of a solar concentrator is limited by the finite size of the solar image as well as by errors in fabrication and tracking. Because of these errors there is invariably a focal spread, instead of a point or line focus, and this results in a reduction in the energy captured by the absorber located in the focal region. While nothing can be done about the first limitation, attempts to overcome the others result in higher costs. Hence the use of second stage concentrators has found frequent favour in literature as an alternative.

THE SECOND STAGE GEOMETRIES

Nonimaging optical profiles are ideally suited for second stage concentrators. Winston and coworkers, who pioneered this class of geometries, first proposed the CPC (Rabl, 1976), then the CEC (Rabl and Winston, 1976), and finally the Trumpet (Winston and coworkers, 1981). Meinel (1976) had suggested both the CPC as well as the Trombe-Meinel Cusp, as possible candidates for application as field lens mirror. Baranov (1977), who had independently invented the CPC collector, had also advocated its use as a secondary concentrator, and had given it the name "parabolotoric focone". The present authors had proposed the use of a three-dimensional version of the Trombe-Meinel Cusp, and had suggested the name "Involute Focone" for the same (Menon, Inayatullah and Rohatgi, 1982). The first three belong to the class of ideal concentrators, as they approach the thermodynamic limit of concentration for a given angle of acceptance. The full CPC has a rather large mirror surface area. This can be offset by truncation. The CEC is particularly adapted for use with finite sources. By restricting the exit angle, grazing angles of incidence on the absorber can be avoided. In fact, the CPC emerges as the limiting case of the CEC, when the exit angle is allowed to approach $\pi/2$. Both these devices are to be placed behind the focal plane of the primary and hence do not cause any interference to the converging beam (Fig. 1). But, since the position of the receiver also has to be correspondingly moved back, these cannot be readily retrofitted on existing systems. Another disadvantage is that many of the rays, which would have directly hit the receiver, are now reflected and rerouted via the secondary. For long focal ratio systems ($f/D > 1.0$), these devices can raise the overall concentration ratio up to about 90% of the thermodynamic limit. For short focus systems the effectiveness may be limited to about 50% of the limit (O'Gallagher and Winston, 1986).

The trumpet is also an "ideal" concentrator, and can be added to an existing system without disturbing the receiver. Another advantage is that the secondary does not interfere with most of the rays which would have directly hit the receiver. Shading is a problem, but can be minimised by judicious truncation. The achievable concentration for optimized hybrid systems approaches 75% of the ideal limit for a given error budget (Winston and coworkers, 1981).

The Trombe-Meinel Cusp has 180^{0} acceptance range, but is not an optical concentrator in the usual sense of the term. If 'd' is the diameter of the receiver, the involute is to be generated from a circle of the same diameter. The aperture is then πd, which is also the circumference of the receiver (Fig.2), making the geometric concentration ratio equal to unity. However, in the three dimensional version, obtained by rotating the involute cusp about its central axis, this ratio becomes $\pi^2/4$ if a spherical receiver is employed. Even if the receiver is a cylinder, with diameter and height equal to 'd', this ratio is still $\pi^2/6$. The fundamental difference is that here the natural shape for the absorber is three dimensional. In the earlier geometries the absorber was assumed to be disc shaped. There are a number of applications, where the cylindrical absorber is a natural choice.

High flux concentration is essential to generate very high tempersture s. But in applications like steam generation, the energy captured is usually the limiting factor, and not the temperature alone. Hence a compromise is possible by allowing the flux to be distributed over a larger absorber area. For example, in the cavity type central receiver, the highly concentrated beam at the cavity aperture is allowed to spread over the heat transfer coils lining the cavity. The involute focone provides an alternative approach by offering about 60% more concentration, and directly allows the concentrated beam to spread over the evaporator coils, which can form a cylindrical absorber surface. The important advantage here is that the secondary does not interfere with any ray which would have directly hit the receiver. Thus unnecessary reflection losses are avoided. Since all the external area is utilized for heat absorption, back-surface losses are also avoided. The internal hollow of the cylindrical coil can function as a cavity absorber.

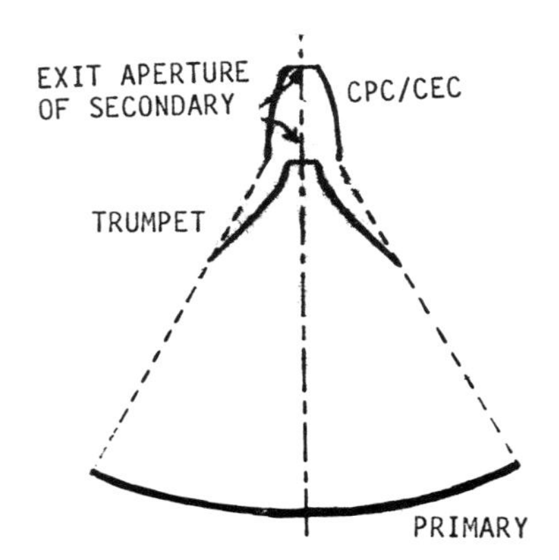

Fig. 1. Secondary configuration using the CPC, CEC, and the Trumpet

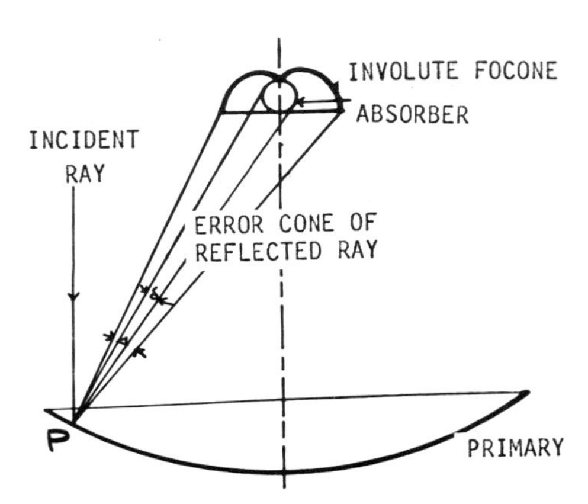

Fig. 2. Involute Focone as secondary

The Coefficient Of Advantage (COA)

The advantage of using a secondary is clear from Fig.2. The probability of any incident ray, which has been reflected from the primary at P, reaching the receiver, is increased by the presence of the secondary. If the direction of the incident ray were perfectly random, this improvement would be Δ / δ, where Δ and δ are the solid angles subtended at P by the bare absorber and the secondary aperture respectively. The value of this ratio is a function of the local radius R. This ratio has to be averaged over the entire range of R. But the flux associated with each R is proportional to the area of the concentric annular strip of radius R. Thus the integration is done using a weighting factor. With some further simplifications, the following approximate expression is analytically obtained for the averaged value of Δ / δ which really expresses the overall advantage of using the secondary concentrator, and hence is named the Coefficient Of Advantage.

$$COA = \pi^2 \Big[\ln Y - Y\Big]_0^{\beta_{max}} \Big/ \Big[Y\Big]_0^{\beta_{max}} \tag{1}$$

$$\text{where} \quad Y = 1/(1 + \cos\beta) \tag{2}$$

The above mentioned simplifying assumptions have introduced some errors in the result. An error analysis showed that these are of the order of $(1/f^2)$ where f is the focal length of the system. Hence, the larger the system, the smaller the error involved. This is also borne out by the exact numerical solution obtained through a digital computer. The following correction formula is, therefore, suggested to compensate for the omissions in the first approximation:

$$COA_{corrected} = COA_{approx.}/(1 + \epsilon) \tag{3}$$

where,

$$\epsilon = \Big[(1 - \pi(R \sin\beta_{max}/\rho)) \quad - 1\Big] \tag{4}$$

ρ being the distance of P from the focus of the paraboloid. A serious limitation of the above analysis is the assumption of randomness of the reflected ray at P. It is much more reasonable to assume it to be a Gaussian distribution about the designed direction of the reflected ray. A refinement of the analysis along this line is being attempted.

EXPERIMENTAL RESULTS

Experiments were conducted on a paraboloid dish solar concentrator to assess the improvement in heat gain consequent to the incorporation of an involute focone secondary in the system. The paraboloid had an f/D ratio of 0.4 and an aperture diameter of 150cm. The reflecting surface was a metallized PVC film stuck on a FRP base. Tracking was manual. The involute focone, generated from a 10cm diameter circle, was spun out of aluminium sheet. The reflecting surface was electroplated. The absorber was a coiled copper tube which formed a cylindrical receiver with diameter = height = 10cm. The heat gain was measured by circulating water through the coil at a constant head. The inlet and outlet temperatures were measured using thermocouples. Readings were obtained with and without the secondary in position. The insolation was monitored using a Pyrheliometer and the heat gain value was normalised to a direct beam insolation of 1000 watts/m^2. The rate of flow of water was measured physically. The Reynolds number was computed and found to be just around the critical value for transition. Higher flow rates were not attempted due to experimental limitations. The system parameters and a summary of the results are presented in Tables 1 and 2.

TABLE 1 System Parameters

PRIMARY	
(Paraboloid dish with metallised PVC film reflecting surface)	
Aperture diameter, 2R	= 150 cm
Rim angle, β_{max}	= 64°
Focal ratio	= f/0.4
Approximate focal spot diameter	= 32 cm
Geometric concentration ratio	= 22
Reflectivity of primary surface	= 0.76
SECONDARY	
(Involute Focone: Spun aluminium sheet with electroplated surface)	
Aperture diameter	= 32 cm
Generating circle diameter	= 10 cm
Receiver dimensions	= 10cm dia. x 10cm ht.
Geometric concentration ratio	= 1.7
Reflectivity of secondary surface	= 0.82
COMBINED SYSTEM	
Geometric concentration ratio	= 37.4
Percentage shading	= 4.55
COA (theoretical)	= 6.75

TABLE 2 Summary of Performance Test Results

	Without Focone	With Focone
Maximum heat gain, with the input normalised to a direct beam insolation of 1000 W/m^2	643 watts	901 watts
Maximum overall efficiency of the system	35.25%	49.42%
Gain in overall efficiency		40.20%

SUMMARY AND CONCLUSIONS

The 40% improvement in overall efficiency obtained with the incorporation of the involute focone secondary compares favourably with the 30% gain observed by O'Gallagher and Winston (1986) for the Trumpet. But it falls short of the geometric concentration of the focone(1.7) and far behind the approximate computation for the COA ($\sim$6.75). The reasons for the large discrepancy in the predicted value for COA have been pointed out earlier. The refined technique for the computation may give a closer approximation, it is hoped.

The thermal efficiency of the system declined with increasing outlet temperatures. This is apparently due to the higher losses from the receiver. It is observed that the presence of the secondary as a "roof" over the receiver suppresses free convection heat losses. The connected heat transfer studies are being separately reported.

In conclusion, it can be stated that the involute focone, when used as a secondary, produces heat gains comparable to those of the trumpet or CPC, causes minimal shading, can easily be retrofitted on existing systems, supresses free convection losses from the receiver, avoids unnecessary optical losses due to additional reflections, and can be easily adapted for use with cylindrical coil receivers for steam generation.

REFERENCES

Baranov,V.K. (1977). Parabolotoric focone as secondary solar energy concentrator. Geliotekhnika, 13:5, 18-23.

Meinel,A.B. and M.P.Meinel (1976). Applied solar energy, Addison-Wesley.

Menon,R.V.G., G.Inayatullah and P.K.Rohatgi (1982). Solar furnace using secondary concentrator. Proc. National Solar Energy Convention of the Solar Energy Society of India (New Delhi), 3.004-3.007.

O'Gallagher,J. and R.Winston (1986). Test of a "trumpet" secondary concentrator with a paraboloidal dish primary. Solar Energy, 36:1, 37-44.

Rabl,A. (1976). Comparison of solar concentrators. Solar Energy, 18, 93-111.

Rabl,A. and R.Winston (1976). Ideal concentrators for finite sources and restricted exit angles. Applied Optics, 15, 2880-2883.

Winston,R., E.K.Kritchman, J.O'Gallagher and P.Greenman (1981). Nonimaging second stage concentrators for point focus systems. Proc. of the ISES Congress (Brighton, England), 3022-3026.

SOLAR POWER PLANT WITH A MEMBRANE CONCAVE MIRROR

*Jörg Schlaich *Rainer Benz

*Schlaich und Partner, Hohenzollernstr. 1
7000 Stuttgart 1

Abstract: A dish / Stirling system with a special fabrication method is described at the example of a pilot plant in Saudi-Arabia.

Keywords: Concentrator, Stirling, Membrane, Dish

As a means of generating electricity from solar energy, high temperature energy conversion with concentrating systems has a very promising future. A large hollow reflector is suspended and supported on rails in such a way that it can track the sun. The reflector has an energy converter which converts the concentrated solar heat into electricity suspended at its focal point.
The new and special feature of the power plant described here is the construction method, which makes a very large concentrator possible. The concentrator is a hollow membrane of thin sheet steel to which mirror glass is bonded. The membrane is plastically deformed to the desired paraboloid shape by air pressure. When the concentrator is in operation the shape of the membrane is kept constant by a partial vacuum in the interior of the concentrator, i.e. between the reflector membrane, the rear membrane and the reflector ring.
The energy conversion system (ECS) consists of a Stirling engine with a receiver located at the focal point of the reflector, the reflected solar rays heat the working gas (hydrogen) of the engine. There is a generator coupled directly to the engine. The ECS has its own monitoring and control system which monitors essential operational data. A central control system calculates and controls tracking, monitors the Stirling control system, processes wind data and indicates malfunctions.

Plant near Riyadh, Saudi-Arabia.
In operation since 1985.

Since each unit is capable of fully independent operation, as many concentrators as desired can be operated in conjunction, according to requirements. They can be operated both in grid connection mode or "stand-alone" mode with storage devices (batteries, pumped storage etc). The prototypes on which this brochure is based are to be tested with both operating modes. Power plants with reflector membranes are capable of an overall efficiency (defined as the ratio of the output usable electricity to the solar irradiation over the reflector surface) of up to 27 %. This has never been achieved with other types of solar plants. As the membrane construction method used for the reflector is relatively inexpensive they also make economic electricity generation a real possibility. The output of the energy converter depends on the accuracy of the beam path. The reflector membrane satisfies this requirement, though only a simple technology is needed for its fabrication.
The pilot plant in Saudi-Arabia is in operation since 1985 and is being evaluated in terms of performance, reliability, operating and maintenance need to gain important experiences for future development. Different structural designs, operation modes, Stirling engines and power levels have been and are examined to get a wide range of systems for specific user needs, which contradict sometimes each other, such as:

fully automatic - semi-automatic operation
stand alone - grid connected mode
production in the user country - in Germany
sophisticated - simple technique
field - single installations

Cost per electrical kWh depend strongly on the cost of the energy converter and therefore on the number of systems produced per anno. With a production rate comparable to a car engine, cost per kWh is supposed to be in the range of conventional electrical power even in western industrialised areas.

Compared to the cost of electrical power from Diesel stations in remote locations, especially in low-income countries without own conventional resources, power from dish/Stirling systems comes into a competitive range already in the very near future.

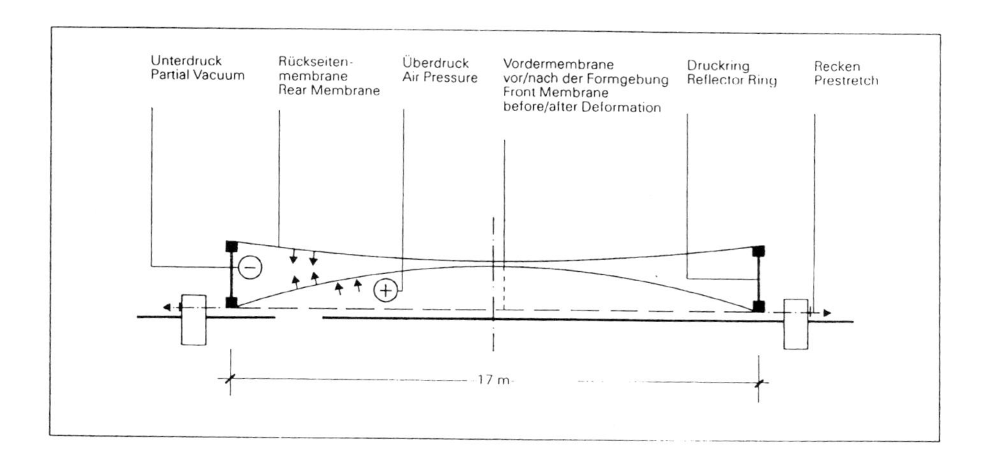

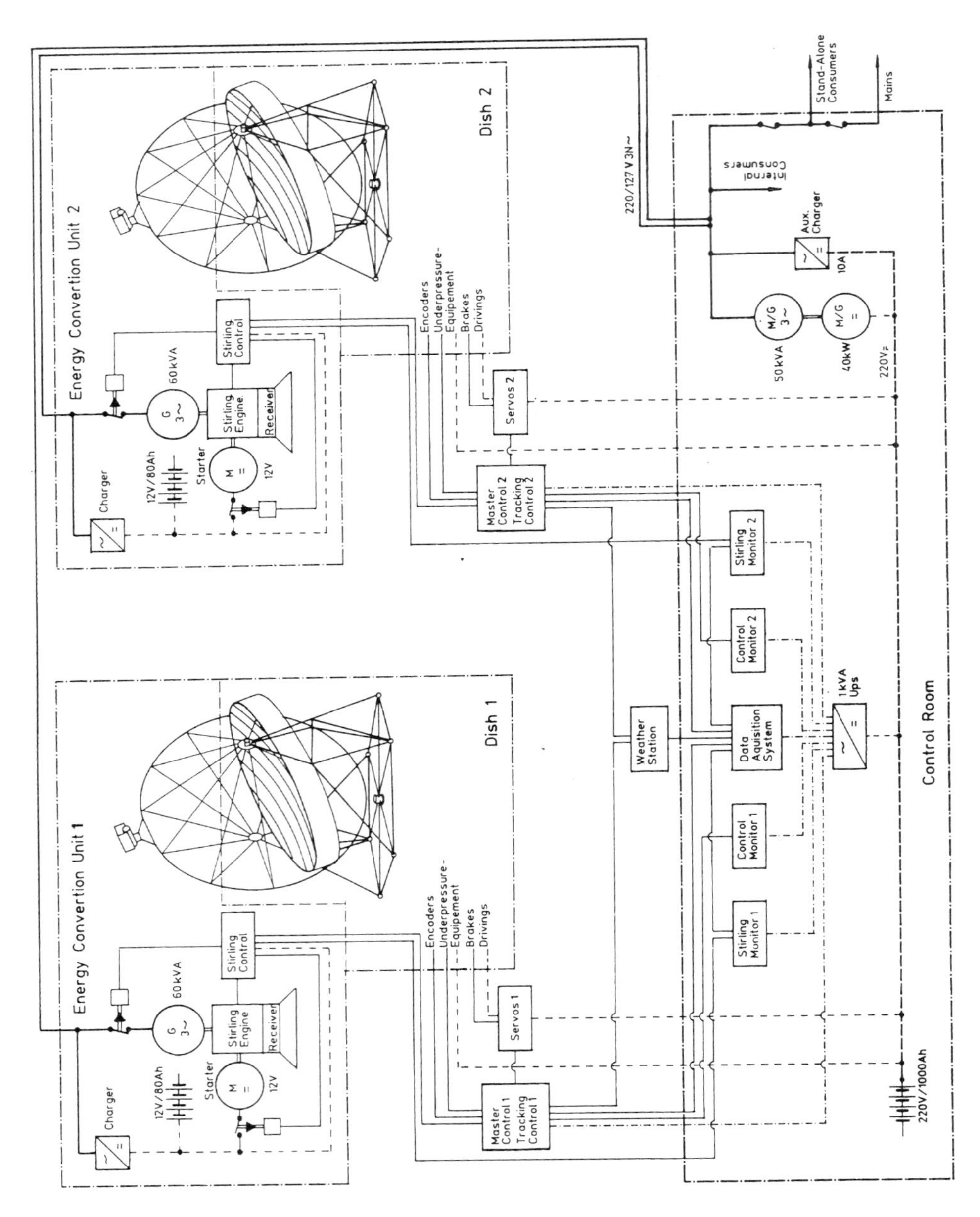

Schematic Diagram:
Pilot plant near Riyadh with two dishes

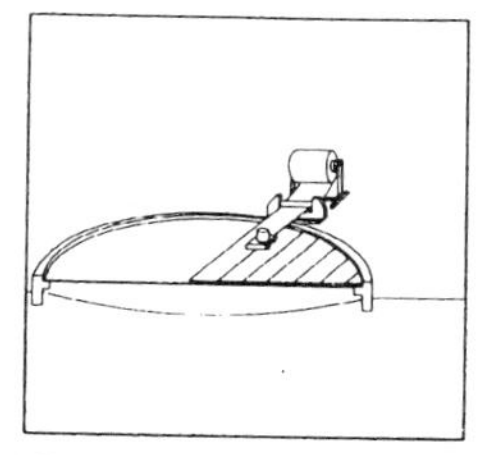

Welding together of membranes from individual 0,5 mm thin sheetmetal strips.

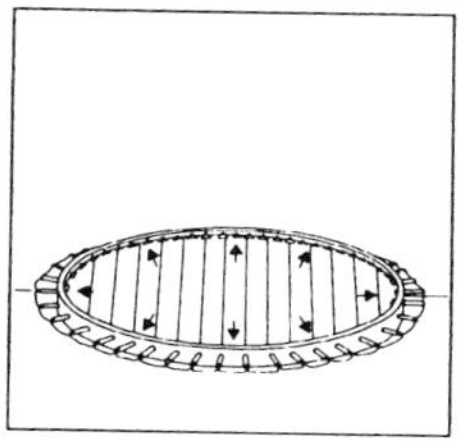

Stretching of membranes in shaping ring.

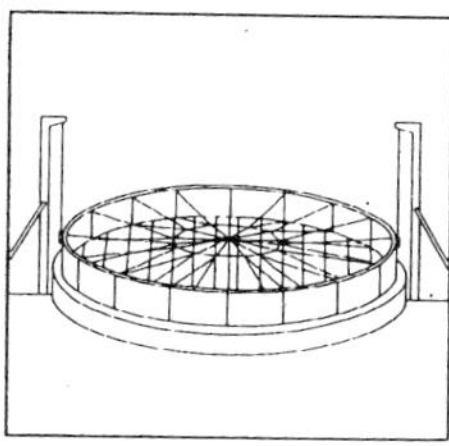

Attachment of rear membrane to concentrator ring.

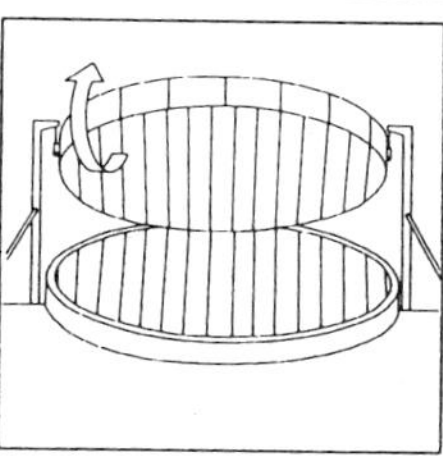

Turning of concentrator ring and placing on stretched front membrane.

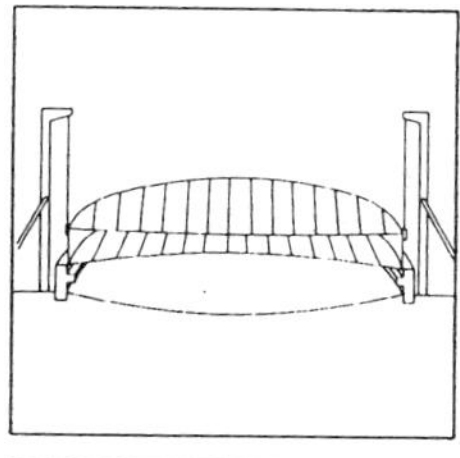

Shaping of front membrane by application of pressure.

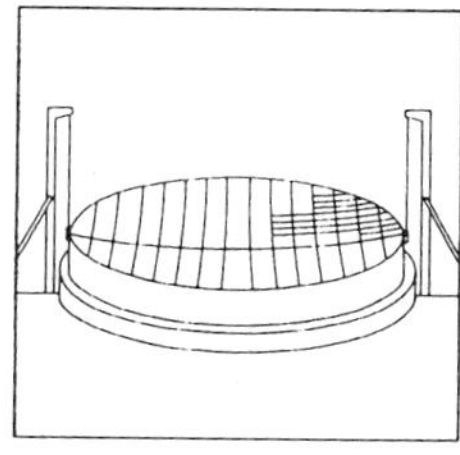

Stabilization of concentrator with partial vacuum, and bonding of glass mirrors on the front membrane.

OPTIMIZATION PROBLEMS IN THERMAL SOLAR PLANTS

A. ADELL, Laboratoire de Chimie-Physique, Université des Sciences et Techniques du Languedoc, 34060, Montpellier, France

ABSTRACT

The aim of this work is to answer such elementary questions as "what is an optimization problem and how to solve it?" The case of thermal solar plants has received a special attention because of the obvious importance of optimization for their industrial development. The principal concepts of the literature have been presented while introducing some new ones such as "quality factors".

KEYWORDS

Optimization problems, solar plants, Thermoeconomics, quality factors, systematic decrease of exergy.

Plant optimization means a complex set of mathematical and technical operations in order to satisfy some predetermined criteria such as maximizing the return on the financial investment, minimizing pollution and so on.

OPTIMIZATION CRITERIA

The optimization criteria concerning the apparatuses of transformation of energy are generally classified into three groups (Maczeck, 1980): thermodynamic socio-economic and technical.
The *technical optimization criteria* proceed from the conditions imposed by technical considerations like the minimum bulkiness, the best means of detecting the escapes of working fluid, the minimum of corrosion... .
The *economic optimization criterion* is the most important for the owner of the machine who wishes to maximize the benefit of his investment. But when the decision is to be taken by, for example, a public body, other socio-economic criteria must be taken into account, e.g. minimizing foreign currency cost, maximizing the number of created jobs,minimizing the energy and materials inputs... .
The *thermodynamic optimization criterion* has to be considered when the decision is to be made by a heat engineer whose only concern is the thermo-energetic behaviour of the system.

GLOBAL OPTIMIZATION

In practice several criteria intervene simultaneously. This is treated in the so-called *global optimization* (Le Goff, 1979). In global optimization some criteria can be contradictory such as minimizing the selling price and maximizing the reliability. It is the politic decider's responsability in important socio-economic projects, to asses the respective weight that must be given to each criterion for establishing priorities. In order to achieve global optimization, three types of consideration must be taken into account:

1- complying with the *technical specifications* of the project. The technical specifications fix the objectives to be reached and indicate the value or the constraints that must be verified by certain parameters; they also contain, more or less implicitly, some optimization criteria with eventually some indications on their respective weight.

2- choosing a particular kind of apparatus the working principle of which allows the satisfaction of the objectives of the technical specifications, and considering some hypotheses concerning the expected behaviour of the system. These hypotheses may be technical in nature which allows the modelling of the operational behaviour of the real machine in a theoretical or empirical way, e.g. the nature of irreversibilities, the life-time of the machine parts ... . These hypotheses may also have a socio-economic nature, e.g. national or world-wide evolution of material or energy costs; then they correspond to the choice of an economic model. All these hypotheses constitute "bets" which all the knowledge of "the man of the art" will tend to minimize the risk of.

3- taking into account additional constraints imposed by the environment of the system; the word "environment" been taken in a wide sense: can the designer realize the parts of the machine at the required dimensions himself or must he buy standard industrial elements? what are the chemical nature and the degree of purity of the available constituents? what are the ambient meteorological conditions?... .

Thus the global optimization can be done and discussed within scenarios which involve various choices: technological choices concerning the type of machine, technico-economic choices linked to the bets considered and political or technocratic choices of the respective weight that should be given to the criteria.

OPERATIONAL, CONSTRUCTIONAL AND DESIGN PARAMETERS

The practice of optimization can be done by two sorts of operations: a mathematical treatment of an a priori problem, allowing the solution of the optimum value for certain parameters of the system, or technical operations which must be done on the machine during its working,among which automatic control constitutes the principal element.

The parameters that should be taken into account in order to achieve the optimization are either those which define the conditions of the working and of the environment from a physical as well as a socio-economic viewpoint, or those that characterize the machine constituents (elements or working fluids): chemical nature, geometry, cost... .

The behaviour of these parameters may be continuous, discrete or simply qualitative; they may also be time-dependent in non stationary systems such as solar energy apparatuses. The knowledge of the qualitative parameters -- chemical nature and degree of purity of fluids, chemical nature and quality of the surface or type of manufacture of pipes... -- is equivalent to the implicit knowledge of a great number of coefficients among which some play an important part in the optimization according to the hypotheses considered. These coefficients generally have a discrete value,as in the case of viscosity and heat transfer coefficients and for the cost of the type of thermal insulator, for instance. The usual method to proceed to the optimization of these qualitative parameters consists of an a priori discussion of the result expected for different initial choices. The inverse method is of course theoretically possible and may lead to interesting information delimiting the possible choices, but generally does not directly result in a concrete solution. So the choices of the qualitative parameters are most often included in the planned scenario.

The set of parameters have been classified into three categories in the literature: operational, constructional, and design parameters (Maczeck, 1980).

DECISION VARIABLES AND OBJECTIVE FUNCTIONS

Optimization problems arise when we have to answer some essential questions concerning a project; these questions relate to the determination of extrema: which machine is cheaper? how to improve an existing machine? what is the best technological channel? Among these questions the principal is: what value should be given to the parameters of the system in order to realize the best plant?

To solve mathematically an optimization problem we must first determine which are the parameters of the system that have to be known to answer the asked question. These parameters constitute the *decision variables* of the questions. In order to formulate mathematically the questions of the optimization problem a quantity called the *objective function* (o.f.) is associated with each criterion contained in the questions. These o.f. must satisfy an extremum condition. The o.f. are explicitly expressed by means of physical or economic laws that apply to the model of the scenario, as a function of certain parameters which are the decision variables of the criterion.

Beside these o.f. associated with the optimization criteria, there are other o.f. with no extremum conditions: they are the parameters constrained by the technical specifications or by the environment. These quantities must in their turn constitute or depend on new variables which, together with the decision variables of the criteria, form the *primary decision variables* also called *essential variables. These essential variables are necessary to answer fundamental points such as: how to fabricate the machine and what are the conditions for its operation?*

In order to achieve the global optimization, that is to satisfy simultaneously all the questions while respecting the objectives and the constraints of the project, a set of coupled mathematical relations must be solved. These relations contain a great number of interdependent variables. This interdependence entails a great complexity for optimization problems. furthermore, as has been seen, some variables can have a discrete value which varies with the scenario and so have to be optimized case by case; this increases the length of calculations, complicates the discussions but affords an opportunity for future innovations.

PARTIAL OPTIMIZATION

Partial optimization is characterized by a limited number of mathematical relations with which is associated a limited number of variables. Partial optimization may be envisaged either to treat separately a few questions the answer of which offers an intrinsic interest, or to bring about a simplification of the problem of global optimization. In the latter case, partial optimization becomes a method of uncoupling the mathematical relations of the problem.

Among the simplifying methods, the well-known method using what is called "*zone decomposition*" (El Sayed, 1970) allows the reduction of the problem of global optimization of a complex machine to the problem of partial optimization in each element or functional zone of the machine. This method becomes of great effectiveness when the concept of *available energy* -- also called exergy -- is used. However this method leads to an approximate general solution because it neglects the interdependence existing between certain variables of each zone. One of its main advantages is to legitimize the use of finer thermodynamic models in each zone, contrary to the global optimization which more often can only be done using coarse or over-simplified models, e.g. " quasi-perfect machine", "basic cycle"... . This is what Tribus, initiator of the method, has justified by saying: "It is much more important to be able to survey the set of possible systems approximately than to examine the wrong system exactly".

Thermodynamic optimization is another well-known example of partial optimization involving the use of only one criterion and a limited number of decision variables.

WORKING PARAMETERS, QUALITY FACTORS, DEGREE OF FREEDOM

Two types of situation occur during the working of machines of transformation of energy:-- thermodynamic equilibrium states (or steady flows), for which the passage from one state to another defines the transformation between these two states.
-- mass and energy transfers which characterize an evolution between these two equilibrium states.
These two situations concern the whole system, that is the machine itself (internal situation) as well as its thermodynamic environment (external situation). THE *working parameters* are used to describe these two situations. These paramaters are composed of the variables of equilibrium states and of the transfer variables. They are interconnected by two sorts of relations:

.fundamental laws that directly link the parameters together;

.phenomenological laws that may have a theoretical or empirical basis and that involve various coefficients depending on the experimental conditions, according to the hypotheses considered in the scenario.

In the case of a machine for the transformation of energy, the exergetic efficiency is the principal optimizable o.f. because contrary to the energetic efficiency, it has an intrinsic physical meaning independently of the temperature of the heat exchanges.
In simple cycle flow machines, efficiency is affected by two types of decrease: on the one hand the *systematic decrease of exergy* and on the other hand the *occasional decrease of exergy*; decrease meaning either a loss of exergy with a corresponding decrease of energy, or a dissipation of exergy with conservation of energy and creation of an equal amount of anergy. These decreases may be internal or external.
The decreases of exergy are associated with what we have called the *quality factors*. These quality factors depend on two kinds of parameters of intensive or extensive nature.

Exergetic Dissipations: if one considers that the irreversible phenomena responsible for dissipations are in their linear domain, then the quality factors connected to the exergetic dissipations depend on the intensive phenomenological coefficients that link the cause of the phenomena (generalised affinities) to their effects (generalised fluxes), e.g. thermal conductivity, diffusion, viscosity... coefficients. The quality factors also depend on extensive coefficients such as the coefficients of dimension or shape, e.g. heat exchangers surfaces, diameter and surface state of pipes, shape of ajutages... .

Exergetic losses: they can also be connected with quality factors depending on two kinds of intensive or extensive coefficients; for instance in the case of the use of heat conveying fluid, these coefficients would be the specific heat and the mass of this fluid.

A constituent of a machine is said to have perfect quality if during the working of the machine no dissipation or loss of exergy due to this constituent intervenes. This situation can be reached by assigning certain conditions to only one of the two components of the quality factors.
The *systematic decreases of exergy* derive from the choices made in the scenario; effectively these choices exercise constraints on some quality factors e.g. choice of a technology (real machine with no perfect Carnot cycle...), choice of certain qualitative parameters (nature of working fluid, type of condenser, particular ajutage...). It must be noted that it is the existence of these systematic decreases of exergy that gives justification to the thermodynamic optimization.
The *occasional decreases of exergy* come from the existence of the non constrained quality factors.
The maximizing of the exergetic effiency as a function of the non constrained quality factors is a trivial operation that consists in taking the value of the quality factor corresponding to an infinite quality. So the non constrained quality factors do not constitute real decision variables for the

thermodynamic optimization.
The primary decision variables which allow the complete satisfaction of all the energetic objectives, are the *thermodynamic decision variables*. The knowledge of these variables allows us to answer the two essential questions: how to fabricate the machine and how to make it work in order to satisfy all the energetic objectives that can be demanded from this type of machine?
The thermodynamic degree of freedom is the number of independent thermodynamic decision variables. More precisely, it is the minimum number of the primary decision variables that must be known to make it possible to fabricate a machine and to make it work so that this machine should satisfy all the energetic objectives and this within a scenario corresponding to a technological choice with no occasional degradations of exergy and without any environmental constraints. Thus the thermodynamic optimization consists in finding the particular value of each of the thermodynamic decision variables corresponding to the greatest exergetic efficiency of the machine.

ECONOMIC OPTIMIZATION OF THERMAL SOLAR PLANTS

Economic optimization -- the term thermo-economic would be more appropriate-- consists in optimizing the actual cost of the project and the actual total financial value of the production, simultaneously. If the decision variables of the economic optimization intervene in both criteria -- as is the case for most of them -- then the unique criterion to be considered will be the maximizing of the benefit. This is accomplished by seeking the maximum value for the following o.f.: actual value of the production / actual cost of the project.

Thermodynamic optimization -- the term thermo-energetic would be more appropriated -- of a machine for the transformation of energy, is done by taking the maximizing of energetic production as the optimization criterion. As already discussed, this thermodynamic optimization is a partial optimization that can present an intrinsic interest, but that can also be done with the aim of simplifying the calculus of the global economic optimization of the project. It has also been seen that the thermodynamic optimization consists in determining the energetically optimum value of the thermodynamic decision variables within a particular scenario.
In order to carry out the thermodynamic optimization, one must take for the non constrained quality factors a value that corresponds to an infinite quality -- that is only systematic decreases of exergy have to be considered. Then, to achieve the global optimization of the system, the value of the quality factors responsible for the occasional decreases of exergy have to be sought. This is accomplished by a technico-economic optimization; but, and that is the interest of the method, the value found by the thermodynamic optimization can be taken as the value of the thermodynamic decision variables. The calculations may next be continued by an iteration repeating the first calculus of the thermodynamic optimization and using the new value found for the quality factors. *That amounts in effect to a transfer from occasional to systematic decreases of exergy*. The justification of this method is based on the fact that generally the variation of the thermodynamic decision variables has little influence on the costs and much on the efficiency, while the variation of the quality factors has little influence on the efficiency and much on the costs.

In the case of thermal solar plants, the thermodynamic optimization presents an increased interest: according to the "gratiutousness" of the solar energy, the financial cost of a solar plant only covers the investment and maintenance costs of each of its elements among which the solar collector takes a very important part. To a first approximation, it can be said that the total cost of a project is proportional to the collecting area S. Besides, the selling price of the energetic production of the machine which is proportional to this production, is then proportional to the product $\overline{N}.S$, where $\overline{N}$ is the mean value of the global solar efficiency during the irradiation. The o.f. linked to the

economic criterion is then proportional to $\bar{N}$: o.f. proportional to $\bar{N}.S/S = \bar{N}$
Thus, except in the vicinity of the optimum economic solution,each variation of a decision variable that increases the exergetic efficiency also increases the economic efficiency, so that the search for the optimum conditions of the exergetic efficiency tends to the global economic optimization of the project.

THERMOECONOMIC ANALYSIS

The thermoeconomic analysis of a machine for the transformation of energy consists in finding out where and how loss and degradation of energy occur,how they depend on the choices of the parameters of the system and how they influence one another. The economic analysis consists in determining where and how much material and energy come into orout the system and how much it costs. Considering the interdependence between the energetic and economic fluxes leads to the so-called *thermoeconomic analysis* (also called *Thermoeconomics*)which has been introduced by Tribus, Evans and El Sayed (1962 ,1970), and has become of frequent use in the study of industrial processes.
The utilization of exergy for studying and optimizing the transformation of energy processes is now a well-known and effective tool(Auracher, 1984). This concept has been used in solar technologies in particular for collectors (Scholten, 1984), solar thermal engines (Vokaerd, 1981) and for solar refrigeration systems like liquid absorption machines (Anan, 1984) or solid adsorption machines (Adell, 1983). Tsatsaronis (1984) has proposed the term *Exergoeconomics* as better than Thermoeconomics to characterize the combination of economic and exergetic analyses.

CONCLUSION

Some concepts of the literature have been presented in order to answer such elementary questions as: what is an optimization problem? how can we solve them? According to the diversity of the problems which contain an optimization condition, it is often difficult to give definitions general enough to be applicable to all situations. The principal domain of application that we have constantly kept in mind is the optimization of the machines for the transformation of energy and essentially those that use renewable sources of energy.

We have emphasized the interest of thermodynamic optimization as a partial optimization which permits the simplification of the global economic optimization in the case of thermal solar plants.

REFERENCES

Adell, A. (1983) Optimisation du fonctionnement des systèmes de réfrigération solaires à adsorption solide. International Congress on Refrigeration,Paris

Anan, D.K. (1984)Second law analysis of solar powered absorption cooling systems. A.S.M.E. J. of Solar Energy Engineering, 106, 291-298.

Auracher, H. (1984). Fundamental aspects of exergy applications to the analysis and optimization of energy processes.Heat Recovery Systems, 4, 323-27.

El Sayed, Y.M. and R.B. Evans (1970); Thermoeconomics and the design of heat systems. J. of Engin. for Power, 92, 27-35.

Le Goff, P.(1979) Energetique Industrielle, 1 et 2 ,Techn. et Doc., Paris

Maczeck, K. (1980) Application of various criteria for optimizing refrigerating plants. In Saving of Energy,International Institute of Refrigeration, Paris

Scholten, W.B. (1984) A comparizon of exergy delivery capablities of solar collectors. A.S.M.E. J. of Solar Energy Engineering, 106, 490-493.

Tribus, M. and R.B. Evans (1962) The Thermoeconomics of sea water conversion. University of California, Los Angeles Report N°62-63.

Tsatsaronis, G. (1984) Combination of exergetic and economic analysis in energy conversion processes.Eur. Cong. on En. Economics and man. Ind.,Algarve,Port.

Vokaerd, D.and J. Bougard (1981) Solar refrigeration Engines, I.S.E.S.Brighton

OPTIMIZATION OF STEAM BASED ENERGY TRANSPORT IN DISTRIBUTED SOLAR SYSTEMS

P.O. Carden and P.K. Bansal

Department of Engineering Physics,
Research School of Physical Sciences, Institute of Advanced Studies,
The Australian National University,
Canberra, Australia.

ABSTRACT

In the present paper, we describe a unique computer software package for finding out the most cost effective layout of the steam based energy transport system for distributed solar thermal power stations. The analysis has been carried out in three optimization procedures: minimization of energy loss, minimization of energy transportation cost and the computer software to find the optimum pattern of interconnections for energy flow within the array of collectors and to seek the optimum array shape. Power station sizes up to 10,000 collectors each of $300m^2$ aperture area have been studied.

KEYWORDS

steam; energy transportation; optimization; solar thermal; power stations; computer software

INTRODUCTION

In solar thermal power stations having distributed paraboloidal collectors, heat is transported as dry saturated steam in insulated pipelines (called the energy transportation system), from individual collectors to the single centrally located power generation unit. The pipelines are made up of straight segments (which never cross each other), each of which begin and end at a collector. The flow in the outgoing segment always equals the sum of the flows of the incoming segments plus the contribution to the flow made by the interconnected collector. The cost of a segment is assumed to be proportional to its length and to a function of the design flow through the segment. The pipelines converge and join together, ultimately forming a single supply line for the power generator.

In order to determine the range of power sizes for which this form of energy transport is economical, equations are developed to yield optimum engineering design parameters for steam based energy transport in pipelines. The cost of thermal insulation has been identified as the dominent component over the cost of piping, circulating pumps, thermal energy loss, frictional energy loss etc. A computer programme has been developed to find out the most cost effective layout for the energy transportation system.

ANALYSIS

Following Carden and Bansal (1987a), the analysis has been carried out in three parts.

Minimization of Energy Loss

For a given energy flow and for a given quantity of thermal insulation, the design of a pipe segment has been optimized to yield the minimum energy loss, comprising frictional and thermal components.

The thermal and fluid friction energy loss per unit length of insulated pipe are respectively given by

$$W_{th} = -\frac{2\pi k\Delta T}{\ln \alpha} \tag{1}$$

and

$$W_f = K\left(\frac{n^3}{V^{5/2}}\right).\left(\frac{1-\alpha^2}{\alpha^2}\right)^{5/2} \tag{2a}$$

where

$$K = \frac{\pi^{0.5}\rho f\beta^5 q^3}{4\varepsilon}\ ;\ \beta = (d_o/d_i) = 1.1;\ \alpha = (\beta d_i/D_o) \tag{2b}$$

k, ρ, f, n, V, ε, d_i, d_o, D_o are respectively the thermal conductivity, density, friction factor, number of flows, volume of insulation (per unit length), heat to work conversion factor, inside pipe diameter, outside pipe diameter, outside insulation diameter; and q is the "frictional mean flow" from a single collector, i.e. it is the flow which, if steady over the power station "on time" Δt, would cause the same frictional loss as the actual varying flows do over the same period.

The minimum energy loss for a given n and V, designated by $W_o(n,V)$ may be obtained by plotting the sum of Eqs. (1) and (2) as a function of α. Evidently as the quantity $(n^3|V^{5/2})$ is now varied, the point of minimum energy will trace a locus. Since for each point on the locus, there exists a unique set of values α_o, W_o and $(n^3/V^{5/2})$; one may also plot the inverse functions. One of these, shown in Fig. 1, forms the basis of a second optimization procedure.

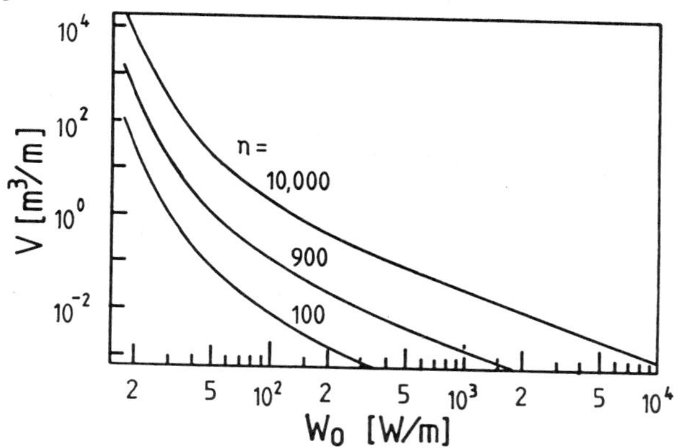

Fig. 1: Volume of insulation of a segment vs. minimum energy loss for various segment sizes.

Minimization of Cost

Total cost of energy transportation is the sum of the costs of lost energy and the thermal insulation. The first depends on the selling price of energy, P_e and the "on" time per day 'Δt'. Thus

$$\$W = W_o J \ (\$/m.yr) \quad (3a)$$

where $$J = (365.\Delta t)x10^{-3} \ P_e \ \varepsilon(\$/W.yr) \quad (3b)$$

The cost of thermal insulation is

$$\$V = V[P_v(inv + dep + O\&M)]; \quad (\$/m.yr.) \quad (4)$$

where P_v is the price of insulation which also includes the price for forming, transportation, installation and other accessories. The sum of these two costs may be expressed (in non-dollar terms) by dividing by J. Thus

$$C = W_o + zV \ (W/m) \quad (5a)$$

where $$z = \frac{P_V(inv + dep + O\&M)^1}{(365.\Delta t)x10^{-3} x(P_e.\varepsilon)} \quad (5b)$$

is a commercial factor that allows several non-engineering parameters such as prices and depreciation allowance to be resolved into engineering units. As explained in Fig. 1, V can be expressed as an inverse function of n and W_o. Therefore, using Eq.(5a), C can be plotted as a function of W_o for various values of n, revealing the minimum $C_o(n,z)$ as shown in Fig. 2. The numerical calculations are based on the following set of relevant parameters:

i) effective area of each collector = 300m²
ii) steam conditions: pressure = 6.8MPa; temperature (T_s) = 280°C; quality = 100%
iii) ambient temperature (T_a) = 30°C and therefore $\Delta T = T_s - T_a$ = 250°C.
iv) q = 0.00198m³/s; k = 0.05W/m°K; ρ = 0.031 kg/m³; f = 0.025; a = ratio of NS to EW spacing of collectors = 0.667.
v) for commercial factor z: inv+dep+O&M = 0.25; ε = 0.25, P_v = \$3000/m³; P_e = 0.1 \$/kWh; Δt = 8hrs; then z = 10,274 and so, 10,000 is selected as baseline value

However, the cost data presented in Fig. 2, over a wide range of z (5000 to 20,000) will cover the geographic and economic factors associated with most solar power station sites. The variation of other optimised parameters as a function of number of flows (n) is exhibited in Fig. 3.

Computer Software

The detailed computer analysis using various empirical methods for finding out the most cost effective energy transport layout has been discussed elsewhere by Carden and Bansal (1987b). That study, gave us an understanding of the lowest-cost states and consequently led us to the discovery of a class of Optimum Typical States which we found to be always at or very near a cost minimum. A family of typical states, denoted by integers l_1, l_2 (including zero) and l_3 (taking successive non-zero values) is exhibited in Fig. 4. The corresponding array sizes N_1, N_2, ... N_i and costs C_1, C_2, ... C_i can be used to define a continuous curve which will yield an actual or hypothetical value of C for every N. For each N, there is an optimum family - the one giving the lowest value of C. In Fig. 5, we show the envelope $C_o(N)$ of all these families which we contend, closely approximates the lowest-possible cost state.

FEATURES OF COMPUTER SOFTWARE OUTPUT

For most practical cases, a designer will nominate the size of the power station (in MW or GW) to be installed, its geographic location and the

1. (inv+dep+O&M) is a factor accounting for investment, depreciation and operation and maintenance.

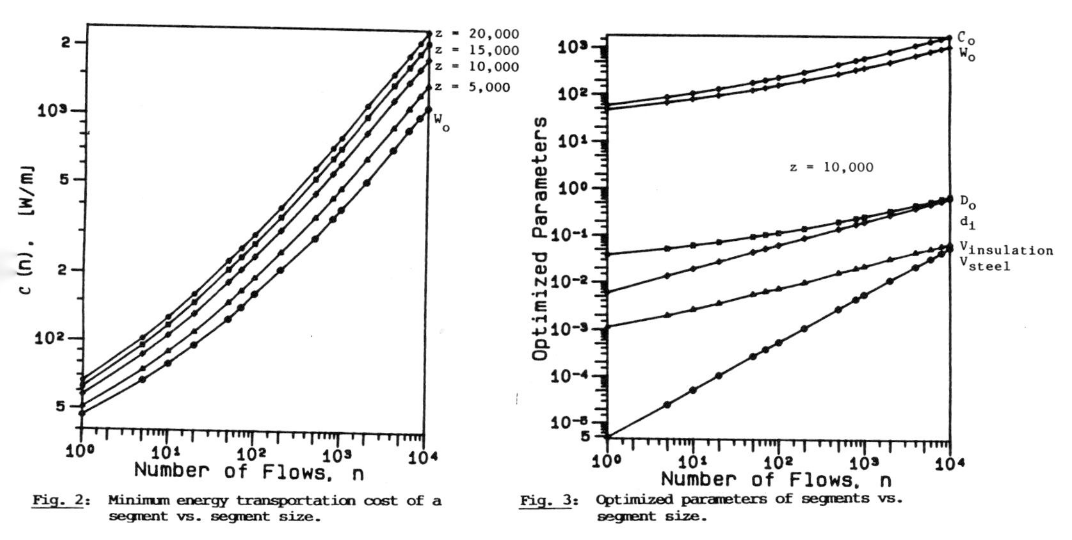

Fig. 2: Minimum energy transportation cost of a segment vs. segment size.

Fig. 3: Optimized parameters of segments vs. segment size.

hourly-daily data of the meteorological parameters (viz. solar direct radiation, ambient temperature, pressure, wind speed etc.) over a few years. To cater for this situation, a computer program has been written which gives the following output:

- Number of collectors to be installed, N
- Printed array layout - optimum shape; example is shown in Fig. 4
- Ratio of volume of steel to insulation, VRatio
- size of each pipe/insulation
- Percentage of maximum pressure drop to inlet pressure, % Ploss
- Percentage of energy loss to collection, % Eloss
- Cost of energy transportation, $C_o(N)$

REFERENCES

Carden, P.O., and P.K. Bansal (1987a). Optimization of steam based energy transport in distributed solar systems - submitted to J. Solar Energy.

Carden, P.O., and P.K. Bansal (1987b). Computer Software for optimum energy transport layouts for distributed solar systems. Proc. Australian-New Zealand Solar Energy Society Annual Conf., Canberra, Australia, Nov. 1987.

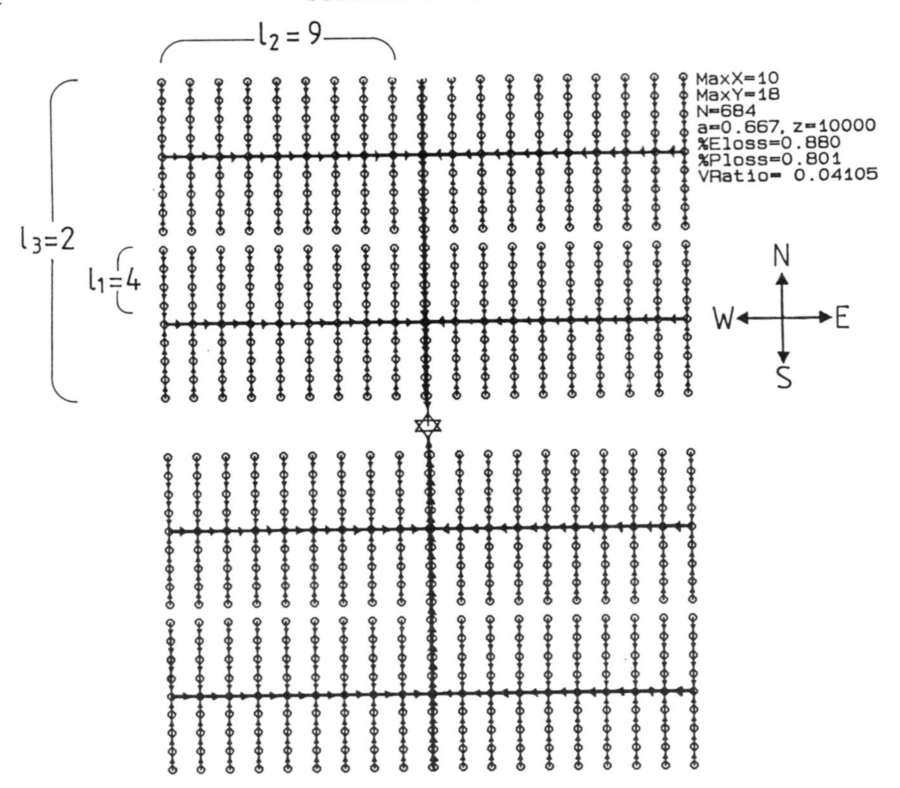

Fig. 4: An example of a typical layout of a collector array, illustrating definitions of l_1, l_2 and l_3

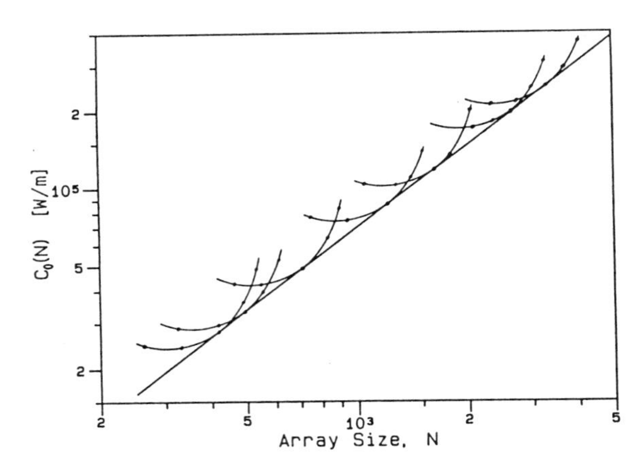

Fig. 5: Minimum energy transportation cost of an array vs. array size. Derivation as an envelop of family cost curves.

ON THE THERMAL REGULATION OF A SOLAR PARABOLOIDAL COLLECTOR

G.Stavrakakis*, G.Bitsoris**, M.Santamouris***

* Control and Automation Group, Electrical Engineering Dept., NTUA, 15773 Zographou, Athens, Greece.

** Automatic Control Lab., Electrical Engineering Dept., University of Patras, Patras, Greece.

*** Society of Appropriate Technology Ltd., Mavromihali 39, Athens, Greece.

ABSTRACT

This paper deals with the problem of the thermal regulation of a solar paraboloidal collector. A short description of the experimental process is given and former modelisation results are presented. Then, based on a bilinear model for the collector receiver a thermal regulator is established. The role of this regulator is to define an equilibrium point for the output temperature of the heat transfer fluid and, in addition, to eliminate the deviations which are due to external perturbations.

KEYWORDS

Solar paraboloidal collector, receiver thermal model, bilinear model, open loop control, stability.

SYSTEM DESCRIPTION

Two couples of paraboloidal solar concentrators have been build on the experimental field of Saint Chamas in Marseille-France. These THEK concentrators (for THErmal-Kilowatt) are thermal concentrating solar collectors intended for operation up to mean output temperature of 300°C and for delivering a power of 30 KW [1]. The THEK collector is permanently pointed towards the sun and concentrates solar energy on a boiler in which circulates a heat transfer fluid, Fig. 1,2.

The boiler is a steel spirally wound tube placed on the focal plane of the reflector. In the receiver, shown in Fig. 2, the concentrated solar energy is transformed to thermal energy transported by heat transfer fluid (Gilo - therm oil) which circulates in the boiler.

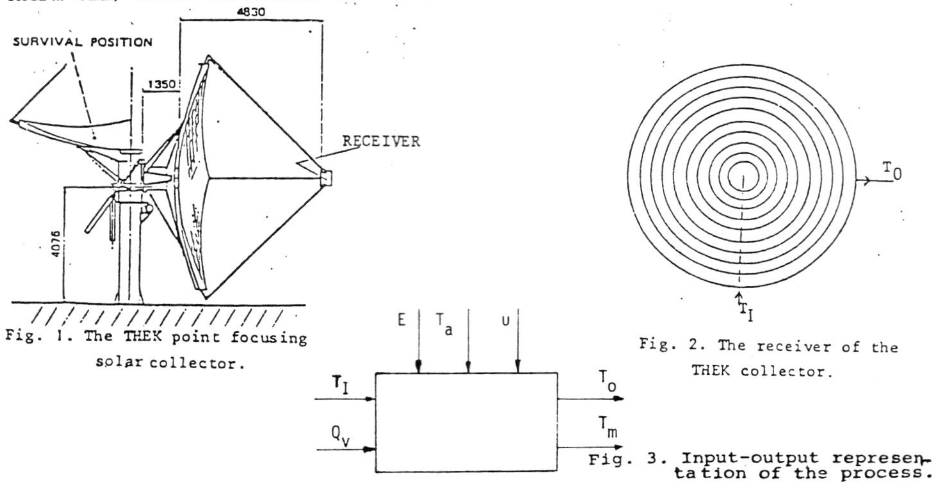

Fig. 1. The THEK point focusing solar collector.

Fig. 2. The receiver of the THEK collector.

Fig. 3. Input-output representation of the process.

The operation of the THEK collector is perturbed by the variations of the solar flux, the ambient temperature and the wind speed. So the problem of designing of a regulator of the output temperature of the heat transfer fluid is arised. To this end, based upon physical considerations, a bilinear state model is established for the determination of an input-output description of the THEK boiler. Then, an open loop regulator is obtained by applying techniques of the stability theory.

A BILINEAR INPUT-OUTPUT MODEL

Since the purpose of the modeling of the system is rather the implementation of a control unit than a simple description of the behaviour of the process, we must derive an input-output description of the process.

The solar energy flux E, the ambience temperature T_a and the wind speed υ are uncontrollable but measurable inputs. The input temperature of the heat transfer T_I is measurable but not directly controllable. The volume flow rate Q_v is directly controllable and could be considered as the control variable of the system. The other inputs are considered as measurable perturbations of the system.

The outputs of the system considered also as the state variables of the system, are the average metal temperature T_m and the fluid output temperature T_o. This input-output representation of the system is shown in Fig. 3.

A sensitivity analysis of the non-linear thermal model for the collector receiver derived in [6] shows that all the optical collector parameters, as the fluid and metal thermal parameters can be joined together in a small number of new constant parameters which can be estimated. An estimation of these parameters at this stage would produce model which would be nonlinear with respect to the control variable because of the relation $h_{mf}=KQ_v^{0.9}$ for the convective heat transfer coefficient between metal and fluid. The analysis of this model would be complicated and the resulting control law could not be easily implemented to the process. For this reasons a simpler bilinear model shall be established.

Let us pose $Q_v(t)=Q_{v_o}+\Delta Q_v(t)$ where Q_{v_o} is a fixed value of the volume flow rate which will be determined in the next section. We express the elements $f_{ij}(Q_v)$ of the matrices in the non-linear model for the receiver given in [6] by their first order approximations

$$f_{ij}(Q_v)=f_{ij}(Q_{v_o})+f_{ij}(Q_{v_o})\Delta Q_v$$

Insertion of this expressions to the differential equations of the non-linear model yields the equation [5,6]

$$\dot{x} = A(t)x+B(t)xu+C(t)wu+D(t)w \quad (1)$$

where $x = [T_m \;\; T_o]^T$, $u = \Delta Q_v$, $w = [T_I \;\; T_a \;\; E]^T$

Equation (1) is a bilinear dynamical thermal model of the THEK receiver with $x(t) = [T_m(t) \;\; T_o(t)]^T$ as state vector and $u(t) = \Delta Q_v(t)$ as control variable. The vector $w(t) = [T_I(t) \; T_a(t) \; E(t)]^T$ represents a measurable uncontrollable input. All the unknown parameters of the non-linear model are joined together in only five parameters θ_i which must be estimated. These are also the unknown parameters of the bilinear system (1) and can be considered as constant.

Since the parameters θ_i $i = 1,2,...,5$ do not depend on the value of Q_{v_o} [5,6] for the estimation of θ_i we choose the value of Q_{v_o} in the middle of the control variable range (from 150 to 1200 lt/h) that is Q_{v_o} = 576 lt/h . With this value of Q_{v_o} a least square estimation algorithm based on Kalman's filter equations has been applied [3,5,6], using real data.

THERMAL REGULATION

The control problem for the receiver consists in the regulation of the output temperature T_o of the heat transfer fluid vis a vis the variations of the ambient temperature, the wind speed, the input temperature of the heat transfer fluid and the solar energy received by the boiler.

For the regulation of the fluid output temperature T_o we must determine a control law $Q_v(t)$ or, equivalently after the choice of Q_{v_o}, a control law $u(t) = \Delta Q_v(t)$, which under normal operation maintains the output temperature equal to a desired value T_{od} and, in addition, in the case of an accidental perturbation it brings the output temperature back to the desired value.

The behaviour of the output temperature T_o is described by the equation

$$\dot{T}_o = a_{21}T_m+a_{22}T_o+(b_{21}T_m+b_{22}T_o)u+(c_{21}T_I+c_{22}T_a+c_{23}E)u+$$
$$+d_{21}T_I+d_{22}T_a+d_{23}E \quad (2)$$

where a_{2j}, b_{2j}, c_{2j} and d_{2j} are the corresponding elements of the matrices A, B, C, D defined in the previous section. This equation can be written in the form

$$\dot{T}_o = a^*T_o+b^*T_ou+c^*(t)u+d^*(t) \quad (3)$$

The parameters a^*, b^*, $c^*(t)$ and $d^*(t)$ are known functions of the estimated parameters θ_i and of the variables $\upsilon(t)$, $T_I(t)$, $T_a(t)$, $E(t)$ and $T_m(t)$. The input variables $\upsilon(t)$, $T_I(t)$, $T_a(t)$ and $E(t)$ are measurable at any instant t. The average temperature $T_m(t)$ of the receiver wall is not measurable but can be determined by the first state equation

$$\dot{T}_m = (a_{11}T_m + a_{12}T_o) + (b_{11}T_m + b_{12}T_o)u + (c_{11}T_I + c_{12}T_a + c_{13}E)u + d_{11}T_I + d_{12}T_a + d_{13}E \tag{4}$$

which is considered as a model of a dynamical observer of the state variable T_m. So, all the parameters of the state equation (3) are known at any ins - tant t.

The control problem of the process can be stated as follows:

"Given a set of admissible controls Ω, determine a control law $u^*(t) \in \Omega$ such that

a) The state T_{od} is an equilibrium state of system (3)

b) The state T_{od} is asymptotically stable with a desired exponential rate of convergence".

We shall consider control laws of the form

$$u(t) = u_d(t) + u_s(t, T_o(t) - T_{od}) \tag{5}$$

where $u_s(t,0) \equiv 0$.

Insertion of (5) into (3) yields the equation

$$\dot{T}_o = a^*(t)T_o + \left[b^*T_o + c^*(t)\right]\left[u_d(t) + u_s(t, T_o - T_{od})\right] + d^*(t) \tag{6}$$

The state T_{od} is an equilibrium state of the controlled system (3) if and only if

$$a^*(t)T_{od} + \left[b^*T_{od} + c^*(t)\right]\left[u_d(t) + u_s(t, T_{od} - T_{od})\right] + d^*(t) = 0$$

or equivalently if and only if

$$u_d(t) = -\frac{a^*T_{od} + d^*(t)}{b^*T_{od} + c^*(t)} \tag{7}$$

Therefore, the first term $u_d(t)$ of the control law $u(t)$ given by relation (7) makes T_{od} an equilibrium state of system (3). The second term $u_s(t, T_o - T_{od})$ would be determined so that the equilibrium T_{od} is asymptotically stable with a desired exponential stability rate.

Let $y(t)$ denote the deviation of the output fluid temperature $T_o(t)$ from the desired value T_{od}, that is $y(t) = T_o(t) - T_{od}$. Then, for the controlled system, from (6) and (7) it follows that

$$\dot{y} = p(t)y + qyu_s + r(t)u_s \tag{8}$$

where

$$p(t) = a^* + b^*u_d(t), \quad q = b^*, \quad r(t) = b^*T_{od} + c^*(t)$$

The regulation problem is now reduced to the determination of the control law $u_s(t, y(t))$ for the equilibrium state $y=0$ of system (8) to be asymptotically stable with a desired exponential stability rate.

We shall consider the trivial case where $u_s(t,y) \equiv 0$. In this case equation (8) becomes

$$\dot{y} = p(t)y \tag{9}$$

and the equilibrium $y = 0$ is exponentially stable if there exists a positive real number ε such that $p(t) \leqslant -\varepsilon$ for all $t \in [0,\infty]$.

We can write a^*, b^*, c^*, d^*, by inspection of their analytic expressions [3,5,6] as:

$$a^* = -a_v Q_{v_o}, \quad b^* = -b_v, \quad c^*(t) = c_v(t), \; d^*(t) = d_v(t) Q_{v_o}$$

Then, the exponential stability condition $p(t) < -\varepsilon$ for system (9) can be written equivalently in the form

$$\frac{\dfrac{d_v(t)}{a_v} - T_{od}}{\dfrac{c_v(t)}{b_v} - T_{od}} \leqslant 1-\varepsilon^* \quad \text{and} \tag{10}$$

$$-\frac{a_v}{b_v} Q_{v_o}(1-\varepsilon^*) \leqslant u_d(t) \leqslant 0 \quad \text{using } (7)$$

where ε^* is a small positive real number related to ε by $\varepsilon^* = \dfrac{\varepsilon}{a_v Q_{v_o}}$.

If condition (10) is satisfied, then the equilibrium $y=0$ of (9) is exponentially stable, i.e.

$$|y(t)| = |y(t_o)| \exp[-\varepsilon(t-t_o)] \quad \text{for all } t>t_o \text{ where } \varepsilon = a_v Q_{v_o} \varepsilon^*$$

Therefore, if (10) is satisfied, then with $u_s=0$ and $u_d(t)$ given by (7') the equilibrium state T_{od} of (3) is exponentially stable, i.e.

$$|T_o(t)-T_{od}| = |T_o(t)-T_{od}| \exp[-\varepsilon(t-t_o)], \quad \forall t>t_o$$

By definition, the control law is admissible if and only if:

$$-Q_{v_o} \leqslant u_d(t) < Q_{max} - Q_{v_o} \tag{11}$$

where $Q_{max} = 3.333.10^{-4}\ m^3/sec$ is imposed by the process specifications. A good choice of the value Q_{v_o} which satisfied both (10) and (11) is

$$Q_{v_o} = \frac{b_v}{a_v} \frac{1}{(1-\varepsilon^*)} Q_{max}$$

SIMULATION RESULTS

In order to evaluate the performance of the control law established in the previous section, we have simulated the process when it is perturbated by the uncontrollable inputs E (solar flux), T_a (ambient air temperature) and υ (wind speed) which are shown in the Fig. 4. The desired output temperature T_{od} has been chosen 300° C. The results obtained when the initial output temperature is $T_o(0) = 265°$ C are shown in Fig. 5. In Fig. 5.c is shown the evolution of the efficiency η_r of the collector.

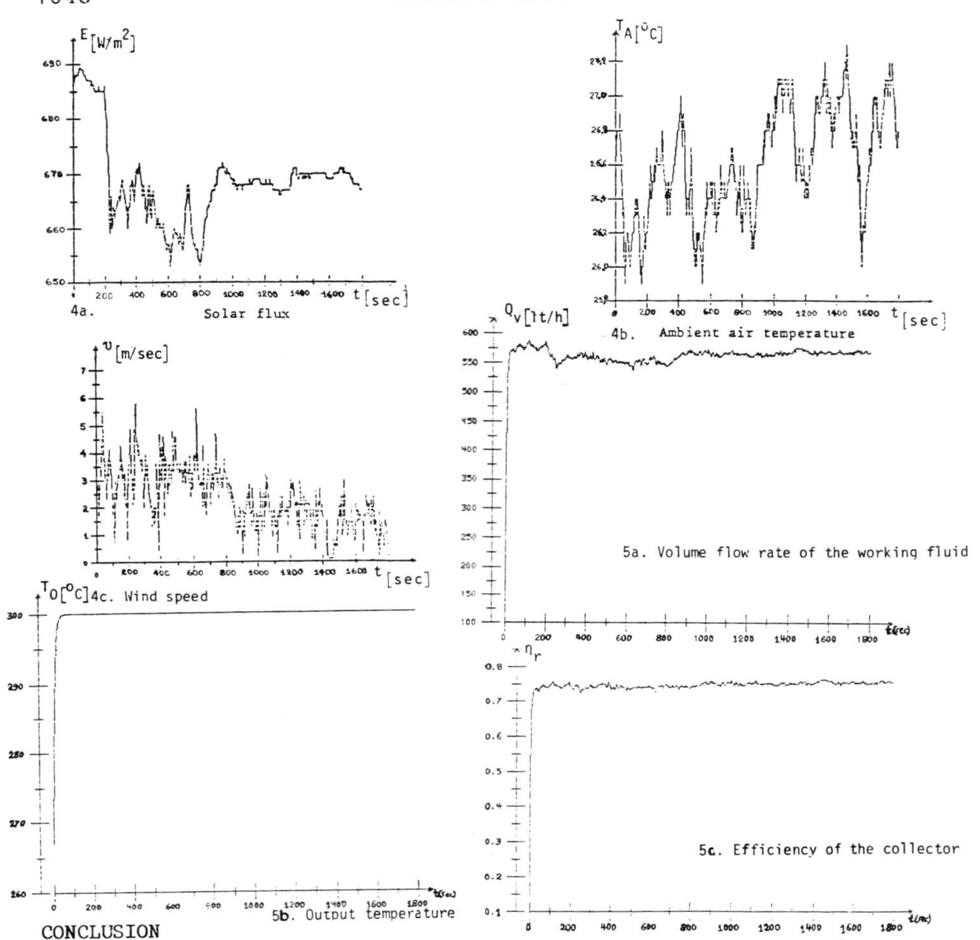

4a. Solar flux

4b. Ambient air temperature

4c. Wind speed

5a. Volume flow rate of the working fluid

5b. Output temperature

5c. Efficiency of the collector

CONCLUSION

We have studied the problem of the thermal regulation of a solar paraboloidal collector. The analysis of the proposed bilinear state model has shown that with an open loop controller we can fix a desired equilibrium state and in addition to eliminate exponentially any deviation caused by the variations of the uncoltrollable inputs as the solar flux, the wind speed and the ambient air temperature.

REFERENCES

1. M. AUDIBERT. La centrale solaire experimentale de production de chaleur industrielle THEK 2. Colloques Internationaux du C.N.R.S. No. 306.
3. J. MIGNOT, G. STAVRAKAKIS, 1983. On modelling and indentifying the thermal behaviour of a solar paraboloidal collector. EEE 83 Symposium,Athens.
5. G. STAVRAKAKIS, 1984. Modelisation et commande du fonctionnement d'un capteur solaire du type THEK., Thèse de Docteur Ingenieur, Université Paul Sabatier, Toulouse, France.
6. G. STAVRAKAKIS, I. AGUILAR-MARTIN, 1986. A Bilinear Model for the thermal behavior of a point focusing solar collector. International Journal of Energy Research, Vol. 10, pp. 291-300.

OVERALL THERMAL LOSS COEFFICIENT AND INCIDENT ANGLE MODIFIER FOR AN ACUREX 3001 DCS: APPLICATION TO THE CONTROL SYSTEM

E. Zarza and J.I. Ajona

CIEMAT-Instituto de Energías Renovables,
Plataforma Solar de Almería (Spain)

ABSTRACT

This report presents the actual incident angle modifer and overall thermal loss coefficient for the ACUREX 3001 distributed collector field installed on the Plataforma Solar de Almería, Spain.

Calculation was based on a database of over 3400 sets of values gathered during the wide temperature range test campaign performed during 1985-86 (Sánchez, 1987). These data were statistically processed and the incident angle modifier function, $K(\theta)$, calculated as a fourth-order function of the incident angle, θ:

$$K(\theta) = 0.991 + 4.455E\text{-}3*\theta - 5.48E\text{-}4*\theta^2 + 1.426E\text{-}5*\theta^3 - 1.252E\text{-}7*\theta^4 \quad (1)$$

Results of Equation (1) are compared with those previously calculated for just one collector (Dudley, 1981).

The overall thermal loss coefficient, U_L, has been calculated for the whole collector field as a first-order function of the difference between the oil inlet temperature, T_i, and the ambient temperature, T_a:

$$U_L = U_{L1} + U_{L2}\,(T_i - T_a) \quad (2)$$

Determination of $K(\theta)$ and U_L has improved the control system through more accurate knowledge of the collector field's characteristic parameters.

KEYWORDS

Mid-temperature collectors; parabolic trough; optical parameters; thermal parameters; incident angle modifier; thermal loss coefficient.

INTRODUCTION

Main technical characteristics of the ACUREX 3001 DCS field under study are:
° Type of collector : ACUREX model 3001.

- Rim angle : 90°.
- Tracking system : horizontal one-axis tracking system, oriented East-West.
- Geometrical concentration ratio : 18.2.
- Mirrors : silver-coated steel reflector, protected by thin glass (0.6 mm).
- Absorber : selective black chrome-coated steel tube receiver.
- Field layout : 10 loops of 2 rows; 2 collectors/row; 12 modules/collector.
- Aperture area of collector : 66.8 m^2.
- Total aperture area : 2672 m^2.
- Land-use factor : 0.37.
- Working fluid : 3M Santotherm 55 thermal oil.
- Storage system capacity : 140 m^3.

The collector field was considered as one collector; parameters calculated and presented are overall parameters to consider when studying the whole collector field behavior.

Small standard deviations obtained from statistical data processing prove that this procedure is reliable. This reliability is more evident taking into account the large amount of data processed.

CALCULATION PROCEDURE

Simple, steady-state, instantaneous energy balance of a solar collector is given by the expression:

$$Q_u = F_r \cdot (K(\theta) \cdot \eta_o)_{max} \cdot I_b \cdot \cos(\theta) \cdot A_t \cdot FIT) - F_r \cdot U_L \cdot (T_i - T_a) \quad (3)$$

where Q_u = thermal energy gained by oil circulating through collectors (W); F_r = heat removal factor of the collector; θ = incident angle (°); $K(\theta)$ = incident angle modifier; $\eta_o)_{max}$ = optical efficiency when θ=0°; I_b = beam solar irradiance (W/m^2); A_t = total aperture area of collector field; FIT = overall geometrical loss factor; U_L = overall thermal loss coefficient (W/°C); T_i = field inlet temperature (°C); and T_a = ambient temperature (°C).

The FIT accounts for diverse geometrical losses in the collector field: losses due to the sheet metal located on the collector sides; when any collector is shadowed by another; or the fact that some reflected sunlight at the end of the collector does not reach the absorber tube. The FIT factor was determined using software, as a function of the julianne day and solar time.

Calculation of U_L and $K(\theta)$ was based on Equation (3). When collectors are defocussed, available beam irradiance on the collector aperture plane, I_a = $I_b \cdot \cos(\theta)$, is 0. In this case Equation (3) becomes:

$$Q_u = m_c \int_{T_i}^{T_o} C_p \cdot dT = \text{thermal losses} = -F_r \cdot U_L (T_i - T_a) \quad (4)$$

where m_c = mass oil flowrate circulating through collector field (kg/s); T_o = field outlet temperature (°C); and C_p = specific oil heat (depending on the oil temperature) (J/kg°C).

Thermal losses can be calculated as the difference between the thermal power carried by the oil at the field inlet and outlet under steady conditions. Assuming that thermal losses do not depend on the reflected solar radiation reaching the absorber, an error $<$ 4 or 5% is introduced. Thus, overall thermal losses were measured circulating oil through the collector field with the collectors defocussed, maintaining constant flow and oil inlet temperature, applying Equation (4). Since F_r depends on the oil mass flow which was not the same in all thermal loss tests, proper heat transfer correlations were used to determine F_r as a function of the oil flow.

Evaluation of Equation (4) for different steady states gave a set of points (U_L, T_i-T_a). Coefficients U_{L1} and U_{L2} in Equation (2) were determined adjusting this set of points to a first-order function using the least square method.

With regard to optical losses, these are usualy determined by circulating thermal fluid through the collector at ambient temperature, obtaining null thermal losses. However, this method cannot be used for the ACUREX 3001 collector field: high viscosity of the working fluid at ambient temperature makes the oil flow too low.

To overcome this problem, data from the wide range temperature test campaign and previously calculated coefficient U_L were used to determine the incident angle modifer $K(\theta)$, applying Equation (3). These tests were made at constant inlet and outlet temperatures. Prior to each test, temperature in the storage tank containing 140 m^3 thermal oil was homogeneous. The collector field was fed from this tank so that the inlet temperature was constant throughout the test. Furthermore, the field control system varied the flowrate and thus maintained constant the outlet temperature.

Steady conditions were maintained at the field inlet and outlet for long periods during each test due to the reliability and accuracy displayed by the control system. This contributed to the large quantity of data (more than 6000 sets of data) available for evaluation purposes.

In addition, only the most accurate data were used (over 3400). These data belong to three different groups of tests, depending on the oil temperature at the field inlet/outlet throughout the test: low temperature tests (100/180°C), medium temperature tests (160/230°C), and high temperature tests (225/295°C).

RESULTS AND CONCLUSIONS

The overall thermal loss coefficient, U_L, obtained is:

$$U_L = U_{L1} + U_{L2}(T_i-T_a);\ U_{L1} = 1034.06\ W/°C,\ U_{L2} = 1.878\ W/°C \quad (5)$$

These values were obtained based on tests performed on 11 different days under diverse steady conditions (oil temperatures and flowrates), according to Equation (5).

Figure 1 shows how U_L evolves versus the temperature difference (T_i-T_a).

Table 1 contains mean values of the incident angle modifier, $K(\theta)$, calculated for each 2° interval of the incident angle. Typical deviation and number of samples statistically processed are also presented. The value of $\eta_{o})_{max}$ obtained is 0.608 for an average field mirror reflectivity of 0.88 and typical deviation of 0.0314.

Evolution of $K(\theta)$ versus θ is shown in Fig. 2, together with the $K(\theta)$ function calculated by Sandia Laboratories for one collector.

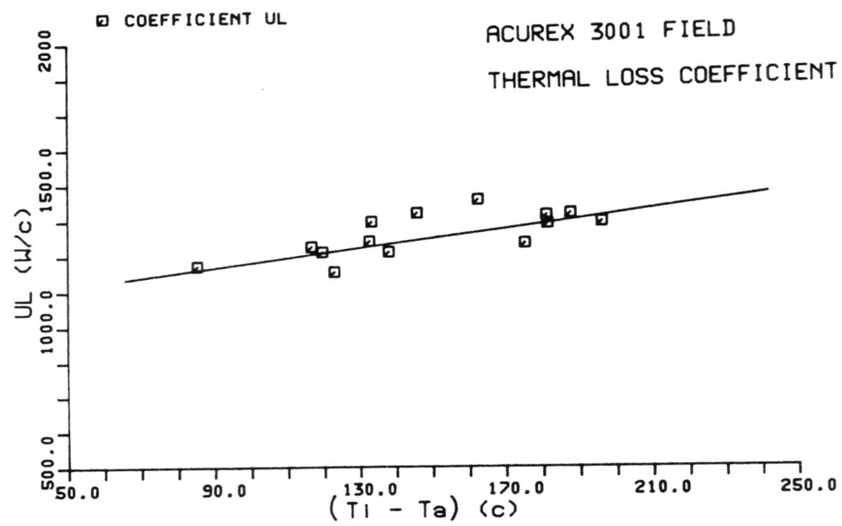

Fig. 1 Overall thermal loss coefficient, U_L

TABLE 1 Instantaneous values of the incident angle modifier K(θ)

(θ)	K(θ)	Desviation	Numb. of Samples
59.	0.778	0.0483	27
57.	0.786	0.0511	36
55.	0.807	0.0791	39
49.	0.823	0.0876	66
47.	0.852	0.0565	82
45.	0.829	0.0614	94
43.	0.861	0.0429	118
41.	0.878	0.0345	125
39.	0.877	0.0456	130
37.	0.900	0.0498	132
35.	0.910	0.0278	143
33.	0.909	0.0415	142
31.	0.923	0.0355	152
29.	0.929	0.0287	133
27.	0.938	0.0408	144
25.	0.942	0.0377	142
23.	0.948	0.0290	150
21.	0.946	0.0497	132
19.	0.949	0.0575	147
17.	0.970	0.0609	148
15.	0.962	0.0637	140
13.	0.987	0.0399	145
11.	0.986	0.0459	144
9.	0.990	0.0439	142
7.	0.994	0.0350	146
5.	0.996	0.0300	141
3.	1.000	0.0281	139
1.	1.000	0.0314	144

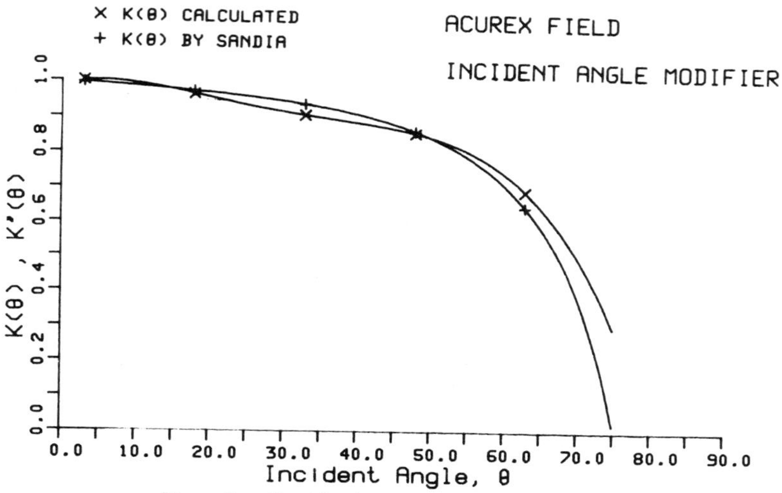

Fig. 2. Incident angle modifier, K(θ)

Small typical deviations obtained prove the accuracy and homogeneity of the data set processed, due to the large amount of data used and the previous calibration campaign, and prove the evaluation procedure reliable for the whole collector field.

The incident angle modifier evaluated by Sandia National Laboratories (Dudley, 1981) used only one ACUREX 3001 collector. Though few data were used by Sandia, values of K(θ) are quite similar (Fig. 2). Differences for $\theta > 65$ are caused by the lack of processed data within this range.

The thermal loss coefficient value, U_L, calculated at present does not differ much from that used since the first thermal loss tests of 1983, despite the fact that it was calculated based on a reduced number of data and with no previous calibration campaign.

Computer simulation was performed of thermal losses of absorber tubes similar to those of the ACUREX field. Calculated and simulated thermal losses are the same if emissivity of the selective surface is 0.35 at 200°C. This value is higher than the nominal value given by the manufacturer in 1981 ($\varepsilon = 0.27$), justified by degraded selective surface properties. Calculation of the actual thermal loss coefficient has made evaluation of this degradation possible, and can be used to evaluate future degradation.

Knowing actual K(θ) and U_L coefficients allows accurate simulation of collector field behavior under different operating conditions, and improves the control system. This control maintains the oil outlet temperature constant changing the oil flow in the field.

REFERENCES

Dudley, V.E., end R.M. Workhaven (1981). Performance testing of the ACUREX solar collector model 3001-03'. SAND80-0872, final draft.

Sánchez, M., R. Carmona, and E. Zarza (1987). Behavior of DCS fields in a wide temperature range. IEA/SSPS Technical Report 1/86.

A DIRECT ABSORPTION SOLAR CENTRAL RECEIVER GAS-TURBINE POWER PLANT

P. De Laquil III, S. M. Schoenung, T. Nakamura, S. M. Patel

Bechtel National, Inc., San Francisco, CA

ABSTRACT

This paper summarizes the results of a feasibility study of an innovative high temperature direct absorption central receiver concept - the particle injection receiver - in which heat from concentrated sunlight is absorbed into the receiver fluid (air) without exposing receiver tubes to the solar flux. Submicron-sized carbon particles suspended in air act as the energy exchange medium. A very small number of these suspended particles are injected into an air stream, flowing through a windowless receiver cavity, where they absorb solar radiation from a focused heliostat field and simultaneously heat the air. The resulting heated air stream passes into a heat exchange region where a pressurized air stream on the secondary side of a ceramic-tube heat exchanger is heated for use in a gas turbine. The turbine exhaust is utilized as the air stream which is darkened with particles and injected into the receiver cavity.

KEYWORDS

Solar thermal central receiver, solar direct absorption, solar Brayton cycle, submicron-sized particle radiation heating, ceramic heat exchanger, intercooled gas turbine.

INTRODUCTION

Since the mid-1970s, the solar central receiver concept has been under development as a potentially competitive, utility-scale power generation technology for the future. Central receiver development activities for steam Rankine-cycle power plants have included a 10 MWe pilot plant, near Barstow, California, the 0.5 MWe SSPS CRS facility and the Cesa Uno plant at Almeria Spain, and the Themis facility in the French Pyrenees. These plants have demonstrated the technical feasibility and environmental acceptability of the solar central receiver concept. In the United States solar central receiver technology development activities are currently being pursued by several Southwest U.S. utilities in a cooperatively funded program with the U.S. Department of Energy. The near-term central receiver technology of choice uses nitrate salt as the receiver heat transfer fluid and storage media and employs a metal-tubed receiver operating at temperatures below 600°C. The nitrate salt is used to generate steam for a Rankine-cycle power plant.

The particle injection receiver concept is a long-term, high potential central receiver option. In the feasibility study, it was examined in a Brayton-electric application because the high temperatures that appear achievable with this receiver concept allow use of modern gas-turbine systems that have good thermal-to-electric conversion efficiencies. In addition, compared to solar Rankine-cycle systems, solar gas-turbine systems have the following advantages that in the long-term should make them attractive to both U.S. utilities and developing countries:

- They are simple power conversion systems with higher reliability and lower operating and maintenance costs.
- They can be commercially attractive at smaller sizes.
- Their inherent low water requirements increase the siting potential for central receivers in arid regions.
- They are more easily adaptable to fossil hybrid operation which increases the plant availability.

BACKGROUND

The basic concept of using concentrated sunlight to heat an air stream darkened by a suspension of submicron-sized particles was first proposed by Hunt (1981, 1982) and a windowed, forced-flow receiver concept was successfully tested in 1982 (Hunt, 1983a 1983b). Hunt's receiver concept has not received significant attention for large-scale systems because of the cost, fabrication difficulties, and operational concerns associated with high temperature windows 5 to 10 m in diameter. The particle injection receiver described here is a new receiver configuration, which also employs submicron-sized particles as heat absorbers, but eliminates the need for an aperture window. This simplifies the design of particle injection receivers for utility-scale power plants.

It appears that very high air temperatures can be achieved with the submicron-sized particle heat absorption process. In theory, temperatures in excess of 2000°C are achievable. Because heat absorption in the darkened air stream takes place at solar flux levels greater than 2 MW/m^2, the size of the heat absorption region is small compared to conventional receiver concepts, and the area subject to reradiation and reflection losses is minimized. Because heat absorption takes place in the air stream and away from the receiver walls, these surfaces are cooler (by more than 100°C) than the air temperature. This further reduces the radiative heat losses and greatly simplifies receiver wall material selection. Finally, the pyrolyzed carbon particles have a high solar absorptivity but a low emissivity in the infrared region; therefore, they act as selective absorbers. This helps to further reduce receiver radiative losses.

Fortunately, the mass of particles required to effectively heat the air is quite small, less than 0.2 weight percent. The absorber particles injected into the heat absorption region can be either reactive with air, such as pyrolyzed carbon, or non-reactive, such as silicon carbide or another ceramic material. However, pyrolyzed carbon particles were selected for the feasibility study because:

- They have a high intrinsic absorptivity.
- They can be readily generated on site at a reasonable cost.
- They eventually oxidize leaving clean air to enter the heat exchanger.

The oxidation of the carbon particles is also the factor that ultimately limits the air temperatures which can be achieved with this receiver concept. Once the particles have completely oxidized, the heat absorption process stops.

PARTICLE INJECTION RECEIVER DESCRIPTION

The particle injection receiver is similar to other solar central receivers in that the receiver itself is located atop a tower, and solar radiation is focused into the aperture of the receiver cavity by a field of heliostats. However, in the particle injection receiver, the concentrated solar energy is absorbed in an air stream darkened by a suspension of submicron-sized carbon particles. Fig. 1 is a schematic illustration of the concept as configured for a Brayton-electric application.

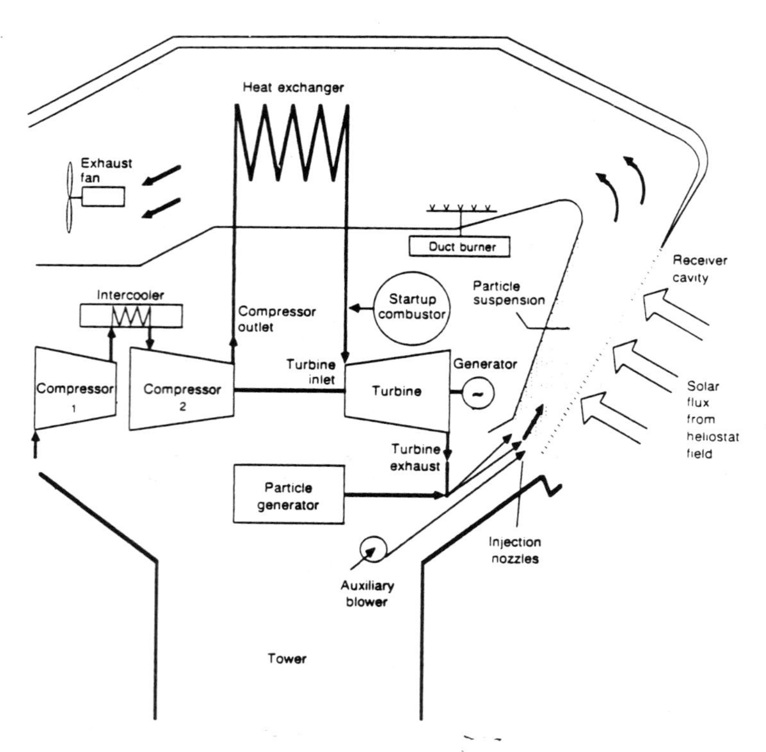

Fig. 1-1. Particle Injection Receiver - Schematic Diagram

The heat exchanger takes the place of the normal gas turbine combustor, and both the heat exchanger and the turbine-generator equipment are located atop the receiver tower.

As shown in Fig. 1, the turbine exhaust gas is darkened with particles and then injected through a series of nozzles to flow upward through the receiver cavity and the solar flux. The receiver cavity is oriented such that the direction of flow is normal to the direction of the solar flux.

Particles are generated on-line from a hydrocarbon oil using a particle generator. The design of this generator is based on process equipment currently used to manufacture carbon black. Because of their small size and intrinsic absorptivity, the carbon particles effectively absorb the concentrated sunlight with very little scattering of the energy back out the receiver aperture. The suspended particles move at the velocity of the air. Therefore, there is no convective heat transfer, only conduction from the particles to the air. The high surface-to-volume ratio of the particles allows them to act as very efficient heat exchangers, and the air temperature closely tracks the particle temperature (Hunt, 1986). The temperature of the particle/air suspension increases as it advances through the solar flux. The particles oxidize at a temperature-dependent rate, disappearing as the air leaves the receiver cavity and flows on to the heat exchanger.

The atmospheric pressure heated air is induced to flow through the heat exchange region by an exhaust fan that overcomes the pressure drop. Because the heat exchanger is convectively heated, a relatively standard design can be used in which the heat exchanger tube bundle fills the entire volume of the heat exchange region. As compared with typical metal or ceramic tube solar receiver concepts in which the heat absorbing surface is arranged around the interior of a large cavity, the result is a relatively compact receiver design. Furthermore, because the heat exchanger tubes receive no direct solar flux, tube stresses due to non-uniform circumferential heating are reduced, and tube temperatures are limited to the maximum air temperature. This simplifies heat exchanger design and allows direct application of the research and development on heat exchangers for industrial heat recovery and for indirect-fired gas turbines.

STUDY RESULTS

The objective of the feasibility study was to develop a pre-conceptual design of the particle injection receiver concept, to evaluate its ability to efficiently achieve high temperatures, and to estimate its cost and performance.

Cycle Configuration

An open, intercooled Brayton cycle with a turbine inlet temperature of 1205°C (2200°F) was selected from the cycle configuration analyses. The optimum cycle includes intercooling because the cycle efficiency is directly affected by the temperature of the atmospheric air exiting the heat exchanger. Intercooling reduces the compressor work and lowers the temperature of the pressurized air entering the heat exchanger. This helps to reduce the energy content in the exiting atmospheric air stream. In effect, the configuration is a recuperated cycle because the energy content of the turbine exhaust gas is recovered by injecting it into the receiver aperture where the cycle heat addition takes place.

Solar Plant Geometry/Receiver Flux Distribution

Utilizing the computer program DELSOL2 (Dellin, 1981), the collector field size, tower height, aperture size and orientation, and field efficiency were optimized for a 40 MWe plant. DELSOL2 was also used to calculate the solar flux distribution throughout the heat absorption region. The peak design point flux calculated for this 40 MWe plant is about 4.5 MW/m^2. The submicron- sized particle heat absorption process can accommodate higher peak fluxes. However, to achieve uniform air heating, a relatively uniform flux distribution across the receiver aperture is desired.

Receiver Fluid Mechanics

The particle/air suspension flowing through the receiver cavity was analyzed to understand the influence that external winds may exert on the air stream and particle distribution. Utilizing potential theory and boundary layer theory, the amount of heated air displaced by cold ambient air was calculated for pressure head effects and viscous mixing effects. Maintaining a high receiver air stream velocity (and thus momentum) gave the most protection against wind influences. Losses of heated air due to a 11.2 m/s (25 mph) wind were calculated to be less than 5 percent at the design receiver flow velocity of 40 m/s (89.5 mph).

Particle/Air Heat Absorption Modeling

The modeling of the heat absorption process was based on theoretical and experimental knowledge of the interaction of very small absorbing particles within a solar radiation field and an understanding of how they couple with the air in which they are entrained. Analysis of the particle heating process began with the development of a one-dimensional computer model that included the influences of scattering, absorption, reradiation, and particle oxidation. The model helped to identify which design parameters significantly affected the maximum air temperatures that could be achieved. In addition, it provided confidence that temperatures in the range of 1400°C could be achieved using pyrolyzed carbon particles. The model was reformulated into a two-dimensional (2-D) computer model that was representative of the final receiver geometry (Schoenung, 1987).

Using the 2-D model, sensitivity analyses were performed to determine the influence of particle size, particle oxidation rate, mass loading, solar flux intensity, air stream velocity, and other parameters affecting the maximum achievable air temperature. Particle oxidation rate has by far the most significant influence on the maximum achievable receiver outlet temperature. The results of the sensitivity analyses were used to select the optimum receiver parameters which yield the design outlet temperature. The calculated design point receiver efficiency was 90%.

Receiver Cost and Performance Estimate

The results from the previous tasks were used to finalize the receiver size and configuration, to design the heat exchange and particle generation equipment, and to select the turbine machinery. This preconceptual plant design provided the basis for developing the cost and performance estimates of the particle injection receiver concept. All the components selected for this design are either existing technology or are currently under development. The turbine machinery selection dictated the size of the plant. A 20 MWe Allison 1220 B1 intercooled, regenerated machine best fit the optimum cycle from Task 1, and two machines were selected that gave the 40 MWe plant size. The ceramic heat exchanger was based on design information received from both Solar Turbines, Inc. and Babcock & Wilcox. The calculation of the annual energy generated by the plant was performed using the computer model SOLERGY (Stoddard, 1987).

Cost of Energy

The 40 MWe particle injection receiver power plant analyzed in this study was calculated to produce almost 92,000 MWhe of electricity annually. This corresponds to an annual capacity factor of 26.2 percent without any additional fossil fuel firing during cloudy weather. The plant annual solar-to-electric conversion efficiency is 20 percent.

The near-term capital cost is $2,595 kWe, and the levelized energy cost, based on the new tax laws, is 13.9¢/kWhe in constant 1986 dollars. This is a significant improvement over estimates for current central receiver technology based upon metal-tubed, cavity receiver systems using a Rankine-cycle power plant.

The potential long-term system cost for the particle injection receiver power plant is $1,357/kWe, and the levelized energy cost is 7.5¢/kWhe in constant 1986 dollars. This is close to the DOE Solar Thermal Program long-term goal for central receiver technology, which when adjusted for the new tax laws is 6.5¢/kWhe. This projected particle injection receiver system cost was calculated by assuming the DOE long-term goal for collector system costs and by adjusting the remaining near-term system costs for the effects of learning, experience, and technology improvements generated by building successive power plants.

CONCLUSIONS

This feasibility study, shows that the particle injection receiver concept is a technically feasible approach for heating air to temperatures in the range of 1400°C at solar receiver efficiencies approaching 90 percent. This is significantly higher than that calculated for other high temperature receiver concepts evaluated to date (De Laquil, 1984). The concept couples well with an intercooled, gas turbine cycle, and a modular 40 MWe power plant was defined using current and near-term technology.

REFERENCES

De Laquil, P. and J. V. Anderson (1984). The Performance of High Temperature Central Receiver Systems. SAND 84-8233. Sandia National Laboratories.

Dellin, T.A., M. J. Fish, C. L. Yang (1981). A User's Manual for DELSO2: A Computer Code for Calculating the Optical Performance and Optimal System Design for Solar Thermal Central Receiver Plants. SAND81-8237, Sandia National Laboratories.

Hunt, A. J., and D. B. Evans (1981). The Design and Construction of a High Temperature Gas Receiver Utilizing Small Particle as the Heat Exchanger (SPHER). LBL-13755, Lawrence Berkeley Laboratory, University of California.

Hunt, A. J. (1982), Solar Radiant Heating of Small Particle Suspension. AIChE 1982 Winter Meeting Symposium Series. "Fundamentals of Solar Energy," Vol. 3.

Hunt, A. J. and C. T. Brown (1983a). Solar Testing of the Small Particle Heat Exchanger Receiver (SPHER). LBL-15807, Lawrence Berkeley Laboratory, University of California.

Hunt, A. J. and C. T. Brown (1983b). Solar Test Results of an Advanced Direct Absorption High Temperature Gas Receiver. Proceedings of the 1983 Solar World Congress, Perth, Australia.

Hunt, A. J., J. Ayer, P. Hull, R. McLaughlin, F. Miller, J. E. Noring, R. Russo, and W. Yuen (1986). Solar Radiant Heating of Gas-Particle Mixtures. FY 1984 Summary Report. Lawrence Berkeley Laboratory, University of California.

Stoddard, M. C., S. E. Faas, C. J. Chaing, and J. A. Dirks (1987). SOLERGY - A Computer Code for Calculating the Annual Energy from Central Receiver Power Plants. Sandia National Laboratories.

Schoenung, S. M., De Laquil, P. and Loyd, R. J. (1987). Particle Suspension Heat Transfer in a Solar Central Receiver. Proceedings of the ASME/JSME Solar Energy Conference, Vol. 2, Honolulu, Hawaii.

SIMULATED OPERATION OF A SOLAR POWER PLANT
A SIMULATION DESK FOR "THEMIS"

Alain Ferrière

A.F.M.E. - C.N.R.S. Centrale Themis 66120 Targasonne - France

ABSTRACT

We present a tool which makes it possible to simulate the Themis power plant exploitation in all its running modes, production as well as break-time ones. The field of use extends from the evaluation of the energy balance to the development of automatisms and operating rules.

KEYWORDS

Solar power plant ; Solar central receiver ; Power plant operation ; Model ; Control.

INTRODUCTION

The 3 years of Themis experimental operation have shown that the improvement of the production balance needs the development of efficient control laws and reliable operating rules. In order to set free from the frequent breakdowns of the installation and to use operating procedures less conservative than those actually used in Themis, we have realized a model which represents a reliable plant. This two-fold paper presents both parts of our study : the knowledge model supplied with an operating software, and its field of use.

I - MODEL AND OPERATING SOFTWARE

Our goal is to account for the behaviour of the plant, not only under nominal conditions, but in every running mode.

1. The Simulation Desk.

More concerned by proving the feasibility rather than beeing exhaustive, we have restricted our study to the main heat transfer chain, constituted by the primary molten salt loop and the secondary thermodynamical cycle. The unnegligible weight of break modes in the losses and the consumptions of the plant had led us to give as much importance to the piping as to the receiver and the steam generator. In a purposely simple way, we describe the state of the system with a minimum set of state variables, temperatures or fluid masses. The hypothesis made in line with those choices do not affect the validity of the model. Figure 1 shows the modelized system and the main variables.

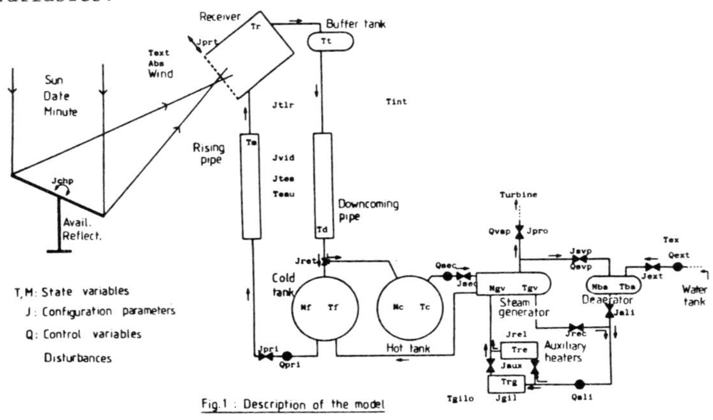

Fig.1 : Description of the model

The energy and mass balances are expressed as differential equations, whose appoximate computer resolution governs the evolutions of the state variables. The model was field proven by measurements during specific running modes, and the identification of the parameters representative of the installation.

2. The Operating Software.

Thanks to a set of configuration parameters, it selects the running mode of the loops, applying the operating rules with respect to the safety rules. The values of the control variables - fluid massflows - are set by simple laws, which reproduce as well as possible the real operation. The systematically acquired meteorological disturbances - flux, wind speed, external temperature - are read from a tape. Other disturbances - reflectivity and availability of the field - are entered in the computer by the user.

3. Validity : comparison between real and simulated results.

We present (see Fig. 2) the detailed energy balances of september 9, 1984, established from measurements first, and then from results given by the simulation. The sunlight was excellent, and therefore the quality of the control laws and of the operating rules had very little influence on the performances of the plant, whose running was satisfactory.

The gaps observed in any point do not exceed 1 %. Thus our model is very representative of the real installation, when relieved of its mechanical breakdowns. It takes a few minutes to our software to simulate a whole day running of the reliable plant.

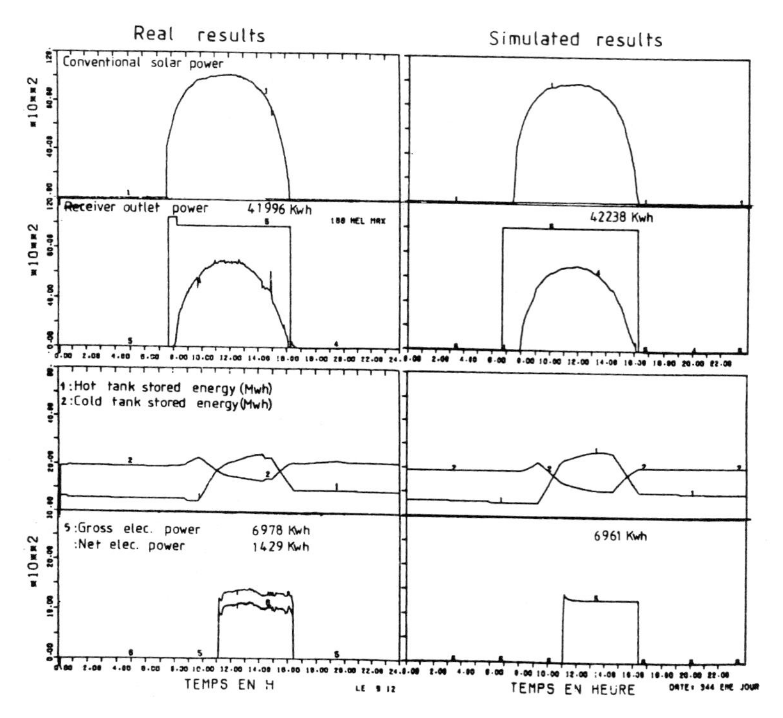

Fig. 2 : Real and simulated compared balances

II - APPLICATIONS

Here follow two illustrations of the use of our simulation desk, the first one as an evaluation tool, and the second one as a testing-bench for a control law. Due to the lack of room, we do not present other interesting uses, such as the test of new procedures, which could be hazardous for the real installation. Let us point out that the field of use of this model would get wider if it would be extended to the whole plant and if the operating software would allow to simulate accidental running modes.

1. Evaluation of the secondary loop conditionning cost.

Three strategies are considered :

- warm-keeping, by intermittent salt flows.
- cooling, and preheating with the gilotherm heater.
- cooling, and preheating with the electrical heater.

The behaviours of the secondary loop have been simulated according to the 3 strategies, during break times as far as 20 consecutive days. We have figured the detailed conditionning costs (see Fig. 3). The electrical consumptions have been multiplied by 4 to take into account the 0,25 rate of thermal/electrical energy conversion.

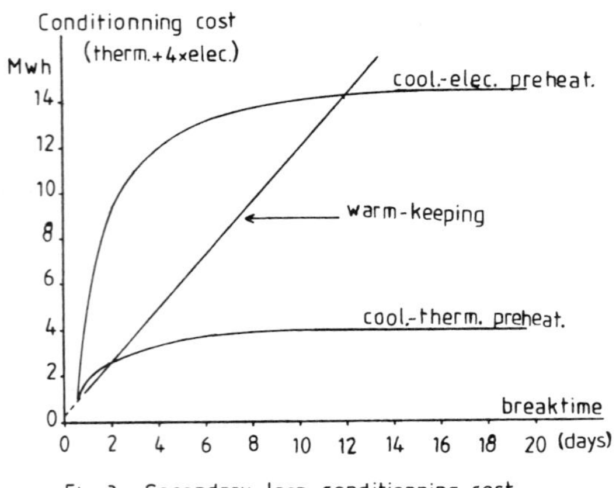

Fig. 3 : Secondary loop conditionning cost

For a break time shorter than 48 hours, warm-keeping is the most economical strategy. Beyond 48 hours, cooling-preheating becomes more interesting, on the condition that we use a thermal heater. Electrical preheating pays off for breaktimes longer than 12 days.

Remark :

A similar study has been made to compare the primary loop conditionning costs according to warm-keeping or draining strategy. The choices of the strategies selected to compute the annual productibility of the plant, are based on the results of these studies. (1)

2. Control law of the molten salt massflow in the receiver.

During the tracking periods, the outlet temperature regulation needs a control law, to govern the salt massflow. The temperature must raise to the preset value as fast as possible when the flux appears, and stick to this value during the tracking. The restrictive conditions are the following ones :

- Minimum massflow of 3.2 l/s, to insure a correct exchange between the salt and the thin pipes of the receiver.
- The preset temperature is 430 °C. A ±5 °C variation is tolerated.
- An overtaking greater than 20 °C beyond the preset value is considered as dangereous.

2.1 Control law proposed in the model : the cancellation of the dynamic term (dT_{out}/dt) in the state equation of the receiver, leads to the expression of the massflow : $Q_{salt} = f\ (T_{pre}, T_{in}, \text{Flux})$, the outlet temperature being taken equal to the preset one.

If applied instantaneously, this law leads to an excellent stability, without overtaking, but the very slow rising of the temperature is a drawback in case of irregular sunlight. The introduction of a delay, proportionnal to the difference between the outlet and the preset temperature, reduces the rising time. A delay of 1.6 sec/°C involves reasonable overtakings (see Fig. 4).

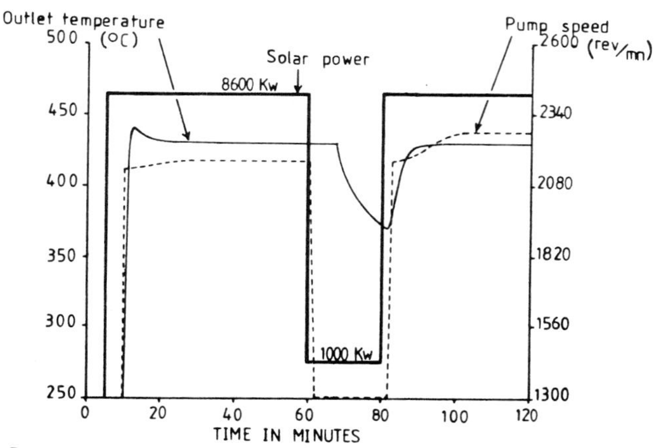

Fig.4 : Salt flow control _ Outlet temperature simulated

2.2 Real control : only few regulation tests have been conducted during the last days before the definitive close-down of Themis. No significant result has been obtained. Major problems were encountered in the consideration of the received solar power, which is not measurable in the receiver. The computing of the reflected flux would be helpful, but it needs further development.

CONCLUSION

The feasibility of a simple but reliable simulation desk is demonstrated. It is a precious tool in the evaluation and development tasks. The extension of the model to the whole plant and the consideration of accidental running modes, together with the adoption of a language derived from the expert systems for the operating rules, constitute the next step in the improvement of the operation and the automation of this kind of plants.

REFERENCES

(1) - Boutin V. (1986). Productibilité des centrales solaires à tour. Thèse de l'université de Nice.

PROSPECTIVE EVALUATION OF MOLTEN SALT SOLAR PLANTS BASED ON THEMIS RESULTS

J.J. Bezian[*], B. Bonduelle[**], B. Rivoire[**]

[*] A.F.M.E. - VALBONNE - FRANCE
[**] I.M.P. - C.N.R.S. - 66120 ODEILLO, FRANCE

ABSTRACT

An evaluation of the potential of solar central receiver plants using molten salt is given, based on Themis testing results.

KEYWORDS

Solar thermal plants ; central receivers ; Themis pilot plants ; power generation .

INTRODUCTION

For 3 years, from 1983 to june 1986, the central receiver solar plant Themis has been running and tested. Its main characteristics are the primary fluid (molten salt HITEC) and two storage tanks. It is now useful to evaluate the yearly potential production of such power plants, based on actual results rather than on design prospects.

1. THEMIS RESULTS .

First, Themis proved that electricity can be produced with central receiver solar plants. It also pointed out the advantages of the receiver and the storage system using molten salt : rapid start-up, flexible operation and a good disconnection between primary and secondary loops.

Then, as far as the power efficiency is concerned, Themis is very close to the specifications : at nominal conditions, we reached a net efficiency of 0.17 instead of 0.19 predicted.

But when considering the annual production, we can get no general and significant lessons out of the direct measurements. As an experimental prototype, Themis encountered numerous periods of breakdown due to failures and its operating conditions were far from industrial productive operations.

Table 1 gives the measured results obtained during the operation of Themis.

	1983 6 months	1984	1985	1986 6 months
GROSS PRODUCTION (MWH)	45.3	574	766	544
PARASITICS/GROSS PRODUCTION	16.4	4.7	2.85	2.08
INSOLATION TIME (HOUR)	1252	2306	2555	1084
GRID CONNECTION TIME (HR)	51	470	526	389

Table 1. Themis gross energy production measured.

It can be seen that during this period, we never stopped increasing the gross production and reducing the parasitics.
But till the end, the production was auto-consumed for different reasons :

. the shutdowns long duration (more than one year because of failures on standard equipment-transformer, alternator, turbine, pumps-and bad meteo conditions) ;

. the non optimized auxiliary consumptions, during shutdowns.

For these reasons, the measured production is not representative of the potential of molten salt plants. So we build up a method to estimate this potential in various configurations.

2. WHAT WOULD BE THEMIS PRODUCTION ?

In a first step, we want to estimate the energy production of a plant designed the same way as Themis, but operated in an optimized manner and having a good reliability : the only modifications are simple technical solutions to correct the minor defects observed. This plant, named Themis A, is defined hereunder.

2.1 Optimized operating strategy

The frequent shutdowns due to the day-night variations are the main problem when managing the primary loop. For molten salt plants, the problem is complicated by the presence of a trace-heating system used to preheat the pipes and to prevent plugs.

In short, we balance between two strategies : on the one hand loops are drained and let cooling down during the breaktime, and on the other hand loops are kept warm.

The first strategy main advantage is to minimize thermal losses. But there are many objections : it increases thermal cyclings which induce stresses and failures ; it needs an efficient tracing system which is used quite daily ; and, because it takes a long time to re-heat pipes, there is a loss of primary loop availability.

The second strategy increases thermal losses. But it reduces stresses, it simplifies the operation and gives a total availability. The auxiliary consumptions are not increased when keeping warm with intermittent flows.

One important result of Themis testing is the proof, (1), that the best strategy is the warm keeping one, which saves more energy and which is easier to operate.

As for the secondary loop, the problem is less important but the conclusion is the same.

2.2 Technical recommendations to increase reliability

Most failures occured in the standard portion of the plant (adjustment of controllers, lacks in protections of the transformer or the turbine, ...) : their analysis always give solutions which are known and they are not described here.

The other failures resulted from the solar prototype new design. They were never serious and they all had solutions, (2). We do not detail them but indicate the major points to which attention must be paid :

. collector field : lightning protection, reliability of encoders or position sensors, batteries maintenance (or design of another safety system).

. insulation and supports of the riser and of the downconner in the tower.

. thermal loops : for their design, we have to stress on the importance of the frequent thermal cycles in solar plants.

. the electric trace-heating system has to be mounted carefully.

. the long shafts of cantilever salt pumps have to be tightened in order to avoid vibrations but to allow dilatation.

. stuffing boxes of salt valves have to be put in vertical position for a good draining.

. simplification of the loops to decrease unavailability (shorten the loops, avoid as many valves as possible).

Taking care of all these points would give a high reliability.

2.3 Annual energy estimate

The annual production is estimated assuming that the losses of Themis A components vary the same way as measured in Themis. The simulation gives a net annual electric production of 787 Mwh (73 Kwh/m^2). This is the possible production of Themis, as it is designed. It is positive, but not yet interesting.

3. TOWARDS AN INDUSTRIAL PLANT .

The losses analysis and the experiments show us that the design of Themis (or Themis A) is not optimized. It incites us to bring slight modifications in the design to increase the energy balance and to simplify the plant. In this second step, we want to calculate the possible gain from these improvements.

The first improvement is to design the auxiliaries for reducing their consumptions during stoppings. As a matter of fact, solar plant designs are usually studied for nominal production states, but they are more often in overnight shut down than in production. For instance, in Themis the turbine turning gear consumes 27 kW : that seems almost negligible in nominal conditions, but as it runs more than 7.500^h/year, its consumption reaches 210 MWhe, 1/3 of Themis A production ! With a 4kW turning gear, which is sufficient, 175 MWhe would have been saved. In the same way, the salt flow during nights should be generated by small specific circulators.

Other important losses are the field reflective losses: the best and thinner mirrors must be installed.

After all, we must emphasize the loops simplification to decrease the thermal losses which occured all year long as for the parasities. In a future plant, we would eliminate the recirculating loop at the receiver, and lay the vents and drains of the receiver in its insulation. It is also possible to suppress the duplication of many auxiliaries because discontinuous operation allows their maintenance during the breaks.

From these improvements, we defined another plant, Themis B, which has the same characteristics as Themis but with a modified design. Its production becomes 1700 MWh/yr (or 158 KWh/m^2.yr).

The gain is as much noticeable as the modifications tend to simplify the design, i.e. to decrease costs. But to know the potential of molten salt plants like Themis, we need to study the influence of the plant scale and of its site.

4. EXPECTED POTENTIAL OF MOLTEN SALT PLANTS .

The plant size has a large influence in the production, because parasities increase less than the size while the group efficiency grows.

For the simulation, we consider plants which are homothetic of Themis. Their sizes are characterised by the collector field size. Results are given by the lower curve of Fig. 1. The annual electric production of a 25 MWe plant (10 x Themis), located in the same site as Themis, is 26425 MWh (245 KWh/m^2.yr).

If Themis were located under a sunnier climate, the production would of course increase. For the simulation, we use the meteo data of Barstow, where is built SOLAR-1. For a direct insolation of 2380 KWh/m^2.yr, the production of a plant of 107.600 m^2 of mirrors (10 x Themis) reacnes 35508 MWh/yr (330 KWh/m^2.yr), as showed by the upper curve of Fig. 1. The net overall efficiency is nearly 14 %.

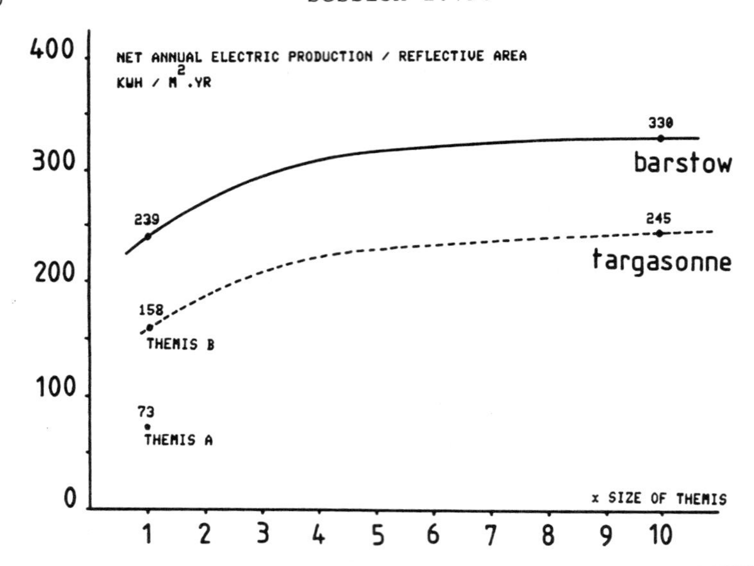

Fig. 1. POTENTIAL ESTIMATION OF MOLTEN SALT SOLAR PLANTS BASED ON THEMIS CONCEPT

5. CONCLUSIONS .

Molten salt solar plants like Themis may have an important production which leads to a cost price of nearly 2 FF per KWhe (30 cts/KWhe). This can be interesting for sunny and isolated sites.

The estimation here mentionned is based on Themis results and on some realistic and feasible modifications. Further improvements have to be investigated, which will still bring substantial gains :

. single vertical storage tank in the tower to shorten the primary loop.

. overheating steam in the thermodynamic cycle.

. use another fluid to raise the receiver outlet temperature.

. find a high-temperature thermal fluid which remains liquid at ambient temperature.

REFERENCES

(1) : "The Up-Keeping of Themis primary and secondary loops : power consumption and procedures optimization."
A. Ferrière, V. Boutin. Proceedings of the 3rd int. Workshop on solar thermal central receiver systems - juin 86 - Springer - Verlag p. 317 - 326.

(2) : "Themis solar plant technology : heliostat field and molten salt primary loop". B. Bonduelle, B. Rivoire. id. (1) - p. 307 - 316.

DISTRIBUTED COLLECTOR AND CENTRAL RECEIVER SYSTEMS ACHIEVEMENTS AT THE PLATAFORMA SOLAR DE ALMERIA

A. Sevilla and R. Carmona

Plataforma Solar de Almería, Apartado 22, Tabernas (Almería), Spain

ABSTRACT

The Plataforma Solar is a research and development center devoted to solar thermal conversion at medium and high temperatures (Fig. 1).

Since 1981, different thermal power plant prototypes have been applied and tested. Ample information regarding distributed collector and central receiver systems has been gathered. Also subcomponent improvements have been achieved on heliostats, receivers, controls, and thermal storage.

Future investigation goals are inlcuded in the fields of electricity production, chemicals and Hydrogen, technology, and diversification.

Fig. 1 Aerial view of the Plataforma Solar

KEYWORDS

Plataforma Solar de Almería; solar energy research and development.

PRESENT MANAGEMENT STATUS

The Plataforma Solar installations are managed by the Instituto de Energías Renovables of the CIEMAT, according to a Cooperation Agreement with the Deutsche Forschungs- und Versuchsanstalt für Luft- und Raumfahrt e.V. (DFVLR) of Germany, based on the principle of equally shared operation and maintenance costs by both institutions.

Control of the implementation of this solar thermal research cooperation agreement at the Plataforma Solar is vested in the Plataforma Solar Steering Committee, which approves and supervises all joint programs and projects. Projects may also be submitted by external clients to the Steering Committee, which then fixes fees and priorities.

POWER PLANT OPERATION RESULTS

Collector Field Behavior

° Normal operation - Daily average 300 W/m^2
  - 3 hours 600 W/m^2
° Operation hours - 2200 hours
° Reflection and concentration efficiency - 60-70% (at normal operation)
° Yearly maintenance relative to investment - 2.5%

Collector Field Economics

° Cost - 500-800 DM/m^2
° Fuel compared cost - 40-70 DM/MWh
° Non-renewable equivalent cost -

	Fuel cost DM/MWh	Additional cost environment DM/MWh
- heavy oil	55	?
- light oil	65	?
- gas	50	?
- coal	25-35	5-10
- uranium	5-10	5-30

Energy Conversion (Receivers)

° Availability and operational reliability - close to 100%
° Performance at design level - 90%

Energy Transport and Handling

° Solar specific components and subsystems met almost all expectations.
° The complete power plant system had operability problems under transient conditions.
° Conventional subsystems were not well matched with the rest of the plant, with too high an internal consumption.
° Clever energy management during operation must be established.

Generation of Electricity

System	Conversion efficiency Beam radiation/electricity Daily	Yearly
DCS	2.5-9	5-6
CRS	8- 20	10-15

° Advanced system design would potentially guarantee competitiveness of solar generated kWh's for CRS plants with conventional power plants.

SUBCOMPONENT IMPROVEMENTS

Heliostats

° Improve mirror construction to avoid corrosion of helisotat facets.

° Modify heliostat shape to reduce blocking and shading with the same land use factor.

° Improve local heliostat control.

° Unite helisotat foundation and pedestal, reducing installation cost.

° Level the heliostat field using software.

° Modify heliostat accuracy, according to an economic point of view, dictated by the influence of receiver losses on the overall plant performance.

° Reduce heliostat cost of existing models by a factor of 2, on a 12 unit series.

Control

A new adaptive control has been designed and implemented on a conventional microcomputer. This control operates the collector field automatically, adjusting the pump flowrate to maintain the field outlet temperature constant. No manpower is necessary for normal operation, and due to the accuracy and component reliability of this control, the collector field operates under strict industrial constraints. The same microcomputer houses the data acquisition system. All these features support a new control philosophy in a modern, flexible, economic approach enabling automatic operation of any DCS.

The structure used in the adaptive control corresponds to a self-tuning regulator (STR), used to calculate the regulator's parameters, assuming that plant parameters are defined and permanent, and substituting them with estimates through an identification algorithm.

An important part of the self-tuning controllers is the parameter identification--this should be recursive, as it must work in real time with the process.

Receivers

The basic receiver technologies tested have been water-steam, sodium, and air.

Two sodium receivers have been successfully operated. The so-called Advanced Sodium Receiver was operated at a flux density of 2.7 MW/m^2. Hence, smaller and lighter receivers could be constructed.

Three different air receivers have been tested. The metallic receiver, a tube concept, was operated at 800°C and 10 bar with complete success.

In 1987, a similar concept of ceramic material has started its testing phase, where an operating temperature of 1000°C and 10 bar is foreseen.

A completely different air concept has also been installed and tested during 1987--the volumetric receiver. This receiver heats air as a volume and not inside the tubes. This concept promises to lead into very cheap receivers.

Basic operation strategy has been gathered concerning all types of receivers and future improvements are foreseen.

CONCLUSIONS

The Plataforma Solar de Almería has been operating since 1981. Its facilities include diverse projects pertaining to both central receiver system and distributed collector system technologies.

Ample experience and knowledge has been obtained, predominantly in the field of electricity production. The feasibility of using solar radiation to generate electrical power was demonstrated. At present, the behavior of conventional components working under solar conditions is studied, and components of new design are tested before proceeding with commercial applications.

Since the suitability and economic viability of solar plants for electricity generation is questioned because of fuel cost evolution, research is now made towards developing diverse industrial heat process applications at medium and high temperatures.

Varied storage and control systems have been tested and studied to increase the solar energy availability.

In addition, the Plataforma Solar de Almería is a research center with human and technical support, enabling it to work with the latest solar technologies, e.g., chemical production, synthetic fuels, and high temperature studies.

REFERENCES

Becker, M., Ellgering, H., Stahl, D. (1983). Construction experience report for the central receiver system (CRS). DFVLR, Cologne, SSPS-SR2.
Bücher, W. (1984). SSPS-CRS first period of operation. Preliminary operation results, experiences and events. DFVLR, Cologne, SSPS-SR4.
Hansen, J. (1984). SSPS-CRS advanced sodium receiver. Construction experience report. DFVLR, Cologne, SSPS-SR5.

Hansen, J. (1984). SSPS-DCS supplement. Construction experience report. DFVLR, Cologne, SSPS-SR6.
Grasse, W. (1985). SSPS - results of test and opration 1981-1984. DFVLR, Cologne, SSPS-SR7.
Kalt, A., Loosme, M. and Dehne, H. (1982). Distributed collector system plant (DCS) construction report. DFVLR, Cologne, SSPS-SR1.
Kalt, A. (1983). SSPS-DCS first year of operation. Verification with design expectations. DFVLR, Cologne, SSPS-SR3.

RESULTS OF THE CESA-1 MOLTEN SALTS STORAGE SYSTEM EVALUATION

F. Rosa and J.M. Andújar

CIEMAT-Instituto de Energías Renovables,
Plataforma Solar de Almería, Apartado 22, Tabernas (Almería), Spain.

ABSTRACT

The CESA-1 molten salts storage system has been evaluated since November 1986. This evaluation provides useful experimental data for system modelling and characterization by determining: charge/discharge efficiencies, and hot/cold tank thermal losses.

It was performed systemmatically, measuring inlet/outlet temperatures of steam/water/salt in the charge/discharge circuit, using thermocouples, steam/water inlet/outlet pressure (pressure transmissor), steam/water and salt flow (venturi-type flowmeters), and hot/cold salt temperature variation with time. Thirteen charge/discharge and thermal loss tests were performed, with the following results:

- After three years of operation, storage salt may degrade, yet it maintains storage capabilities.
- Charge/discharge efficiencies of the heat exchangers appear to be reasonably good (90% for the charge system and 91% for the discharge system), and an overall 24 h round-trip efficiency of 72%.
- Thermal losses of the storage tanks are low: respective values of the overall heat transfer coefficient for the cold and hot tanks are U_c = 0.265 and U_h = 0.327 W/m^2°C, respectively.
- No major problems or outages have been detected.

Conclusions drawn from this evaluation are that the molten salts thermal storage system is efficient, practical, and reliable (Andújar, 1987).

KEYWORDS

Molten salts, storage, charge/discharge efficiencies, thermal losses.

INTRODUCTION

The CESA-1 thermal storage facility was designed for 3 h/d plant operation, with an output of 875 kW$_e$ (Muñoz, 1983) in discharge mode. To ensure this output, 4165 kW_{th} must enter the steam generator where steam is produced at 330°C and 16 bar. For this, the system is charged by steam at 520°C and 100 kg/cm^2 pressure.

A total of 260 Tn storage fluid is contained in the facility, with a C_p=1.3039 + 0.6066 x 10^{-3}T kJ/kg°C (T in °C) and a maximum storage capacity of 12772.5 kWh_t or 2685 kWh_e. Each tank has 200 m^3 capacity.

This system, located outside the machine hall between levels 0.2 and 14 m, consists in 2 storage tanks (hot and cold), 3 chemical pumps to drive the salt from one tank to another, 2 water pumps (at the charge and discharge circuit), and 6 heat exchangers (3 in each circuit).

The thermal storage fluid is an euthetic mixture of sodium nitrite ($NaNO_2$) 40%, sodium nitrate ($NaNO_3$) 7%, and potassium nitrate (KNO_3) 53% (design composition).

THERMAL STORAGE SYSTEM EVALUATION

Tests will be referred to in the following as DDMMYY, i.e., the test performed on 13/12/86 becomes 131286.

Charge Tests

In these tests, the salt flowrate was varied to obtain data regarding system efficiency under different operational conditions. Table 1 summarizes results.

Results in Table 1 show that, within the flowrates investigated, overall instantaneous efficiency (η_c) of the system does not depend on the salt/steam flow. Higher efficiencies obtained at lower flowrates can be explained due to the fact that the mean test temperature in the charge system is the lowest obtained. Mean overall efficiency of this subsystem is acceptable: 90.2%.

TABLE 1 Efficiencies and Effectiveness of the Charge System

Test DDMMYY	Salt flow (m^3/h)	Steam flow (kg/h)	Instantaneous efficiency (%)	Effectiveness (%)
050687	42	4330	89.14	82.60
140687	35	4240	88.86	81.03
130687	34	4330	91.44	79.23
120687	35	3410	91.54	85.29

Discharge Tests

In these tests, salt flowrates did not meet nominal outputs because the subsystem was operated without the turbine venting the steam produced to the ambient, and not enough steam sucking was obtained. Test 130687 was performed to check the behavior of the secondary evaporator. Table 2 summarizes results of these tests.

TABLE 2 Efficiencies and Effectiveness of the Discharge System

Test DDMMYY	Salt flow (m^3/h)	Steam flow (kg/h)	Instantaneous efficiency (%)	Effectiveness (%)
070587	28.5	1080	89.69	71.95
130687	34.5	1820	89.78	70.31
230687	35.5	1960	92.44	79.29

Cold and Hot Tank Losses

Table 3 presents operational conditions and results of cooling tests.

TABLE 3 Cooling Test Operational Conditions and Results

Test DDMMYY	Duration(h)	Salt mass (kg)	Init.T (°C)	Final T(°C)	P_s[1] (kW)	P_m (kW)	P_i (kW)	P_t[2] (kW)	R (°C/W)	Heat losses (kW)[3]
				Cold tank						
131286	35	231669	201.4	196.4	11.8	0.4	0.2	12.3	0.012	11.55
211286	312	235464	203.2	174.0	8.7	0.3	0.6	9.1	0.017	10.60
131286	190	234072	193.3	176.6	8.1	0.3	0.4	8.5	0.019	8.73
				Hot tank						
150687	47.5	241300	293.1	283.8	18.9	0.5	0.3	19.7	0.013	20.9
170687	71.5	241300	283.7	272.4	15.3	0.5	0.3	16.0	0.016	16.5
200687	47.5	241300	272.1	264.6	16.5	0.5	0.3	17.3	0.014	18.1

Heat losses of the thermal storage tanks were determined measuring the cooling rates dT/dt. Assuming an identical cooling rate for the salt vessel and insulation, total power lost by the tanks is:

$$P_t = (M_s C_{ps} + M_m C_{pm} + M_i C_{pi})\ dT/dt \qquad (1)$$

where M_m, M_s, and M_i are tank metal, insulation, and salt masses; and C_{ps}, C_{pm}, and C_{pi} are specific heats of salt, metal, and insulation. Cooling rates were used in Equation (1) to evaluate material, insulation, salt, and total tank heat losses, and are summarized in Table 3. Note that both the vessel and insulation contribute equally ($\leq 3\%$) to the total heat loss.

The tank temperature time dependence can be expressed as:

$$T = T_a + (T_i = T_a)\ \exp\ (-t/RC) \qquad (2)$$

where T_a is the ambient temperature; T_i, the initial tank temperature; t, the time; R, the thermal resistance; and $C=M_s C_{ps}+M_m C_{pm}+M_i C_{pi}$. Values of R are sumarized in Table 3. Total heat power lost by the tank is then:

$$P_t = (T-T_a)/R \qquad (3)$$

[1] P_s, P_m, P_i, P_t refer to power lost by salt, metal, insulation, and total.
[2] Determined through application of Equation (1).
[3] Deterrined through application of Equation (3).

P_t values calculated applying Equation (3) are presented in Table 3. A comparison of the values in Table 3 shows that the heat loss derived from experimental results through Equations (1) and (3) obtain similar values. Average values of the tank thermal resistance are: R_h = 0.014°C/W (where R_h is the thermal resistance of the hot tank); and R_c = 0.017 °C/W (where R_c is the thermal resistance of the cold tank).

Considering that both tanks are identical with an external sruface of 213 m^2, values the overall heat transfer coefficient, U, are 0.327 W/m^2°C and 0.265 W/m^2°C, respectively for the hot and cold tanks. These values, together with Equation (3), will serve to evaluate the tank heat losses at design temperature (T=340°C).

THERMAL STORAGE SYSTEM EFFICIENCY

Overall system efficiency can be expressed as:

$$\eta = \eta_c \eta_d - (\eta_d C_1(T_{i1} - T_a)(1 - e^{-t/R_h C_1}) + C_2(T_{i2} - T_a)(1 - e^{-t/R_d C_2}) + E)/Q_i \quad (4)$$

where C_1 and C_2 are the corresponding values of C for the hot and cold tank; T_{i1} and T_{i2}, the initial hot and cold tank temperatures; η_c and η_d, efficiencies of the charge and discharge circuits; Q_i, the heat input; and E is the equivalent thermal energy required to drive the system.

The following assumptions are made to calculate : (1) thermal storage is charged and discharged daily at full capacity; (2) T_i of the hot and cold tank are 340°C and 220°C, respectively; (3) efficiency of the charge and discharge systems are 90.2% and 91.1%, respectively; and (4) equivalent thermal energy for salt pumping is 809.5 kWh_t (Andujar, 1987). Under these conditions, thermal storage efficiency is calculated applying Equation (3) as η = 72%.

Figure 2 presents main losses and overall efficiency of the thermal storage system.

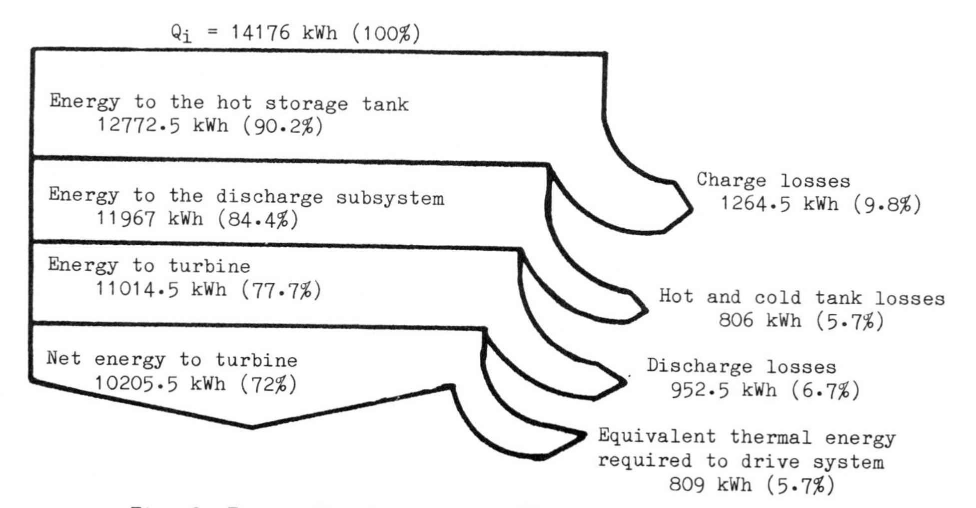

Fig. 2 Energy flowchart of the CESA-1 thermal storage system

MAINTENANCE EXPERIENCE AND CORROSION

After pertinent modifications were made of the thermal storage system (August 1987), general behavior of the system has been very satisfactory. Very few maintenance interventions were necessary to maintain the loop operable (Andújar, 1987).

With regard to corrosion, the type of steel sheet used in the salt equipment of the plant appears appropriate: during the last three years, no evidence of any serious corrosion have been reported by CESA-1 operation and maintenance teams.

CONCLUSIONS

On the basis of this evaluation, it can be concluded that the molten salt thermal storage concept has proven an efficient, practical, and relatively reliable solution.

This conclusion takes into account the following:

° A hot tank storage temperature of 300°C was tested with no problems. Higher temperatures (450°C) are advisable (Amri, 1985) if an arrangement other than the CESA-1 charging system's heat exchangers is foreseen.

° Heat losses of the thermal storage tanks, heat exchangers, and piping are reasonably low leading to 72% 24 hour round-trip overall efficiency. This efficiency refers to a conplete daily charge/discharge mode. It must be noted that tests were performed operating the system at less than nominal conditions of steam/salt flows where the relative losses are higher and efficiency is subsequently lower than the maximum obtainable.

° The thermal storage concept has been designed to decouple completely between sunlight and electricity production. This results in high adaptability to a large variety of meteorological and electricity production conditions.

° During a year's operation, no storage failure was detected and few maintenance problems reported. No evidence of serious corrosion problems or salt pump aging were recorded.

REFERENCES

Andújar, J.M., and Rosa, F. (1987). CESA-1 thermal storage evaluation. Plataforma Solar Task IV proceedings.

Amri, A., Izygon, M., Tedjiza, B., and Etievant, C. (1985). Themis thermal storage subsystem evaluation. Sandia National Laboratories.

Muñoz Torralbo, A., Hernández González, C., Avellaner Lacal, J., Ortíz Rosas, C., Sánchez Sudón, F., Caso Neira, B., and Navarro Asenjo, L. (1983). Descripción general de la Central Electroslar de Almería, CESA-1. Ministerio de Industria y Energía, Centro de Estudios de la Energía.

RESULTS OF THE TECHNOLOGY PROGRAM "GAST"

P. Wehowsky
Interatom GmbH
P.O. Box, D-5060 Bergisch Gladbach 1
Federal Republic of Germany

ABSTRACT

Within the frame of the bilateral German-Spanish Technology Program GAST (1981-1987) solar-specific design methods/computer codes for techno-economical system optimization for GAS-cooled Solar Tower plants have been developed, system analyses performed and backed-up by construction and testing of relevant plant components in full scale like cavity type solar receiver modules/hot gas piping as well as suitable high concentrating heliostats incl. field control systems. These components have been tested at the Spanish Solar Platform near Almeria utilizing the existing tower plants SSPS-CRS and CESA-1 as test facilities. A survey of the results of this experimental as well as the GAST system work is presented.

KEYWORDS

Solar thermal tower plant, GAST, high temperature gas/air-cooling, central cavity receiver, high accurate heliostat field, full scale component tests, Plataforma Solar de Almeria, system analyses and optimization, computer codes, power/process heat co-generation, techno-economical investigations

INTRODUCTION

In October 1978 the GAS-cooled Solar Tower project GAST has been started in Germany. The German Federal Minister for Research and Technology (BMFT) is since that time promoting this program: design and development of GAST system concepts as well as construction and testing of its solar-specific components. Since summer 1981 the project has been continued as a Technology Program in bilateral German-Spanish co-operation by Interatom and the Spanish research association Asinel in Madrid, under further participation of the German companies MAN-Technologie, MBB and Dornier System as well as of Spanish subcontractors (Casa, Initec, etc.).

A first 20 MWe GAST power plant reference concept served as a basis for the Technology Program work - consisting of a mirror field of appr. 2000 heliostats of 52/55 m² mirror surface. Air has been selected as heat transfer medium which is heated within two cavity receivers on top of an about

170 m high tower to more than 800 °C for high-efficient Brayton cycle demonstration. An open KWU gas turbine cycle with a bottoming steam turbine circuit is being provided for optimized energy conversion. In place of a passive energy storage system, this plant concept contains a supplementary fossil-fuel fired heating system allowing electric power and/or process heat generation even during periods with insufficient or no incident solar radiation as well as plant operation in grid connection or within an insulated net-work.

Up to the end of 1987 solar-specific GAST components will be developed, qualified and tested under realistic solar conditions using the existing solar tower plants at the "Plataforma Solar de Almeria" - the Spanish CESA-1 and the IEA/SSPS-CRS - as test facilities. In parallel system analyses have been performed investigating the best possible applications of the GAST system. Thus, all key questions concerning GAST system design should be clarified and the essential soft- and hardware technologies developed and available for a possibly following system demonstration phase. More details about the GAST project organization, the 20 MWe pilot plant concept, and the intermediate results of the Technology Program are reported e.g. in (1,2).

TECHNOLOGY PROGRAM "GAST": OBJECTIVES, RESULTS

At present the major experimental phase with the qualification/testing of the metallic and ceramic tube heat exchanger receiver panels followed by a hot gas piping test rig under 800 resp. 1000 °C panel air outlet temperature as well as two small heliostat test fields in Almeria has nearly and successfully finished. With respect to the program organization the activities are structured as follows:

Heliostat/Heliostat Field

The aim has been development of cost-effective heliostats ready for mass-production, suitable for slant ranges up to 600 m and solar concentration factors >1000 in large fields including the construction of German and Spanish heliostat prototypes and small test-series fields as well as their qualification and testing in Almeria in cooperation with specially developed heliostat field control systems (2).

For the relatively small aperture size of the GAST cavity receiver (appr. 30 m^2) an admissible deviation of the reflected beam of 2 mrad was originally required, later on extended to 3 and 4 mrad specified total error budget (1σ-value/RMS, beam quality and pointing/tracking accuracy). On German side MBB developed and constructed two 52 m^2 heliostat protoypes for 2 resp. 3 mrad, the second one under reduced requirements regarding acceptable wind velocities for survival and operation limits. The Almeria test results showed full accordance with the requirements by measured total errors of appr. 2 mrad for both units; for the second one cost and weight reductions of appr. 30 % were realized.

Based on the second prototype, the components of the 30 German test-series heliostats were manufactured by German and Spanish companies in 1983/84 and afterwards assembled behind the original SSPS-CRS field. The follow-up test program concerning single heliostat, group and complete field tests showed again excellent results: measured total errors of 2.2 to 2.6 mrad. For the total field performance tests the CRS sodium cooled receiver was used as a "cooled target". The test program is being completed by permanent life cycle tests of 2 heliostats as well as field stand-by tracking showing a high heliostat system availability and good component condition (e.g. tracking electronic, gear boxes, double glass facets).

Following a heliostat field strategy to arrange more cost-effective heliostats of lower image quality nearest to the tower, a Spanish prototype of 4 mrad total error budget and 55 m² mirror surface was specified, manufactured and at least succesfully tested by Asinel and Casa: measured total error < 3 mrad. Taking advantage of the error budget reserve, Asinel consequently has enlarged the mirror surface to 65 m² and 12 test-series heliostats assembled behind the SSPS-CRS field, too. The tests are on-going, however first measuring results show again total errors of appr. 3 mrad. (Regarding "Analysis of GAST Heliostat Image Measurements" compare with Poster/Paper No. 2.9.10).

For both, the German and Spanish test fields different field control systems have been developed, installed and tested by Dornier System/MBB and Asinel, able to simulate large heliostat field handling. The Spanish one avoids individual heliostat controllers; up to 32 heliostats will be controlled by group controllers which are largely independent from the central field computer.

Having started in 1981 with a total heliostat cost figure of 1,250.-- DM/m² mirror surface, the 52/55 m² heliostat types of the two different error budget levels were offered in 1985 for 520.-- DM/m² (Asinel) to 620.-- DM/m² (MBB) if 10,000 heliostats p. a. will be manufactured (1984 cost, heliostat ready for operation). After supply and assembly of 65 units of the 55 m² heliostat to the Israel Weizman Institute during 1986/87, Asinel today offers cost of 620,-- DM/m² (65 m² units, 1987 cost basis).

GAST Hot Gas Component Test Facility Almeria

In order to test the developed full scale GAST components "receiver panel/ hot gas piping" in Almeria, a complete closed-loop gas supply system was designed and placed by Interatom/Asinel at the free upper part of the CESA-1 tower using its heliostat field for producing the required radiation flux profiles at the receiver test panels for the following max. test conditions: 1000 °C, 10 bar, 3.2 kg/s for air as heat transfer medium. The necessary gas cycle components incl. the instrumentation and control as well as the electrical systems were mainly supplied by Spanish companies (1,2). Due to the extensive instrumentation of the a. m. test components a special data acquisition system was developed and supplied by Dornier System.

Receiver

The aim has been development of a gas-cooled solar cavity receiver for air outlet temperatures of min. 800 °C including tests of heat exchanger tube panels in real scale in Almeria using metallic tubes for test temperatures of 800 °C (MAN) resp. ceramic tubes for 1000 °C (Dornier System).

With respect to this new high-temperature receiver application, extensive thermo-dynamic and stress analysis work was performed using specially developed computer codes, and in addition enlarged metallic and ceramic material and component laboratory tests in newly erected test facilities were realized in these companies. Furthermore, receiver cavity materials, panel supporting and insulating structures were tested in Asinel's test facilities having used a high-concentrating Lajet parabolic dish.

From March 85 until June 86 the original sized high-temperature metallic tube receiver test panel (1.5 m width, 8 m irradiated tube length, 18 tubes of 42/37.8 mm ø each) was succesfully tested in Almeria as an "external receiver" (although designed for cavity use) in solar operation mode under the design conditions: 625/800 °C in-/outlet temperature, 9.3 bar system pressure, 2.5 kg/s mass flow. (Details compare with (1,2) and esp. with Poster/Paper No. 2.9.13 for test results.)

Because of severe difficulties regarding the soldering technique of the SiSiC ceramic tube material to realize the originally planned 6 m irradiated tube length (panel width 1.3 m, 10 tubes of 42/32 mm ø each), the project partners Interatom, MAN and Asinel decided in 1985/86 to qualify and test a interim solution of appr. 4.5 m tube length with a mechanical thread connection of the already existing ceramic tube material using a ceramic sleeve. This panel has been under test in Almeria since August of this year by gradual increase of the air outlet temperature - and the design value of 1000 °C could be obtained and kept first of all for 1.5 hours coinciding with the "ISES Solar World Congress, Hamburg 1987" as reported there. The complete test program includes solar mode operation examples and will be finished by end of October this year.

Hot Gas Piping (HGP)

For air temperatures of 800 °C the development and test of hot gas pipe components were performed at Interatom and the nuclear research center KFA Jülich, using their experience and test facilities from the high-temp. nuclear reactor field (HTR). All essential HGP components were installed as a test rig on top of the CESA-1 tower - being part of the a.m. gas supply system (1,2) - and succesfully tested in connection with the receiver test panels. For the 1000 °C air outlet temp. of the ceramic panel a water cooler was added to meet the at the time being allowed 800 °C HGP inlet temperature.

System Engineering and Optimization

In addition to the experimental part of the GAST Technology Program, sub-system/system analyses and plant optimization work were performed for definition, performance and evaluation of the tests incl. development/use of special computer codes (1), also for investigations regarding the best technical and economical applications of the GAST system. Examples of computer codes:

- **HFLCAL** optimizes the system configuration of heliostat field, tower and receiver for the nominal design conditions following the criterion "maximum produced thermal/electric annual energy per heliostat" and considering the part load characteristics of receiver and gas/steam turbine cycles (no consideration of cost data).

- **GASBIE** uses HFLCAL results, simulates and analyses the GAST systems' and subsystems' operational behaviour, and calculates the plant's energy production for site realistic solar or solar/fossil fuel buffered operation without and with consideration of clouding, taking into account real measured and statistically evaluated annual meteorological site conditions.

  Example for the **20 MWe GAST** Reference System (1980 heliostats, 200 m tower, real weather, without fuel support) for sites Almeria and Barstow:

MWhrs/y \ site	thermal energy to field	receiver	cycle	el.en. outp. by gas cycle	steam cycle	para-sitics	net el. energy output
Almeria	200,053[1)]	112,667	92,257	15,622	9,558	5,405	19,775
Barstow	270,820[2)]	155,024	126,912	22,087	13,451	5,975	29,563

1) $\hat{=}$ 62.4 %, 2) $\hat{=}$ 85.8 % of the undisturbed site insolation

Influence of clouding shown by comparison of ratio "**B**arstow/**A**lmeria" for

$$\text{Insolation } \frac{B}{A} = 1.35; \text{ Energy output } \frac{B}{A} = 1.50$$

-**ASPOC** serves for system site and cost related optimization using GASBIE formalisms as well as all plant investment and operation cost data - that means modifying the until now physically based system/subsystem configuration by weighting it with subsystem/component cost figures - and delivers at least energy production cost: thus, two 20 MWe GAST plants - either optimized for the concerned design point - show cost of 31 ¢/kWhr for site Almeria and 19 ¢/kWhr for site Barstow (real weather, without fuel support, 1984/85 cost data, 2.80 DM/$).

-**DYNAG** simulates the behaviour of the GAST plant gas-turbine cycle for analyzing its non-steady state operation modes e.g. normal and emergency shut-down, cold or warm start-up/ restart, short-/long-term clouding without/with fossil fuel firing.

Within an extensive "Analysis of the Potential of the GAST System" (3) the GAST power plant concept was compared with other solar tower systems/heat transfer media (sodium, salt); the economical results have been found to be similar: 11 to 13.5 ¢/kWhr for 100 MWe plants of approx. 40 % capacity factor.

Beside pure electrical power generation, alternative advanced GAST plant concepts were studied within (3), utilizing the GAST potential for industrial process heat production in power/heat co-generation mode as well as for high-temperature heat applications e. g. for "solar fuels and chemicals" like Methane steam reforming, Hydrogen production, etc. (see also Poster/Paper 2.19.12).

-For example, a **50 MWe GAST-SOLFOS** solar-fossil hybrid fired plant (fuel-saver concept) supplies electricity of 53 MW from the gas-turbogenerator set for a fixed price of 5.5 ¢/kWhr and steam of two conditions (454 °C/ 74 bar, 144 °C/4 bar produced from the gas-turbine waste heat) for a mixed cost of 12.5 ¢/ton to compare with conventionally produced steam of 20.5 ¢/ton (all cost data 1984/85, 2.80 DM/$, real present value method for 1990-2010, 80 % plant capacity factor).

Due to the importance and availability of necessary real meteorological site data Interatom implemented and installed 1982 a well equipped GAST Meteo Station in Almeria, and measured and analyzed the till 1986 collected data using the related developed evaluation software (compare Poster/Paper 2.19.11). This station and/or further available Small Environmental Data Acquisition Systems (8 to 16 signals, incl. special software) are movable as well as site- and, the last one, grid-independently operable (power supply by a Pv-System).

CONCLUSION

After successful termination of the Technology Program by the end of 1987 the continuation of the GAST project is at the time being somewhat uncertain. However, within the frame of the Spanish-German 5-years contract of CIEMAT/IER and DFVLR regarding their common operation and maintenance of the Solar Test Center Almeria, there exist advanced plans to extend the GAST test facility at CESA-1 able to qualify and test a small ceramic tube cavity receiver (2-3 MWt, 1100-1200 °C) as well as e.g. volumetric receiver-, hot gas storage- and methane-steam reforming modules, all that in collaboration with related SSPS tasks. GAST results, the component, subsystem and system analysis experience and the design codes are finally used by those GAST firms engaged in the running European 30 MWe PHOEBUS tower feasibility study.

REFERENCES

1. P. Wehowsky/D. Stahl/J. de Marcos/L. Crespo, The Gas-cooled Solar Tower Project GAST; Proceedings of 2nd Int. Workshop on Design, Construction and Operation of Solar Central Receiver Projects, June 1984, Varese/Italy

2. P. Wehowsky/J. de Marcos, GAST-Status of Work: Tests in Almeria, System Analyses - and Future Prospects; Proceedings of 3rd Intern. Workshop...(see (1)) June 1986, Konstanz/F.R. Germany

3. Technologieprogramm "Gasgekühltes Sonnenturm-Kraftwerk" (GAST) - Analyse des Potentials, BMFT-Forschungsbericht T 86 - 087, ed. by M. Becker and others, 1986

TRANSIENT MODELS FOR THE CESA-I SIMULATION PROGRAM

M. Castro *, J. Cabañas **, J. Peire ** and P. Martínez **.

* Electronic and Control Engineering Department
Universidad Nacional de Educación a Distancia
Ciudad Universitaria, s/n. 28040 Madrid - SPAIN
Phone number : 34-1-2439127
** Electronic Engineering Department / Universidad Politécnica
c/ José Gutierrez Abascal, 2. 28006 Madrid - SPAIN
Phone number : 34-1-4117517

ABSTRACT

In this paper the inclusion of several models to the CESA-I Simulation Program is described.

The receiver model allows to simulate the transient response of the Solar Receiver, least-squares adjusted to the data of the operation done at the CESA-I. The storage charging and discharging models, include in only two models the five existent heat exchangers, with piping and associated losses.

Inclusion of these three new models, in addition of the Receiver efficiency exponential model adjusted to the obtained operating values, allows to extend the simulation program to the plant start-up and stop performances and the cloud transients.

KEYWORDS

Solar energy; central receiver plant; mathematic model; computer simulation; receiver transient response.

CESA-I SIMULATION PROGRAM DESCRIPTION

The CESA-I Project (Central Electrosolar de Almería) is a 1.2 MWe Central Solar Receiver Plant, located in " La Plataforma Solar de Almería " near Tabernas in the southeast of Spain, (CEE, 1983; Martínez, 1983; Castro, 1987). This plant has a 300 heliostat field with a total of 12,000 m2 total reflectance area, focusing beam radiation onto a cavity receiver with an aperture size of 11.5 m2.

The power conversion system used is a water/steam Rankine regenerating cycle, obtaining a gross electrical power output of 1.2 MWe at design conditions. The thermal storage system is made up of two 18 MWth tanks being the working fluid molten salt. Tanks" storage temperatures are 220 and 340 C.

According to which plant operating strategy (operation mode) is in use, steam at the receiver outlet can be sent to the main turbine inlet, to the heat storage system, or to both of them, as it can be seen in the figure 1, where is displayed the general block diagram of the central simulation. Operating conditions define the plant operation modes. The operation mode is selected according to weather conditions, hot storage tank level and plant operating strategies.

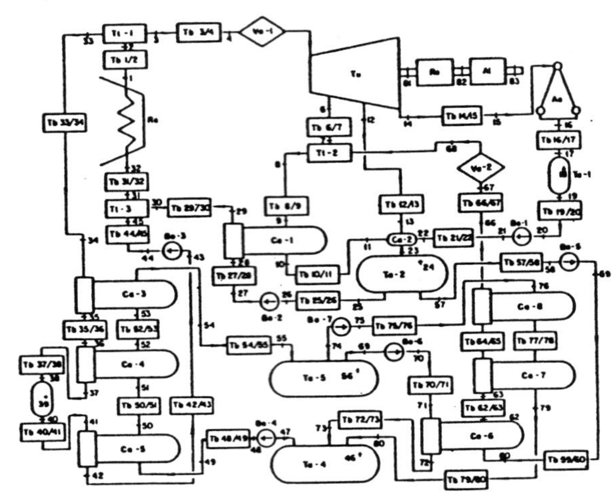

Fig. 1. CESA-I simulation block diagram.

The CESA-I simulation program allows to simulate the steady-state response of all the plant components, and includes the heliostat field, receiver, hot and cold storage vessels, charging heat exchangers, discharging boiling and superheater heat exchangers, turbine, reducer, alternator, aerocondenser, deaerator, pumps, switchs and interconnecting piping, (fig. 1). Continuous data for a typical meteorological year for Almeria in a cloudless-sky are used with any time increment, (Castro, 1985).

RECEIVER MODEL

Firstly, the receiver model has been modified in its efficiency equation, which has been fitted to the CESA-I receiver data operation, (IDAE, 1984). The curve obtained is an exponential approximation as

$$Nrec\ (o/oo) = 0.91 - 1.413 \times e^{-0.00085884 . Win(kW)} \qquad (1)$$

where Nrec is the receiver efficiency and Win the thermal input power, having a mean square error of 8.72 %, as it can be shown in the figure 2.

The other receiver topic modeled is the start-up and stop performances. These effects have a high influence in the global efficiency of the plant and its operation.

The start-up process in the cavity type receivers (as in the CESA-I one) is slower than in the open ones, but they have the possibility to close the external doors and to bottle the steam in the intermediate boiler, allowing a faster start-up sequence having higher initial conditions of pressure and temperature. The sequence of operation after an insolation transient (as in a cloud passing) or after the night stop is to close the doors and bottle the receiver. After this moment, the receiver does not have fluid flowing through it, and the pressure and temperature are decreasing with the time. These values have been least-squares fitted to the operation values of the receiver, (IDAE, 1984; CESA, 1985), (fig. 3).

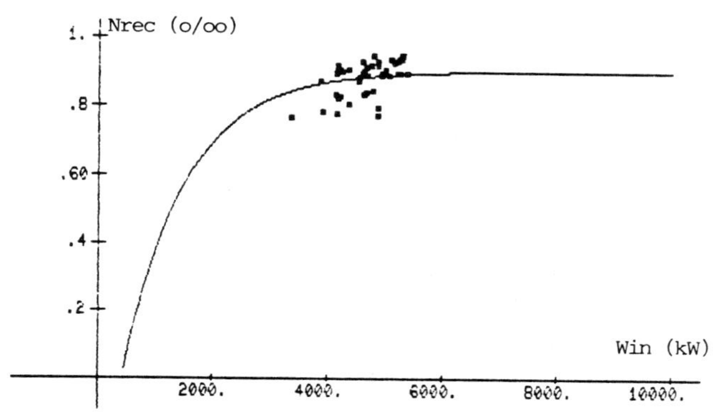

Fig. 2. CESA-I receiver efficiency.

$$P/Pref = 102.95 - 5.41.t + 0.109.t^2 - 0.00096.t^3 + 0.00000295.t^4 \quad (2)$$

$$T/Tref = 314.61 - 5.01.t - 0.000161.t^2 + 0.046.t^3 \quad (3)$$

being Pref = 110 bar abs and Tref = 550 C.

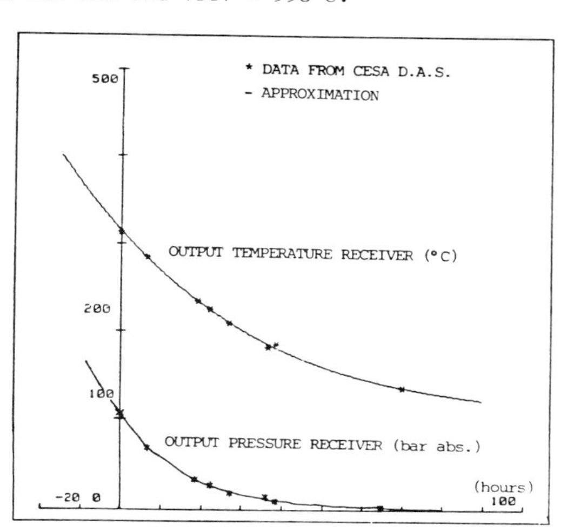

Fig. 3. Receiver output pressure and temperature fitted in stop plant operation.

When the bottling time is over begins the start-up process, which have been modeled from the receiver data operation. The start-up equations have been least-squares fitted to these curves, (CESA, 1985), assuming that the time needed to obtain staedy-state conditions only depends on the pressure and temperature after the bottling time.

$$P/Pref = 1.467 + 13.867.t - 3.467.t^2 + 24.133.t^3 \quad (4)$$

$$T/Tref = 59.53 + 614.229.t - 285.959.t^2 + 50.789.t^3 \quad (5)$$

RECEIVER OPERATION RESULT

As a result of this model it has been simulated a day variation of the receiver and its variables, P, (output pressure, bar abs), T, (output temperature, C) and I (solar radiation, W/m2). The figure 4 shows the output receiver variations with the input solar radiation and the previous conditions, (normal operation, start-up and bottle). This figure shows the reduction of the receiver values due to the transient conditions.

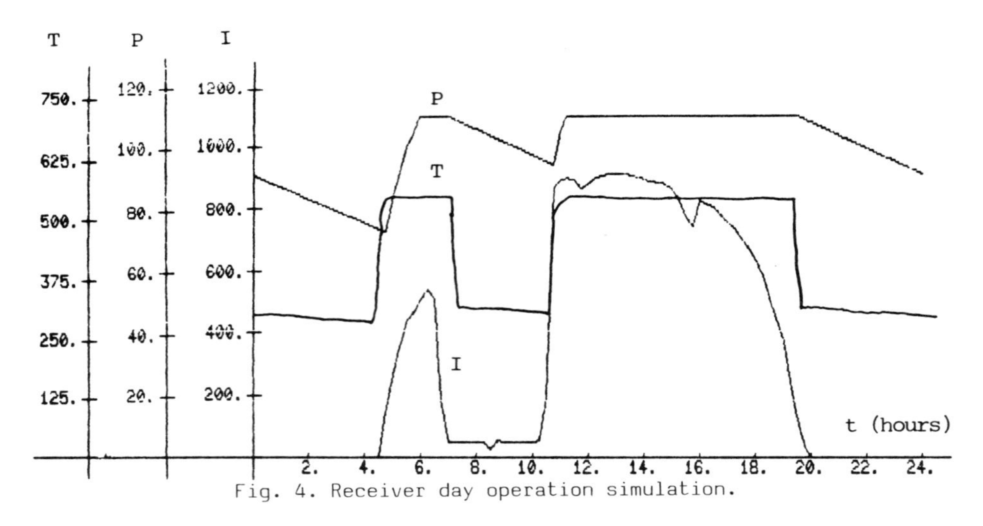

Fig. 4. Receiver day operation simulation.

CHARGING AND DISCHARGING MODELS

These models are the most expensive ones in computer running time due to its complexity. For this reason these models have been simplified obtaining a higher computer response within an acceptable error. The new block diagram is shown in the figure 5, where is possible to compare this one with the figure 1 to appreciate the number of blocks saved and the circuit complexity avoided. The storage charging and discharging models include in only two models the five existent heat exchangers, with piping and its associated losses.

The model of each heat exchanger is modeled fitting the temperature terminal differences of each one to its operational values, (IDAE, 1984), with a mean square error of 0.1 %.

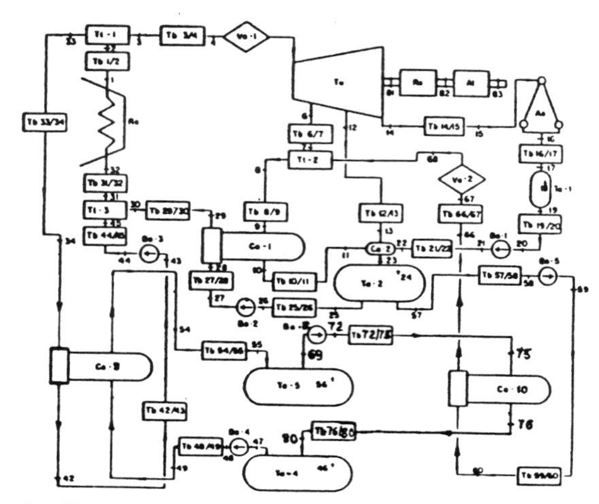

Fig. 5. CESA-I simulation block diagram simplified.

CONCLUSIONS

The new models added to the CESA-I simulation program have been developed in order to include the dynamic performance of the receiver and to reduce the program running time, simplifying the charge and discharge storage models.

This R&D project is developed as a part of the works carried out in the " Solar Central Receiver Joint Program " under the auspicies of the U.S.A.-Spain Joint Committe for Technological and Scientific Cooperation.

ACKNOWLEDGEMENTS

The authors would like to express their gratitude to the personnel of the " Plataforma Solar de Almeria ", who collaborated in the data acquisition and to the Sandia National Laboratories Central Receiver team.

REFERENCES

Castro, M., Peire, J. and Martínez, P., (1985). A final real approach to CESA-I Simulation, Proc. ISES, INTERSOL"85, Montreal, 1403-1407, Pergamon Press.

Castro, M., Peire, J. and Martínez, P., (1987). Five-year CESA-I simulation program review. Solar Energy, 38, 415-424.

CEE, Centro de Estudios de la Energía, (193). Proyecto CESA-I. Descripción General. Ministerio de Industria y Energía, Madrid.

CESA-I, Plataforma Solar de Almería, (1985). Informe de la primera fase de operación y ensayos. Ministerio de Industria y Energía, Madrid.

IDAE, Instituto para la Diversificación y el Ahorro de la Energía, (1984). Datos de la operación de la Central Electrosolar de Almería. Ministerio de Industria y Energía, Madrid.

Martínez, P., Muñoz, M., Peire, J., Prieto, J. and Castro, M., (1983). CESA-I Simulation, Proc. ISES, First Arab International Solar Energy Conference, Kuwait, 372-376, Pergamon Press.

RECEIVER TRANSFER FUNCTIONS OF THE GAST METALLIC PANEL

F. Rosa Iglesias (*) and J. Guerra Macho (**)

(*) Plataforma Solar de Almería, CIEMAT-IER

(**) C. Termotecnia, E.S.Ingenieros Industriales, Universidad Sevilla

ABSTRACT

Three transfer functions of the GAST metallic panel are proposed, based on a theoretical dynamic model. Each function relates inlet temperature, mass flow and radiant solar flux to the outlet temperature. Parameters for these transfer functions are suggest in three operation ranges and the model is validated using real operation values recorded by the Data Acquisition System (DAS)

KEYWORDS

Solar thermal plant; receiver system; transfer function; theoretical model; test method; model validation.

INTRODUCTION

In a receiver system, the controls have to respond to variations in heat flux in order to maintain constant the outlet temperature. This state is achieved by modifying the coolant mass flow. The receiver transient response may be characterized in terms of the ratio between variations in the outlet temperature and variations in the flowrate, radiant solar flux and inlet temperature.

To study the receiver behavior, a dynamic model must be developed. Introduction of steady-state and perturbation variables in this model facilitates the definition of receiver transfer function. Since the receiver is not a linear system, time constant and residence times of the transfer function are operation range functions. Therefore, several test will be performed to determine these parameters about equilibrium points.

GENERAL CHARACTERISTIC OF THE GAST PANEL

The GAST (Gas-cooled Solar Tower) metallic panel is an air-cooled receiver composed of 18 cylindrical tubes in parallel, placed vertically on the same plane. The tubes measure 8.4 m long, 42 mm outer diameter and the walls are 2.1 mm thick. A transient model schematic is shown in Fig. 1.

The nominal outlet temperature, pressure and mass flow are 800 ºC, 9.5 bar and 2.45 Kg/s, respectively (Wehowsky, 1984).

RECEIVER TRANSFER FUNCTION

The receiver is divided in three subsystems: radiated pipes, inlet and outlet headers. Each subsystem is represented by an associated set of differential equations.

The following basic assumptions were made:

a) The incident radiant flux is supossed equal to the average value on the receiver and only as a time function.
b) Air physical properties are considered as temperature functions, but are assumed constant in each subsystem and operation range.
c) The radial temperature variation and the axial heat conduction in air and tube are neglected.
d) Viscous dissipation is neglected.
e) Heat transfer coefficients are constant in each operation range.
f) Radiated pipes have been considered as distributed parameter one dimensional system. Headers have been considered as a single lump where perfect mixing is assumed.
g) Air flow rate is considered only as a time function and, a every instant, is the same in each radiated tube.

For the radiated tubes, the energy equations for the pipe metal and the fluid have been written in the following way:

$$\rho_m C_m A_m \, \partial T_m/\partial t = \alpha Q - h_o P_o (T_m - T_a) - h_i P_i (T_m - T_f) \tag{1}$$

$$\rho_f C_f A_f \, \partial T_f/\partial t + \dot{m} C_f \, \partial T_f/\partial x = h_i P_i (T_m - T_f) \tag{2}$$

where ρ is the density, C the specific heat capacity, Tm the metal average temperature, Tf the fluif temperature, Ta the ambient temperature, m the air flowrate, αQ the absorbed power per unit length, h_i and h_o the inner and outer heat transfer coefficient, Pi and Po the inner and outer tube perimeter, Am the tube metal cross section and Af the air flow cross section.

The equations (1) and (2), can be normalized by introducing the following dimensionless variables: length $(X=x/L)$, time $(\tau=t/t_r)$, heat transfer coefficient $(h=h_oP_o/h_iP_i)$, heat flux input $(H(t)=Q(t)/(\rho_f C_f A_f V(T_{fo}(L)-T_a)))$, flow rate $(W(t)=m(t)/m_o)$, temperature $(\theta_s(x,t)=(T_s(x,t)-T_a)/(T_{fo}(L)-T_a))$ and time constant $(\tau_{s1}=\tau'_{s1}/t_r)$.

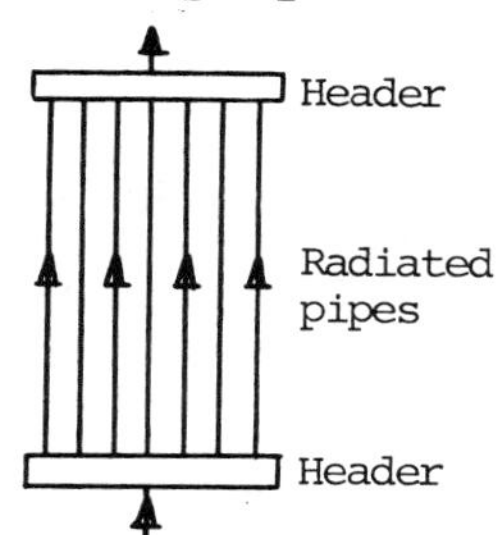

Fig. 1. GAST receiver flow diagram

TABLE 1 Proposed Parameters

TEST	W	τ_{m1}	τ_{f1}	h	α	$\partial\theta_{fs}(x,0)/\partial X$	M	τ_{m2}	τ_{f2}
7110	1	60	0.209	0.079	-	0.1974	7×10^{-4}	1074	45.5
7210	0.8	60	0.209	0.101	-	0.1915	1×10^{-3}	1074	45.5
7310	0.6	87	0.209	0.130	-	0.1974	2×10^{-3}	1074	45.5
7120	1	61.38	0.227	0.088	0.92	-	2.75×10^{-3}	1074	45.5
7220	0.8	70.36	0.227	0.109	0.92	-	2.75×10^{-3}	1074	45.5
7320	0.6	89	0.296	0.133	0.92	-	4×10^{-3}	1074	45.5
7410	1	59	0.190	0.093	-	-	4×10^{-3}	1074	45.5
7420	1	76.20	0.290	0.083	-	-	4×10^{-3}	1074	45.5
7430	1	85.50	0.430	0.089	-	-	4×10^{-3}	1074	45.5

In the previous expressions the dummy subscript s is replaced by either m (pipe metal) or f (fluid). t_r is the fluid residence time ($t_r=L/V$), τ'_{s1} is a time constant defined by $\rho_s C_s A_s / h_i P_i$, L is a characteristic length, m_o and V are a characteristic fluid flowrate and velocity, and $T_{fo}(L)$ is a reference outlet temperature.

Linearization of normalized energy equations about an equilibrium point yields (Doebelin, 1980),

$$\tau_{m1}\ \partial\theta_{mp}/\partial\tau = \tau_{f1}\alpha H_p + \theta_{fp} - (1+h)\theta_{mp} \tag{3}$$

$$\tau_{f1}\ \partial\theta_{fp}/\partial\tau + \tau_{f1}W_s\ \partial\theta_{fp}/\partial X + \tau_{f1}W_p\ \partial\theta_{fs}/\partial X = \theta_{mp} - \theta_{fp} \tag{4}$$

where each dimensionless variable θ_m, θ_f, H and W is defined as the sum of an initial steady-state value (subscript s) and a dynamic perturbation (subscrip p).

L-transforming equations (3) and (4) and solving the resulting ordinary differential equation, we obtaining the radiated pipes transfer functions between the outlet temperature and the air mass flow $G_1(s)$, the solar flux $G_2(s)$ and the inlet temperature $G_3(s)$,

$$G_1(s) = \theta_{fp}(1,s)/W_p(s) = (\partial\theta_{fs}(X,0)/\partial X)\ (\exp(-\phi(s)/W_s(0))-1)/\phi(s) \tag{5}$$

$$G_2(s) = \theta_{fp}(1,s)/H_p(s) = \alpha(1-\exp(-\phi(s)/W_s(0)))/(\phi(s)(1+h+s\tau_{m1})) \tag{6}$$

$$G_3(s) = \theta_{fp}(1,s)/\theta_{fp}(0,s) = \exp(-\phi(s)/W_s(0)) \tag{7}$$

where $\phi(s) = s+(h+s\tau_{m1})/(\tau_{f1}(1+h+s\tau_{m1})$

Through a similar processes, is possible to obtain the header (considered as a single lump) transfer function between the outlet and inlet temperature,

$$G_4(s) = \theta_{op}(s)/\theta_{ip}(s) = MW_s(0)/(1+MW_s(0)+s\tau_{f2}-1/(1+s\tau_{m2})) \tag{8}$$

where M is a dimensionless heat capacity equal to $m_o C_f / h_i S_i$ and the others dimensionless variables have a similar definition of the radiated pipes parameters. S_i represents the internal header surface.

Taking into account the cascade connection of the headers and the radiated pipes in the receiver, the global transfer functions between receiver outlet temperature and air flowrate, radiant solar flux and receiver inlet temperature are the following,

$$K_1(s) = \theta_{rop}(s)/W_p(s) = G_4(s)_o\ G_1(s)\ 0.5/(s+0.5) \tag{9}$$

$$K_2(s) = \theta_{rop}(s)/H_p(s) = G_4(s)_o\ G_2(s)\ 0.5/(s+0.5) \tag{10}$$

$$K_3(s) = \theta_{rop}(s)/\theta_{rip}(s) = G_4(s)_o\ G_3(s)\ G_4(s)_i\ 0.5/(s+0.5) \tag{11}$$

where the subscript o and i on $G_4(s)$ indicates the outlet and inlet header respectively. The factor $0.5/(s+0.5)$ is the response of a thermocouple with a time constant of 2 seconds. Table 1 presents the proposed parameters for every transfer function in several operation ranges (Sánchez, 1985).

MODEL VALIDATION

Nine tests were performed under different operation conditions to validate the proposed model. A first group of test (7110, 7210, and 7310) was carried

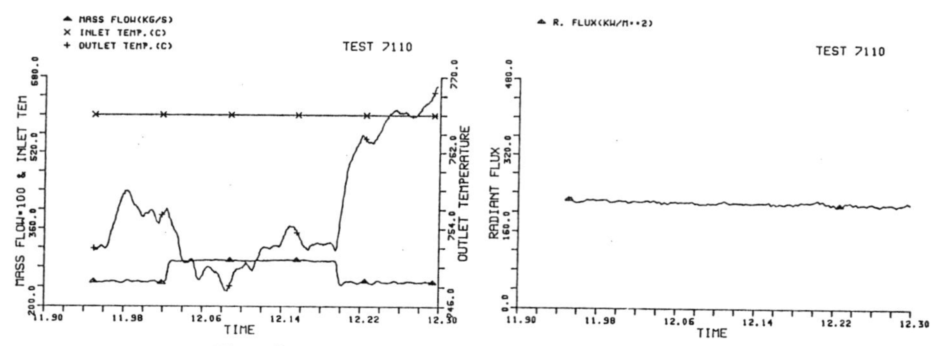

Fig. 2. Operational conditions in test 7110

out, producing a step change in the mass flow while the incident power and the inlet temperature remained constant. These test validate equation (9) relating outlet temperature to mass flow. Figure 2 shows the operation conditions in test 7110, where a step change in mass flow from 2.5 to 3.0 Kg/s was made.

The second group (test 7120, 7220 and 7320) validates equation (10), where outlet temperature is related to radiant solar flux. During these test, a step change in incident power was made while the mass flow and the inlet temperature remained constant.

The third group (test 7410, 7420 and 7430) validates equation (11), where the outlet and inlet temperature are related. A step change in inlet temperature was made manipulating the air heaters, while the radiant solar flux an mass flow were constant.

To compare the model's results with DAS values, the input signal has been achieved in a nondimensional form, adjusted to a polynomial expression, and the L transform of that expresion calculated. Simulated outlet temperature response is obtained through the inverse transformation of the product of the L transform by equation (9), (10) or (11), using the Bromwick's integral to inverse L transform (Levinson, 1975). Figures 3 to 5 show the simulated and experimental outlet temperatures for test 7110, 7210 and 7410.

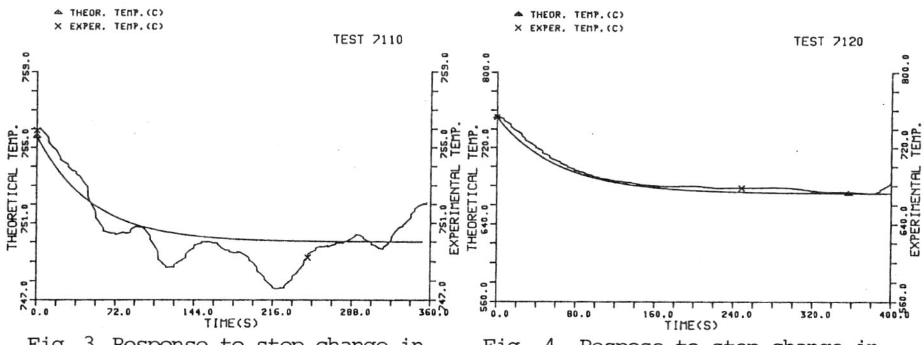

Fig. 3. Response to step change in mass flow.

Fig. 4. Respose to step change in radiant solar flux.

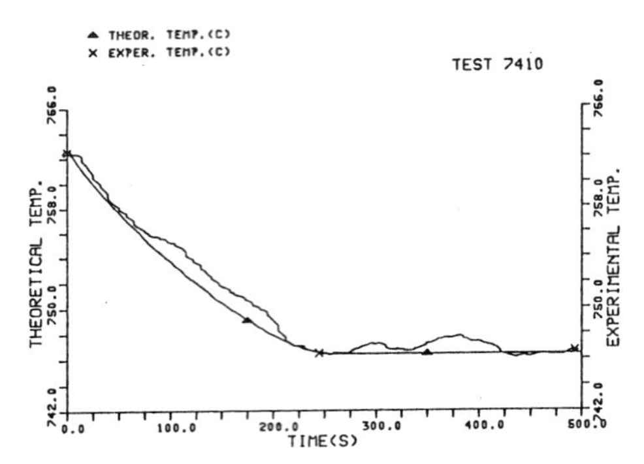

Fig. 5. Response to step change in inlet temperature.

CONCLUSIONS

Three transfer functions of the GAST metallic receiver panel have been obtained to relate outlet temperature to mass flow, radiant solar flux and inlet temperature. Nine test were realized under different operation conditions to validate the proposed model.

The transfer functions obtained represent the receiver response accurately. These functions can be used to predict what receiver outlet temperature will occur for arbitrary changes in mass flow, radiant solar flux and inlet temperature. These functions are also useful to study the stability of the metallic panel.

REFERENCES

Wehowsky, P., Stahl, D., de Marcos J. and L. Crespo (1984). The Gas-cooled Solar Tower Project GAST, Proceedings of 2nd Int. Workshop on Design, Construction and Operation of Solar Central Receiver Projects, June 1984, Varese/Italy.

Doebelin, E.O. (1980). System modelling and response, J. Wiley and Sons, pp. 437-447.

Sánchez, M. (1985). Receiver Dynamic Response, GAST-1/85 MS.

Levinson, N. and R.M. Redheffer (1975). Curso de variable compleja, Ed. Reverte pp. 176-257.

ANALYSIS OF "GAST" HELIOSTAT IMAGE MEASUREMENTS

M. Kiera
Interatom GmbH
P. O. Box, D-5060 Bergisch Gladbach 1
Federal Republic of Germany

ABSTRACT

Within the GAST project highly focussing heliostats have been designed and developed in order to reflect the irradiated sun light into small cavity apertures required by gas-cooled receivers at high temperature. For testing purposes the sun image produced by one or more heliostats are measured with distant camera systems (HERMES, MBB) in terms of its instantaneous intensity distribution reflected from a target plane. This paper describes how this distribution may be interpreted by means of the incoming radiation (sunshape), the structure of the reflecting heliostat surface (curvature, glass waviness) and the actual alignement of the heliostat modules (astigmatic aberration, canting). Each contribution is considered separately and parametrised by an appropriate distribution function. Finally they are composed to a theoretical image on the target plane using the convolution prescription. The parameters which the model functions depend on have a specific physical meaning (appearant sun's radius, focal length etc.) and are determined by adapting to the measured intensity distribution. In this way reflecting systems may be investigated in detail. The optical property without the time dependent astigmatism is usually expressed by the beam quality, i. e. the standard deviation of the angular dispersion of the reflected ray cone. The results of several experiments involving GAST heliostats are:

- module of prototype Pl.1: 0.83 mrad
- prototype Pl.1, off axis canted: 1.50 mrad
- prototype Pl.1, on axis canted: 1.27 mrad
- pretestserie heliostat Pl.3, on axis canted: 1.35 mrad
- 30 test heliostats, on axis canted: 1.30 - 1.52 mrad

In addition from the experiment with a group of 8 test field heliostats the pointing accuracy is estimated to about 0.8 mrad. If compared with the standard deviation of the original sunshape found in the range 2.25 $\pm$ 0.05 mrad the good optical performance of the GAST heliostats is demonstrated.

KEYWORDS

central receiver systems; heliostat; optics; beam quality; pointing accuracy; sunshape; canting; Fourier transform; convolution.

MEASUREMENT DEVICE

All the heliostats analysed are manufactured by MBB having about 52 sqm reflecting area. Each of them consists of 16 mirror modules canted on the supporting structure to a focal length which is equal to the distance between heliostat and target. Both on and off axis canting modes were realised in the experiments. The heliostats were erected at the Plataforma Solar near Almeria (Spain) in the rear part of the CESA-1 and SSPS-CRS heliostat field respectively. They are tracked on a 8 x 8 sqm flat target facing to north and being mounted at the corresponding tower shaft. The target distance of the heliostats is about 350 m. In roughly the same distance a CCD camera system is installed which measures the sun image reflected from the target in a square mesh of 256 x 256 equidistant points corresponding to a spacing of about 3 cm on the target plane. Because there is no absolute calibration provided only the relative intensities are detected given in digital units ranging from 0 to 255. Subsequently the heliostat is moved in stand-by position, and the skylight contribution on the target is measured to be subtracted from the heliostat image. The corrected data normalised to unity enter the analysis. Figure 1 shows schematically the basic experimental configuration of the system HERMES whereas Fig. 2 presents graphically the image distribution detected by the camera.

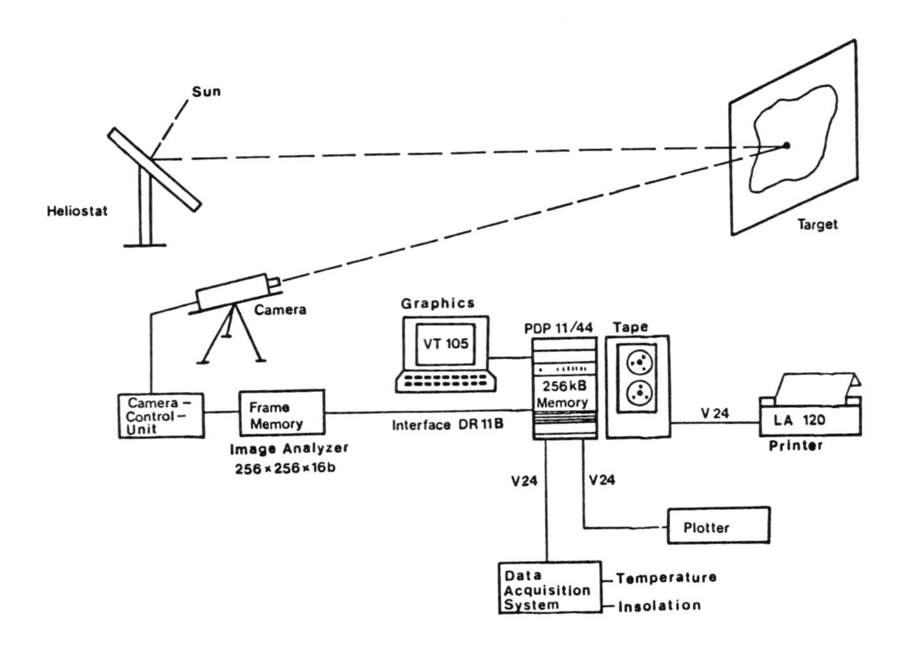

Fig. 1 Physical configuration of the measurement device HERMES

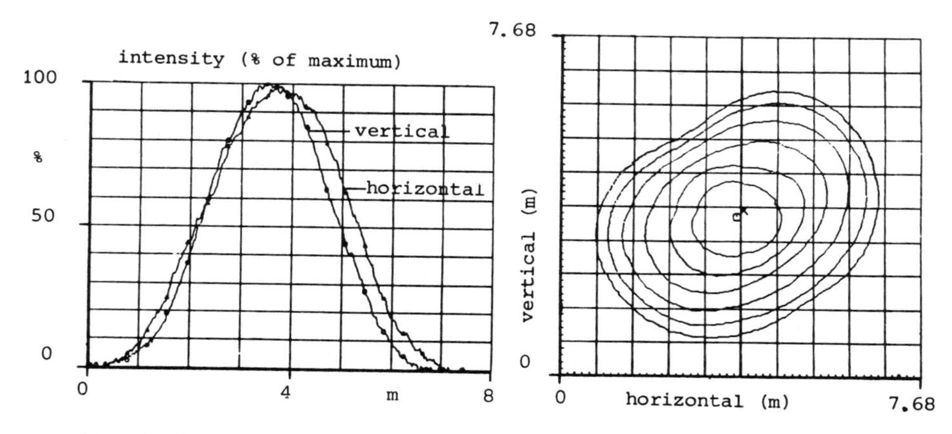

Fig. 2. Target cross sections **Isointensity lines on target**

THEORETICAL APPROACH

The analysis is preferably performed in Fourier space because the Fourier transform of the measured intensity distribution separates the low frequency components from the high frequency ones to which mainly the short range fluctuations contribute due to the finite resolution of the camera. So the analysis may be restricted to a much smaller grid in Fourier space (e. g. 9 x 9) preserving the physically interesting feature of the image and reducing considerably the computational amount. Secondly the Fourier transform of the model image function has a much simpler form than the convolution integral itself and may be calculated in closed analytical form. The basic tool in constructing the model image is the reflection law in the general vectorial form. If $\vec{s}$ denotes the unit vector of the sunposition, $\vec{x}$ the vector of any point on the reflecting surface relative to its centre and $\vec{n}(\vec{x})$ the corresponding surface normal then the direction $\vec{z}(\vec{x})$ of the reflected ray is given by:

$$\vec{z}(\vec{x}) = -\vec{s} + 2(\vec{n}(\vec{x}) \cdot \vec{s})\,\vec{n}(\vec{x}) \qquad (1)$$

from which the intersection point $\vec{r}(\vec{z})$ in the target plane will be determined. The intensity distribution over the incoming sunshape is modeled by a circular symmetric function which drops toward the sun's limb according to a power law as proposed by Pierce and Slaughter (1977). The appearant sun radius and the exponent enter as fitting parameters to be determined in every image experiment. The surface structure of the module area is assumed to be the superposition of the design curvature and a cosine like glass waviness. Therefrom the normal $\vec{n}(\vec{x})$ at any surface point is calculated depending on the actual module focal length, wave length and amplitude of the wavy structure. $\vec{n}(\vec{x})$ further depends on the alignement of the module which may deviate from the designed one (on/off axis) due to canting errors. These deviations provided for each module serve as additional fit parameters. Now following the ray of each point of the sunshape, reflected at any point of the heliostat surface the intensity distribution on the target can be constructed by superposition which is mathematically expressed by the convolution integral. The procedure has been described by several authors (Burkhard and Shealy, 1973; Walzel and co-workers, 1977).

DATA EVALUATION AND RESULTS

In order to determine the 44 parameters required for the analysis of a complete GAST heliostat a general program is applied which fits iteratively a given discrete data set by a function of arbitrary shape according to the least chisquare method the relevant subroutines being given by Bevington (1969). For physical reasons many of the parameters may be estimated whose values are used to start the iteration procedure and to speed up the convergence towards a stable parameter configuration. This final set serves to calculate the standard deviations of the single effects (sunshape, etc.) in order to show up how strongly they contribute to the heliostat image dispersion. In this way the deviation of the optical performance of the heliostat from the designed one will be quantified. Finally the inverse Fourier transform of the fitted model image is computed allowing for a direct comparison with the measurement in the original 256 x 256 grid in terms of the overall accuracy and graphical representations of selected intensity cross sections. In this way 30 experiments involving GAST heliostats have been analysed up to now. Figure 3 visualises the good agreement between the fit result (smooth curves) and the experiment having been performed with the pretest serie heliostat P1.3 in off axis canting mode at 10/19/1983, 15 h 17 min solar time and also been shown in Fig. 2.

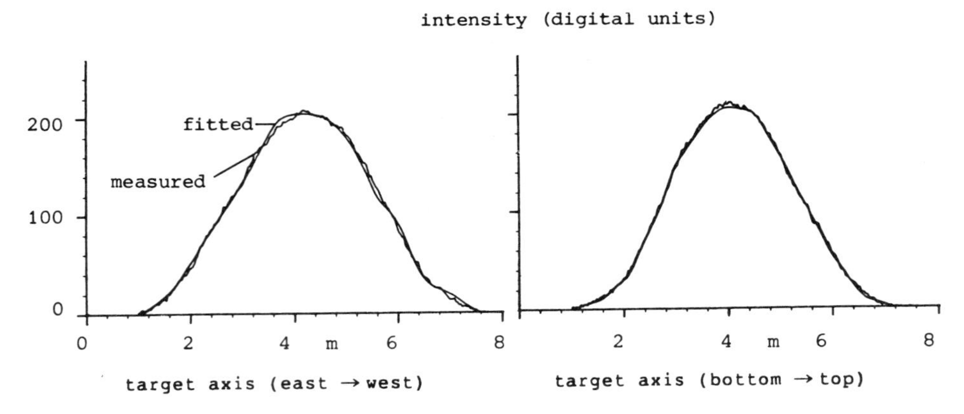

Fig. 3. Intensity cross sections through centre of gravity

It should be noted that all cross sections show similar agreements as in Fig. 3. Expressed in angular standard deviations the following contributions were found in this experiment:

- module curvature: 0.24 mrad
- glass waviness: 0.64 mrad
- module canting: 1.16 mrad
- astigmatism: 2.11 mrad
- sunshape: 2.23 mrad (appearant sun radius: 4.46 mrad)

beam quality: 1.35 mrad

The mean square deviation between experiment and fit is 2.5 digital units being comparable with the camera resolution of ± 1 digital unit. Of big importance for the image size is the canting mode which influences strongly the astigmatic aberration, i. e. the image dispersion due to the slant incidence of the sunlight. At the time when the present experiment was performed off axis canting (realised at summer solstice noon) leads to large aberration inducing image distortions as visualised in Fig. 2. Another experiment with a heliostat canted on axis at about the same time yields an aberration value of only 0.8 mrad demonstrating that the on axis mode should be realised in solar tower plants.

REFERENCES

Barthelmess, W. (1983). Prototyp-Heliostat: Ergebnisse der Qualifikation und Standorterprobung. GAST-IAS-BT-300000-093.

Barthelmess, W. and H. Wessel (1984). Results of long-term testing of MBB-prototype heliostat Pl.1 in Almeria. GAST-IAS-BT-300000-185.

Barthelmess, W. and H. Wessel (1985). Results of long-term testing of MBB-pretestserie heliostat Pl.3 in Almeria. GAST-IAS-BT-300000-186.

Bevington, P. R. (1969). Data reduction and error analysis for the physical sciences. Academic Press, New York.

Burkhard, D. G. and D. L. Shealy (1973). Optica Acta, 20, 287-301.

Kiera, M. (1985). On the interpretation of measured heliostat images; GAST-IAS-BT-200000-071.

Pierce, A. K. and L. D. Slaughter (1977). Solar limb darkening. Solar Physics, 51, 25-41.

Schiel, W. (1983). Bericht über die Vermessung des MBB-Prototypheliostaten in Almeria im Herbst 1982. GAST-IAS-BT-300000-091

Schiel, W. (1983). HERMES Measurements. In M. Becker (Ed.) SSPS-CRS mid term workshop. SSPS Technical Report 4/83, 326-348.

Walzel M. D., F. W. Lipps and L. L. Vant-Hull (1977). Solar Energy, 19, 239-253.

Wessel, H. (1986). Summary of test results - test series heliostats (Oct. 1984 and June 1985). GAST-IAS-BT-300000-194.

METEO STATION "GAST": MEASURING EQUIPMENT AND DATA ANALYSIS

M. Kiera, E. Sauermann
Interatom GmbH
P. O. Box, D-5060 Bergisch Gladbach 1
Federal Republic of Germany

ABSTRACT

The basic devices of the meteorological station GAST are described measuring important parameters involved in solar energy applications like insolations, speed and direction of wind, ambient temperature, relative humidity and atmospheric pressure. The data collected during almost three years (1982 - 1985) at the Plataforma Solar (Almeria/Spain) have been evaluated with respect to the average daily behaviour and frequency distribution providing for a comprehensive qualification of the site. From that experience some general algorithms for reconstructing the mean daily characteristics of insolations and ambient temperature were derived allowing for the extrapolation to different sites if only reduced information is available, e. g. monthly averages. Additional measurement campaigns were devoted to questions which arise at central receiver systems involving heliostat fields like field shadow transients due to cloud passages, atmospheric attenuation of the reflected sun light between heliostat and receiver and the turbulent structure of the wind behind heliostats.

KEYWORDS

meteorology; solar energy; central receiver systems; insolation; wind; temperature; humidity; cloud transients; atmospheric transmittance.

MEASURING EQUIPMENT AND DATA EVALUATION

The meteorological station GAST was designed to measure the most important parameters for solar energy applications in high time resolution. The data aquisition is installed within a container being surrounded by the sensors. Up to 30 independent channels may be recorded at a time rate of 10 sec/5 min during the day/night. The data recording system consists of the data aquisition unit (HP 3497A), the process computer (HP 1000 Mod. 9 with RTE-XL operating system and 512 KByte storage capacity), two floppy disc stations, the terminal with integrated printer (HP 2623A Grafik) and the magnetic tape unit (HP 7970 E). This system together with the sensors has been erected at the Plataforma Solar near Almeria/Spain and has operated from November 1982 until July 1985 with a data availability of about 80 %. The corresponding evaluation software provides for graphical presentations, consistency checks and several procedures involving averages and frequency distributions in order to get a comprehensive survey about the climatological parameters and their interrelations. Based on this experience a small measuring station with reduced size (8 - 16 channels) has been developed allowing the optional power supply by means of PV cells and consequently high mobility and autonomous operation. The system was already used sucessfully at various sites, e. g. in the Sahara. The following list presents the continuously measured parameters and the corresponding sensors applied:

- direct insolation: pyrheliometers (one in redundancy)
- global insolation: pyranometer
- diffuse insolation: pyranometer with shadow ring
- wind speed in 2 and 10 m height: cup anemometers
- wind direction in 10 m height: wind vane
- ambient temperature in 2 and 10 m height: resistance thermometers
- relative humidity in 10 m height: psychrometer (CHH-Sensor)
- atmospheric pressure: aneroid barometer

RESULTS OF CONTINUOUS MEASUREMENTS

The data are presented preferably in total monthly averages and maxima revealing their seasonal variation and in monthly averages and maxima at fixed solar time showing up the mean daily behaviour as illustrated in Fig. 1 for the direct insolation, the wind speed at 10 m and the ambient temperature:

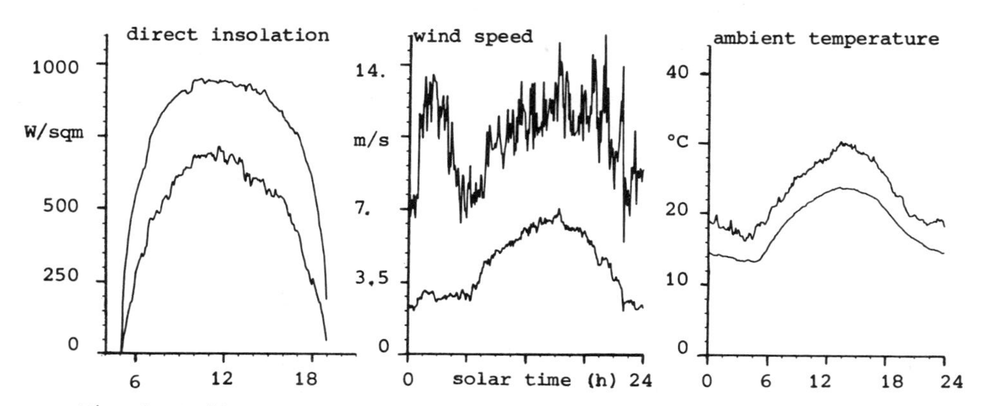

Fig. 1. Daily behaviour of monthly maximum and average (June 1984)

The whole measuring period of almost three years is characterised most favourably in terms of frequency distributions as it is shown in Fig. 2 in the cases of the direct insolation and 10 m wind speed. To summarise the Plataforma Solar represents a site where about 40 % of the insolation at clear days is lost by cloudiness. In terms of annual sums and averages we found:

. direct insolation (kWh/sqm/a)	clear:	3266	real: 1958 ≙ 60.0 %
. global insolation (kWh/sqm/a)	ex. terr:	2965	real: 1777 ≙ 59.9 %
. diffuse insolation (kWh/sqm/a)			real: 612
. sunshine hours (direct insol. > 300 W/sqm)			real: 2760
. wind speed (m/s) at 2 and 10 m			mean: 2.6/3.7
. ambient temperature (°C)	min: -6.7	max: 35.4	mean: 14.0
. relative humidity (%)			mean: 49.3
. atmospheric pressure (hpa)			mean: 945

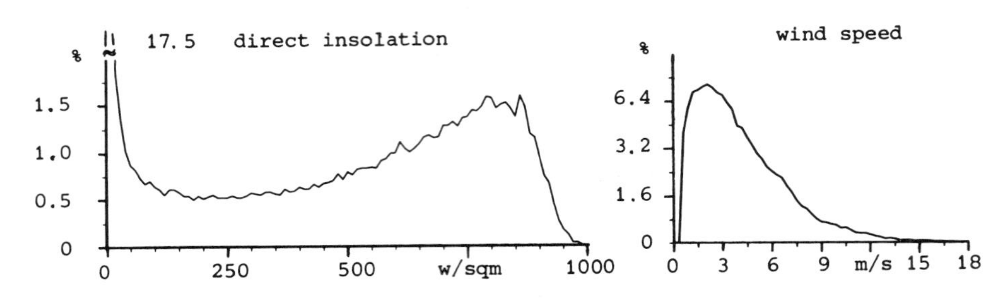

Fig. 2. Frequency distributions (all data)

The structures of the direct insolation and ambient temperature as shown in Fig. 1 and the annual results open the possibility to a semiempirical approach in order to predict their mean daily course if only reduced data material is availabe. Given the total monthly averages of global and diffuse radiation the average daily behaviour per month of the three radiation channels may be computed using the radiation balance on the horizontal surface and the following approach of the direct insolation:

DIR = DIR (EXTRATERR.) * CLEARNESS INDEX * EXP (-EXT*ATM. MASS)

From the input the unknown parameter EXT is determined. The results are illustrated in Fig. 3.

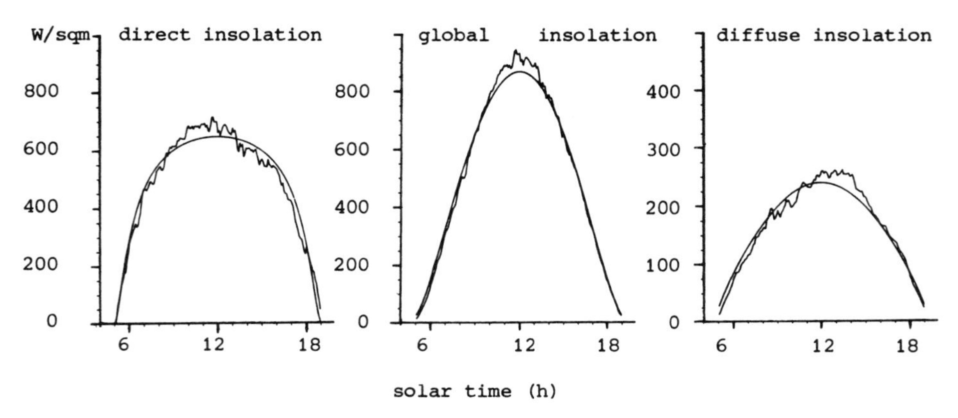

Fig. 3. Measured and computed insolations (June 1984)

The annual direct sum is predicted within an error of about 2 %.
Given the minimal, maximal and average temperature per month a similar approach of the ambient temperature may be performed using a cosine like behaviour during the day and an exponential decay during the night as it is indicated in Fig. 1.

RESULTS OF SPECIAL MEASUREMENT CAMPAIGNS

To study shadow transients on large fields 16 equidistant solar cells were installed in a square grid of 600 m x 600 m size. During 3 weeks in spring 1984 the global insolation incident on every cell was measured in time steps of 10 sec. The correlations of the signals in time and space are used to estimate the speed, direction and diameter of the moving cloud formations as well as characteristic shadow cycles. Figure 4 presents an event when a cloud is approaching the field, covering the field for a while and is disappearing again. The stars on top of each other indicate the number of shadowed cells of the total field and of the 4 sectors (NW, NE, SW, SE).

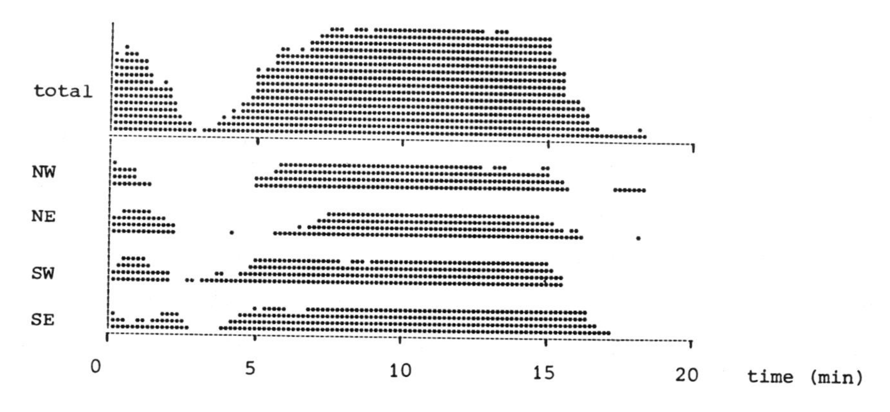

Fig. 4. Shadow sequence of solar cells at day 55 (1984)

All cloud transients analysed reveal changes of the field covering degree within ranges of several minutes which would induce rapid variations of the receiver load in central receiver systems. The second campaign concerns the transmittance of the low atmospheric layers for reflected sunlight. The measuring device consists of an emitter (Xenon spark lamp) installed on top of the CESA-1 tower and of an 500 m distant absorber (Selen photocell) at ground level. The campaign comprised about 3 weeks in summer 1985. The measured signals show characteristic variations during the day according to the changing aerosol content of the atmosphere. Figure 5 presents the averaged daily behaviour and the frequency distribution of the transmittance.
The average of 86.6 % corresponds to a visual range of about 10.5 km and indicates high aerosol densities which are typical for the site during the summer when also reduced clear insolations are observed. The third campaign investigates the turbulent structure of the wind behind a heliostat which might lead to undesired resonances of the supporting structures of the heliostats in its neighborhood. Due to the design specifications resonance frequencies above 5 Hz are expected such that the portable two axis wind velocity sensor TSI with fast response was used. It consists of a capsule with an ion source inside and measures the deflection of the ionic ray due to the wind entering the capsule through slits. The subsequent Fourier analysis provides for the wind velocity spectra where significant gust periods may be read off which are linked to the mean alignement error of the heliostat due to wind influences. Figure 6 shows both the spectral densities for free wind and wind 10 m behind the heliostat according to a measuring time of 2 hours at day 176 (1985).

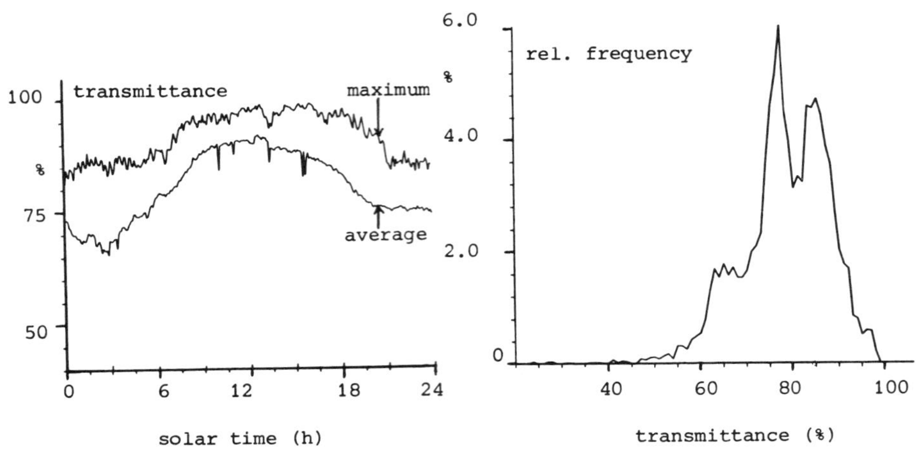

Fig. 5. Daily course and frequency distribution of transmittance

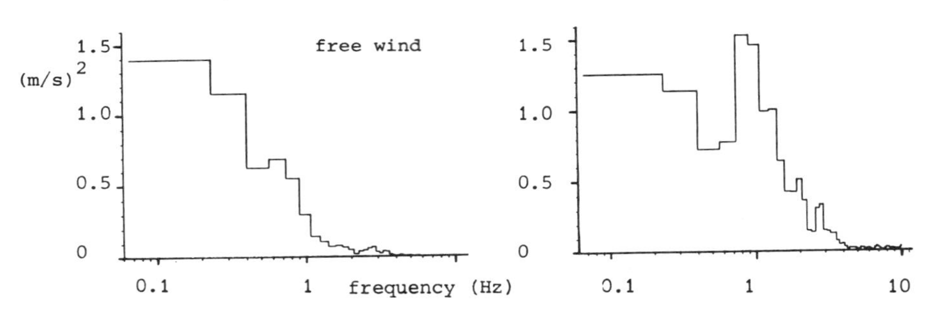

Fig. 6. Spectrum of free wind and wind 10 m behind heliostat

Even though the short measurement time may not be representative the free wind spectrum shows the characteristic drop meaning that turbulent structures occur only at frequencies below 1 Hz. The wind behind the heliostat contains additional periodic gusts around 1 Hz, but not significant ones which may affect neighboring heliostats.

REFERENCES

Imig, G, W. Pinterowitsch and E. Sauermann (1983). Aufbau und Betrieb der meterologischen Meßstation. GAST-IAS-BT-100100-053.

Kiera, M. (1986). Meteostation GAST: Analysis of the meteodata in 1983/84 at Plataforma Solar, Almeria (Spain). GAST-IAS-BT-200000-072.

Kiera, M. (1986). Meteostation GAST: Analyse der Transmittometer- und Windfrequenzmessungen. GAST-IAS-BT-200000-073.

Kiera, M. (1986). Meteostation GAST: Analyse der Bewölkungstransienten (Meßkampagne Februar 1984). GAST-IAS-BT-200000-076.

Sauermann, E. (1984). Beschreibung einer meteorologischen Kleinmeßstation. GAST-IAS-BT-703000-012.

Sauermann, E. (1986). Erfahrungsbericht über den Testeinsatz der meteorologischen Kleinmeßstation. GAST-IAS-BT-703000-009.

"GAST" TECHNICAL AND ECONOMIC INVESTIGATIONS OF DIFFERENT POWER/PROCESS HEAT APPLICATIONS

W. Meinecke, P. Wehowsky
INTERATOM GmbH
P. O. Box, D-5060 Bergisch Gladbach 1
Federal Republic of Germany

ABSTRACT

Within the framework of an analysis of the potential of gas-cooled solar tower plants carried out in parallel to the **GAS**-cooled **S**olar **T**ower project **GAST** [1], selected short term and future-oriented applications of the **GAST** technology were investigated initially from a technical and an economic point of view assuming the insolation model of Barstow/USA. The results indicate that there exist advantegeous short term GAST applications, in particular concerning the co-generation and the solar/fossil hybrid power generation. The GAST technology required for this is already available today.

KEYWORDS

Solar thermal tower plant, high temperature, gas/air-cooling, central cavity receiver system, power generation, co-generation, process heat, technical investigations, economic investigations.

INTRODUCTION

In parallel to the GAST Technology Programme, an analysis of the potential of the gas-cooled solar tower power plant (GAST) was carried out by joint cooperation of DFVLR, Sharan Engineering, Interatom, Asinel, M.A.N, MBB und Dornier System in 1984/1985 [1].
Within the framework of this analysis, following GAST applications were selected and investigated, the purpose of which was to define suitable advanced GAST conceptual designs in more detail and to get an idea about their profitability:

a) Short term applications:

(1) GAST SOLFOS power plant (solar/fossil hybrid operation), flow diagramme see Fig. 1

(2) GAST co-generation plant, power and process heat (hot gas), flow diagramme see Fig. 2

(3) GAST co-generation plant, power and process heat (steam), flow diagramme see Fig. 3

b) Future applications:

(4) GAST synthesis gas production plant (steam reforming of methane or of natural gas), flow diagramme see Fig. 4

(5) GAST hydrogen production plant (HOT ELLY of Dornier System), flow diagramme see Fig. 5.

The 20 MW_e GAST power plant reference design comprises a KWU open gas turbine cycle (gas temperature of max. 800 °C at the turbine inlet) with bottoming steam turbine circuit.

The investigated plants operate in so-called "fuel saver" operation mode, i. e,. approx. 22-31 % of the annual energy put into the turbine cycle at rated power operation is supplied by solar energy.

RESULTS

Table 1 gives a preliminary economic summary of the cost analysis.

The essential input data for the calculations made by DFVLR (plants (1) to (3)) and by Interatom (plants (4) and (5)) are listed in the Tables 2 and 3.

All cost data are applicable for the time the economic investigations were made in 1984.

The plants considered herewith are pilot plants (i. e. no series plants). But development costs, which are still necessary, were not taken into account.

Due to the still uncertain cost estimates for the above mentioned future-oriented plants (4) and (5), estimated cost data must be considered as approximate values.

Returns from power generation by co-generation plants were assumed to be 15 Dpf/kWh (1984).

Regarding the economic viability of the projects, the following attributable cost ("market price") can be taken as a basis (1984 DM):

- for power: 10 Dpf/kWh (grid connection)
- for hot gas: 44 DM/MWh (price of heat generated from natural gas)
- for process steam: 37 DM/t (price of steam generated by a hard coal-fired boiler for GAST steam conditions, cost of coal 260 DM/t)

- for methanol: 500 DM/t

- for hydrogen: 1,500 DM/t.

CONCLUSIONS

In summary, the results of the initial technical and economic studies indicate that there exist advantgeous GAST applications from a technical and an economic point of view.

The SOLFOS power plant version resulted in relatively favourable power generating cost because of the currently low fuel price and its low price escalation, but also because of the assumed high number of yearly hours of full load operation.

The future-oriented plants designed to produce synthesis gas and hydrogen by solar/fossil energy have not yet been able to show appealing economic prospects although, generally speaking, the direct application of high-temperature solar heat to chemical processes using the GAST technology looks promising from a technical point of view.

In contrast, GAST co-generation plants (simultaneous generation of power and process heat), have a favourable economic starting base. These plants are virtually ready for sale on the market and are good examples of the generation of solar process heat.

REFERENCE

[1] Becker, M., Dunker, H. and Sharan, H. (Ed.), (1986). Technologieprogramm Gasgekühltes Sonnenturm-Kraftwerk (GAST) - Analyse des Potentials. Bundesministerium für Forschung und Technologie (BMFT), Forschungsbericht T86-087, p. 372.

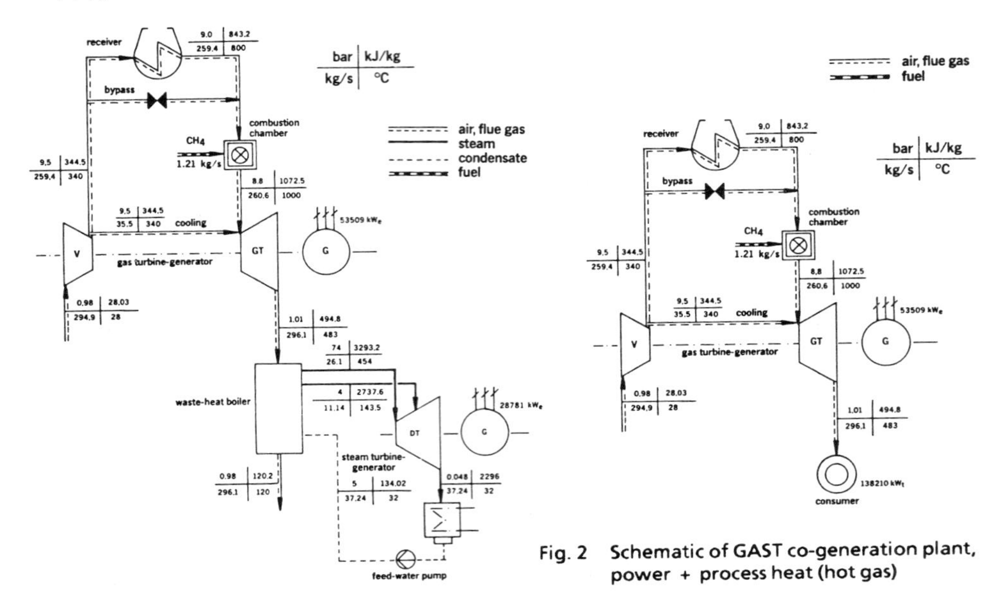

Fig. 1 Schematic of GAST SOLFOS power plant, 82 MWe (gross)

Fig. 2 Schematic of GAST co-generation plant, power + process heat (hot gas)

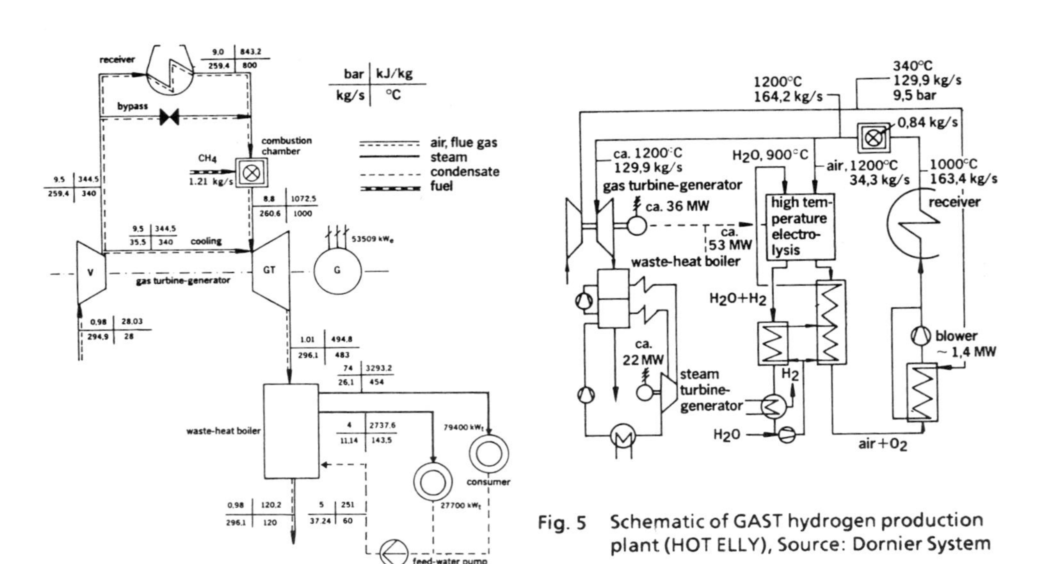

Fig. 3 Schematic of GAST co-generation plant, power + process heat (process steam)

Fig. 5 Schematic of GAST hydrogen production plant (HOT ELLY), Source: Dornier System

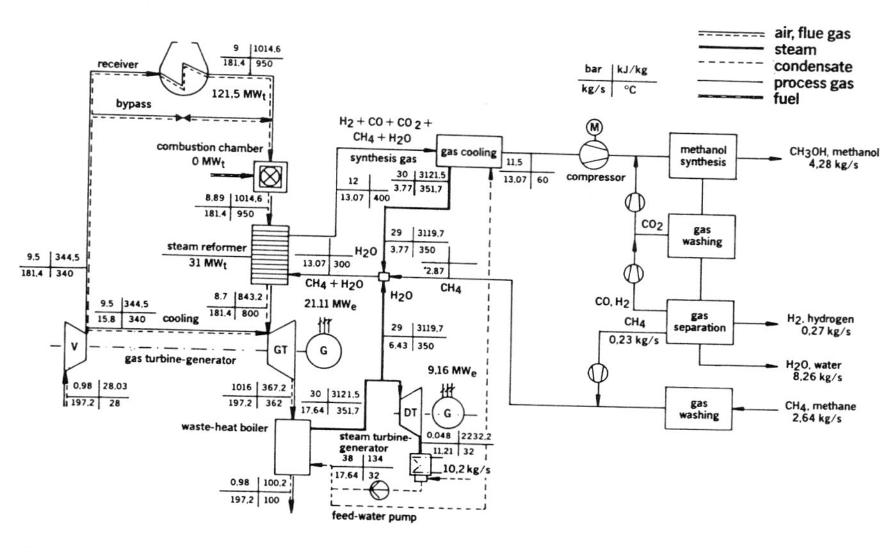

Fig. 4 Schematic of GAST synthesis gas production plant (steam reforming of methane)

TABLE 1 Preliminary Economic Summary of Advanced GAST Plants, 1984 DM
Source: DFVLR (1) to (3), Interatom (4) and (5)

Type of plant	GAST hybrid power plant (solar/fossil)				GAST hybrid plants (solar/fossil)							
					power + process heat production				chemicals production			
Economic summary	(1) GAST-SOLFOS power plant				(2) GAST combi plant		(3) GAST combi plant		(4) GAST synthesis gas production plant		(5) GAST hydrogen production plant (HOT ELLY)	
Main product(s)	power				power + hot gas		power + proces s steam		methanol		hydrogen	
Main by-product(s)	-				-		-		hydrogen, power		-	
Market price (1984)	power: ≈ 10 Dpf/kWh				hot gas: ≈ 44 DM/MWh		process steam: ≈ 37 DM/t		methanol: ≈ 500 DM/t hydrogen: ≈ 1,500 DM/t		hydrogen: ≈ 1,500 DM/t	
Generating return Dpf/kWh	-				15		15		15		-	
Price of burned fuel/raw material	fuel oil/- 75 DM/MWh		natural gas/- 44 DM/MWh		natural gas/- 44 DM/MWh		natural gas/- 44 DM/MWh		methane/methane 44 DM/MWh		natural gas/- 44 DM/MWh	
Price escalation rate of natural gas, methane, fuel, oil, power %/a real	2	3	2	3	2	3	2	3	2	3	2	3
Price of generated energy (start-up 1990)	power Dpf/kWh		power Dpf/kWh		process heat DM/MWh		process steam DM/t		methanol DM/t		hydrogen DM/t	
annuity method	29.2	30.1	22.3	22.8	50.0	49.7	59.3	58.7	1,218	1,241	10,800	10,900
real present value method	28.6	31.9	20.3	22.2	30.1	29.3	39.3	34.8	1,089	1,159	9,600	10,200
nominal present value method	44.0	48.9	31.1	34.0	47.5	45.1	62.1	61.6	1,592	1,695	14,000	14,900

TABLE 2 Performance Data and Cost Estimate Summary of Advanced GAST Plants, 1984 DM

Type of plant / Plant characteristics	(1) GAST-SOLFOS power plant	(2) GAST combi plant, power + hot gas	(3) GAST combi plant, power + process steam	(4) GAST synthesis gas production plant	(5) GAST hydrogen production plant (HOT ELLY)
1 annuall full-load hours [h/a] (capacity factor [%])	7,000 (80)	7,000 (80)	7,000 (80)	7,000 (80)	7,000 (80)
2 rated power output, gross/net [MWe]	82.3/75.7	53.5/50.3	53.5/50.3	30.3/18.3	approx. 58/ -
3 annual power output, net [MWe /h]	529,947	352,100	352,100	128,240	-
4 annual output of process heat [MWh /a]	-	967,470	749,700	-	-
5 annual output of chemical products [10^3 kg/a]	-	-	-	methane: 107,870 hydrogen: 6,804	hydrogen: 12,700
6 annual consumption of natural gas [10^3 kg/a]	74,342	74,342	74,342	combustion chambers: 41,982 combined process: 66,500	62,762
7 total plant efficiency annual average [%] (heat utilization)	32	79	66	29	?

Type of plant / Cost estimate summary	(1) GAST-SOLFOS power plant	(2) GAST combi plant, power + hot gas	(3) GAST combi plant, power + process steam	(4) GAST synthesis gas production plant	(5) GAST hydrogen production plant (HOT ELLY)
Total plant installlation cost (1984) 10^6 DM	551	452	476	643	689
Cost of maintenance and repair 10^6 DM/a	4.5	4.5	4.5	6.8	7.7
Operating staff expenses 10^6 DM/a	3.6	3.0	3.0	6.2	4.4

TABLE 3 General Economic Assumptions of GAST Cost Analysis

-	Prices based on	year	1984
-	Year of start-up	year	1990
-	Depreciation time = general service life	years	20
-	Erection/starting time	years	3
-	Destruction time	years	0
-	Rate of interest	%/a	8
-	Taxe rate	%/a	0
-	Insurance rate	%/a	1
-	Rate of general price escalation	%/a	5
-	Price of fuel oil (pessimistic assumption)	DM/MWh	75
-	Price of natural gas	DM/MWh	44
-	Escalation rate of fuel prices (in real terms)	%/a	2 (3)

Test of a High-Temperature Heat Exchanger Unit for the Gas-cooled Solar Tower GAST

E. Melchior, M. Dogigli
MAN Technologie, Munich, West Germany

ABSTRACT

The test and the test results of a metallic high-temperature heat exchanger unit on the tower of the CESA 1 plant, Spain, are described.

KEYWORDS

Solar tower, Solar receiver, High-temperature heat exchanger unit (Panel), Special measuring devices

INTRODUCTION

With the aid of advanced reliable analytical design methods in thermodynamics and stress analyses we have designed, built and tested a metallic high-temperature heat exchanger unit (so-called panel, Fig. 1, Table 1).

TABLE 1 Main Data of Metallic Test Panel (Design Point)

Inlet temperature	625°C
Inlet pressure	9,3 bar
Outlet temperature	800°C
Mass flow	2,45 kg/s
Rated output	489 kW
Irradiated length	8 m
Outer diameter of tubes	42 mm
Wall thickness of tubes	2,1 mm
Spacing ratio	2,0
Number of tubes	18
Material of tubes: X10 NiCrAlTi3220H	

Fig. 1 Originally scaled high-temperature gas-cooled receiver test panel

DESIGN METHODS

DESIGN LIFE:	10 YEARS
ENVIRONMENT:	AIR AND WORKING MEDIUM
TEMPERATURES	
GAS OUTPUT:	800°C
METALLIC PIPE:	700 - 900°C
FRONT-TO-BACK GRADIENTS:	60 - 120°C
THERMAL TRANSIENTS: EVENTS	ABOUT 10.000 REFERENCE CYCLES; USUALLY WITH HIGH CHANGE RATE (CLOUDS, DAY CYCLE)
DWELL TIME:	20.000 HOURS IN TOTAL WITH CYCLE TIME 1 TO 6 HOURS

Fig. 2 Bounds on panel design and working conditions

The design of components for conventional plants has taken place up to now essentially on the basis of proven rules and standards, such as TRD (Technische Regeln für Dampfkessel), BS (British Standard) and ASME (The American Society of Mechanical Engineers). This method of proceeding is completely satisfactory, provided that the temperature limits (600-700°C) laid down in the rules are not exceeded (Fig. 2). Besides the high material temperatures up to 900°C these components, unlike those in conventional power plants, are also exposed to frequent and fast temperature change cycles which result from the solar operation itself. The main requirements are summarized and give a first impression on the difficult - and in particular unique - operating conditions.

TEST ON THE PLATAFORMA SOLAR

From March 85 until June 86 the original sized GAST panel was tested in solar operational mode on the 'Plataforma Solar de Almeria', Spain. The panel, mounted on the tower of the CESA 1 plant at a height of 80 metres in front of a wall of high-temperature-resistant ceramic-fibre blocks (Fig. 3), was exposed to radiation from a heliostat field with 300 heliostats.
The panel was operated as an external receiver, although it is designed for use in a cavity. This was taken into account in the test program by the selection of cavity-relevant operating conditions. The objectives of the tests were

- to prove the operational reliability of the components under realistic solar conditions, especially during unsteady operational modes
- to gather experience in design, construction and operational behaviour of high temperature heat exchangers and their additional equipment
- to reach conclusions on the validity of the model assumptions used as a basis for the calculations

Fig. 3 CESA 1 tower plant equipped with the test panel

In the panel sector gas temperatures, temperatures of tube-walls, headers, back wall and structure were measured, as were pressures, mass flow, radiation and panel movements (Fig. 4 and 5). Other measuring points were installed in the area of the hot-gas line and on the tower, where insolation and other environmental parameters were measured.
For that purpose special measuring devices were developed by MAN:
Special thermocouple probes (marked 1-8) installed in three tubes in the panel make it possible to measure the gas temperature and the tube wall temperatures at the circumference of the tube.

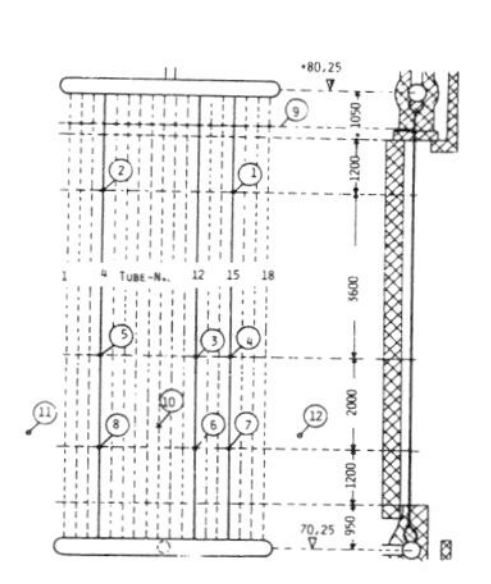

1 - 8 thermocouples: gas temperature and tube temperatures

9 thermocouples: gas outlet-temperatures

10 thermocouples: rear wall temperatures

11,12 radiometers: insolation

Fig. 4 Location of measurement probes at the panel area

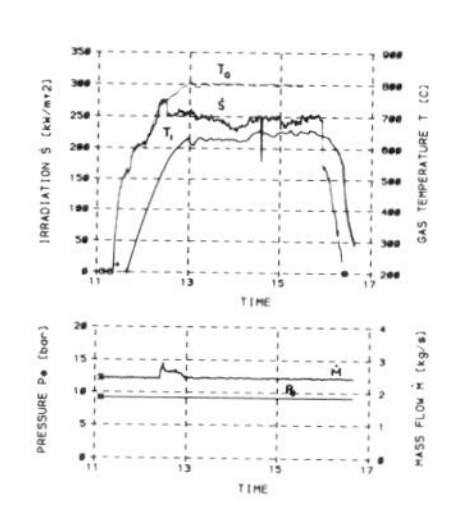

Fig. 5 Concentrated radiation on the panel ($\dot{S}$), panel inlet and outlet temperatures (T_i, T_o), pressure and mass flow during panel operation on 11.01.1986

The VISIR measuring system is envisaged for measurement of the surface temperatures and the insolation. This system records from the ground via a mirror telescope the radiation (temperature) emitted in the infrared (IR) range of the spectrum and the radiation (insolation) reflected in the visible (VIS) range. A redundant flux analysing system OSIRIS from the Spanish company ASINEL is installed. The test program consisted of test groups with different test specimens (Fig. 6). It comprised tests at various temperature levels under steady-state operation, tests under unsteady operating conditions, load-changing tests and tests of a hot-gas line. During one test 163 measured data were recorded and stored by the data acquisition system (DAS), most of them being recorded at intervals of 10 seconds. The main results of computations are spatial and temporal fields in the part of the tube walls exposed to radiation. Two selected investigations covering steady-state operation and unsteady operating conditions respectively give an insight into the test results:

With the solar flux measuring equipments (OSIRIS, VISIR) the insolation was measured after steady-state conditions had been reached. Fig. 7 shows the insolation as designed and measured along the tube length. The deviation of the measured curve to the theoretically designed curve results from the possibilities of heliostat field control and from effects at the panel ends caused by the panels supporting construction. The mentioned effects are: shadowing of the tube's inlet area by a protecting insulation wall and additional irradiation from the high-reflection insulation material towards the tube's end.
Due to the deviations in the measured insolation the calculated tube material temperatures, as well as the gas temperature, differ from the design values. Fig. 8 shows the maximum tube front and gas temperature, calculated with the actual irradiation of the experiment and, in comparison to that, the corresponding curves for the theoretical design irradiation. The temperatures measured by the thermocouples of tube 15 are also included in the figure and marked with a symbol.

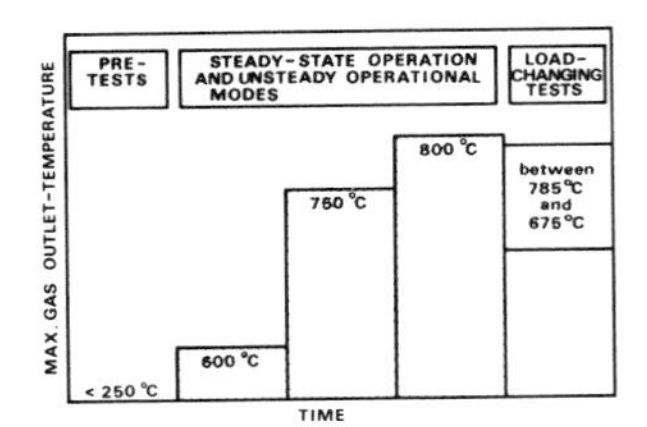

Fig. 6 Test program, schematic graph

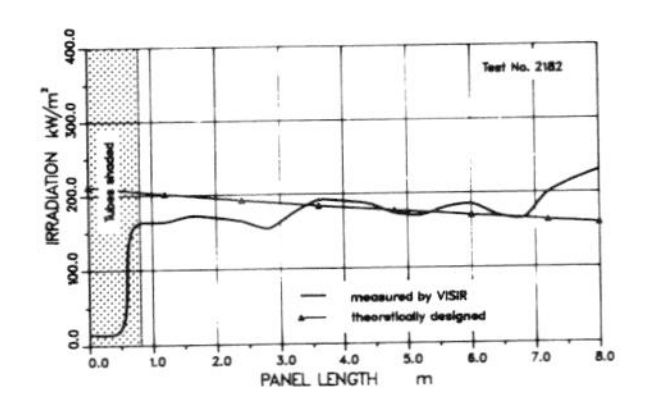

Fig. 7 Irradiation on the panel area versus panel height

Unsteady operational modes were investigated during start-up procedures, cloudy periods, defined decrease and increase of irradiation by the heliostat field control and controlled reduction of panel mass flow. Fig. 9 shows the quick reduction of the irradiation on the panel in 3 steps by heliostat control; it drops during a period of 50 seconds from 100 % to 25 %. Fig. 10 shows the panel characterized by the temperature course in a cross-section of one tube, on the front and back sides of the tube and in the gas.

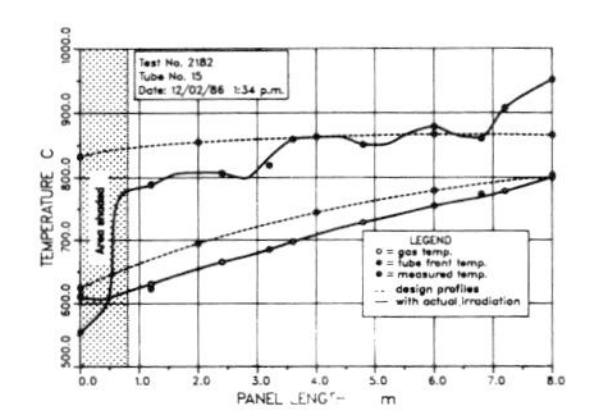

Fig. 8 Gas and max. tube temperature versus tube length

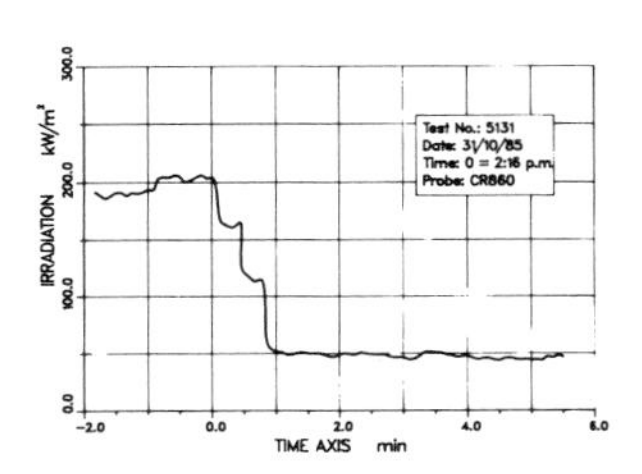

Fig. 9 Quick reduction of panel irradiation from 100 % to 25 %

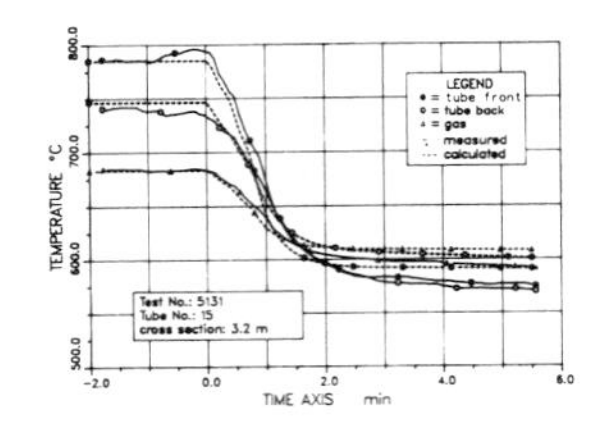

Fig. 10 Panel response on quick reduction of irradiation from 100 % to 25 % level

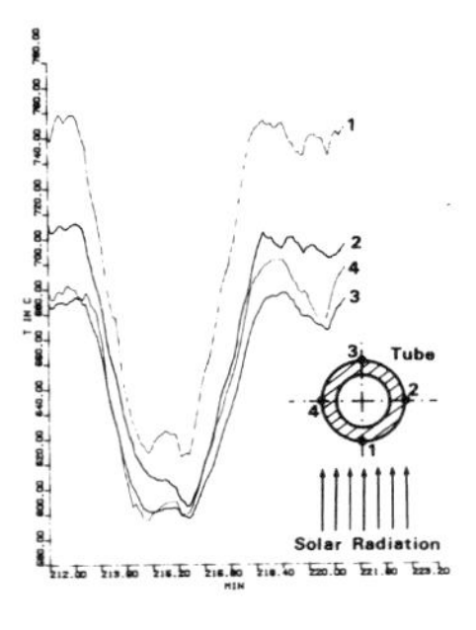

Fig. 11 Tube wall temperatures during "cloude cycle" test

Only 3 minutes after the reduction of the irradiation on the panel a new steady-state is reached. The measured data are in good accordance with the calculated, especially during the transient period.
Every time the tubes are heated up and cooled down greater thermal stresses are induced by greater temperature differences than at steady-state conditions (Fig. 11). This results in accelerated material damage.

SUMMARY

Fig. 12 Test panel in solar operation

Our complete test program has been carried out with great success:

- Perfect functioning under all tested operational modes was demonstrated.
- More than 650 hours of solar operation have been run (Fig. 12).
- Solar radiation intensities up to 260 kW/m² have been realized with gas temperatures up to 800°C and material temperatures of max. 900°C.
- The maximum temperature gradients measured (tube front/tube back) lie between 60 - 120 K.
- Besides the total number of 200 different tests more than 100 cycles simulating the load case "cloud covers and leaves the heliostat field" have been carried out by heliostat field-control.

Three detailed inspections of the complete panel during the panel tests did not show any signs of damage or distortion. The developed advanced design methods have thus been well verified for use in the design of high-temperature solar receiver components and may generally help to solve similar problems on other hot components.

ACKNOWLEDGEMENTS

This work has been undertaken within the GAST-project, which is performed under the contract of Interatom and sponsored by the German Federal Ministry for Research and Technology (BMFT).
The authors thank Mr. Roos from our department of thermodynamics and Dr. Agatonovic from our department of stress analyses for their helpful assistance.

THE IEA/SSPS HIGH FLUX EXPERIMENT

Testing the Advanced Sodium Receiver

up to Heat Fluxes of 2.5 MW/m²

W.J. Schiel, M.A. Geyer

Deutsche Forschungs- und Versuchsanstalt

für Luft- und Raumfahrt e.V. (DFVLR)

Pfaffenwaldring 38-40, D-7000 Stuttgart 80, West Germany

ABSTRACT

At the solar test facility Plataforma Solar de Almeria/Spain a Central Receiver (CRS) High Flux Experiment (HFE) was carried out with the Advanced Sodium Receiver (ASR). This Experiment demonstrated the feasibility of high power and high flux levels with the ASR concept by increasing the design heat flux from 1.4 MW/m² to 2.5 MW/m². Receiver efficiency was measured in various tests in a range from 0.8 MW to 3.5 MW with sodium inlet and outlet temperatures being 200/400, 270/530, 280/560 °C. The incoming flux and the receiver surface temperature was measured. These data were used to verify different thermodynamic models. The experiment showed that current design upper flux, power and temperature limits have not yet been reached and that there is a promising potential for increasing annual thermal energy output by up to 25% with a high flux sodium receiver.

KEYWORDS

Central receiver system, sodium, advanced sodium receiver, high flux, front surface temperature measurement, heliostat, SSPS-CRS

INTRODUCTION

Several Central Receiver System (CRS) Projects have shown that it is feasible to concentrate solar radiation and convert it into high temperature heat at promising efficiencies. The sodium heat transfer system at the *Small Solar Power Systems Project of the International Energy Agency* (IEA-SSPS) in spain has yielded the best receiver performance. The Advanced Sodium Receiver (ASR) was originally designed by Franco Tosi to absorb 2.5 MW heat with a 7.3 kg/s sodium massflow and 270/530°C inlet/outlet temperature, and to withstand a peak heat flux of 1.38 MW/m² and an average heat flux of 350 kW/m². The more detailed description is given by Grasse, 1985.

Before the High Flux Experiment, the ASR had been operating for almost 500 days with accumulated operating time of approx. 900 hours (Grasse, 1985). Maximum peak flux reached during this period was approx. 1.1 MW/m². However, at the designed 1380 kW/m², the SSPS Advanced Sodium Receiver's peak flux limitation has still not been reached. Stress analysis indicated that maximum incident fluxes on the order of 3000 kW/m² may be achieved. This would cut the receiver area required in half. Such a sodium receiver would have the following advantages over air or steam receivers: Have a better yearly efficiency because of reduced thermal losses, reduced thermal inertia, be more economical because much less material is required and therefore be of a simpler construction. This paper summarizes the HFE results. The detailed report is being published by Schiel et all, 1987.

EXPERIMENTAL RESULTS

Incident Flux Distribution

In order to achieve the desired high flux levels, the original 3-aiming-point strategy of the heliostat field was abandoned and a new 1-aiming-point strategy was implemented, in which all heliostats are focussing on the fifth panel. Figure 1 shows the two different flux distributions (1-aiming- and 3-aiming-point strategies) across the vertical centerline of the ASR-Receiver as calculated by HELIOS for solar noon at equinox for the 93 MMC heliostats with an irradiance of 920 W/m² and a design reflectivity of 91%. The drastic change in peak flux from 1.4 MW/m², the ASR design level, to nearly 2.5 MW/m² is clearly shown, while the total incident power stays constant at the ASR-design level of 2.8 MW.

Flux distribution was experimentally measured with the HFD-bar (Becker et all, 1981). This vertical bar is equipped with 11 water cooled HYCAL radiometers, that had been calibrated before the test by the manufacturer. For each measurement, the bar runs from West to East with eleven stops in between, taking a total of approximately 120 seconds for the round trip.

All peak/power combinations attained in the course of the High Flux Experiment are presented in Fig. 2. The receiver was operated at power levels ranging from 0.8 to 3.5 MW. Peak flux levels varied from 0.4 to 2.5 MW/m². The figure indicates a linear relationship between incident power and corresponding peak flux for both aiming point strategies. This linearity confirms that the flux distribution stays constant over the full range of incident power.

When changing from the 3-aiming-point to the 1-aiming-point strategy, the individual load of each panel changes significantly. Figure 3 shows the total power absorbed by the individual panels. While the joint contribution of the two outer panels drops from 15% to 9% when changing from the 3-aiming-point to the 1-aiming-point strategy, the fifth panel's share increases from 30 to 45%. This leads to high vertical surface temperature gradients on the fifth panel and horizontal on the third and fourth panels.

Temperature Distribution

Temperature distributions were evaluated for the receiver as a whole and for each individual panel with the THERESA and HOTREC receiver models. The infrared imaging system, HERMES (Schiel et all, 1986) was used to verify the ASR front surface temperatures calculated. This system resolves temperature differences of 1 K. Assuming 1.5% uncertainty in the emissivity of the Pyromark receiver coating and 1.5% in its reflectivity, the measured temperature uncertainty factor would then be 11.3 K. At 800 °C measured maximum surface temperature, this corresponds to a 1.4% error.

Depending on the a.m. flux concentration, a rise in the maximum receiver surface temperature from 630 °C (1.4 MW/m²) to 770 °C (2.5 MW/m²) can be observed. The thermodynamic calculations with HOTREC show that at this location the temperature drops from 770 °C on the Pyromark coating to the local bulk sodium temperature of 520 °C by 250 °C. This represents a temperature gradient of almost 100 °C across the 45μm Pyromark layer. The corresponding maximum gradient along the circumference of the central tube was calculated with THERESA and amounts to 132 °C (Rosa, 1986).

The average panel surface and sodium bulk temperatures were calculated by HOTREC (Lemperle, 1986) for two different peak fluxes (i.e. 1.4 MW/m² and 1.9 MW/m²) at a constant incident power level of 2.8 MW. A 20 °C rise in the average surface temperature can be observed on the central panel, while the average surface temperature of the other four panels drops with the flux concentration. Since the drop on these four panels outweighs the rise in the fifth panel, the average total receiver surface temperature drops from approx. 400 °C to 380 °C. This is confirmed by infrared measurement to a 3% tolerance. As the average surface temperature in the outer panels is reduced, the average sodium bulk temperature also drops there, leading to a decrease in the sodium inlet temperature in the fifth panel. Therefore, the average sodium bulk temperature also drops in the fifth panel, (from 472 °C to 463 °C), although its average surface temperature rises. The average temperature gradient between the front surface and the sodium bulk temperature increases from the original 68 °C to 93 °C.

Thermal Losses

An increase in average surface temperature leads to an increase in thermal losses. Several independent loss tests were carried out during the HFE in order to find out how these losses change with flux concentration.

The ASR average surface temperature measured with the HERMES infrared imaging system for the one-aiming-point tests are shown in Fig. 4. The thermal losses then have to be calculated as follows:

1. The *infrared losses* have been evaluated with Planck's law for every pixel of the infrared image and have been integrated over the whole receiver aperture area (Brinner, 1986).
2. The *convection losses* have been calculated with the HOTREC thermodynamic model by overlaying the free and forced convection (Lemperle, 1986) for the design wind velocity of $v_w = 5m/s$.
3. The *conduction losses* have been assumed to be 5% of the total thermal losses.

The calculated ASR infrared reradiation and convection losses are also shown in this figure. The average ASR surface temperature drops by changing from the three- to the one-aiming-point strategy. The total thermal ASR losses drop accordingly leading to a rise in efficiency. For the one-aiming-point strategy, the average ASR surface temperature $\bar{T}_R$ can be approximated with an accuracy of $\pm 3\%$ from the incident power P_R with the following linear relation:

$$\bar{T}_R[^{\circ}C] = 329.3 + 0.021 \cdot P_R[kW] \quad (1)$$

Advanced Sodium Receiver Efficiencies

Due to diurnal variation in solar insolation, the partload behaviour of the receiver is of utmost importance for the daily and yearly energy output of a solar thermal plant. An important goal of the High Flux Experiment, therefore, consisted in finding the receiver efficiency curve as a function of incident power and peak flux. A reliable way to obtain the receiver efficiency is the analysis of receiver losses Φ_L:

$$\eta_R = \frac{P_\alpha}{(P_\alpha + \Phi_L)} \quad (2)$$

In this relationship for receiver efficiency, the error in measurement of absorbed power P_R, nearly balances out. For the determination of the partload efficiency curve, it is of greater importance to measure relative changes in absorbed power than its absolute values. The efficiencies so obtained from all tests conducted are presented in Fig. 5 and compared with the theoretical efficiency curve obtained with HOTREC.

Agreement between measurements and theory may be affirmed. No significant change in receiver efficiency is observed when the aiming strategy is changed. Since neither increase in power nor flux concentration causes a significant rise in the average receiver surface temperature due to a consequent rise in thermal loss, receiver efficiency does not drop when operating the receiver under high peak and power levels, but even rises slightly over its original design value of approx. 90%. Furthermore, no dependence of receiver efficiency on variation in sodium inlet and outlet temperatures can be found.

Additional information for the design of future high flux sodium cooled external receivers can be obtained from an analysis of the highly irradiated center receiver panel no. 5. At the same incident power levels, panel 5 shows the lowest individual efficiencies. The reason again lies in its higher average surface temperature and the consequently higher thermal losses.

CONCLUSIONS

The High Flux Experiment demonstrated that sodium receivers can absorb much higher power and withstand much higher fluxes than it was designed for. This was shown in a wide temperature range from 200/480 °C to 280/560 °C sodium inlet/outlet temperatures. During the tests performed, the designed peak flux of 1.4 MW/m² could nearly be doubled. The total incoming power could be increased from the designed 2.7 MW (averaged flux of 340 kW/m²) to 3.4 MW (averaged flux of 430 kW/m²) by making use of the 30 MBB Heliostats that have been added to the original Martin Marietta heliostat field.

The results presented indicate that external sodium receivers can be operated under much higher flux and power levels than they are presently designed for. This shows that current design upper flux, power and temperature level limits have not yet been reached, and that there is promising potential for improvement. The following conservative estimate assesses this potential: In order to attain flux conditions similar to those attained in the High Flux Experiment in which peak flux was approx. 2.5 MW/m², with the existing receiver absorbing area, the heliostat field area would have to be twice that used with the current 3-aiming-point strategy. The receiver would then operate in a more favourable partload region of its efficiency curve. The corresponding shift of the operating range from the original 0.6-2.8 MW range to the high flux 2.5-5.0 MW range is illustrated in Fig. 6. Apart from a 5% improvement in average daily efficiency, early startup becomes possible, since with the increased solar multiple, minimum radiation of 200 W/m², instead of the originally required 300 W/m², appear to be sufficient for a positive thermal energy gain. Thus even though receiver absorbing area and heliostat field area were to be increased with the original "low" flux levels, the proposed high flux operation would still increase the amount of thermal energy collected annually by up to 25%.

REFERENCES

M.Becker, J.Bäte, F.Diessner: Device for the measurement of heat flux distributions (HFD) near the receiver aperture plane of the Almeria CRS Solar Power Station. IEA-SSPS Technical Report 5/81. Cologne 1981

A.Brinner: Test of the ASR with an IR-camera system during the High Flux Experiment. Internal DFVLR-Report IB444 002/86. Stuttgart 1986

W.Grasse: IEA-SSPS Results of Test and Operation 1981-1984. IEA-SSPS SR7. DFVLR Cologne. May 1985

G.Lemperle: Numerische und experimentelle Untersuchung der instationären Temperaturverteilung in Strahlungsabsorbern. DFVLR Forschungsbericht DFVLR-FB 86-43 Köln 1986

F.Rosa: Thermodynamic simulation of the ASR using 'THERESA'. IEA-SSPS Internal Report R-36/86FR.Almeria 1986

W.Schiel, M.Geyer, R.Carmona: The IEA-SSPS High Flux Experiment - Testing the Advanced Sodium Receiver at Heat Fluxes up to 2.5 MW/m². Springer-Verlag Berlin Heidelberg New York. October 1987

W.Schiel, G.Lemperle: Surface Temperature Measurements of a Sodium Cooled Solar Central Receiver using an Infrared Imaging System. Paper No. A86-54 in the Proceedings of The Metallurgical Society of AIME. New Orleans 1986

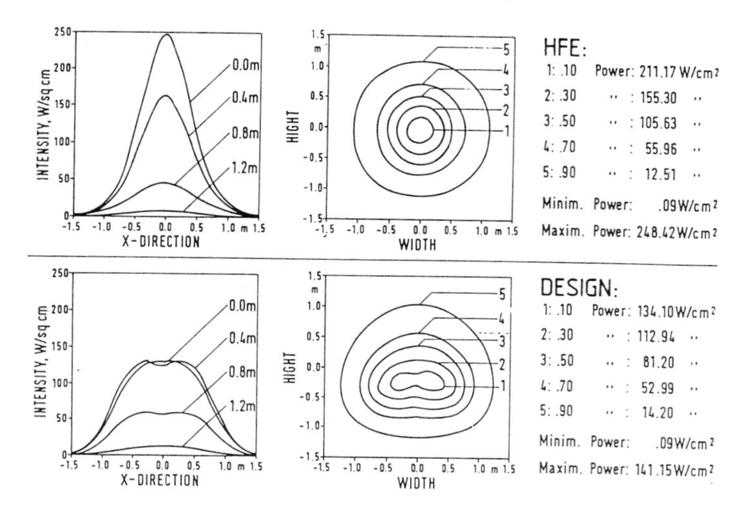

Figure 1. **Flux across the vertical center and Iso-contour lines on the ASR plane** as calculated by HELIOS for the 3-aiming- and 1-aiming-point strategies (solar noon day 80, 93 MMC heliostats, 920 W/m² irradiance).

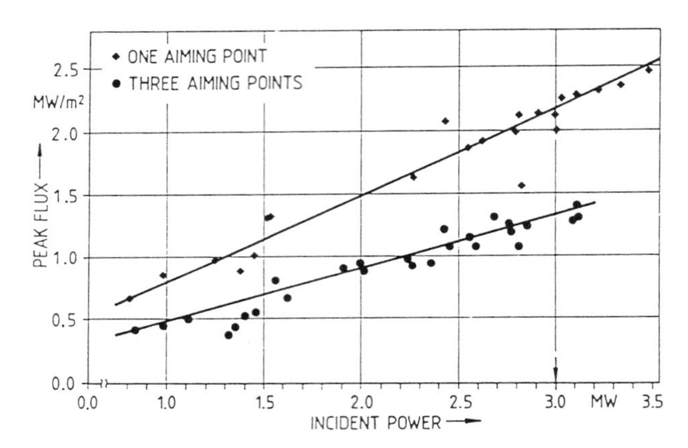

Figure 2. Peak/power levels attained in the High Flux Experiment

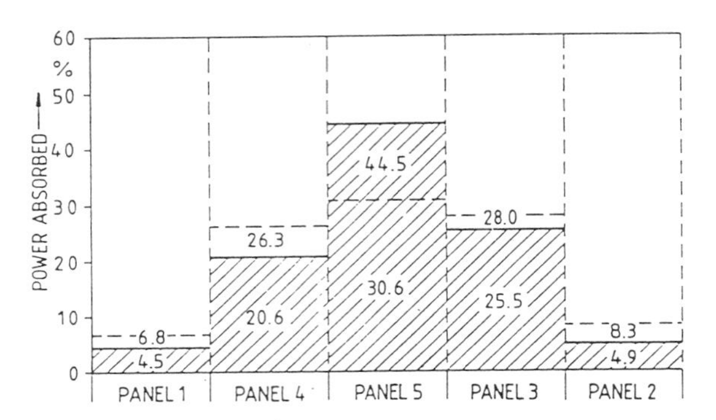

Figure 3. Average power absorbed by the individual panels with both aiming-point strategies

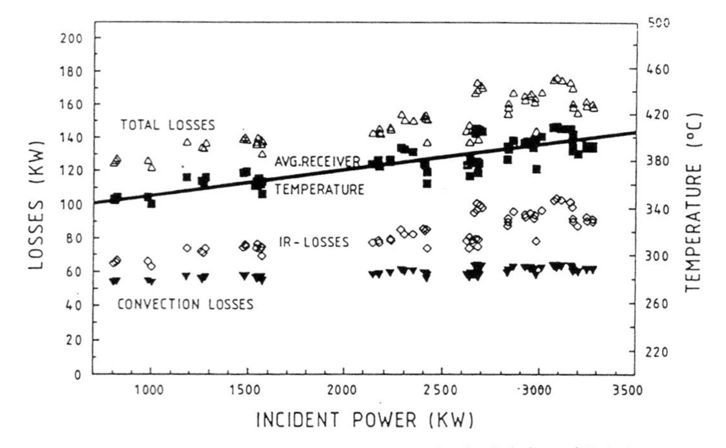

Figure 4. ASR losses and surface temperature for the 1-aiming-point strategy

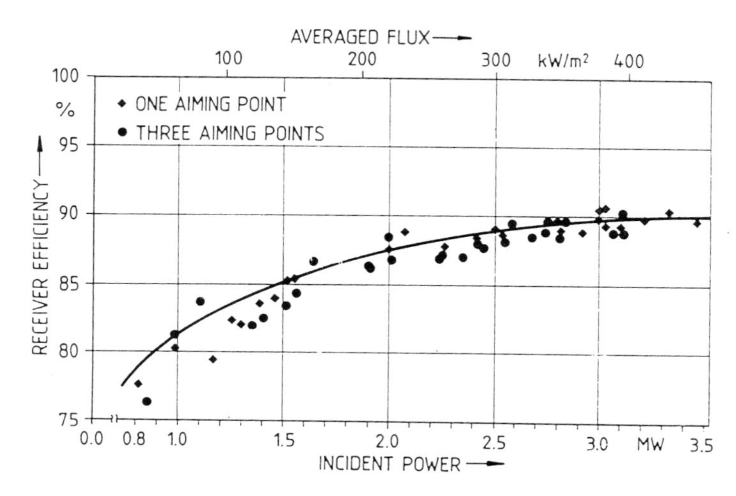

Figure 5. **Receiver efficiencies obtained from HFE loss tests**

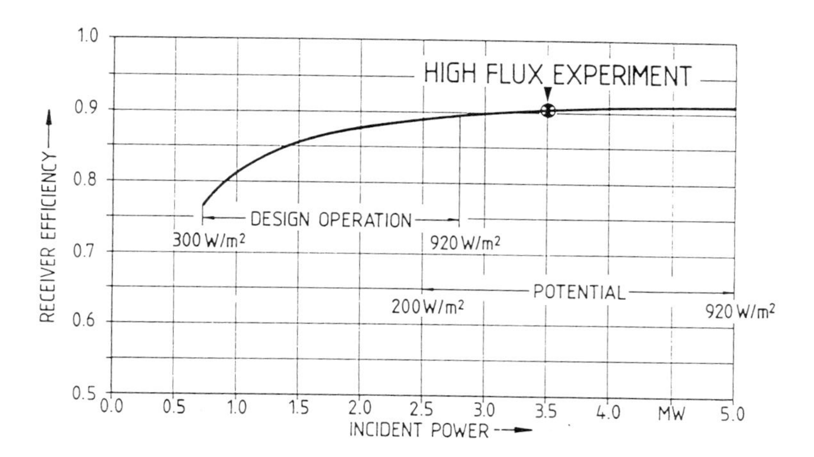

Figure 6. **Potential performance of a future high flux sodium receiver** compared to ASR design performance

DETERMINATION OF THE FEASIBLE POWER/PEAK REGION AND THE HELIOSTAT FIELD CONFIGURATION TO ACHIEVE THE DESIRED POWER/PEAK LEVELS

R. Carmona, M. Blanco, and M. Silva

CIEMAT-Instituto de Energías Renovables,
Plataforma Solar de Almería (Spain)

ABSTRACT

Two computer programs are presented in this paper: one implements a method of determining feasible power/peak levels for a central receiver system under given conditions; the other selects a heliostat field configuration to produce the desired power/peak level within a defined margin of error.

KEYWORDS

Heat-flux distribution; incident thermal power; peak heat flux; power-peak region; heliostat field configuration; heliostat aiming strategy.

INTRODUCTION

For the measurement campaign of the Advanced Sodium Receiver High Flux Experiment (ASR-HFE) (Schiel, 1987), evaluators agreed to define working points in terms of four parameters; two are the coolant temperatures at the receiver inlet and outlet. The other two refer to the flux incident on the ASR plane--they are: (1) the peak of the flux density distribution, and (2) the total power incident on the receiver surface. Not all flux distributions are available under given conditions.

A computer program, FERE, implements a method to determine which power/peak levels are achievable. Once a working point has been selected, another computer program, HELSEL, provides a heliostat field configuration producing the desired power/peak level, within a user-defined margin error. Results of the application of these programs during the HFE prove their utility.

DETERMINATION OF THE FEASIBLE POWER/PEAK LEVELS

Within certain limits, incident thermal power and heat flux peak on the receiver surface can be chosen by selecting a specific heliostat field configuration for a given aiming strategy.

Limits are determined by two curves: the first maximizes the ratio 'peak flux/thermal power'; the second maximizes the inverse relation 'thermal power/peak flux'. These two curves depend on meteorological (irradiance and Sun position) and operational (aiming strategy, heliostat availability, and heliostat mirror reflectivity) factors.

A computer program, FERE, was developed to determine the limits of the feasible power/peak levels. The input data are: HELIOS output file for a given day, time, heliostat aiming strategy, reflectivity=1, irradiance-1000 W/m^2; identification numbers of the non-available heliostats; actual average field reflectivity; and actual (predicted) irradiance at the time of the test.

The program produces a graphic output (Fig. 1). Feasible power/peak levels for the selected aiming strategy are plotted in the power-peak plane. If a given power/peak level is within limits of the feasible power/peak region, that working point is available within a margin of error, given the discrete nature of the heliostat field within operating and meteorological conditions given to the program.

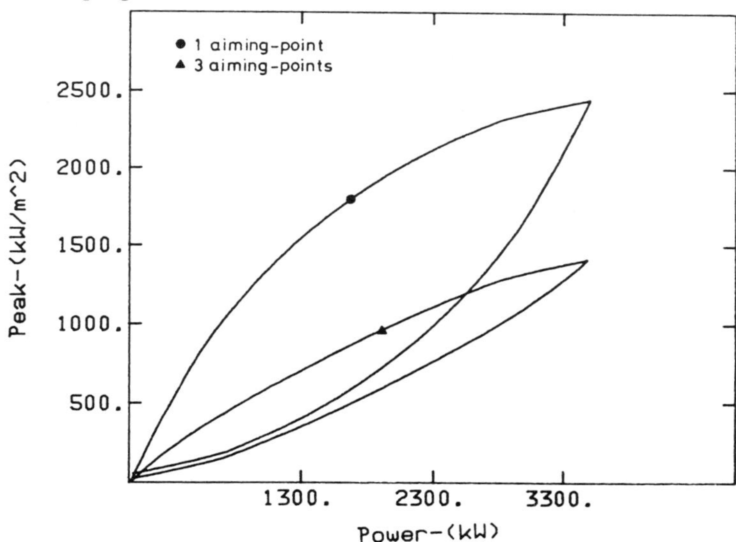

Fig. 1 Feasible power-peak regions calculated by FERE for day 80 at noon: irradiance=920 W/m^2, reflectivity=0.911, 1- and 3-aiming points

Parameters which Affect the Feasible Power-Peak Region

Irradiance. When all other parameters are constant, irradiance is a multiplicative factor which does not affect the shape of flux distribution on the receiver surface, but its size.

Sun position. This is a function of the day of the year and solar time. It affects the heliostat field geometrical factor, thus modifying the capacity of the field to produce a given flux distribution on the receiver surface.

Heliostat aiming strategy. An increased number of aiming points produces the following effects: a small decrease in incident thermal power due to increased spillage; and reduced peak flux for a given thermal power. Therefore, feasible peak/power levels obtained by 1-aiming point strategy will be higher than those produced with 3-aiming points in the peak/power plane.

Heliostat availability. Decreasing the number of available heliostats decreases the maximum feasible incident thermal power. Moreover, flux distributions produced by different heliostats differ, so that heliostats placed in the first row produce a higher flux peak and incident thermal power than those at the bottom of the field. It is not enough to know how many, but which heliostats are unavailable.

Reflectivity of the heliostat mirrors. Reflectivity, similar to irradiance, works as a multiplicative factor. Therefore, it has the same effect.

DETERMINATION OF THE HELIOSTAT FIELD CONFIGURATION TO ACHIEVE THE DESIRED POWER/PEAK LEVELS

Once a feasible power/peak level has been chosen, one must select which heliostats have to be placed in track to achieve it. This selection depends on meteorological and plant conditions at the moment of the test.

HELSEL has been developed to perform the heliostat selection. It is based on artificial intelligence techniques, and written in PASCAL language. PASCAL was used because some of its features make it particularly appropriate for this kind of application.

Input Data

HELSEL needs two groups of data: Group A supplies information relating to the operating point to be achieved; Group B supplies information about the heliostat field status. Table 1 shows the input data of each group.

TABLE 1 Input Data Necessary for HELSEL Groups A and B

GROUP A		GROUP B	
Datum	Definition	Datum	Definition
SPEAK	Flux power on ASR plane (kW/m^2)	IRRAD	Direct irradiance (W/m^2)
SPOWER	Incident power at ASR (kW)	INFILE	File name containing HELIOS info. about normalized flux combination of heliostats
SERROR	Maximum combined error allowed on peak and power (%)		
NODES	Maximum number of nodes to explore	HELOUT	heliostats out of service
		REF	aver. heliostat reflectivity

In Group A, SPOWER and SPEAK are the power and peak levels defining the operating point to be achieved. SERROR may need some clarification. The discrete nature of the heliostat field (i.e., only complete heliostats may be put in or out of track), makes obtaining an exact operating point impossible. Desired peak and power values will differ from those which could be supplied for any selected group of heliostats. However, it does not make sense to search all possible heliostat sets for those minimizing these differences. Instead, the margin of error allowed for a given operating point is defined and the search stops when a set fulfills this condition; SERROR specifies this margin. If a solution is obtained, the final set of heliostats selected by HELSEL satisfies the following expression:

$$SERROR = ERRPK^2 + ERRPW^2 \qquad (1)$$

where ERRPK and ERRPW are, respectively, the relative differences (%) between calculated peak and power levels for selected heliostat sets and desired levels.

In Group B, actual attainment of a required operating point depends strongly on the accuracy of the input data in this group. For each heliostat in service, HELSEL must estimate its possible flux combination at the receiver at time TS, when the operating point must be achieved. The possible flux combination for a heliostat is characterized in the code by HMATRIX(N) and HPOWER(N) where: (1) HMATRIX(N), an 11 x 11 matrix defining the flux distribution produced by heliostat N at the receiver at TS, if the heliostat is in track; and (2) HPOWER(N), the integral of this flux distribution, or incident power that the heliostat could supply to the receiver.

To estimate flux combinations, HELSEL uses direct irradiance data (IRRAD), and the reflectivity (REF), and information contained in the input file INFILE. INFILE data are extracted from HELIOS output. For each heliostat, these data are: (1) identification name HLABEL(N); (2) normalized incident power (kW) NPOWER(N); and (3) 11 x 11 normalized flux matrix distribution (kW/cm^2) NHMATRIX.

Normalized incident power and flux matrix distribution are obtained running HELIOS (for a given time and sunshape) with irradiance equal to 1000 W/m^2 and reflectivity equal to 1.

Expressions used in HELSEL to calculate HMATRIX(N) and HPOWER(N) for a given heliostat are:

$$\text{HPOWER(N)} = \text{IRRAD.REF.NHPOWER(N)} \quad (2)$$

$$\text{HMATRIX(N,I,J)} = \text{IRRAD.REF.NHMATRIX(N,I,J,)} \quad (3)$$

where REF is the reflectivity.

Output Data

With HMATRIX and HPOWER for each heliostat in service, the code uses a heuristic search procedure to find a set of heliostats satisfying conditions imposed by the parameter of Input Group A. If a solution is found, the following message is displayed on the screen:

SOLUTION FOUND AT NODE NUMBER 5
SET OF HELIOSTATS RECORDED ON FILE: OUTFILE.OUT

OUTFILE.OUT contains the identification number of selected heliostats and values of parameters IRRAD, REF, SPOWER, and SPEAK, used in running HELSEL.

If no solution is found within the number of explored nodes specified by NNODES, the following message will be displayed on the screen:

SOLUTION NOT FOUND

and no OUTFILE.OUT will be created.

RESULTS

Table 2 lists results obtained running HELSEL for different operating points during the HFE campaign.

The first two columns give dates and times of actual measurements. The third column, PIRR, lists irradiance predicted by extrapolating from irradiance measurements made between 8:00 and 9:00 on each given day. Predictions were made for 11:00 or 12:00, according to the chosen operating point. The fourth column, OIRR, gives the actual irradiance reading pyrheliometer at the time of the test. The next column, PKA, gives the peak flux at the ASR plane calculated from measurements taken by the HFD bar, and the PVA column lists total power incident on the receiver, calculated from these measurements.

TABLE 2 HELSEL Results

Operating points: PEAK, POWER	Date	Time	PIRR	OIRR	PKA	PWA
500, 1300	2308	11:10	625	663	542	1424
	2808	10:51	870	905	536	1409
	2808	11:03	870	907	542	1417
	2808	10:26	870	905	541	1444
	2908	11:11	905	907	548	1359
	2908	10:22	905	907	561	1400
	3008	11:47	720	773	493	1363
1000, 2300	2008	12:49	905	878	1046	2279
	2808	13:03	905	881	1004	2229
	2808	13:19	905	867	1026	2243
	2908	12:52	870	900	1124	2422
	2908	13:00	870	883	1091	2382
	2908	13:08	870	883	1153	2410
	3008	13:32	720	***	1006	2115
1000, 1300	3008	11:46	***	***	935	1328

REFERENCES

Biggs, F., Vittitoe, Ch. (1979). The HELIOS model for the optical behavior of reflecting solar concentrators. Sandia National Laboratories, Albuquerque, New Mexico.

Schiel, W. and others (1987). The IEA/SSPS high flux experiment, Springer-Verlag.

THEORETICAL INVENSTIGATIONS ON THE REFLECTING AND FOCUSING SURFACE OF A FIXED MIRROR LINE FOCUS SYSTEM

G. Atagündüz*, M. Eltez*

*Solar Energy Institute, University Ege
Bornova, İzmir, Turkey

ABSTRACT

Research and development studies for the production of electrical energy from solar energy are undergoing in many countries as an alternative to conventional and nuclear power plants. The Solar Power Plant with Concentrating Mirrors is one of the methods of conversion of solar energy in electricity. It has been proven that the major investment cost percentage of solar plants with tracking heliostats is the Heliostat Structure which consists mainly of reflector surfaces, heliostat support constructions and tracking mechanism. The investigations have been performed for minimizing the number of mobile parts in Reflecting and Focusing area for a low-cost alternative to the heliostat systems, in Ege University Solar Energy Institute.

KEYWORDS

Solar power plants, fixed mirror system, line focusing, reflecting and focusing fixed space surface, solar concentration.

INTRODUCTION

Different techniques have been used to concentrate the solar energy. The oldest concentration technique is the one axis tracking parabolic mirros. Another technique is the so called "power tower which consists mainly one central reciver and two axis tracking heliostats. The concentration of the solar energy using a receiver at the top of a tower can be obtained also through fixed mirrors. There are two main systems of the fixed mirrors ;

Fixed Mirror Solar Concentrator and
Fixed Mirror Distributed Focus-System.

Solar Energy Institute of the Universtiy EGE has attempted to establish a new concept of fixed mirror systems, namely, Reflecting and Focusing Space Surface which should fulfill the sun tracking function through single, double and even triple reflection characteristics of the choosen space surface.

REFLECTING AND FOCUSING SPACE SURFACE CONCEPT OF THE SOLAR ENERGY INSTITUTE

The basic idea to design a three dimensional space surface which should follow the sun was to use single, double even triple reflections. The geometrical place of focal points of the reflected solar rays from this surface should be a line. In order to realize this idea a space surface will be estimated and the calculations of the focus points of the estimated reflecting and focusing surface will give an estimated focus line. If the estimation of the space surface were correct, then the calculated focusing points should fit to the focusing line of the existing tower with the height of 22 m, Fig.1; if not a new reflecting and focusing area must be estimeted to repeat the whole calculations. The iteration will last until the difference between the calculated focus line and the actual focus line of the existing tower will be near zero, Fig.1. For this purpose a computer program has been developed which analys s the following points :

1- A surface equation is estimated. The coefficients of this equation are loaded to the computer as unknown quantities,
2- This estimated surface is put in parametric form,
3- To start the calculation, a certain day and the local time are put in the program,
4- The coordinates of the reflecting point A are given,
5- The local latitute angle is given to calculate necessary angles,
6- Direction cosines of the incident solar rays, are calculated for the given local time,
7- Direction cosines of the surface normal at point A is determined in terms of surface coefficients,
8- Direction cosines of the first reflection ray, are determined in terms of surface coefficients through reflection laws,
9- The beam of the first reflection strikes the fixed reflecting focusing surface at the point, B ,
10-Determination of the direction cosines of the surface element at the point B in terms of surface coefficients,
11-Establishment of the relations for the beam of the second reflection which should strike on the receiver,
12-Going back to the 4th analysis point a new point A will be choosen,
13-Repeat the calculation between the steps 4. and 12. for the each point of the whole reflecting and focusing surface,
14-Going back to the 3rd analysis point an other local time, will be choosen and the whole calculations will be repeated,
15-After having these analysis compute the surface coefficients,
16-If the error matrix is out of the usual limits, repeat the whole calculations for a new estimated reflecting and focusing surface.

CONCLUSION

A general approach is made to the reflecting and focusing surfaces and present technologies are reiewed. Development studies should be maintained at this subject but preferance should be on fixed surfance reflecting and focusing surfaces instaed of heliostat technology in Turkey.

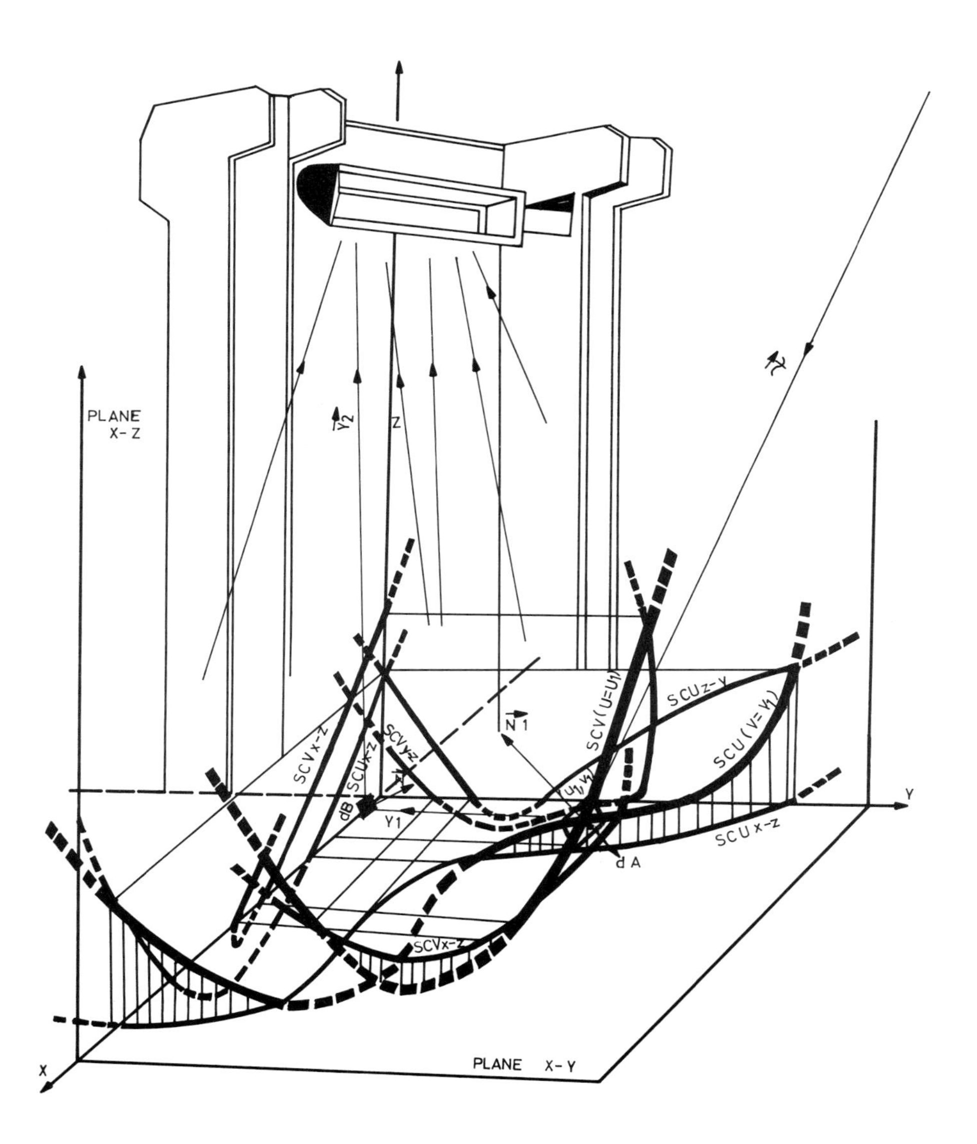

Fig.1. Reflecting and focusing space surface concept of the Solar Energy Institute.

Results of the solar power plant using fixed mirror technology will be evaluated and used at the commercial solar power plant stage. Results of the theoretical investigations on the reflecting and focusing surface of fixed mirror line focus system show that there is a chance to decrease the total direct cost of a solar power plant.

REFERENCES

Eltez, M. (1986). Sabit Yansıtıcılı-Çizgisel Odaklı Kule Projesinde Yansıtıcı Odaklayıcı Yüzeyin Şekillendirilmesi. Ph.D. Thesis, Ege University Solar Energy Institute, İzmir.

Selçuk, M.K. (1978). Fundamentals and collectors of the past and present. JPL., California Institute of Technology, Pasadena.

Hearing. (1976). The Economics of Solar Energy. Joint Economic Comittee of the United States 94th Congress.

Backus, C.E. (1975). Fixed Mirror Solar Concentration for Electrical Generation. Arizona State University, Tempe.

Ortabaşı, U. (1980). An internal cusp reflector for an evacuated tubular heat pipe solar thermal collector. Solar Energy, Vol.25, 67-78.

Umarov, G. (1976). Radiant vector distribution in the radiant field of a parabolcylindiric concentrator. Geliotekhnika, Vol.12, No.1, 27-32.

ESTIMATION OF REFLECTION LOSSES FOR A HORIZONTAL AND INCLINED HELIOSTAT FIELD

S. D. Oda N. A . Dawood

Solar Energy Research Center
P.O.Box 13026 ,Baghdad ,Iraq

ABSTRACT

In this study an analytical comparison was carried out between horizontal and inclined heliostat fields.The required equations were derived for the two dimensional case for two identical heliostats placed in line. Shading, blocking and cosine reflection angle losses were determined daily and throughout the year

KEYWORDS

shading; blocking; central receiver; inclined field; flat field

INTRODUCTION

The design of solar central receiver plant(SCRP) depends on four major principles: heliostat design, the performance of the central receiver(heat transfer losses and flux densities), the tracking accuracy of the heliostat and the distribution of heliostats in the field. The later is studied by considering the distribution of heliostats around the central receiver to give an optimum design with minimum optical(reflection) losses. These losses consist of shading,(the interception of the incident rays by the front heliostat) blocking (the interception of the reflected rays from the rear heliostat by the front one) and cosine losses (due to the incident ray is not normal to the heliostat). Optical losses are variable during the daily hours and throughout the year. The reduction of these losses will not only have an effect on the plant performance, but also on the initial cost by reducing the area covered by the heliostats. Flat fields are widely used in SCRP s and many studies (Abatut, 1978; Aparisi,1980; Elsayed,1986; Vant-Hull,1974)have been made to estimate shading, blocking losses and to optimize the heliostat field area . In this work , a preliminary study was carried out on the inclined field type to find out the effect of slope angle on receiver height for the perpose of reducing capital cost .

ANALYSIS OF THE RAYS GEOMETRY

In this study the required equations were derived for the two dimensional case. The receiver, incident and reflected rays and the heliostats were taken in one plane passing through a north-south line. The receiver position was taken for several points on the east-west line. The required equations for the estimation of solar angles were adapted from Threlkeld(1962). The location of the heliostats was taken for a latittude angle of 33.3° . Initial field features were selected from Eurelios (1984) (suchas receiver height 50m,distance between two heliostats and heliostat size 5m). Figure 1 shows a general view of the two types of field (I and II)with the main angles involved. The distance(R) from the receiver was taken equal in both types of field. The tilt angle(Z) is estimated as:

$$Z = 90 - Z' \quad \text{(Field I)} \qquad (1)$$
$$Z = 90 - Z' - S \quad \text{(Field II)} \qquad (2)$$

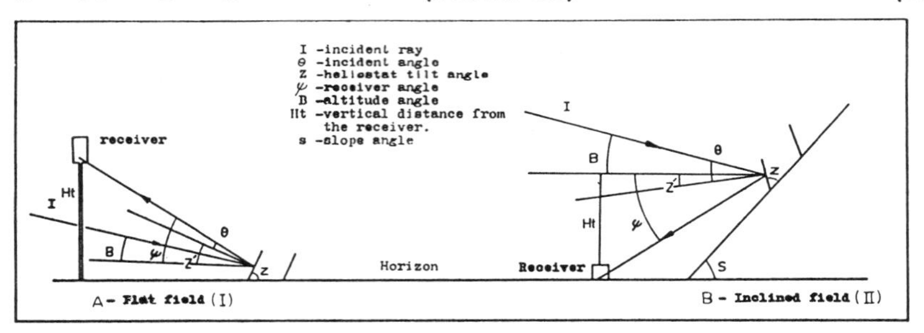

Fig. 1. Field angles

Figure 2 illustrates the angles between two heliostats. In this figure there are two conditions to estimate the angle (Vs) :

I- When the angle U > 90

$$Vs = 180 - \cos^{-1}((L^2 + R^2) - W^2)/2LR - Z \qquad (3)$$

II-When the angle U <= 90

$$Vs = 180 - \tan^{-1}(Rv / Rs) - Z \qquad (4)$$

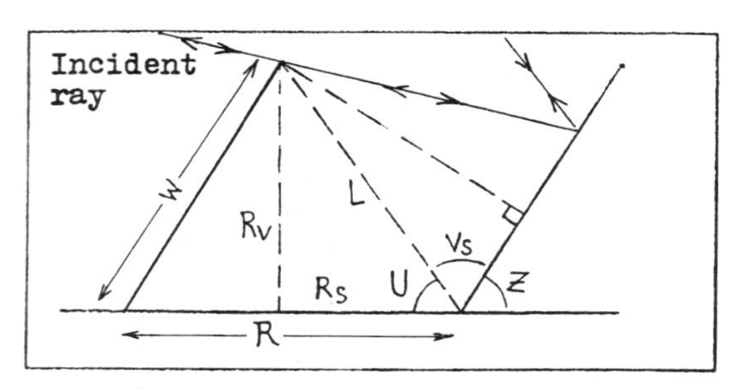

Fig. 2. Heliostat angles

The value of the angle (Vs) restricts the type of shading (Sh) and blocking (Bl). Thus there are two cases:

I- When Vs<= 90

a- $Sh = L (\cos Vs - \sin Vs / \tan V)$ (5)

Equation(5) represents the shading for both fields. It should be noted that in the flat field, when the altittude angle is less than receiver angle(ψ), the negative sign is changed to positive.

b- $Bl = L (\cos Vs + \sin Vs / \tan V)$ (6)

Equation (6) represents the blocking for both fields. In the flat field and when altittude angle is less than receiver angle (ψ) , blocking will have no effect (the reflected rays are not intercepted by the front heliostat).

II- When Vs > 90

This condition appears when the distance between heliostats is greater than the heliostat width (W) and the value of tilt angle (Z) is small.

a- $Sh = L (\sin (180 - Vs)/\tan V - \cos (180 - Vs))$ (7)

Equation (7) is valid for field I only. For field II this case is impossible since the incident ray will be under the horizon (see fig.1 & 2).

b- $Bl = L (\sin(180 - Vs)/\tan V - \cos (180 - Vs))$ (8)

Equation (8) is used for both fields, and it is looks like eq.(7).

A computer programme was constructed to estimate the losses hourly and throughout the year.

DISCUSSION AND CONCLUSION

The results of the above analysis can be summarized in the following remarks:

1-The distance from the receiver does not have the same effect on the two fields . As shown by fig.3 total losses (shading ,

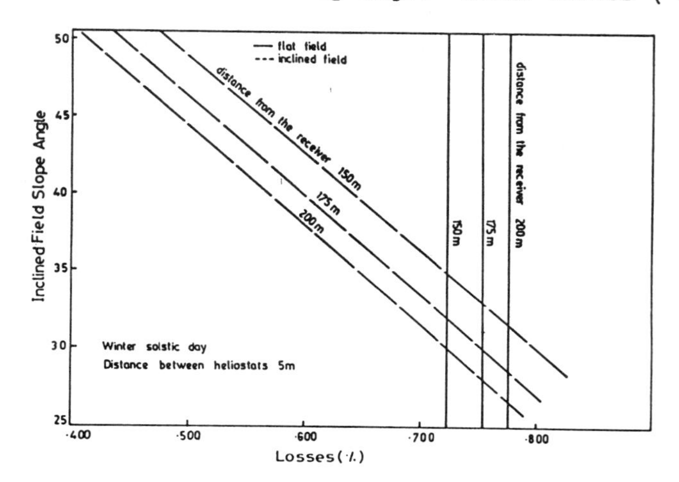

Fig. 3. The effect of slope angle on the reflection losses .

blocking and cosine losses) increase when the distance from the receiver increases in field I.This is opposite to the effect on field II whereas the total losses change by slope angle value.

2- The behaviour of field losses is shown in fig.4 which reveals the component of optical losses for winter solstic day.For both fields shading is decreasing upto solar noon and then starts increasing .On the other hand blocking in both fields increases with daily hours upto solar noon. This behaviour of blocking is not constant for the inclined field during the daily hours throughout the year. An opposit shape of losses occurs in summer solstic as shown by fig.5.

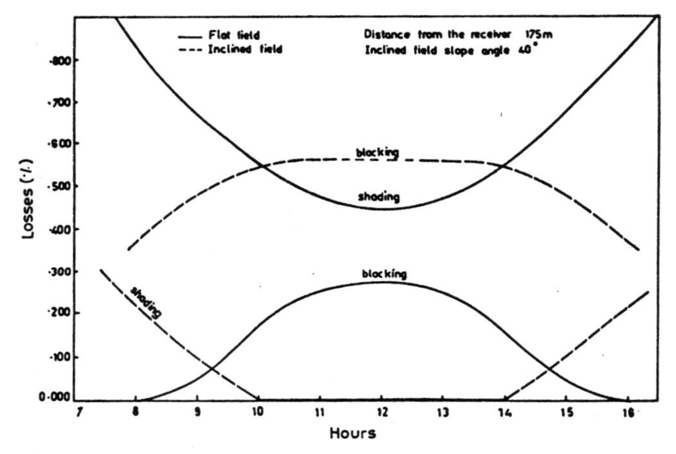

Fig. 4. Reflection losses for winter solstic.

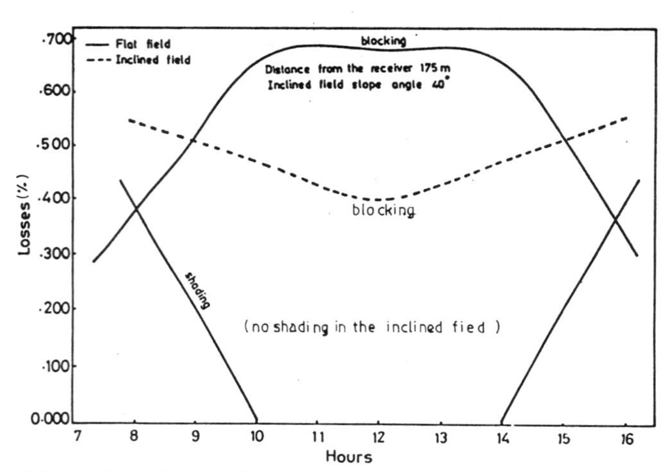

Fig. 5. Reflection losses for summer solstic.

3- Representation of these losses throughtout the year is shown by fig.6. For the condition shown by this figure , the inclined field has no shading losses throughout the year while for the flat type shading does not occur only in the period between day number 101 and 241 . Figure 6 reveals the small effect of cosine losses in the flat field while it is considerable in the inclined one .

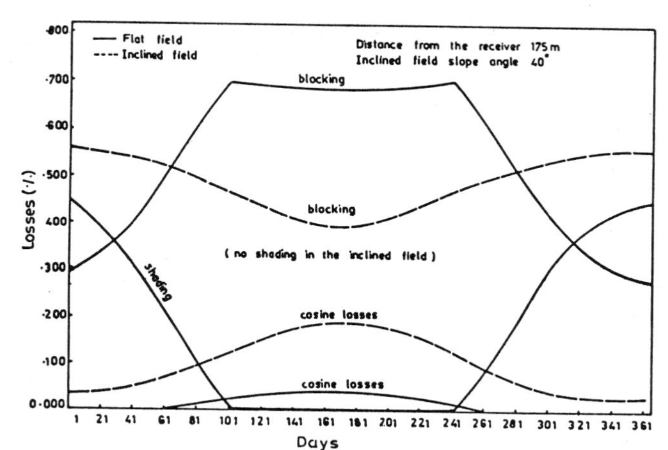

Fig. 6. Reflection losses for solar noon throughout the year .

As a conclusion , the application of inclined field is recommended as longe as there is a relatively large distance between the heliostats and the receiver. The reduction of its losses is enhanced by the increament of slope angle.

REFERENCES

Abatut,J.L., and A.Achaibou (1978). Analysis and design of a field of heliostat for a solar power plant.Sol.Energy ,vol.21 453-463.

Aparisi,R.R. ,D.I.Teplyakov, and B.G.Khantsis(1980). Geometry and kinematic characteristics of heliostats for a tower-type solar power plant.Gelioteknika ,vol.16,1, 19-25

Elsayed,M.M., and K.A.Fathalah (1986).Estimation of percentage useful area of a heliostat when considering shading and blocking.sol.&wind Tech.,Vol.3,3,199-205.

EURELIOS (1984). The 1MWel experimental solar thermal electric power plant of the european community,seriesB,vol.1

Lipps,F.W., and L.L.Vant-Hull(1974).Shading and design geometry for a solar tower concentrator with rectangular mirrors. ASME papar 74-wa/sol-11. New York.

Threlkeld J .L. (1962) . Thermal Enviromental engineering prentice-Hall,U.S.A.

ACHIEVEMENTS OF HIGH AND LOW RADIATION FLUX RECEIVER DEVELOPMENT

M. Becker and M. Boehmer
DFVLR, RF-ET
D-5000 Koeln 90

Abstract

The present state of the solar thermal receiver development is described. Cavity and external concepts are explained; volumetric and direct absorption receivers are presently under research.

Keywords: **Solar Thermal Energy, Solar Tower Test Facilities, Receiver, Cavity, External, Volumetric, Direct Absorption**

Within the last decade, the solar thermal technologies were developed from concepts to demonstration units by various research institutions and industrial companies. The systems under consideration are the centralized tower with temperatures from 500 ^{o}C to more than 1000 ^{o}C, the parabolic dish with similar temperatures and the parabolic trough farm with temperatures below 350 ^{o}C. Essential parameters are also the power ranges (100 kW to 100 MW) and the applications of the thermal energy (e.g. electricity production; utilization in chemical processes to drive endothermic reactions).

The central receivers of the tower system display a spectrum of universal application and are therefore highly attractive also for research and economical view points. Such a receiver has the capacity to heat various heat transfer media by large modules to high temperatures and achieve high and efficient performances. The topic of the present paper, therefore, deals with the investigations on receivers of different concepts for high, medium and low flux densities.

The following table describes the possibilities of using different heat transfer fluids as water/steam, molten salt, sodium and gas, especially air. The characteristic features of these selections and the test receivers operated at solar facilities in the USA, Spain, France, Italy and Japan are specified.

Fluid	Characteristics	Test Receivers
Water / Steam	- Available Technology - High Pressure - Steamturbine	- Solar 1 (Barstow / USA) - CESA 1 (Almeria / Spain) - Eurelios (Adranon / Italy) - Sunshine (Nio / Japan)
Molten Salt	- High Heat Flux Density - Good Thermal Conductivity - Best Suitability for Storage - Corrosion Problems	- CRTF (Albuquerque / USA) - Themis (Targasonne / France)
Sodium	- Best Heat Flux Density - Best Thermal Conductivity - Safety Requirements - Best Receiver Response	- IEA - SSPS (Almeria / Spain) - CRTF (Albuquerque / USA)
Gas	- Possibility of Open Cycles - Gasturbine - Low Thermal Conductivity - High Temperatures (> 1300 K)	- GAST (Almeria - Spain) - CRTF (Albuquerque / USA)

Most of the facilities used water/steam since this technique has the advantage to operate within one cycle receiver and steam turbine. Besides, this is the conventional way of electricity production in present power stations. Molten salts have good thermal properties (heat conductivities) and are very suitable to be integrated with storage systems to enable and secure the operation of a large-scale tower power station.

The best thermal performances (heat conductivities) can be expected from sodium since so the operational temperature range spreads from 200 to 600 ^{o}C. However, safety regulations cause problems because sodium reacts with water and burns in case of leakages.

Up to now, air has been used in the investigations of simplified panels (characteristic receiver elements) in Almeria, Spain and Albuquerque, USA. This concept offers the advantage of using a gas turbine within the receiver cycle or feeding a high temperature chemical process.

Fig. 1 shows the schematic sketch of a solar tower facility. The heliostat field concentrates the incident solar radiation, the receiver transfers the radiation into thermal energy. The upper part of Fig. 1 shortly describes the set-up of electricity generation for different flow media, the lower part presents other possibilities of thermal heat utilization as for process heat or for the production of fuels and chemicals.

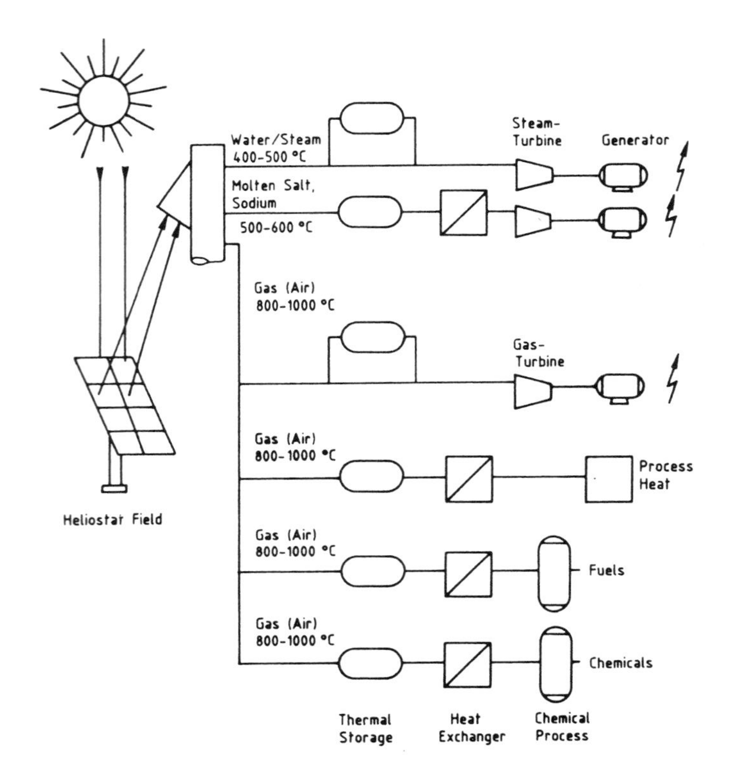

Fig. 1
Schematics of Solar Tower Facilities

The choice of heat transfer fluid determines the range of heat flux densities applicable in the needed materials respectively the containments. From these considerations, consequently the geometric receiver concept has to be derived. This can be characterized by the following ways:

- Cavity Tube Receiver (<u>Fig. 2</u>)
  The piping is mounted in a closed room that has only a minimum opening towards the heliostat field for reasons of protection against thermal losses. This is needed, if the tolerable heat flux density is small (e.g. gases).

- External Tube Receiver (<u>Fig. 3</u>)
  The heat transfer area is designed as light and small as possible and therefore needs no protection because the losses relate to the exposed area. So, the heat flux densities have to be developed to maximum values (e.g. sodium and salt).

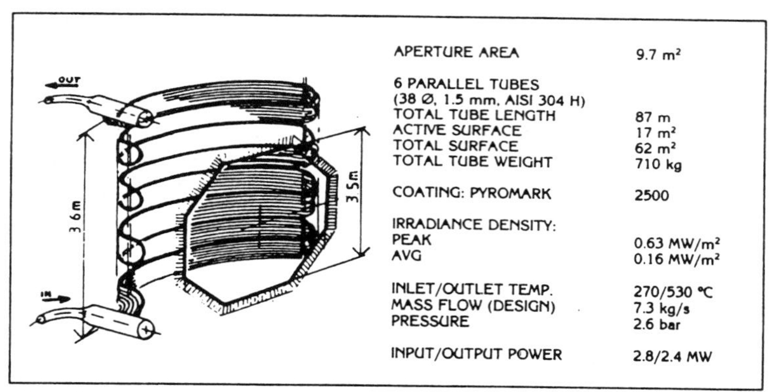

OUT
IN
3.6 m
3.5 m
APERTURE AREA 9.7 m²
6 PARALLEL TUBES
(38 Ø, 1.5 mm, AISI 304 H)
TOTAL TUBE LENGTH 87 m
ACTIVE SURFACE 17 m²
TOTAL SURFACE 62 m²
TOTAL TUBE WEIGHT 710 kg
COATING: PYROMARK 2500
IRRADIANCE DENSITY:
PEAK 0.63 MW/m²
AVG 0.16 MW/m²
INLET/OUTLET TEMP. 270/530 °C
MASS FLOW (DESIGN) 7.3 kg/s
PRESSURE 2.6 bar
INPUT/OUTPUT POWER 2.8/2.4 MW

Fig. 2
Typical Cavity Receiver (Sulzer, Switzerland)
SSPS, Almeria 1983/84

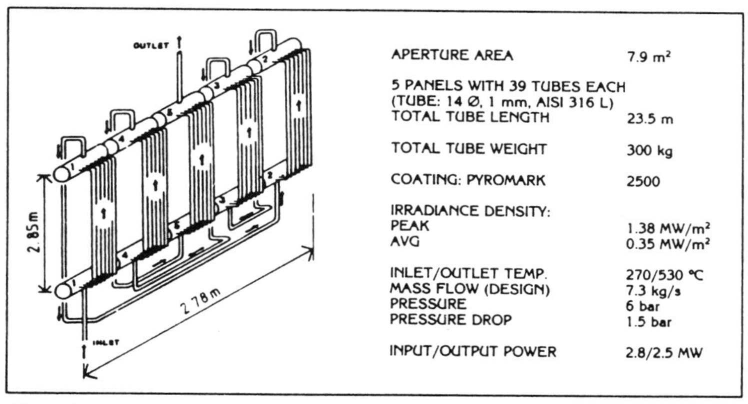

OUTLET
INLET
2.85 m
2.78 m
APERTURE AREA 7.9 m²
5 PANELS WITH 39 TUBES EACH
(TUBE: 14 Ø, 1 mm, AISI 316 L)
TOTAL TUBE LENGTH 23.5 m
TOTAL TUBE WEIGHT 300 kg
COATING: PYROMARK 2500
IRRADIANCE DENSITY:
PEAK 1.38 MW/m²
AVG 0.35 MW/m²
INLET/OUTLET TEMP. 270/530 °C
MASS FLOW (DESIGN) 7.3 kg/s
PRESSURE 6 bar
PRESSURE DROP 1.5 bar
INPUT/OUTPUT POWER 2.8/2.5 MW

Fig. 3
Typical External Receiver (Franco Tosi/Agip, Italy)
SSPS, Almeria 1985/86

- Volumetric Receiver
  The gaseous heat transfer fluid is convectively heated in a volume of suitable geometric arrangements (e.g. wire net or foils) which are irradiated to the wanted temperature range. An advantage here is the non-tube shape; the necessity of a window separation hints for further intensive development.

- Direct Absorption Receiver
  The heat transfer fluid of the reactant receives directly the energy radiatively. The receiving fluid may be a liquid or a particle suspension. This is the most attractive solution, if the chemical process with substantial energy need can be coupled. The state of the art is early research.

The main technical results are often summarized in the determination of efficiency, though there are also other points of interest. Fig. 4 is a collection of the presently available data on this issue. It must be stated that this information is drawn from the first operational results. Concept optimizations are topics of the future. It is apparent from this figure that efficiencies between 80 and 90 % are clearly achievable both for cavity and external receiver concepts. However, the following preliminary test results can be reported:

Molten Salt External	71 - 75 %	(Albuquerque)
Molten Salt Cavity	81 - 93 %	(Albuquerque)
Gas Panel	65 - 85 %	(Almeria).

Volumetric and direct absorption receivers are too early in development to be represented here.

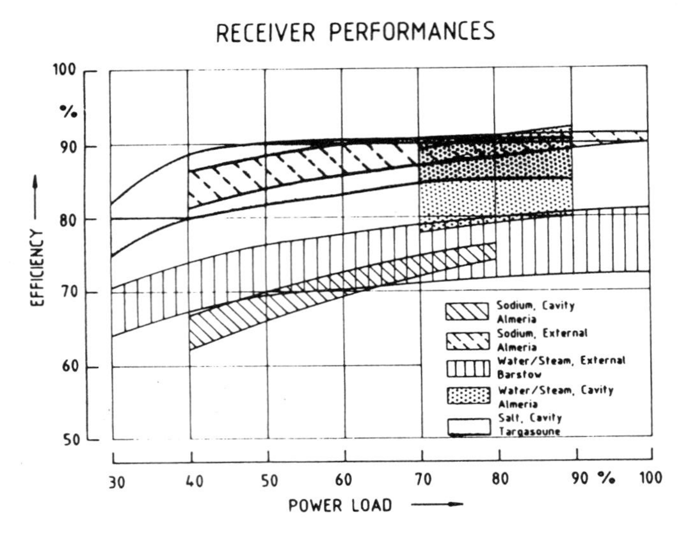

Fig. 4
Efficiencies of
Different Receiver
Performances

Fig. 5 shows the results of incident peak flux densities measured at the external sodium receiver in Almeria. Parameter is the aiming strategy to define one or three points for the whole heliostat field.

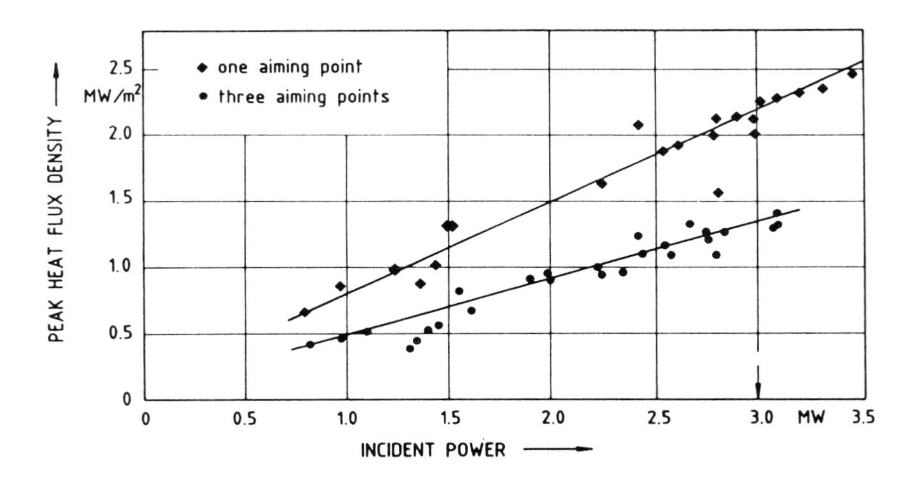

Fig. 5
Peak Heat Flux Densities at the External Sodium Receiver, Almeria 1985/86

For the different heat transfer media, the following maximum values are achievable:

Sodium	2.5 MW/m^2
Molten Salt	1.0 MW/m^2
Water/Steam	0.7 MW/m^2
Gas	0.3 MW/m^2

Fig. 6 gives the development of temperatures with air as heat transfer medium. The gas and tube front material temperatures are displayed as calculated and measured. The mean exit temperature of this test series was 800 ^{o}C (metallic). Presently, ceramic material tests with 1000 ^{o}C mean exit gas temperatures are in progress. These receiver versions can be combined with high efficiency gas turbines or with chemical processes that need high temperature energy.

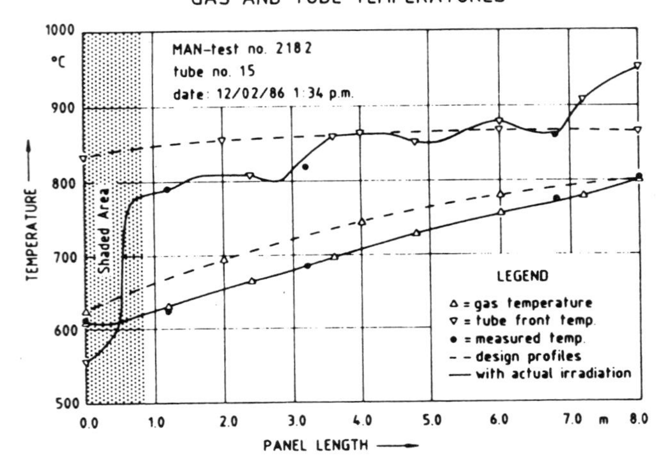

Fig. 6
Gas and Tube Temperatures of the Metallic Panel Test Within the GAST Technology Program, Almeria 1986

References

1. Grasse, W.
SSPS, Results of Test and Operation 1981-1984, SSPS-SR 7, IEA-SSPS Operating Agent DFVLR, Cologne, Germany, May 1985

2. Kuntz Falcone, P.
A Handbook for Solar Central Receiver Design, Sandia Report, SAND 86-8009, UC-62a, Livermore, USA, Dec. 1986

3. Becker, M. (Editor)
Solar Thermal Central Receiver Systems, Proceedings of the Third International Workshop in Konstanz, June 23-27, Germany, ISBN 3-540-17052-9, Springer Verlag, Oct. 1986

4. Radosevich, L. G.
Final Report on the Experimental Test and Evaluation Phase of the 10 MWe Solar Thermal Central Receiver Pilot Plant, Sandia Report, SAND 85-8015, UC-62d, Livermore, USA, Sept. 1985

5. Kesselring, P., Selvage, C. S. (Editors)
The IEA/SSPS Solar Thermal Power Plants - Facts and Figures -, Final Report of the International Test and Evaluation Team (ITET), ISBN 3-540-16145-7, Springer Verlag, 1986.

SECONDARY CONCENTRATORS FOR A CENTRAL RECEIVER SYSTEM

Manuel Blanco Muriel* and Jose G. Martin**
* E.S.I.I de Sevilla, Sevilla, Spain.
** University of Lowell, Lowell, MA 01854 USA

ABSTRACT

In future central receiver systems, secondary concentrators may make it possible to reach high temperature and to relax design requirements on heliostats. In the course of evaluating the high-flux receiver at the Almeria CRS plant, it was proposed that a terminal concentrator would allow the use of the plant for high flux experiments and at the same time serve to evaluate the concept. When designing such a concentrator, it is necessary to compare the optical behavior of different configurations. To aid this effort, a software package has been developed at the University of Lowell to analyze the optics of secondaries. The main features of the package, which is based on raytracing techniques, are its flexibility and modularity. The program determines the distributin of power absorbed by the reflector walls and incident at the receiver. Preliminary results show that it is possible to increase the concentration ratio by a factor of 2.6 with a considerable flattening of the power profile and an efficiency of 80%. The results are encouraging.

KEYWORDS

Solar thermal plants, central receivers, secondary concentrators.

INTRODUCTION

Two major lessons learned from the experience with central receivers are that the cost of the generated power is high and that their main appeal lies in the high temperature applications (Carmona, 1985). Second-stage concentration addresses these two issues.

The work performed in configurations where the primary concentrator is a dish is formidable. Although configurations where a heliostat field is the primary have not received similar attention, recent work on flat secondary reflectors (Chuah, 1987) indicates that a two-dimensional terminal concentrator may reduce the aperture size while producing a minimal effect on the intercept factor.

Several configurations based on nonimaging optics have been proposed for secondaries. These concentrators can reach the thermodynamic limit of concentration in 2-D or cylindrical geometry and can approach it in 3-D or axisymmetric geometry. The

CPC (Winston, 1975) consists of parabolic reflectors inclined with respect to the optical axis. In the CEC (Rabl, 1976) the profile curves of the reflector are portions of distinct ellipses. The trumpet (Winston, 1980) conforms to a hyperbolic shape .

In large plants, the primary reflector field will subtend a large angle about the receiver. Several concentrators may be combined to form a 'fly-eye' terminal concentrator. There have been suggestions for compound mirrors with flat surfaces, based on the optics of macrurean crustacean eyes (Martin, 1984) and for mirrors reflecting light onto a ground-based receiver (Kesselring, 1983).

When designing a secondary, the designer must compare the optical behavior of different configurations. There are computer codes that simulate receiver optics, but they do not offer adequate built-in options to account for reconcentration. It is not realistic to consider the heliostat field as a Lambertian source. Ray tracing techniques are a suitable approach.

OPTICAL ANALYSIS

A software package has ben developed at the University of Lowell to analize the optics reconcentrators. The main features of the package, which is based on raytracing techniques, are its flexibility (different geometries do not require changes in the software) and modularity (sophisticated treatments may be acomplished by the addition of software modules).

For a given geometry and flux distribution at the input plane, the program determines the distribution of power absorbed by the reflector walls and incident at the receiver. To do this, each surface is discretized. Rays are generated at established points of the input plane. The number, origin, power and direction of rays are selected to reproduce the incoming flux radiation. Each ray is followed (traced) through its reflections: the process ends when it reaches the output or escapes.

To determine the power distribution on each one of the surfaces (including the input and output planes) using raytracing techniques, the following steps have been executed:

1. The reconcentrator surfaces are divided into small area elements.

2. Rays are generated at established points of the input plane. Each ray is associated with a specific direction and amount of power. The origin, power and direction are selected to reproduce the incoming flux.

3. Ray tracing techniques are used to follow each ray. When a reflection occurs, the surface element hit by the ray is determined and an update its the power balance is carried out. The process ends when the ray reaches the output or input plane. The element of the plane hit by the ray is determined and its power updated. In some geometries, a ray may become trapped: the calculation is ended when the ray energy becomes less than some threshold.

4. From the power calculated on each reconcentrator element, one estimates the total power and the power density distribution on each surface.

Much computational effort is devoted to estimating where the rays intersect surfaces. For a concentrator with a bounded cylindrical geometry, it is possible to reduce this effort from one in three dimensions to one in two (Van Wijk, 1978).

For flexibity the definition of the reconcentrator geometry and the generation of rays (incoming flux) have been considered as external tasks. The package is written in C and implemented on a VAX 11/780.

APPLICATION TO THE ALMERIA CRS PLANT

The original purpose of the IEA Small Solar Power Systems plants in Almeria, Spain, was to compare different thermal systems. At the Central Receiver System (CRS), 135 heliostats reflect enough power onto the receiver that about 4 Mwth were absorbed. The 8 m2 advanced billboard receiver, cooled by sodium, operated at high flux densities and tolerated transients. This Advanced Sodium Receiver (ASR) could operate - and was operated - at 2.5 Mw/m2 peak power fluxes. This limit was set by the field configuration and beam quality, when all the heliostats were aimed at the center of the aperture. The high flux allowed the receiver to operate at high efficiencies (>=90%) at relatively high temperatures (530°C).

Earlier, a sodium receiver of the cavity type had been operated at the site. The motivation to build the ASR was to compare the performance of a high-flux billboard and a cavity receiver. The project did verify that high efficiencies could be achieved at high fluxes. For other receivers operating at higher temperatures, acceptable efficiencies will be achieved only if the flux density at the aperture is higher.

The program developed at the University has been used to analyze different CPC trough reconcentrators for the CRS plant. The input plane was a square of 3 x 3 meters with center at the actual receiver center and tilted -28° in the vertical direction. The reflectivity of the surfaces was taken as 0.90. The incoming flux from the heliostats was calculated at design conditions (equinox, noon, 920 W/m2 of direct irradiance, and heliostat reflectivity equal to 0.911). The Helios program (Biggs, 1979) has been used to estimate the incoming flux distribution at the entrance of the reconcentrator. The area of the parabolic mirrors for the 18° CPC configuration is 45.1 m2.

The results are tabulated in Table 1 for the CPC's with half acceptance angles of 16 and 18 degrees. Even though the field subtends a half-angle of 16 degrees in the vertical direction, the CPC with the larger acceptance angle has a larger collection efficiency (80% vs. 73%). (The field subtends 84.4° horizontally. No rays escape for this configuration, and the average number of reflections is slightly lower.

TABLE 1:EFFECT OF TROUGH CPC RECONCENTRATORS

(Reflectivity = 0.9, Input Apertura = 9.0 m2; width = 3.0 m; Input power=2895.4 kW)

		16°	18°
Half acceptance angle		16°	18°
Aperture area, output, m^2		2.48	2.78
Depth, m		6.67	6.04
% absorbed power,	top surface	6.1	5.3
	bottom surface	2.0	1.5
	left side	8.7	6.7
	right side	8.4	6.5
	output	72.6	80.0
% 'lost' power, front		2.3	0
On output plane: Power, kW		2102	2317
maximum flux density, kW/m2		1940	1840
minimum "		39	9
average "		830	806
standard deviation kW/m2		420	406

Figures 1 and 2 show the flux distribution in different parts of the reconcentrator for the CPC reconcentrator with the 18 degree aperture. The average flux at the outplut plane is approximately 800 KW/m2, with a standard deviation of 405 KW/m2. Without the reconcentrator, under the same conditions, the average flux is 304 kW/m2, with a standard deviation of 498 kW/m2. The reconcentrator succeeds in increasing the concentration by a factor of 2.66 while flattening the power density profile.

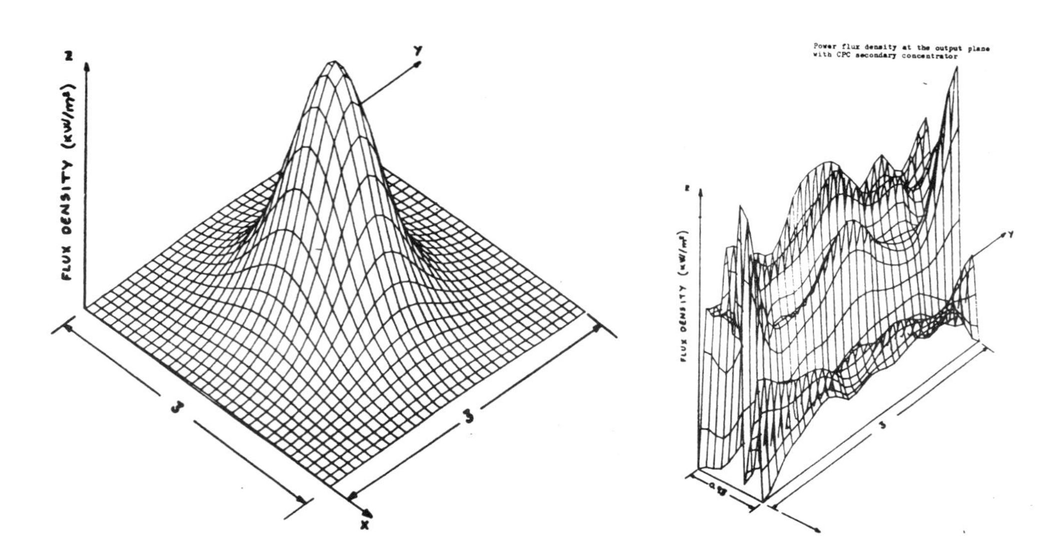

Figure 1: Flux without concentrator.

Figure 2: Flux at the CPC output plane

Due to the asymmetrical distribution of the heliostats in the field, the amount of power absorbed by the upper reflector surface is approximately four times greater than that absorbed by the lower reflector surface.

DISCUSSION

Analytical methods are a valuable tool to understand the principles and the limits of concentration, but they must be complemented by careful computational techniques to estimate performance data for a central receiver system. For the CRS system, these techniques establish that it is possible to increase the concentration ratio substantially while flattening the power profile and achieving an output/input efficiency of 80%. These results are preliminary but encouraging. Other shapes and configurations remain to be analyzed.

Many challenges remain. The surface of a secondary concentrator should have a large, and preferably specular, reflectivity. However, that concentrator will be subject to harsh conditions. High power fluxes from the primary and the receiver will be incident on the reflecting surfaces, which will likely be exposed to the outside and the receiver environments. They will be heated. Even if reflecting surfaces are identified which can operate at high temperatures in the relevant media, refrigeration may be required. There will be large temperature gradients on a system which is subject to cycling. There is a need to learn about high solar flux effects on materials, and these challenges introduce another motivation for the work on reconcentrators.

REFERENCES

Biggs, F., C.W. Vittitoe (1979). The Helios Model for the Optical Behavior of Reflecting Solar Concentrators. Sandia National Laboratories. SAND76-0347.

Carmona, R., J.G., Martin (1985). Central Receiver Systems - Potential for Improvements. IEA/SSPS Final Report, DFVLR, Cologne.

Chuah, Y.K., J.V. Anderson, T. Wendelin (1987). Terminal Concentrators for Solar Central Receiver Applications. Paper submitted to the ASME Journal.

Kesselring, P. (1983). The Secondary Mirror Concept. Proceedings of the Intl. Seminar on Solar Fuels and Chemicals. University of Stuttgart.

Martin, J.G. (1984). A Secondary Solar Collector. Proceedings of the International Solar Forum (Berlin). DGS-Sonnenergie Verlags, Munich.

Rabl, A., R. Winston (1976). Ideal Concentrators for Finite Sources and Restricted Exit Angles. Applied Optics, Vol. 15.

Van Wijk, J.J. (1978). Ray tracing objects defined by sweeping planar cubic splines. Delft Univ. of Technology, ACM Trans. on Graphics.

Winston, R., H. Hinterberger (1975). Principles of Cylindrical Concentrators for Solar Energy. Solar Energy, Vol. 17.

Winston, R., W.T. Welford (1980). Design of non-imaging concentrators as second stages in tandem with image-forming first stage concentrators. Applied Optics, Vol. 19.

NEW MULTILAYER SEMI-TRANSPARENT PACKED BED HIGH TEMPERATURE SOLAR RECEIVER

T. Menigault*, G. Flamant*, D. Gauthier*, P. Crausse**

* Institut de Science et de Génie des Matériaux et Procédés
C.N.R.S. BP 5 - 66120 ODEILLO, FRANCE.

** Institut de Mécanique des Fluides, I.N.P. - C.N.R.S.
2 rue Camichel - 31071 TOULOUSE, FRANCE.

ABSTRACT

A new high temperature solar absorber consisting in two spectrally dissimilar layers of particles is studied. Experimental results related to a bed composed of glass beads and silicon carbide particles are described. These data are discussed using a two-flux based heat transfer model.

KEYWORDS

Solar Thermal conversion, packed bed, high temperature receiver, combined heat transfer, bands model.

INTRODUCTION

Direct absorption of concentrated solar energy using packed beds is a very simple way to heat gas up to temperatures higher than 500°C. Previous computed and experimental data show evidence of the solid overheating near the irradiated surface (Flamant and Olalde, 1983). In this paper, the theoretical and experimental study of a new packed bed absorber consisting in two layers of different particles with air flowing through it is presented.

The two layers properties are :
Layer 1 - Solar radiation transparent and infrared radiation absorbent.
Layer 2 - Solar and infrared radiations absorbent.

The physical basis of this system may be expressed as follows :

Let us consider two slabs composed of particles whose scattering and absorption coefficients are different in two spectral ranges, for example solar range ($0.2 < \lambda < 1.5$ µm) and infrared range ($1.5 < \lambda < 20$ µm). Two absorption and scattering coefficients must be defined (a_s, a_r, σ_s', σ_r' respectively) for both slabs. Let suppose that slab 1, situated above slab 2 and submitted to concentrated solar radiation, is transparent in the solar band but absorbent in the infrared band. If slab 2 absorbs solar radiation, its temperature increases but the backward I.R. emission is

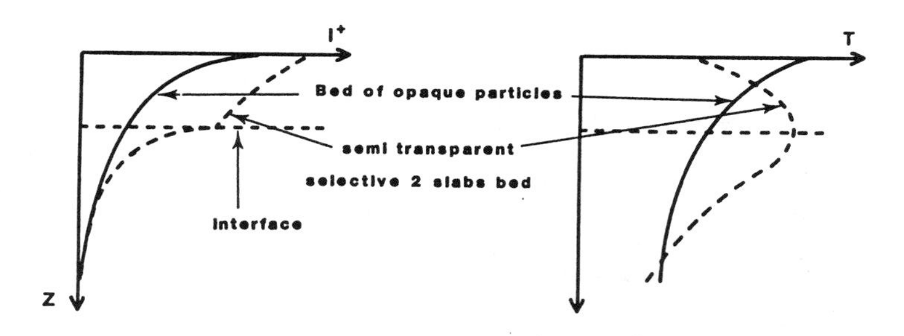

Fig. 1. Physical concept

absorbed by slab 1. Therefore, there exists a temperature maximum inside the bed. To explain this phenomenon, a qualitative comparison of two situations is described on Fig. 1. Solar flux and temperature distributions are shown for a bed composed of solar absorbent particles and a two-slab packed bed defined by previous properties. The main difference is the surface temperature reduction and consequently a large decrease of infrared emission losses.

Air is supposed to flow through the packed beds Since one of our objectives is gas heating using solar energy.

EXPERIMENTAL

The experimental device is schemed on Fig. 2. This receiver is set at the focus of the 6.5 kW solar concentrator described by Arnaud et al (1982).

Particles are set inside a 80 mm I.D. alumina tube placed in a stainless steel shell. Cold air is introduced at the upper part of the bed, between its irradiated surface and the transparent window. The 6 mm thick window transmissivity in the solar band was determined by Pierrot (1987) : $\tau_s = 0.94$. Temperatures are measured using 6 bare K thermocouples placed inside the bed and 2 suction pyrometers. The latter are alumina shielded thermocouples indicating the inlet and outlet gas temperatures : (θ^i, θ°).

The receiver efficiency is defined as :

$$\eta_g = \int_{\theta^i}^{\theta^\circ} F_g \, C_g(\theta) \, d\theta / P_i$$

Where F_g and P_i are the air mass flowrate and the incident solar power at the focus respectively.

Experiments were run with a two-layer packed bed composed of glass beads (slab 1) and silicon carbide particles (slab 2). The thicknesses and particles diameters are Z_1 = 25 mm, d_1 = 2 mm and Z_2 = 55 mm, d_2 = 2.36 mm respectively. The receiver efficiency is plotted versus gas outlet temperature on Fig. 3. for a SiC packed bed (SiC P.B.) and a two-slab packed bed (2-S.P.B.)

A large decrease of η_g is observed for the SiC P.B. whereas the 2-S.P.B. efficiency is roughly constant. The following explanation

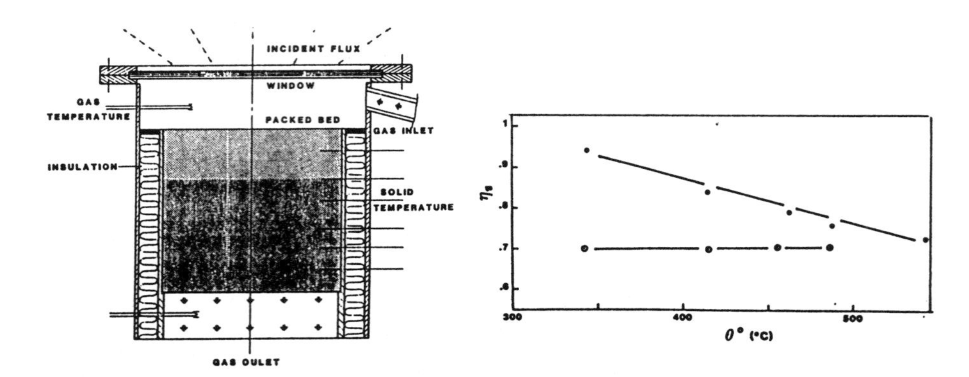

Fig. 2. Experimental device

Fig. 3. Efficiency versus outlet air temperature (•) : SiC, (⊖): two-slab bed

may be proposed : At low temperature, the 2-S.P.B. efficiency is limited by solar radiation scattering on glass beads. When the gas temperature rises, infrared emission increases ; Fig. 4. shows a large decrease of surface temperature when using the selective absorber. Note that the black body emission decreases 7.5- fold when the surface temperature varies from 1000 °C to 500 °C. All experiments were run with temperatures lower than 500 °C, to avoid any risk of glass melting. Nevertheless, extrapolating the curves on Fig. 3, the 2-S.P.B. efficiency seems to become larger than the SiC-P.B. one for temperatures above 550 °C. Thus, we may propose quartz for high temperature operations.

THEORETICAL AND DISCUSSION

The model main assumptions are the following :

- steady state conditions
- one dimensionnal plane parallel slabs
- each slab is composed of identical isotropic spherical particles, scattering is isotropic on each hemisphere (2-flux approximation)
- the fluid is incompressible and non absorbent. The conductive heat transfer in the fluid may be neglected in comparison with the convective one.
- two spectral bands are considered : solar band (subscript s), $0.2\ \mu m < \lambda < 1.5\ \mu m$, and infrared band (subscript r), $\lambda > 1.5\ \mu m$ The infrared emission in the solar band is neglected.

Accounting for conduction, convection and radiation, the 6 governing equations of transfer for a slab i are (I is the radiant heat flux) :

$$1/2(dI_s^+/dZ) = -(a_{s,i} + B_s\sigma'_{s,i})I_s^+ + B_s\sigma'_{s,i}I_s^- \quad (1)$$

$$1/2(dI_s^-/dZ) = (a_{s,i} + B_s\sigma'_{s,i})I_s^- - B_s\sigma'_{s,i}I_s^+ \quad (2)$$

$$1/2(dI_r^+/dZ) = -(a_{r,i} + B_r\sigma'_{r,i})I_r^+ + B_r\sigma'_{r,i}I_r^- + a_{r,i}\sigma T^4 \quad (3)$$

$$1/2(dI_r^-/dZ) = (a_{r,i} + B_r\sigma'_{r,i})I_r^- - B_r\sigma'_{r,i}I_r^+ - a_{r,i}\sigma T^4 \quad (4)$$

$$\rho_f C_f \xi_i v(d\theta/dZ) = [6(1 - \xi_i)/d_i]h_i(T - \theta) \quad (5)$$

$$(1-\xi_i)\lambda^*_i(T)(d^2T/dZ^2) + 2a_{s,i}(I_s^+ + I_s^-) + 2a_{r,i}(I_r^+ + I_r^-) = [6(1-\xi_i)/d_i]h_i(T - \theta) + 4a_{r,i}\sigma T^4 \quad (6)$$

B_s and B_r are the back-scatter fractions in the solar and infrared bands, defined as :

$$B = 1/2\int_0^1\int_{-1}^0 p(\mu,\mu')d\mu'd\mu \quad (7)$$

where $p(\mu,\mu')$ is the scattering phase function.

T and θ are the particles and gas temperatures ; ξ, V, d are the mean bed porosity, the fluid average velocity in the axial direction and the particles diameter.

The equivalent thermal conductivity, $\lambda^*_i(T)$, and the gas solid heat exchange coefficient, h_i, are defined according to Baddour, Yoon (1960) and Donnadieu (1961), respectively.

Boundary conditions are based on the following considerations :
- Bed surface (Z=0) : conservation of solar and I.R. intensities.
- Boundary between slab 1 and slab 2 ($Z=Z_1$) : conservation of solar and I.R. intensities, solid and gas temperatures are equal.
- Bottom ($Z=Z_1 + Z_2$) : extinction of solar intensity, no temperature gradients.

Fig. 5. shows a good agreement between experimental and theoretical data (parameters values are listed on Table 1).

The gas outlet temperature difference is less than 10 K and the temperature distribution is well-fitted inside slab 2 ; whereas the thermocouple

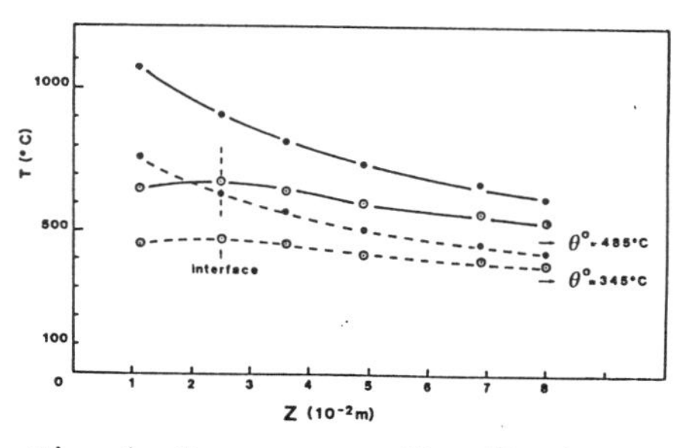

Fig. 4. Temperature distributions inside the SiC P.B. ● and the 2-S.P.B. ⊙ for two outlet gas temperatures

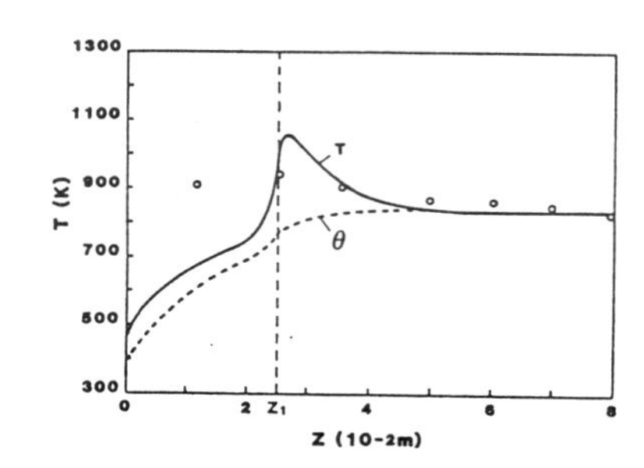

Fig. 5. Comparison between theoretical and experimental data. Solar intensity : 600 kW/m^2

TABLE 1. Parameters for computed data

Slab	d	ε	a_s	a_r	σ'_s	σ'_r
1	2	0.42	15	700	85	100
2	2.36	0.54	155	175	40	500

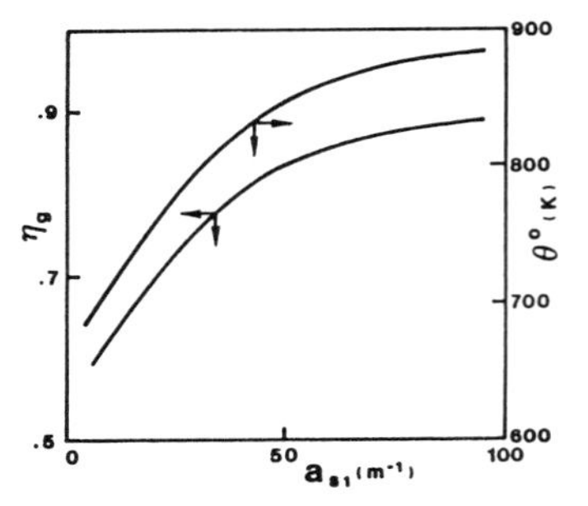

Fig. 6. Efficiency and gas outlet temperature versus slab 1 absorption coefficient

in slab 1 overestimates the temperature since it is irradiated by solar flux.

Another application of the model is illustrated on Fig. 6. The receiver efficiency may be improved by increasing the slab 1 absorption coefficient (reduction of the scattering coefficient since the extinction coefficient is supposed to be constant). Such a phenomenon may be approached using colored glass.

REFERENCES

- Arnaud G., Flamant G., Olalde G., Robert J.F., (1981), "Les fours solaires de recherche du L.E.S. d'Odeillo". Entropie, 97, 139-146.

- Baddour R.T. and Yoon C.Y., (1960), "Local radial effective conductivity and wall effect in packed beds". Chem. Engng. Prog. Symp. Series, 32, 35.

- Donnadieu G., (1961), "Transmission de chaleur dans les milieux granulaires. Etude du lit fixe et du lit fluidisé". Revue de l'Institut Français du Pétrole, 16, n° 11, 1330.

- Flamant G., Olalde G., (1983), "High temperature solar gas heating. Comparison between packed and fluidized bed receivers." Solar energy, 31, n° 5, 463-471.

- Pierrot A., (1987) "Contribution à l'étude des transferts de chaleur à haute température dans les milieux alvéolaires". Thesis, Perpignan University, 05.15.1987.

HIGH TEMPERATURE HONEYCOMB SOLAR RECEIVER FOR GAS HEATING

G.Olalde, A.Pierrot

Institut de Science et de Génie des Matériaux et Procédés
C.N.R.S. BP 5 - 66120 ODEILLO, FRANCE

ABSTRACT

This paper presents the study of a solar honeycomb receiver system with a gas flowing through it. A theoretical model expressing the gas-matrix heat transfer is developped. It takes into account the exchanges by conduction, convection and radiation, the solar radiation penetration into the matrix cells and the variations with temperature of the gas and solid physical properties. Also, this model considers the possible presence of a quartz window.Experimental results related to silicon carbide and mullite honeycomb receivers are described.
Then, theoretical and experimental results (temperature profiles, solar radiation penetration into the matrix) are compared. Also, heat balances over the receiver are presented and its thermal efficiency is calculated.

KEYWORDS

Solar thermal conversion, honeycomb exchanger, high temperature receiver.

INTRODUCTION

High temperature gas heating (up to 1000°C) may be used for ore drying (phosphates...); for thermodynamic conversion of energy (Brayton and Stirling cycles) or for several chemical reactions. Radiative sources, such as solar concentrators, I.R. radiative furnaces and plasma torches are probably the most promising sources to reach high temperatures. The use of honeycomb matrices presents many advantages :

- cavity effect for better radiation absorption
- important wall area (about $1000m^2/m^3$)
- good ratio weight/exchange surface ($0.5kg/m^2$)
- good temperature standing qualities (up to 1300°C or even more than 2000°C depending on the material)
- good mechanical strength
- setting up and operating easiness (no moving part)

THEORETICAL

The honeycomb matrix is taken as a particular porous medium.
Let assume :
- steady state
- gas transparent and chemically inert
- uniform density of the source radiation
- one-dimension flow
- running parameters are the same for a whole section
- the gas conductive contribution to the heat transfer is negligible

The energy equations are :
for the solid :

$$(1-\xi)\frac{\partial}{\partial x^+}\left(\lambda_s \frac{\partial T}{\partial x^+}\right) - A\ D_H\ Nu\ \lambda_F\ (T-\theta) - D_H \frac{\partial}{\partial x^+}(Q_{IR} + Q_R) = o$$

for the gas (1)

$$\xi\frac{\partial}{\partial x^+}\left(\lambda_F \frac{\partial \theta}{\partial x^+}\right) - \xi\ R_\theta\ P_r\ \lambda_F \frac{\partial \theta}{\partial x^+} + A\ Nu\ \lambda_u\ \lambda_F\ D_H\ (T-\theta) = o$$

Where Q_R is the solar radiation which penetrates inside the matrix and Q_{IR} is the infrared radiation inside the matrix. Q_R is experimentally evaluated with an optical fiber system.

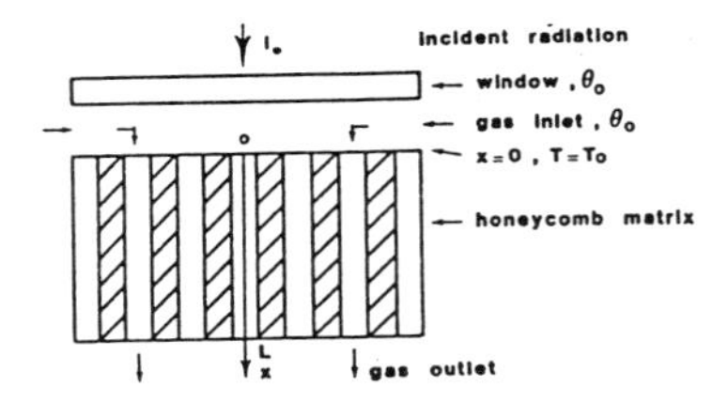

Fig.1 : Schematic representation of the system

Boundary conditions

. 3 local conditions

$\theta = \theta_o$ at $x^+ = o$ $x^+ = x/D$

$$\frac{\partial T}{\partial x^+} = o \text{ at } x^+ = L^+$$

$$\frac{\partial \theta}{\partial x^+} = o \text{ at } x^+ = L^+$$

. 1 global condition : the overall energetic balance

PI		PG		PP
total incident power	=	gas enthalpy increase	+	total losses

with

$$PG = \dot{m} \int_{\theta_o}^{\theta_f} C_p \, d\theta$$

$$PP = \left[1-\tau_s + \frac{\tau_s^2 \, (1-\xi)(1-\alpha_s)}{1-(1-\tau_s)\,(1-\alpha_s)\,(1-\xi)} \right] I_o + |Q_{IR}\,(0)| +$$

$$(1-\xi)\,\sigma\,\epsilon_{IR}\left(T_o^4-T_e^4\right)$$

Nusselt's number

The local Nusselt's number Nu (x^+) is a very important parameter in the energy equations system (1). We consider according to Pierrot (1987) the expressions :

$$Nu_x = Nu_d\left(1 + 0.0401\,(x^+ + \delta)^{-0.601}\right) \qquad x^+ < .004$$

$$Nu_x = Nu_d\left(1 + 0.002444\,{x^+}^{-1.108}\right) \qquad x^+ \geqslant .004$$

The systems of equations (1) is solved using Patankar's numerical method (1980)

EXPERIMENTAL

The honeycomb receiver was tested at the I.M.P'.s 6.5kW solar concentrator at ODEILLO.
It is a vertical cylindrical cavity closed by a transparent quartz window (Fig.2). the honeycomb matrix is maintained inside by a metallic support. A mineral insulating wool surrounds the matrix to reduce radial heat losses. Gas flow is supplied through a pressure regulator system. It enters the matrix at ambient temperature (293K).
Inside the matrix a set of K thermocouples has been located at different levels along the honeycomb matrix, all of them immersed in a cell. These thermocouples indicate the solid temperature at these levels. Inlet and outlet gas temperature are measured by two suction pyrometers, composed of a K thermocouple and two alumina radiation shields. The solar radiation penetration into the matrix has been recorded by means of mobile optical fiber, which is connected with a photodetector. This device allows measurement of the radiant heat flux (Q_R) variation as a function of the

distance (x^+) from the irradiated surface.

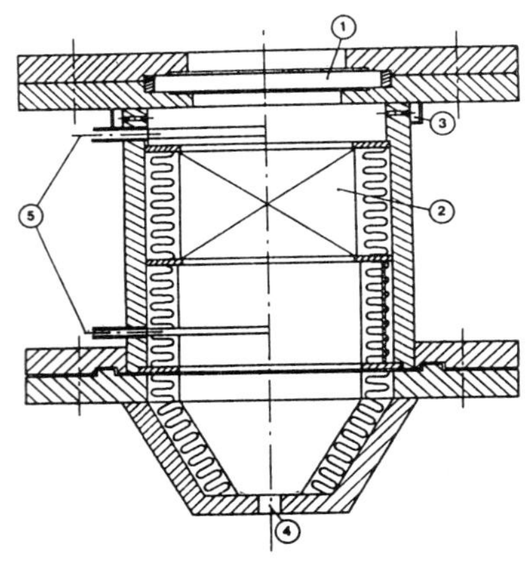

Fig.2 : Receiver scheme : 1/quartz window, 2/honeycomb matrix, 3/gas inlet, 4/gas outlet, 5/suction pyrometers.

Silicon carbide (SiC) and Mullite ($Al_2O_3.SiO_2$) honeycomb matrices were used. The main physical properties of these matrices are given in Table 1. The working fluid is dry air.

TABLE 1 Properties of the matrices

Material	SiC	Al_2O_3-SiO_2
porosity	0.76	0.76
D_H (mm)	1.35	1.35
Cell geometry	Square	Square
A	2250m^2/m^3	2250m^2/m^3

RESULTS

Temperature profiles.

Fig.3 shows the theoretical and experimental solid temperature (T) profiles and the theoretical gas temperature (θ) profilesince the gas temperature could not be measured inside the cells.

Thermal efficiency

As shown in Fig.4, the thermal efficiency decreases when the temperature increases. Notice, for every temperature (θ_f) of the gas, thermal efficiency is higher with silicon carbide than with mullite since the matrix temperature near the surface is higher in the late case, therefore creating larger radiation losses.

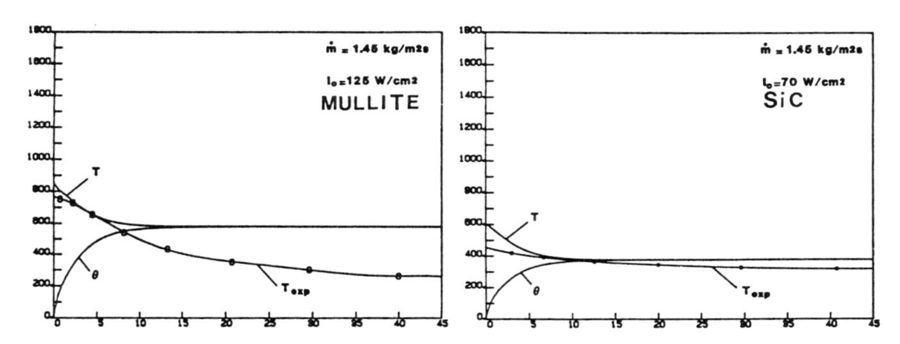

Fig.3 : Temperature profiles Vs x^+ for two materials (SiC and mullite)

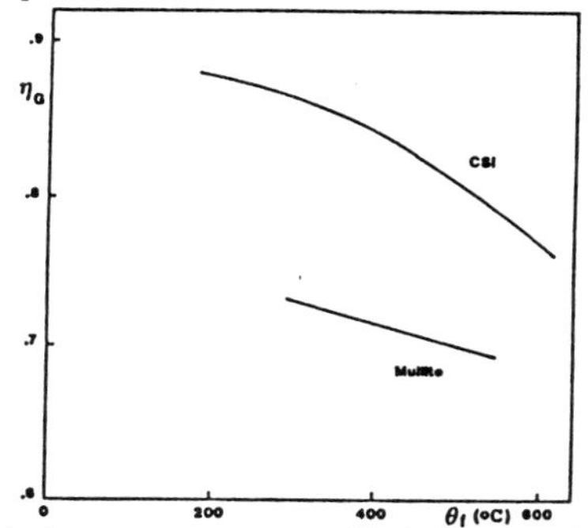

Fig.4 : Thermal efficiency vs. gas outlet temperature

$\left(\dot{m} = 0.96 \text{ kg/m}^2\text{s}\right)$

REFERENCES

- Patankar S, "Numerical heat transfer and fluid flow" Mc Graw Hill Ed, New York (1980)

- Pierrot A, (1987) "Contributions à l'étude des transferts de chaleur à haute température dans les milieux alvéolaires". Thesis, Perpignan University, 05.15.1987

3

Wind Energy

ACHIEVEMENTS AND CHALLANGES FOR THE NEXT FUTURE IN WIND ENERGY TECHNOLOGIES AND APPLICATIONS

P.B.S. Lissaman
Aero Vironment Inc., 825 Myrtle Avenue
Monrovia, California 91016, USA

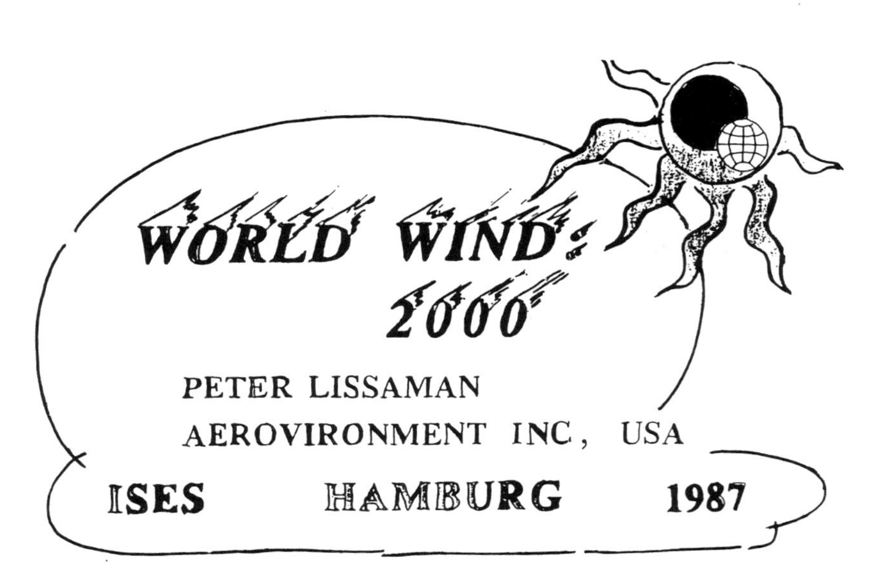

WORLD WIND:
2000
PETER LISSAMAN
AEROVIRONMENT INC, USA
ISES HAMBURG 1987

A RUSHING, MIGHTY WIND
The first energy to replace
man & animal power.
1

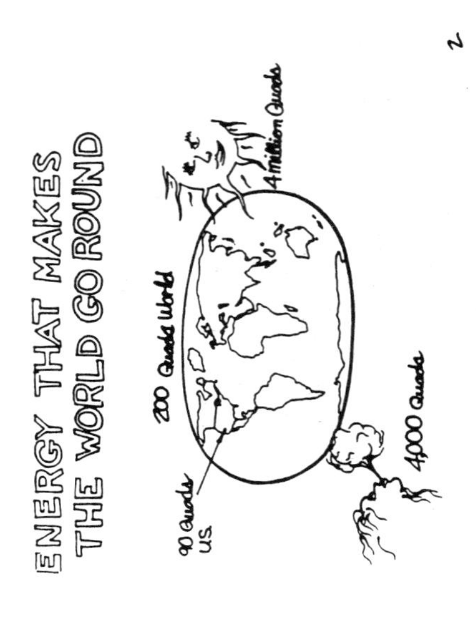
ENERGY THAT MAKES
THE WORLD GO ROUND
200 Quads World
4 Million Quads
90 Quads
U.S.
4,000 Quads
2

ON VERACITY
There are three kinds of lies,
lies
damned lies
&
Statistics
Benjamin Disraeli
1804-1881
3

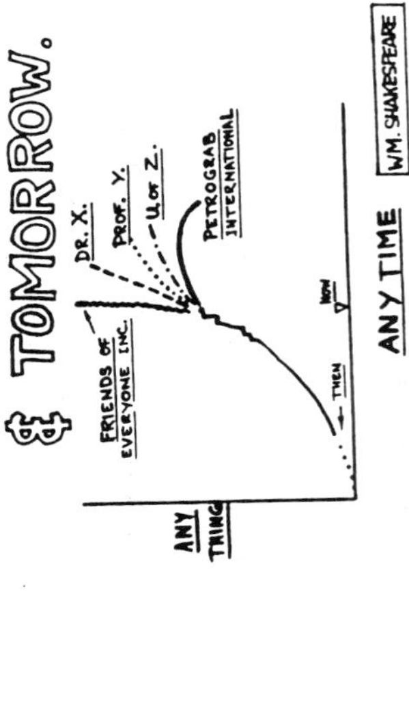
TOMORROW, & TOMORROW
& TOMORROW.
DR. X.
PROF. Y.
U. of Z.
FRIENDS OF
EVERYONE INC.
PETROGRAB
INTERNATIONAL
ANY
THING
THEN
NOW
ANY TIME
WM. SHAKESPEARE
4

ON PRECISION

THE AVOIDANCE OF STRICT ACCURACY IS A GENERAL SIGN OF GOOD BREEDING

EDCRATEZ 400 BC

5

PRICE & PETRO-POLITICS

85 U.S.$ BBL

IRAN/IRAQ

U.S. DECONTROL

ARABS/ISRAEL

1970 1980 1990

Year

REPORT OF U.S. DOE TO THE PRESIDENT '87

6

THE PHASES OF U.S. WIND

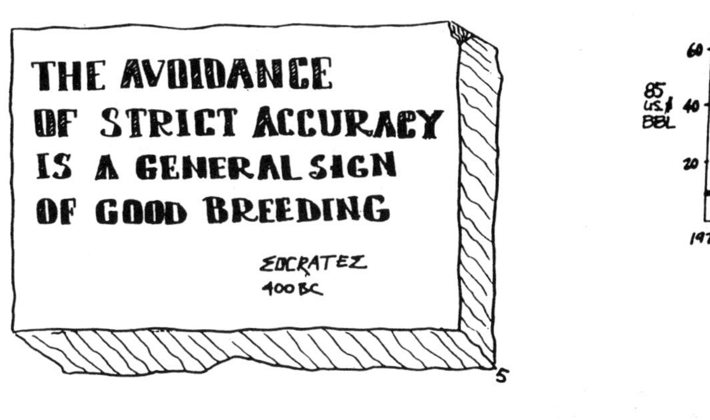

7

MEGABUCKS BREED MEGAWATTS

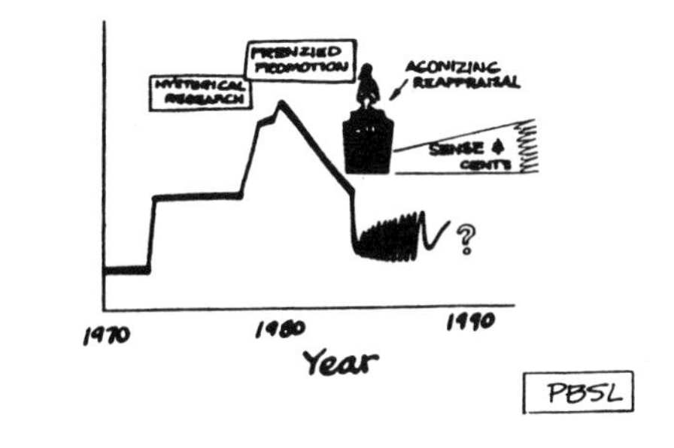

8

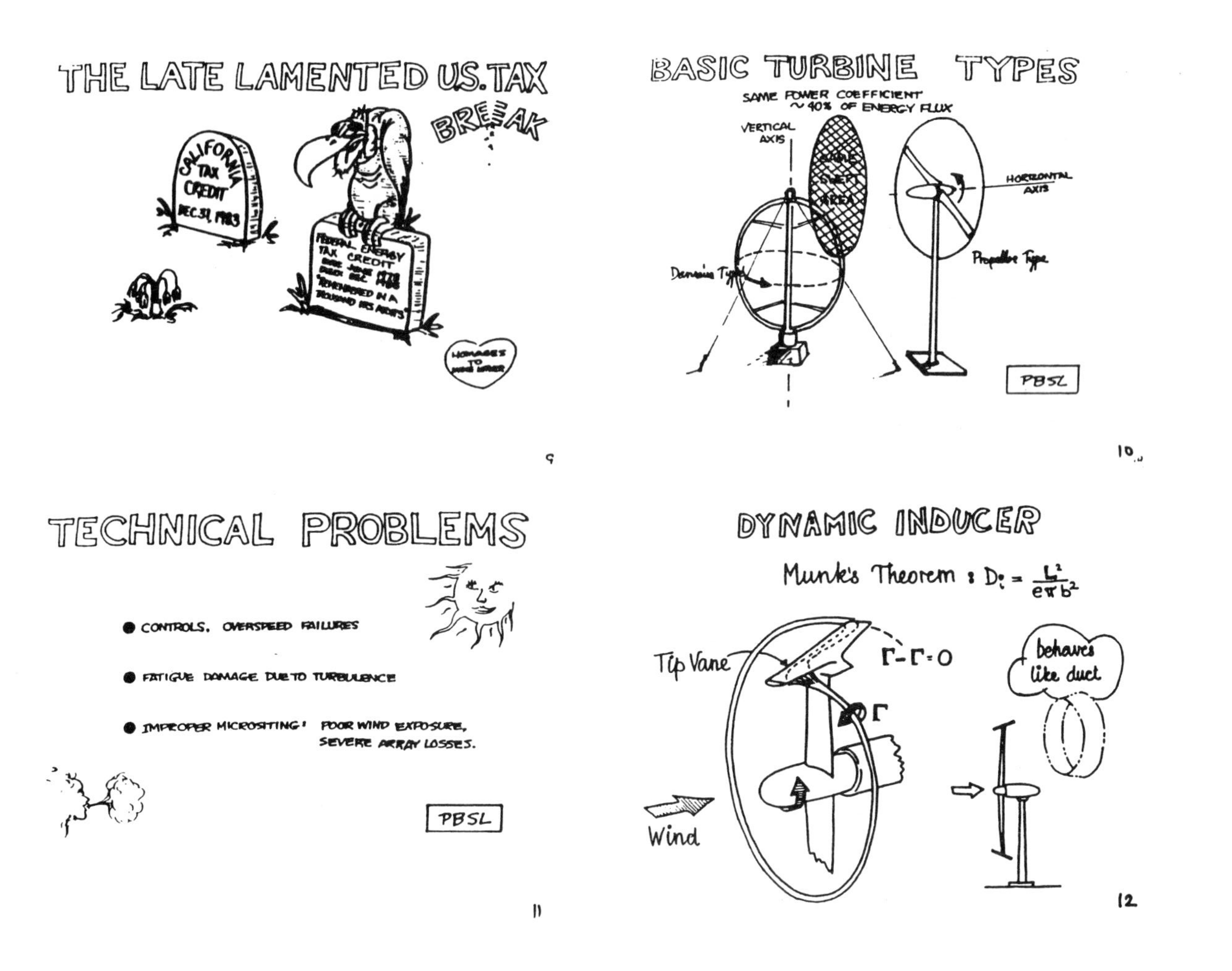
THE LATE LAMENTED US. TAX BREAK
CALIFORNIA TAX CREDIT DEC.31, 1983
FEDERAL ENERGY TAX CREDIT
9
BASIC TURBINE TYPES
SAME POWER COEFFICIENT
~40% OF ENERGY FLUX
VERTICAL AXIS
HORIZONTAL AXIS
Propeller Type
PBSL
10
TECHNICAL PROBLEMS
CONTROLS, OVERSPEED FAILURES
FATIGUE DAMAGE DUE TO TURBULENCE
IMPROPER MICROSITING: POOR WIND EXPOSURE, SEVERE ARRAY LOSSES.
PBSL
11
DYNAMIC INDUCER
Munk's Theorem : $D_i = \frac{L^2}{e\pi b^2}$
Tip Vane
$\Gamma - \Gamma = 0$
Γ
behaves like duct
Wind
12

WINGLETS DOWN-UNDER

Rule limits keel depth, d;
not effective depth, √e d.

13

TECHNICAL PROSPECTS

ADVANCED CONFIGURATIONS:	X NO BREAK THRU LIKELY.
ADVANCED MATERIALS:	X PROBABLY NOT COST EFFECTIVE.
MICRO PROCESSOR CONTROLS:	✓ NOT MUCH PERFORMANCE GAIN BUT GOOD FOR O & M.
SAFE LIFE REPLACEABLE COMPONENTS:	✓ GOOD PROSPECTS SELF VALIDATING.
LIGHTER STRUCTURE LOWER DESIGN LOADS:	✓✓ GOOD PROSPECTS, HIGH RISK.
MICRO-COMPUTER SITING/ECONOMICS:	✓✓ FUNDED BY RISØ & US DOE.

14

AGRI VERSUS AERO TECH

	OLD	TODAY	TOMORROW	
RATING : SPEED	65:12.5	250:17.4	750:15.0	kW:m/s
DIAMETER	15.2	20.6	45	m
TOTAL WEIGHT	11,400	11,400	51,000	kg
ROTOR WT/AREA	6.6	1.9	2.0	kg/m²
TOT. WT/AREA	63	34	32	kg/m²
POWER/WT	3.8	14.1	14.7	W/kg

*ALL AT 15.0 m/s

Europe — USA — MID-ATLANTIC!

STODDARD & PBSL

15

ENERGY PRODUCTION EFFECTIVENESS

	FOSSIL	60%	U.S. AVERAGE
	NUCLEAR	52%	INCLUDING SHUTDOWNS
	WIND	30% 20%	8 m/s 6.5 m/s
	PHOTO VOLTAIC	23%	FIXED TILT SUNNY CLIME

PBSL

16

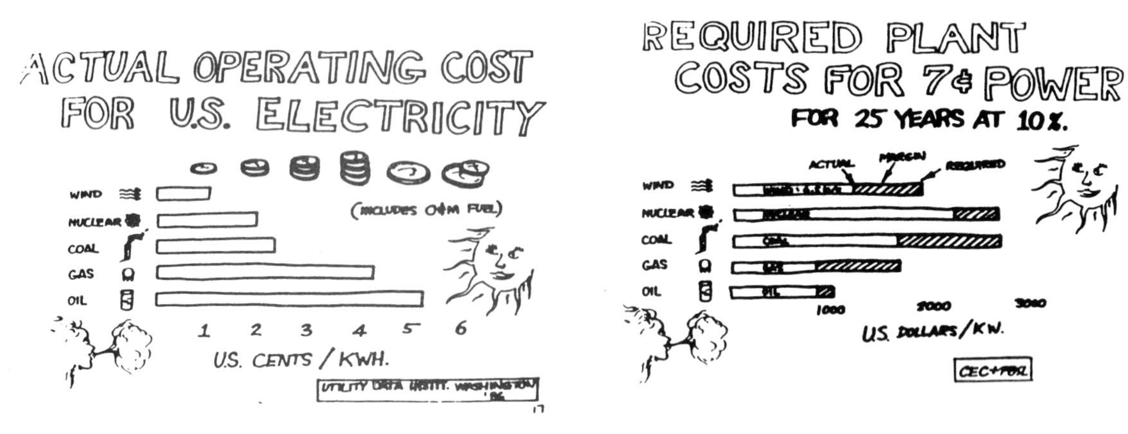

The address contained a number of

photographic slides not reproduced here.

These included shots of :

* Earth from Space
* Various California wind farms
* Dynamic Inducer advanced wind turbine rotor
* The SOLAR CHALLENGER photo-voltaic powered aircraft for Du Pont
* The SUNRAYCER photo-voltaic trans-continental racer for General Motors

19

SOLAR VOYAGES

SOLAR CHALLENGER '81
350 KM BY AIR
AEROVIRONMENT
SUNRAYCER '87
3500 KM BY LAND.

20

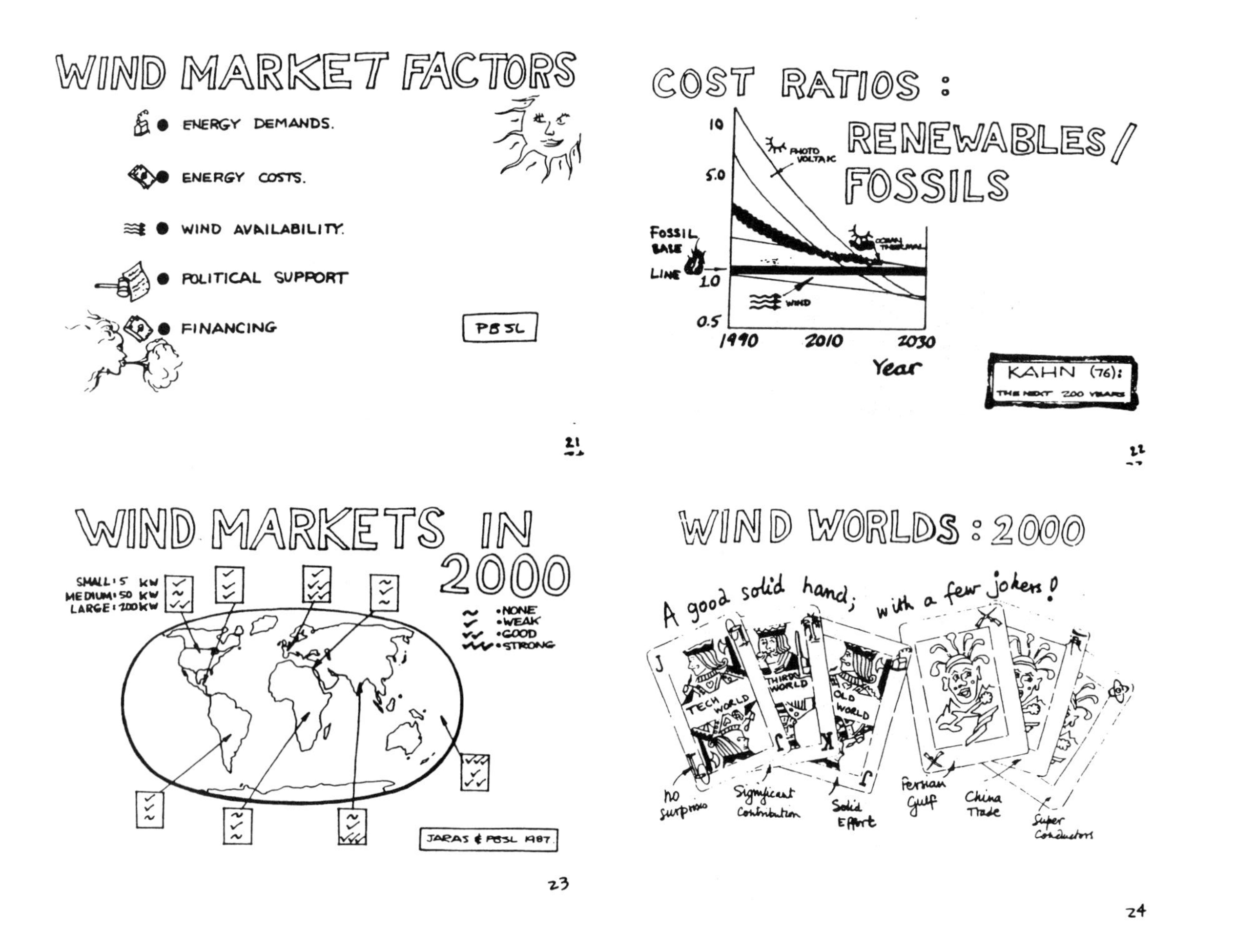
WIND MARKET FACTORS
ENERGY DEMANDS.
ENERGY COSTS.
WIND AVAILABILITY.
POLITICAL SUPPORT
FINANCING
PBSL
21
COST RATIOS :
RENEWABLES /
FOSSILS
10
5.0
1.0
0.5
PHOTO VOLTAIC
FOSSIL BASE LINE
WIND
1990
2010
2030
Year
KAHN (76):
THE NEXT 200 YEARS
22
WIND MARKETS IN 2000
SMALL: 5 KW
MEDIUM: 50 KW
LARGE: 200 KW
• NONE
• WEAK
• GOOD
• STRONG
JARAS & PBSL 1987.
23
WIND WORLDS : 2000
A good solid hand; with a few jokers!
TECH WORLD
THIRD WORLD
OLD WORLD
no surprises
Significant Contribution
Solid Effort
Persian Gulf
China Trade
Super Conductors
24

UN BEL DI'

Some day, after mastering
The wind, the waves, the tides & gravity
We shall harness for God the Energies of Love
And then
For the second time in the history of the World
Man will discover Fire.

Teilhard de Chardin

25

WORLD WIND: 2000

PETER LISSAMAN
AEROVIRONMENT INC, USA
GRAPHIC DESIGN C.J. Haase
California

26

WIND ENERGY UTILIZATION: STATUS OF RESEARCH AND DEVELOPMENT IN THE FEDERAL REPUBLIC OF GERMANY

W.Stahl

Federal Ministry of Research and Technology,
P.O. Box 20 07 06, D-5300 Bonn 2

N. Stump

Project Management for Biology, Ecology and Energy Research, Nuclear Research Establishment Jülich,
P.O. Box 1913, D-5170 Jülich

ABSTRACT

Most of the R&D activities in the field of wind energy in the Federal Republic of Germany are being carried out in projects supported by the Federal Minister of Research and Technology (BMFT). They follow the general objective to develop reliable wind energy conversion systems (WECS) and to stimulate the commercialization of wind power, particularly through demonstration projects.

Activities range from the generation of electricity using WECS of less than 50 kW up to several megawatts to the modification of small and medium-sized machines for special applications. Other projects are concerned with detailed measurements of wind characteristics, basic research and the technical optimization of turbines.

The following report mainly describes the support measures and policies of BMFT and concludes with a review of future funding policies for wind energy programmes in Germany.

KEYWORDS

Wind energy; research; development; Federal Republic of Germany

INTRODUCTION

In the middle of the nineteen seventies, the problems of dependency on imports, exhaustion of resources and pollution connected with the utilization of fossil sources of energy, instigated more interest in regenerative sources of energy in the Federal Republic of Germany. Today, as a result of accidents in nuclear power plants, further arguments have been added in favour of research into the use of these energy sources. With a wind energy supply of 4 000 kWh/m²a on the northern coast, the utilization of wind energy could theoretically make a considerable contribution to the production of electricity in the Federal Republic of Germany.

Since 1973, the Federal Ministry for Research and Technology (BMFT) has been supporting R&D in the field of wind energy converters (WECS) and up to now has spent about 200 mill. DM for this purpose. Although initially basic investigations were in the foreground of interest, today German industry and electricity utilities are increasingly contributing financially towards the development of wind energy technologies (Table 1).

Table 1: Wind Energy Projects since 1975; Costs in mill. DM

Year	BMFT Contributions	Total Costs
1975	0.1	0.1
1976	0.2	0.2
1977	4.1	5.1
1978	9.3	11.5
1979	17.8	18.9
1980	32.4	33.9
1981	42.2	43.8
1982	31.0	32.3
1983	16.0	17.9
1984	8.9	10.9
1985	10.2	11.7
1986	12.1	16.6
1987	17.2	27.4

More public money is flowing into the task of making wind energy usable through tax credits and grants:

- Commercial companies: 7.5 % tax-free allowance according to § 4a of a law on investment allowances
- Tax payers: according to § 82 of a tax regulation 10 % of the costs incurred can be claimed against tax for 10 years
- Agricultural concerns: either § 4a or § 82 above; in addition there are other possibilities included in the programme "Improvement to the structure of agriculture and protection of the coastline"
- Favourable loans are available from the Kreditanstalt für Wiederaufbau for the protection of the environment
- Support can be given within the framework of programmes of the Federal State Governments.

In addition, grants from the European Community are also being made available for German projects.

By the end of 1988, the installed power of WECS in R&D and D projects in the Federal Republic will have reached about 8 MW. However, commercialization is still in its early stages.

Initially, in addition to the purely scientific basic investigations into wind energy and demonstration projects for small-scale WECS, R&D concentrated on large-scale plants. At the turn of the seventies, the Federal Republic of Germany undertook one of the most comprehensive tests worldwide on LS WECS with its 3 MW GROWIAN plant.

In particular, GROWIAN produced the following scientific and technical results:

- a very good control of the overall plant
- a good control of vibrations and the overall dynamics
- rotor blades: in principle, even very large blades can be manufactured with high aerodynamic efficiency, in particular for part-loads
- excellent auxiliary automatic devices
- measurement of meteorological phenomena, previously unknown.

The problems leading to the decision to dismantle the plant, due to take place imminently, are found in the hub (pendulum frame and traverse), where all forces and moments are transmitted from the rotor blades to the hub. Cracks occurred here. Unfortunately, the exceptional size of the GROWIAN construction means that repairs cannot be carried out in spite of the justifiable costs of the repair measures themselves. For this reason, this first large-scale wind experiment cannot be used as a test power plant as it is unsuitable for continuous operation. In spite of this negative result, the positive experiences mentioned above, will be used in our present and future programme described below.

TODAY'S STATUS

Today the support of research within the different areas of wind energy technology presents a well-balanced picture (Table 2).

Table 2 : Contributions for Wind Energy

Category	mill. DM	No. of projects
Large scale WECS	99	4
Medium-sized WECS	46	3
Small WECS	38	27
Other projects	22	26
Windpower in the Third World	20	< 10 *

* not all purely wind projects

We would like to describe the following points in more detail.

In the near future, as a result of a support programme for WECS up to 250 kW, almost all manufacturers of wind energy plants in the Federal Republic will each be able to set up 5 plants with a maximum public grant of 400 DM per m^2 of area swept out by the rotors (Figure 1). This will be accompanied by a 5-year measuring programme to give essential information on the production of electricity, availability and maintenance.

Another 55 small wind power plants will be going into operation in two wind parks, each of 1 MW, shortly to be completed. Operating data specific to wind parks, which is of particular interest to electricity utilities, will be collected in special measuring programmes (Figure 2).

In the field of LS WECS, a 3-blade 1.2 MW plant (WKA 60) is being built on the North Sea Island of Helgoland, based on experiences made with GROWIAN. A similar plant (AWEC 60) is being set up in Spain in a Spanish-German co-operation.

One development which is only being pursued in Germany is the single-blade wind turbine MONOPTEROS. A 400 kW demonstration plant is being operated near Bremerhaven (Figure 2). Three further-developed single-bladed plants (rated power 640 kW) will be erected shortly.

German technology was chosen for the manufacture of the rotor blades for the 2 MW LS WECS "Aeolus" in Sweden. This plant is definitely one of the most successful first plants internationally.

Fig.1: Support Program for WECS up to 250 kW

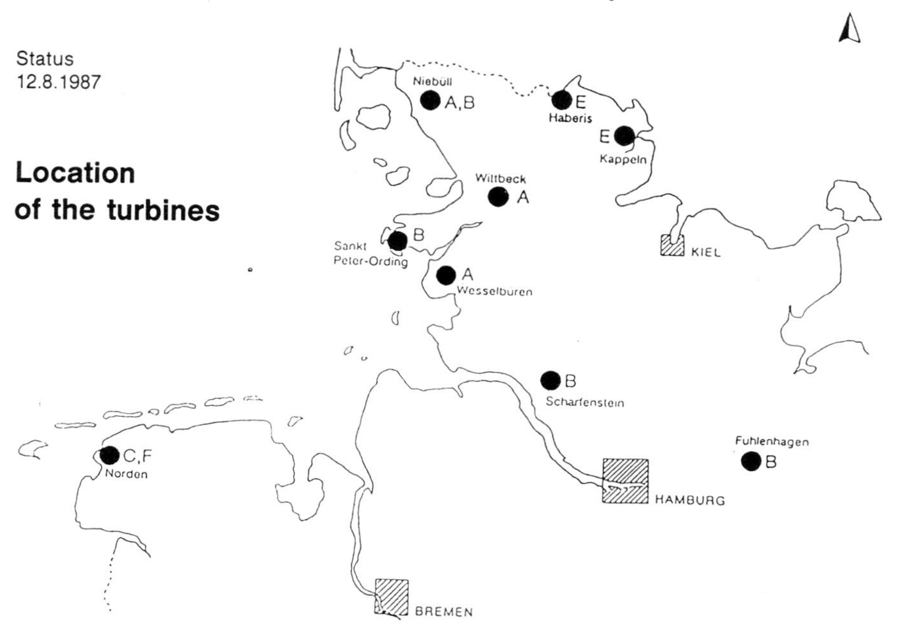

	Type Manufacturer	Power kW	Location
A	ADLER Köster, Heide	165	2260 Niebüll 2244 Wesselbüren 2251 Wittbeck
B	aeolus 12,5 Aerodyn, Damendorf	33	2053 Fuhlenhagen 2053 Fuhlenhagen 2211 Scharfenstein 2252 Sankt Peter-Ording 2260 Niebüll
C	Enercon 16 ENERCON, Aurich	55	2980 Norden 6349 Mademühlen (2960 Aurich) (2904 Kirchhatten) (4592 Lindern)
E	elektrOmat Windkraft-Zentrale Kappeln	25	2340 Kappeln 2391 Haberis
F	NOAH Schönball, Bonn	90	2980 Norden

Fig. 2: Locations of medium and large scale wind turbines and wind farms in northern Germany

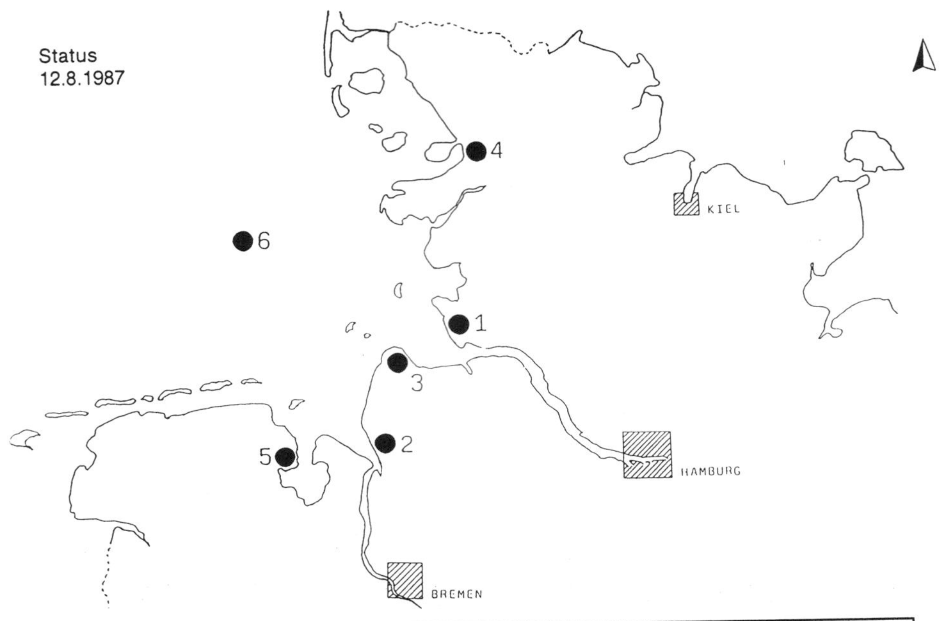

	Location	Type	Power	Status
1.	Kaiser Wilhelm-Koog	GROWIAN Windfarm: • Aeroman • ENERCON • elektrOmat	3 MW 1 MW: • 20 à 30 kW • 5 à 55 kW • 5 à 25 kW	Finished August 87 First rotation August 87
2.	Weddewarden	MONOPTEROS	400 kW	Finished summer 88
3.	Near Cuxhaven	Windfarm • MONOPTEROS • ENERCON	1 MW • 15 à 30 kW • 10 à 55 kW	First rotation winter 87/88
4.	Husum	HSW 200	200 kW	Turbine in operation
5.	Wilhelmshaven	MONOPTEROS 50 MONOPTEROS 30	3 à 640 kW 200 kW	First rotation June 88 First rotation late 88
6.	Helgoland Cabo Villano (coop. with Asinel)	WKA 60 AWEC 60	1,2 MW 1 MW	First rotation spring 89 First rotation spring 89
7.	N.N.	2 LS WECS	MW Class	Planning stage

In addition to the development, construction, test and demonstration of wind turbines, the BMFT is also supporting scientific and technical projects necessary for the utilization of wind energy. These include wind measurements at the height of the rotor of large plants, the formulation of guidelines for testing, commissioning and monitoring wind power plants and possible licensing tests for the different types. The Fördergesellschaft Windenergie e.V., a group of industrial companies and institutes interested in wind, is being given support for its projects on environmental aspects (e.g. noise emissions) and on the measurement and assessment of plants.

A series of co-operative projects are being carried out to test wind plants in developing countries. Some of these are being undertaken in close association with the Federal Ministry for Economic Co-operation or with the German Agency for Technical Co-operation (GTZ). For example, a project of the Kreditanstalt für Wiederaufbau has the goal to erect 45 wind plants each of 30 kW in small wind parks in various countries of the Third World. Local people will be trained in the operation and maintenance of the plants. In the Federal Republic of Germany, much attention is being given to wind-diesel systems. Field tests are being carried out outside Germany, for instance, on the Island of Kythnos in Greece or on Clear Island, Ireland.

FUTURE ACTIVITIES

A great deal of importance is being attached to further research into wind energy and its utilization. It is planned to raise the annual public funds available from the present amount of just under 20 mill. DM. In the field of basic research, new materials will not only help to improve the costs but also the lifetime of wind power plants. For outputs over 250 kW there is still some development work to be carried out as very little reliable data is available both on the technological as well as on the economic side.

Research and development is also being continued for LS WECS stimulated by increased interest being shown by the electricity utilities. Several LS WECS of more than 1 MW will go into operation in the near future to produce reliable data on the availability, capacity effect and electricity production costs.

In many fields a certain conclusion to R&D work is expected. Research is no longer the central point of certain areas, particularly those involving smaller plants. Information important to industry, such as reliability, maintenance, lifetime, annual electricity production and installation costs are being pushed into the foreground of interest.

With the steps towards series production which are necessary for the forecasted wide utilization of wind energy, BMFT will no longer be able to provide financial support. A noteworthy contribution of wind energy towards electricity production will require an extensive use of mature plants. In order to build these today help from other institutions is still imperative.

At the Ministry of Research and Technology there is still a generous expansive attitude towards innovative research and technology in the field of wind energy. Whether or not advantage will be taken of this will depend on the quality of new proposals for projects.

STATUS OF VERTICAL AXIS WEC'S

A. Fritzsche, Dornier System GmbH, Friedrichshafen, FRG

ABSTRACT

The Darrieus-type VAWT gained increasing importance in recent years. The technical performance attained a level comparable to HAWTs. In many industrial and developing countries there is a growing interest in its relatively simple design features and manufacturing technology. Series production in the USA and in Canada covers a range of 20 to 500 kW. Several hundred units are installed in the Californian windfarms. A Canadian manufacturer just recently commissioned a 4 MW VAWT.
The design data of some prototype and series VAWTs have been compiled. Recent results of VAWTs development, as well as the potential for large size units confirm that the Darrieus type VAWT has to be considered as a valuable alternative for all wind regions, possibly to be even more economical with respect to investment and operation costs.

KEYWORDS

Vertical axis wind turbine; Darrieus rotor; performance of VAWTs; development potential; economics of VAWT.

INTRODUCTION

It is worthwhile to start with a brief comparison of the quality of solar and wind energy at a conference dominated by solar technology. A solar radiation of 800 W/m^2 corresponds to 11 m/s wind speed (at sea level and normal temperature). The maximum value of solar intensity outside the atmosphere is 1.37 kW/m^2 and equivalent to 13 m/s. Maximum wind speeds of more than 65 m/s have been observed (typhoons and extreme speeds in Patagonia e.g.) equivalent to an energy density of 165 kW/m^2.
The energy conversion efficiency of solar thermal processes amounts to 10 to 30 % depending on concentration ratio and process temperature. Photovoltaic conversion provides less than 17 % including the most advanced technology but disregarding the efficiency of harnessing and power conditioning steps. Physical and mechanical constraints leave from a 800 W/m^2 wind energy density about 280 W/m^2 per rotor swept area, of which only 5 to 30 % are covered materially by the rotor blades. This results in an average efficiency of 35 % for electricity production by exploiting the energy of the wind. Thus it is comprehensible that an economic analysis provides a ratio of the cost of electricity of about 1:12 in favour of wind energy compared to photovoltaics.
Both principles of wind energy utilization, the horizontal (HAWT) and the vertical axis turbines (VAWT) have made good progress. One aim of both types is to promote clean energy. Many configurations have been proposed and investigated to convert the kinetic energy of a more or less horizontal wind flow into torque of a vertically rotating axis. Of the many varieties of VAWTs only those based on the Darrieus type including some design variations reached a technological level suitable for a wide range of applications. Air drag VAWTs gained some interest in the very low power range and for water pumping. They will not be considered here.
Darrieus earned his patent in 1926, it was reinvented by South and Rangi

(NRC Ottawa) in 1973. From there on this striking configuration has been subject of considerable interest and activities. It has to be mentioned that Darrieus' patent claimed also the H- and giromill rotor type. The VAWT made good progress in energy output in the recent years, see Fig. 1. The theoretical inferiority of the aerodynamic efficiency has been overcome. The independence of wind directions is not only an energetical but also a design advantage.

A very schematic overview of VAWT-activities all over the world is summarized in Fig. 2, making no difference between test, prototype and series VAWTs. It is visible that operational experiences and test data now are available for a power range from less than 1 to nearly 4000 kW. Besides Canada and USA many countries appreciate the technical and economic potential of the VAWT.

SMALL AND MEDIUM SIZE VAWTS

The first larger prototype was commissioned in 1977 on the Magdalen Islands in the Gulf of St. Lawrence, Canada. The 230 kW unit has been developed and built by the NRC Ottawa, DAF and Hydro Quebec. After some modifications and extensive testing the grid connected plant is still in operation (rotor diameter 24 m, rotor height 36 m, design wind speed 15 m/s, 2 extruded aluminium blades).

Based on this technology DAF, today Indal Technologies Inc., scaled down and up. The 50 and the 500 kW type have been designed for grid connected operation. A 30 kW VAWT/100 kW diesel hybrid power plant is in operation at the Atlantic wind test site in Canada. The 50 kW VAWT near Churchill, Manitoba, attained almost 20,000 operating hours. 15 units of that type have been sold, the 100 kW type is under test in Canada and the USA.

Another manufacturer of medium size VAWT in Canada is ADECON. A 150 kW prototype has been tested successfully. A characteristic feature is a truss support substituting the guy wires. The low weight rotor blades are assembled from 3 different extruded aluminium sections. 25 units have been ordered. Ex-works costs of about 1,000 US $/kW have been announced for future series production.

In the USA Sandia Laboratories, envolved in VAWT development since the late seventies have still a leading position. A 34 m diameter research oriented wind turbine is in operation since 1986 with step tapered blades, specially designed for natural laminar flow thus considerably increasing energy output under part load operation conditions [1,2]. A reduction of blade flatwise bending stresses is expected from a slight modification of the troposkien blade shape. The hight to diameter ratio of the rotor was reduced to 1.25 without any losses of structural integrity.

Flowind Corporation calculated an increase of annual energy output using the Sandia profile SNLA 0018/50 by 49 % at 6.3 m/s, and by 29 % at 7.2 m/s decreasing to 4 % at 9 m/s. Existing Flowind 19 M VAWT was the reference and a Rayleigh wind frequency was assumed. Flowind operates 511 turbines at Tehachapi and Altamont wind farms, 309 of the 17 M VAWT type (153 kW), 200 of the 19 M VAWT type (300 kW) and 2 of the 25 M VAWT type (500 kW) in April 1987. Electric output of the turbines was 110 GWh until April 1987 at Tehachapi and 66 GWh until February 1987 at Altamont.

Flowind is completing the design and tests of structural retrofits for the 17 and 19 M VAWTs. The goal is a 0.20 $/kWh ratio of installed cost versus annual energy output for wind farm conditions in California. (Multiplying this ratio by the fixed charge rate provides the cost of electricity in cents/kWh).

The effect of clean and dirty blade surfaces measured by Flowind merits special attention of theorists and economists, see Fig. 3. A loss of up to 7 %

may by envisaged at high wind speed sites.
VAWT-Power Inc. operates 30 VAWTs at Palm Springs (150 kW).

An overview of VAWT-activities in Europe should start in Darrieus' native country, in France. CENG, Centre d'Etudes Nucleaires de Grenoble, developed a 2 and 3 blade version of a 5 kW turbine for heating and grid connection, now in operation at 6 sites in France. A 30 kW turbine, supported by a 12 m high structure was tested in 1986.
Special design features can be found at the Darrieus turbines developed in Belgium and in the Netherlands. The 30 kW turbine, STRO-V10 of the Vrije Universiteit of Bruxelles for electric heating is a 4 blade rotor of tropo-skien shape (profile 0018, 10 m diameter, 10 m rotor height). The STRO-V6, a 3 kW turbine is designed for water pumping equipped with 3 blades of a large cord length (950 mm) for low wind speed aerodynamic starting.
Polymarin B.V., Netherlands, in 1983 developed a cantilevered VAWT, the Pioneer I, thus saving space at the site, see Fig. 4. Omitting the guy wires of that 40 kW turbine leads to a heigher weight of about 250 kg/kW, being more than twice the average of Darrieus type VAWTs.
Dornier built VAWT test plants in Germany for technology transfer to less developed countries. The 20 kW wind turbine installed in 1981 at Comodoro Rivadavia, Southern Argentina, and tested until 1983 has been taken over by a local oil company and is in operation again. A 30 kW turbine on the Höchsten near Dornier premises serves for training and testing at different operation modes, Fig. 5. Operation at constant tip speed ratio is part of an adjustable generator control, including stall control limiting power output at high wind speeds. The AC-DC-AC link to the grid soon will be removed by the installation of a high efficient double fed generator. A hybrid wind-diesel plant is under preparation. Minimum fuel consumption and increased life time of the diesel will be achieved by a microprocessor controlled variable speed operation mode [3]. A turbine of the same size is in operation in the P.R. of China, another one is on the way to be comissioned in India.
Cesen and Ansaldo Impianti built the first VAWT in Italy. The tests of the 5 kW turbine have been performed in 1983. Tema, a company associated with the ENI group, developed a variable geometry rotor according to a patent held by VAWT Ltd., London. The 4 blades of the Musgrove type rotor are mounted in pairs inclining gradually in order to reduce the swept area above 11 m/s wind speed (max. power 40 kW). The variable speed generator with fixed frequency output has been developed using a patent filed by Terma (rotating 'stator').
Different configurations and sizes of VAWTs have also been constructed and tested in Romania. Technical details have been presented in 1985 at the 7th BWEA conference [4]. A 100 kW turbine has been developed corresponding to informations received in 1986. Any other VAWT activities in Eastern Europe are unknown to the author.
Alpha Real, a company in Switzerland, designed a 160 kW VAWT, which was commissioned recently at a site near Martigny in the valley of the Rhone. The main design features of the predecessor type have been retained. A low weight construction (63 kg/kW) was achieved by mounting the rotor directly on top of the gearbox, thus saving any ground support structure. The nose and tail section of the blades were welded together (722 mm total cord length). The turbine (4/6 pole-asynchronous generator for grid connection) operates on microprocessor control.
A short report on the Musgrove-type VAWT development in the U.K. concludes the European overview [5]. The basic driver was the aim to reduce dynamic loads of a VAWT operating in the cyclic stall mode, which result in fatigue problems. The variable geometry principle, a modification of the H-type Darrieus, was proposed by Musgrove in 1975. Blade inclination provides an

effective mean of altering the blade angle of attack thus limiting power output in strong winds without relying on cyclic stall. This principle may open the way for large units at off-shore sites.
A 130 kW turbine (25 m diameter, 2 blades, 18 m long, total height 34 m) was commissioned in late 1986 at Carmarthen Bay, South Wales, Fig. 6. There have been no periods of unscheduled downtime due to component failures.
In 1986 a 14 m diameter, 40 kW VAWT has been installed in Sardinia, only recently a 100 kW turbine at St. Mary's, Isles of Scilly. VAWT Ltd. is undertaking a concept study for a 1.2 MW cantilevered turbine, funded by UK Department of Energy and EEC.
Some research activities of industry and research institutions in Asia have to be mentioned briefly. In the P.R. of China a multiple of small genuine and modified Darrieus type turbines have been installed at Li Shan and Padaling test station. Troposkien, H-shaped and double blade rotors have been developed for battery charging, and autonomous power supply. A German-Chinese cooperation programme (Dornier System and Electric Dept. of Tsinghua University, Beijing) supports the efforts of Chinese manufacturers to go into batch and series production. A similar implementation programme is performed by BHEL, Hyderabad, India, supported by Indian Institute of Technology, Madras and Dornier System (funded by the German Ministry for Research and Technology and the Department for Non-Conventional Energy Sources, Delhi).
At Hokkaido University, Japan, selfstarting of VAWTs was studied in 1978/1979. In 1984 Oriental Kiden Co. Ltd. offered, 8 and 45 kW H-type turbines for heating and electricity supply.

THE FIRST MW-SIZE VAWT

The project Eole started in 1981 in cooperation of the National Research Council of Canada, Hydro Quebec and Lavalin-Shawinigan (at that time subcontractor) with a total budget Can. $ 35.2 millions funded by the Canadian government. Commissioning and testing began in 1987. At the time when the key design data were specified only the basic aerodynamic and aeroelastic theories of VAWTs where available, proven by some operational experiences in a power range not exceeding 230 kW. Lavalin-Shawinigan is now the project leader. Electricity output is sold to Hydro Quebec for a special rate of 0.20 Can $/kWh, thus financing the test campaign.
Eole has been erected at Cap Chat, north-east of Quebec, right coast St. Lawrence river. The table summarizes the main design data [6], Fig. 7 gives an impression of the size of the VAWT.

Table: Main Design Data of Eole, 4 MW-VAWT

nominal turbine power rating	4 MW
rotor diameter and height	64/96 m
swept area	4000 m²
number of blades and airfoil profile	2/NACA 0018
cord length of the blade	2.4 m
annual mean wind speed at the site	9.2 m/s
cut-in wind speed	4.5 m/s
cut-out wind speed	62.0 m/s
(all values at 10 m above ground level)	
rated rotational speed	14.24 rpm
annual energy output (Weibull factor k = 1.56 and shear coefficient 0.11)	
mode A constant speed	1.83 MWH/M2
mode B variable speed	1.85 MWH/M2
mode C variable speed with power limit	2.07 MWH/M2

The middle section of the blade is a cylindrical steel construction, tail and nose are made from GFRP. Aerodynamic spoilers (3 at each blade) and 2 disk brakes (3 calipers each) guarantee a very reliable overspeed protection. The power train, in principle very simple, demonstrates one of the main charac teristic features of the VAWTs, Fig. 8. Rotor mast, couplings, brake-system and generator are mounted in-line, all major and heavy components for electric energy conversion and security are located on the ground, easily protected, air conditioned and accessible. The 162 pole synchronous generator (20 Hz) has been developed and built by General Electric, Montreal. The static exitation system is fed from the grid. Eole is started by means of a motor mode of the generator.
No serious problems occured during commissioning. Eole will go in automatic operation in October, controlled by 5 engineers for the next 6 months. Then the design of this possibly overprotected first prototype will be assessed. Any fundamental problems of any kind have not been discovered, 6 and even 8 MW VAWTs seem to be feasible.

CONCLUSIONS

In spite of the simplicity of the aerodynamic and mechanical design principle of the Darrieus type VAWT the problems of aeroelastics, structural dynamics and operation control required considerable effort. But now the appropriate design tools are available, as well as the modern power electronics for an optimum wind energy utilization in a wide range of wind speeds. The Darrieus is not longer only a high wind speed WEC. Some modified versions have been proposed also for off-shore sites [7]. Based on the good test and operational results in a wide power range from 10 kW to 4 MW, the trade-off of the most economic size for large grid connected plants is open again. 1200 to 1500 $/kW are considered to be realistic figures achievable in the next years in batch production at sites comparable to the Canadian conditions.
Further improvements in aerodynamics, electrics, and operational reliability accompanied by a decrease in manufacturing costs will enhance the attractive offer of VAWTs for a worldwide market.

ACKNOWLEDGEMENT

This up-to-date overview was only possible by the support of manufactures and research institutions in Europe, Canada and the USA, particularly by Mr. Quraeshi, Lavalin-Shawinigan, who was originally envited to present this paper.

References

[1] P.C. Klimas, Tailored Airfoils for Vertical Axis Wind Turbines. SANDIA Report 1062, 1984.

[2] D.J. Malcolm, High Production Darrieus Rotors. EWEC '86, pp. 605-608.

[3] A. Fritzsche, Entwicklungsprobleme autonomer Wind-Diesel-Anlagen. Tagung Windenergie der DGW, Universität Oldenburg, März 1987.

[4] G.H. Voicu, C.R. Tantareanu, M. Tudor, Perspectives of wind power utilization in Romania. Proceedings of the 7th BWEA Wind Energy Conference, 1985, pp. 29 - 35.

[5] I.D. Mays, R. Clare, The U.K. Vertical Axis Wind Turbine Programme, Experiences and Initial Results. EWEC' 86, pp. 183 - 178.

[6] B. Richards, S. Quraeshi, The Design, Fabrication and Installation of Project Eole, 4 MW Prototype Vertical Axis Wind Turbine Generator. Proceedings EWEC' 86, pp. 609-614.

[7] O. Ljungström, Some Innovative Concepts in Axial and Cross Flow Wind Turbine Systems. Proceedings EWEC '86, pp. 657 - 668.

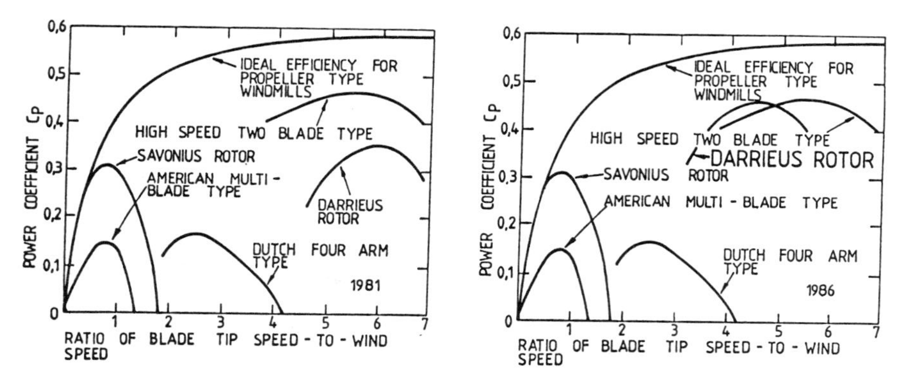

Fig. 1: Typical performance of several wind machines, status of Darrieus rotor 1981 and 1986 (DOE)

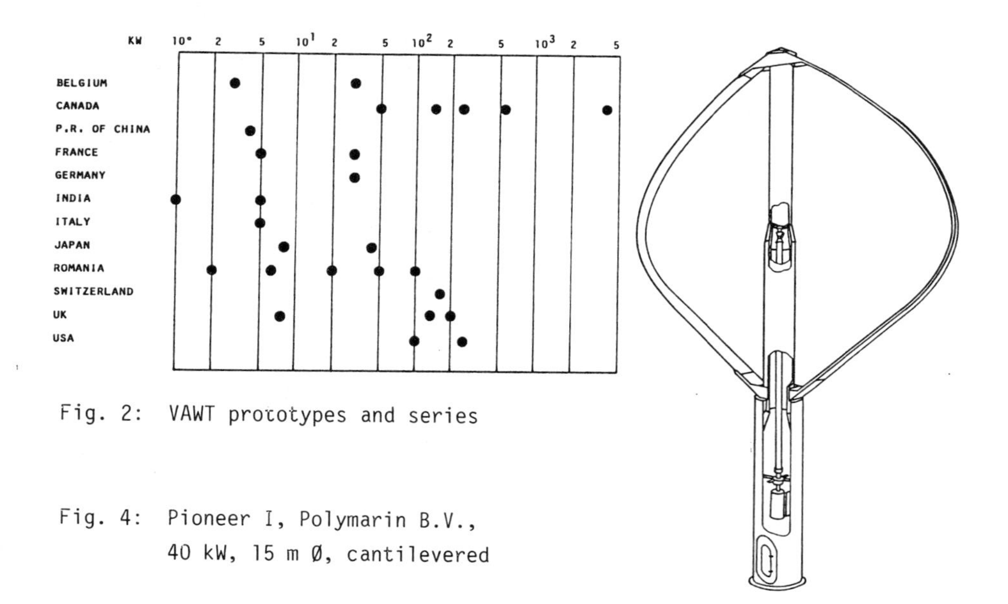

Fig. 2: VAWT prototypes and series

Fig. 4: Pioneer I, Polymarin B.V., 40 kW, 15 m Ø, cantilevered

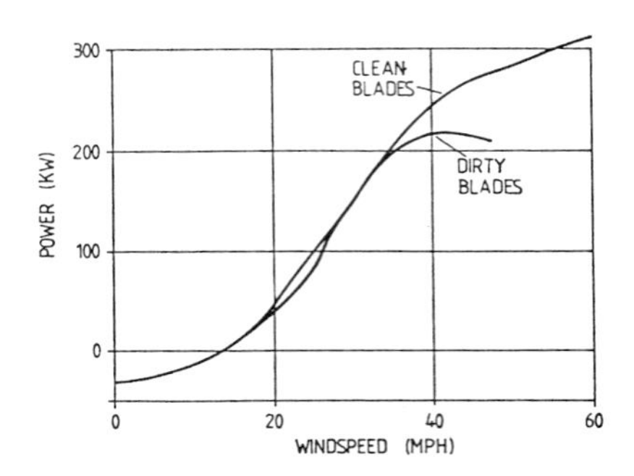

Fig. 3: Effect of blade washing on turbine performance (Flowind 19 M)

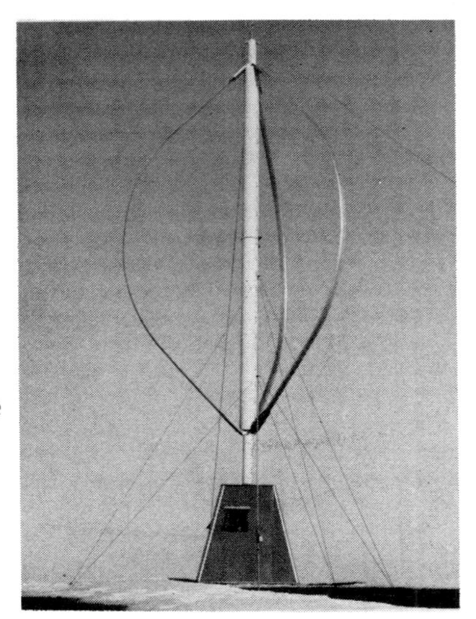

Fig. 5: Dornier test plant, 30 kW, 12 m Ø, variable speed

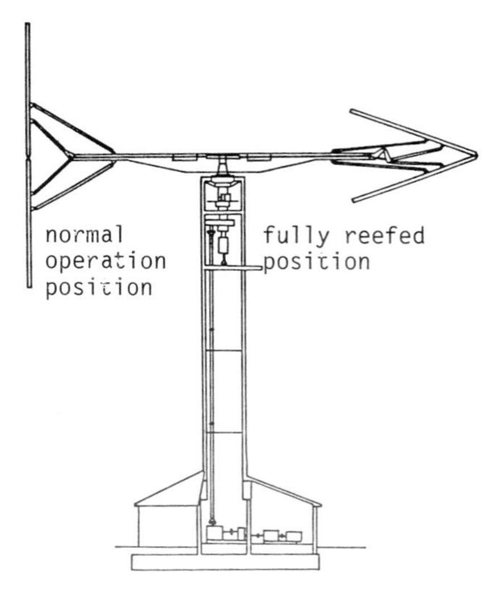

Fig. 6: Musgrove type, VAWT, research prototype, 120 kW, 25 m Ø

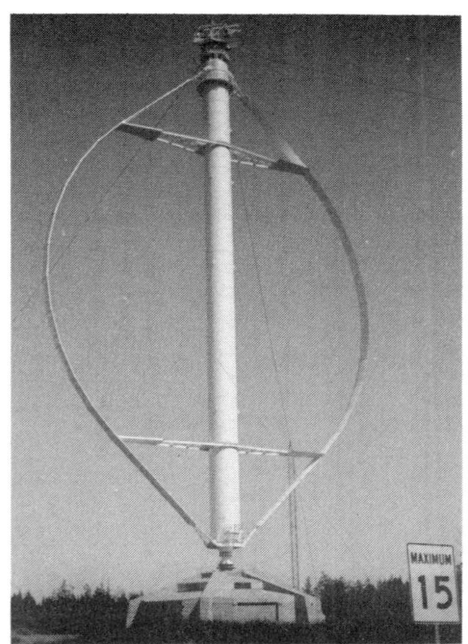

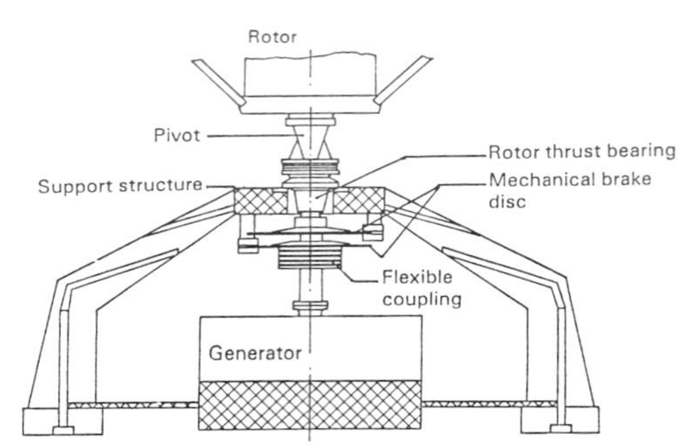

Fig. 7: Eole, 4 MW, 64 m Ø

Fig. 8: Eole, power train

ESSENTIAL SYSTEM FEATURES OF HORIZONTAL AXIS WECS

E. Hau *, W. Kleinkauf **

* MAN Technologie GmbH, München, W. Germany

** Universität Gh-Kassel, W. Germany

ABSTRACT

Proceeding on the structure of a horizontal axis wind energy converter, the first part states fundamental considerations concerning aerodynamics and mechanics. The second part is focussed on electrical requirements and resulting design variants. Finally, general trends are pointed out appearing for the design of large and small plants.

KEYWORDS

Horizontal axis WEC; Rotor configurations; Tower conceptions; System design; Generator variants; Isolated and grid-connected operation; Speed-variable plants.

1. INTRODUCTION

The horizontal axis type is by far the predominant conception for wind energy converters up to now. Its dominance is caused by some fundamental virtues:

- The horizontal axis rotor optimally utilizes the aerodynamical lift forces, and therefore has a very high aerodynamical power coefficient.
- The control of aerodynamical power input and rotor speed can be obtained comparatively easy and effective with a propeller rotor.
- Not least the technological development advantage of this type is an importand argument for its broad application.

The basic design of a horizontal axis wind energy converter is shown in fig. 1. The rotor consists of the rotor blades and the rotor hub, which contains the blade pitch mechanism in larger and even some small plants. The mechanical drive train consists of the rotating mechanical components (rotor shaft, rotor brake, transmission gear, and generator drive shaft with bearings, couplings, etc.).

The electrical system consists of the electric generator, the control system, and the grid connection resp. the connection of special electrical consumers. The mechanical drive train and the greater part of the electrical system are installed in the nacelle. It follows the wind direction by means of the yaw drive. Rotor and nacelle are carried by the tower.

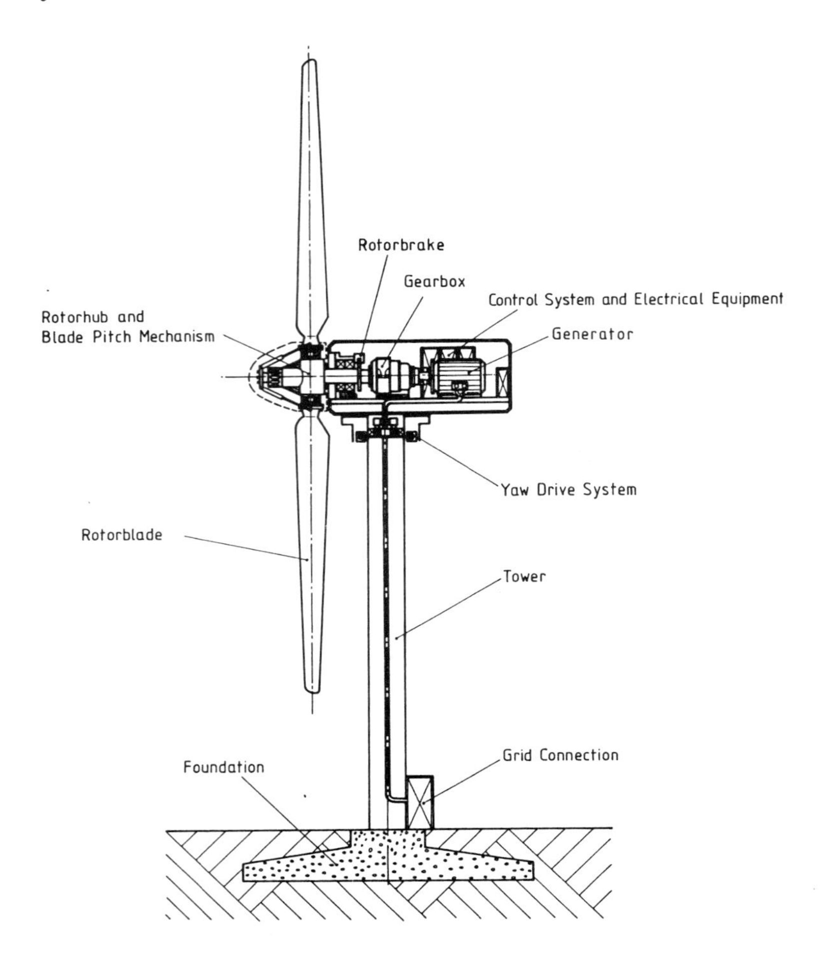

Fig. 1 Fundamental design of a horizontal axis WEC.

In the frame of this fundamental design of a horizontal axis wind energy converter, the mentioned sub-systems are realized with largely differing functional and constructional characteristics. These system features characterize a horizontal axis WEC regarding to its technical conception and to its application feasibilities.

2. AERODYNAMICAL AND MECHANICAL SYSTEM CHARACTERISTICS

The aerodynamical and mechanical system characteristics of a wind energy converter mainly determine the construction expenditure, i.e. the construction weight and the mechanical complexity. With regard especially to the series production the manufacturing cost of the WEC is affected by this.

2.1 ROTOR CONCEPTION

The most important technical options for rotor design are shown in fig. 2. To discuss the advantages and disadvantages bound to these in detail is beyond the scope of this survey. Two fundamental tendencies can be distinguished. On the one hand rotors with medium speed (usually 3 rotor blades), with hingeless, stiff structure (stiff, heavy rotor blades and hingeless hub), and with passive aerodynamical power limitation ("stall"), and on the other rotors with high speed (one or two-blade), load balancing rotor hub hinges, and active blade pitch setting. Whilst the first conception leads to high construction weight, the second reduces it considerably, but leads to higher mechanical and dynamical sensibility and mechanical complexity.

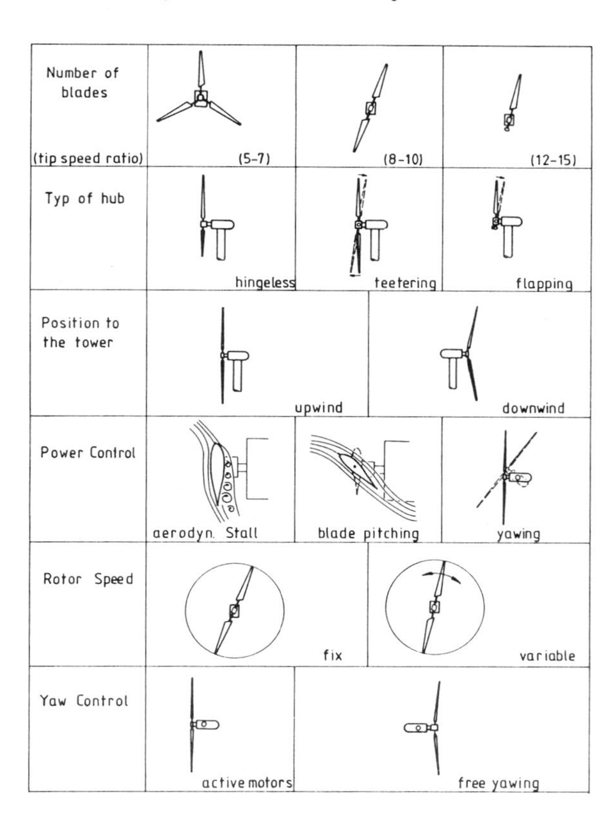

Fig. 2.1 Main options for rotor design.

2.2 MECHANICAL DRIVE TRAIN AND STATIC CONCEPTION OF THE NACELLE

The arrangement of the mechanical components and the static conception of the nacelle are closely connected. The most frequent type at present has the components of mechanical drive and electric generator mounted "in series" on a load carrying bedplate. The fairing has no statical function (fig. 3).

Weight and volume are reduced when the mechanical components are partly or fully integrated in the statical function of the load carrying bedplate. The gearbox carries all loads from rotor weight, and is directly mounted to the tower. This type is less expensive due to the reduced mass, but makes necessary higher costs for development.

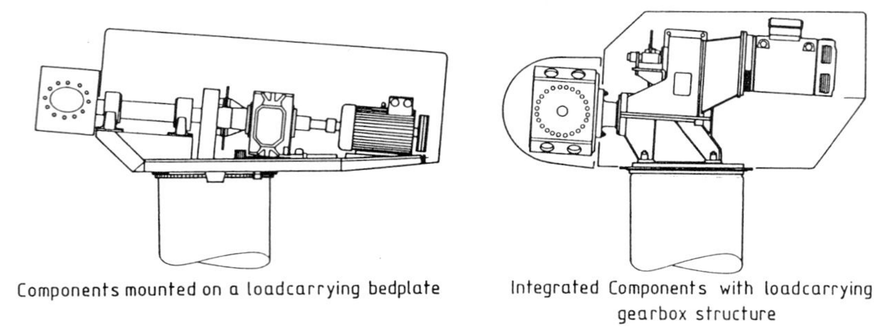

Components mounted on a loadcarrying bedplate
Integrated Components with loadcarrying gearbox structure

Fig. 2.2 Typical arrangements of drive train and nacelle.

2.3 TOWER CONSTRUCTION AND STIFFNESS DESIGN OF WECS

The main tower configurations for horizontal axis WECs are:
- free-standing steel tubes
- free-standing lattice masts
- free-standing concrete towers.

The selection of the material, the geometrical shape, and the static conception are determined by height, stiffness requirements, and mounting costs. The selection of tower height in relation to rotor diameter cannot always be determined as economic optimum by increased power yield and increased building cost. Different traditions evidently play an important role. The stiffness design of the tower is of great importance regarding plant vibrations (coupled vibrations of rotor and tower). The location of the towers first natural bending frequency relative to the dynamical excitation of the rotor is the essential criterion. Fig. 4 shows the different options for stiffness design using a two-blade rotor as example.

Stiff designs with a first natural bending frequency above the aerodynamical excitation (double rotor frequency 2p for a two-blade rotor) lead to a considerable material expenditure (see fig. 5). In this case lattice masts or concrete towers are prefered configurations; but there is an increasing tendency towards so-called "soft" tower constructions with a first natural bending frequency below the aerodynamical, in some cases even below the mass-dynamical excitation (1p). Free-standing lattice masts fullfil this requirement best.

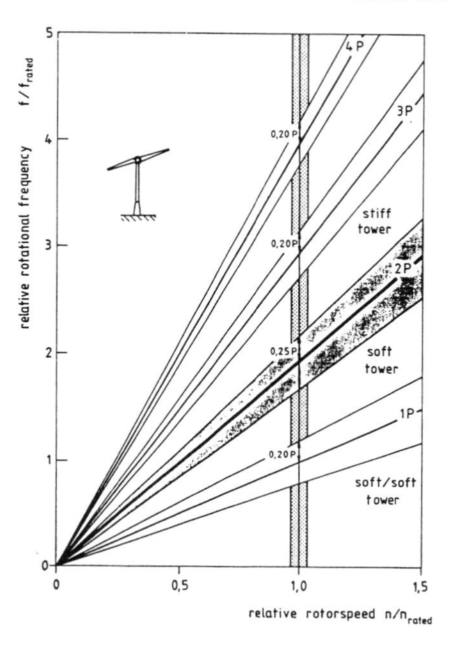

Fig. 2.3

Options for tower stiffness design for a two-blade WEC.

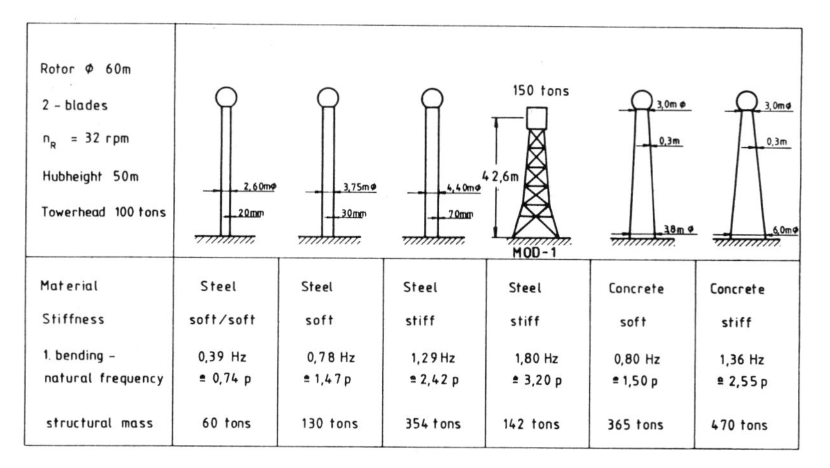

Material	Steel	Steel	Steel	Steel	Concrete	Concrete
Stiffness	soft/soft	soft	stiff	stiff	soft	stiff
1. bending - natural frequency	0,39 Hz ≙ 0,74 p	0,78 Hz ≙ 1,47 p	1,29 Hz ≙ 2,42 p	1,80 Hz ≙ 3,20 p	0,80 Hz ≙ 1,50 p	1,36 Hz ≙ 2,55 p
structural mass	60 tons	130 tons	354 tons	142 tons	365 tons	470 tons

Fig. 2.4 Tower concepts with various stiffness for a large WEC.

2.4 OVERALL CONFIGURATIONS

The discussed system characteristics can be found in different variations and combinations in present horizontal axis WECs. In this variety two basic conceptions appear, which can be regarded as corner-points of the design variety (fig. 6).

One basic conception is mainly represented by the small and medium converters of Danish type. They have the following features:
- 3-blade rotor with low tip speed ratios (5-6)
- hingeless rotor hub
- upwind rotor
- power limitation without blade pitch setting by the aerodynamical stall
- constant rotor speed
- drive train on load carrying bedplate
- relatively stiff tower design.

This design is mechanically and dynamically uncomplicated, but relatively heavy. It is limited on a power range up to 200-300 kW resp. 20-30 m rotor diameter.

The other conception is the light construction prefered by American and European manufactures:
- two-blade or one-blade rotor with high speed coefficient
- teetering hub
- upwind or downwind rotor
- power and speed control with blade pitch setting
- constant or variable rotor speed
- component integrated drive train design
- soft tower design.

This design is more and more prefered for small plants. For greater plants it is nearly a must to keep the building expenditure within economic limits.

Figure 7 shows the characteristics of the discussed basic conceptions regarding the construction weights presently resulting by means of the specific mass of rotor and nacelle (tower head mass) relative to the swept rotor area. The statistics also show the increase of specific mass with increasing plant size. In this connection, though, it must not be forgotten that the specific power yield of the swept rotor area also increases with increasing plant size (hub height of the rotor). Furthermore, it is to be considered that the mentioned masses are not only influenced by the described system features; e.g. the installed generator power or different load assumptions and building regulations play some role. In spite of this, the influence of the system characteristics remains dominant.

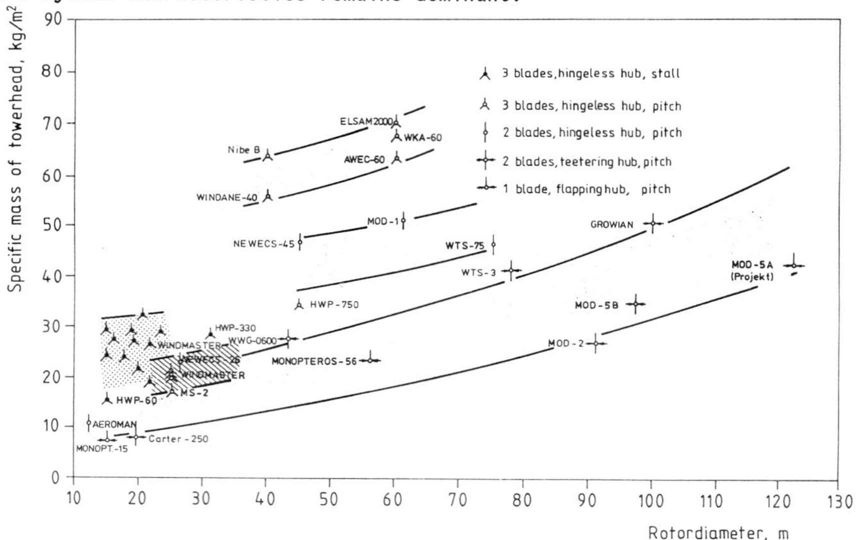

Fig. 2.5 Specific towerhead mass of horizontal axis WECs.

3. ELECTRICAL PLANT DESIGN
3.1 APPLICATIONS AND REQUIREMENTS

A fundamental distinction for the application of electrical power supply plants is between isolated and grid connected operation. Isolated operation means the use of an electrical generator for the supply of one or more consumers without being coupled with other supply devices. Grid-connected operation of a plant is in contrast to this characterized by the supply of electrical energy to an existing AC supply grid. Systems for DC power supply only play a subordinate role. The terms isolated grid and interconnected grid are used to characterize design and features of grids. The interconnected grid is stable regarding load variations. Typical for isolated, weak grids is a geographical situation making expensive or impossible the connection to an interconnected grid. The ratio of maximum grid power and single consumer power is very much smaller here than in an interconnected grid.

The processes in electric grids can be described by means of the grid quantities frequency f, voltage U, active power P, and reactive power Q. Considering a distributed grid in stationary state, local voltage differences may occur due to different line losses. Therefore the voltage can be influenced even locally - i.e. decentrally - by feeding power to the grid. In contrast, the frequency is equal everywhere in the grid. It is of central importance.

When using rotating mechano-electrical energy converters, e.g. synchronous generators, frequency depends on the driving torques. Active and reactive power output are coupled by means of the load and exciter current caused rotor displacement angle ϑ (fig. 3.1). Desired relations between active power and frequency, or between reactive power and voltage (statics), can be reached by respective control circuit design.

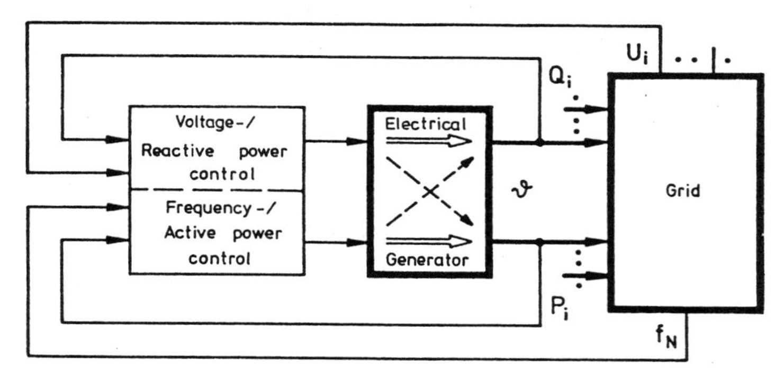

Fig. 3.1 Synchronous generator in grid-connected operation.

Fundamentally different conditions occur when the grid is formed by static converters, due to the missing relation between rotating mass and frequency.

Electrical plants and devices can be distributed to the following classes regarding their active power balance: Consumers ($P < 0$), Suppliers ($P > 0$), and Compensation Units ($P \approx 0$). Storages (e.g. batteries) may be consumers or suppliers. Compensation units make possible the supply of reactive currents. The control of frequency and voltages is carried out by means of controllable supplies of active and reactive power. Units being able to do this in a defined range are called grid forming elements.

The consumer demands are substantial for the possible tolerance ranges of the grid quantities. The rating of consumers to be connected to the grid is based on an allowed range for grid voltage und frequency, and an admissible harmonic content. The conditions specified in VDE 0160 (power electronics equipment) may serve as general rule. Following this, voltage variations in a range from 90 % to 110 % of nominal voltage are admissible. The undisturbed operation of devices and plants is required for frequency variations up to ±1 % of the nominal value. The majority of consumers allows frequency variations up to ±5 % of nominal frequency, and voltage variations from -15 % to +10 % of the nominal voltage. The fundamental content should be more than 99,5 % (distortion factor $k \leq 10$ %). Own experience has shown that isolated grids (e.g. the supply stations on Malta or the Greek island of Kythnos) often do not keep the tolerances, but still fulfil their supply task.

Many loads need a specific energy amount only during a certain period, e.g. interruptable mechanical drives like water pumps. A design oriented by the consumers characteristics makes it possible to run wind energy converters in isolated operation. Here the dynamical problems (starting procedures, energy offer variations, etc.) occur in full acuteness due to the direct coupling of consumers and wind energy converters. Thats why the conception has to be carried out considering the single case.

3.2 CHARACTERISTICS OF MECHANO-ELECTRICAL ENERGY CONVERTERS

The generator is the connection between the mechanical part of the WEC and the grid resp. consumer Therefore it has to comply with the requirements of both sides, if possible. The loadings of mechanical plant components, i.e. the configuration of the speed-torque relations, stand in the foreground on the one side, whilst conditions under which electrical power should be supplied (voltage, frequency, active power, reactive power, harmonic content) are on the other. Basically it must be distinguished whether electrical converters make possible

1. a de-coupling of mechanical speed and electrical frequency, or
2. supply reactive power besides active power.

Figure 3.2 shows 5 design variants with very different characteristics. As well asynchronous as synchronous generators may be used as real mechano-electrical converters. The very simply constructed asynchronous generators with cage rotor have become the most successful machines for grid-connected operation up to the range of some 100 kW. They have essential advantages compared to synchronous generators, because they make possible better operational stability at fast altering driving torques, a power-dependent speed variation in the range of admissible slip values, and a simplification of grid connecting devices.

The essential disadvantage in using an asynchronous generator is that the inductive reactive power necessary for the excitation of and asynchronous generator must be supplied by the grid or by a compensation device. In most cases a fixed reactive power compensation is sufficient. Figure 3.3 shows the fundamental electrical design of a wind energy converter suitable for interconnected operation. Figure 3.4 shows the appertaining control structure of such a plant. During normal operation the plant works grid-connected. To avoid overloading due to the frequency being preset then the power input has to be limited. This power control is superposed by a speed control engaging in the operation states starting, shutting down, and support of the plant during grid failures. The plant type with blade pitch setting used here has advantages compared with converters using the stall for the limitation of maximum power, especially when used in weak grids, due to the possibility to limit output power continuously.

conversion system	characteristics	supplementary components	suitable for
A. Asynchronous generator with cage rotor (n, ASG, f)	$n \approx (1-s)\,f$ $-s \approx 0 \ldots 0{,}08$ speed dependent inductive reactive power consumer	compensation, blade pitch setting device	grid-connected operation weak grid connected operation
B. Synchronous generator with excitation (n, SG, f)	$n = f$ controllable reactive power supply	vibration damper	isolated operation grid connected operation
C. Synchronous generator with DC link (n, SG, DC, f)	$n \approx 0{,}5 \ldots 1{,}2\,f$ controllable inductive reactive power consumer	harmonics filter compensation	grid-connected operation interconnected operation (coupling on DC-side) weak grid connected operation
D. Oversynchronous converter cascade (n, ASG, f)	$n \approx 1 \ldots 1{,}3\,f$ controllable inductive reactive power consumer	harmonics filter compensation	grid-connected operation
E. Double-fed asynchrongenerator (n, ASG, f)	$n \approx 0{,}8 \ldots 1{,}2\,f$ controllable controllable reactive power supply	harmonics filter blade pitch setting device	grid-connected operation great plants

Fig. 3.2 Mechano-electrical power conversion systems
n..mechanical speed, f..electrical frequency, s..slip.

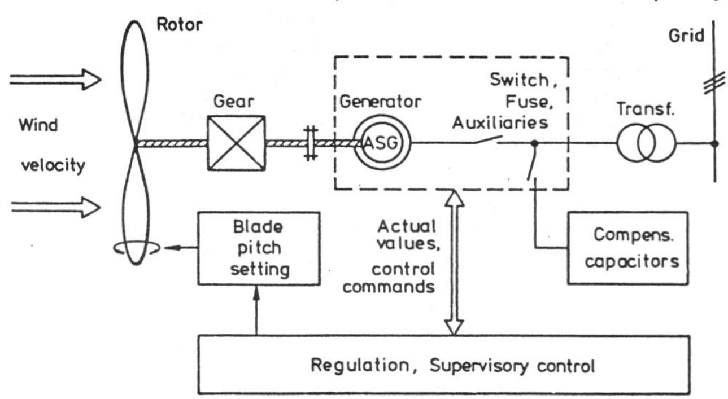

Fig. 3.3 Basic design for the operation of a wind energy converter with asynchronous generator connected to the grid.

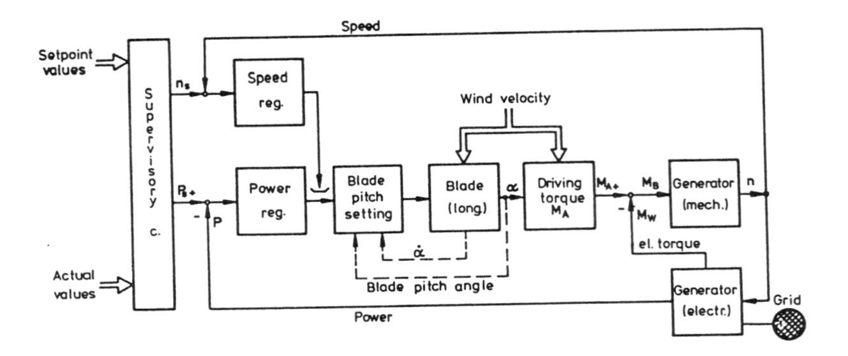

Fig. 3.4 Basic block diagram for the control of a wind energy converter with asynchronous generator in grid-connected operation.

Synchronous generators with controllable exciter units have a fix frequency-speed coupling, but are able to supply controlled reactive power, and therefore they are suitable for isolated operation.

The essential technical advantages of a plant with speed-variable coupling of rotor and electrical output compared with a fix coupling are:
- Speed/power output ratio is continuously controllable.
- Mechanical parts can be unloaded by the variation of speed.
- Blade pitch setting actions can be reduced.
- Rotor speed setting favourable for power conversion is possible.
- Electrical output is smoothed using kinetic energy (flywheel principle).
- Better control facilities improve interconnected operation.

Speed-frequency variable generator systems which need inductive reactive power are shown in fig. 3.2 as variants C and D. The speed variation range of variant D (e.g. AWEC 60, Spanish 1.2 MW plant) is smaller than that of system C (e.g. WKA 60, German 1.2 MW plant) due to their equipment characteristics. Generator conception D allows as well speed-variable converter operation as controlled supply of reactive power. The realization (e.g. 3 MW GROWIAN plant, 5 MW MOD5 plant) is expensive so that this solution is used only for very great plants.

Particularly difficult to estimate are the effects of speed-frequency variable generator system on the design of mechanical plant components and in this way on the complete plants cost.

4. FINAL CONSIDERATIONS

The future development of wind energy converters will more and more be determined by the necessity to reduce the manufacturing costs in series production. Weight reducing system characteristics have therefore become necessary. Flexible, light constructions, speed-variable generator systems, and "intelligent" control methods can do their part to avoid material fatiguing load states. Speed-variable generator/converter systems might help to make smaller plants lighter even without blade pitch setting devices, and to control difficult isolated operation states. For great plants there is no alternative to light construction and high-standard dynamical control on rotor and on generator side in the near future.

FIVE YEARS OF OPERATIONAL EXPERIENCE WITH TWO LARGE WECS IN SWEDEN

Staffan Engström
National Energy Administration, Sweden
S-117 87 Stockholm, Sweden

Göran Olsson
Sydkraft AB
S-217 01 Malmö, Sweden

Lars Möller
Karlskronavarvet AB
Box 1008
S-371 24 Karlskrona, Sweden

Pär Svenkvist
Nohab KMW Turbin AB
Box 1005
S-681 01 Kristinehamn, Sweden

Per Swenzén
Vattenfall
S-162 87 Vällingby, Sweden

ABSTRACT

Operational experiences from the Maglarp 3 MW and Näsudden 2 MW wind turbines is described, covering five years since the first rotation in 1982. The accumulated time of operation is 10 500 and 8 700 hours respectively, with a total generation of 15.3 and 9.5 GWh (August 1987).

KEYWORDS

Wind power, Energy conversion and storage, Natural resources, Power production, Operational experience, Availability, Maintenance, Environmental impact, Lightning protection, TV-interference.

THE SWEDISH WIND ENERGY PROGRAM

The two large WECS prototypes at Maglarp and Näsudden constitute a major effort in the Swedish wind energy program, initiated in 1975 and financed and coordinated by the National Energy Administration (before 1983 the National Swedish Board for Energy Source Development). Objectives for the board to order the units in 1979 were to develop and demonstrate the technology needed, to gain operational experience, to evaluate different design solutions and, in general, to evaluate whether wind energy is a viable energy option for the future taking economical and environmental factors in consideration. In this paper we will concentrate on the operational experiences.

NÄSUDDEN

The plant

The Näsudden 2 MW Wind Energy Conversion System (WECS) prototype is located on the southwestern coast of the island of Gotland in the Baltic. Main data is presented in Table 1. The plant is a turn-key delivery from Nohab-KMW Turbin AB (former Boving KMW Turbin AB). MBB (former ERNO) of Western Germany acted as a subcontractor for the blade delivery. The turbine machinery was mounted on the concrete tower in mid 1982, first rotation occured in February 1983 and formal delivery in August 1983.

Table 1. Main data for the Näsudden and Maglarp WECS prototypes.

Turbine	Näsudden	Maglarp
Number of blades	2	2
Diameter	75.0 m	78.2 m
Speed	25 rpm	25 rpm
Position relative to tower	upwind	downwind
Type of hub	rigid	teethered
Power regulation	variable pitch	variable pitch
Cone angle	0°	6°
Tilt angle	10°	0°
Blade		
Material	steel with GRP leading/trailing edges	GRP
Weight/blade	21 t	13,5 t
Twist	15°	13°
Airfoil	NACA 64-4xx	NACA 23xx
Tower		
Hub height	77 m	80 m
Material	concrete	steel
Diameter top/base	4.5/10.4 m	3.8/3.8 m
Weight	1500 t	281 t
Concrete foundation	1900 t	1600 t
Machinery		
Gear type	planetary with bevelled endstage	planetary
Gear ratio	60:1	60:1
Generator type	induction	synchronous
Rating	2 MW	3 MW
Cut-in windspeed	7.0 m/s	7,8 m/s
Rated windspeed	12.5 m/s	14.0 m/s
Cut-off windspeed	21 m/s	21 m/s

Support organisation

Vattenfall (the Swedish State Power Board) was contracted to supervise the delivery on behalf of the customer and to bear the main responsibility for the evaluation program. Vattenfall is the largest utility in Sweden and produced 67 TWh of electricity in 1986, roughly half of the country´s needs.

Vattenfall Östsverige, a division operating on Gotland and in the south eastern parts of Sweden, is contracted to perform operation and maintenance for the unit. It also buys the electricity produced. Their Gotland centre for operation and maintenance is the Ygne rectifier station, which is the connection point for the HVDC links between Gotland and Swedish mainland. The station is located some 70 km north of the WECS. Ygne is manned only during normal working hours, total number of employed eight, with one person at stand-by for emergencies at other occasions.

At Ygne system monitoring and manual control can be performed of the normally unmanned WECS. Before May 1986 only alarms could be transmitted. After the expiration of the warranty period the services of the manufacturer have been secured through a service contract. Maintenance is performed by the utility organisation,with use of local contractors for e.g. computer and hydraulic work. Maintenance of the blades and evaluation are handled by the central Vattenfall project group that was responsible during the delivery phase.

Operational history

As a final test before the unit was left in an unattended mode of operation, a 100 hour-test was performed during the last weeks of October 1983. The first day after the reliability test a loss of grid occured. One man who, by chance, was in the control room when the turbine stopped heard a heavy noise from the nacelle. An inspection of the gearbox revealed that a toothed coupling inside the gearbox was broken. The investigation showed that the reason for the damage was an insufficient design of the coupling. The high fluctuating torque in the drive train following the grid loss, were therefore not the actual cause for the damage but made it happen somewhat earlier than would otherwise have been the case.

The damaged parts were removed from the gearbox after dismantling it inside the limited space of the nacelle. The damaged parts were lowered to ground level inside the tower. The weight of the heaviest piece was 5 tons. After redesign the gearbox was reassembled during the spring of 1984.On May 1 the turbine resumed operation, which means a standstill of 6 months.

In December 1985 a crack was discovered in the 14 mm steel plate at one of the numerous bolt holes for attaching the leading and trailing edge GRP panels to the main steel spar. The crack was located near the most highly fatigue stressed zones of the steel spar, but in a way that further crack growth was not expected to influence the repair procedure. Thus it was decided to continue operation with short intervals for inspections, to gain experience from watching the crack growth. In May 1986 the crack had grown to 50 mm length and was succesfully repaired by TIG-welding. The plant was out of operation for six weeks.

In 1986 a modification program was carried through in order to increase the availability and reduce the cost for operation. The installation of a monitoring and remote control system made it possible to interprete the cause of an unexpected shut-down. If the reason indicated is not vital, the unit is

restarted which reveals if it was just a temporary fault.
Another modification introduced was a change in the charging procedure of the emergency stop hydraulic accumulators in the hub. This considerably reduced start-up time and thus increased operation. An organisational change gave better excess to personell to inspect the unit after un-expected shut-down when the remote system does not allow restart.

During early 1987 the unit operated for several hundred hours with a malfunctioning wind vane (the vane plate had disappeared!), causing continous operation at skew wind of as much as 50°. No immediate damage occured although it is a rigid hub unit, and the incident resulted in abundant data about the relationship between performance and wind incident angle, which proves to be valuable for the research concerning skew wind operation, for power regulation. However, it also clearly demonstrates the need for sufficient routines for regular inspections.

General experiences from the operation

The choice of an induction generator and the fairly weak electrical grid in the area necessitated a special solution to keep the voltage variations within acceptable limits. It ended up with exciting the generator with capacitators and then synchronizing it to the grid utilizing conventionel synchronization equipment. This has performed quite well with voltage and frequency variations within acceptable limits.

The main electricity supply to Gotland is via the HVDC links, which means that the inertial masses in the system are quite small. Thus a disturbance of the link may cause very fast frequency variations, 20 Hz/s has been registered. This causes rapid power fluctuations in the generator. Fortunately there is a "brake pin" arrangement built into the secondary shaft. It disconnects the generator from the gear before the stresses get too high. The design has worked according to expectations.

At two occasions lightnings have hit the blades, indicated by small (5x5 cm) areas of burnt paint. Also a damage to the computerized data acquisition system may be due to a lightning strike.

The blade inspection platform, running on rails on the tower, has proved to be efficient and easy to use for blade inspection and repair.

A 5-10 mm ice layer on the nacelle and hub has been observed twice. No ice has ever been seen on the blades.

The ice warning system was far too sensible and has not given any useful information.

Power production

Including August 1987 the Näsudden WECS has produced 9.5 GWh during 8700 hours.

The calculated power output versus wind speed is in good correspondence with the actually measured power curve, according to the IEA-method. Assuming 100 % availability and Näsudden mean wind statistics this would result in an annual net production of 6.0 GWh. Actual production since May 1984 has been 2.9 GWh/year or 48 % of the calculated value. On the whole this is explained by the availability losses and by yearly variations of available wind energy (around ± 15 %). A more detailed investigation will be performed during the final evaluation.

The auxiliary electrical power needs during operation are around 50 kW (note that the main hydraulic pumps are driven mechanically) and at stand-by 30 kW. About half is due to the data acquisition system and other prototype dependent needs

Availability

The availability is calculated as the fraction of the total time during which the plant has been technically available, i.e. operation or in a stand-by mode due to high or (mostly) low wind speeds.

Causes of unavailability are mainly scheduled and unscheduled maintenance. The latter includes time to get a maintenance man to the site, to inspect, to get spare-parts and to get the unit in order. In Table 2 the unit has been supposed to be available also during shut-downs due to external reasons (mostly losses of grid)

Table 2. Operational statistics for the Näsudden WECS

Year	1984 May-Dec	1985	1986	1987 Jan-June
Operation	24 %	25 %	33 %	31 %
Stand-by	25	28	24	37
External reasons	5	7	3	2
Scheduled maintenance	5	5	4	16
Unscheduled maintenance	41	35	36	14
Availability	54 %	60 %	60 %	70 %

When studying the table it is evident that the unscheduled maintenance is the major cause of the unavailabilities before the 1986 installation of a remote control, which drasticly reduced the need for time-consuming on-site inspections. Thus the availability increased from around 60 to 70 %. So far ten months have been better than 80 %.

The unscheduled maintenance is distributed among the various systems as follows from Table 3.

Table 3. Distribution of down-time due to unscheduled maintenance for the Näsudden WECS.

	Year		
	1985	1986	1987 (Jan-June)
down-time	3127 h	3176 h	580 h
System:			
Turbine blades	44,1 %	36,6 %	4,3 %
Blade pitch system	7,8	8,8	41,1
Control system	10,9	38,4	7,1
Slipring	13,1	10,4	7,8
Electrical system	5,0	0,7	2,1
Gearbox lubrication	1,7	0,4	25,0
Other	17,4	4,7	12,6

Design verification

The wind environment at Näsudden, especially the turbulence and wind gradients, is somewhat more benign than was stated in the technical specification for the tender.

The variations of the blade root edgewise bending moments are somewhat higher than calculated while the flapwise moments are well predicted. The nacelle thrust force variations are also well predicted, but the horizontal transverse force variations are roughly ten times higher than predicted. These deviations are mainly due to a different and more complex dynamic behaviour than foreseen. E.g. the phase angle of the maximum flap load differs from what was assumed.

Thus the blade fatigue loads are generally higher than predicted, which means that the 10^{-5} probability of failure in the steel regulation is slightly increased, although not to an unacceptable level. The regular ultrasonic inspections, at intervals determined by fatigue crack propagation calculations, are judged to give a sufficient level of safety.

Environmental influence

The noise from the turbine and, to less extent, the machinery is audible above the background noise in the rural area up to a distance of ca 1 km at low wind speeds. The extensive noise measurements demonstrate low levels of infrasound that can not cause any discomfort. The closest neighbours (0.7 km) do not consider the sound to be uncomfortable.

The measurements of the impact on TV reception indicates less disturbance than expected. This is mainly due to a small difference in field intensity between ground level and hub height. There have not been any complaints from the neighbours.

Investigations of nesting birds 1979-83 did not reveal any significant changes due to the power plant. At a few instances birds have been killed due to collisions with the tower.

MAGLARP

The plant

The Maglarp prototype WECS is located near the coast 22 km south of Malmö, close to Trelleborg, in an area dominated by farming. The plant is a turn-key delivery from Karlskronavarvet AB. Hamilton Standard from the US was subcontractor for turbine technology and blades. A similar unit, WTS-4, is situated at Medicine Bow, Wyoming. The wind turbine was erected and first rotated in mid 1982 and formally delivered in September 1983.

Support organisation

Sydkraft performs the same functions for Maglarp as Vattenfall for Näsudden. Sydkraft is the second largest utility in Sweden, with a turnover of 20 TWh/year (1986).

The wind-power plant is operated unmanned, and data on wind speed, power production an operational status are transmitted to a control centre in Malmö. This is permanently manned and monitors power production and distribution of about 1500 MW. The control centre receives detailed error messages from the wind turbine.

If a shutdown, requiring inspection on site, occurs, then a maintenance man will normally be sent to the plant within an hour. This duty maintenance man, however, has only limited training for wind-turbine servicing. His job is merely to check the cause of the shutdown, and, if possible, restart the turbine while observing the procedure.

Sceduled maintenance is carried out on part-time basis by a Sydkraft staff of six and, through a service contract, by seven Karlskronavarvet employees.

Operational history

Since the delivery in 1983 the plant has generally worked well, with no dramatic events. Except for 1987, extensive blade maintenance has been carried out every summer. It will be described later.

The spinner, which is a secondary but large structure of welded aluminum, has been subjected to cracks at several occasions, which have needed some down-time for repair and modification.

In mid 1985 the center section of the gear-box casing was exchanged due to cracks at the oil outlet.

The so far longest down-time period occured in November 1986 - June 1987 due to a replacement of a teeter bearing. It is described later.

Minor modifications of e.g. lubrication, blade maintenance platform and control program have been introduced successively.

General experiences from the operation

Electrical connection to the power grid is done very smoothly. Power control normally maintains alternator output to rated power plus or minus 10-20% when the wind speed exceeds the rated wind speed of 14 m/s. At wind speeds exceeding about 18 m/s high turbulence sometimes (3-5 times/year) causes the plant to shut down, owing to power peaks exceedeing 140% of rated power. However, this does not cause any problems other than the necessity of a restart on site.

Lightning hits the plant several times a year, but normally only causes damage to the blade protection tape, as intended. Only once there has been damage to the engineering-data-collection equipment. The control system has been given a more extensive protection and has never been damaged.

Ice formation on the blades was expected to be a problem, but has not been known to occur any time. Instead, however, ice and wet snow a few times every winter disturb the wind speed and direction sensors. This causes no remaining damage, but cleaning on site has to be done.

At an early design stage, it was believed that the wind turbine would keep in alignment with the wind direction without power assistance. Later, model tests made this less likely. To reduce the wear on the yaw bearing cog race, a second hydraulic yaw drive had to be included in the design. Full-scale tests confirmed that there was no stable yaw condition, and the yaw drive is constantly in operation. After some running-in control problems it now works quite adequately.

Power production

Including August 1987 the Maglarp wind turbine has produced 15.3 GWh during 10 500 operational hours.

The power production capacity as a function of wind speed exceeds the calculated values at all wind speeds below rated. At very high wind speeds the mean produced power is slightly lower than rated, due to a cutback in power control at combinations of high wind speed and high turbulence.

The annual energy production, assuming 100 % availability, would be 8.4 GWh using Maglarp mean wind statistics. Actual average power production since September 1983 is 4.3 GWh/year, which equals 51 % of the theoretical capacity. The reasons for availability losses will be discussed in more detail below.

The auxiliary power requirements for the plant is 50-60 kW during operation, including 10-15 kW required for the control building and the data acquisition system. At standby, the power requirement amounts to about 20 kW.

Availability

Availability has been calculated in the same manner as for Näsudden, except that "external" reasons for down-time are not specified separately. Maintenance is carried out in a scheduled manner, which does not allow for the use of low wind periods especially for maintenance activities. However, major maintenance tasks are carried out in the summer, which makes availability high during the high-wind winter period and low in the summer.

Total availability during the October 1983 - September 1986 period has been 61%, thirteen months have been better than 80%. The decline in availability the third year may be due to the change in routines in that no duty maintenance man is sent to the plant until office hours if it is not likely to be enough wind for production. As can be seen in Table 4, the decrease has mostly affected the standby-time.

The best results sofar were achieved in July 1987, after the teeter bearing modification period, with 93%. The main cause of unavailability has been scheduled maintenance. Normally the turbine is shut off once a month for planned maintenance. Duration varies from 1-3 days for monthly routine checks and maintenance up to major overhaul done every summer with a duration of 4-7 weeks. Table 4 indicates shutdown time for scheduled and unscheduled maintenance, given as a percentage of the total time.

Table 4. Operational statistics for the Maglarp WECS

	Oct 83 - Sep 84	Oct 84 - Sep 85	Oct 85 - Sep 86
Operation	31%	35%	32%
Stand-by	30	30	24
Scheduled maintenance (including blades, modification, training)	30	29	32
Unscheduled maint.	9	6	12
Availability	61%	65%	56%

Fig. 1. The blade maintenance platform at Maglarp WECS is considered necessary to carry out blade work.

From the first year of operation to the third, the amount of mandays for Sydkraft personnel decreased from 605 to 460. The amount of scheduled maintenance is now revised according to the experience from several years of operation. The intention is that this will increase the availability and decrease the operational costs. It is the opinion of the authors that for a series-produced wind turbine which is located in groups of 10 or more, the maintenance labour will be cut down to less than 25 % of the present values.

Design verification

Deviations from the original design are the earlier mentioned yaw behaviour and power control at high wind speeds. Consequences of these deviations are minor from loads point of view, due to the limited yaw moment which somtimes fails to align the nacelle with the wind direction and that energy is lost at high wind speeds due to a decreased power reference.

The loads in the main components such as blades, nacelle and tower as a whole agree with design loads. It can be concluded that the philosophy of the "soft" design has been fulfilled, which also includes the drive train.

Blade maintenance

The blades are equipped with conductive tape glued on the blade in a grid pattern to protect the GRP structure and sensor cables from damages by lightning current. Strokes by lightning occur about 5-15 times a year and result in small burn marks on the tape, about 1-5 cm in diameter. In all about 40 lightning impacts have been observed on the blades.

Due to the excessive costs associated with the maintenance of the lightning protection and to the negative influence on tv-reception (although solved at this very plant), it was decided to remove the protection outside 85 % radius, i.e. outside the strain gauge installations. This was performed in the summer of 1986. Since then some more lightnings have hit the blades, but only the remaining protected areas. Thus it seems possible to remove the rest of the lightning protection when the measuring program is finished. That means that all cables etc also have to be removed.

Another blade problem is a lot of small surface cracks in the surface layer of the GRP. This carries no load and the purpose is only go give a smooth surface

EUROPEAN WINDFARM PROJECTS

Ir.G.G.Piepers

Past Chairman EWEA

Het Vierkant 2
1852 RA HEILOO
Netherlands

ABSTRACT

Windpower could make a significant contribution to meeting future energy needs in European countries with adequate wind resources.Experts predict that 10 to 20 per cent of their current electricity demand could be supplied by windturbines.The most effective way to obtain these goals is grouping windturbines in clusters to produce electricity that is fed into the utility grid.These windpower stations or windfarms may consist of a few to more than 100 units of different types and sizes. A very rapid development of windfarms took place in California where mid 1987 more than 15000 windturbines were installed with a total capacity of 1200 MW.In Europe the development of windfarms is occurring at a much slower pace for various reasons.In this paper an overview is given of the present status of windfarming in a number of European coumtries,including the political,technical,economical and environmental restraints that have to be overcome in order to get permission to install large numbers of windturbines.

KEYWORDS

Windfarms;windpowerstations;windturbines for windfarms;windfarm experience;output and control;integration into the public grid;environmental restraints;international developments.

INTRODUCTION

The wind energy resource in Western Europe is a very large one and could provide a significant contribution to meeting future energy needs.An assessment study made by the Commission of the European Communities(CEC) has shown that even after allowing for all likely siting restrictions there are approximately 400.000 sites available for multi-megawatt windturbines which could provide 340 TWh of electricity annually.This potential,

equal to 780 million tonnes of oil equivalent(m.t.o.e.) annually,is about three times larger than the present electricity consumption within the European communities.Though it would be impractical to provide all European communities electricity needs from the wind,systems integration studies have shown that despite its variability in European countries with adequate wind resources 10 to 20 per cent of the current electricity demand could be supplied by windturbines,before adverse effects on the system operation become significant.
The most effective way to realise this contribution is grouping windturbines in clusters to produce electricity that is fed into the public electricity network.No additional storage facilities are required since periods of calms can be met by regulation of the output of the conventional mix of power plants. These windpower stations or windfarms may consist of some 5 to more than 100 units of different types and sizes.
A very rapid development of windfarms took place in California where by this time more than 15000 windturbines are installed with a total rated capacity of 1200 MW.This big boom is mainly due to a system of generous tax credits and other financial incentives in favour of investers in windfarm projects and not so much the result of a sound technical development.
In spite of the promising prospects the introduction of windpower on a large scale in Europe is occurring at a rather slow pace.The main reason is the fact that the present generation of windturbines is still too expensive and is not able to compete with the depressed oil and coal prices.A serious obstacle to progress in some countries also appears to be the environmental issue,which leads to a very tedious procurement of getting building permits.In the following chapters the present situation in a number of European countries concerning windfarming is briefly outlined and discussed.

DENMARK

Traditionaly Danish windmills have been erected as individual units. During the last 10 years about 1000 electricity producing windmills have been installed and connected to the national grid.Since 1984/1985,however,about 16 MW windpower has been installed in real windfarms.These windfarms are distributed all over the country and the majority of them are owned by groups of private persons who sell the surplus of produced electricity to the utilities.
There is,however, a growing tendency under pressure of the utilities to discourage the further expansion of privately owned windfarms.At the end of 1985 an agreement was reached between the Ministry of Energy and the Danish Electricity Supply Undertakings to install 100 MW in windfarms.These windfarms will be completely financed by the utilities and should be completed before 1990.It is expected that any future installation of windfarms will become an almost hundred per cent utility activity.
Studies have shown that 10% of the Danish electricity demand in the year 2000 could be covered by wind energy.Some doubts, however are arising if this figure will ever be realised because of the many hindrances already experienced in allocating sites for the 100 MW plan.

The first off-shore windfarm in the world is planned for construction 2,5 km from the coastline near Aarhus.A total of 28 windturbines with a rated capacity of 200-300 kW each will be installed on seperate foundations and controlled and monitored from a land-based control centre.The originators of the project believe that the off-shore windfarm will promote the utilization of the large off-shore wind energy potential,especially in countries where the availability of suitable land-based sites is restricted.

Location	Owner	Capacity	Windmill Manufacturer
Oddesund Syd	Dansk Vindmølle	13x100 kW	Bonus
Oddesund Nord	Park A/S	20x55 kW	Bonus
Tønder Tekniske Skole	Public Institution	9 units of various types	
Ebeltoft	Municipality	16x55 kW	Nordtank
		1x100 kW	Nordtank
Aerø	Co-operation	11x55 kW	Vestas
Fanø	"	6x55 kW	Vestas
Nordby	"	6x55 kW	Vestas
Aale	"	10x75 kW	Windmatic
Ranum	"	14x75 kW	Vestas
Sydvestmors	"	10x75 kW	Vestas
Masnedø	Utility	5x750 kW	DWT
Hundested	Utility	3x300 kW	DWT
Hasle	Municipality	10x100 kW	Smedemester
Taendpibe	Co-operation	30x75 kW	Vestas

Survey of Danish Windfarms

SWEDEN

Sweden is one of the first countries to investigate the possibilities of the large scale utilization of wind power.The interest is focused on big machines to be installed in windfarms. Two 3 MW/80m prototypes have been developed and are still in the testing stage.In the meantime siting studies are being carried out,based on the following production of electricity: 10 TWh/yr on land and 20 TWh/yr off-shore.These studies should be completed by the end of 1987.
The two prototypes after initial problems are operating satisfactorily,but so far no concrete plans for building windfarms have been released.First the results of the prototype testing and the siting studies will be evaluated.Probably the Energy Bill to be released in 1990 will include windfarm projects. The utilities have already founded a new company called"The Utilities Wind Power Company",to deal with future wind energy utilization.

NORWAY

Norway has an abundant off-shore supply of oil and natural gas, much more than necessary for its own use.A substantial amount is exported.The government is therefore only indirectly interested in the utilization of wind energy.A modest programme is being executed,mainly to gather background information to support the decision if a programme should be initiated that could lead to a large scale application of windpower.The decision depends largely on wind energy being able to compete with other renewable or conventional energy sources.

Windfarm at Ebeltoft,Denmark

WEST GERMANY

Until recently the Federal Government and the Utilities have not shown much interest in the utilization of wind energy.It has been looked on mainly as a technology for export to the Third World,and its reputation has not been improved by the performance of the prototype windgenerator Growian.
After Chernobyl,however,the attitude seems to have changed and West German energy planners are looking again at the prospects for wind power.According to the Ministry of Technology,wind energy in Germany has a useful potential of 3,4% of the total primary energy use.
Regional governments like Schleswig-Holstein,Hessen and

Nieder Sachsen and local utilities are showing now quite some interest.A contract has been signed between the Wilhelmshaven Utility and MBB for the building of a small windfarm near Wilhelmshaven consisting of 3 one-bladed Monopteros 50 type windturbines at a cost of 16 million DM.The project is due for completion by 1989.Discussions are taking place concerning the construction of up to three windfarms in the north of the country,each consisting of 20 to 30 smallscale windturbines. The Federal Government would contribute to the cost of these projects.

Location	Owner	Capacity	Windmill Manufacturer
Westcoast at Growian site	Windpark Westküste GmbH	20x30 kW 5x25 kW 5x55 kW	Man Windkraft-Zen ENERCON
Cuxhaven	Überlandwerk Nord-Hannover AG (ONH)	20x25 kW 10x55 kW	MBB ENERCON
Husum		?x200 kW	?
Wilhelmshaven	ONH	3x640 kW	MBB

Survey of planned windfarms in Germany

BELGIUM

In Belgium 60% of the electricity consumption is produced by nuclear power.Government and Utilities do not show much interest in wind energy.Nevertheless Belgium has an important windturbine manufacturer HMZ,fabricating the successful 200 kW Windmaster.Many of them are operating in Californian windfarms in the USA and in several other countries.
In the outer harbour of Zeebrugge a windfarm consisting of 23 Windmasters has been built and is operating satisfactorily since February 1987.It is a project of the Administration for Electricity and Electromechanics,financed by the government. The windfarm is primarely intended as a showroom for the promotion of the Belgian windmill industry.

NETHERLANDS

From 1976 on successive R&D programmes on wind energy financed by the government are carried out.The initial positive results led to the release in 1982 of an official government standpoint stating that as a substantial part of the future electricity supply system 2000 MW of wind power should be installed by the year 2000.In 1984,however, this target was reduced to 1000 MW,because the utilities foresaw regulation problems of the combined production system when more than 1000 MW had to be integrated.In order to get a better insight into the problems involved in the large scale application of wind energy it was decided to build an experimental windpower station.The Co-operating Electricity Producers(SEP) would design and operate the plant.The windfarm comprises 18 windturbines of 300 kW each and after some delay will be completed before the end of 1987.
The national programmes have resulted in much know-how at the research institutes but not given birth of a flourishing Dutch

windmill industry.The market for WECS in the Netherlands has been quite stagnant due to the current generation of windmills still being too expensive to compete with conventional power plants.In order to stimulate the Dutch windmill industry the Minister of Economic Affairs announced in May 1986 the "Integral Programme Windenergy"(IPW).This 130 million Dutch guilders programme calls for the installation of medium and large scale windturbines with a total capacity of 100 to 150 MW over a five years period(1987-1991).A sum of 60 million will be spent on research and development of a "cost-effective" windturbine and 70 million on subsidies to users to cover the difference between the actual and the economically justified cost of WECS. The subsidy for users is designed to decrease annually from 700 guilders per installed kW to 100 guilders per kW in 1991. From then on no state subsidy will be provided.

Decisive factors for becoming eligible for subsidy are the approval of a business plan to be submitted by the windmill manufacturer and especially the certification by ECN of his turbine models. The announcement of the IPW started a real race for subsidy. Mid 1987 more than 150 project have been submitted for the maximum sales subsidy with a total capacity of 155 MW by industries,non-profit organisations,utilities and by far the most by individuals.For 1987,however,funds are made available corresponding to only 30 MW.Besides not all of the 13 windmill manufacturers involved have got their business plan approved and so far only three of them have been granted a certificate for one of their machines.Below a survey is presented of the windfarms which are among the proposed projects and are likely to meet the requirements for realisation.(the last one is not included in the IPW).

Location	Owner	Capacity	Windmill manufacturer
Friesland	Utility PEB	40x250 kW	Bouma
Along dike of N.O.-polder	Utility IJsselcentrale	25x300 kW	HMZ
N.-Brabant	Local utility	16x250 kW	Newinco
N.-Holland	Utility PEN	18x160 kW	Bouma
Delfzijl	Private organisation	22x160 kW	Bouma
Sexbierum (not in IPW)	Board of utilities(SEP)	18x300 kW	Holec/Polenko

Survey of Dutch windfarms

A very serious obstacle appears to be the procurement of building permits.The environmental issue is a strong barrier to the implementation of windfarms.Pollution of nature,killing birds, noise,etc. are the arguments used against installing of windturbines.All selected sites so far raised objections.It is evident that without effective measures enacted by the central and local authorities even the 100 MW in 1991 will never be realised,let alone the 1000 MW in 2000.

Nevertheless at least three windfarms(Sexbierum,N.O.-polder and Delfzijl) are likely to be completed by the end of 1987. Furthermore a group called "Megawind" has been founded by three provincial utilities to promote the development of a large "cost-effective" windturbine based on the sussessful 1 Mw prototype NEWECS 45 designed and built by FDO/WES.Plans for the installation of 18 units at various locations are making good progress.

UNITED KINGDOM

The wind energy activities in the UK started somewhat later than in Denmark,Sweden,W-Germany and the Netherlands.Government officials and the Utility Board (CEGB) were quite sceptical about the prospects of windpower.Thanks to the persistent efforts of the British Wind Energy Association (BWEA) those responsible for securing energy supply are now recognising that wind power can make asubstantial contribution to meeting future energy needs.The Departement of Energy,the Generating Boards, research councils and industries are these days co-operating well in research and demonstration programmes,so that many promising projects are under way.Experts are predicting that up to 20% of the current electricity demand could be supplied by land-based windturbines and that an even greater contribution will be possible by installing off-shore arrays of windturbines. The windmill manufacturers James Howden,Glasgow and Wind Energy Group Ltd.(WEG,a joint company of British Aerospace and Taylor Woodrow Construction)are active in developing and testing prototypes of medium and large machines suited for application in windfarms.James Howden has built the world's most powerful windfarm in California,consisting of 75x330 kW units.Because of blade fixing problems this farm,however,is out of operation for more than one year,WEG has also built a windfarm in California,consisting of twenty 250 kW/25m machines.The windturbines were all commissioned and available for operation before the end of 1986.
WEG recently applied for planning permission to erect a windfarm on the Isle of Man.The proposed windfarm consist of ten 250 kW units similar to those installed in California.WEG initially received planning consent but the Manx Museum and National Trust appealed against the permission on the grounds that the site was near a place of historical interest and that there would be a loss of visual amenity.This appeal was up held on the basis that the environmental "cost" did not outweigh the economic benefits at current oil prices.

IRELAND

Ireland imports about 80% of its total primary energy needs, most in the form of oil.The cost of energy imports represents a substantial drain on the economy.The Irish Departement of Industry and Energy therefor shows considerable interest in the potential of windpower.The country experiences a very good wind regime,with annual average wind speeds in the coastal regions of 6 m/sec and even higher.Financing of wind energy projects,however,is a big problem.So far the activities have been limited to assessment studies and the testing of small windturbines from various countries to compare which machines are best suited for use under Irish circumstances.

FRANCE

France has plenty of relatively cheap hydro- and nuclear power, so that government and utilities are not much interested in the utilization of wind energy.There are some possibilities at isolated locations on the main land,on Corsica and on the French islands in the Pacific and the Caribbean.A modest programme for the development of WECS is co-ordinated by the "Agence Française pour la Maîtise de l'Energie".As part of this programme a windfarm with 10 kW machines has been set up in the south of France for demonstration purposes.

SPAIN

At the beginning of 1984 the Institute for Energy Diversification and Savings(IDEA) was founded within the Ministry for Industry and Energy.One of the principal aims of this Institute is the promotion of the utilization of renewable energy sources.In 1985 IDEA has published several aspects of the "Plan for Wind Energy Utilization"(PAOLO).During the period 1986 to 1992 the installation of 35 to 50 MW windpower has been planned.Spain has already in operation two small windfarms.
A feasibility study is currently being carried out on the possibility of building two windpower stations on Gran Canaria and near Tenerife as part of the wind energy plan PECAN.

ITALY

Italy is almost completely dependet on the import of fossile fuel for its energy supply.Consequently there has been increasing interest in the potential of windpower.Though the wind regime in Italy is generally less favourable than in most Europe,there are many areas in the south,in the Apennines, the Alps and on Sardinia and Sicily,where wind speeds are high enough for an economically justified use of windpower.There is particularly interesr in the application of larger machines to supply electricity to the grid system.The Italian Corporation for Electric Energy(ENEL) and the Italian Committee for R&D of Nuclear and Alternative Energy Sources(ENEA) have recently decided to have three 1 MW windturbines built by Aeritalia/Fiat.The first prototype should be completed in 1988. On Sardinia a pilot plant windfarm at Alta Nurra,comprising 10 50 kW ENEL-Fiat windturbines has been set up.

GREECE

Greece has excellent wind conditions on its numerous islands. The government and the Utility PPC have announced plans to install windfarms on the largest islands to save oil consumed by diesel generator sets.Progress,however, is slow since Greece has only a very modest windmill industry and no foreign exchange to buy windturbines abroad.On a few islands individual Danish windmills have been erected.Some years ago MAN,Germany has installed a small windfarm consisting of 5 Aeromans on the isle of Kythnos.On the island of Limnos a windfarm consisting of 6x100 kW units is planned at the end of 1987.
The Hellenic Agency for Local Development and the local government in collaboration with the Technical University of Athens recently decided to undertake a study for the installation of a 30 MW windfarm on the island of Creta.

ASSESSMENT OF WIND ENERGY FOR BAHRAIN

N.S. AL-Baharna* and G. Fregeh**

*Gulf Polytechnic, P.O.Box 32038, Isa Town - Bahrain
**Bahrain Center for Research & Studies
P.O.Box 496, Manama - Bahrain

ABSTRACT

The wind speed records at different stations in Bahrain is compiled to assess the potential of local wind resources for power generation. The application of wind energy for water desalination and ice making in rural areas of Bahrain is also investigated.

KEYWORDS

Wind energy conversion system; capacity factor; rated capacity; rated speed; cut in speed; cut out speed

INTRODUCTION

Oil and gas are the main sources of energy in Bahrain. While these are high grade sources, they are non-renewable, and particularly for Bahrain which has quite small reservoirs, exhaustion of the supply within the next twenty years is probable. It is inevitable then that substitute energy sources must be sought. One substitute which should be considered is wind energy.

The most important application of wind as an energy source is in the generation of electricity. However, electrical power must be generated constantly and abundantly, this presents the first problem, as wind, by its very nature, is not constant and does not prevail at a steady rate. There is also the variable nature of wind speed which must be taken into account, since the speed of wind is subject to enormous changes in short time periods. The quantity of energy which can be extracted from wind is depenbdent directly on this speed.

WIND ENERGY POTENTIAL IN BAHRAIN

It is essential to determine the wind speed acurately at a particular site for determing the wind turbine size and energy output. The data given in a form of tables of wind speed intervals versus the time duration on daily, monthly or yearly basis will be converted to mid point speeds, V_n, versus the annual percentage duration, T_n. Accordingly the wind speed, and the power density can be calculated as:

$$V = \sum_n (V_n \times T_n) \tag{1}$$

$$P = 1/2\rho \sum_n (V_n^3 \times T_n) \tag{2}$$

Assuming the air density to be 1.225 kg/m³ at 15 C° and atmospheric pressure, then eq. 2 becomes:

$$P = 0.61 \sum_n (V_n^3 \times T_n) \tag{3}$$

The term capacity factor, F, was defined as the ratio of the total energy generated by the wind energy conversion system (WECS) per year under the wind conditions at that site to the energy generated per year if the WECS is operating at its rated capacity all the times. The value of F, is a function of the wind characteristics and the turbine power curve which is governed by its cut-in, rated and cut-out speeds, and efficiency. The capacity factor can be computed from the following equation:

$$F = \sum_n \left\{ \left(\frac{V_n}{V_r} \right)^3 \times T_n \right\} + T \tag{4}$$

Figure 1, shows the wind speed for three different locations in Bahrain.

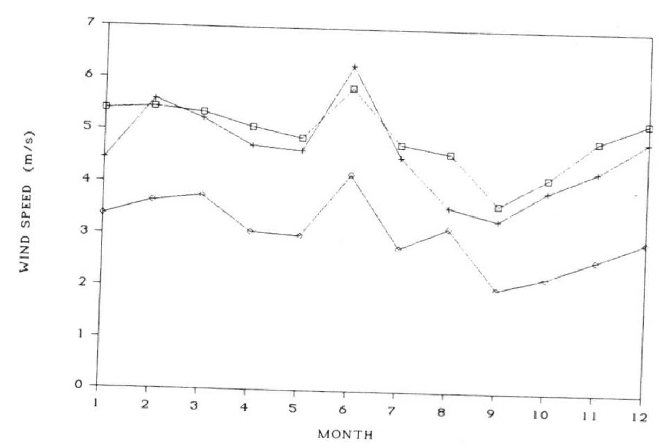

Fig. 1 The Average Monthly Wind Speeds for Three Different Locations in Bahrain

Figure 2, shows the power density for the same locations.

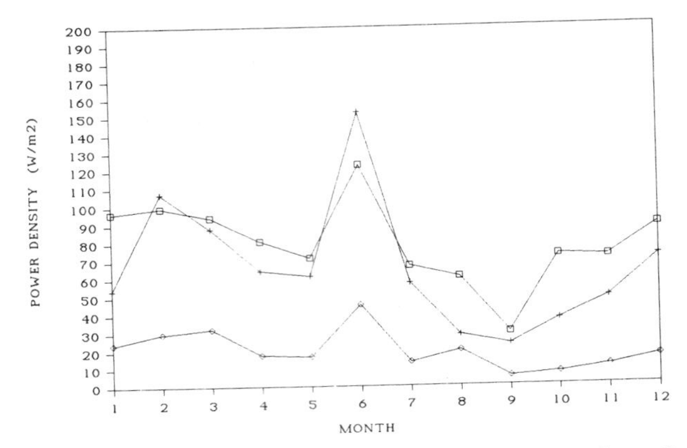

Fig. 2 The Average Monthly Power Density for Three Different Locations in Bahrain

The capacity factors as a function of the rated speed are computed for different sites in Bahrain as shown in Fig. 3.

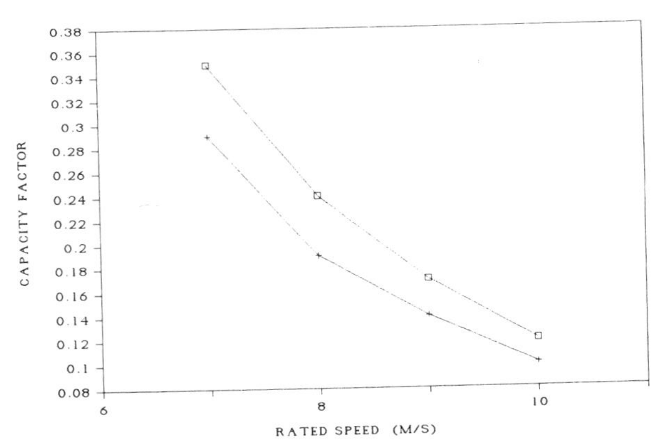

Fig. 3 The Capacity Factor as a Function of the Rated Speed at Different Locations in Bahrain

APPLICATION OF WIND ENERGY

The basic energy requirements for the people in rural areas are mainly those covering food preparation, domestic water supply, lighting, and agriculture. Wind energy can be used for sea or brackish water desalination and also for ice making. The amount of energy required for fresh water production depends on the salinity of water and its amount.

Water Desalination

The desalination plants for remote areas have to be reliable, easy to maintain and simple to operate by local people . Accordingly a reverse-osmosis wind powered desalination plant will be used. The energy required to produce one cubic meter of fresh water can be approximated as :

$$E = H/(25 \times R) \tag{5}$$

where, H is the working pressure, and R is the recovery ratio.

Ice Making

Several of the fishing villages are scattered along the coast of Bahrain. Wind energy can be utilized to drive an ice-making plant producing ice from sea-water, which can be used by the fishermen to preserve their fish.

Energy consumption for producing ice varies from 65 kWh /ton for small plants (3 tons/day) to 40-50 kWh /ton for larger plants (6 tons/day) .

Economic Consideration

The cost "C" of one kWh in mills generated by WECS with the assumption of expected life firm of 15 years can be expressed as:

$$C = I(0.084r + 0.11m + 0.0066)/F \tag{6}$$

where I is the total capital investment in the WECS in US dollar per installed Kilowatt, r is the annual interest rate per unit, and m is the fraction of the capital investment needed per year for the operation and maintenance of the WECS.

The daily cost for the capital investment can be calculated from the folllowing simplified equation:

$$C = B(1/N + r/2) \tag{7}$$

where B is the price of the unit, and N is the expected life of service of the unit.

DISCUSSION AND CONCLUSION

For the sake of calculating the cost of producing one kWh , a WECS is selected with a rated power output of 2.5 KW, a cut in speed of 3.5m/s, and a rated speed of 7.0 m/s. The cost of the system is $6600. From Fig. 3 the capacity factor is 0.35. Using eq. 6 with r=0.1 and m=0.1, the cost per kWh is $0.20.

To get an idea about the cost to produce one ton of ice, the capital investment for an ice maker to produce 3 tons per day is $24,000. Accordingly assuming a life period of 15 years and an interest rate of 10% the daily capital cost per ton will be obtained by eq. 7 to be $2.5. The energy required to produce one ton is 65 kWh. Therefore the total cost to produce one ton of ice will be (2.5 + 65 x 0.2 =) $15.5.

It is clear from the above that wind energy is more expensive than non renewable energy ,however it should be noted that wind energy will be used for rural areas with specific applications.

NOMENCLATURE

B Price of the unit ($)
C Cost of the capital investment ($)
E Energy per cubic meter (kWh/m^3)
F Capacity factor
H Working pressure (atm.)
I Total cost of WECS per installed kW ($)
N Expected life service of the unit
P Power density (W/m^2)
R Recovery ratio
T_n Number of hours corresponding to the n speed interval divided by the total no. of hours for which the wind measurements have been recorded
T The no. of hours for which the wind speed is larger than the rated speed and smaller than or equal to the cut-out speed divided by the total no. of hours for which measurements have been recorded
V_n Mid-point speed for the n speed interval in the range between the cut-in speed and the rated speed (m/s)
V_r Rated speed (m/s)
ρ Density
m Fraction of capital investment per year for operation and maintenance of WECS
r Annual interest rate

REFERENCES

1. Saleh, M. A. (1986). Wind energy for rural and remote areas in ESCWA region. UNESCWA Document No. E/ESCWA/NR/86/WG.1/3.

2. Saad, M. (1982). RO supplies look to wind and sum to power rural system. World water.

MICROSCALE FLOW VARIATIONS OVER THE SOUTH FREMANTLE WIND TURBINE TEST SITE

S.J. Dear, M.F. Steketee and T.J. Lyons

Environmental Science, Murdoch University,
Murdoch, W.A. 6150, Australia.

ABSTRACT

A comparison between observed and modelled wind profiles at an operational wind turbine site highlights the inability of a neutral flow model to accurately predict the flow under low to moderate wind speeds. The importance of these flow conditions to power generation is demonstrated.

KEYWORDS

Wind profiles, wind energy, wind flow models, atmospheric stability.

INTRODUCTION

The South Fremantle wind turbine test site is located south of Perth (32 S, 116 E), Western Australia, in moderately rough terrain, approximately 800 m inland on a 50 m high north-south ridge running parallel to the coast (see Fig.1). On site are three wind turbines with rated power output of 60, 30 and 20 kW. The site has an instrumented 54 m tower located in the midst of the three machines, the nearest machine being 1.7 rotor diameters from the tower. The tower is instrumented every 8 m up to 48 m and wind speed and direction data have been collected from October 1986 to April 1987, under normal turbine operation. During April 1987, an additional experiment was conducted with the turbines shut down.

In conjunction with the field programme, the wind field over the turbine site and surrounds was calculated using the neutral flow (MS3DJH/3R) model of Walmsley, Taylor and Keith (1986). An analytical approximation to the terrain was employed in the model runs as shown in Fig. 2. This approximate terrain reproduces the following features of the actual terrain:
(i) the maximum height and slope of the ridge,
(ii) the 10-20 m undulations along the ridge line, and
(iii) the major surface roughness transitions.

The present paper presents the profile results both observed and predicted and draws conclusions with regard to the suitability of the MS3DJH/3R model (and neutral flow models in general) for estimating power output at hub heights significantly greater than the original monitoring height.

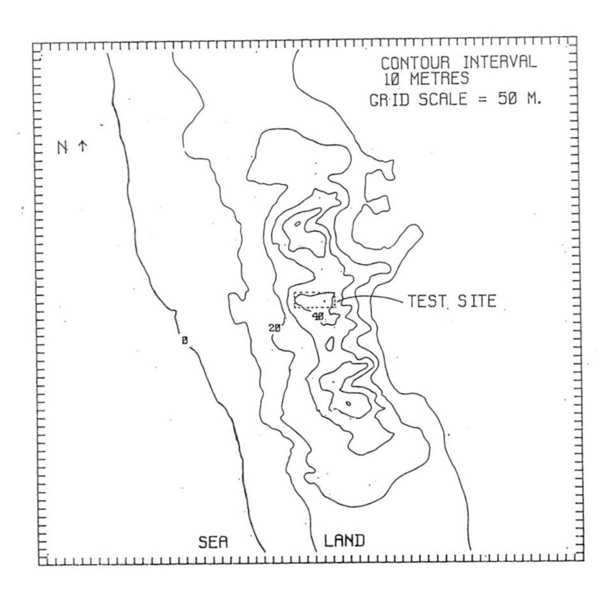

Fig. 1. Actual terrain surrounding the South Fremantle wind turbine test site.

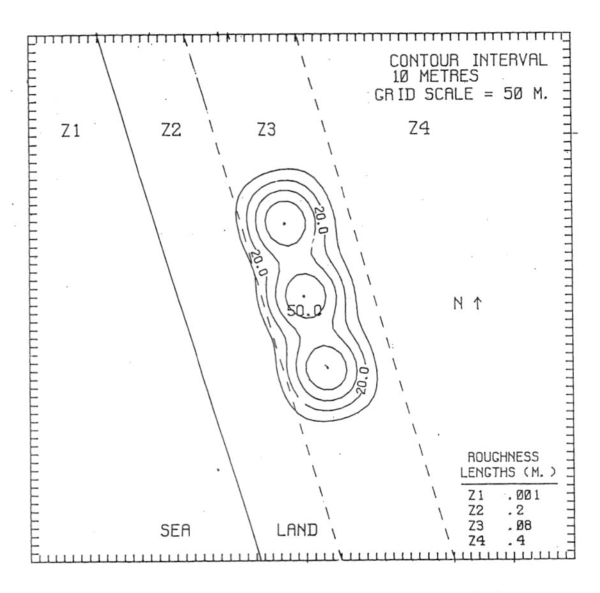

Fig. 2. Analytical terrain used for model runs.

FIELD AND MODEL RESULTS

In order to demonstrate the dependence of the wind speed profile on measured wind speed, the ratio of the wind speed at 48 m (hereafter V48) to that at 16 m (V16) was calculated and plotted against V16. This process was repeated for each of eight 45 degree wind direction sectors. As the terrain is strongly two dimensional with the ridge line and coast line running 15 degrees west of north, the sector analysis was done with respect to this terrain axis rather than true north.

As the instrumented tower is placed amongst the site turbines, considerable wake effects exist at the 16 m anemometer. Thus it is necessary to exclude data from certain wind directions in order to collect unperturbed profile data. In accordance with IEA guidelines (Anon., 1982), a wake extent with semi-angle 60 degrees was attributed to each machine and data deleted accordingly. This data base was supplemented with observations taken during significant turbine outage periods. During the April experiments, none of the turbines were operated, guaranteeing wake free data.

Plots of V48/V16 versus V16 are shown for four sectors in Fig. 3. The data was reduced by a binning procedure (V16 classified into 1 m/s bins) with a bin mean and standard deviation being depicted. The south-west data (sea fetch) shows a relatively constant V48/V16 ratio converging to presumably a neutral flow profile at wind speeds greater than 8-10 m/s. The easterly profiles however increase significantly at lower wind speeds and only appear to converge to a neutral flow limit at around 10 m/s. Similar behaviour is displayed by other land sectors.

As wind profiles illustrate a marked dependence on atmospheric stability (Irwin, 1979), the south sector data have been split into day and night profiles. This illustrates that increases in V48/V16 are not limited to a particular stability, but are observed in both stable and unstable flow.

The MS3DJH/3R model was run on a 64x64 grid with a grid spacing of 50 m, for each of the eight wind sectors. The predicted value of V48/V16 for each sector is included in Fig. 3. As the MS3DJH/3R model is a neutral flow model, the ratio is independent of wind speed and the different values for each sector illustrate the influence of roughness variations between sectors.

DISCUSSION

In general there is good agreement between the model predicted values of V48/V16 and the observed high wind speed limit of the ratio. However for the east, south and north-east sectors there are significant increases in V48/V16 as the wind speed drops below 10 m/s. For the south-western sector, the increases in V48/V16 are less and the ratio appears to approach the limiting value at lower wind speeds than the land sectors.

To illustrate the impact of these steeper moderate wind speed profiles on wind energy output, calculations were made of estimated power output at 48 m based on a Rayleigh wind speed distribution at 16 m with a given mean value. Extrapolation of this distribution to 48 m was achieved either by using (i) the MS3DJH/3R predictions of V48/V16 for the east sector or (ii) the observed values of V48/V16 for the east sector (for wind speeds above 10 m/s the MS3DJH/3R value is assumed).

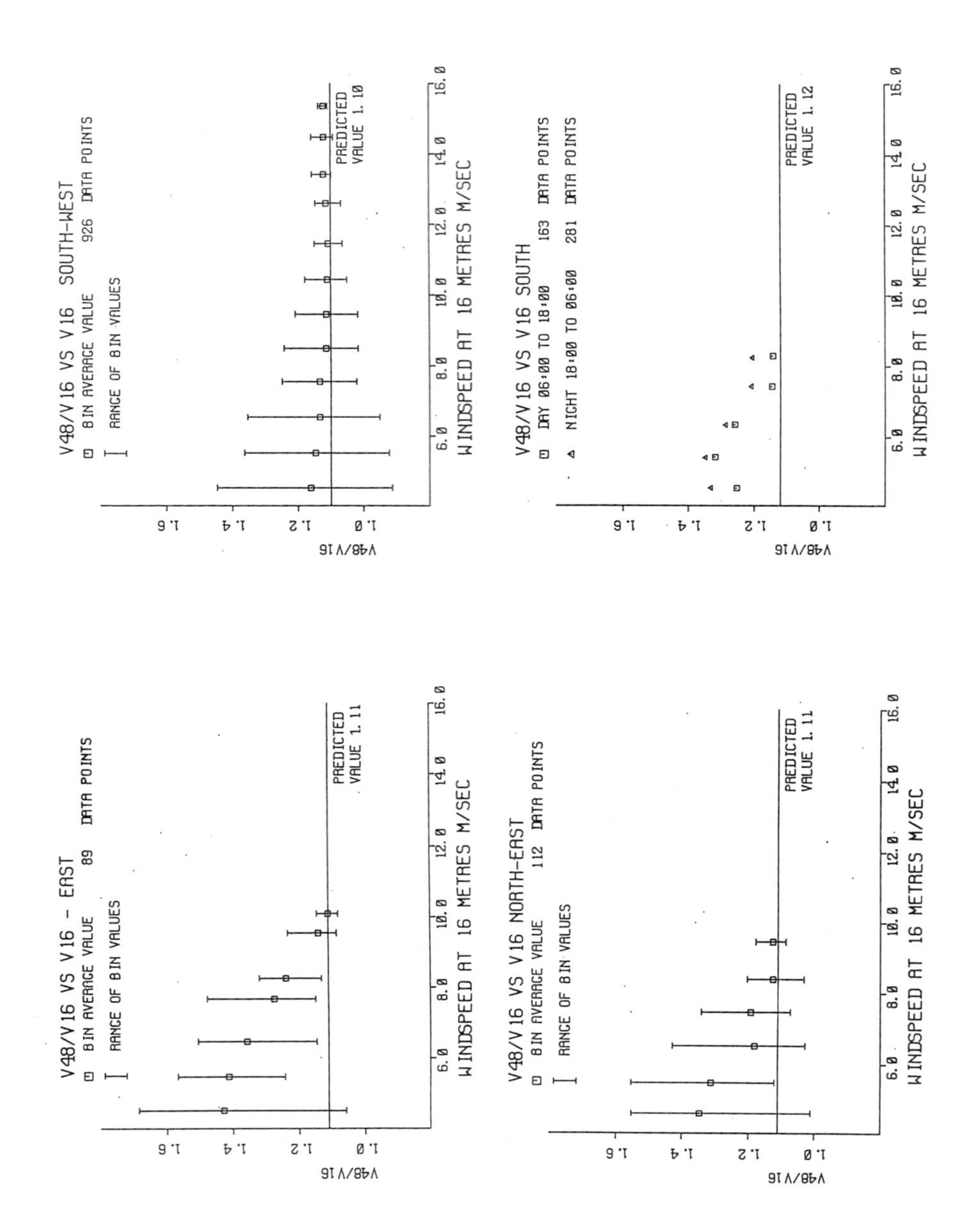

Fig. 3. Variation of V48/V16 as a function of wind speed at 16 m for the sectors shown.

The resulting calculated wind speed distribution at 48 m was combined with the power output curve for the Westwind 55 kW turbine and average power output at 48 m calculated. The results of this calculation are shown in Table 1 and indicate that power estimates based on the observed wind profiles can be 20 to 50 % larger than those based on neutral flow assumptions. The largest percentage increase is to be found for lower wind speed sites. This is to be expected as such sites produce a greater percentage of energy from moderate speed winds where the observed departures from neutral profiles are greatest.

CONCLUSION

An analysis of profile data from the South Fremantle wind turbine test site illustrates the importance of low to moderate wind speeds on power production. Under these conditions the flow illustrates a marked departure from that predicted by a neutral flow model and highlights the need for a better understanding of stability and roughness effects, particularly in the low wind speed regimes encountered at many inland sites.

ACKNOWLEDGEMENTS

Monitoring at South Fremantle has been funded under the auspices of the Solar Energy Research Institute of Western Australia. Throughout this programme, one of the authors (SJD) was in receipt of a Commonwealth Postgraduate Scholarship. All of this assistance is gratefully acknowledged.

REFERENCES

Anonymous (1982). Recommended practices for wind turbine testing and evaluation. 1. Power performance testing. Executive Committee of the International Energy Agency Programme for Research and Development on Wind Energy Conversion Systems, International Energy Agency, 16pp.

Irwin, J.A. (1979). A theoretical variation of the wind profile power law exponent as a function of surface roughness and stability. Atmos. Environ., 13, 191-194.

Walmsley, J.L., P.A. Taylor and T. Keith (1986). A simple model of neutrally stratified boundary-layer flow over complex terrain with surface roughness modulations (MS3DJH/3R). Bound.-layer Meteorol., 36, 157-186.

Average Wind Speed at 16 Metres m/s	Average Power Output at 48 Metres kW	
	MS3DJH Profile	Observed Profile
4.5	8.4	12.5
5.0	10.9	15.2
5.5	13.6	17.8
6.0	16.2	20.3
7.0	21.1	25.0
8.0	25.5	28.9

Table 1. Estimated wind power production at 48 m.

ANALYSIS OF DATA FROM THE DFVLR 100-kW TURBINE DEBRA-25

J. P. Molly, H. Seifert

German Aerospace Research Establishment (DFVLR)
Pfaffenwaldring 38-40, D-7000 Stuttgart 80
Federal Republic of Germany

ABSTRACT

The DEBRA-25 wind energy converter, developed by DFVLR, is in an unattended operation since nearly two years. Data from various measurement campaigns and long term operation were recorded and evaluated. More than 6000 grid connected hours at the test site in an average wind speed of 4.2 m/s proved the special design properties for low wind speed areas. All interesting data like power curve, availability, generated energy, load factor and auxiliary energy need are given in the report. Experiences gained from test operation resulted in a slightly changed tower design and in a change of the hydraulic pitch control system. The loads measured on the rotor and their combination with the load cycles during one year show the fatigue relevant load spectra of the WEC. The power fluctuation coefficient is given as a function of wind speed and demonstrates the quality of the pitch control system.

KEYWORDS

Power curve, power duration curve, power control, power fluctuations, availability, fatigue loads.

INTRODUCTION

The DEBRA-25 wind energy converter (Fig. 1, Table 1) has been developed by DFVLR and went into operation at the end of 1984. To meet optimum operation properties at low and high average wind velocities the three-bladed rotor of 25 m diameter drives two generators of together 100 kW rated power. One of the generators can be switched from 1000 and 1500 rpm by changing the number of activated poles. Thus the rotor speed varies from 33 rpm (Mode I) using one generator to 50 rpm (Mode II) with both generators. Referring to the operation loads and transportation weight emphasis was laid on a light weight design (Molly, 1985). The application of aeronautical design experiences together with the lower loads caused by blade pitch control resulted in a rotor blade mass of only 320 kg which is about 1/3 of comparable rotor blades used in stall controlled WEC (Hald, Kensche, 1985). Also the total WEC mass of only 14,500 kg, tower included, is much less than WECs of same size.

After the WEC had been erected on the DFVLR test site "Ulrich Hütter", a longer period of test operations followed. In 1985 the WEC was 500 h in operation mainly for measurement campaigns and optimization tests concerning the blade pitch control system and the general operation supervisory system.

Beginning 1986 the DEBRA-25 went into unattended continuous operation. Only 637 hours of operation were achieved during the first half of 1986 caused by many shut downs released by false alarms. After the reasons were found the on-line operation time considerably rised to 6008 hours and an energy production of 158 MWh until end of August 1987. The good test results and the proven reliability of the DEBRA-25 convinced the company Friedrich Köster to take a licence of the WEC which now is offered as Adler-25.

POWER PRODUCTION

The DEBRA-25 wind turbine has been designed for application in remote low and medium wind speed areas. Therefore the expected wind velocities of the test site (long term average 4.6 m/s) are suitable for testing the WEC under design conditions.

The power curve given in Fig. 2 was taken from 7888 10-min-averages measured in the undisturbed wind directions. The comparison with the calculated power curve shows higher outputs below 80 kW and worse values above this level. The reason is that any pitch control influence had been neglected in the calculations. As Fig. 2 shows pitch control is at least partly in action for all average wind speeds higher than 8.5 m/s. Average rated power of 100 kW is not reached because the power limit had been kept constant at 100 kW for the whole operation period considered. The difference of about 4 kW in rated power can be explained by the response time of the control system and by the fact that also in high wind speed averages wind velocities of less than rated occur. Therefore a simple change of the power control set-point to 104kW would lead to the desired rated output. The maximum overall efficiency is reached at the design wind speed of 8.0 m/s with 0.36 which results in an aerodynamic power coefficient of about cp = 0.42 as average for all weather conditions, including 50 percent power loss in heavy rain and icing periods.

To get a reference year the DEBRA-25 operated continuously from July 1986 to June 1987. 11 percent of that time the WEC was not available for operation due to an improvement of the hydraulic pitch control system and various measurement campaigns. Maintenance time caused about 10 hours loss in operation. In Table 2 the main items of the operation time statistics are given.

The relatively long time lost by automatic stops of the DEBRA-25 can be explained by the unmanned test site and the 80 km distance from the office which didn´t allow to switch on again the WEC immediately after an automatic stop. In all stop cases only a simple reset of the control system had to be done to start operation again. In nine of ten cases the shut-down was caused by a hypercritical sensitivity of the control sensor inputs.

Figure 3 shows the monthly energy generation and operation time statistics for the reference year. Approximately 35,000 kWh or 34 percent of the total generated energy was consumed on the test site itself, the rest was sold to the utility. The average power factor for the whole period was 0.93

achieved by a partly compensation of the reactive power.

In Fig. 4 three power duration curves of the respective year are given. A comparison of the curves for average, minimum and maximum monthly mean wind velocities shows big differences in the energy output but only small changes in the total operation time which is an effect of the two rotor speeds used. These result in a higher availability of low power outputs compared to a single speed operation. The estimated gain for a two-speed operation on the test site is 1750 h per year, equivalent to an increase from 36 percent operation time to 56 percent of the considered year. The average wind velocities in Fig. 4 are calculated from the wind velocities during the technical availability of the WEC and therefore are different from those given in Fig. 3.

IMPROVEMENTS

During the test period two important technical improvements of the DEBRA-25 had been done. The first speed-up tests already showed a too high natural bending frequency of the tower. Instead of having the calculated frequency of 1.4 Hz the measured was 1.55 Hz. In the operation Mode I the exciting frequency for the tower due to the three rotor blades is 1.67 Hz. The small difference of the two frequencies therefore didn´t allow a vibration free power operation in Mode I. A simple shortening of the stay poles at half length of the tower diminished the natural tower frequency to the desired 1.4 Hz. In Fig. 5 the resonance diagram shows the final measured vibration modes of tower and rotor blades.

The driving mechanism of the blade pitch control is a medium pressure hydraulic system. Until November 1986 a control valve with a relatively high leakage flux was used which forced the electric pump to refill the pressure reservoir every 10 seconds (running time 1 s) even at stand still of the DEBRA-25. To avoid the very often switching of the hydraulic pump the control valve had been changed for a one of less leakage flux. Together with the installation of a small mechanically driven pump which should replace the leakage flux continuously the electric pump didn´t switch on any more during operation of the WEC. The auxiliary electric energy consumption fell down from an average of 410 Watts to 260 Watts including the whole energy needed for the control unit. Now the yearly auxiliary energy consumption is about 2,300 kWh.

POWER CONTROL

The power control of the DEBRA-25 is done by a total span blade pitch control with a theoretical maximum change speed of 15 deg/s. Normally the pitch change speed is far away from that limit. Nevertheless certain operation conditions need fast pitch changes, e.g. to avoid overspeed during shut-down. To improve the control behaviour the controller performs an automatic adaption of the amplification factors related to the actual wind speeds. A measure for the quality of the power control had been proposed by Stam (1986). The definition of the Power Fluctuation Coefficient PFC is given in Fig. 6 together with the measured PFC as a function of the average wind speed. The curve passes through 1.0 when the power coefficient is best and the small cp-deviations left and right of the maximum don´t affect the third power dependence of the PFC-value. The graph shows for high wind speeds a constant value of less than 0.2 which means that only twenty per cent of the power content of the fluctuations is taken by the rotor.

LOADS DURING LIFETIME

Several measurement campaigns were performed to get load data from the rotor during operation. Loads from all operation conditions were put together in Fig. 7 to allow an easy comparison of the flapwise bending moment levels. Negative loads are bending the rotor blade against the wind. The built-in cone angle of 7 degrees of the rotor blades unloads the blade with increasing power output and leads to the highest loads at rated rotor speed and zero power condition. In general all transitions where the WEC is not grid connected are combined with higher loads than in grid operation. The maximum flapwise bending moment at the blade root appears during emergency shut down with -45 kNm or 45 percent of the design load j = 1. The normal transition from Mode II to Mode I leads to the second highest bending moment of about -35 kNm. Both average powers in Mode I and Mode II result in nearly the same bending moment of about -3.5 kNm due to the different rotor speeds. Only near rated power the rotor blade bends in the positive direction showing that a better balance between positive and negative bending moments could be achieved by a smaller coning angle. To get an information about the number of load cycles all operation conditions were registered for one year and are written together with the respective loads in Fig. 7.

REFERENCES

Hald, H. and Ch. Kensche. (1985). Development and tests of a light weight GRP rotor blade for the DFVLR 100 kW WEC. Windpower `85, San Francisco, 506-512.

Molly, J. P. (1985). Design and operation of the DFVLR 100 kW wind energy converter. Windpower `85, San Francisco, 451-456.

Stam, W. J. and N. J. C. M. van der Borg. (1986). Fluctuations in the electric power from wind turbines. Contribution for the IMTS 1986, Schnittlingen, ECN Wind Turbine Test Station, The Netherlands.

ROTOR	
Blade number	3
Diameter	25 m
Rotational speed	33 and 50 rpm
Position	downwind
Cone angle	7 deg
BLADE	
Profiles	FX-77W-XXX + NACA 44XX
Mass	320 kg
TOWER	
Type	guyed steel tube
Height	22 m
Erection	tilting tower

CONTROL SYSTEM	
Supervisory syst.	microprocessor
Power control	full span pitch
Max. change speed	15 deg/s
Power	400 W
PERFORMANCE	
Rated power	100 kW
Cut-in wind speed	3.4 m/s
Rated wind speed	10.5 m/s
MASSES	
Rotor	1,400 kg
Nacelle	6,000 kg
Tower	7,100 kg

Tab. 1 Main data of the DEBRA-25

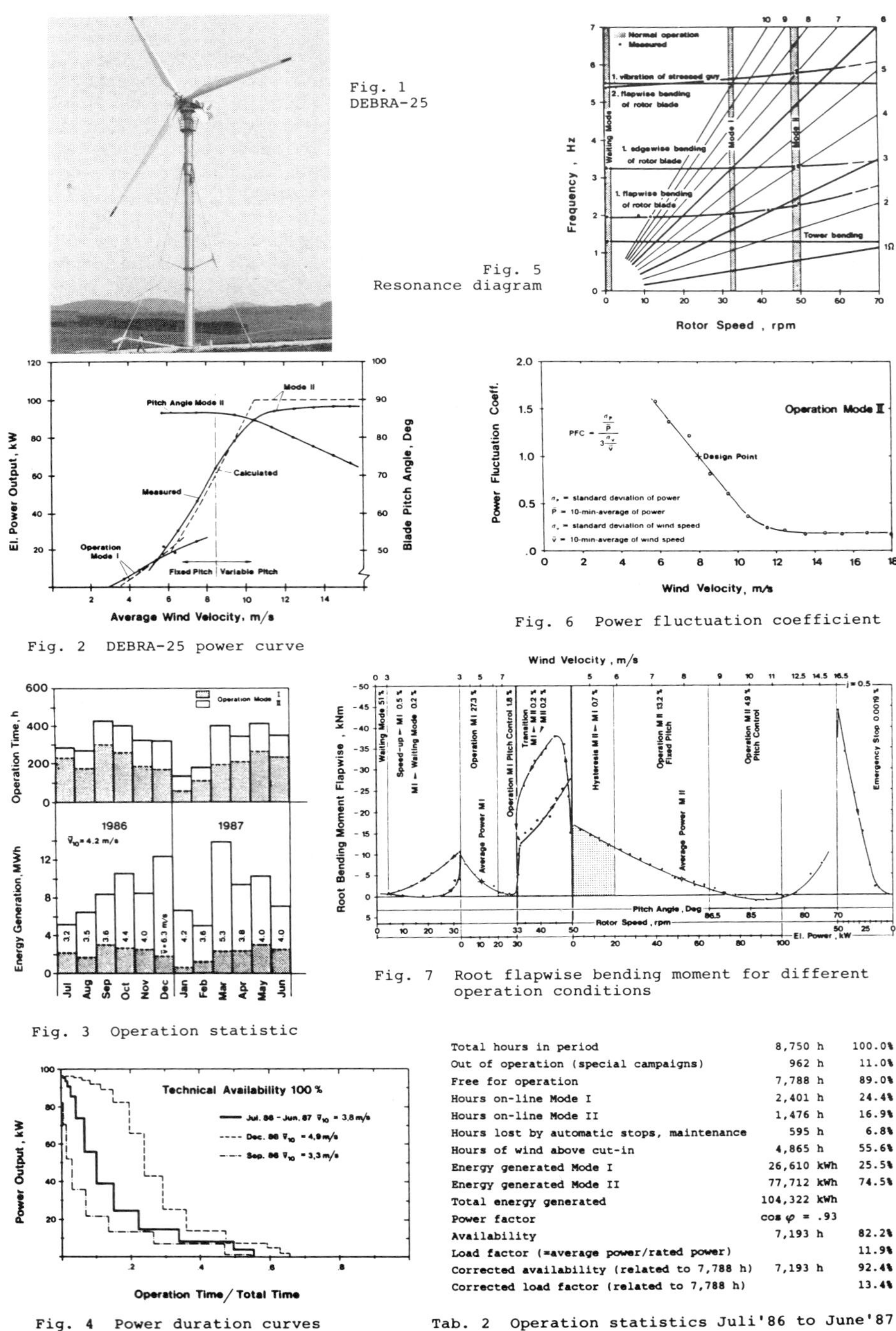

Fig. 1 DEBRA-25

Fig. 5 Resonance diagram

Fig. 2 DEBRA-25 power curve

Fig. 6 Power fluctuation coefficient

Fig. 3 Operation statistic

Fig. 7 Root flapwise bending moment for different operation conditions

Fig. 4 Power duration curves

Total hours in period	8,750 h	100.0%
Out of operation (special campaigns)	962 h	11.0%
Free for operation	7,788 h	89.0%
Hours on-line Mode I	2,401 h	24.4%
Hours on-line Mode II	1,476 h	16.9%
Hours lost by automatic stops, maintenance	595 h	6.8%
Hours of wind above cut-in	4,865 h	55.6%
Energy generated Mode I	26,610 kWh	25.5%
Energy generated Mode II	77,712 kWh	74.5%
Total energy generated	104,322 kWh	
Power factor	$\cos\varphi$ = .93	
Availability	7,193 h	82.2%
Load factor (=average power/rated power)		11.9%
Corrected availability (related to 7,788 h)	7,193 h	92.4%
Corrected load factor (related to 7,788 h)		13.4%

Tab. 2 Operation statistics Juli'86 to June'87

Wind Direction and WEC Yawing

Körber, F., Besel, G.*

Results within the scope of the GROWIAN Test Program

*) Dr.-Ing., Dipl.-Ing., MAN Technologie GmbH, Dachauer Strasse 667, 8000 München 50

Abstract

This paper presents results of the German 3 MW WEC GROWIAN concerning the fluctuation of the wind direction and the WEC's yawing.

For a 6 and a 39 months' period, wind and machine data were evaluated at the GROWIAN site in the coastal area of North Germany.

The short time evaluations which covered 12,000 yawing processes proved a very accurate function of the system. The nacelle misaglignment was less than ± 10 deg. in stable wind conditions and less than ± 20 deg. in heavily fluctuating winds. The operational time of the yawing drive was found to be about 5% of the WEC's operating time.

In the 39 months' period 35 full clockwise revolutions of the wind vector were recorded at wind speeds above 4 m/s. A certain annual periodicity with an increased number of revolutions from April to September was noticed.

Yaw Drive

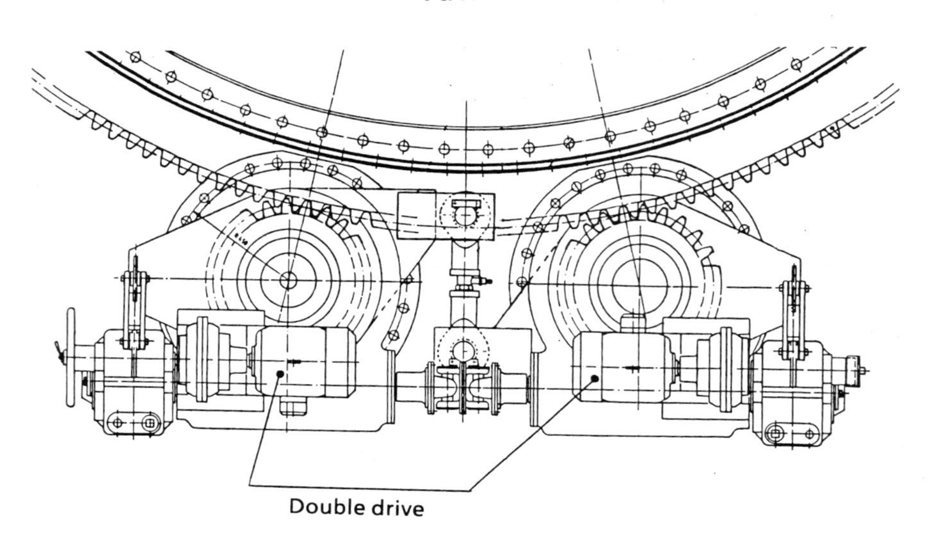

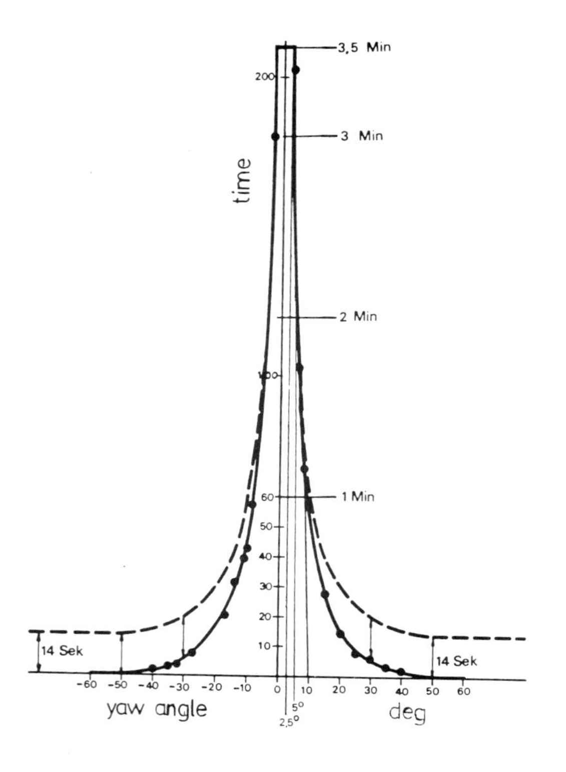

Yaw Control Strategy

The start of the yaw drive depends on the integral of the yaw misalignment and the duration time of that misalignment. The full line shows that correlation in theory. In reality, a time delay of 14 s (dotted line) is caused by the initial period of the yaw drive. A misalignment below 5 deg. starts the yawing after 3.5 minutes. The wind direction is a 12 s average value.

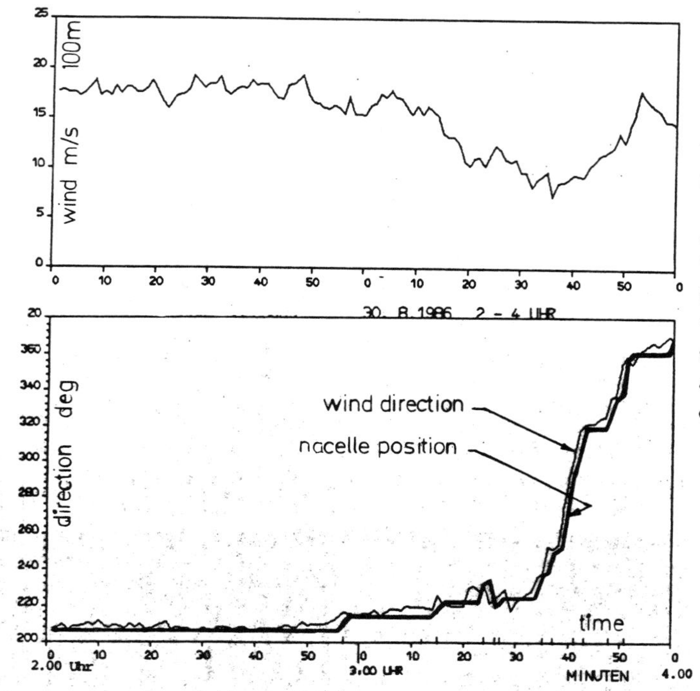

Record of the Yaw Process

This figure shows a 2 hours' record of wind speed in the 100 m level, wind direction and nacelle position. In this "normal" wind conditions the misalignment is less than 10 deg. in average. More varying wind causes a misalignment of less than 20 deg.

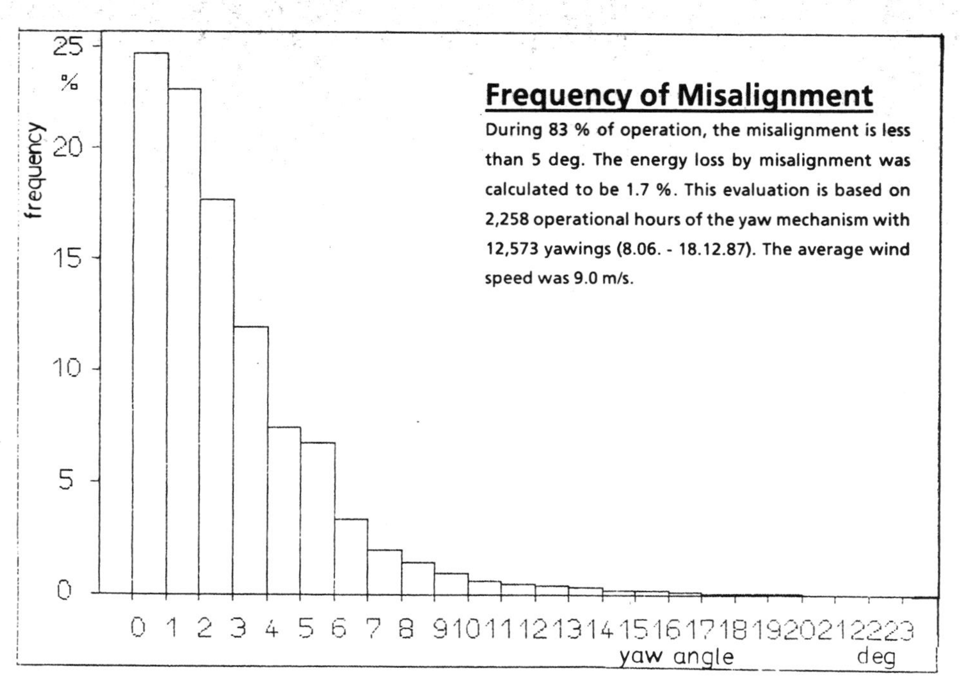

Frequency of Misalignment

During 83 % of operation, the misalignment is less than 5 deg. The energy loss by misalignment was calculated to be 1.7 %. This evaluation is based on 2,258 operational hours of the yaw mechanism with 12,573 yawings (8.06. - 18.12.87). The average wind speed was 9.0 m/s.

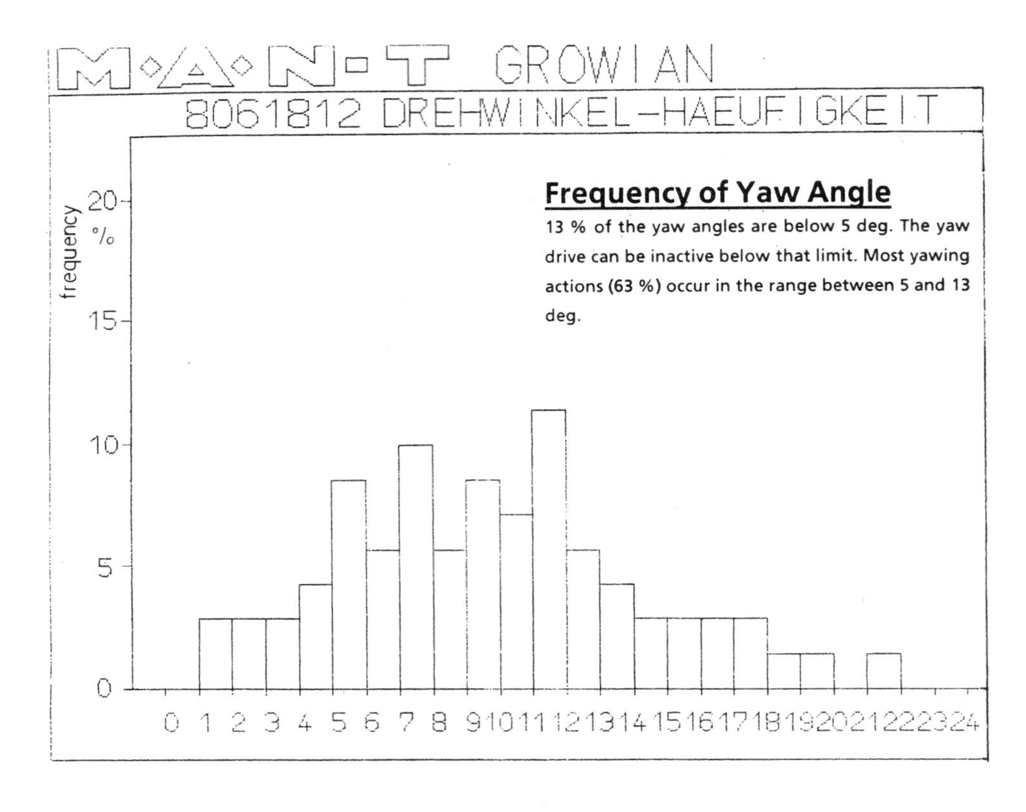
M·A·N-T GROWIAN
8061812 DREHWINKEL-HAEUFIGKEIT
Frequency of Yaw Angle
13 % of the yaw angles are below 5 deg. The yaw drive can be inactive below that limit. Most yawing actions (63 %) occur in the range between 5 and 13 deg.
frequency
%
20
15
10
5
0
0 1 2 3 4 5 6 7 8 9 10 11 12 13 14 15 16 17 18 19 20 21 22 23 24

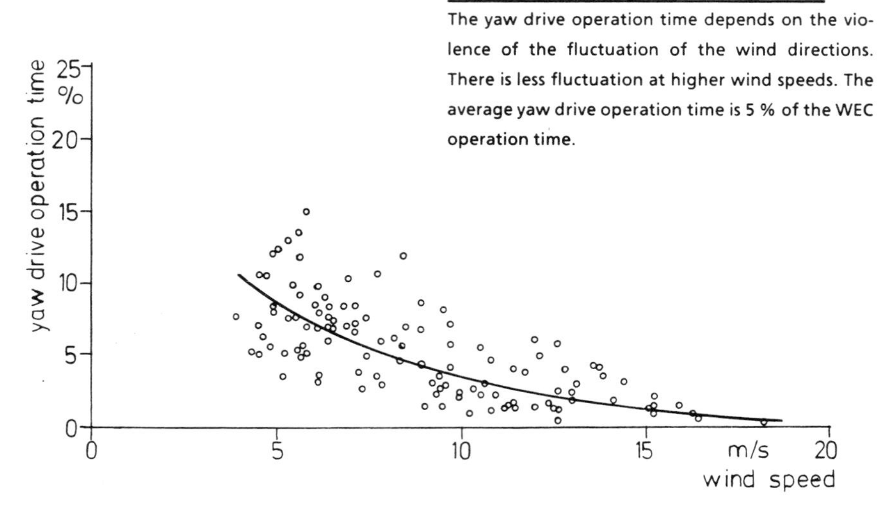
Yaw Drive Operation Time
The yaw drive operation time depends on the violence of the fluctuation of the wind directions. There is less fluctuation at higher wind speeds. The average yaw drive operation time is 5 % of the WEC operation time.
yaw drive operation time
%
25
20
15
10
5
0
0
5
10
15
m/s
20
wind speed

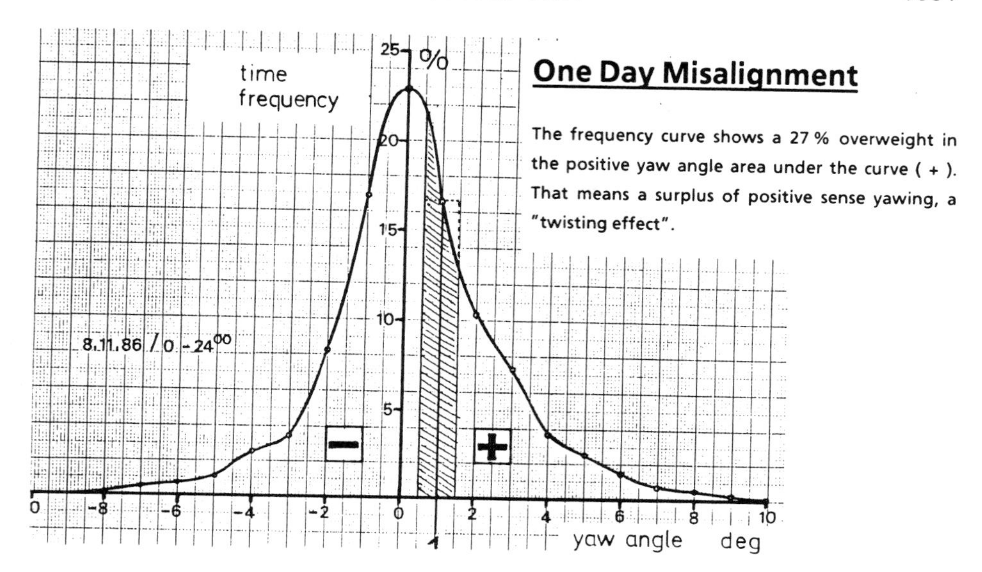

One Day Misalignment

The frequency curve shows a 27 % overweight in the positive yaw angle area under the curve (+). That means a surplus of positive sense yawing, a "twisting effect".

Longtime Yaw Angle of GROWIAN

This printout shows about 4 months of GROWIAN yawing. A tendency to more positive turns is observed. There are some days missing because of computer breakdown. The overall result should be valid because the missing values are at random.

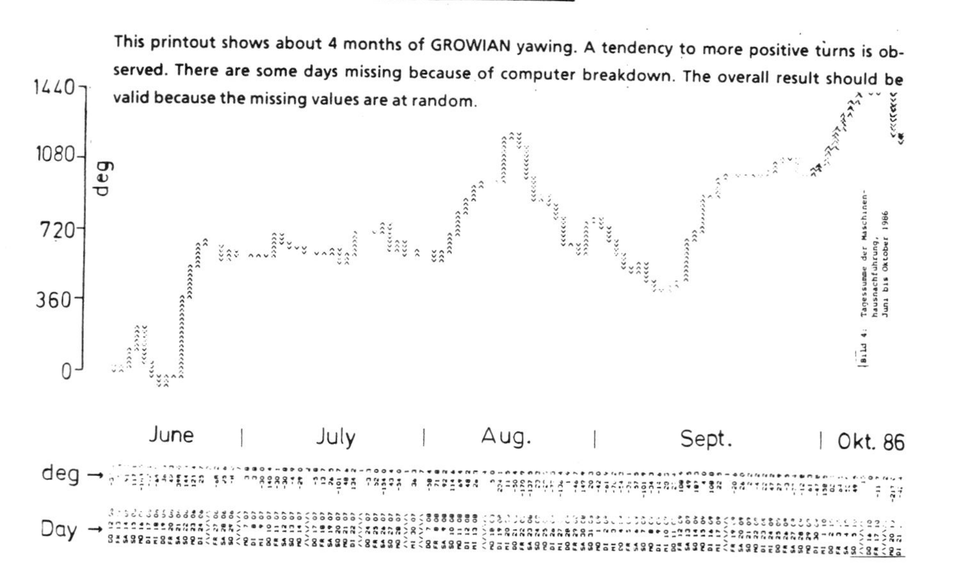

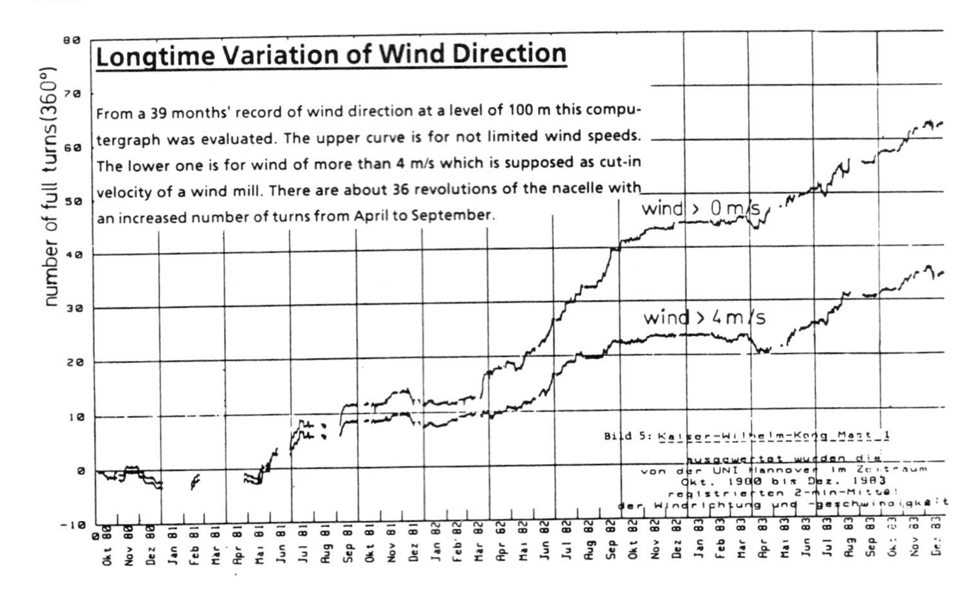

Longtime Variation of Wind Direction
From a 39 months' record of wind direction at a level of 100 m this computergraph was evaluated. The upper curve is for not limited wind speeds. The lower one is for wind of more than 4 m/s which is supposed as cut-in velocity of a wind mill. There are about 36 revolutions of the nacelle with an increased number of turns from April to September.
number of full turns (360°)
wind > 0 m/s
wind > 4 m/s

Annex:

GROWIAN SYSTEM DATA

Performance

- Rated capacity	3 MW
- Mean annual energy output	12 GWh
- Power-to-area ratio	380 W/m2
- Rotor speed	18.5 rpm ± 15 %

Wind

- Rated wind speed	12 m/s
- Cut-in speed	6 m/s
- Cut-out speed	24 m/s
- Maximum speed	60 m/s

Dimensions

- Rotor diameter	100 m
- Hub height	100 m
- Mass of nacelle with rotor	400 t

THE OPERATIONAL STUDY OF A DARRIEUS WIND TURBINE GENERATOR FOR APPLICATION DEMONSTRATION

Wang Cheng xu

Dept. of Electrical Engineering, Tsinghua Univ.
Beijing, China

ABSTRACT

This paper describes an application demonstration study of a Darrieus wind turbine generator which can be connected to the local network or operated in stand alone condition. The study includes system layout, control strategy and performance test.

KEYWORDS

Darrieus wind turbine generator; system layout; grid connected operation; stand alone operation; control strategy; AC/AC converter; load regulator; performance test.

INTRODUCTION

In recent years, the wind energy utilization have been paid more attention in China, up to now most of the islands along the seashore and in the remote countryside, grassland where are no electricity at all or lack of it.On the other hand, in these districts the renewable energy, including solar and wind energy, are rich, so using wind energy for electricity generation is getting more and more attractive.

Based on this situation a joint study of wind energy conversion system have been carried out between Dornier System GmbH (Federal Germany)and Tsinghua University (China).

The study shows that using AC/AC converter or using load regulator both two different operation system can be realized in steady and safety by adopting suitable control strategies.

The performance testing shows a satisfactory results. Based on the performance data and technical-economic evaluation, the application prospect of such Darrieus wind turbine generator in windy areas is promising.

CONSTRUCTION OF WTG

The Darrieus WTG's construction is shown in Fig.1.

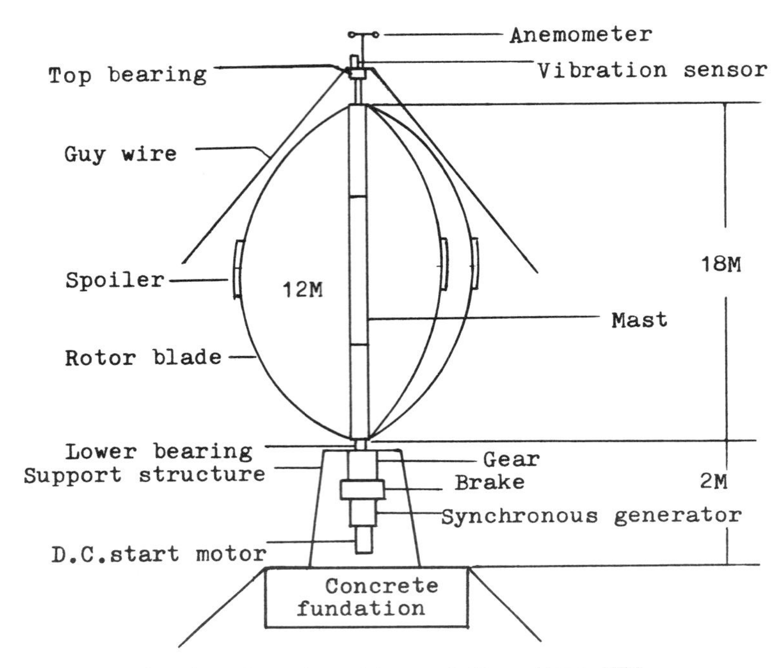

Fig.1. Construction of Darrieus WTG

The Darrieus rotor has three aluminium extruded blades. The swept area of the rotor is 140mm². The profile of blade is 634-021 and the cord length of the blade is 300mm. The gearbox, synchronous generator, disc brake, lower bearing and starting motor are all assembled inside a steel made support structure which are fixed above the concrete foundation. The rated revolving speed of the alternator is 1500 rpm. The gear ratio between the alternator and Darrieus rotor is 16.78.

An anemometer is installed on the top of the mast, it provides the signals to start the rotor. A vibration sensor is also installed on the top of the mast. When the vibration of the mast becomes severe, the sensor will give out a signal to stop the WTG automatically.

In order to limit the overload and overspeed of the alternator, three spoilers had been assembled on three blades to absorb the extra wind energy during high wind speed running period.

Three guy wires are fixed on three concrete foundations.The prestress of each guy wire is about 3KN. The total weight of the WTG is about 3 tons.

OPERATION MODE

The operation mode scheme is shown in Fig. 2.

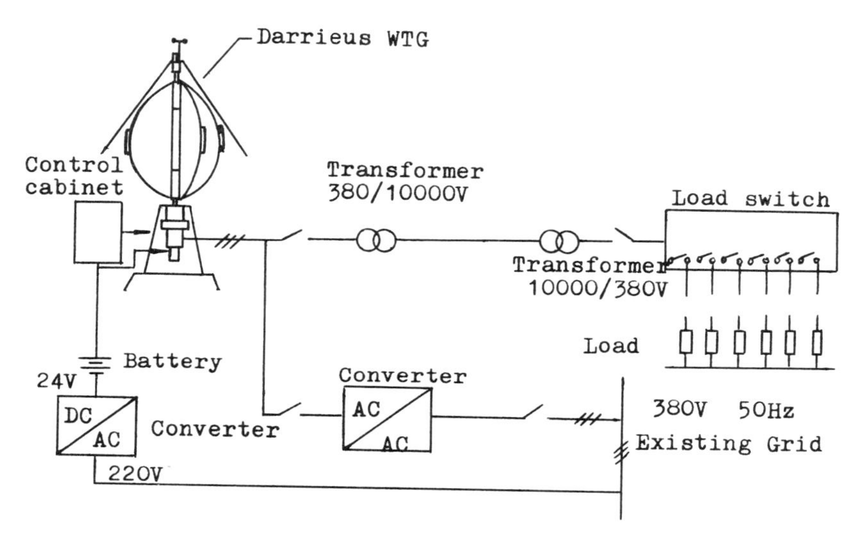

Fig.2. Operation Mode Scheme

Stand Alone Operation Mode

In this operation mode, the load switch could be switched in with load according to the frequency. The frequency and voltage of the alternator are proportional to its revolving speed, i.e, to wind speed. For experimental purpose at first stage a set of 20kw electric heaters which was divided into 2,4,8,12, 16,20kw steps were installed and tested.

To minimize the variation of the frequency while the WTG is running with load, another load regulator for stand alone operation had been developed and tested. It consists of SCR, frequency/ voltage converter, consumer's load, dump load, alternator and wind turbine simulator. All these components form a speed-frequency-power close loop control system. In this system, the regulation process is automatically continuous and the frequency could be kept in constant.

Grid Connected Operation Mode

In this operation mode, the A.C. power with variable frequency from alternator goes into the converter, and the A.C. power with constant frequency can be obtained at the output side of the converter. This means it permits to extract more energy from the wind. In addition, using AC/AC converter between the alternator and network makes the alternator independent of the grid, so it alleviates the synchronisation problem.

The shortcoming of this operation mode is the relatively high cost of the AC/AC converter. In spite of this fact, it is still attractive while considering the advantage above mentioned and the direct drive capability of using a low speed alternator.

CONTROL STRATEGY

The control strategy of the WTG is shown in Fig. 3.

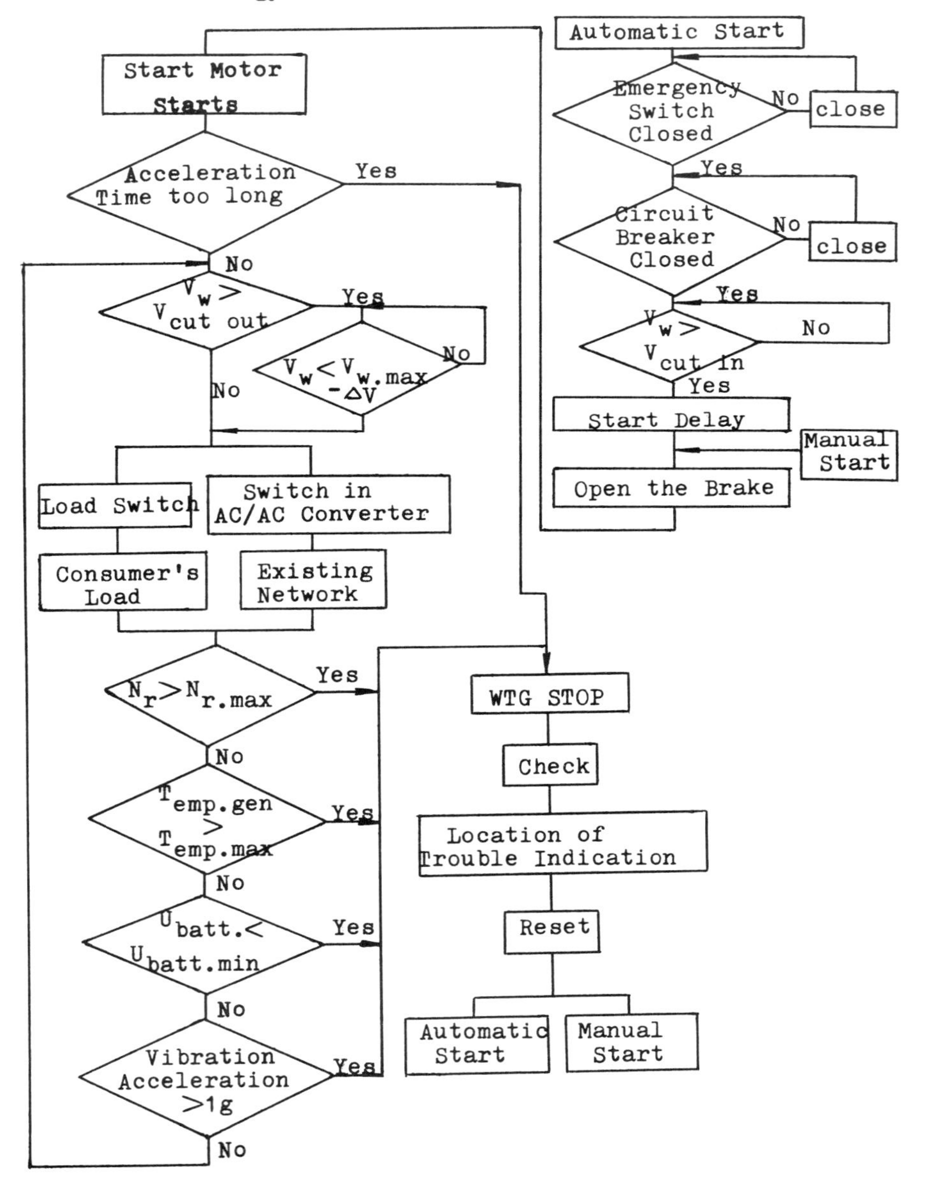

Fig.3. Control Logic Flow Chart

OPERATIONAL PERFORMANCE

According to the design data, the cut in wind speed is 4.5m/s, the rated wind speed is 10m/s, corresponding to 20kw electric power output and the cut out wind speed is 15m/s.

Fig.4. is the operational characteristic which represents the electric power of the WTG as a function of wind speed. From the measured points it is obvious that the expectant effect had been reached.

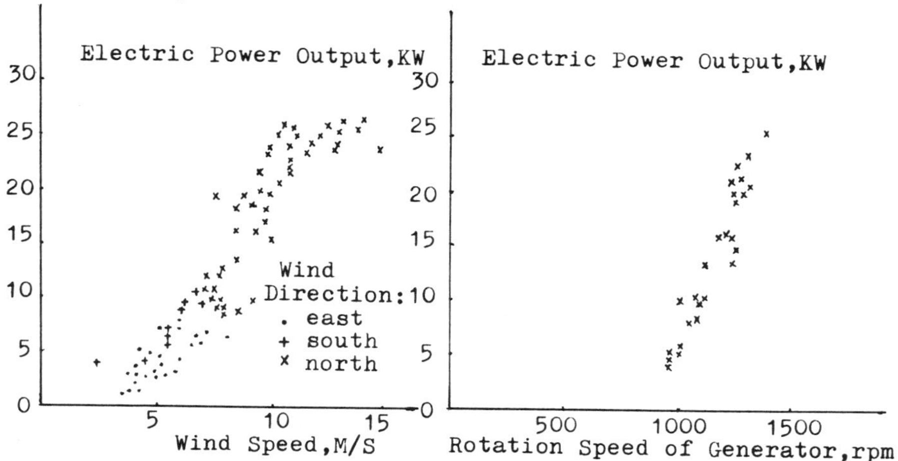

Fig.4. Electric Power Output at Different Wind Speed

Fig.5. Characteristic of Electric Power Versus Rotating Speed of the Alternator

The maximum electric power output depends on the rated power of the alternator and its ability for short time overloads. The rated power of the alternator in this wind energy conversion system is 20KW. To limit the power output within 20KW and keep this power output as the wind speed goes up and higher than 10m/s, the spoilers open at 1300 rpm of the alternator. This based on the characteric of electric power versus rotating speed of the alternator, see Fig.5.

REFERENCE

DANIEL M.SLMMONS.(1975). Wind Power. NOYES DATA CORPORATION.

J.P.MOLLY.(1978). Windenergie in Theorie und Praxis, Verlag C.F. Müller Karlsruhe.

V.D.HUNT.(1981). Wind Power. VAN NOSTRAND REINHOLD COMPANY.

R.J. TEMPLIN and R.S.RANGI. Vertical-Axis Wind Turbine Development in Canada, IEE PROCEEDINGS, Vol.130,Pt.A,No.9,December 1983

A.FRETZSCHE,K.SPEIDEL, WANG CHENG XU. Technical-Economic Evaluation of Vertical Axis Wind Turbine of 30-50kw, Report, Dornier System, Tsinghua University, 1986.

SOLAR CHIMNEYS

SOLAR ELECTRICITY FROM SOLAR RADIATION

J. SCHLAICH * K. FRIEDRICH *

*Schlaich und Partner, Beratende Ingenieure im Bauwesen
Hohenzollernstr. 1, 7000 Stuttgart 1

Abstract:

Solar chimneys convert solar radiation into electrical energy by combining in a novel way the well known principles of the greenhouse, the chimney and the wind turbine generator.

Keywords:

updraught, solar chimney, greenhouse, translucent roof, solar radiation.

Prototype solar chimney in Manzanares, Spain

The guyed chimney is made from thin corrugated sheet.

Height of chimney	194.6 m
Radius of chimney	5.08 m
Mean radius of collect.	122.00 m
Height of collector roof	1.85 m
Number of rotor blades	4
Diameter of rotor	10.00 m
Design speed of turb.	100 rpm
Solar radiation	1000 W/m²
Increase in temp.	20 K
Ambient air temp.	302 K
Efficiency of collector	32 %
Efficiency of turbine	60 %
Coefficient of friction	0.2
Updraught - under load	9 m/s
Updraught - on release	15 m/s
Design power output	50 kW

The solar chminey combines three well-known physical principles - those of the greenhouse, the chimney, and the wind turbine - in a novel way. Solar radiation, absorbed under a large and translucent horizontal roof, heats the air. The warm air flows to the chimney und rises, i.e., the difference in temperature between the heated air and the fresh air is transformed into kinetic energy, an updraught via the chimney. This updraught is converted into mechanical energy by a wind turbine and then into electricity by a conventional generator.
Apart from its simple principle of operation the solar chimney has a number of other advantages: the collector exploits not only direct radiation, but also diffuse light when the sky is partially or completely overcast. (Other technologies, e.g., parabolic dishes or heliostates, only exploit direct radiation). The turbine-transmisson-generator unit is the only moving part of the plant which has to sustain a permanent mechanical load. Hence, two distinctive features of the solar chimney as compared to other solar power plants are its high level of availability for power generation and its extremely low susceptibility to malfunctions.
The soil beneath the collector roof acts as a natural energy storage reservoir. During the first hours of daylight it absorbs a portion of the radiated energy; in the second half of the day, and into the night, it releases the energy into the collector air. On the one hand this helps to balance fluctuations of supply resulting from changes in cloud cover, and on the other hand it transfers some energy production in a natural way to the night hours. Large-scale plants produce 10-12% of their energy during hours of complete darkness. Though during the day strong side winds are undesirable, as they increase the inevitable losses of the collector due to convection, they are welcome at night: having an effect which is in principle similar to that of a water-jet pump, the side winds cause the air pressure at the chimney outlet to be reduced. As a result, cold fresh air is pumped trough the plant.
It has to be realized that in a large solar chimney, and depending on operating conditions, it takes a single molecule of air between 10 and 30 minutes to pass through the plant. Therefore during operation a computer model simulates the performance of the plant on the basis of meteorological data and plant characteristics, and transmits the appropriate control signals to the turbine and generator at the right moment.
Solar chimneys need collectors with very large surface areas. However, since they can in any case only operate economically in sunny countries, this does not represent a major disadvantage, because these are generally the very countries where there is sufficient dessert land available.
The materials needed to build large-scale plants - concrete, steel, and glass, are available in most Third-World countries. This means that in contrast to other solar technologies, the construction of solar chimneys does not involve the user country in heavy expenditure of foreign currency. Instead, local resources are exploited and jobs are created.
The plant can be operated in either grid connection mode or stand-alone mode (with storage batteries or to produce hydrogen). In addition, the size is such that it has enough space for a number of conceivable complementary functions, e.g., biomass production, drying, irrigation, or even desalination. A broad field of research is beginning to open up here, because many of the factors relevant to dual-purpose or multipurpose utilization of solar chimneys have so far only been dealt with superficially.

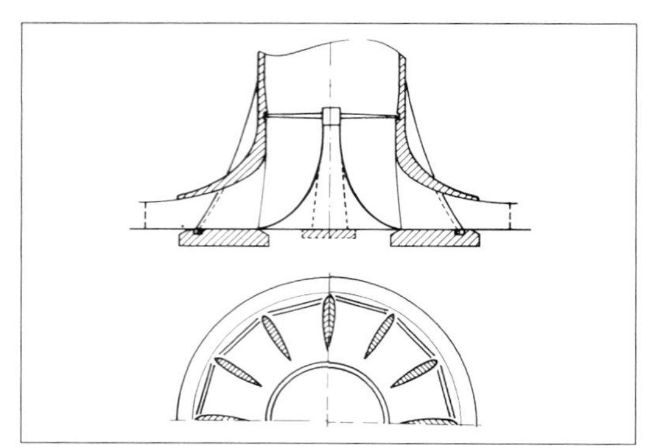

Usually the chimney will be made from reinforced concrete to secure a good durability and long lifetime. Concrete also suits to shape the transistion zone between collector and chimney. This area - the chimney inlet - has to be designed and built with great care to minimize friction losses resulting from deflection of the air flow. The wind turbine, which converts the kinetic energy into mechanical rotation energy, is located in the inlet area.

Drawing of 100 MW solar chimneys with a light weight membrane structure, suitable for earthquake endangered regions. They could produce hydrogen - the energy source of the future - in deserts by large-scale hydrogen electrolysis.

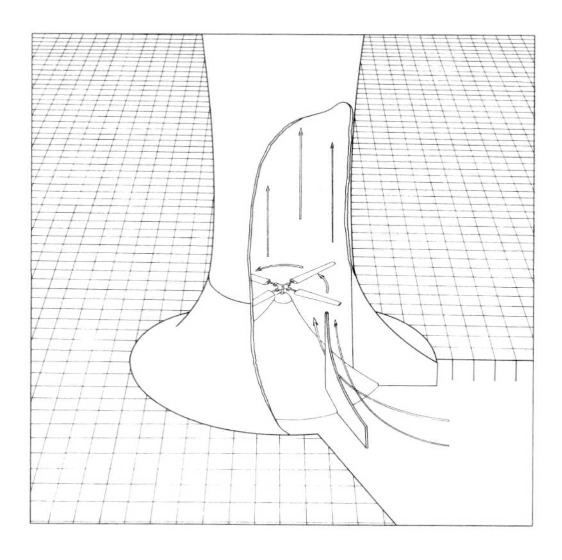

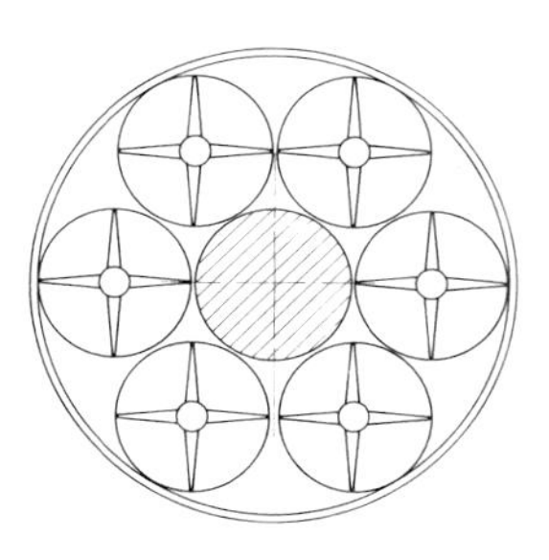

While smaller units can operate with a single wind turbine, larger power plants would work better with several turbines arranged in a circular configuration. The rotor blades, made of glass fibre-reinforced polyester, are specially designed to withstand the permanent updraught load. During operation automatically controlled setting gears adjust the blades to the optimal face impact angle.

Glassroof of the prototype at Manzanares, Spain
It resisted heavy storms for many years without harm and proved to be self cleaning thanks to the occasional rain showers.

Plastic membrane roof in Spain; installed there for comparison with the glass roof. It was cheaper than glass, however plastic gets brittle with time and thus tends to tear. The membranes are clamped to a frame and stressed down to the earth at the centre by use of a plate with drain holes.

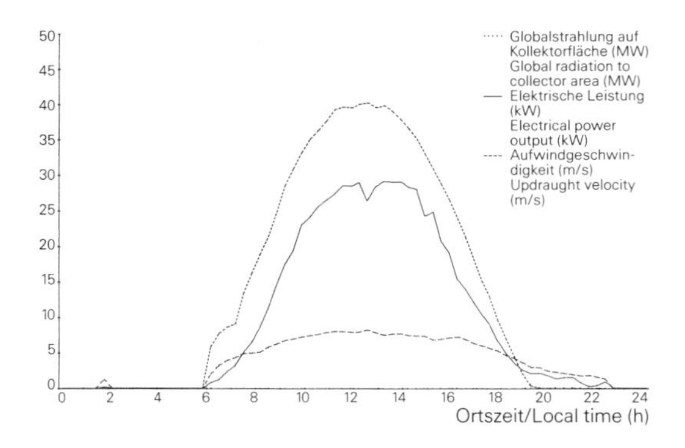

The electricity generated by the power plant basically depends on the amount of solar radiation. Under optimal conditions the collector of the prototype plant, with solar radiation of 1000 W/sq.m., increases the temperature by 20 K above the ambient temperature. The updraught then reaches velocities of 15 m/s when the turbine is switched off and approx. 10 m/s under load.

Basically the solar chimneys posses a low degree of efficiency in comparison with other types of solar power plants, but this is offset by low collector costs. Their efficiency increases sharply with the absolute size of the plant.

The costs of the produced energy (DM/kWh) even of small units around 1 to 5 MW are so slow, that in remote areas they are not only competitive but superior to any other energy generation. For larger units around 50 or 100 MW the comparison with oil or nuclear plants indicates specific costs in the same range, and this already today, which means, that also and mainly larger solar chimneys are economically superior to other energy generation, when taking the long-term fuel availability and costs into account. These plants can produce electrical energy for DM 0,20/kWh.

Solar chimneys are the ideal source of electrical energy for countries with high solar radiation and sufficient free land.

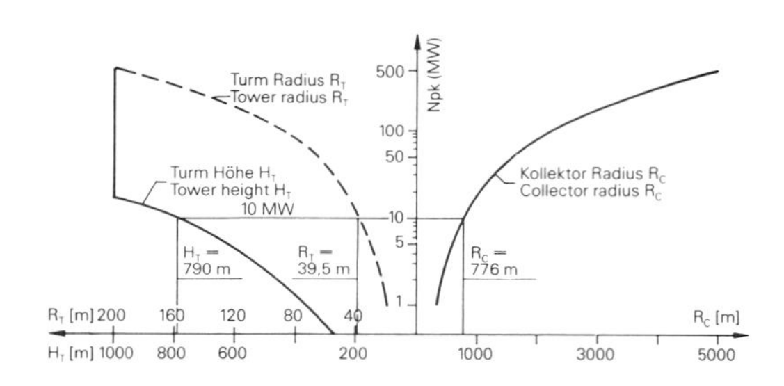

Peak output Npk as a function of the dimensions

Institutions and companies involved in the prototype plant project:

Interatom GmbH, Bergisch-Gladbach
Institut für Aerodynamik und Gasdynamik, Stuttgart University
Balcke-Dürr, Ratingen
SEG, Kempen
Maurer Söhne, Munich
IST, Wollbach-Kandern
Hoechst AG, Gendorf
DuPont, Geneva
SIC, Hamburg

Inquiries should be addressed to the authors:
Schlaich und Partner
Civil Engineering
Consultants
Hohenzollernstr. 1
D-7000 Stuttgart 1
Telephone: (0711) 6491016
Telefax: (0711) 6405970
Telex: 721401 sps

Sponsors:

Federal Ministry of Research and Technology Bonn (BMFT)
Union Electrica Fenosa S.A., Madrid

Development and Project Management:

Schlaich und Partner
Civil Engineering Consultants
Prof. Dr.-Ing. Jörg Schlaich
Dipl.-Ing. Rudolf Bergermann, Stuttgart

Adviser:

Prof. Dr.-Ing. Drs. h.c. mult. Fritz Leonhardt

Examinations of 20 Wind Energy Converters AEROMAN 20 kW in Power to Grid Operation in Germany

M.Dams, W.Hauber
Forschungsstelle für Energiewirtschaft
D-8000 München, Am Blütenanger 71

(Research Institute for Energy Utilisation and Technology, Munich)

This project "application of small wind energy converters (WEC)" was set up in order to get more information about the use of WEC in direct to grid connected operation mode. For this purpose 20 AEROMAN 12 ø/20 kW windturbines have been installed between 1984 and 1986 in the northern part of Germany. Different types of users: farmers, private househoulds, restaurants, camping sites, shops etc. were selected in order to gather more experience how a WEC can be used by non scientist owners. Besides the annual electrical demand and the annual mean wind speed had to meet predefined requirements.

A special measuring program, fund by the Federal German Ministery of Research and Technology, the manufacturer MAN (Munich) and the utility SCHLESWAG (Rendsburg), was started to analyse the generation- and demand-situation. For this purpose electrical consumption meters were installed. These consumption meters deliver pulses to a Data Acquisition System (DAS). This DAS records all accumulated values every 15-minutes and resets the counters. Via telephone link the data can be read out without interrupting the long term data acquisition. The build-in storage capacity of 32 KByte RAM allows periods of approximatly four weeks until a reading access becomes necessary.

List of measured values:

1. Power Generation
2. Power Delivery to Grid
3. Power Supply from Grid
4. Signal High/Low price category
5. Windspeed

This measuring program was started in april 1987 and will be finished in september 1988, although the whole project is still in progress we want to present some first results.

The map in Figure 1 gives an overview of the distribution of the WECs in the northern part of Germany. The site numbers refer to all the other figures presented in this report.

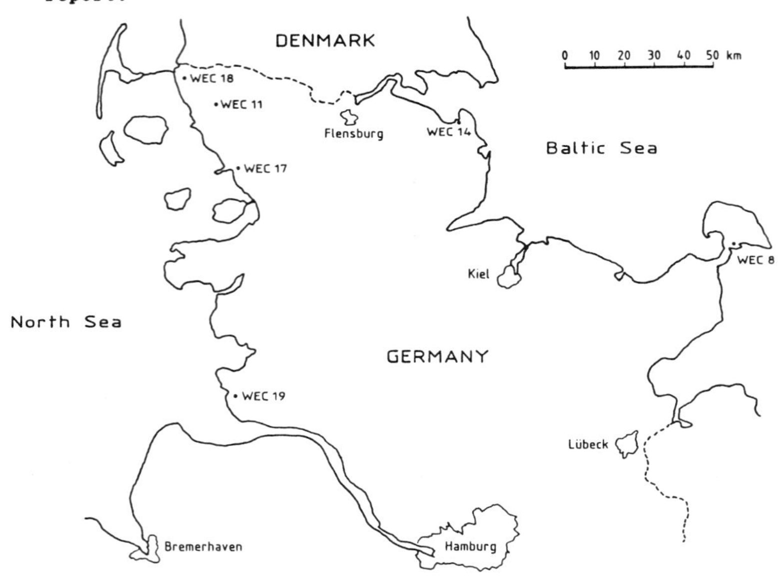

Fig. 1: Map of Northern Germany - Location of WEC's

Figure 2 shows a plot of the power generation and demand situation of a farm (WEC 18). The timescale is one week. The first line contains the wind speed [m/sec] taken from an anemometer. The second line contains the power generation shown in kWh. As already mentioned the intervall-time was programmed to 15 minutes, so 4 kWh stands for a power output of 16 kW. The third line shows the part delivered to grid. Below that the power supply from the grid is shown. In the bottom line the calculated power demand is shown according the formula:

$$\text{Power Demand} = P_{gen} - P_{to\ grid} + P_{from\ grid}$$

The peaks are caused by a bruising mill need to prepare the feed for the pigs. This unit alone needs about 16 kW power, for that reason even on the last day of that week, when the WEC delivered up to 20 kW, there was still a remainder supplied by the grid. If this bruising mill would be a smaller unit with a storage place and controlled by an energy management system, the farmer would have reduced his grid-supply by about 100 kWh in that week.

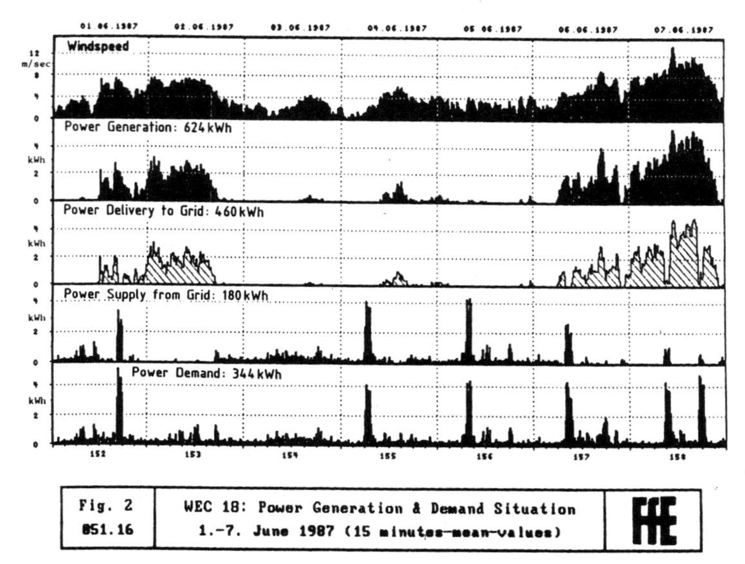

Fig. 2 051.16	WEC 18: Power Generation & Demand Situation 1.-7. June 1987 (15 minutes-mean-values)	FfE

These measured values lead to the energy balance for that particular week shown in Figure 3.

The power generation of 624 kWh per week is not typical for that site. The annual mean value could be exspected in the range of approximatly 900-1000 kWh/week.

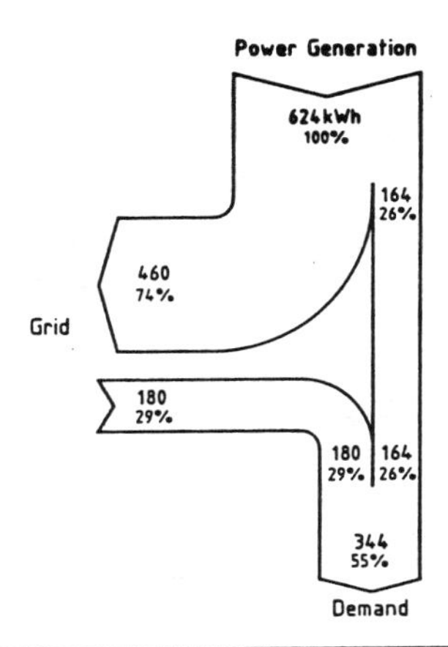

Fig. 3 051.16	WEC 18: Sankey Diagram 1.-7. June 1987 (15 minutes-mean-values)	FfE

AEROMAN Wind Turbine technical data:

Upwind two blade rotor, diameter:12 m
Asynchron Generator, rating: 20 kW
Steel pipe tower, hub heigth: 16 m
Rotor-Speed and Generator-Output: pitch controlled

The used DAS is a small stand alone battery backed up system especially designed for long term measurement periods. Via MODEM the system can be fully controlled via a standard telephone line. For that reason the clocks can be adjusted from time to time. Due to these possibilities we can offer the following plot, which shows data of the week 24.-30.8.1987. Figure 4 contains the Power Generation of four WECs. In the first line the data of WEC18, located in the upper left corner of that area. Beneath that WEC14 located in the upper right corner, followed by WEC8 located on the island Fehmarn. The last one is WEC19 located not far away from the test site of GROWIAN, the probably well known 3 MW german windturbine. The bottom line contains the accumulated sum of these four wind turbines. These data will give an idea about the power cycles which a number of distributed WECs would have delivered to the grid. In reality the real power is to be reduced by the local power consumption.

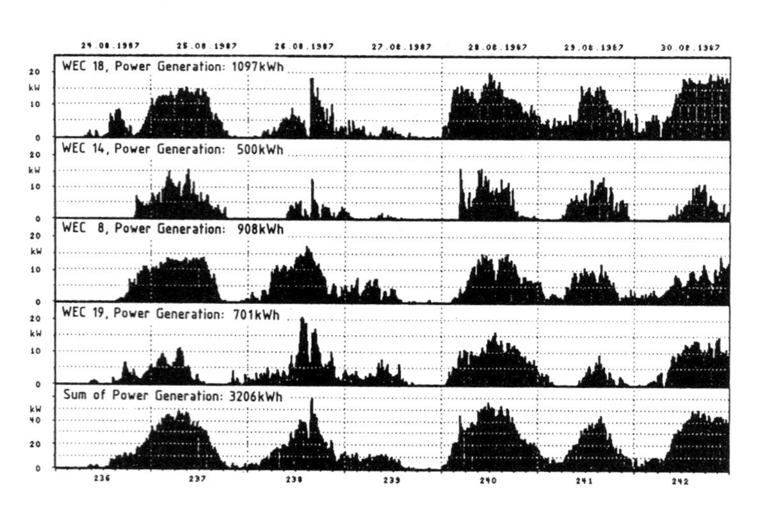

Fig. 4 051.16	WEC 18/14/8/19: Power Generation and sum 24.-30.8.1987 (15 minutes-mean-values)	FfE

The duration time graph in Figure 5 gives a first overview on the effect adding four WECs over a period of one week. More sophisticated statistical analyse will be done when more data over a longer period are available. In this particular week the WEC18 is dominant. All WECs together had been in operation in 95 % of the time, which was reached as well from WEC18 alone. Contrary to that the operation time of WEC14 was only 78 % of the whole period.

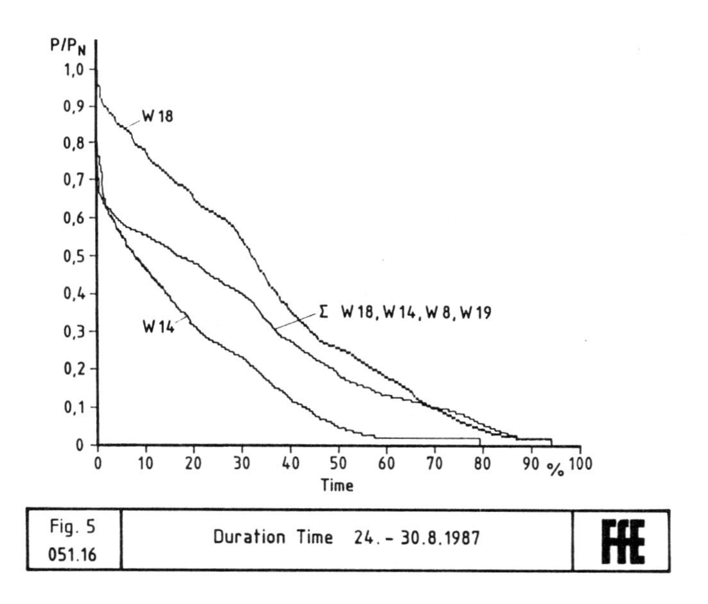

Fig. 5 051.16 Duration Time 24.- 30.8.1987

Conclusion:

This examination, when finished in the end of 1988, will offer severe information concerning the aspects of connecting local distributed WECs to the public grid.

Contrary to theoretical examinations using calculated power output values based on measured wind data , using a typical Power-Wind-Curve, this time real produced Power outputs can be analysed. Due to the remote controlled mode of the DAS the Time intervall can be set to other values, for example down to 1 minute, in order to get a better resolution.

KANO-ROTOR-30kW: DESIGN, CONSTRUCTION & FATIGU ANALYSIS OF A 4-BLADED HORIZONTAL-AXIS WINDTURBINE AFTER NEARLY ONE YEAR OF OPERATION

M. Lührs

KANO ROTOR, Kähler Maschinenbau GmbH
2246 Norderheistedt, Federal Republic of Germany

General overview

KANO-ROTOR is part of the house KÄHLER-MASCHINENBAU-GmbH in Norderheistedt at the countryside of Schleswig-Holstein, 15 miles besides the Northern Sea and so "near by the wind".

KANO-ROTOR developed WEC`s since 1983. In November 1984 a prototype with 30 kW output has been installed in Norderheistedt.

The experience with this windturbine is realized in the new type, the KANO-ROTOR 30 kW machine.

The design of the KANO-ROTOR 30 kW meets the guidelines and rules for wind energy converters set forth by German Authorities called the "Typenprüfung".

The "Typenprüfung" is a quality and safety control given by the "Landesamt für Baustatik", an independant government institution.

Also the technical design is coordinated with the "Germanischer Lloyd Hamburg", a worldwide wellknown independent expert organisation.

The intention of KANO-ROTOR is to produce aerogenerators with less costs, less maintainance and designed in a way, taht the machine is easy to build and to repair.

There a some more eminent points to mention about this machine:

The KANO-ROTOR is designed nearly maintainanceless. Only once a year a inspection is prescribed and maintainance is easy realized because of the well-ordered construction. Many structural parts are approved in other machines, so the brake system for example is a heavy lorry design.

To predict a rated lifetime of more than 20 years and to confirm the calculated loads of the "Typenprüfung" the machine is equipped with an extensive but relative simple measuring system.

The KANO-ROTOR 30 kW was erected on December 23 in the year 1986 and was running nearly 4.000 hours in test- and most of the time in continous operation and a whole host of measuring data achieved us and our customers the certainty to produce a young but matured wind energy converter.

So if you are interested in the results of our efforts contact

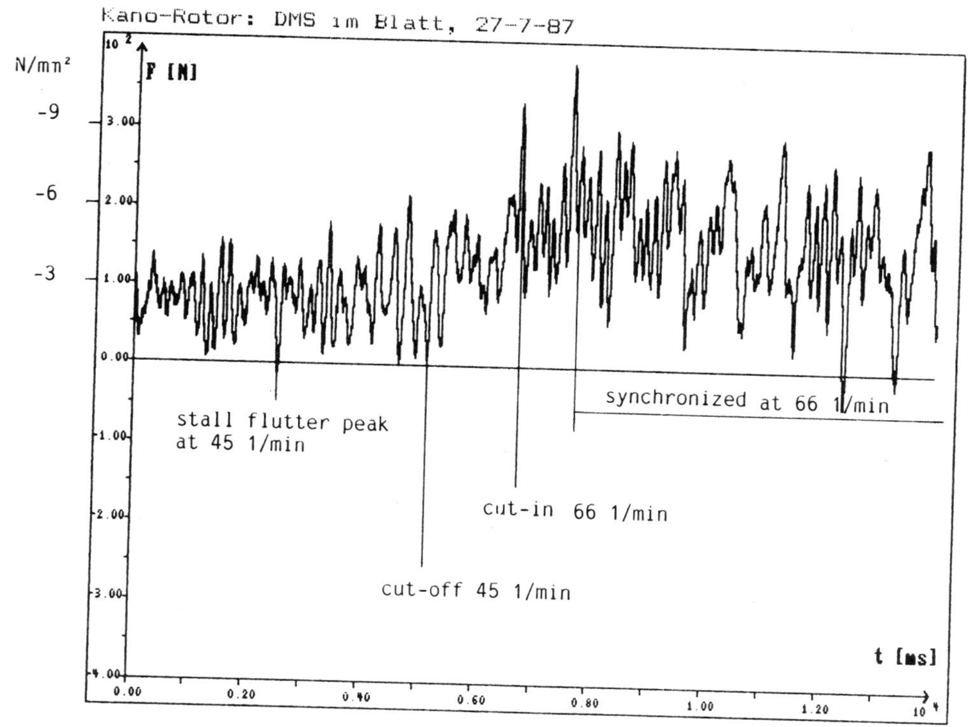

Switchung up from 45 1/min to 66 1/min, windspeed 9 m/s

Achtern Diek Elektronik GmbH

Blade strain gage at r = 4,45 m

Kano-Rotor: DMS im Blatt, 4-8-87

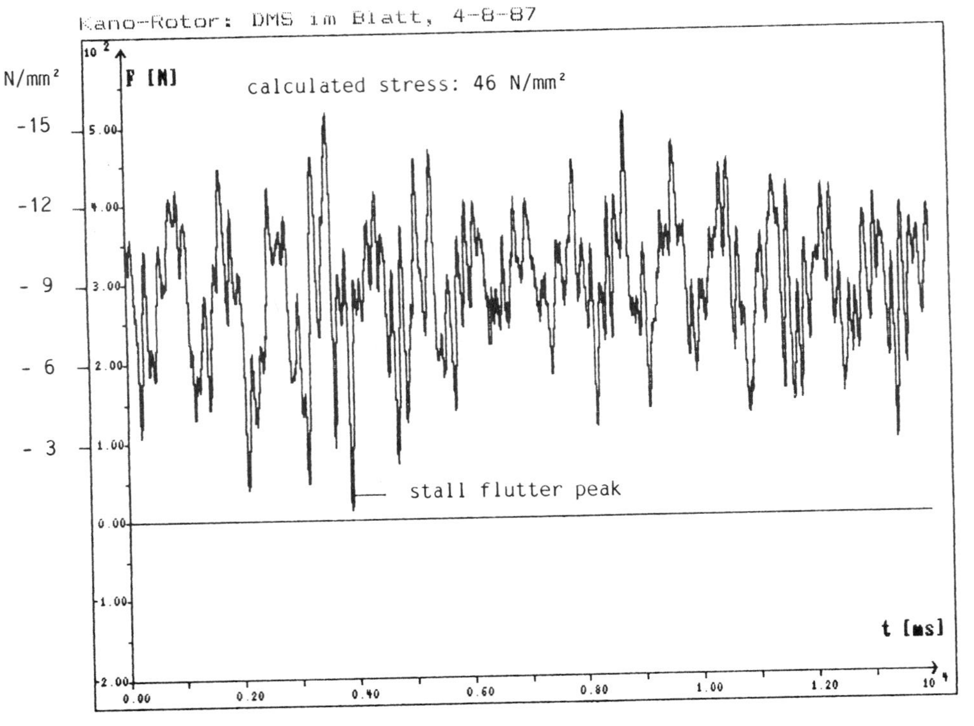

Böenwerte bis 18 m/s, Z:280-284

Gusts up to 18 m/s, stall flutter

Achtern Diek Elektronik GmbH

Blade strain gage at r = 4,45 m

Kano-Rotor: DMS im Blatt, 27-7-87

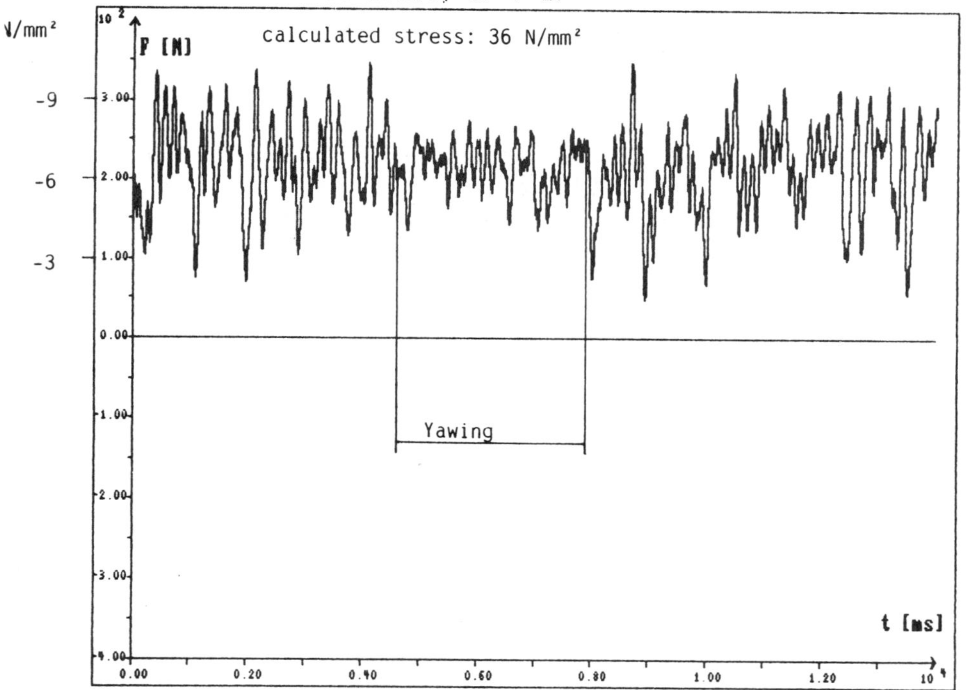

ca 20-25 KW bei 10-11,5 m/s, Z:052-055

Yawing at 20 - 25 kW output, wind-speed 10 - 11,5 m/s

Achtern Diek Elektronik GmbH

Blade strain gage at r = 4,45 m

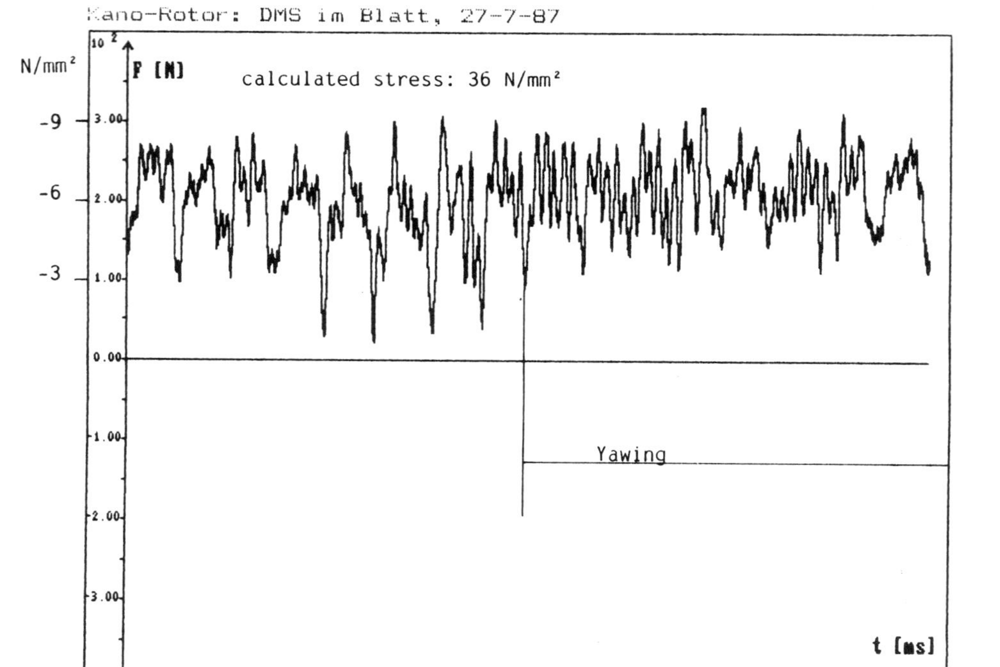

Windnachfuehrung bei 10 m/s, Z:043-046

Yawing at 10 m/s wind-speed

AUTONOMOUS DECENTRAL ELECTRICAL POWER SUPPLY SYSTEM

Dipl.-Ing. G. Cramer

SMA-Regelsysteme GmbH
Hannoversche Str. 3, 3501 Niestetal 1
Federal Republic of Germany

ABSTRACT

A modular system for the supply of remote electrical consumers was developed, which makes possible a variable integration of wind energy and photovoltaic plants in connection with a diesel engine and a battery storage.

The wind energy converters, equipped with asynchronous generators, and a fast pitch control, work parallel with a synchronous generator. The generator is driven by a diesel engine by means of an overrunning clutch, or started by a small DC-motor. If the diesel is off, or the starting process by the DC-motor is finished, the synchronous generator works as a rotating phase-shifter and takes over voltage control and supply of reactive power.

The speed versus power control of the wind energy converters does not only make possible an optimized parallel operation with the diesel-generator unit, but also directly takes over frequency control when operated singly without diesel engine. The location of the wind energy plants does not depend on the site of the diesel engine, because control cables are not necessary.

The PV-generator is connected to the 3-phase AC-bus at any location by means of a converter with MPP-controller.

To avoid too frequent starting of the diesel, a storage battery is installed. A line-commutated converter is incorporated to charge the batteries, and to feed back energy into the grid. It stands out for simple construction and low expense.

The report describes the electrical and control technical design on principle, as well as the experience with the following, already built plants:

- Coupling of two wind energy plants for the supply of water irrigation pumps.
- Combination of two wind energy plants with a short-time battery storage and a diesel power unit for the supply of the Irish island of Cape Clear.
- Combination of two wind energy plants with a photovoltaic generator and a battery storage for the supply of a village in Jordan.

INTRODUCTION

During the last years, the installation of wind energy converters for the supply of remote electric consumers not connected to a public grid, as well as the combination of wind energy converters with photovoltaic generators, battery storages, and diesel power sets, has proved to be a very interesting application for small wind energy plants.

Due to the high cost of generating electric power in decentral power supply systems using diesel generator units, an economical operation of small electric power supply systems using the power of wind is already possible nowadays.

Based on six years of experience in developing, manufacturing, and operating stand alone systems, SMA has developed several Advanced Autonomous Electrical Power Supply Systems, which, in the meantime, are successfully installed in several versions.

WIND PARK IN ISOLATED OPERATION

The parallel operation of several wind energy converters for the supply of greater consumer power should, if possible, be carried out with wind energy converters having a fast speed and power control by a variation of blade pitch angle.

Besides an increase of availability of the compound system, the installation of wind parks has the great advantage, that the short-time variations of output power due to wind energy offer variations (in the seconds range) are better balanced.

Examples for the coupling of two wind energy plants in isolated operation are the project "Small Stand Alone Wind Farm for Water Pumping", which shall be installed by SMA on order of the RSS in Jordan in September 1987, and a project on the Cape Verde Islands.

The wind parks are designed for two wind energy converters. A third converter may be installed later if required. The wind energy converters are standard-type versions for grid operation with asynchronous generators. The rated power of each converter is 20 kW resp. 30 kW. Each plant has a speed and power control by means of a fast-working pitch control. Voltage control is carried out by a rotating phase-shifter with a rated power of 40 kVA, which stands on the floor. For starting after dead calm periods, the phase-shifter is driven up to nominal speed by means of a small DC motor. In this way it can build the grid conditions necessary for the wind energy converters, so that they can start up. If one of the plants has connected its generator to the grid, the DC motor is switched off and mechanically disconnected by means of an overrunning clutch.

The commands for the connection of the several pumps and for starting the rotating phase-shifter are given by a consumer control. The logic is based on a compact microprocessor system, so that all switching criteria may be altered even at the plant's location.

The system is shown in Fig. 1.

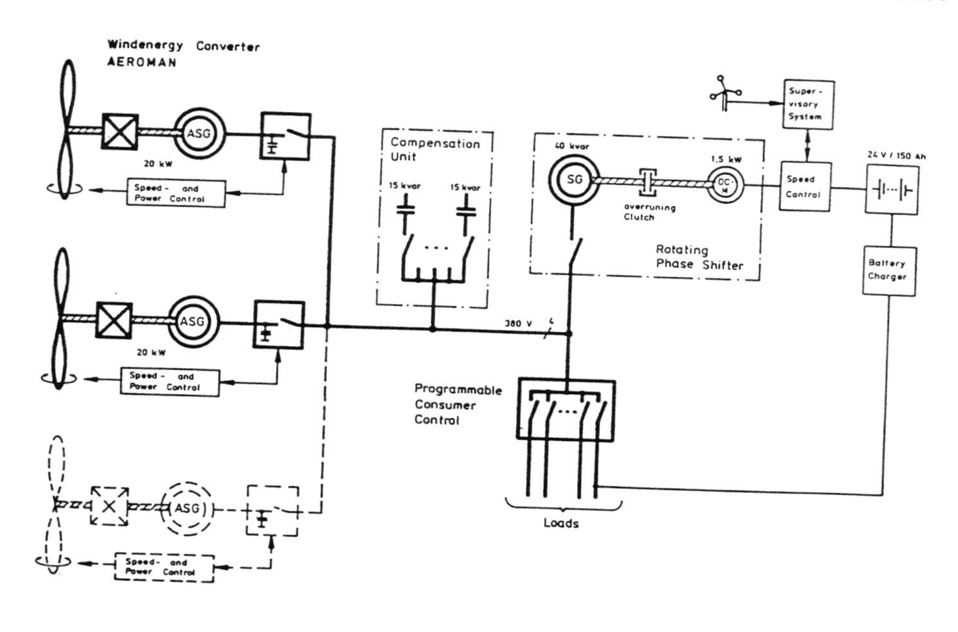

Fig. 1: Block diagram of a wind park in isolated operation (e.g. for water pumping)

WIND/PHOTOVOLTAIC/BATTERY-COMBINATION

The enlargement of the wind park described before by a solar generator and a battery storage is useful, if a supply of essential consumers shall be maintained even during dead calm periods.

The small village Jurf El-Daraweesh situated in the desert between Amman and Aqaba shall be supplied by this system.

The supply system consists of 2 wind energy converters of the AEROMAN-type with a nominal power of 25 kW each, a photovoltaic generator with a nominal power of 15 kW_p, and a battery system with a capacity of about 280 kWh, which is coupled by means of a converter with the 3-phase AC-bus (v. Fig. 2).

WIND/DIESEL/BATTERY-SYSTEM

If a safe supply shall be maintained, a diesel engine and a battery storage are necessary for the supply during dead calm periods. The use of a battery for long-time storage is not economical, that's why the battery system in the following wind-diesel-conception is installed only as short-time storage.

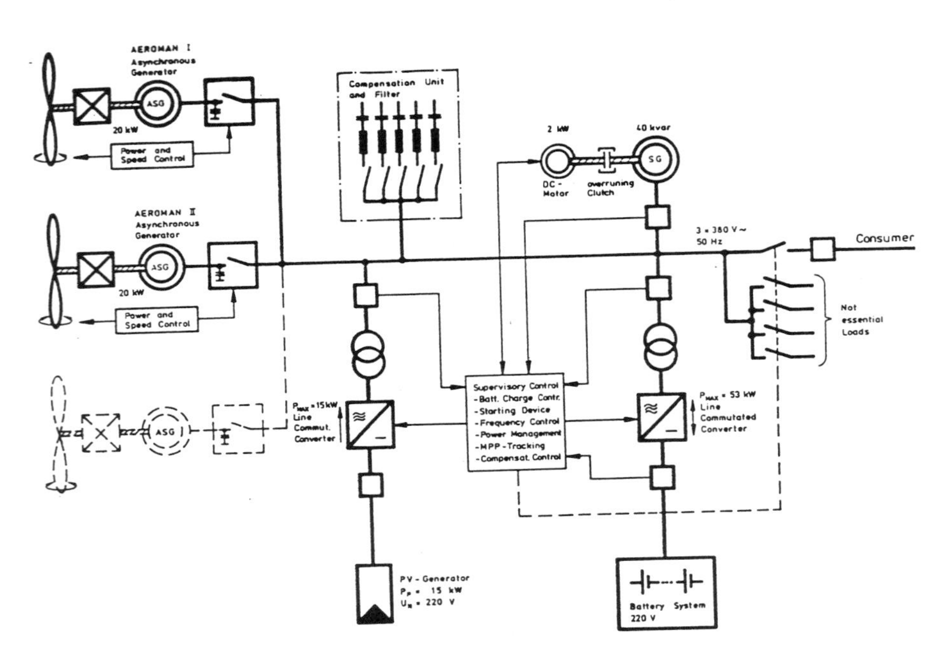

Fig. 2: Block diagram of the wind/photovoltaic/battery-combination for a village supply system

This design makes possible a supply even without running diesel engine during good wind conditions or during low-load periods. The diesel engine is equipped with an electrically controllable clutch or a mechanical over-running clutch instead of a fixed coupling of motor and generator, so that the synchronous generator works as a rotating phase-shifter when the diesel is off, and takes over voltage control.

To avoid an unnecssary frequent starting of the diesel engine caused by changes of wind-speed or consumption, the system is equipped with a battery storage. The capacity of this storage is small (appr. rated power for about 30 minutes), and depends on the local conditions and the consumer's demands. For the charging of batteries and to supply the isolated grid, a line (machine)-commutated converter is installed, which stands out for simple construction and low price.

Furthermore, the installation of a storage unit makes possible operation of the diesel engine in more suitable power ranges for a longer time, because e.g. load peaks, which would make necessary starting of the diesel, can be checked by the storage unit. Because load peaks exceeding the rated power of the diesel normally occur only for short times, the diesel engine may be dimensioned smaller. The load peaks are checked by the battery storage then.

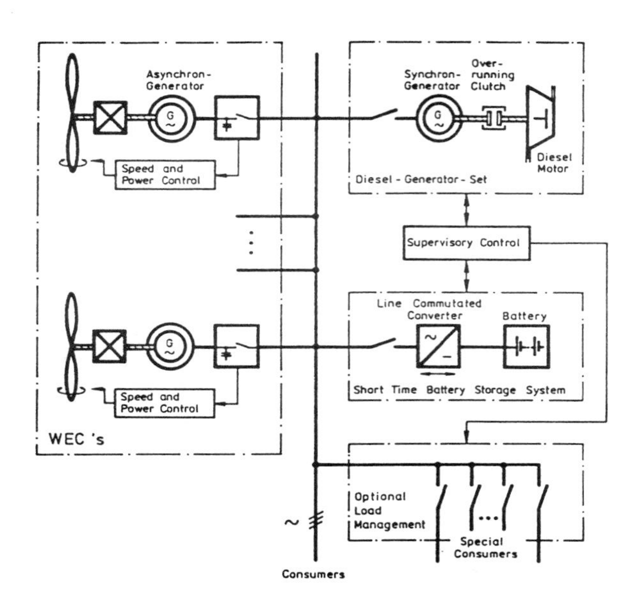

Fig. 3: Simplified block diagram of the modular system for an autonomous electrical power supply (with static converter)

The Cape Clear Project

The feasibility and economy of the plant should be demonstrated by the installation of the Modular Autonomous Electrical Power Supply System on the small Irish Island of Cape Clear. This project is carried out in the frame of a demonstration project of the Commission of the European Communities, and is additionally supported by the German Federal Ministry for Science and Technology. The Irish Department of Energy finances all infrastructural measures on the island. The project management on German side is taken over by SMA Regelsysteme GmbH, whilst the project coordination in Ireland is carried out by the National Board for Science and Technology. The supply system is operated by the local Coop organization on behalf of the Irish Electricity Supply Board.

The system consists of two wind energy plants AEROMAN, manufactured by MAN Technology, with a nominal power of 30 kW each, a MAN diesel-generator set with a power of 85 kVA, and a battery storage manufactured by Hagen Batterie AG with a capacity of 100 kWh.

The island of Cape Clear is located south of Ireland; it has about 300 inhabitants. The wind conditions are very good with an annual average wind velocity of 7.8 m/s.

Formerly electrical power was generated by two diesel generators with a nominal power output of about 100 kW each, and was distributed by means of a 3-phase 10 kV overhead line grid. The yearly power consumption is approximately 250 MWh; the daily load curve has its minimum at about 15 kW in the early morning, and its maximum at 80 to 120 kW in the evening.

The two wind energy plants were installed on the 162 m high hill of the island, because here the power yield is several times larger than on sea level. The distance between the diesel power station and the wind energy plant is about 2.5 km. The diesel generator set and the battery storage with inverter are built up near the existing diesel power station in the island's harbour.

The complete system is in operation since August 1987.

Fig. 4: View of the wind energy converters AEROMAN on Cape Clear

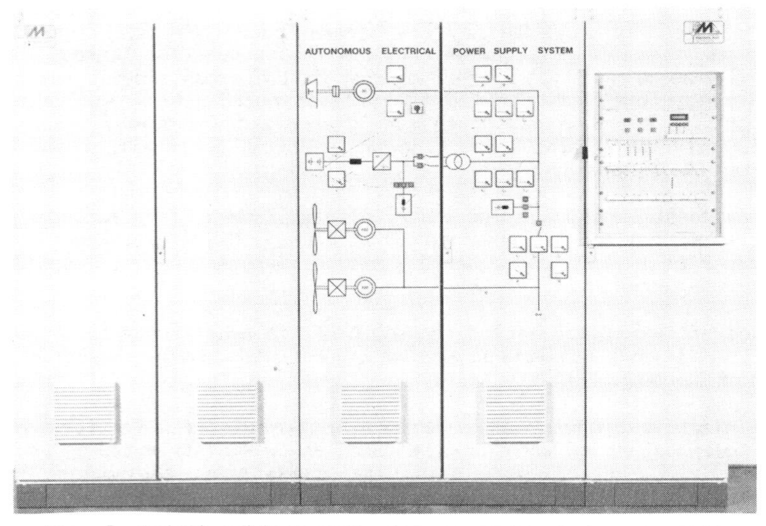

Fig. 5: Switch cabinet unit with converter and supervision

WIND ENERGY WATER PUMPING IN JORDAN

RIZEQ TA'ANI

HEAD OF THE WIND ENERGY SECTION
SOLAR ENERGY RESEARCH CENTER
ROYAL SCIENTIFIC SOCIETY
P.O. BOX 925819 / AMMAN - JORDAN

ABSTRACT

In this paper; Jordan experience in introducing wind energy technology for water pumping from different types of wells is described.

Different wind pumping systems are utilized to transfer the technology, to modify it according to the specific needs of Jordan, and to gain technical know how to enable Jordan to build its industrial capabilities in this field. Preliminary results will be presented.

KEY WORDS

Mechanical wind pumping system, Electrical wind pumping system, coupling of two or more wind converters, submersible pumps.

1. INTRODUCTION

In the frame work of a Research and Development Programme between the Royal Scientific Society (RSS), Jordan Water Authority (WA), and the Federal Ministry of Economic Cooperation (BMZ) through the German Agency for Technical Cooperation (GTZ), a project for the introduction of wind energy technology is defined.

The main objectives of this project are:

1. Introducing wind energy technology for water pumping.
2. Determination of the potential applications of such technology in pumping from shallow and deep wells.

3. Test, modification, and evaluation of different wind energy pumping systems.
4. Determination of the technical and socio-economical aspects of such technology.

To fullfill the objectives of this project, different pumping systems were acquired, installed and tested. The main tangable result of this project is the development of a local made mechanical wind converter for water pumping. In this paper the main results are presented.

2. WATER RESOURCES

RSS group [1] conducted a survey to identify the characteristics of water wells in Jordan. The survey showed that Jordan has water wells possessing a variaty of depths and yields. Looking at the appropriate technologies, it is clear that most of Jordanian wells can be equipped-one way or another-with solar and wind energy systems.

The total minimum power that is required for pumping at this point is estimated to be about 10 MW.

3. WIND ENERGY RESOURCES

An extensive study was made by RSS to evaluate the wind data available by Jordanian Institutions [2]. This study determine roughly the wind energy potentials. It highlighted furthermore the necessaty of wind speed measurements at 10-m Heights for different locations representing various regions and topography. These data will be augumented by installing a battery powered data logger capable for determing the 10-minute average values of wind speeds and thus a complete historical and wind distribution data will be available for 16 stations at the end of 1987.

4. DEVELOPMENT OF A TEST PROGRAMME

A programme for all test activities for the research and development phases was prepared. This programme includes test procedures for the mechanical as well as for electrical wind pumping systems.

Recommended practices for wind turbine testing and evaluation encompassed all aspects of a wind energy conversion system ranging from energy production, quality of power, reliability, durability and safety through cost effectiveness, noise characteristic, etc. are developed from expert groups and presented in [3].

5. WIND ENERGY PUMPING SYSTEMS

In this section the experience gained and the results obtained from the different pumping systems will be presented as follows:

A. Mechanical wind pumping system (KIJITO - KENYA).

The system consists of the KIJITO wind energy converter, 55 m^3 storage tank, and the necessary piping system. The functional test showed that this pump is functionally accepted.

Tests were made to obtain the cut-in wind speed, the Q-V performance curve, and the long term behaviour. Ta'ani et.al reported detailed results in [3]. Fig. (1) shows the experimentally determined Q-V curve. It is obvious from this figure that the rated wind speed is 8.5 m/sec and the absolute maximum 10-minute average value for the flowrate was 5.895 m^3/h.

Figure (2) presents the correlation between the monthly mean wind speed and the annual totals of pumped and consumed water. The annual pumped water was 14933.5 m^3, while the annual total of consumed water was 16702 m^3.

B. Bruemmer Medium Technology Electrical Wind Converter

The system consists of the Bruemmer wind converter, 55 m^3 storage tank, 3 submersible pumps, consumer control and the necessary piping system. The 10 kW wind converter is a down wind machine with a mechanical pitch control for speed regulation. The functional test showed that the wind converter had problems in the turn-table design and the adjusting mechanism of the blades.

C. M.A.N. High Technology Wind Converter

The system consists of the M.A.N. AEROMAN 12.5/14 kW wind converter, 55 m^3 storage tank, 3 submersible pumps (Grundfos SP 8-15, SP 16-8 and SP 16-8), consumer control and the necessary piping system.

The wind energy converter has a rotor diameter of 12.5m. Speed and power are controlled by a fast rotor blade adjusting device (pitch control) with a synchronous generator for stand alone operation mode. The wind energy converter is connected to a consumer control, which switches the loads (in our case the pumps) according to the availability of wind speed.

The functional test showed that the wind converter is functionally acceptable. Only minor problems concerning oil leakage, voltage control cards and some switches in the electro hydraulic control units were recognized. The RSS team and MAN personnel are working together to solve these problems.

Tests were made to obtain the power curve, Q-V curve and other relevant information. Figure (3) shows the experimentally determined P-V curve, and the Q-V curve. It is seen from this figure that the rated wind speed is 13.5 m/sec, the absolute maximum power equals to 12.5 kW, the absolute maximum flowrate is 32.5 m^3/h. Figure (4) presents the 10 minute mean value parameters for one day operation. It is concluded from the results, that this machine is of high technology, efficient, and is suitable for areas with high wind profile.

D. Development of a Local Mechanical Wind Pumping System

Based on the experience gained from the Research and Development Programme, a mechanical wind pumping system was developed at the RSS. The system utilizes materials and technical know how available in Jordan. The wind converter has a rotor diameter of 7.75 m and is now under extensive tests to develop it according to the wind regime available in the country.

ACKNOWLEDGEMENT

The Royal Scientific Society and the Author express their sincere gratitude to the Federal Ministry for Economic Cooperation (BMZ), and to the German Agency for Technical Cooperation for all their help and support throughout this endeavour without which this work could not have been possible.

REFERENCES

1. Ta'ani, R. et. al (1983). Assessment and Analysis of Basic Energy Needs to be supplied by Solar Energy: Royal Scientific Society publication (3) 83(31).

2. Mulki, H. et.al (1983). Assessment and Analysis of Available Energy Resources: Royal Scientific Society publication (3) 83(32).

3. Frandsen, S. (1982). "Recommended practices for wind turbine testing and evaluation", Riso National Laboratory, DK 4000 Rokilde, Denmark.

Figures

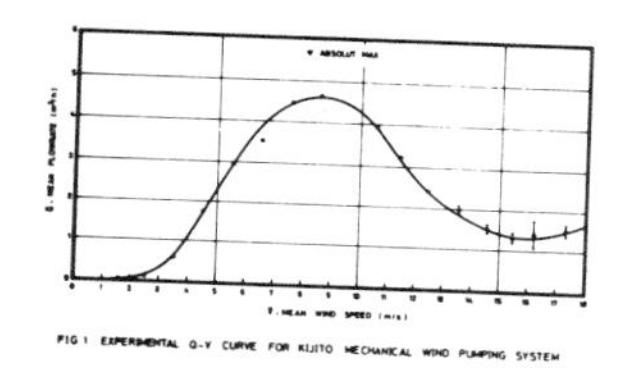

FIG 1 EXPERIMENTAL Q-V CURVE FOR KIJITO MECHANICAL WIND PUMPING SYSTEM

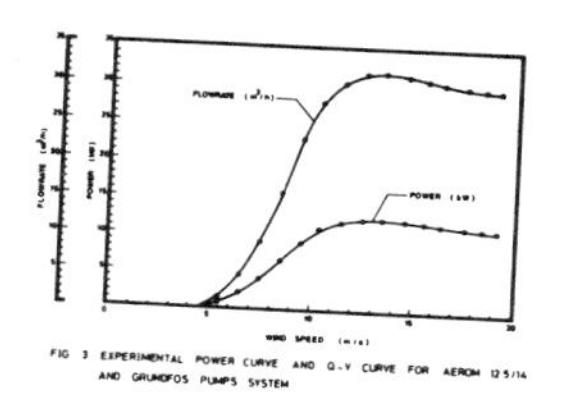

FIG 3 EXPERIMENTAL POWER CURVE AND Q-V CURVE FOR AEROM 12.5/14 AND GRUNDFOS PUMPS SYSTEM

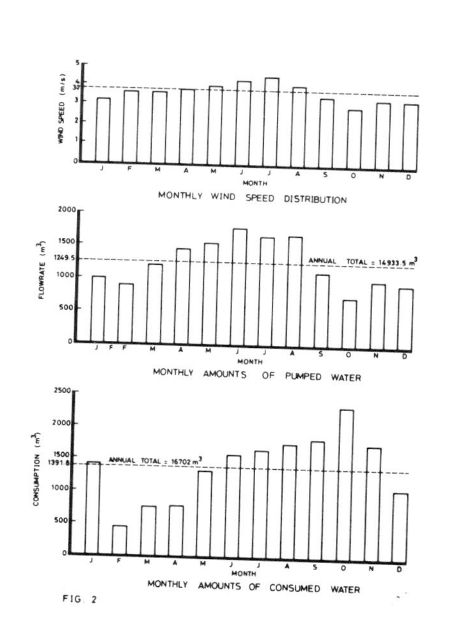

FIG 2

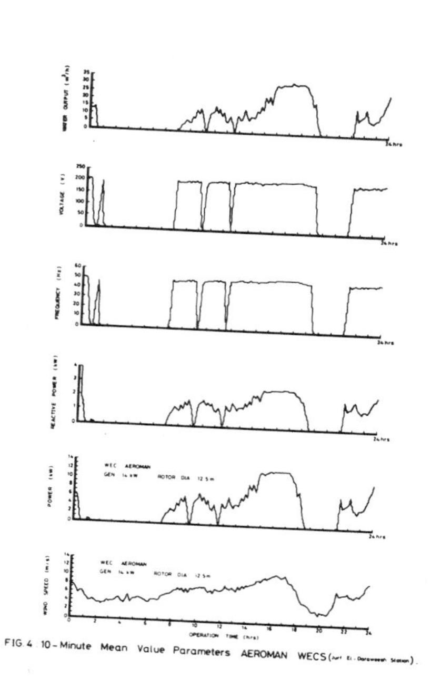

FIG 4 10-Minute Mean Value Parameters AEROMAN WECS (...).

COST ASPECTS TO BE CONSIDERED WHEN PLANNING WIND ENERGY PARKS

Wolfgang Maier
Manag. Director

FICHTNER Development Engineering
Sarweystr. 3 D- 7000 Stuttgart 1
Tel. 0711/8995-225 Tlx: 72 2360 2

ABSTRACT
Two different types of cost have to be considered when planning a wind park: Proper costs of the wind energy converter and costs of the peripherical installations. The paper discusses possibilities for cost reductionof both types. For cost comparison, specific costs should be related to energy yield rather than to rated power.

KEY WORDS

Optimal configuration of wind parks; fraction of peripherical cost; cost optimization related to annual energy yield.

An investigation, initiated by the German Federal Ministry of Economics gives an estimate of the potential for wind energy converters in the coastal region of West Germany up to the year 2000. The study anticipates an installed capacity of about 1700 MW and an annual energy yield of approximately 3800 GWh/a which is an average value between the best (4600 GWh/a) and the worst (3200 GWh/a) scenario assumed in the study.

Although this represents only around 1.1 % of the electricity generation the Federal Republic in 1985 it is still a large number, especially when considering the positive side-effects on employment in the coastal region and on the reduction of pollution.

Yet the yields quoted seem rather optimistic. The question arises as to what can be done to ensure that the assumptions are certainly optimistic but not utopian.

In the first place this is a question of cost reduction. When dealing with the costs of electricity generation from wind energy, two groups of cost have to be considered: the costs of the wind energy converter itself, and the costs of the infrastructure and supplementary measures. This second group amounts to as much as 50 to 70 % of the costs of the wind converters proper. It comprises construction works (both below-ground and surface), connection to the grid and control facilities. An important additional aspect is the granting of offical administrative approvals and permits for erection. Some possibilities for reducing the costs of this second group may be mentioned, i.e.

- reduction of the infrastructure to the minimum required for satisfactory operation
- the use of standardized elements in the design
- exploitation of steps which have to be repeated when erecting plants for different operators at various locations, e.g. the stress analysis calculations
- clear definitions of interfaces, unequivocal demarcation of competencies
- setting up of an organisation bearing overall responsibility for the effective teamwork of all participants
- quality assurance in all phases.

Particular attention is drawn to the learning effect when repeating permit application procedures. The establishment and cultivation of contacts and the exchange of experiences between regulatory authorities with responsiblities for various technical aspects and for different locations are among the long-term objectives for achieving a certain degree of standardization of the procedure. This applies, for instance, to safety aspects, for which it is not necessary to start from scratch for each new location, if they have already been treated thoroughly on a sound technical basis and with proper corroboration elsewhere.

It is clear that this represents an important field of activity for engineering consultants, and one which also benefits considerably the implementation of such projects. Such an organisation will accumulate engineering, construction and operational experience from many installations, will generalize this know-how and will then apply it further over a wide range of projects.

At first sight, optimization of the costs of the wind converter itself seems to be exclusively the domain of the manufacturer. This, however, is not quite true. Of the wind parks in which our company - FICHTNER Development Engineering - is involved as consultant, two solutions which are completely opposite in character may be mentioned as examples. These are the Wilhelmshaven Wind Park operated by Jade Windenergie Wilhelmshaven GmbH, and the Cuxhaven Wind Park operated by Überlandwerke Nordhannover AG.

The essential difference becomes apparent just by considering the following site plans - Fig. 1. It shows up even more in the technical data given in Fig. 2.

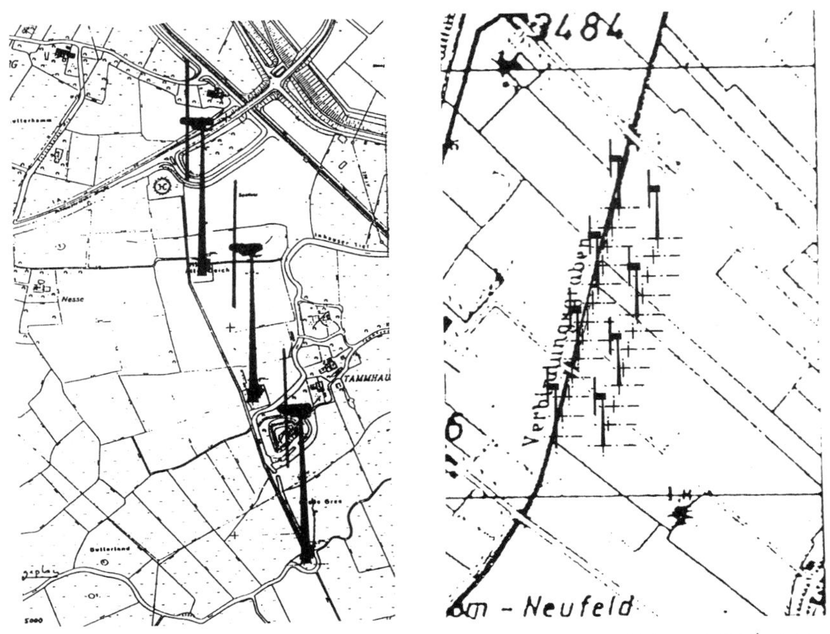

Fig. 1 Site plans of Windpark Wilhelmshaven (left) and Cuxhaven

Name of Windpark	JWE Wilhelmshaven	ÖNH Cuxhaven	
Total capacity (MW)	1,92	1	
Number of WEC's rated power per unit	3 x 640 kW	25 15 x 30 kW + 10 x 55 kW	
Rotor diameter (m)	56	15	16
Height of hub (m)	60	20	22
Speed of rotation (r.p.m.)	32 - 43	117	20 - 48
Wind velocity start (m/sec) (m/sec)	5	5	3
Wind velocity shut-down	16	16	30
Grid conncection	Synchron. Generator (frequency converter	Induction Generator	Synchron. Generator
Theoretical annual energy yield (MWh/a)	5850	1130	870

Fig. 2 Technical data of Windparks Wilhelmshaven and Cuxhaven

It is indeed a rather difficult optimization problem to decide on the number and type of converters for cost minimization, or rather cost optimization, for a specified total capacity.

The task is rendered even more difficult due to the fact that it seems inevitable that converters with a higher rated power output in the range of 0.8 to 2 MW will be developed, to replace those of 50 to 200 kW which is today's state of the art. Such big converters do, in fact, already exist, but so far only in a very small number which does not yet provide a sufficient basis for either long-term operating experience or cost reduction by series manufacture.

It has further to be kept in mind that for a wind energy converter, costs cannot be expressed with reference to rated power or maximum capacity.

The frequency distribution of wind velocity at a given site has to be considered as well. This defines, together with the technical properties of the converter, the maximum annual energy yield, and specific costs have to be seen in relation to this figure rather than to the rated power.

Thus the factors determining the optimum costs are the power output as a function of wind velocity, minimum and maximum wind velocity for startup and shutdown and the possibilities for optimizing energy yield during periods of frequent changes of wind velocities. As an example, curves from two different converters are shown in Fig. 3.

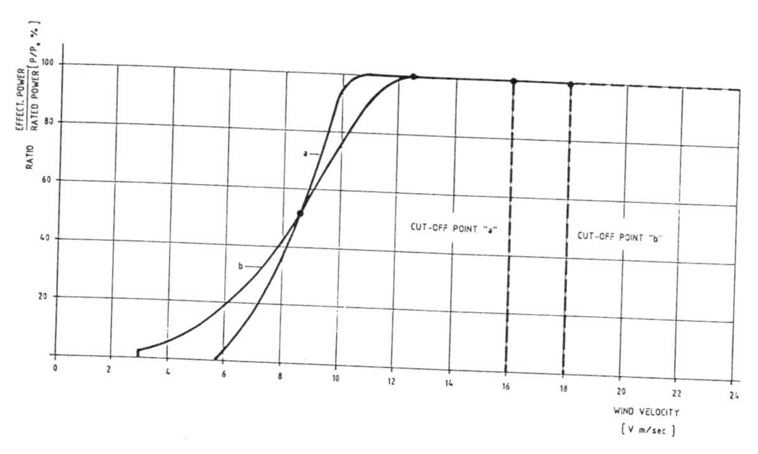

Fig. 3 Energy yield as function of wind velocity for 2 different types of converters

It can be easily imagined, that, assuming equal costs per kW for the two converters in question, in areas where high wind velocities are more frequent, curve "a" gives the greater energy yield per year, and consequently a more economic performance, whereas curve "b" gives better results in areas where low and medium wind velocities are more frequent.

So far for the two parks mentioned above only theoretical calculations performed on a computer are available to answer the question of the maximum annual energy yield.

It will, however, be of considerable interest to gain a stock of practical experience on the basis of measurement programm, once the different configurations of parks come on stream. I am sure, that in due time a report on the results of such comparative long-term measurements will be presented.

A COMPARATIVE TECHNOECONOMICS OF STANDA-LONE WIND ELECTRIC CONVERTERS (WECs) EXISTING IN INDIA

J.V. KAUDINYA

BIRLA ENGINEERING SERVICES
3RD FLOOR, GURU ANGAD BHAVAN
71, NEHRU PLACE, NEW DELHI
INDIA - PIN 110 019

1. ABSTRACT: The poor progress due to scarcity of electricity in most of populated and windy parts of India spurred the interest of the Indian Govt. to develop wind farms in potential areas. Out of so many windy areas of the country, DNES and State Governments have set-up at a number of places, wind electric converters, connected to respective State Grid as well as also in stand-alone operation. The stand-alone WECs supplement a certain local demand by their own individual power conditioning units. No doubt, this makes the machine a bit costlier but altogether increases the degree of its usability and hence the importance. The present communication deals with a comparative technoeconomics of all the three existing (so far) stand-alone WECs in India. These are from Windane of Denmark, Wind Power of Holland and W.E.D. of West Germany. All the three, in the range of 20-30 KW rated, have been compared with one of the most competitive WECs monopteros-15 from MBB of West Germany of almost same power range. The analysis highlights the comparative technical and economic considerations of monopteros-15 of MBB to make it a useful introduction to every body interested.

KEY-WORDS: Technoeconomics, Technical features, CIF cost, power curve, cost of energy per unit (CEPU).

2 INTRODUCTION

The overall technoeconomic viability and its effective operational cost are the major governing factors to the success of marketing any durable engineering system. Similarly for the wind electric converters the following specific factors are more deciding in worth of the system as a whole :

1. Cut in wind velocity must be as low as possible.
2. Rated wind velocity must also be optimally low but should enable the system to extract maximum power out of it.
3. Design of turbine must work efficiently in a wide range of wind velocity.
4. System must be least bulky but durable and easy in installation.
5. Technology must be simple to indeginise.
6. Power production of WEC must be as high as possible.
7. Total effective cost (capital + operational) of the entire system in running condition should be satisfactorily low.
8. A reduction in efficiency should be quite acceptable if the resultant reduction in capital cost is greater than the resultant loss in power.

All the above conditions have been accounted in the present assessment of the existing stand alone WECs and form Table-1, as one tool, which shows the comparative technical features and capital CIF costs of the machines. The second tool to make a comparison is, the power curves of these machines which has been shown by fig.1. These two tools (Table 1 and Fig.1) lead this study to the following comparative analysis.

3 COMPARATIVE ANALYSIS

Table 1 shows that all these stand-alone WECs are almost in the same power range, but with different technical features & weight. All the first three machines are relatively quite heavier than the fourth,Monopteros-15. All are microprocessor based automatic units with synchronous generators but may be with different reliability. The Windane-12 machine has been working satisfactorily at Mandvi since 1984. While the WPH-15-3-25 and WE1501 have just been introduced.

Fig.1 shows that although WPH and WED machines are of a higher rated power (30KW), but MBB's Monopteros-15 works better at a lower band of wind velocities. The curves show that certainly

Monopteros has an edge over the Windane machine with a lower rated wind velocity.

Table 2 depicts the same result as that from Fig.1. At 10 m/s, the annual power generated by monopteros and Windane are almost same but at a low velocity like 4 m/s, the annual power generated by former is about 3½ times more than the latter. Similarly at other mostly existing wind velocities, the former performs better than latter, but eventually bows before the other two machines of 30 KW rated.

With the help of Table-2 and Table 3, the cost of energy per unit (CEPU) KWH for each machine has been derived and figured in table 4. This table depicts that the CEPU with Windane is three times higher than that with Monopteros at 4 m/s. Surprisingly at 5 m/s again it is lesser than all of the rest three. On and after 6 m/s, all these generate a lesser CEPU than the Monopteros.

The analysis is based on the available data in the pamphlets and cost of machines of the respective companies and can only be considered a simple comparison.

4 CONCLUSION

The present study concludes that Monopteros-15 certainly performs better than Windane of the same order at only wind velocities between 4 to 5.5 m/s, which is the most existing wind velocity at most of the windy parts of India. The other two machines perform better than the rest two because of their higher rated power. The apparent thing with Monopteros which disfavours it, is its higher cost than all the three existing machines. But, even then, its attractive power curve makes it a competitive choice among all these. Moreover, its lesser weight and robust design add in its bank of assets and make it a technoeconomically viable system to India.

ACKNOWLEDGEMENT

The author is grateful to Mr. S.P. Singh and Birla Engineering Services for the financial support extended to him to carry out this study and to enable him to present this paper in the Solar World Congress 1987, Hamburg, West Germany.

Author is very much thankful to Dr. Raja Ahamar, DNES, Govt. of India for providing the sufficient time to the useful discussions at various occasions.

TABLE 1 COMPARATIVE FEATURES OF STAND-ALONE WECs WITH SYNCHRONOUS GENERATOR

S. No	Mnfrs' Name & Machine Model	Executing Agency & Work Site in India	Machine Rated Power KW	Rotor Dia. Mtr	Hub height	Performing Wind velocity, M/s			Weight of machine Kg				Apprx CIF cost* in Rs. for 1 m/c (Base Aug.87)
						Cut-in	Rated	cut-off	Rotor	Tower	Nacelle	Total	
1	Windane Windance-12	Gujrat Wind Farm Ltd Mandvi	22	12	18	4	13	none	500	1500	750	2750	405000/-
2	Wind Power Holland WPH-15-3-25 SW-500-110	Himachal Pradesh Electricity Board	30	15	18	4	10	25	-	-	350	2000	466700/-
3	W.E.D. West Germany WE-1501	Mizoram State Govt. Power Deptt, (to be erected)	30	15	15	4.5	10	-	-	-	-	-	455820/-
4	M.B.B. West Germany Monopteros-15	-	25	12.5	15	3.5	12	20	40	800	700	1540	479700/-

*The sources of the respective costs are the representatives of the principal companies, which is still could not be authenticated because of various reasons. Notably these identical costs do not include any customs/import duties.

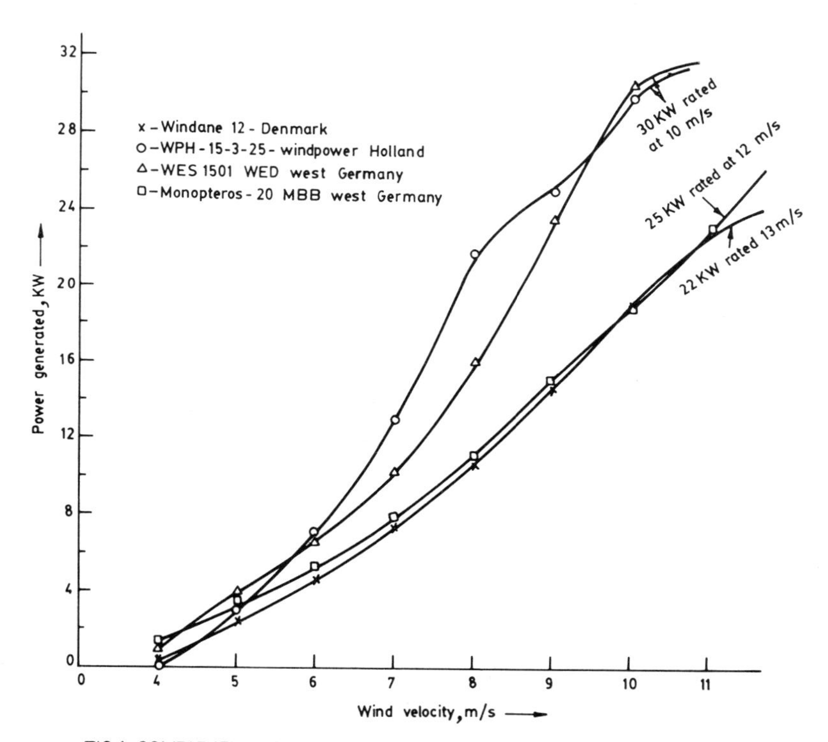

FIG.1.COMPARATIVE PERFORMANCE OF STANDALONE-WEC$_S$

TABLE 2 DIURNAL AND ANNUAL POWER GENERATED IN KWH AT DIFFERENT WIND VELOCITIES

Power Generation	S. No	Machine Model & Rated Capacity	Hourly Averaged Wind Velocities, m/s						
			4	5	6	7	8	9	10
Diurnal	1	Windane-12 Danmark, 22KW	0.4	2.4	4.6	7.4	10.6	14.5	19.0
	2	WPH-15-325 Holland, 30KW	0	2.9	7.0	13.0	20.8	25.0	30.0
	3	WES-1501, West Germany, 30KW	0.8	2.8	6.6	10.1	16.0	23.7	30.4
	4	Monopteros-15 MBB, West Germany 25KW	1.4	3.4	5.2	7.8	11.0	15.0	19.0
Annual	1	Windane-12 22 KW	3504	21024	40296	64824	92856	127020	166440
	2	WPH-15-3-25	0	25404	61320	113880	182208	219000	262800
	3	WES-1501	7008	24528	57816	88476	140160	207612	266304
	4	Monopteros-15	12264	29784	45552	68328	96360	131400	166440

TABLE 3 OWNERSHIP & OPERATING COST OF WECs PER YEAR IN RUPEES

S.No	Machine Model & Capacity	Capital Cost	Interest 8.8%	Depreciation 5%	Operational cost 3%	Maintenance charges 3%	Total cost of m/c per yr.	Ownership cost per year	Oprtg. costper year
1	Windane-12 22KW	405000	35640	20250	12150	12150	80190	55890	24300
2	WPH-15-3-25 30KW	466700	41069	23335	14001	14001	92406	64404	28002
3	WES-1501 30KW	455820	40112	22791	13675	13675	90253	62903	27350
4	Monopteros -15, 25KW	479700	42213	23985	14391	14391	94980	66198	28782

TABLE 4 COST OF ENERGY GENERATION PER UNIT (CEPU)KWH IN RUPEES AT DIFFERENT WIND VELOCITY

S No	Machine Model & Capacity	Hourly averaged wind velocity, m/s						
		4	5	6	7	8	9	10
1	Windane-12, 22KW	22.88	3.81	1.99	1.24	0.86	0.63	0.48
2	WPH-15-3-25 30KW	-	3.63	1.51	0.81	0.51	0.42	0.35
3	WES-1501, 30KW	12.88	3.67	1.56	1.02	0.64	0.43	0.34
4	Monopteros-15,25KW	7.74	3.18	2.08	1.39	0.98	0,72	0.57

MONOPTEROS 15 WIND-REVOS
WINDDRIVEN SEAWATER DESALINATION UNIT

Dipl.-Phys. Gerhard Huppmann
MBB, P.O. Box 801109, D-8000 München 80

Harnessing the Wind for Drinking Water in Egypt

MBB's wind energy converter MONOPTEROS 15 has been in operation since the end of last year, pumping water in Marsa Matrouh, a small coastal town in the north of Egypt.
This smallest member of the MBB wind energy converter family has a rated output of 25 kilowatt and exploits the wind, a renewable power source, to generate electricity.
According to the actual power requirements of the connected consumers a diesel generator gets support automatically the WEC to ensure continuous service security. If enough wind is available the diesel generator stops.

This WIND DIESEL POWER GENERATING SET MONOPTEROS 15 has been connected in June, 87 with an adapted seawater desalination unit on the basis of reverse osmosis, called WIND-REVOS, by Kraftanlagen Heidelberg AG.

The MONOPTEROS 15 WIND-REVOS operates in a fully automatical way and needs only chemicals for water pretreatment and finishing.

The WIND-REVOS produces an amount of approximately 25 cubicmeters fresh water per day from seawater under continuous operation and the expected annual fraction of its energy supply by wind will be under the windsituation of Marsa Matrouh approximately 40%. The windfraction will increase, if a discontinuous operation according to the actual available windspeeds is allowed by implementation of a storage tank for potable water.

The flow scheme shows the principle of the winddriven desalination unit WIND-REVOS.

The diesel supported WEC feeds the R.O. unit with continuous and stable (in frequency, voltage and power) electrical energy. This is ensured by the Master Control Unit.

The power cogeneration is controlled only by measurement of the frequency of output power. It can be described as follows:

Assuming increasing windspeed the WEC disloads the dieselgenerator. The dieselengine increases its speed, but the speedcontrol fixes the speed by controlling the injection pump; vice versa for decreasing windspeed.

In the past there have been some experimental plants of such kind of wind/diesel power cogeneration; but most of them had the disadvantage that diesel has to be of a higher performance as the WEC and the diesel has to work continuously for controlling output power frequency and the rotational speed of the WEC. We eliminated this disadvantage by using a synchronous generator at the diesel and an asynchronous generator at the WEC. The synchronous generator is coupled to the diesel with an overspeed clutch, which allows to the generator to rotate with a higher speed than the diesel, but not vice versa. In this way the synchronous generator operates as a phase lifter, if the wind energy converter produces enough energy to fulfill the requirements of the user (here the R.O. unit). The diesel can be stopped in this situation. In order to minimize the start-stop frequency of the diesel, a time delay between the point of feeding the consumer by windenergy alone and stop of the diesel is implemented.

As shown in the figure "Performance Distribution" the windfraction under the windregime of Marsa Matrouh will increase to approx. 50% by using the advanced MONOPTEROS 15, pitch controlled.

The additional energy fraction can be used for other purposes, e.g., pumping or heating water etc.

This system represents the solution of the wind/diesel power cogeneration, if continuous power supply is required. This system also can be used for feeding small grids by windfarms or single WEC's with a performance up to approx. two times of the windfraction.

Our second step in this field will be the design and erection of a stand-alone version of a winddriven desalination unit without a diesel or a gridconnection. For this system we use a desalination unit, which is able to increase and decrease its production due to the variable energy feed from the WEC.

It is planned to collect empirical data by mid 1988 with the MONOPTEROS 15 WIND-REVOS about the drinking water amount obtained with this type of plant to find out the optimum between increasing investment cost and decreasing fuel consumption and to minimize the specific cost of potable water production.

Because of the high prices of water supply in remote regions like this of Marsa Matrouh, the MONOPTEROS 15 WIND-REVOS operates already economically and it can be assumed that the specific water price can be decreased by using an advanced type of MONOPTEROS 15 - pitch controlled - which reaches an approx. 1.3 times higher annual energy yield compared to the used stall controlled MONOPTEROS 15 .

Upon successful completion of this demonstration project, it is planned to set up larger wind energy converters from the MONOPTEROS family in the 200 to 800 kilowatt power range combined with larger desalination plants. It is also planned to install WEC's of the AEOLUS type in the power range of 2 - 3 Megawatt. These converters have been designed primarily to feed the public grid.

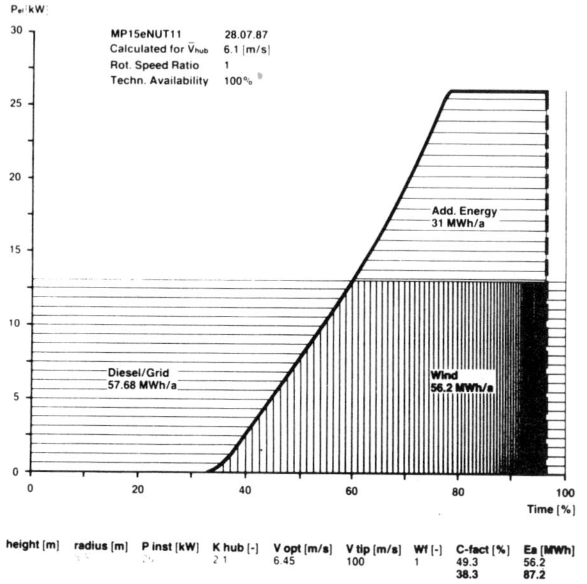

Performance Distribution, MARSA MATROUH, EGYPT
advanced MONOPTEROS 15, continuous pitch controlled
P_{el} [kW]
MP15eNUT11 28.07.87
Calculated for $\bar{V}_{hub}$ 6.1 [m/s]
Rot. Speed Ratio 1
Techn. Availability 100%
Add. Energy 31 MWh/a
Diesel/Grid 57.68 MWh/a
Wind 56.2 MWh/a
Time [%]
height [m] radius [m] P inst [kW] K hub [-] V opt [m/s] V tip [m/s] Wf [-] C-fact [%] Ea [MWh]
2 1 6.45 100 1 49.3 56.2
38.3 87.2

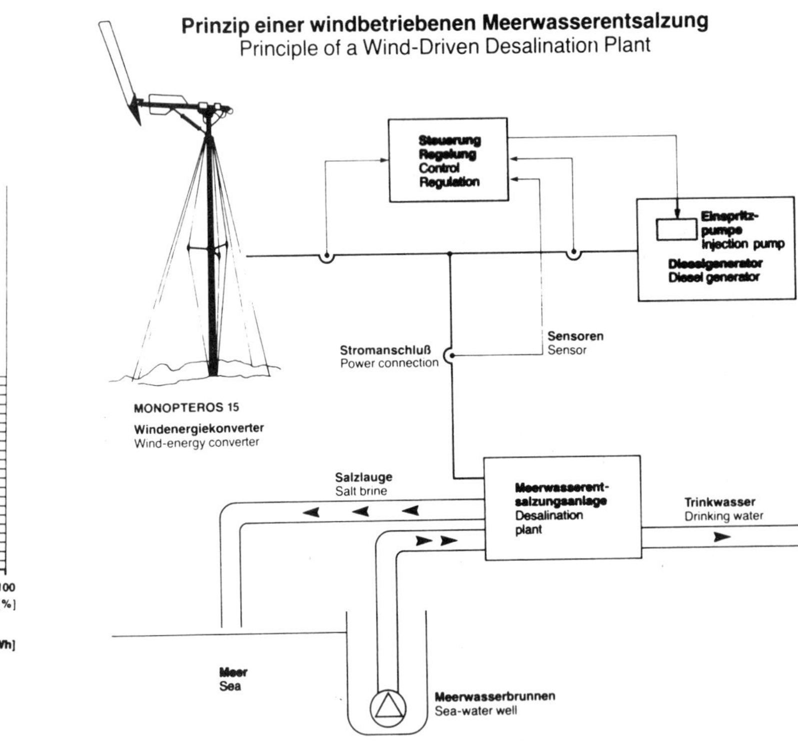

Prinzip einer windbetriebenen Meerwasserentsalzung
Principle of a Wind-Driven Desalination Plant
Steuerung
Regelung
Control
Regulation
Einspritz-pumpe
Injection pump
Dieselgenerator
Diesel generator
Stromanschluß
Power connection
Sensoren
Sensor
MONOPTEROS 15
Windenergiekonverter
Wind-energy converter
Salzlauge
Salt brine
Meerwasserent-salzungsanlage
Desalination plant
Trinkwasser
Drinking water
Meer
Sea
Meerwasserbrunnen
Sea-water well

ECONOMY OF WIND ENERGY CONVERTERS

Messerschmitt-Bölkow-Blohm-GmbH, Hoykenkamp, West Germany
now with: Ludwig-Bölkow-Systemtechnik GmbH, München,
West Germany

Horst Selzer

Daimlerstraße 15
D - 8012 Ottobrunn

ABSTRACT

The economy is being determined on one side by the investment cost including machine costs, infrastructure, land costs, customs, taxes, provisions, operational costs for maintenance and repair, operation, insurance, taxes and on the other side by the income due to the electricity generated and the tax depreciation. In general figures, the machine costs (ex works) represent about 50% of the overall investment costs for small WECs and 80% to 90% in case of large WECs. The operational costs range in the order of 2,5% of the costs for hardware and infrastructure.

Instead of the income - depending on the specific tariffs and tax laws - the generating costs are widely used. These can be calculated using the energy -specific investment costs of the diagram. The values have to be multiplied by the assumed cost factor for the infrastructure and then by the annuity due to the interest rate and economic lifetime.

KEYWORDS

Wind energy converters, machine costs, infrastructure, electricity generating costs

INTRODUCTION

Although the technical viability of wind energy converters is no longer questioned in Europe, the hesitation by many utilities and also governments to decide on utilisations of large WECs comes from low experiences and questions of economy. As large WECs have not yet been built in bigger quantities, the failure of single units is taken as characteristics even for the good design. The same applies to the costs, as prototype costs significantly differ from quantity production prices. The aim of this paper is to analyse the present situation of those WECs being under series production and to extend the findings to wind farms and large WECs for large scale electricity generation. For the purpose of better comparison commercial data related to German conditions are used.

TECHNICAL STATE OF THE ART OF WECs

Worldwide there exist about 110 companies offering approximately 210 commercial models of WECs ranging between 10kW and 4000 kW of rated power. Although more than 20.000 units have already been installed exceeding 1400 MW of installed capacity, only 19 models are being produced in large quantities, about 50 models in small series with less than 100 units per batch. Among these models in small quantity production, there are 2 vertical axis rotors, 1 single bladed rotor, 24% 2-bladed and 75% 3-bladed rotors. In large series production the relevant figures are 1 vertical axis, 16% 2-bladed and 84% 3-bladed rotors.

The share within all commercially available WECs is:

- vertical axis rotors : 8,8%
- 1-bladed rotors : 1,5%
- 2-bladed rotors : 29,0%
- 3-bladed rotors : 68,0%

showing a strong dominance of the 3-bladed rotors.

According to the rotor size, the following distribution can be recognised:

- 11% between 5 - 9 m diameter
- 51% "" 10 - 19 m "
- 31% " 20 - 34 m "
- 3% " 35 - 59 m "
- 4% " 60 - 100 m "

and according to the powering range:

- 11% between 10 - 19 kW
- 19% " 20 - 49 kW
- 34% " 50 - 150 kW
- 26% " 150 - 400 kW
- 5% " 401 - 1499 kW
- 3% " 1500 - 5000 kW

The accumulation of certain rotor sizes shows how experience was made, i. e. a concentration of 93% of WECs with diameters smaller than 34 m and 92% with rated power lower than 400 kW. Starting with WECs of 10 and 20 kW, power and size were increased step by step with a strong accent around 15 - 20 m in diameter and 50 - 150 kW. Now the marketing of the 250 kW-class takes place in a large scale. Contrary to the small and medium sized rotors, the large WECs have been developed in experimental units or prototypes only, thus resulting in a limited quantity of only 9 units above 1000 kW (5 units above 2 MW) being suited for the next step of small quantity production.
Despite of this limited experience with large WECs the technology was proven in general assuming 30 years lifetime and about 1% of initial investment for annual maintenance and repair. The corresponding figures are 20 years and 1,5% M & R for the small and medium WECs.

COSTS OF WIND TURBINE SYSTEMS

This chapter deals with costs based on series production and on installations for large scale electricity generation. One must carefully separate series production costs from prototype costs being twice or three times as high. The same applies to windfarms if they are installed as demonstration plants. Another misunderstanding results from not precisely defined costs.

Three types of costs have to be considered:

- machine costs "ex works"
- machine costs "installed", including transport, setting-up and assembly
- total costs, including additional infrastructure of the windfarm, i. e. roads, net connection, control buildings, fences etc. Land costs are sometimes included, too.

As the costs "ex works" for many WECs are already shown in catalogues , they give a rather reliable basis for comparison between most of the Danish and Dutch WECs, as shown in Fig. 1. The catalogue used also includes the cost basis "installed"; this means an average increase of about 30% compared to "ex works". Most WEC types in the 100 - 300 kW range are not yet included in price lists, these prices are based on personal findings from statements of the main manufactures in Europe. The price bases on series production.

The total cost of windfarms result from the following analysis:

- power range 20 - 50 kW: additional costs of 100% on "ex works costs as verified by Californian windfarms,

- power range about 250 kW: additional costs between 60% - 100% due to statements of interviewed manufacturers

- power range Megawatts: additional costs of 10% - 20% due to publications of Danish and Swedish utilities.

Of course, these total costs may differ according to the special situation and the needs of the customer or sponsor. Nevertheless, the above mentioned enables a general evaluation of the generating costs and provides a basis for comparison.

Plotting the power-specific prices (Fig. 2) and the energy-specific prices (Fig. 3), the findings of an earlier EEC-study can be confirmed: the simpler design of the small and medium WECs mostly stall-controlled offer a cost minimum in the range of 200 - 400 kW compared to the more expensive pitch-controlled large rotors.

It must be mentioned that:

- the prices of the small and medium WECs are based on the production of large quantities. The prices of the 2/3 MW unit relate to the Swedish-German WTS AEOLUS and were taken from a published ceiling offer to the Swedish government for the 10th respectively 100th unit.

- the dots of the plot represent prices influenced by the companies marketing policy. As bankruptcy already happened, a further decrease in the price minimum is very unlikely

- the very simple reason for some of the low values "DM/kW" is: a high power rating was selected for this rotor diameter in order to provide a good sales argument. Therefore, the performance of a rotor can be judged more objectively from the specific costs for the energy production. These values, of course, can only be determined on site.

While the prices "installed" range between 1500 - 2000 DM/kW as a minimum (6000 DM/kW for the prototype of LWECs and 3000 - 4000 DM/kW for line produced units), the energy-specific investments are levelling at 0,80 DM/(kWh)$_1$, for the medium WECs and 1,20 DM/(kWh)$_1$, for large WECs, a site provided at the German North Sea. Assuming an interest rate of 5% p.a. and 15 years commercial lifetime, i. e. an annuity of 9,6%, the capital costs will be 0,077 DM/kWh respectively 0,115 DM/kWh, if only the machine costs and installation are considered. Quite often, these values are used by mistake to show, how cheap electricity from wind turbines is, however, neglecting the various additional cost influences.

ELECTRICITY GENERATING COSTS

Therefore, some tabulated examples will be presented indicating the major cost parameters for large scale utilization of wind turbines.

In Germany, a profit making company has to pay various taxes, approx. 70%. Assuming an amortization period of 10 years, taxes of about 54% of the initial investment can be saved. As this tax saving possibility depends on the situation and structure of the company, the generating costs of the following table are calculated for both cases, full and zero amortization. Furthermore, 7,5% of investment subsidy can be claimed legally.

A major cost factor for windfarms might come up by the land costs. Due to the wind shadow effect, the turbines should not be placed closer than 8 - 10 rotor diameters, otherwise the efficiency will be reduced. For small and medium sized rotors this means that the land in between cannot be used the same was as before - or with less profit. Therefore, the land for a windfarm must be bought or rented trying to compensate the profit loss.

In the case of large WECs, the distance between the towers will be in the range of 600 - 1000 m, hence no disturbance to the land used will occur and only the land due to safety requirements must be bought, which is in the order of the rotor swept area.

The next cost term is due to the annuity. Here, the commercial lifetime was assumed to be 50% of the technical lifetime. Maintenance and repair reported to be 1,5% of the initial investment for medium WECs and 1% for large WECs, including land care. Operational costs might be 1% on the average. The amount of annual electricity generation at a site near the German North Sea will be 400 MWh (40% availability) for a 200 kW-turbine and 7,100 MWh for a 3 MW-turbine. Hence, the generating costs will come to 0,17 DM/kWh without tax depreciation and 0,10 (0,09) DM/kWh including depreciation.

Without touching the aspect of fuel/capacity credit the generating costs have to be compared with those of conventional power stations. In many cases, today's electricity costs are compared to future wind energy costs. The comparison must of course be done on the same time basis. In this sense, Fig. 4 was taken from sources not favouring renewable energies. The generating costs of wind power stations range between 0,15 and 0,40 DM/kWh depending on fuel prices and annual time of operation.

CONCLUSION

The commercial values presented indicate the economic competability of wind energy with future coal and even nuclear-power stations. The opposing arguments of influencing the total costs of a power station mix in a grid was proven not to be valid, if the wind energy penetration is kept below 10% of the annual energy consumption of the grid.

Due to the series production of the small and medium WECs, the 200 - 300 kW class offers the lowest costs today. But experience is available that large WECs of future small quantity production will match with these costs. Land consumption taken into account, favours the application of large WECs if large scale wind energy utilisation is considered in Central Europe.

LITERATURE

1. Jaras, T.F. (1987). Windenergy 1987, Wind shipments and applications Stadia Inc. Great Falls, Virginia, ISBN 0-944 038-00-X

2. Simpson, P.B. (1987). Product Guide-Wind Turbines. Modern Power Systems Sept. 87, Vol. 7 ISSUE 9, ISSN 0260-7840, pp. 73-95

3. Selzer, H. (1986). The economy of wind energy in centralized and decentralized applications, Int'l. Symp. on Appl. of Solar and Ren. Energy, Cairo

4. Selzer, H. (1986). Wind Energy, Potential of wind energy in the European Community, an assessment study: Reidel Publ. Comp., Dodrecht, ISBN 90-277-2205-6

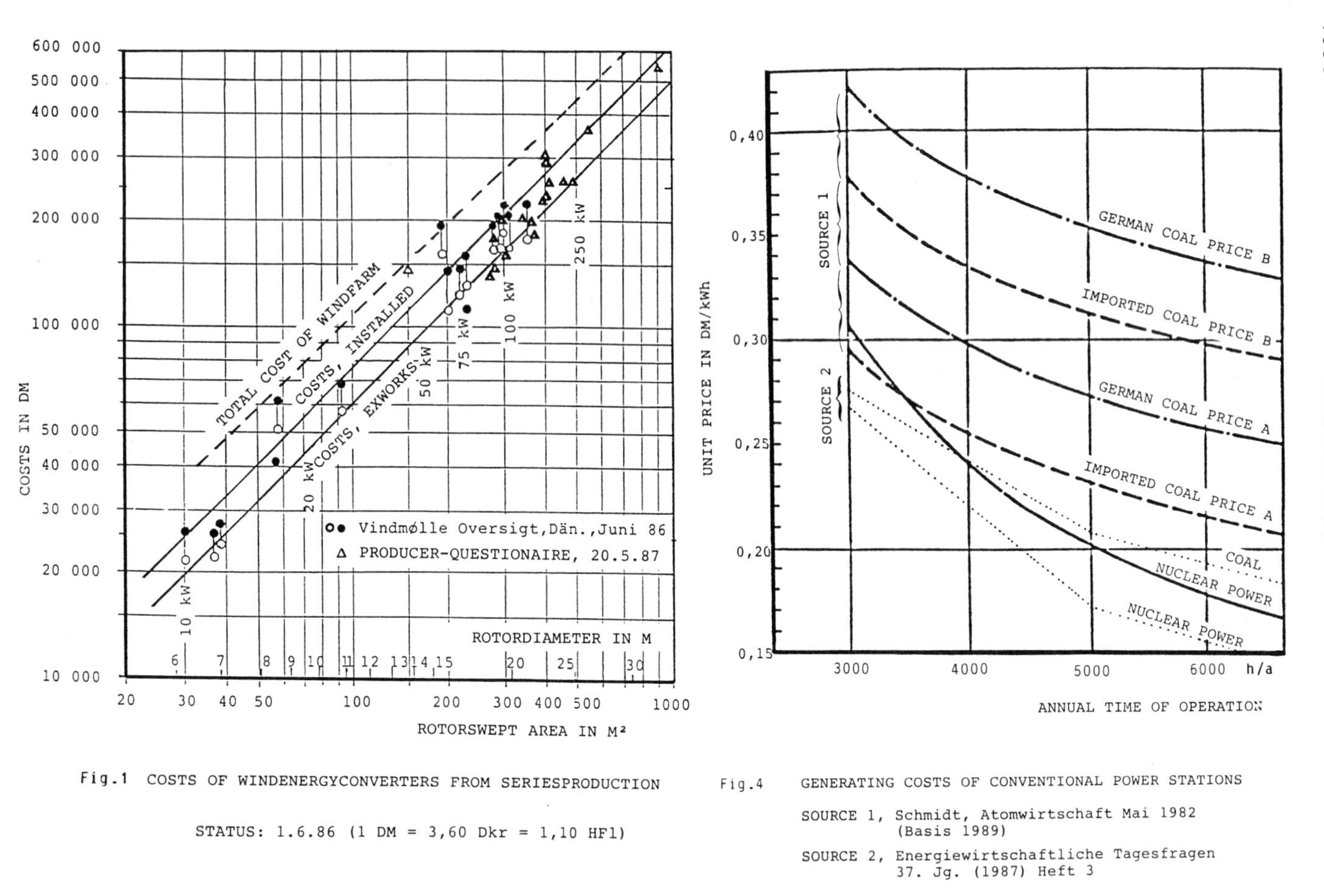

Fig.1 COSTS OF WINDENERGYCONVERTERS FROM SERIESPRODUCTION

STATUS: 1.6.86 (1 DM = 3,60 Dkr = 1,10 HFl)

Fig.4 GENERATING COSTS OF CONVENTIONAL POWER STATIONS

SOURCE 1, Schmidt, Atomwirtschaft Mai 1982 (Basis 1989)

SOURCE 2, Energiewirtschaftliche Tagesfragen 37. Jg. (1987) Heft 3

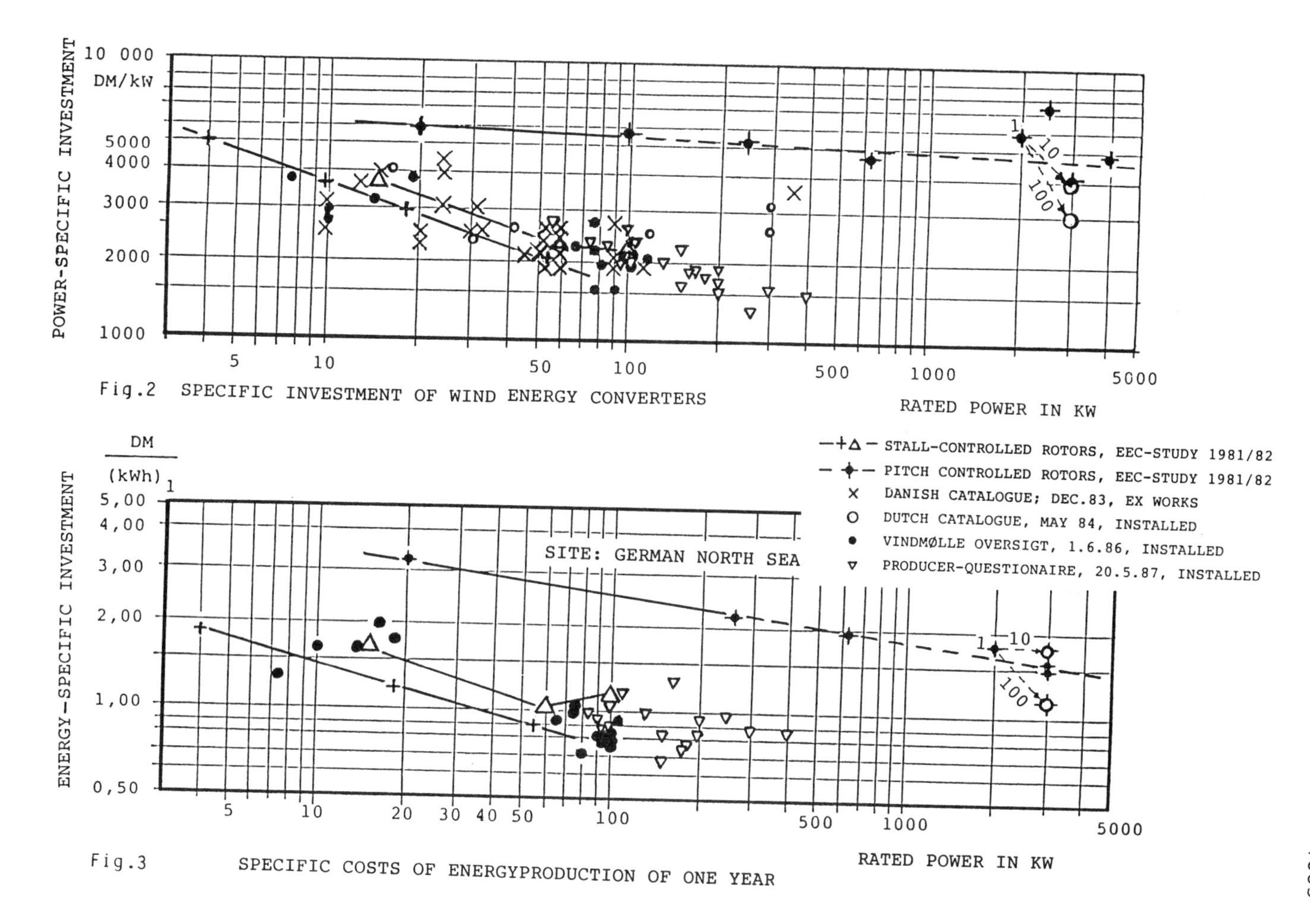

Fig.2 SPECIFIC INVESTMENT OF WIND ENERGY CONVERTERS

Fig.3 SPECIFIC COSTS OF ENERGYPRODUCTION OF ONE YEAR

Electricity generating cost of series-produced WECs

	200 KW		3 MW		
Cost(installed)					
Small-series (20)			10.500.000	10.500.000	DM
Large-series (100)	320.000	320.000			DM
Infrastructur					DM
+ 30 %	96.000	96.000			DM
+ 10 %			1.050.000	1.050.000	DM
	416.000	416.000	11.550.000	11.550.000	DM
./.Invest-Subsidy 7,5%	- 31.200	- 31.200	- 866.250	- 866.250	DM
./.Depreciation					
0%	- 0		- 0		
54%		- 224.640		- 6.237.000	DM
	384.800	160.160	10.683.750	4.446.750	DM
Real Estate					
40000 qm à 1,50 DM/qm	+ 60.000	+ 60.000			
5600 qm à 1,50 DM/qm			8.400	8.400	DM
Total Costs	444.800	220.160	10.692.150	4.455.150	DM
Cost per year					
1. Capitalcost					
Annuity 0,13 (10J./5 % Interests)	57.824	28.620			DM
Annuity 0,096 (15J./5 % Interests)			1.026.446	427.694	DM
Maint./Repair					
1,5 %	7.140	7.140			DM
1,0 %			115.584	115.584	DM
Operation/Insurance 1,0 %	4.760	4.760	115.584	115.584	DM
Costs/year	69.724	40.520	1.257.614	658.862	DM/a
Electricity generation per year	400.000	400.000	7.100.000	7.100.000	kWh/a
Generating Costs	0,174	0,101	0,177	0,093	DM/kWh

COUPLING GEOGRAPHICALLY DISPERSED WIND CONVERTER SYSTEMS TO THE MAINS GRID

Joachim Luther, Robert Steinberger-Willms
Department of Physics, Universität Oldenburg,
D-2900 Oldenburg

ABSTRACT

Geographical diversity is one strategy to reduce the fluctuations in energy output of solar and wind converter systems. We report here on an attempt to evaluate the effect of geographical dispersion on the power output of wind energy converter systems. We have performed simulation calculations for the Northwest region of the Federal Republic of Germany.

KEYWORDS

Wind energy systems, grid coupling, geographical diversity

INTRODUCTION

The power output of single renewable energy converter systems (photovoltaics, wind) is generally highly fluctuating. This may be a restriction when renewable energy converters are added to a utility grid, since power commitment in the utility's power station system could become complex and considerable reserve capacity may have to be kept in spinning reserve.

One simple measure to improve system characteristics is to distribute converters over distances of some one hundred kilometres in order to obtain a relatively high diversity in wind speeds experienced by the system as a whole. It is generally acknowledged that the system reliability should improve in such a grid connected system.

We report here on attempts to evaluate the performance of a system of two (hypothetical) wind converters situated about 120 km apart on the Northwest German coast. The weather data utilized were recorded at Norderney and Bremerhaven weather stations as hourly mean values of wind speed and direction (refer to map).

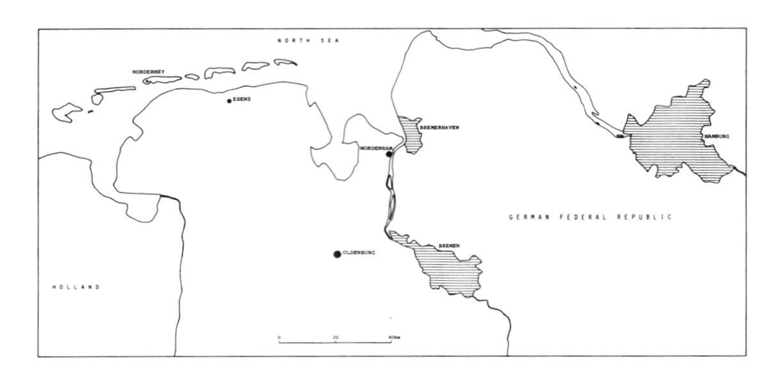

SYSTEM ANALYSIS

The analysis was carried out using simulation calculations at time steps of one hour. In order to achieve normalized results the wind data of the Bremerhaven site were scaled up to yield the same annual mean wind speed as the Norderney site. This roughly corresponds to greater hub height at the less favourable site. Power output characteristics of the system were evaluated using the power duration function as a criterion. Improved system characteristics are expected to show as an increase in availability below the system mean power output (about one third of rated power for the sites investigated). The availability of high power output should decrease, since the probability of the system of two converters supplying rated power is smaller than that of a single converter supplying maximum power.

Figure 1 shows a calculated power duration curve. Output power P is normalized by the respective system rated power P_{inst}. The duration curve reveals only marginal effects in power availability, though a five percent increase is discernible at low power levels. These results must be explained by a rather high correlation in wind data utilized. For 1985 data a correlation coefficient of 0.75 was calculated, 1982 data yielded 0.79. Though correlation is distinct in the hourly mean values of the meteorological data, shorter averaging time spans should show less correlation.

In order to further investigate into these results, auto- and cross-correlation functions for 1985 data were evaluated and are shown in Fig. 2. Though some effect is obvious for short time spans, there is no remarkable uncoupling of the power output in the cross-correlation as compared to the auto-correlation of the Norderney site.

The coincidence in weather changes and occurrence of high winds may also be noted in the sample graph of wind power output in Fig. 3 (fourth week of January 1985). A shift of one to four hours is observable in the arrival of strong winds in the middle section. Some anti-correlation is also apparent, but the general tendency is that of a rather close coupling of the output of the single converters and the coincidence of output peaks and lulls.

To cross-check the assumption that the unsatisfactory results originated from too small a distance of the converter sites we ran simulation calculations with practically uncorrelated data sets, i.e. Norderney data for 1985 and 1986 (correlation coefficient 0.008). These data show no

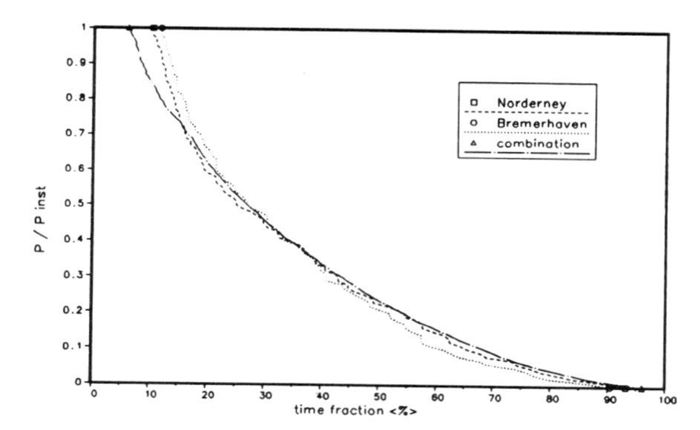

Fig. 1: Power duration curve for Norderney and Bremerhaven sites and for a combination of the two (1985 wind data).

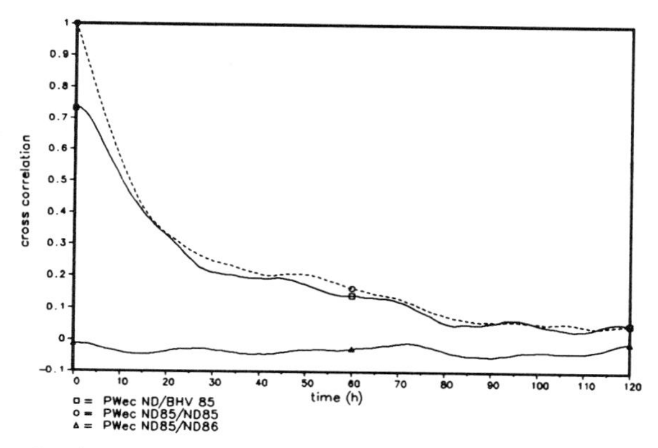

Fig. 2: Cross-correlation of power output from Norderney (ND) and Bremerhaven (BHV) sites (1985), auto-correlation of the Norderney output data (1985), and cross-correlation between 1985 and 1986 Norderney data.

short-term correlation, although a seasonal trend remains. The cross-correlation function is also shown in Fig. 2 as lower graph. The power duration curve in Fig. 4 resulted and shows the more favourable characteristics originally expected.

CONCLUSIONS

The results presented do not seem sufficient to satisfyingly clarify the effect of a spacing of little more than one hundred kilometres on grid coupled wind energy systems on the German coast. Improvements in system behaviour should be detectable if shorter time steps were used, since short-term meteorological variations and turbulence effects would play an increasing role.

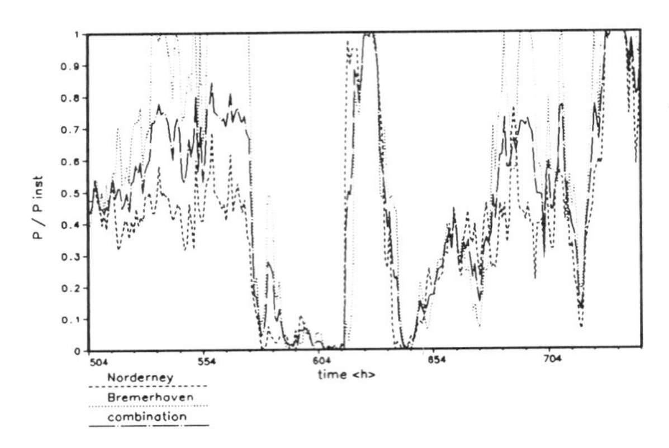

Fig. 3: Sample graph of power output, Norderney site, fourth week of January 1985.

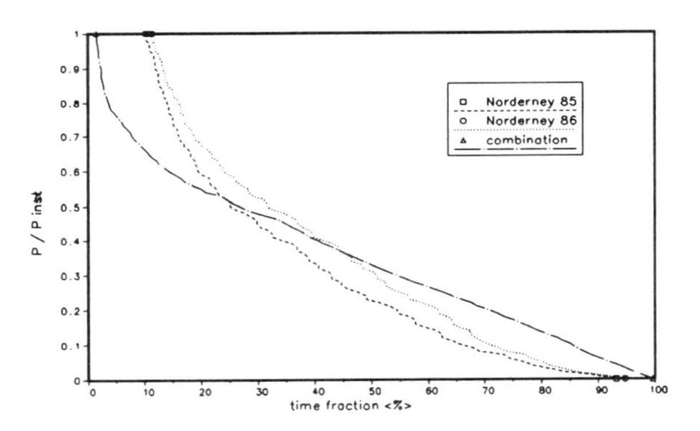

Fig. 4: Power duration curve for uncorrelated data sets Norderney 1985 and 1986.

Future work will now concentrate on data taken at time intervals shorter than one hour (e.g. 15 minutes) in order to take short-term meteorological variations into account. Four recording stations have recently become available to us at about 100 km mutual distance.

Thanks are due to the Deutsche Wetterdienst (DWD) for supplying meteorological data and to the Energieversorgung Weser-Ems for cooperation in recording weather data.

4

Heat Pumps

AMBIENT ENERGY UTILISATION WITH HEAT PUMPS - ACHIEVEMENTS AND CHALLENGES -

Dr.-Ing. Bernd Stoy

Rheinisch-Westfälisches Elektrizitätswerk (RWE), Head Office,
4300 Essen 1, Federal Republic of Germany

ABSTRACT

The activities of Californian utilities are very impressive in terms of size, capacity and development state. However they cannot be transferred to Western Europe. Nevertheless electricity utilities of many European countries have been making efforts to advance and test the utilisation of renewable energy sources.
Apart from solar radiation and wind energy there is another important energy source among the renewable energy sources in question, that is ambient heat.
Ambient heat presents the biggest potential in quantitive and capacity terms in the moderate zone of the earth. Today the heat pump capacity installed worldwide is estimated at 50,000 MW. Against this background wind energy or photovoltaics are still relatively unimportant.
In countries where air conditioning or cooling is the most important application (e. g. in the US and Japan) the purchase of a heat pump is relatively independent of the economic efficiency. This means that in those countries, particularly in Western Europe, where heating is virtually the sole application, the future market development will mainly depend on the future energy prices for fuel oil, gas and electricity.
The heat pump technology harnessing ambient energy exclusively will assert itself worldwide.

KEYWORDS

Definition of ambient heat; potential of renewable energy sources; development of energy prices; heat pump market situation.

Strictly speaking, I am supposed to give an introductory presentation on the heat pump workshops. The ISES programme committee asked me, however, to include some information on the solar activities of European electricity utilities. If I may, I would like to do so at the beginning.

The activities of Californian utilities described by Mr. Carl Weinberg are very impressive in term of size, capacity and development state. However, they cannot be transferred to Western Europe also because, or mainly because, of the climate prevailing there and the topological conditions.
With the exception of the large-scale solar thermal project at Almeria sponsored in Spain by the International Energy Agency (IEA) there are no similar plants in Western Europe to harness solar radiation and wind energy. However, there are major plants for the utilisation of ambient heat in Europe.

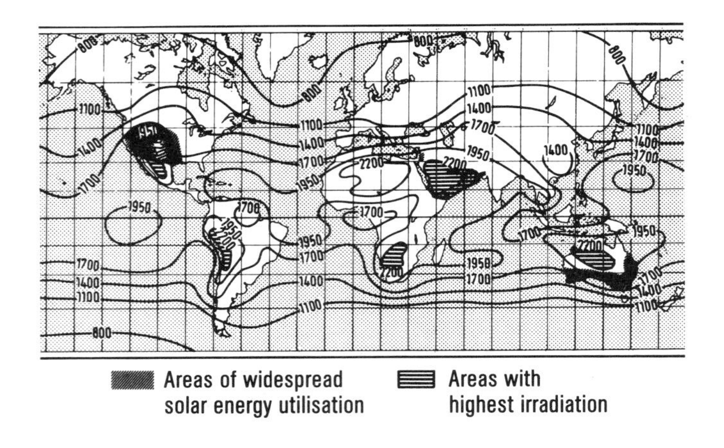

Fig. 1 Solar and sky radiation map of the earth (average annual insolation on to a horizontal area in kWh/m²)

An illustration (see Fig. 1) of the insolation and the latitudes of the regions in question shows quite clearly that these different concepts are well justified.
Nor is there in Western Europe any cold ocean fed by the Arctic Ocean such as the Pacific, but exactly the opposite, a relatively warm ocean fed by the Gulf Stream and no desert region behind a 500 m high and 1,000 km long mountain range such as the Altamont Mountains. Both are prerequisites for the daily and regular wind energy supply in California. The isochronism of electricity demand and electricity generation in wind energy or solar thermal plants does not exist either. On the contrary: at times of annual peak load the availability of wind energy is severely restricted and that of solar radiation is virtually nil in countries of Western and Northern Europe.

Nevertheless, Swedish, Danish, English, Dutch, French, Spanish, Italian, Swiss, Austrian and not least German electricity utilities have been making efforts of varying intensity for years to advance and test the utilisation of the energy of the wind and solar radiation. As for time constraints I cannot present the diversity of the projects here. Anyway, they come partly up to the megawatt range and will be dealt with at this congress in the respective workshops.

The lively interest of the electricity utilities in these subjects is also underlined by the composition of the audience at this world congress. For the Federal Republic of Germany we find that approximately 15 % of all the German participants are employees of electricity utilities.

If we ignore hydroelectric power for the moment, there is apart from solar radiation and wind energy another important energy source among the renewable energy sources in question, that is ambient heat.

Having been a committed advocate of this inexhaustible and particularly low-polluting energy source for years I would therefore like to express my great pleasure that the area "ambient heat" will for the first time be given the same priority at an ISES world congress as hitherto the other areas of renewable energy utilisation.

This is also to be welcomed because ambient heat presents the biggest potential in quantitative and capacity terms in the moderate zone of the earth.

First a linguistic definition of ambient heat (see Fig. 2):

SPRACHE	LANGUAGE	
deutsch	german	Umweltwärme
englisch	english	Ambient heat
französisch	french	Chaleur ambiente
italienisch	italian	Calore ambientale
spanisch	spanish	Calor ambiental
schwedisch	swedish	Omgivningsvärme
russisch	russian	Циркуляционное тепло
arabisch	arabic	حرارة البيئة
japanisch	japanese	周囲環境熱
chinesisch	chinese	环境热

Fig. 2 Linguistic definition of ambient heat

A technical definition of the term ambient heat to follow the linguistic one:
The so-called ambient heat, one of the renewable energy forms, is the solar energy temporarily stored in the mass and atmosphere of the earth as well as in flora and fauna. However, ambient heat is peculiar in that it can be harnessed only indirectly by means of suitable energy exchangers (heat pumps) as a result of the temperature level. In addition, there is the so-called anthropogenic ambient heat generated by man because all energy conversions ultimately end in low-temperature heat and therefore come on top of solar-generated ambient heat.
In the Federal Republic of Germany a densely populated country with a very high specific energy turnover, the anthropogenic proportion amounts to approximately 2 % of the ambient heat resulting from solar energy. Related to the entire globe this percentage is only a negligibly small quantity, i.e. 0.004 %.
The potential of ambient heat in comparison with other forms of renewable energy sources is generally underestimated. The pertinent literature does obviously not contain any comparative data on the energy content. Therefore, a proposal for such an estimation is to be submitted here.
The theoretical potential of all renewable energy sources is virtually inestimable, defies the imagination and goes beyond the potential which could ever be harnessed technically and economically.
If you want to estimate the capacity and quantity potentials of renewable energy extraction, you can only approach the problem by human standards, which means that this estimation should be subdivided into:

- o amount of energy which can be technically/theoretically extracted (e. g. kWh/m^2 and year of active exchanger surface)
- o thereof measures which can be economically implemented and
- o thereof measures which can be implemented under landscape, political or mentality-related aspects.

Only the comparison of the technical/theoretical energy extraction per year will be feasible, although here, too, just an idea can be conveyed because of varying parameters to what extent the renewable energy sources differ in terms of their efficiency.
Any approach is based on the respective active exchanger surface. An amount of energy in kWh which can theoretically be obtained per m^2 and year is taken as a basis.

o Biomass	1 kWh/m^2•a	(area under cultivation)
o Photovoltaics	100 kWh/m^2•a	(active solar cell surface)
o Wind energy	1,000 kWh/m^2•a	(area passed over by the rotor)
o Ambient heat	10,000 kWh/m^2•a to 100,000 kWh/m^2•a	(active evaporator surface of an air/water heat pump)

Of course, in terms of exergy, you cannot compare these figures.
This comparision makes it clear that the potential for the utilisation of ambient heat is very large indeed in every country since the energy sources air/groundwater/surfacewater/ground etc. are virtually abundant. The actually extractable amount of ambient heat is therefore rather a question of economic efficiency than one of the potential.
Among all the forms of renewable energy sources ambient heat is more than any other dependent on auxiliary energy (e. g. electrical energy or fuels) for its utilisation according to the thermal dynamic principle of the heat pump.
With electric heat pumps approximately one third, with gas heat pumps approximately two thirds of the delivered useful energy are generally accounted for by auxiliary energy, i. e. a substantial proportion of the heating output is the result of electricity or fuels.

This is why not only the system manufacturers and installation companies are interested in this technology but also the energy utilities, above all the electricity supply undertakings since given the relatively high prices of daytime electricity (compared with the heat content of gas or oil) the best way to use electricity economically in the heat market is to combine it with free ambient heat through the use of heat pumps.
Since the operation of heat pumps has already reached economic efficiency in many countries today, in particular in combination with cooling or air conditioning in summer, there has been for years a broad range of applications and fully developed technologies worldwide.
Therefore, the harnessing of ambient heat by heat pumps represents today jointly with the solar collector technology the most advanced application for the utilisation of renewable energy in quantitative and capacity terms.
Just two figures in this respect: the water power potential installed worldwide, which is generally assumed to represent the biggest share of renewable energy utilisation, amounts to 550,000 MW related to power station capacity. Today already, the heat pump capacity installed worldwide is estimated at 50,000 MW:
Against this background wind energy or photovoltaics are still relatively unimportant.

I have been asked to present some statistical information on the worldwide development of heat pumps. Reliable data, however, were not available and, come to that, you are certainly aware from your own experience that "no statistics should be trusted unless falsified by yourself".

In the context of this general presentation it is not possible to go into technical details. To add a little bit of humour I should like to mention a few superlatives:

- The first and hence oldest heat pump in the world, at the time still used to harness waste heat, was set up in 1857 in Upper Austria by Peter Ritter von Rittinger with a capacity of approximately 80 kW.
- The currently biggest heat pump installtion in the world is operated in Sweden with a thermal capacity of 150 MW. The heat is fed into the district heating grid of Stockholm.
- The most frequently mentioned heat pump in Europe is the system commissioned in 1938 in Zürich to heat the townhall and whose cold energy source is the river Limmat.
- To our knowledge the most mobile heat pump in the world is also found in Europe, again in Sweden. This is a 100 MWth heat pump floating on a pontoon and currently feeding into a district heat network north of Stockholm. After the completion of a cogeneration station it will swim 20 km across the Malaren Lake and supply the city of Varby with heat.
- The most distinguisted heat pump in the world is also operated in Sweden, namely in the castle of the Swedish King, in Drottningholm.

Given the sharp decline in fuel prices the current competitive situation is rather unfavourable for heat pump applications especially in Europe if just heating energy is compared and no use is made for cooling purposes.
The cartoon showing renewable energies being drowned in a glut of oil is a better description than one of many words (see Fig. 3).

Fig. 3

However, this picture is misleading. In actual fact mankind is moving on the globe with undiminished speed towards a shortage of fossil energies and an increase in the pollution of the atmosphere. The following cartoon is quite a humorous description of this situation (see Fig. 4).

Fig. 4

As far as the generation of low-temperature heat, i. e. useful energy with a low energy content, by heat pumps is concerned, one can nevertheless be optimistic: the system prices are largely calculable today. Should increasing numbers of items be produced in future, they will go decreasing just as installation costs will fall as experience grows.
And in countries where air conditioning or cooling is the most important application (e. g. in the US and Japan) the purchase of a heat pump is relatively independent of the economic efficiency. The heating task is a desirable byproduct as it were.

This means that in those countries, particularly in Western Europe, where heating is virtually the sole application, the future market development will mainly depend on the future energy prices for fuel oil, gas and electricity. In other words, as soon as fuel prices start soaring again, there will be an increase in heat pump applications in Western Europe, too.

Nonetheless there are bright people who plan on a longterm basis and operate a heat pump now already - in most cases alongside the oil-fired heating system - to be prepared for all contingencies.
The following figure for the ten-year period 1976 to 1986 (see Fig. 5) shows that the price of electricity, the auxiliary energy of electric heat pumps, has been surprisingly stable.

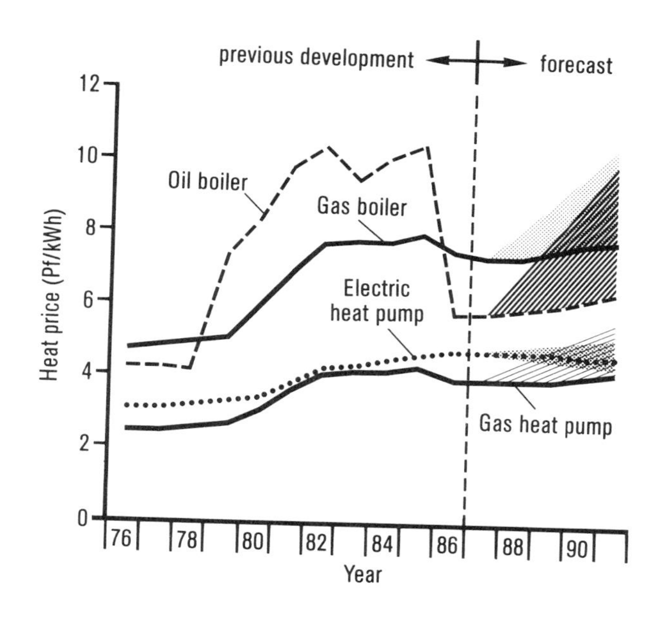

Fig. 5 Development of average energy and heat prices for room heating (FRG 1976 - 1991)

What sort of development does the future hold?
As far as oil and gas prices are concerned, forecasts are largely a matter of conjecture. For electricity prices a concrete outlook is easier.

For the Federal Republic of Germany it can be assumed that by 1991 electricity prices will show a falling trend or will remain stagnant since the growth rates of electricity consumption are declining heavily and no new power stations will be required in the next five to ten years. This is true of quite a number of mainly European countries.
Since most of the plants for electricity generation and transmission will furthermore largely be written off, electricity prices are likely not to rise further but will rather drop during this period. Oil and gas prices,

however, will rather show an upward trend. This would in the final analysis have a favourable impact on the application of e.g. gasdriven heat pumps.
Therefore, we expect for the future a positive trend for heat pump applications, although at present the mere running costs of electric heat pumps are frequently higher than those of oil and gas heating systems.
In sum, the following conclusions can be drawn:

- o The potential of ambient heat is extremely large, exists in every country (unlike solar radiation) and is available for manifold applications depending on the ambient heat source (air, groundwater, riverwater, ground etc.).
- o With respect to both the heat source and the heat sink the possible range of applications is wide, very small capacity ranges are available (some 100 W) but also very big ones (some 100 MW) and the required technology is fully developed.
- o The current competitive situation in Western Europe is likely to be temporary in nature.
  The heat pump technology which can harness ambiet energy, i. e. anergy, exclusively will assert itself worldwide.

REFERENCES

Stoy, Dr.-Ing. B. (1980). Wunschenergie Sonne. 3. überarbeitete Auflage.
Fasholz, J. (1986). Die Nutzung regenerativer Energieträger. 1. Auflage.

HEAT PUMPS FOR RESIDENTIAL USE IN THE USA

James M. Calm

Technical Manager, IEA Heat Pump Center, FIZ – Karlsruhe,
D-7514 Eggenstein-Leopoldshafen 2, Federal Republic of Germany

Consulting Engineer, 4580 Alpine Road, Portola Valley, CA 94025,
United States of America

ABSTRACT

More than a quarter of all new houses built in the United States in the last decade are heated with solar energy by means of heat pumps. Including replacements and retrofits, the residential market is approaching one-million heat pumps per year, or approximately three-quarters of the country's heat pump sales. The quantity installed and average efficiency are increasing despite the dramatic decline in world oil prices, but penetration has fallen slightly after a period of steady growth. Installations typically use air-to-air, dual-mode (heating and cooling), electrically-powered heat pumps with heating capacities of 6-12 kW. This paper presents the equipment used, market situation, and outlook for these energy saving, economic, and environmentally attractive systems. Factors influencing the unique status of heat pumps in the United States, including climate and energy prices, are briefly addressed.

KEYWORDS

Heat pumps, heating, air conditioning, building energy systems, residential, market analysis, forecast

EQUIPMENT USED

The most widely used heat pump in residential applications is the unitary[1] type. Several configurations are available, of which the most common is the split system with an outdoor heat exchanger and compressor unit and an indoor blower and heat exchanger unit. These assemblies are connected by refrigerant tubing, often precharged, with quick-connect fittings to simplify installation. The heat pump can be fully installed for $1,900-3,700 in the common 6-12 kW_t heating capacities. Virtually all residential systems are dual-mode, heating and cooling, using R-22 (chlorodifluoro-

---

[1] The *unitary* designation is defined by the performance rating standard and certification program established by the manufacturers' trade association [1-3]. In practice, the unitary class of equipment refers to heat pumps (and air conditioners) that are manufactured in one, two, or sometimes three assemblies designed to be used together, to form a complete heat pump (or air conditioner), with a minimum of application engineering.

methane) as refrigerant and air as both the thermal source (sink for the air conditioning mode) and indoor distribution medium. Supplemental and backup heat are commonly provided by electrical resistance heaters in one to three stages of 4-5 kW_e each.

Other popular configurations include air-to-air single-package heat pumps and packaged water-to-air units for incorporation into groundwater, ground-coupled, and

Table 1: Heat pump equipment groups and typical usage

	Equipment group					
		Space heating			Space and process heating	
Market sector	Domestic water heating	Consumer appliance	Unitary	Applied package	Large unitary[a]	Field[a] engineered
Residential						
single family detached and multi-family low rise	heat pump water heater (HPWH)	window	split system, single package	water-source[b] heat pump (WSHP)	–	–
multifamily high rise	HPWH[c]	–	–	packaged terminal and water-loop heat pump (PTHP and WLHP)	air-to-water (a-w) and water-to-water (w-w)	a-w, w-w[c]
Commercial and institutional						
light/small	HPWH	window, thru-the-wall	single package, split system	WSHP	–	–
heavy/large	[c]	–	–	PTHP, WLHP	w-w[c]	a-w, w-w[f]
Industrial	–	–	–	–	w-w	a-a, a-w, w-w, w-a, mvr, other
1986 sales (thousands)	7	~60	884	289	[d]	[d]
Representing[e] trade association	GAMA	AHAM	ARI	ARI	ARI	ARI[f]

[a] While equipment specifically produced for heat pump use is available, equipment and components manufactured as or for air-conditioning chillers are often adapted.
[b] Water-source equipment is frequently applied in groundwater, ground-coupled, surface-water, and (less commonly) city-water systems; aggregate installations were approximately 14,000 in 1986.
[c] Various forms of heat reclaim devices are also used to recover heat as a cooling byproduct.
[d] Data not available (combined sales of large unitary and field engineered heat pumps in 1985 were estimated at 5,000 units).
[e] Trade associations: Gas Appliance Manufacturers Association (GAMA), Association of Home Appliance Manufacturers (AHAM), and Air-Conditioning and Refrigeration Institute (ARI)
[f] ARI is the trade association for many of the products and components, but numerous other associations are involved.

surface-water systems; applications using city water as a source (and/or sink) are rather uncommon due to the cost and disposal requirements for the water. Packaged terminal and water-loop (closed-loop water-source) heat pumps as well as central air-to-water and water-to-water heat pumps are also used for multifamily residences, and particularly in high-rise and mixed-occupancy complexes (e.g., apartments over commercial space).

The wide array of heat pumps marketed in the United States is summarized in Table 1; references 4-10 provide a more detailed introduction.

HEAT PUMP MARKET

Annual sales of heat pumps in the USA grew to approximately 1.25 million units in 1986. In simple quantities, the U.S. market is the second largest in the world, following that in Japan; it is by far the largest – for either new or existing installations – in terms of capacities [11]. The overall market for residential heat pumps increased again in 1986, with declines only in the water-heating and residential groundwater categories.

The sales growth for unitary heat pumps is depicted in Figure 1. This figure and related shipment statistics, however, do not reveal strong undercurrents that are likely to affect market progression. Indeed, a misleading perception of sustained expansion is given by the relatively steady increases in annual sales since 1982.

CONSTRUCTION RATES

Figure 2 compares unitary heat pump sales to residential construction[2] rates. Single-family detached dwellings are the most popular, and represent nearly two-thirds of the total in most years. Such houses and low-rise multifamily buildings are the primary markets for unitary heat pumps, although retrofit and replacement sales of the latter are also significant. It is, nevertheless, interesting to note that annual sales of unitary heat pumps have matched approximately half the total residential construction rate consistently since 1980. Figure 2 also reveals a cyclic pattern of construction peaks and recessions, the next of which is imminent according to economic forecasts. Heat pump sales can be expected to temporarily decline – by as much as 30% – for two to three years when the construction slump occurs.

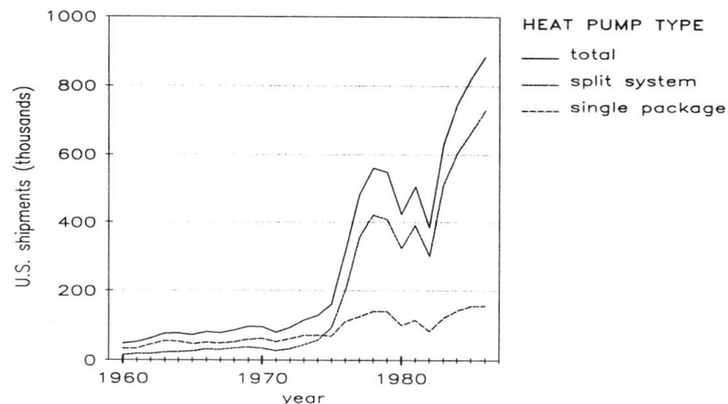

Fig. 1. Unitary heat pump use (based on data from reference 12)

[2] Figure 2 includes only privately-owned houses; the public share (e.g., for military or special purposes) is negligible as a fraction of the total.

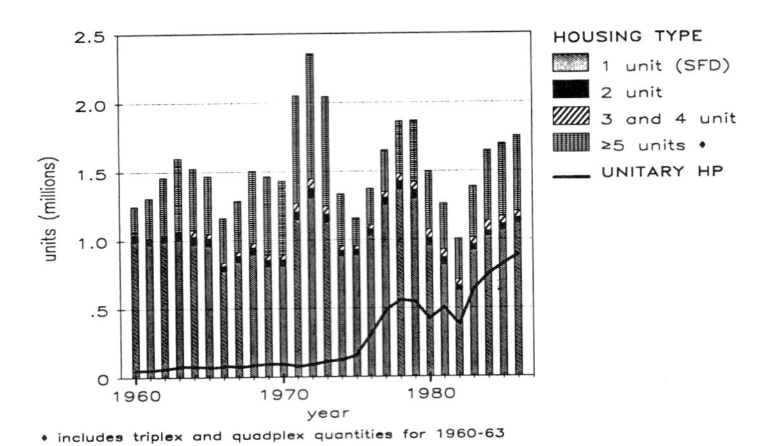

Fig. 2. Comparison of residential construction and unitary heat pump use (based on data from references 12 and 13)

HEATING FUELS

Further analysis of residential heat pump use requires examination of the fuels used for heating. The 1950s were marked by intense competition between the electric and gas utilities. Strategic marketing by the two groups has resumed, with both generally having adequate capacity – barring disruptions – for the near future. The electric utilities are projected to require more new capacity for the mid- and late-1990s, but the shortfall is generally for summer cooling. High-efficiency air conditioners and heat pumps are therefore being promoted, the latter to build a more even annual load profile. The gas industry, meanwhile, is promoting more efficient furnaces and other gas-fired systems to increase its competitiveness; it is also aggressively seeking to develop viable gas-fired engine-driven and absorption-cycle heat pumps. The gas heat pump option would be especially attractive for its summer load, a low-sales period for gas utilities. Although coal is used for more than half the nation's electric generation, it is no longer a significant fuel for building heating except in a relatively small number of district and central plants. Oil is competitive only in several areas, primarily New England, certain areas in the Midwest on the Great Lakes, a few west coast locations, Hawaii, and parts of Alaska.

Figures 3 and 4 show the dominance of gas and electricity as well as the displacement and limited resurgence of oil as a heating fuel for single- and multifamily residential construction, respectively. The gas share has generally increased for the last decade, though total sendout has declined with conservation measures. The electric fraction has, for the same period, declined slightly. Overlaid on figures 3 and 4, are the shares of heat pump penetration. By contrast to figures 1 and 2, small declines can be seen in the last year for single-family and in the last two years for multifamily residences. It is also apparent, in both cases, that heat pumps represent an increasing share of the electric systems, displacing resistance furnaces and boilers as well as zonal resistance and radiant systems; all unitary and virtually all larger heat pumps in current residential use are electrically powered.

REGIONAL PENETRATION

Having started with a perception of rapid growth and moved to one of a slight decline, a regional analysis provides still a different view of heat pump application trends.

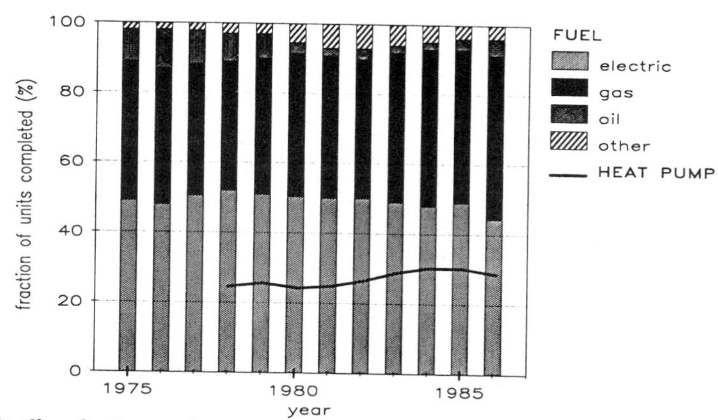

Fig. 3. Heating fuels used and heat pump penetration in single-family detached houses (based on data from reference 13)

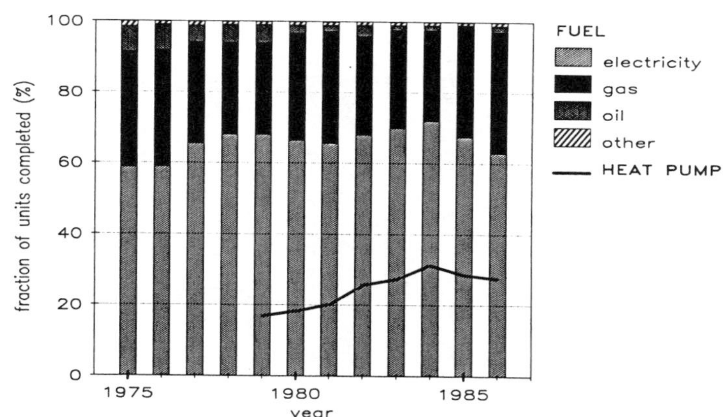

Fig. 4. Heating fuels used and heat pump penetration in multifamily residences (based on data from reference 13)

Because of its sheer size, the climatic variations in the United States far exceed those of the European countries individually or even collectively. Using the four regions for which econometric data are commonly grouped, figure 5 shows the very different regional patterns of heat pump acceptance. Penetration is highest in the South, where warmer climates typically result in the highest seasonal performance for air-source heat pumps; use in this region has increased appreciably since 1981. By contrast, heat pump popularity rose and fell over the past decade in the West. Except for the last year, steady growth – and nearly a doubling – has occurred in the Northeast with the opposite result in the Midwest; the last five years have been somewhat erratic in the latter, host to the most intense utility competition.

Figure 6 adds some perspective to the regional examination by showing the construction rates for single-family detached dwellings. Comparing the regional penetration of heat pumps with construction, it is evident that the recent decline in heat pump penetration on a national basis, also shown in figure 5, is more heavily influenced by demographic shifts than declines in heat pump acceptance.

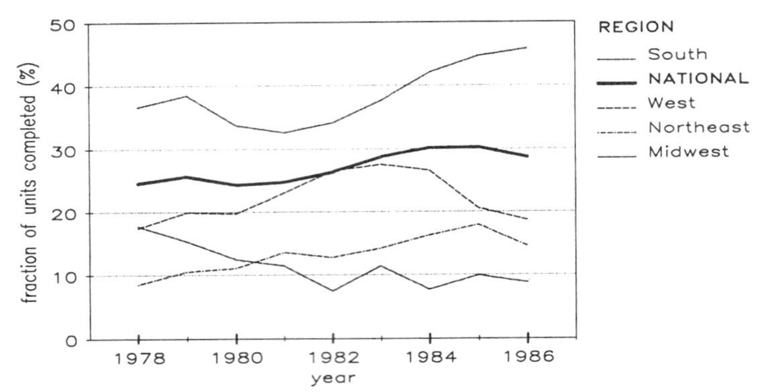

Fig. 5. Regional heat pump penetration in new single-family detached houses (based on data from reference 13)

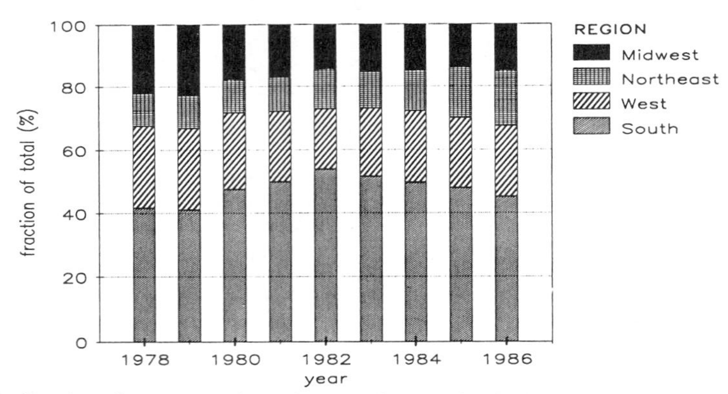

Fig. 6. Regional construction shares of new single-family detached houses (based on data from reference 13)

COMPARISON TO EUROPEAN AND JAPANESE HEAT PUMPS

The typical residential heat pump in the United States differs starkly from its European counterpart both because it is dual-mode (heating and cooling) and because it uses air as the indoor distribution (delivery) medium. Single-mode machines are quite rare; nearly 70% of all new single-family and 90% of multifamily residences in the USA are air conditioned as shown in figure 7. These fractions are actually higher if air-conditioners added after initial construction are considered; more than 2.8 million room (mostly window) air conditioners were purchased in 1986, a small decrease from preceding years, and many homes are also retrofit with central systems. Warm air – usually filtered and sometimes humidified – is used for heating; it is generally ducted to individual rooms or zones of the house. Cool, dehumidified, and filtered air is used in the air-conditioning mode. Beyond the physical differences, the total installed cost of a heat pump

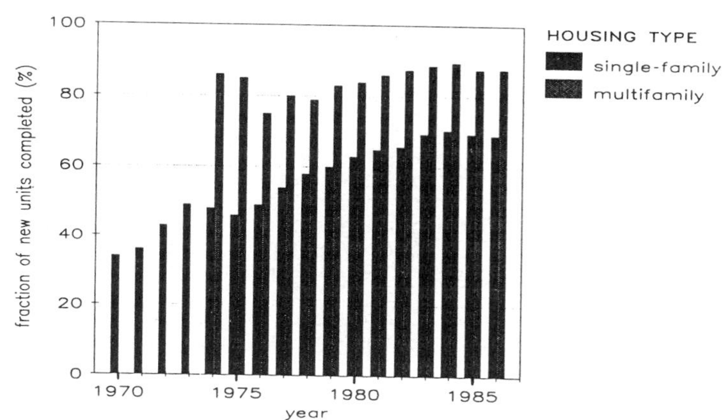

Fig. 7. Air-conditioning use in residences (based on data from references 12 and 13)

system for a single-family residence is commonly a factor of 2-4 times less expensive than the heating-only, hydronic-distribution, European counterpart. The price difference is due more to volume production and competitive marketing than physical or even application differences.

Contrasted to the dominant Japanese heat pump – the *room air conditioner heat pump* – the typical U.S. product is of a higher-capacity and is applied in a central (to the building) system. Adjustable-speed unitary heat pumps, using inverters, have been commercially available in the United States since 1979, but relatively few have been sold; two-speed and dual-compressor products have been available even longer, again with low penetration. The two largest manufacturers of unitary heat pumps will have begun sales of inverter-driven heat pumps by the time this paper is published. By contrast, the split-system, room air-conditioner heat pump with inverter drive has displaced its single-speed counterpart as the norm in Japan. Another difference lies in the ratio of air conditioner (cooling only) to heat pump (heating and cooling) sales. Although this ratio has fluctuated significantly in the last decade [6], it has generally exceeded 3 in the USA. By contract, ratios less than 1 have been common in Japan in recent years. The availability – or more correctly lack of widespread availability – of pipeline gas for furnaces, higher fraction of multifamily high-rise residences, and warmer climates in the populous regions explain this difference.

European use of heat pumps in block centrals and district heating far exceeds that in the United States on a per capita or even absolute basis. Several factors are at play, foremost of which are higher thermal demand density, due both to climate and housing styles (preponderance of large, clustered multifamily buildings), and institutional differences in the municipal planning and construction infrastructures.

OUTLOOK

The consensus of future projections for residential heat pumps in the United States is very favorable (see references 6, 8, 14, and 15). Nevertheless, a near-term decline in sales is probable, stemming from projected construction reductions and demographic shifts resulting in lower building rates where heat pumps have been most heavily applied; neither of these causes is expected to endure.

Several factors point to a longer-term expansion of the heat pump market. Product improvements, regulatory actions, utility incentives, and consumer reactions to energy prices have resulted in sustained increases in heat pump seasonal efficiencies. As shown in figure 8, the improvement in seasonal performance has approximated 2.5%/year for the last decade. This rate is likely to increase due to intensive research and development efforts initiated several years ago during the period of rising energy process. Furthermore, expected penetration of inverter-driven heat pumps is projected to cut both off-peak cycling losses and peak use of supplemental heat. Regulatory actions, including a national minimum efficiency standard that becomes effective in 1992, will also contribute. Expanding foreign competition in the domestic market as well as cost competition with high-efficiency furnaces, and to a lesser extent boilers, insure that installed heat pump costs will be maintained at competitive levels.

An important long-term influence will be the price ratio of electricity to gas, the two dominant heating *fuels*. As shown in figure 9, this ratio has declined fairly steadily

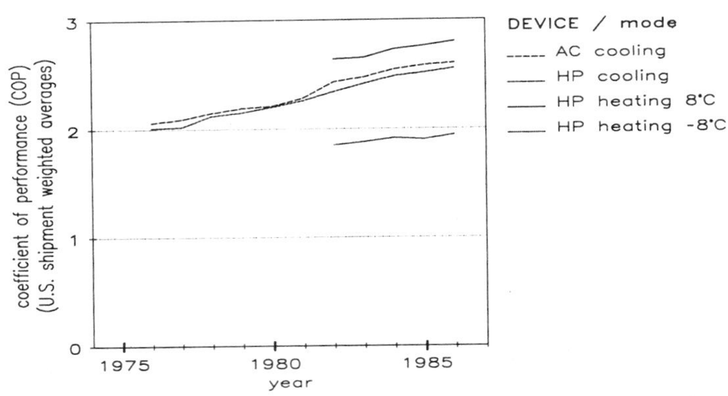

Fig. 8. Trends for shipment weighted average efficiencies for unitary heat pumps (the cooling performance shown is the industry standard energy efficiency ratio, EER, through 1980 and seasonal energy efficiency ratio, SEER, thereafter)

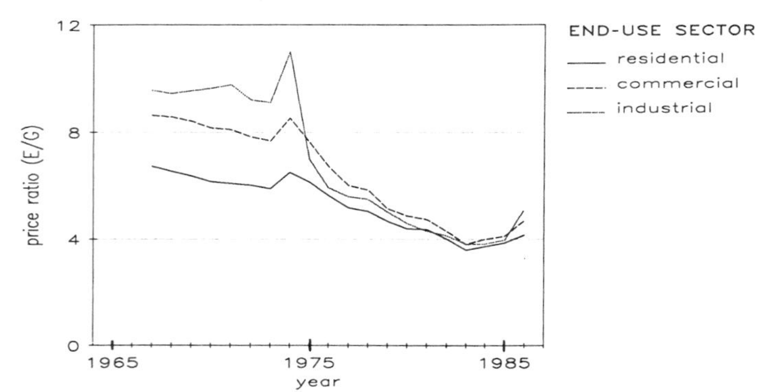

Fig. 9. Historical electric to gas price ratios (E/G) (based on data from reference 16)

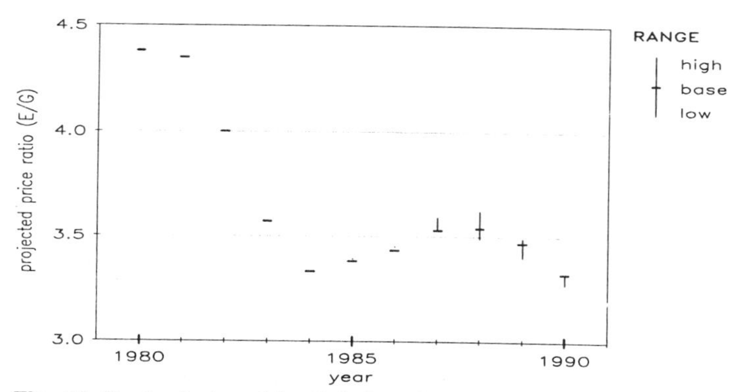

Fig. 10 Projected residential electric to gas price ratios (E/G) (based on data from reference 17)

(except during the 1973-4 interval) over the preceding twenty years [16]. The trend reversed sharply in the last several years, adversely impacting electric heat pumps relative to competing gas furnaces. Electric to gas price ratios are, however, projected to fall again within the next several years as shown in figure 10.

Anticipated efforts to commercialize gas-fired heat pumps in the 1990s will have two influences on the heat pump market. First, any penetration will increase aggregate heat pump use. Second, such efforts will necessitate restraint in the aggressive campaign, by the gas industry, against electric heat pumps as the negative claims are likely to also affect the gas heat pump image by association.

The heat pump outlook could be impacted by current deliberations on chlorofluorocarbon (CFC) refrigerant curtailments. R-22 (the dominant residential heat pump working fluid) is not, however, regarded among the CFCs of greatest concern. Moreover, the use of residential heat pumps is clearly beneficial in terms of other pressing environmental concerns.

CONCLUSIONS

Heat pumps are being used to heat and air condition nearly 30% of new residences in the United States and in nearly two-thirds of of those heated electrically. New records have been established for heat pump sales in each of the last four years. While the longer term outlook for expanded use is very favorable, declining sales are likely after 1987 for a short interval.

REFERENCES

1. **Unitary Air-Source Heat Pump Equipment**, Standard 240-81, Air-Conditioning and Refrigeration Institute, Arlington, VA, USA, 1981.
2. **Unitary Air-Conditioning and Air-Source Heat Pump Equipment**, Standard 210/240-84, Air-Conditioning and Refrigeration Institute, Arlington, VA, USA, 1984.
3. **Directory of Certified Unitary Products**, Air-Conditioning and Refrigeration Institute, Arlington, VA, USA, August 1987.

4. **ASHRAE Handbook: Equipment**, American Society of Heating, Refrigerating, and Air-Conditioning Engineers, Atlanta, GA, USA, chapter 44, pages 44.1-44.6, 1983 (revision in publication).

5. **ASHRAE Handbook: HVAC Systems and Applications**, American Society of Heating, Refrigerating, and Air-Conditioning Engineers, Atlanta, GA, USA, chapters 9 and 17, pages 9.1-9.16 and 17.1-17.6, 1987.

6. J. M. Calm, *Heat Pumps in the United States*, 13th Congress of the World Energy Conference, Cannes, France, October 1986; republished in the **International Journal of Refrigeration**, volume 10, pages 190-196, July 1987.

7. J. M. Calm, *Electric Heat Pump Trends in the United States*, **Proceedings of the International Energy Agency Heat Pumps Conference, Current Situation and Future Prospects**, Verlag für die Technische Universität Graz, Graz, Austria, volume 1, pages 15-26, May 1984.

8. J. G. Crawford, *Heat Pumps in the United States: A Status Report and View to the Future*, **Proceedings of the International Energy Agency Heat Pump Conference, Current Situation and Future Prospects**, Verlag für die Technische Universität Graz, Graz, Austria, volume 1, pages 27-28, May 1984.

9. **The Electric Heat Pump Option**, Edison Electric Institute, Washington, DC, USA, August 1983.

10. **Heat Pump Manual**, report EM-4110-SR, Electric Power Research Institute, Palo Alto, CA, USA, and National Rural Electric Cooperative Association, Washington, DC, USA, August 1985.

11. J. M. Calm, *Estimating International Heat Pump Use*, **Prospects in Heat Pump Technology and Marketing**, Proceedings of the International Energy Agency Heat Pump Conference, edited by Kay H. Zimmerman, Lewis Publishers, Chelsea, MI, USA, chapter 42, pages 553-565, April 1987.

12. **Statistical Profile of the Air-Conditioning, Refrigeration, and Heating Industry**, Air-Conditioning and Refrigeration Institute, Arlington, VA, USA, July 1986.

13. Bureau of the Census, U.S. Department of Commerce, Washington, DC, USA.

14. T. C. Gilles, *Factors Affecting the Marketing of Heat Pumps in North America*, **Prospects in Heat Pump Technology and Marketing**, Proceedings of the International Energy Agency Heat Pump Conference, edited by Kay H. Zimmerman, Lewis Publishers, Chelsea, MI, USA, chapter 34, pages 465-478, April 1987.

15. G. C. Groff and R. E. Ertinger, *Heat Pumps in the USA Projections for the Future*, **Proceedings of the International Energy Agency Heat Pump Conference Current Situation and Future Prospects**, Verlag für die Technische Universität Graz, Graz, Austria, volume 1, pages 207-217, May 1984.

16. **Annual Energy Review 1986**, report DOE/EIA-0384(86), U.S. Department of Energy, Washington, DC, USA, May 1987.

17. **Annual Energy Outlook 1986**, report DOE/EIA-0383(86), U.S. Department of Energy, Washington, DC, USA, February 1987.

GAS HEAT PUMPS
TECHNICAL AND COMMERCIAL STATUS REPORT

Dr. Joachim Paul
SABROE Kältetechnik GmbH
P.O. Box 2653, Ochsenweg 73
D-2390 Flensburg

SUMMARY
The report deals with the status of presently applied systems, highlights the main issues for systems, comments shortly present applications and sums up new developments.

1. Presently Applied Systems are:

- closed cycle systems
- open cycle systems and
- Carnot systems

After the last years´ depression the heat pump never really gained speed again, which is mainly due to badly planned, badly executed and poorly matching systems. Too many hoped to make "quick money", the loosers are both end-users and producers.

Common issues for closed cycle systems show an interest to go from Carnot to Lorenz cycles, partly by using multistep machines or nonazeotropic mixtures. Whereas domestic heating has only marginal importance, high temperature applications have been introduced with temperatures up to 120 deg. C, mainly in industrial applications. The same goes for pinch-point considerations when designing industrial heatpumps by using Linhoff´s theory on heat transport feasibility.

New heat sources have been applied, one example later will give a comment on a so-called vacuum-ice system.

In the light of the ongoing CFC discussions traditional heatpump systems have to be changed to use non-hazardous refrigerants, however, these substitutes are technically not applicable yet.

When it comes to specific heatpumps like gas-driven heatpumps, the modular concept offers both a "threat and opportunity". Some producers have made a standard, modular design with the threat, that host processes can not fully be matched and there has to be found a compromise between standard equipment and individual needs. The opportunity behind the modular concept is, however, that standard designs with the lower price and the proven performance and reliability could increase the total quality of the system and by this recover some of the confidence lost before. During the last few years, catalytic converters have been introduced, by the same time also considerable improvements on gasmotors have been made to mainly NOx. Only few Sterling machines have been introduced, however, they were not commercially successful yet and it needs a certain market volume to justify costly developments. The last gas-driven issue is the boosting of engines with ORC-cycles (Organic Rankine Cycles), where waste heat from a gas-driven motor is not used as heat but converted into mechanical power, thus increasing the overall power without the need of managing power and heat at the same time.

Open cycle systems have in some countries (UK) become increasingly important, among the developments, the improved designs for Mechanical Vapour Recompression plants (MVR) must be mentioned. Both screw compressors and turbines have been applied, the aim is to have at reasonable price the water vapour resistant machine, these applications point mainly in the direction of industrial applications.

The success of sorption systems cannot be accounted for by a number of installations but by the increasingly sophisticated design. Without further comments some key-words out of this environment: single-stage, double-effect, double-lift absorption systems or heat transformers. Investigations on working fluids are uncompleted, there have been serious corrosion problems, on the other side solid absorption systems (salts, Zeolith) or pediodical machines have been introduced respectively presented. Just in order not to forget, chemical heat storage should be mentioned to complete the overview.

2. Main Issues for Systems

Over the last few years, discussions and publications aim mainly for the goal, how to cope with conditions of the host process. It is not the heatpump itself but the incorporation of the heatpump in a thermal system which created many of the well-known problems, this resulted mainly from incorrect sizing and consequent under-utilization of the system. Furthermore, the cycle configuration has become increasingly important and it is rather the software idea which counts than the component design.

Furthermore, passive systems are more and more included and it has never been understood by the author, that heat recovery measures together with the design of heatpumps has not been taken into account to a larger extent. But the awareness is now increasing. Alltogether, the aim is for maximum reliability and efficiency with the goal, to keep promises once made.

3. Applications

As already mentioned, applications of heatpumps are becoming increasingly important in industrial and commercial environment and ecological aspects have added to the purely economical ones. During the last few years many examples of successful applications have been reported and published, among them should be mentioned:

- fish farming
- greenhouses
- timber driers
- cooling of effluents
- evaporation and distillation processes

4. New developments

There has not really been that terribly much in terms of new dedevelopments, among the most important are the following:

- a new look on heat sources (ground coupled, air coolers, water coolers)
- multiple-effect MVR-systems
- conversion of heat into electrical/mechanical power
- extention of temperature and operating pressure limits
- study on "new" refrigerants for substitution on CFCs

When looking on this list one can state, that the industry has taken the initiative by really digging into the problems and the small troubles are on the way out. It is certainly very expensive, to do research and development on the mentioned issues and this work has to be performed by producers and cannot be done by installers.

Among these new developments, the author wants to highlight a new development, which sails under the name of "vacuum-ice" and presents both a heat source and the chiller/ice maker.

The production of vacuum ice means, to freeze water at triple point conditions (0 deg. C, 6 mbar for pure water). The beauty of the system is, that the vacuum-ice machine works both as a water cooler and as ice-maker, depending on load conditions. As a direct-contact system, a vacuum-ice maker can extract heat even from water under 0 deg. C temperature and there is no risk of frosting and therefore no needs for defrosting. As a direct-contact system, the vacuum-ice maker can use any kind of water with any kind of pollution or contamination, as long as a pump can transport the feed water. The vacuum-ice maker is a continuous process and is able to generate ice which is pumpable and which can be used for thermal energy storage, industrial applications or the like resp. disposed off without any hazard to the environment. Fig. 1 shows the triple point of pure water which is the idea behind the whole issue.

What makes vacuum-ice so interesting? It is the fact, that this machine works with the highest possible COP of all ice makers and is even better than normal water chillers. This can be seen in Fig. 2, where the generation temperatures for ice is shown on the right hand side on a "thermometer", the consequential COP is shown on the left hand side of fig. 2. One can express consequence of this also in energy demand, which is shown in Table 3. All other ice-makers have an energy demand which is between 25 and more than 100% and higher than the one of vacum-ice. At the same time it must again be stated, that with the same high evaporating temperature the vacuum-ice maker can also make chilled water without any risk of freezing. Therefore the minimum efficiency of a vacuum-ice maker is at least as good as from any water chiller.

The vacuum-ice maker itself is shown in Fig. 4 and is a very simple system. The main part is a vacuum-tank, almost empty inside, where feedwater partly evaporates and as a consequence cools down another part of the feed water resp. freezes feed water. The water vapour is extracted by a vapour compressor with very little power consumption and condensed on the evaporator of a normal conventional refrigeration system (heat-pump). The frozen ice crystal drops into the sump and forms a kind of ice-slurry which can be extracted and either returned to the heat source (e.g. river water, ground water, cooling water), or be used in any kind of cooling process or even to generate pure ice. For this purpose, an ice concentrator can be used to make any concentration between 0 and 100%.

Looking on the cost and efficiency of such a system, investment is slightly higher than the one of a conventional chiller, the flexible and troublefree adoption and ability to generate both cold water and ice makes it a very interesting alternative for water-heat-source heatpumps and is by all means superior to any ice-making equipment presently applied.

To conclude the vacuum-ice system it can be stated, that it is

- heat source for heatpumps under even harsh conditions
- thermal energy storage for accumulation of "cold" energy i.e. off-peak tariffs to avoid demand charges or to reduce the installed capacity of cooling equipment
- production of large quantities of ice (for 100 tons/day and much more in one unit) for industry and airconditioning
- chiller for cooling tunnels, shafts and mines
- ice maker for chemical industry and civil engineering such as concrete cooling

Another new application is the Organic Rankine Cycle system, Fig. 5 shows one application to make use of the useful heat of a gas engine. Many plants lack from the possibility to control both mechanical resp. electrical power generated and the heat. For district heating systems, heating demand is up and down over the year, the process is mainly governed by only one energy (electricity or heat) and a surplus or shortage of the other energy is the consequence. In other installations, there is not need for heat at all or heat distribution will be very costly.

Therefore it would be nice, to reduce the control parameters to one, i.e. mechanical power resp. electricity. Fig. 5 shows, how the useful heat from exhaust and jacket of the gas engine can be used by feeding an ORC system. The useful heat evaporates a working fluid at high pressure, this working fluid is taken into an expander, where it generates mechanical power. The expander delivers mechanical heat which can be fed on the shaft of the gas engine and by this increases the yield of power generated resp. reduces the gas consumption at same power. The expanded working fluid is condensed, put on high pressure with a feed pump and the cycle starts again.

By these means, the efficiency of the gas engine can be increased by 20%, the example gives very conservative figures for presently marketed machines, where the power at the shaft is increased from 30 to 36%. Taking these ideas into account, the operation of a gas engine in a co-generation plant or in an ORC-booster becomes interesting. Figure 6 shows the energy flow diagrams. Instead of producing mechanical/electrical power and heat at the same time at the same place, one could boost the same gas engine with an ORC-plant and distribute the electricity to decentralized electrical heat pumps. There will be cost reductions for extensive district heating pipework, efficiency and load of the gas engine can be always at optimum and in times, when no heat is needed, the power consumption of the motor can be reduced resp. its generated power can be used without any considerations to heat management. Fig. 7 shows such an ORC system with decentralized electric heat pumps and one can see the simplicity.

Of course, ORC systems can also be used for converting low-grade heat into mechanical/electrical power in other applications, but this is not topic of this paper. To conclude, one can make the following statements for ORC-systems:

- ORC-systems make use of waste heat at temperature levels between 90 deg. C and 200 deg. C
- efficiencies are between 10 to 15%
- waste heat emission is reduced
- there exists environmental benefits (e.g. replacing river water cooling)
- ORC-systems are possible as integrated part of machinery and plants
- electric/mechanical power can be generated costfree from waste heat
- ORC-systems can be used as booster for combustion engines and turbines.

5. Conclusion

The importance of the heatpump has suffered over the last years considerably, especially for heating only and domestic applications heatpumps have dropped drastically. However, in industrial and large scale application, heat pumps are still alive today and with some new development have a good chance to contribute to the energy scenario in our today´s world.

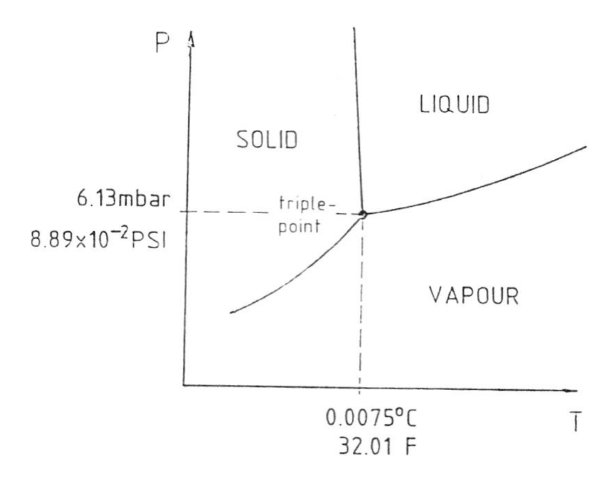

1 TRIPLE-POINT OF PURE WATER

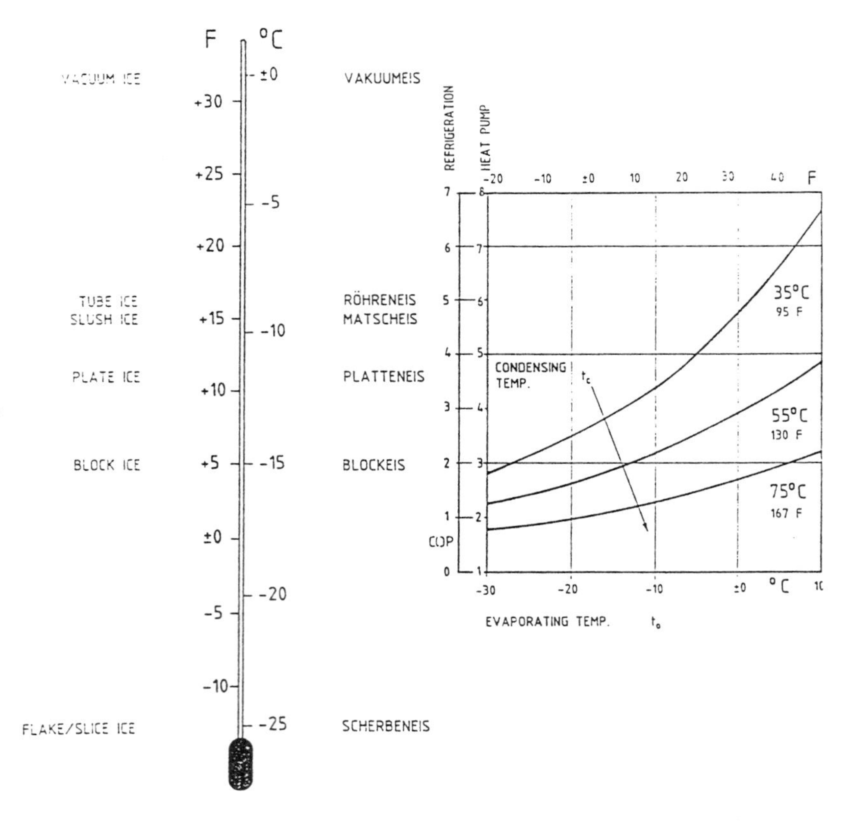

2 TEMPERATURES AND C.O.P. FOR GENERATING ICE WITH DIFFERENT METHODS

	real energy demand	
vacuum - ice	100 %	=+ 0 %
tube - ice	125 %	=+ 25 %
slush - ice	133 %	=+ 33 %
plate - ice	150 %	=+ 50 %
block - ice	166 %	=+ 66 %
flake/slice - ice	210 %	=+110 %

3 ENERGY-DEMAND FOR PRODUCTION OF ICE

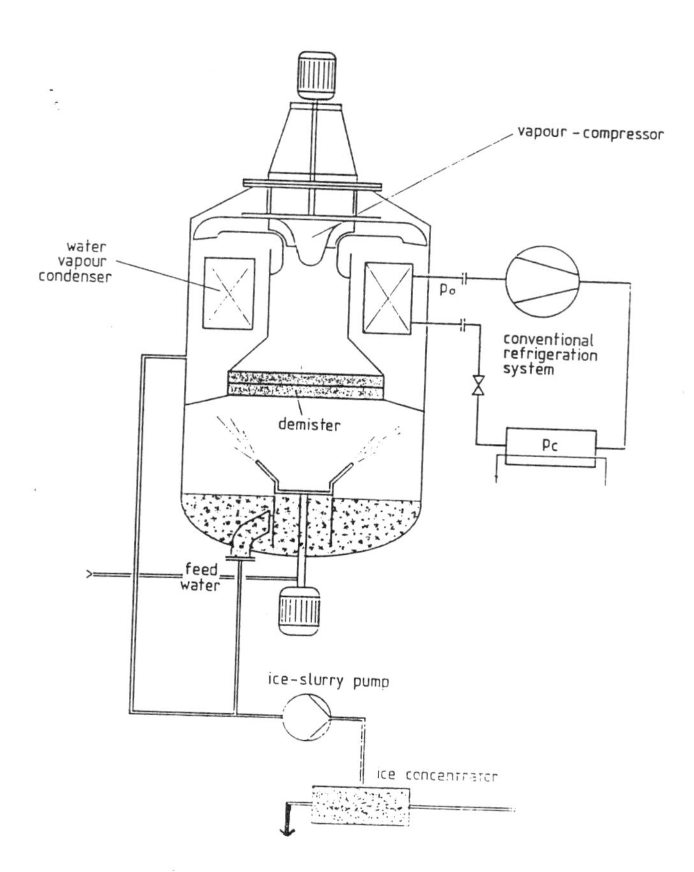

4 VACUUM-ICE MAKER

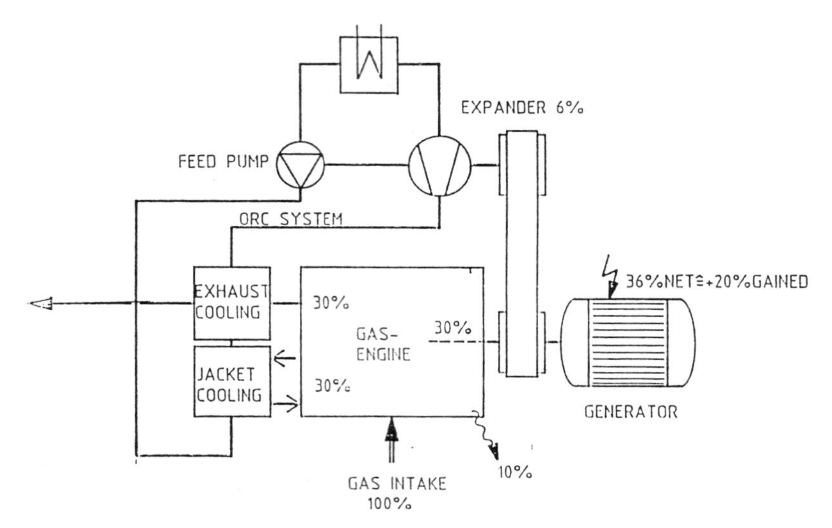

5 BOOSTING A GAS-ENGINE WITH AN ORC-SYSTEM

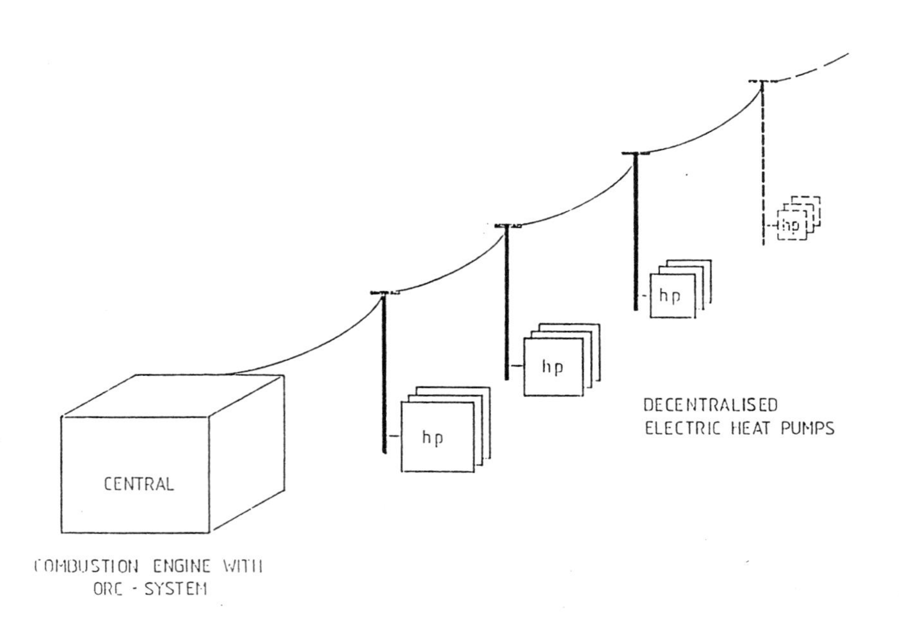

7 DECENTRALIZED HEAT PUMP SYSTEM WITH ORC BOOSTER FOR COMBUSTION ENGINE

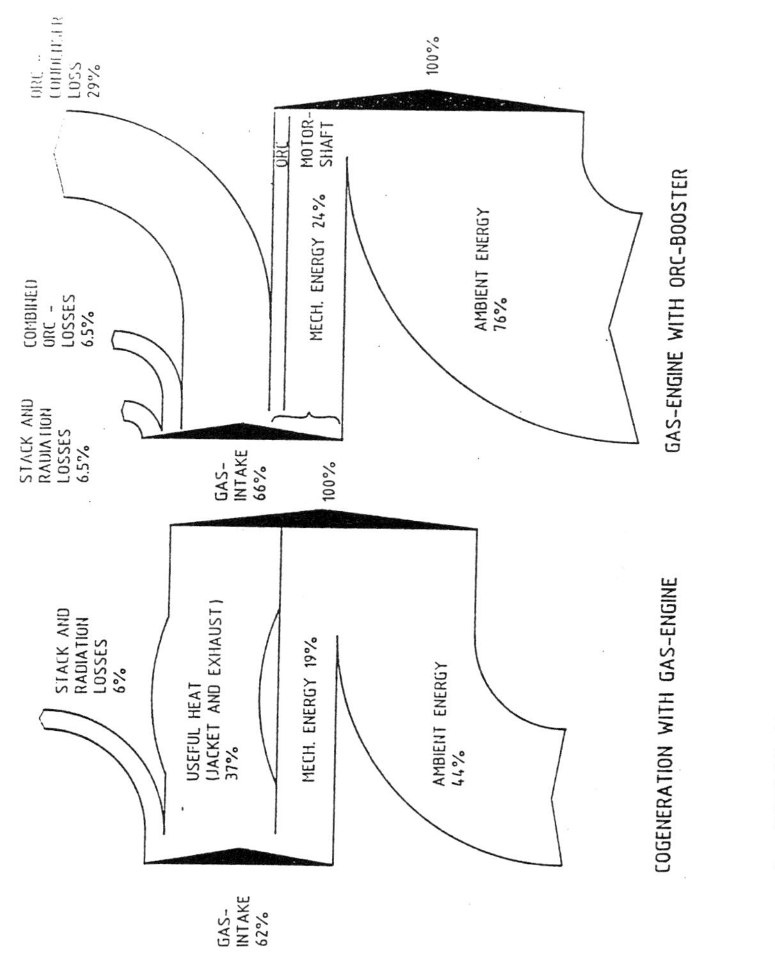

6 ENERGY FLOW CHART FOR GAS HEAT PUMP AND ORC-BOOSTED GAS MOTOR SYSTEM WITH DECENTRALIZED ELECTRIC HEAT PUMPS

Block Central Applications for Electric and Gas Heat Pumps

Hans-Gerhard Rumpf

Rheinisch-Westfälisches Elektrizitätswerk AG (RWE), Head Office
4300 Essen 1, Federal Republic of Germany

ABSTRACT

The block central heat pump with gas or electric motor is used for heating single large buildings or groups of buildings and smaller settlements. In the Federal Repulic of Germany, it is almost exclusively operated in the dual source mode (bivalent mode). Five plants which have been tested in practical terms for some years manifest the numerous and manifold possibilities of block central heat pumps. Heat pumps dimensioned to meet 30 to 45 % of the maximum heat requirements of the consumer are apt to satisfy 65 to 90 % of the annual heat demand. The heat sources power station waste heat, groundwater, river water and outdoor air challenge the conception of the block central heat pump in various ways. The structure of the buildings to be heated is the decisive factor in finding the most suitable arrangement for the components heat source system, heat pump, boiler plant and heating water distribution system.

KEYWORDS

Power station waste heat; cold district heat; industrial waste heat; bivalent operation; heat pump share; heating water network; waste heat utilization for refrigeration.

THE BLOCK CENTRAL HEAT PUMP IN BETWEEN DISTRICT HEAT AND SMALL STANDARD HEAT PUMP

For one- and two-family houses the decision for the heat pump is taken by the owner who occupies his house himself. This decision for the heating technology is taken under very long-term and personal aspects, that also applied to the decision for his own home. Short-term economic benefits are of secondary importance.
District heating is up to large companies and is strongly influenced by the government in most countries. Depending on the situation of the respective country, aspects such as priority given to domestic fuels, expansion of cogeneration, displacement of independent fireplaces from an

agglomeration or other arguments play a decisive role. The use of large heat pumps within district heating systems also depends heavily on considerations of economic policy apart from purely economic ones.
The gas-driven or electric block central heat pump has its application in between these two markets. In a region whose centre is supplied with district heat, e. g. from cogeneration (CHP) it is the outskirts of the agglomeration that are supplied in this way. In smaller towns it is the numerous local authority buildings such as schools, administrative buildings, swimming pools and sport centres that can be supplied by block central heat pumps. Other possible applications are residential areas on the outskirts of cities and areas in the vicinity of econimic centres where a lot of people have come to live in recent years.
Several projects of this kind were implemented in the last few years. It was often the use of an attractive heat source that was at the origin and the centre of the activities. The conditions on the heat sink side, i. e. on the side of the heating systems in the buildings and the related problems, however, are of decisive importance for the feasibility and the economic efficiency. Five examples are to provide an overview of the numerous and manifold possibilities but also problems that exist for the supply of residential and administrative buildings by block central heat pumps.

THE ARZBERG EXAMPLE - POWER STATION WASTE HEAT

The first example is an object at Arzberg in the region of Upper Franconia in the Federal Republic of Germany. On the outskirts of the town of Arzberg, a coal-fired power station is in operation. A partial flow of some 120 m^3/h is taken from the cooling water circuit of this power station and is transferred via a water pipe of about 1,500 m length to the heat pump plants in several buildings of the town. Since water is in ample supply and with a temperature of only 15 °C to 35 °C represents low-grade heat energy, pipes of asbestos cement without thermal insulation have been used.
Afer cooling in the heat pumps' evaporators, the water is discharged into a brook (Flitterbach) flowing through the town. Upstream the power station, this brook runs into the river (Röslau) feeding water to the cooling circuit of the power station.

Cold District Heat

The concept of combining several block central heat pumps with one heat source system, i. e. developing one single extraordinary heat source has so far been rarely put into practice. This method makes it possible to utilize even the waste heat from industrial plants, refuse incinerators, sewage water treatment installations etc., if the temperature is not high enough to allow direct heating. Such a concept is also referred to as "cold district heat" and with the use of a one-pipe system is considered the most favourable kind of industrial and power station waste heat utilization.
The interconnection of several block central heat pumps with one heat source is also feasible in connection with groundwater, occasionally even with surface water. It is a reasonable option in both technical and economic terms if the expenditure for developing and pretreating the water heat source is high and the capital cost is largely uninfluenced by the size of the plant.

The plant at Arzberg was planned as early as 1978 and was completed in 1979. In the winter months from December to March, cooling water temperatures vary between 15 °C and 30 °C, depending on power station output, air temperature and air humidity.
Up to 1987, a school, an outdoor swimming pool and twelve other objects, even though much smaller, have been connected to the cooling water long-distance pipe with one electric heat pump each. This development goes far beyond the planned scale of 1980 (Fig. 1). Further projects are scheduled. Due to the favourable temperature level the heat pumps' coefficient of performance range between 4.0 and 5.5 for heating the buildings and attain values from 5.5 to 6.5 for the outdoor swimming pool. These figures do not take into account the pumping energy of 25 kW which is needed for the long-distance pipe. This energy is made available by the power station without any costs for the consumers.

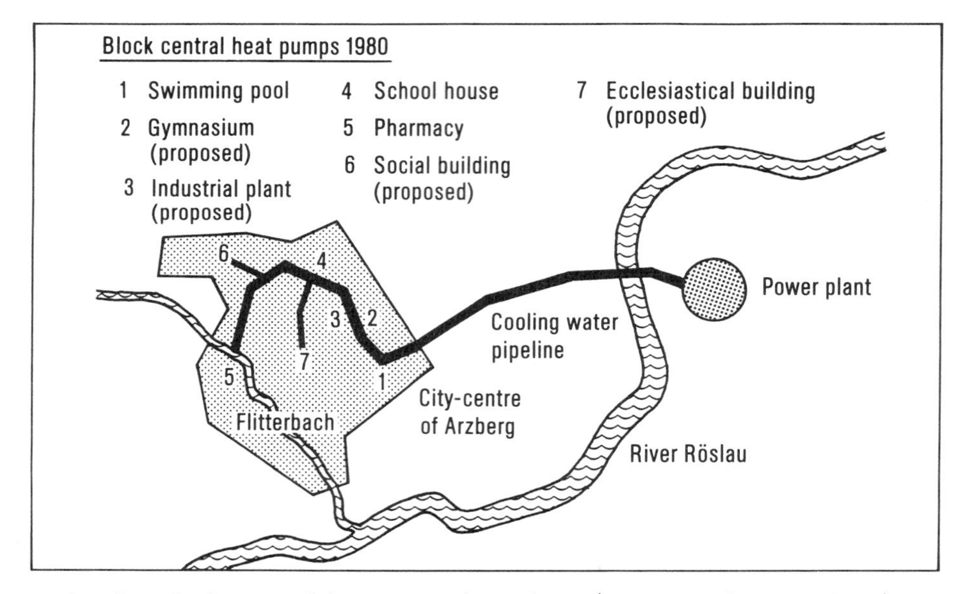

Fig. 1 Block central heat pumps in Arzberg (power station wast heat); "cold district heat supply"

The outdoor swimming pool is supplied by a heat pump which is designed for a boiler capacity of about 50 %. It almost completely satisfies the annual energy requirements. The oil-fired boiler continues to be available for start-up heating in spring and during outages.
The school has a maximum heat demand of 2.6 MW. The condenser capacity of the heat pump is 1.2 MW. This division of capacities between boiler plant and heat pump as well as the temperatures of the heat distribution system in the school result in the typical operating conditions of the bivalent heat pump as encountered in a multitude of block central heating stations.

Bivalent Mode of Operation

Bivalent operation means that up to a certain operating point represented by the outdoor temperature the heat pumps works on its own. At the Arzberg school this point is about -2 °C. When this temperature is reached supplementary heating by a boiler is required in order to attain the necessary heating water temperature and to meet the heating demand of the building. From a second operating point onwards, the boiler heats alone. In the present example the outdoor temperature is -6 °C. This operating condition is reached when the heat distribution system of the building attains a return temperature that can no longer be handled by the heat pump. It is particularly the point at which the heat pump is switched off that is subject to highly varying influences and thus requires particular attention in every plant. As far as the design of the heat distribution system can be influenced this second switching point can be further reduced or made superfluous. Such measures, however, are not compulsory and even under economic aspects are not of urgent necessity since the amount of heat used on such extremely cold days accounts for less than 10 % of the annual heating energy consumption.
It is an economic necessity to restrict the heat pump's efficiency to at most 50 % of the maximum heat requirements. If a heat pump whose only task was to heat a building would be designed such that it is capable of meeting the heat demand of the coldest winter day on its own, its rate of utilization is too low and the high cost of investment is inefficiently used. This important statement applies to all the projects presented in this paper and is independent of whether the heat pump is driven by a gas or electric motor.

THE DORSTEN EXAMPLE - RIVER WATER

The second example of block central heat pumps is a plant where the three components heat source system, heat pump and boiler plant are concentrated in one location and the connected buildings are supplied via heating water pipes. The project is named "Maria Lindenhof" and was realised in Dorsten on the northern edge of the Ruhr valley in 1983. The heat is tapped from the river Lippe whose temperature never, not even in winter, falls below 8 °C. This phenomenon is due to cooling water discharges by industrial plants.
The block central heat pump consists of two gas-fired boilers with a heating capacity of about 5.8 MW and three gas-driven heat pumps which at a water temperature of 10 °C are capable of delivering a heat output of about 2,100 kW, about 1,500 kW of which are condenser capacity. The refrigeration machine of a nearby ice stadium is also integrated into the heat source system. Thus, it is possible to use the waste heat from the refrigeration process for the heat pumps.
The heating water pipe supplies buildings with a maximum heat demand of 5.6 MW. The buildings which were erected before 1983 already are equipped with conventional heating systems with heating water temperatures of 90/70 °C. In the newly constructed buildings the heating systems are designed for 90/30 °C. As the returning heating water flows mix, the heat pump's range of application is improved. Its limit temperatures is set at 65 °C. A major component of the heat pump is a storage vessel which can be charged with 40 m^3 of heating water. Thus, the heating energy of a one-hour heat pump operation can be taken up to equalize considerably the operation of the heat pumps.

It is planned to extend the block central heat pump in line with projected housing development. Planning already takes account of a third boiler and a fourth heat pump.

THE WESEL EXAMPLE - GROUNDWATER

The third example for a block central heat pump is a plant using groundwater as heat source. In the district administration building of Wesel an oil boiler plant with a heating capacity of 2.2 MW and an electric heat pump plant comprising three units of 280 kW (altogether 840 kW) heating capacity each have been installed.
Since the winter of 1984/85 the district administration building (1,650 kW) and from 1987/88 the police station (550 kW) have been and will be supplied by this central heating installation. The second building is connected by an underground heating water pipe (Fig. 2). The development of the heat source involved considerable expenditure, as illustrated in the figure. The maximum amount of water of 120 m³/h taken from the production well had to be distributed to four well drains. Moreover, a balancing pool was required since the well drains are arranged at different heights. One of the three heat pumps is used to cool the district administration building in summer. The resulting waste heat is in turn used to heat the police station which has a low heating demand in summer, too.

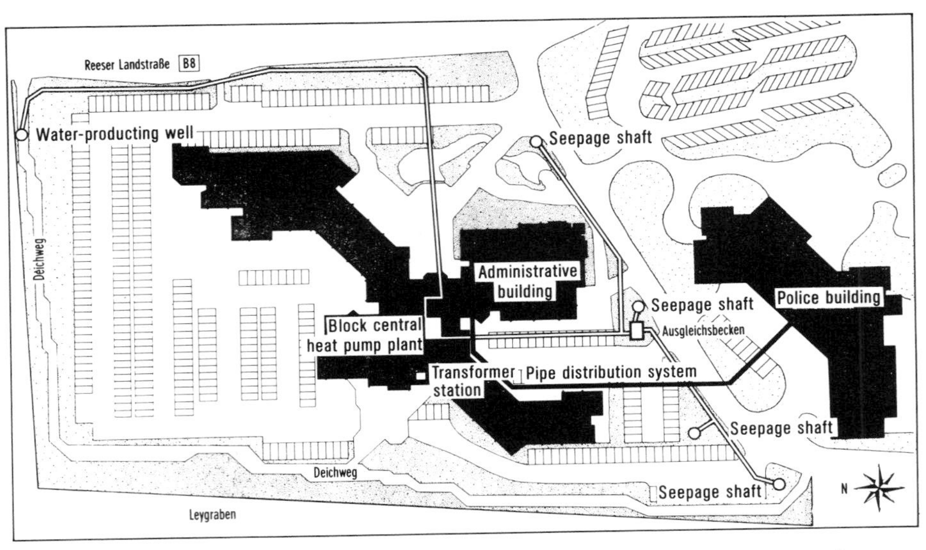

Fig. 2 Layout of the Wesel block central heat pump (groundwater)

The results achieved by this bivalent plant for heating the district administration building in the winter of 85/86 are illustrated in Fig. 3. The heat pump heating systems with a condenser capacity of 560 kW intended for the district administration building have supplied more than 90 % of the annual heating energy consumption. The coefficient of performance was 3.1 and considers all peripheral consumers including the losses of the 400 kVA transformer. The police station will extend energy supplies of the block central heat pump by about 50 %. Details on this project can be taken from the poster 4.3.6 that will be shown in the poster session.

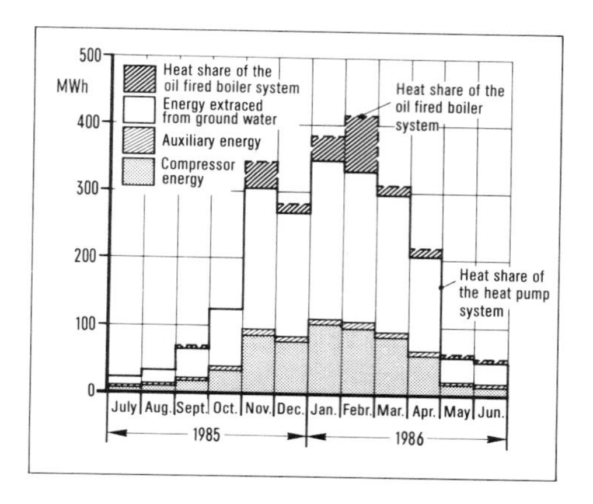

Fig. 3 Monthly heat supplies

THE ESSEN AND GELSENKIRCHEN EXAMPLES - OUTDOOR AIR

The examples 4 and 5 illustrate block central heat pumps using outdoor air as heat source. In comparison with industrial or power station waste heat, in comparison with river water and groundwater this heat source seems an ordinary one. In fact, it offers a lower energy density and poor temperatures in winter. Nevertheless, outdoor air opens up the greatest possibilities as a heat source especially for the large number of small standard heat pumps. As far as the smaller number of block central heat pumps is concerned, which is discussed in the present paper, growing condenser capcities increase the chances of the "special" heat sources. As the capacity of the block central heat pump improves, the larger expenditure for the development of the special heat sources loses in importance. The problems relating to air quantities, air control and sound insulation, however, gain in significance as the efficiency of the heat pump grows.

A settlement comprising 584 flats in 28 four-storey buildings was redeveloped and modernized in 1985 in Essen. The heat requirements of the floor space consisting of 41,500 m² was reduced by 38 %. Thus, the temperatures of the heat distribution system automatically favoured the heat pump. The new gas boiler plant has a capacity of 3.0 MW. The gas-driven heat pump has a capacitiy of 1,032 kW at an outdoor air temperature of +5 °C. The share of the condenser heat in the heating efficiency of the heat pump amounts to 65 %, that is 670 kW. The heat pump supplies approximately 65 % of the annual heating energy for the settlement although its share in the heating capacity of the central installation accounts for only 34 %. The heat distribution network in the settlement, which consists of underground pipes, existed right from the beginning. This has proved a considerable economic advantage for this block central heat pump.

In the neighbouring town of Gelsenkirchen another block central heat pump using ambient air as the heat source was erected. The five four-storey buildings were completed in 1982. The heat distribution in the building system has been designed for 68/53 °C, thus accomodating both the needs of the heat pump in terms of moderate temperatures and the builder's wish for low investments. The 104 flats are supplied via two block central heat pumps (Fig. 4). This division was the most favourable option to keep the costs for the manufacture of the heating water pipes and the structures as low as possible. The two boilers of the bivalent plant are located in one residential building each. The electric heat pumps are housed in standard garages with improved sound insulation. For 8,250 m² of floor space with a theoretical heating demand of 504 kW (61 W/m²) a boiler capacity of 460 kW has been installed. The heating efficiency of the heat pump is 228 kW, that accounts for 45 % of the heat requirements. In 1986 the share in the annual heating energy consumption was 72 %, the annual coefficient of performance of the entire plant was 2.8.

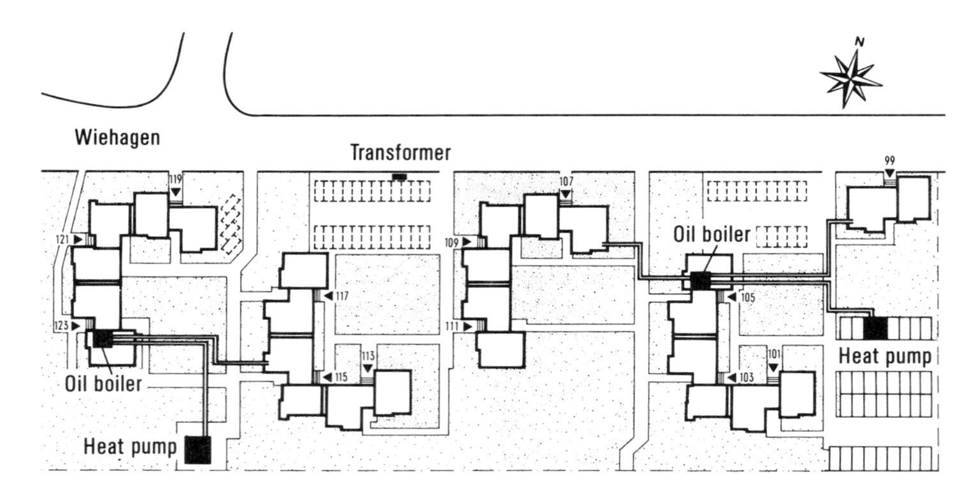

Fig. 4 Layout of the Gelsenkirchen block central heat pumps (ambient air)

REFERENCES

Haase, J. (1981). Nahwärmeversorgung des zentralstädtischen Gebietes "Maria Lindenhof". ASUE-Schriftenreihe, Nr. 9, 20-28

Müller, L., and E. Wegmann (1981). Nutzung von Kraftwerksabwärme zur Heizung. Kraftwerk und Umwelt, 1981, 173-179.

Müller-Schnick, H. (1985). Nahwärmeversorgung der Humboldtsiedlung. ASUE-Schriftenreihe,, Nr. 9, 38-45

Rumpf, H.-G. (1985). Wärmeversorgung von Wohn- und Verwaltungsgebäuden mit Wärmepumpenheizzentralen. Fachtagung Industrielle Elektroprozeßwärme, Nürnberg 1985, 12.1-12.14

OPTIMIZATION OF THE REFRIGERATION CYCLE WITH THE ELECTRONICALLY CONTROLLED EXPANSION VALVES

H. Itoh
Toshiba Fuji Works, 336 Tadehara
Fuji City, 416, Japan

This paper introduces an electronically controlled expansion valve newly developed with the high electronics technology,and its application to heat pump room air conditioners. Fig.1 shows the market of room air conditioners in Japan. Heat pump room air conditioner,especialy with the inverter,has rapidly increased. Last year, over 2 million heat pump room air conditioners,including more than 1 million inverter heat pumps are sold. Fig.2 shows the total sales of cars and that of ICs in Japan. From this chart,we can easily understand how rapidly the production of ICs has increased. In this electronics age,the application of micro-electronics technology has become to be widely used in every field of industry,and our heat pumps are no exception. Electronics revolution has come to our heat pump room air conditioners. The compresoor,the heat exchanger and the expansion device are the 3 major essencial areas of the heat pump engineering. In 1978, the fan control was applied to regulate the capacity of the heat-exchanger. In 1982,the compressor has become to be controllable by the inverter. As for the expansion device,the electronic expansion valve was put into practical use in 1984. Thus these all items of heat pump RACs,have become to be controllable by the micro-processor.

In Japan,2 types of electronic expansion valve are generally used.(See Fig.3) The revolution of the pulse-motor is lowered down by the gear,and transmitted to the driver. The driver is threaded,and the motor rotation is converted into vertical movement of the driver. By this up-and-down movement,the valve opening degree can be optionally changed.(See the picture on the left side in Fig.3) In the other type(the picture on the right side),the rotor of the pulse-motor is in the refrigerant circuit. The motor is the canned type,that is,the stator and the rotor are parted by the can. So the valve is directly driven by the rotor. The pulse command to the motor is optionally decided by the refrigeration system designer. In Fig.4,the electronic expansion devices are compared with the conventional expansion devices. The electronics expansion valve is superior in control freedom,control range and cycle response. The cost is slightly high,but the reversibility of the refrigerant flow,is very convenient for heat pump,and the refrigeration cycle can be simplified. Fig.5 shows an example of the application to the residential room air conditioner which is most popular in Japan. The expansion valve is generally controlled by the superheat,but in this case the electronic expansion valve is controlled by the micro-processor with the discharge temperature of the rotary compressor and the operation frequency of the inverter. The control by the discharge temperature has the following merits.

1. The compressor reliability increases,because the compressor temperature can be kept at our desired degree.
2. The refrigeration cycle is simple and efficient,because the liquid injection cooling is not necessary.
3. Low cost; One senser covers the whole control. The superheat control method needs at least 3 sensers for cooling and heating.

In addition, the valve is controlled by the micro-processor. Therefore the problem of hunting that sometime goes with the thermostatic expansion valve can be easily solved.

Next, I describe how to control the valve with the discharge temperature and the operation friquency and how to keep the maximum capacity. Fig.6 shows the correlation between the discharge temperature and the cooling capacity. If the compressor is operated at 30 Hz,the cooling capacity becomes maximum when the discharge temperature is 68℃. If the compressor is operated at 60 Hz,the cooling capacity becomes maximum when the discharge temperature is 95℃. Fig.7 shows the correlation between these compressor operation frequency and the cooling capacity.

If we put these data into the ROM(read only memory), we can get and always keep the maximum cooling capacity.

This electronic expansion valve is usable not only for the purpose just mentioned above,that is,to keep the maximum capacity but also for other various purposes,for the pulse commands can be optionally decided by the designer. The valve can be fully opened or closed or kept to some fixed opening degree for some fixed time by using the information of the temperature,pressure,time and the operation mode of the refrigeration cycle. For example if we want to keep the compressor temperature below a certain level,120℃ for example,for the long life of the compressor,we only have to put the command into the built-in program of the micro-processor. I'll show by example that the defrosting time is shortened by using this valve and the inverter. Fig.8 shows the refrigeration cycle. The 2-way valve is newly added.When the frost accumulate on the outdoor coil and the micro-processor judges the defrosting is needed,the preparation for defrosting starts. The electronic expansion valve is closed slightly,and the revolution of the compressor is raised to it's maximum speed by the inverter. Then the suction gas is superheated and the compressor temperature rises in a short time. After that, the compressor temperature is checked by the discharge temperature and when it reaches the fixed degree,the 2-way valve is opened and at the same time,electronic expansion valve is fully opened.Then the very hot discharge gas is provided into the outdoor coil through the 2-way valve. The electronic expansion valve is fully opened during defrosting. Therefore the refrigerant stagnation in the indoor coil is prevented and the suction pressure can be kept high. Accordingly the maximum input of the compressor is kept. In this way the defrosting time is reduced very much due to high compressor temperature before defrosting, and the maximum compressor input while defrosting. Fig.9 shows the comparison of the energy for the defrosting with the 1HP-class residential heat pump room air conditioner. The energy necessary for defrosting is about 150 Whs in the conventional reverse cycle method,and the defrosting operation takes 6 min. The energy consumed in 6 min is 60 Whs,as the compressor input is 600Ws. The energy received from the compressor is very little. On the contrary in the new defrosting method,defrosting is finished in only 3 min. The energy consumed in 3 min is 70 Whs as the compressor input is 1400 Ws. I'll show another application; the refrigeration system controlled by the electronic expansion valve and the inverter. The system in Fig.10,we call super multi-system,was marketted in this January and we enjoy a good reputation.This kind of multi-system is getting to be popular in the commercial use in Japan. This system consists of 3 parts; the outdoor unit,the control unit and the indoor units.In the cooling mode the discharged gas goes through the reversing valve,the outdoor coil,the piping,the electronic valve and the indoor coil,and return to the compressor. The refrigerant is distributed with the electronic expansion valve to each indoor coil. There are many types and sizes in the indoor units,and 2 or 3 indoor models,ceiling cassette type with 2HP or ceiling duct type with 3HP for example,can be selected and connected to this system,and operated individually. Therefore it is very important to control the compressor output and refrigerant flow to keep the refrigeration cycle and the capacity best. For this purpose,each indoor micro-processor sends the command according to the load to the micro-processor in the control unit. The control unit micro-processor computes the total load and commands the outdoor inverter micro-processor to operate the compressor with the optimal frequency. The control unit micro-processor also controls the on-off of the 2-way valve and the opening degree of each electronic expansion valve to keep the refrigeration cycle best. Next I'll explain another system which I now call the flame source heat pump system. The energy for the flame is the gas. So it has the strong merit to keep the constant heating capacity against the lower outdoor temperature. The cooling mode has the conventional refrigeration cycle.The discharged gas is condensed in the outdoor coil and vaporized in the indoor coil. But in the heating mode,the liquid refrigerant is heated with the flame and vaporized in the heat exchanger. The discharged refrigerant vapor is liquified in the indoor coil,and it's flow is controlled by the electronic expansion valve. The refrigerant flow control is very important,because if the flow control is wrong,the system is strongly damaged from the deterioration of oil and refrigerant. Electronic expansion valve is very convenient for it's quick response,wide control range and especially complete cotrollability by the micro-processor.

CONCLUSION

The electronics technology has come to be used for the heat pump engineering,and the application of the inverter to the heat pump made the energy saving performance,and the quality of comfort more and more excellent.Furthermore,with the application of the electronic expansion valve,the heat pump room air conditioner has additional merits as follows: (1) The defrosting time is shortened,because the compressor input and the compressor temperature are controllable. (2) The compressor reliability is increased,because the overheating and overcooling trouble of the compressor is preventable. (3) The refrigeration cycle is simplified because of the reversible flow of the expansion valve and no injection cooling. (4) The compressor efficiency is improved,because the injection cooling is not necessary. (5) The annual performance factor is improved,because the refrigeration cycle has always maximum efficiency. At last,in this electronics age,these electronic control expansion valve to assure the free and accurate control with the micro-processor can also be used for other purposes. The system with high reliability and performance can be made by using this new valve and the inverter.

ROOM AIR CONDITIONERS
MARKET IN JAPAN

(6,000~20,000 BTU/H)

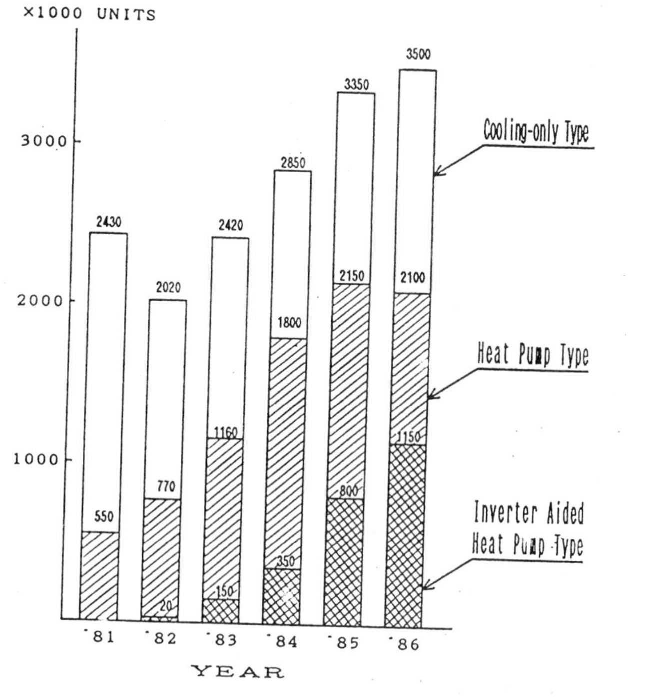
×1000 UNITS
3000
2000
1000
2430
550
2020
770
20
2420
1160
150
2850
1800
350
3350
2150
800
3500
2100
1150
Cooling-only Type
Heat Pump Type
Inverter Aided
Heat Pump Type
'81 '82 '83 '84 '85 '86
YEAR

Fig. 1

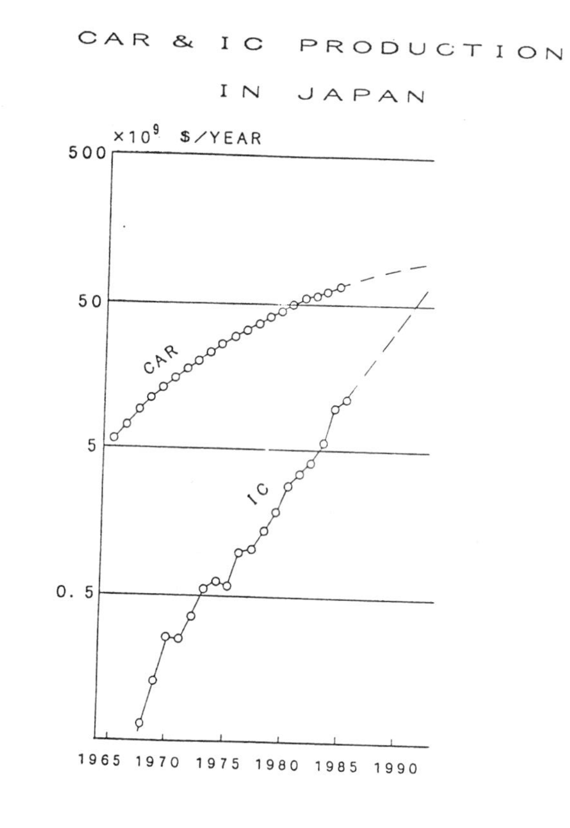
CAR & IC PRODUCTION
IN JAPAN
×10⁹ $/YEAR
500
50
5
0.5
CAR
IC
1965 1970 1975 1980 1985 1990
YEAR

Fig. 2

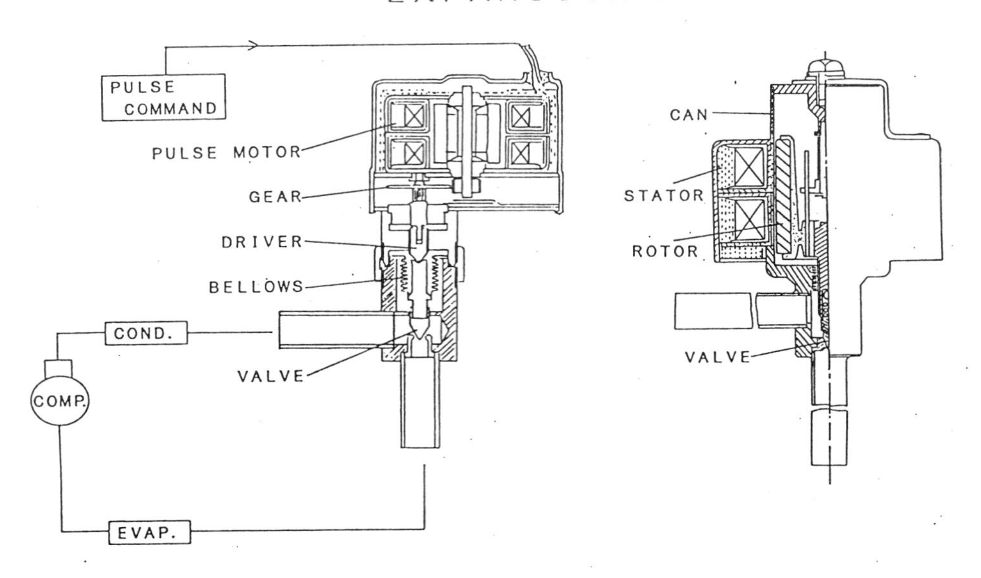

Fig. 3

COMPARISON OF EXPANSION DEVICES

	ELECTRONIC CONTROL EXPANSION VALVE	THERMOSTATIC EXPANSION VALVE	CAPILLARY TUBE
CONTROL FREEDOM	◎	×	×
CONTROL RANGE	◎	○	△
REVERSIBILITY	○	×	○
CYCLE RESPONSE	◎	△	○
ADDITIONAL FUNCTION	◎	×	×
COST	×	△	◎

◎: EXCELLENT ○: GOOD △: AVERAGE ×: INFERIOR

Fig. 4

EXAMPLE OF REFRIGERATION CYCLE WITH ELECTRONIC EXP. V.

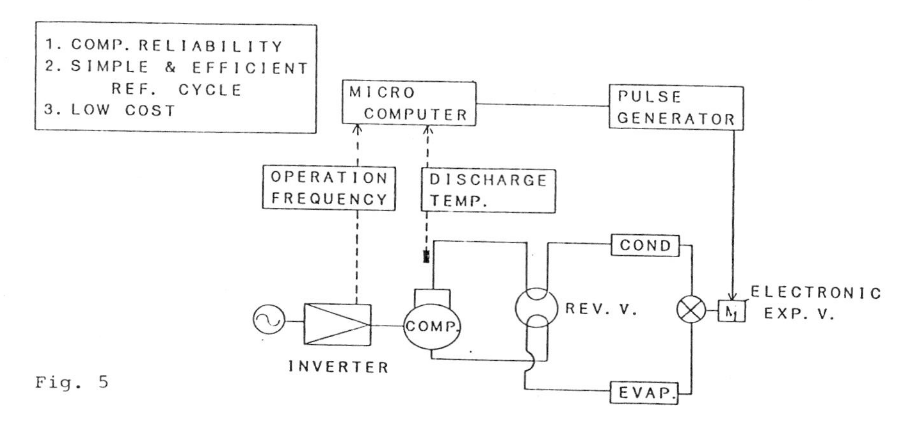

Fig. 5

HOW TO KEEP MAX. CAPACITY

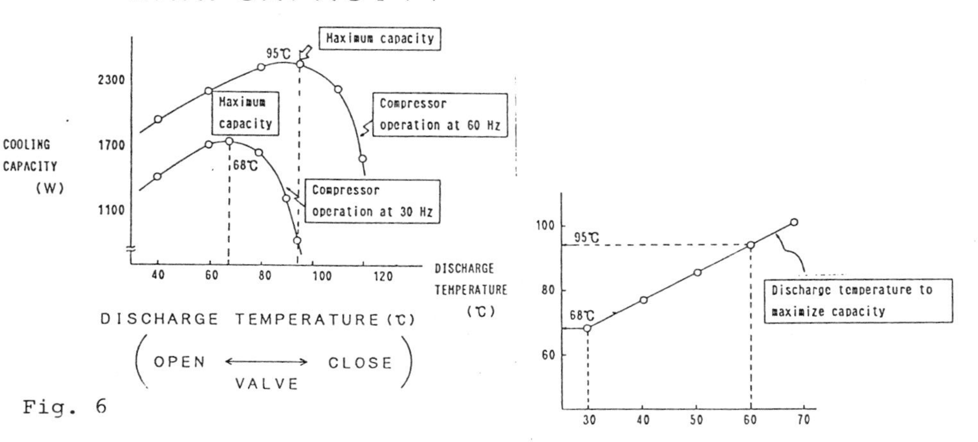

Fig. 6

COMPRESSOR OPERATION FREQUENCY (HZ)

Fig. 7

MINIMIZATION OF DEFROSTING TIME

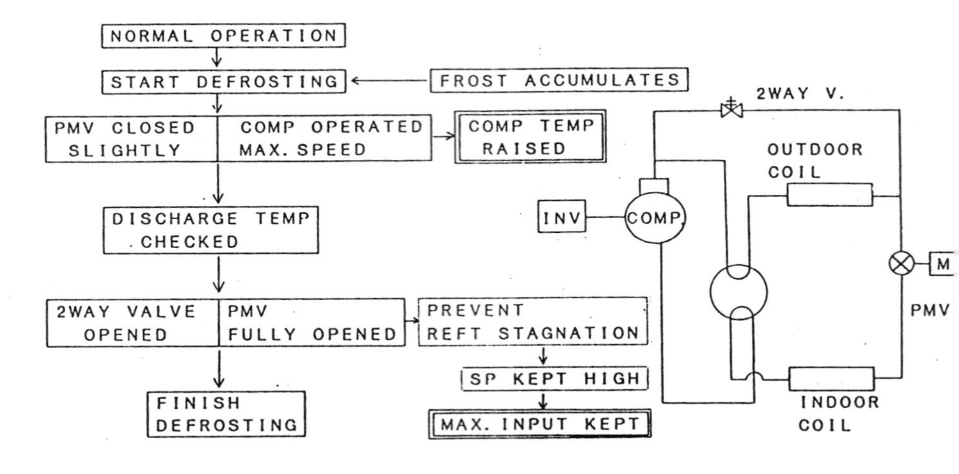

Fig. 8

COMPARISON OF ENERGY FOR DEFROSTING

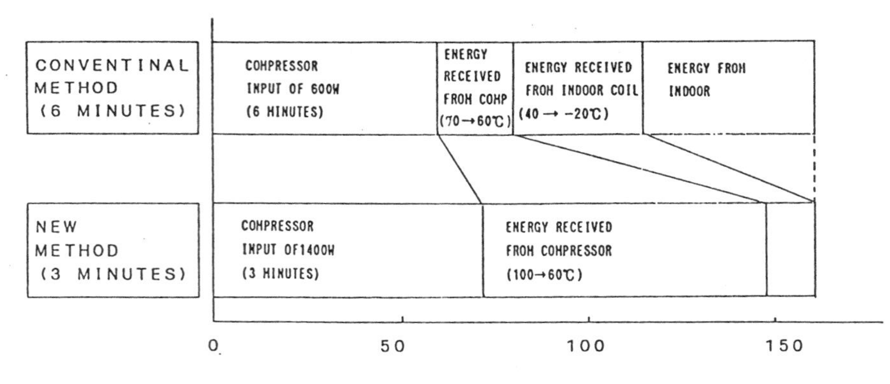

Fig. 9

SUPER MULTI-SYSTEM

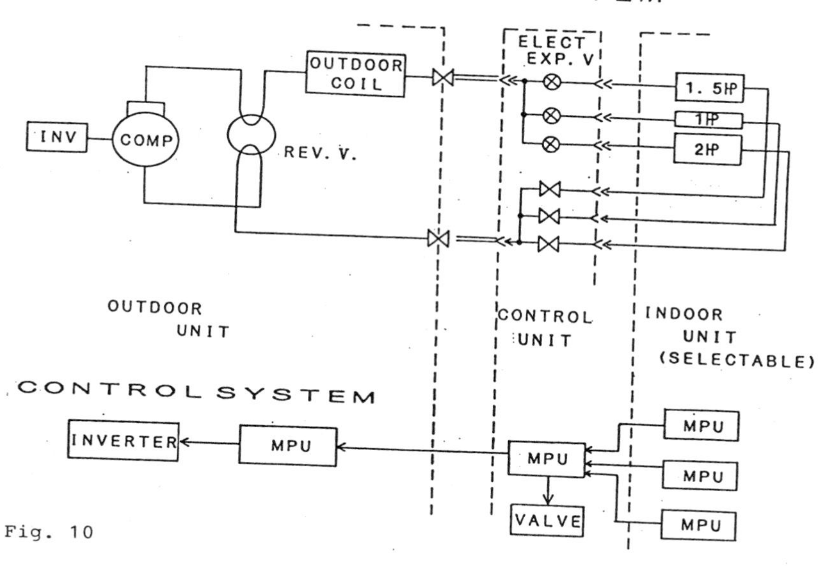

Fig. 10

FLAME SOURCE HEAT PUMP SYSTEM

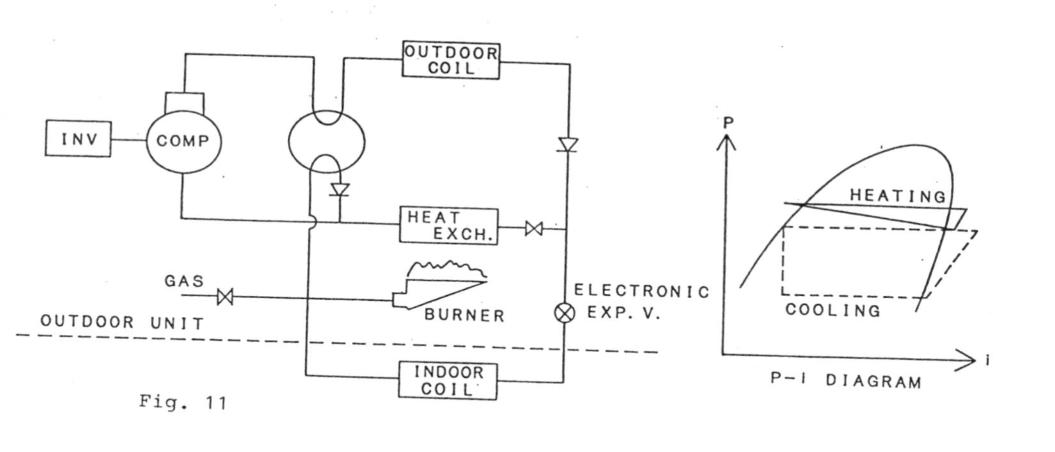

Fig. 11

AN ANALYSIS OF SOLAR ASSISTED HEAT PUMP SYSTEM FOR APARTMENT BUILDING HEATING IN POLISH CLIMATIC CONDITIONS

D. Chwieduk

Inst.of Fund.Tech.Research Polish Academy of Sciences
00049 Warsaw,Poland

ABSTRACT

The text is dealing with an analysis of possibilities of utilizing series solar assisted heat pump (SAHP) system for apartment building heating in Poland. An analysis of system operation modes for providing solar energy is described. The calculations of the system are made through computer simulation using mathematical model based on a "first-cut" method. The results: load profiles, useful rate of energy from solar collectors, storage temperature and thermal performance factor of the SAHP system are presented for typical Polish heating season.

KEYWORDS

Solar energy; useful energy gain from solar collectors; solar assisted heat pump system; building heating; performance factor; storage temperature.

INTRODUCTION

In order to identify when solar energy could be used for apartment building heating it is necessary to have an analysis of energy consumption of the building. In the following text an example of the typical Polish apartment building of standard design is selected for investigation of the series solar assisted heat pump system. The heating load of the apartment building depends on the construction of the building, climatic region, the indoor and outdoor conditions and other factors. The analysis is done to determine the possibilities of satisfying the energy requirements for the chosen building via the SAHP system. This analysis does not address itself to the technical nor the economic details of the problem but does provide an answer as to its possibility.

THE POSSIBILITY OF PROVIDING SOLAR ENERGY FOR APARTMENT BUILDING HEATING

The typical Polish apartment building has: four floor, 32 apartments each one about 60 m^2 (150 m^3). The external walls and ceiling at last floor are insulated with styrofoam and mineral wool respectively. Based on various assumed design conditions the average hourly heating loads are calculated using the standard accepted procedure (Polish Standard PN-82/B-02020). It is assumed, that heating requirements are equal to total heat losses from the building, i.e. heat transfer losses and ventilation losses, which are expressed in a function of ambient temperature.

In Poland nearly 80 percent of total annual solar radiation occurs during the warmer months from April to the end of September. In winter average available insolation is low (see Table 1, data from Publication of the Inst. of Geophysics (1983)). This obviously limits the possibile utilization of solar energy. Thus because of climate any standard design building could not be provided with adequate space heating from solar energy alone. The efficiency of the solar system must be improved. It can be done by applaying a heat pump.

TABLE 1 Mean Monthly Global Horizontal Insolation and Mean Monthly Ambient Temperature Averaged over 30yr Period of Records at Warsaw

Item	Jan	Feb	Mar	Apr	May	June	July	Aug	Sept	Oct	Nov	Dec
Mean inso. MJ/m^2	62	106	248	356	501	573	547	463	305	165	62	42
Diffuse %	67	65,8	53,8	49,8	47,4	43,8	45,3	45,2	46,7	54,8	68,6	74,4
Mean temp. °C	-3,5	-2,6	1,2	7,8	13,8	17,3	19,1	18,2	13,9	8,1	3,0	-0,6

Description of the Solar Assisted Heat Pump System

A simple series solar assisted heat pump system is chosen for analysis. In this system solar energy is collected by flat plate solar collectors and stored in a water storage tank. Energy from storage tank can be used directly or via a heat pump. In SAHP system heat pump is dependent on the solar heating system. Since the heat pump evaporator requires relatively low temperatures, the temperature of the storage can be also relatively low. If the solar heating system directly or via a heat pump can not meet the load an auxiliary heater would supply necessary energy.
The system is analysed for a variety of solar collectors area A_c and storage volume V using simulation program. Primary design parameters are constant during system operation,

Mathematical Model of the SAHP System

The calculations of the system are made through computer using simple mathematical model based on a "first-cut" method, given by Estes and Kahan (1978). It is assumed, that storage is fully mixed and characterized by a single temperature T_s. An energy balance of storage is also fundamental balance equation of the system and is given by:

$$(\dot{m}c_{pl})\frac{dT_s(t)}{dt}=\begin{cases}0 & (1)\\ Q_u(T_s(t),t) & (2)\end{cases}-\begin{cases}Q_h(t)+Q_s(T_s(t)) & (a)\\ Q_{hp}(T_s(t),t)+Q_s(T_s(t)) & (b)\\ Q_s(T_s(t)) & (c)\end{cases}$$

(1) there is no insolation or T_s is above the allowable maximum

(2) insolation is available; Q_u -rate of useful energy gain of collectors

(a) direct solar heating; Q_h - heating load

(b) heating via a heat pump; Q_{hp} - rate at which heat is extracted by the heat pump evaporator ($Q_{hp} = f(COP(T_s(t)), Q_h(t))$, COP - coefficient of performance)

(c) auxiliary heating; Q_s - rate of energy loss of storage

The energy balance equation is a linear first order differential equation and if all coefficients are constant it can be solved analytically. In order to obtain the constant coefficients, the weather data are represented by periodical waves with constant values over each time step. Daily ambient temperature fluctuations are represented by sinusoidal function. Temperature amplitude is assumed to be constant over heating season (3,5 °C). Maximum and minimum temperatures have been determined using mean monthly temperature averaged over 30yr period (Table 1). The mean hourly radiation sums (direct and diffused) on horizontal surfaces of mean day of each month during the heating season are averaged over the 20yr records period and represented by hourly rectangular waves. The calculation of the system are made for Polish heating season: October - April. The hourly process is repeated for each day of each month over a whole heating period.

RESULTS AND CONCLUSIONS

Figure 1 shows the mean daily fluctuations of useful rate of solar energy gain and partly of heat demand for four months of heating season for various collector area. It is obvious that with the increase of collector area the magnitude of net heat derived from collectors also increase. It can be seen that the curves of heating load are opposite to curves of useful solar energy, the curves of heating loads reach a maximum before sunrise when there is no insolation. The difference between course of these two curves especially high in winter illustrate how it is difficult to utilize solar energy in Polish climatic conditions. During early autumn and spring (spring months are not shown in Fig. 1.) when curves of Q_h intersect the curves of Q_u solar energy could be effectively utilize and stored for night time.

Figure 2 presents daily typical fluctuation of storage temperature plotted vs time for four considered values of collector area. This hourly variation is then repeated for each day during each month. When T_s is higer then 10°C, solar heating via a heat pump is possible. From November (not all months are represented in Fig. 2.) to January storage temperature does not reach high values, due to low insolation level, short duration of solar radiation and low ambient temperature. In February and especially in March, April and October much better insolation conditions couse the storage temperatures increase. During this time solar heating system can operate longer. In spring and early autumn, an excess of solar energy stored in the storage tank in a form of sensible heat can be provided to the heating system during night.

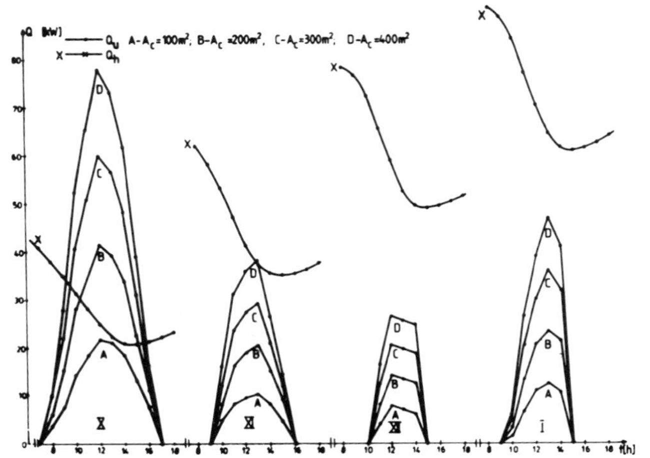

Fig. 1. Daily distribution of useful rate of solar energy gain.

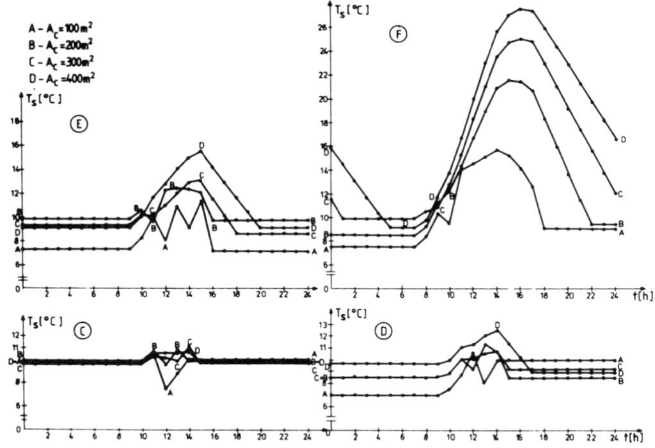

Fig. 2. Daily typical profiles of storage temperature vs collector area. Letters in circles indicate months: C - December, D - January, E - February, F - March.

In the presented study the performance factor of the SAHP system,F, is equivalent to the percentage fraction of solar energy meeting heat demand directly or via a heat pump. Resulting data of the fraction of the monthly total heating load supplied by solar energy are shown in schematic diagram - Fig.3. It is evident, that solar energy fraction QF is dependent on variations in load and solar availability. In winter SAHP system with collectors area of

100 and 400 m^2 can provide only about 5-10 or 15-30 % of heat load,but in spring 26-74 or 74-100 % respectively. It is noticable that solar heating of the building is mainly accomplished via a heat pump. Direct solar heating is possible, when large collector area and high insolation level meet low loads. For increasing collector area from 100 to 400 m^2 following values of seasonal performance factor are achieved: 0,16; 0,26; 0,37; 0,45 (see Fig. 4.).

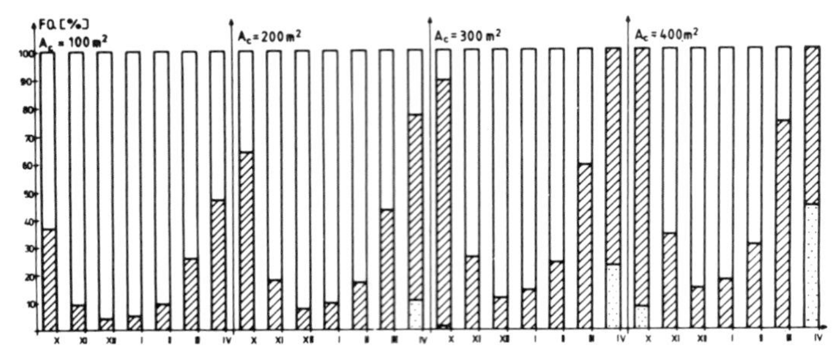

Fig. 3. Monthly fraction of solar energy meeting heat load. Dots indicate direct solar heating, lines heating via a heat pump.

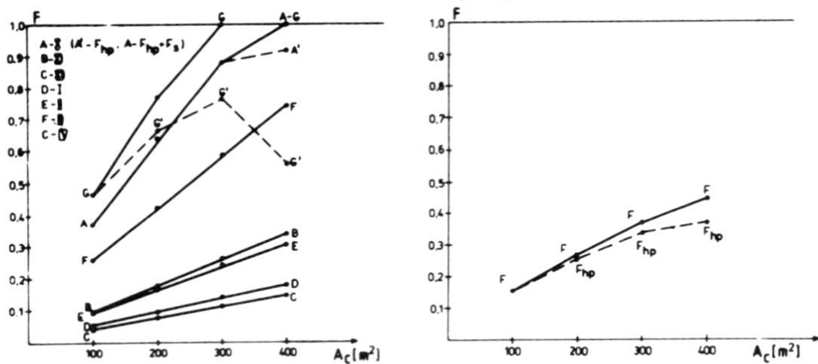

Fig. 4. Monthly (on the left) and seasonal (right) performance factor F, vs collector area (F_{hp}-for heat pump).

Reffering to thermal performance factors of the SAHP system and all obtained results, it can be said, that considered system with large collector area can be rather useful for apartment building heating in Poland. As monthly performance factors in winter decrease, due to difficult climatic conditions, the standard design of apartment building must be modified. Improving the insulation quality of the building is important since it can significantly decrease the energy losses. For example a new design of apartment building must require that the building be constructed in such a manner that the thermal resistance values of walls floor and ceiling will be higher. Thus the heating load of such building will be lower then that conventional one.

REFERENCES

D-20/78/1983. (1983). Solar Radiation. Publication of the Institute of Geophysics. Polish Academy of Sciences.

Estes, R. C., W. Kahan, (1978). Analytical selection of marketable SAHP systems.In T. N. Veziroglu (Ed.). Solar Energy and Conservation. Vol. 1, Pergamon Press, New York. pp. 455-475.

THE LUEBECK HEAT PUMP ASSISTED SOLAR HEATING SYSTEM FIVE YEARS OF OPERATION

H. Weik and J. Plagge
Fachbereich Angewandte Naturwissenschaften
Fachhochschule Lübeck
D-2400 Lübeck, Fed. Rep. of Germany

ABSTRACT

A double-storage, heat pump-assisted solar space heating concept, realized in a two-story, 130 m^2 one-family home inhabited by a real family, has proved its reliability during a five-year test as an all-year heating system even in the North-German climate with its inherent low winter insolation. During the long and rather cold years 1985/86 and '86/87, with about 12 % more Kelvin days than average, a 60 % solar fraction was obtained with a 28 m^2 collector field, a 18/3.2 m^3 water double-storage unit and a 2 kW heat pump. The system-COP averaged 30 %, and the auxiliary energy consumption amounted to 5.8 MWh/a in '85/86 and 4.9 MWh/a in '86/87.

KEYWORDS

Dual-storage solar heating concept; heat pump assisted solar heating system; active solar space heating; seasonal storage; latent heat storage; coefficient of performance ; solar fraction.

INTRODUCTION

At the ISES Conference in Brighton, UK, Weik, Plagge and Rosche (1982) presented experimental data demonstrating the feasibility of using the latent heat of the water-ice phase transition as an integral part of a new solar heating concept for space heating of dwellings in middle- and north-European climates.

This solar heating concept, that can be characterized as «heat pump assisted dual-storage solar space heating», was to be tested under real conditions in a two-story one-family home inhabited by a real (i.e. not-computerized) family, then under construction on the site of the Lübeck Fachhochschule.

After the building was completed and the family had moved in, in summer 1982 a long-term test period could be started. The goal was to demonstrate the merits of the dual-storage solar heating concept and to collect data on its performance under real living and climatic conditions of Northern Europe with its inherently poor winter insolation.

DUAL-STORAGE SOLAR HEATING CONCEPT

High latitude locations are characterized by an extreme unbalance between the main solar energy supply and the space heat demand. In Lübeck (54° N latitude), for example, 65 to 70 % of the year's total insolation occurs between April and September, while 85 % of the total space heat demand for a well insulated one-family home is needed during the October-till-March period. If one wants to attain the most efficient use of solar thermal energy in our climate, only a double-storage concept appears economically meaningful, with the storage tanks being at two different temperature levels during the winter months and having different functions (Fig. 1).

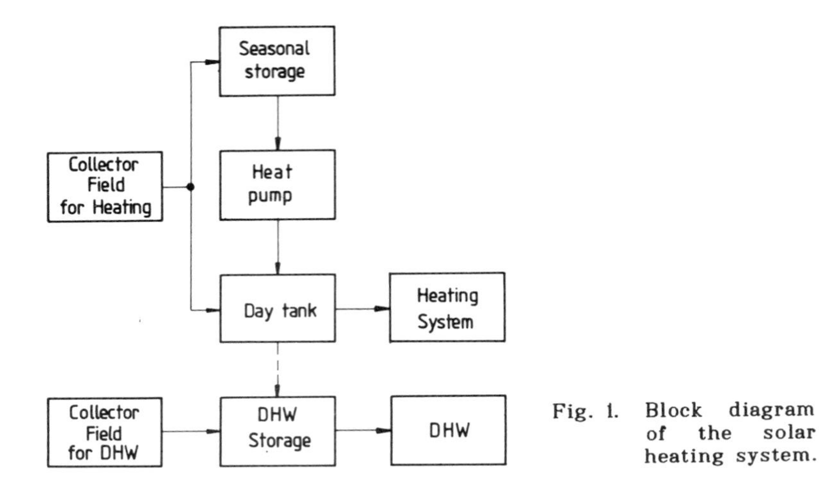

Fig. 1. Block diagram of the solar heating system.

The small tank ('day tank'; about 3.5 m^3 in volume for a heated area of 120 to 130 m^2) takes care of the daily heat requirements of the house and therefore is kept at a temperature between 35 and 50 °C, depending upon the ambient temperature. (Hereby a low-temperature heating system is a necessary condition.) The large tank ('seasonal tank', about 20 m^3 in volume, mainly depending upon the level of the winter insolation), on the other hand, serves as heat reservoir for the day tank, either by direct heat transfer to the heating system at the beginning of the heating season, or - later in winter - as heat source for a small heat pump, for which a driving power of only 2 kW is sufficient.

The solar heat for the domestic hot water supply may be furnished independently, for instance by coupling it to the main storage unit through a heat pipe connection, as indicated in the block diagram.

HEATING/STORAGE CYCLE

For the first 2 or 3 weeks after the beginning of the heating period (in Lübeck around 5 September), heat gains from the collector field will generally compensate the heat consumption of the house and the heat losses of the storage, which, at that time has a temperature of about 60 to 65 °C. Later on, around the beginning of October, first the day tank and then the seasonal tank will be discharged, the latter also by direct heat exchange to the heating system of the house (not shown in Fig. 1).

After all useful heat (so-called 'exergy') of the storage unit has been consumed, the rest of the sensible heat of the large tank will be extracted by means of the heat pump and thus be made available for heating via the day tank. This goes on even during the solidification of the water in the tank, which now acts as a phase change (latent heat) storage.

Starting end of April and/or beginning of May, generally a solar surplus occurs. It will be used to liquify the ice in the large tank, which will subsequently be heated up during the summer months to its initial 60 to 65 °C temperature for the new heating period.

REALIZATION OF THE DOUBLE-STORAGE CONCEPT

In order to test the dual-storage solar heating concept under real inhabitat conditions, a one-family home was constructed. The building complex (Fig. 2) consists of the two-story house of about 130 m^2 living area, and a one-story laboratory annex, the latter being equipped with a 'conventional' single-storage solar system with gas as auxiliary heat source.

Fig. 2. The Lübeck experimental solar one-family house. (South roof 60° inclined)

The house was designed with a mean U-value of 0.35 W/m^2K and an enclosure surface of 450 m^2. Taking the long-term averages of the ambient temperature of Lübeck (5.3 °C) during the 260 days of the heating period , and of the number of degree days for an average room temperature of 20/17 °C (day/night), 3580 Kd, the total heat demand of the dwelling house for space heat (i.e. transmission and ventilation) is roughly 18.5 MWh/a; hereby, no heat demand during the summer months is assumed (Weik and Plagge, 1986).

Subtracting about 3.5 MWh for internally produced heat by four persons and appliances during the winter, and 2.5 MWh for passive solar gains through the (relatively small) south windows, a net winter heat demand of about 12.5 MWh remains. This was supplied by an active-solar system consisting of 28 m^2 qualitatively good solar collectors, a 2 kW heat pump and a water-double-storage unit of 18/3.2 m^3 volume (Fig. 3 on next page).

SYSTEM PERFORMANCE

Already during the first heating period (1982/83), the double-storage solar system worked to our and to our tenant's satisfaction. In the following years, various steps were taken to optimize the system and to minimize the auxiliary energy consumption: various shortcomings and deficiencies in the thermal insulation of the building were removed, some of the piping altered, and the double-storage unit rearranged (cf. Weik and co-workers, 1982).

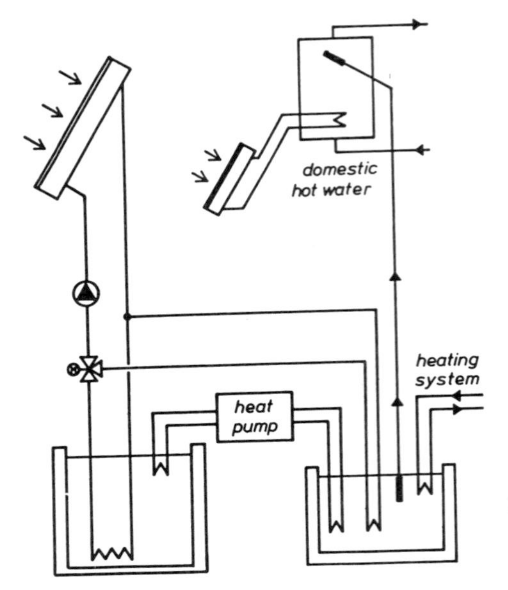

Fig. 3. Simplified solar system lay-out of the dwelling house.

Most representative of the technical maturity of the dual-storage solar system are the results of the last two (consecutive) heating periods, 1985/86 and 1986/87, which were extremely cold and long and preceded by unusually chilly summers. Figure 4 gives the monthly energy/heat gains and demands, respectively, and Table 1 the yearly performance data (Weik and Plagge, 1987).

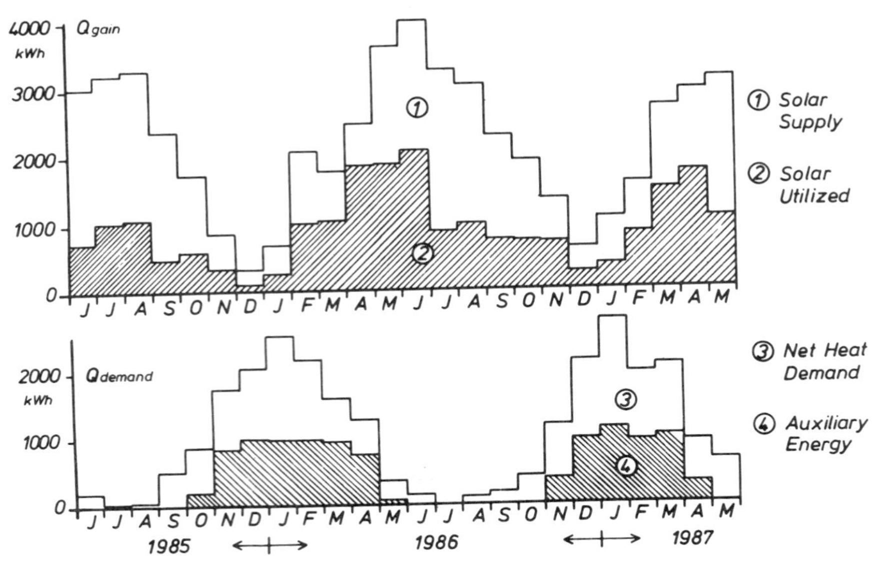

Fig. 4. Energy/heat balance of the solar heating system from June '85 til May '87.

Table 1. Performance Data of the Dual-Storage System

		1985/86	1986/87
Yearly degree days, 20/17 °C	(Kd/a)	4210	4220
Mean ambient temperature Sep-May	(°C)	4.7	4.5
Insolation onto collector plane	(MWh/a)	25.4	28.1
Insolation stored in system	(MWh/a)	10.4	11.6
Net heat demand of dwelling house	(MWh/a)	13.5	12.9
Energy consumption of heat pump	(MWh/a)	5.8	4.9
COP (collector)	(%)	41	41
COP (storage unit)	(%)	75	69
COP (system)	(%)	31	29
Solar fraction	(%)	57	62

CONCLUSIONS AND FUTURE INVESTIGATIONS

As the performance data indicate, the dual-storage solar system is capable of fully supplying the space heat for a standard size one-family home. Furthermore, it proved to be a reliable heat-generating installation.

From a technical standpoint, the solar system worked with a relatively high system-COP, that was considerably higher than was obtained by earlier investigators at a comparable object (Bruno and co-workers, 1978).

During the coming months, the Lübeck solar house will be converted into a demonstration object for the combination of passive and active thermal solar applications by adding sun spaces to the south fassade of the house. At the same time, changes in the collector field will be made: while thus far, a variety of collector types has been installed in order to test their individual long-standing efficiencies, only the solar collector type with the best COP and the best price-to-efficiency ratio will remain on the roof.

Both measures will increase the solar fraction to an estimated 70 %, and reduce the auxiliary energy demand correspondingly. Besides, the effective collector area for heating will be reduced from now 28 m^2 to then 24 m^2.

In addition, part of the roof will be covered with solar generators to provide (at least partly) the electric energy for the heat pump. This will further reduce the auxiliary energy taken thus far from the public electric supplier.

REFERENCES

Bruno, R., W. Hermann, H. Hörster, R. Kersten, and K. Klinkenberg (1978). The Philips experimental house: a system's performance study. In *Proceedings of the CCMS/ISES Intern. Conf. « Performance of Solar Heating and Cooling Systems »* CCMS Report No. 85, Düsseldorf. pp. 249-263.

Weik, H., J. Plagge, and G. Rosche (1982). Latent heat storage utilizing the water-ice transition. In *Proceedings Int. Solar Energy Soc. Congr. Brighton, UK,* Vol. 1, Pergamon Press, Oxford. pp. 731-735.

Weik, H., and J. Plagge (1986). The experimental solar house of the Lübeck Technical Institute. Results of four years of operation. In *Proceedings « North Sun '86 », Int. Conf. Copenhagen.* pp. 224-229.

Weik, H., and J. Plagge (1987). Das Lübecker Solarheiz-Projekt. Bilanz nach fünf Betriebsjahren. In preparation for publication.

EXPERIMENTAL STUDY ON A SOLAR-HEAT PUMP SYSTEM

K. KANAYAMA and H. BABA

Department of Mechanical Engineering
Kitami Institute of Technology
165 Koen-Cho Kitami, Hokkaido, 090 Japan

ABSTRACT

The renewable energy laboratory at Kitami Institute of Technology is an experimental facility in cold regions whose heat for the space heating and domestic hot water is supplied by solar and underground water energies using a solar-heat pump system. The laboratory consists of two-storied house No. 1 and one-storied house No. 2. The house No. 1 on which 24-solar collector is fixed has a study room, a measurement room, and a machine room in which the heat pump, tanks and auxiliary boiler are settled. The house No. 2 has three experimental rooms required to heat. The hot water obtained by the solar-heat pump system is fed to the houses No. 1 and No. 2 to heat the space and to supply the domestic hot water. By changing the combination and control methods in several modes, the performance of the system is examined on the basis of the system COP in connection with heat load. Consequently, the parallel solar-heat pump system whose system COP is 4.20 is more effective than the series solar-heat pump system whose system COP is 3.70.

KEYWORDS

Solar energy; Underground water energy; Solar system; Heat pump; Series solar-heat pump system; Parallel solar-heat pump system; System COP.

INTRODUCTION

To heat the space of a house and to supply domestic hot water using solar energy and environmental energies such as underground water and atmospheric air, several analytical and experimental studies for a solar-heat pump system which is combined a solar system with a heat pump have been done at the foreign countries hitherto; (Freeman, 1979), (Anderson, 1980), (Svard, 1981), (Chandrashekar, 1982), (Manton, 1982), and so on.
In Japan, the result of measurements for the well-known Yanagimachi solar house (Yanagimachi, 1977) was formerly reported in detail. The renewable energy laboratory at Kitami Institute of Technology is an experimental facility for heating, cooling and hot water supplying by means of a hybrid system of solar-heat pump using mainly solar energy and additionally under-

ground water energy. The authors (1986, 1986, and 1987) have published the results of the measurements and the analyses of its performance already. In this paper, the results of performance tests for this experimental facility during two winter seasons are summerized; for the series connection of the solar system and the heat pump on the only house No. 1 from December, 1982 to March, 1983, and for the parallel connection of the solar system and heat pump on both houses No. 1 and No. 2 from November 1983 to February 1984.

Symbols and Definition

t_{CT} = temperature of collecting tank, °C
t_{ST} = temperature of storage tank, °C; $t_{CT} = t_{ST}$ for the parallel system
t_R = room temperature, °C; t_a = ambient temperature, °C
t_1 = temperature of heat reservoir, °C; t_2 = temperature of heat sink, °C
Δt = temperature difference between the reservoir and sink, °C
H_{ND} = normal-direct solar radiation, MJ/m^2 period
Q_J = total solar radiation on a surface, MJ
Q_C = collected solar energy, MJ
Q_{CE} = electric power for collecting solar energy, MJ
Q_{TL} = total heat load (= heat leaving the house), MJ
Q'_{TL} = total heat supply (= heat entering the house), MJ
Q_L = heat load for space heating, MJ
Q_{HW} = heat for domestic hot water, MJ
Q_{MAC} = heat load calculated for machine room, MJ
Q_S = heat absorbed by heat pump from collected energy, MJ
Q_E = electric power of heat pump, MJ
Q_W = heat of underground water energy, MJ
Q_H = heat rejected by heat pump = $Q_E + Q_W$, MJ
Q'_H = directly measured heat rejected by heat pump, MJ
Q_B = heat supplied by auxiliary boiler, MJ
Q_ℓ = heat loss from the system, MJ
ΔQ_{ST} = heat stored in the system, MJ; $\Delta Q_{ST} = 0$ for a long period of measurement
η_C = collection efficiency = Q_C/Q_J
η_{CE} = rate of electric power for collecting pump = Q_{CE}/Q_C
F_S = fraction of solar energy for the system = Q_C/Q_{TL}
F'_S = fraction of solar energy through heat pump = Q_S/Q_{TL}
$F'_S = F_S$ for the parallel system
F_W = fraction of heat of underground water = Q_W/Q_{TL}
F_{NE} = fraction of renewable energy for the system = $(Q_C + Q_W)/Q_{TL}$
$= F_S + F_W$
F'_{NE} = fraction of renewable energy through heat pump = $(Q_S + Q_W)/Q_{TL}$
$= F'_S + F_W$; $F'_{NE} = F_{NE}$ for the parallel system
F_E = fraction of electric power of heat pump for the system = Q_E/Q_{TL}
$= 1 - F_{NE}$
F'_E = fraction of electric power of heat pump = $Q_E/(Q_S + Q_W + Q_E)$
$= 1 - F'_{NE}$; $F'_E = F_E$ for the parallel system
F_B = fraction of heat supplied by auxiliary boiler = Q_B/Q_{TL}
COP = coefficient of performance of heat pump
$(COP)_S$ = coefficient of performance for the system
$(COP)_{SB}$ = coefficient of performance for the system including auxiliary boiler heat

SYSTEM OUTLINE

The total plan of the houses No. 1 and No. 2 is shown in Fig. 1. The house

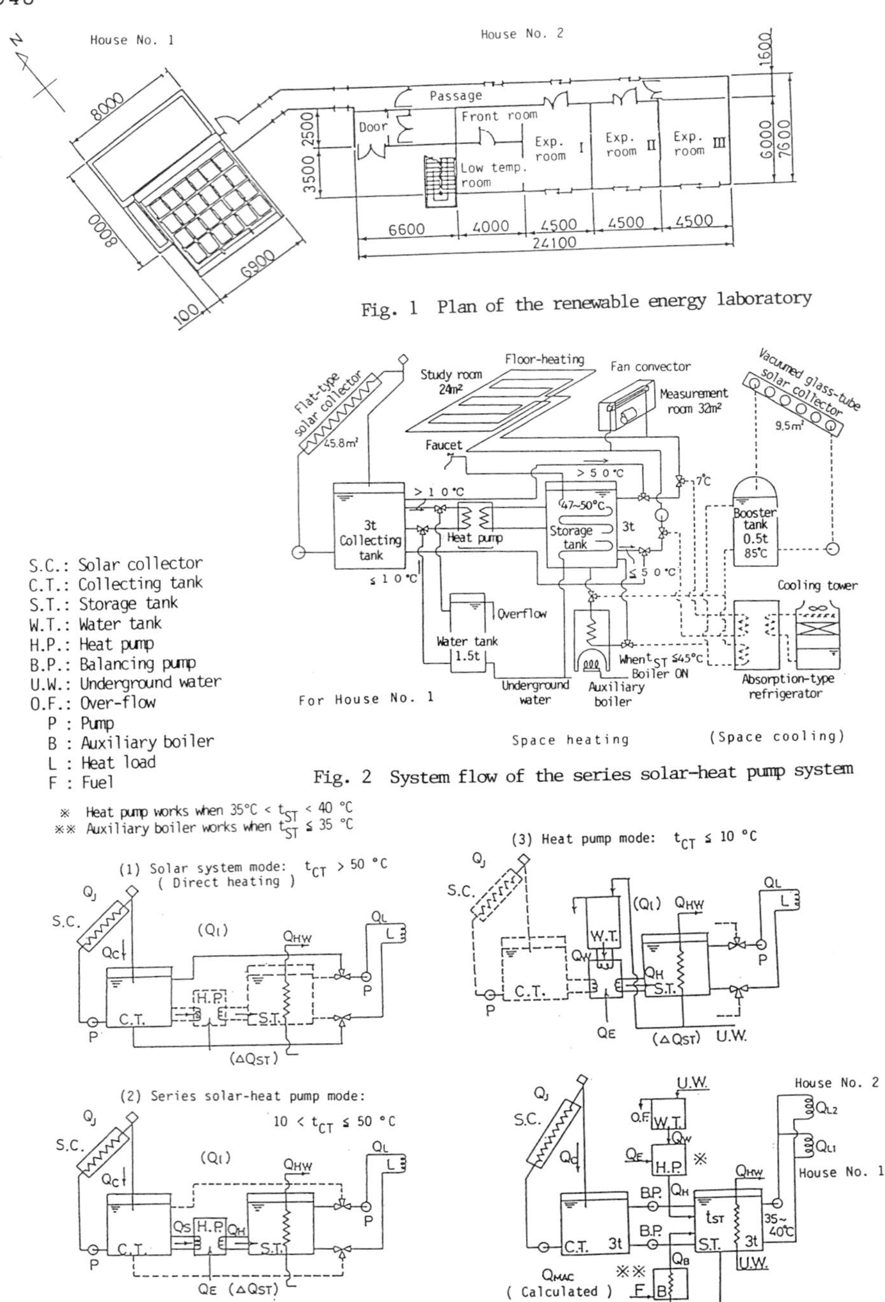

Fig. 1 Plan of the renewable energy laboratory

Fig. 2 System flow of the series solar-heat pump system

Figs. 3 (1) ∿ (3) Working modes of the series solar-heat pump system

Fig. 4 Working mode of the parallel solar-heat pump system

No. 1 is two-storied building made of steel-frame and ALC construction whose floor area is 96 m^2, the measurement room and the machine room are arranged on the first floor and the study room is arranged on the second floor. The wall of the building is well insulated with 100 mm thick glass-wool, and the window is air-tight with a triple glass-plate and covered by an insulation shutter on it. On the south side of the house, 24-flat-type solar collector (collecting area is 45.8 m^2 in total) is settled in the way of 6-row and 4-step facing due south with inclination angle of 60°. Machineries and equipments, such as the heat pump, tanks, auxiliary boiler and so on, are installed in the machine room. One-storied house No.2 whose floor area is 150 m^2 consists of three rooms, the wall is insulated with 50 ∿ 100 mm thick glass-wool. The experimental rooms I, II and III are heated by hot water circulated from the house No. 1.
The solar system and the heat pump are combined in two ways of series and parallel connections. In the former it forms normally the series solar-heat pump system, and sometimes it is switched to the stand-alone solar or heat pump system by controlling with temperature in the storage tank depending on the relation between solar radiation and heat load. In the latter, the space heating and hot water supply are first done by the solar system, the most of the shortage of required energy is provided by the heat pump which works always parallel to the solar system and the rest of that is provided by the auxiliary boiler. System flow of the series solar-heat pump system and its working mode are illustrated in Fig. 2 and Figs. 3 (1) ∿ (3). Working mode of the parallel solar-heat pump system is illustrated in Fig. 4.

MEASURED RESULTS

The measured results are shown by the total, average and fractional values for a period of the temperature at each part, heat quantity, required electric power, solar radiation and ambient temperature. Figure 5 and Table 1 show the results for the series solar-heat pump system. Figure 6 and Table 2 show the results for the parallel solar-heat pump system. Since the equation of coefficient of the heat pump (1.7 kW) used in this experiment is described as Eq. (1) for the ordinary temperature range, the COP of the stand-along heat pump is about 2.8 between the temperatures of the sink and the reservoir to be 10 and 50°C respectively.

$$COP = 3.427 + 0.423\ \ln(t_2/(t_1\Delta t)^{1/2}) \qquad (1)$$

CONCLUSIONS

Since the total heat supply for the parallel solar-heat pump system increased by the factor of about 1.6 in comparison with that for the series solar-heat pump system, the fraction of solar energy for the former decreased, while the fraction of underground water energy and the fraction of heat rejected by heat pump increased. Moreover, a little quantity of the required heat was supplied by the auxiliary boiler in order to assist the shortage of the heat supply. Judging from the coefficient of performance for the system, $(COP)_S$, the parallel solar-heat pump system had a more excellent thermal performance than the series solar-heat pump system had, in spite of increase of the heat load.

References

Anderson, J.V., J. W. Mitchell, and W. A. Beckman (1980). Solar Energy, 25, 155-163.

Chandrashekar, M., N. T. Le, H. F. Sullivan, and K. G. T. Hollands (1982).

Solar Energy, 28, 217-226.

Freeman, T. L., J. W. Mitchell, and T. E. Audit (1979). Solar Energy, 22, 125-135.

Kanayama, K., H. Baba, and T. Fukuda (1986). J. of JSES., 12, No. 2, 13-20.

Kanayama, K., H. Baba, and M. Yamamoto (1986). Trans. of JSME., B, 52, 4046-4052.

Kanayama, K., H. Baba, and T. Fukuda (1987). J. of JSES., 13, No. 5, 12-17.

Manton, B. E., and J. W. Mitchell (1982). Trans. of ASME., J. of Solar Energy Eng., 104, Aug., 158-164.

Svard, C. D., J. W. Mitchell, and W. A. Beckman (1981). Trans. of ASME., J. of Solar Energy Eng., 103, 135-143.

Yanagimachi, M. (1977). J. of JSES., 3, No. 1, 27-31.

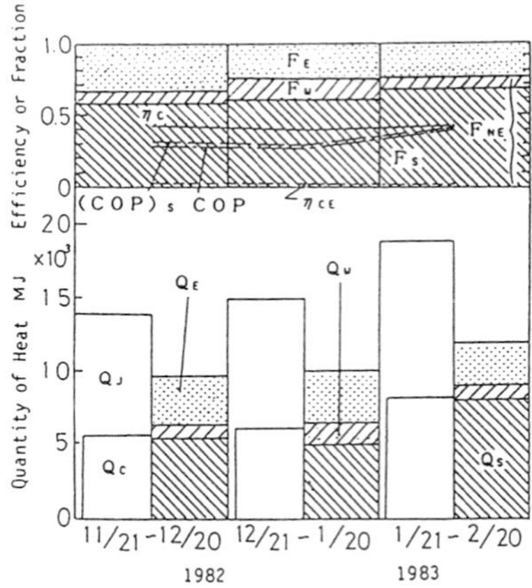

Fig. 5 Experimental results of the series solar-heat pump system

Fig. 6 Experimental results of the parallel solar- heat pump system

Table 1 EXperimental Results of the Series Solar-Heat Pump System

Items \ Period		12. 10. '82 – 1. 9. '83	1. 10. '83 – 2. 9. '83	2. 10. '83 – 3. 9. '83	Sum total Average
H_{ND}	MJ/m²	9760.4	9609.6	13369.4	32739.4
	Rate ※	0.864	0.720	0.722	0.769
t_a ℃	max	1.1	-1.8	-2.7	-1.1
	min	-10.5	-12.8	-15.4	-12.9
	mean	-4.7	-7.3	-8.5	-6.9
t_R ℃	max	19.6	18.9	18.5	19.0
	min	14.3	12.8	11.9	13.0
	mean	17.5	16.7	15.9	16.7
t_{CT}, ℃		20.5	19.4	27.8	22.6
Q_J, MJ		13816.8	14859.9	18817.4	47494.1
Q_C, MJ		5810.3	5963.7	8027.5	19801.5
Q_{CE}, MJ		192.13	176.67	127.48	538.27
Q_{TL}, MJ		9889.3	9993.2	11802.3	31684.8
Q_E, MJ		3403.8	3632.2	2953.8	9989.6
Q_W, MJ		933.7	1491.4	916.1	3341.2
Q_S, MJ		5560.9	4869.8	7932.4	18363.1
η_C		0.421	0.401	0.427	0.416
η_{CE}		0.033	0.030	0.016	0.027
F_S		0.588	0.597	0.680	0.625
F'_S		0.561	0.487	0.672	0.580
F_W		0.094	0.149	0.078	0.105
F_{NE}		0.682	0.746	0.758	0.730
F'_{NE}		0.656	0.637	0.750	0.685
F_E		0.318	0.254	0.242	0.270
F'_E		0.344	0.363	0.250	0.315
COP		2.91	2.75	4.00	3.18
(COP)$_S$		3.14	2.94	4.13	3.70

※ Rate of solar radiation to the average of solar radiation from 1978 to its preceding year

Table 2 Experimental Results of the Parallel Solar-Heat Pump System

Items \ Period	11. 21. '83 – 12. 20. '83	12. 21. '83 – 1. 20. '84	1. 21. '84 – 2. 20. '84	Sum total Average
t_a, ℃	-2.6	-6.8	-8.8	-6.07
t_R, ℃	16.8	16.9	16.2	16.63
t_{CT}, ℃	38.6	37.8	37.6	38.0
Q_J, MJ	16192.437	14070.802	17272.348	47535.587
Q_C, MJ	5240.199	3696.108	6017.690	14953.997
Q_{CE}, MJ	128.862	86.307	110.975	326.144
Q_{TL}, MJ	15213.3	17396.1	19372.0	51981.4
Q'_{TL}, MJ	15008.473	16668.491	18451.844	50128.808
Q_H, MJ	9668.911	12626.625	11446.557	33742.093
Q_E, MJ	3124.641	3955.393	3594.610	10674.644
Q_W, MJ	6544.268	8671.232	7851.947	23067.447
Q_B, MJ	99.382	352.991	983.099	1435.472
η_C	0.324	0.261	0.348	0.315
η_{CE}	0.024	0.024	0.021	0.022
F_S	0.349	0.222	0.322	0.298
F_W	0.437	0.520	0.428	0.460
F_{NE}	0.786	0.742	0.750	0.759
F_E	0.208	0.238	0.196	0.213
F_B	0.006	0.021	0.054	0.029
COP	2.707	2.816	2.787	2.780
(COP)$_S$	4.305	3.771	4.906	4.198
(COP)$_{SB}$	4.208	3.497	4.091	3.747

DYNAMIC OPTIMISATION OF SOLAR THERMAL ENERGY SYSTEMS WITH A HEAT PUMP AND SEASONAL STORAGE

A.G.E.P. van Delft and A.R. Logtenberg

Eindhoven University of Technology, System and Control Engineering Group,
Building W&S 1.38, P.O.Box 513, 5600 MB EINDHOVEN, Netherlands

ABSTRACT

This paper describes the dynamic optimal control strategies for solar energy systems incorporating a heat pump and seasonal storage.
Dynamic optimisation reveals how the control variables (fluid flows between components, rotation speed of the heat pump compressor) have to be varied as a function of time, to ensure the best performance of the system.
In 2 examples it is shown that dynamic optimal control establishes a significantly better system performance, owing to a better exploitation of the components in the system, as well as a better anticipating on daily or seasonal changes in heat demand patterns. Furthermore, this approach leads to a better system design, on the one hand because bottle-necks are revealed and system dimensions can be optimised in the same procedure, on the other hand because the results suggest more cost-effective lay-outs and configurations.

KEYWORDS

Solar Energy, Heat Pumps, Storage, Optimal Control, Design.

INTRODUCTION

The System and Control Engineering Group of the Eindhoven University of Technology is investigating solar assisted heat pump systems with seasonal storage, for residential heating.
The aim of the work is to find the dynamic optimal control strategy. This strategy reveals how the control variables (fluid flows between components, rotation speed of the heat pump compressor) have to be varied as a function of time to ensure the best performance of the system over a given period.

In the research on solar energy and heat pump systems the control strategy is often a neglected subject. Some authors have reported optimal control strategies for relatively simple systems under restrictive conditions.
Much attention has been paid, however, to the design of these systems, for instance: how to choose heat pump capacity or collector area to obtain the best system (under conventional control).
We propose a new design method for solar energy and heat pump systems. The predominant characteristic of this method is that optimally controlled, instead of conventionally controlled, design alternatives are compared. This selection based on optimal control leads to a

better system design. Moreover the optimal control strategy of the best design alternative serves as a guideline for deriving sub-optimal robust control strategies.
In this paper we will focus on the first step in the new design method: the calculation of the optimal control of a design alternative. The next chapter deals with the optimisation method and gives a brief description of the systems considered. In chapter 3 two examples are presented, that clearly illustrate some typical characteristics of dynamic optimal control.

SYSTEM DESCRIPTION AND OPTIMISATION METHOD

Figure 1 gives a general view of the systems considered in this paper.

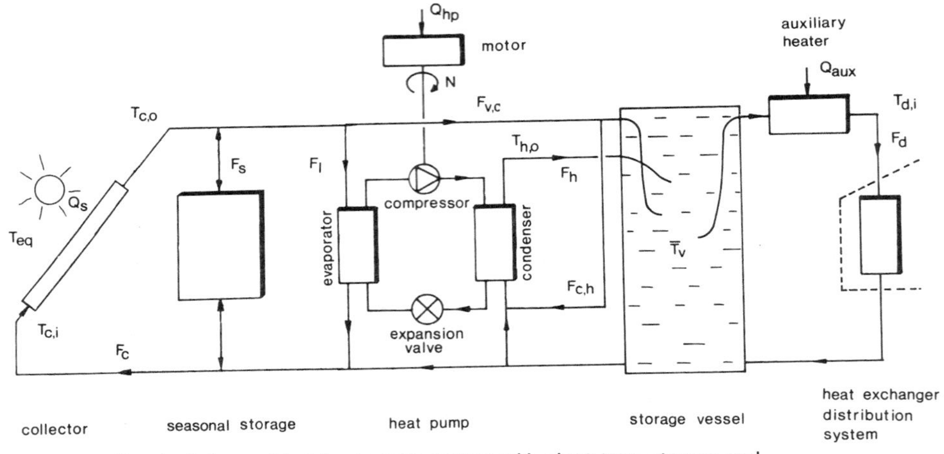

Fig. 1 Solar assisted heat pump system with short term storage and an optional seasonal storage.

The symbols T, F and Q stand for temperature, fluid flow and energy flow. Teq is the equivalent ambient temperature, combining the ambient temperature and solar irradiance.
Hot water from the solar collectors can be used as a heat source for the heat pump, or directly stored in the short term vessel.
The presence of a seasonal storage provides an alternative for injection or extraction of heat.
The space heating demand is met by the short term storage vessel (with thermal stratification).
The auxiliary heater comes into operation if the storage temperature is below the demand temperature.
An interesting operational alternative is provided by this system: A part of the collector heat can be used to preheat the flow through the heat pump condenser, which appears to be a clever mode under certain circumstances.
To calculate the behaviour of the system, experimentally verified simulation models are used, with a typical time step of one hour. The control variables are the fluid flows between the components (Fl, Fh, Fv,c, Fc,h, Fs) and the rotation speed of the heat pump compressor (N). The performance of the system is expressed in total energy consumption of heat pump, auxiliary heater and circulation pumps over the optimisation interval (resp. IQhp, IQaux, IQpump). In this paper the optimisation interval is restricted to 24 hours, for illustrative purposes.
The optimisation problem can be formulated as follows:
Consider the system of Fig. 1 and a typical weather pattern.
Find the optimal values of the control variables $\underline{U}$ in every time step that minimize a performance index PI, in this case the total energy consumption over the optimisation interval {tb,te} is:

$$\min_{\underline{U}} PI = \min_{\underline{U}} \sum_{tb}^{te} (Qhp + Qaux + Qpump).\Delta t \tag{1}$$

Calculating the optimal control strategy is an iterative procedure. Starting with a chosen control strategy, its components are adapted in the opposite direction of the gradient to reduce PI as much as possible. This leads to another control strategy, the gradient is calculated again and the whole procedure is repeated until the control strategy converges to the optimal strategy. Full details of the optimisation method are given by Van Delft (1987).

The computerprograms for simulation of the components and calculation of the optimal control strategy form the most important parts of the SYNTES program-package (SYNthesis of Thermal Energy Systems) that is currently being developed in the System and Control Engineering Group. The SYNTES-package will allow for a selection of the best design alternatives, based on optimal control.

It is important to stress the difference between dynamic, static and momentary optimisation. In static optimisation, each control variable can only have one value over the entire optimisation interval. In momentary optimisation, only the effect of the control variables on the momentary energy consumption is taken into account, irrespective of the possible future consequences.

RESULTS

In this chapter two examples of optimal control are dealt with, based on the system of Fig. 1, with and without seasonal storage. Example 1 illustrates the difference between dynamic optimal and conventional control of a system without seasonal storage, over a 24 hour period and with a typical weather pattern. For an extensive discussion of other results with this system in our group, we refer to Slenders (1987) and Bottram (1984). At this point we confine ourselves to a treatment of the predominant effects.

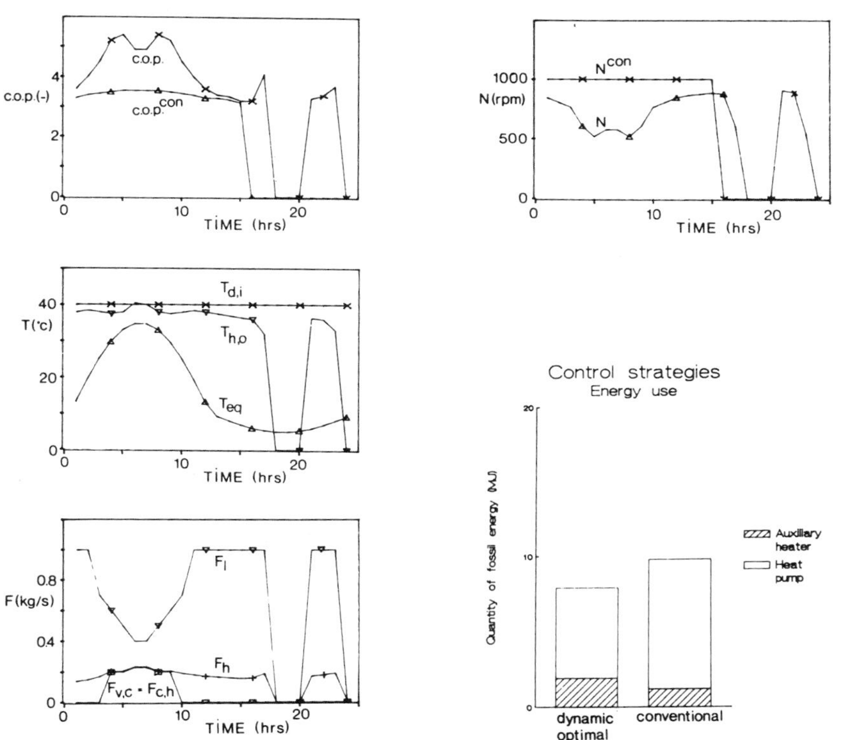

Fig. 2 Dynamic optimal vs conventional control.

Figure 2 gives the weather and demand pattern, the heat flows and the control variables for both cases. The superscript 'con' refers to the conventional strategy. In the conventional control strategy N has a constant value if the heat pump is in operation, Fl is at a maximum value and Fh equals the demand flow Fd.
Figure 2 shows some interesting effects. The dynamic optimal control strategy reduces the total energy consumption by 16 %. It seems rather surprising that this is achieved by a decreasing energy consumption of the heat pump, combined with an increase of the auxiliary energy. This illustrates a phenomenon that is often encountered when applying dynamic optimisation to these systems: maximising the contribution of the (high efficiency) heat pump, or minimising the consumption of the (low efficiency) auxiliary heater, is not the same as minimising the total energy consumption. As is shown in Fig. 2, the output temperature of the heat pump is somewhat below the demand temperature. The heat pump, however, could have produced the required heat if it was operated at a higher rotation speed. But this would reduce the heat pump efficiency and would perhaps have future repercussions because the storage temperature is increased, resulting in a higher collector inlet temperature and thus a lower collector yield.
The special operational mode introduced by Fc,h, as dicussed before, is used if the collector output temperature is rather high, but not high enough for direct storage without the heat pump.

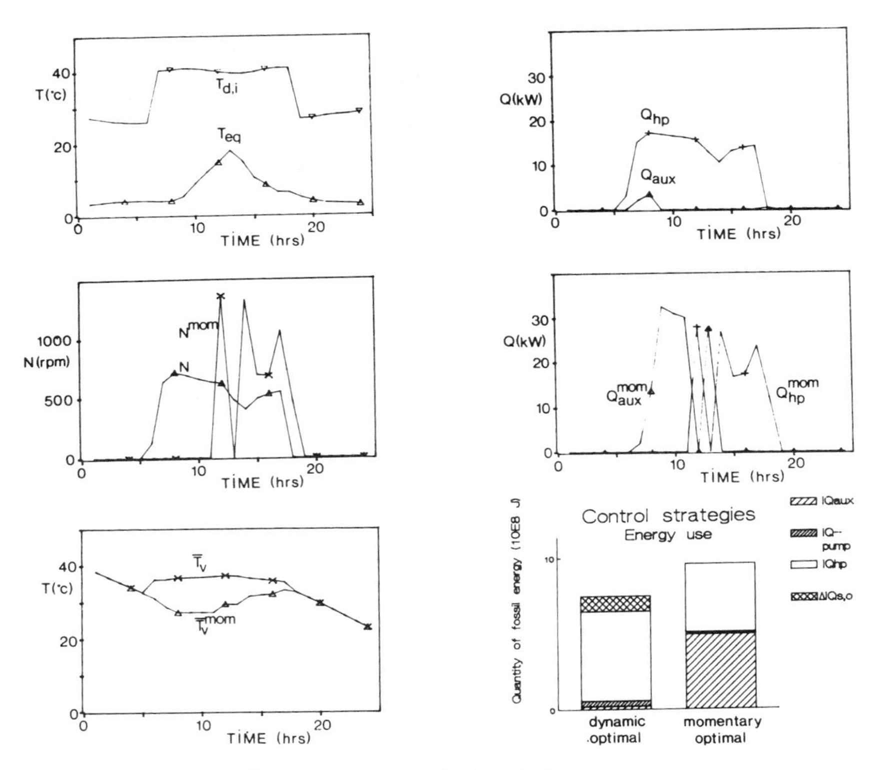

Fig. 3 Dynamic vs momentary optimal control.

In example 2 the difference between momentary and dynamic optimisation is illustrated for the system of Fig. 1 with seasonal storage. Figure 3 gives weather and heat demand patterns, heat flows and control variables. The superscript 'mom' refers to the momentary optimal case. Only N is shown, the fluid flows have a similar pattern. The dynamic optimal control anticipates on the sharp increase in heat demand after 6 hours. The short term storage is kept at the required temperature level, shown by the curve of the average storage temperature $\overline{T}v$. This is not the case with momentary optimal control. This control attaches no value to storing heat, resulting in a large amount of auxiliary energy between 6 and 12 hours.
The total net energy consumption (the use of heat from the seasonal storage, ΔIQs,o being taken into account) is reduced by 20 % for this typical day.

CONCLUSIONS

Dynamic optimal control establishes a significantly better system performance, owing to a better exploitation of the components in the system, as well as a better anticipating on daily or seasonal changes in heat demand patterns. The two examples discussed briefly in this paper show that:

- The energy consumption of a dynamic optimally controlled collector-heat pump system is reduced by 16 % as compared to a conventional control strategy.
- A collector-heat pump-ground storage system shows a reduction in energy consumption of about 20 % as compared to a momentary optimal control strategy.

Dynamic optimisation has also proved succesfull in solar heating systems with seasonal storage but without a heat pump (Logtenberg, 1987). Finding the dynamic optimal control is the first step in the new design method proposed by the authors. The SYNTES program package will provide a helpful tool for the selection of the best design alternative.

REFERENCES

Bottram, A.M.M., and Slenders, P.J.F., (1984). The control of thermal solar energy systems for space heating of houses and heating of tap-water. Journal A, 25, 4, 215-222.

Delft, A.G.E.P. van, and Meerman, J.J., (1987). Dynamic optimisation of thermal energy systems: Methodology and Results. ServoBode (Journal of the System and Control Engineering Group), 33, 20-36.

Logtenberg, A.R., and Delft, A.G.E.P. van, (1987). Dynamic optimization of control for designing the optimal geometry of a seasonal ground storage. Submitted for publication this conference.

Slenders, P.J.F., and Delft, A.G.E.P. van, (1987). Dynamic-optimal control of a heat pump system with an energy-roof. Submitted for publication in ASME Journal of Solar Energy Engineering.

HEAT PUMPS WITH SEASONAL STORAGE - ECONOMIC MARGIN

Stefan Aronsson, Per-Erik Nilsson

Department of Building Services Engineering
Chalmers University of Technology
S-412 96 Gothenburg
Sweden

ABSTRACT

Heat stores combined with heat pumps can be justified only if they result in economically viable operation of the heating system. Both heat sources and heat stores are therefore constrained to lie within given economic limits, determined by the costs of the alternatives available. If these economic limits are exceeded, the alternatives will usually be applied.

In the following, these economic frameworks are discussed for heat pumps with seasonal storage. Other types of heat store, such as buffer stores and short-term stores, are assumed to be included in the system costs, quoted for heat pumps.

This paper begins with a detailed description of the reasoning behind the work and defines the various, more important concepts that are employed. Then there is an example of how to present the results and how to use them in practice.

KEYWORDS

Heat pump, heat store, seasonal store, economic model, marginal cost, cost limit.

METHOD OF ASSESSMENT

Limiting Cost and Upper Cost Limit

It is possible to assess the prospects for some given new technology in an even simpler way by estimating what would theoretically be possible under ideal conditions in the present-day energy context. It then becomes possible to make an approximate estimate, by how much the system concerned can be reduced in price and increased in efficiency, and to calculate the kWh cost for an ideal plant of this type, which then represents the lowest possible cost of energy from it. If this ideal limiting cost lies below a present-day acceptable level, the technology can be regarded as having an acceptable development potential. A limiting cost consideration of this type has been applied in paper. When assessing a limiting cost, assumptions should be as optimistic as possible without being unrealistic. The earlier a particular technology is in its development stage, the more optimistic the assumptions should be, in connection with determination of the limiting cost.

Assessment of development potential has therefore been carried out in this report by checking to see whether the limiting cost is acceptably low. It is important to decide on what is an acceptable level. This has been done here on the basis of the assumption that no heating system can be expected to establish any significant natural market if the cost of the heat that it supplies is significantly more expensive than that of heat from simple conventional heat sources. With this reasoning as a background, it is possible to establish an upper cost limit which thus indicates the level at which any system would definitely be eliminated by alternative systems.

Marginal Cost and Alternative Systems

If a heat pump system is chosen to be designed, we can then go further and attempt to choose between two alternative heat pump designs, of which one is more expensive than the other, but also has a higher efficiency. When deciding which of these two heat pumps to select, it is natural to weigh the difference in capital cost against the additional energy output. This results in a cost of the additional energy which is referred to as its marginal cost. This marginal cost must at least be less than the corresponding energy cost of heat from gas oil if the choice of the more advanced heat pump is to be justified. If every consideration involves increased investment, such a marginal cost analysis must be employed.

In the following, we have assumed that the marginal cost must lie below the upper limiting cost, which therefore is normally equivalent to the cost which would be obtained by the use of gas oil. This reasoning can be illustrated by a heat pump system working with a low-temperature ground heat store.

The prerequisite to be able to talk about a marginal cost is that it must be possible to define the increased capital cost resulting in a future saving or increase in production. In practice, this means a comparison between two systems with different capital costs. When analysing a system in one particular version, it is necessary to be able to compare it with a postulated alternative system, which must be selected in such a way that the incremental part to be investigated can be identified and separated.

If, for example, we want to compare the value of a heat store connected to a heat pump system, we must have a comparison system which makes it possible to identify and isolate the actual effect of the heat store on heat production. In this case, the natural comparison system is an air heat pump having the same compressor power. The difference between the quantity of heat produced by the heat pump system with the heat store and that produced by the compared heat pump system without a heat store represents the marginal gain in energy output resulting from incorporation of the heat store. To this must also possibly be added a further gain in output resulting from a higher coefficent of performance. The difference in the capital cost between the two alternatives represents the expenditure necessary to obtain the marginal increase in heat output. From this, a cost can be calculated for the marginal heat which can be related directly to the performance improvement resulting from the heat store. This marginal cost (SEK/kWh) must not exceed the upper cost limit, i.e. the cost of heat produced from gas oil.

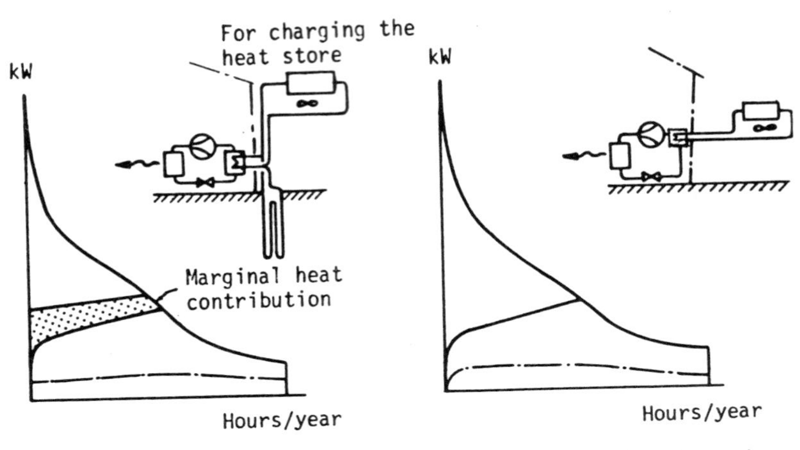

Fig. 1 The left-hand duration curve is that of a planned heat pump with a low-temperature ground heat store. Both curves indicate the output from the respective heat pumps and the electrical energy input necessary for powering the heat pump.

Bearing in mind what has previously been said concerning the upper cost limit, this is the highest marginal cost which can be accepted for a preliminary assessment of the realism of a particular design.

Economic Margin

When heat stores are being considered, it is possible to arrive at a permissible maximum cost for a store in a given system if the relevant marginal cost criteria are to be met. This means that the economic margin for various technologies must be calculated, starting from the system in which the technology is to be incorporated. This can continue by checking to what extent various types of heat store can be contained within the economic margin. The limiting cost at least, must lie within this economic margin if the particular type of heat store concerned is to be economically justifiable and thus have a realistic chance of competing with other alternatives.

Low-Temperatue Seasonal Store- Example

- Low-temperature heat store, charged with heat directly from outdoor air.
- Store working temperature: +5 °C to +15 °C.
- The heat pump utilizes the store as its heat source throughout the year.
- The reference installation is an outdoor air heat pump.

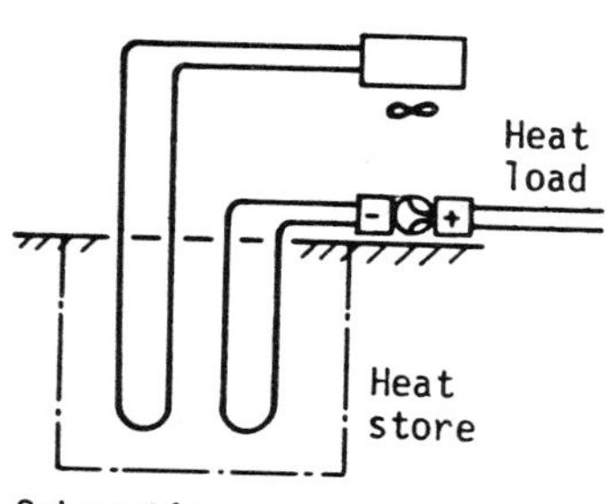

Schematic arrangement of heat pump and heat store

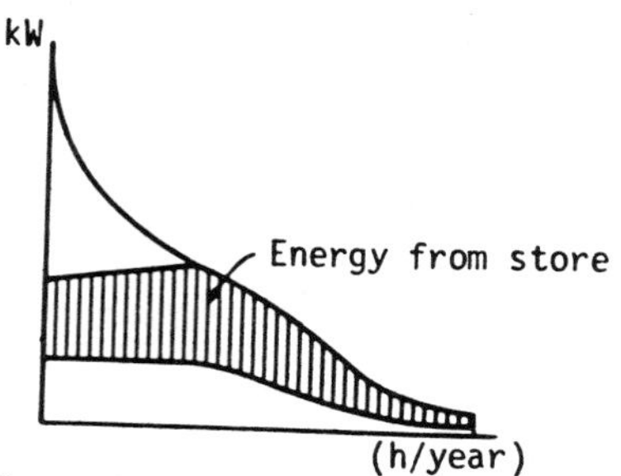

Power duration diagram for heat pump with heat store. The shaded area indicates the energy supplied by the store.

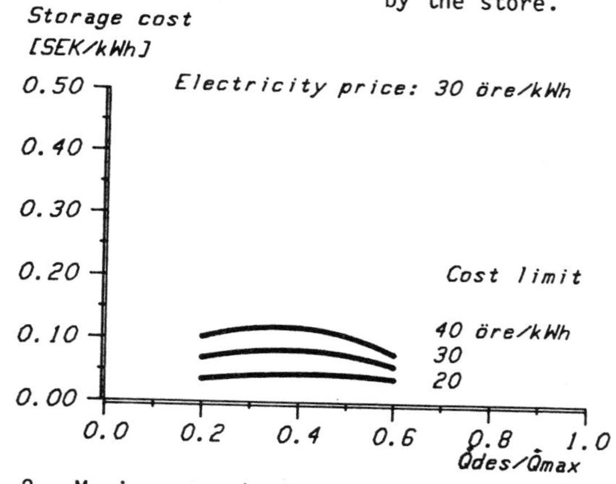

Fig. 2. Maximum permissible cost of heat supplied by the store.

This cost has been calculated on the assumption that the marginal cost of the additional heat supplied by virtue of the presence of the store does not exceed the specified cost limit, which is the highest marginal cost that can be accepted without choosing to provide the heat in some other manner. The reference heat source/system chosen here is gas oil; three alternative oil prices have been used:

40 öre/kWh (= SEK 3.000/m^3)
30 öre/kWh (= SEK 2.300/m^3)
20 öre/kWh (= SEK 1.500/m^3)

The maximum permissible cost includes not only the cost of the store itself but also all costs for necessary connections on the supply and discharge side, as well as additional costs involved for the heat absorber and other system parts as a result of incorporating the storage function.

A prerequisite for the diagram is that the cost of the total heat energy supplied by the system is less than the cost limit.

Comparison with the Costs of Heat Stores in Sweden

The storage costs presented in this comparison can be regarded as limiting costs, i.e. very low initial costs.

The range in both specific thermal storage capacity and specific cost means that the cost of energy from stores will fall within the range over a relatively wide interval.

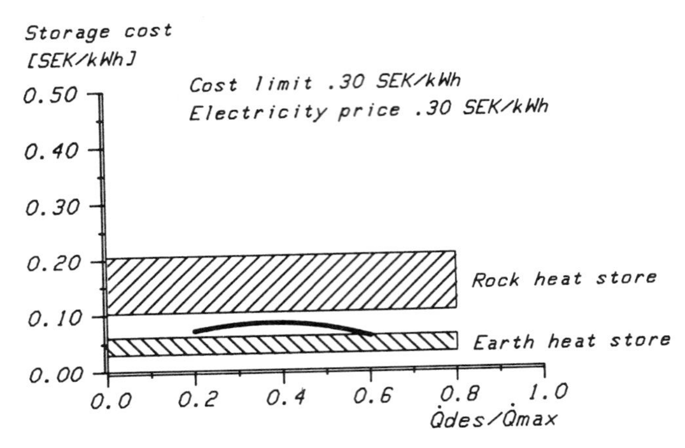

Fig. 3. Current costs of heat stores superimposed on the maximum permissible cost of a low-temperature heat store charged with heat from outdoor air.

CONCLUSIONS

It is important to have access to some evaluation yardstick, which objectively quantifies the costs of any heat-producing installation. The model reasoning must be sufficiently simple and easy to understand to allow it to be easily checked. The marginal cost model with valuation against cost limits, presented here, is aimed to fulfil these demands. The economic margin of seasonal store combined with heat pumps, is given by the model as a maximum permissible cost of heat supplied by the store. If the given cost is exceeded, other alternative installations will be chosen on strictly economical bases.

REFERENCES

Abel, E., Aronsson, S., Dalenbäck, J-O., and Nilsson, P-E (1987). The economic margin for alternatives in new energy technology. Swedish Council for Building Research, D4:1987, Stockholm.

COMPARISON OF ENERGY AND EXERGY ANALYSES OF A SOLAR-ASSISTED HEAT PUMP WITH SEASONAL STORAGE.

L. Mazzarella, E. Pedrocchi

Dip. di Energetica, Politecnico di Milano
P. Leonardo da Vinci 32 - 20133 Milano - Italy

ABSTRACT

This report outlines the exergy performance of a complex heating-plant consisting of solar-assisted heat pumps with ground-coupled seasonal storage. The system has been monitored for some years enabling us to quantify the energy balance. Since heat is supplied using a mix of different energies, the energy balance is not wholly satisfactory. For this reason, we have also prepared an exergy balance. The energy and exergy balances are compared. They reveal a great difference. The exergy balance allows us to highlight the most important irreversibilities of the process.

KEYWORDS

Exergy analysis; heat-pump; energy-exergy comparison; solar-assisted heat-pump; heat-ground storage.

PURPOSE OF THE WORK

We have been monitoring and analysing the data of a heating-plant consisting of solar-assisted heat-pumps with ground-coupled seasonal storage for some years. The energy balance of the system has already been presented (Mazzarella and Pedrocchi, 1986). We summarize the main findings:

- Climate data of the site (Treviglio near Milan) (Table 1).
- Graphic of the system overall (Fig. 1).
- Graphic of the heating-plant (Fig. 2).

Table 1 Concise Climate Data in Treviglio

Treviglio is located 40 Km North-east of Milan
(longitude 9°30' east, latitude 45°31', 120 m a.s.)

Average June temperature	19.7	°C
Average December temperature	3.3	°C
Yearly solar radiation on horiz. surf.	4.7	GJ/m^2
Solar radiation on horiz. surf. in June	684	MJ/m^2
Solar radiation on horiz. surf. in December	96	MJ/m^2
Period of heating (from 15 Oct until 14 Apr)	180	days
Degree - days (at 19°C)	2350	°C d

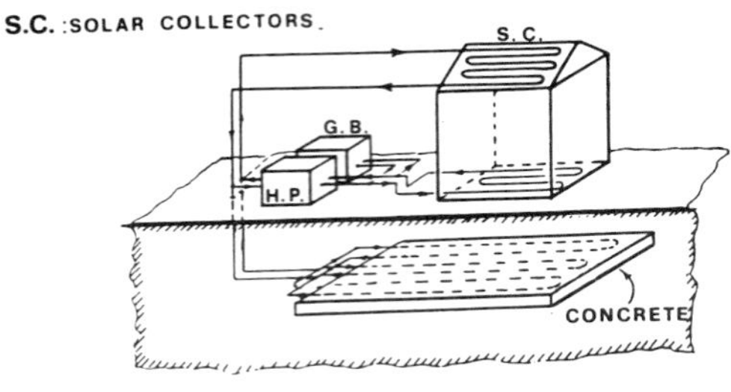

Fig. 1 System graphic

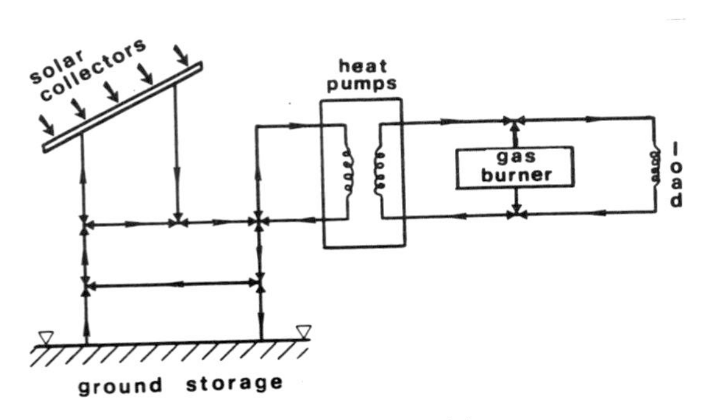

Fig. 2 Plant graphic

The heating plant can operate in four typical different modes depending on external conditions:

1. There is no building load and conditions are such that the collectors can charge the ground storage (this occurs mainly in Summer).
2. There is heat demand and solar radiation is sufficient to provide energy for the heat pumps (this occurs mainly in Autumn and Spring).
3. There is heat demand and solar collectors are not able to provide the necessary amount of energy for the heat-pumps. In this case, the latter utilizes the ground storage as cold source (typically in Winter and at

night-time).

4. The building load is greater than the heat pumps' capacity and the gas-burner provides the heat demand (this should only be an emergency mode).

A typical energy balance for a whole year is schematically represented in Fig. 3 (the gas-burner never functioned).

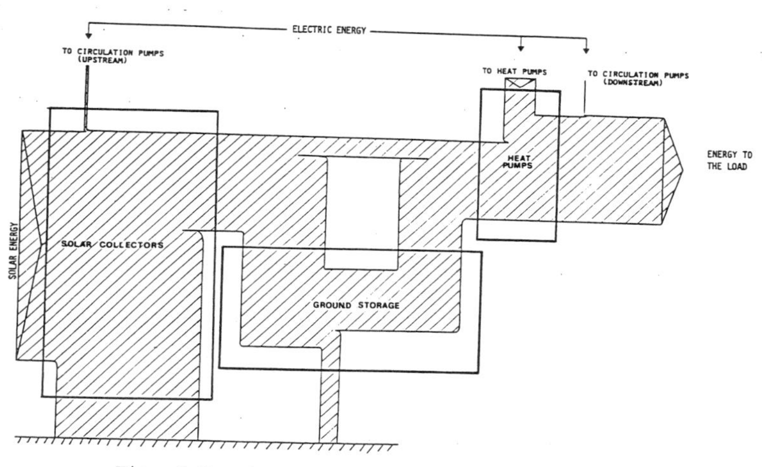

Fig. 3 Yearly energy balance (├──┤ 400 GJ)

We also conducted an economic analysis of the system; the synthesis is shown in Table 2.

Table 2 Initial Approximate Costs Analisys

	Conventional system	This system
Construction costs for a 100 m^2 apartment	100 MLit	110 MLit
Heating costs for one year (all inclusive)	0.6 MLit	0.4 MLit
Pay-back period without discount rate	$\frac{10 \text{ MLit}}{0.2 \text{ MLit/y}} = 50\text{y}$	

We would now like to describe an exergetic analysis of the system to assess the thermodynamic quality of the process and to illustrate the more important exergy losses and relative irreversibilities.

THEORETICAL APPROACH

To carry out the exergetic analysis of the whole plant, we have divided it into four sections (subsystems):

1. Solar-collectors subsystem

2. Ground storage subsystem
3. Heat-pumps subsystem
4. Heating distribution subsystem

The circulation pumps located upstream of the heat-pumps have been enclosed within the solar collectors subsystem; the circulation pumps located downstream of the heat-pumps have been enclosed within the heating distribution subsystem. We have calculated the exergetic balance for each subsystem using the method proposed by Prof. L. Borel, (1984). This is the most systematic and suitable for our work.

So as to avoid being obliged to deal with a non-stationary state with variations of exergy in the ground storage, we refer to a whole year presupposing that the ground storage after such a period reverts to the same start conditions: in this way we also avoid any difficulty in discovering the mass ground involved and its temperature distribution. This method does not permit separation of the exergetic losses during charge and discharge periods. We have found it suitable to make use of the daily medium temperature of the air as temperature of the dead state.

One important matter is calculating the solar radiation exergy. Although it is known (Gribik and Osterle, 1984) that the value of the exergy is in theory slightly less than its energy, (for T_a=300K approx. 93%) in reality the maximum exergy obtainable with solar collectors without concentration is 0.54 of solar energy when the ambient temperature is 300K (Haught, 1984). Further reductions are due to the scattering and absorption by the atmosphere. In our case, due to the particularly cheap type of collectors used, we have preferred to refer to the exergy of the solar radiation corresponding to a Carnot factor $(1-T_a/T)$ where T is the daily maximum value of the stagnation temperature for this type of collector: this ranges from 20°C to 120°C.

The exergy given to the building for the heating has been calculated assuming a constant thermal level equal to 20°C.

RESULTS AND CONCLUSIONS

The results of the exergetic analyses are schematically represented in Fig.4. Due to the low thermal level, it should be borne in mind that exergy flowing from the collectors and exergy flowing out and into storage, depends heavily on the choice of dead state. A variation in the latter could radically modify that part of the exergy diagram located between the collectors' outlet and heat-pumps' inlet. However, the insignificant quantities of the exergetic fluxes relative to this area of the plant would not substantially modify the exergetic scheme overall.

In conclusion, the following findings may be summarized:

1. A significant exergetic loss relative to the solar-collectors' subsystem; since we assumed the exergy of solar radiation corresponding to the stagnation temperature, this loss only takes function into account.
2. The exergy flowing into the heat-pumps' evaporators is very small in comparison to the exergy of incoming electric energy (this consideration is valid independently of any choice as to dead state).
3. A heavy, possibly unexpected exergetic loss is the heat-pumps' subsystem; this is due to the low COP value compared to the theoretical

one corresponding to the working limit temperatures. These heat-pumps are in fact designed to produce hot water at 50-55°C; in the case under consideration, hot water would suffice at 30-35°C with a possible increase in COP.

4. The only subsystem of the plant with good exergetic efficiency is the distribution of heat in the building.

We have found a total exergetic efficiency of ~7% (please remember we assumed a low value for the solar radiation exergy). This total exergetic efficiency may be compared to the one of the traditional heating system (~5%) or to the one of the electric heating system (~7%) (Borel, 1980).

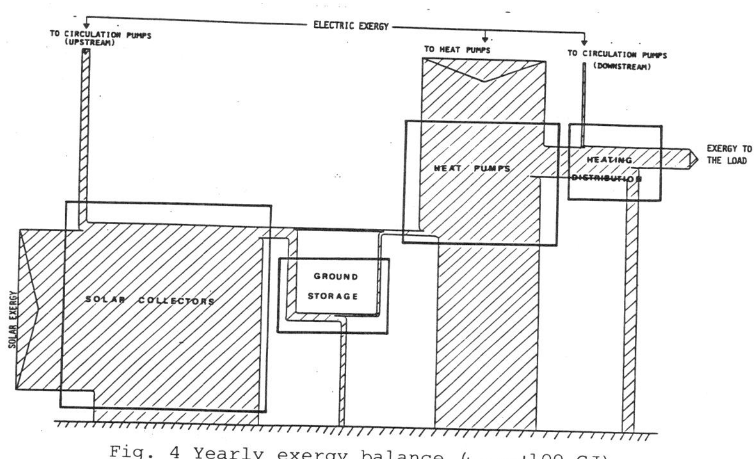

Fig. 4 Yearly exergy balance (├──┤100 GJ)

Only a heating system utilizing a high-efficiency heat-pump with heat distribution at low temperatures would provide better exergetic efficiency.

REFERENCES

Borel, L. (1980). Economie énergétique et exergie. Ec. Pol.Fed. de Lausanne 1980, IV.

Borel, L. (1984). Thermodynamique et énergétique. Presses Polytechniques Romandes.

Gribik, J.A. and J. F. Osterle, (1984). The Second Law Efficiency of Solar Energy Conversion. Jour. of Solar Energy Engineering Vol. 106/17.

Haught, A.F. (1984). Physics Considerations of Solar Energy Conversion. Jour. of Solar Energy Engineering Vol. 106/3.

Mazzarella, L. and E. Pedrocchi, (1986). Evaluation of the Treviglio Project. Task VII of Solar Heating and Coding Programme IEA.

Mazzarella, L. and E. Pedrocchi, (1986). Monitoring Solar-assisted Heat Pumps with seasonal Ground-coupled Storage: Analyses of the first results. Euroclima 86 Brussels.

A PASSIVE SOLAR HEATING SYSTEM COMBINED WITH A HEATPUMP AND A LONG TERM HEAT STORAGE

D. van Hattem, R. Colombo and P. Actis-Dato

Commission of the European Communities
Joint Research Centre, Ispra Establishment
I-21020 Ispra (VA), Italy

ABSTRACT

This paper describes the design and the first preliminary performance results of a sunspace attached to an existent building, combined with a heatpump and a long term heat storage. The aim of the project is to study the possibility of storing the excess heat of the passive system in a low temperature storage, which is used as cold source for a heatpump. The advantages of the presented system are that the energy flows in the passive solar system can be controlled and that a rather high solar fraction can be obtained (around .7 to .8 in the climate of Ispra).

KEYWORDS

Passive solar energy, heat pump, heat storage

INTRODUCTION

In general passive solar systems are not well suited to store the collected energy for more than about twelve hours (Shea 1984). But if the energy could be stored over longer periods of time, much higher solar fractions could be obtained.

Fig. 1. Solar laboratory of the CEC Joint Research Centre in Ispra, after rebuilding.

The long term storage of heat from passive solar systems can be realized by dumping the excess heat during hours of over heating, in a low temperature heat storage. This storage is at the same time the cold-source of a heatpump. With this heatpump the house is heated during the periods with insufficient sunshine. The heat can be extracted from the building with fancoil units connected to the heat storage.

The storage can be a large uninsulated underground water vessel or the ground itself with a buried heat exchanger. The temperature of the storage must be in the range form 5 to 15 ^oC. Lower temperatures would result in a bad heatpump performance and higher temperatures could make the heat extraction from the building difficult.

The purpose of this project is to assess if a heating system based on this concept is a viable solution from a technical economical point of view.

SYSTEM DESIGN

The building to which the sunspace is attached, is the Solar Laboratory of the Joint Research Centre in Ispra, which is used as an office. The floor area is about 160 m^2. Under the building there is a large heat storage of 50 m^3 of water which has been used for various experiments in the past, especially in combination with a heatpump (see van Hattem and others 1982 and 1984).

A large sunspace has been added to the building and the area of the South facing windows of has been increased. The building has now 15 m^2 of windows orientated at the South, with an inclination of 60 degrees, which give a direct solar gain. In Fig. 1, a view of the Solar Laboratory with the sunspace after completion of the building works is given.

The design of the sunspace is adapted to the rather warm summers in Ispra. Therefore, only vertical glazings have been used, since these can be shaded rather easily in summertime. Also, glazings oriented to the East or West have been avoided because of overheating. The South facing windows of the sunspace

Fig. 2. Shading system on the windows of the sunspace

and the main building are very well protected against the penetration of direct solar radiation in summer. Fixed shading devices are used for the sunspace, and moving blinds on the main building (see Fig. 2). The sunspace has a floor area of 90 m^2 and a South facing glazing of 60 m^2. Ten fancoil units are placed inside for heat extraction.

The heatpump is an electrical water to water heatpump, with a heating power of 16 kW. The heatpump and the fancoil units are used in summer time for the cooling of the building. A cooling tower is used for the heat rejection.

SYSTEM OPERATION AND CONTROL

A schematic view of the heating system is given in Fig. 3. It consists essentially of:

- the sunspace with a floor heating system and fancoil units
- building with a floorheating system and fancoil units
- large storage system (50 m^3 water)
- electrical water to water heatpump (16 kW thermal)

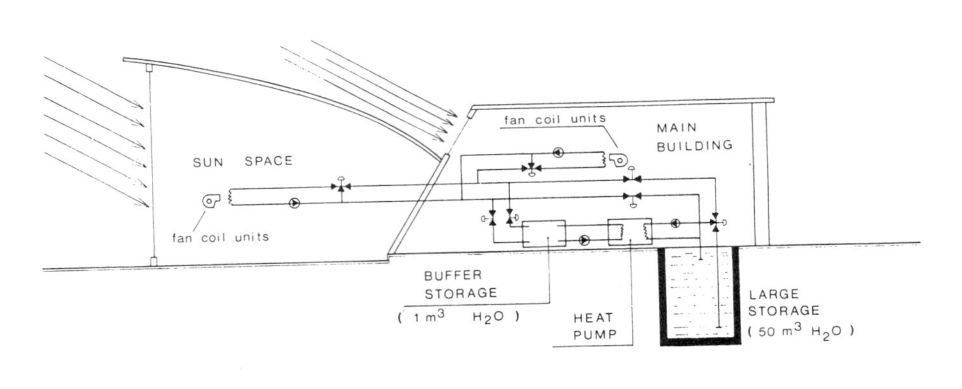

Fig. 3. Schematic view of the heating system.

In wintertime, the system operates as follows:

During the day the building and the sunspace are kept at a minimum temperature of 20 °C with the heatpump or the direct solar gain or both. If the temperature in one of the rooms rises higher than 22 °C, the fancoil units in that room will switch on and cool the room with the water from the large storage. The heat put into the large storage is used later through the heatpump to provide heating during the night or evening for the building. A schematic diagram illustrating these energy flows is given in figure 4. When, in spring time, the temperature in the large storage rises too much to cool the rooms, the shading devices on the windows will stop the influx of solar radiation.

When the temperature in the large storage drops below the critical storage temperature, that is the lowest temperature at which the heatpump can be

operated, an auxiliary heating takes over. However, this should happen only on a few days each year.
The sunspace is heated only during office hours (8-18 hrs). This heating is done with the underfloor heating system. In the morning when the sunspace must be heated up quickly, the fancoil units are also used for the heating.
In summertime the heatpump keeps the large storage at a fixed temperature (about 14 °C). The fancoil units are used for the climatization of the building and the sunspace. The operation of the heatpump operation is limited to night time only.

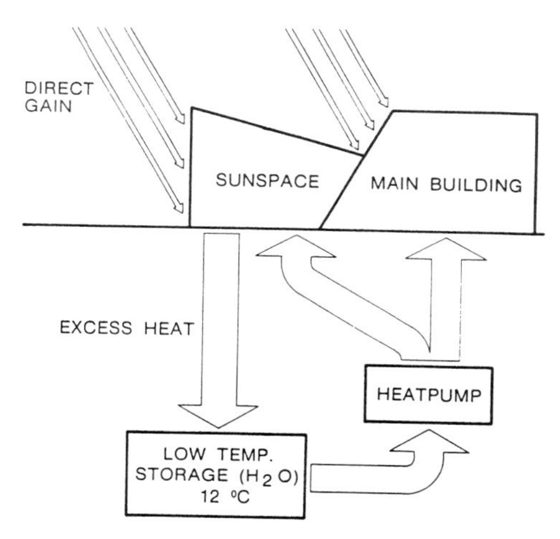

Fig. 4. Schematic view of the energy flows.

PERFORMANCE

Winter performance

Due to delays in the construction of the sunspace the planned detailed monitoring has been delayed as well. However some temperatures have been recorded and the the electricity of the heatpump was measured as well. The temperature recordings were mainly used to tune the control system.
The results presented here refer to the period 23-12-1986 to 5-4-1987. At that time some construction works were still going on in the building, so no undisturbed performance was obtained. Also the control was not completely satisfactory yet. An other problem was air in the hydrolic parts of the system, which did reduce certain flows significantly at certain days. Though these problems have been solved now, the results presented here should be regarded as preliminary results.

During the period 23-12-1986 to 5-4-1987 the estimated building load was 22000 kWh. The heatpump used, in this period, 2832 kWh of electricity. The auxiliary heating used during the same period was 2240 kWh. The energy used for lighting and instruments was almost nihil: 140 kWh. This because the building was not used yet and was hardly instrumented.

These preliminary results seem to confirm the high solar fractions which were expected on basis of the simulations (van Hattem and others 1986).

Summer performance

Also during the summer some measurements were done in order to get a first impression of the summer performance of the system. The results refer to the period 25-5-1987 to 7-9-1987 (105 days). In this period the heatpump, now used for the cooling of the building, used about 2500 kWh electricity in order to keep the building and the sunspace in comfort conditions (= 24 kWh/day). The electricity used for lights and instruments was about 900 kWh. If one assumes a COP of the heatpump of 3.5, the cooling load of the building was 6300 kWh (= 60 kWh/day). Not all the data have been treated yet so only these global figures can be given. It was observed that the shading devices worked well.

CONCLUSIONS

Though the preliminary results presented here look promissing one should not draw any definitive conclusions yet, concerning the viability of these kind of systems. More performance data are required. A detailed monitoring in going on now and next summer more extensive performance data will be available. During the monitoring emphasis will be put on the comfort conditions created by the system.

The economics and operational aspects will also be taken into account for the final system assessment. At the actual local economic conditions a simple pay back time of about 18 years is expected. However, one should realise that the economics are strongly influenced by the the costs of electricity and the kind of heat storage which is used. This will change from case to case. The construction of the sunspace is finished now. Further modelling studies will be done, especially the use of different types of heat storage systems will be studied, for example the use of multi layer horizontal ground coils.

REFERENCES

Shea, M. (1984), Monitoring the Thermal Performance of Passive Solar Buildings, Commission of the European Communities, EUR 9417 EN

van Hattem, D., (1984), Performance of a Solar Cooling System Using Different Types of Evacuated Tubular Collectors, Commission of the European Communities, EUR 9711 EN

van Hattem, D., E. Aranovitch and P. Actis-Dato, (1982), Performance of a Hybrid Solar Heating System of the Solar Laboratory of the JRC-Ispra, IEA-EC Workshop on Solar Assisted Heatpump Systems with Ground Coupled Storage, Commission of the European Communities SA.1.04.D2.84.08

van Hattem and R. Colombo, (1986), Design and analysis of a passive solar heating system with a long term heat storage for an office building. Conseil International du Batiment 10ème Congrès, Washington D.C. 1986.

HEATING PUMP-SOLAR ENERGY MIXED USING A COVERED SWIMMING POOL AS COLD FOCUS IN WINTER

F. Hernãndez and J.L. Esteban

Instituto de la Construcciõn y del Cemento "Eduardo Torroja", C.S.I.C. 28033 Madrid, Spain

ABSTRACT

Basically, it is about serie circuit where, in winter, the traditional cold focus (reservoir) is substituted by a swimming pool covered. This system is suitable in climates where frost is not frecuent (Mediterráneo coasts), heating in winter a detached house with swimming pool, and offers besides the following advantages:

1.- In summer, the heating-pump cools the house transfering power to the condenser; in this case, the swimming pool dissipate it partially by evaporation, radiaton and convection. The solar system is shutdown.

2.- The working period is extendable to spring and autumn. During those seasons, the heating-pump do not works, being only a work the solar installation in order to heat the swimming pool.

KEYWORDS

Heating-pump; solar energy; swimming pool; collectors; heat pump COP; solar energy fraction.

INTRODUCTION

The purpose behind replacing a conventional system with a heating pump-solar energy mixed system is to obtain a sufficiently large proportion of free solar energy to make it competitive economically or, if the period of amortization is too long to make investment in the project seem worthwhile, to provide year-round complementary services that could not be supplied by a conventional facility.

In the case of a heating pump-solar energy mixed system that uses a swimming pool covered with a transparent plastic sheet as a cold focus in winter, and excluding the swimming pool from the fixed overhead for the installation, such a mixed system can provide particularly attractive additional services for a one-family dwelling in a location that enjoys an equable climate (such as a Mediterranean climate). In addition to the basic service of heating in winter, such services include air conditioning in summer and heating for the swimming pool in spring and autumn.

HEATING PUMP-SOLAR ENERGY MIXED SYSTEM USING A COVERED SWIMMING POOL AS COLD FOCUS IN WINTER

Figure 1 presents a diagramme of the operation of the system. Basically the system consists of a mixed circuit set up in series, in which the solar collector is replaced by a swimming pool covered with a trasparent plastic sheet in winter in order to prevent evaporation losses and to reduce convection losses.

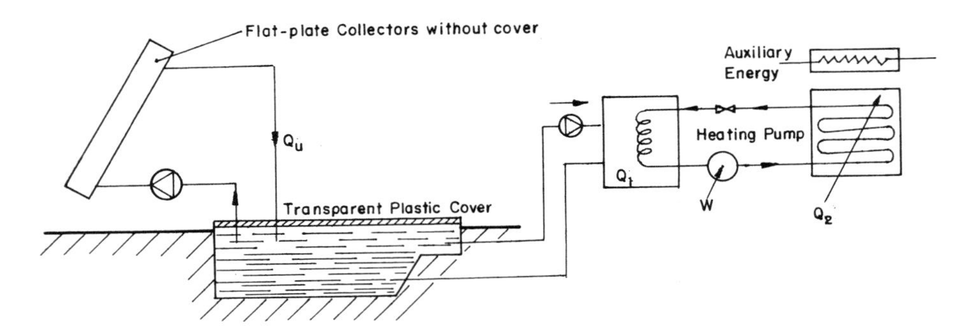

Fig. 1. Heating pump-solar energy mixed system using a covered swimming pool as cold focus in winter

In the wintertime, the heat pump's evaporator excheanges heat with the covered swimming pool and the dwelling's condenser. The heat pump is situated between the solar energy systems and the thermal load and is in operation only when the swimming pool temperature is higher than the minimum operating temperature of the heat pump. This temperature is attained by means of the solar energy absorbed as radiation together with the energy supplied by uncovered flat-plate collectors. When the swimming pool temperature is below the minimum, auxiliary energy is used.

In the summertime, the heat pump cools the dwelling by means of the evaporator and supplies the condenser (the swimming pool), with energy, partially dissipated by evaporation, convection, and radiation. The uncovered flat-plate solar collectors are not used during this period.

In the spring and autumn, the heat pump is not in operation, and just the solar energy subsystem is used to heat the swimming pool.

As a result, a mixed system of this kind is in operation all year long; just the heat pump in summer, the heat pump subsystem in spring and autumn, and all system components in winter.

SIMULATION OF SYTEM OPERATION

A computer was used to simulate the behaviour of the mixed heat pump-solar energy system. A system of equations obtained by establishing the thermal balance of each system component yielded a recursive analytical model providing swimming pool water temperature at any given time (h) based on the temperature at the time minus one hour.

$$T\ (h + 1) = T(h) + Qu\ .\ Ac - Q - Q_1)/V$$

where Qu is the usable heat delivered by the collectors in function of collector performance (Duffie and Beckman, 1980), Ac is collector surface area, Q is the sum of all the swimming pool's heat losses (convection + radiation, disregarding evaporation, which is negligible through the use of the plastic sheet, less direct solar energy accumulated during the hours of daylight), Q_1 is the heat obtained from the evaporator, which is a function of the thermal load and the heat pump's COP (variable in function of swimming pool water temperature), V is swimming pool volume, and T(h) and T(h + 1) are swimming pool temperature at times h and h + 1. Figure 2 depicts this energy balance graphically.

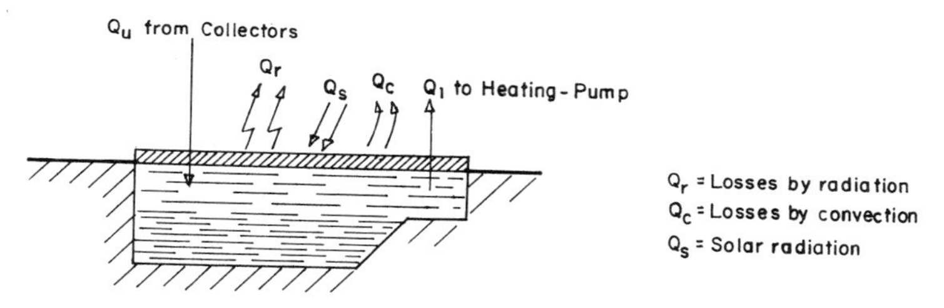

Fig. 2. Energy balance in the swimming-pool

EFFECT OF CERTAIN SIZE VARIABLES ON THE SOLAR ENERGY FRACTION

The solar energy fraction, $F = (Qu\ .\ Ac)/Q_2$, where Q_2 is the thermal load of the dwelling. Optimum size of system components must be determined by analyzing the variation in each of the variables with respect to changes in the solar energy fraction F. This parameter provides us with the required information on the economic suitability of the mixed system.

Each of the following figures reflects the behaviour of certain variables with respect to F, using the least favourable, most extreme conditions for the system in comparison with other Mediterranean localities as the input data for simulating the climatic conditions for a typical year in the province of Madrid.

Collector Area

Figure 3.1 presents the trend in the solar energy fraction F in fuction of collector area Ac. The lower curve represents an uncovered swimming pool, while the upper curve represents a swimming pool ocvered with a transparent sheet of plastic.

The two curves are rather flat, and the increase in F in the simulation with respect to the uncovered pool achieved by using the plastic sheet is practically constant (between 8 and 14%). Such an increase more than offsets the cost of the cover.

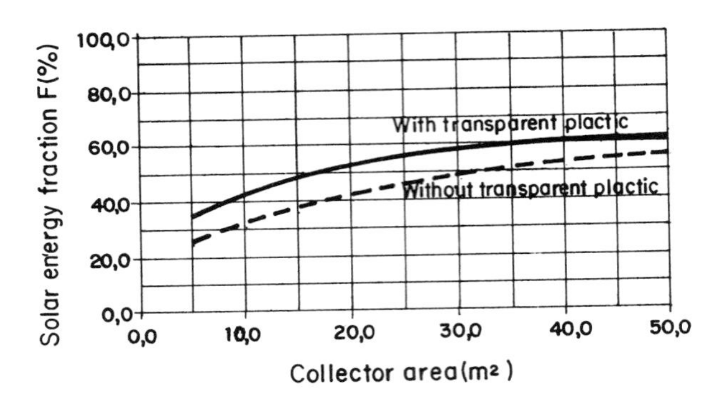

Fig. 3.1. Solar energy fraction versus collector area

Minimum Starting Temperature For The Heat Pump

Heat pump operation is normally bounded by two temperature limits:

- the minimum evaporator temperature at which the heat pump can commence operating safely such that the return water will not freeze (related to pressure), and
- the maximum evaporator temperature, related to the minimum temperature, determined by the nature of the coolant (González, 1975).

Figure 3.2 represents the case in which the operating temperature of the heat pump ranges between 1.5 and 10°C.

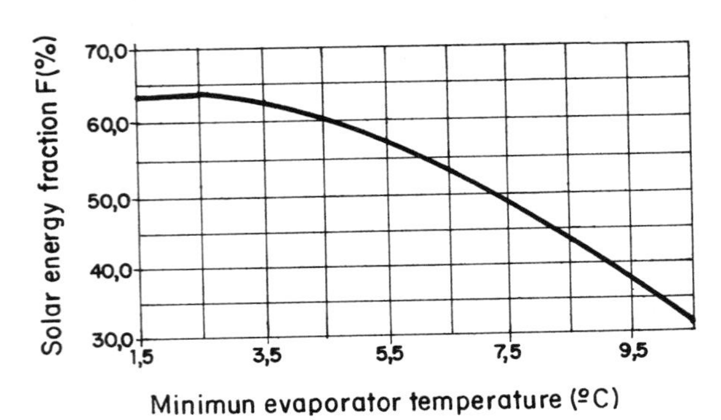

Fig. 3.2. Solar energy fraction versus minimum starting temperature

F decreases as the temperature approaches the minimum, although the function decreases hardly at all until 4°C. No adjustment of the heat pump to an operating temperature of below 4°C is warranted. The maximum fluctuation in F. however, in the range between 1.5 and 10°C is 32.5%.

CONCLUSIONS

The mixed heat pump-solar energy system using a swimming pool covered with a trasparent plastic sheet as a cold focus affords the following advantages over other mixed systems:

- the period of system operation can be extended through spring and autumn for the purpose of heating the swimming pool;

- in view of the low operating temperatures of the heat pump's solar collector-swimming pool-evaporator subsystem, the performance of the uncovered flat-plate solar collectors, which are inexpensive, is improved, and no insulation of piping is required. The mean temperature of the swimming pool during the operating hours of the system was 13,5°C;

- the sizeable heat losses occurring in the pool limit the area of application of this type of system to climatc regions with mild winters, e.g., the Mediterranean region;

- it would be interesting to design a heat pump specifically for such a mixed system, e.g., one with large heat exchanger surfaces, particularly for the evaporator.

LIST OF REFERENCES

Duffie, J.A. and Beckman, W.A. (1980), Theory of flat-plate collector. In J. Wiley & Sons (Ed), Solar Engineering of Thermal Processes, New York, pp. 246-247.

González, F. (1975). Aplicaciones del efecto frigorífico a la calefacción. Memoria del Centro Experimental del Frío (C.S.I.C.), pp. 17-24.

Harpiter, H.E. and Eltimsahy, A.N. (1978). A heat pump driven optimal heating system. I.E.E.E. Transactions on Industry Applications, Vol. IA, 14, July-August, pp. 112-123.

LARGEST HEATPUMP OF GERMANY

Peter Bailer
Manager International Sales
Centrifugal Refrigeration Units and Heatpumps
Sulzer Bros. Ltd.
CH - 8401 Winterthur, Switzerland

Heatpump for the district heating network in Kiel.

Customer:
Preussen Elektra / Stadtwerke Kiel

Heatpump-Supplier:
Sulzer Escher Wyss Lindau, FRG

Heatpump Type:
Sulzer Unitop 34 FY with a two stage centrifugal compressor Type Uniturbo 34 FY. The heatpump is operating in two modes :

1. Cooling water of the powerplant as heatsource with a temperature of 16°C, at this heatsource level and district heating water temperatures of in/out 55/85°C a heatcapacity of 9 MW with a COP of 3,1 can be achieved.

2. In case the powerplant is not in operation the heatpump is designed to use seawater at 3°C. With district heating water of in/out 65/78°C a heatcapacity of 7 MW with a COP of 2,9 can be achieved.

In order to be able to use seawater of this temperature level a plate heatexchanger has been used. Refrigerant is evaporating inside the plates, water is distributed equally to flow as a film on the outside of the plates. This design allows a water outlet temperature close to 0°C without danger of freezing.

With the same heatexchanger design Sulzer has built the largest heatpump installation in the world in Stockholm - Värtan with a total heatcapacity of 180 MW - split into 6 units with 30 MW each.

The heatpump in Kiel has been put officially in operation by the Ministerpresident of Schleswig Holstein, Dr. Barschel, during an inauguration on 28.10.1986 and has been successful in operation since that time.

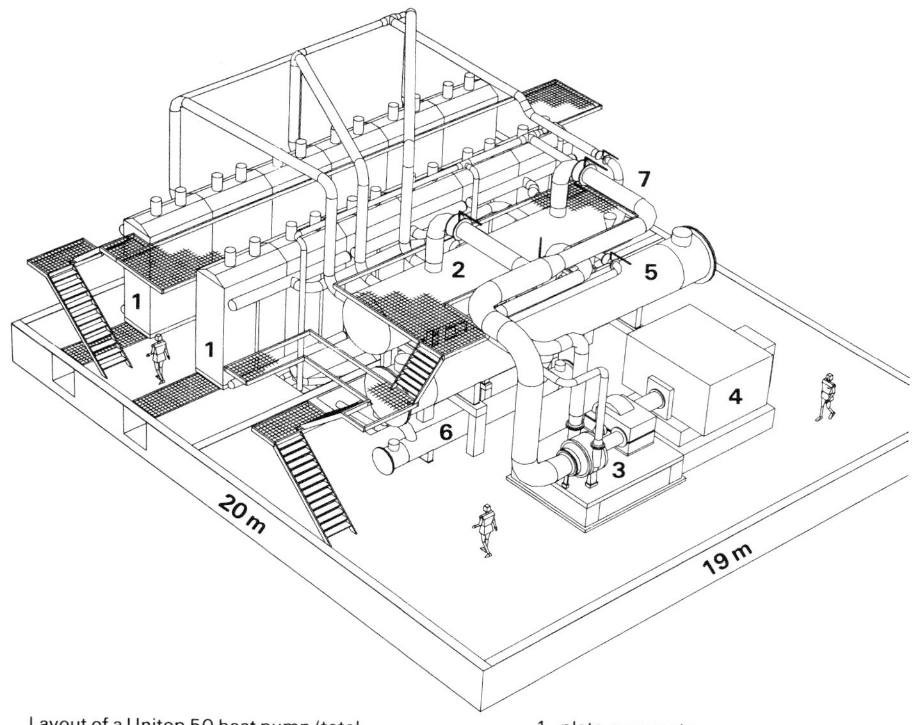

Layout of a Unitop 50 heat pump (total of 3) at the Värtan-Ropsten heatpump station with a heating capacity of about 28000 kW, engineered and supplied by Sulzer for the district heating installation Stockholm Värtan, Sweden.

1 plate evaporator
2 separator
3 Uniturbo 50 FY compressor package
4 electric motor compressor drive
5 condenser
6 sub-cooler
7 intermediate vessel (hidden by 2)

MATHEMATICAL MODELLING AND EXPERIMENTAL VALIDATION OF HEAT AND MASS TRANSFER TO UNFROSTED WIND CONVECTORS

D.P. Finn*, P.F. Monaghan* and P.H. Oosthuizen**

* Department of Mechanical Engineering, University College, Galway, Ireland

** Department of Mechanical Engineering, Queen's University, Kingston, Ontario, Canada K7L 3N6

ABSTRACT

Wind convectors may be used to replace conventional fan-assisted evaporators for heat pumps. These outdoor heat exchangers transfer heat from the environment without the use of fans. Heat transfer is multimode and is due to (a) convection (b) condensation (c) longwave radiation and (d) solar radiation. This paper describes the development of a mathematical model that predicts the wind convector heat transfer when there is no frost or rainfall on the convector surfaces. The model is incorporated in a general simulation program and validated against experimental tests.

KEYWORDS

Wind convectors, heat pumps, heat transfer, computer model, experimental validation.

INTRODUCTION

Conventional air-source heat pumps use fans to create forced convection on the air-side of the evaporator. Wind convectors are outdoor heat exchangers which rely on wind driven and free convection to transfer heat to an anti-freeze solution (a brine of methanol and water). Limited work is available however, which investigates wind convector heat transfer from a comprehensive theoretical and/or experimental basis and this is seen as a barrier to the exploitation of this innovative heat pump development. This paper describes a mathematical model that predicts the heat transfer to a convector based on long cylindrical tubes (see Fig. 1) when there is no frost or rainfall on the heat transfer surfaces and when forced convection conditions due to wind prevail.

Heat transfer is due to the following mechanisms in series (Finn and Monaghan, 1987):
1. Multimode heat transfer to the convector outer surfaces.
2. Conduction heat transfer through the convector walls (thick-walled cylinder).
3. Convection heat transfer from the convector inner walls to the brine.

On the outside surfaces the following heat transfer processes take place in parallel:
1.1 Wind driven convective heat transfer from the ambient air.
1.2 Latent heat transfer due to condensation of air-borne water vapour.
1.3 Longwave radiation from sky, ground and buildings.
1.4 Solar radiation heat transfer.

The convector overall heat transfer coefficient U is affected by 3 sets of independent variables:
1. Climatic variables: air temperature, air humidity, wind speed, wind direction, solar radiation.
2. Design parameters: brine flowrate and properties, convector design specifications.
3. Heat pump operating condition: brine inlet temperature.

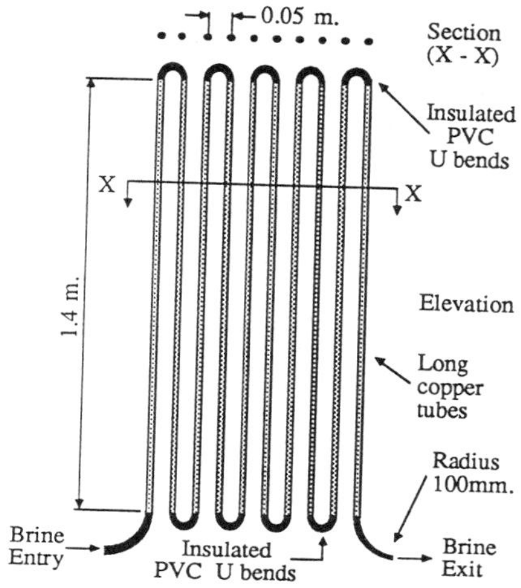

Fig. 1 Wind convector based on vertical cylindrical copper tubes.

Previous work examining heat transfer to wind convectors is limited and consists of long term seasonal/economic analysis or pure experimental work. The mathematical model developed here is incorporated in a convector simulation program and predicts convector performance due the combined effects of the above heat transfer processes as a function of the independent variables.

MODEL

Heat transfer to a convector is calculated as follows (Finn and Monaghan, 1987): (a) determine the external heat transfer convection coefficients for convector geometry (b) calculate heat transfer due to longwave and shortwave radiation (c) calculate internal convection coefficients due to brine flow and (d) use energy balance methods to calculate the heat transfer to the brine, taking account of thermal resistance of the tube wall and condensate film.

The external convection coefficients are determined using the following correlations:

1. Single cylinder in crossflow: Churchill and Bernstein (1977), Morgan (1975), and Zhukauskas and Zhiugzdha (1985).
2. Tube rows in crossflow: Grimison (1985) and Zhukauskas (1972).

The mass transfer coefficients are determined using the heat and mass transfer analogy, (Incropera and DeWitt, 1981).

For longwave radiation modelling (Finn and Monaghan, 1987) treat a convector as a small body surrounded by two infinitely large bodies (i.e. sky and buildings/ground). The sky effective temperature is estimated by Swinbank's formula (Duffie and Beckman, 1980). The ground and buildings are assumed to be gray bodies and their surface temperatures are let equal to air temperature. Solar radiation is calculated using techniques described in Duffie and Beckman (1980). For these tests the plane of the convector was oriented in the north-south direction.

Internal convection coefficients between the convector walls and the brine are estimated using the following assumptions: (a) smooth transition between the PVC and copper tubes (b) all pipes up to the entry region of the convector are insulated and (c) brine flow is either laminar or turbulent. Correlations based on a thermal entry length attributed to Hausen (1983) and Gnielinski (1976) are used to calculate an average convection coefficient for the convector tube.

The solution procedure for the mathematical model analyses the cylindrical convector as a single tube of length 14 m. Heat transfer is calculated by assuming steady state over a 5 minute period. This time step is selected on the basis of the experimental data (Monaghan and others, 1987). An energy balance is carried out at the convector surface so that the heat transfer due to convection, condensation, longwave and solar radiation is equal to the heat conducted through the convector wall. Solution is achieved by iteration which uses an initial guess of convector exterior surface temperature and converges when the energy balance is satisfied.

RESULTS

Validation of the simulation models against experimental data is carried as follows:

1. A correlated relationship between wind speed and U is obtained from the experimental data (Monaghan and others, 1987). Simulation results using the different Nusselt correlations are compared with the experimental data. The most suitable correlations are selected.
2. The simulation predictions are compared against two 24 hour time based tests.

Fig. 2 shows the experimental relationship between wind speed and U. This relationship is obtained from approximately 400 hours of data collected at the wind convector test facility. Data processing constrains the other independent variables within the limits shown, so that a relationship between wind speed and U is obtained. For these tests, no relationship was evident between U and wind direction or air temperature, therefore they are not constrained. Solar radiation effects are eliminated by using night time data. In Fig. 2, a series of curves generated by the simulation program using each of the above external convection correlations are plotted. The correlations attributed to Grimison (1985) and Zhukauskas (1972) give the closest results.

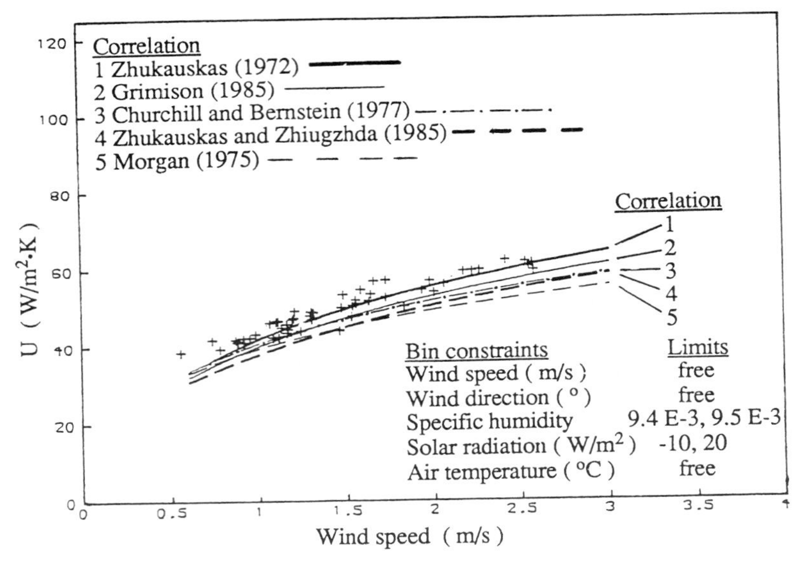

Fig. 2 U plotted against wind speed - A comparison of experimental data (+) (Monaghan and others, 1987) with simulation model using five different external Nusselt number correlations.

Fig. 3 shows a 24 hour test for laminar brine flow conditions, the data input to the simulation programs are collected at the wind convector test facility and U is calculated from this data. Specific humidity was found to be almost constant over these tests and therefore is not plotted. The following points are noted (a) at wind speeds of 1.5 to 3.0 m/s solar radiation relative to forced convection has a small influence on U, (b) the simulated results using Zhukauskas

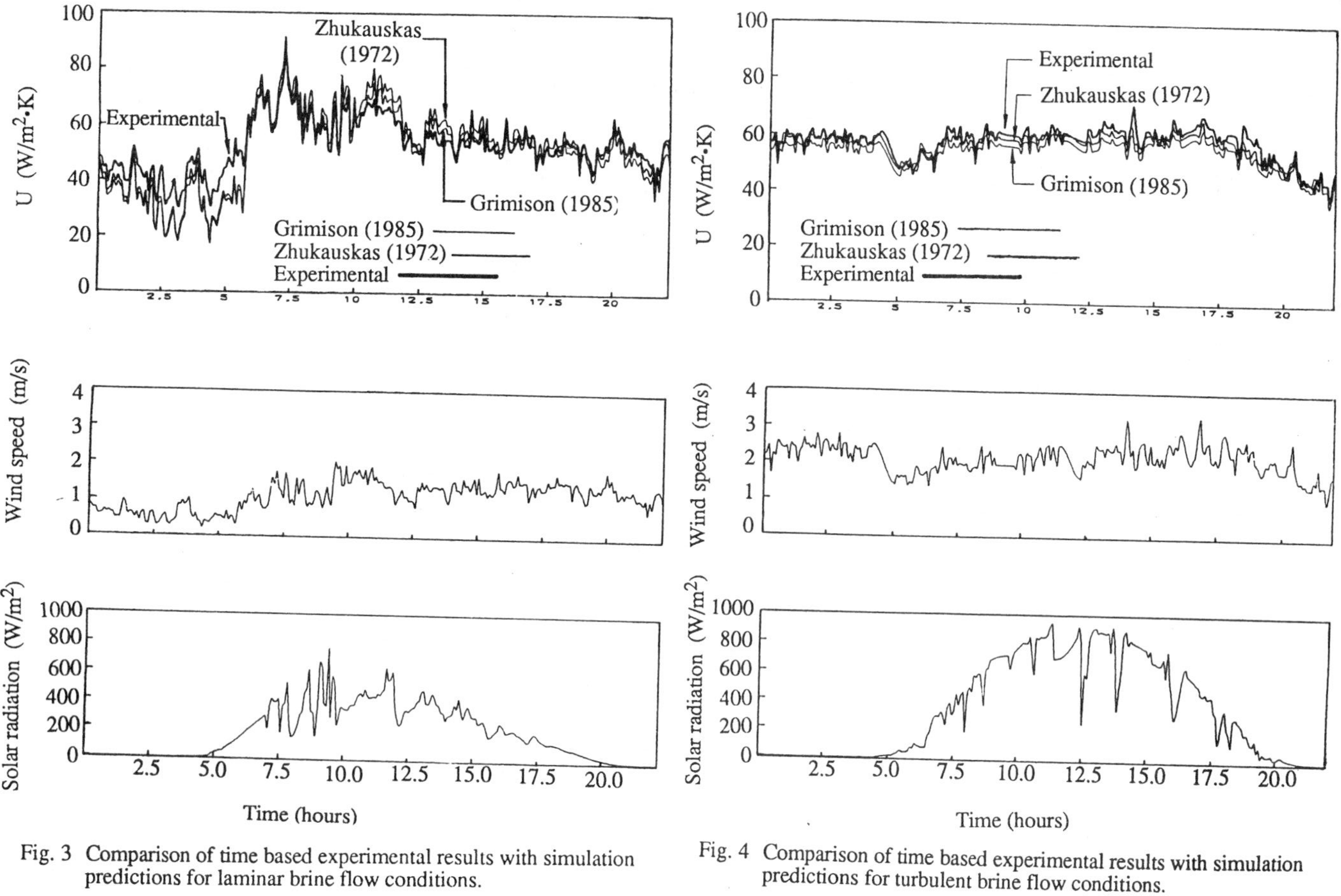

Fig. 3 Comparison of time based experimental results with simulation predictions for laminar brine flow conditions.

Fig. 4 Comparison of time based experimental results with simulation predictions for turbulent brine flow conditions.

(1972) gives the closest results. Fig. 4 shows a similar test to Fig. 3 with turbulent brine flow. At low wind speeds (less than 1 m/s) poorer agreement is found between the experimental and theoretical results. Possible reasons for discrepancies between experimental and model results are: (a) errors in the input data to the simulation programs arising from errors associated with the various climatic measuring instruments at the test facility, (b) error in U (experimental) estimated at ±5%, (Monaghan and others, 1987) and (c) errors at low wind speeds are possibly due to mixed convection effects, or greater anemometer inaccuracies at wind speeds below 1 m/s.

Results from studies using the validated model indicate the following, (a) at night, 95% of total heat transfer is due to convection and 5% due to longwave radiation, (b) at night, condensation contributes between 35% and 50% of total convective heat transfer when relative humidity levels vary from 70% to 100%, (c) solar radiation contributes up to 25% of total convector heat transfer under optimum solar conditions (i.e. early morning and evening periods, when radiation received by a horizontal surface is approximately 50-400 W/m^2), (d) thermal resistance due to conduction through the condensate film and metal wall is less than 1% of total resistance, (e) the percentage of total resistance caused by the brine/wall interface is less than 5% for turbulent flow and 20% to 40 % for laminar flow.

CONCLUSIONS

A mathematical model is developed that predicts the heat transfer to wind convectors for conditions when there is no frost or rainfall on the convector surfaces. The model is incorporated in a general simulation program and is validated against experimental test results. Satisfactory agreement is found between the model and experimental results.

ACKNOWLEDGEMENTS

This project was financially assisted by Erinova Industries, Dublin, National Board of Science and Technology, Dublin and Commission of European Communities, Brussels. We also thank technical staff at UCG Mechanical Engineering Department and Buildings Office staff at UCG.

REFERENCES

Churchill, S., and M. Bernstein (1977). A Correlating Equation for Forced Convection from Gases and Liquids to a Circular Cylinder in Crossflow. *ASME J. of H. T.*, *99*, 300-306.

Duffie, J.A., and W.A. Beckman (1980). *Solar Engineering of Thermal Processes,* 2nd ed. Wiley, New York. Chaps. 1-4, pp. 1-110.

Finn, D.P., and P.F. Monaghan (1987). *Mathematical Modelling and Experimental Validation of the Heat and Mass Transfer Associated with Cylindrical Wind Convector*, Report WC-5, Department of Mechanical Engineering, University College, Galway, Ireland

Gnielinski, V. (1976). New Equations for Heat and Mass Transfer in Turbulent Pipe and Channel Flow. *Int. Chem. Eng.*, *16*, 359-368.

Grimison, E.D. (1985). External flow. In F.P. Incropera and D.P. DeWitt. *Fundamentals of Heat and Mass Transfer*, 1st ed. Wiley, New York. Chap. 7.

Hausen, H. (1983). Laminar heat transfer in ducts. In L.C. Burmeister. *Convective Heat Transfer*, 1st ed. Wiley, New York. Chap. 6, pp. 245-249.

Incropera F.P. and D.P. DeWitt (1981). Introduction to Convection. In *Fundamentals of Heat and Mass Transfer*, 1st. ed. Wiley, New York. Chap. 6, pp. 247-320.

Monaghan, P.F., D.P. Finn, P.H. Oosthuizen, and B.J. McDermott (1987). Design and Testing of the UCG Wind Convector Test Facility. Paper No. 4.2.02, *Proc. of 1987 Solar World Congress.*

Morgan, V.T. (1975). The Overall Convective Heat Transfer from Smooth Cylinders. In T. Irvine, Jr. and J. Hartnett (Eds.), *Advances in H. T.*, *11*, Academic Pr., N.Y. pp. 199-264

Zhukauskas, A. (1972). Heat Transfer from Tubes in Crossflow. In J.P. Hartnett and T.F. Irvine, Jr. (Eds.), *Advances in Heat Transfer*, *8*, 93-160, Academic Press, New York.

Zhukauskas, A., and J. Zhiugzhda (1985). Average heat transfer. In *Heat Transfer of a Cylinder in Crossflow*, 1st ed. Hemisphere, Washington. Chap 7, pp. 129-149.

DESIGN AND TESTING OF THE UNIVERSITY COLLEGE GALWAY WIND CONVECTOR TEST FACILITY

P.F. Monaghan*, D.P. Finn*, P.H. Oosthuizen** and B.J. McDermott*.

* Department of Mechanical Engineering, University College, Galway, Ireland

** Department of Mechanical Engineering, Queen's University, Kingston, Ontario, Canada K7L 3N6

ABSTRACT

This paper deals with measurement of heat transfer performance of wind convectors. An automatically controlled and monitored outdoor wind convector test facility that is capable of measuring heat transfer rates and overall heat transfer coefficients to within ±5% measurement uncertainty has been designed, built and tested. Data on wind speed and direction, air temperature and humidity and solar radiation are simultaneously collected. The choice of measurement technique for each variable and an error analysis for each measurement are discussed. Typical graphs of test results are presented.

KEYWORDS

Heat pumps, wind convectors, energy absorbers, heat transfer, experimental.

INTRODUCTION

A wind convector is an alternative evaporator system for air-source heat pumps. Conventional air-source heat pumps use fans while wind convectors are metal or plastic outdoor heat exchangers which rely on wind-driven or free convection to transfer heat from outdoor air to an anti-freeze solution. From the viewpoint of energy performance, capital cost and reliability, wind convectors may have certain economic and operational advantages over conventional fan-assisted evaporators. Limited data on the relationships between climate and wind convector performance are available and this is seen as a barrier to the practical use of wind convectors.

This paper deals with design, error analysis and results processing of a wind convector test facility. The purposes of the facility are (a) to provide a data-base on the performance of wind convectors for a range of weather conditions and (b) to provide convector performance data suitable for heat pump designers. The operation of the test facility is demonstrated by the testing of two types of wind convectors when there is no rainfall and no frost growth on the convector surfaces. Under these conditions, heat transfer from the environment to the convector heat transfer fluid (a brine, consisting of methanol and water) can be due to convection, condensation longwave radiation and solar radiation. Overall wind convector heat transfer rate, Q, and overall heat transfer coefficient, U, are defined by the following expressions:

$$Q = \rho * V * c_p * [T_{m,o} - T_{m,i}] \quad (1)$$

$$U = Q / [A * \Delta T_{ba}] \quad (2)$$

Wind convector heat transfer performance may be related to the following measured quantities:

1. Dependent variables: brine temperature difference between inlet and outlet.
2. Design parameters: flowrate, inlet temperature, density and specific heat of brine.
3. Independent variables: wind speed and direction, air temperature and humidity, solar radiation.

EXPERIMENTAL APPROACH

The outdoor wind convector test rig consists of (a) a brine circuit including a heat pump, wind convectors and a temperature control system (Fig. 1) and (b) monitoring instrumentation and a data collection system (Fig. 2). Convectors may be simultaneously tested by installation on outdoor support structures. Brine flowrate may be independently varied. The temperature control system consists of a heat pump, electric resistance heaters and a storage tank and maintains the inlet brine to within ±0.5 °C of a pre-set temperature. The sensors are connected to a programmable monitor/ controller and this, in turn, is interfaced to a microcomputer and mainframe computers to allow data analysis. The wind convectors' urban location provides air flow patterns at the test site that are, to some extent, similar to those experienced by wind convectors in urban/suburban environments.

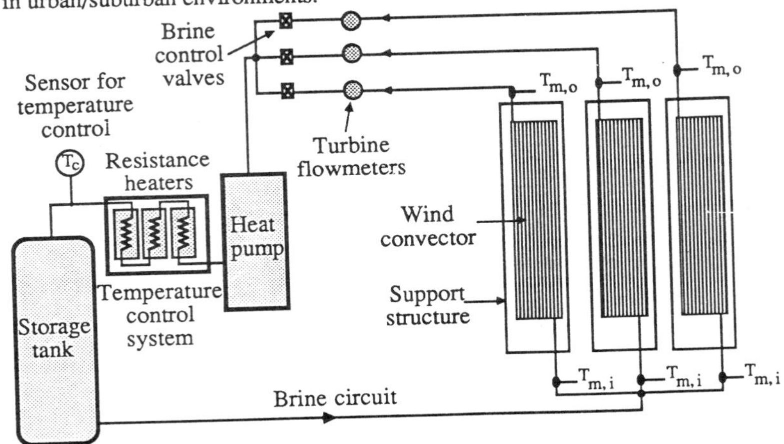

Fig. 1 Schematic diagram of the heat transfer fluid circuit.

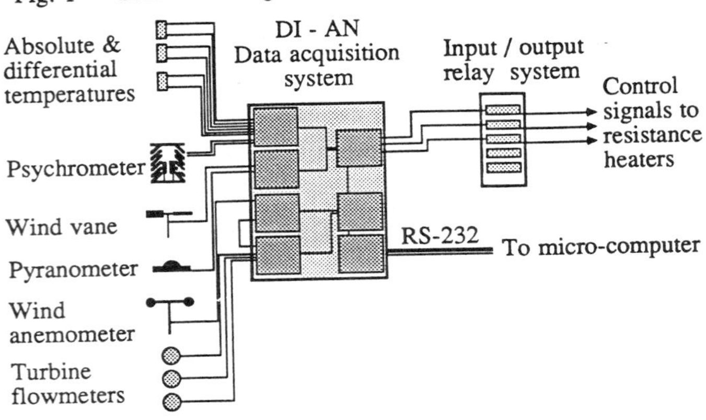

Fig. 2 Wind convector test facility data collection system.

MEASUREMENT TECHNIQUES

Finn and Monaghan (1987) give further details for each sensor. The results are summarised here. With the exception of brine density and specific heat, all dependent, independent and design variables discussed earlier, are automatically logged on the data collection system.

The brine temperature difference across a convector depends on the operating conditions and the convector being tested. Its value may be as small as 0.25 °C for parallel tube convectors (brine flowrate of 3000 l/hr) or up to 5 °C for simpler convectors based on a series arrangement of cylindrical tubes (brine flowrate of 125 l/hr). This temperature difference is measured by a pair of individually calibrated 1/10 DIN Platinum Resistance Thermometers (PRTs) with a differential temperature conditioner which gives a common excitation to each PRT. The outputs from the differential conditioner are the temperature difference and the inlet temperature, thus the temperature difference is itself directly measured. Other methods using separate PRTs and thermopiles were investigated and rejected. Air temperature and humidity levels are measured by using an aspirated psychrometer, which uses PRTs as wet and dry bulb sensors. The wet bulb depression is measured using a differential conditioner which also outputs dry bulb temperature. The above approach is preferred to polymer film humidity devices because these devices are not suited to relative humidity levels above 90%. Wind speed is measured by a wind cup anemometer placed at mean convector height. Hot wire anemometry is not used for this project because of its lack of robustness and inaccuracies associated with rainfall and changing wind direction. An analogue wind vane measures wind direction and a pyranometer is used for solar radiation. Brine flowrate is measured by three turbine flowmeters, which cover a wide range of flowrates. Brine density was measured regularly at 20 °C using laboratory scales and pipette and was found to be constant. This value is used to find the water/methanol mass ratio. Brine specific heat is calculated using the superposition of water and methanol specific heats.

ERROR ANALYSIS

Collet and Hope (1983) indicate that component errors for each measurement may be combined by addition in quadrature to give an overall uncertainty of measurement for each sensed variable, within a 99.7% confidence interval. This approach is used for each of the measurements. Errors were considered under the following broad classifications (a) systematic sensor/transducer and data acquisition errors, (b) random sensor/transducer and data acquisition errors. Total uncertainty of measurement depends on the actual value of the variable being measured and on post-processing of data. As discussed in the next section, data is averaged over a 5 minute period. This averaging process is assumed to effectively eliminate random errors and the total uncertainty associated with these averaged data is comprised only of systematic uncertainties. Based on the above approach, measurement uncertainties are:

Wind speed	± 0.05 m/s	(based on 1 m/s)
Wind direction	± 10.3 degrees	
Solar radiation	± 18.2W/m^2	(based on 1000 W/m^2)
Air temperature	± 0.081 °C	(based on 10 °C)
Wet bulb depression	± 0.027 °C	(based on 1 °C)
Air-brine differential	± 0.06 °C	(based on 10 °C, i.e., 0.6%)
Brine temperature diff.	± 0.009 °C	(based on 0.25 °C, i.e., 3.6%)
Brine temperature diff.	± 0.015 °C	(based on 2.0 °C, i.e., 0.8%)
Brine temperature diff.	± 0.026 °C	(based on 4.0 °C, i.e., 0.6%)
Brine Flowmeter 1	± 2.8 l/hr	(based on 125 l/hr, i.e., 2.2%)
Brine Flowmeter 2	± 5.5 l/hr	(based on 400 l/hr, i.e., 1.8%)

The total measurement uncertainties for brine density and specific heat are calculated to be ±1.8 kg/m^3 (based on 965 kg/m^3, ±0.19%) and ±8.6 J/kg•K (based on 3906 J/kg•K, ±0.22%).

The absolute error made in calculating U is determined by expanding Equation 2 using a Taylor series (Doebelin, 1984). The uncertainty figures given above indicate that the largest errors arise when either brine flowrate or temperature differential is small. Other variables occurring in

Equations 1 and 2 contribute a total uncertainty of about ±1%. Examples of the absolute error in calculating U for the cylindrical and the commercial (WRE) wind convectors are given here:

Convector	*Cylindrical*	*Commercial*
Flowrate	125 l/hr	400 l/hr
Temp. Diff.	4 °C	2 °C
Flowrate Error	±2.2%	±1.8%
Temp. Diff. Error	±0.6%	±0.8%
Overall U Error	±3.9%	±3.1%

The conclusions to this analysis are (a) the uncertainty in calculating U is better than ±5% if it is assumed that random uncertainties are made negligibly small by time-averaging and (b) error analysis for particular tests may show measurement uncertainty to be less than this.

RESULTS PROCESSING

For each sensor, the sampling strategy is follows. For the wind anemometer and the brine flowmeters, a cumulative pulse count is taken for 6 seconds and there is a 2 second pause. For the wind vane and pyranometer, a single analogue sample is taken every 2 seconds and every 4 samples are averaged. For the psychrometer and brine temperature PRTs, one analogue sample is taken every 8 seconds. Therefore, one set of data is available every 8 seconds. Results processing is carried out in two stages (a) time-averaging and (b) *bin* sorting.

Time-averaging is used to average out short term rapid fluctuations in wind speed and solar radiation and to neutralise the effects of time lag in the wind convector. After preliminary tests of 1, 5, and 10 minute averaging periods, the 5 minute period was chosen as the one which reduced scatter without loss of data. Fig. 3 shows unaveraged and 5 minute averaged data. Bin-sorting allows the effect of each climatic variable on convector heat transfer to be examined by sorting all data into *bins*. Within a bin, variations in all other climatic conditions are confined to a small range. The number of bins assigned to each weather variable was as follows.

Variable	Wind speed	Wind direction	Solar radiation	Specific humidity	Air temperature
No. of Bins	20	3	6	40	11
Total Range	0.2 - 3.8 m/s	0 - 90 °	-10 - 1153 W/m^2	0.0048 - 0.0125	6.5 - 20.7 °C

TYPICAL RESULTS

The results of a typical test run for the cylindrical convector are shown in Fig. 4. The results show:

1. Temperature and specific humidity of air change very slowly with time.
2. Peaks and valleys on the U and wind speed plots seem well correlated.
3. A relationship between U and solar radiation is not obvious.
4. Wind direction appears to vary significantly at consecutive data points.

Establishing the relationship between U and climate will be the subject of future papers.

CONCLUSIONS

An automatically controlled and monitored outdoor wind convector test facility that is capable of measuring heat transfer rates and overall heat transfer coefficients to within ±5% measurement uncertainty has been designed, built and tested. Continuing work deals with extension of the facility's capabilities to allow the effects of rain and frost on convector performance to be measured. Empirical and theoretical models of convector performance are also being developed.

ACKNOWLEDGEMENTS

This project was financially assisted by Erinova Industries, Dublin, National Board of Science and Technology, Dublin and Commission of European Communities, Brussels. We also thank technical staff at UCG Mechanical Engineering Department and Buildings Office staff at UCG.

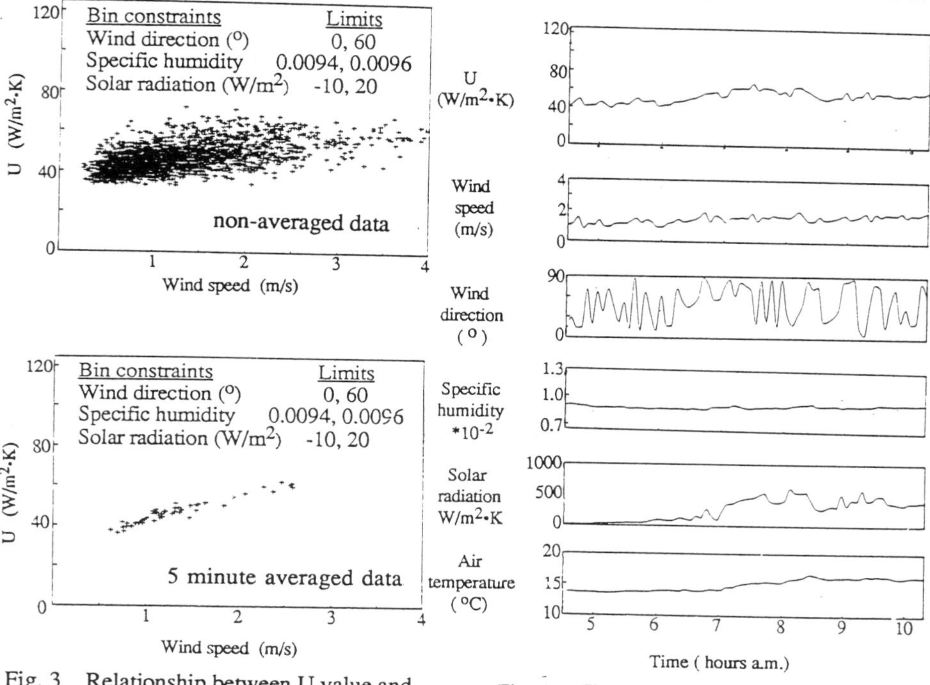

Fig. 3 Relationship between U value and wind speed for non-averaged data and 5 minute averaged data.

Fig. 4 Five minute averaged values of U and weather variables plotted against time for a 6 hour test run.

REFERENCES

Collet, C.V., and A.D. Hope (1983). Traceability, repeatibility, accuracy. In *Engineering Measurements*, Pitman, London. Chap. 2, pp. 32-53.

Doebelin, E.O. (1984). Generalised performance characteristics of instruments. In *Measurement Systems Applications and Design*, 3rd ed. McGraw-Hill, Japan. Chap. 3, pp. 37-207.

Finn, D.P., and P.F. Monaghan (1987). *Design and Testing of the UCG Wind Convector Test Facility*, Report WC-4, Department of Mechanical Engineering, University College, Galway, Ireland.

NOMENCLATURE

A	=	convector outside surface area, m^2
c_p	=	brine specific heat, J/kg·K
Q	=	total wind convector heat transfer rate, W
ΔT_{ba}	=	air-brine log mean temperature difference, °C
$T_{m,i}$	=	inlet temperature of brine to convector, °C
$T_{m,o}$	=	outlet temperature of brine from convector, °C
U	=	overall heat transfer coefficient referred to outer surface area, $W/m^2 \cdot K$
V	=	volumetric flowrate of brine through convector, m^3/s
ρ	=	brine density, kg/m^3

MATHEMATICAL MODELING OF PERFORMANCE AND OPERATIONAL MODE OF ENERGY-ABSORBERS

W. Krumm, F.N. Fett and H. Bradke

Institute of Energy Technology; University of Siegen, Federal Republic of Germany

ABSTRACT

A mathematical model is presented which describes the operational mode and the performance characteristics of energy-absorbers. The simulation of different types of energy-absorbers shows the dependence of the daily and annual average value of specific energy gain and its composition due to different energy sources from the environment. Furthermore the annual average value of the specific energy gain as a function of the temperature difference between environmental and working fluid and as a function of the working fluid flow rate is shown for different types of energy-absorbers. The climatic input values for the calculation are generated by a weather model. Finally suggestions are given for the optimum operation of energy absorbers.

KEYWORDS

Energy-absorbers, mathematical model, performance characteristics, simulation

INTRODUCTION

Energy-absorbers are usually uncovered collectors that utilize solar energy as well as energy from the ambient through convection, condensation of humidity and longwave radiation. They are used as heat source devices for heat pumps. In order to design such a system it is useful to utilize a mathematical model that determines the energy-absorber characteristics. In this paper a general mathematical model for energy-absorbers and simulation results for five different types of energy-absorbers are presented.

MATHEMATICAL DESCRIPTION

Despite of different configurations of the energy-absorbers a mathematical model of general validity can be developed. The first step in the model is to subdivide the energy-absorber into several volume cells. For each of these volume cells the transient mass and energy balances are formulated. I.e. for the energy-absorber

$$m_A \cdot c_A \cdot \frac{\partial T_A}{\partial t} = \alpha_a \cdot A \cdot (T_L - T_A) - \alpha_I \cdot A \cdot (T_A - T_S) \tag{1}$$

and for the working fluid (= antifreeze water mixture (= AWM))

$$\frac{\partial T_S}{\partial t} + v \frac{\partial T_S}{\partial x} = a \frac{\partial^2 T_S}{\partial x^2} + \frac{\alpha_I \cdot A}{m_S \cdot c_S} \cdot (T_A - T_S) \tag{2}$$

This set of differential equations is solved simultanously by a modified Newton-Euler method. The energy gain is the sum of the heat fluxes, which occur due to convection, condensation of humidity, hoar-frost generation that means solidification of the condensed humidity in case of sub-zero (°C) absorber temperatures and short and longwave radiation (Maßmeyer, Posorski, 1983). The leaving working fluid temperature is given by

$$\vartheta_{S,aus} = \vartheta_L - (\vartheta_L - \vartheta_{S,ein}) \cdot \exp(-NTU) \tag{3}$$

where NTU is the number of transfer units (Schlünder and others, 1983)

$$NTU = \frac{k \cdot A}{\dot{V}_S \cdot c_S \cdot \rho_S} \tag{4}$$

and k is the overall heat transfer coefficient, which is determined by

$$k = 1 / \left(\frac{1}{\alpha_a} + \frac{\delta_A}{\lambda_A} + \frac{\delta_R}{\lambda_R} \cdot \psi + \frac{1}{\alpha_I} \right) \tag{5}$$

The correction factor ψ takes into account the surface structure of a possible hoar-frost layer of thickness δ_R. A synopsis of the equivalent heat transfer coefficients due to short and longwave radiation, due convection and condensation is given by Krumm (1985).
The simulation results agree well with experimental data, see Krumm (1985).

DESCRIPTION OF THE INVESTIGATED ENERGY-ABSORBERS

Three of the five investigated energy-absorbers are carried out in plate configuration, the other two are compact configurations. The two plate energy-absorbers consist of two point welded steel sheets with integrated ducts. One of the energy-absorber types is insulated on the backside. The third type of plate energy-absorber is composed of two PVC mats which are kept on distance by a special textile. The first compact energy-absorber is a rib pipe system. The second compact energy-absorber consists of several parallel flow absorber plates.

SIMULATION RESULTS

A weather model (Krumm, 1985) generates identical climatic input values for all energy-absorber calculations. The working fluid (AWM) flow rate and the forcing temperature difference between the working fluid and the environment are constant for each calculation. Fig. 1 shows the simulation results for the energy gain of a compact energy-absorber. The energy gain is composed of the different amount of the above mentioned energy sources.
The simulation results for a plate energy-absorber are shown in Fig. 2. Due to solar radiation there is an increased energy-gain during the summer. During the whole year an energy loss due to longwave radiation exchange with atmosphere can be recognized.
The average values of the energy gain of the different types of energy-absorbers are presented in Fig. 3. They can be achieved if the energy-

absorber would be operated throughout the whole year. In Fig. 4 and Fig. 5 the specific energy gain is shown as a function of the temperature difference between working fluid and environment and of the working fluid flow rate.

The following conclusions can be made:

- The energy gain is approximately a linear function of solar radiation, humidity and temperature difference between environment and working fluid.
- The energy gain is a non-linear decreasing function of wind velocity and flow rate of the working fluid.
- The energy gain is strongly dependent upon the cloud cover, that means the atmospheric radiation causes an increase or reduction of the energy gain depending on the atmospheric temperature.
- The length of the plate absorbers determines the energy gain as a function of the absorber arrangement in a complete absorber system. A series arrangement of several absorbers or a long single plate absorber are not recommended, because in this case there is virtually no energy gain in the last part of the absorbers.
- The optimum plate absorber orientation in an area with little sunshine is installed north facing and vertically.
- A compact absorber should be installed on areas exposed to wind, for example on the top of the roof.

REFERENCES

Krumm, W. (1985). "Mathematische Modellierung des Betriebsverhaltens und der Leistungsfähigkeit von Energieabsorber-Wärmepumpenanlagen zur Gebäudebeheizung". Dissertation Uni Siegen.

Maßmeyer, K. and R. Posorski (1983). "Wärmeübergänge am Energieabsorber und deren Abhängigkeit von meteorologischen Parametern". Jül.-Spez. 184.

Schlünder, E. U. and others (1983). "Hochschulkurs Wärmeübertragung I, II". TH Karlsruhe.

NOMENCLATURE

Symbol	Quantity	Units	Symbol	Quantity	Units
A	area, surface	m^2	q	specific heat	kWh/m^2
a	temperature conductivity	m^2/s	T	thermodynamic temperature	K
c	specific heat capacity	kJ/kgK	t	time	s
k	overall heat transfer coefficient	W/m^2K	$\dot{V}$	volume flow rate	m^3/h
m	mass	kg	v	velocity	m/s
			x	locus	m
α	heat transfer coefficient	W/m^2K	ρ	density	kg/m^3
Δ	difference		ϑ	temperature	°C
δ	thickness	mm	ψ	correction function	
λ	heat conductivity	W/mK	∂	differential	

Subscripts

A	energy-absorber	i	inside
a	outside	L	air, ambient, environment
aus	outlet	R	hoar-frost
ein	inlet	S	antifreeze water mixture(AWM)

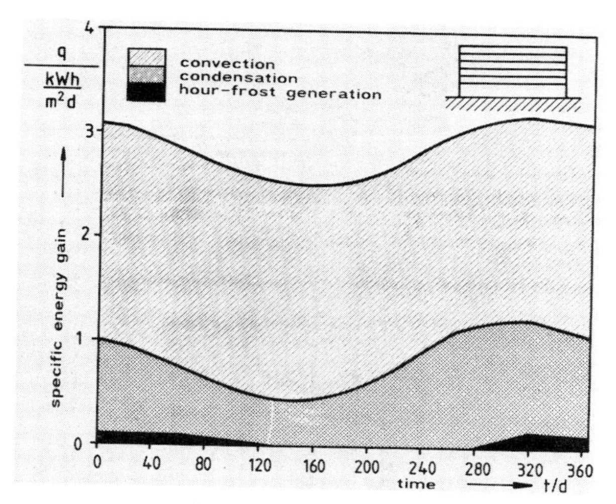

Fig. 1 Daily average volume of the specific energy gain and its composition due to different energy sources as a function of time. Absorbertype: compact energy-absorber: "energy block". AWM volume flow rate: $\dot{V}_S$ = 3 m³/h; temperature difference between environment and AWM: $\Delta\vartheta$ = 10 K (AWM: antifreeze water mixture)

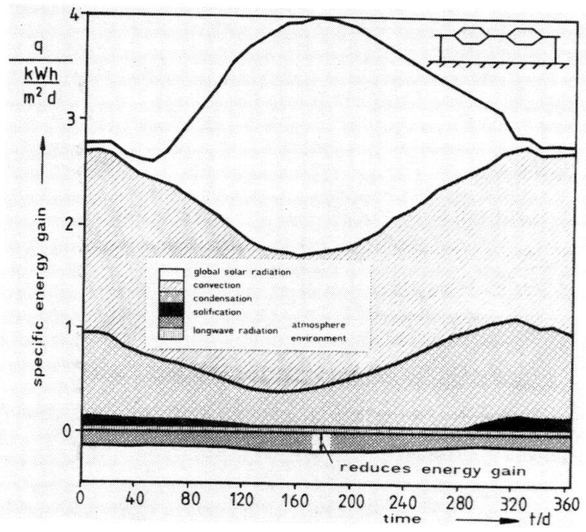

Fig. 2 Daily average value of the specific energy gain and its composition due to different energy sources as a function of time. Absorbertype: Plate energy-absorber with integrated AWM-ducts, backside ventilated. AWM volume flow rate: $\dot{V}_S$=3 m³/h temperature difference between environment and AWM: $\Delta\vartheta$= 10 °C Absorber orientation: horizontal (AWM: antifreeze water mixture)

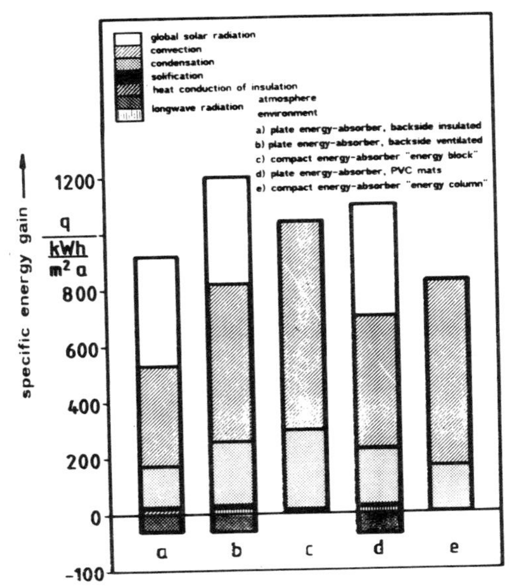

Fig. 3 Annual average value of the specific energy gain and its composition due to different energy sources. $\dot{V}_S$ = 3 m³/h

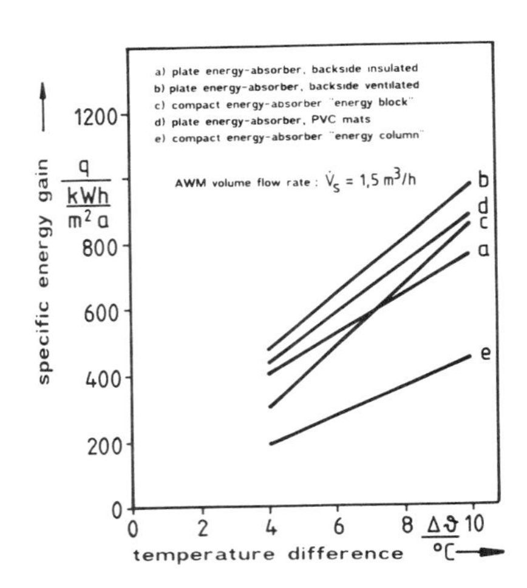

Fig. 4 Annual average value of the specific energy gain as a function of the temperature difference between environment and AWM for AWM volume flow rate of $\dot{V}_S$ =1,5m³/h (AWM = antifreeze water mixture)

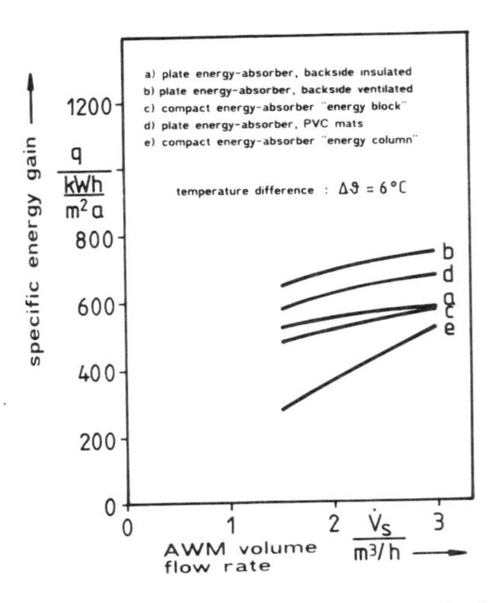

Fig. 5 Annual average value of the specific energy gain as a function of the AWM volume flow rate for Δϑ = 6°C (AWM = antifreeze water mixture)

THEORETICAL AND EXPERIMENTAL INVESTIGATION OF HOAR-FROST GENERATION AND THAW OFF PROCESS ON A COMPACT ENERGY-ABSORBER

H. Bradke, W. Krumm and F.N. Fett

Institute of Energy Technology; University of Siegen, Federal Republic of Germany

ABSTRACT

A mathematical model is presented which describes the hoar-frost generation and the thaw off process on a compact energy-absorber. Simulation results which show the influence of a hoar-frost layer on the energy gain are discussed. The simulation results also show the dependence of the energy gain and the energy demand of the thaw off process under different operation and weather conditions. Finally suggestions are given for the optimum operation of an energy absorber.

KEYWORDS

Energy-absorber, hoar-frost generation, thaw off process, simulation

INTRODUCTION

During the operation of energy-absorbers in form of uncovered collectors, it is possible that in case of sub-zero degree centigrade absorber temperatures the total absorber surface is covered with frozen condensed humidity or hoar-frost. There are a lot of open questions concerning the influence of this hoar-frost layer on the operational mode and performance characteristics of an energy-absorber.

MATHEMATICAL DESCRIPTION

The compact energy-absorber investigated consists of ten parallel flow absorber plates. Each of these absorber plates is produced by point welding two steel sheets with integrated flow ducts. Fig. 1 shows a longitudinal sketch of a flow duct. The first step of the modeling is the subdivision of the absorber plates into volume cells. An energy balance for each volume cell is set up. The most important heat fluxes for compact energy-absorbers from the ambient to the working fluid, an antifreeze water mixture (= AWM), are caused by convection and condensation (Krumm,1985). The heat transfer coefficient has to take into account the increased mass transfer due to hoar-frost generation as well as the rib efficiency of the hoar-frost needles. A correction factor takes into account the surface structure of the hoar-frost layer. In case of hoar-frost generation, an exact analogy between

heat and mass transfer does not exist. Therefore the analogous Sherwood number has to be multiplied by a factor which is a function of the hoar frost density. The generated mass flow rate of hoar-frost is a function of the mass transfer coefficient, a correction factor for mass transfer, the difference of the water vapor partial pressure between environment and absorber surface, the gas constant of water vapor and the mean thermodynamic temperature.
The change of the thickness of the hoar-frost layer results from the generated mass flow rate and the hoar frost density.
Lotz (1968) found by a lot of experimental data the interrelation between the hoar-frost density, the hoar-frost layer thickness and the time. The specific energy amount which is necessary for the thaw off process has to be delivered either by the working fluid or by the weather. An energy balance determines the gradient of the hoar-frost layer thickness during thawing off. A synopsis of all equations is given by (Bradke, Krumm, Fett, 1985).

SIMULATION RESULTS

Outside laboratory experiments with the compact energy-absorber show that the simulation results agree well with the experimental data.

Fig. 2 shows the calculated specific energy gain and the thickness of the hoar-frost layer as a function of time. The forcing temperature difference between the ambient and the entering working fluid temperature is varied from 5 to 15 K. During the operation of an energy-absorber with a temperature difference of 5 K a very compact hoar-frost layer with a smooth surface is generated. At the beginning of the hoar frost generation increasing temperature differences cause a porous hoar frost layer which increases the external surface. So the energy gain increases due to the increased surface. In course of time the density of the hoar-frost increases, and the energy-gain decreases due to an increased insulation effect. These facts cause a maximum energy gain within the first operation period of the energy-absorber.
A similar effect is caused by the wind velocities, see Fig. 3 .
The required energy amount and the necessary time are of interest when thawing off the energy-absorber by a heated working fluid . Fig. 4 shows the influence of the inlet working fluid temperature on the thaw off process. The lower the entering working fluid temperature, the longer the thaw off time and the lower the energy amount. The reduction of the working fluid flow rate also leads to an extension of the thawing off time. At the same time the required energy amount for the thaw off process increases considerably, Fig. 5.
If the energy-absorber can not gain sufficient energy from the ambient due to the resistances of heat conductivity of a hoar-frost layer, it has to be thawed off. A regular thawing off of the energy-absorber can be recommended, if the total energy-balance is positive.
The simulation results show that under normal operational conditions the thaw off process of the energy-absorber is not recommended, because the amount of melting energy is greater than the increased energy gain later on. That means the particular absorber plates have to be installed in such a way that an air current between the plates is not prevented by a hoar-frost layer.
The following conclusions can be made:

- Wind exposed installation of energy-absorber is recommended. High wind velocities lead to increased energy gain. Furthermore they cause very dense hoar-frost layers with a high heat conductivity and low layer thicknesses. In case of temperatures higher zero degree centigrade the thawing off times are shortened considerably.

- Short current length leads to an increased heat transfer coefficient.
- High working fluid flow rates shorten the thaw off process. The thawing off times are reduced considerably and the total energy input is decreased.
- Low working fluid flow rates in case of normal energy-absorber operation have positive influence. High flow rates cause a higher electrical power consumption for the pump without achieving a reasonable increase in energy gain.
- Low temperature differences between environment and entering working fluid are favourable. High temperature differences cause very thick hoar-frost layers with a high resistance of heat conductivity.
- High inlet temperatures of the working fluid shorten the thawing off time and usually require a higher melting energy amount.
- To avoid heat losses the thaw off process has to be stopped immediately after the disappearance of the hoar-frost layer.
- A vertical position of the absorber plates improves the thaw off process. Due to gravity the hoar-frost layer can be removed by its dead weight.

REFERENCES

Bradke, H., W. Krumm and F.N. Fett (1985). "Theoretische und experimentelle Untersuchung des Bereifungs- und Abtauprozesses bei einem Kompaktenergieabsorber". Fortschr.-Ber. VDI Reihe 6 Nr. 183. VDI-Verlag, Düsseldorf.

Krumm, W. (1985). "Mathematische Modellierung des Betriebsverhaltens und der Leistungsfähigkeit von Energieabsorber-Wärmepumpenanlagen zur Gebäudebeheizung". Dissertation Uni Siegen.

Lotz, H. (1968) "Wärme- und Stoffaustauschvorgänge in bereifenden Lamellenrippenluftkühlern im Zusammenhang mit deren Betriebsverhalten". Dissertation TH Aachen.

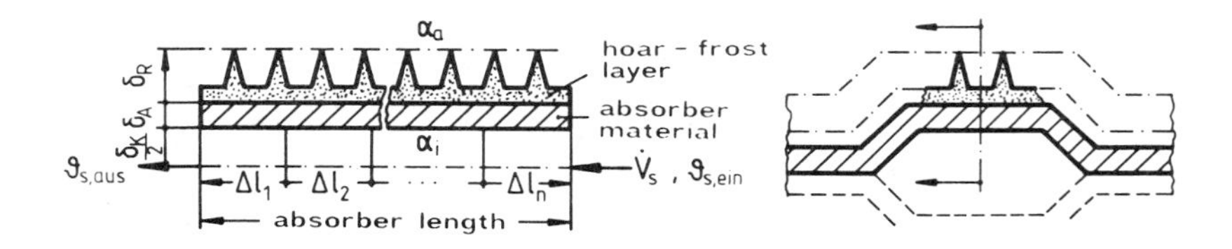

Fig. 1 Longitudinal sketch of a flow duct

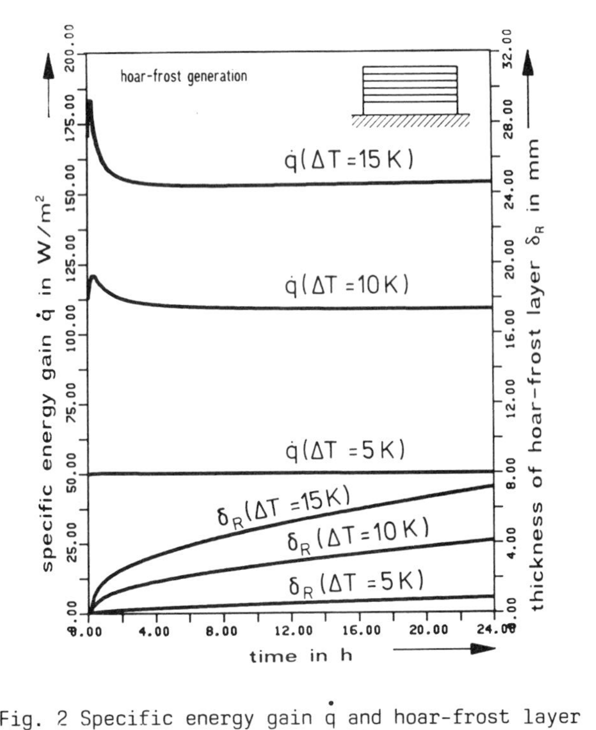

Fig. 2 Specific energy gain $\dot{q}$ and hoar-frost layer thickness δ_R as a function of time.
Parameter: forcing temperature difference Δ_T between environment and entering working fluid
Constant input values: ambient temperature ϑ_U = 0 °C; relative humidity φ = 80 % ; wind velocity v_W = 2 m/s ; working fluid flow rate $\dot{V}_S$ = 3,5 m^3/h ; temperature difference ΔT = 5/10/15 K ; calculated mean specific energy gain $\bar{\dot{q}}$ = 72/137/174 W/m^2

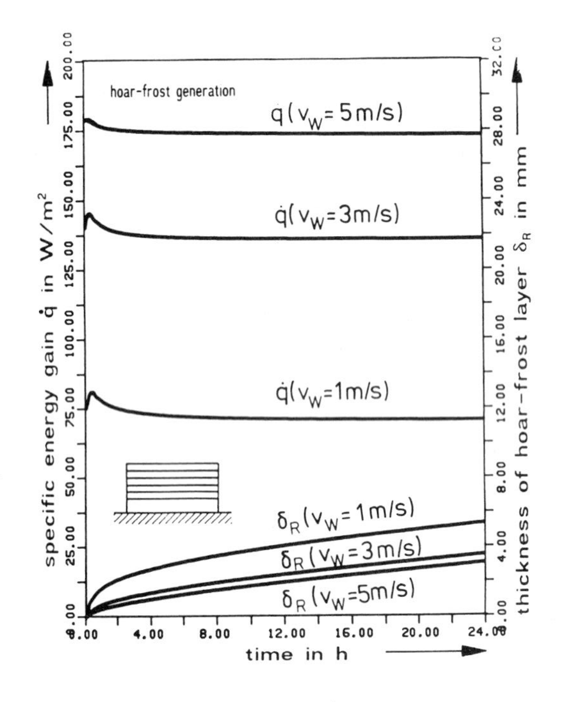

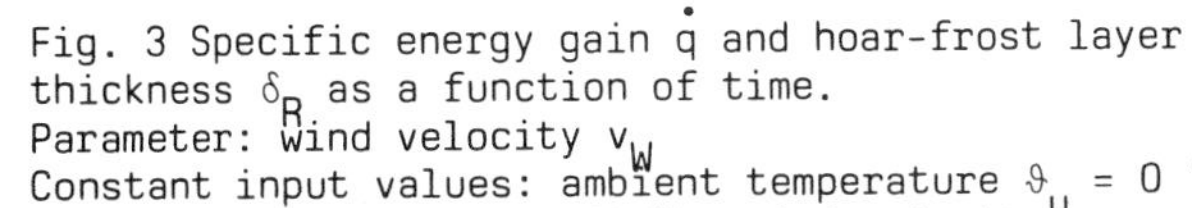
Fig. 3 Specific energy gain $\dot{q}$ and hoar-frost layer thickness δ_R as a function of time.
Parameter: wind velocity v_W
Constant input values: ambient temperature ϑ_U = 0 °C; relative humidity φ = 80 % ; wind velocity v_W = 1/3/5 m/s ; working fluid volume flow rate $\dot{V}_S$ = 3,5 m^3/h ; temperature difference ΔT = 10 K
calculated mean specific energy gain $\bar{\dot{q}}$ = 72/137/174 W/m

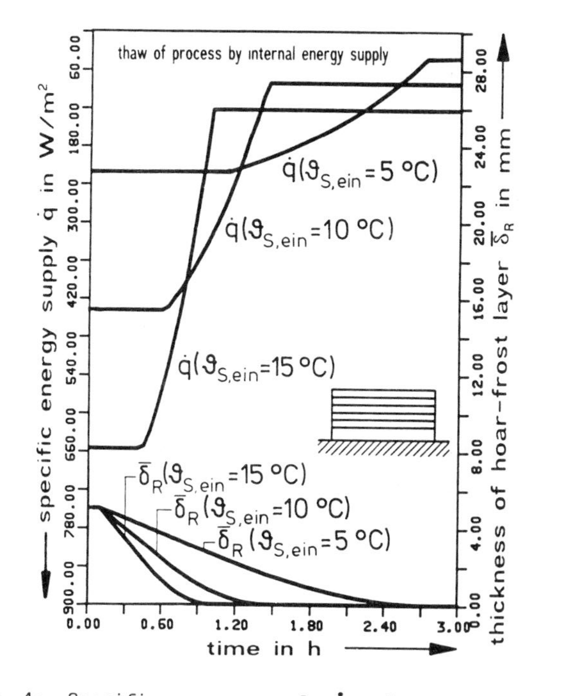

Fig. 4: Specific energy supply $\dot{q}$ and mean thickness of hoar-frost layer δ_R as a function of time.
Parameter: inlet temperature of working fluid $\vartheta_{S,ein}$
Constant input values: ambient temperature $\vartheta_U = 0\,°C$; relative humidity φ = 80 % ; wind velocity v_W = 2 m/s ; working fluid volume flow rate $\dot{V}_S$ = 3,5 m /h ; entering temperature of working fluid $\vartheta_{S,ein}$ = 5 / 10 / 15 °C ; thickness of hoar-frost layer $\delta_{R,auf}$ = 5 mm ; density of hoar-frost ρ_R = 917 kg/m³;
Calculated melting energy q = 561 / 674 / 775 Wh/m²

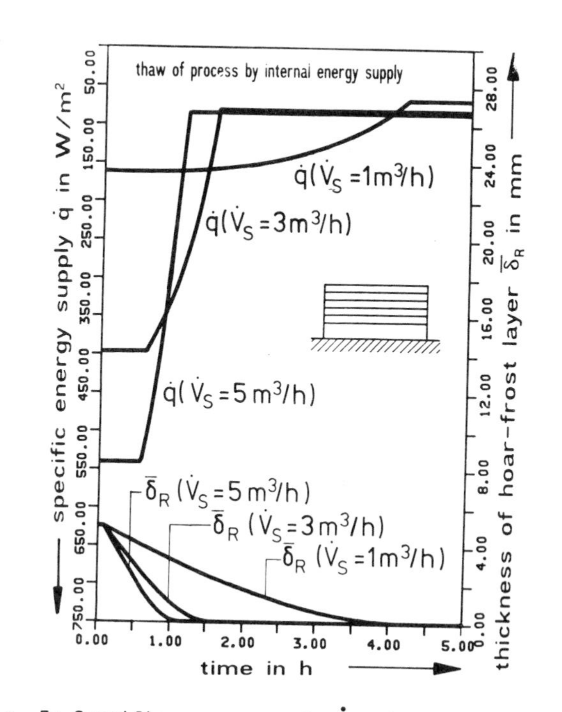

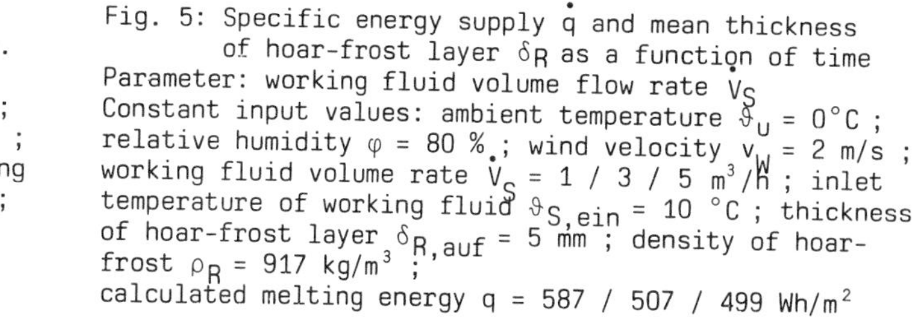

Fig. 5: Specific energy supply $\dot{q}$ and mean thickness of hoar-frost layer δ_R as a function of time
Parameter: working fluid volume flow rate $\dot{V}_S$
Constant input values: ambient temperature $\vartheta_U = 0\,°C$; relative humidity φ = 80 % ; wind velocity v_W = 2 m/s ; working fluid volume rate $\dot{V}_S$ = 1 / 3 / 5 m³/h ; inlet temperature of working fluid $\vartheta_{S,ein}$ = 10 °C ; thickness of hoar-frost layer $\delta_{R,auf}$ = 5 mm ; density of hoar-frost ρ_R = 917 kg/m³ ;
calculated melting energy q = 587 / 507 / 499 Wh/m²

HEAT PUMP SYSTEM WITH COMPACT ENERGY ABSORBER FOR HOME ASSEMBLY

W. Kimpenhaus, G. Picken

Rheinisch-Westfälisches Elektrizitätswerk AG (RWE), Head Office
4300 Essen 1, Federal Republic of Germany

ABSTRACT

In addition to flat absorbers such as energy roof and energy facade, the so-called compact absorbers have been making inroads into the energy absorber market in the last few years. Due to the compact design of these absorbers the utilisation of solar energy plays a minor role, but this deficiency is made up for by such advantages as easy production and installation, low space requirements and favourable capital costs.
Jointly with Lahmeyer AG in Mechernich, RWE's Department of Applied Technology has developed a self-contained heat pump heating system with energy absorber, which makes consistent use of the above described advantages.
The heat pump can be operated down to a minimum outside temperature of about -3 °C and is suitable for supplementing oil-fired boilers. With an electric load of about 2 kW and depending on the house size, the heat pump covers between 75 % and 40 % of the annual heating energy consumption. This is equivalent to a saving of 1,500 to 2,500 litres of fuel oil.
The combined system of heat pump and energy cone has been tested successfully by an independent control body. The DIN test mark and PI certification guarantee the maturity and high technical standard of the heat pump system.

KEYWORDS

Complete heat-pump system; compact energy absorber; brine/water heat pump; bivalent operation; possible home assembly; low installation costs; high efficiency; tested by the German Technical Control Association (TÜV).

INTRODUCTION

In addition to flat absorbers such as energy roof and energy facade, the so-called compact absorbers have been making inroads into the energy absorber market in the last few years. Due to the compact design of these absorbers the utilisation of solar radiation plays a minor role, but this deficiency is made up for by such advantages as easy production and installation, low space requirements and favourable capital costs.
Jointly with Lahmeyer AG in Mechernich, RWE's Department of Applied Technology has developed a self-contained heat pump heating system with an energy absorber, which makes consistent use of the above described advantages.
The heat pump system with the compact energy absorber - also called energy cone because of its appearance - has been especially designed for the application in single- or two-family houses.
The heat pump heating output of approximately 7 kW is mainly suited for the dual-source heating operation in combination with a fossil-fired boiler.

SYSTEM DESCRIPTION

The heat pump system with the compact energy absorber is designed as a complete system. It was built up of approved plant components.
The complete system consists of:

- brine/water heat pump as compact unit
- energy absorber, cone-shaped
- compact installation of the brine circuit
- compact installation of the heating circuit
- pre-fabricated connecting hoses
- antifreeze (ethylene glycol)
- control devices

The advantage of the complete system is to be found in the use of harmonized components. In this way, the simple connection of the energy absorber to the heat pump and the trouble-free integration of the heat pump system into the existing heating system are guaranteed.
That means a very low installation expenditure by the professional installer and hence low installation costs.
Furthermore, a house-owner with a handcraft skill has the possibility to do himself a part of the installation and assembly work - especially when installing the energy cone. This leads to favourable overall costs.

HEAT PUMP

The heat pump is a compact brine/water heat pump. The compressor, the heat exchangers and all internal safety, switching and control devices are installed in a galvanised, powder-coated and well sound-insulated metal case.
The heat exchangers made of refined steel guarantee a high protection against corrosion and frost damages. Favourable output values are attained by an optimum construction and large-scale dimensioning of the heat exchanger surfaces.
The very low pressure loss on the heating water and brine sides and the low and still admissible minimum heating water flow rate make a direct integration of the heat pump into the heating system possible. In general, the available circulation pump of the heating system may be used further.

The heat pump case has very favourable dimensions and a space requirement of only approximately 0.3 m². Therefore, its installation in narrow heating rooms is also possible without any problems.

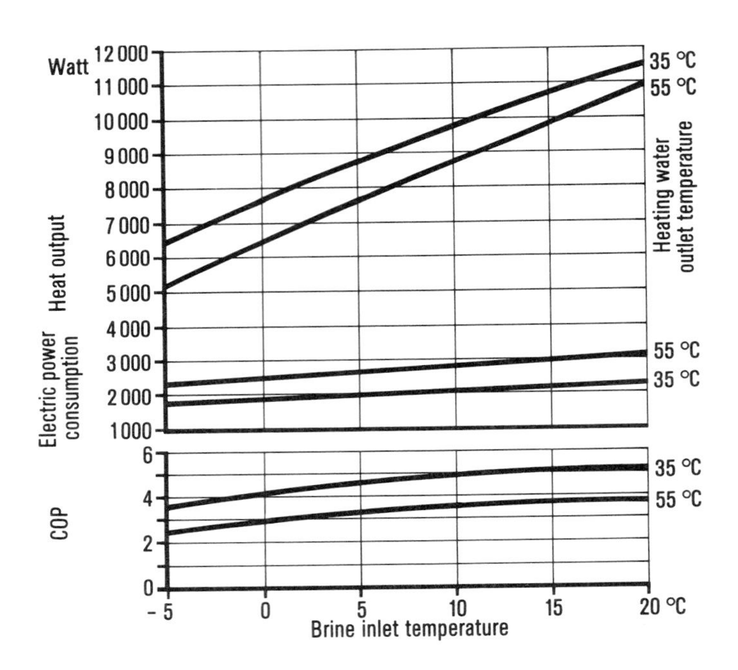

Fig. 1 Heat output, electric power consumption and coefficent of performance (COP) of the heat pump

COMPACT ENERGY ABSORBER (ENERGY CONE)

The energy cone is assembled on site from prefabricated parts and installed outside the building.
The supporting frame with integrated brine distributor is made of galvanised and powder-coated metal pieces. To connect the different parts of the supporting frame, screws made of stainless steel are used. The frame has a black colour.
The heat exchanger consists of continuously weathering plastic pipes of polyethylene (PE) material.
It is divided in eight pipe circuits totalling some 400 m in length. These are spirally rolled round the supporting frame in identic pipe spacings.
The connections of the pipe circuits to the brine distributor are made via self-sealing brass screwings.

Brine, a mixture of water and ethylene glycol, is used as heat transfer medium.
In a concentration of 35 % of ethylene glycol (Antifrogen N) frost protection is guaranteed up to -20 °C.
The ethylene glycol is delivered in two tanks. After filling them up with water, 2 x 60 l of brine are available; that is sufficient for applications with usual lenghts of pipes.

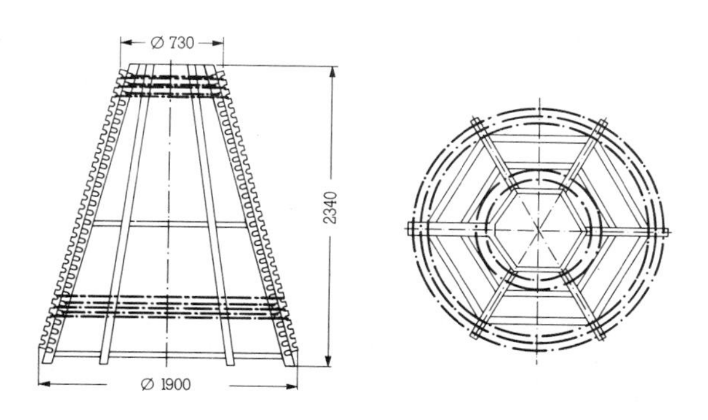

Fig. 2 Construction details and dimensions of the energy cone

COMPACT INSTALLATION

The integration of the heat pump into the heating and absorber circuits is done by means of pre-assembled building units (compact installation) and flexible heat-insulated special hoses.
This has the advantage that errors during installations are avoided, that the heat pump is integrated into the existing heating system in accordance with general installation rules, and that the installation expenditure in situ remains low.
The connection of the energy cone to the evaporator of the heat pump and the integration of the heat pump condenser into the heating circuit is performed via two compact installations. These are equipped with the required safety, isolating and ventilation devices as well as with the appropriate thermometers. The compact installation for the brine circuit is equipped additionally with an expansion tank and the necessary special circulation pump for the brine.

CONTROL

Control of the heat pump is done via an adjustable thermostat in combination with a sensor in the compact installation of the heating circuit, or alternatively with a regulator operated depending on the outdoor temperature that may be integrated into the heat pump case.
The heat pump is equipped with the safety switching elements prescribed and at the lower limit of application (outdoor temperature of approximately -3 °C) the heat pump guarantees an automatic switch-off and a new switch-on respectively.

The delayed relay for a new switch-on and the limitation of heat pump starts required to fulfil the regulations of the electricitiy supply utilities, are available in the series units. It is no problem to install other controls in addition to those available.

CONCLUSION

The heat pump and the energy cone, as a system, have been tested with success by an independent research centre. The DIN designation and the declaration in the product information (PI) for the final consumer ensure a sophisticated product with a high technical standard. The benefits of the complete system reduce the installation expenditure for the heat pump heating system and enable the house owner to do himself a part of the required work.
Due to the concept and the mode of operating the system, a considerable share of the annual heating heat consumption is made available by the heat pump heating system.
Depending upon the weather character, the heat pump alone cannot provide a sufficient heating output in case of outdoor temperatures below approximately 8 °C. Therefore, the system should be operated in a partly parallel mode of operation, that means that the heat pump continues its operation together with the oil-fuelled boiler. Only when the outdoor temperatures fall below approximately -3 °C, the oil-fuelled boiler alone will meet the requirement of heating heat.
Owing to the partly parallel operation of both heat generators, between 75 % and 40 % of the annual consumption of heating heat, depending upon the size of the house, may be covered by the heat pump. In this way, between 1,500 and 2,500 l of fuel oil are saved per year.

REFERENCES

Rheinisch-Westfälisches Elektrizitätswerk AG, Abt. Anwendungstechnik (1981). Stand und Trend der Entwicklung von Wärmeabsorber-Systemen. RWE informiert 176.
Rheinisch-Westfälisches Elektrizitätswerk AG, Abt. Anwendungstechnik (1985). Wärmepumpen-Kompaktinstallation. RWE-Fachberatung.
Rheinisch-Westfälisches Elektrizitätswerk AG, Abt. Anwendungstechnik (1986). Einsatzmöglichkeiten von Wärmepumpen mit Energieabsorbern. RWE informiert 216.
Lahmeyer Aktiengesellschaft für Energiewirtschaft (1987). Umweltwärme nutzen. VT 284/2/87

NEW ROOF ABSORBER HEATING PLANT WITH DIGITAL ENERGY MANAGEMENT

E. Bollin*, J. Schmid*, E. Schäfer**

**Kraftwerk Laufenburg, 7887 Laufenburg

*Fraunhofer Institut für Solare Energiesysteme,
Oltmannstraße 22
7800 Freiburg/FRG

ABSTRACT

New roof absorber elements have been developed and tested. In 1984 a bivalent heating plant with 500m² of roof absorber was put into operation. An automatic data aquisition enables a detailed control of the operating behaviour of single components as well as the survey of the control strategy. Furthermore, control strategies were developed which are transferable to other similar multivalent heating plants.

KEYWORDS

New roof absorber elements, 500 m² pilot plant, digital energy management, optimal heat pump control, temperature stratification in heatstorage.

INTRODUCTION

At the Institute for Solar Energy Systems (ISE) in Freiburg, FRG, a new roof absorber conception was developed and tested in a heating plant with 500m² of roof absorber. This report will explain the roof absorber and the heating plant conception including the control and regulation conception as well as the operating behaviour of single components plus control and regulating strategy.

1) The New Roof Absorber Conception

An aluminum absorber element with rear ventilation was developed at the ISE for installation on aluminum industrial roofs. The element is produced using the rollbond process.
In fig. 1 the cross section and the side view of an absorber element are illustrated.
Construction costs can be cut by half in comparison with conventional roof absorber constructions.The absorber elements have been bench-tested on the roof of the ISE.
Fig. 2 shows the heat transfer coefficient α_a, as a function of wind speed for the absorber. A design performance of 120 W/m² of absorber surface is

achieved.

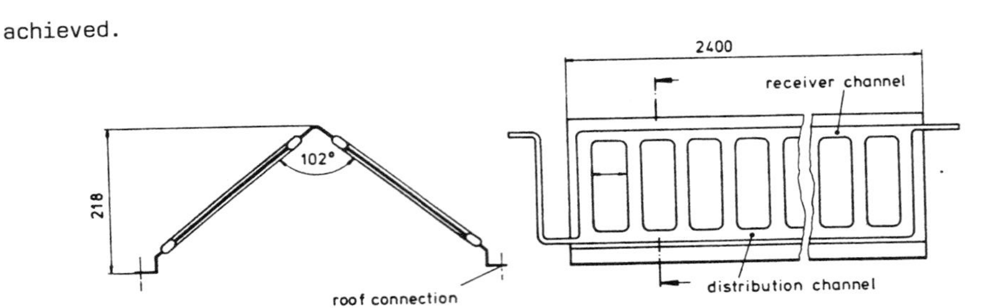

Fig. 1) Cross section and side view of an absorber element.

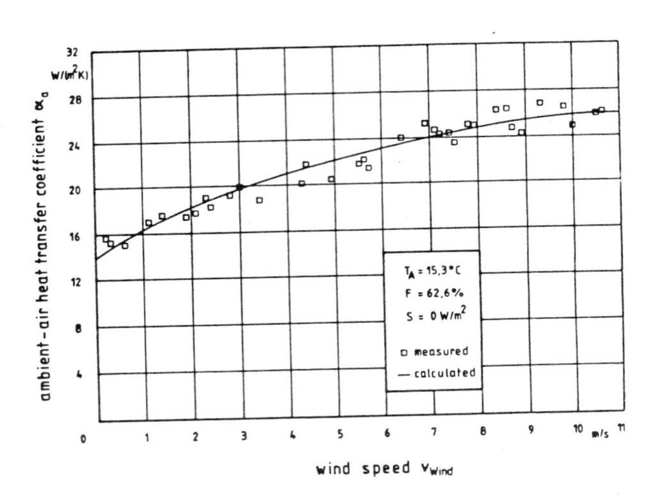

Fig. 2) Outdoor heat transfer coefficient for an absorber element as a function of wind speed.

2) Description Of The Heating Plant In Engen

A bivalent heat pump plant with roof absorbers was designed for two buildings operated by the Kraftwerk Laufenburg utility: 500 m² office building and 788 m² automobile shed and warehouse. On the roof surface inclined to the south, 300 absorber elements with 500 m² of surface area and a design performance of 70 kW were installed.
Apart from the function to heat the electric utility Kraftwerk Laufenburg in Engen the plant should also contribute to gain some experience in handling components for the utilization of renewable energy. Therefore the plant was outlined in such a way that the most operating variants generally known in practice can be run. These reflections resulted in the plant conception as shown in fig.3.

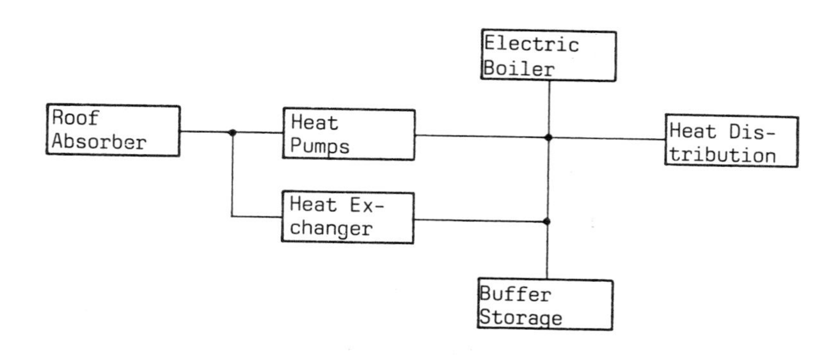

Fig. 3) The flow chart of the Roof Absorber heating plant in Engen.

The essential components of the plant are:

*Two electric-motor-driven heat pumps with a total heating capacity of 93 kW at a brine inlet of 0°C and heating supply temperature of 55°C. Both heat pumps can be operated at 64% of design capacity by means of cylinder unloading.

*An electric boiler with a design capacity of 175 kW, available for meeting peak loads.

*A 25 m³ water tank for heat storage. It is an underground cylindrical steel tank.

The great variety of these operating possibilities overstrain a conventional control and regulation system. Therefore, the plant was equiped with a microcomputer-controlled system which does not only control and regulate but also supervise the plant and register completely as well as demonstrate all important data of the plant.
This conception has proved to be particularly good and allows to control the plant not only to direct criterions (temperature) but also to consider dynamically calculated factors such as the present factor of the stored energy.

3) Operating Behaviour Of The Heating Plant And Of Single Components

3.1 Roof Absorber: The roof absorber construction has totally proved to be good both under the extreme climatic conditions at the site in Engen and as a supplier of energy for the two heat pumps. The design performance of about 120 W/m² could be proved.

3.2 Buffer Storage: One component of the plant, the buffer storage, plays a central role in a plant equiped with heat pumps and electric boiler. It can easily equalize the fluctuations in offer and demand and this "neutralizes thermically" the heat pumps and the heat distributing system. However, the storage of energy is mostly subject to loss and it is necessary to predetermine precisely the heat demand required. In cold seasons the buffer storage will be charged with the heat pumps and the electric boiler on the base of low tariff during the night and the other day it will be discharged when operating on the base of "heating with storage". In this case it is also interesting to know whether the temperature stratification due to differences in the volumetric weight continues or whether a mixture occurs. In order to avoid this mixture during the charging and discharging procedures the inflow and outflow tubes of the lying 25 m³ earth storage were distributed over the total length of the storage. Additionnally, the microcomputer regulates the charging and discharging in such a way that cold water will always be drained off at the bottom and warm water, on the contrary, always on the top. The charging will be regulated in the same way. Fig.4 shows the temperature stratification inside the storage and the influence of the charging and discharging procedures to this stratification within a 24 hours' period.

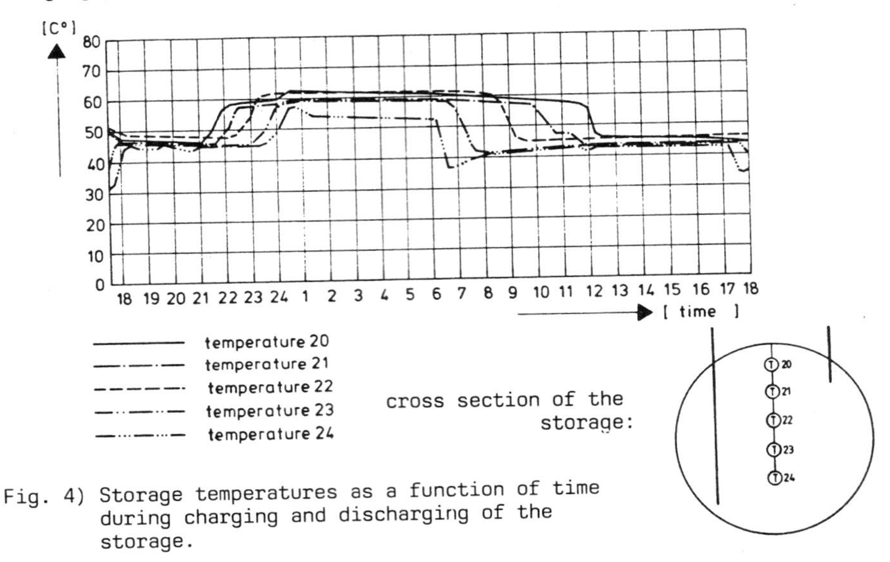

Fig. 4) Storage temperatures as a function of time during charging and discharging of the storage.

3.3 Heat Pumps: The C O P of the heat pumps has been registered over a period of 5 months. The C O P of the small heat pumps which is 3.16 is clearly above the big ones with 2.79. This is due to the more favorable operating conditions at high outside temperatures.

4) Examination Of Various Strategies

First, a so-called economically optimized control was installed. That means a control which, according to the present tariffs for electricity (low tariff =0.5 x high tariff), does only put the heat pumps into operation if there are favorable conditions during the day.
In this way the heating is mainly effected by the buffer storage which is charged at low tariff. The daily energy balance of this control modus is shown in fig.5 b).
Due to the charging of the storage at high supply temperatures unfavorable C O Ps of the heat pumps (around 2.4) and the whole plant (1.2) result.

Escaping from this fixed economical structure by giving total priority to the heat pumps an energy-optimized plant control will be the result, i.e. the heat pumps have priority in cold seasons and in the warm seasons one dispenses totally with the charging of the storage during the night.

Due to this modus operandi C O P of the heat pumps ameliorates clearly and the C O P of the plant achieves 2.1 (fig.5 a))

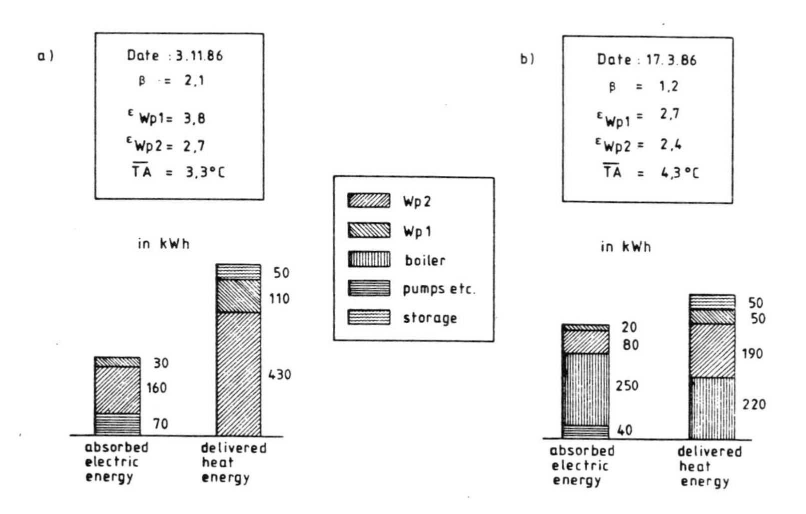

Fig. 5) Daily energy balances of the plant with different modi operandi.

ENERGY ABSORBER HEAT PUMP SYSTEM TO SUPPLEMENT HEAT RECOVERY SYSTEMS IN AN INDOOR SWIMMING POOL

K. Leisen

Rheinisch-Westfälisches Elektrizitätswerk AG (RWE)
Head Office; D-4300 Essen; Federal Republic of Germany

ABSTRACT

Compared with convontional indoor swimming pools with traditional plant engineering, the Schwalmtal indoor swimming pool has a final energy consumption of just 40 %. This low consumption is achieved by improved insulation of the building's enveloping surface, through the operation of systems for the recovery of heat from drain water and waste air as well as by the operation of a heat pump system to gain ambient heat. The decentralised heat recovery systems met between 40 and 80 % of the heat requirements in the supply areas where they were used. The electric heat pump system, which is operated in the bivalent mode in parallel to a heating boiler, could generate 75 % of the heat provided by the central heating circuit to meet the residual heat requirements.
The report illustrates the structure of the residual heat requirements of the central heating circuit. A description is given of the measured coefficients of performance of the brine/water heat pump connected by a brine circuit with two different energy absorber types - energy stack and energy roof. Finally, the ambient energy gained with the absorbers is broken down into the various kinds of heat gains from radiation, convection, condensation etc.

KEYWORDS

Energy absorber; energy stack; energy roof; heat pump; heat recovery systems; indoor swimming pool; energy engineering concept.

ENERGY ENGINEERING CONCEPT

The heat requirement of the Schwalmtal indoor swimming pool is
- reduced by improved thermal insulation of the enveloping surface of the buildings
- reduced by orienting the window areas of the pool hall, which has to be heated to a higher temperature, mainly to the south
- met primarily by the recovery of waste heat with independent systems

in decentralised arrangement
- met secondarily by the extraction of ambient heat for the central heating circuit.

The following systems were installed to meet the heat requirements:
- a system for the recovery of heat from waste water of the shower, pool and filter flushing drain water, which consists of a recuperative heat exchanger and a downstream small heat pump system (see poster no. 4.3.09);
- a system for the recovery of heat from the humid exit air of the pool hall and the shower rooms with a cross-flow heat exchanger and an air/air heat pump system;
- three systems for the recovery of heat from the exit air of the adjacent rooms with cross-flow heat exchangers or vertical storage blocks;
- a heat pump system with energy stack and energy roof for the extraction of ambient heat.

DESCRIPTION OF THE CENTRAL HEATING CIRCUIT

The heat pump system is operated in the bivalent mode in parallel with an external heating (Fig. 1). It consists of an electric brine/water heat pump which is connected by a brine circuit with an energy stack and an energy roof.

The technical specifications of the heat pump are:
- heating output (3 compressors, brineinlet 0°C, heating water outlet 55°C) 81 kW
- driving power (see above) 30 kW
- maximum heating water outlet temperature 55°C
- number of individually switchable compressors 3

The energy roof consists of 204 absorber panels which were mounted with the associated brine distribution pipes as an on roof system (supports with a spacing of 15 cm above the roof covering of the pool hall). Its entire surface of 807 m² is involved in the transfer of ambient heat to brine.

The energy stack was placed on the flat roof of a stair well. Eight blocks with 24 absorber panels each and a heat transfer surface of 793 m² were put into a steel frame structure.

The heat pump system can be operated at brine temperatures above -15°C. It heats up the shower water to temperatures between 45 and 50°C. The temperature of the heating water is determined as a function of the outdoor air temperature (55°C at 0°C outdoor air temperature, 35°C at 20°C outdoor air temperature). The controller R1 switches on the desired number of compressors according to this heating curve.

If required the shower water and the heating water are heated by the external heating additionally.

TOTAL ENERGY BALANCE

The final energy consumption of the Schwalmtal indoor swimming pool was reduced to 40 % of the consumption of a comparable modern indoor swimming pool with conventional plant engineering (Fig. 2).

An increased consumption of electrical energy of 200 MWh/a contrasts with a saving of conventional energy sources for heat production of 1200 MWh/a. In that way the primary energy consumption was reduced by 30 %, too.

The shares of the renewable energy sources and the energy saving systems in meeting the heat requirements were as follows:

- heat recovery in heat requirement of the supply areas — 39 % to 58 %
- decentralised heat recovery systems in heat requirement of the supply areas — 40 % to 80 %
- gain of ambient energy in the remaining heat requirement of the central heating circuit — 44 %
- heat pump system in the remaining heat requirement of the central heating circuit — 73 %

ENERGY BALANCE OF THE CENTRAL HEATING CIRCUIT

By the operation of the decentral heat recovery systems the heat consumption of the different consumers from the central heating circuit was considerably reduced (Fig. 3). Concerning the monthly additional heat requirement the consumers showed a different behaviour

- the shower water was regularly subsequently heated
- the heat requirements of the pool hall and the adjacent rooms were very low in the summer months and increased in the winter months
- the heat requirement of the pool water circuit increased in the winter months and especially in November, when heating up the fresh pool water filled after major cleaning

The additional heat requirement of the consumers (219 MWh/a) was met in the summer months with the heat pump only. In January the share of the external heating was the highest with 59 %. The heat pump fed 194 MWh/a of heat and the external heating 70 MWh/a of heat into the central heating circuit. For the operation of the heat pump 23 MWh/a of electricity and for the operation of the auxiliary equipment and circulation pumps 26 MWh/a of electricity were required.

HEAT PUMP AND ENERGY ABSORBERS

The compressors of the heat pump reached an annual running time of 1450, 1665 and 2240 hours. The monthly average of the COP of the heat pump was between 2.24 in January and 2.76 in July and August; the annual average of the COP amounted to 2.52.

Fig. 4 shows typical curves of essential values of a week in November, which were measured or calculated by simulation (S):

- from Tuesday evening to Thursday noon an increasing amount of heat was fed into the pool water circuit and the pool water temperature was raised from 28° to 30°C
- at outdoor air temperatures below about 3°C heat was additionally supplied by the external heating
- the gain of ambient energy with the energy roof can be broken down for this week into the various kinds of heat gains (S): global radiation 29 %, convection 52 % , condensation 19 %
- if operated with an energy stack the following breakdown would have been expected (S): global radiation 2 %, convection 74 %, condensation 24 %

CONCLUSION

This energy engineering concept

- replaces the usually high consumption of conventional energy sources by recovered heat, ambient heat and the heat equivalent of the electrical energy to drive the energy saving systems
- is transferrable to the engineering systems of existing indoor swimming pools
- can be implemented step by step in an indoor swimming pool
- renders possible the installation of industrially manufactured, tested and completely developed standard sets, which can be connected quickly and be put into service at once.

REFERENCES

Leisen, K. (1987): Central heat pump installation in an indoor swimming pool with decentralised heat recovery systems. Translation from German from "elektrowärme international Edition A, 3/1987, pp. 115 - 118. RWE, Department AT-RE, Essen.

Biasin, K. and Leisen, K. (1986): Small indoor swimming pool Schwalmtal. Results of measurements for an efficient use of energy. **Final report 178.86.** Forschungsstelle für Energiewirtschaft, Munich.

Biasin, K. und Leisen, K. (1986): Ergebnisse energietechnischer Untersuchungen in einem Hallenbad. **RWE-informiert Nr. 218.**

Leisen, K. (1984): Small indoor swimming pool Schwalmtal. Heating system with heat pump and absorbers. **2nd interim report 178.86.** Forschungsstelle für Energiewirtschaft, Munich.

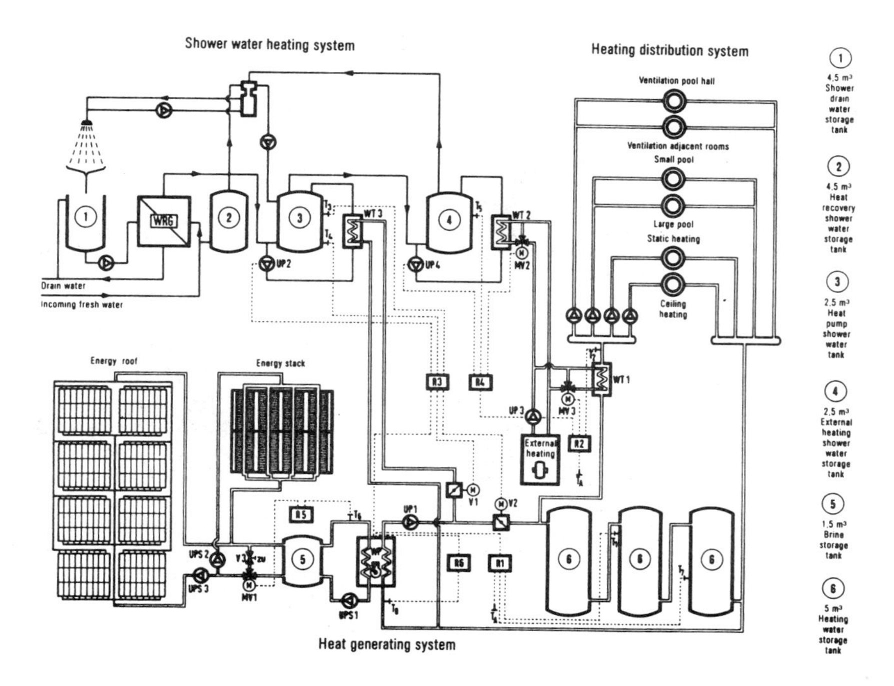

Fig. 1: Diagram of the heating and shower water circuit

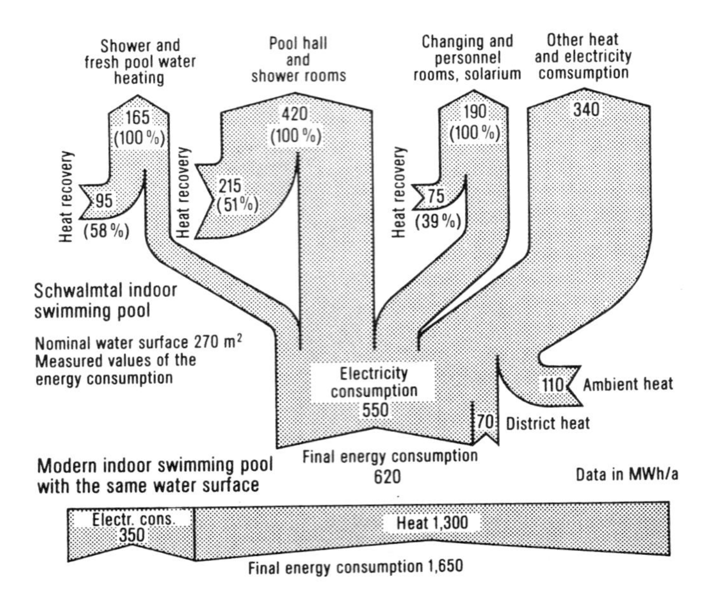

Fig. 2: Measured values of energy consumption (01-03-86 to 28-02-87)

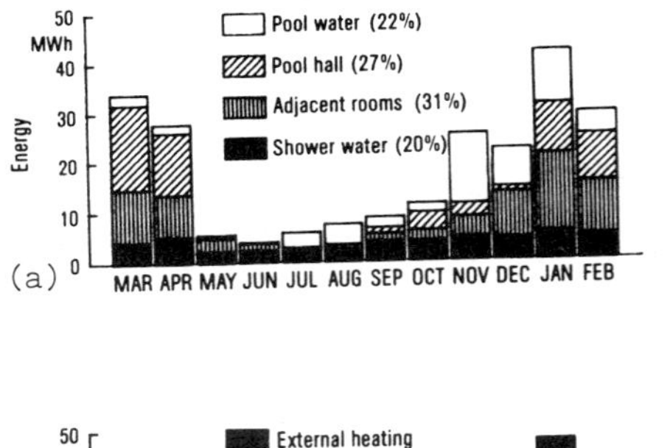

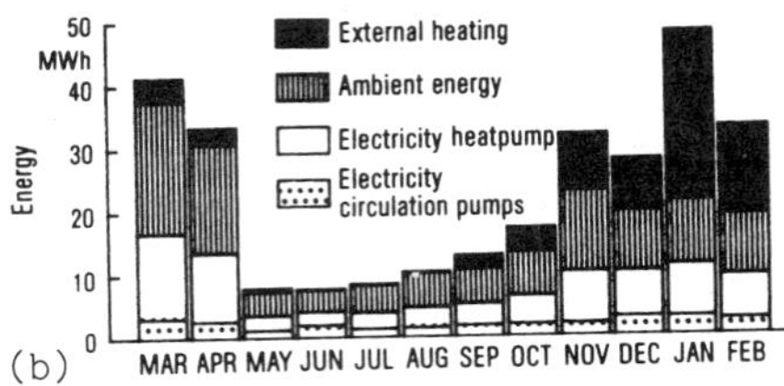

Fig. 3: Heat taken by the consumers from the central heating circuit (a)
Energy to meet the remaining heating and shower water heat requirements (b)

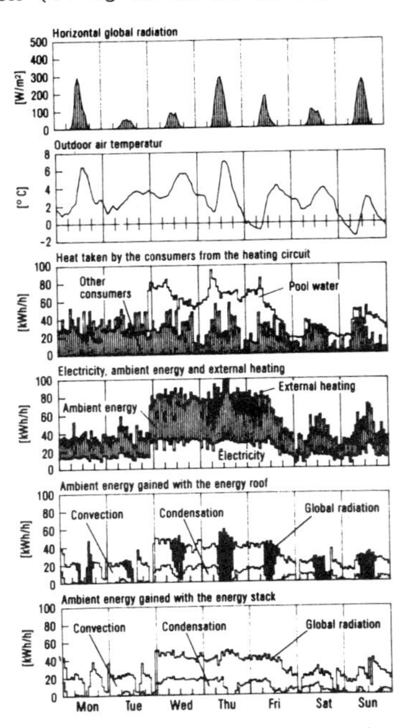

Fig. 4: Hourly values in a November week

SOLAR RECHARGING OF MULTIPLE WELL SYSTEMS FOR GROUND COUPLED HEAT PUMPS.

L. Spante and M. Larsson
Swedish State Power Board, Alvkarleby Laboratory,
S-810 71 Älvkarleby, Sweden

ABSTRACT

The Swedish State Power Board (SSPB) is actively encouraging the introduction of heat pumps in Sweden. A demonstration programme carried out by SSPB includes 30 ground coupled heat pumps with vertical pipes in a multiple well system as the heat source. In order to decrease the required bore-hole depths some of the systems are equipped with simple air convectors or solar collectors for recharging of the wells.

In this paper results from one of these ground coupled heat pumps with recharging equipment are presented. The plant shows good annual performance and is more cost-effective than a corresponding system with electric boilers. Results from measurements and computer simulations indicate that it is more cost-effective to design a multiple well system with optimal bore-hole depth, than to design it with recharging.

KEYWORDS

Heat pump; ground coupled; multiple well system; recharging; measurements.

INTRODUCTION

The Swedish State Power Board (SSPB) is actively encouraging the introduction of heat pumps for heating and hot water production in Sweden. The goals are a reduction of the use of oil and a more efficient use of electrical energy.

The programme carried out by SSPB includes today 450 demonstrations plants of which 30 are ground coupled heat pumps for apartment and school buildings, with vertical pipes in a multiple well system as the heat source. In order to decrease the required bore-hole depth and thereby the costs for drilling some of the systems are equipped with simple air convectors or solar collectors for recharging of the wells.

In this paper the Kragstalund plant in Vallentuna, just outside Stockholm, is presented. The plant consists of three 100 kW heat pumps installed in identical new apartment buildings. Bore-hole depths and degree of recharging differ between the three systems. The heat pump systems are equipped with meters for power supply, operating time, heat delivery and auxiliary heat. In addition the temperatures in the well and on the brine circuit are recorded.

The main purpose of the evaluation is to find out how the recharging influences the annual mean and the winter minimum temperature of the multiple well system and to estimate its economic value.

DESCRIPTION OF THE PLANT

The three heat pumps were installed during the autumn of 1983 in a newly built district with apartment buildings in Vallentuna, 16 km north of Stockholm. Each heat pump is placed in a separate heat station which supplies 50 apartments (Fig. 1). The heat pump, TETAB 18L, has a heating capacity of 93 kW at a brine temperature of -1°C and a condensing temperature of 55°C. Auxiliary heating during cold periods is produced with an electric boiler with a heating capacity of 72 kW in each heat station. The annual heat requirements are about 470 MWh for each group of apartments (A,J,N).

The multiple well system consists of 8 bore-holes inclined 20° to the vertical direction with a closed-loop system of polyethylene pipes. The distance between the bore-holes at the ground level is 0.7 metre. In this way a large volume of rock is used but the occupied surface area is kept at a minimum. The energy roof is of low temperature type and consists of aluminium strips with copper tubes placed under the tile roof (Fig. 1). The brine is circulated through the copper tubes when the air temperature under the tile roof is higher than the temperature in the well system. The annual heat output from the energy roof is about 470 kWh/m² strips surface.

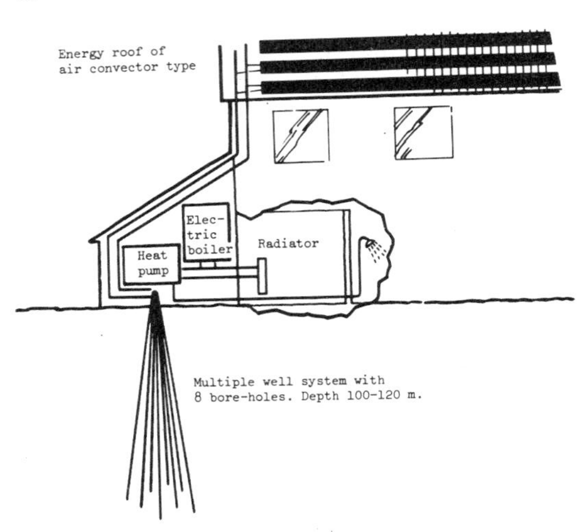

Fig. 1. The heat pump is connected to a multiple well system with 8 boreholes. The well is recharged with an energy roof and exhaust air recovery.

Energy from the exhaust air system is continuously delivered either to the inlet brine circuit or to the well system. About 25 MWh/year is in this way recharged from the exhaust air to each heat pump system. The difference in bore-hole depth and recharging equipment between the three heat pump systems is described in Table 1.

TABLE 1. Description of the Three Heat Pump Systems

House	A	J	N
Bore-hole configuration	8*120 m	8*100 m	8*120 m
Effective bore-hole depth	909 m	748 m	925 m
Recharging system	Exhaust air and energy roof 100 m²	Exhaust air and energy roof 100 m²	Exhaust air only

The investment for each heat station, including heat pump system with recharging devices and electric boiler, is about 600 000 SEK (1 US$=6.3 SEK, August 1987). The whole plant with three heat pump systems is more cost-effective than a corresponding system with electric boilers.

RESULTS AND DISCUSSION

Depending on varying performance and availability for the heat pumps, the heat extraction from the well system differs a lot for the three systems. It is therefore difficult to analyse the effect of recharging and varying bore-hole depths from measured data only. Because of this the analysis is based on measured data from the A-house system and on computer simulations with the SBMA-program (Eskilsson,1985).

The heat production during the period March-84 to Feb-85 for the A-house heat station is presented in Fig. 2. The heat load in the system was 473 MWh of which the heat pump delivered 408 MWh (86 %) with a seasonal performance factor of 2.6. The total heat to the evaporator is 253 MWh/year. The recharging system produced about 72 MWh/year during the first two years of operation. This is less than expected.

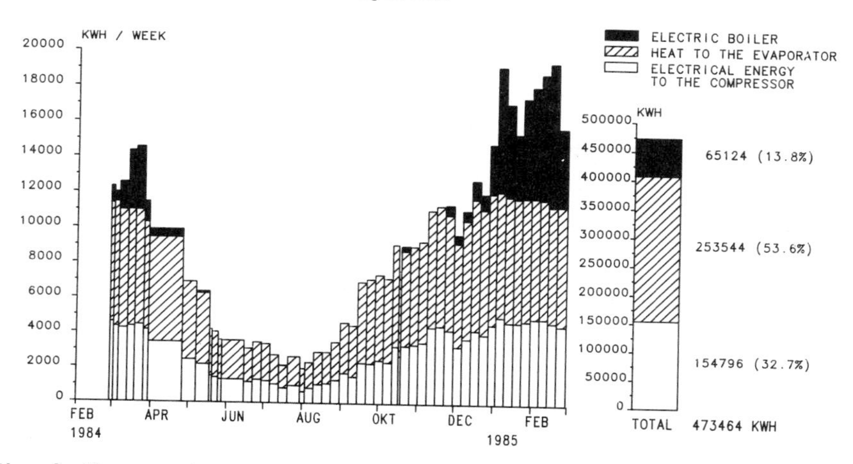

Fig. 2 Heat production during the period 1984/85 for the heat station (A).

In Fig. 3 recorded mean temperatures in the well water during the first two years of operation are shown, together with results from the computer simulations. The simulated result is a mean temperature at the bore-hole wall. The temperature difference between the wall and the water is estimated to be about 1°C when the water is frozen to ice which will be valid for the period November to April. Taking this into account the computer simulation is seen to be in good agreement with observed temperatures. This has also been the result from other analyses with the SBMA-programme (Larsson, 1985). Results for the A-house system without recharging are also presented in Fig. 3. As expected the temperature is lower in this case.

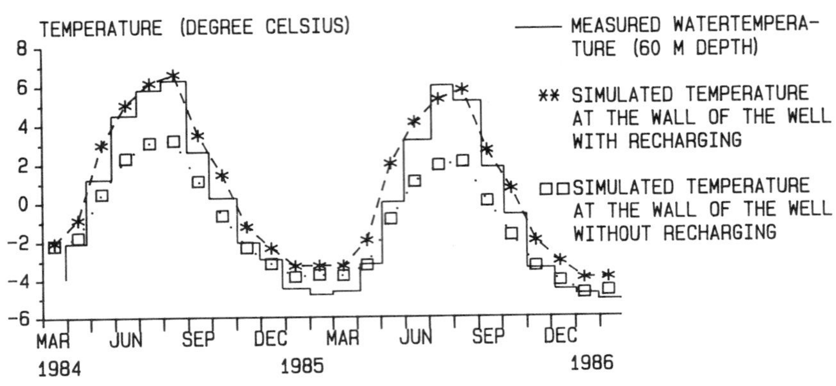

Fig. 3. Measured and simulated monthly mean temperatures in the well system during the first two years of operation.

The long term temperature development for the A-house system is simulated and the results during the 15th year of operation are presented in Table 2 (Case 1). The monthly minimum temperature in the well will be about 3 °C lower the 15th year than the second year of operation. This means that the evaporation temperature for the heat pump will be very low during winter periods, which could cause operation disturbances.

Two ways of solving this problem have been analysed with computer simulations.

Case 2. Two extra bore-holes and recharging with the exhaust air system only. Total active bore-hole depth H=1120 metres. Recharged energy 25 MWh/year.

Case 3. A new recharging system with fan-driven ambient-air heat exchangers and the existing exhaust air system. Total active bore-hole depth H=909 metres and recharged energy 125 MWh/year.

The results are presented together with Case 1 in Table 2 as minimum temperatures and energy weighted annual mean temperatures year 15. The simulated results show that the difference in the well temperature during the heating season between Case 2 and 3 is small. Compared to the existing system (Case 1) the minimum temperature will be about 2 °C higher after 15 years of operation if the Case 2-solution is selected. This will decrease the risk of too low evaporation temperatures and also increase the heat pump performance. The saving in the total electrical energy consumption will be in the range of 5 - 10 MWh/year compared to the existing system.

The investment for two extra bore-holes with pipes will be 45000 SEK. The solution with fan-driven ambient-air heat exchangers will cost more than 45000 SEK. This means that the best way of improving this system is to increase the total bore-hole depth (Case 2). As the heat pump performance will be almost the same for Case 2 and 3, the lower investment cost and simplified system solution of Case 2 is conclusive in making this solution the better choice.

TABLE 2. Simulated Bore-hole Temperatures after 15 Years of Operation for Three Different System Solutions of the A-house Heat Pump System.

Case	1 No change of the well or recharge system.	2 Two extra bore-holes. No energy roof system.	3 Recharging with fandriven air heat exchangers
Active bore-hole depth:	H= 909 m	H= 1120 m	H= 909 m
Annual recharging:	Q= 72 MWh	Q= 25 MWh	Q= 125 MWh
Annual mean heat extraction, W/metre bore-hole:	23	23	16
Monthly minimum temp. :	-6.0 °C	-4.2 °C	-4.6 °C
Energy weighted annual mean temperature:	-3.7 °C	-2.9 °C	-1.5 °C

CONCLUSIONS

The Kragstalund plant shows good annual performance. This installation demonstrates that the ground coupled heat pump, in the range 50-200 kW, with a multiple well system as heat source is a competitive option for a large number of apartment buildings in Sweden, both for new buildings and for retrofit.

Results from measurements and computer analyses presented in this paper indicate that it is more favourable to add bore-length than to add recharging equipment. This will give a more cost-effective system with a simpler and more reliable system design.

REFERENCES

Eskilson, P. (1985). Superposition Bore-hole Model (SBMA). Simulation model for oblique or vertical bore-holes. University of Lund, Sweden.

Larsson M., and L. Spante (1985). Solar charging of energy wells for small heat pump systems. Proceedings of the ISES Congress INTERSOL 85, Montreal, Canada.

GROUND-COUPLED HEAT PUMP SYSTEMS FOR RESIDENCES

Georg Milborn

University of Innsbruck, Institut für Bauphysik
A-6020 Innsbruck, Technikerstr. 13, Austria

1. ABSTRACT:

This study treats an analysis of ground-coupled heat pump systems for residences. Different installations in western Austria are described and measuring data is compared with computer calculations. Low-temperature heat is drawn out of heat exchanger pipes installed vertically in the ground. The ground is actively cooled down by a liquid to liquid electrical compressor heat pump unit which feeds its energy to the heating system and hot water supply of residential houses.

Keywords: Heat-pumps, Ground-coupled, Low-temperature ground collectors, Vertically drilled holes as low temperature source.

2. DESCRIPTION OF THE SYSTEM:

Two types of ground-coupled systems are known:
a) The flat collector of polyethylene pipe in a serpentine array 1.5 - 2 m deep, with 0.5 to 1 m spacing between the pipes. The thermal load depends on the type of ground and varies between 10 to 30 W/m2 for gravel to swamp ground. One variant of the flat collector is the trench collector where the pipes are arranged in a long ditch 3 m deep. The cost per m^2 is about 20 DM.
b) Vertically drilled holes, 60 to 120 m deep, also with polyethylene pipes. In this case the thermal load varies between 30 and 50 W/m. The thermal load varies with the type of ground. The optimal ground is swamp ground, saturated with water or compact rock with high thermal conductivity. The worst ground is loose, dry gravel. The standard situation in the valleys of the alps is a mixed ground with different layers, the upper layers earth, clay, gravel; at the bottom under 40 to 60 m is rock.

In some cases it is possible to combine the low-temperature ground heat source with solar collectors. During the daytime, solar energy can be used to raise the temperature level in the ground a number of degrees. Some amount of heat can be stored in the ground, especially for short periods, for example from day- to night-time. If the low-temperature heat is to be stored for longer periods, e.g. from Sommer to Winter, this combination is only useful when the ground can store heat without disturbing the heatflow by drying out. The best ground for this purpose is solid rock.

The hole is made by a rotary drilling machine. It is fitted with four pipes, each 26 mm in diameter, or with a coaxial pipe having 9 cm outer diameter. The space between the pipes is filled with drilling mud with high thermal conductivity. One drilling down to 65 m, which is the standard deepness, is done in one to two days, hence the whole installation for a residential house is finished in one week. The cost of drilling and installation of 1 m is 100DM.

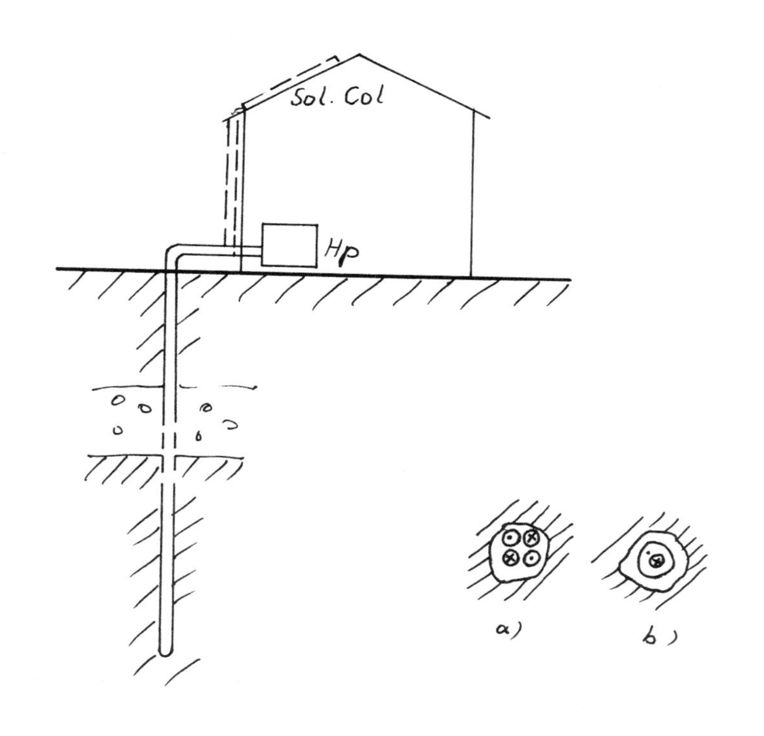

Ground-coupled heat-pump systems for residences with optional solar connection. The two types of heat exchanger pipes.

3. ADVANTAGES AND DISADVANTAGES:

The main advantage of ground-coupled heat-pump systems compared with ground water systems is that the low-temperature circuit of the heat-pump funktions in a closed circuit. Chocking with sand or electro-chemical problems cannot arise.

a) Flat-collector: cheap and easy to install in the case of a new installation; it requires a big land area and it is dependent on the ambient temperature level; a retrofit is nearly impossible.

b) Drilling in the ground: very little or nearly no land is needed, retrofit is possible, high cost, small heat- exchanging surface.

4. THERMAL AND ENERGETIC PROPERTIES:

The thermal load of vertical heat exchangers in the ground is 30 to 50 W/m. The annual amount of low temperature heat per meter withdrawn from a vertical drilling varies between 70 and 100 kWh/m*a. This results in an average decrease of the temperature by 4.5 K; the maximum is about 10 K. The downgoing and the raiser pipes are very close to each other and they have a relatively good thermal contact. This results in the effect that the upper layers are cooled more than the lower ones.

5. RESULTS:

In the three last years 8 systems as described were installed in different locations in western Austria, mainly in Vorarlberg. The ground varies between swamp-ground in the Rhein-valley and dry rock in Lech at an altitude of 1800 m above sea level. All installations are working well and the results are good.

The first system was installed was done in 1984 in Hörbranz, Vorarlberg. The residential house "Boch" is a little house with 120 m^2 floor area and a heat load of maximal 10 kW. The house is equipped with a heat-pump for space heating only. The heating capacity of the heatpump is 8 kW. Two vertical drillings down to 65 m were made. The ground is watersaturated rock (Flysch). The undisturbed ground temperature was 12.1 C. During the winter of 1985/86 this temperature dropped to the minimum of 3.1 C. In the following summer the temperaure rose to 11.6 C. The complete heating system, heat-pump and pumps for the circuits had a consumtion of electricity of 5477 kWh. The efficency of the heat-pump, cop, was 3.42, for the whole system including pumps 2.84. The measuring equipment included PT-100 temperature sensors and meters for the low- and high-temperature heat and electricity.

OUTPUT OF A SUPERCALC CALCULATION PROGRAM

Mon	T-amb C	DD 23/12	End kWh	Heat-P kWh	Th-Ld W/m	T-grd C	Ht-Pp h/M	Elec kWh	cop	cop +Pumps
Aug	17.6	10	0	0	.0	10.4	0	0	–	–
Sep	15.3	0	0	0	.0	10.4	0	0	–	–
Okt	9.6	295	913	902	6.8	10.0	78	255	4.0	3.58
Nov	2.3	579	2226	2118	15.8	7.8	217	751	3.5	2.96
Dec	3.4	575	2391	2324	16.5	6.5	259	798	3.4	3.00
Jan	1.5	639	2663	2596	18.1	5.4	301	917	3.3	2.90
Feb	-4.4	762	3089	2988	22.0	3.3	407	1164	3.0	2.65
Mar	3.6	545	2176	2065	14.5	5.1	280	773	3.4	2.81
Apr	7.8	389	1392	1327	10.0	7.0	165	459	3.6	3.03
May	15.2	49	217	206	1.5	8.9	37	68	3.9	3.20
Jun	16.6	91	154	149	1.2	9.7	11	45	4.0	3.39
Jul	18.0	0	0	0	.0	10.2	0	0	–	–
Sum	8.9	3934	15221	14674	13.3	7.9	1755	5232	3.5	2.91

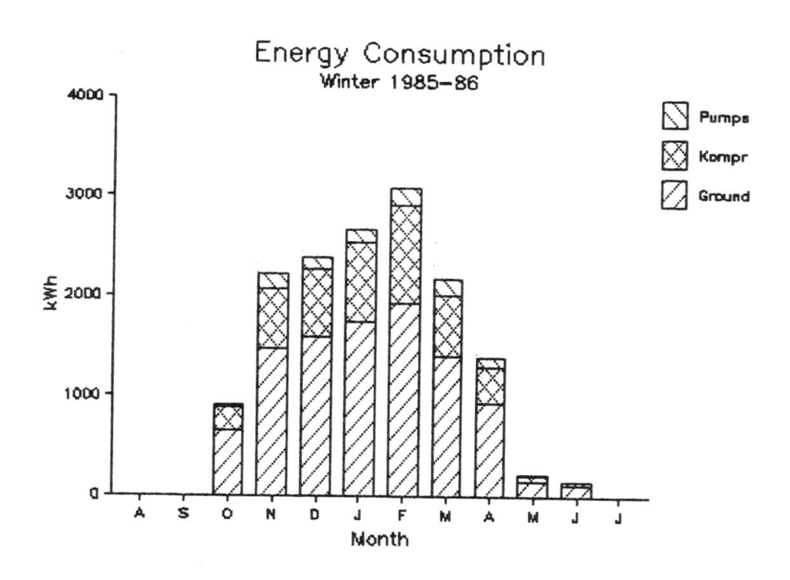

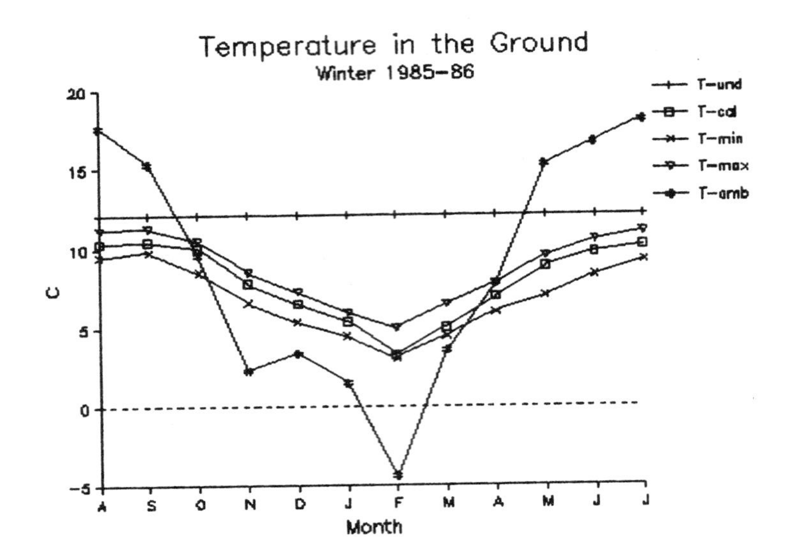
Temperature in the Ground
Winter 1985-86
T-und
T-cal
T-min
T-max
T-amb
C
Month
20
15
10
5
0
-5
A
S
O
N
D
J
F
M
A
M
J
J

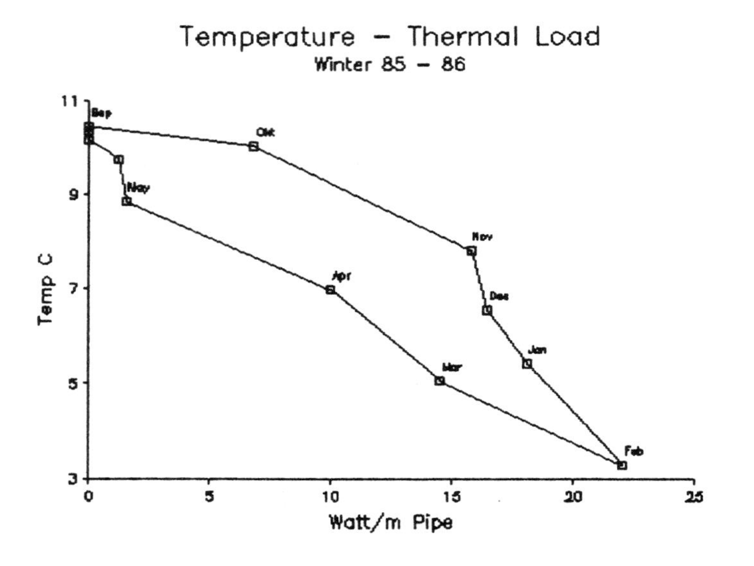
Temperature - Thermal Load
Winter 85 - 86
Temp C
Watt/m Pipe
Sep
Okt
Nov
Des
Jan
Feb
Mar
Apr
May
11
9
7
5
3
0
5
10
15
20
25

PERFORMANCE TEST OF A DIRECT EXPANSION HEAT PUMP SYSTEM

J. A. Edwards*, P. Safemazandarani*, R. R. Johnson *
and Y. Mohammad-zadeh*

Applied Energy Research Laboratory, North Carolina State
University, School of Engineering,
Ragleigh, NC 27695-7910, USA

ABSTRACT

In this paper results are given for a direct expansion heat pump system which is uniquely coupled to the earth through a heat transfer concentrator (GLHXC-Ground Loop Heat Exchanger Concentrator). In effect the system reduces several of the thermal resistances associated with the heat flow paths in other types of ground coupled heat pump systems. These reductions result from (1) letting the refrigerant circulate through a heat exchanger in the earth, thus eliminating the resistance associated with the water-refrigerant heat exchanger used with a conventional water source heat pump, and then by (2) letting the refrigerant coil in the earth first transfer heat by natural convection to water and then having this heat transferred to the earth through the large surface area of the water-earth interface. The GLHXC system avoids the most critical thermal resistances encountered with buried pipe systems.

Performance results for the GLHXC system are given for both cooling and heating operations based on a one year test period. The results are compared to experimental data obtained from four other types of earth coupled configurations (horizontal pipe, pebble bed, surface tank, and a closed loop vertical well) and an air source heat pump tested at AERL.

KEYWORDS

Heat pump; heating; cooling; ground coupled.

INTRODUCTION

Ground-coupling of residential heat pumps seeks to take advantage of two aspects that are not available to conventional air-to-air heat pump systems. First, the slow response of the ground to changing ambient temperatures, because of high thermal mass, results in source or sink temperatures which are not subject to large extremes. Second, the solar insolation absorbed by the ground in winter acts as a heat source. Because of these features, a properly designed ground coupled system will match or outperform an air-to-air system in cooling (summer), and show significant improvement in terms of the seasonal coefficient of performance in heating (winter). During winter, the most troublesome times for conventional heat pump operation are during cold snaps when the outdoor temperature falls so low as to reduce the COP and capacity of the heat pump to the point that back-up resistance heat is needed. Ground-coupling can carry the system through such transients without a drastic drop in performance.

Experience with different ground-coupling configurations has shown that one of the key parameters affecting performance is the thermal resistance between the working fluid in the outdoor loop and the surrounding ground (Edwards, 1985). For one of the most common ground-coupling configurations, the horizontal buried pipe, the thermal resistance is critically governed by the outside surface area of the pipe and the layer of insulative ground adjacent to that surface. Designs for such systems therefore call for a considerable length of buried pipe in order to provide the needed heat transfer surface area. As a means of reducing the overall thermal resistance we have tested several systems which are designed to balance the resistances in the critical heat flow path (Johnson, 1984 and 1985).

In this paper results are given for a direct expansion system which is uniquely coupled to the earth through a heat transfer concentrator (GLHXC-Ground Loop Heat Exchanger Concentrator). In effect the system reduces several of the thermal resistances associated with the heat flow paths in other types of ground-coupled heat pump systems. Allowing the refrigerant to circulate through a heat exchanger in the earth eliminates the resistance associated with the water-refrigerant heat exchanger used with a conventional water source heat pump. There is also a reduction in resistance by having the refrigerant coil in the earth first transfer heat by natural convection to water where the effective convective coefficient ranges from 30 to 100 Btu/hr-ft^2-^{o}F (170 to 600 W/m^2 ^{o}C) as opposed to an effective value of 3 to 7 Btu/hr-ft^2-^{o}F (17 to 39.7 W/m^2 ^{o}C) for a pipe to earth, and then having this heat transferred to the earth through the large surface area of the water-earth interface.

SYSTEM DESCRIPTION

Figures 1 and 2 show the general flow arrangement of the GLHXC system and give some details of the earth coupling. Referring to Fig. 1, it is noted that the overall configuration of the system is similar in many respects to any heat pump system. Starting with the compressor the high pressure refrigerant vapor flows through an oil separator to a reversing valve. When operating in the cooling mode, the reversing valve routes the refrigerant to the earth coupled coil which serves as the condenser. After leaving the condenser the refrigerant flows to an expansion valve and the inside coil (evaporator for cooling). After leaving the inside coil it is returned as a vapor to the compressor. When operating in the heating mode, the reversing valve sends the refrigerant to the inside coil first. The bypass arrangement around the expansion valves is accomplished with ball check valves and requires no manual attention.

The refrigerant pressure is measured at four locations and its temperature at six locations. These measurements, along with a refrigerant flow measurement, are used to do heat balances on the evaporator and condenser. The electrical consumption of the system is measured with a kW-hr meter in addition to a watt reading obtained from an in-line watt transducer. Other temperatures are also measured in the thermal storage tank, the earth-coupling device, and the surrounding earth.

In Fig. 2 a cross-sectional view of the earth-coupling arrangement is shown along with other descriptive information. Essentially, the earth-coupled heat exchanger consists of 170 ft (31.8 m) of 7/8 in (2.22 cm) OD copper tubing which is immersed in a water bath. The water bath is formed by using a semi-cylindrical section of metal culvert. Gravel is placed around the culvert up to a depth of 3 ft (0.91 m). Soil is placed on top of the gravel. A vapor liner is placed in the excavated trench to retain the water. The heat exchanger is buried in such a fashion that it is not noticeable from the earth's surface and the area above the heat exchanger can be used for a lawn, garden or left bare. In over two years of operation we have not experienced any difficulties with the earth-coupled heat exchanger. Any loss of water due to evaporation of leakage has been replenished via rain.

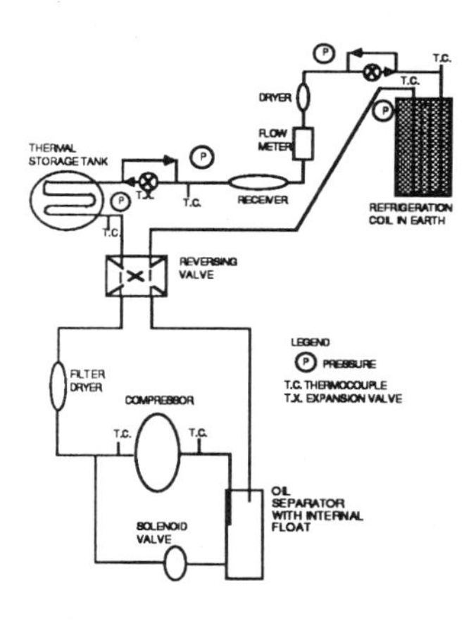

Fig. 1. Flow Diagram.

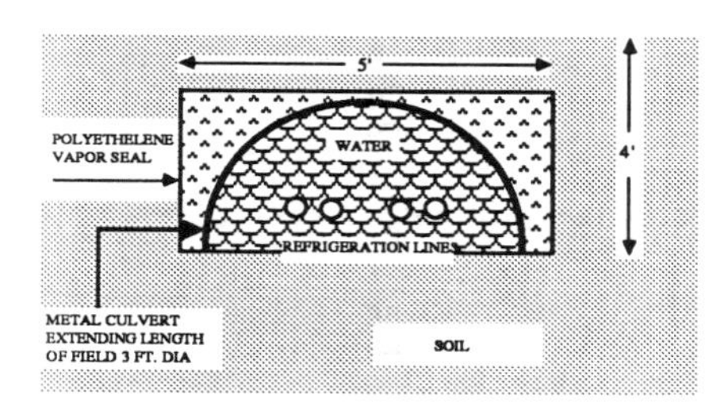

Fig. 2. Cross-sectional view of the earth coupling.

RESULTS

The cooling mode results are presented in three groupings: (1) random nonsteady operation, June 19 to July 19, (2) continuous operation (heat pump never shut off) for July 19 to August 25, and (3) operation with a 4 hour on - 4 hour off cycle from August 26 to October 22, 1984. The results are presented as daily average coefficients of performance and daily average heat transfer rates for the evaporator. During the initial period (6/19 to 7/19) the cooling COP ranged from a high of 3.8 to a low of 2.9. The cooling effect ranged from 15000 Btu/hr (4.39 kW) to 11,700 Btu/hr (3.43 kW). The maximum daily cooling effect was 300,000 Btu/day (87.9 kWh/day). Starting on July 19 the system was operated on a continuous basis (24 hours/day) until August 25. This type of schedule was used to stress the system to its maximum capacity and do so during the hottest portion of summer. Even under these severely stressed conditions the cooling COP ranged from a high of 3.1 to a low of 2.3, with an average value of 2.55. The cooling effect during this period ranged from a high of 312,000 Btu/day (91.2 kWh/day) to a low of 256,800 Btu/day (75.12 kWh/day), with a daily average of 272,000 Btu/day (79.68 kWh/day). The heat dumped through the ground coupling ranged from a high of 400,000 Btu/day (117.2 kWh/day) to a low of 317,000 Btu/day (92.9 kWh/day) with a daily average of 336,000 Btu/day (98.4 kWh/day). After returning to cyclic operation on August 26 (4 hr on/4 hr off), the cooling COP started to rise again and reached a maximum value of 3.5. During this period the maximum daily average cooling effect was 14,600 Btu/hr (4.28 kW) (based on 12 hours of operating time per day), and the maximum daily average amount of heat discharged through the ground coupled device was 18,000 Btu/hr (5.27 kW). The average cooling effect was 13,200 Btu/hr (3.87 kW) and the heat rejection rate was 15,900 Btu/hr (4.66 kW). This gives an average daily cooling load of 158,000 Btu/day (46.32 kWh/day) and an average heat rejection rate of 191,000 Btu/day (55.92 kWh/day).

The heating results are presented as daily average COP and daily average heat transfer rates for the condenser. In Fig. 4, we show the average COP obtained on a daily basis during the test period (12/6 to 3/6). During this period the heating COP ranged from a high of 5.0 to a low of 3.42. The heating effect ranged from 14999 Btu/hr (4.10 kW) to 18750 Btu/hr (5.49 kW). The maximum daily heating effect was 225000 Btu/day (65.9 kWh/day). On a monthly basis the COP and heat effect was:

Month	Avg. Heating COP	Avg. Daily Heating Effect Btu/day	kWh/day
December - 1984	4.34	193,000	56.40
January - 1985	3.91	167,700	45.96
February - 1985	4.51	193,640	56.64

The above results were obtained with a duty cycle of four hours on following by four hours off.

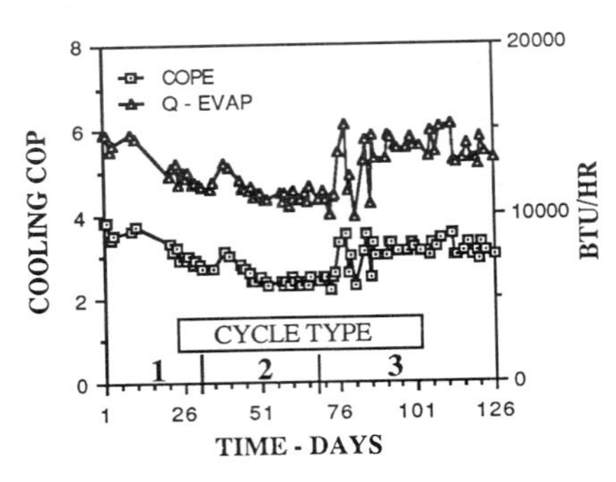

Fig. 3. Cooling season test results

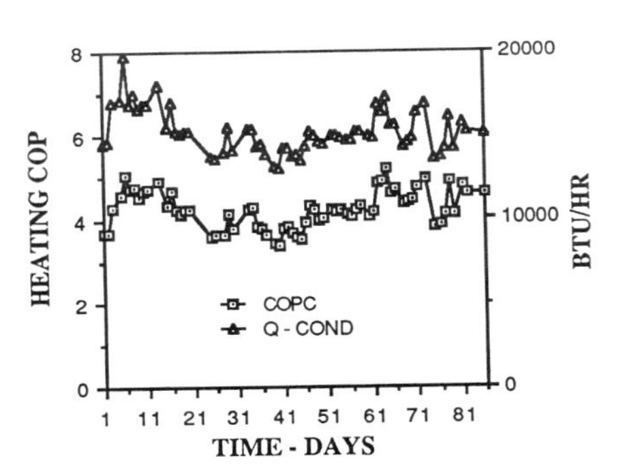

Fig. 4. Heating season test results

Based on comparative test results (Edwards, 1985; Johnson, 1985; Johnson, 1984; Mohammad-zadeh, 1987) with other heat pump systems tested at AERL under similar load and cycle conditions, the GLHXC system showed the following energy savings:

Percentage less electricity cooling	heating	when Compared to
7	15	Closed loop vertical well*
13	28	Pebble bed*
20	37	Surface tank*
50	28	Horizontal pipe*
37	34	Air source heat pump

* Earth-coupled to water source heat pumps and tested at AERL.

SUMMARY

1. A direct expansion earth-coupled heat pump using the newly devised earth-coupling technique described herein provides an effective system for heating and cooling residential buildings.

2. The system tested, which had a compressor rated at one ton capacity, can heat and cool a residence with an overall UA of 280 Btu/hr - °F (147 W/°C) and a heating design temperature differential of 50°F (27.7 °C) without the need of back-up heating.

3. The earth-coupling is not complicated and only requires a land area of 5 x 50 ft (1.52 x 15.2 m) per ton of cooling capacity.

4. The system only uses commercially available components and can be assembled by a competent refrigeration mechanic.

5. The only moving part, and energy consuming part, of the primary heat transfer system is the compressor. Of course, like any other heat pump it requires an inside blower unit.

REFERENCES

Edwards, J. A. (1985). Heat transfer from earth-coupled heat exchangers - experimental and analytical results. ASHRAE Transactions, Vol. 91, Part 2.

Johnson, R. R., Edwards, J. A., and Mulligan, J. C. (1984). Heat pump data acquisition project. Applied Energy Research Laboratory Reports AERL-9, AERL-13, AERL-17 and AERL-30, Department of Mechanical and Aerospace Engineering, North Carolina State University.

Johnson, R.R., Edwards, J. A., and Mulligan, J. C. (1985). Comparative seasonal operation of ground-coupled heat pumps. Proceedings of the ASHRAE/REHVA World Congress on Heating, Ventilating and Air-conditioning, CLIMA 2000, Copenhaven, Denmark, Vol. 6, pp. 135-140.

Mohammad-zadeh, H., Johnson, R. R., Edwards, J. A., and Safemazandarani, P. (1987). Experimental evaluation of an open loop pebble-bed method of ground-coupling for residential heat pumps. Proceedings of 22nd Inter-society Energy Conversion Engineering Conference. Vol. 1, pp. 1138-1143.

HEATING A SCHOOL WITH GROUNDWATER HEAT PUMPS

H. Jänich

HASTRA, Humboldt Straße 33, 3000 Hannover 1
Federal Republic of Germany

ABSTRACT

The school heating system comprises a groundwater heat pump installation. The environmental heat acquisition piping is located approx. 1.5 m deep under the parking lot, the school yard and the green spaces. Two heat pumps heat the water to the required input temperature. The school has under-floor water heating with the advantage of a very low input temperature. The overall system combines reliability with low energy consumption.

KEYWORDS

Heating system; heat source; heat pumps; operating results.

INTRODUCTION

The primary school in Nordstemmen near Hanover has a heated floor area of 1586 m^2 (1896.8 square yards) heated by two compact heat pumps. These two units are from series production, have a heating capacity of 57 kW each and can meet the heat requirement of 94 kW (or 60 W/m^2). The system works as a groundwater heat pump.

HEATING SYSTEM

All rooms have natural lighting, providing sufficient window ventilation, and negating the need for air conditioning. The client preferred a heating system with static heating surfaces. Furthermore, the generation of heat was to be characterized by low energy costs, require no supervision and little maintenance work. For these reasons it was decided to install a warm water under-floor heating system, the pipes of which were laid in the still moist concrete floor.

To provide good performance efficiency even at low outdoor temperatures and to obtain good operating figures throughout the year, the floor heating is designed for a maximum in-flow temperature of only 42°C with a spread of 3 K. Since the heating pipes are covered with 50 mm (1.96 in) of concrete floor, the total thickness of the floor is 80 mm (3.14 in), which provides for heat accumulation and helps reduce energy costs.

Fig. 1 View of the school

The individual heating circuits are not longer than 100 m (109.6 yards). In total, 132 individual circuits were laid in the floor, supplied by 25 heating-circuit manifolds. On average, 6.8 m (7.43 yards) of polyethylene pipes are installed per square meter of heated area. The pipes have a diameter of 20 mm (0.78 in) and are spaced about 150 mm (5.9 in) apart.

SOURCE OF HEAT

To extract the ground heat, a system of pipes is laid at a depth of about 1.5 m (1.6 yards). The piping is installed underneath the paved school-yard, under the car parking area and the green spaces. It was decided to employ PE pipes with an outside diameter of 20 mm (0.78 in), of which 6,000 m (6,564 yards) is laid in an area of 2,900 m^2 (3,172 yards) -- i.e. nearly double the heated area. The pipe length is subdivided in heating circuits of 100 m (109.6 yards) each. The individual heating circuits are connected to a collector and a distributor, which in turn are installed in shafts outside the building. A mixture of water and antifreeze, with a freezing point below -12°C (10.4°F), circulates through the piping system to provide heat absorption and transport.

The heat absorption capacity required is 7 W/m or 15 W/m^2. In the moist clayey soil, which contains a certain amount of sand, these values are easily attainable at any outside temperature. Two pumps rated at only 75 kW each are installed to provide for circulation of the brine. This indicates that the diameter and the length of the pipes in the heating system and the condenser of the heat pump are extremely well matched.

HEAT PUMPS

Two standard compact brine-water heat pumps with a heating capacity of 57.4 kW each at 35°C (95°F) flow temperature and 0°C (32°F) heat-source temperature are installed to cover the demands of both heating and ventilation (94.5 kW). Each of the pumps has a power requirement of 15.2 kW. Consequently, the condensing capacity of each pump is 42.2 kW.

Fig. 2 Heat pump system

The temperature of the heating water is raised by approx. 3 K in the heat pump and is equivalent to the cooling effect of the floor heating, making coupled-storage systems superfluous. On the other hand, the storage capacity of the concrete floor is large enough to make buffer storage systems unnecessary. The outdoor temperature and a thermometer in the floor of a pilot room determine the regulation of the flow temperature. Thus, long flow intervals are ensured, preventing frequent switching of the compressor. An automatic timer adapts the operating times of the heat pump to the times the school is being used.

From 10 p.m. to 6 a.m. the electricity to run the heat pumps is available at a particularly low, off-peak night rate. For this reason the heat pump control system raises the flow temperature towards morning above the set-point value. In conjunction with the floor heating, which, due to its design, delays the transfer of the heat from the excessively high inflow temperature into the room, it is possible to generate the heat necessary a) to compensate for the night temperature drop and b) for first morning hours using the cheap off-peak power.

OPERATING RESULTS

During five years of operation the system has not only proved its reliability, it has also fulfilled all expectations.

Table 1 shows the consumption of electricity for the heating from 1983 to 1986. The high proportion of cheap off-peak power was achieved, as already explained, by slightly raising the input temperature of the floor heating in the early morning hours towards the end of the night-rate period and because of the fact that the school has basically only a single day shift.

TABLE 1 Electricity Consumption of Heat Pumps

Year	Standard Rate kWh	Cheap Rate kWh
1983	12,422	42,162
1984	18,039	48,336
1985	19,770	43,562
1986	19,370	50,622

FINNED AIR COOLERS

Are they as efficient as the manufacturers promise?

A.Imhof, J.Keller
Swiss Federal Institute for Reactor Research (EIR)
CH-5303 Wuerenlingen / Switzerland

ABSTRACT

The objective of this investigation was to check the ability of various manufacturers of finned air coolers to design properly their product with the blower mounted at the exchanger's case. For this purpose, five air coolers available on the Swiss market were bought. Each of them was sized by the manufacturer according to a given set of specifications. On the one hand, these air coolers have been tested on the outdoor test facility of the EIR, the conditions of operation being similar to those in practice (in-situ measurements). On the other hand, three of them were investigated in an indoor test facility where a uniform air flow was established (the cooler's blower being removed).

Only two of the five manufacturers were able to predict the thermal performance of their product with an accuracy better than 10%, this for conditions without as well as with condensation. Regarding the seasonal performance (obtained by simulation over a winter period), the highest value is acheived by the cooler equipped with the strongest blower. Compared to this value the product with the poorest performance yields only 10 to 60%, depending on the operation mode. However, the cooler with the highest thermal performance is able to deliver only 2 to 4 times the energy consumed by the blower whereas for the remainig products this ratio is of the order of 10 to 15.

KEYWORDS

Heat exchangers, air coolers, air/water heat exchangers, heat pumps, heat recovery systems, performance check.

INTRODUCTION

Air coolers used in industrial processes, heat recovery or heat pump systems are often built as finned heat exchangers integrated in a case with an axial blower mounted on. As there are very few data about the reliability of the performance data of air coolers with case-mounted blowers, five finned air coolers from different manufacturers or sales reperesentatives have been investigated. Air coolers in heat pump systems are mostly designed as refrigerant evaporators. Because of experimental reasons (e.g. phase change only on one side of the heat exchanger, existing test facility) the air coolers have been designed and tested with a (33% ethyleneglycol, 67% water)-mixture as heat transfer medium. They had to be sized by the manufacturers according to the following specifications:

thermal power $\dot{Q}_{th}$ at $t_{a,i}$ = 10 °C, r.h. = 80 %, $t_{gl,i}$ = -5 °C, $\dot{V}_{gl}$ = 1200 l/h : $\dot{Q}_{th}$ = 7.5 kW.

The quantities $t_{a,i}$, $t_{gl,i}$, r.h. and $\dot{V}_{gl}$ are the air and the glycol inlet temperature, the relative humidity and the glycol flow rate, respectively. Besides these basic data, the manufacturers were asked to provide the thermal performance for three additional points of operation, one of them being a state without condensation.

The objectives of this investigation are:
1) Verification of the thermal performance (manufacturer's data vs experimental data)
2) Influence of the blower on the thermal performance
3) Seasonal performance under operating conditions in practice

The program was financially supported by the Federal Office for Economic Policy (BfK) and the Swiss Federal Institute for Reactor Research (EIR).

OUTDOOR EXPERIMENT

The air coolers to be tested were installed on the modified outdoor test facility for solar collectors at the EIR (Fig. 1). A ventilated tent roof protected the experiment from unwanted solar radiation. All exchangers were tested at equal inlet temperatures and flow rates. For the measurement of the air flow through the air cooler, a settling down tube with a reference anemometer was mounted at the air inlet (without affecting the point of operation).

Over the day the following quantities were measured: glycol flow rate $\dot{V}_{gl}$, glycol inlet temperature $t_{gl,i}$ and outlet temperature $t_{gl,o}$, air inlet temperature $t_{a,i}$, air temperature drop Δt_a, reference air velocity v_{ref} (as a measure of the air flow rate V_a), electric power of the blower P_{el}.

DATA EVALUATION METHOD

The Mollier-diagram t(X) depicted in Fig. 2 shows the states of the air during its cooling down. Due to the difference of the two inlet temperatures Dt = $t_{a,i} - t_{gl,i}$ the air is cooled down to its outlet temperature $t_{a,o}$. If the air temperature falls below the dew point, a certain amount of water vapor DX condenses. The specific enthalpy of the air will be decreased by DH. Using the quantities shown in Fig. 2, the heat transfer of the air cooler is characterized by the following quantities:

- enthalpy gradient : DH / Dt
- heat transfer coefficient (multiplied by the h.ex.surface) : k*A
- heat exchanger effectiveness (referring to air) : $\Phi_a = \Delta t_a$ / Dt
- heat exchanger effectiveness (referring to glycol) : $\Phi_{gl} = \Delta t_{gl}$ / Dt

These quantities as well as the air flow rate V_a are dependent on humidity (water droplets reduce the free cross section of the heat exchanger leading to an enhanced air pressure drop). It turned out that they can be written as functions of the quotient DX/Dt.

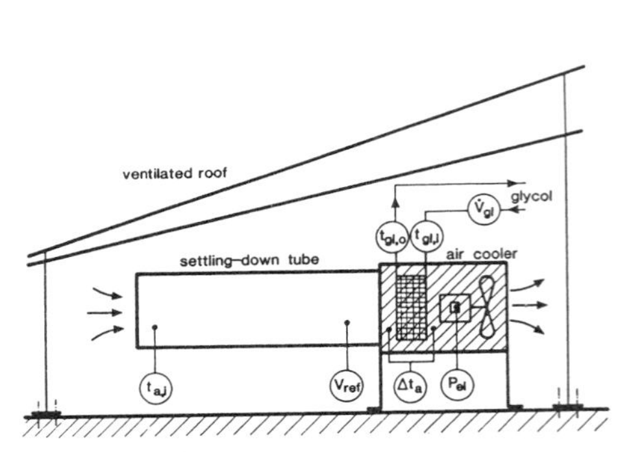

Fig. 1: Experimental setup of the outdoor test facility

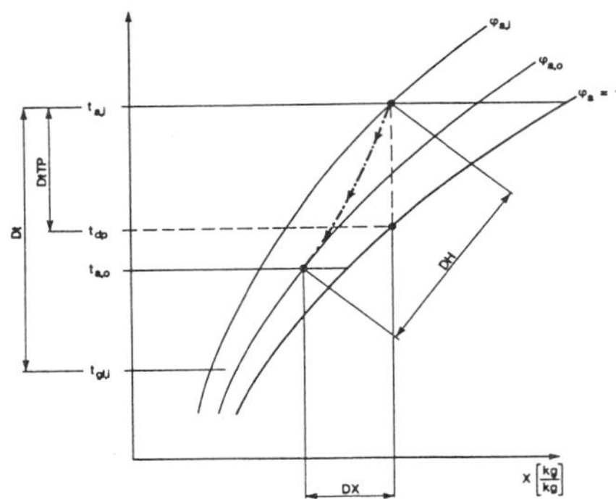

Fig. 2: Air states during cooling, depicted in the Mollier-diagram t(X)

On the one hand, the enthalpy gradient and the air flow rate can be combined to yield the thermal power output of the heat exchanger as a function of the argument DX/Dt:

$$(1) \qquad \dot{Q}_{th,1} = DH * \dot{V}_a * \rho_a = \dot{Q}_{th,1}(DX / Dt)$$

On the other hand the thermal power may be determined by means of the heat transfer coefficient:

$$(2) \qquad \dot{Q}_{th,2} = (k*A) * \Delta t_{log} = \dot{Q}_{th,2}(DX / Dt) \qquad ,$$

where the log-mean temperature difference Δt_{log} is obtained by combining the two heat exchanger effectivenesses Φ_a and Φ_{gl}. If the inlet temperatures of both the air and the glycol, the air pressure and the dew point are given, an iterative solution of the relation $\dot{Q}_{th,1} = \dot{Q}_{th,2}$ yields the point of operation of the air cooler. Based on this procedure it was possible to calculate the thermal power for the specified data given to the manufacturers in order to check the performance promised by the companies. Moreover, using meteo data from the test facility recorded during a winter season, the seasonal energy gain SEG and the seasonal electricity consumption SEC at given glycol inlet temperatures were calculated.

RESULTS

Sizing Check

In order to quantify the reliability of the manufacturer's data the so-called sizing factor SF is defined as

$$(3) \qquad SF = \dot{Q}_{situ} / \dot{Q}_{spec}$$

where $\dot{Q}_{situ}$ and $\dot{Q}_{spec}$ are the thermal powers based on the outdoor test and on the manufacturer's data respectively. The sizing factors of the air coolers were determined at the design point (Fig. 3) as well as under three additional operating conditions. In Fig. 4, in dark hatching the sizings factor for outdoor operating conditions without condensation are depicted. In addition, the light hatched bars show the sizing factors of three air coolers tested indoors under uniform air flow conditions, the blowers being removed. The air flow rate of the test facility was adjusted to the value specified by the manufacturer. In order to quantify the effect of the blower, the so-called blower factor is defined as

(4) $BF = \dot{Q}_{situ} / \dot{Q}_{lab}$,

$\dot{Q}_{lab}$ being the thermal power obtained from the indoor test.

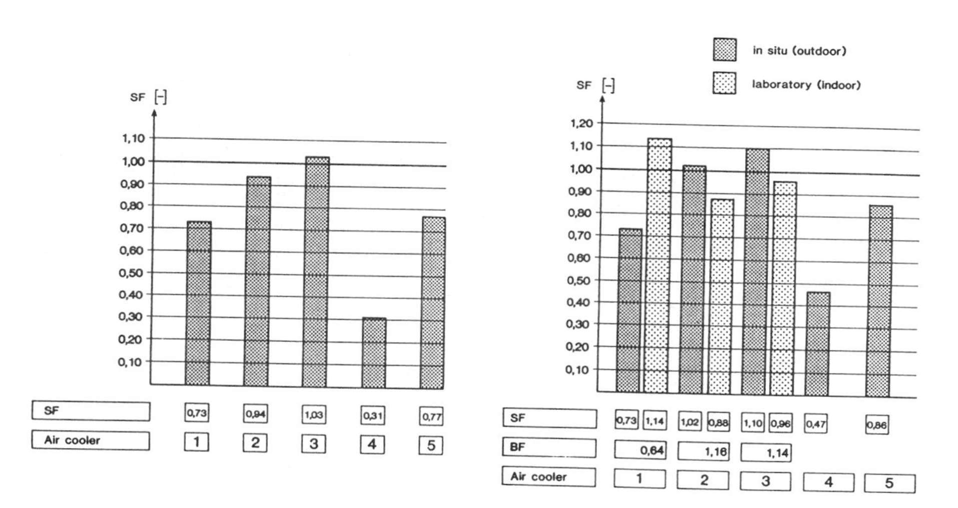

Fig. 3: Sizing factors SF of the air coolers at the design point.

Fig. 4: Sizing factors SF of the air coolers under the conditions of no condensation. Dark hatched bars: outdoor measurements, light hatched bars: indoor measurements (uniform air flow). BF: blower factor.

Seasonal Energy Gain

Two important quantities characterizing the seasonal performance of an air cooler at given input data are the seasonal energy gain SEG and the seasonal electricity consumption SEC. In Fig. 5 the seasonal energy gains for $t_{gl,i} = -5\ ^0C$ are given. As a bar diagram the SEG values normalized to the SEG of the cooler having the highest one are depicted. The basis of the simulation was the set of meteo data obtained during the outdoor tests from January to May 1986. The simulation model includes the formation of ice in the following way: the thermal power values are only summed up if the air outlet temperature $t_{a,o}$ is above 0 ^{0}C. It can be shown that air coolers with small fin gaps or unsufficiently sized blowers are substantially susceptible to ice formation, thus having small SEGs (e.g. cooler 1).

If an air cooler is used as the energy collecting component of a heat pump, the ratio of the seasonal energy gain SEG to the seasonal energy consumption SEC of the blower is of great importance. In Fig. 6 the so-called energy ratio ER is depicted, which is defined by

(5) $$ER = SEG / SEC$$

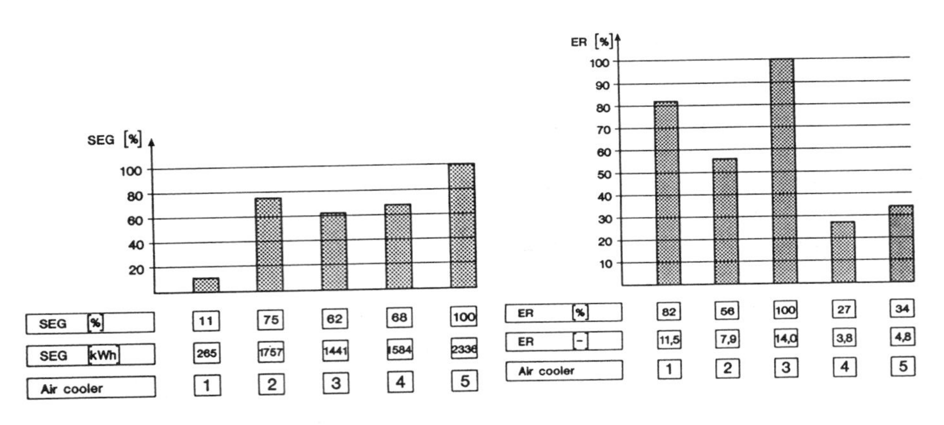

Fig. 5: Seasonal energy gain SEG for $t_{gl,i}$ = -5 ^{0}C

Fig. 6: Energy ratio ER for $t_{gl,i}$ = -5 ^{0}C

These figures show that the air cooler with the highest seasonal energy gain is provided with the most powerful blower leading to a poor energy ratio ER.

CONCLUSIONS

- Only 40% of the tested heat exchangers meet the specifications within ±10%. One cooler does not even reach 50% of the thermal power promised by the manufacturer.
- The decrease of the thermal performance due to an unsufficient air flow rate was 2 to 3 times greater than the variations due to a non-uniform air velocity profile.
- In regard to a high energy ratio the sizing of the heat exchanger and of the blower is much more important than the mode of operation of the air cooler.

These investigations gives a strong evidence that the sizing of air coolers (in particular for conditions where condensation may occur) is still a business for which not every manufacturer is skill enough. We hope that our results are helpful for a better optimisation of air coolers for heat pump and heat recovery systems as well as for industrial processes.

MATHEMATICAL MODELING OF AIR-REFRIGERANT EVAPORATORS AIMING AT DESIGN OPTIMIZATION

S.M. Ordones*, M. Fortes*, R.N.N. Koury* and A. Cordier**

*Thermal Engineering Department, Federal University of Minas Gerais.
Av. Antônio Carlos, 6.627 - Campus UFMG - CEP. 31.270
Belo Horizonte - Minas Gerais - Brazil

**Laboratory of Solar Energy - Paul Sabatier University, 118
Route Narbonne - Toulouse - Cedex - France

ABSTRACT

A full physical-mathematical modeling of air-refrigerant evaporators, with spatial discretization, aiming at optimized design, is presented, along a methodology for solving the resulting set of non-linear equations. The computer results show that the performance of evaporator relies heavily on all the input parameters (refrigerant mass flow, quality and evaporation pressure and air mass flowrate, relative humidity and dry bulb temperature)and the desired refrigerant superheating level. The relationship among input and output parameters is highly non-linear and, thus, precludes the usage of interpolating polynomials: - the full set of equations must be solved for the desired condition.
Environmental conditions affect, in a sensible way, the evaporator performance: - refrigerant or air massflow controllers are required.

KEYWORDS

Mathematical modeling; air-refrigerant evaporators; heat exchangers; heat pumps.

INTRODUCTION

In the reviewed literature (Ordones, 1986), no information was obtained on a detailed spatially-discretized modeling of finned tube evaporators.
The objectives of this paper are to present a methodology for a spatially-discretized modeling of air-refrigerant evaporators and analyze the parameters which affect their performance.

MATHEMATICAL MODELING

The modeling to be presented is based on steady state conditions. Air and R12 thermodynamic and transport properties can be found in Ordones (1986). A diagram of a plane serpentine, rectangular finned evaporators is shown in fig. 1. The modeling neglects heat losses and elbow pressure drops, as suggested by Shah (1981). The rectangular fins were considered as circular fins with a rectangular profile and equivalent area, as proposed by Sauer (1983). The refrigerant enters the evaporator as humid saturated vapor and leaves as

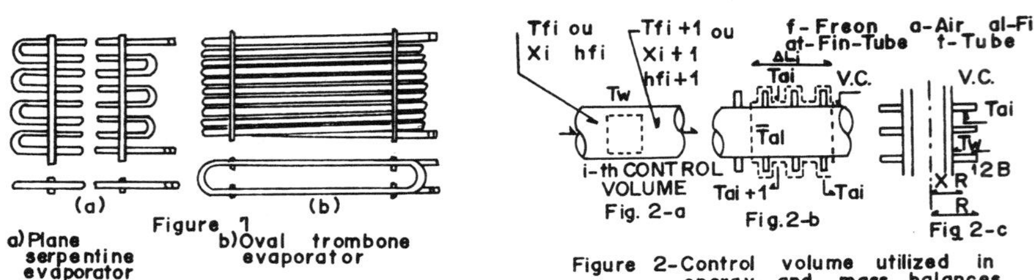

Figure 1

Figure 2-Control volume utilized in energy and mass balances

overheated vapor. It was assumed that the air entering a row does not interact with the air in adjacent rows and that air mass flowrate is constant in all tube rows.
The physical-mathematical modeling equations, based on mass and energy conservation equations are presented in table 1. The associated control volumes are shown in fig. 2.

METHOD OF SOLUTION

The solution of the high non-linear coupled set of balance equations presented in table 1 gives the full profiles of R12 temperature and quality and air temperature and relative humidity for any air or R12 inlet conditions. These non-linear equations were solved by Newton-Raphson's method and other iterative procedures.
An important feature of the solution is that, in even rows, the R12 properties should be iteratively corrected so that outlet R12 conditions in that row are equivalent to the ones in the previous row. If a phase change starts in even rows, the control volume, where the phase change occurred, should be determined and a correction of the R12 inlet values should be made. After convergence is reached, a balance is made on the row to determine outlet values of temperature and relative humidity.
It should be noted that the algorithm allows the relative values to reach values higher than 100%. When this happens, the saturated refrigerant enthalpy should be evaluated from the given quality and its overheating temperature from the just calculated enthalpy value. If the overheated R12 temperature reaches a value which is higher than the inlet air temperature, the equilibrium temperature between the R12 and the air should be calculated and this equilibrium condition should remain constant for the rest of row.

RESULTS AND DISCUSSION

Based on the physical-mathematical modeling, evaporators with 2 and 3 horizontal rows and 4 tubes per row, were analyzed. For simplicity, inlet quality was fixed at 20,45% and evaporation pressure at 3.86×10^5 Pa.
Fig. 3-a,b and 4-a,b show the effect of the simultaneons variation of inlet air temperature, mass flowrate and relative humidity on the evaporator performance. These results indicate that:

1) Two and three row evaporators behave differently under equivalent input conditions and their difference in behavior is not amenable to simple physical explanations.

TABLE 1 - ENERGY BALANCE EQUATIONS

EVAPORATOR REGION WHERE THERE IS:	EXPRESSION	COMMENTS	FIGURE
SATURATED R12	$T_{EV} = T_w + M(X_i + X_{i+1})$	$M = \dfrac{\dot{m}_f \, h_{gf}}{h_{\overline{fe}} \, \pi d_1 \Delta L_i}$	1. a
OVERHEATED R12	$T_f = T_w + P(h_{fi} - h_{fi+1})$	$P = \dot{m}_f / (h_{\overline{se}} \, \pi d_1 \Delta L_i)$; $T_f = (T_{fi} + T_{fi+1})/2$	1. a
FIN	$\bar{T}_{al} = \bar{T}_a + (\bar{T}_w - \bar{T}_a)\eta_{al}$	$\eta_{al} = \dfrac{2X}{N^2(1-X)^2} \cdot \dfrac{I_1(N)K_1(XN) - I_1(XN)K_1(N)}{I_1(N)(K_0(XN) + I_0(XN)K_1(N)}$ $\bar{T}_a = (T_{ai} + T_{ai+1})$, $N = h_{\overline{a\varkappa}} R^2 / b.K$ I and K are 1st and 2nd order modified Bessel Functions	1. b
DRY AIR ($T_{ai} > T_{dp}$)	$T_a = T_{at} + \dfrac{\dot{m}_{as}}{h_{\overline{a\varkappa}} \cdot A\phi} \cdot (h_i - h_{ar\ i+1})$	$A = 2\pi R (X\Delta L_i + NA\ R(1 - X^2))$ in which NA is the number of fins per control volume	1. c
MOIST AIR ($T_{ai} < T_{dp}$)	$h_{ar,00} = h_{ar\ sa,w} + \dfrac{\dot{m}_{as}}{H_t \cdot A\phi} h_{ari} - h_{ar\ i+1} - (W_i - W_{i+1}) \cdot h_{we}$	$m_{as} = m_a/(1 + W)$; $T_{at} = \dfrac{m_{al} C_{al} T_{al} + m_T C_T T_W}{m_{al} C_{al} + m_T C_T}$	
R12 SATURATED -AIR	$X_{i+1} = X + \dfrac{H_t}{S}(h_{ar,00} - h_{ar\ sa,w})$	$S = \dot{m}_f \, h_{fg} / A\phi$	1. c
R12 OVERHEATING-AIR	$T_a = T_{at} + \dfrac{Z}{h_{\overline{a}}} \cdot (h_{fi+1} - h_{fi})$	$A = \dot{m}_f / A\phi$	1. c

2) There is, for each geometric and environmental conditions, an allowed range of air mass flowrate. Outside this range, the refrigerant will become excessively overheated (overheating > 5°C) or saturated at the outlet.
3) An increase in environmental air temperature, in the range of 25 to 31°C, has little influence on the delivered power (power variation is at most 10% for $\Delta T = 6^{\circ}C$). On the other hand, an increase in inlet air temperature broadens the allowed air mass flowrate ranges for 2-row evaporators; in the case of 3-row evaporators the air mass flowrate range is shifted to a lower flowrate range and may reach a point where no further increase in air temperature is allowed.
4) An increase in inlet air relative humidity causes a small reduction in the delivered power and a shift of the allowed air mass flowrates to regions of lower flowrates. Fig. 4-a shows that a 3-row evaporator designed to work with an inlet air temperature of 31°C and 62% relative humidity does not work properly when the relative humidity is increased to 82%.
5) The highy non-linear curves associated with allowed flowrate ranges precludes the utilization of polynomial statistical fit. It should be mentioned that the number of independent parameters is 8. The outlet refrigerant temperature is limited by its evaporation temperature and inlet air temperature. The refrigerant temperature range is reduced by an increase in air temperature and is shifted to lower values for 3-row evaporators. On the other hand, this range is broadened with an increase in inlet air temperature, in the case of 2-row evaporators.
Air relative humidity changes affect sensibly the R12 outlet temperature.

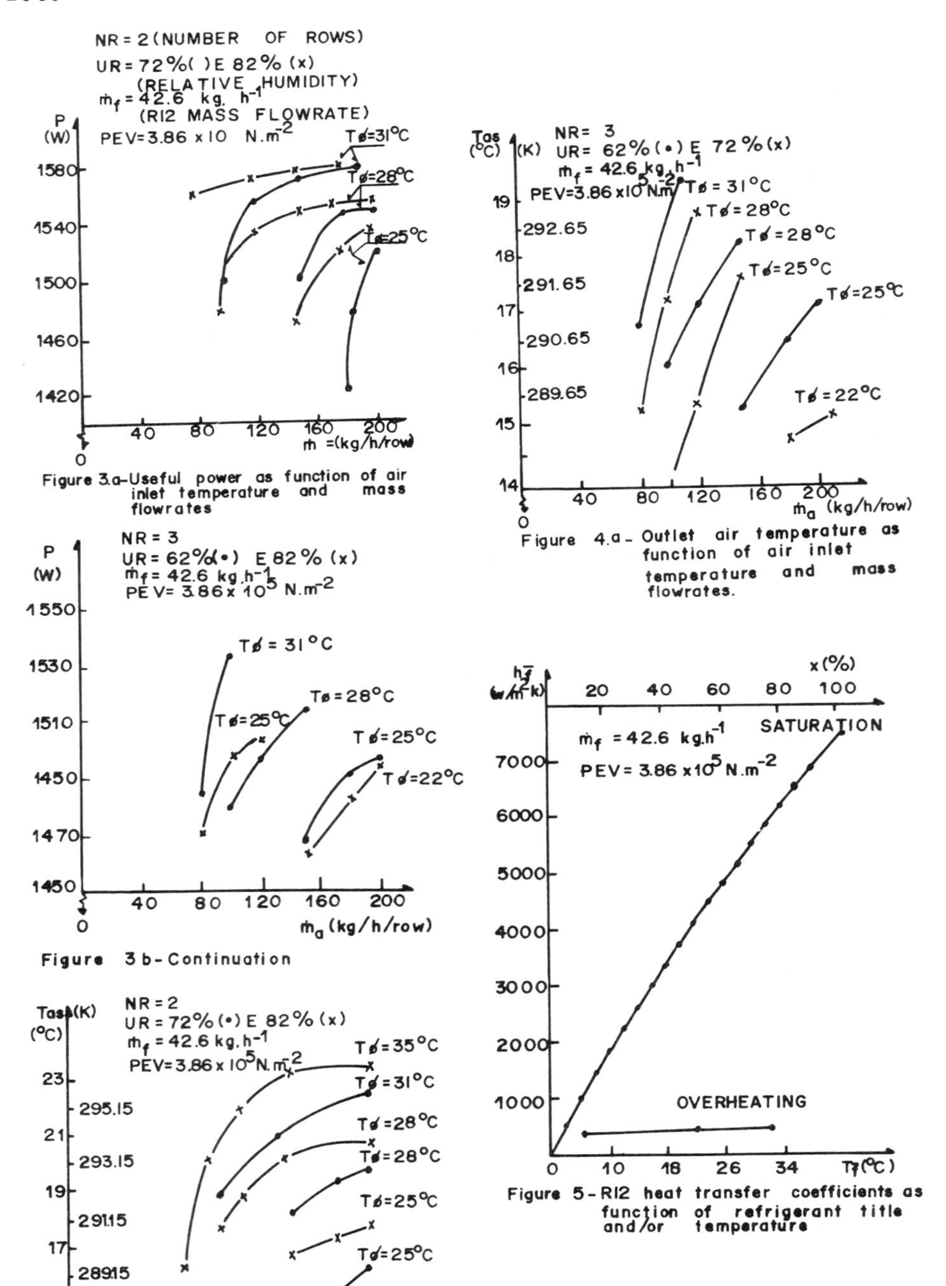

Figure 3.a-Useful power as function of air inlet temperature and mass flowrates

Figure 4.a - Outlet air temperature as function of air inlet temperature and mass flowrates.

Figure 3 b - Continuation

Figure 5 - R12 heat transfer coefficients as function of refrigerant title and/or temperature

Figure 4.b - Continuation

Comments are analogous to the ones just made. R12 phase change heat transfer coefficient is highly dependent on quality (x) and at x = 0,5 (50%) it is ten times higher than the overheated R12 heat transfer coefficient (420 $W.M^{-2}.K^{-1}$), fig. 5. These values reflect the simulated result that evaporation occurs in about 75% of the evaporator lenght.
Air heat transfer coefficient is practically independent of temperature (under environmental conditions and heat pump applications) and shows a mild dependence on relative humidity, fig. 5. Air mass flowrate (and, thus, Reynolds number) is the parameter which affects heavily the heat transfer coefficient.

CONCLUSIONS

A physical-mathematical modeling analysis of air-refrigerant evaporators allows to conclude that:

1) Inlet evaporator conditions (refrigerant mass flowrate, quality and evaporation pressure and air mass flowrate, relative humidity and dry bulb temperature) and the desired degree of R12 overheating affect, in a non-linear way, the evaporator performance;
2) Response curves to changing inlet conditions depends heavily on the number of rows;
3) Results were presented for 2 and 3-row evaporators; these results show that the delivered power is not affected sensibly by the inlet air relative humidity. However, small changes in inlet air temperature may cause a change in evaporator operating conditions;
4) The existing non-linearity among inlet and outlet parameters and the existence of allowed operating ranges suggest that a rational design can be accomplished solely by the solution of the balance equations, associated to thermodynamic and transport property equations. Polynomial fits would become extremely complex to describe the involved physical phenomena;
5) An extremely important result is the assessment of the high sensitivity of evaporators performance to small variations in inlet air conditions. A controller is required;
6) The presented modeling can be extended to the analysis of any air-refrigerant evaporators.

REFERENCES

ORDONES, S.M. Modelagem com discretização espacial e análise de condensadores e evaporadores a serem utilizados em bombas de calor ar-ar. M. Sc. Thesis, UFMG, 1986, Brazil.

SAUER, E.J.; HOWELL, R.H. Heat pump systems. New York, John Wiley And Sons Inc., 1983.

SHAH, M. M. Heat transfer during film condensation in tubes and annuli: a review of the literature, ASHRAE Transaction, 87(1): 1086-1104, 1981.

WILHELM, L. R. Numerical calculation of psychrometric properties, Anual meeting of ASAE, Paper nº 75-4019, June, 22-25, 1975.

MATHEMATICAL MODELING WITH SPATIAL DISCRETIZATION OF AIR-REFRIGERANT CONDENSERS

S.M. Ordones*, M. Fortes*, R.N.N. Koury*, and A. Cordier**

*Thermal Engineering Department, Federal University of Minas Gerais.
Av. Antônio Carlos, 6.627 - Campus UFMG - CEP. 31.270
Belo Horizonte - Minas Gerais - Brazil

**Laboratory of Solar Energy, Paul Sabatier University, 118
Route de Narbonne, Toulouse, Cedex, France

ABSTRACT

A methodology for computer-assisted analysis of air-refrigerant condensers is presented. The developed physical-mathematical model comprises spatially discretized energy balances of finned-tubes bank condensers. The results, presented in several graphs, show the existing non-linearity between inlet parameters (air temperature and mass flowrate and condensation pressure) and outlet parameters (air and refrigerant temperatures, released thermal power, and the number of rows and tubes per row required for a given subcooling). The effect of utilizing innacurate expressions for the condensation heat transfer coefficient is discussed. The presented modeling and analysis should be useful in design and optimization studies of heat pumps.

KEYWORDS

Mathematical modeling; air-refrigerant condensers; heat pumps; heat exchangers.

INTRODUCTION

Design and modeling equations for shell and tube air-water heat exchangers are found in Almeida (1982) and for finned air-R12 evaporators and condensers, in James (1976). Correlations for heat transfer parameters and friction factors in finned tubes were presented by McQuiston (1981). ASHRAE Standards for testing air-air serpentine coil heat exchangers, with dehumidification, are given by Mueller and Co-Workers (1981). Other types of heat exchangers, such as rotative, open cycle and coil loop run-around exchangers analysed by Sauer (1981). Kays and London (1958) present equations and graphs to be used in the evaluation of crossed-flow finned heat exchangers efficiency. In the reviewed literature, no information was available on a detailed and discretized modeling of finned tube heat exchanger banks. This is the objective of this work.

MATHEMATICAL MODELING

A plane serpentine finned condenser was modelled. Heat losses and pressure drops were neglected (Shah, 1981) and the serpentine was taken to be a

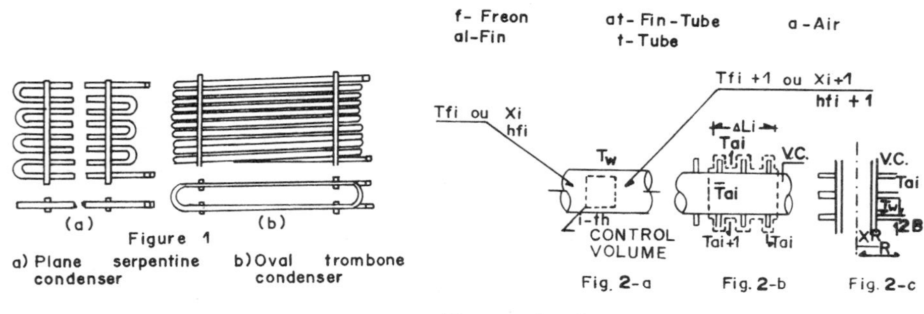

Figure 1
a) Plane serpentine condenser b) Oval trombone condenser

Figure 2 - Control volumes utilized in energy and mass balances

straight tube, fig. 1. Air passing in one row was assumed not to interact with air in the other rows; air flowrate was taken as constant in all tube rows.
Air and R12 thermodynamic properties expressions are presented by Ordones (1986). Air exit temperature was considered to be the average of the air temperatures at all condenser rows.
The physical-mathematical modeling which included energy and mass balances, led to the equations presented in table 1. Fig. 2 shows the control volumes utilized.

SOLUTION METHOD

The solution of the energy equations led to the full R12 temperature and quality profiles and air temperatures and moisture ratios, for any given air and R12 inlet conditions. The equations are non-linear and their solution by the Newton-Raphson method required a careful handling of the iterative processes in the control volumes. Process modeling in even rows required an iterative attribution of values for the R12 properties at the inlet of the first control volume of the row under consideration; this iterative procedure ceased when the outlet condition of this row was the same as the outlet condition of the previous (upper) row. When a phase change occurred in the even rows, it was necessary to determine the control volume where this phase change occurred, in each iterative step. Since in each iterative step, the R12 condition were changed, the exit air temperature at the control volume where the phase change occurred and the row exit temperature had to be changed.

RESULTS AND DISCUSSION

Air relative humidity at the condenser inlet does not affect significantly the exit air temperature and the useful power delivered by the condenser. Fig. 3 shows the curves of the exit air temperature and useful delivered power as functions of air mass flowrate and inlet temperature. The condenser under study, consists of 4 horizontal rows with 4 tubes in each row.
The simulation results show that the useful power increases non-linearly with the air mass flowrate, for a fixed inlet air temperature and decreases with increasing inlet air temperature, for a fixed air mass flowrate. Exit air temperature decreases with air mass flowrate for a given inlet air temperature.
As there are many parameters which affect the performance of the condenser, it is not feasible to establish a polynomial which may substitute the set of modeling equations.
At low air mass flowrates, an increase in environmental air temperature may

TABLE 1 - Energy balance equations

Condenser region where there is:	Expression	Comments	Figure
Overheated R12	$T_f = T_w + A(h_{fi} - H_{fi+1})$	$A = \dot{m}_f/(h_{ci} - \pi d_1 \Delta Li)$; $T_f = (T_{fi} + T_{fi+1})$	2.a
Saturated R12	$TSAT = T_w + G(X_i - X_{i+1})$	$G = \dot{m}_f h_{ev}/(h_{mf} - \pi d_1 \Delta Li)$	2.a
Subcooled R12	$T_f = T_w + C(h_{fi} - h_{fi+1})$	$C = \dot{m}_f/(h_r - \pi d_1 \Delta Li)$	2.a
Fin	$T_{al} = \frac{T_a + 2X(T_w - T_a)}{N(1 - X^2)} \cdot AA$ $n_{al} = \frac{2X}{N(1 - X^2)} \cdot AA$	$T_a = (T_a + T_{a+1})/2$; $N = h_a R^2/(BK)$ $AA = \left[\frac{I_1(N) K_1(XN) - I_1(XN) K_1(N)}{K_1(N) I_0(XN) + K_0(XN) I_1(N)}\right]$ I and K are 1st and 2nd order modified Bessel Functions	2.c
Air	$\bar{T}_a = T_{at} + \frac{\dot{m}_{as}}{h - A\phi}$ $Cpa(Tai - Tai+1) + W(h_{wi} - h_{wi+1})$	$Ao = 2\pi R(X\Delta Li + NA\ R(1 - X^2))$ In which NA is the number of fins per control volume $\dot{m}_{as} = \dot{m}_a/(1 + w_c)$ $T_{at} = \frac{\dot{m}_{al} C_{al} \bar{T}_{al}\quad m_t\quad C_t\quad T_w}{m_{al} C_{al} + m_t C}$	2.b
Air-Overheated R12	$TSAT = T_w + D(\bar{T}_{at} - \bar{T}_a)$	$D = h_a - A\phi/(h_{ci} - \pi d \Delta Li)$	2.b
Air-Saturated R12	$TSAT = T_w + E(\bar{T}_{at} - \bar{T}_a)$	$E = h_a -/h_{mf}$	2.b
Air-Subcooled R12	$T_f = T_w + F(\bar{T}_{at} - \bar{T}_a)$	$F = h_a -/h_{\bar{r}}$	2.b

cause a partial condensation of the R12 refrigerant. Thus, air mass flowrate is a limiting parameter in condenser design and operation. For the given condenser geometry air mass flowrate cannot be less than 100 kg/h per row. The useful delivered power will remain the same if equal increments occur in the environmental (inlet) air temperature and air mass flowrate (fig.3). Fig. 4, 5 and 6 show the effect of air and R12 mass flowrate on the delivered power and exit air and R12 temperatures,while the other parameters conditions are kept constant.

Summarizing and supplementing, the following comments should be added, with respect to fig. 2 to 5:

1) The curves reflect the non-linearity of the equation which generated them.
2) Unless high order regression polynomials with four or more independent variables are utilized, a condenser should be modeled and analyzed from the conservation equations.
3) The curves are complete in themselves; that is, extrapolations would lead to regions of overheated or excessively subcooled refrigerant and, thus, outside adequate operating conditions.
4) This model is useful to analyze air-refrigerant cross-flow condensers, with any number of rows or control volumes per row.
5) The operating condition of a condenser should be well established and is determined, for a given geometric configuration, not only by the mass flowrat_, but also by the environmental conditions and required air temperature. It is possible that the environmental conditions impose operating restrictions on air mass flowrate in a given day.
6) The optimization of operating conditions will require, in mathematical

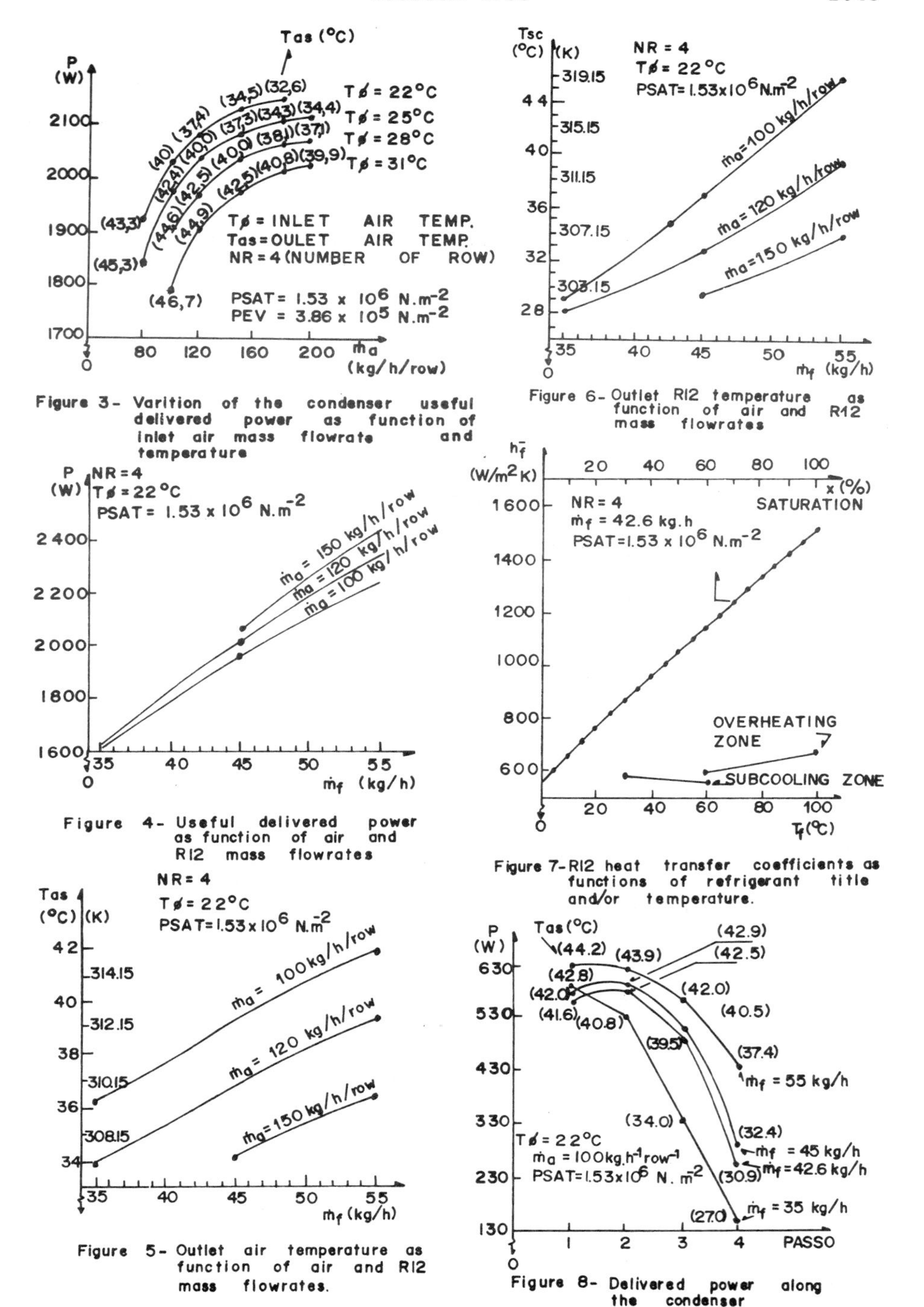

Figure 3 - Varition of the condenser useful delivered power as function of inlet air mass flowrate and temperature

Figure 6 - Outlet R12 temperature as function of air and R12 mass flowrates

Figure 4 - Useful delivered power as function of air and R12 mass flowrates

Figure 7 - R12 heat transfer coefficients as functions of refrigerant title and/or temperature.

Figure 5 - Outlet air temperature as function of air and R12 mass flowrates.

Figure 8 - Delivered power along the condenser

terms, the solution of semi-stationary conservation equations, under climatic restrictions, and taking into account an air flowrate controller.

Selected effects of some physical parameters associated to the balance equations can be appreciated in fig. 7 and 8. The following points should be emphasized:

a) Phase change heat transfer coefficient of the refrigerant changes considerably with quality (550 $W.M^{-2}.K^{-1}$ at a quality of 0% to 1530 $W.M^{-2}.K^{-1}$ at 100%). Heat transfer coefficients of the overheated vapor and compressed liquid are approximately 630 and 570 $W.M^{-2}.K^{-1}$, respectively, for the stated operating conditions. Phase change occurs in 60% of the condenser and this fact shows the importance of the phase change heat transfer coefficient.
b) Air heat transfer coefficient shows a variation inferior to 3% along the condenser.
c) Fig. 7 shows the delivered power and air temperature variation in each condenser row. The intersection of the curves and its concavity are amenable to a simple physical reasoning: the global heat transfer coefficient in the second row can be higher than in the first one, due to the presence of saturated R12. This figure also shows the effect of an R12 mass flowrate variation on the delivered power and outlet air temperature.
d) Knowledge of the model sensitivity to poor estimates of the phase change heat transfer coefficient is essential to the engineer. By taking that the R12 mass flowrate to be 42,6 kg/h (fig. 7), the simulation results showed that a variation of 30% in the phase change heat transfer coefficient causes a 5% variation in the delivered power and 2% change in the outlet air temperature.

CONCLUSIONS

The physical-mathematical modeling analysis of the air-R12 condensers allowed to infer the following conclusions, stated briefly:

1) Air mass flowrate and inlet temperature and refrigerant mass flowrate, saturation pressure and inlet temperature affect the condenser performance in a non linear way. The effect of the air relative humidity is negligible.
2) Changes of +30°C in inlet air temperature lead to aproximately +4% variation in the delivered power and +2,5°C in the outlet air temperature, if the other operating conditions are kept constant.
3) The number of rows and tubes per row are parameters of main importance in the analysis of condensers. If an analysis is made for a certain geometric configuration, it may not be useful when extrapolated. The presented results refer to condensers with 4 rows and 4 tubes per row.
4) A variation of 20% in R12 mass flowrate may cause a 6% change in the delivered power and outlet air temperature.
5) Overheated and undercooled R12 heat transfer coefficients are practically constant and are worth 630 $W.M^{-2}.K^{-1}$ and 550 $W.M^{-2}.K^{-1}$ respectively. On the other hand, the phase change heat transfer coefficient ranges from 550 to 1540 $W.M^{-2}.K^{-1}$. A 30% error in the evaluation of the phase change heat transfer coefficient may cause a 5% evaluation error of the delivered power and 2% of the outlet air temperature.
6) Environmental conditions should be taken into consideration.
7) The presented model can be utilized in the analysis of any air-refrigerant, plane serpentine condensers and in optimization studies.

REFERENCES

ALMEIDA, M.S.V. Modelamento Matemático de um sistema de refrigeração por compressão em regime permanente e transitório, com análise comparativa de 2 tipos de controle de temperatura. M.S. Thesis, UNICAMP, 1982, Brazil.

JAMES, R.W and Co-Workers. The heat pump as a mean of utilising low grade heat energy. Building Services Engineer, 43: 202-206, 1976.

KAYS, W.M.; LONDON, A.L. Compact heat exchangers. California, McGraw-Hill Book Company, Inc., 1958.

MCQUISTON, F.C. Finned tube heat exchangers: state of the art for the air side. ASHRAE Transactions, 87(1): 1077-1085, 1981.

MUELLER, M.A. and Co-Workers. Development of a standard test facility for evaluation of all types of air-to-air energy recovery systems. ASHRAE Transactions, 87(1): 183-198, 1981.

ORDONES, S.M. Modelagem com discretização espacial e análise de condensadores e evaporadores a serem utilizados em bombas de calor. M. S. Thesis. UFMG, 1986, Brazil.

SAUER, H.J.; HOWELL, R.H. Promise and potential of air-to-air energy recovery systems. ASHRAE Transactions, 87(1): 167-182, 1981.

SHAH, M.M. Heat transfer during film condensation in tubes and annuli: a review of the literature. ASHRAE Transactions, 87(1): 1086-1104, 1981.

HEATPUMPS IN CHINA

Wu Zhi-jian Li Zong-zhe He Rong-zhi Lin Yi-cheng

Guangzhou Institute of Energy Conversion
Chinese Academy of Sciences
P.O.Box 1254 Guangzhou China

ABSTRACT

Heat pumps development in China is still in an early stage. Number of heat pumps installed is not more that 5000 units,mainly for air conditioning, hot water supply, drying, evaporation, concentration and heat recovery. Although heat pumps have an enormous potential domestic market, owning to the shortage of electricity suply at present and the irrational fuel-to-electricity price ratio (relatively high electricity price), heat pump application would be "saving in energy but not in money" thus not so attractive for enterprises. Chinese refrigeration industry has had a relatively solid basis and this, plus the importance attached by the government to energy conservation, will enable heat pumps to have a rapid development provided there is a encouraging policy in question.

KEYWORDS

Heat pump development; Energy conservation; Electricity shortage; Market potential; Compression and Absorption heat pumps

INTRODUCTION

Heat pumps was begun in application research early in 1950s. However heat pump development had been very slow because of the weak industrial basis at that time and irrational fuel (coal)-to-electricity price ratio. Since late 1970s, owing to the rapid economic development and the resulting improved living standard accompanied with the "open to the world" policy, energy-consuming products such as air-conditioners and electric heaters are used in large amount. This has stimulated the Chinese refrigeration-air conditioning industry to develop in higher speed and provided a favourable environment for heat pump development.

Furthermore, since energy production can not meet the need by the rapid developing economy and thus energy pinching is quite apparent totay in China, energy conservation is attached high importance. These have provided favourable conditions for heat pump development. Nevertheless, so far application of heat pumps is still very limited in number due to the electricity shortage, the high electricity rate and insufficient propaganda of significance of heat pump in energy saving and the unfamiliarity of heat pump to the public. Only about 4000 units of air conditioning heat pump (including imported units) are in operation, plus less than 100 units of other types.

STATUS QUO

At present about 30 factories manufacture air-conditioners and refrigeration equipment in China, among which about 10 manufacture heat pumps at a small annual production (Fig.1) but very steep tendency of development. The heat pumps manufactured were mostly compression type used in air-conditioning, drying, evaporation & concentration, hot water supply, and waste heat recorvery. Small air-conditioning types constitute the majority. Consumers were hotels, hospitals, department stores, laboratories, offices, exhibition centers, etc. As for household heat pumps, owing to the electricity shortage, the high electricity rate and the high capital cost, heat pumps can not be expected to enter common families in large quantity in the near future as refrigerators did.

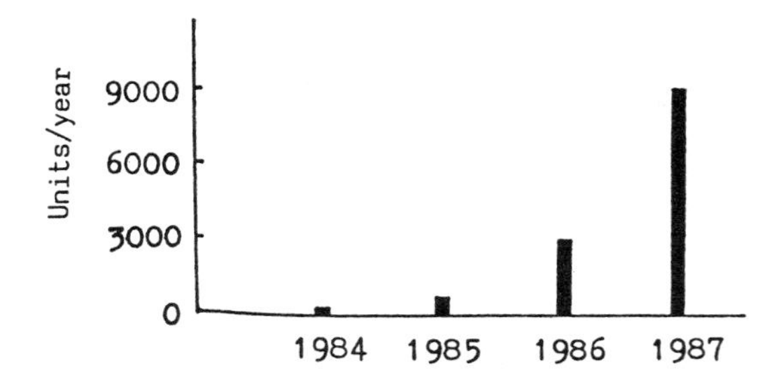

Fig. 1 Number of Heat Pumps Manufactured Annually

Thanks to the policies pursuited by the Chinese government to reenforce energy conservation and encourage use of advancd energy saving equipment, plus the lower electricity rate for industry and agriculture (about 1/2 to 1/3 of the non-productive electricity rate), development, application and popularization of heat pumps are under way with appreciable results in some areas of production fields.

An arts and crafts center in Shanghai undertook trial application of production scale in an earlier time (1980) where a compression heat pump installation was and is used for cooling in summer and heating in winter with capacity of one million kcal/hr. The evaporation facility using a compression heat pump, developed by the Peijing atomic energy institute saves 38-41% of energy originally consumed when used on paper pulp concentrating and consumes 90% less of coal when used on concentration of corn milk [2]. In Guangdong and Zhejiang provinces heat pumps are used on drying of tea and wood. In drying of tea the cooling and heating capacities of the heat pump can be fully utilized resulting in a reduction in total cost by about 20% and improved product quality; in drying wood the energy consumption ranges only about 30-40% of the conventional steam drying [4]. The mechanicl vapor compression heat pump developed by Shengyang Chemical Machinery Plant exhibited good waste heat recovery when used in a NaOH concentration column and is planned to be used on concentration of glycerine later this year. Four sets of propylene-propane separation system, using heat pump, for petro-chemical factories are under design and construction. One of them, which is planned to be used at a petrochemical factory in Guangzhou, has a compressor of 1147 kw and is designed to reduce the original energy consumption by about 44% with a two year pay back time. The Guangzhou Institute of Energy Conversion developed a water/water low temperature heat pump which, when combined with solar greenhouse, yielded appreciable results in water heating for swimming pool and aquaculture, with a COP of 4 to 5. The heat pump-solar heating system is shown in Fig.s 2 and 3.

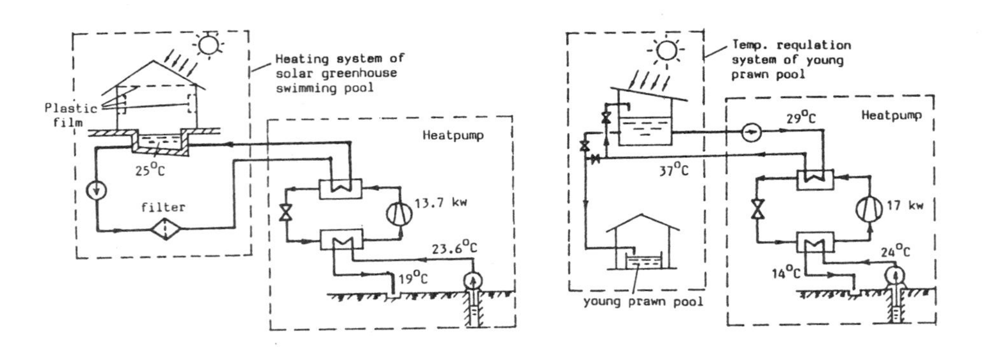

Fig. 2 Heat pump-Solar energy swimming pool

Fig. 3 Heat pump system for prawn breeding

RESERACH AND DEVELOPMENT

In general heat pump application in China is still in the preliminary stage. But heat pump technology is attached importance in the technical circle. Tens of universities, research institutes, designing institutes and manufacturers have proceeded with heat pump development and application research. This R & D includes two aspects, one is to increase COP through the improvement of component parts and the research in heat pump new cycle; the other is the engineering application research for optimization as well as development of equipment. So far heat pumps used in engineering fields are principally compresion types and the rest are ejection type and absorption type. The composition of energy consumed in China is roughly as indicated in the following table.

TABLE 1 Energy Comsumed by Different Fields

Industry	Transportation	Agriculture	Commercial & civil	Others
69.4%	5.7%	4.9%	17.2%	2.8%

It can be seen from the above table that industry, commerce and civil application consumed major portion of the total energy. Moreover energy efficiency appears quite low in China (national average 30%). Therefore heat pumps will find their use in the fields of high energy consumption, and priority will be given to application where ample waste heat is present and where both heating and cooling are required. Besides consideration should be given to the replacing with heat pumps those direct electric heaters to which required temperature is not high. Recently industrial applications of absorption heat pumps have received much importance on account of the electricity shortage in China, the high electricity rate and the low coal price. At present about 10 units have undertaken researches in this respect.

MARKET POTENTIAL

Heat pumps are in urgent need in China where energy pinching has been perceived during economic construction and energy efficiency apears not so high. China has enormous industrial and agricultural systems which require improvement in energy efficiencty,and millions of family as well as the third trade which require space heating and hot water supply. These constitute an enormous potential market for heat pumps. Heat pumps will also find their market in waste heat recovery and low temperature heat explotation in chemical, textile, paper making, medicine industries, etc. Although in the above fields only some individual factories have practically used heat pumps and some are still planning or undertaking design and installation, industrial application of heat pump will have a breakthrough once the energy saving benefit of heat pumps is proved through actual operation demonstration.

In consideration of the climate conditions in China, the vast areas along Yellow river and Changjiang river are the most promising areas for heat pump heating. In these areas, winter is long but not seriously cold, and the days of winter and summer added together are quite a number, totalling about 8 months. This will ensure a very high equipment service rate when supplying heat in winter and cooling in summer. In south China, heat pump development will be centered on air-conditioning plus hot water making because of the year long high ambient temperature.

OBSTRUCTION

For heat pump application in China, although of great potential, there would be many difficulties. Firstly, there is at present not yet favoured policy for popularization of energy saving heat pump installations although Chinese government regards energy conservation important. Secondly, development of electric industry in China can not catch up with the development of industry and agriculture and the need by the increasing living standard. In consideration that electricity shortage of 15-30% is often seen in Chinese cities, electricity utilities have taken measures to restrict use of electricity. Thirdly, because of high price of electricity and low price of coal, use of heat pumps would appear "saving in energy but not in money", which will not be attractive for enterprises.

CONCLUSION

Energy pinching, government's encouragement of energy saving, plus a relatively solid basis of refrigeration industry have provided favourable conditions for development of heat pump industry.

Market potential of heat pump in China is enormous but heat pump development had been slow owing to the electric shortage, high electricity price, high capital cost and the want of awarding policy for heat pump energy saving installations. However, onece the energy saving benefits of heat pumps are recognized by the social public, they will enjoy rapid development. Considering that Chinese refrigeration industry which is based on compression type has already a relatively solid basis, judging from the technological establishedness, heat pump development in China will has the priority on compression type. Besides, absorption types will also enjoy a relatively rapid growth when electricity shortage is considered.

REFERENCE

1. "National Reports on the Status of Heat Pumps", 13th Congress of the world Energy Conference, Cannes, 5/11 October 1986
2. Li Cheng-chun, Han Wei: "Heat pump evaporation-efficient energy-saving technology" Science and technology Literature publisher, Jan 1986 (in Chinese)
3. Pang He-ding etal:"Efficient energy-saving heatpump technology" Atomic Energy Publisher May 1985 (in Chinese)
4. Lin Yi-cheng, He Rong-zhi, Li Song-zhe : "A Survey on Energy Saving Technologies for low Temperature WAste Heat Utilization" Guangzhou Institute of Energy Conversion, Chinese Academy of Science. April 1987 (in Chinese)

3 MW OUTDOOR AIR HEAT PUMP EVALUATION

V Olving

Department of Building Services Engineering
Chalmers University of Technology
S-412 96 Gothenburg
Sweden

ABSTRACT

This paper reports on results and experiences obtained from two years of operation and evaluation of a 3,2 MW outdoor air heat pump plant.

Originally the heat pump was installed on purely economical grounds in order to decrease the amount of oil used in an existing oilfired boiler plant.

At that time, in 1984, there where few, if any, outdoor air heat pumps of this size. Thus, it was decided that a comprehensive evaluation programme should be carried out. The studies include a continuous advanced measurement programme, a follow up of the practical operation and a technical and economical evaluation of the plant as a whole.

An obvious risk with an outdoor air heat pump of this size is that its noiselevel will be unacceptable. In this case this is no problem. On the contrary, the plant has been almost noiseless and the cost for obtaining this has been negligible.

The saving of oil has been about 1200 m^3/year. Yet this is not enough to give a fully acceptable plant economy due to the present low oil price. Gradually the availability of the plant has increased to about 70 % and can still be improved quite easily.

KEYWORDS

Outdoor air; large heat pump; bivalent heating plant; oil saving; energy saving; operation experiences; evaluation.

INTRODUCTION

Outdoor air has for many years, been used as the heat source for heat pumps in Sweden. In 1984 a very large outdoor air heat pump was connected to an existing central oil boiler plant in Kungälv just north of Gothenburg. The heating plant provides space heating and domestic hot water, via a district heating network to some 2000 apartments in multifamily houses, schools and shopping precincts.

The existing heat load was estimated at some 34 TWh/year with a winter peak load around 12 MW. In Fig. 1, the flow diagram is shown.

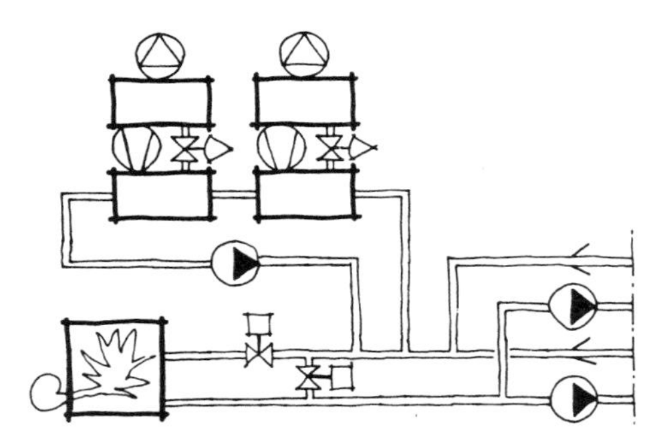

Fig. 1. Flow diagram

The heat pump was dimensioned to meet the entire load down to +8 degrees C outdoor air temperature which required a heat output in the order of 3,2 MW, which would cover nearly 40 % of the entire heating load, that is to say 13 TWh/year. The seasonal performance factor was estimated to be 2,6.

The heat pump building has two floors. The heat pump consists of two separate heat pump systems. The evaporators (five per each system) are situated on the upper floor.

The heat pumps are assembled on site and the systems are filled with 12 tons R12.

The heat pump compressors are of the screw type. Consequently the oil system is more complex than in other compressor types. The heat pump has also got a new type of defrosting system.

To increase the running time of the heat pump in the summer, when only domestic hot water is needed, accumulators were installed in every consumer service unit.

To keep the oil warm, also during the summer when the oil boilers do not operate, an electric boiler was installed in the central boiler plant.

THE MEASURED RESULTS

Figure 2 shows that the measured total energy is smaller than predicted and that the energy produced from the heat pump is also smaller than predicted, in spite of the fact that the measured years were colder than a normal year.

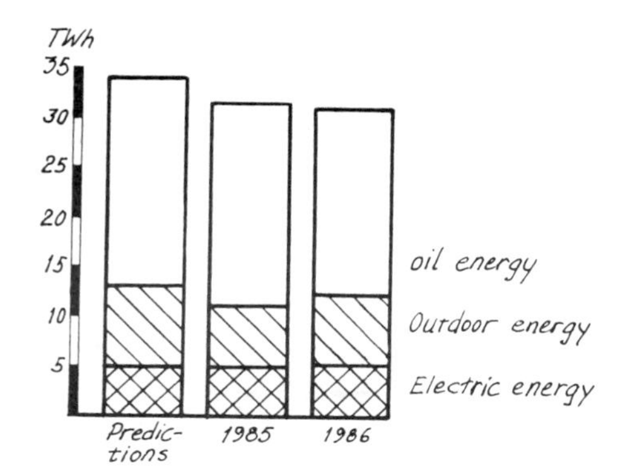

Fig. 2. The measured energy division in comparison with the predictions.

Figure 3 shows two different SPF:s from the measured values. The largest value is the result of SPF when only electrical energy for the heat pump is considered. The other value expresses all additional electric energy related to the heat pump installation, that is to say, energy for air conditioning, warming the heat pump building in wintertime and warming the oil in summertime.

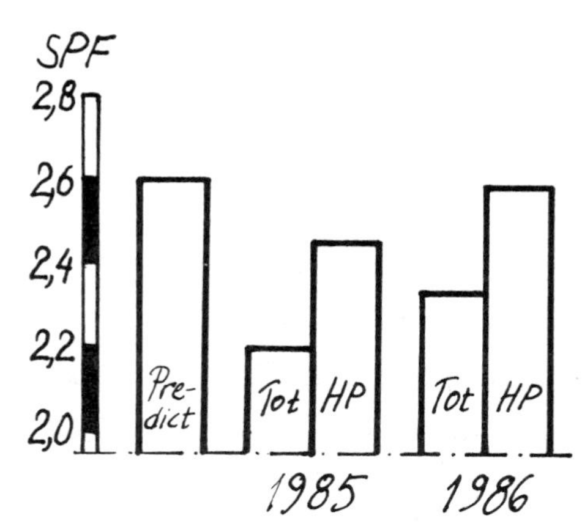

Fig. 3. The SPF:s for the two years in comparison with predictions.

The following diagrams, Fig. 4, show that the peak load is lower than the predicted 12 MW. One reason for this is that the hot water flow was too small and because of this the heat plant could not produce the maximum effect during the coldest days. The diagrams also show that the heat pump did not produce as much energy as it could, if the production had been trouble-free.

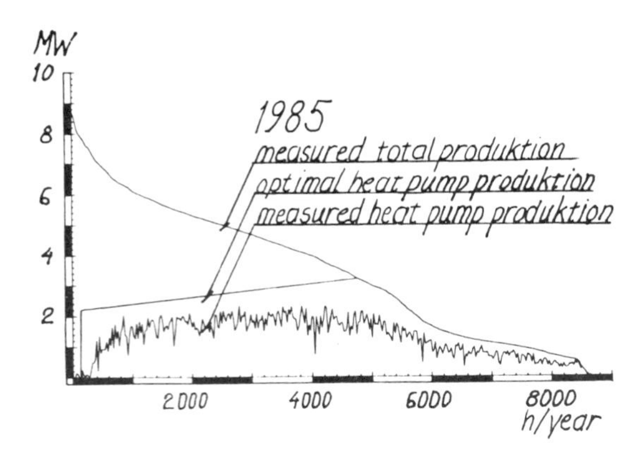

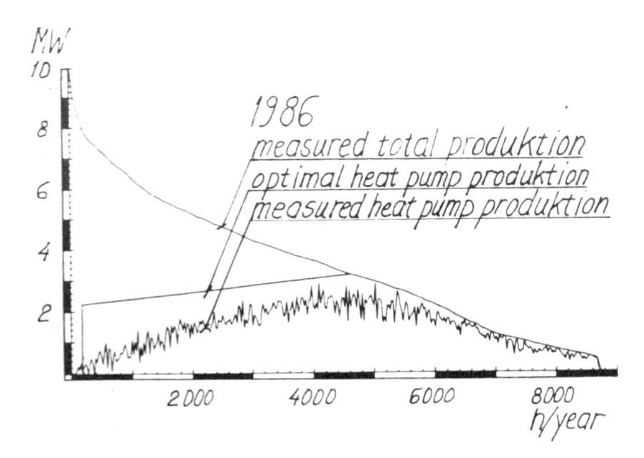

Fig. 4. The duration diagrams for 1985 and 1986.

The reasons for production trouble are:

- The new type of defrosting system.
- Contamination in the piping has caused some shut-downs.
- Problems with leakage of refrigerant. The leakage throughout the two first years, was about 3000 tons.
- Trouble with keeping the oil pressure at the right level has caused some shut-downs as well. The oil system is more complex in this type than in other compressor types.
- The heat pump has also had some problems in delivering energy due to a too high temperature of the water returning to the condensors. Because of complaints from the tenants it was sometimes necessary to raise the outlet hot water temperature. This problem was caused by bad regulation in the consumer service units.

However, the greatest part of the shut-down-time is accounted for by the delay between a breakdown and the repairing of it. Thus, an easy way to increase the utilization time would be through a better heat pump service.

The usual problem with loud noise from the evaporator fans has been reduced very much, with a very small investment only.

ECONOMY

Assuming a heat pump effect of 3,2 MW will give a total cost for the heat pump installation of 5300 SEK/kW heat output. (1 SEK = 0,16 US $).

The following Fig. 5 shows the production costs in relation to the utilization time of the heat pump.

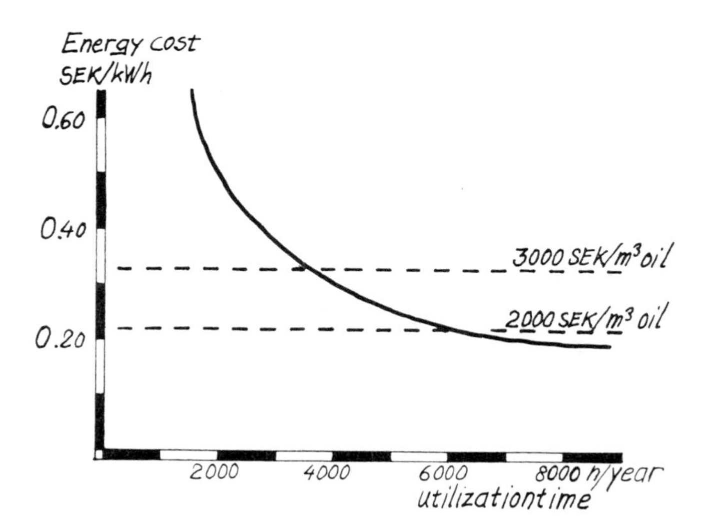

Fig. 5. The operating costs as a function of utilization time.

The calculations were made with an assumed lifetime of 15 years and at a 6 % rate of interest and with actual operation and maintainance costs.

The diagram shows that:

- The energy cost for the heat pump was 0,33 SEK/kWh in 1985, equivalent to 3000 SEK/m^3 oil for operating the boilers instead of the heat pump.

- In 1986 the energy cost was 0,30 SEK/kWh because of a longer utilization time, that is equvivalent to 2700 SEK/m^3 oil.

- If the heat pump could operate without any disruptions at all the utilization time could be 5500 h/year and the total operating cost could be reduced to 0,24 SEK/kWh, corresponding to 2200 SEK/m^3 oil.

BLOCK CENTRAL HEAT PUMP FOR THE DELIVERY OF SHORT-DISTANCE HEAT

Hans-Gerhard Rumpf

Rheinisch-Westfälisches Elektrizitätswerk (RWE), Head Office
4300 Essen 1, Federal Republic of Germany

ABSTRACT

The administrative building of the Wesel rural distric (completed in 1984) is supplied with heating and cooling by the local electricity utility RWE. Generally all technical equipment has been planned and installed by RWE up to the interface with the distribution grids for warm water and in this example also for cold water. The heating and cooling centre in the basement of the building is owned by RWE. The heat price is a market price. The economic risks of the capital-intensive energy saving technology are borne by the short-distance heat supplier. The heat supply contract between both parties is valid for about ten years. During that the RWE also takes care of the technical servicing for which there is no adequately qualified staff available, particularly in buildings of smaller local authorities and block of flats.
The operating results of the first three heating seasons have shown a heat pump share of 85 % in the aforementioned annual heating energy as well as a coefficient of performance of 3.1 for the heat pump system.

KEYWORDS

Groundwater heat pump; dual-mode heating system; energy saving technology; short-distance heat supplier.

TECHNICAL DESCRIPTION OF THE DUAL-MODE HEATING SYSTEM

The block central heat pump in the basement of the administrative building consists of three groundwater heat pumps of equal capacity with altogether 840 kW of heating output. A heat pump unit has been equipped with a plate heat exchanger to meet the cooling regirements in summer, so that the priority operating mode of this unit can be changed over hydraulically from heating to cooling, depending on the requirements. The heat optained from the cooling mode is largely used for heating purposes. The maximum cooling output is 320 kW. The heat source system consists of a water-producing well with three submerged pumps, each of which delivers 40 m³ of water per hour from a depth of approximately 25 m, about 400 m long cold-water pipes and four seepage wells which ooze the cool water away via a distribution shaft. On top of the heat pump system, two heating boilers with a total heating output of 2,200 kW have been installed. the object of the boiler system is to meet peak demand in parallel with the heat pump system and, moreover, to take over the heat supply of the building at very low outdoor temperatures, e. g. from -5 °C.

MONTHLY HEAT SUPPLIES OF THE HEAT PUMP AND BOILER SYSTEM FOR THE 1985/86 HEATING SEASON

The bar chart provides a comprehensive overview of the monnthly heat supplies to the district townhall. The heat supplies delivered by the heat pumps are made up of the electrical energy required for the compressors and the auxiliary drives such as groundwater and condenser booster pumps as well as of the ambient heat extracted from the groundwater.
The total amounts of energy optained for this period were the following:

593.27 MWh/a compressor energy
64.73 MWh/a auxiliary energy
1,429.60 MWh/a energy extracted from the groundwater
corresponding altogether to
2,087.60 MWh/a heat share of heat pump system
211.30 MWh/a heat share of oil-fired boiler system
2,298.90 MWh/a total amount of heat.

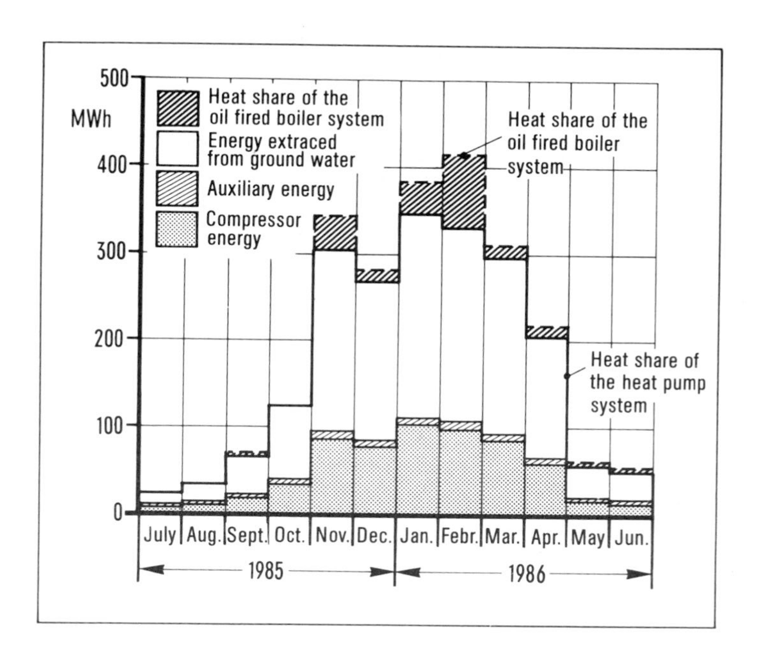

Fig. 1 Monthly heat supplies

DESCRIPTION OF THE ENERGY FLOWS OF THE DUAL-MODE BLOCK CENTRAL HEAT PUMP IN THE WESEL TOWNHALL (HEATING AND COOLING)

101 parts of primary energy, thereof 14 parts of imported crude oil and 87 parts of domestic coal, are processed, transported or converted into electricity.
The remaining 42 parts of final energy from the conversion chain are broken down into 29 parts of electricity and 13 parts of fuel oil.
For the 100 parts of useful heat of the building, 22 parts of the electricity path and 9 parts of the fuel oil path as well as 61 parts of ambient heat and 1 part of reject heat from the basement are required. 2 parts of the aforementioned useful heat are accounted for by the hot water production of the kitchen.
The cooling with 2 useful parts is ensured by 1 part of electricity for the operation of the cooling unit.

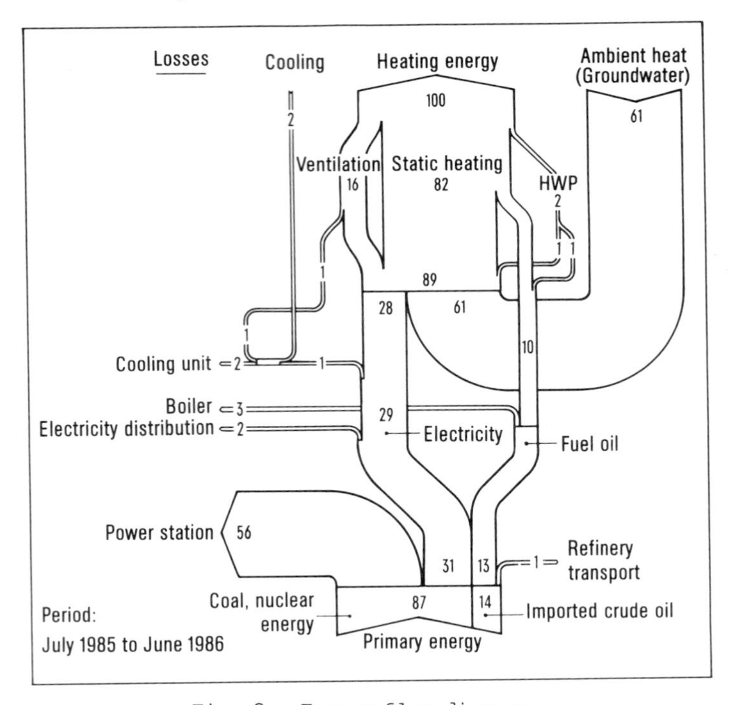

Fig. 2 Energy flow diagram

CONCLUSION

The operating results of three heating seasons have shown that the annual heat requirements of the administrative building total on average 2,350 MWh/a. If these consumption values are compared with a purely oil-fired heating system with an annual utilisation rate of 78 % for the boiler system, approximately 300,000 l of fuel oil would have been burnt per year. The actual consumption totals on average 48,000 l/a; this corresponds to an

oil substitution of 252,000 l/a. This means furthermore that through the use of 610 MWh/a of electrictiy 1,388 MWh/a of ambient energy - groundwater - are harnessed.
The decision of the district administration to also connect the new police building to the energy concept rounds off the positive result.

HEATING AN OUTDOOR SWIMMING POOL WITH A RIVER WATER HEAT PUMP

H. Jänich

HASTRA, Humboldt Straße 33, 3000 Hannover 1
Federal Republic of Germany

ABSTRACT

An outdoor swimming pool is heated from river water. The system centerpiece is a large heat pump. Additional compact heat pumps provide heat for changing rooms and hot water. The installation has proved its reliability and economy over many years of service.

KEYWORDS

Heat pump - river water extraction; compact heat pumps; room heating; operating results.

INTRODUCTION

A public outdoor swimming pool was built in Wendeburg, district of Peine, Lower Saxony, and designed to cater for recreation and sport. The catchment area of this town, which has about 3,000 inhabitants, was too small in itself for a modern outdoor swimming pool. However, it was thought -- provided the swimming pool could be made attractive enough -- that it would be possible to include part of the city of Braunschweig into the planning as a potential catchment area.

Based on these premises, the decision was taken to build an outdoor swimming pool suitable for sports and equipped with a diving platform, a warm-up hall and heated changing and recreation rooms. The combined pools have a total surface area of 1,220 m^2 (1,459 square yards). For these reasons the pool water temperature was fixed at the unusually high temperature of 28°C (82°F).

HEATING SYSTEM

Heating demands

The heating system has to fulfill the following demands:
1. Heating the pool water up to 28°C (82°F),
2. dehumidifying the air in the warm-up hall, and
3. supplying the heat for the heating of changing and recreation rooms.

Fig. 1 Outdoor swimming pool Wendeburg

Meeting the heating demands

A heat pump which obtains heat from river water was installed to provide heating for the pool. Its heating capacity is 765 kW when the temperature of the river water is 12°C (53°F) and the temperature of the heating water is 30°C (86°F).

An inlet gate was built to draw water from the river. It is provided with screens and bar screens; the river water pump is also installed here. A second construction, about 50 m (54 yards) downstream from the inlet, is for the return flow of the water, which is cooled down by about 5 K in the heat pump.

The heat pump was delivered as a ready fabricated unit. Two 70 kW, electrically driven piston compressors ensure the compression of the refrigerant.

On site only the water and electricity supplies had to be connected. The installation of a refrigeration plant was not necessary.

The refrigerating capacity (evaporative capacity) is 640 kW when the flow rate of river water is 110 m^3/h (143.88 cubic yards/h); the compressor power amounts to 600 kW when the flow rate of pool water is 165 m^3/h (215.82 cubic yards/h).

In the warm-up hall the pool surface area of 30 m^2 (36 square yards) gives off water vapor continuously to the air. In order to keep the relative humidity at an agreeable level and to maintain the temperature of 30°C (86°F), a ventilation system with heat recovery was installed.

Fig. 2 Heat pump of the outdoor swimming pool in Wendeburg. Source of heat: river water.

In the ventilation unit an air-water heat pump with a refrigerating capacity of 9.5 kW dehumidifies the hall air. The heat obtained in this manner in the condenser -- condensing capacity: 12 kW -- is passed to the pool water.

Hot-water under-floor heating is installed as heating system in the changing and recreation rooms. Over an area of 445 m^2 (532 square yards), a system of plastic pipes was laid in the still moist concrete. The consumption of heat amounts to around 20 kW. The heater for the floor heating is a water-water heat pump, which has a capacity of 46 kW at a flow temperature of 40°C (104°F), a reflux temperature of 36°C (96°F) and a temperature of the heat source of 28°C (82°F). The river water was not taken as a source of heat since inlet and outlet pipes would have been extremely long and an additional river water pump would have been necessary. This heating system also supplies the heat necessary for heating the hall air.

OPERATING RESULTS

During ten years of operation the system has proved its reliability. The desired temperatures have been maintained at all times. The system has provided faultless mechanical functioning throughout. Energy costs have always remained at the predicted level. Environmental energy has been used to a satisfactory extent. Owing to the heat pump, the high pool water temperature of 28°C (82°F), which was demanded for the system, can be maintained at very low energy costs.

FUNDAMENTALS OF SIZING HEAT PUMPS FOR AIR DEHUMIDIFICATION
- EVAPORATION RATE FROM INDOOR WATER SURFACES -

K. Biasin*, F.D. Heidt**

*Rheinisch-Westfälisches Elektrizitätswerk AG
RWE-Anwendungstechnik, 4300 Essen 1

**Universität Siegen, Fachbereich Physik
5900 Siegen

ABSTRACT

Measurements of the evaporation rate of smooth water surfaces as well as of the air condition and the water surface temperature were made in three indoor swimming pools. The results cover a wide range of values. They can be explained by simple considerations on the basis of the theory of free and forced convection. This results in an estimation formula which calculates the measuring results with sufficient accuracy from the ambient conditions. In general, the estimation formula for evaporation rates yields lower mass flow rates than have so far been taken as a basis for design calculations of heat recovery systems. In future, this permits a better adjustment of the size of heat pumps for air dehumidification. Other applications of the formula are e.g. the determination of the required air exchange rate to keep up an air humidity condition, the examination of the heat requirements due to evaporation or the estimation of the effect of the water and air temperature on the air exchange rate or the heating requirements.

KEYWORDS

Measurements of evaporation rate, smooth water surface, evaporation due to forced and free convection, theoretical modelling, comparison of theory with measurements.

PROBLEM STATEMENT

Smooth water surfaces exist inter alia in houses in the living, sanitary and kitchen area as well as in public and private indoor swimming pools outside opening hours. These surfaces are potential sources of a permanent production of water vapour which can be important in three respects:

(1) The evaporating body cools down.

(2) The relative humidity rises or is kept at a constant maximum value by increased air exchange.

(3) Heat can be recovered by means of condensation from exhaust air.

For the optimum sizing of a heat pump with respect to the volume flow rate and thermal output, e.g. in case (3), the knowledge of the evaporation rate must be as detailed as possible:

If the heat pump has been designed too large, it is not only too expensive but it also works in a part-load range which is unfavourable in terms of energy efficiency. If it is too small, not all the recoverable energy can be harnessed. Hence, there must be an optimum.

A physically founded formula for the evaporation rate is developed by the analysis of many measurements in three indoor swimming pools with a smooth water surface. It is based upon the laws of convection and on the analogy between heat and mass transfer.

MEASUREMENT OF EVAPORATION RATES

The evaporating water quantities as well as the associated variables were measured in three swimming pool halls with smooth water surfaces of 340 m^2, 64 m^2 and 38.5 m^2 (Biasin, 1987). For this purpose a heat pump was used which dehumidifies the pool hall air. The air exchange with the outside air was cut off, and the water vapour condensed by the heat pump in the circulating air mode was measured over a period of 4 to 8 hours. The surface-specific water vapour flow rate $\mathring{m}$ was calculated from these data. This is only permissible if no water vapour condenses on the walls in the swimming pool hall or evaporates from there. This condition was (approximately) fulfilled in all the measurements.

Moreover, the water surface temperature T_0 and the air condition (temperature T_1, relative humidity RH) were recorded with standard thermometers and an air humidity measuring device as stationary mean values. The measuring point of the air condition was in two cases 1.72 m and in one case 3.00 m above the water surface. 51 data sets (T_0, T_1, RH, $\mathring{m}$) were obtained altogether which are compiled in Table 1. The water surface temperatures vary between 24°C and 30°C, the air temperatures between 22°C and 32°C, the relative humidities between 40% and 77%. This results in mass flow rates between 17 and 163 $gm^{-2}h^{-1}$.

DISCUSSION OF MEASURED RESULTS

For the parameter values for T_0, T_1 and RH, which can be regarded as representative, the water vapour flow rate $\mathring{m}$ shows a variation of almost one order of magnitude. The question arises how $\mathring{m}$ depends on these values in quantitative terms.

The reasons for the existence of an evaporation rate are depicted schematically in Fig. 1. It shows a vertical section through the interface between water and air. For the sake of simplicity it is assumed that the temperatures T_0 and T_1 of water surface and air are the same (profile 1). Immediately at the water surface the relative humidity is always 100%. Within a boundary layer it decreases to the value RH in the room air (profile 2). This involves a difference $\Delta\ p_W$ of the water vapour pressure (profile 3). It is calculated from

$$\Delta\, p_W = p_S(T_o) - RH \cdot p_S(T_1) \tag{1}$$

where $p_S(T)$ is the saturation vapour pressure of the water as a function of the temperature. Since humid air has a lower density than dry air (ρ_A)

$$\rho = \rho_A(T)(1 - 0.38 \cdot RH \cdot p_S(T)/p) \tag{2}$$

ρ grows in the described case as the height increases (profile 4). Therefore, an unstable situation exists which results in a free convective flow. Additionally, air velocity boundary layers are possible above the water surface (profile 5) which can give rise to a mass transfer as a result of forced convection.

Because of Henry's Law an increase of $\dot{m}$ is expected as $\Delta\, p_W$ grows. Although this is approximately true as shown in Fig. 2, the mass flow rates fluctuate too irregularly around this trend to give a satisfactory representation of the results. This is due to the fact that e.g. the impact of the free convection is not at all expressed in this representation. If the mass flow rate (related to the non-dimensional pressure difference $\Delta\, p_W/p$) is plotted across the non-dimensional density difference $\Delta\rho/\rho$ (Fig. 3) a much more compact picture is obtained. The mass flow rate is obviously increased by higher buoyancy. Even when there is no free convection or when the layers are stable ($\Delta\rho \leqslant 0$), however, a mass transfer exists which can only be brought about by the imposed velocity field in the boundary layer. A theoretical description of the mass flows must therefore comprise the joint effect of free and forced convection.

THEORETICAL MODELLING

The mass flow $\dot{m}$ between two areas 1 and 0 of different (non-dimensional) concentrations c_1 and c_0 is normally described with a mass-transfer coefficient β:

$$\dot{m} = \beta(c_1 - c_0) \tag{3}$$

The non-dimensional concentration c is the ratio of the density of the water vapour ρ_W to that of humid air:

$$c = \rho_W/\rho \tag{4}$$

With the ideal-gas equation of state for water vapour and equation (2) follows

$$c_1 - c_0 = 0.62 \cdot \Delta\, p_W/p \tag{5}$$

The mass-transfer coefficient β depends via the Lewis relation (Eckert, 1966) on the surface conductance α. With the definitions for the Nusselt number Nu and the thermal diffusivity a the following equations hold:

$$\beta = \alpha/c_p, \quad Nu = \alpha \cdot \ell/\lambda\ , \quad a = \lambda/(\rho \cdot c_p) \tag{6}$$

and combined:

$$\beta = a\rho \cdot Nu/\ell \tag{7}$$

where ℓ represents a characteristic length of the transfer process. The following formulation is proposed as relation for the Nusselt number for combined convection:

$$Nu = A(Gr^{+} + (B/A)^3 \cdot Re^2)^{1/3} \tag{8}$$

Here, A and B are numerical constants, Gr^+ the Grashof number ($= l^3 g \, \Delta \rho / \rho \cdot \nu^{-2}$) calculated only for positive $\Delta \rho$ and Re the Reynolds number ($= u l \, \nu^{-1}$).

Lack of air velocity u or lack of buoyancy ($\Delta \rho = 0$) result in the borderline cases

$$Nu = A \cdot Gr^{1/3} \quad \text{and} \quad Nu = B \cdot Re^{2/3}$$

which are well known (VDI, 1981). Since the quantity Gr/Re^2 is a measure for the ratio of the force induced by buoyancy to the inertial force, the linear combination of Gr and Re^2 in eq. (8) describes the joint effect of both forces on the heat or mass transfer.

With the definition of Gr^+ as well as the equations (8), (7), (5) and (3) the mass flow rate $\mathring{m}$ is finally calculated from

$$\mathring{m} = A(g \cdot \nu^{-2} \, \Delta \rho/\rho + (B/A)^3 \cdot Re^2/l^3)^{1/3} \cdot \Delta p_w \cdot 0.62 \, p^{-1} \, \rho \, a \tag{9}$$

For the physical quantities appearing in this equation the values given in Table 2 are inserted and the mass flow rate $\mathring{m}$ is given in $gm^{-2}h^{-1}$. From this follows:

$$(\mathring{m}/\Delta p_w)^3 = 7.68 \cdot (A^3 \cdot \Delta\rho/\rho + B^3 \cdot u^2 l^{-1} g^{-1}) \left(\frac{gm^{-2}h^{-1}}{Nm^{-2}}\right)^3 \tag{10}$$

COMPARISON OF THEORY WITH MEASUREMENTS

The described relationship of the variables $\mathring{m}$, Δp_w, $\Delta\rho/\rho$ and u in equation (10) is only partly verifiable with the experimental data since the velocities u were not measured and the reference length l can only be estimated. Nonetheless, a linear relationship

$$(\mathring{m}/\Delta p_w)^3 = A_0 + A_1 \cdot \Delta\rho/\rho \tag{11}$$

can be postulated for the measured data from which the parameters A_0 and A_1 can be determined numerically. By comparison with equation (10) follows from $A_1 = (2.33 \pm 0.18) \cdot 10^{-2}$ for the parameter A characterising the free convection:

$$A = (A_1/7.68)^{1/3} = 0.145 \pm 0.003 \tag{12}$$

It is approx. 10% higher than would be expected when applying the relationship $Nu = 0.15 \cdot (PrGr)^{1/3}$ ($Pr = 0.707$) which is known from literature. Parameter $A_0 = (2.08 \pm 1.13) \cdot 10^{-5}$ yields only collective information on B^3u^2/l which is not to be discussed any further here. It is regarded as a typical mean value affected by some fluctuation and characterising the convective exchange forced by the flow of room air. The estimation formula for $\mathring{m}$ therefore reads:

$$\mathring{m} = \Delta p_w \cdot (2.08 \cdot 10^{-5} + 2.33 \cdot 10^{-2} \, \Delta\rho/\rho)^{1/3} \tag{13}$$

If the mass flow rates $\mathring{m}$ calculated by means of equation (11) with the parameter A_0 and A_1 are compared with the measured values (Fig. 4), acceptable agreement is obtained. The standard deviation of the measured mass flow rates from the computed ones is $\pm$ 11.4 $gm^{-2}h^{-1}$, and the mean value of the relative error is smaller than 14%. In light of possible measuring inaccuracies and systematic errors these deviations are fairly realistic.

REFERENCES

Biasin, K. (1987): Die energetischen Vorgänge an Freibadbecken mit Abdeckeinrichtungen / Die Wasserverdunstung nicht benutzter Becken im Hallenbad. VDI-Verlag GmbH, Reihe 19, Nr. 17, Düsseldorf.

Eckert, E.R.G. (1966): Einführung in den Wärme- und Stoffaustausch. Springer-Verlag, Berlin, S. 277.

VDI-Wärmeatlas (1981), VDI-Verlag, Düsseldorf. S. A10, A11.

NOMENCLATURE

a	thermal diffusivity	m^2s^{-1}
A	numerical constants	-, $(gm^{-2}h^{-1}/Nm^{-2})^3$
B	numerical constant	-
c	specific heat, concentration	$Nm\ kg^{-1}K^{-1}$, -
g	acceleration of gravity	ms^{-2}
Gr	Grashof-number	-
ℓ	characteristic length	m
$\dot{m}$	evaporation rate	$kg\ m^{-2}s^{-1}$, $gm^{-2}h^{-1}$
Nu	Nusselt-number	-
p	pressure	Nm^{-2}
Re	Reynolds-number	-
RH	relative humidity	-
T	temperature	K, °C
u	velocity	ms^{-1}

Greek symbols

α	surface conductance	$Wm^{-2}K^{-1}$
β	mass transfer coefficient	$kg\ m^{-2}s^{-1}$
Δ	difference	-
λ	thermal conductivity	$Wm^{-1}\ K^{-1}$
ν	kinematic viscosity	m^2s^{-1}
ρ	density	$kg\ m^{-3}$

Subscripts:

0,1	numbers
p	constant pressure
s	saturation
W	water

TABLE 1: Summary of measured data

No.	T_0 °C	T_1 °C	RH -	$\dot{m}$ $\frac{g}{m^2h}$	No.	T_0 °C	T_1 °C	RH -	$\dot{m}$ $\frac{g}{m^2h}$	No.	T_0 °C	T_1 °C	RH -	$\dot{m}$ $\frac{g}{m^2h}$
1	28,2	28,3	0,54	81	18	28,2	28,3	0,52	95	35	28,6	26,3	0,58	163
2	28,1	28,0	0,57	69	19	28,3	30,1	0,45	79	36	28,7	28,5	0,49	140
3	28,1	30,0	0,48	69	20	30,2	32,3	0,45	84	37	28,2	28,3	0,43	125
4	28,2	30,0	0,49	72	21	30,3	32,2	0,46	75	38	23,0	21,1	0,72	63
5	28,0	32,0	0,44	53	22	30,2	30,4	0,50	109	39	23,0	22,9	0,64	42
6	28,9	26,1	0,60	123	23	30,0	30,5	0,48	104	40	24,1	26,0	0,58	22
7	29,7	30,2	0,49	124	24	26,7	28,0	0,51	60	41	24,7	26,9	0,49	46
8	30,1	27,9	0,57	154	25	26,2	28,2	0,51	45	42	25,6	25,9	0,56	76
9	28,2	29,8	0,46	87	26	25,1	26,5	0,58	37	43	25,9	24,4	0,58	109
10	26,5	29,8	0,47	38	27	25,1	28,2	0,51	27	44	26,8	24,5	0,60	122,6
11	25,5	28,0	0,56	25	28	25,2	24,5	0,64	55	45	27,9	26,0	0,53	132,4
12	26,0	24,4	0,67	75	29	25,1	26,1	0,57	41	46	28,5	29,8	0,55	71,8
13	25,3	25,7	0,60	58	30	24,1	21,9	0,77	64	47	28,5	29,7	0,54	66,8
14	24,3	25,5	0,58	40	31	24,1	24,2	0,65	42	48	29,0	29,8	0,55	69,1
15	24,2	27,9	0,51	17	32	24,5	25,6	0,58	43	49	29,0	30,0	0,55	72,9
16	28,0	30,0	0,52	41	33	27,7	24,8	0,64	134	50	28,5	29,5	0,54	65,9
17	28,3	28,4	0,52	97	34	27,4	27,1	0,57	103	51	29,0	29,7	0,55	117,6

TABLE 2: Values of variables used to evaluate eq. (9)

$\rho = 1.16$ kg m^{-3}	
$a = 2.22 \cdot 10^{-5}$ m^2s^{-1}	$g = 10$ ms^{-2}
$\nu = 1.57 \cdot 10^{-5}$ m^2s^{-1}	$p = 10^5$ Nm^{-2}

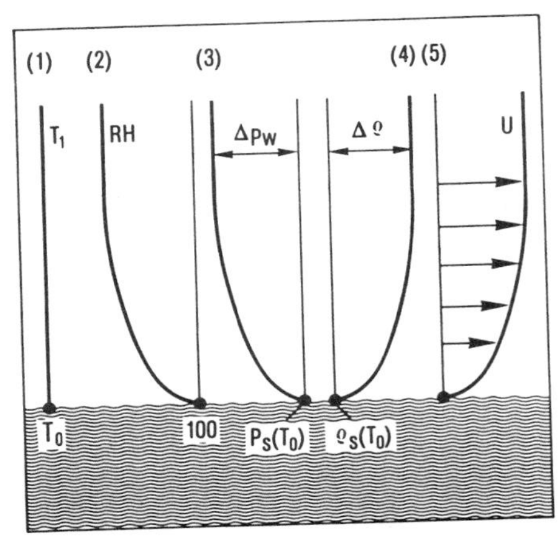

Fig. 1: Typical boundary layer profiles above an air-water interface for the case of identical temperatures of water surface and air

Fig. 2: Measured evaporation rates $\dot{m}$ from three indoor swimming pools vs. vapour pressure difference Δp_W between water surface and air

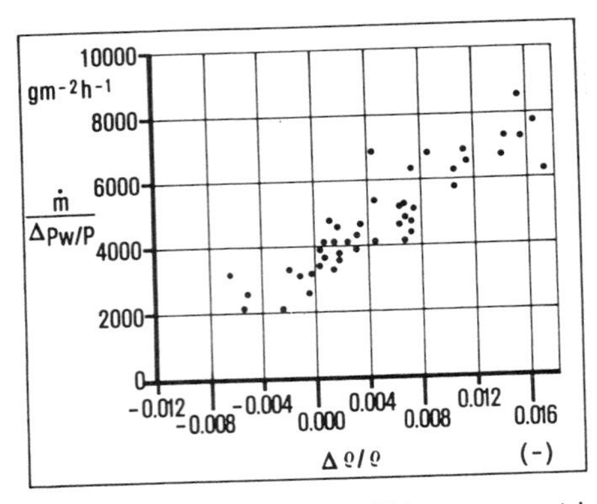

Fig. 3: Pressure-specific evaporation rate $\dot{m}$ / ($\Delta p_W/p$) vs. dimensionless air density difference $\Delta\rho/\rho$

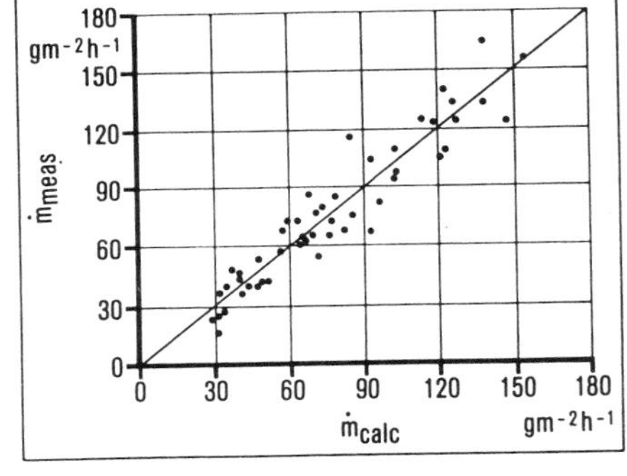

Fig. 4: Measured vs. calculated evaporation rates $\dot{m}$

HEAT RECOVERY FROM WASTE WATER BY MEANS OF A RECUPERATIVE HEAT EXCHANGER AND A HEAT PUMP

*K. Biasin, **F. D. Heidt

*Rheinisch-Westfälisches Elektrizitätswerk AG,
RWE-Anwendungstechnik, 4300 Essen 1, FRG

**University of Siegen, Department of Physics,
5900 Siegen, FRG

ABSTRACT

The useful heat of warm waste water is generally transferred to cold water using a recuperative heat exchanger. Depending on its design, the heat exchanger is able to utilise up to 90% of the waste heat potential available. The electric energy needed to operate such a system is more than compensated for by an approximately 50-fold gain of useful heat.
To increase substantially the waste heat potential available and the amount of heat recovered, the system for recuperative heat exchange can be complemented by a heat pump. Such a heat recovery system on the basis of waste water is being operated in a public indoor swimming pool. Here the recuperative heat exchanger accounts for about 60%, the heat pump for about 40% of the toal heat reclaimed. The system consumes only 1 kWh of electric energy to supply 8 kWh of useful heat. In this way the useful heat of 8 kWh is compensated for by the low consumption of primary energy of 2.8 kWh. Due to the installation of an automatic cleaning device, the heat transfer surfaces on the waste water side avoid deposits so that the troublesome maintenance work required in other cases on the heat exchangers is not required.

KEYWORDS

Shower drain water, recuperative heat recovery, heat recovery by means of a heat pump, combination of both types of heat recovery, automatic cleaning device for the heat exchangers, ratio of useful heat supply vs. electric energy consumption, economic consideration.

INTRODUCTION

The consumption of conventional primary energies can by reduced considerably in public indoor swimming pools. This reduction is achieved by the application of systems for heat recovery and for using environmental heat. By the example of the Schwalmtal indoor swimming pool it is to be demonstrated to what extent the consumption of conventional primary energie may be reduced by heat recovery from waste water.

This indoor swimming pool makes use of a heat recovery system from waste water. The heat recovery device itself is a "heat pump recuperator device" of Menerga Ltd., Type 43.12.01. year of construction 1984. It has been put into service in January 1985. In this device, the waste water is at first cooled down recuperatively from 30 to 16°C and then from 16 to approx. 7°C in the evaporator of a heat pump. Due to the extensive heat extraction,the waste water draining away has a temperature that is always 2 to 3 K below the temperature of the incoming fresh water.
Schwalmtal public indoor swimming pool was visited by 87.000 persons in 1985. Since every bather used 40 l of shower water, and 30 l of pool water per bather were exchanged for fresh water, a waste water volume of 6.100 m^3 were supplied into the heat recovery system. During the opening time of the indoor swimming pool the heat recovery system is exlusively fed by shower drain water. When the daily opening period of the indoor swimming pool has expired, heat extraction of the pool drain water and the heat-up of the pool fresh water can start.

CONFIGURATION AND MODE OF OPERATION

Configuration: The heat recovery system a diagrammatic view of which is shown in Figure 1, consists essentially of the heat recovery device, a shower drain water storage tank and a fresh water storage tank. The most important nominal details of the system are summarized in Table 1. A recuperative heat exchanger and a small heat pump are installed in the heat recovery device. Figure 2 shows the arrangement of the heat exchangers in flow direction of the waste water and fresh water respectively.
The cleaning device is a special characteristic of the heat recovery device for the transfer surface on the outlet side: At one and a half hours intervall, cleaning substances are trasported by the waste water flow through the recuperative heat exchanger and the evaporator of the heat pump (Figure 3). This automatic and periodical cleaning procedure avoids deposits on the heat transfer surfaces of the waste water side during long periods.

Mode of operation: During the opening period of the indoor swimming pool the heat recovery device is switched on and off via the level control in the waste water storage tank. The level difference between the switch-on and switch-off contact corresponds to a waste water volume of approx. 400 l. Since the waste water flow and the fresh water flow respectively are constant and amount to 1,2 m^3/h each, an operation time of at least 20 minutes is allocated to every switch-on procedure of the heat recovery device.
Heat is extracted recuperatively from the waste water by 14 K and in the evaporator of the heat pump by 9 K (Figure 2, at the bottom). On the other hand, a temperature rise by 12 K is attained in the condenser of the heat pump (Figure 2, at the top). The resulting temperature of the preheated fresh water of 36°C is heated to the required value of approx. 45°C via a subsequent heating unit.

OPERATING EXPERIENCE

The heat recovery system has been operating since January 1985. The periodical cleaning of the heat transfer surfaces on the outlet side led todate to steady conditions for heat transfer. With a view to the condition of these heat transfer surfaces it may be expected that additional maintenance work on the heat exchangers will not become necessary if the

cleaning device of the heat recovery device functions properly.
The primary filter in front of the waste water pump has proved to be essential for trouble-free operation (Figure 1). It should be cleaned twice per week. The expenditure of time is low because the overall cleaning procedure takes only approx. 5 minutes. If this filter is not cleaned, the waste water flow can be reduced in such a way that the heat pump is switched off via the control facilities available.
During the operating time from January 85 through August 87, the heat recovery system has fulfilled the expectations concerning the availability, the heat recovery and a low maintenance expenditure.

MEASURED RESULTS

Operation of the system and shower water consumption: The low number of visitors during 7 and 15 h and the low waste water volumes lead to a discontinuous operation of the heat recovery system (Figure 4). From 15 h the system turns to a comparably continuous operation. Approximately an hour after the bathing activity has stopped, heat is extracted from the shower drain water and the exchange of pool water fro fresh water begins. It may be directly seen from Figure 4 for the quasistationary operation of the heat recovery system that the output of useful heat of the system amounting to 8 kWh/0.25 h is compensated for by consumption of electric energy of 1 kWh/ 0.25 h.
The consumption of shower water per bather was 40 l. This value was determined during the public opening hours excluding school sports. This figure corresponds to a shower period of 3.3 minutes at a water flow of 12 l per shower and minute. The consumption of shower water per pupil visiting the indoor swimming pool during school sports, was 30 l on average and the shower duration was 2.5 minutes. Both shower periods are applicable to other indoor swimming pools as approximative values.
The temperature of the shower drain water was between 29 and 33°C. It depends upon the number of visitors and increases with the number of visitors increasing. The mean waste water temperature of 30°C ascertained for a fourmonth measuring period may also be transferred to other indoor swimming pools as an approximative value.

Useful heat recovered and electric energy consumed: The combined coefficent of performance (CCP) represents the ratio of useful heat output by the heat recovery system vs. electric energy consumption for driving the system. Here the useful heat ouput consists of the useful heat of the recuperative heat exchanger Q_R and the useful heat of the heat pump Q_{HP}. When dividing the electric energy accordingly in E_P, E_{HP} and the other electric energy E_C (control, etc.) the following eqution applies for the combined coefficient of performance

$$CCP = \frac{Q_R + Q_{HP}}{E_R + E_{HP} + E_C} .$$

Biasin, Heidt and Leisen (1985) show how to represent the CCP by means of the equations for recuperative heat exchangers and heat pumps where such equations are common in techniques.
In the period from 1 March to 30 June the output of useful heat of the heat recovery system and the consumption of electric energy were measured. As shown in Table 2, the CCP for the total period amounted to approx. 7.9. It varied between the monthly average of 8.9 in March and 6.7 in June. The different values of the CCP may be explained, to a large extent, by the fresh water temperature which was 10°C in March and 14°C in June. For the period of one year, a CCP value of eight and a heat-up of the fresh water

to approx. 36°C may be supposed according to the measurements.
Figure 5 is based on the use of primary energy in the power plant and describes the energy flow until the output of useful heat of the heat recovery device. The energy flow of the heat recovery device is based on a CCP value of eight. For this value it can be seen from Figure 5 that an output of useful heat of 100 energy units is compensated for by a use of primary energy of only 35 units.

ECONOMIC CONSIDERATION

The installation of a heat recovery system had been included in the planning of the indoor swimming pool. By means of favourably selected sites for the individual devices the costs of the connecting pipes and the installation costs were low. The capital cost of the heat recovery system amounted to 48.000 DM. For a waste water volume of 6,100 m³ in 1985 the heat recovery system had an output of useful heat of 149 MWh and consumed an electric energy of 18.6 MWh. The total annual costs that are equal to the sum of the maintenance, energy and capital costs, amount to approx. 10.000 DM per year. From these costs and the useful heat output of 149 MWh per year, the specific costs of the useful heat output of about 6.7 Pf per kWh are calculated. These specific costs are relatively low. For an economic appraisal of the system, these costs would have to be compared with the coresponding costs of competing heating systems.
The annual waste water volume has a considerable effect on the specific costs of the useful heat output. If the number of bathers coming to the Schwalmtal indoor swimming pool would be 120.000 instead of 90.000 per year, the specific costs would be approx. 5.7 Pf per kWh of useful heat output for the system design presented.

OUTLOOK

The representation of the heat recovery system was aimed at showing the considerable technical performance standard available today in the field of heat recovery from waste water. The system represented is an industrially manufactured series device that may be installed as a complete device in an indoor swimming pool. If it is connected to the electric network and to the different pipes, it may be directly put into service.
For the operator of an indoor swimming pool such a system has the advantage that it may be evaluated independently of the additional energy engineering equipment and may be installed at any time in his indoor swimming pool. According to the result of the measurements and investigations carried out in the Schwalmtal indoor swimming pool, systems for heat recovery from waste water in accordance with the construction described, are very well suited for being used in indoor swimming pools.

REFERENCES

Biasin, K., F.D. Heidt, K. Leisen, (1985). Wärmepumpen-Rekuperatoren für die rationelle Energienutzung Bericht der RWE-Anwendungstechnik, Essen.
Leisen, K. (1986). Small Indoor Swimming Pool Schwalmtal - Final Report. Forschungsstelle für Energiewirtschaft, München.
Biasin, K., K. Leisen, (1986). Ergebnisse energietechnischer Untersuchungen in einem Hallenbad. RWE informiert Nr. 218. RWE-Anwendungstechnik, Essen.

Heat recovery device		Water storage tank	
waste water flow	1.2 m³/h	shower drain water storage	5 m³
fresh water flow	1.2 m³/h	shower fresh water storage	5 m³
useful heat output	33,5 kW	flushing water storage	(30 m³)
electric energie consumption	4 kW	The storage was designed for additional filters (outdoor swimming pool).	
combined coeffizient of performance	8,3		

Table 1: Design data

parameter considered	period considered				
	March	April	Mai	June	Sum and mean value respect.
heat recovered kWh	10,300	9,000	7,900	5,100	32,300
electric energy kWh	1,310	1,230	1,220	,900	4,660
CCP	8.86	8.32	7.48	6.67	7.93

Table 2: Measured results

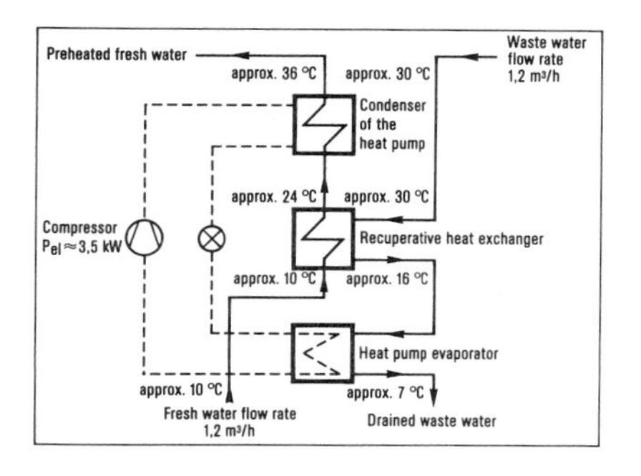

Figure 2: Arrangement of the heat exchangers

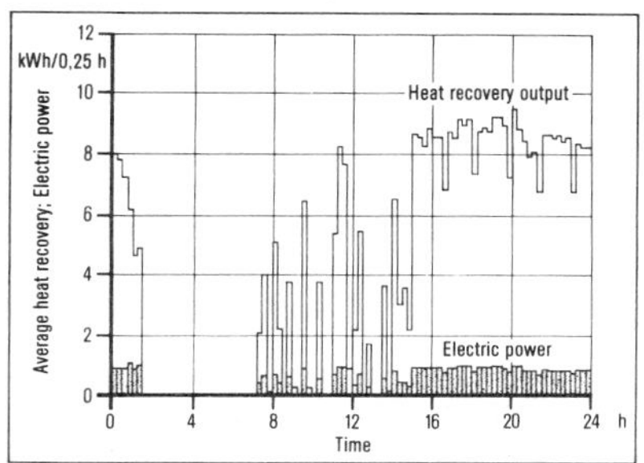

Figure 4: Development of the emission of useful heat in terms of time

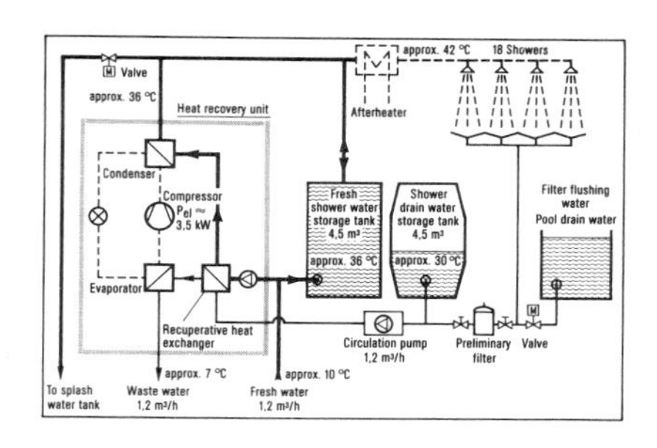

Figure 1: Diagram of the heat recovery system

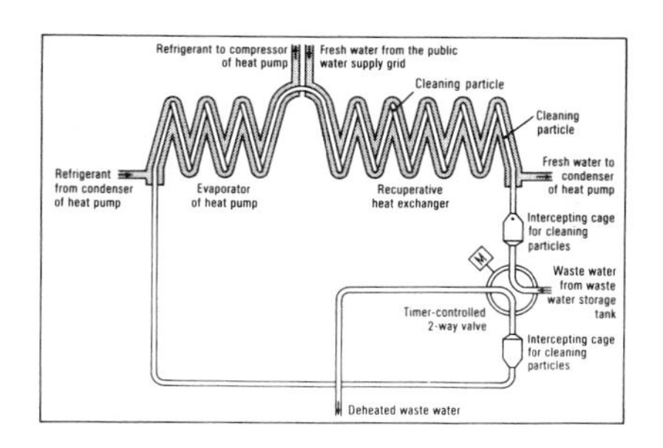

Figure 3: Diagram of the cleaning device

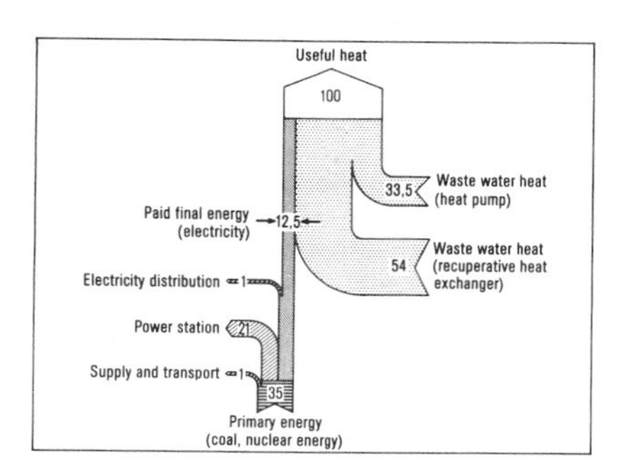

Figure 5: Energy flow in the heat recovery device

THERMAL PARAMETERS CONTROL IN AN AQUACULTURE PLANT.

M. Izquierdo[x]; M. Carrillo[xx]; J.A. Carrasco[xxx]; D. Tinaut[x].

[x]Instituto de Optica (C.S.I.C.). Serrano, 121. 28006 Madrid.
[xx]Instituto de Acuicultura (C.S.I.C.). Torre la Sal. Castellón.
[xxx]Instituto del Frío (C.S.I.C.). Ciudad Universitaria. 28003 Madrid.

ABSTRACT

The paper shows a technical solution to obtain optimal temperatures in an Aquaculture Plant by sea water treatment. A simplified calculation method for water -water heat pump, determining the global profitability of the plant with static energy recuperator is described.

This thermal generation system is compared with others of present use, and the net price of thermal Kwh, either by using liquid, gas fuels or heat pump are calculated.

KEYWORDS

Aquaculture plant; thermal generation system; temperature control; heat recovery; refrigeration machine; static recuperator.

INTRODUCTION

Sea water temperature is one of the main parameters conditionating biological development of marine organisms. Operating cost to keep optimal temperature in an aquaculture plant are high because energy is consumed by the heating or the cooling system according to seasonal needs.

In the present study, the objectives are the heat production to keep optimal temperature in an hatchery and the recovery of the residual heat.

In figure 1 we can see the major components of a gas-oil installation utilised for sea water heating which possess

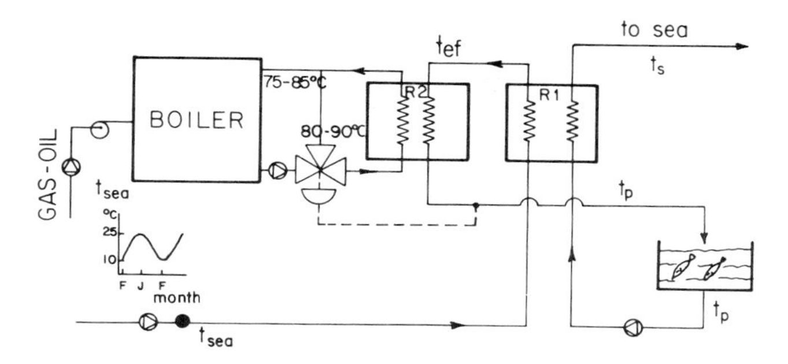

FIG. 1.- Gas-oil installation for sea water heating with heat recuperator.

a heat recuperator.

A plant designed for extreme conditions acts as an oversized installation the year around due to the seasonal variations of sea water temperature. Furthermore, as the generator temperature stays near 80ºC, increased control requeriments will be necessary to operate within ± 1ºC as required by biologists.

REFRIGERATION MACHINE INTEGRATED IN AN AQUACULTURE PLANT.

In the system depicted in Fig. 2 the boiler has been substituted for a frigorific machine which carries sea water to process temperature and in second step cools return water.

The level of the heat source is determined by the process temperature. This suppose a decrease of at least of 40ºC with regard to the traditional system. Thus control difficulties when working in out of design conditions are smaller. Return water of production tanks is used as cold source, so evaporation temperature is maintained practically constant.

The remainder main section of the plant is: Static recuperator R_1 with recovery capacity of 50-60% of the total thermal load. Dinamic recuperator R_2 which through a second recuperation acts as cold source. Thermal storage unit Q. Provides thermal inertia during cleanliness time of production tank, acts as heat source when the frigorific machine stops.

Attached devices like pumps, valves and controls are alike to those of conventional plant.

CALCULATION METHOD

1. Design global Thermal load

$$Q_d = m\, C_p (t_p - t_{mm}) \qquad (1)$$

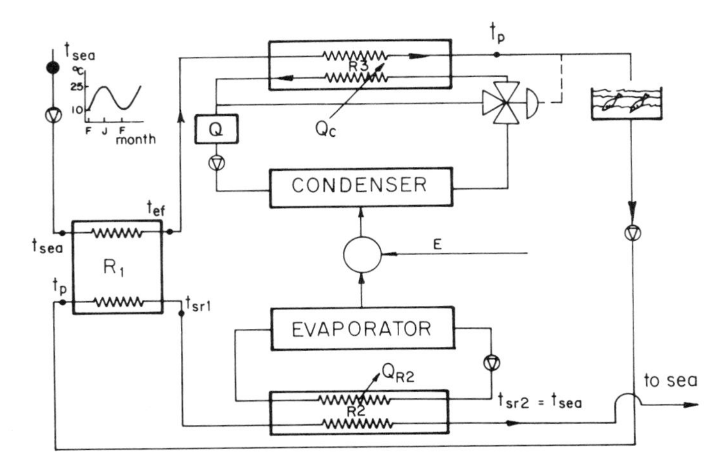

FIG. 2.- Installation for sea water heating with frigorific machine.

2. Recuperated heat at R_1

$$Q_{R1} = 0.5\ Q_d \tag{2}$$

3. Recuperated heat at R_2

$$Q_{R2} = Q_e = \text{evaporation heat}$$

4. Supplied heat at R_3

$$Q_{R3} = Q_c = \text{condensation heat}$$

$$Q_c = Q_d - Q_{R1} = 0.5\ Q_d \tag{3}$$

5. Frigorific machine

E = Absorbed energy at compressor

$$Q_c = Q_{R3} = Q_{R2} + E \tag{4}$$

$$Q_c = (COP)_c . E \tag{5}$$

From Eqs (3) and (5):

$$E\ (COP)_c = 0.5\ Q_d \tag{6}$$

$$E = 0.5\ Q_d/(COP)_c \tag{7}$$

Further substitution in Eq (4), gives

$$Q_{R2} = 0.5\ Q_d\ (1 - 1/(COP)_c) \tag{8}$$

To determine numerical value of the above equations we need calculate the heating coefficient of performance of the frigórific machine. Howell (1987) /1/.

6. Calculation of $(COP)_c$

The frigorific machine performance is determined by the real thermodynamic cycle of refrigerant shown in Fig. 3. corresponding efficiencies are calculated. Calorific efficiency is obtained from Eq. (9).

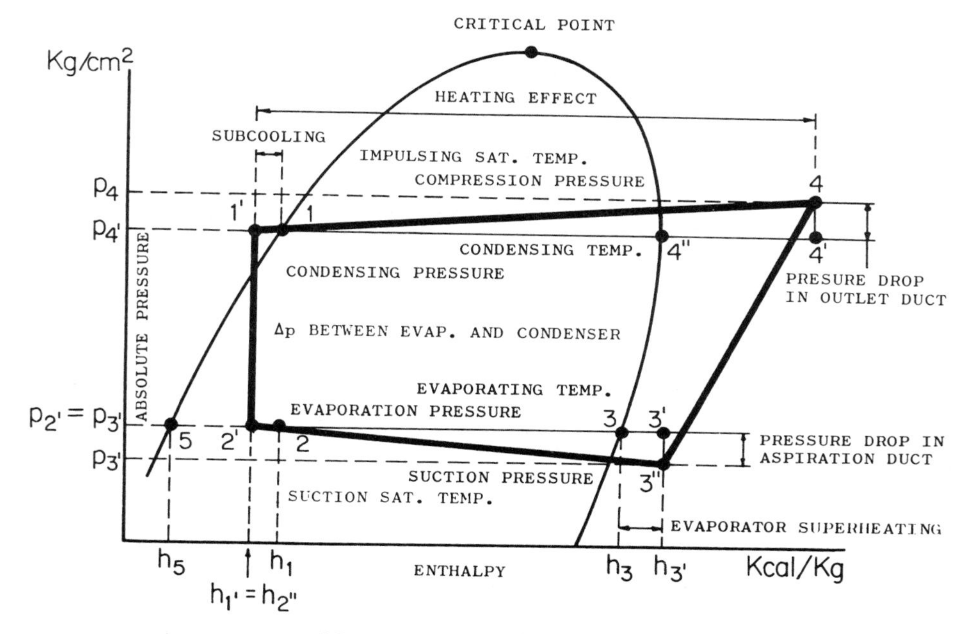

FIG. 3.- Real cooling cycle.

$$(COP)_c = \eta_i \ \eta_m \ \eta_e \ \frac{h_4 - h'_1}{h_4 - h_{3''}} \qquad (9)$$

Indicated efficiency η_i is given by the compression ratio $P_4/P_{3''}$ plotted in Fig. (4). J. Bernier (1979).

7. Global instantaneous efficiency η_c

We define the ratio between producted and absorbed energy

$$\eta_G = (Q_{R1} + Q_{R2} + E)/E =$$
$$\eta_G = 1 + (Q_{R1} + Q_{R2})/E \qquad (10)$$

ECONOMIC COMPARISON

A comparison of gas-oil, natural gas and frigorific machine system is given. Economic analyses is carried out considering cost of energy delivered. Other factors as rate of scalation of fuel price, inflation rate, and so, for difficult prediction reasons have not been taken into consideration for comparison purposes.

Cost of combustible per unit of producted energy C:

$$\text{Gas-oil: } C_{GO} = 0.5\ P_{GO}/\ \eta_{GO}\cdot(PCI/860) \qquad (11)$$
$$\text{Natural gas: } C_{GN} = 0.86.0.5\ P_{GN}/\eta_{GN} \qquad (12)$$

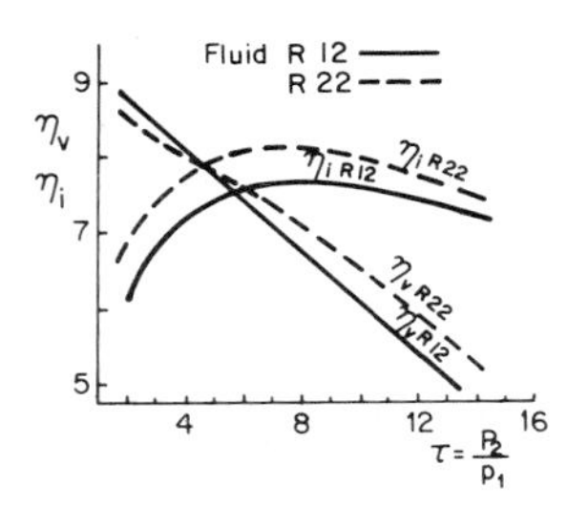

FIG. 4.- Volumetric and indicated efficiency of reciprocating compressor.

Frigorific machine: (Cost of electricity per unit of producted energy)

$$C_{MF} = 0.5 \; Pel/(COP)_c \qquad (13)$$

DEMONSTRATION CASE

The specification read:

$$Q_D = 10^6 \text{ Kcal/h}$$

	η_{comb}	P.C.I.	P	C
GAS OIL	0.65	10.000 Kcal/Kg	46 Pts/Kg	3.04 Pts/KWHt
NAT. GAS	0.75	THERM	2 Pts/Th	1.15 Pts/KWHt
FIRG. MACH: $(COP)_c$=3.19			7.5Pts/KWHe	1.18 Pts/KWHt

Frigorific machine:

Refrigerant R22
Condens. temp. t_c=47ºC
Evap. temp. t_e = 7ºC
Vapor superheating = 3ºC

Liquid subcooling = 5ºC
Condens. loss of pres. =0.5 Kg/cm^2
Evap. loss of pres. = 0.5 Kg/cm^2
Elec. and mech. eff.: 0.85

From these data we determine by the thermodynamic diagram the following values:

discharge pressure: 19.5 Kg/cm^2;
suction pressure: 5.1 Kg/cm^2

$h_1' = 113.5$ Kcal/Kg
$h_3'' = 150.2$ Kcal/Kg
$h_4 = 157.7$ Kcal/Kg

Compression rate: 3.39

From curves plotted in Fig. 4 we obtain an indicated efficiency $\eta_i = 0.75$.

$(COP)_c = 3.19$

$Q_{R1} = 0.5.10^6$ Kcal/h

$Q_c = 0.5.10^6$ Kcal/h

Q_{R2}= 343.000 Kcal/h

η_G : 6.38

E = 182.25 KWe

Cost of the aquaculture plant with frigorific machine can be estimated as two or three times higher than the cost of a plant with fossil combustible.

CONCLUSION

A simplified calculation method for designing an aquaculture installation using a frigorific machine has been developed.

A substancial energy saving can be achieved, and control requirements for thermal parameters in these installations are reduced.

Cost of energy supplied is equivalent to that corresponding to installations using natural gas and 2.5 times lower to those with gas-oil.

NOMENCLATURE

m sea water mass flow rate, Kg/h.
c_p specific heat, Kcal/Kg ºC.
P_{GO} gas oil price, Pts/Kg.
P_{GN} natural gas price, Pts/Th
P_E electricity price, Pts/Kg
η_i η_m η_e indicated, mechanical and electric efficiency, %.
η_G global efficiency.
P.C.I. inferior calorific power.
t_p process temperature, ºC.
$(COP)_c$ heating performance coefficient.
h enthalphy.
t_{mm} min. temp. of sea water.

REFERENCES

Howell, J.R. (1987). Active hybrid solar cooling systems. Solar Energy Utilization. NATO ASI Series. Serie E nº 129.
Bernier, J. (1979). La pompe de chaleur. Mode d'emploi. Vol. 1. PYC Ed.

A NEW ICE-MAKER HEAT PUMP

M.A. Paradis, Ph.D., Professor
G. Faucher, Ph.D., Professor
Dept Génie mécanique
Université Laval
Sainte-Foy, Qc, Canada, G1K 7P4

ABSTRACT

A new type of ice-maker heat pump based on water supercooling is being developed. A prototype has been built and tested. It has been possible to operate the water-to-air heat pump with entering water temperatures of about 0,5 °C and exit liquid (supercooled) water temperatures of about -1,5 °C. The supercooled water changes to ice outside of the evaporator, in the storage reservoir.

KEY WORDS

Heat pump; Ice maker heat pump; water supercooling; ice making.

INTRODUCTION

An Ice-Maker Heat Pump (IMHP) is basically a vapor compression heat pump unit that extracts energy from water by freezing the liquid to ice. A large number of references exist on the subject and the problems associated with their use are rather well known. Indeed, the IMHP technology has not changed much during the last 30 or 40 years. The two basic types are still the coil-in-bin arrangement, either direct or indirect, and the plate-above-bin, for which periodic ice harvesting is necessary.

All these systems have a relatively poor efficiency, are complex and costly and they can only be justified economically for some large commercial applications. Thus, it seems that a major breakthrough is needed to make the smaller system attractive, especially for northern climates.

That breakthrough could now be in the making, however. For example, the intensive research and development performed on clathrate systems could result in major improvements to existing systems. Another alternative is the supercooling IMHP being developed by the authors.

WATER SUPERCOOLING

It has been observed that phase changes are often accompanied by an "instability" by which a substance tends to keep its initial phase for temperatures well beyond phase equilibrium temperatures. Water supercooling is one of those "instabilities": water stays in the liquid phase down to temperatures well below the so-called "freezing-point". Actually, it seems that instead of talking about instabilities, one should rather differentiate between two different notions: the nucleation temperature (that at which an initial ice crystal is formed within the water) and the phase equilibrium temperature of 0 °C. For absolutely pure water, the nucleation temperature seems to be below -20 °C, maybe even something like -40 °C (Dorsey,1948; Gilpin,1977).

The only fool-proof method for initiating freezing of pure supercooled water (at any temperature above its natural nucleation point) is the introduction of an ice crystal in the liquid. Freezing will then begin immediately at the surface of that ice crystal and propagate quickly until enough latent energy has been liberated to increase the water temperature to 0 °C. The ice thus formed is not of the solid type normally seen. It grows instead as dendritic ice, i.e. as thin plate-like crystals interspersed in the water.

Freezing can also be initiated by a number of methods: by mechanical or thermal shocks, by the introduction of nucleants (like silver iodide), etc. In ordinary static tap water, the multitude of impurities present act as nucleants and nucleation will normally occur at temperatures of the order of -5 °C (Gilpin, 1977). Other parameters will also affect the nucleation temperature: heat transfer rate, type of container, movement, etc.

If freezing supercooled water is easy, preventing it from freezing is more difficult. Water is more readily supercooled (Mousson, 1858) if it is

- in the form of fine droplets
- squeezed between two glass plates well clamped together
- in a fine capillary

Thus, it would seem that everything that impedes a rearrangement of the molecules will facilitate supercooling. For the case of water at rest, supercooling has been studied by a large number of researchers, mainly in connection with water pipe freezing and with water droplets freezing in the atmosphere. For the case of water flowing in pipes, however, information is rather scarce.

The study most pertinent to the present work is the one by Mukushi and Takahashi (1982) in which the nucleation temperature of tap water flowing in small diameter pipes was studied experimentally as a function of the Reynolds number and pipe diameter. The main conclusions were that supercooling is possible in capillary tubes, that the exact nucleation temperature is difficult to predict but also that it is easy to determine with some certainty the conditions for which freezing always occurs and those for which it seldom occurs. The parameter used to define those regions was L/d, the ratio of tube length to internal diameter, as shown in fig. 1. The Reynolds number is important in determining the maximum value of the supercooling ΔT_s possible in the pipe. As shown in fig. 2, values of the nucleation temperature as low as -5 °C can be obtained (or $\Delta T_{s,max}$ = 5 °C) at low Reynolds numbers. The IMHP described below is based on the fact that supercooling can be

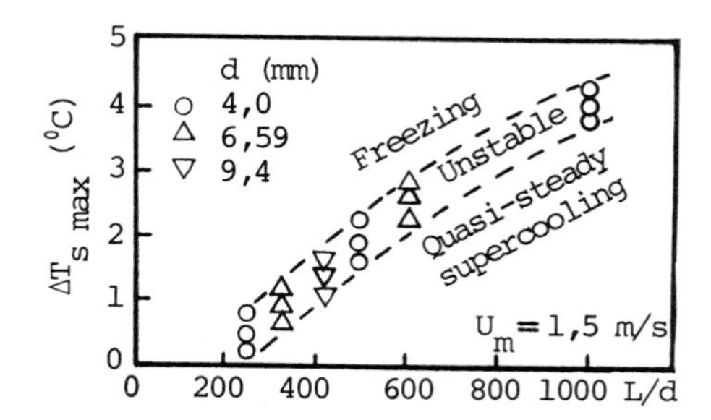

Fig. 1. Maximum supercooling as a function of L/d.

quasi-stable for water flowing in small diameter pipes.

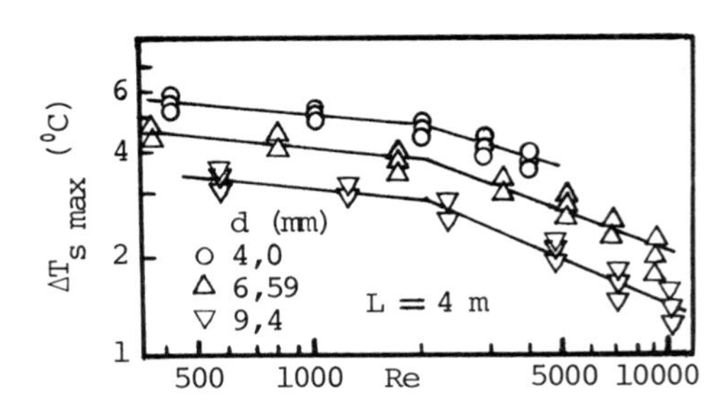

Fig. 2. Maximum supercooling as a function of Re.

THE SUPERCOOLING IMHP

The main characteristic of an IMHP based on water supercooling is its capacity to extract energy from liquid water entering the evaporator at temperatures approaching 0 °C. When coming out of the evaporator, the water is still in the liquid phase and at a temperature well below the phase equilibrium temperature, i.e. supercooled. If this water is fed to a reservoir, the contact of the jet with any part of the reservoir (mechanical shock) will normally result in a phase change: water will separate in a liquid and a solid part, separation which will put the mixture back to 0 °C. Considering the value of the phase change energy, one can anticipate 1,25 % of solid particles per degree of supercooling.

Figure 3 shows schematically a solar-assisted IMHP using water supercooling. The usual components can be seen: compressor, condenser, expansion valve and evaporator, reservoir and solar collectors. The design of the evaporator, however, is somewhat unusual: water flows through small diameter tubes before falling in the storage reservoir where the liquid-solid separation occurs. In figure 3, solar energy is used to melt the floating ice crystals.

As presented in fig. 2, the results of Mukushi and Takahachi are difficult to use for designing a supercooling evaporator. All the results of fig. 2, however, can be represented with little error by the equation

$$1000\, \Delta T_s\, (L/d^2)^{-0,25} = 310 - 5\, Re^{0,4} \qquad (1)$$

where L and d are in meters. With it, one can easily verify if a certain choice of dimensions, based on the usual heat transfer calculations, will result in the desired value of supercooling.

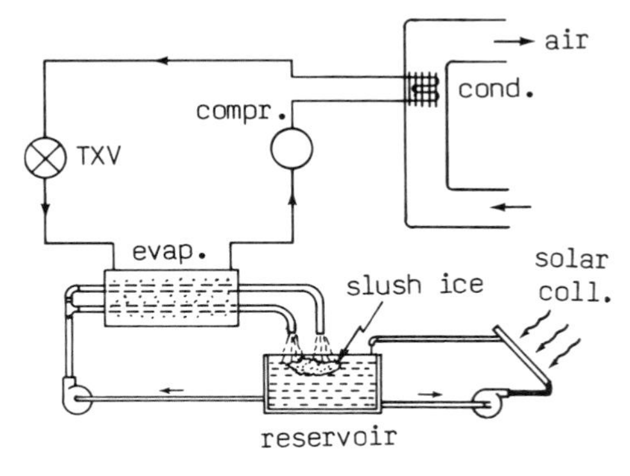

Fig. 3 Solar assisted supercooling heat pump

EXPERIMENTAL APPARATUS

An experimental rig has been constructed in order to verify if an IMHP based on water supercooling would work. After taking into account the heat pump components already available in the laboratory as well as the usual evaporator design criteria (desired capacity, mass flows, heat transfer rates, pressure losses, etc.) and the new design constraints (desired Re, L/d^2 and

ΔT_S), it was decided to use 20 capillaries with a 2,93 mm internal diameter and 1,9 m effective length for water circulation. Using equation (1), one finds a maximum value for ΔT_S of about 4 °C for such a case, which is more than sufficient and should leave a good safety margin.

The evaporator was vertical and consisted of an exterior polycarbonate tube (38 mm I.D.) for the refrigerant flow and, inside, a bundle of 20 capillary tubes for the circulation of water. A central steel rod kept the refrigerant flow area under control and ensured a good rigidity. The water (10 to 20 L/min) entered at the top and was distributed uniformly to each capillary tube. The refrigerant entered at the bottom and the type of flow present (bubbly, slug, dry, etc) could be easily determined by looking through the transparent outer tube.

The compressor was of the open type: the advantage was that the compression power could easily be varied (from 0,5 to about 1,5 hp); but the resulting COP was lower than with a hermetic compressor. The storage reservoir contained approximately 1000 L of water during the experiments. Temperatures were measured throughout with carefully calibrated thermistors. A more thourough description of the system has been given by Mercier (1986).

PRELIMINARY RESULTS

As a water-to-air heat pump, the system worked well from the beginning. But when the reservoir temperature dropped below 2 °C approximately, the evaporator froze solid. After weeks of minor modifications all aimed at eliminating mechanical and thermal shocks on the water side, it was possible to obtain water supercooling. The slush ice formed in the reservoir was very soft to the touch.

Values for some of the variables in a typical test were as follows:

- water temp.,	evap. inlet	0,56 °C
	evap. exit	-1,48 °C
- refrig. temp.,	evap. inlet	-4,12 °C
	evap. exit	-1,71 °C
	cond. exit	23,86 °C
	valve inlet	18,36 °C
- water flowrate in evaporator		0,291 kg/s
- evap. capacity		2,42 kW
- cond. heat rejection		2,87 kW
- global COP		2,2

At those conditions, the supercooling, and ice fabrication, was quite stable. But when the reservoir temperature dropped below 0,5 °C approximately, there was a tendency for the evaporator to freeze, and the system had to be stopped. The minimum nucleation temperature obtained during the test period was -2,12 °C.

Figure 4 shows the variation of capacity of the evaporator as a function of the evaporator entrance temperatures (water and refrigerant sides) for 4 typical tests. The small temperature difference between water and refrigerant should be noticed: 4 °C approximately. In conventional systems, this difference is often more than twice as large because of the ice buildup on the evaporator walls. In fig. 5, which shows the approximate evolution of the temperatures along the evaporator, only the inlet and outlet values were obtained directly from measurements. The saturation temperature of the refrigerant was deduced from pressure measurements. The shape of the curves were deduced from calculations.

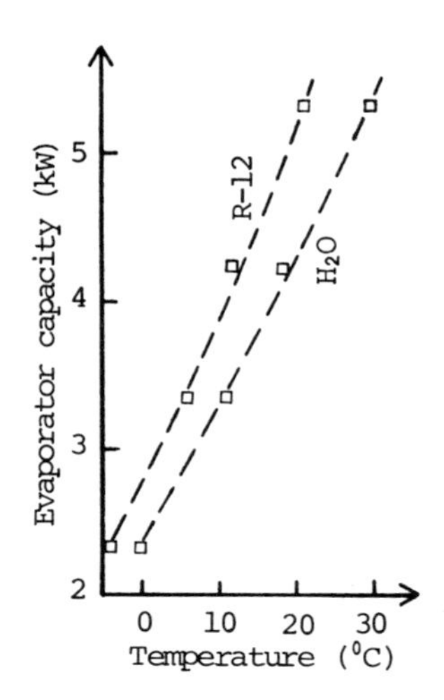

Fig. 4. Evaporator capacity versus water and refrigerant temperatures.

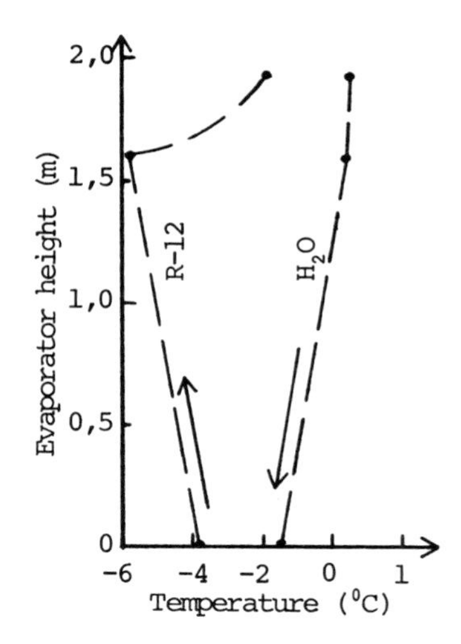

Fig. 5. Approximate temperature evolution in the evaporator for test #74.

CONCLUSION

Preliminary experiments have been performed on a prototype of a supercooling IMHP and it has proved possible to operate it in a stable way within a restricted range of operation. Preliminary results are encouraging and the new heat pump concept seems promising.

The next step in the development of the new IMHP will be to try increasing the range of stable operation of the system. A new version is now being constructed and should be tested within the coming year. Low temperature solar collectors of very simple construction are also being developed which will efficiently melt the ice formed by the heat pump.

If it can be fully developed, this type of heat pump could have many applications other than the one envisaged at the beginning of the project. For example, water to air heat pumps operating with river, lake or ocean water could work all year long very efficiently.

REFERENCES

Dorsey, N.E. (1948). The freezing of supercooled water. Trans. Am. Phil. Soc. 38, 245.

Gilpin, R.R. (1977). The effects of dendritic ice formation in water pipes. Int. J. Heat Transfer, 20, 693-699. Pergamon Press.

Mercier C. (1986). Simulation et expérimentation d'un évaporateur à glace opérant en régime continu. MSc thesis, Dept Mech. Eng., Université Laval, Québec, Canada.

Mousson, A. (1858). Ann. d. Physik (Pogg.) 105, 161-174.

Mukushi, T., Takahachi, S. (1982). Supercooling of liquids in forced flow in pipes and the growth of ice crystals. Hitachi Research Laboratory, Hitachi Ltd, Scripta Publ. Co.

VACUUM-ICE
THERMAL ENERGY STORAGE, HEAT SOURCE FOR HEAT PUMPS AND MEANS TO IMPROVE THE CAPACITY OF COOLING

Joachim Paul

SABROE GmbH - SAREF-Division
P.O.Box 2653
D-2390 Flensburg / West-Germany

ABSTRACT

Generation of ice at the triple-point of water yields the highest possible COP among all methods of ice-production. Together with unique energetic efficiency goes a range of properties for handling, storing and transporting the "vacuum-ice". Therefore this kind of ice offers exciting possibilities for applications in

- thermal energy storage (off-peak tariffs, avoiding demand charges)
- supply of cold water at constant temperatures (freezing point), therefore steady conditions and reduced heat transfer surfaces
- heat source for heat pumps using water at freezing point without danger of freezing-up or need for defrosting
- generation of ice at any mass-fraction (0-100 %) for application in industry and civil engineering
- cooling of tunnels, shafts and mines

Vacuum-ice is a perfect method to produce ice in large quantities and simple design. Standard capacities range between 100 and 1.000 metric tons per day. The first full-scale plant has been installed last year to extract energy from sea-water for a heat pump in a district heating plant.

KEYWORDS

ice, vacuum-ice, ice-generator, energy storage, thermal energy storage, heat source, heat pump, ice-bank, energy accumulator

INTRODUCTION

Conventional ice-makers operate a fairly low temperatures, as through the surface of a wall heat is extracted from water. This requires not only more energy, but also mechanical installations to harvest the ice from the surface. Furthermore the capacity per ice-maker unit is limited, which makes multiple installations for larger ice-quantities necessary, resulting in high costs and extensive piping and control. All these considerations resulted in the idea, to produce ice at the triple-point of water (0 °C and 6 mbar), called "vacuum-ice". It is a direct process, where no heat transfer surface is needed. Water entering the ice-tank is - by evaporating at triple-point - frozen and forms a slurry in the tank. This means, that part of the water evaporates by extracting energy from water and thus freezing it. As a direct process is unaffected by heat transfer losses trough a wall, the energetic efficiency is highest and the design extremely simple.

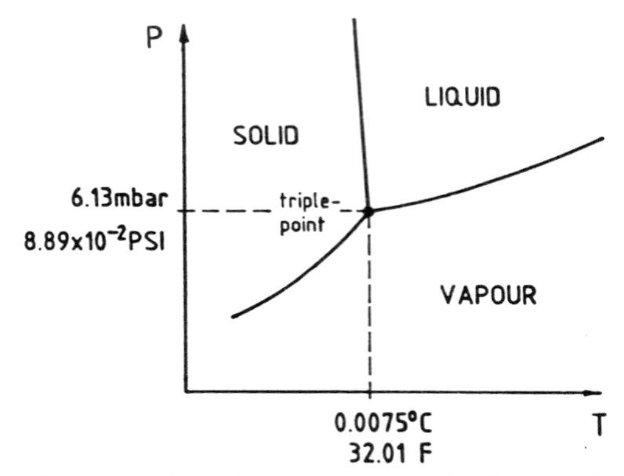

Fig. 1. Configuration of the triple-point of pure water

Figure 1 shows the triple-point of pure water, where the neighbourhood of all three phases is clearly visible.

DESIGN OF A VACCUM-ICE MAKER

The simplicity of the vacuum-ice system is shown in Fig. 2. Into an almost "empty" vessel under triple-point pressure, water is injected. On the top of the tank is a vapour-compressor, removing water vapour generated in the tank and compressing is to abt. 6 mbar. This compressed vapour is condensed by the evaporator of a conventional refrigeration system (heat pump) and is desalinated water rsp. potable water, regardless of the feed water quality. The frozen ice rsp. ice-crystal together with the water in the tank forms an ice-slurry, which can be removed from the tank and concentrated to anything between 0 and 100 % mass-fraction. 0 % concentration indicates already another feature of the ice-maker which is, that this machine also works as a water chiller, if wanted. Vaccum-ice makers therefore serve as chillers and/or ice-generators at the same time with the same equipment.

Due to the simplicity of the ice-maker, any kind of feed water can be used. Sweet- or sea-water, dirty or polluted water, chemically, biologically or mechanically otherwise troublesome waters can also be applied with the potential, that this water to a very large extent is cleaned and desalinated. As a direct process, a contamination from the ice-maker can also be excluded.

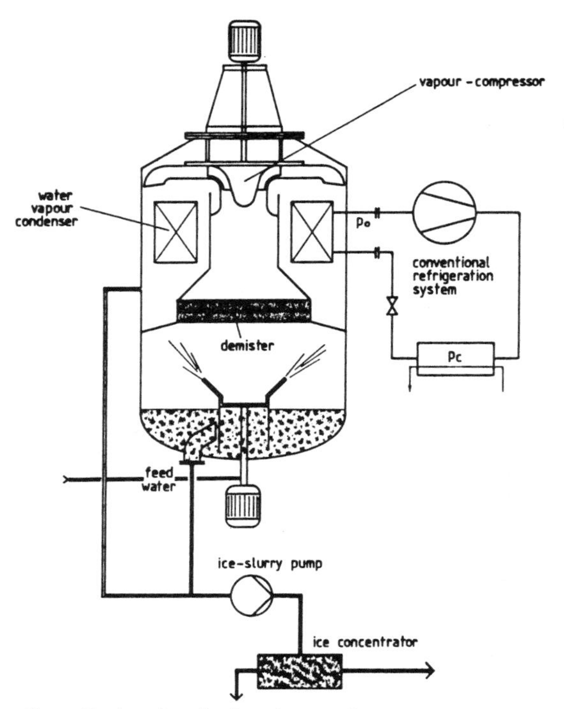

Fig. 2. Principal design of a vacuum-ice maker

ENERGETIC ADVANTAGE OF VACUUM-ICE IN REFRIGERATION- AND HEAT PUMP APPLICATION

As the vaccum-ice maker operates at highest COP af all systems in use today, the energetic performance is superior to all others. Table 1 gives values for ice-making systems with respect to the real energy demand compared to vacuum-ice (value 100 %). The figures include the total energy consumption of all equipment in use, exclude - however - handling, transportation and storage needs. There, vacuum-ice also offers very good conditions, as it can be pumped up to very high mass-fractions.

	real energy demand	
vacuum - ice	100 %	=+ 0 %
tube - ice	125 %	=+ 25 %
slush - ice	133 %	=+ 33 %
plate - ice	150 %	=+ 50 %
block - ice	166 %	=+ 66 %
flake/slice - ice	210 %	=+110 %

TABLE 1 ENERGY DEMAND OF DIFFERENT ICE-MAKING SYSTEMS (VALUE FOR VACUUM-ICE 100 %)

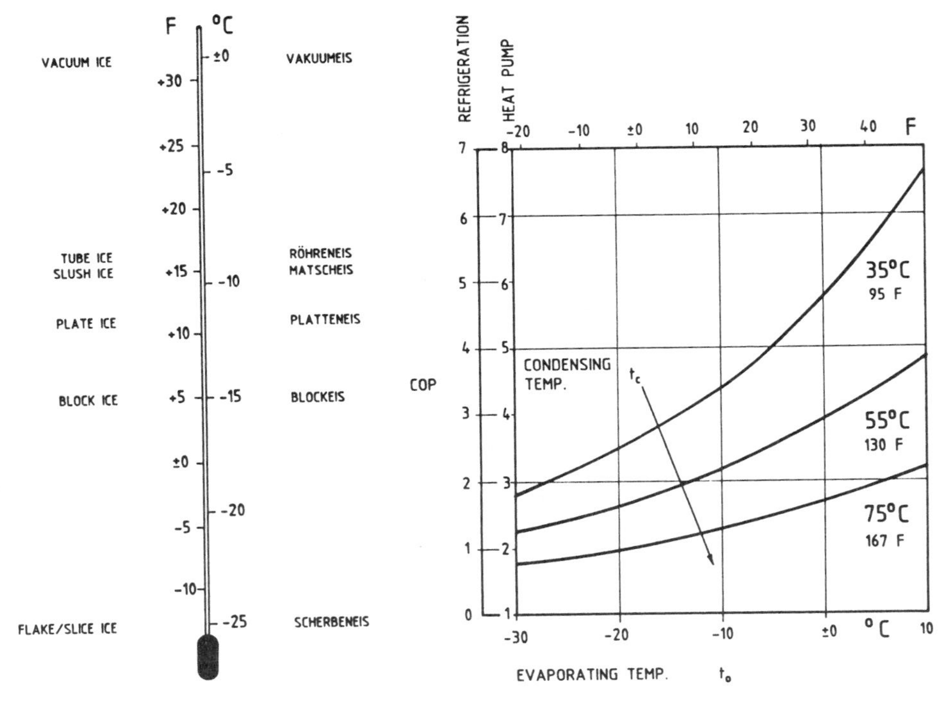

Fig. 3. Temperatures of production of different ice (left) and COP-values for different condensing- and evaporating temperatures

Figure 3 shows the alternative ice-systems with typical operating temperatures. It must be stated, that danger of freezing, energy demand for defrosting, ice-cutting, losses due to discontinuous operation atc. cannot be expressed in a simple graph and are therefore not taken into account. For vacuum-ice, however, no such losses and operational needs are present, making to process therefore simple and easy to control. Continuous operation is a further advantage.

FULL-SCALE VACCUM-ICE MAKER INSTALLED AS HEAT SOURCE FOR A HEAT PUMP IN A DISTRICT HEATING

By the end of 1986, the first full-scale vacuum-ice maker has been started to extract energy from sea water. This water is completely untreated and comes from a harbour at the Baltic Sea. Main operating data are as follows:

- evaporating/condensing-temp. 0 °C / 64 and 75 °C
- total heating capacity 2.700 kW
- total gas consumption 1.600 kW
- energy efficiency ratio 1,68
- ice-capacity (metric tons per day) 175 tons/day
- average sea-water inlet temperature 4 °C

It was considered to be of importance, that untreated water also at freezing point could be used. The generated ice-slurry was released under the surface of the ice-cover in the harbour and due to the water-like behaviour of the slurry did not cause any problems. Because of the vacuum, organisms are killed and this cost-free issue caught the eye of the authorities positively.

MAIN FEATURES AND ADVANTAGES OF VACUUM-ICE

- ice-making system with highest possible COP of all
- pumpability of ice with very high concentrations
- any quality of water can be used - potable water, sweet water, sea water, dirty water, polluted water
- the vacuum-ice process is a water-cleaning process: it desalinates and kills organisms - potable water is a by-product
- using vacuum-ice increases the capacity of existing installations and reduces the diameter of pipework for a given capacity - the energetic density is much higher
- the vacuum-ice process is an ice-maker and/or a water-cooling process (chiller) at the same time
- vacuum-ice can be generated in fractions between 0 and 100 % by mass
- the process works continuously and needs very little space compared to conventional ice-makers
- a vacuum-ice plant - be it heat pump, ice-maker, chiller or thermal energy storage - consists of conventional parts and components, has only few moving parts and is therefore simple and reliable
- vacuum-ice makers can be placed indoors and outdoors
- properties of vacuum-ice are very favourable for all kind of handling, transportation and storage (pumping, pneumatic transportation, agitation)
- vacuum-ice systems do not need exotic or sophisticated equipment for operation and handling

APPLICATION OF VACUUM-ICE+)

Due to the flexible properties of the design, produced ice and applications, there is almost no limitation to apply for numerous purposes. Vacuum-ice is applicable in the following fields:

- heat source for heat pumps (even harsh conditions)
- thermal energy storage for accumulation of "cold" energy, i.e. off-peak tariffs to avoid demand carges or to reduce the installed capacity of cooling equipment
- production of large quantities of ice (100 tons/day and much more in one unit) for industry and air-conditioning
- chiller for cooling tunnels, sahfts and mines
- ice-maker for chemical industry and civil engineering such as concrete cooling

---

+) Vacuum-ice maker design and applications are patented

SEASONAL PERFORMANCE OF HEAT PUMP WATER HEATERS

J. C. Mulligan, R. R. Johnson, and J. A. Edwards

Applied Energy Research Laboratory
North Carolina State University, Raleigh, NC, USA 27695-7910

ABSTRACT

Results are presented for the seasonal performance of two types of heat pump water heaters. First is a heat pump desuperheater in which the heat available in the desuperheater portion of a space heating or cooling heat pump is used to heat water. In winter the desuperheater system converted 11 to 19% of the heat output of the heat pump into hot water at an average coefficient of performance of 1.8. In summer the system recovered 7 to 13% of the electrical energy consumed by the heat pump. Second is a heat pump dedicated solely to making hot water. The heat pump delivered water at 66°C and the average coefficient of performance, including tank and cycle losses, winter penalty, and summer cooling effect, was 1.7. A third, new cycle is proposed but not tested, which could improve performance for cases where the water needs to be heated over a wide range of temperatures.

KEYWORDS

Heat pump water heater; desuperheater water heater; seasonal performance of water heater; pressure gradient condenser.

INTRODUCTION

The use of heat pumps and solar assisted heat pumps for domestic hot water have been advocated in recent years because of the potential energy savings when compared with direct electric heaters. Residential hot water heating is an important component of the central electric utility peak load, especially the early morning winter peak. For this reason, it is important that more efficient techniques to perform this task with electricity receive proper evaluation. Using the heat rejected from compressors (heat pump or air conditioner) is a technique to provide this hot water. The compressed vapor leaving the compressor of a heat pump system is in a superheated condition and at a temperature well suited to making hot water. There are two strategies to take advantage of heat pumps for heating water. The water may be heated in the desuperheater of a system designed for space cooling or heating, or in a heat pump dedicated solely to water heating.

BACKGROUND

Two types of heat pump water heaters available commercially were tested over a period of two years in actual operating circumstances (Mulligan, 1984a, 1984b and 1985). First is the heat-pump desuperheater shown in Fig. 1. This system is a commercial three-component configuration consisting of an outdoor coil unit, air indoor coil unit, and a two-speed compressor unit. The cooling capacity of the system is nominally two tons of cooling at low speed, and three tons at high speed. The speed is controlled by an outdoor thermostat and indoor set point differential. A shell-and-tube desuperheater is incorporated within the refrigerant circuit. A small circulating pump is used to circulate water through the desuperheater.

The system operated within a calorimeter house of 12 m^2 that was designed to simulate the load of a residence of 150 m^2. A drain cycle to simulate a hot water usage pattern was controlled by a series of mechanical timers and a solenoid valve. The total hot water draw was 450 liters per day.

In the winter, the desuperheater system converted from 11 to 19% of the total heat output of the heat pump system into hot water, ranging in temperature from 38°C to 42°C. The resulting 'penalty' to the space heating task was characterized by the fact that the delivered temperature of the space air supply was reduced a maximum of 1.0°C. This temperature reduction is considered insignificant. The total heating coefficient of performance, COP_t, for the low speed setting ranged from 2.8 to 3.0 when the tank storage temperature was high, to 3.6 when the tank storage temperature was low. It needs to be pointed out that at low speed the unit was operating in a high efficiency mode in which the indoor coil was the limiting exchange unit. For high speed operation the heating COP was in the range 1.7 to 2.0, and did not depend as much on the tank temperature.

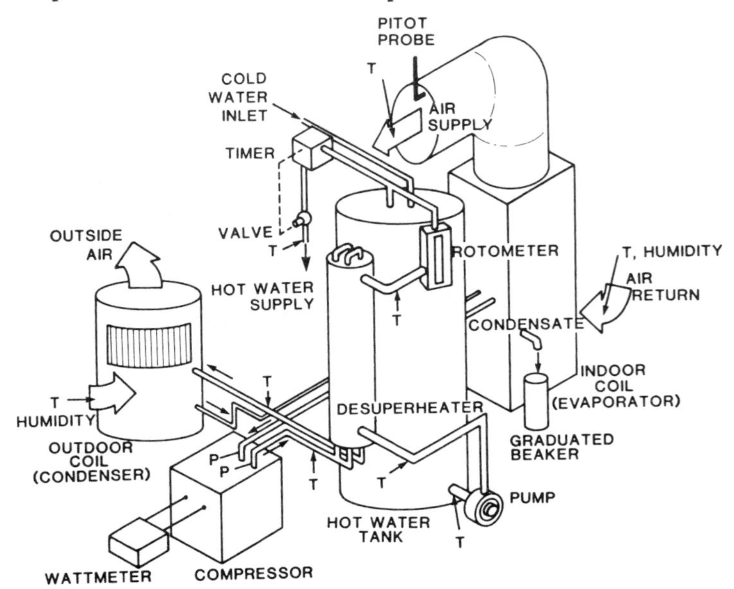

Fig. 1. Desuperheater system of water heating.

In the summer the desuperheater system recovered 7 to 13% of the electrical energy consumed. Because this is usually considered as rejected heat, it increased the cooling COP of 1.6 to 2.1 to a total COP_t of 1.8 to 2.2 for high speed operation of the compressor. The tank temperature ranged from a high of 40.6°C on hot days to 27.2°C on cool days. This wide range in tank temperatures indicates the potential difficulty of the desuperheater system which is operated on a schedule of space heating and cooling, and not on demand for hot water.

The second system tested was a heat pump water heater dedicated solely to making hot water, as shown in Fig. 2. The system tested was a commercial, 1-ton, air-to-water unit with a tube-in-tube condenser, a water circulating pump, and a thermostat in the return water line from the bottom of the hot water storage tank. The unit was connected to a standard 200 liter tank, with disabled heating elements. The system was totally contained within the conditioned space of a calorimeter house, and the hot water draw was 450 liters per day on a schedule designed to simulate a typical hot water usage pattern.

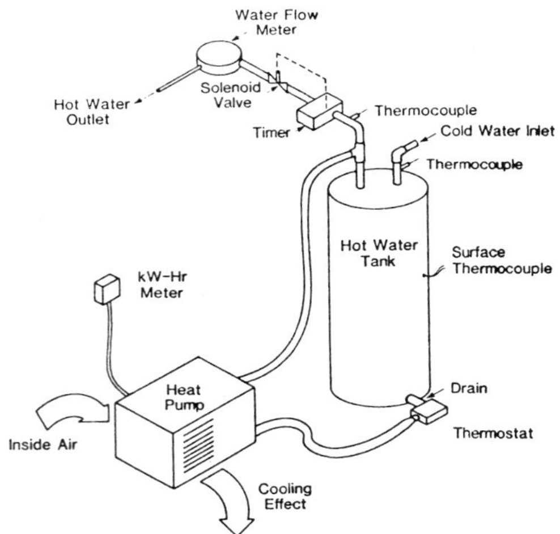

Fig. 2. Heat pump system of water heating.

The heat pump water heater delivered 100% of the daily requirements at 66°C with a COP of approximately 2.0 averaged over a full year of operation, including tank losses, machine cycle losses, and summer cooling effect. The winter cooling penalty, because the unit drew heat from the conditioned space, lowered the winter COP from 2.1 to 1.2. On a yearly average, it is expected that the COP would be of the order of 1.7 including the winter cooling penalty. When compared with 0.8 for an electrical water heater, such a unit would cut the energy cost for hot water by 60%.

Both the desuperheater and heat pump systems show attractive energy savings over the more conventional method of using electric heaters. They do, however, have some drawbacks. The desuperheater is run on a schedule that is primarily set by space conditioning demand and therefore needs to be supplemented at times of water draw when there is little, or no, space heating or cooling. The heat pump system tested here was within the conditioned space. If, however, the air exchanger was outdoors, or the tap water particularly cold, the unit may have had difficulty delivering water at a high enough temperature. The next section of this paper presents a system that has not yet been tested, but holds promise of improving the COP, and can be set to deliver the high water temperatures.

NON-UNIFORM PRESSURE CONDENSER

The proposed device for improving unit efficiency, and for maintaining delivery temperature is a condensing heat exchanger in which the pressure of the condensing refrigerant varies along the length of the heat exchanger. To understand the proposed changes, consider the T-s diagram of the cycle shown in Fig. 3. In the T-s diagram, the cycle 1-2-3-4 represents the conventional heat pump cycle with a maximum temperature T_2

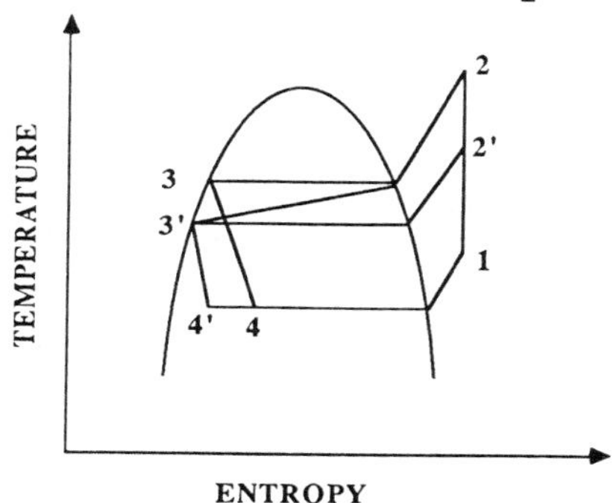

Fig. 3. T-s diagram for vapor compressor cycle.

and constant pressure condenser and evaporator. In order to deliver water at temperatures approaching T_2, the condenser pressure would have to be p_3. The cycle 1-2'-3'-4' is a conventional heat pump cycle with a maximum temperature T_2' and constant pressure condenser and evaporator. Water that entered at temperature just below T_3' would only approach $T_{2'}$ on delivery. The cycle 1-2-3'-4' is the new cycle proposed. The pressure does not remain constant through the condenser from state 2 to state 3', but rather the pressure, and temperature progressively decrease. The ideal cycle, thermal coefficients of performance are given by the following two expressions:

$$COP_{1-2-3-4} = \frac{h_2 - h_1}{h_2 - h_3} ; \qquad COP_{1-2-3'-4'} = \frac{h_2 - h_1}{h_2 - h_{3'}} \tag{1}$$

Shown in Fig. 4 is a plot of the two coefficients of performance given in equations (1) and (2) over a range of differences between inlet and outlet water temperatures. Temperature $T_{3'}$ is assumed to be 10°C hotter than the water inlet temperature. The water outlet temperature is assumed to be 20°C cooler than the peak cycle temperature (T_2 or $T_{2'}$). The nominal value of the evaporator temperature T_5 is -5°C, and the superheat at state 1 is 5°C. The refrigerant is R-12.

The values in Fig. 4 are ideal values and therefore are considerably above the actual values that might be expected in tests similar to those described earlier. The values, however, do indicate the relative advantage of decrease of pressure through the condenser. Of course, some of that

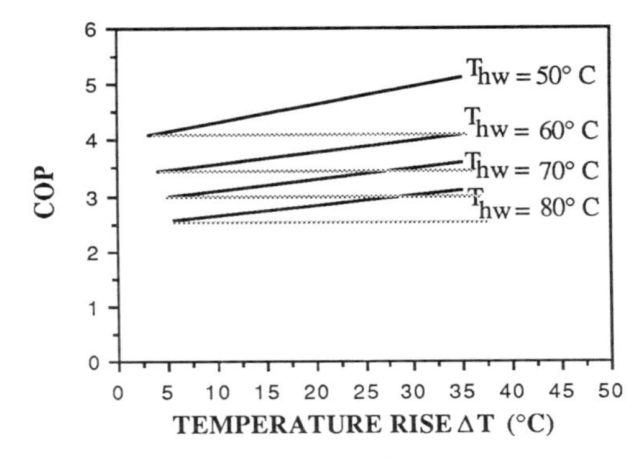

Fig. 4. COP as a function of temperature rise of water. Solid line for non-uniform pressure.

decrease occurs anyway because of the frictional pressure drop, or could also be achieved by subcooling in the condenser. The idea of controlling the pressure gradient in the condenser also suggests a way of building an on-demand heat pump water heater, because cold water could potentially be heated to a suitable temperature in one pass through the heat pump.

SUMMARY

The results are presented for experimental tests on two types of water heaters. In the winter the desuperheater system converted 11 to 19% of the total heat output of the heat pump into hot water ranging in temperature of 38°C to 42°C. For the high speed compressor the total system COP ranged from 1.7 to 2.0. For summer operation the desuperheater system recovered 7 to 13% of the electrical energy consumed. The total system COP ranged from 1.8 to 2.2. The other set of experiments were for a 1-ton heat pump dedicated solely to heating water. The heat pump water delivered 100% of the daily requirements at a temperature 66°C and at a COP of 1.7 averaged over the year and including tank losses, cycle losses, summer cooling effect, and winter penalty. A new cycle is proposed, but not tested, which allows the heating of water through a large temperature rise, with a high coefficient of performance. The proposed cycle develops a pressure drop through the condenser. The cycle is compared with the more conventional heat pump cycles.

REFERENCES

Mulligan, J. C., Edwards, J. A., and Johnson, R. R. (1984a). Test of a Commercially Available Heat Pump Water Heater. Applied Energy Research Lab Report No. AERL 20. North Carolina State University, Mechanical and Aerospace Engineering, Raleigh, NC.

Mulligan, J. C., Edwards, J. A., and Johnson, R. R. (1984b). Effectiveness of Heat Pump Desuperheaters in Producing Domestic Hot Water. Applied Energy REsearch Lab Report No. AERL 19. North Carolina State University, Mechanical and Aerospace Engineering, Raleigh, NC.

Mulligan, J. C., Edwards, J. A., and Johnson, R. R. (1985). Performance Tests of Heat Pump Systems. Proceedings of the ASHRAE/REHVA World Congress on Heating, Ventilating and Air-Conditioning. CLIMA 2000, Copenhagen, Denmark, Vol. 6, pp. 135-140.

POWER ELECTRONICS FOR AIR CONDITIONING

O. G. Hancock* and J. F. Schaefer**

* Florida Solar Energy Center
300 State Road 401, Cape Canaveral, FL 32920

** Department of Electrical Engineering
The Citadel, Charleston, SC

ABSTRACT

The Florida Solar Energy Center has established a Power Electronics for Air Conditioning Program. The results of one of the first tasks, to develop a performance simulation model, are presented. Characteristic curves describing operational parameters are given for single-speed, two-speed and variable-speed heat pumps; and, drive motor choices are discussed. Cooling season and annual performance results are also presented and analyzed.

KEYWORDS

Power electronics; space cooling; space heating; air conditioning; simulation

DISCUSSION

The Florida Solar Energy Center (FSEC), an Institution of the State University System of Florida, was created by the Florida legislature in 1974 to perform solar energy research and development. Over the years, FSEC has expanded its scope of interests and is today performing major research in hydrogen energy production from renewables, photovoltaics, buildings energy use, the use of heat pipes in space conditioning, and power electronics. More than thirty percent of all electricity sold in the state is used for space cooling. Therefore, much interest was found to support research leading to reduced energy consumption in space cooling and heating.

Semiconductor devices which can modify or control tens of kilowatts of electrical power are now available off-the-shelf. Current and potential use of these devices covers a broad spectrum of applications, and the term "power electronics" (PE) has been coined to cover such use. If one were to couple such devices with microprocessors properly programmed and having appropriate sensor inputs, the resulting intelligent controls would be capable of performing many functions which could not practically be done before. This combination offers the possibility of altering the characteristics of appliances so that they perform their function better and, at the same time, use less energy or have less energy demands.

The first task was a study of the potential for PE in air conditioners and heat pumps by evaluating possible energy savings and projecting the effect of PE on an electric utility. After a literature search, it was decided to develop a computer simulation program to predict performance of 3 generic types of heat pumps: single-speed, two-speed and adjustable-speed.

Figures 1, 2 and 3 characterize the heat pump and its thermal loads and Table 1 defines names used in the characterization below and in the program.

- o Thermal loads are assumed to be linear functions of temperature. Loads are defined by the pairs (THSET, LOAD17) in heating mode and (TCSET, LOAD95) in cooling mode.
- o Capacities are assumed to be linear functions of temperature except for discontinuities in the heating mode at T=42°F. Heating capacity in the range from 17°F to 42°F is presumed to be degraded due to defrost cycles. CAP17 and CAP47 define the heating mode, and CAP87 and CAP95 define the cooling mode.
- o Low-speed capacity is assumed to be less than high-speed capacity.
- o Coefficient of Performance (COP) is assumed to be a linear function of temperature, defined by the pairs (COP17, COP47) and (COP87, COP95).
- o Low-speed COP is assumed to be greater than high-speed COP.
- o Resistance heat backup is provided below the heating balance point.

Resultant capacities and COP's are shown by the heavy lines and reflect the degradations discussed above and cycling degradations.

TABLE 1 Definition of Nomenclatures

Steady-state parameter at indicated outdoor db temperatures (°F):	Definition:
CAP17, CAP47, CAP87, CAP95	Capacity of 1-speed pumps.
CAPHI17, CAPHI47, CAPHI87, CAPHI95	Capacity of 2-speed & variable-speed pumps at constant high speed.
CAPLO17, CAPLO47, CAPLO87, CAPLO95	Capacity of 2-speed & variable-speed pumps at constant low speed.
CAPFAC	Ratio of a pump's capacity at minimum speed to its capacity at highest speed.
COP17, COP47, COP87, COP95	COP of 1-speed pumps.
COPHI17, COPHI47, COPHI87, COPHI95	COP of 2-speed & variable-speed pumps at constant high speed.
COPLO17, COPLO47, COPLO87, COPLO95	COP of 2-speed & variable-speed pumps at constant low speed.
LOAD17, LOAD95	Building thermal load at indicated outdoor db temperatures (°F).
SEER	Seasonal Energy Efficiency Ratio.
TAU	Time required for a pump to reach full cooling or heating capacity from an equalized "off" condition.
TCSET, THSET	Cooling setpoint TCSET and heating set-point THSET. (Building thermal load is 0 when outdoor db temperature equals TCSET or THSET.)

Figure 4 shows efficiency curves typical of two types of electric motors which might be considered for use with adjustable-speed drives. These curves use measured data. It is readily apparent that the brushless DC motor is considerably more efficient at slower rotational speeds. Results of cooling season performance as a function of TAU are presented in Fig. 5. The two-speed and adjustable-speed models assume 20 percent COP improvement over full-

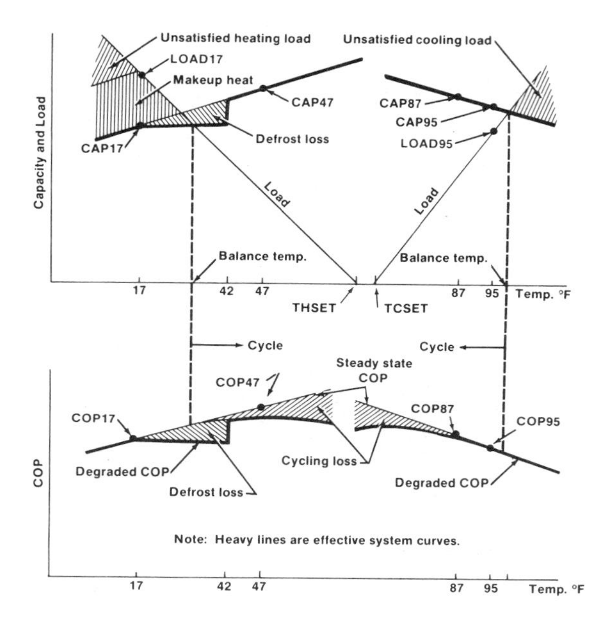

Figure 1. Characteristic curves of single speed heat pump: capacity and COP as a function of temperature.

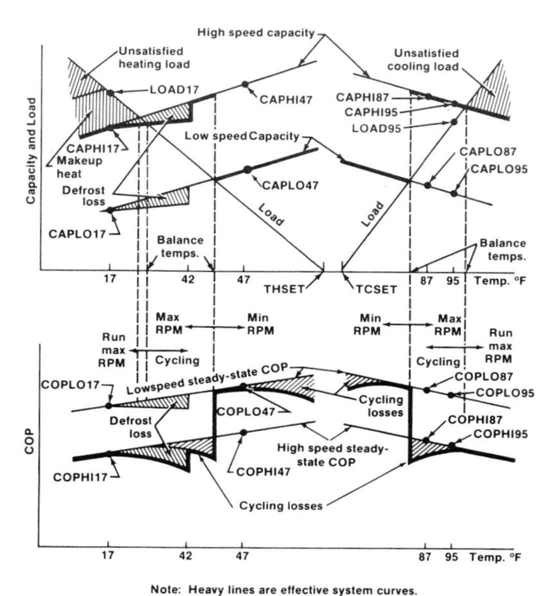

Figure 2. Characteristic curves of 2-speed heat pump: capacity and COP as a function of temperature.

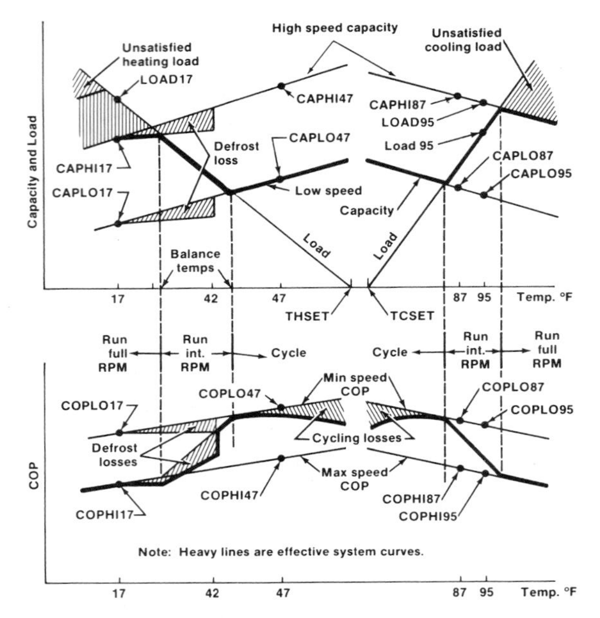

Figure 3. Characteristic curves of a variable-speed heat pump: capacity and COP as a function of temperature.

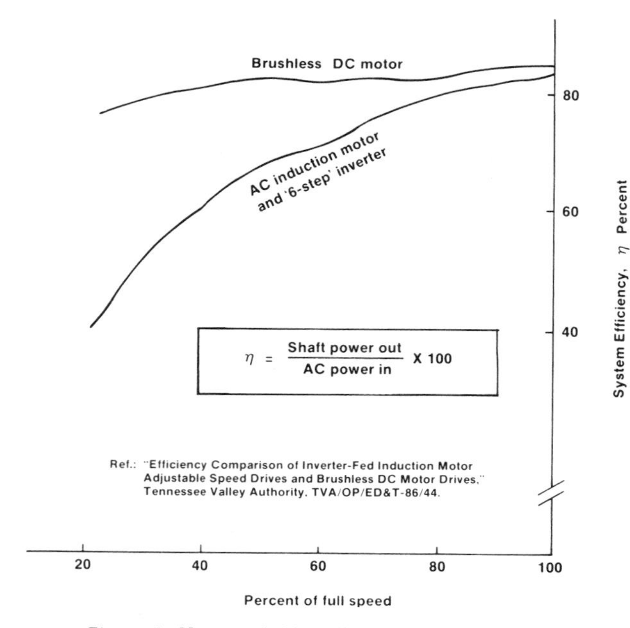

Figure 4. Measured drive efficiencies @ 100% torque.

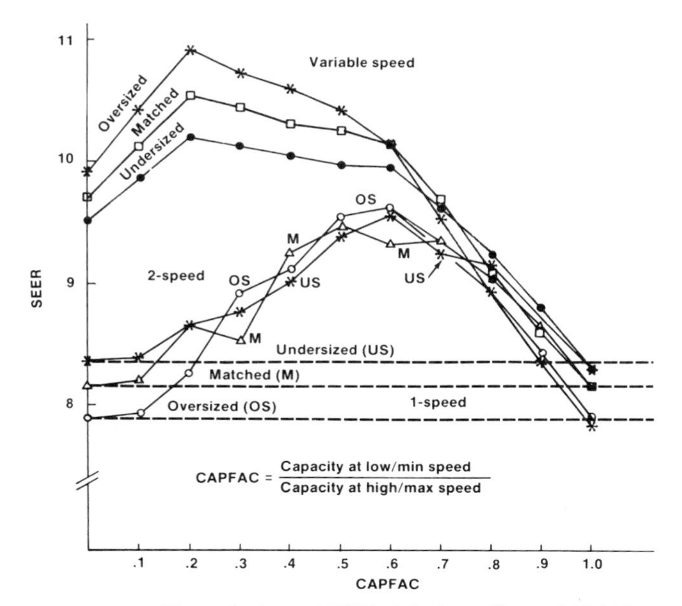

Figure 5. Annual SEER at Jacksonville vs. CAPFAC (capacity factor) for TAU = 100 seconds.

speed COP for CAPFAC ranges from 0.2 to 0.5. From CAPFAC of 0.2 to 0 and from 0.5 to 1.0, COP decreases linearly to full-speed COP.

CONCLUSIONS

Based on many simulation runs of three types of "generic" heat pumps modeled, it was concluded that:

- Both two-speed and adjustable-speed pumps offer efficiency improvement over single-speed pumps. There is little difference in gains of two-speed and adjustable-speed pumps for CAPFAC between 0.7 and 1. For CAPFAC less than 0.7, adjustable-speed pumps are better.
- Oversized adjustable-speed pumps have the highest EER.
- For 1-speed pumps, undersizing is more energy-efficient than oversizing, largely due to less cycling loss.
- For 2-speed pumps, the cycling penalty is less but remains significant while COP increases at low speed. Actual speed ratios and control strategies will be quite important in determining energy efficiency and economics.
- For variable-speed pumps which buildup to full capacity rapidly (tau < 1 minute), an oversized pump yields highest SEER; for tau > 1 minute, cycling losses begin to dominate. Minimum-to-maximum speed ratio and control strategy will be less critical than for 2-speed pumps.
- For all pumps there is a one-point loss in SEER for every minute delay in capacity buildup over the range modeled.
- It is not enough to consider electronic converter efficiency alone. One must consider the combined efficiency of both converter and motor if good efficiency is to be achieved.